Journal of Chromatography Library – Volume 3

# LIQUID COLUMN CHROMATOGRAPHY

A Survey of Modern Techniques and Applications

JOURNAL OF CHROMATOGRAPHY LIBRARY

Volume 1 Chromatography of Antibiotics
by G.H. Wagman and M.J. Weinstein

Volume 2 Extraction Chromatography
edited by T. Braun and G. Ghersini

Volume 3 Liquid Column Chromatography. A Survey of Modern Techniques and Applications
edited by Z. Deyl, K. Macek and J. Janák

Volume 4 Detectors in Gas Chromatography
by J. Ševčík

Journal of Chromatography Library – Volume 3

# LIQUID COLUMN CHROMATOGRAPHY

## A Survey of Modern Techniques and Applications

Editors

**Zdeněk Deyl**
*Physiological Institute, Czechoslovak Academy of Sciences, Prague*

**Karel Macek**
*3rd Medical Department, Medical Faculty, Charles University, Prague*

**Jaroslav Janák**
*Institute of Instrumental Analysis, Czechoslovak Academy of Sciences, Brno*

**ELSEVIER SCIENTIFIC PUBLISHING COMPANY**
AMSTERDAM – OXFORD – NEW YORK 1975

ELSEVIER SCIENTIFIC PUBLISHING COMPANY
335 Jan van Galenstraat
P.O. Box 211, Amsterdam, The Netherlands

AMERICAN ELSEVIER PUBLISHING COMPANY, INC.
52 Vanderbilt Avenue
New York, New York 10017

Library of Congress Card Number: 73-89151

ISBN 0-444-41156-9



Printed in The Netherlands

## Contributors

V. Betina, Department of Microbiology and Biochemistry, Faculty of Chemistry, Slovak Polytechnical University, Bratislava (Czechoslovakia)

K. Čapek, Laboratory of Monosaccharides, Prague Institute of Chemical Technology, Prague (Czechoslovakia)

J. Churáček, Department of Analytical Chemistry, Technical University, Pardubice (Czechoslovakia)

J. Čoupek, Institute for Macromolecular Chemistry, Prague (Czechoslovakia)

J. Davídek, Department of Food Science and Analysis, Prague Institute of Chemical Technology, Prague (Czechoslovakia)

Z. Deyl, Physiological Institute, Czechoslovak Academy of Sciences, Prague-Krč (Czechoslovakia)

J. Dršata, Department of Biochemistry, Faculty of Pharmacy, Charles University, Hradec Králové (Czechoslovakia)

J. Gasparič, Department of Physical Chemistry, Faculty of Pharmacy, Charles University, Hradec Králové (Czechoslovakia)

I. M. Hais, Department of Biochemistry, Faculty of Pharmacy, Charles University, Hradec Králové (Czechoslovakia)

J. G. Heathcote, Department of Chemistry and Applied Chemistry, University of Salford, Salford (Great Britain)

S. Heřmánek, Institute for Nuclear Research, Řež near Prague (Czechoslovakia)

J. Janák, Institute of Instrumental Analysis, Czechoslovak Academy of Sciences, Brno (Czechoslovakia)

M. Janda, Department of Organic Chemistry, Prague Institute of Chemical Technology, Prague (Czechoslovakia)

P. Jandera, Department of Analytical Chemistry, Technical University, Pardubice (Czechoslovakia)

M. Juřicová, Physiological Institute, Czechoslovak Academy of Sciences, Prague-Krč (Czechoslovakia)

F. Jursík, Department of Inorganic Chemistry, Prague Institute of Chemical Technology, Prague (Czechoslovakia)

I. Kluh, Institute of Organic Chemistry and Biochemistry, Czechoslovak Academy of Sciences, Prague (Czechoslovakia)

M. Krejčí, Institute of Instrumental Analysis, Czechoslovak Academy of Sciences, Brno (Czechoslovakia)

M. Kubín, Institute of Macromolecular Chemistry, Czechoslovak Academy of Sciences, Prague (Czechoslovakia)

K. Macek, Third Medical Department, Medical Faculty, Charles University, Prague (Czechoslovakia)

O. Mikeš, Institute of Organic Chemistry and Biochemistry, Czechoslovak Academy of Sciences, Prague (Czechoslovakia)

O. Motl, Institute of Organic Chemistry and Biochemistry, Czechoslovak Academy of Sciences, Prague (Czechoslovakia)

J. Novák, Institute of Instrumental Analysis, Czechoslovak Academy of Sciences, Brno (Czechoslovakia)

Z. Pechan, Department of Biochemistry, University J. E. Purkyně, Brno (Czechoslovakia)

J. Pokorný, Department of Food Science and Analysis, Prague Institute of Chemical Technology, Prague (Czechoslovakia)

Ž. Procházka, Institute of Organic Chemistry and Biochemistry, Czechoslovak Academy of Sciences, Prague (Czechoslovakia)

Z. Prusík, Institute of Organic Chemistry and Biochemistry, Czechoslovak Academy of Sciences, Prague (Czechoslovakia)

Z. Šesták, Institute of Plant Physiology, Czechoslovak Academy of Sciences, Prague (Czechoslovakia)

J. Staněk, Jr., Laboratory of Monosaccharides, Prague Institute of Chemical Technology, Prague (Czechoslovakia)

I. Stibor, Department of Organic Chemistry, Prague Institute of Chemical Technology, Prague (Czechoslovakia)

J. Turková, Institute of Organic Chemistry and Biochemistry, Czechoslovak Academy of Sciences, Prague (Czechoslovakia)

R. Vespalec, Institute of Instrumental Analysis, Czechoslovak Academy of Sciences, Brno (Czechoslovakia)

R. J. Washington, Department of Chemistry and Applied Chemistry, University of Salford, Salford (Great Britain)

S. Wičar, Institute of Instrumental Analysis, Czechoslovak Academy of Sciences, Brno (Czechoslovakia)

J. Zabranský, Research Institute of Food Technology, Czechoslovak Agricultural Academy, Prague (Czechoslovakia)

S. Zadražil, Institute of Organic Chemistry and Biochemistry, Czechoslovak Academy of Sciences, Prague (Czechoslovakia)

Z. J. Zmrhal, Research Institute of Plant Production, Prague-Ruzyně (Czechoslovakia)

# Contents

APPLICATIONS

## Foreword

Contemporary liquid column chromatography is a rapidly expanding analytical technique. In order to ensure its further growth, it is important that periodically all the information available should be summarized in a clear and easily understood manner. Such a compilation serves three purposes. For the novice, it represents the basis on which additional knowledge can be built and practical observations interpreted. For the practical analyst facing different problems, it often allows the possibility of checking whether his "new" problem has already been solved. Finally, for the "expert" – the analyst who is experienced in both the theory and practice of the technique and who may have participated in its development – it permits him to organize his thoughts, to examine the various modifications of the technique in their proper context, and to realize those interrelationships which might have escaped his attention.

In chromatography, the technique that is being discussed here, there are good examples of the influence of such compilations. In classical liquid column chromatography, Zechmeister's book, first published in Europe in the mid-1930s, represented the springboard for the rapid expansion of the technique within a few years and the English translation, published in the early 1940s, contributed significantly to the adaptation of the technique in the United States. The handbook by Stahl on thin-layer chromatography, first published in 1962, served a similar purpose in the expansion of this very practical technique. In gas chromatography, most of us who worked with the technique in its infancy obtained the theoretical basis from Keulemans' book, while the books by Dal Nogare and Juvet, Purnell, and Littlewood, which were published virtually simultaneously in 1962, served as the foundation that enabled many thousands of analysts to increase their practical knowledge.

Modern liquid column chromatography as we know it is only a few years old and was clearly developed from the knowledge accumulated in gas chromatography. However, the technique itself is not young; in fact, it represents the oldest variant of chromatography, and many of those who recently started to apply the "new" technique in their research were not aware of the excellent results obtained by early workers and hence could not apply this old experience to their present work. Also, by restricting their coverage of liquid column chromatography to only the latest results reported in the analytical journals, they may have overlooked the opportunity of examining the technique in its proper context in relationship to other liquid chromatographic techniques that are also usually carried out in columns.

The aim of this book is precisely what was specified above as the proper purpose of a compilation that deals with a rapidly growing analytical technique: to represent the basis on which newcomers to the field can build their future work, to give a rich compendium of the various applications of liquid column chromatography, with adequate references to lead the reader to additional information, and to show the interrelationships of the technique and the common theoretical basis of the individual variants. I believe the Editors have succeeded in achieving these objectives.

The essence of many research and review papers has been reported in this fine volume. To *Liquid Column Chromatography*, its Editors and its readers, my best wishes for success.

*University of Houston,*
*Houston, Texas (U.S.A.)*

ALBERT ZLATKIS

## Preface

In the last 10 years, the advances made in liquid column chromatography have been comparable with those of flat-bed techniques and gas chromatography several years ago. In principle, there are two reasons for this increasing interest in liquid column chromatography. Firstly, the present situation reflects the steadily increasing demands being made upon separation techniques, mainly in biochemistry, drug analysis, the analysis of environmental pollutants, etc. Secondly, the theoretical aspects of liquid column chromatography have been developed substantially, mainly as a result of the application of model situations in gas chromatography which resulted in the integration of the knowledge achieved in the techniques and instrumentation of the different individual chromatographic variants. A deeper understanding of the process of chromatographic separation, especially more complete knowledge about the factors that influence equilibria and the selectivity of separations, showed clearly how far from the optimum are the conditions used in classical liquid column chromatography. These advances resulted in liquid column chromatography becoming a more complete analytical procedure, both qualitatively and quantitatively, and it became a very rapid method, offering numerous ways of automation and a delicate approach to the separation of different substances.

The success in this field is intimately connected with the application of new types of sorbents that allow rapid mass transfer and are suitable for many types of separations. In addition to their chemical properties, the importance of physical properties such as particle size and shape have been stressed during the recent development of the new techniques of separation. The high speed of mass transfer enables the separation time to be shortened substantially and flow-rates to be increased by using high pressures. A high column overpressure is, of course, inevitable when one uses small size particles, which, on the other hand, help the rapid separation to occur. The application of high pressures also necessitates increased investment costs. As the use of high pressures is one of the most striking differences compared with classical techniques, these modern procedures are referred to as high-pressure or, alternatively, high-speed or high-resolution chromatography.

One of the most rapidly developing features of column chromatography lies in the detectors. Instead of slicing the column or using fraction collectors with subsequent laborious assays on each fraction, the column eluate is nowadays analyzed continuously in devices based on different physical and chemical principles. Sometimes these detectors are very specialized and can be applied only to a single type of compound. On the other hand, however, the sensitivity of these detectors does not attain the sensitivity that we expect in gas chromatography. As always, there are a few exceptions from this general rule. The level of sensitivity in liquid column chromatographic techniques is the result of the different nature of the mobile phase compared with gas chromatography. It is to be expected that attempts will be made to increase the detector sensitivity in the near future.

In spite of the advantages mentioned above, high-speed techniques may not provide a universal solution to the many difficulties that one may meet in chromatographic separations. In some fields they already dominate or will dominate in the near future, while in other areas classical techniques will be retained. It is not only the high costs involved in the new techniques, but it might well be the nature of the separation itself which may direct the chemist to make his own choice between the classical and modern techniques.

We are currently witnessing a paradoxical situation. In many laboratories in which classical liquid column chromatographic methods are used, considerable delays have arisen that could have been avoided but for a poor knowledge of the theoretical principles, methods and apparatus involved in the modern version of liquid column chromatography. On the other hand, however, there are a number of laboratory workers who have approached the application of modern liquid column chromatography armed only with theoretical and instrumental experience related to gas chromatography. These workers usually have some gaps in the vast range of experience that has been accumulated in the years of development of classical liquid column chromatography, and there seems to be an undesirable tendency to forget about the past and to re-discover facts already published a long time ago. Other publications on liquid column chromatography have also indicated this situation.

We have tried to amalgamate the existing knowledge on classical and modern liquid column chromatography in the belief that it is the most useful way to stimulate the imagination of those who wish to use this method in the solution of practical problems. In general, we have tried to maintain a balance when preparing the book. We have tried to include sufficient theory to allow an inexperienced worker to acquire an adequate level of understanding of chromatographic separations and possibly to adopt procedures of his own for a particular problem, while on the other hand we have tried to collect reliable separation procedures for individual types of compounds, which are surveyed in the Applications part of the book.

To some, it may seem that the lengths of the different chapters are disproportionate. We believe, however, that this reflects precisely the current status of liquid chromatographic techniques. While separations of hydrophilic substances such as nucleotides, amino acids, proteins, enzymes, etc. are so frequent that there is hardly a paper on enzymes, for instance, which would avoid the use of a liquid chromatographic separation step, with hydrophobic compounds there are far fewer papers but those which have appeared usually involve substantially new procedures. Again, the development is far from proportionate even when judged separately in these two areas.

The system of listing individual chapters in the Applications section follows the general arrangement of organic chemistry, with additional chapters devoted to inorganic separations, radiochemical techniques, dyes and plastics, and is easily comparable with the system of listing references in the Bibliography Section of the *Journal of Chromatography* and in the *Bibliography of Column Chromatography 1967–1970*. In our opinion, this may help to bring everyone rapidly up-to-date without losing too much time in tedious literature searching. However, the rapid expansion of the literature will make parts of this book obsolete even before its appearance in print. It is recommended, therefore, that additional sources providing a fast information service should be followed, *e.g.*, the Bibliography Section of the *Journal of Chromatography*.

In discussing separations of individual types of compounds, we have tried to offer readers as much practical laboratory information as possible for the size of the book, keeping in mind that it is desirable to have a book that can be used directly in the laboratory without the need to study the original literature. In order to be concise, we have included the maximum amount of information in figures and tables and have less frequently described individual methods directly in the text. The latter method was used in situations in which there are widely used separations available for a particular type or series of compounds. The extent to which we have succeeded in achieving the aims outlined above must be judged by the reader. We are, of course, aware that a book of this size could not be perfect and we would therefore appreciate any comments and criticisms in this respect.

It is also a matter of courtesy to acknowledge the help given by collaborators who are neither authors of individual chapters nor Editors of the book. However, we feel that there should be something more than mere courtesy expressed in these acknowledgements: when preparing this manuscript, we realized that technical aid may frequently be more important than the undecipherable original manuscript of the authors, and apparently simple things such as attaching figures and tables to chapters and giving them proper numbers or listing references may easily turn Hercule Poirot's job. We express our deepest gratitude to Miss J. Krausová, M. Ciprová and H. Málková of the Laboratory of Subcellular Structures, Physiological Institute of the Czechoslovak Academy of Sciences, Prague, to Dr. Ž. Procházka of the Institute of Organic Chemistry and Biochemistry, Czechoslovak Academy of Sciences, Prague for translating a substantial part of this book, and, last but not least, to Mrs. R. Diartová, for drawing most of the figures.

*Prague*
*April, 1974*

Z. DEYL
K. MACEK
J. JANÁK

# THEORETICAL ASPECTS OF LIQUID CHROMATOGRAPHY

*Chapter 1*

# Fundamental concepts

J. NOVÁK and J. JANÁK

## CONTENTS



## INTRODUCTION

It is now about 70 years since Tswett (1903, 1906) described his classical experiment on the separation of coloured pigments of leaves and termed the procedure "chromatography". Although earlier experiments by Runge on capillary analysis and observations by Davy on changes in the composition of crude petroleum when in contact with rocks displaying adsorptive activity can now be defined as chromatographic experiments, it was only Tswett who understood and defined his method as a general means of separating substances (including colourless substances) and coined the term "chromatography". This priority is now generally recognized.

In Tswett's original experiment, a light petroleum extract of the colouring substance of leaves was introduced on to a calcium carbonate column in a glass tube under continuous percolation of light petroleum. The result was the separation of the mixture into a series of green and yellow bands (Fig.1.1). This invention came at a time when experimental methods of scientific investigation were still concerned mainly with the mere description of phenomena, and the problems of industrial practice were not considered in detail, and the importance of Tswett's work was not recognized and was revived only in the 1930s by Kuhn *et al.*, in connection with investigations on models of natural substances.

Of major significance in the further development of chromatographic methods was the work by Martin and Synge, the former being a Nobel prize winner; they used a liquid as the sorbent and showed the possibilities deriving from the linear sorption isotherm and copious selection in suitable liquids. This work was a precedent to the development of other chromatographic techniques, such as paper chromatography (Consden *et al.*), counter-current distribution (Craig), ion-exchange chromatography (Mayer and Tompkins, Samuelson), electrophoresis (Haugaard and Kroner), thin-layer chromatography (Ismailov and Shraiber, Kirchner *et al.*), molecular sieve action and, later, gel permeation chromatography (Barrer, Porath and Flodin) and, eventually, gas chromatography (Claesson, James and Martin).

*References p.10*

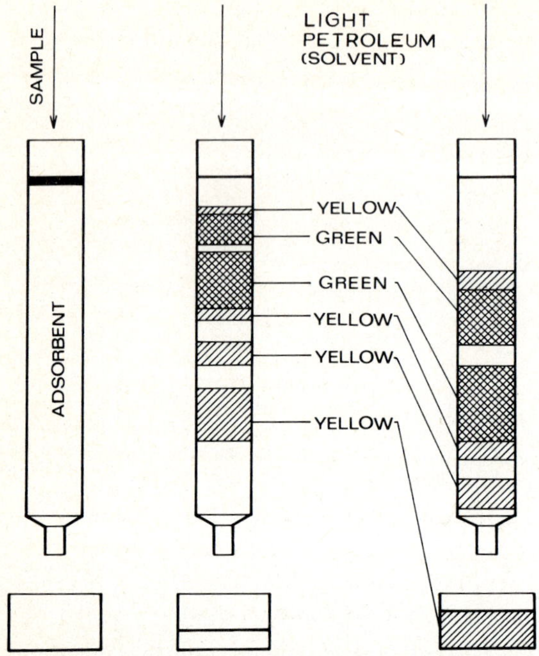


Fig. 1.1. Representation of Tswett's classical chromatographic experiments.

The advent of gas chromatography resulted in major advances in the theory as well as in the instrumentation and methodology of chromatography. It is therefore not surprising that this technique served as the basis for the development of a consistent theory of chromatography, applicable to all or a large group of chromatographic techniques (Giddings). Recently, the application of the theoretical and methodological knowledge obtained in gas chromatography led to a renaissance of the classical Tswett chromatography in the field of column techniques. A significant contribution to the theory of the modern high-speed procedures of liquid chromatography in columns was made by Huber. It appears that liquid chromatography affords both a separation efficiency and time of separation that are comparable with those usual in gas chromatography.

## PRINCIPLE OF CHROMATOGRAPHY

Chromatography can be characterized as a separation method based on the differential migration of solutes through a system of two phases, one of which is mobile. The mechanism proper that determines the regularities of the movement and spreading of the chromatographic zone can be considered as a continual convective disturbance of the equilibrium distribution of a solute between the phases and simultaneously compensating for the deficiencies by diffusion. As only those molecules of the substance chromatographed which occur in the mobile phase are subject to convective transport, the velocity of migration of the zone is proportional to the probability of the occurrence of the solute molecules in the mobile phase. It follows that although the chromatographic process involves inseparably the processes of convective transport, diffusion and sorption equilibration, it is an appropriate difference in the distribution constants that definitively determines the possibility of the differential migration of the chromatographic zones. However, if the mobile phase ceased to move, there would be no migration at all. On the other hand, were it not for diffusion, the differences in distribution constants would lose their significance as there would be no possibility of restoring the disturbed equilibrium. Hence, convective transport, diffusion and differences in the distribution constants are inseparable factors in the chromatographic process, each representing a necessary, but not complete, condition for achieving a chromatographic separation.

The above factors constitute, on the one hand, conditions for separation, but on the other hand, they imply a limitation in the separation efficiency. Geometrical irregularities of the chromatographic bed and the hydrodynamic properties of fluids lead to an uneven flow velocity of the mobile phase and to the formation of velocity profiles. Further, owing to a finite rate of equilibration (finite rate of lateral solute diffusion), the leading (sorption) part of the chromatographic zone during migration contains a certain excess of solute in the mobile phase, as compared with the equilibrium concentration, whereas the rear part of the zone shows a deficiency in the solute concentration in the mobile phase. Finally, longitudinal diffusion proceeds in both phases during chromatography. All of the above factors cause a spreading of the chromatographic zone. As will be shown later, chromatographic separation is feasible owing to the fact that while the spacing between two given zones increases linearly with increasing length of the migration path, the broadening of the zones increases with only the square root of the migration path. Hence, from the viewpoint of separation efficiency, the problem of zone spreading is as important as the problem of selectivity, *i.e.*, the problem of differences in distribution constants.

The above qualitative treatment indicates that a chromatographic separation will be the more effective the greater are the differences in the distribution constants of the solutes chromatographed, the more uniform is the chromatographic bed and the flow of the mobile phase, the more rapid is the establishment of sorption equilibrium, and the slower is the longitudinal diffusion of the solute in both phases. The problems of flow, diffusion and sorption equilibrium are discussed in greater detail in the following sections.

## CHROMATOGRAPHIC SYSTEMS

Chromatographic systems can be classified according to

(i) the state of aggregation of the phases;

(ii) the physical arrangement of the phases;

(iii) the mechanism underlying the distribution equilibrium.

Hence one can combine gaseous, liquid and solid phases to form a total of five two-phase systems, *i.e.*, solid–solid, liquid–solid, liquid–liquid, gas–solid and gas–liquid. Systems typical in liquid chromatography are liquid–solid and liquid–liquid. A solid–solid system is undoubtedly real, but it can hardly be of any use in chromatography as it is impossible in such a system to effect, under the usual chromatographic conditions, the relative

*References p.10*

movement of the phases without the interface being disturbed. Systems involving a gas as the mobile phase are typical in gas chromatography. The further discussion will be centred around liquid–liquid and liquid–solid systems.

As for the physical arrangement of the phases, chromatographic systems can be divided into two large groups – columnar and planar systems – and columnar systems can further be classified with respect to the geometry of the chromatographic bed proper. Thus, we can speak about capillary columns, coated either with a film of a stationary liquid or with a layer of a solid adsorbent adhering to the inner wall, and packed columns, containing a grained solid material that functions either directly as the partitioning phase or merely as a support for a stationary liquid. Columnar systems are much easier to define than planar systems, represented by paper and thin-layer chromatography.

In liquid–liquid systems, regardless of their arrangement, a very important point is the way in which the stationary liquid film is formed. Basically, there are two alternatives; in the first, the liquid film is being created during chromatography by adsorption of components of the solvent, while in the other a defined amount of stationary liquid insoluble in the mobile phase is deposited prior to carrying out the chromatographic run. The latter procedure is obviously preferable if it is desirable to define the amount of the stationary liquid held up within the chromatographic bed. The possibility of defining the amount of the partitioning material is essential whenever relationships between retention data and the parameters of sorption equilibrium are to be expressed.

Differential zone migration occurs by virtue of the differential distribution of solutes between the phases. This distribution can be due to various mechanisms, such as physical interactions of solute molecules with the solvent, formation of chemical bonds between the solute and the solvent, or merely a kind of hindered diffusion of solute molecules in the matrix of the sorbent.

Depending on the nature of the system, the solute–solvent interactions can take place on a two-dimensional or three-dimensional scale, *i.e.*, on the surface or within the bulk phase. It is these two alternatives from which the terms "adsorption" and "partition" chromatography, respectively, have been derived. However, if viewed on a molecular level, both alternatives can involve either of the above mechanisms; in other words, both ordinary physical interaction chromatography and ion-exchange or ligand-exchange chromatography and affinity chromatography can be effected as either a partition or an adsorption process. It is worth mentioning that electrophoresis displays phenomena similar to those of chromatography, but the differential zone migration in electrophoresis is based on a different mechanism and the above concepts of solute distribution between the phases are completely inapplicable. Therefore, this interesting separation method has not been included in this book.

The distribution equilibrium in a solute–sorbent–mobile phase system is characterized by a sorption (distribution) isotherm, representing the relationship between the solute equilibrium concentrations in the coexisting phases. If the distribution isotherm is linear, the distribution constant is a constant independent of the solute concentrations. In this case, we speak about linear chromatography. If the distribution isotherm is curved, the distribution constant is a function of the solute concentrations in the system, which results in non-linear chromatography.

## CHROMATOGRAPHIC TECHNIQUES

Any chromatographic system allows elution or frontal chromatography to be performed. In systems with a solid adsorbent in the role of the sorbent proper, displacement chromatography can also be included.

In the elution technique, a narrow band of the mixture to be chromatographed is introduced intermittently at a starting point (column inlet) on the chromatographic bed. Upon the action of the flowing mobile phase, the zone begins to migrate, the forward velocities of the individual components of the initial zone being inversely proportional to their distribution constants. The mobile phase is assumed to be virtually non-sorbed on the stationary phase in this technique. An elution chromatogram is shown in Fig. 1.2A. The distances of the peak maxima from the starting point are characteristic of the quality of the solute substances, whereas the peak areas are proportional to the total amounts of the substances in the chromatographic zones in elution chromatography.

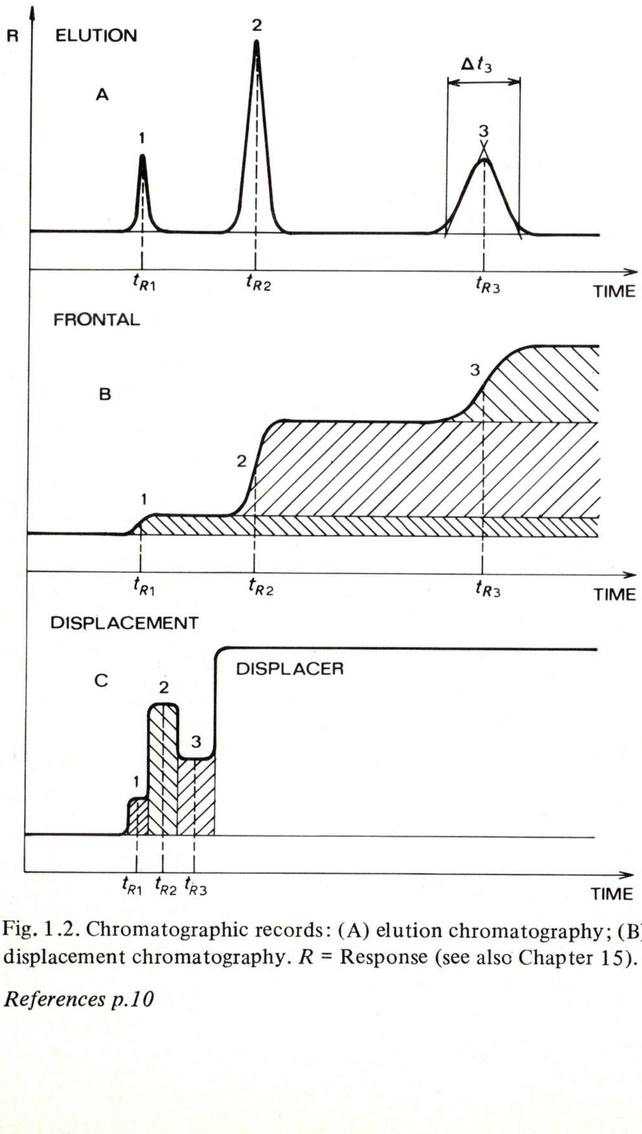


Fig. 1.2. Chromatographic records: (A) elution chromatography; (B) frontal chromatography; (C) displacement chromatography. $R$ = Response (see also Chapter 15).

In frontal chromatography, the mixture to be analyzed enters the chromatographic bed continuously at the start, and hence the mobile phase is actually formed by the mixture analyzed. The resulting course of the concentrations of the solute components in the effluent is shown in Fig. 1.2B. The relative movement and the eventual spacing of the individual steps are controlled by the distribution coefficients in the same manner as quoted for the zones in elution chromatography. The height of the front is proportional to the concentration of the component in the mixture being introduced.

In displacement chromatography, the procedure is similar to that in elution chromatography, but, instead of employing a non-sorbed eluent, the migration of zones is accomplished by displacement with a substance that is sorbed more strongly than any component of the mixture being separated. A displacement chromatogram is illustrated in Fig. 1.2C. The analytical meaning of the position and the area of a chromatographic peak is the same as in elution chromatography. When employing solid adsorbents as chromatographic materials (liquid–solid and gas–solid systems), the displacement effects are operative, to a certain extent, also in elution and frontal chromatography.

The chromatographic zones can either be eluted from the chromatographic bed or the chromatographic run can be stopped before their elution. In the first instance, the volume of the eluent necessary to elute the concentration maximum of the zone is a measure of retention. In the second instance, the distance travelled by the centre of the zone from the starting point serves as a criterion of retention. Planar techniques represent the second alternative. As the elution technique with the complete elution of the zones is of greatest concern in modern liquid chromatography, the discussion will be concentrated only on this version of chromatography.

## BASIC CHROMATOGRAPHIC QUANTITIES

When monitoring the composition of the effluent leaving the column outlet during the elution process, one obtains an elution chromatogram in either a differential (Fig. 1.3A) or an integral form (Fig. 1.3B). The ratio of the net retention time ($t_R - t_m$) of a solute component to the retention time ($t_m$) of a non-sorbed substance is called the capacity ratio and is denoted by $k$:

$$k = (t_R - t_m)/t_m \tag{1.1}$$

where $t_R$ is the total retention time of the solute substance. Actually, the value of $k$ represents the ratio in which a given amount of solute distributes itself between the stationary and the mobile phases in the given sorption system, *i.e.*,

$$k = \frac{\text{amount of solute in the stationary phase}}{\text{amount of solute in the mobile phase}} \tag{1.2}$$

It follows from eqn. 1.2 that

$$k = KV_s/V_m = K\varphi_s/\varphi_m \tag{1.3}$$

where $K$ is the distribution constant, $V_s$ and $V_m$ are the volumes of the stationary and the

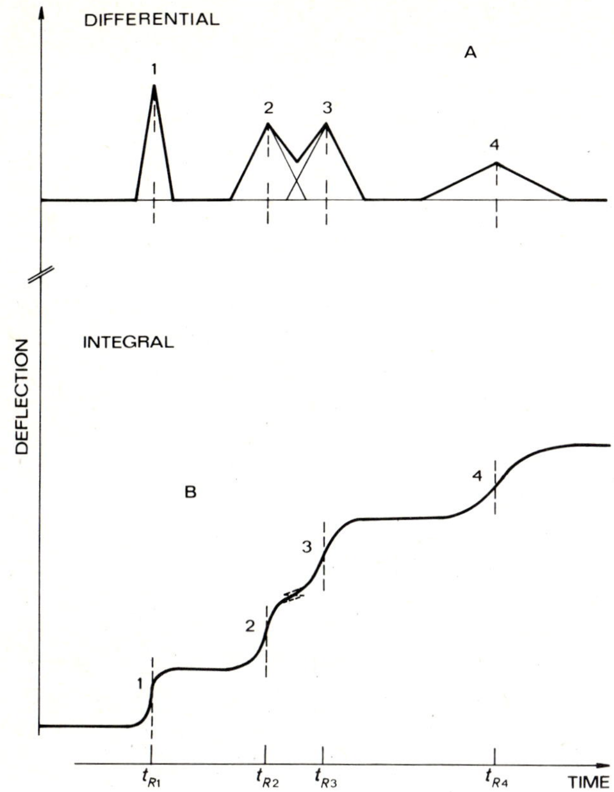


Fig. 1.3. Schematic representation of a differential (elution) chromatogram (A) and the corresponding integral chromatogram (B).

mobile phase, respectively, in the column, and $\varphi_s$ and $\varphi_m$ are the cross-sections of the respective phases. The distribution constant is defined by

$$K = c''/c' \tag{1.4}$$

where $c''$ and $c'$ are the solute equilibrium concentrations (number of moles per unit volume) in the stationary and in the mobile phase, respectively. The multiplication of $t_R$ and $t_m$ in eqn. 1.1 by the volume flow-rate of the mobile phase and combination with eqn. 1.3 gives

$$V_R = V_m\,(1 + k) = V_m + KV_s \tag{1.5}$$

where $V_R$ is the overall retention volume and $V_m$ is equal to the dead retention volume. Hence, for a non-sorbed substance, $k = 0, K = 0, t_R = t_m$ and $V_R = V_m$.

Eqn. 1.5 can be rearranged (Martin and Synge, Phillips) to read as follows:

$$V_m/V_R = V_m/(V_m + KV_s) = \boldsymbol{R} \tag{1.6}$$

where $\boldsymbol{R}$ is the so-called retardation factor. The quantity $\boldsymbol{R}$ is theoretically identical with the quantity $R_r$ employed in planar techniques (LeRosen):

$$R_F = \frac{\text{distance of the solute zone from the start}}{\text{distance of the eluent front from the start}} \tag{1.7}$$

In practice, the equation

$$R = R_F/\theta \tag{1.8}$$

holds (Giddings *et al.*), where $\theta$ is a factor the value of which varies within the range 0.8–0.9.

The extent of the separation of a pair of solutes is expressed empirically by a quantity called "resolution"; the calculation of resolution ($R_s$) from the chromatogram is carried out by using the equation

$$R_s = 2(t_{R2} - t_{R1})/(\Delta t_2 + \Delta t_1) \tag{1.9}$$

where $t_{R2}$ and $t_{R1}$ are the retention times of the two components and $\Delta t_2$ and $\Delta t_1$ are the time-widths of the respective zones. Resolution is dependent on the separation efficiency, expressed by the number of the theoretical plates, $N$:

$$N = 16(t_R/\Delta t)^2 \tag{1.10}$$

where $\Delta t$ is again the time-width of the peak. The number of theoretical plates is given by the column length ($L$) and the height equivalent to a theoretical plate, $H$, the latter quantity being defined by

$$H = \sigma^2/L = L/N \tag{1.11}$$

where $\sigma$ is the standard length deviation of the chromatographic zone.

## REFERENCES


Barrer, R. M., *J. Soc. Chem. Ind., London*, 64 (1945) 130.

Claesson, S., *Ark Kemi Mineral. Geol.* A23, No. 1 (1946).

Consden, R., Gordon, A. H. and Martin, A. J. P., *Biochem. J.*, 38 (1944) 224.

Craig, L. C., *J. Biol. Chem.*, 155 (1944) 519.

Davy, D. T., *Proc. Amer. Phil. Soc.*, 36 (1897) 112.

Giddings, J. C., *Dynamics of Chromatography*, Marcel Dekker, New York, 1965.

Giddings, J. C., Steward, G. H. and Ruoff, A. L., *J. Chromatogr.*, 3 (1960) 239.

Haugaard, G. and Kroner, T. D., *J. Amer. Chem. Soc.*, 70 (1948) 2135.

Huber, J. F. K., in E. Kováts (Editor), *Column Chromatography, Lausanne 1969.* Sauerländer AG, Aarau, 1970, p. 24.

Ismailov, N. A. and Shraiber, M. S., *Farmacia*, 3 (1938) 1; *C.A.*, 34 (1940) 855.

James, A. T. and Martin, A. J. P., *Biochem. J.*, 50 (1952) 679.

Kirchner, I. G., Miller, J. M. and Keller, G. I., *Anal. Chem.*, 23 (1951) 420.

Kuhn, R., Winterstein, A. and Lederer, E., *Hoppe-Seyler's Z. Physiol. Chem.*, 197 (1931) 141.

LeRosen, A. L., *J. Amer. Chem. Soc.*, 67 (1945) 1683.

Martin, A. J. P. and Synge, R. L. M., *Biochem. J.*, 35 (1941) 1358.

Mayer, S. W. and Tompkins, E. R., *J. Amer. Chem. Soc.*, 69 (1947) 2866.

Phillips, C. S. G., *Discuss. Faraday Soc.*, 7 (1949) 241.

Porath, J. and Flodin, P., *Nature (London)*, 183 (1959) 1657.

Runge, F. F., *Farbenchemie*, Band III, Mittler u. Sohn, Berlin, 1850.

Samuelson, O., *Ion Exchange in Analytical Chemistry*, Wiley, New York, 1963.

Tswett, M. S., *Proc. Warsaw Soc. Nat. Sci., Biol. Sect.*, 14, No. 6 (1903).

Tswett, M. S., *Ber. Deut. Bot. Ges.*, 24 (1906) 234, 316 and 384.

*Chapter 2*

# Basic processes in chromatography

J. NOVÁK, J. JANÁK and S. WIČAR

## CONTENTS



## FLOW OF MOBILE PHASE THROUGH A PACKED COLUMN

There are basically two reasons for studying the hydrodynamics of the mobile phase in chromatographic systems. The first is the fact that the direct convective solute mass transport along the stream-lines between separate regions of the column is one of the most important solute transport mechanisms in the column. The second reason is of rather a technical nature and concerns the estimation of the pressure necessary to achieve the required flow-rate of the mobile phase through the column.

The state of a moving liquid within a column is determined by a group of six quantities, namely, by three components of the velocity vector $\vec{u}(\vec{r}, t)$, *i.e.*, by the values of $u_1(\vec{r}, t)$, $u_2(\vec{r}, t)$ and $u_3(\vec{r}, t)$, and by the density, $\rho(\vec{r}, t)$, pressure $p(\vec{r}, t)$ and temperature $T(\vec{r}, t)$, where $\vec{r}$ represents the positional vector and $t$ the time, respectively. When restricting the problem to the study of steady isothermal streaming of an incompressible liquid through a column, the number of determining quantities will be reduced to four: three components of the velocity vector, $u_1(\vec{r})$, $u_2(\vec{r})$ and $u_3(\vec{r})$, and the pressure, $p(\vec{r})$.

The above four independent variables are connected by four relationships:

$$\rho \,\mathbf{div}\, \vec{u} = 0 \tag{2.1}$$

$$\rho\,(\vec{u}\,\mathbf{grad})\,\vec{u} = \vec{F} - \mathbf{grad}\,p + \mu\,\mathbf{div\,grad}\,\vec{u} \tag{2.2}$$

The scalar equation 2.1 expresses the fact that in any region within the column, the mass entering that region equals the mass leaving it. Hence, eqn. 2.1 is a hydrodynamic expression of the mass conservation principle. The vectorial equation 2.2, *i.e.*, the Navier–Stokes equation, expresses the equality between the differentiation with respect to time of the

*References p.23*

momentum of an arbitrarily chosen volume of the streaming liquid (left-hand side) and the resultant of the exterior forces acting on the above element (right-hand side). The force $\vec{F}$ is the resultant of volume forces; in chromatography, these forces are represented by gravity, *i.e.* $\vec{F} = \rho \vec{g}$, where $g$ is the gravitational constant. While in classical liquid column chromatography gravity is usually the main driving force for the flow of the mobile phase, in modern high-speed liquid chromatography the role of gravity is negligible in comparison with the pressure forces represented by the second term of the right-hand side of eqn. 2.2. The last term of the equation stands for the resultant of the shear forces. Eqn. 2.2 is equivalent to three scalar equations corresponding to the individual components of the velocity vector, $u_1$, $u_2$ and $u_3$.

Eqns. 2.1 and 2.2 represent a non-linear system of partial equations of the second order, expressing merely the general regularities of streaming of liquids. In application to a specific problem, it is necessary to solve the respective boundary problem, *i.e.*, to seek such solutions of the system that fit the given boundary conditions. These conditions also involve the geometry of the bed within which the movement of the liquid takes place. Such solutions are known only for some instances of geometrically simple and laminar flow (Schlichting); Hagen–Poisseuille streaming in tubes of generally non-circular cross-sections serves as an example. Thus, the Poisseuille equation holds for steady isothermal streaming of an incompressible liquid through a tube of a circular cross-section:

$$u_z(r) = [(a^2 - r^2)/4]\, \Delta p/\mu L \qquad (2.3)$$

provided that the volume forces are negligible. In eqn. 2.3, $u_z$ is the forward velocity of the streaming liquid along the tube axis at a distance $r$ from the axis, $a$ is the radius of the tube, $\Delta p = p_1 - p_2$ is the difference in the pressures at the inlet and the outlet of the tube, and $\mu$ is the dynamic viscosity of the liquid.

The need to describe analytically the complex structure of a packed column is one of the obstacles to applying eqns. 2.1 and 2.2 directly to the movement of the mobile phase in the chromatographic column. Therefore, the necessary relationships are searched for experimentally and the results are generalized by virtue of the methods of similarity theory. Eqn. 2.2 can be rewritten in a component form while omitting the gravitational term:

$$u_k \cdot \frac{\partial u_i}{\partial x_k} = -\frac{1}{\rho} \cdot \frac{\partial p}{\partial x_i} + \frac{\mu}{\rho} \cdot \frac{\partial^2 u_i}{\partial x_k^2} \qquad (2.4)$$

where $i, k = 1, 2, 3$.

Upon introducing the substitutions $X_i = x_i/l$, $U_i = u_i/w$ and $P = p/\rho w^2$, where $l$ and $w$ are undefined characteristic quantities, eqn. 2.4 will become a system containing the Reynolds number, $Re = wl\rho/\mu$, as a single parameter. This parameter is a dimensionless constant comprising the original constants of the system and the characteristic quantities. Hence, it follows from the similarity theory (Ehrenfest-Afanassjewa, Konakov) that the functions $U_i = U_i(P, X_i)$ for a given value of $Re$ will be identical for geometrically similar systems, provided that the boundary conditions are similar.

The Reynolds number, which characterizes the ratio of the effects of shear and inertial forces in the streaming liquid, is alone a determining criterion of isothermal steady streaming of an incompressible liquid. Under the predominant influence of shear forces, when $Re < Re_0$, the streaming is of laminar (viscous) character and the stream-lines follow more or less the shapes of the bodies past which the liquid is streaming. At higher $Re$ values ($Re > Re_1$), the effect of inertial forces is predominant and one can speak about fully developed turbulent streaming, at which the velocity of streaming acquires locally non-stationary values and only the mean values of the local velocities remain constant. The interval within $Re_0$ and $Re_1$ corresponds to a transitory region in which both types of flow occur. When choosing the mean velocity $u$, determined from the volume flow-rate, for the characteristic velocity $w$, then the following general function for similar systems holds:

$$P = P(Re, \xi_1, \xi_2, \ldots) \qquad (2.5)$$

where $\xi_1$, $\xi_2$, etc., are geometrical simplexes derived from the boundary conditions. However, the actual shape of the function $P$ has to be found experimentally.

The first assumption for the applicability of the similarity theory to the hydrodynamic properties of packed columns is the geometrical similarity of the arrangement of the particles that constitute the packing. The simplest case of a packed bed formed by regularly arranged spheres of equal diameters can be considered first. The spheres can be arranged regularly in a plane to form either a rectangular lattice with the centres of the spheres representing the corners of a square with sides of length equal to the sphere diameter, $d$, or a triangular lattice in which the centres of the spheres form the apexes of an equilateral triangle with sides of length $d$. Spatial structures can now be formed from the above planar arrangements by gradually depositing the layers upon one another and shifting them appropriately. However, only those shifts will be permissible which enable mechanically stable structures to be formed. Geometrical considerations now lead to two limiting spatial structures that differ from each other by the degree of filling the space of the bed, *i.e.*, by the porosity, $\epsilon$, defined by the relationship

$$\epsilon = 1 - (v/V) \qquad (2.6)$$

where $v/V$ is the volume of the fixed phase per unit volume of the bed.

The loosest possible arrangement is represented by square nets deposited upon one another without any shift. In such a bed, each sphere touches six neighbouring spheres, and the entire bed is interwoven with a continuous network of channels. The porosity of such a structure is $\epsilon = 1 - (\pi/6) \approx 0.476$. The other extreme is the arrangement derived from the triangular net; the individual layers are positioned in such a way that the apexes of the triangles of a layer coincide with the centres of the triangles of the neighbouring layers. In this way, the most stable structure is formed, in which each sphere of the bed is in contact with eleven other spheres, and the porosity of such a structure has a value of about 0.26. It should be emphasized that the porosity of a bed composed of spheres of equal dimensions is independent of the size of the spheres, but depends on their arrangement.

The porosity of randomly packed beds, composed of spheres of equal diameters, varies within the range 0.35–0.40 and is therefore substantially higher than that of the expected most stable structure. A possible explanation of this discrepancy may be the breakdown of the bed structure into smaller agglomerates with the minimum porosity and the formation of some bridges between the agglomerates (Giddings, 1965a). In this

*References p.23*

case, more than a quarter of the free volume of the bed would be in the spaces between the agglomerates. Another possibility is a combination of agglomerates of unequal structures and smaller porosities connected with bridges.

The smallest volume unit for which it is meaningful to speak about the porosity defined by eqn. 2.6 is an elementary cube with edges of length $d$. The porosity of this element will be the same as that of the bed of any unconfined size and of the same structure. However, real beds are always bound by the walls of the container, and the structure of the bed has to conform, at least at its outside, to the shape of the wall restricting the bed. Provided that the wall is a planar or cylindrical surface, as is normal in practice, the accommodation of the structure to the container wall is always associated with a sharp increase in porosity in the immediate proximity of the wall. At the surface proper of the wall, the formally understood porosity is unity, as there are only point contacts between the spheres and the wall; at some distance from the wall ($\delta > 0.5$) the porosity remains approximately constant. The course of porosity in the proximity of a wall was described by a semi-empirical equation by Sonntag:

$$\epsilon = 1 - A\delta(1 - \delta) \tag{2.7}$$

which is applicable for the region where $\delta < 0.5$, where $\delta$ is the distance from the wall expressed in multiples of the sphere diameter and $A$ is an empirical constant with a value of 3.2 for loosely packed beds. A similar course of the bed porosity in the proximity of a wall was also found by Schwartz and Smith, who measured radial velocity profiles in a packed column. Hence, the porosity as determined experimentally will always be higher than that of the interior structure of the bed; the following equation holds approximately for cylindrical beds (Sonntag):

$$\epsilon_m = \epsilon_0 + 0.263(1 - \epsilon_0)\,d/R \tag{2.8}$$

where $\epsilon_m$ is the average porosity, $\epsilon_0$ is the porosity of the interior structure and $d/R$ is the ratio of the diameter of the sphere and the radius of the column.

Substantially more complex are beds produced by loosely pouring spheres of different diameters or even particles of irregular shapes yet with the characteristic particle dimensions varying within narrow limits. Although the equality of the porosities of two beds obviously does not necessarily imply that the structures of the beds are similar, the porosity is the only useful representation of the structure of the bed.

It is now useful to consider again a column the packing of which is formed by regularly arranged spheres of the diameter $d$, the length of the bed and the porosity being $L$ and $\epsilon$ respectively. Further, assume that the ratio $d/R$ (*cf.* eqn. 2.8) is sufficiently small and that it is possible to write $\epsilon_m = \epsilon_0 = \epsilon$. The geometry of the bed is characterized by the diameter of the spheres and by the porosity, and the length of the bed is irrelevant to the above problem. Let us now insert in eqn. 2.5 the diameter of the sphere, $d$, as the characteristic dimension $l$ of the bed. Hence, the geometry of the bed will be described by two dimensionless ratios, $\xi_1 = d/L$ and $\xi_2 = \epsilon$, and eqn. 2.5 will become:

$$\frac{\Delta p}{\rho u^2} = f\left(\frac{du\rho}{\mu}, \frac{d}{L}, \epsilon\right) \tag{2.9}$$

It can be seen that $p$ in the variable $P$ has been replaced by the pressure difference, $\Delta p$. A form of the function $f$ describing the regions of both low and high values of $Re$ was given by Ergun:

$$\frac{\Delta p}{\rho u^2} \cdot \frac{d}{L} \cdot \frac{\epsilon^3}{1-\epsilon} = 150 \cdot \frac{1-\epsilon}{Re} + 1.75 \tag{2.10}$$

For $Re/(1-\epsilon) < 10$, eqn. 2.10 becomes the well known Kozeny–Carman equation:

$$\frac{\Delta p}{L} = \text{constant} \cdot \frac{(1-\epsilon)^2}{\epsilon^3} \cdot \frac{\mu u}{d^2} \tag{2.11}$$

Eqn. 2.11 was derived from the concept that the packed bed is a band of parallel capillaries; the proportionality

$$\frac{\Delta p}{L} \approx u \tag{2.12}$$

for small $Re$ values has been known for a long time as the Darcy law.

The applicability of relationship 2.9 was postulated only for beds formed by a random arrangement of spheres of equal diameters. The concept of the mean hydraulic diameter, $d_h$, of the bed particles (Ergun and Orning):

$$d_h = \frac{6\varphi L\,(1-\epsilon)}{S} \tag{2.13}$$

where $\varphi$ is the bed cross-section normal to the direction of flow of the liquid, $L$ is the length of the bed and $S$ is the total geometrical surface area of the particles constituting the bed, makes it possible to extend the applicability of eqns. 2.9–2.12 to beds composed of spheres of unlike diameters as well as particles of irregular shapes. However, the determination of the geometrical surface area of the particles with their own internal porosity is a separate problem.

## DIFFUSION OF SOLUTE WITHIN THE PHASES

### General rules of diffusion

Diffusion is a direct consequence of the tendency of any system to reach a state of minimum potential energy and maximum randomness, as dictated by the second law of thermodynamics. In this respect, diffusion is one of the most universal phenomena of nature, characterized by spontaneous dilution and mixing of matter. Regarding the nature of chromatography, diffusion is a basic constituent of the chromatographic process. The diffusional transport is usually considered to be the result of a concentration gradient. As discussed by Giddings (1965b), this interpretation is essentially incorrect, despite being in accordance with observation. In fact, the only driving force in diffusion is the tendency of the system to attain a minimum free energy, *i.e.*, a state of equilibrium. Diffusion is a random process in which the molecules present in regions of different concentrations undergo their diffusional transport independently of one another. As the number of molecules diffusing out of a given region is proportional to the number of

*References p.23*

molecules present in the region, there will always be a net solute flux towards the more dilute region, which gradually levels the concentration difference.

One-directional isothermal diffusion under steady-state conditions of a substance (solute) through a stationary layer of another substance is expressed by Fick's first law:

$$J = -D \cdot \frac{\partial c}{\partial z} \tag{2.14}$$

where $J$ is the number of moles of solute passing in unit time through a unit area normal to the concentration gradient, $D$ is the diffusion coefficient and $\partial c/\partial z$ is the concentration gradient. At low solute concentrations, the diffusion coefficient is virtually constant.

A case in which the solute concentration can change with time in any region of the solution is described by Fick's second law. The solute mass balance for a region of unit cross-section and of length d$z$ leads to

$$\frac{\partial c}{\partial t} = D \cdot \frac{\partial^2 c}{\partial z^2} \tag{2.15}$$

where $\partial c/\partial t$ is the increase in solute concentration within the above region in unit time. It will be shown in Chapter 3 that eqn. 2.15 is of basic importance in the general description of the chromatographic process. A solution of this equation for an extremely thin initial profile introduced at $z = 0$ (Crank) is

$$c = \frac{n^\circ}{(4\pi Dt)^{1/2}} \cdot \exp\,(-z^2/4Dt) \tag{2.16}$$

where $n^\circ$ is the number of moles of solute introduced per unit cross-section of the concentration profile (column).

Eqns. 2.15 and 2.16 describe diffusion in stationary media. In chromatography, however, diffusion of the solute in flowing fluids usually occurs and, when employing fixed coordinates, this case can be described by

$$\frac{\partial c}{\partial t} = D \cdot \frac{\partial^2 c}{\partial z^2} - u \cdot \frac{\partial c}{\partial z} \tag{2.17}$$

where the term $u(\partial c/\partial z)$ accounts for the convective transport (along the coordinate $z$) at a velocity $u$. Assuming that the same initial and boundary conditions as those mentioned in the solution of eqn. 2.15 apply, and provided that $Dt \ll z^2$, the solute concentration profile can be described in this case by

$$c = \frac{n^\circ}{(4\pi Dt)^{1/2}} \exp\,[-(z - ut)^2/4Dt] \tag{2.18}$$

Eqns. 2.16 and 2.18 represent Gaussian profiles, which are usually characterized by means of the standard deviation, $\sigma$, through the use of Einstein's equation:

$$\sigma^2 = 2Dt \tag{2.19}$$

A more detailed discussion of the role of convective diffusion in chromatography is presented in Chapter 3.

### Diffusion in liquids

The rates of diffusion in liquids are lower by a factor of $10^4$–$10^5$ than in gases. The great difference between liquid and gas diffusivity is very important in many respects when comparing gas and liquid chromatography, and it is necessary to be very cautious if the results obtained in GC are to be applied to the problems of LC. Typical diffusion coefficients in liquids are of the order $10^{-5}$ cm$^2$/sec. One of the main causes of the difference between gas and liquid diffusivity is the difference in the lengths of the random steps travelled by the molecules in their thermal movement. While with gases the length of these steps is approximately equal to the mean free path of the molecules, with liquids this length is similar to the molecular diameter. According to the random walk model (see Chapter 3), the rate of diffusional transport should be proportional to the second power of the step length. Further, as the molecules of a liquid are held together relatively strongly by intermolecular forces, a molecule must accumulate an appreciable amount of thermal energy in order to undergo a displacement step. Hence, a considerable activation energy is associated with diffusion in liquids, thus making the diffusion coefficient strongly dependent on temperature (Frenkel). On the other hand, owing to the relative incompressibility of liquids, diffusivity in liquids is only slightly dependent on pressure. Of numerous equations that have been suggested for calculating diffusion coefficients in liquids (Gambill), only that due to Wilke and Chang will be quoted here. This equation is:

$$D_i = 7.4 \cdot 10^{-8}(F_s M_s)^{1/2}\ T/\mu_s V_i^{0.6} \tag{2.20}$$

where $M_s$ is the molecular weight of the solvent, $T$ is the temperature, $\mu_s$ is the dynamic viscosity coefficient (a function of temperature) of the solvent, $V_i$ is the molar volume of solute and $F_s$ is an association factor ranging from about unity for non-polar compounds to 2.6 for water. This equation proved successful when applied to liquids with molecules of small and medium size. The equation clearly shows the dependence of $D_i$ on the nature of both the solvent and the solute.

It follows from the above discussion that there are significant differences in the diffusion coefficients as well as in their dependences on temperature, pressure and type of substance for liquids and gases. These differences may lead to very different situations when comparing the effects of the operating conditions on the performance of gas and liquid chromatographic systems.

## EQUILIBRATION OF SOLUTE BETWEEN THE PHASES

### Equilibrium in binary two-phase liquid systems

It follows from the definition of chromatography that a chromatographic system of any type must always involve two immiscible phases. The composition of the phases and the physical arrangement of the system are chosen with respect to the given separation problem. Immiscibility is a problem mainly in liquid–liquid systems and will therefore be discussed with reference to the latter only. The phenomenon of immiscibility is closely related to the concept of non-ideal solutions with positive deviations from Raoult's law.

*References p.23*

Increasing positive difference in the cohesive forces between like and unlike molecules will raise the tendency of both the molecules to escape from the mixture. When the above difference is sufficiently large, the liquids will show incomplete miscibility within some temperature range. Azeotropic solutions can be considered as intermediates between systems of complete and partial miscibility.

A thermodynamic condition for two liquids to show incomplete miscibility is that the system composed of these liquids should have a minimum free energy when it is split into two phases. The solubility gap usually (but not always) narrows and eventually decreases upon increasing the temperature. The above situation is shown schematically in Fig. 2.1, which represents a temperature–composition diagram of a binary system composed of substances $M$ and $S$. The symbol $c_S$ represents the molar concentration (number of moles per unit volume) of component $S$, and the points on the line $\overline{MmysS}$ correspond, respectively, to pure component $M$, the $M$-rich phase (just saturated with $S$), a two-phase system, the $S$-rich phase (just saturated with $M$) and pure component $S$. Phases $m$ and $s$ are called conjugated solutions. The curve, called binodal, separates the homogeneous (outer) and the heterogeneous (inner) areas. The point $\alpha$ on the binodal curve is a critical mixing point corresponding to the temperature at which the entire system becomes homogeneous. Lines connecting the points of the composition of the coexisting phases are called tie lines or conjugation lines.

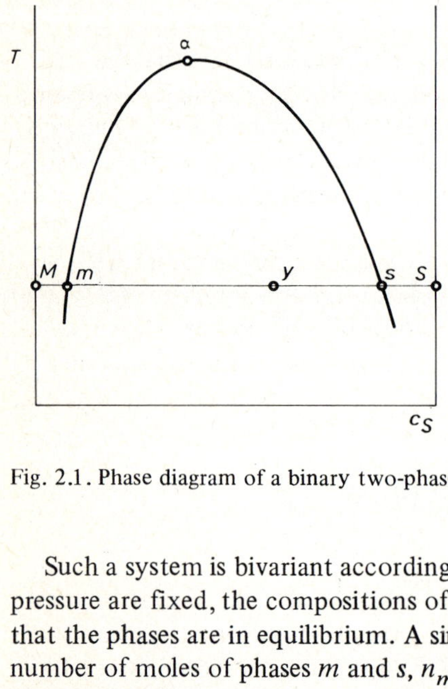


Fig. 2.1. Phase diagram of a binary two-phase liquid system.

Such a system is bivariant according to the Gibbs phase rule; if the temperature and pressure are fixed, the compositions of both conjugated phases are invariable provided that the phases are in equilibrium. A simple mass balance shows that the ratio of the number of moles of phases $m$ and $s$, $n_m/n_s$, corresponding to point $y$ of the heterogeneous area, is given by

$$n_m/n_s = \overline{ys}/\overline{my} \tag{2.21}$$

where $\overline{ys}$ and $\overline{my}$ denote the distances between the respective points in the diagram.

**Equilibrium in ternary two-phase liquid systems**

While the binary systems discussed above can be compared with any liquid–liquid chromatographic system without the solute component, a three-component two-phase liquid system is the simplest representation of a solute zone in a liquid–liquid chromatographic system. Considering a system of three components of which only one pair displays incomplete miscibility, this pair represents the stationary and the mobile phase (components $S$ and $M$) in the corresponding chromatographic system while the third component, which is completely miscible with the other two, represents the solute ($i$).

If it is imagined that component $i$ is gradually added, at constant temperature and pressure, to the initially binary two-phase system of components $S$ and $M$, the solute will distribute itself in certain proportions between the conjugated phases, thus changing both the composition and relative amount of the latter. The change in composition arises not only because the binary conjugated solutions become ternary solutions with an increase in the amount of the third component, but the presence of the third component will alter the equilibrium in such a way that the proportions of components $S$ and $M$ will also gradually be changed to different extents in both phases. This is in accordance with Gibb's phase rule; as a ternary two-phase system is essentially trivariant, the concentration of only one component can be varied independently at a constant temperature and pressure. The effect of the third component in a ternary system such as that discussed above is similar to the effect of temperature in binary two-phase systems. As more solute is added to the ternary system, the compositions of the coexisting phases move closer to each other and, at a certain critical composition (point $\alpha$), the system becomes homogeneous.

Ternary systems can be represented most conveniently by means of triangular diagrams, and such a diagram is shown schematically in Fig. 2.2. The corners of the triangle represent pure components $M$, $S$ and $i$. The concentration of $i$ is plotted along the sides $\overline{iM}$ and $\overline{iS}$ and the concentrations of the other two components are plotted similarly (*cf.*, the dotted lines to point $\alpha$). Hence, on any line drawn parallel to a side bound by two components, the concentration of the third component is constant. In Fig. 2.2, the side $\overline{MS}$ corresponds to a binary system such as that shown in Fig. 2.1. The

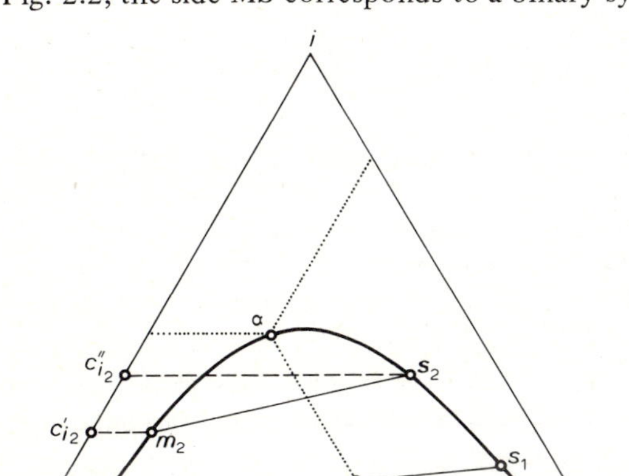


Fig. 2.2. Phase diagram of a ternary two-phase liquid system.

*References p.23*

binodal curve $\overline{m\alpha s}$ binds the heterogeneous area, while the remainder of the diagram (above the curve) corresponds to a homogeneous system. The tie lines $\overline{ms}$, $\overline{m_1 s_1}$ and $\overline{m_2 s_2}$ connect the points representing the compositions of the respective conjugated solutions. As with binary systems, any point on a given tie line represents a system composed of two phases of constant compositions. Provided that the concentrations are expressed in numbers of moles per unit volume, the ratio of the number of moles of the coexisting phases corresponding to point $y_1$ is given by

$$n_{m_1}/n_{s_1} = \overline{y_1 s_1}/\overline{m_1 y_1} \tag{2.22}$$

It is apparent from the diagram that each pair of conjugated points on the binodal curve actually represents data for expressing the distribution constant of component $i$. The distribution constant is usually defined as the ratio of the solute equilibrium concentrations in the sorbent and mobile phases, which can be directly read off on the side $\overline{Mi}$. For example, the distribution constant corresponding to the conjugated pair $m_2$ and $s_2$ is:

$$K = c''_{i_2}/c'_{i_2} \tag{2.23}$$

In some instances, the tie lines may be parallel to the $\overline{MS}$ side or can even show an opposite slope, so that the distribution constant can be either zero or less than unity, respectively.

The diagram indicates that the value of $K$ can be appreciably dependent on the overall composition of the system, *i.e.*, on the solute concentration. However, at very low solute concentrations in the system, the dependence between the concentrations of solute in the coexisting phases can be approximated as linear, thus resulting in a constant distribution constant, which is the situation of interest in chromatography. The situation discussed above concerned isothermal and isobaric conditions. While condensed systems depend only slightly on pressure, changes in temperature can lead to appreciable alterations in the equilibrium concentrations and, consequently, in the distribution constants.

## Equilibrium in liquid–solid systems

Owing to relatively strong intermolecular interactions in liquids, a molecule in the proximity of the liquid surface is pulled towards the interior owing to surface tension. As an increase in the surface area is always associated with the transfer of molecules from the bulk phase to the surface, energy must be used in order to accomplish this increase. The energy necessary to increase the surface of a given amount of a liquid by 1 $cm^2$ is equal to the surface free energy of unit surface area, which is identical with the surface tension of the liquid.

The surfaces of solid substances also display a surface free energy, *i.e.*, surface tension, but solids cannot reduce their surface area as the particles of solids do not undergo translational thermal movement. However, the free energy of the surfaces of solids is a most important factor in the adsorption of gases and liquids on solid adsorbents. The process of adsorption can be understood by discussing separately the following three systems: binary two-phase liquid–solid system; binary single-phase liquid system; and binary liquid mixture–solid adsorbent system.

*Binary two-phase liquid–solid system*

A system is considered in which a liquid ($m$) is in intimate contact with an adsorbent ($s$). Both substances display certain intermolecular cohesion forces and thus certain surface tensions. Hence, at the interface, cohesion forces occur from both sides so that the interfacial tension is lower than the surface tensions. In this event, the interfacial tension, $\sigma_{sm}$, can be expressed as $\sigma_{sm} = \sigma_s - \sigma_m$, where $\sigma_s$ and $\sigma_m$ are the surface tensions of components $s$ and $m$, respectively. In contrast to gas–solid systems, the formation of an adsorption layer of increased density is impossible with liquids, as they are incompressible.

*Binary single-phase liquid system*

This system is simply a mixture of two miscible liquids. Strictly, such a system should be considered as a ternary two-phase system if the liquid is assumed to be in contact with air. As mentioned above, a pure liquid can attain a minimum free surface energy by reducing its surface area to a minimum. A mixture of two or more liquids has yet another possibility of reaching a state of minimum surface free energy. If the components have different surface free energies, the system will try to remove from the surface layer those molecules the presence of which requires a higher surface energy, while accumulating in the layer the molecules of the other components. Thus, there will be a higher concentration of the component of lower surface tension in the surface layer than in the bulk phase at equilibrium. Such a segregation can, of course, occur only in systems that involve a surface-active compound.

*Binary liquid mixture–solid adsorbent system*

In this system, a difference in the solute concentration in the bulk liquid and at the liquid–solid interface can be achieved by virtue of both of the mechanisms described in the sections *Binary two-phase liquid–solid system* and *Binary single-phase liquid system.* It can be inferred from the previous discussion of the problem of adsorption that the presence of a solid adsorbent in a binary liquid mixture can lead to three different situations: in the first two, the chemical nature of the adsorbent can be such that either the solute or the solvent molecules will be preferred in interactions with the solid surface; the third possibility is that there is no preference for either compound. The larger the extent of the interactions, the greater is the effect of decreasing the interfacial tension by the presence of the respective component in the interfacial layer and, consequently, the greater is the enrichment of the layer by this component. If the adsorbent does not discriminate between the solute and solvent in the above respect, some enrichment of the interfacial layer can still occur as a result of the mechanism discussed in the section Binary single-phase liquid system, provided the component to be adsorbed is a surface-active one. In any event, the adsorbent provides a large interfacial area, thus making the total amount of the component accumulated in the adsorbed layer appreciably large.

The above concepts are expressed quantitatively by the Gibbs adsorption isotherm,

*References p.23*

which can be written for dilute solutions in the form

$$\Gamma = -\frac{c'}{RT}\left(\frac{\partial\sigma}{\partial c'}\right)_{T,p} \tag{2.24}$$

where $\Gamma$ is the so-called concentration excess of the solute component in the interfacial region, $c'$ is the molar concentration of this component in the bulk liquid, $R$ is the gas constant, and $\sigma$ is the interfacial tension. The dependence of the interfacial tension on $c'$ can be described by the empirical equation (Szyszkowski)

$$\sigma_0 - \sigma = a\ln(1 + bc') \tag{2.25}$$

where $\sigma_0$ is the interfacial tension in the adsorbent–pure solvent binary system and $a$ and $b$ are constants. The differentiation of $\sigma$ with respect to $c'$ and substitution into eqn. 2.24 gives

$$\Gamma = \frac{a}{RT}\cdot\frac{bc'}{1 + bc'} = a^* \cdot \frac{bc'}{1 + bc'} \tag{2.26}$$

Eqn. 2.26 actually represents the Langmuir adsorption isotherm, where $a/RT = a^*$ is the maximum attainable concentration of solute in the interfacial layer ($c' \gg 1/b$).

The quantity $\Gamma$ can be expressed from the following solute mass balance. Let $n$ be the number of moles of solute component in the initial solute–solvent mixture of volume $V$. After adding an adsorbent to the solution, the solute will distribute itself between the bulk and interfacial phases. At equilibrium, the situation can be described by

$$n = Vc = c'(V - V_a) + V_a c'' \tag{2.27}$$

where $c$ is the solute concentration in the initial solution, $c'$ and $c''$ are the solute concentrations in the bulk phase and in the interfacial adsorption layer of the liquid–solid system, respectively, and $V_a$ is the volume of the interfacial layer. $\Gamma$ is defined as the difference between the number of moles of solute in the initial solution and in the bulk phase after adding a weight amount $w_s$ of the adsorbent, related to 1 g of the adsorbent. Thus, dividing both sides of eqn. 2.27 by $w_s$ and rearranging gives

$$\Gamma = (V/w_s)(c - c') = (V_a/w_s)(c'' - c') \tag{2.28}$$

As $V_a c''/w_s$ can be expressed as $n''/w_s$, *i.e.*, the number of moles of the adsorbed solute per unit weight of the adsorbent, the quantity $\Gamma$ can be expressed by

$$\Gamma = (n''/w_s) - (V_a/w_s)c' \tag{2.29}$$

On combining eqns. 2.29 and 2.26, we obtain

$$\frac{n''}{w_s} = a^* \cdot \frac{bc'}{1 + bc'} + \frac{V_a}{w_s}\cdot c' \tag{2.30}$$

The ratio $V_a/w_s$ in eqn. 2.30 expresses the volume of the interfacial adsorption layer per gram of the adsorbent. This volume is small, and if, in addition, $c'$ is also small, the term can be neglected. Hence, at very low solute concentrations ($c' \ll 1/b$), eqn. 2.30 can be reduced to

$$n''/w_s = a^* bc' \tag{2.31}$$

Eqn. 2.31 is very suitable for expressing the distribution constant. As $c' = n'/V_m$, where $n'$ is the number of moles of solute in a volume $V_m$ of the solvent (mobile phase), we can write

$$K_a = (n''/w_s)/(n'/V_m) = a^* b \tag{2.32}$$

where the subscript $a$ in $K_a$ differentiates between the adsorption distribution constant and that defined for liquid–liquid systems ($K$).

## REFERENCES

Carman, P. C., *Trans. Inst. Chem. Eng.*, 15 (1937) 150.
Crank, J., *The Mathematics of Diffusion*, Oxford Univ. Press, New York, 1956.
Darcy, H., *Les Fontaines Publiques de la Ville de Dijon*, Paris, 1856.
Ehrenfest-Afanassjewa, T., *Math. Ann.*, 77 (1916) 259.
Ergun, S., *Chem. Eng. Progr.*, 48 (1952) 89.
Ergun, S. and Orning, A. A., *Ind. Eng. Chem.*, 41 (1949) 179.
Frenkel, J., *Kinetic Theory of Liquid*, Clarendon Press, Oxford, 1946, Ch. IV.
Gambill, W. R., *Chem. Eng. (London)*, June 30 (1958) 113.
Giddings, J. C., *Dynamics of Chromatography*, Marcel Dekker, New York, 1965a, p.200.
Giddings, J. C., *Dynamics of Chromatography*, Marcel Dekker, New York, 1965b, p.228.
Konakov, P. K., *Izv. Akad. Nauk SSSR, Otd. Tekh. Nauk*, (1949) 240.
Kozeny, J., *Sitzungsber. Akad. Wiss. Wien, Math.-Naturwiss. Kl., Abt. 2B*, 136 (1927) 271.
Navier, M., *Mem. Acad. Sci. Inst. Fr.*, 6 (1827) 389.
Schlichting, H., *Boundary Layer Theory*, McGraw-Hill, New York, 1955, Ch. V.
Schwartz, C. E. and Smith, J. M., *Ind. Eng. Chem.*, 45 (1953) 1209.
Sonntag, G., *Chem.-Ing.-Tech.*, 32 (1960) 317.
Stokes, G. G., *Trans. Cambridge Phil. Soc.*, 8 (1845) 287.
Szyszkowski, B., *Z. Phys. Chem. (Leipzig)*, 64 (1908) 385.
Wilke, C. R. and Chang, P., *AIChE J.*, 1 (1955) 746.

*Chapter 3*

# General description of the chromatographic process

J. NOVÁK, J. JANÁK and S. WIČAR

## CONTENTS



## SOLUTE MASS BALANCE IN THE CHROMATOGRAPHIC SYSTEM

The transport of the solute mass in a chromatographic column takes place by two basic mechanisms – molecular diffusion and convection. The intensity, $\vec{J}$, of both of these transport phenomena can be described by means of the vectors of diffusional and convective flow (subscripts $D_j$ and $C_j$, respectively),

$$\vec{J}_{Dj}(\vec{r}, t) = -D_j \,\mathbf{grad}\, c_j(\vec{r}, t) \tag{3.1}$$

$$\vec{J}_{C_j}(\vec{r}, t) = \vec{u}(\vec{r})\, c_j(\vec{r}, t) \tag{3.2}$$

The magnitudes of the above vectors indicate the velocity of the mass flux through a unit area and the direction of the vectors is identical with that of the mass flux at any point within the column. In eqns. 3.1 and 3.2, $D_j$ is the diffusion coefficient of the $j$th component of the given phase, $\vec{u}$ is the forward velocity of the phase at a point of the position vector $\vec{r}$, and $c_j$ is the concentration of the $j$th component at the point of the position vector $\vec{r}$ and at a time $t$ in the given phase.

Let us choose a volume element $V$ enclosed by a continuous surface $S$ in the interparticle space of the column through which the mobile phase flows. Further, let $(J'_{D_j})_n$ and $(J'_{C_j})_n$

*References p.43*

be the components of the vectors, as defined by eqns. 3.1 and 3.2, normal to a surface element d$S$ (see Fig. 3.1). The integral

$$\int_S [(J'_{D_j})_n + (J'_{C_j})_n] \, \mathrm{d}S \tag{3.3}$$

represents the difference between the rates of the supply and outflow of the mass of the $j$th component of the mobile phase within the element enclosed by surface $S$. It follows from the mass conservation principle that the value of the integral 3.3 must be equal to the rate of accumulation of mass of the component within the element, provided that no chemical reactions take place within the element. Hence,

$$\frac{\mathrm{d}}{\mathrm{d}t} \int_V c'_j \, \mathrm{d}V = \int_V \frac{\partial c'_j}{\partial t} \, \mathrm{d}V = \int_S [(J'_{D_j})_n + (J'_{C_j})_n] \, \mathrm{d}S \tag{3.4}$$

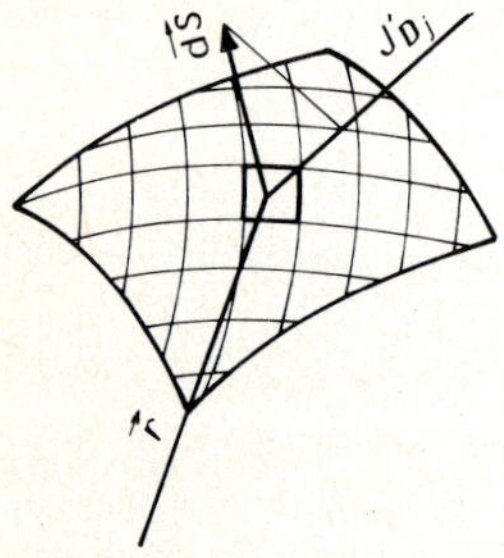


Fig. 3.1. Representation of the solute mass fluxes in the interparticle space of the column.

According to the Gauss theorem, the surface integral on the right-hand side of eqn. 3.4 can be expressed by a volume integral over the volume closed by the surface $S$:

$$\int_V \frac{\partial c'_j}{\partial t} \, \mathrm{d}V = \int_V \mathbf{div} \, (\vec{J}'_{D_j} + \vec{J}'_{C_j}) \, \mathrm{d}V \tag{3.5}$$

As the choice of the volume element in the mobile phase was arbitrary, eqn. 3.5 also holds true in the differential form:

$$\frac{\partial c'_j}{\partial t} = \mathbf{div} \, (\vec{J}'_{D_j} + \vec{J}'_{C_j}) \tag{3.6}$$

The same treatment for a volume element of the stationary phase leads to a mass balance equation for the $j$th component in the stationary phase:

$$\frac{\partial c''_j}{\partial t} = \mathbf{div} \, (\vec{J}''_{D_j} + \vec{J}''_{C_j}) \tag{3.7}$$

where $c''_j$ is the concentration of component $j$ in the stationary phase; the convective flow within the stationary phase can also be brought about by the diffusion fluxes.

The system of eqns. 3.6 and 3.7 holds for any component in both phases, but the solute component is of most concern in the context of the present discussion. At low solute concentrations in the column, the changes in the carrier liquid and stationary phase concentrations can be neglected and the convection in the stationary phase can be omitted. Hence, if the solute concentrations in the mobile and stationary phases are denoted simply by $c'$ and $c''$, respectively, the system of eqns. 3.6 and 3.7 becomes a system of two equations:

$$\frac{\partial c'}{\partial t} = \mathbf{div}\,(-D'\,\mathbf{grad}\,c' + \vec{u}c') \tag{3.6a}$$

$$\frac{\partial c''}{\partial t} = \mathbf{div}\,(-D''\,\mathbf{grad}\,c'') \tag{3.7a}$$

provided that the solute components in both phases behave independently of each other at low concentrations.

The complicated geometry of a packed column does not permit the boundary problem to be formulated analytically, thus hampering the solution of eqns. 3.6a and 3.7a for a packed column. The only means of obtaining approximate analytical solutions is the replacement of the local concentrations $c'(\vec{r}, t)$ and $c''(\vec{r}, t)$ and the local flow velocity $\vec{u}(\vec{r})$ by mean concentrations $< c'(z, t) >$ and $< c''(z, t) >$ and a constant vector $< \vec{u} >$, respectively. The averaging is carried out over the cross-sections normal to the column axis. This substitution will reduce significantly the problems that arise from the complicated pattern of the packed column, but it is necessary to employ a number of empirical coefficients and the results will be only approximate.

Let us imagine a column of a circular cross-section $\varphi$ and length $L$. The entire volume of the column is homogeneously filled either with a solid adsorbent or with a liquid stationary phase deposited on a solid support. The column void volume through which the mobile phase flows is $\epsilon\varphi L$ and the velocity of the latter within the void space is $u = v/\epsilon\varphi$, where $v$ is the volume flow-rate of the mobile phase. Hence, the vector $< \vec{u} >$ is assumed to be always in a direction parallel to the column axis and of a magnitude equal to $u$. The volume occupied within the column by the packing is $(1 - \epsilon)\varphi L$. If the packing is a solid support coated with a stationary liquid, the volume of the latter within the column is $q(1 - \epsilon)\varphi L$, where $q$ is the volume fraction of the stationary liquid of the entire column-packing volume. The discussion below will concern this type of packing.

In the column idealized as shown above, if we consider a volume element bound within two sections normal to the axis and situated at distances $z_1$ and $z_2$ from the column inlet, the volume $V$ of this element is $V = \varphi\,(z_2 - z_1)$. The concept of the mean concentrations and the mean velocity of flow eliminates the problem of radial concentration gradients and radial components of the vector $< \vec{u} >$, and hence there will be no mass flux through the cylindrical surface enclosing the element. With exact models also there is zero mass flux through the column wall owing to the impermeability of the wall, but in the case discussed here, the above fact is part of the boundary conditions.

When substituting mean values consistently into eqns. 3.1 and 3.2 in place of the local values and neglecting the longitudinal diffusion of solute in the stationary phase, as it proceeds on the very small cross-section $q(1 - \epsilon)\varphi$, the difference between the intake and outflow of solute mass in the mobile phase (see Fig. 3.2), as expressed for the above specified element, will acquire an especially simple form (*cf.*, integral 3.3 of the exact model):

$$\epsilon\varphi\left[-D'\left(\frac{\partial<c'(z_2,t)>}{\partial z}-\frac{\partial<c'(z_1,t)>}{\partial z}\right)+u\,(<c'(z_2,t)>-<c'(z_1,t)>)\right]=$$

$$=\epsilon\int_S(-D'\,\mathbf{grad}<c'>+u<c'>)\,\mathrm{d}S \tag{3.8}$$

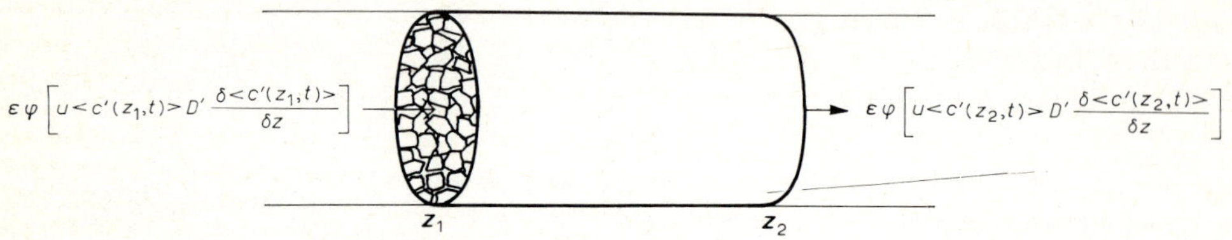


Fig. 3.2. Solute mass balance in an idealized column.

On the other hand, the difference is equal to the rate of increase of the solute mass in the element of the volume $\varphi(z_2 - z_1)$, *i.e.*, in both phases of the element, which is given by the integrals

$$\int_{V_M}\frac{\partial<c'>}{\partial t}\,\mathrm{d}V+\int_{V_S}\frac{\partial<c''>}{\partial t}\,\mathrm{d}V \tag{3.9}$$

where $V_M = \epsilon V$ and $V_S = q(1-\epsilon)V$. The formal difference, as compared with the exact equation 3.4, is already apparent here. Combining eqns. 3.8 and 3.9, employing the Gaussian theorem, and with regard to the arbitrary choice of the volume element, the mass balance for the column is also valid when expressed in a differential form:

$$\frac{\partial<c'>}{\partial t}+\frac{q(1-\epsilon)}{\epsilon}\cdot\frac{\partial<c''>}{\partial t}=D'\frac{\partial^2<c'>}{\partial z^2}-u\frac{\partial<c'>}{\partial z} \tag{3.10}$$

Eqn. 3.10 describes precisely the properties of the idealized column, but the behaviour of a real column of the same mean concentrations and mobile phase volume flow-rate, as with the idealized column, is characterized only approximately. For instance, there is no theoretically substantiated relationship between real and apparent diffusional flow in a packed column, $<D\,\mathbf{grad}\,c(\vec{r},t)>$ and $D\,\mathbf{grad}<c(z,t)>$, or between real and apparent convective flow, $<\vec{u}(\vec{r})\,c(\vec{r},t)>$ and $u<c(z,t)>$.

This fact indicates the necessity for correcting the results by introducing an effective diffusion coefficient in addition to corrections for eddy diffusion. Prager, for instance, derived, for the effective diffusion coefficient in a bed formed by randomly arranged beads, the inequality

$$\frac{D_{\text{eff}}}{D}<\frac{\ll g(\vec{r})\,[\mathbf{grad}\,c(\vec{r})]^2\gg}{\ll\mathbf{grad}\,c\gg^2}$$

where $g(\vec{r})$ is a random function with values of zero in the solid regions and unity in the void regions of the porous medium ($\ll g(\vec{r})\gg=\epsilon$), $D_{\text{eff}}$ is the effective diffusion coefficient and $\ll\gg$ designates the mean value averaged over the volume $V$. Another significant difference between eqns. 3.6a and 3.7a and eqn. 3.10 is that each of the former two contains only one unknown function whereas eqn. 3.10 contains two unknown functions, $<c'>$ and $<c''>$.

## Equations of the chromatographic zone

In order to be able to solve eqn. 3.10, it is necessary to obtain one further relationship between the above mean concentrations. We shall discuss two cases in the further treatment. The first case is described by

$$<c''> = f(<c'>) \tag{3.11}$$

*i.e.*, it will be assumed that the solute concentrations in both phases are equilibrated while the sorption isotherm need not necessarily be linear. The second case is characterized by

$$\frac{\partial <c''>}{\partial t} = k_s <c'> - k_d <c''> \tag{3.12}$$

which implies that there is a finite rate of equilibration in the column. The latter case is analytically described as if a reversible first-order chemical reaction with sorption and desorption rate constants $k_s$ and $k_d$ proceeded between the solute in the mobile and the stationary phase in the column. The dimension of both constants is $(\text{time})^{-1}$.

The boundary conditions characteristic of elution chromatography can be described by the equations

$$<c'(z,0)> = <c''(z,0)> = 0 \tag{3.13}$$

and

$$<c'(0,t)> = c_0\, \delta(t) L/u \tag{3.14}$$

Eqn. 3.13 reflects the fact that there is no solute in the column until the moment the sample charge, of solute mass $n$, is injected at a time $t = 0$. The concentration $c_0$ is then equal to $n/\epsilon\varphi L$, and $\delta(t)$ is the Dirac delta-function with the dimension $(\text{time})^{-1}$. Besides the above conditions, we shall consider, wherever necessary, that the function $<c'(z,t)>$ is confined, *i.e.*, for $z \to \infty$ $<c'(z,t)>$ it remains finite. Through the substitutions $C = \frac{<c'>}{c_0}$, $S = \frac{q(1-\epsilon)}{\epsilon} \cdot \frac{<c''>}{c_0}$, $Pé = \frac{uL}{D}$, $T = \frac{tu}{L}$, $Z = \frac{z}{L}$, $k(C) = \frac{q(1-\epsilon)}{\epsilon} \cdot \frac{f(<c'>)}{c_0}$, $K_s = \frac{q(1-\epsilon)}{\epsilon} \cdot \frac{k_s L}{u}$ and $K_d = \frac{k_d L}{u}$, eqns. 3.10–3.14 will be transformed into dimensionless forms:

$$\frac{\partial C}{\partial T} + \frac{\partial S}{\partial T} = \frac{1}{Pé} \cdot \frac{\partial^2 C}{\partial Z^2} - \frac{\partial C}{\partial Z} \tag{3.10a}$$

$$S = k(C) \tag{3.11a}$$

$$\frac{\partial S}{\partial T} = K_s C - K_d S \tag{3.12a}$$

$$C(Z, 0) = 0 \tag{3.13a}$$

$$C(0, T) = \delta(T) \tag{3.14a}$$

The combination of eqns. 3.10a and 3.11a for $1/Pé \to 0$ gives

$$\left(1 + \frac{dk(C)}{dC}\right)\frac{\partial C}{\partial T} + \frac{\partial C}{\partial Z} = 0 \tag{3.15}$$

which describes the behaviour of an ideal chromatographic column. A general solution can be expressed only in an implicit form (Courant):

$$F(C) = T - (1 + \frac{dk(C)}{dC})\, Z \tag{3.16}$$

A concrete form of the function $F$ depends on the boundary conditions. In the special case when $f(<c'>) = K<c'> = \frac{\epsilon}{q(1-\epsilon)}k<c'>$ (linear isotherm), the solution to eqn. 3.15 can be expressed explicitly by

$$C = F_0\ [T - (1 + k)\, Z] \tag{3.17}$$

Then, for the boundary conditions expressed by eqns. 3.13a and 3.14a and for $Z = 1$ (column outlet):

$$C(1, T) = \delta\ [T - (1 + k)] \tag{3.17a}$$

Hence, one obtains at the column outlet a pulse of the same shape as that of the initial pulse, but with a time delay which characterizes the ideal retention time, $T = 1 + k$. The equations for ideal linear chromatography are discussed in greater detail later.

Eqns. 3.10a and 3.11a for $1/Pé = \alpha\,(\alpha > 0)$ have been solved analytically for the special case of a linear isotherm; then, for $Z = 1$ under the conditions expressed by eqns. 3.13a and 3.14a:

$$C(1, \tau) = \frac{1}{2\tau(\pi\alpha\tau)^{1/2}} \cdot \exp\left(-\frac{(\tau-1)^2}{4\alpha\tau}\right) \tag{3.18}$$

where $\tau = T/(1 + k)$ and $\alpha = 1/Pé$. The result is a curve approaching a Gaussian with the variance $\sigma_\tau^2 = 2\alpha$. If the variance is expressed in the coordinates $T$, as is usual in column chromatography, one obtains $\sigma_T^2 = 2\alpha\,(1 + k)$. The coordinate of the maximum of the above curve, $\tau_{max} = (1 - 9\alpha^2)^{1/2} - 3\alpha$, differs from the ideal retention time $\tau^* = 1$, *i.e.*, $t_R = (1 + k)L/u$, by about $3\alpha$.

Eqns. 3.10a and 3.12a for $1/Pé \to 0$ describe the behaviour of a chromatographic column with negligible longitudinal diffusion, while respecting a finite rate of interphase solute equilibration in the column. Together with the conditions 3.13a and 3.14a, the above equations lead to the solution

$$C(1, T) = \mathbf{1}(T-1)e^{K_s}\{\delta(T-1) + e^{K_d(T-1)}\sqrt{K_sK_d/(T-1)}\ \cdot$$

$$\cdot I_1\ [2\sqrt{K_sK_d(T-1)}]\ \} \tag{3.19}$$

where $\mathbf{1}(t)$ is the Haeviside unit step function and $I_1(x)$ is the modified Bessel function of the first order. At sufficiently large values of the rate constants $K_s$ and $K_d$, the above solution can also be converted into a form approaching a Gaussian curve with the variance $\sigma_T^2 = 2k/K_s$, where $k = K_s/K_d$. In this case, the coordinate of the maximum of the curve, $T_{max} \approx 1 + k - (2k/K_s) + (k/K_s^2)$, will also differ from the ideal retention time, the difference again being a function of the determining criteria $K_s$ and $K_d$ (Wičar *et al.*). In both cases, the variance of the quasi-Gaussian approximations, $\sigma_T^2$, is proportional to $k$.

A complete solution of eqns. 3.10a and 3.12a, without any restricting assumptions, was presented by Lapidus and Amundson for the boundary conditions of frontal chromatography, $C(Z, 0) = 0$ and $C(0, T) = \mathbf{1}(T)$. From this result, Van Deemter *et al.* derived a solution for a narrow concentration pulse, which, for conditions 3.13a and 3.14a, can be written in the form

$$C(Z, T) = e^{-K_s T} \cdot \frac{Z}{2T\sqrt{\pi\alpha T}} \cdot \exp\left(-\frac{(T-Z)^2}{4\alpha T}\right) + \int_0^T \left\{ e^{-K_s X - K_d(T-X)} \cdot \right.$$

$$\left. \cdot \frac{Z}{2X\sqrt{\pi\alpha X}} \cdot \exp\left(-\frac{(X-Z)^2}{4\alpha X}\right) \sqrt{\frac{K_s K_d X}{T-X}} \cdot I_1 \left[2\sqrt{K_s K_d X(T-X)}\right] \right\} dX \quad (3.20)$$

Hence, the complete solution is a convolution of the limiting solutions 3.18 and 3.19.

## CONCEPT OF IDEAL LINEAR CHROMATOGRAPHY

The concept of ideal linear chromatography is based on the most simplified model of the chromatographic process. This model is characterized by:

(i) infinitely rapid equilibration of solute between the phases;

(ii) negligibly slow longitudinal solute diffusion;

(iii) constant partition coefficient, independent of the concentration of solute in the phases;

(iv) piston flow of the mobile phase.

Despite being hypothetical, this model is very useful in that it provides a simple and reasonable characterization of chromatographic retention. It is possible to arrive directly at the differential equation of ideal linear chromatography by omitting the terms that refer to the non-ideal effects in the equations of real models (for instance, see eqns. 3.10–3.12), but it may nevertheless help in comprehending the problem of retention if ideal linear chromatography is discussed independently.

Let us consider a chromatographic column with an overall cross-section $\varphi$, of which the fractions occupied by the mobile and by the stationary phases are $\varphi_m$ and $\varphi_s$, respectively; the column is assumed to contain a migrating solute zone. Further, let us imagine a thin slab bound perpendicular to the column axis somewhere in the region of the zone and observe the change in the total number of moles of solute (in both phases) with time within the slab, $\partial n/\partial t$. Under the conditions of ideal chromatography, this change is given simply by the difference in the number of moles of solute brought into and out of the controlled space by the flow of mobile phase. Thus, provided that the column axis is oriented in the direction of coordinate $z$ and the gradient of solute

*References p.43*

concentration in the zone is positive with respect to the mobile phase velocity ($u$), the above situation can be described by the following simple mass balance:

$$\frac{\partial n}{\partial t} = -u\frac{\partial n'}{\partial z} \tag{3.21}$$

where $\partial n'/\partial z$ is the instantaneous gradient of the number of moles of solute in the mobile phase within the controlled slab.

In any region of the zone and at any instant, the total number of moles of solute, $n$, is distributed between the mobile and the stationary phases, and the respective solute concentrations are always in equilibrium. This equilibrium can be characterized by the conventional distribution constant. Hence, the equation

$$n = n' + n'' \tag{3.22}$$

holds, where

$$n'' = Kn'V_s/V_m \tag{3.23}$$

$n'$ and $n''$ being the numbers of moles of solute present in the mobile and stationary phases and $K$ the distribution constant, defined as the equilibrium ratio of the numbers of moles of solute per unit volumes of the stationary and the mobile phases, *i.e.*, $K = (n''/V_s)/(n'/V_m)$, where $V_s$ and $V_m$ are the respective volumes of the phases. Substituting $n''$ from eqn. 3.23 into eqn. 3.22 and $n$ from eqn. 3.22 into eqn. 3.21 and expressing the ratio $V_s/V_m$ by $\varphi_s/\varphi_m$, and rearranging, we obtain

$$\frac{\partial n'}{\partial t}\left(1 + K\frac{\varphi_s}{\varphi_m}\right) = -u\frac{\partial n'}{\partial z} \tag{3.24}$$

where the term $K(\varphi_s/\varphi_m)$ represents the so-called capacity ratio, $k$. If both sides of eqn. 3.24 are divided by the volume of the mobile phase contained in the controlled slab ($\varphi_m\,\mathrm{d}z$), one obtains finally

$$\frac{\partial c'}{\partial t}(1 + k) = -u\frac{\partial c'}{\partial z} \tag{3.25}$$

where $c'$ is the solute concentration in the mobile phase.

Eqn. 3.25 can actually be considered as a definition of ideal linear chromatography. A solution of this equation leads to the basic retention equation

$$t_R = L(1 + k)/u \tag{3.26}$$

where $t_R$ is the retention time and $L$ is the column length. Eqn. 3.26 can be rewritten into several forms. As the velocity of zone migration, $u_i$, is $L/t_R$, one can write

$$u_i/u = \boldsymbol{R} = 1/(1 + k) \tag{3.27}$$

where $\boldsymbol{R}$ is the so-called retardation factor (related to the $R_F$ value used in chromatography in flat-bed systems). Further, as $L/u$ is the mobile phase hold-up time (dead retention time), termed $t_m$, eqn. 3.26 can be expressed in the form

$$t_R = t_m(1 + k) \tag{3.28}$$

By multiplying both sides of eqn. 3.28 by the volume flow-rate of the mobile phase, $v$, an expression in terms of the retention volume, $V_R$, is obtained:

$$V_R = V_m(1 + k) \tag{3.29}$$

where $V_m$ is the actual mobile phase hold-up of the column (dead retention volume). As $\varphi_s/\varphi_m = V_s/V_m$ with homogeneous columns, $V_s$ being the total volume of stationary phase contained in the column, eqn. 3.29 can be simplified to

$$V_R = V_m + KV_s \tag{3.30}$$

In many instances, both the stationary and the mobile phase can be mixtures of several components. Also, the process of sorption in a given liquid phase can be accompanied by adsorption effects exerted by the solid support and chromatographic retention is therefore the result of several simultaneous sorption processes in such instances. These situations can still be described by retention eqns. 3.26–3.30 provided that $K$ is considered to be an effective overall distribution constant.

None of the above equations can be used to predict zone-broadening effects because no assumptions concerning this phenomenon have been involved in the model of ideal linear chromatography used. Actually, the above assumptions imply that an initial solute band of any concentration profile would travel through the column unskewed, *i.e.*, all parts of the band would migrate at the same velocity as given by eqn. 3.27. The most simplified yet real model is that of non-ideal linear chromatography. In this case, the zone spreading due to non-ideal effects, such as finite rate of solute equilibration and appreciable longitudinal diffusion of solute, can be assumed to be symmetrical with respect to the centre of the zone. Hence the centre of the zone can be taken as a region very near to sorption equilibrium, and the velocity of migration of the centre of the zone should approach the velocity expected in ideal linear chromatography.

## CONCEPT OF THE THEORETICAL PLATE

The model of the theoretical plate was introduced into chromatography by Martin and Synge. In this concept, the chromatographic column is considered to be composed of a series of hypothetical layers (perpendicular to the direction of zone migration ) in which the solute concentrations in the participating phases are assumed to be equilibrated. The chromatographic column is thus comparable with a series of mixing vessels the length of each of which is identical with the thickness of the above layers. This thickness represents, in the plate theory, one height equivalent to a theoretical plate (HETP, $H$). The model discussed is illustrated schematically in Fig. 3.3, where the space between two vertical lines represents one HETP and the dashed curve shows the profile of the overall solute concentration, $c$, in the column. The column axis is oriented along the coordinate $z$ and the mobile phase flows in the positive direction of this coordinate.

The discussion presented below follows Glueckauf's approach, leading to a definition of the HETP in terms of effective diffusion; a change of the amount of solute in an infinitesimally short time interval within the plate centred around the point $z$ (see Fig.

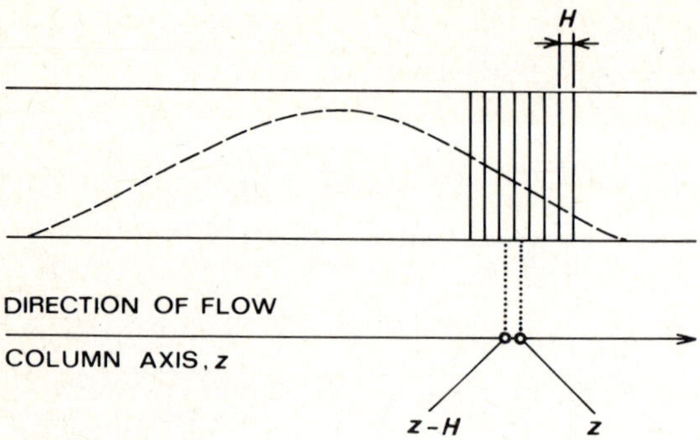


Fig. 3.3. Schematic representation of the plate model.

3.3) can be expressed by the following mass balance:

$$Ru\varphi(c_{z-H} - c_z)\,dt = H\varphi(c_{t+dt} - c_t) \tag{3.31}$$

where $\varphi$ is the overall cross-section of the column and $R$ is the retardation factor. If the HETP is sufficiently small in comparison with the width of the chromatographic zone, the term $c_{z-H}$ can be expanded into a Taylor's series about $z$ and eqn. 3.31 can be expressed in the form

$$\frac{\partial c}{\partial t} = \frac{RuH}{2} \cdot \frac{\partial^2 c}{\partial z^2} - Ru\frac{\partial c}{\partial z} \tag{3.32}$$

The broadening of the zone can be considered to be due to apparent longitudinal diffusion, controlled by an overall effective diffusion coefficient $D_{\text{eff}}$. According to eqn. 2.17, the situation in which a zone is simultaneously subject to longitudinal diffusion and convective migration at a forward velocity $u_i$ is described by

$$\frac{\partial c}{\partial t} = D_{\text{eff}} \cdot \frac{\partial^2 c}{\partial z^2} - u_i \cdot \frac{\partial c}{\partial z} \tag{3.33}$$

Comparison of the coefficients in eqns. 3.32 and 3.33 gives $u_i = Ru$ and $D_{\text{eff}} = RuH/2$. Employing Einstein's equation, the HETP can be expressed by the relationship

$$H = \sigma_L^2 / Rut_R = \sigma_L^2/L \approx (\Delta z)^2/16L \tag{3.34}$$

where $\sigma_L$ and $\Delta z$ represent the length standard deviation and the approximate width of the chromatographic zone as measured at the column outlet, respectively, $t_R$ is the retention time and $L$ is the column length.

Note that for a Gaussian profile, $\sigma_L \approx \Delta z/4$. In some instances, it may be more appropriate to express the standard deviation in time or volume units, *i.e.*, $\sigma_t$ or $\sigma_V$, respectively. As

$$\sigma_L = \sigma_t Ru = \sigma_V Ru/v \tag{3.35}$$

where $v$ is the volume flow-rate of the mobile phase, the plate height can also be expressed by

$$H = \sigma_t^2 Ru/t_R = \sigma_V^2 Ru/V_R v \tag{3.36}$$

where $V_R$ is the retention volume ($vt_R$). The ratio $u/v$ is actually the inverse of the flow-through cross-section of the column. The equations $\sigma_t \approx \Delta t/4$ and $\sigma_V \approx \Delta V/4$ also hold in these cases, where $\Delta t$ and $\Delta V$ are the time and volume of the mobile phase, respectively, necessary to shift the zone in the column by a length $\Delta z$.

The number of the theoretical plates, $N$, is given by

$$N = L/H \tag{3.37}$$

However,

$$L = Rut_R = RuV_R/v \tag{3.38}$$

so that $N$ can be calculated from the following equations:

$$N = (L/\sigma_L)^2 = (t_R/\sigma_t)^2 = (V_R/\sigma_V)^2 \tag{3.39}$$

It follows that although the plate model is unrealistic, it serves as a useful criterion of the separation efficiency of the chromatographic column. However, this criterion itself can give no information about the mechanism of zone development or the role of the individual spreading factors. In addition to the formal definition of the HETP, eqn. 3.34 indicates that the extent to which a zone broadens on its migration down the column is proportional to the square root of the length of the migration path:

$$\Delta z \approx 4\,(HL)^{1/2} \tag{3.40}$$

It is imperative, however, that the parameter $H$ should be regarded as a quantity dependent on the nature of the chromatographic system and on the operating conditions employed, particularly on the mobile phase velocity.

## DYNAMICS OF ZONE SPREADING

An exact solution of the system of differential equations that describes a real chromatographic model has not yet been found even for as simple a system as the capillary column. Therefore, there is full justification for approaching the problem of zone broadening in a manner in which the individual contributions to zone spreading are treated separately. Although such procedures are not exact from the physico-mathematical point of view, they are very useful in that they give quantitative criteria of the separation efficiency and relate them to the factors that control the mechanism of chromatography. The most significant contribution to this rational approach is Giddings' (1959) generalized non-equilibrium theory. In this section, the problem of zone broadening is discussed briefly in terms of the "random walk model", which approach was also first applied in chromatography by Giddings (1958, 1965).

The basic factors responsible for zone broadening are:
(i) non-uniformity of mobile phase flow;
(ii) longitudinal diffusion of the solute;
(iii) deviations from sorption equilibrium.

Longitudinal diffusion of the solute proceeds in both the mobile and stationary phases. With deviations from equilibrium, at least three regions have to be considered, *viz.*, the

*References p.43*

sorbent proper, interparticle (moving) mobile phase and intraparticle (stagnant) mobile phase. Hence there are six factors, each of which has associated with it a certain variance contributing the total variance of the zone. As all of the above factors are essentially controlled by the laws of statistics (Chandrasekhar), there is some reason to suppose that the total variance is the sum of the partial variances. In addition, the random nature of the processes that determine zone broadening implies a Gaussian concentration profile of the zone, which is in reasonable agreement with experience. However, the above additivity of the partial variances is not applicable without any limitations. As shown by Giddings (1961a), the additivity rule does not hold for the variances incidental to mass transfer processes that are mutually competitive. Such situations are fairly common in liquid chromatography because of the important role of mass transfer in the mobile phase.

The variance resulting from a multi-step random walk of a large number of molecules can be described by

$$\sigma^2 = l^2 n \tag{3.41}$$

where $l$ is the average length of the random step and $n$ is the number of steps taken by a molecule. The application of eqn. 3.41 enables the relationships discussed below to be formulated for the variances due to the individual zone spreading factors.

## Non-uniformity of flow

The overall flow of the mobile phase through a packed column consists of a large number of stream-lines, the velocities of which fluctuate widely around the mean velocity. If $t_e$ is the time for which a molecule persists in a stream-line of velocity $u^*$, it will depart by a distance $t_e(u^* - u)$ from the molecules migrating at the mean velocity $u$. This distance can be considered to be the length of a random step. The number of steps taken is apparently $n = L/u^*t_e$, where $L$ is the column length. Provided that $u^*t_e$ is proportional to the particle diameter, $d_p$, of the packing and $u^* - u$ is proportional to $u$, the variance due to non-uniformity of flow, $\sigma^2(A)$, is proportional to $d_pL$. This proportionality is usually written in the form

$$\sigma^2(A) = 2\lambda d_p L \tag{3.42}$$

where $\lambda$ is the eddy diffusion coefficient.

## Longitudinal diffusion in the mobile phase

The chromatographic zone consists of a part formed by the mobile phase and a part fixed in the stationary phase. In any position in the column and at any instant, the mean forward velocity of the mobile phase part of the zone is equal to the mobile phase velocity $u$. Hence, all molecules of any solute component spend, on average, the same time in the mobile phase during their residence in the column while migrating down it, no matter what the extent of retention may be. This time is simply $t_m = L/u$ (dead retention time). When combining this relationship with Einstein's equation for the diffusional variance and taking into account the fact that diffusional transport is hindered by the presence of the packing particles, the variance due to longitudinal diffusion in the mobile phase, $\sigma^2(B_m)$, is given by

$$\sigma^2(B_m) = 2\psi' D' L/u \tag{3.43}$$

where $\psi'$ is the obstructive factor for the mobile phase.

### Longitudinal diffusion in the stationary phase

Solute molecules are subject to diffusion in the stationary phase only during the time they are present in this phase, $t_s$, and it can easily be shown that $t_s$ is related to $t_m$, $u$ and $u_i$ by the equation $t_m/(t_m + t_s) = u_i/u = R$. Thus, when substituting $t_s$ from the above relationship into Einstein's equation and expressing $t_m$ by $L/u$, the variance due to longitudinal diffusion in the stationary phase, $\sigma^2(B_s)$, can be expressed by

$$\sigma^2(B_s) = 2\psi'' D''(1 - R)L/Ru \tag{3.44}$$

where $\psi''$ is the obstructive factor for the stationary phase.

### Non-equilibrium in the sorbent

*One-site adsorption kinetics (adsorption chromatography)*

The process of adsorption can be compared to a first-order reaction with $t_a = 1/k_a$, where $t_a$ is the time necessary for a solute molecule to be adsorbed and $k_a$ is the corresponding reaction rate constant. During the time $t_a$, the molecule travels at the velocity $u$ while the velocity of the centre of the zone is only $Ru$. Thus, the distance between the molecule and the centre of the zone that develops during the time $t_a$ is $(1 - R)ut_a$ and represents the length of a random step. As each adsorption step is followed by a desorption step, the number of steps, $n$, is $2L/ut_a$. Hence, the variation due to non-equilibrium in the sorbent in adsorption chromatography, $\sigma^2(C_{sa})$, is given by

$$\sigma^2(C_{sa}) = 2(1 - R)^2 Lu/k_a \tag{3.45}$$

The above variance can also be expressed by means of the desorption rate constant, $k_d$. As $t_a/t_d = R/(1 - R)$, where $t_d$ is the time taken by a desorption step, there holds for the variance the equation

$$\sigma^2(C_{sa}) = 2R(1 - R)Lut_d = 2R(1 - R)Lu/k_d \tag{3.46}$$

*Diffusion-controlled kinetics (partition chromatography)*

In this case, the expression for the variance can be obtained by a procedure similar to that for adsorption. The desorption time, $t_d$, is the time necessary for a solute molecule to diffuse through a film of stationary phase of effective thickness $d_f$. According to

*References p.43*

Einstein's equation, this time is given by $t_d = d_f^2/2D''$, where $D''$ is the diffusion coefficient in the stationary phase. The substitution of $t_d$ from this equation into eqn. 3.46 yields

$$\sigma^2(C_{sp}) = R(1 - R)d_f^2Lu/D'' \tag{3.47}$$

for the variance due to non-equilibrium in the sorbent in partition chromatography. A more detailed theory (Giddings, 1959) showed that the right-hand side of eqn. 3.46 should contain a geometrical factor ($q$) to account for the internal geometry of the sorption bed (for example, $q = 2/3$ for a uniform liquid film).

**Non-equilibrium in the interparticle mobile phase**

With chromatographic beds of porous particles, it is expedient to distinguish between the void volume dislocated within the particles and that between the particles. The interparticle fraction of the total void volume is actually the only space available for the solvent to flow through the bed, *i.e.*, to function as a true mobile phase. The solvent bound within the pores of the particles is stationary, and permits only the diffusion of solute into and out of the pores. This situation has some important implications in liquid chromatography and constitutes the basis of separation in gel permeation chromatography.

If the mass transfer in the mobile phase is considered to be a diffusion-controlled process, the basic parameter will be the length of the diffusion path. Provided that this length is proportional to $d_p$, it follows from Einstein's equation that the time $t_e$ for which a solute molecule persists in a stream-line of velocity $u^*$ is $t_e \approx d_p^2/2D'$. This relationship can now be combined with those arrived at in the discussion of the variance due to the non-uniformity of mobile phase flow (p.36). Thus, if again $l = t_e(u^* - u)$, $n = L/u^*t_e$ and $u^* - u \approx u$, the variance due to non-equilibrium in the "mobile" mobile phase can be expressed by

$$\sigma^2(C_m) = \omega d_p^2 Lu/2D' \tag{3.48}$$

where $\omega$ is again a geometrical factor characterizing the complex structure of the packing.

**Non-equilibrium in the intraparticle mobile phase**

As discussed above, the mobile phase present in the intraparticle space is essentially immobile. It was shown by Giddings (1961b) that the variance due to non-equilibrium in the "stationary" mobile phase is

$$\sigma^2(C_m^*) = [(1 - \varphi_m^* R)^2/30\psi_m^*(1 - \varphi_m)]\, d_p^2 Lu/D' \tag{3.49}$$

for a spherical particle, $\varphi_m^*$ and $\psi_m^*$ denoting the fraction of the mobile phase present in the interparticle void volume and the corresponding obstructive factor, respectively. Provided that the intraparticle space is completely filled with a stationary liquid sorbent, the respective non-equilibrium variance will be

$$\sigma^2(C_{sp}^*) = [R(1 - R)/30\psi_s^*]d_p^2Lu/D'' \tag{3.50}$$

where $\psi_m^*$ is the obstructive factor for solute diffusion in the stationary liquid bound in the pores. A comparison of eqns. 3.47 and 3.50 shows that the expression in square brackets in eqn. 3.50 actually represents the geometrical factor $q$ for this particular case, $d_f$ being effectively equal to $d_p$. A more complete discussion of the problem of zone spreading associated with intraparticle mass transfer, involving also instances such as porous layer open tubes, porous layer beads, and non-adsorbing and adsorbing gel filter particles, was given by Hawkes.

## Combination of plate height contributions

With respect to the definition of the plate height ($H = \sigma_L^2/L$), eqns. 3.42–3.50 can be rewritten as

$$H(A) = 2\lambda d_p = A \tag{3.51}$$

$$H(B_m) = 2\psi' D'/u = B_m/u \tag{3.52}$$

$$H(B_s) = 2\psi'' D''(1 - R)/Ru = B_s/u \tag{3.53}$$

$$H(C_{sa}) = 2R(1 - R)u/k_d = C_{sa}u \tag{3.54}$$

$$H(C_{sp}) = qR(1 - R)d_f^2 u/D'' = C_{sp}u \tag{3.55}$$

$$H(C_m) = \omega d_p^2 u/2D' = C_m u \tag{3.56}$$

$$H(C_m^*) = [(1 - \varphi_m^* R)^2/30\psi_m^*(1 - \varphi_m^*)]\, d_p^2 u/D' = C_m^* u \tag{3.57}$$

$$H(C_{sp}^*) = [R(1 - R)/30\psi_s^*]d_p^2 u/D'' = C_{sp}^* u \tag{3.58}$$

According to the classical theory, the total value of the HETP would be given simply by the sum of the plate height contributions relevant to the processes that take place in a particular system. However, this simple concept fails in a number of cases. A typical illustrative example is the situation described by the coupling theory of eddy diffusion (Giddings and Robison). The time $t_e$ for which a molecule stays in a stream-line of given velocity ($u^*$) can be drastically shortened by diffusion in the mobile phase. On the other hand, the path of diffusion can be substantially shortened owing to mixing effects of eddy flow. In this way, the plate height contributions due to both factors are decreased by their mutual effect on each other to an extent proportional to the significance of either of them. This is of great importance in liquid chromatography owing to the low rates of diffusion that occur in liquids. According to the coupling theory, the effective plate height contribution due to inequality of velocities and mobile phase non-equilibrium effects is given by (*cf.* eqns. 3.51 and 3.56):

$$H(A, C_m) = 1/[(1/A) + (1/C_m u)] \tag{3.59}$$

This effect is very favourable because the $H(A, C_m)$ value will approach the value of $A$ upon increasing the flow velocity rather than increasing proportionally to the latter. In order to show more closely the rules that govern the combination of the individual $H$ contributions, some typical systems are quoted overleaf.

*References p.43*

(I) In liquid–liquid chromatography, with a uniform stationary liquid film on a macroporous solid support:

$$H = 1/[(1/A) + (1/C_m u)] + B_m/u + B_s/u + C_{sp}u \tag{3.60}$$

(II) In liquid–liquid chromatography, with microporous support particles and the pores filled completely with the stationary liquid:

$$H = 1/[(1/A) + (1/C_m u)] + B_m/u + B_s/u + C^*_{sp}u \tag{3.61}$$

(III) In liquid–solid chromatography, with microporous adsorbent particles and the pores filled completely with stagnant mobile phase:

$$H = 1/[(1/A) + (1/C_m u)] + B_m/u + B_s/u + C^*_m u + C_{sa}X_a u \tag{3.62}$$

where $X_a$ is the fraction of the retained solute which is adsorbed.

The longitudinal diffusion terms for both phases are generally negligible in modern high-speed liquid chromatography.

## CHROMATOGRAPHIC RESOLUTION

It was shown earlier in this chapter (p. 35) that while the distance between the centres of two zones increases linearly with the average length of their migration paths, the widths of the zones increase proportionally only to the square root of the paths. This effect is the basis of chromatographic separations and is dealt with in some detail in this section. From the relationships presented earlier, the time standard deviation can be expressed by either

$$\sigma_t = (HL)^{1/2}/Ru \tag{3.63}$$

or

$$\sigma_t = t_R/N^{1/2} \tag{3.64}$$

Eqn. 3.63 reveals that for a given component, the time width of the zone increases proportionally to the square root of the column length, whereas eqn. 3.64 shows that on a given column the time widths of zones of different solutes increase proportionally to their retention times.

Chromatographic resolution is usually defined as the spacing between the maxima of two zones, expressed in the units of the mean standard deviation. Thus, the resolution, $R_s$, of components 1 and 2 with $t_{R2} > t_{R1}$ is given by

$$R_s = 2(t_{R_2} - t_{R_1})/(\sigma_{t_2} + \sigma_{t_1}) = (t_{R_2} - t_{R_1})/\bar{\sigma}_t \tag{3.65}$$

where $\bar{\sigma}$ represents the mean standard deviation. Thus, a virtually complete separation is attained if $t_{R_2} - t_{R_1} = 4\bar{\sigma}_t$, which corresponds to $R_s = 4$. As discussed earlier:

$$t_R = L/Ru = (L/u)\,[1 + K(V_s/V_m)] \tag{3.66}$$

so that

$$t_{R2} - t_{R1} = \frac{L}{u}\left(\frac{1}{R_2} - \frac{1}{R_1}\right) = \frac{L}{u} \cdot \frac{V_s}{V_m}\ (K_2 - K_1) \tag{3.67}$$

The substitution of $t_R$ from eqn. 3.66 into eqn. 3.64 gives eqn. 3.63 and the equation

$$\sigma_t = \frac{(HL)^{1/2}}{u} \cdot \left(K \cdot \frac{V_s}{V_m} + 1\right) \tag{3.68}$$

which results in

$$\bar{\sigma}_t = \frac{\sigma_{t2} + \sigma_{t1}}{2} = \frac{(HL)^{1/2}}{2u}\left(\frac{1}{R_2} + \frac{1}{R_1}\right) = \frac{(HL)^{1/2}}{2u} \cdot \frac{V_s}{V_m}\left(K_2 + K_1 + 2\frac{V_m}{V_s}\right) \tag{3.69}$$

Eqns. 3.67 and 3.69 show explicitly the dependence of both $t_{R2} - t_{R1}$ and $\bar{\sigma}_t$ on the basic characteristics of the column. Thus, when plotting $t_{R2} - t_{R1}$ and $4\bar{\sigma}_t$ calculated from eqns. 3.67 and 3.69 against $L$, a graph similar to that shown in Fig. 3.4 is obtained, where the intersection of the two curves determines the column length at which $R_s = 4$. The combination of eqns. 3.65, 3.67 and 3.69 and the solution for $L/H = N$ yields either

$$N = \left(\frac{R_s}{2} \cdot \frac{R_1 + R_2}{R_1 - R_2}\right)^2 \tag{3.70}$$

or

$$N = \left[\frac{R_s}{2} \cdot \frac{K_2 + K_1 + 2(V_m/V_s)}{K_2 - K_1}\right]^2 \tag{3.71}$$

where $N$ is the number of theoretical plates necessary to achieve the required resolution $R_s$.

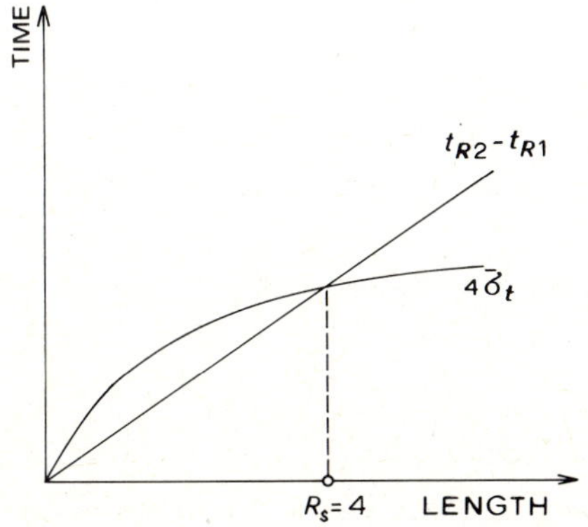


Fig. 3.4. Plots of the mean time width and the difference in the retention times of two adjacent zones *versus* the length of the path of chromatographic migration.

The resolution as defined by eqn. 3.65 obviously relates to the separation of two components. However, this fact does not imply any serious limitations to the applicability of the criterion, as problems in the separation of multi-component mixtures can be reduced to those of resolving the pairs of components that are most difficult to separate.

Resolution problems involving a larger number of components can be treated in terms of the "peak capacity", introduced by Giddings (1967) and further discussed in detail by Scott. The peak capacity is defined as the number of peaks that can be placed in the chromatogram between the peak of a non-sorbed component and that of a final component, provided that the resolution of all the peaks is $R_s = 4$. The quantitative expression relating the peak capacity ($\Pi$) to the column parameters is

$$\Pi = 1 + 0.6N^{1/2}\log(1 + k) \tag{3.72}$$

where $k$ refers to the final component in the chromatogram.

As $L = NH$ (see eqn. 3.37), eqn. 3.66 can be rewritten as

$$t_R = (NH/u)(1 + k) \tag{3.73}$$

Hence, if $N$ for the required $R_s$ and $K$ for the more retained component are substituted into eqn. 3.73, the time and column length necessary to achieve the required extent of separation in a particular sorption system can be calculated, provided that the values of $H/u$ and $V_s/V_m$ are also known. Eqns. 3.70–3.73 and the considerations on the necessary pressure drop (*cf.* Chapter 2) are the basis for the optimization of the time of separation (Huber and Hulsman). Eqn. 3.73 also enables the rate of plate generation to be expressed as follows:

$$N/t = (u/H)/(1 + k) \tag{3.74}$$

The expressions of resolution can be simplified by introducing the term "effective plate number", $N_{\text{eff}}$ (Desty *et al.*), defined by

$$N_{\text{eff}} = [(t_R - t_m)/\sigma_t]^2 = [k/(1 + k)]^2 N \tag{3.75}$$

## Extra-column zone broadening

There are several factors that add to the broadening which the zone suffers on its passage through the chromatographic column alone. This additional broadening is mainly due to non-zero volumes of:

(a) sample charge;
(b) injection port–column inlet connecting line;
(c) column outlet–detector inlet connecting line;
(d) sensing space of the detector.

Provided that the variances incidental to the individual spreading factors are expressed in volume units, they can be assumed to be additive and the total variance, $\sigma^2$, is given by

$$\sigma^2 = \sigma^2_{\text{col}} + \sigma^2_{\text{ex},a} + \ldots + \sigma^2_{\text{ex},d} = \sigma^2_{\text{col}} + \Sigma\sigma^2_{\text{ex}} \tag{3.76}$$

where $\sigma^2_{\text{col}}$ is the variance due to the column alone, $\sigma^2_{\text{ex},a}$, $\sigma^2_{\text{ex},b}$, etc., are the extra-column variances due to the above factors, and $\Sigma\sigma^2_{\text{ex}}$ is the sum of the extra-column variances. As only volume variances are considered in this discussion, the subscript $V$ is omitted for brevity. It follows from eqn. 3.76 that

$$\sigma^2 = \sigma^2_{\text{col}}[1 + (\Sigma\sigma^2_{\text{ex}}/\sigma^2_{\text{col}})] \tag{3.77}$$

Now, $\sigma^2_{col}$ can be expressed as

$$\sigma^2_{col} = V_R^2/N \tag{3.78}$$

where $V_R$ is the retention volume as measured in the column alone, *i.e.*, without involving any extra-column hold-up volumes, and $N$ is the number of the theoretical plates in the column. Hence, eqn. 3.77 can be rewritten as

$$\sigma^2 = \sigma^2_{col}[1 + (N\Sigma\sigma^2_{ex}/V_R^2)] \tag{3.79}$$

As the plate height is generally given by $H = \sigma_V^2 L/V_R^2$, the apparent height of a theoretical plate, $H_{app}$, can be expressed as

$$H_{app} = H_{col}[1 + (N\Sigma\sigma^2_{ex}/V_R^2)] \tag{3.80}$$

In eqn. 3.80, $H_{app}$ is the plate height as calculated by the conventional method from the plate number and column length without taking into account the extra-column effects, whereas $H_{col}$ is the true column plate height. Note that $H_{app}$ is always larger than $H_{col}$.

Eqn. 3.80 can be further rearranged by expressing $V_R$ in terms of the column parameters. As $V_R = (vL/u)(1 + k)$ and $v/u = \varphi_m$, which can further be expressed by $\varphi_m = \varphi\epsilon = \pi r^2 \epsilon$, where $\varphi$, $\epsilon$ and $r$ are the cross-section of the empty column, total porosity of the column packing and the radius of the column, respectively, the apparent plate height can be expressed as

$$H_{app} = H_{col}\left\{1 + \Sigma\sigma^2_{ex} \cdot \frac{N}{[\pi r^2 L\epsilon(1 + k)]^2}\right\} \tag{3.81}$$

Eqn. 3.81 shows directly how the deleterious role of $\Sigma\sigma^2_{ex}$ depends on the basic parameters of the column. It is evident from this equation that the radius of the column is the most significant parameter in this respect.

## REFERENCES

Chandrasekhar, S., *Rev. Mod. Phys.*, 15 (1943) 1.
Courant, R., *Partial Differential Equations*, Interscience, New York, 1962, Ch. I, §7.
Desty, D. H., Goldup, A. and Swanton, W. T., *ISA Proc.*, (1961) 83.
Giddings, J. C., *J. Chem. Educ.*, 35 (1958) 588.
Giddings, J. C., *J. Chem. Phys.*, 31 (1959) 1462.
Giddings, J. C., *J. Chromatogr.*, 5 (1961a) 46.
Giddings, J. C., *Anal. Chem.*, 33 (1961b) 962.
Giddings, J. C., *Dynamics of Chromatography*, Marcel Dekker, New York, 1965, p.29.
Giddings, J. C., *Anal. Chem.*, 39 (1967) 1027.
Giddings, J. C. and Robison, R. A., *Anal. Chem.*, 34 (1962) 885.
Glueckauf, E., *Trans. Faraday Soc.*, 51 (1955) 34.
Hawkes, S. J., *J. Chromatogr.*, 68 (1972) 1.
Huber, J. F. K. and Hulsman, J. A. R. J., *Anal. Chim. Acta*, 38 (1967) 305.
Lapidus, L. and Amundson, N. R., *J. Phys. Chem.*, 56 (1952) 984.
Martin, A. J. P. and Synge, R. L. M., *Biochem. J.*, 35 (1941) 1358.
Prager, S., *Physica*, 29 (1963) 129.
Scott, R. P. W., *J. Chromatogr. Sci.*, 9 (1971) 449.
Van Deemter, J. J., Zuiderweg, F. J. and Klinkenberg, A., *Chem. Eng. Sci.*, 5 (1956) 271.
Wičar, S., Novák, J. and Ruseva-Rakshieva, N., *Anal. Chem.*, 43 (1971) 1945.

*Chapter 4*

# Physico-chemical basis of chromatographic retention in liquid – liquid and liquid – solid systems

J. NOVÁK

## CONTENTS



## INTERACTION OF SOLUTE WITH THE PHASES

### Nature of the distribution constant

It has been pointed out on several occasions in the preceding chapters that it is primarily the difference in the distribution constants that leads to the differential migration of solute zones in a chromatographic system. It is now appropriate to recall that the absolute chromatographic retention in linear chromatography is given by the equation $t_R = (L/u)[1 + K(\varphi_s/\varphi_m)]$, which readily shows that except for $K$ all the variables on the right-hand side are parameters unrelated to the chemical nature of the system. Within certain limits, these parameters can be adjusted by an appropriate choice of the working conditions and the arrangement of the sorption system. The intrinsic sorption properties of the system are given by the chemical nature of the phases and are represented by the distribution constant. In this respect, the distribution constant can be considered as a most important chromatographic quantity.

The distribution constant reflects an equilibrium between the escaping tendencies which the solute exerts in the media of the stationary and the mobile phases. The escaping tendencies are determined by the disparities in solute–solute, solvent–solvent and solute–solvent intermolecular interactions and can be characterized quantitatively in terms of the excess Gibbs free energy. As there is a simple relationship between the partial molar excess Gibbs free energy of the solute in a given solute–solvent mixture and the respective Raoult's law activity coefficient, the latter is also a suitable quantitative measure of the escaping tendency of the solute. Hence it follows that there is a relationship between the chromatographic distribution constant and the thermodynamic properties of the

*References p.56*

sorption system, but it should be noted that the distribution constant itself is not an unequivocally defined thermodynamic quantity. This important problem is dealt with in detail later in this chapter.

In context with the above considerations, solute–solvent intermolecular interactions will be discussed with reference to the Raoult's law activity coefficient, $\gamma^0$, defined by $\Delta G^E = RT \ln \gamma^0$, where $\Delta G^E$ is the partial molar excess Gibbs free energy and $R$ and $T$ are the perfect gas constant and absolute temperature, respectively. In chromatography, the situation is substantially simplified by the fact that extremely dilute solutions are almost always involved with which the $\gamma^0$ values are practically independent of solute concentration. Hence, all considerations concerning the activity coefficient will subsequently always refer to extremely dilute solutions.

Actually, the above activity coefficient accounts for the differences in the type and strength of the intermolecular forces acting between the molecules of pure solute and solvent alone, and also for the differences in the sizes of the solute and solvent molecules. These two aspects can be considered as the mechanism that underlies the non-ideality of solutions in view of Raoult's law ($\gamma^0$ deviating from unity). A detailed discussion of the activity coefficient exceeds the scope of this chapter. Of the original work concerning this problem, Guggenheim's theoretical paper, Hildebrand and Scott's book and the chromatographically oriented papers by Martire (1956, 1961) and Martire and Pollara can be referred to. Much information can be found in papers on empirical correlations between the activity coefficient and the structure of the solute and solvent molecules (Pierotti *et al.*).

## Types of interaction

By the term "interaction", mutual attraction or repulsion of molecules or ions, accompanied by a release or consumption of energy, can be understood. The amount of interaction energy exchanged is a measure of the strength of interaction and can serve as a criterion of whether the interaction is of a physical or a chemical nature. In chromatography, reversible interactions are of primary importance. Although most of the interactions that are of interest from the chromatographic viewpoint are physical interactions, chemical interactions cannot be excluded from consideration. Physical interactions, in terms of both their quality and intensity, are determined by the molecular constitution, which is fixed by the species of the atoms present and by the nature of the bonds between them (metallic, ionic, covalent, coordination).

Typical physical interactions are those that arise from dispersion forces and electrostatic forces (dipole–dipole and ion–dipole interactions, formation of hydrogen bonds). Dispersion (London) forces are of a non-polar nature and act as the only type of interaction between molecules of hydrocarbons. The physical nature of dispersion forces has been studied and explained on a quantum mechanical level by Lennard-Jones and by London.

Dipole–dipole interactions, usually termed orientation interactions, can occur only between molecules that possess permanent dipole moments. It follows from the studies of Keesom (1921a, b) that the significance of orientation interactions decreases strongly with increasing temperature. Induction (Debye) forces can arise between molecules with permanent dipoles, or ions, and non-polar but polarizable molecules, *i.e.*, those which have a system of easily shiftable electrons. Interactions between ions and dipoles cause solvation of ions to occur, and they are sometimes strong enough to produce stoichiometrically defined formations.

Another type of polar interaction that is very important in chromatography is hydrogen bonding. A hydrogen bond can be formed between a molecule that contains hydrogen attached to atoms of high affinity towards electrons (fluorine, oxygen, nitrogen) and a molecule that possesses unshared pairs of electrons. The interactions arising through hydrogen bonding are relatively strong; in some instances, the interaction energy approaches that of a weak chemical bond. Hydrogen bonds can be formed both between molecules and within a molecule. In the latter instance, hydrogen bonding entails a depolarization of the molecule, *i.e.*, the saturation of its tendency towards intermolecular interactions. Similar effects are encountered in the formation of dimers or oligomers through hydrogen bonding. Ewell *et al.* developed a scheme for predicting the formation of azeotropes due to hydrogen bonding; this scheme was also a subject of discussion in the field of chromatography (Keulemans). In addition to interactions of a pronouncedly physical nature, interactions also occur in chromatography that fall into the area of chemistry. For example, such interactions are those of the ion–ion type (ion-exchange chromatography, precipitation chromatography) and the formation of complexes.

Interactions of the above type contribute to that component of the activity coefficient which is related to the interaction energy between the solute and solvent molecules. In addition to this interaction component, the overall activity coefficient comprises an entropic component that accounts for the differences in the sizes of the solute and solvent molecules [a system of molecules of different sizes is characterized by a lower degree of ordering, *i.e.*, such a system displays a larger number of distinguishable configurational arrangements compared with systems of molecules of equal sizes (*cf.*, Ashworth and Everett, Denbigh)]. This statistical factor always presents a negative contribution; the equation $\ln \gamma^0 = (\Delta H^E/RT) - \Delta S^E/R$ applies, where $\Delta H^E$ and $\Delta S^E$ are the partial molar excess enthalpy and entropy.

Intermolecular interactions are significantly affected by steric effects. These effects can either be of a merely spatial character or may bear upon energetic factors. In the former case, the steric effects are approximately comparable with those of the size of molecules. In the latter case, the role of steric effects is more varied. Thus, for example, the position of a polar group in the molecule determines the magnitude of the permanent dipole moment. Further, the mutual configuration of polar and non-polar groups in a molecule determines the accessibility of the polar groups to be involved in interactions, thus determining the possibility of whether intermolecular association can occur (hindrance effects). Finally, the mutual configuration of polar groups influences both the magnitude of the permanent dipole moment and the possibility of the potential for intermolecular interactions to be saturated by intramolecular associations (intramolecular hydrogen bonding, for instance). The steric effects were discussed in detail by Green and McHale with reference to the deviations from Martin's postulate on the additivity of $\Delta R_M$ values.

The above considerations concern interactions in liquid solutions. Systems that involve solid adsorbents present different situations, which are more difficult to describe. In

*References p.56*

liquid solutions, the contacts between molecules of the solute and solvent are necessarily accompanied by some cancelling of the contacts among the solvent molecules, so that the solute–solvent interactions take place on account of the solvent–solvent interactions. With solid adsorbents, the rigid pattern does not allow any solute molecules to enter the interior of the bulk adsorbent and, in addition, keeps the active sites on the surface sufficiently far apart to prevent mutual interactions between them. Hence, the surface solute–sorbent interactions take place without any effect on the cohesion forces within the adsorbent. However, in systems typical of liquid chromatography, the surface of the adsorbent is saturated with the eluent and, under these conditions, solute–sorbent interactions are accompanied by the displacement of the eluent molecules from the surface, *i.e.*, by breaking of eluent–sorbent bonds. There is a significant difference between the mechanisms that underlie the partition of solute between two liquid phases and the adsorption of a solute from a liquid mixture on a solid adsorbent: while in partition the solute–solvent interactions that take place within one phase are virtually unaffected by the other phase, the solute molecules adsorbed on the solid surface are exposed to simultaneous interactions with the interfacial eluent layer.

## THERMODYNAMICS OF SORPTION EQUILIBRIUM

Let us consider the process of sorption as a reversible reaction in which 1 mole of solute is transferred from the mobile phase into the stationary sorbent under isothermal and isobaric conditions. Denoting the solute ($i$) in the mobile and stationary phases by $i(m)$ and $i(s)$, respectively, the reaction can be represented by

$$i(m) \rightarrow i(s)$$

This process has associated with it a change in the solute partial molar Gibbs free energy, which will be termed the Gibbs free energy of sorption and denoted by $\Delta G_s$. Now,

$$\Delta G_s = \mu_{is} - \mu_{im} \tag{4.1}$$

where $\mu_{is}$ and $\mu_{im}$ are the chemical potentials of the solute in the sorbent and the mobile phase, respectively. The chemical potentials can further be expressed by

$$\mu_{is} = \mu_{is}^0 + \ln a'' \tag{4.2}$$

$$\mu_{im} = \mu_{im}^0 + \ln a' \tag{4.3}$$

where $\mu_{is}^0$, $\mu_{im}^0$, $a''$ and $a'$ are the standard chemical potentials and the activities of solute in the respective phases. Hence, $\Delta G_s$ is given by

$$\Delta G_s = \mu_{is}^0 - \mu_{im}^0 + RT \ln (a''/a') \tag{4.4}$$

where the difference $\mu_{is}^0 - \mu_{im}^0$ is the standard Gibbs free energy of sorption and will be denoted by $\Delta G_s^0$.

Eqn. 4.4 also holds for systems that are out of sorption equilibrium (*e.g.*, the leading or trailing part of the chromatographic zone), and the negative value of $\Delta G_s$ is actually the driving force of sorption. At equilibrium (approximately in the centre of the zone),

$\Delta G_s = 0$ and the standard Gibbs free energy of sorption is

$$\Delta G_s^0 = -RT \ln (a''/a')_{eq.} \tag{4.5}$$

where the subscript eq. indicates that equilibrium solute activities are being considered. The ratio $(a''/a')_{eq.}$ can be considered as a thermodynamic distribution constant. It is readily apparent that this thermodynamic distribution constant is not *a priori* identical with the conventional chromatographic distribution constant.

For the activity ($a$) of the solute in either phase,

$$a = f/f^0 \tag{4.6}$$

where $f$ is the actual fugacity of the solute and $f^0$ is the fugacity of the solute in a chosen standard state. Further, in order to define a relationship between the activity and concentration of the solute in the given phase, a reference state (the state of unit activity coefficient) has to be chosen. Hence, the chromatographic and thermodynamic distribution constants can be related to each other only through the appropriate choice of the standard and reference states for the solute in both phases. It should be noted that the numerical values of the thermodynamic distribution constant and, consequently, of the $\Delta G_s^0$ values calculated from it depend on the choice of the standard and reference states. It is therefore necessary that these states be specified unequivocally whenever numerical data are presented on the thermodynamic properties dependent on the above choice. Examples are presented later in this chapter. It is evident from the above discussion that only *standard* thermodynamic properties of sorption can be calculated from retention data.

The temperature and pressure dependence of the Gibbs free energy and also data on the enthalpy ($H$) and entropy ($S$) can be obtained through the use of the general thermodynamic definitions, such as:

$$(\partial G/\partial T)_p = -S \tag{4.7}$$

$$[\partial(G/T)/\partial T]_p = -H/T^2 \tag{4.8}$$

$$(\partial G/\partial p)_T = V \tag{4.9}$$

$$G = H - TS \tag{4.10}$$

where $p$, $T$ and $V$ denote pressure, absolute temperature and volume, respectively.

**Distribution constant in liquid–liquid systems**

In order to distinguish between the thermodynamic and chromatographic distribution constants, the former will be denoted by $K_{th}$. Further, systems with completely immiscible liquids as the chromatographic phases, giving binary solute–solvent solutions, will be considered. In the case of partial miscibility of the liquids, the solute–solvent mixtures can be considered as pseudo-binary solutions and the quantities describing the properties of such systems conceived accordingly.

Let us make the following choice of the standard and reference states: for solute standard states in both phases, pure solute at $T$ and $p$ of the system; and for reference states for the solute in both phases, an infinitely dilute solution of solute at $T$ and $p$ of the system (quantities referring to this reference state will be denoted with an asterisk).

We can then write

$$f_{is} = \gamma^*_{is} h'' x'' \tag{4.11a}$$

$$f_{im} = \gamma^*_{im} h' x' \tag{4.11b}$$

$$f^0_{is} = h'' \tag{4.12a}$$

$$f^0_{im} = h' \tag{4.12b}$$

$$K^*_{\text{th}} = (\gamma^*_{is} x'' / \gamma^*_{im} x')_{\text{eq.}} \tag{4.13}$$

where $\gamma^*$ is the Henry's law activity coefficient ($\gamma^* \rightarrow 1$ for $x \rightarrow 0$), $h$ is the Henry's law constant, and $x$ is the mole fraction of the solute in the respective solution. Under the usual chromatographic conditions, both $x''$ and $x'$ are extremely small, so that the actual values of $\gamma^*_{is}$ and $\gamma^*_{im}$ approach unity. Hence, the thermodynamic distribution constant and the standard molar sorption Gibbs free energy expressed in terms of the above convention are

$$K^*_{\text{th}} = x''/x' \tag{4.14}$$

and

$$\Delta G^*_{\text{s}} = -RT \ln (x''/x') \tag{4.15}$$

(the subscript eq. will be omitted henceforth).

The chromatographic distribution constant is given by

$$K = \frac{V_R - V_m}{V_s} = \frac{n''/V_s}{n'/V_m} = \frac{x'' M_m \rho_s}{x' M_s \rho_m} \tag{4.16}$$

where $M_s$ and $M_m$ are the molecular weights and $\rho_s$ and $\rho_m$ are the densities of the sorbent and the mobile phase, respectively; for the other symbols, see eqns. 3.23 and 3.30. Hence,

$$K = K^*_{\text{th}} M_m \rho_s / M_s \rho_m \tag{4.17}$$

and

$$\Delta G^*_{\text{s}} = -RT \ln (K M_s \rho_m / M_m \rho_s) \tag{4.18}$$

Let us now change the reference states for those of the pure substance at $T$ and $p$ of the system, leaving the standard states the same as before. Thus, the standard and reference states will be identical (the quantities referring to this reference state will be denoted with a superscript zero). With this choice, Raoult's law is applicable so that it is possible to write:

$$f_{is} = \gamma^0_{is} f^0_i x'' \tag{4.19a}$$

$$f_{im} = \gamma^0_{im} f^0_i x' \tag{4.19b}$$

$$f^0_{is} = f^0_i \tag{4.20a}$$

$$f_{im}^0 = f_i^0 \tag{4.20b}$$

$$K_{th}^0 = \gamma_{is}^0 x''/\gamma_{im}^0 x' \tag{4.21}$$

where $\gamma^0$ is the Raoult's law activity coefficient ($\gamma^0 \to 1$ for $x \to 1$) and $f_i^0$ is the fugacity of the pure solute. In contrast to the previous choice of the reference states, the $\gamma^0$ values show maximum deviation from unity at values of $x$ approaching zero. The standard molar Gibbs free energy of sorption can be expressed in this case by

$$\Delta G_s^0 = -RT \ln (\gamma_{is}^0 x''/\gamma_{im}^0 x') \tag{4.22}$$

However, as $\Delta G_s^0$ refers to solute transfer between the same states, it follows that $\Delta G_s^0 = 0$, which gives $K_{th}^0 = 1$, and one can write with respect to eqn. 4.14:

$$x''/x' = \gamma_{im}^0/\gamma_{is}^0 = K_{th}^* \tag{4.23}$$

The substitution of $x''/x'$ from eqn. 4.23 into eqns. 4.15 and 4.16 gives

$$\Delta G_s^* = -RT \ln (\gamma_{im}^0/\gamma_{is}^0) \tag{4.24}$$

and

$$K = \gamma_{im}^0 M_m \rho_s/\gamma_{is}^0 M_s \rho_m \tag{4.25}$$

respectively.

From eqn. 4.24 $\Delta G_s^*$ can readily be expressed in terms of the excess Gibbs free energy. As

$$\Delta G^E = RT \ln \gamma^0 \tag{4.26}$$

eqn. 4.24 can be rewritten as

$$\Delta G_s^* = \Delta G_{is}^E - \Delta G_{im}^E \tag{4.27}$$

where $\Delta G_{is}^E$ and $\Delta G_{im}^E$ are the partial molar excess Gibbs free energies of the solute in the sorbent and in the mobile phase, respectively.

Eqn. 4.27 reveals that there is a simple correlation between the properties of a liquid–liquid system and the two gas–liquid systems in which either of the above liquid phases serves as the sorbent, provided that all the systems are compared at the same temperature. If the liquid–liquid system and the two corresponding gas–liquid systems are compared on a consistent standard and reference states basis, it is possible to write

$$\Delta G_s^* (\text{LL}) = \Delta G_s^* (\text{GL})_s - \Delta G_s^* (\text{GL})_m \tag{4.28}$$

where $\Delta G_s^*$ (LL) is the standard molar Gibbs free energy of sorption of a solute in the liquid–liquid system and $\Delta G_s^* (\text{GL})_s$ and $\Delta G_s^* (\text{GL})_m$ are the standard molar Gibbs free energies of sorption of the same solute in the gas–liquid systems with the sorbents of the stationary and the mobile phase of the liquid–liquid system, respectively. Therefore,

$$\Delta G_s^* (\text{GL})_s = \Delta G_{is}^E + \Delta G_c \tag{4.29}$$

and

$$\Delta G_s^* (\text{GL})_m = \Delta G_{im}^E + \Delta G_c \tag{4.30}$$

where $\Delta G_c$ is the standard molar Gibbs free energy of condensation of solute vapour to pure liquid solute. As $\Delta G_c$ is the same in both cases under the above conditions, eqns. 4.29, 4.30 and 4.27 give eqn. 4.28. Similarly, it can easily be shown that the chromatographic distribution constant of the liquid–liquid system is equal to the ratio of the corresponding gas–liquid distribution constants, *i.e.*,

$$K(\mathrm{LL}) = K(\mathrm{GL})_s / K(\mathrm{GL})_m \tag{4.31}$$

where the subscripts $s$ and $m$ refer to the gas–liquid systems with the sorbents of the stationary and the mobile phase of the corresponding liquid–liquid system, respectively. The above correlations, of course, are precisely valid only if the two liquids are completely immiscible.

*Dependence of the distribution constant on temperature and pressure*

Eqn. 4.25 indicates that $K$ is dependent on temperature and pressure through the temperature and pressure dependence of the activity coefficients and densities. In order to discuss these problems, it is appropriate to start with the following relationships:

$$\left(\frac{\partial \ln K}{\partial T}\right)_p = \left[\frac{\partial \ln (\gamma_{im}^0/\gamma_{is}^0)}{\partial T}\right]_p + \left[\frac{\partial \ln (\rho_s/\rho_m)}{\partial T}\right]_p \tag{4.32}$$

and

$$\left(\frac{\partial \ln K}{\partial p}\right)_T = \left[\frac{\partial \ln (\gamma_{im}^0/\gamma_{is}^0)}{\partial p}\right]_T + \left[\frac{\partial \ln (\rho_s/\rho_m)}{\partial p}\right]_T \tag{4.33}$$

With respect to eqns. 4.8 and 4.26, the temperature dependence of $\gamma^0$ is given by

$$\left(\frac{\partial \ln K}{\partial T}\right)_p = -\frac{\Delta H^E}{RT^2} = -\frac{\bar{H} - H^0}{RT^2} \tag{4.34}$$

where $\Delta H^E$ is the solute partial molar excess enthalpy, $\bar{H}$ is the solute partial molar enthalpy and $H^0$ is the molar enthalpy of pure solute. Further, as

$$\left(\frac{\partial \ln \rho}{\partial T}\right)_p = -\frac{1}{V}\left(\frac{\partial V}{\partial T}\right)_p = -\alpha \tag{4.35}$$

where $V$ is the volume and $\alpha$ is the coefficient of thermal expansion of the liquid, it is possible to write

$$\left(\frac{\partial \ln K}{\partial T}\right)_p = \frac{\bar{H}' - \bar{H}''}{RT^2} + \alpha_m - \alpha_s \tag{4.36}$$

where $\bar{H}'$ and $\bar{H}''$ are the solute partial molar enthalpies in the mobile and stationary phase, respectively, and $\alpha_m$ and $\alpha_s$ are the thermal expansion coefficients of the mobile and the stationary phase, respectively. Upon integration within the temperature limits $T_1$ and $T_2$, assuming $\bar{H}' - \bar{H}''$ to be constant within the temperature interval, we obtain

$$\frac{K_{T_2}}{K_{T_1}} = \exp\left[-\frac{\bar{H}' - \bar{H}''}{R}\left(\frac{1}{T_2} - \frac{1}{T_1}\right) + (\alpha_m - \alpha_s)(T_2 - T_1)\right] \tag{4.37}$$

where $K_{T_2}$ and $K_{T_1}$ are the distribution coefficients at temperatures $T_2$ and $T_1$, respectively.

It can be expected that the values of $\bar{H}'$ and $\bar{H}''$ and also of $\alpha_m$ and $\alpha_s$ will not differ very much from each other in liquid–liquid systems, which indicates that the temperature dependence of $K$ will be rather small in most instances. This implies that the ratio $K_{T2}/K_{T1}$ can acquire values very near to unity. For example, a temperature dependence of $K$ only about half as high as that encountered in GC can be inferred from Locke's data, although the latter were measured in systems that afforded the highest dependences of $K$ on temperature (alcohols chromatographed in a glycerol–hydrocarbon system).

The pressure dependence of $\gamma^0$ can be formulated by virtue of eqns. 4.9 and 4.27:

$$\left(\frac{\partial \ln \gamma^0}{\partial p}\right)_T = V^E = \bar{V} - V^0 \tag{4.38}$$

where $V^E$, $\bar{V}$ and $V^0$ are the solute partial molar excess volume, partial molar volume and molar volume of pure solute, respectively. The pressure dependence of the density is expressed by

$$\left(\frac{\partial \ln \rho}{\partial p}\right)_T = -\frac{1}{V}\left(\frac{\partial V}{\partial p}\right)_T = \beta \tag{4.39}$$

where $\beta$ is the compressibility coefficient of the liquid. Hence,

$$\left(\frac{\partial \ln K}{\partial p}\right)_T = \frac{\bar{V}' - \bar{V}''}{RT} + \beta_s - \beta_m \tag{4.40}$$

where $\bar{V}'$ and $\bar{V}''$ are the solute partial molar volumes in the mobile and in the stationary phase, and $\beta_m$ and $\beta_s$ are the compressibility coefficients of the mobile and of the stationary phase, respectively. Integration within the limits $p_1$ and $p_2$ gives

$$\frac{K_{p2}}{K_{p1}} = \exp\left\{\left[\frac{\bar{V}' - \bar{V}''}{RT} + (\beta_s - \beta_m)\right](p_2 - p_1)\right\} \tag{4.41}$$

provided that $\bar{V}' - \bar{V}''$ is constant.

It can again be assumed that the values of $\bar{V}'$ and $\bar{V}''$ and also of $\beta_s$ and $\beta_m$ will be similar, so that the pressure dependence of $K$ will be inappreciable in most instances.

It should be noted that the temperature and pressure dependence of $K$ could be arrived at directly through the respective dependences of $\Delta G_s^*$. Thus, eqn. 4.18 can be rewritten to read

$$\ln K = -(\Delta G_s^*/RT) + \ln(\rho_s/\rho_m) + \ln(M_m/M_s) \tag{4.42}$$

where $\Delta G_s^* = \mu_{is}^* - \mu_{im}^*$. As the $\mu^*$ terms actually represent the solute partial molar Gibbs free energies at infinite dilution in the respective phases, it follows with respect to eqns. 4.8 and 4.9 that

$$\frac{1}{R}\left[\frac{\partial(\Delta G_s^*/T)}{\partial T}\right]_p = \frac{1}{R}\left\{\frac{\partial[(\mu_{is}^* - \mu_{im}^*)/T]}{\partial T}\right\}_p = \frac{\bar{H}' - \bar{H}''}{RT^2} \tag{4.43}$$

and

$$\frac{1}{RT}\left(\frac{\partial \Delta G_s^*}{\partial p}\right)_T = \frac{1}{RT}\left[\frac{\partial(\mu_{is}^* - \mu_{im}^*)}{\partial p}\right]_T = -\frac{\bar{V}' - \bar{V}''}{RT} \tag{4.44}$$

This direct approach will be applied in the discussion of the distribution constant in liquid–solid systems.

**Distribution constant in liquid–solid systems**

As pointed out above, we cannot speak about solute–sorbent solutions when considering systems of a solid adsorbent and a liquid and it is therefore impossible to define the concentration or activity of the solute in the stationary phase in terms of the mole fraction. With respect to the definition of the liquid–solid distribution constant as presented in Chapter 2 (p. 20), it is appropriate to express the above concentration in terms of number of moles of the solute adsorbed per gram of the adsorbent. This concentration unit will be denoted by $\boldsymbol{m}$, and the chromatographic partition constant, defined by the ratio $(n''/w_s)/(n'/V_m)$, is denoted by $K_a$. The activity of the solute in the stationary phase will be expressed by the use of a practical activity coefficient defined by

$$a'' = \gamma_{is}^{+} \boldsymbol{m}'' \tag{4.45}$$

while the situation in the mobile phase will be described in the same manner as with LLC. All quantities referring to the above convention of the practical activity coefficient will be designated by a superscript$^{+}$. Provided that the specific surface area of the adsorbent is known, all of the quantities referred to the concentration units of $\boldsymbol{m}$ can readily be converted into those referred to unit surface area.

The following choice of the standard and reference states will be made in this instance: for the solute standard state in the stationary phase, one mole of the solute per gram of the adsorbent ($\boldsymbol{m}'' = 1$) at $T$ and $p$ of the system; for the solute standard state in the mobile phase, pure solute at $T$ and $p$ of the system; for the reference state for the solute in the stationary phase, infinitely low deposition of solute on the adsorbent surface ($\gamma_{is}^{+} \rightarrow 1$ for $\boldsymbol{m}'' \rightarrow 0$) at $T$ and $p$ of the system; and for the reference state for the solute in the mobile phase, infinitely dilute solution of solute ($\gamma_{im}^{*} \rightarrow 1$ for $x' \rightarrow 0$) at $T$ and $p$ of the system.

Thus,

$$f_{is} = \gamma_{is}^{+} \kappa \boldsymbol{m}'' \tag{4.46a}$$

$$f_{im} = \gamma_{im}^{*} h' x' \tag{4.46b}$$

$$f_{is}^{0} = \kappa \tag{4.47a}$$

$$f_{im}^{0} = h' \tag{4.47b}$$

$$K_{\text{th}}^{+} = \gamma_{is}^{+} \boldsymbol{m}'' / \gamma_{im}^{*} x' \tag{4.48}$$

where $\kappa$ is a proportionality constant. As in liquid–liquid chromatography, both $\gamma_{is}^{+}$ and $\gamma_{im}^{*}$ approach unity under usual conditions, so that

$$K_{\text{th}}^{+} = \boldsymbol{m}''/x' \tag{4.49}$$

and

$$\Delta G_{\text{s}}^{+} = -RT \ln (\boldsymbol{m}''/x') \tag{4.50}$$

The chromatographic distribution constant in LSC is given by

$$K_a = m''/(n'/V_m) \qquad (4.51)$$

As $n'/V_m = x'\rho_m/M_m$, it can readily be shown that

$$K_a = m''M_m/x'\rho_m \qquad (4.52)$$

and

$$\Delta G_s^+ = -\ RT \ln (K_a\rho_m/M_m) \qquad (4.53)$$

Owing to the great influence of the mobile phase on the sorption properties of adsorbents, there is no analogy between LSC and LLC regarding the correlation between data obtained in a LSC system and the corresponding GSC systems, *i.e.*, eqn. 4.31 is inapplicable in liquid–solid chromatography.

*Dependence of the distribution constant on temperature and pressure*

The solution of eqn. 4.53 for $\ln K_a$ gives

$$\ln K_a = -(\Delta G_s^+/RT) - \ln \rho_m + \ln M_m \qquad (4.54)$$

where $\Delta G_s^+ = \mu_{is}^+ - \mu_{im}^*$. Hence, employing the concepts mentioned with eqns. 4.43 and 4.44, it is possible to write with regard to eqns. 4.35 and 4.39:

$$\left(\frac{\partial \ln K_a}{\partial T}\right)_p = -\frac{\bar{H}' - \bar{H}''}{RT^2} + \alpha_m \qquad (4.55)$$

and

$$\left(\frac{\partial \ln K_a}{\partial p}\right)_T = \frac{\bar{V}' - \bar{V}''}{RT} - \beta_m \qquad (4.56)$$

Upon integration, eqns. 4.55 and 4.56 will obviously give relationships 4.37 and 4.41 with $\alpha_s$ and $\beta_s$, respectively, equal to zero.

If either $\ln K$ or $\ln K_a$ is differentiated with respect to $1/T$ rather than $T$, then

$$\left[\frac{\partial \ln K}{\partial \left(\frac{1}{T}\right)}\right]_p = \frac{\bar{H}' - \bar{H}''}{R} - T^2(\alpha_m - \alpha_s) \qquad (4.57)$$

and

$$\left[\frac{\partial \ln K_a}{\partial \left(\frac{1}{T}\right)}\right]_p = \frac{\bar{H}' - \bar{H}''}{R} - \alpha_m T^2 \qquad (4.58)$$

Hence, if the $\alpha$ terms are negligible in comparison with the $\bar{H}$ term, plots of $\log K$ or $\log K_a$ against $1/T$ should be straight lines within certain temperature limits, but with slopes appreciably less than those of analogous plots in gas chromatography.

## REFERENCES

Ashworth, A. J. and Everett, D. H., *Trans. Faraday Soc.*, 56 (1960) 1609.

Denbigh, K. G., *Principles of Chemical Equilibrium,* Cambridge Univ. Press, London, New York, 1957, pp. 242–246.

Ewell, R. N., Harrison, J. H. and Berg, L., *Ind. Eng. Chem.*, 36 (1944) 871.

Green, J. and McHale, D., *Advan. Chromatogr.*, 2 (1966) 99.

Guggenheim, E. A., *Proc. Roy. Soc., Ser. A,* 203 (1944) 183.

Hildebrand, J. H. and Scott, R. L., *Regular Solutions*, Prentice-Hall, Engelwood Cliffs, N.J., 1962.

Keesom, W. H., *Phys. Z.*, 221 (1921a) 126.

Keesom, W. H., *Phys. Z.*, 221 (1921b) 643.

Keulemans, A. I. M., *Gas Chromatography*, Reinhold, New York, 2nd ed., 1959, p.168.

Lennard-Jones, J. E., *Proc. Phys. Soc., London*, 43 (1937) 461.

Locke, D. C., *J. Gas Chromatogr.*, 5 (1967) 202.

London, F., *Trans. Faraday Soc.*, 331 (1937) 8.

Martin, A. J. P., *Biochem. Soc. Symp.*, 3 (1950) 4.

Martire, D. E., in A. B. Littlewood (Editor), *Sixth Int. Symp. on Gas Chromatography*, Institute of Petroleum, London, 1956, Paper 5.

Martire, D. E., *Anal. Chem.*, 33 (1961) 1143.

Martire, D. E. and Pollara, L. Z., *Advan. Chromatogr.*, 1 (1965) 335.

Pierotti, G. J., Deal, C. H., Derr, E. L. and Porser, P., *J. Amer. Chem. Soc.*, 78 (1956) 2989.

*Chapter 5*

# Gel permeation chromatography

M. KUBÍN

## CONTENTS



## INTRODUCTION

Gel permeation chromatography (GPC) is a column separation technique based on a non-ionic molecular-sieve effect and, in contrast to other chromatographic methods, separates solutes exclusively according to their size. The origin of the method dates back to the early 1950s (Lindqvist and Storgårds, Synge); however, the main incentive for its rapid development was the publication of a fundamental paper by Porath and Flodin. Gel filtration in aqueous solutions with the well known Sephadex gels soon became a widely used technique in biochemistry (see the review articles by Determann, Gelotte, and Morris and Morris for references). The first attempts to fractionate synthetic high polymers by this method were made by Brewer (1960) and Vaughan; however, it was not until 1964, when Moore published his work on the preparation of GPC packings based on highly cross-linked polystyrenes with controlled pore size, that a rapid increase of research activity in this field began. The early history of GPC was summarized in a review by Altgelt and Moore.

## PRINCIPLES OF GEL PERMEATION CHROMATOGRAPHY

A chromatographic column is packed with particles that have a specified maximum pore size and a specified pore size distribution, and the voids between the particles and the pores are filled with a suitable solvent. A small sample is introduced as a dilute solution on to the top of the column and the solvent is continually passed through, the solute concentration in the effluent being monitored with a suitable sensitive detector. The detailed mechanism of the separation is not yet fully understood and several theoretical models that have been proposed will be discussed later, but the simplest explanation is that the molecules of the solute are partitioned between the stationary solvent in the pores and the moving solvent outside.

*References p.66*

In terms of this simple model, the volume of the solvent in the column, $V_t$, consists of the volume of all available pores, $V_i$, and the void volume, $V_0$:

$$V_t = V_0 + V_i \tag{5.1}$$

The largest molecules are completely excluded from the packing and will be eluted at the volume $V_0$; small molecules have access to all pores and elute at $V_t$, whereas a certain fraction of the pore volume is available for the intermediate-sized solutes and they will appear between $V_0$ and $V_t$. Thus,

$$V_R = V_0 + K_d V_i \tag{5.2}$$

where $V_R$ is the retention volume of a given solute, characterized by a certain value of the volume partition coefficient, $K_d$, which is a measure of the apparent fractional permeation of the pores by this solute. If no specific interaction (adsorption) takes place between the solutes and the matrix of the packing, the partition coefficient, $K_d$, will lie between zero and unity. For a specific polymer-solvent system on a given column, there exists an unequivocal relationship between the molecular size (molecular weight) and retention volume; a calibration curve (usually plotted as log $M$ *versus* $V_R$) can be established by determining the elution pattern of a series of narrow fractions with molecular weights measured by an independent method. A typical calibration curve of a GPC column is plotted schematically in Fig. 5.1.

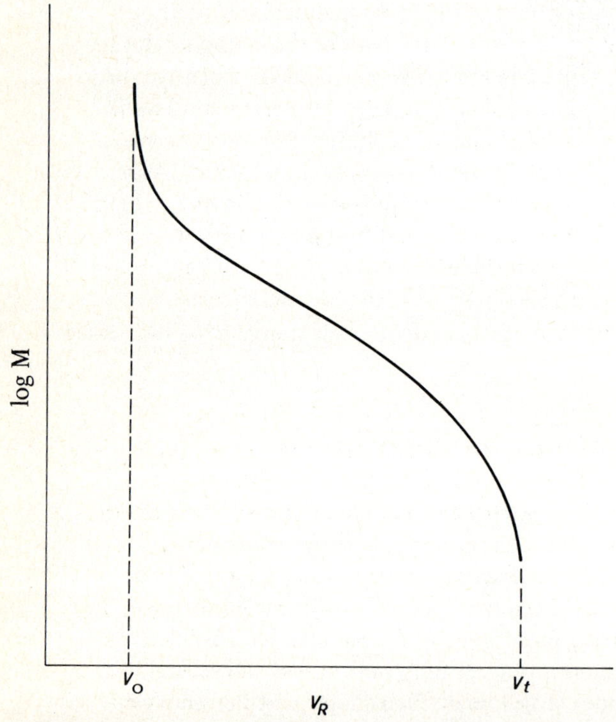


Fig. 5.1. Schematic representation of the molecular weight–retention volume dependence of a single gel permeation column.

## PHYSICAL BASIS OF THE SEPARATION PROCESS

Gel permeation chromatography (the most popular of several suggested names for the method, *e.g.*, gel filtration, gel chromatography and exclusion chromatography) can be classified as a type of liquid–solid chromatography, although it might be preferable to consider it as a special case of liquid–liquid chromatography in which the solvent immobilized in the "pores" of the packing acts as a stationary phase. However, in spite of the popularity of the technique and its great success in solving many practical separation problems, a comprehensive theory describing the physical basis of separations by GPC has not yet been formulated.

### Theories of the partition coefficient with regard to gel permeation

As pointed out by Casassa (1971), all of the theories concerned with the relative velocity with which a given solute travels through the column can be expressed in the form of eqn. 5.2; their real content is condensed into the explicit form that they predict for the partition coefficient, $K_d$, and its dependence on the relevant characteristics of a particular packing–solvent–solute system and on the operating parameters of the GPC apparatus.

The equilibrium theories start with the simple assumption that the retention volume of a polymeric solute is determined by the equilibrium partitioning between the liquid in the interstitial phase and the liquid entrapped in the micropores of the packing, so that $K_d$ is the ratio of solute concentrations in the two phases at equilibrium. All the early theories of GPC fall into this category. Thus, Porath (1963) considered a simple model of spherical solutes characterized by their effective hydrodynamic radius in conical pores of equal size. Laurent and Killander, starting from the results published by Ogston, proposed a model of infinitely long, randomly distributed straight rods for the structure of the porous packing. Le Page *et al.* (1968) and, independently, Cantow and Johnson derived a relationship between the pore-size distribution of the beads and the calibration curve. The exclusion of rigid particles of various shapes from pores of geometrically simple forms was described by Giddings *et al.*

Casassa (1967) and later Casassa and Tagami ascribed the change in free energy that accompanies the transfer of a flexible molecule from the outer solution to the limited space within a pore entirely to the reduction of conformational entropy and calculated the partition coefficient of a polymer between the bulk phase and porous packing for different assumed geometries of cavities (spheres, cylinders, slabs). The results can be expressed in the form of the equation

$$K_d = Z \sum_{m=1}^{\infty} \alpha_m^{-2} \exp\left[-\beta_m^2 (R/a)^2\right] \qquad (5.3)$$

where $R$ is the root-mean-square radius of gyration of the (unconfined) flexible polymer chain and $a$ is a characteristic dimension of the pore; the parameters $Z$, $\alpha_m$ and $\beta_m$ have different forms depending on the geometry of the cavity. This theory predicts that the separation in GPC depends mainly on the effective hydrodynamic volume of the chains,

*References p.66*

regardless of whether they are linear or branched. Yau and Malone (1971) published the results of an experimental study in support of this treatment. Casassa's theory was further generalized by Pouchlý (1963, 1970) to include the effect of adsorption forces near the partially permeable phase boundary.

Carmichael extended the stochastic theory of chromatography developed by Giddings and Eyring and by McQuarrie to GPC.

The equilibrium theories, apart from their basic simplicity, have the additional advantage that the $K_d$ values are considered to be true equilibrium constants and can be directly compared with static sorption experiments. Few such comparisons have been attempted and their results are rather conflicting. For example, Yau *et al.* (1968) found excellent agreement between static and elution experiments performed with narrow polystyrene fractions on a porous glass packing, but with high-molecular-weight polystyrene samples on cross-linked polystyrene beads marked deviations were observed. Similarly, Ackers, working with Sephadex gels and proteins, found that on the less swollen gels similar results were obtained in static and elution partitioning experiments, but large differences occurred with very highly swollen, slightly cross-linked packings. Haller's results, obtained with porous glass with a very uniform pore-size distribution, indicated that whereas for both very large (tobacco mosaic virus) and small molecules (benzyl alcohol), the steric exclusion principle is valid, the behaviour of an intermediate-sized solute does not fit the equilibrium steric exclusion theory.

The concept that the retention volume is determined solely by equilibrium partitioning is generally accepted in other types of chromatography, which usually deal with small molecules; its validity in the GPC of macromolecules is less apparent owing to the low diffusion coefficients of such molecules and to the relative inaccessibility of micropores with complicated shape deep within the beads. Another mechanism which might be operative is that of restricted diffusion. Ackers combined this mechanism with the steric exclusion of larger particles from narrow pores. Yau and Malone (1967) considering the effect of axial diffusion on the retention volume, derived the equation

$$V_R = V_0 + V_i\,[(\gamma_1/U^{1/2})(1 - e^{-U\gamma_2}) + \mathrm{erfc}\,(U^{1/2}\gamma_3)] \quad (5.4)^*$$

where $U$ is the solvent velocity (cm/sec) in the interstitial volume and the parameters $\gamma_i$ depend on the diffusion coefficient of the solute in the gel phase. However, their results showed a strong dependence of the retention volume on flow-rate, which was never observed experimentally in GPC. Detailed theoretical treatments of longitudinal and axial diffusion on the elution behaviour (Giddings, 1965; Hermans; Kubín; Vink) have shown that they have no effect on the retention volume and influence only the width and shape of the elution band.

A different mechanism of separation, due to the Poiseuille flow of the solution through a system of narrow capillaries, has been proposed by DiMarcio and Guttman (1969, 1970), Guttman and DiMarcio and Verhoff and Sylvester. These models were analyzed in detail by Casassa (1971), who showed that they are in effect equivalent and, moreover, that they predict retention volumes (apart from a vanishingly small additional term) which coincide with the values predicted by equilibrium steric exclusion theories.

---

*erfc = Error function complement.

**Column efficiency**

The fact that $K_d$ in eqn. 5.2 satisfies the inequality $0 \leqslant K_d \leqslant 1$ has a serious consequence for the separating ability of GPC: the total working region of retention volumes in GPC is confined to the interval of about $(0.4–0.9) \cdot V_{total}$, where $V_{total}$ is the total volume of the (empty) column. This is in distinct contrast to the other types of chromatography, where the retention volumes can be very high, limited only by practical considerations such as finite detector sensitivity, time and cost of analysis. Therefore, much emphasis must be laid on the quality of GPC packings and on the efficient and balanced design of the whole GPC system and it is necessary to operate the instruments near the optimum values of all operational variables.

In 1961, Flodin studied experimentally the influence of gel type, particle mesh size, solvent type and flow-rate on the efficiency of Sephadex columns. Giddings and Mallik modified the well known Van Deemter equation for the specific case of GPC. Giddings (1967) calculated the maximum number, $n_{max.}$, of components that can be resolved (to baseline separation) within the volume working interval of a GPC column:

$$n_{max.} \approx 1 + 0.2\, N^{1/2} \tag{5.5}$$

where $N$ is the number of theoretical plates. Later, Giddings (1968) analyzed the resolution and optimization of operational variables in GPC on the basis of the general theory of chromatography and found that the highest resolution is to be expected with long, narrow columns packed with fine particles.

Several workers studied theoretically and experimentally the influence of operational variables on peak spreading in GPC. Hendrickson resolved the zone spreading into components due to axial and longitudinal diffusion, interstitial volume, sample polydispersity and extra-column effects, and concluded that longitudinal diffusion played a minor role. With high polymers, the axial diffusion in the beads is mostly responsible for the peak broadening. Billmeyer *et al.*, Billmeyer and Kelley and Kelley and Billmeyer (1969, 1970a, 1970b) also found that longitudinal molecular diffusion has a negligible effect and that the zone broadening of permeating solutes is diffusion-controlled. Biesenberger and Ouano and Ouano and Biesenberger examined the extra-column effects and found that the connecting tubing and the detector cell can contribute significantly to the width and skewness of elution peaks in GPC, especially for high-molecular-weight solutes. The increase in column efficiency effected by decreasing the flow-rate and decreasing the particle size has been confirmed by several investigators (Čoupek and Heitz, Le Page *et al.*, Smith and Kollmansberger).

The analysis of the molecular-weight dependence of peak spreading is complicated by the fact that the fractions of high polymers used as standards are not monodisperse. In order to isolate the real peak spreading from the effect of sample polydispersity, Tung *et al.* proposed a reversed-flow technique: when a polydisperse sample has been allowed to proceed half way through the column and then the direction of flow is suddenly reversed, the molecular-weight separation process is also reversed and thus compensated for, whereas the zone spreading process continues to broaden the peak. The real zone spreading factor can thus be determined for the two halves of a column separately. The arrangement used in these experiments is shown schematically in Fig. 5.2. Two four-port

*References p.66*

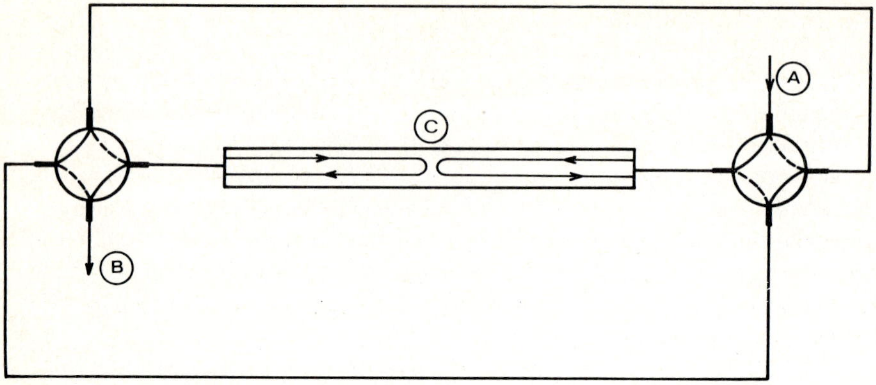


Fig. 5.2. Schematic representation of the experimental arrangement for determining the spreading factor, $h$, by the reversed-flow technique. The solvent is delivered at A, the detector is connected at B and C is the calibrated column. The arrows indicate the direction of movement of the sample in the two halves of the column.

valves are inserted into the stream of solvent just before and after the calibrated column. For example, in order to determine the spreading factor for the right-hand half of the column, the sample is introduced with the two valves in the first position (full lines) and, after a volume of the solvent corresponding to half of the pre-determined elution volume of the sample has passed through, both valves are simultaneously turned to the other position (dashed lines). Yau *et al.* (1970) concluded, on the basis of these experiments, that the hydrodynamic separation mechanism proposed by DiMarcio and Guttman (1969, 1970) is probably not operating under normal conditions in GPC.

In contrast to the other types of chromatography, in the GPC of polymers the concepts of baseline separation, number of theoretical plates and resolution are of relatively minor importance; the main problem is to calculate the true molecular-weight distribution of a given polymer from the GPC record, which is usually the sum of excessively overlapping unresolved peaks of individual components (mers)*. In solving this problem, the chromatogram must be corrected for imperfect resolution of the column, otherwise the calculated molecular-weight distribution will be distorted and will contain end-sections that are non-existent in the real sample, due only to the spreading of the zone.

It has been shown by Tung (1966a) that the correction for imperfect resolution consists in the solution of an integral (Tung's) equation:

$$f(v) = \int_{V_0}^{V_t} w(y)\, W(v, y) \cdot \mathrm{d}y \qquad (5.6)$$

where $f(v)$ is the observed chromatogram, $w(y)$ is the function sought (chromatogram corrected for imperfect resolution) and the function $W(v, y)$ (this kernel of the integral

*Nevertheless, the HETP measured with some suitable low-molecular-weight solute is often used to characterize the efficiency of GPC columns; however, Bly (1968) has shown that the HETP determined in this manner can be very misleading.

equation is often called the spreading function) is characteristic of the given column arrangement and can be regarded as its response to an extremely narrow input of a sample having a retention volume $y$; in most instances, this function has the so-called convolution form $W(v - y)$; further, for high-resolution columns, it can be represented by a Gaussian curve:

$$W(v - y) = (h/\pi)^{1/2} \exp\left[-h(v - y)^2\right] \quad (5.7)$$

The information on the resolving power of the column is now contained in the parameter $h$ (which may or may not depend on the elution volume); its relation to the variance $\sigma^2$ of a single-component peak centred at $y$ is

$$h = 1/2\,\sigma^2 \quad (5.8)$$

Methods proposed for solving eqn. 5.6 are discussed in more detail in Chapter 3; in the present context, it is of interest that the resolution factor, $h$, is usually an increasing function of elution volume (May and Knight; Smit *et al.*; Tung, 1966b; Tung *et al.*, 1966), although sometimes a minimum is observed (Hendrickson; Tung and Runyon; Yau *et al.*, 1970).

**Calibration of GPC column systems**

The hopes that GPC with well characterized porous packings could be developed into an absolute method for determining molecular weights and their distribution (Beau *et al.*) have not been substantiated, and calibration of GPC columns with narrow fractions of known molecular weight is necessary. This calibration is carried out simply by injecting a series of well defined fractions. The logarithms of their molecular weights are then plotted against the elution volumes measured at the peak maximum and the best line is drawn through the points either by hand or by means of some standard statistical procedure.

It has been ascertained for a wide variety of polymers on various packings and with different solvents that a linear relationship of the form

$$\log M = A - B\,V_R \quad (5.9)$$

is approximately valid. The unique properties and advantages of a linear semi-logarithmic calibration curve have been discussed by Bly (1969). It is usually necessary to combine several columns (capable of separating polymers of different maximum molecular weights) in series in order to obtain good linearity of the calibration curve. The effect of some operational variables on the reliability of calibration was studied by several workers. Boni *et al.* (1968a), Lambert, Meyerhoff (1965a) and Rudin examined the effect of sample size on retention volume and discussed methods of minimizing it. It is recommended that the calibration should be carried out at the same values of the operational variables (flow-rate, temperature) as used in the subsequent analyses of the given polymer. In this manner, it is possible to obtain molecular weights and distributions of good accuracy from GPC data, and the calibration curve remains unchanged for a long time, although some deterioration of column efficiency has been reported (Hazell *et al.*).

*References p.66*

The necessity to have, for each analyzed polymer, a series of narrow fractions of known molecular weights for calibration is a serious disadvantage. Attempts to construct calibration curves on the basis of polymers with broad molecular-weight distributions have been reported by Cantow *et al.*, Frank *et al.*, Purdon and Mate and Weiss and Cohn-Ginsberg, but the results seem to be less accurate than those from narrow fractions. Many workers have therefore proposed methods for the correlation of calibration curves for different polymers or polymer–solvent systems. The solution of this problem would permit calibrations to be made with polystyrene or polypropylene oxide fractions, which are commercially available, and then the calibration curve thus obtained to be used for other polymers of interest. Thus, Moore and Hendrickson proposed the plot of extended

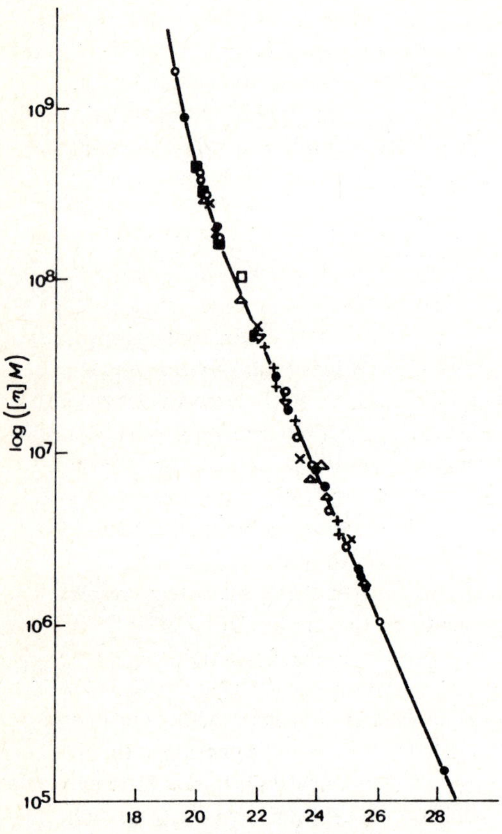


Fig. 5.3. Universal calibration in GPC (Grubisic *et al.*) expressed as the dependence of the hydrodynamic volume parameter, $[\eta]M$, on elution volume for several polymers and copolymers with different structures. •, Polystyrene; ○, "comb-type" branched polystyrene; +, "star-shaped" branched polystyrene; ×, poly(methyl methacrylate); ◑, poly(vinyl chloride); ▪, "ladder-type" polyphenylsiloxane; □, polybutadiene; △, graft styrene–methyl methacrylate copolymer; △, heterograft (block) copolymer polystyrene–poly(methyl methacrylate)–polystyrene. Results obtained on four Styragel columns in tetrahydrofurane (1 ml/min) at room temperature.

chain-length *versus* retention volume as the basis of such a universal calibration method, but it soon became apparent that significant differences between individual polymers persist (Salovey and Hellman). Meyerhoff (1965b) used the product of the effective hydrodynamic radius and the square root of the molecular weight as a universal parameter, but the suggestion of Benoit *et al.* and Grubisic *et al.* that the effective hydrodynamic volume is a decisive factor in the elution behaviour of different polymers on GPC columns was soon widely accepted. This universal calibration parameter has a sound physical basis and is easy to apply, as the hydrodynamic volume of a polymer at a given temperature in a certain solvent is proportional to the product $[\eta] \cdot M$, and the intrinsic viscosity, $[\eta]$, can easily be measured. Fig. 5.3 shows the excellent agreement of calibration curves plotted as log $([\eta]M)$ *versus* $V_R$ for several very different polymers, including linear and branched polystyrenes, two grafted styrene—methyl methacrylate copolymers with different structures, polybutadiene, poly(vinyl chloride), poly(methyl methacrylate) and stiff-chain ladder polyphenylsiloxane.

These results have been confirmed by other workers (Boni *et al.*, 1968b; Dawkins and Hemming; Le Page *et al.*; Rohn; Wild and Guliana). Some conflicting evidence has also been published (see Crouzet *et al.*; Meyerhoff, 1965a,b; Swenson *et al.*). Recently, Rudin and Hoegy attributed part of this discrepancy to the fact that the hydrodynamic volume of the solvated polymer *at infinite dilution* is implicitly introduced into the universal calibration parameter, $[\eta]M$, whereas the calibration is performed at finite concentrations of the solute, and they proposed a modified procedure in order to remove the apparent discrepancies.

If it is desirable to correct the chromatogram for imperfect resolution prior to converting it into the molecular-weight distribution, the columns must also be calibrated with respect to the resolution factor, $h$ (see eqn. 5.7); this calibration is carried out either by the already mentioned reversed-flow technique (Tung, 1966b; Tung *et al.*), by a method using the leading edge of chromatographic peaks (Tung and Runyon) or by means of the additivity of variance method proposed by Hendrickson (see also May and Knight). Balke and Hamielec also proposed a method that obviated the use of the tedious and time-consuming reversed-flow technique. A method that is particularly attractive for its simplicity has been proposed by Kendrick: he assumed the validity of the calibration curve according to eqn. 5.9 and further assumed that the molecular-weight distribution of all fractions can be adequately approximated by the log-normal distribution (this latter approximation is certainly valid for the narrow fractions that are normally employed for calibration purposes). Under these conditions, an analytical solution of eqn. 5.6 exists (Tung, 1966a) and the following relationship can serve for determining the spreading factor, $h$, from the chromatogram of a fraction and from its known polydispersity:

$$f(v_{\max.}) / \int_{V_0}^{V_t} f(y)\mathrm{d}y = [\pi\,(1/h + \beta^2/B^2)]^{-\frac{1}{2}} \tag{5.10}$$

where the left-hand side is the height of the peak divided by the total area of the chromatogram, $\beta = 2 \ln (\bar{M}_w/\bar{M}_n)$ and $B$ is the slope of the calibration curve, defined by eqn. 5.9; $\bar{M}_w$ and $\bar{M}_n$ are the weight- and number-average molecular weights of the fraction, respectively.

*References p.66*

## REFERENCES


Ackers, W., *Biochemistry*, 3 (1964) 723.
Algelt, K. H. and Moore, J. C., in M. J. R. Cantow (Editor), *Polymer Fractionation,* Academic Press, New York, London, 1967, p. 123.
Balke, S. T. and Hamielec, A. E., *J. Appl. Polym. Sci.*, 13 (1969) 1381.
Beau, R., Le Page, M. and De Vries, A. J., *Appl. Polym. Symp.*, 8 (1969) 137.
Benoit, H., Grubisic, Z., Rempp, P., Decker, D. and Zilliox, J., *J. Chim. Phys.*, 63 (1966) 1507.
Biesenberger, J. A. and Ouano, A., *J. Appl. Polym. Sci.*, 14 (1970) 471.
Billmeyer, Jr., F. W., Johnson, G. W. and Kelley, R. N., *J. Chromatogr.*, 34 (1968) 316.
Billmeyer, Jr., F. W. and Kelley, R. N., *J. Chromatogr.*, 34 (1968) 322.
Bly, D. D., *J. Polym. Sci., Part C*, 21 (1968) 13.
Bly, D. D., *Anal. Chem.*, 41 (1969) 477.
Boni, K. A., Sliemers, F. A. and Stickney, P. B., *J. Polym. Sci., Part A-2*, 6 (1968a) 567.
Boni, K. A., Sliemers, F. A. and Stickney, P. B., *J. Polym. Sci., Part A-2*, 6 (1968b) 579.
Brewer, P. J., *Nature (London)*, 188 (1960) 934.
Cantow, M. J. R. and Johnson, J. F., *J. Polym. Sci., Part A-1*, 5 (1967) 2835.
Cantow, M. J. R., Porter, R. S. and Johnson, J. F., *J. Polym. Sci., Part A-1*, 5 (1967) 1391.
Carmichael, J. B., *J. Polym. Sci., Part A-2*, 6 (1968) 517.
Casassa, E. F., *J. Polym. Sci., Part B*, 5 (1967) 773.
Casassa, E. F., *J. Phys. Chem.*, 75 (1971) 3929.
Casassa, E. F. and Tagami, Y., *Macromolecules*, 2 (1969) 14.
Čoupek, J. and Heitz, W., *Makromol. Chem.*, 112 (1968) 286.
Crouzet, P., Martens, A. and Mangin, P., *J. Chromatogr. Sci.*, 9 (1971) 525.
Dawkins, J. V. and Hemming, M., *Makromol. Chem.*, 155 (1972) 75.
Determann, H., *Angew. Chem.*, 76 (1969) 635.
DiMarcio, E. A. and Guttman, C. M., *J. Polym. Sci., Part B,* 7(1969) 267.
DiMarcio, E. A. and Guttman, C. M., *Macromolecules,* 3 (1970) 131.
Flodin, P., *J. Chromatogr.*, 5 (1961) 103.
Frank, F. C., Ward, I. M. and Williams, T., *J. Polym. Sci., Part A-2*, 6 (1968) 1357.
Gelotte, B., in A. T. James and L. J. Morris (Editors), *New Biochemical Separations,* Van Nostrand, Princeton, N. J., 1962, p.93.
Giddings, J. C., *Dynamics of Chromatography. Part 1: Principles and Theory*, Marcel Dekker, New York, 1965.
Giddings, J. C., *Anal. Chem.*, 39 (1967) 1027.
Giddings, J. C., *Anal. Chem.*, 40 (1968) 2143.
Giddings, J. C. and Eyring, H., *J. Phys. Chem.*, 59 (1955) 416.
Giddings, J. C., Kucera, E., Russell, C. P. and Myers, M. N., *J. Phys. Chem.*, 72 (1968) 4397.
Giddings, J. C. and Mallik, K. L., *Anal. Chem.*, 38 (1966) 997.
Grubisic, Z., Rempp, P. and Benoit, H., *J. Polym. Sci., Part B*, 5 (1967) 753.
Guttman, C. M. and DiMarcio, E. A., *Macromolecules,* 3 (1970) 681.
Haller, W. J., *J. Chromatogr.*, 32 (1968) 676.
Hazell, J. E., Prince, L. A. and Stapelfeldt, H. E., *J. Polym. Sci., Part C*, 21 (1968) 43.
Hendrickson, J. G., *J. Polym. Sci., Part A-2,* 6 (1968) 1903.
Hermans, J. J., *J. Polym. Sci., Part A-2*, 6 (1968) 1217.
Kelley, R. N. and Billmeyer, Jr., F. W., *Anal. Chem.*, 41 (1969) 874.
Kelley, R. N. and Billmeyer, Jr., F. W., *Anal. Chem.*, 42 (1970a) 399.
Kelley, R. N. and Billmeyer, Jr., F. W., *Separ. Sci.*, 5 (1970b) 291.
Kendrick, T. C., *J. Polym. Sci., Part A-2*, 7 (1969) 297.
Kubín, M., *Collect. Czech. Chem. Commun.*, 30 (1965) 1104 and 2900.
Lambert, A., *Polymer*, 10 (1969) 213.
Laurent, T. C. and Killander, J., *J. Chromatogr.*, 14 (1964) 317.
Le Page, M., Beau, R. and De Vries, A. J., *J. Polym. Sci., Part C*, 21 (1968) 119.

Lindqvist, B. and Storgårds, T., *Nature (London)*, 175 (1955) 511.
McQuarrie, D. A., *J. Chem. Phys.*, 38 (1963) 437.
May, Jr., J. A. and Knight, G. W., *J. Chromatogr.*, 55 (1971) 111.
Meyerhoff, G., *Ber. Bunsenges. Phys. Chem.*, 69 (1965a) 866.
Meyerhoff, G., *Makromol. Chem.*, 86 (1965b) 282.
Moore, J. C., *J. Polym. Sci., Part A*, 2 (1964) 835.
Moore, J. C. and Hendrickson, J. G., *J. Polym. Sci., Part C*, 8 (1965) 233.
Morris, J. C. O. R. and Morris, P., *Separation Methods in Biochemistry*, Wiley-Interscience, New York, 1963.
Ogston, A. G., *Trans. Faraday Soc.*, 54 (1958) 1754.
Ouano, A. and Biesenberger, J. A., *J. Appl. Polym. Sci.*, 14 (1970) 483.
Porath, J., *J. Pure Appl. Chem.*, 6 (1963) 233.
Porath, J. and Flodin, P., *Nature (London)*, 183 (1959) 1657.
Pouchlý, J., *Collect. Czech. Chem. Commun.*, 28 (1963) 1804.
Pouchlý, J., *J. Chem. Phys.*, 52 (1970) 2567.
Purdon, Jr., J. R. and Mate, R. D., *J. Polym. Sci., Part A-1*, 6 (1968) 243.
Rohn, C. L., *J. Polym. Sci., Part A-2*, 5 (1967) 547.
Rudin, A., *J. Polym. Sci., Part A-1*, 9 (1971) 2587.
Rudin, A. and Hoegy, H. L. W., *J. Polym. Sci., Part A-1*, 10 (1972) 217.
Salovey, R. and Hellman, M. Y., *J. Polym. Sci., Part A-2*, 5 (1967) 333.
Smit, J. A. M., Hoogervorst, C. J. P and Staverman, A. J., *J. Appl. Polym. Sci.*, 15 (1971) 1479.
Smith, W. B. and Kollmansberger, A., *J. Phys. Chem.*, 69 (1965) 4157.
Swenson, H. A., Kaustinen, H. M. and Almin, K.-E., *J. Polym. Sci., Part B*, 9 (1971) 261.
Synge, R. L. M., *Inst. Int. Chim. Solvay, Cons. Chim. (Rapp. Discuss.)*, 9 (1953) 163.
Tung, L. H., *J. Appl. Polym. Sci.*, 10 (1966a) 375.
Tung, L. H., *J. Appl. Polym. Sci.*, 10 (1966b) 1271.
Tung, L. H., Moore, J. C. and Knight, G. W., *J. Appl. Polym. Sci.*, 10 (1966) 1261.
Tung, L. H. and Runyon, J. R., *J. Appl. Polym. Sci.*, 13 (1969) 2397.
Vaughan, M. F., *Nature (London)*, 188 (1960) 55.
Verhoff, F. M. and Sylvester, N. D., *J. Macromol. Sci., Chem.*, 4 (1970) 979.
Vink, H., *J. Chromatogr.*, 52 (1970) 205.
Weiss, A. R. and Cohn-Ginsberg, E., *J. Polym. Sci., Part A-2*, 8 (1970) 148.
Wild, L. and Guliana, R., *J. Polym. Sci., Part A-2*, 5 (1967) 1087.
Yau, W. W. and Malone, C. P., *J. Polym. Sci., Part B*, 5 (1967) 663.
Yau, W. W. and Malone, C. P., *Polym. Prepr., Amer. Chem. Soc., Div. Polym. Chem.*, 12 (1971) 797.
Yau, W. W., Malone, C. P. and Fleming, S. W., *J. Polym. Sci., Part B*, 6 (1968) 803.
Yau, W. W., Malone, C. P. and Suchan, J. L., *Separ. Sci.*, 5 (1970) 259.

*Chapter 6*

# Fundamentals of ion-exchange chromatography

O. MIKEŠ

## CONTENTS



## PRINCIPLES AND TERMINOLOGY

Ion exchangers can be defined as polyvalent materials that are insoluble in water, contain bound ionogenic groups and are capable of dissociating and exchanging ions in solution. Sometimes the shortened term ionex is used instead of ion exchanger. In spite of the fact that there are natural ion exchangers, most ion exchangers have been prepared synthetically. Natural and synthetic ion exchangers may consist of inorganic or organic materials and are usually solid substances, but liquid ion exchangers are used in special circumstances.

Ion exchangers and their properties have been described in thousands of papers and tens of monographs, and many of these articles also deal with chromatographic aspects. The most important publications in this field since 1960 are those by Dorfner (1963a, b), Helfferich (1962a), Hering, Inczedy, Marinsky, Osborn, Paterson, Reuter, Saldadze *et al.* and Samuelson. Griesbach (now Reuter) is producing a comprehensive German work in numerous volumes. There are specialized chapters on ion exchange in monographs on chromatographic methods by Flaschka and Barnard, Genge, Kunin, Mikeš, Morris and Morris, and Walton (1967). Systematic reviews have been published every other year, the last three being by Walton (1968, 1970, 1972).

In a typical ion exchanger (see the schematic representation in Fig. 6.1), there are two main components: a porous matrix (or network) and electrically charged, covalently bound functional ionogenic groups. Ion exchangers can be divided into four main groups, depending on the composition of the matrix:

(1) inorganic exchangers, based on aluminium silicates and other suitable minerals;
(2) synthetic resins of many types;
(3) ion-exchange cellulose;
(4) ion-exchange polydextran.

*References p.86*

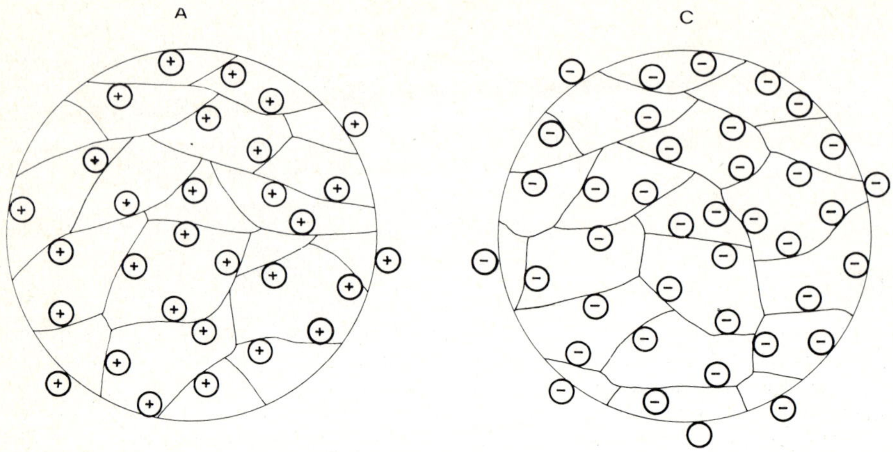


Fig. 6.1. Schematic representation of resinous ion-exchange particles. A = Anion exchanger; C = cation exchanger. The lines illustrate the polymer chains and cross-linking of the matrix, and the charges in circles the functional ionogenic groups. Counter-ions have been omitted.

There are five principal classes of functional groups present in ion exchangers, and hence exchangers can be classified on this basis as follows:

(1) cation exchangers;
(2) anion exchangers;
(3) amphoteric and dipolar ion exchangers;
(4) chelating ion exchangers;
(5) selective (or specific) ion exchangers.

Cation exchangers are high-molecular-weight polyvalent insoluble anions (polyvalent acids), the ionogenic groups of which are saturated with individual soluble cations. These are able to dissociate when the exchanger comes into contact with an aqueous solution, and thus may be exchanged, *e.g.*:

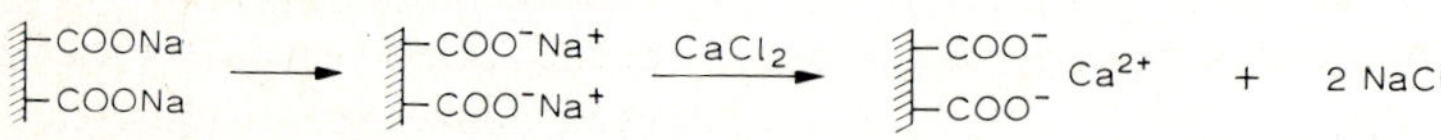


The shortened term catex is also used instead of cation exchanger.

An anion exchanger (anex) is a high-molecular-weight polyvalent insoluble cation (polyvalent base), which is able to exchange electrostatically bound soluble anions, *e.g.*:

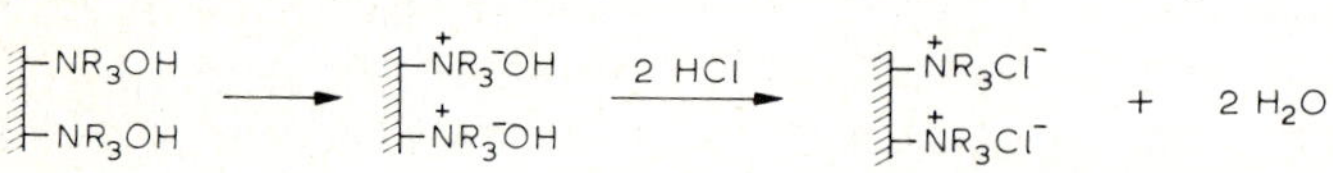


Amphoteric ion exchangers are polyvalent insoluble "zwitter-ions" (polyvalent "inner salts"), which dissociate in aqueous solution without the release of ions into the solution. However, they are then able to bind dissociated salts from the solution, cations to anionic groups and anions to cationic groups, *e.g.*:

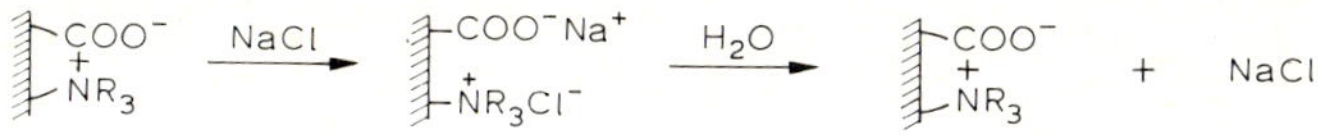

This binding represents a dynamic equilibrium. After washing with a large amount of water, the original internally neutralized form of the exchanger is regenerated and the salt is released. It is difficult to prepare these exchangers with exact equivalence of ionogenic groups. Dipolar ion exchangers can be prepared by binding amino acids (*e.g.*, arginine via amino group) to the matrix. These dipolar ion exchangers are used advantageously for the separation of biopolymers.

Chelating ion exchangers are resins prepared by incorporating complex-forming groups, *e.g.*, iminodiacetic acid. These ion exchangers retain only specific ionic groups (*e.g.*, heavy metals or alkaline earths) and are therefore more selective than cation exchangers. An example is given below, in which $M^{2+}$ represents, for example, $Cu^{2+}$, $Ni^{2+}$, $Zn^{2+}$, $Co^{2+}$ or $UO_2^{2+}$:

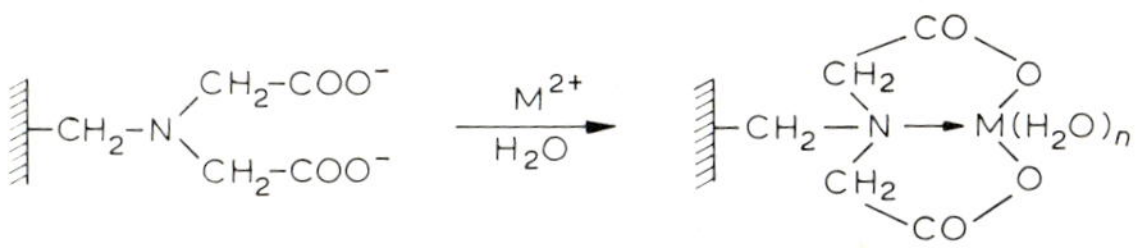

Selective (or specific) ion exchangers are synthesized experimentally for a specific purpose. They contain functional groups that are able to retain only one type of ion or a very limited number of types. For example, Skogseid synthesized a resin with trinitrophenyliminodinitrophenyl groups. These groups specifically bind potassium ions:

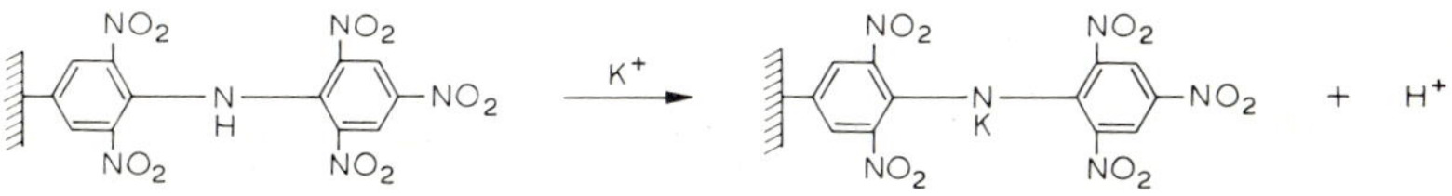

This type of ion binding cannot be considered as a true ion exchange, but is more closely related to potassium precipitation by hexanitrodiphenylamine (dipicrylamine). The principle of selective ion exchange resembles that of affinity chromatography.

The homogeneity of functional groups is very important in ion-exchange chromatography. Exchangers can be classified as monofunctional if they contain only one type of ionogenic or chelating group; these are sometimes called homoionic exchangers, and are most suitable for chromatography. Cheaper polyfunctional exchangers are used only for certain technical purposes. The presence of different kinds of ionogenic (or chelating) groups leads to a loss of resolution in the chromatography.

The terminology of ions should now be explained. Let us consider a homoionic exchanger, in which there is one type of charged functional group covalently bound to the matrix (*cf.* Fig. 6.2). In every case, dissociable counter-ions with an opposite charge are bound to it by electrostatic forces and thus form part of the ion exchanger. Counter-ions are mentioned in the literature and in commercial pamphlets when the ionic form of an ion exchanger is illustrated. Co-ions play an important role in the mechanism of ion-exchange chromatography and are ions with the same type of charge as the functional group, but they are soluble and capable of forming salts, acids, bases or water with counter-ions. The co-ions compete with the functional group in attracting counter-ions,

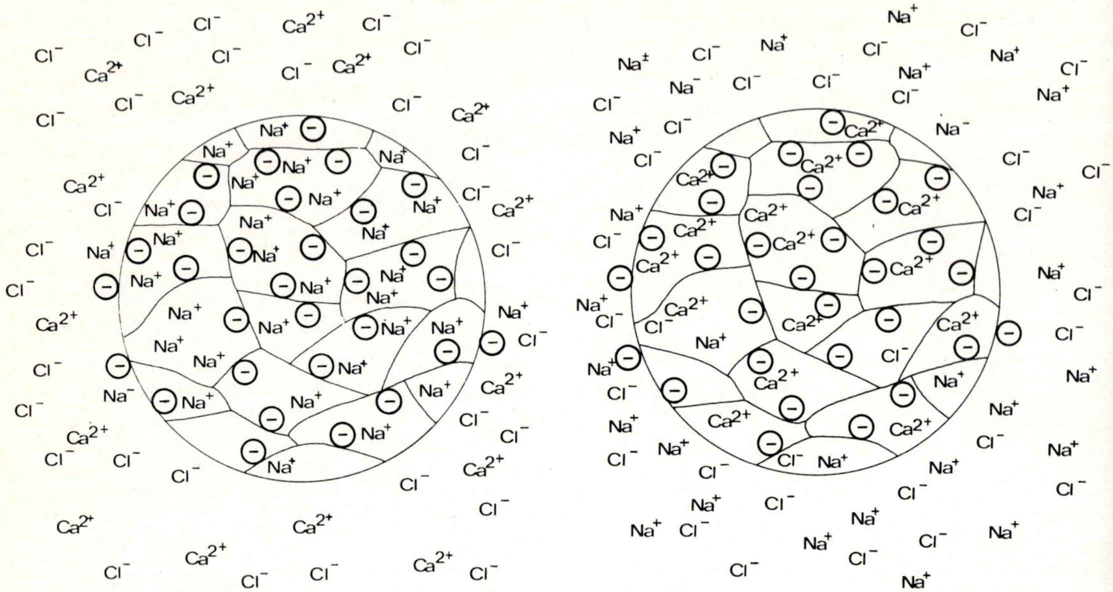

Fig. 6.2. Scheme of cation exchange. Left: bead of cation exchanger in the $Na^+$ form (the negative circles represent carboxylic functional groups, $Na^+$ the counter-ions in the ion-exchanger, $Ca^{2+}$ the counter-ions in solution and $Cl^-$ the co-ions); right: bead after exchange. The process can be expressed by the equation

$$2(R{-}COO^-Na^+) + CaCl_2 \rightleftharpoons (R{-}COO^-)_2\,Ca^{2+} + 2\ NaCl$$

where R = resin, $Ca^{2+}$ are the counter-ions in the ion-exchanger, $Na^+$ the counter-ions in solution and $Cl^-$ the co-ions.

thereby facilitating the exchange of the desired counter-ion. Examples to illustrate this terminology by Helfferich (1962a) are as follows:

| *Type of exchanger* | *Functional group* | *Counter-ion in ion exchanger (ionic form of exchanger)* | *Co-ion in solution* | *Counter-ion in solution (to be exchanged)* |
|---|---|---|---|---|
| Cation | $-COO^-$ | $Na^+$ | $Cl^-$ | $Ca^{2+}$ |
| Cation | $-SO_3^-$ | $H^+$ | $Cl^-$ | $K^+$ |
| Anion | $-\overset{+}{N}(CH_3)_3$ | $Cl^-$ | $Na^+$ | $OH^-$ |
| Anion | $-\overset{+}{N}(CH_3)_3$ | $OH^-$ | $Na^+$ | $CH_3COO^-$ |

The first of these four examples is represented schematically in Fig. 6.2.

The complete change of counter-ions in an ion exchanger is called cycling, and sometimes the term "cycle" is used instead of "ionic form". Cycling at the end of an ion-exchange operation to produce the ion exchanger in its original form is called regeneration.

## CHARACTERIZATION OF ION EXCHANGERS

Information describing the properties of individual ion exchangers is necessary before they can be used for chromatography. It is usually presented in the form of tables (*cf.* Chapter 13). First the type of exchanger must be known. Both cation and anion exchangers are classified according to the nature of the active groups, as shown below:

| *Ion exchanger* | *Type* | *Usual functional group* |
|---|---|---|
| Cation exchanger | Strongly acidic | Sulphonic |
| | Medium acidic | Phosphonic |
| | Weakly acidic | Carboxylic |
| Anion exchanger | Strongly basic | Quaternary ammonium |
| | Medium basic | Mixture of tertiary and quaternary ammonium groups |
| | Weakly basic | Amines, polyamines |

The ionic form of a commercial ion exchanger is usually indicated by the manufacturer. Cation exchangers are produced in the acidic form, designated $H^+$ (or H*), or in the salt from ($Na^+$, $Li^+$ and others). Anion exchangers are produced as the free base, designated $OH^-$ (or B*), or as salts (usually $Cl^-$).

The type of matrix (lattice) of an ion exchanger must be considered carefully. Polystyrene or polyacrylic types find a wide application, while phenolic types are, in general, not suitable for chromatography, because phenolic R–OH groups dissociate in alkaline media and thus form an additional cation exchanger group, $R{-}O^-$, with different properties from those of the main functional group. For the chromatography of many biopolymers, polystyrene or other aromatic matrices are not suitable because of their denaturating effect. Polyacrylic types are better, but a cellulose or a polydextran matrix is usually the best.

The degree of cross-linking of the matrix is very important in chromatography, and defines the average porosity of exchangers. The symbol X, accepted as a measure of the degree of cross-linking, represents the percentage of divinylbenzene in the styrene polymerization mixture used to prepare this type of resin (*cf.* Chapter 9, Ion-exchange materials). The process of cross-linking is easily controlled and therefore it is possible to produce a resin with a porosity suitable for a given purpose, and in commercial resins it varies from X1 to 16, X2 to X9 being most often used. The more cross-links present, the less an exchanger swells. The functional groups of individual exchangers have the same electric charge and therefore they have the tendency to extend the network to a maximum, and this process is accompanied by the hydration of functional groups and bound ions. The strength of the repulsing force is influenced by the type of functional group, ionic strength, pH and the nature of any bound ions, but it is strictly limited by the cross-linking. Swelling is reversible and can be considered as a state of balance between the tension of the elastic network and the osmotic pressure of the inside solution, arising from the presence of counter-ions. Volume changes in swollen resins, due to changes in the composition of the surrounding solution, sometimes disturb the chromatographic operation, and these difficulties occur more often with resins with a low degree of cross-linking. However, the lower cross-linked resins change ions more rapidly, but they are less selective. As a result of variations in the cross-linking reactions, Pepper *et al.* recommended the use of water regain values (W.R.) to characterize ion exchangers instead of the X values. The water regain is defined as the maximum weight in grams of water taken up to 1 g of completely dry ion exchanger. The relationship between the X and W.R. values for strongly acidic cation exchangers is shown below:

| *X* | *W.R.* | *X* | *W.R.* |
|---|---|---|---|
| 2 | 3.45 | 10 | 0.83 |
| 4 | 1.92 | 15 | 0.59 |
| 6 | 1.36 | 20 | 0.48 |
| 8 | 1.04 | 25 | 0.38 |

The cross-linking and porosity are equivalent terms used for ion-exchange resins, but this terminology is not valid in the case of macroreticular resins (*cf.*, Chapter 9). The cross-linking defines here the composition of the matrix only; very large pores of macroreticular resins are best expressed by the maximal molecular weight of substances penetrating the beads.

The capacity of an ion exchanger is a measure of the total amount of ions the resin is able to bind and is usually expressed as milliequivalents (mequiv.) per gram of dry resin (in the $H^+$ or $Cl^-$ form) or as milliequivalents per millilitre of fully swollen wet resin (in the $H^+$ or $Cl^-$ form) packed in the bed. This value, which is a measure of all charged groups present, is not usually achieved in practice, but is lower for various reasons, *e.g.*, operating under non-equilibrium conditions. Factors that influence the available capacity are concentration and ionic strength of the eluent, pH, temperature, the accessibility of functional groups and the nature of the counter-ions. The dependence of the ion-exchange capacity on pH is illustrated by the titration curves in Fig. 6.3. It can be seen that for strongly acidic and strongly basic exchangers, the capacity is virtually independent of pH and they can therefore be used over a wide pH range. However, the capacity of weakly acidic cation exchangers and of weakly basic anion exchangers is strongly dependent on pH, so the use of weakly acidic exchangers is therefore limited in media of low pH and weakly basic anion exchangers are not very efficient in alkaline media.

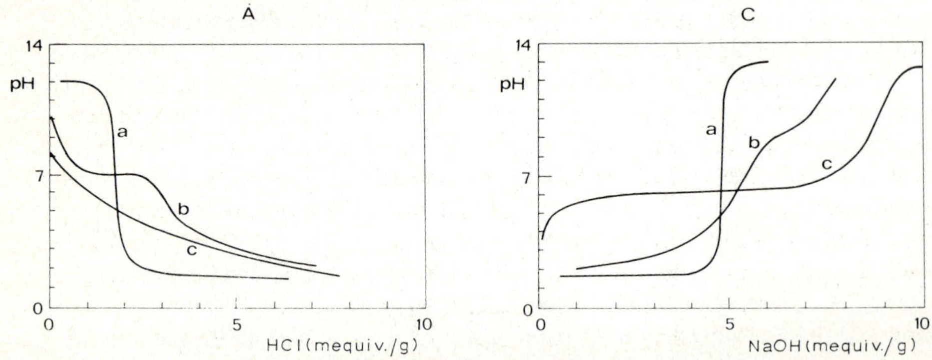


Fig. 6.3. Titration curves of ion exchangers in a dilute solution of neutral salts. A = anion exchangers; C = cation exchangers; a = strongly basic anion exchanger or strongly acidic cation exchanger; b = medium basic or acidic exchanger; c = weakly basic anion exchanger (amine-type) or weakly acidic (carboxylic) cation exchanger.

Particle size and particle form are important characteristics of ion exchangers. The particle size determines how quickly equilibrium is established and hence influences the sharpness of a chromatographic separation: the smaller the particle size, the sharper is the separation. If the particles are too small, the flow resistance of the chromatographic column is increased and higher pressures are required for elution and care must therefore be taken when choosing a suitable grain size. The particle form is also an important factor. Ion exchangers are delivered either in the form of grains prepared by grinding the resin gel (these have irregular shapes) or as uniform beads (spheres), prepared by polymerization in an emulsion. Generally, the latter are better for chromatography, because they do not pack the column so tightly and consequently there is a lower flow resistance. The bead form possesses better mechanical properties and hence the losses caused by friction are lower. The particle size is usually expressed in terms of the size range of dry copolymer beads before any ionic groups are attached, and is measured by standard mesh screens. Sometimes the particle size is expressed as the wet mesh range after maximal swelling*. The wet mesh size depends, of course, on the many factors mentioned in the preceding paragraphs. Millimetres (mm) and microns ($\mu$m) are also used to measure grain size. For the conversion table for U.S. standard mesh screens, see Table 11.1 (p.286). For fine chromatographic separations, it is important that the particle size should be as uniform as possible. The narrower the variation in grain size, the sharper is the separation obtained.

## REACTIONS, AFFINITY AND SELECTIVITY IN ION EXCHANGE

Exchange reactions of simple ions are best described in terms of an ionic redistribution between the ion-exchange gel and the aqueous phase. These reactions are always stoichiometric because the electroneutrality of the resin must be maintained. As no covalent bonds are formed or broken during this process, there is little heat evolution or absorption accompanying ion exchange. The only exceptions are neutralization reactions involving a cation exchanger in the $H^+$ form or an anion exchanger in the $OH^-$ form, in which the formation of low-dissociated water is the source of heat.

Ion exchange is, in general, a reversible process and therefore an equilibrium is obtained, *e.g.*:

$$2(R{-}SO_3^- Na^+) + Ca^{2+} \rightleftharpoons (R{-}SO_3^-)_2\, Ca^{2+} + 2\, Na^+$$

where R = resin. This equilibrium depends not only on the relative affinities of ions for the exchanger, but also on the relative ionic concentrations. Therefore, ions with a low affinity for the exchanger can regenerate it and replace ions of a greater affinity, if the former are present at a higher concentration. In practice, this is made use of in the regeneration of water softeners: $Ca^{2+}$ and $Mg^{2+}$ are present in natural hard water at relatively low concentrations. Because they have a higher affinity for the cation exchanger (in the $Na^+$ form), they exchange with $Na^+$. A large amount of water can be treated in this way, to produce water that contains only monovalent ions which do not precipitate

*American manufacturers use U.S. standard mesh screens as their standard, and British manufacturers B.S.S. standard mesh screens.

soap. When the ion exchanger is exhausted, a small amount of concentrated sodium chloride solution quickly regenerates the ion exchanger to the original $Na^+$ form.

The affinity of ions for an ion exchanger is sometimes called the "ion-exchange potential" and in dilute aqueous solutions it increases with the size of the ionic charge. Polyvalent ions are more strongly bound than monovalent ions, the affinity being proportional to the charge. For ions of the same charge, the exchange potentials are inversely proportional to the radius of the hydrated ions. Because the radii of many cations are inversely proportional to their atomic weights, the affinities of these cations can be arranged in order of atomic weights. The exchange potentials of cations are similar to or identical with the so-called lyotropic series. The affinity of anions is governed by similar rules. In addition, the exchange potentials increase with the ability of anions to polarize. Examples (*cf.* Kunin and Myers, and Nachod) of typical affinity sequences are given below.

Composite affinity sequence for cations: $(CH_3)_4N^+ < Li^+ < Na^+ < NH_4^+ < K^+ < Rb^+ < Cs^+ < Tl^+ < Ag^+$; $Mg^{2+} < Ca^{2+} < Sr^{2+} < Ba^{2+}$; $Fe^{2+} < Co^{2+} < Ni^{2+} < Cu^{2+} < Zn^{2+}$; $Al^{3+} < Sc^{3+} < Lu^{3+} < Yb^{3+} < Tm^{3+} < Er^{3+} < Ho^{3+} < Y^{3+} < Dy^{3+} < Tb^{3+} < Gd^{3+} < Eu^{3+} < Sm^{3+} < Pm^{3+} < Nd^{3+} < Pr^{3+} < Ce^{3+} < La^{3+}$. The position of $H^+$ on strongly acidic cation exchangers is nearly the same as that of $Li^+$, and on weakly acidic (carboxylic) cation exchangers it is about the same as that of $Ba^{2+}$.

Composite affinity sequence for anions: fluoride < acetate < formate < chloride < bromide < chromate < molybdate < phosphate < arsenate < nitrate < iodide < tartrate < sulphate < citrate. The position of $OH^-$ on strongly basic anion exchangers lies between those of fluoride and acetate, while on weakly basic anion exchangers it has a much higher affinity.

These series vary slightly, depending on individual ion exchangers and differing conditions.

The difference in affinity between individual ions is called selectivity. Other factors that influence the selectivity are ionic concentration, the quality of the solvent and the nature of the exchanger. In principle, the selectivity of an ion-exchange process increases with the degree of cross-linking in the resin and is little affected by temperature (except where Van der Waals forces play a role). Of course, the pores in the network have to be large enough to allow the penetration of ions into the grain, but if the pores are too large, then the differences in the effective radii of various ions entering the grain will have only a small influence, which will lead to a decrease in ionic selectivity. The exchange selectivity for ions of similar size and charge may be increased by the addition of complexing agents.

In the chromatography of biochemical substances, the affinities of amphoteric ions (zwitter-ions) for ion exchangers are important. For example, a typical amino acid, $H_2N{-}CH(R){-}COOH$, in a neutral aqueous solution exists in the completely ionized form, $H_3\overset{+}{N}{-}CH(R){-}COO^-$. In acidic solution on a cation exchanger, the carboxyl group interacts with a proton to give the unionized moiety. Therefore, when a mixture of amino acids is chromatographed on a cation exchanger in an acidic medium, the amino acids should emerge as a function of their $pK_1$ values. The dissociation of the carboxyl group determines the affinity. In a basic solution on an anion exchanger, the ammonium group reacts with a hydroxyl group and loses its proton without any change on the carboxyl.

Therefore, the dissociation constant of the ammonium group, p$K_2$, determines the affinity. The behaviour of amphoteric ions is illustrated below (the decisive dissociation constant is indicated with an asterisk):

| *Ion exchanger* | *Medium* | *Ionic form* | *Formula* | *Dissociation constant* |
|---|---|---|---|---|
| Cation exchanger | Acidic | Cation | $H_3N^+$–CH(R)–COOH | p$K_2$; p$K_1^*$ |
| Aqueous solution | Neutral | Zwitter-ion | $H_3N^+$–CH(R)–$COO^-$ | p$K_2$; p$K_1$ |
| Anion exchanger | Basic | Anion | $H_2N$–CH(R)–$COO^-$ | p$K_2^*$; p$K_1$ |

The sequence of emergence of some amino acids in practice was found to differ from the theoretical sequence based on p$K$ values in an ion-exchange experiment, owing to adsorption due to Van der Waals forces. The hydrophobic side-chains of amino acids are adsorbed on the aromatic network of the resin, and this process can be decreased by increasing the temperature. Sometimes these interactions can help in chromatographic separations.

These rules are valid for the chromatography of low-molecular zwitter-ions (*e.g.*, amino acids and smaller peptides). The behaviour of high-molecular-weight amphoteric substances and of their fragments (*e.g.*, proteins and large polypeptides) follows other rules (Porath and Fryklund). They usually exist only in two alternative states in contact with ion exchangers: either completely sorbed or not sorbed at all. The state depends on the conditions in the solution (pH and ionic strength). The choice of a suitable composition of the buffer used for the sorption and for the selective desorption of proteins is explained in Chapter 10 (p.264). Certain valuable results on the resolution of proteins and large peptide fragments have been reported by Novotný and by Novotný *et al.*

## ION-EXCHANGE EQUILIBRIA AND KINETICS

Rigid inorganic exchangers contain water-filled pores and the ions are bound in specific positions. A better model describing elastic resinous exchangers immersed in water is a two-phase system. A swollen grain of the resin resembles an aqueous electrolyte solution. Both the functional groups and counter-ions are dissolved in absorbed water, and the counter-ions are not fixed to individual functional groups, but are present as a cloud containing mutually repulsive ions distributed throughout the whole resin volume. This cloud is attracted by the oppositely charged bound functional groups. The counter-ion cloud extends partly to the grain surface and thus an electric double-layer is formed.

If the grain is placed in an electrolyte solution containing the same type of counter-ion, $A^-$, the penetration of the electrolyte into the grain is not complete and an equilibrium is reached. Initially, this can be expressed by the Donnan equation:

$$(a_{A^-})_r \cdot (a_{B^+})_r = (a_{A^-})_s \cdot (a_{B^+})_s$$

where $a$ is the activity of the counter-ion $A^-$ or co-ion $B^+$ in the resin (subscript $r$) and in the solution (subscript $s$). If the ion exchanger (anion exchanger) is in the $A^-$ form, there is a higher concentration of $A^-$ and a lower concentration of $B^+$ in the ion exchanger phase relative to the solution. If $A^-$ begins to diffuse out of the ion exchanger, following the concentration gradient, and $B^+$ into it from the solution, a net positive charge will result on the grain and a net negative charge in the solution and this Donnan potential will stop any further diffusion. This Donnan equilibria will affect not only the ion-exchange process, but also the swelling of the resin.

If the grain in the N form is placed in an electrolyte solution containing different counter-ions, M, the Donnan potential also limits the entry of ions of the same charge, but the re-distribution of different counter-ions between both phases begins and continues until equilibrium is reached. Because electrical neutrality between the grain and the solution must be maintained, the exchange is strictly stoichiometric. According to Samuelson, the selectivity coefficient, $k_{\mathrm{N}}^{\mathrm{M}}$, describing the equilibria can be defined by the expression

$$k_{\mathrm{N}}^{\mathrm{M}} = \frac{[\mathrm{M}]_r^n \cdot [\mathrm{N}]_s^m}{[\mathrm{N}]_r^m \cdot [\mathrm{M}]_s^n}$$

where M and N are the ions being exchanged and $m$ and $n$ are the absolute values of their charges. The brackets represent the concentration in the external solution (subscript $s$) or in the resin phase (subscript $r$). If the selectivity is calculated from experimental data, the total amount of M and N, including their concentration due to co-ions, must be considered in the expressions $[\mathrm{M}]_r$ and $[\mathrm{N}]_r$.

For an understanding of ion-exchange chromatography, the rate at which equilibrium is reached is very important and must be considered in addition to the equilibrium itself. Therefore, the basic principles of ion exchange kinetics are now mentioned briefly in accordance with the general theory of Boyd *et al.*

Let us consider the exchange of $N^+$ (present in the exchanger) for $M^+$ (present in the external solution; *cf.*, Fig. 6.4): this can be expressed by the equation

$$\mathrm{NR} + \mathrm{M}^+ \rightleftharpoons \mathrm{MR} + \mathrm{N}^+$$

where R represents the resin. Now, assume that the exchanger is suspended in a mixed solution or fixed in a chromatographic column through which the solution is flowing. The exchange process can be divided into five individual steps: (1) the transport of $M^+$ from the solution to the surface of a bead of ion exchanger; (2) the diffusion of $M^+$ through the matrix to the functional group; (3) the chemical exchange inside the particle expressed by the above equation; (4) the diffusion of the exchanged $N^+$ from the functional group to the surface of the exchanger; (5) the transport of $N^+$ into the external solution. It is clear that the slowest of these processes should be rate controlling.

These individual steps will now be discussed briefly. The bead of exchanger, in spite of the fact that it is situated in a mixed liquid, is surrounded by a still, thin layer of solution. This layer is not mixed with the external solution (or the mixing is not perfect) and it moves with the particle. This relatively stable Nernst film is formed for thermodynamic reasons and the ions can move through it only by diffusion, so for steps (1) and (5) the rate of diffusion through this film is therefore limiting. For steps (2) and (4), the rate of diffusion through the gel of the swollen ion-exchanger particle is limiting. The process of actual ion exchange, step (3), in the swollen gel of exchanger can be considered to be instantaneous, because the ions are not firmly bound to the functional groups but are attracted to them only in a dynamic equilibrium. Only in special cases does the velocity of

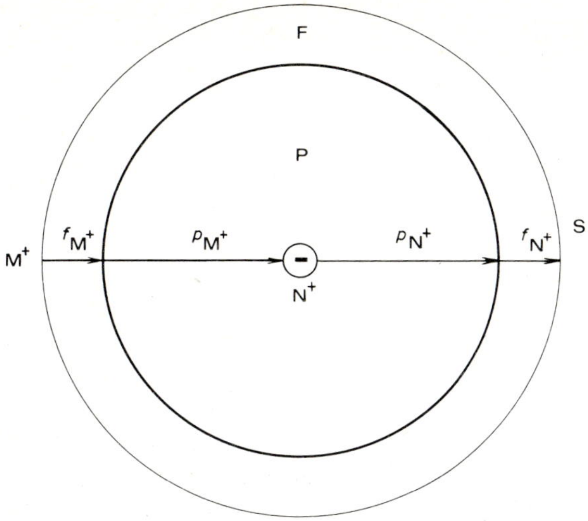


Fig. 6.4. Schematic representation of the fundamental terms in ion-exchange kinetics. $M^+$ and $N^+$ = exchanging ions; P = ion-exchange particle; F = Nernst film; S = external solution; ⊖= functional group; $f$ = rate of diffusion of ions through the film; $p$ = rate of diffusion of ions through the particle.

the chemical reaction in step (3) play a role. From this discussion, it can be seen that there are two main rate-limiting steps in ion-exchange kinetics: "film diffusion" and "particle diffusion".

Film diffusion is enhanced by small particles and particle diffusion by large particles, which can be explained by the fact that film diffusion is inversely proportional to the diameter of the ion-exchanger particle, whereas particle diffusion is inversely proportional to the square of the diameter. Film diffusion is rate limiting at low concentrations of exchanging ions, and particle diffusion at high concentrations, which is understandable when it is realized that higher concentrations of ions in solution increase the rate of film diffusion but do not affect particle diffusion. The exchangeable ions have to jump from the electrostatic field of one charged functional group to another, which requires a considerable activation energy. The film diffusion velocity is obviously independent of resin cross-linking. However, when particle diffusion is rate determining, the exchange rate is decreased substantially by an increase in cross-linking. An increase in temperature causes a greater increase in the rate of diffusion within the particle in comparison with the increase in film diffusion and therefore the rate-controlling influence of film diffusion is enhanced.

Kinetic data can be expressed in two ways: (1) by the time necessary for complete ion exchange under defined conditions (the time necessary to reach equilibrium), or (2) by the time of half-exchange. Both methods have been used in the literature, and the data published by Kressman are given as an example. He measured the time for half-exchange on a strongly acidic sulphonated cation exchanger (polystyrene type of resin) in the $NH_4^+$ form under comparable conditions and found the following values: $Na^+$, 1.25 min; $\overset{+}{N}(C_2H_5)_4$, 3.0 min; $\overset{+}{N}(CH_3)_4$, 1.75 min; and $C_6H_5-\overset{+}{N}(CH_3)_2CH_2-C_6H_5$, 1 week. These results illustrate the influence of ionic size. The relationship between full exchange and the time required for half-exchange was illustrated by Reichenberg and Wall, who measured the kinetics on a similar type of exchanger (X17):

| | |
|---|---|
| complete exchange: | $Na^+/H^+$, 2–10 min; |
| half-exchange: | $Na^+/H^+$, few seconds to 1–2 min. |

These effects have been studied mostly on strongly acidic resins, and strongly basic resins have not been studied to the same extent. The kinetics in weakly acidic and weakly basic exchangers depends on the degree of ionization of their groups and if this is high enough, the parameters are comparable with strongly acidic and strongly basic exchangers. When there is a low dissociation of functional groups, the rate of exchange is decreased and in every case controlled by particle diffusion. The exchange of $H^+$ in weakly acidic cation exchangers and of $OH^-$ in weakly basic anion exchangers is always very slow and is also controlled by particle diffusion, even in very dilute solutions.

In most chromatographic columns, particle diffusion limits the rate of exchange. An increased flow-rate through the column will decrease the thickness of the Nernst film and thus will favour mass-transfer control. The theoretical treatment of ion-exchange kinetics is very complicated. When the mobilities of exchanging ions are different ($p_{M^+} \neq p_{N^+}$ in Fig. 6.4), an electrical potential gradient is formed which affects the diffusion of both types of counter-ions. The faster ion will be held back, while the slower ion will be accelerated. However, the condition of electrical neutrality must be maintained. The ion leaving the exchanger plays the most important part in determining the rate ($p_{N^+}$ in Fig. 6.4). The reader who requires more detailed information concerning ion-exchange kinetics should consult more specialized publications, *e.g.*, Helfferich (1966).

## COLUMN OPERATION AND ION-EXCHANGE CHROMATOGRAPHY

A very efficient means of achieving ion exchange is to pack the beads into a column and then slowly filter the electrolyte through it. The solution entering the column is called the influent and that leaving the column the effluent. By the term capacity of the column is meant the total number of exchangeable groups (in milliequivalents) in the column. When the column is being saturated with a new electrolyte, a small part of the sorbed substance reaches the bottom of the column before the total capacity of the column is exhausted. The break-through capacity of the column is defined by the sorption ability up to the appearance of the sorbed substance in the effluent, and depends on the particle size of the sorbent, flow-rate and composition of the influent. When the sorption process is continued after the break-through, the concentration of the sorbed substance in the effluent eventually reaches that of the influent. This process is usually expressed by break-through curves, the slopes of which define the quality of the sorption and show the chromatographic efficiency of the column under the operating conditions. The steeper the break-through curve, the better is the sorption process.

When the influent contains two or more ions, the ion exchanger in the column can act as a chromatographic agent. As the ions differ in their affinity for the exchanger, they are sorbed to different extents, or they are sorbed completely at the top of column and then eluted gradually using different conditions. Ion-exchange chromatography is the dynamic replacement of zones of bound ions by ions newly entering the column, accompanied by their separation. The separation depends not only on the difference in exchange potential (in dissociation constants) between the substances being separated, but also on their adsorptivity on the network of the exchanger.

All three general types of chromatography can be realized on ion exchangers: frontal analysis, displacement chromatography and elution chromatography.

Frontal analysis is the most simple technique. A solution containing the counter-ions to be separated, *e.g.*, A, B and C, is pumped to the top of the column, which is in the M form (M having lowest affinity for the exchanger). At first, only M ions are found in the effluent, indicating the exchange A → M, B → M and C → M. Then A ions emerge, these having the least affinity in the influent mixture, which indicates that the capacity of the column is exhausted. The column is now in the A form and the exchange is limited to B → A and C → A. A ions can be isolated in a pure form from the effluent only during this stage. After some time, A ions begin to be accompanied by B ions, the effluent having the composition A + B, which indicates that the column is in the B form, and ion exchange is limited to the process C → B. Finally, the whole column is in the C form and no separation takes place, the effluent now having the same composition, A, B and C, as the influent. Frontal analysis in ion exchange has little practical value for preparative purposes and today it also has little importance for analytical use.

Displacement chromatography differs from frontal analysis in the experimental procedure required. Let us suppose the same mixture of counter-ions, A, B and C, is applied to the column of exchanger in the M form. The relative affinities are the same. However, only a limited amount of solution with the mixture to be resolved is applied to the top of the column, then a new solution containing the counter-ions Y, having the highest affinity for the exchanger, is used as the influent. During the flow through the column, the ions are arranged in the sequence of their affinities. M is eluted first, followed by A, which is displaced by B, and the following C displaces B. All three types of ions are forced down the column by Y ions, the function of which may be compared with a piston. The yield of pure components separated by this procedure is not high, as mixing occurs across the zones of separation giving an effluent composition of A, A + B, B, B + C, C, C + Y and Y. Displacement chromatography has limited analytical applications, but is sometimes used for preparative purposes as the column capacity is better exploited in comparison with elution chromatography, described below.

The most important method is elution chromatography, in which a much smaller amount of material can be applied to the column compared with displacement chromatography. However, the separation of components is often complete and this procedure is therefore very valuable for many analytical applications and for those preparative purposes where the complete or very effective separation of components is required. In order to describe the principle of the method, let us assume that the same mixture of counter-ions, A, B and C, is used and that the column is in the M form, M having the lowest affinity for the exchanger. A dilute solution containing a small amount of A, B and C is applied to the column, so that all the ions are sorbed in a zone at the top of the column. Then an eluting solution containing the counter-ion M, at a low concentration, is used as the influent, which slowly releases the ions A, B and C from their positions at the top of the column and washes them down. Because they have higher affinities than the ion M for the

*References p.86*

exchanger, a relatively large amount of the eluting solution is required in order to effect a short movement of zones A, B and C. The important fact here is that during elution, the movement of ions A, B and C is governed solely by ion-exchange equilibria:

$$RA + M \rightleftharpoons RM + A$$

$$RB + M \rightleftharpoons RM + B$$

$$RC + M \rightleftharpoons RM + C$$

where R = resin. This depends on the different affinities of A, B and C, the mobilities of these ions being independent. Therefore, they can emerge in the effluent as individual peaks separated by pure eluting solution.

Because of the importance of this technique, elution ion-exchange chromatography will be discussed further in this section. Martin and Synge were the first to apply the idea of a theoretical plate (*cf.*, Chapter 3) to column chromatography and Mayer and Tompkins were the first to extend it to ion-exchange columns. It was found that the particle size has the most important influence on the efficiency of ion-exchange chromatography. If column elution is slow enough to reach equilibrium, then the height equivalent of a theoretical plate (HETP) is approximately equal to the diameter of the particle. Such low values are seldom obtained in practice, as channelling in the column (due to irregularities in the packing) causes this value to be several times higher. Glueckauf (1955a, b) derived formulae from which the plate height can be calculated under non-equilibrium conditions, such conditions usually being found in practice. The HETP, $H$, is divided into three terms:

(1) $H_{\text{particle size}} = 1.64\, r$ (corresponds to the conditions of equilibrium);

(2) $$H_{\text{particle diffusion}} = \frac{D_v}{(D_v + \epsilon)^2} \cdot \frac{0.142 \cdot r^2 \cdot F}{D_S};$$

(3) $$H_{\text{film diffusion}} = \left(\frac{D_v}{D_v + \epsilon}\right)^2 \cdot \frac{0.266 \cdot r^2 \cdot F}{D_L(1 + 70\, r \cdot F)}$$

where $r$ = particle radius, $D_v$ = volume distribution coefficient (amount of sorbed solute per ml of the column divided by the amount of solute per ml of solution), $F$ = linear flow-rate of solution in the column above the resin bed (cm/sec or (ml of solution/cm$^2$ of column)/sec), $\epsilon$ = void fraction of column and $D_S$ and $D_L$ are diffusion constants (cm$^2$/sec) for the solute in the resin and in the interstitial volume, respectively. The HETP increases under non-equilibrium conditions, the value being given by:

$$H = H_{\text{particle size}} + H_{\text{particle diffusion}} + H_{\text{film diffusion}}$$

Fig. 6.5, according to Glueckauf (1955a), shows the factors that control the HETP according to the operating conditions. Usually, ion-exchange chromatography is carried out in the particle diffusion area of this graph. With a moderate flow-rate and a high distribution coefficient, film diffusion is the main factor. At very slow flow-rates, equilibrium is reached. Because the distribution coefficient can be changed by varying the concentration of the eluting solution, according to this theory it is possible to operate in any part of the diagram. Sharp peaks with an acceptable rate of elution are obtained when operating in the vicinity of point B in the diagram.

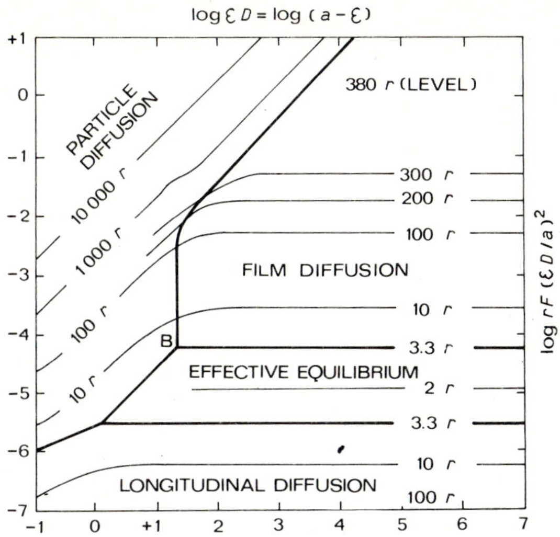


Fig. 6.5. Theoretical plate height as a function of operating conditions (after Glueckauf, 1955a). $r$ = radius of resin beads; $F$ = linear flow velocity (cm/sec); $\epsilon$ = void fraction of the column; $D$ = distribution coefficient (amount of solute per gram of resin divided by amount of solute per cubic centimetre of the solution); $a$ = distribution coefficient (total solute per cubic centimetre of the column divided by dissolved solute per cubic centimetre of the solution).

According to the theoretical plate concept, it is possible to predict the length of column necessary for a particular separation. Selected examples of such calculations for inorganic ions have been given in monographs (Helfferich, 1962a; Samuelson; Walton, 1967). These calculations are valid if it is assumed that there is a linear exchange isotherm, small and nearly equal amounts of separating substances, not exceeding a few per cent of the column capacity, and perfect packing of the column. For most laboratory separations, the necessary data for these calculations are not available. Therefore, orienting experiments based on published data must substitute for a mathematical treatment and consequently the examples given in the Chapters 32 and 34–37 in this book are very important.

## ION EXCLUSION, ION RETARDATION, THE ION-SIEVE PROCESS AND PARTITION CHROMATOGRAPHY ON ION EXCHANGERS

The principle of ion exclusion can be explained simply by considering the Donnan equilibria that affect the ion-exchange process. Let us suppose that one bead of cation exchanger in the $K^+$ form comes into contact with a solution of a salt. For simplification, let us consider a salt containing the same cation, *e.g.*, KCl, and in addition some non-ionic soluble substance (*e.g.*, ethylene glycol). The non-electrolyte diffuses into the bead without hindrance, but the diffusion of both ions of the salt is greatly affected. Because of the high concentration of $K^+$ originally present in the bead, the entry of additional ions of the same charge is limited and this factor also influences the penetration of $Cl^-$. In order to maintain a neutral charge, the total number of $Cl^-$ ions penetrating must equal the number of $K^+$ ions that penetrate the bead. After equilibrium has been reached, the relative concentrations of salt and non-electrolyte in the bead therefore differ from the relative concentration of both components in the solution. This shift is repeated within the chromatography of the mixture of both components on a column of ion exchanger and results in their separation. The peak of salt is excluded before the peak of non-electrolyte and therefore this process is called "ion exclusion".

In effect, ion retardation is the opposite of ion exclusion. Let us consider the same example as above (potassium chloride and ethylene glycol), the only difference in this case being the use of a special amphoteric ion exchanger called a "snake-cage" resin, which contains two types of functional groups, both cationic and anionic, rotating on the same matrix. These groups are internally neutralized and when the internal salt linkages have been split by the penetration of a salt solution, the functional groups are capable of binding both cations and anions. In the example given above, the $K^+$ ions of the salt are bound to acidic groups and the $Cl^-$ ions to basic groups. After equilibrium has been reached, the salt is concentrated in the grains of the ion exchanger. Molecules of the glycol penetrate the grain freely and diffuse out without being affected. The chromatographic effect appears when the mixture of salt and non-electrolyte flows slowly through a column prepared from this type of resin. The glycol peak in our example is eluted first because the salt is bound. However, the ions of the salt are in dynamic equilibrium with the functional groups and the flow of water through the column causes this equilibrium to be slowly shifted in the direction of flow. After the resin has been thoroughly washed with water, the functional groups will again form "inner salts", and the result of this process is the retardation of the electrolytes, which are then eluted as a second separate peak. Therefore this process is called "ion retardation".

There are two main advantages that ion-exclusion and ion-retardation chromatography possess over conventional ion-exchange methods: (1) electrolytes and non-electrolytes can be separated and (2) the column does not require regeneration after use. Simple elution of both peaks is sufficient, the column then being ready for the next separation. The disadvantage of ion exclusion is the dilution of non-electrolyte.

In some instances, ion exchangers can be used to effect chromatography based on other principles. Sometimes this is regarded as a special form of partition chromatography (*cf.*, Reichenberg). The use of ion sieves permits the separation of large ions from small ions; a suitable degree of cross-linking allows the penetration of the small ions, while the large ions are excluded. The same principle is valid for the separation of non-electrolytes. For example, D-glucose can be separated from methanol using X8 resin. In other cases, Van der Waals forces play a role, *e.g.*, a mixture of acetic acid and *n*-butyric acid can be separated on X5 sulphonated polystyrene resin with water as eluent, the *n*-butyric acid being retarded. Several separations are known that indicate complicated polar interactions, which influence the chromatographic process, between solute molecules and the functional groups of the exchanger.

## LIGAND-EXCHANGE CHROMATOGRAPHY

The complex-forming ability of some ions plays an important role in chromatography, because it often improves the selectivity of ion exchange. The interaction of ions with buffers is assumed to occur in this case and the methods describing the exchange of ions in complex form are treated briefly in Chapter 10 (p.267). In this section, another method will be mentioned, namely the so-called ligand-exchange or ligand chromatography. In this case, a metal ion capable of forming complexes (*e.g.*, $Ag^{+}$, $Cu^{2+}$, $Ni^{2+}$ or $Fe^{3+}$) is attached firmly to the exchanger and the ligands are exchanged (*i.e.*, the molecules coordinated to the metal ion).

Sjöström attached $Fe^{3+}$ to a cation exchanger and β-diketones were sorbed selectively on such a column. According to Helfferich (1961, 1962b), amines are sorbed on a cation-exchange column loaded with a nickel–ammonia complex and displace ammonia, which can be used in the next step as an eluting agent. Wuster *et al.* sorbed esters of unsaturated acids on a resin loaded with silver (which forms complexes with π-electrons of double bonds; the so-called argentation chromatography should be mentioned here, *cf.*, Morris and Nichols). Latterel and Walton and Shinomura *et al.* separated amines, and Hill *et al.* compared the selectivities of the complex-forming $Ni^{2+}$ bound to organic resins and to inorganic exchangers (zirconium phosphate). Seigel and Degens used the commercial chelating resin Chelex 100 ($Cu^{2+}$) for the isolation of amino acids from sea-water and thus extended the use of ion-exchange chromatography to solutions of high ionic strength (saline, brines, etc.). Buist and O'Brien separated peptides from amino acids in urine by the same procedure. Goldstein isolated nucleotides, nucleosides and nucleic acid bases by this method.

## ION EXCHANGE IN NON-AQUEOUS SOLUTIONS

True ion exchange is also possible in organic solvents and in mixed solutions and alcohols, acetone and other solvents have been used in the pure form and in mixtures with water. The necessary condition is the partial dissociation of the solute. The variation of swelling due to changes in solvent composition often makes chromatography very difficult, especially when stepwise or gradient elution is used. The resin requires preliminary conditioning with the solvent before application of the sample. The distribution of organic solvents between the resin phase and the outer solution is complicated and depends on the polarity of the organic solvent, its concentration in the solution, the ionic form of the exchanger and the composition of the matrix. London forces between the solvent and the matrix often play an important role, and the chemical similarity of the network and the solvent facilitates penetration and swelling.

In some cases, the ion-exchange potentials of particular ions are increased in mixtures with water and organic solvents, but the rate of exchange is slower in non-aqueous systems and diminishes with decreasing polarity of the solvent. Under these conditions, gel diffusion is usually the main factor that determines the rate of exchange. Macroreticular resins are very useful for this type of chromatography owing to the presence of macro-pores inside the matrix; the separation is much faster.

*References p.86*

The capacity of the exchanger is not usually exhausted in non-aqueous or mixed solutions, but in some instances neutral adsorption of electrolytes causes a large uptake of ions. It is also possible to achieve some special ion exchanges in water-free ammonia.

Generally, chromatography in non-aqueous or mixed solutions is used very seldom compared with chromatography in aqueous solutions. In special instances it may offer some advantage, but usually there are difficulties. It is not possible to give a general approach for specific problems of this type. Samuelson published a short survey of this type of chromatography.

## REFERENCES

Boyd, G. E., Adamson, A. W. and Myers, Jr., L. S., *J. Amer. Chem. Soc.*, 69 (1947) 2836.
Buist, N. M. R. and O'Brien, D., *J. Chromatogr.*, 29 (1967) 398.
Dorfner, K., *Ionenaustauscher: Eigenschaften und Anwendungen*, De Gruyter, Berlin, 1963a.
Dorfner, K., *Ionenaustauschchromatographie*, Akademie-Verlag, Berlin, 1963b.
Flaschka, H. A. and Barnard, J. R., *Chelates in Analytical Chemistry*, Marcel Dekker, New York, 1967.
Genge, J. A. R., in D. R. Browning (Editor), *Chromatography*, McGraw-Hill, London, 1969.
Glueckauf, E., *Ion Exchange and Its Application*, Society of Chemical Industry, London, 1955a.
Glueckauf, E., *Trans. Faraday Soc.*, 51 (1955b) 34.
Goldstein, G., *Anal. Biochem.*, 20 (1967) 447.
Griesbach, R., *Ionenaustauscher in Einzeldarstellungen*, Akademie-Verlag, Berlin, 1957.
Helfferich, F. G., *Nature (London)*, 189 (1961) 1001.
Helfferich, F. G., *Ion Exchange*, McGraw-Hill, New York, 1962a.
Helfferich, F. G., *J. Amer. Chem. Soc.*, 84 (1962b) 3237 and 3242.
Helfferich, F. G., in J. A. Marinsky (Editor), *Ion Exchange*, Marcel Dekker, New York, 1966, p. 65.
Hering, R., *Chelatbildende Ionenaustauscher*, Akademie-Verlag, Berlin, 1967.
Hill, A. G., Sedgley, D. and Walton, H. F., *Anal. Chim. Acta*, 33 (1965) 84.
Inczedy, J., *Analytical Applications of Ion Exchangers*, Pergamon Press, New York, 1966.
Kressman, T. R. E., *J. Phys. Chem.*, 56 (1952) 118.
Kunin, R., in E. Heftmann (Editor), *Chromatography*, Reinhold, New York, 1961, p. 315.
Kunin, R. and Myers, R. J., *Ion Exchange Resins*, Wiley, New York, 1952.
Latterel, J. J. and Walton, H. F., *Anal. Chim. Acta*, 32 (1965) 101.
Marinsky, J. A. (Editor), *Ion Exchange*, Marcel Dekker, New York, 1966.
Martin, A. J. P. and Synge, R. L. M., *Biochem. J.*, 35 (1941) 1358.
Mayer, S. W. and Tompkins, E. R., *J. Amer. Chem. Soc.*, 69 (1947) 2866.
Mikeš, O., in O. Mikeš (Editor), *Laboratory Handbook of Chromatographic Methods*, Van Nostrand, London, 1964, p. 247.
Morris, C. J. O. R. and Morris, P., *Separation Methods in Biochemistry*, Pitman, London, 1964.
Morris, L. J. and Nichols, B. W., in E. Heftmann (Editor), *Chromatography*, Reinhold, New York, 2nd ed., 1967, p. 466.
Nachod, F. C., *Ion Exchange*, Academic Press, New York, 1949.
Novotný, J., *FEBS Lett.*, 14 (1971) 7.
Novotný, J., Franěk, F. and Šorm, F., *Eur. J. Biochem.*, 16 (1970) 278.
Osborn, G. H., *Synthetic Ion Exchangers*, Chapman and Hall, London, 2nd ed., 1961.
Paterson, R., *An Introduction to Ion Exchange*, Heyden and Sons, Philadelphia, 1970.
Pepper, K. W., Reichenberg, D. and Halle, D. K., *J. Chem. Soc.*, (1952) 3219.
Porath, J. and Fryklund, L., *Nature (London)*, 226 (1970) 1169.
Reichenberg, D., in C. Calmon and T. R. E. Kressman (Editors), *Ion Exchangers in Organic and Biochemistry*, Interscience, New York, 1957, p. 178.
Reichenberg, D. and Wall, W. F., *J. Chem. Soc.*, (1956) 3364.

Reuter, H., *Kunstharzionenaustauscher (Symposiumbericht)*, Akademie-Verlag, Berlin, 1970.
Saldadze, K. M., Pachkov, A. S. and Titov, V. S., *Ionoobmennye Vysokomolekularnye Soedineniya*, Goskhim. Izdat., Moscow, 1960.
Samuelson, O., *Ion Exchange Separation in Analytical Chemistry*, Wiley, New York, 1963.
Seigel, A. and Degens, E. T., *Science*, 151 (1966) 1098.
Shinomura, K., Dickson, L. and Walton, H. F., *Anal. Chim. Acta*, 37 (1967) 102.
Sjöström, E., *Trans. Chalmers Univ. Technol., Gothenburg*, 136 (1953) 7.
Skogseid, A., *Dissertation*, Norges Techniske Hogskole, Trondheim, 1946.
Walton, H. F., in E. Heftmann (Editor), *Chromatography*, Reinhold, New York, 2nd ed., 1967, pp. 287 and 325.
Walton, H. F., *Anal. Chem.*, 40 (1968) 51R.
Walton, H. F., *Anal. Chem.*, 42 (1970) 86R.
Walton, H. F., *Anal. Chem.*, 44 (1972) 256R.
Wuster, C. F., Copenhauer, J. H. and Shafer, P. R., *J. Amer. Oil Chem. Soc.*, 40 (1963) 513.

*Chapter 7*

# Affinity chromatography *

J. TURKOVÁ

## CONTENTS



## PRINCIPLES OF AFFINITY CHROMATOGRAPHY

Affinity chromatography is a technique for the isolation of biologically active substances, making use of their exceptional property of selective and reversible binding of other substances, for which Reiner and Walch introduced the term "affinant". An affinant can therefore be considered as a special case of a ligand. If an insoluble affinant is prepared, usually by covalently attaching to a solid support, and if the extract containing the biologically active component to be purified is passed through this column containing the above material, then all substances which possess no affinity for the given affinant pass directly through the column with the eluate. Substances that have an affinity for the bound affinant are retarded in proportion to their affinities under the experimental conditions used. The specifically adsorbed substances can then be eluted either by a soluble affinant or by changing the solvent composition in such a way as to cause the dissociation of the complex of the isolated substance with the bound affinant. Most commonly, changes in pH, ionic strength or temperature or the addition of reagents that cause the dissociation of the complex are used.

As an example of affinity chromatography, the isolation of active chymotrypsin and trypsin from a pancreatic extract is shown in Fig. 7.1. The isolation was carried out on a column of 6% agarose with bound soya bean trypsin inhibitor (Porath).

In alkaline medium, chymotrypsin and trypsin form a complex with the bound trypsin inhibitor, thus separating from the inactive material which passes through the column unretained. By making use of a pH gradient, it is possible to separate the complex of enzymes with the insoluble inhibitor. At pH 4.5, chymotrypsin is eluted first, while for the dissociation of the trypsin complex a lower pH should be used. For the release of chymotrypsin and trypsin from the complex with the insoluble inhibitor, instead of a change in pH, elution with specific inhibitors at a constant pH can also be employed, as

*Obviously the term "affinity chromatography" is incorrect, as chromatographic separations in general are based on differences in affinities. Therefore, the more recently introduced name for this technique, bioaffinity chromatography, seems to be more suitable. However, as this term is not very common yet, "affinity chromatography" has been used throughout this book.

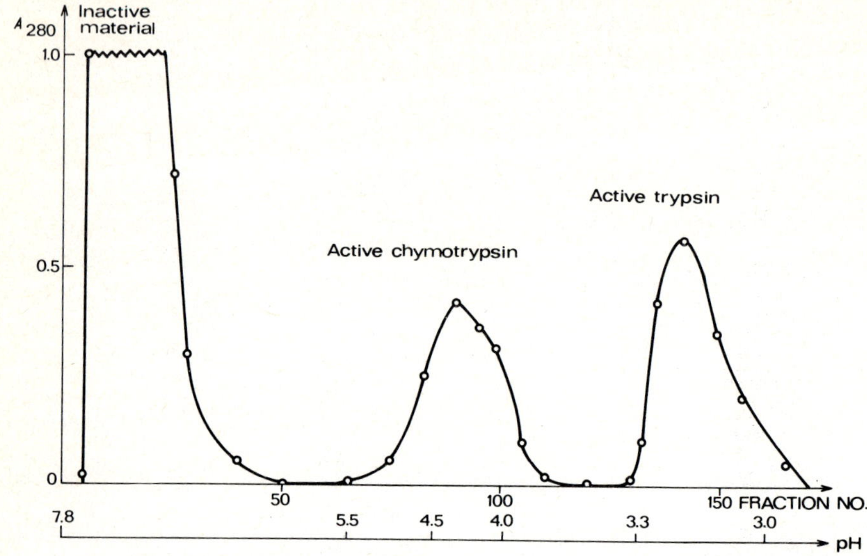


Fig. 7.1. Affinity chromatography of a pancreatic extract on a column of 6% agarose with bound soya bean trypsin inhibitor (Porath). Volume of column, 30 ml. The chromatogram was developed with a solvent system with decreasing pH, as indicated.

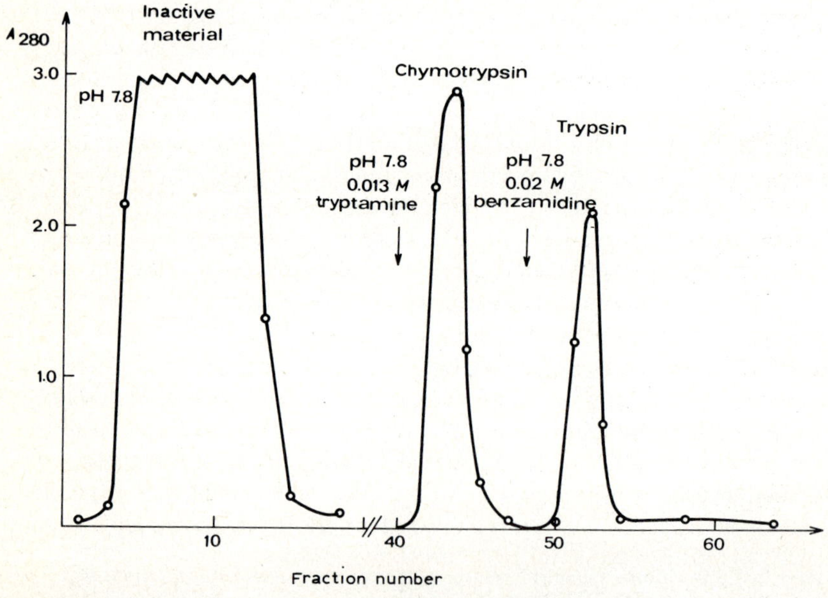


Fig. 7.2. Affinity chromatography of a pancreatic extract on a column of 6% agarose with bound soya bean trypsin inhibitor (Porath). Stepwise elution was accomplished with specific inhibitors.

shown in Fig. 7.2 (Porath). A solution of tryptamine, which is a specific inhibitor for chymotrypsin, releases only chymotrypsin from the column of the insoluble trypsin inhibitor, while trypsin can be liberated by using a solution of benzamidine. The principle of affinity chromatography is often made use of in a very simplified manner when the biologically active substance to be isolated is sorbed on to an insoluble affinant, from which it can be desorbed in a single peak by a convenient procedure. As an example, the separation of active pancreatic trypsin inhibitor from inactive ballast on a column of hydroxyalkyl methacrylate gel with bound chymotrypsin is shown in Fig. 7.3 (Turková *et al.*). If the affinity of the isolated substance for the bound affinant is high, the batch process can be employed with advantage. The isolation procedure can then be considered to be a precipitation rather than chromatography.

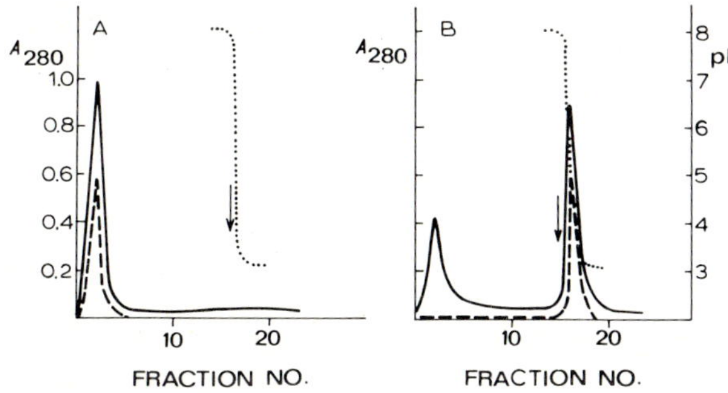


Fig. 7.3. Affinity sorption of lung trypsin inhibitor on a column (10 × 1 cm) of hydroxyalkyl methacrylate gel with bound chymotrypsin (B), compared to chromatography on unmodified gel (A) (Turková *et al.*). Vertical arrow, elution buffer changes from pH 8.1 (0.05 *M* Tris–hydrochloric acid buffer) to pH 3.1 (approximately 0.1 *M* acetic acid). Solid line, absorbance at 280 nm; broken line, inhibitor activity; dotted line, pH.

The principle of affinity chromatography has been known for more than 20 years. In 1951, Campbell *et al.* used it for the first time for the isolation of antibodies by means of a cellulose column with covalently bonded antigen. For the isolation of an enzyme, affinity chromatography was used for the first time in 1953 by Lerman, who isolated tyrosinase on a column of cellulose with ether-bonded resorcinol residues. Subsequently, this method has been only seldom used (Silman and Katchalski), evidently because of the nature of the insoluble carriers of affinants, which did not permit sufficient freedom for the formation of a complex between the isolated substance and the bonded affinant. Major developments of the method have occurred in recent years, the most significant step in the utilization of affinity chromatography being the introduction of the method of affinant binding on to cyanogen bromide-activated Sepharose (Axén *et al.*, Porath *et al.*). It was demonstrated by Cuatrecasas and Anfinsen (1971a) that Sepharose fulfils almost all of the requirements for an ideal carrier of the bound affinant. Thus, in 1968, using a Sepharose-bound affinant, Cuatrecasas *et al.* achieved excellent results in their first work on the affinity chromatographic isolation of nuclease, chymotrypsin and carboxypeptidase. That report, in which affinity chromatography was defined for the first time, led to the wide use of this method in the isolation of enzymes, their inhibitors, antibodies and antigens, nucleic acids, transfer and repressor proteins, hormones and their receptors, and many other substances.

*References p.96*

## CHOICE OF BOUND AFFINANT

In general, any compound is suitable for the isolation of biologically active substances that bind it specifically, firmly and reversibly. Because of the widely varying character of biologically active substances, chemically affinants are also of very diverse types and their classification is therefore based on their biochemical function rather than on their chemical structure.

For the isolation of enzymes, competitive inhibitors can be utilized as affinants (Baker and Siebeneick; Berman and Young; Blumberg *et al.*; Cuatrecasas, 1970a, b; Cuatrecasas *et al.*; Steers *et al.*) or substrates (Baggio *et al.*, Chua and Bushuk). In these instances, the affinant is bound to the active site of the enzyme and becomes more selective with increasing binding specificity. Further, effectors can also be used as affinants (Chan and Takahashi, Sprossler and Lingens). In this instance, the bonding does not take place at the enzyme active site, and it is also possible to isolate enzymes of the same class and similar substrate specificity if the differences occur at the site for the binding of the effector. Enzymes, from which apoenzymes can be prepared by the elimination of coenzymes, can be isolated on carriers with bound coenzyme (Lowe and Dean, Weibel *et al.*). A disadvantage of this method is that in the presence of several apoenzymes that have the same coenzyme, a mixture is often obtained.

For the isolation of inhibitors, the corresponding enzymes serve as affinants (Feinstein, Fritz *et al.*). For proteins binding vitamins, the corresponding vitamin serves as the affinant, for example biotin for the isolation of avidin (Cuatrecasas and Wilchek, McCormick). Thyroxine-binding globulin can be isolated using thyroxine as the affinant (Pensky and Marshall), and for estradiol receptor protein, estradiol serves the same purpose (Cuatrecasas, 1970a). For the isolation of polysaccharides or glycoproteins that contain glycosyl as the terminal group, concanavalin A is a suitable affinant (Aspberg and Porath, Edelman *et al.*, Lloyd, Yariv *et al.*). For the isolation of SH-proteins, *p*-aminophenyl mercuriacetate is a useful affinant (Sluyterman and Wijdenes).

Synthetic ribonuclease S-peptide was purified by means of ribonuclease S-protein bound to agarose (Kato and Anfinsen), the peptide of the active site from the tryptic hydrolyzate of the inhibited nuclease was purified using nuclease bound to agarose (Wilchek), and the peptides containing modified amino acid residues using insoluble antibodies (Wilchek *et al.*). Blasi and Goldberger (see Cuatrecasas and Anfinsen, 1971b) have shown that histidyl-tRNA synthetase is a suitable affinant for the isolation of histidyl-tRNA. DNA is a suitable affinant for the isolation of gene-specific mRNA (Bautz and Reilly, Nyggard and Hall). Insoluble oligonucleotides and polynucleotides of cellulose were employed for the isolation, fractionation and structure determination of various nucleic acids (Erhan *et al.*, Gilham and Robinson, Sander *et al.*) and DNA polymerases (Jovin and Kornberg, Litman). For the isolation of DNA polymerases, adsorption on DNA-agarose was found to be more suitable (Poonian *et al.*).

For the isolation of antigen, the corresponding antibodies can be used as affinants, while for the isolation of antibodies, the corresponding antigens are employed (Akanuma *et al.*; Cuatrecasas, 1969; Goetzl and Metzger; Omenn *et al.*; Silman and Katchalski; Weintraub; Wofsy and Burr). Hapten can also serve as an affinant for the isolation of whole cells producing antibodies against the bound hapten (Truffa-Bachi and Wofsy).

The thermodynamic character of the bond between the affinant and the isolated biologically active substance (the enzyme) was discussed in a review by Reiner and Walch.

The bond between the monomeric affinant A and the enzyme E is expressed by the equilibrium constant of the reaction, $K_A$, on the supposition that this exists in a single tertiary structure form:

$$E + A \underset{k_2}{\overset{k_1}{\rightleftharpoons}} [E{\cdot}A]$$

$$K_A = \frac{C_{[E]} \cdot C_{[A]}}{C_{[E \cdot A]}} \text{ ; } \Delta G_A = -RT \ln K_A$$

After the binding of the affinant to the solid support, the equilibrium constant $K_A$ is affected to a certain extent. An increase in $K_A$ results in modification of the affinant by binding to the matrix, and the steric accessibility of the affinant is limited as a consequence of this binding. On the other hand, a decrease in $K_A$ causes non-specific adsorption of the enzyme to the solid support and to the molecules of the already adsorbed enzyme. On the assumption that a single enzyme of the crude protein has an affinity for the matrix, the equilibrium between the bound affinant and the isolated enzyme is given by the equation:

$$E + A' \underset{k'_2}{\overset{k'_1}{\rightleftharpoons}} [E{\cdot}A']$$

$$K'_A = \frac{k'_2}{k'_1} = \frac{G_{[E]} \cdot C_{[A']}}{C_{[E \cdot A']}} \text{ ; } \Delta G'_A = -RT \ln K'_A$$

For the successful isolation of an enzyme by affinity chromatography, $K_A$ or $K'_A$ should be very small for the desired enzyme and should be much smaller than any dissociation constant for adsorption between the protein and the matrix surface (*i.e.*, non-specific adsorption). We can estimate the maximum $K'_A$ value as follows. Starting from a $10^{-3}$ mole/l concentration of inhibitor in the insoluble affinant and the requirement of a 99% retention of the enzyme from the raw material, which contains about $10^{-5}$ mole/l of enzyme in the three-fold volume of the insoluble affinant, we obtain $10^{-4}$ mole/l as the upper limit of the $K_A$ value of an effective affinant. In a 3% protein solution, where the active enzyme constitutes 10% of the total protein, which should have a molecular weight of $10^5$, *ca.* 10% of the matrix capacity is utilized under the above conditions.

From this estimate, it further follows that in view of the bond that can be formed between the inhibitor and the enzyme, the whole purification process should be considered to be precipitation rather than chromatography. This can also be shown by means of the adsorption isotherm for affinity chromatography shown in Fig. 7.4, from which it is evident that the gross adsorption isotherm (curve 3) can be defined as the sum of the specific (curve 1) and non-specific (curve 2) adsorption isotherms. The specific adsorption isotherm characterizes the ideal specific adsorption, when the adsorption energy, $\Delta G'$, for all adsorbed particles is constant and relatively large. Adsorption ceases when all accessible "affinant" sites are occupied. The non-specific adsorption isotherm characterizes the adsorption of proteins on non-specific sites of the matrix and on already adsorbed protein.

*References p.96*

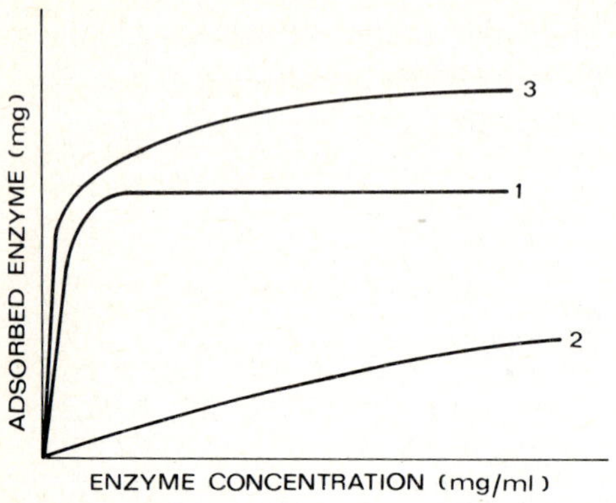


Fig. 7.4. Adsorption isotherm for affinity chromatography (Reiner and Walch). Adsorption isotherms: 1, specific; 2, non-specific; 3, gross.

The $\Delta G'_A$ value is the sum of $\Delta G_A$ and $\Delta G_{\text{non-sp.}}$, where $\Delta G_{\text{non-sp.}}$ is the reaction energy of non-specific complexing and hindrance. Inserting a mean value of $10^{-5}$ mole/l for $K_A$ gives a value of *ca.* 7 kcal/mole for $\Delta G_A$.

The adsorption energy for non-specific adsorption, $\Delta G_{\text{non-sp.}}$, results from the hydrophobic, hydrophilic or even ionic interactions and is comparable with the adsorption energy in normal chromatography. It depends greatly on the nature of the solid carrier and the protein. $\Delta G_{\text{non-sp.}}$ should be as low as possible because it also comprises the adsorption of molecules forming non-specific complexes with the affinant.

However, instances may occur in which the crude protein contains two or more enzymes that display an affinity for the bound affinant. If the equilibrium constant of the reaction of the second enzyme $K_A$(II) is greater than $10^{-3}$ mole/l, then only minute amounts of the second enzyme will be retained together with the enzyme sought. If $K_A$(II) is less than or equal to $10^{-4}$ mole/l, then a mixture of both enzymes will be adsorbed, even though the $K_A$ value of the desired enzyme might be much less than $K_A$(II). This follows from the specific form of the adsorption isotherm for affinity chromatography, because the heat of adsorption is extremely high under chromatographic conditions. If $K_A$(II) differs from $K_A$ by more than 50–100, a separation can still be achieved if differential elution is applied, for example, or if the isolation is carried out by the batch process, using an amount of the insoluble affinant that corresponds exactly to the more intimately binding enzyme, or if chromatography during which equilibrium between [E(II) · A] and [E(I) · A] must be attained is very slow.

## GENERAL ASPECTS OF THE AFFINANT–SORBENT BOND

At the beginning of this chapter, it was indicated how important a role is played by the properties of the solid support in affinity chromatography. The nature of these properties and the extent to which they might affect the results of affinity chromatography are discussed in Chapter 10.

The difference between $\Delta G_A$ and $\Delta G'_A$ is due not only to the nature of the matrix and the modification of the affinant, but also to the change in its steric accessibility. In view of the different structures of the isolated substances, no general rule exists on the minimum distance between the affinant and the surface of the solid support. However, the affinant should be located at such a distance from the carrier surface that the bond would not require the deformation of the isolated substances. The effect of the distance of the affinant 3′-(4-aminophenylphosphoryl) deoxythymidine-5′-phosphate from the solid support surface (both Sepharose 4B and Bio-Gel P-300) on the capacity of the gel in the chromatography of staphylococcal nuclease (Cuatrecasas, 1970a) is shown in Table 7.1. In type A, the inhibitor is bound directly to the matrix, and in other types a chain of varying length is inserted between it and the carrier surface.

TABLE 7.1
CAPACITY OF INSOLUBLE AFFINANTS PREPARED BY BINDING 3′-(4-AMINOPHENYLPHOSPHORYL) DEOXYTHYMIDINE-5′-PHOSPHATE ON SEPHAROSE 4B AND BIO-GEL P-300 DERIVATIVES IN THE AFFINITY CHROMATOGRAPHY OF STAPHYLOCOCCAL NUCLEASE (CUATRECASAS, 1970a)

| Type of inhibitor bound on matrix | Structure | Capacity (mg of nuclease/ml of gel) on derivative of | |
|---|---|---|---|
| | | Sepharose 4B | Bio-Gel P-300 |
| A | 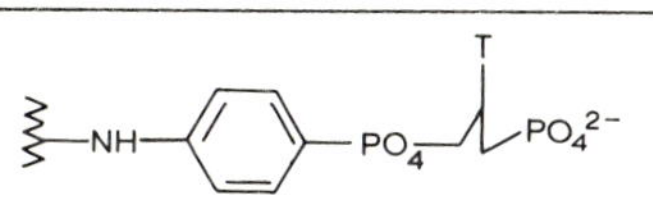 | 2 | 0.6 |
| B | 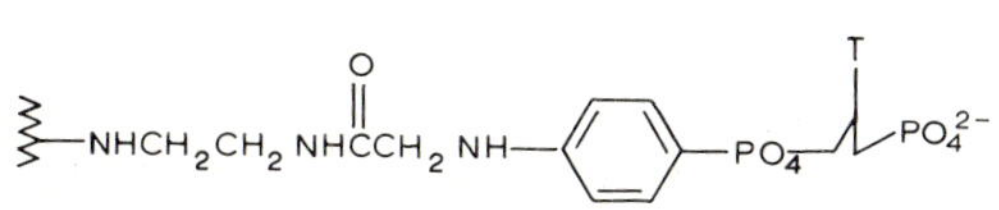 | 8 | 2 |
| C | 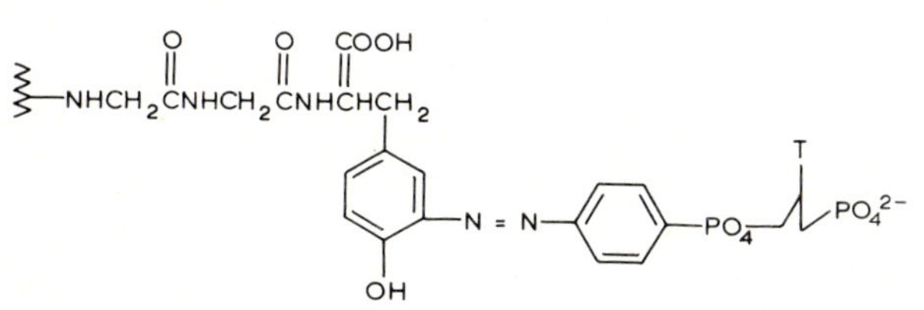 | 8 | 3 |
| D | 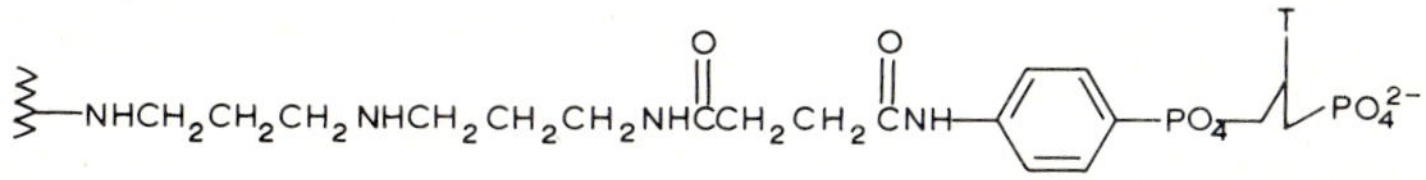 | 10 | – |

The course of affinity chromatography is also affected by the concentration of affinant on the matrix. When Steers *et al.* isolated β-galactosidase on agarose containing a small amount of *p*-aminophenyl-β-D-thiogalactopyranoside, the adsorbed enzyme could be eluted with buffers containing the substrate. At a high concentration of the bound affinant, the adsorbed $\beta$-galactosidase could not be eluted with substrate-containing buffers; on the contrary, even after binding, it remained active and the substrate present in the buffer was cleaved by it during its passage through the column. During the binding of chymotrypsin on various agarose derivatives, Axén and Ernback found that with larger amounts of the chymotrypsin bound per millilitre of the carrier the relative proteolytic activity decreased. Kalderon *et al.* found that on increasing the concentration of the bound affinant [N-($\epsilon$-aminocaproyl)-*p*-aminophenyl] trimethylammonium bromide above 0.16 $\mu$mole per millilitre of the support, the adsorbent specificity for the binding of acetylcholinesterase decreased. This might be explained by the non-specific sorption on the adsorbent which acquired ion-exchanging properties through the increase in the content of ammonium groups. It had originally been hoped that the enzyme would be selectively adsorbed on to the insoluble affinant at high ionic strengths at which non-specific electrostatic interactions would be avoided. However, the decreased affinity of acetylcholinesterase inhibitors for the enzyme at high ionic strength precluded the use of this approach.

The capacity of the bound affinant is further influenced by preserving its original conformation with as little change as possible. Cuatrecasas (1970a) isolated insulin on columns of Sepharose with an antibody against hog insulin bound at both pH 6.5 and pH 9.5. As will be shown in Chapter 10, protein is bound on cyanogen bromide-activated Sepharose by its non-protonated forms of amino groups. On decreasing the pH, a decrease in the number of binding groups also takes place and the result of the pH difference was that the first derivative was able to bind almost 80% of the theoretically possible amount of insulin, while the second derivative, prepared by binding at pH 9.5, took only 7% of the capacity for insulin. As the total content of the bound affinant was identical in both instances, the second derivative must have contained immunoglobulin, which is unable to bind antigen effectively. In the case of a large number of bound amino groups, disturbance of the native tertiary structure evidently occurred. Even at a low pH, adsorbents can be obtained that contain a large amount of active protein bound to Sepharose if the amount of cyanogen bromide is increased during the activation and the amount of protein during the binding. The importance of preservation of the original conformation of the affinant after binding on solid support was also shown by Lowe and Dean for the binding of the cofactor.

## REFERENCES


Akanuma, Y., Kuzuya, T., Hayashi, M., Ide, T. and Kuzuya, N., *Biochem. Biophys. Res. Commun.*, 38 (1970) 947.

Aspberg, K. and Porath, J., *Acta Chem. Scand.*, 24 (1970) 1839.

Axén, R. and Ernback, S., *Eur. J. Biochem.*, 18 (1971) 351.

Axén, R., Porath, J. and Ernback, S., *Nature (London)*, 214 (1967) 1302.

Baggio, B., Pinna, L. A., Morel, V. and Siliprandi, N., *Biochim. Biophys. Acta*, 212 (1970) 515.

Baker, B. R. and Siebeneick, H. U., *J. Med. Chem.*, 14 (1971) 799.

Bautz, E. K. F. and Reilly, E., *Science*, 151 (1966) 328.

Berman, J. D. and Young, M., *Proc. Nat. Acad. Sci. U.S.*, 68 (1971) 395.

Blumberg, S., Schechter, I. and Berger, A., *Eur. J. Biochem.*, 15 (1970) 97.
Campbell, D. H., Luescher, E. L. and Lerman, L. S., *Proc. Nat. Acad. Sci. U.S.*, 37 (1951) 575.
Chan, W. W. C. and Takahashi, M., *Biochem. Biophys. Res. Commun.*, 37 (1969) 272.
Chua, G. K. and Bushuk, W., *Biochem. Biophys. Res. Commun.*, 37 (1969) 545.
Cuatrecasas, P., *Biochem. Biophys. Res. Commun.*, 35 (1969) 531.
Cuatrecasas, P., *J. Biol. Chem.*, 245 (1970a) 3059.
Cuatrecasas, P., *Nature (London)*, 228 (1970b) 1327.
Cuatrecasas, P. and Anfinsen, C. B., *Methods Enzymol.*, 22 (1971a) 345.
Cuatrecasas, P. and Anfinsen, C. B., *Annu. Rev. Biochem.*, 40 (1971b) 259.
Cuatrecasas, P. and Wilchek, M., *Biochem. Biophys. Res Commun.*, 33 (1968) 235.
Cuatrecasas, P., Wilchek, M. and Anfinsen, C. B., *Proc. Nat. Acad. Sci. U.S.*, 61 (1968) 636.
Edelman, G. M., Rutishauser, U. and Millette, C. F., *Proc. Nat. Acad. Sci. U.S.*, 68 (1971) 2153.
Erhan, S. L., Northrup, L. G. and Leach, F. R., *Proc. Nat. Acad. Sci. U.S.*, 53 (1965) 646.
Feinstein, G., *Biochim. Biophys. Acta*, 236 (1971) 73.
Fritz, H., Gebhardt, M., Mester, R., Illchmann, K. and Hochstrasser, K., *Hoppe-Seyler's Z. Physiol. Chem.*, 351 (1970) 571.
Gilham, P. T. and Robinson, W. E., *J. Amer. Chem. Soc.*, 86 (1964) 4985.
Goetzl, E. J. and Metzger, H., *Biochemistry*, 9 (1970) 1267.
Jovin, T. M. and Kornberg, A., *J. Biol. Chem.*, 243 (1968) 250.
Kalderon, N., Silman, I., Blumberg, S. and Dudai, Y., *Biochim. Biophys. Acta*, 207 (1970) 560.
Kato, I. and Anfinsen, C. B., *J. Biol. Chem.*, 244 (1969) 5849.
Lerman, L. S., *Proc. Nat. Acad. Sci. U.S.*, 39 (1953) 232.
Litman, R. M., *J. Biol. Chem.*, 243 (1968) 6222.
Lloyd, K. O., *Arch. Biochem. Biophys.*, 137 (1970) 460.
Lowe, C. R. and Dean, P. D. G., *FEBS Lett.*, 14 (1971) 313.
McCormick, D. B., *Anal. Biochem.*, 13 (1965) 194.
Nyggard, A. P. and Hall, B. D., *Biochem. Biophys. Res. Commun.*, 12 (1963) 98.
Omenn, G., Ontjes, D. A. and Anfinsen, C. B., *Nature (London)*, 225 (1970) 189.
Pensky, J. and Marshall, J. S., *Arch. Biochem. Biophys.*, 135 (1969) 304.
Poonian, M. S., Schlabach, A. J. and Weissbach, A., *Biochemistry*, 10 (1971) 424.
Porath, J., *Biotechnol. Bioeng., Symp.*, No. 3 (1972) 145.
Porath, J., Axén, R. and Ernback, S., *Nature (London)*, 215 (1967) 1491.
Reiner, R. H. and Walch, A., *Chromatographia*, 4 (1971) 578.
Sander, E. G., McCormick, D. B. and Wright, L. D., *J. Chromatogr.*, 21 (1966) 419.
Silman, I. H. and Katchalski, E., *Annu. Rev. Biochem.*, 35 (1966) 873.
Sluyterman, L. A. Æ. and Wijdenes, J., *Biochem. Biophys. Acta*, 200 (1970) 593.
Sprossler, B. and Lingens, F., *FEBS Lett.*, 6 (1970) 232.
Steers, E., Cuatrecasas, P. and Pollard, H., *J. Biol. Chem.*, 246 (1971) 196.
Truffa-Bachi, P. and Wofsy, L., *Proc. Nat. Acad. Sci. U.S.*, 66 (1970) 685.
Turková, J., Hubálková, O., Křiváková, M. and Čoupek, J., *Biochim. Biophys. Acta*, 322 (1973) 1.
Weibel, M. K., Weetall, H. H. and Bright, A. J., *Biochem. Biophys. Res. Commun.*, 44 (1971) 347.
Weintraub, B. D., *Biochem. Biophys. Res. Commun.*, 39 (1970) 83.
Wilchek, M., *FEBS Lett.*, 7 (1970) 161.
Wilchek, M., Bocchini, V., Becker, M. and Givol, D., *Biochemistry*, 10 (1971) 2828.
Wofsy, L. and Burr, B., *J. Immunol.*, 103 (1969) 380.
Yariv, J., Kalb, A. J. and Levitzki, A., *Biochim. Biophys. Acta*, 165 (1968) 303.

# TECHNIQUES OF LIQUID CHROMATOGRAPHY

*Chapter 8*

# Instrumentation for liquid chromatography

M. KREJČÍ, Z. PECHAN and Z. DEYL

## CONTENTS

## CLASSICAL INSTRUMENTATION FOR LIQUID CHROMATOGRAPHY

### Introduction

The development of chemistry in this century, and especially the rapid development of the chemistry of natural products after the Second World War, is directly connected with the utilization of chromatographic separation methods. The techniques of liquid chromatography which are now considered to be classical played an important role in this development and today form the basis of high-efficiency liquid chromatography. The extensive use of liquid column chromatography in the past was evidently the result not only of the extension of the applicability of separation, analytical and preparative methods but also of the simplicity and cheapness of chromatographic equipment. These properties are to a great extent common to both classical liquid column and paper chromatography, and later also to thin-layer chromatography.

TABLE 8.1
EXPERIMENTAL DIFFERENCES BETWEEN CLASSICAL AND HIGH-EFFICIENCY LIQUID CHROMATOGRAPHY

| Characteristics | Classical chromatography. | High-efficiency chromatography |
|---|---|---|
| Efficiency (HETP) (mm) | *ca.* 0.5 (0.1–5)* | *ca.* 0.1 |
| Output (number of plates per second) | 0.02 | 5 |
| Pressure | Fraction to *ca.* 5 kp/cm$^2$ | > 5, usually 20–300 kp/cm$^2$ |
| Flow-rate (mm/min) | 5–50 | 600 |
| Time of experiment | Hours to days | Minutes to hours |
| Equipment | Simple; column and accessories | Chromatograph |
| Orientation (purpose) | Predominantly preparative | Predominantly analytical |
| Detection of separation | Usually analytical, in fractions; sometimes continuously with a detector | Always with a detector |
| Scale | μg to kg (even tons) | ng to μg |

*Gel chromatography.

At present, the classical procedures of column chromatography are still used predominantly and cannot be considered to have been surpassed from the methodical point of view. Even in the future many cases will remain for which their application will be advantageous and useful. Most of the analytical procedures described in the literature today are based on the classical technique of liquid chromatography, and the costs of installing a modern liquid chromatograph considerably exceed those necessary for the application of the classical technique.

Although a separate section is devoted to the differences between classical and high-efficiency liquid chromatography (p.123), it will still be useful to indicate briefly at this point the chromatographic techniques that are considered as classical and those which are of high efficiency. Characteristic differences are given in Table 8.1. It is evident that such a differentiation is formal to a certain extent, but it is indispensable for the classification of the material in this chapter.

Dunnill and Lilly analyzed statistically the preparative papers published on the isolation of microbial enzymes. From these statistics, it follows that more than 90% of papers in this field are based on techniques that are indicated here as classical, and the situation is almost certainly similar in other areas. This can be substantiated by the material from other chapters in this book or by the review by Zweig and Sherma.

In this chapter we shall devote attention to the preparation of columns for analytical and preparative purposes, to questions of the elution technique, and to the corresponding equipment, such as fraction collectors.

It can be seen from Table 8.1 that the amount of the worked-up sample is appreciable. In this section, we deal with the current scale used in laboratory techniques, say from micrograms to kilograms, and a special section is devoted to the problems of industrial-scale liquid chromatography. We have purposely omitted the special techniques involved in ultramicro-scale liquid chromatography because we consider that this is the domain of other types of chromatography: paper, thin-layer, gas and, last but not least, high-efficiency liquid chromatography.

## Columns and accessories

### *Columns*

In classical liquid chromatography, glass tubes (polyacrylate tubes are also commercially available) are almost always used in the preparation of columns. Their dimensions are dictated primarily by the amount of the sample: laboratory columns usually have a diameter of 2–70 mm and a length of 15–150 cm. The amount of sorbent necessary is given by its bulk density. Various materials differ to such an extent in their bulk densities and swelling properties that no universal rule for the weight of sorbent to be used for a given tube volume can be given. For alumina, Reichstein and Shoppee proposed a standard series of tubes of ten dimensions; for example, for 1 g of alumina, a 110 × 8 mm tube is suitable, for 250 g a 600 × 50 mm tube, etc. The calculation of the necessary volume or, conversely, of the diameter and the length of the tube, is much facilitated by the nomogram proposed by Kočent (Fig. 8.1).

*References p.168*

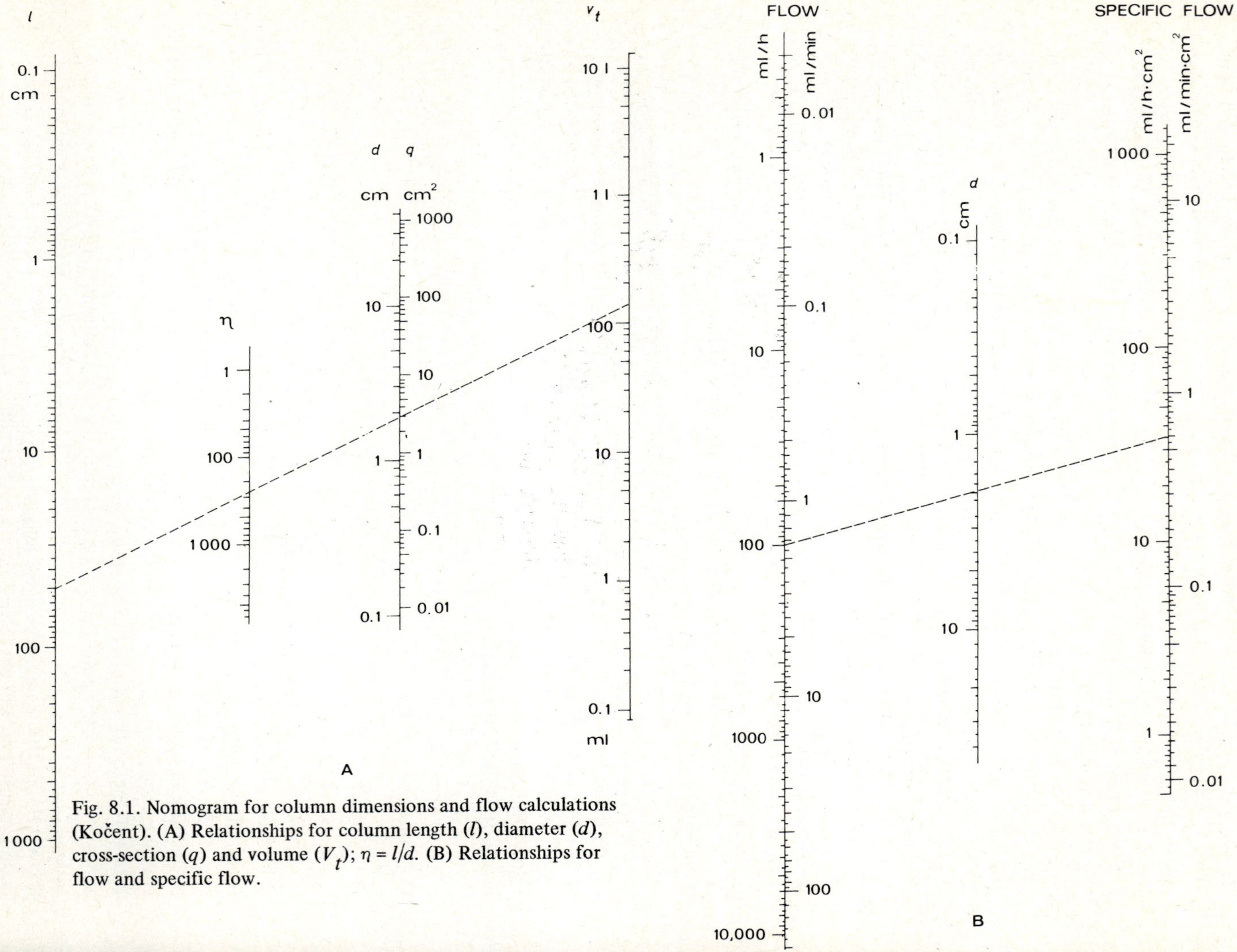


Fig. 8.1. Nomogram for column dimensions and flow calculations (Kočent). (A) Relationships for column length (*l*), diameter (*d*), cross-section (*q*) and volume ($V_t$); $\eta = l/d$. (B) Relationships for flow and specific flow.

The ratio of the column diameter to its length cannot be recommended generally. As the homogeneity of the bed is of prime importance for efficiency, a rather wide range of ratios of diameter to length may be tolerated. The efficiency of the separation and the complexity of the separated mixture should be taken into account in order to have a sufficiently, but not excessively, long column: the elution of a too strongly sorbed component could prolong the experiment too much and the separation effect would deteriorate. Most workers choose a ratio of diameter to length between 1:10 and 1:100.

Some common types of columns are shown in Fig. 8.2. The upper end is either rounded by melting or provided with a ground-glass joint for the connection of the inlet tube from the mobile phase reservoir. A tube narrowed at the lower end is most commonly used; if a rubber or plastic tube is fitted on it, a stopcock is not necessary, and the flow-rate, if no pump is used, can be regulated by the height of its orifice. By raising it above the level of the sorbent, a safety syphon is obtained that prevents the sorbent from drying out. Above the sorbent bed there should be a space for the mobile phase. The volume of the tube used is adequate if it corresponds to 10–30% of the adsorbent column (bed) length. This is, of course, valid under the assumption that the elution takes place by gravity and with a single solvent. In other instances, especially in continuous gradient elution (see below), this space represents a dead volume and must be decreased to the minimum.

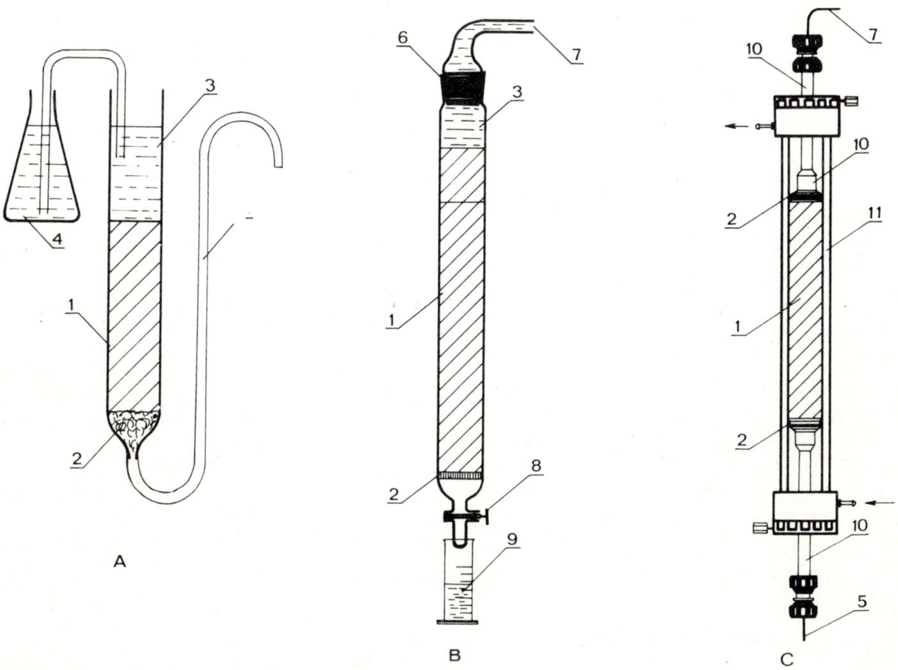


Fig. 8.2. Columns: (A) simple column without stopcock; (B) column with fritted support and adaptor; (C) piston column with thermostatting jacket. 1 = Sorbent; 2 = bed support; 3 = mobile phase; 4 = eluent reservoir; 5 = outlet tube; 6 = connecting adaptor; 7 = connection to reservoir (pump); 8 = stopcock; 9 = collecting vessel; 10 = piston flow adaptor; 11 = thermostatting jacket.

*References p.168*

TABLE 8.2
PRODUCERS OF CHROMATOGRAPHIC EQUIPMENT FOR CLASSICAL LIQUID CHROMATOGRAPHY

C = Columns; A = accessories; FC = fraction collectors; P = pumps; M = monitors

| Company | Equipment available | | | | |
|---|---|---|---|---|---|
| *Ace Glass*, 1430 N. West Blvd., Vineland, N.J. 08360, U.S.A. | C | | FC | | |
| *American Instrument Co.*, Div. of Travenol Labs., 8030 Georgia Ave., Silver Springs, Md. 20910, U.S.A. | | | | P | M |
| *Applied Science Labs.*, P.O. Box 440, State College, Pa. 16801, U.S.A. | C | | | | |
| *Baird and Tatlock*, Freshwater Rd., Chadwell Heath, Essex, Great Britain | | | FC | | |
| *Beckman Instruments*, 2500 Harbor Blvd., Fullerton, Calif. 92634, U.S.A. | | A | FC | P | M |
| *Boehringer Mannheim GmbH*, 6800 Mannheim 31, G.F.R. | C | | | | |
| *Brinkmann Instruments*, Cantiague Rd., Westbury, N.Y. 11590, U.S.A. | C | | FC | P | M |
| *Buchler Instruments*, Div. of Nuclear-Chicago Corp., 1327 17th St., Fort Lee, N.J. 07024, U.S.A. | C | A | FC | P | |
| *Carlo Erba*, Scientific Instruments Div., P.O. Box 4342, 20 100 Milan, Italy | | A | FC | P | M |
| *Chemical Data Systems*, Rd. 2, Box 74, Oxford, Pa. 19363, U.S.A. | | | FC | | |
| *Chromatronix*, 2743 9th St., Berkeley, Calif. 94710, U.S.A. | | A | | P | M |
| *Corning Glass Works* (N.Y.), Houghton Pk., Corning, N.Y. 14830, U.S.A. | C | | | | |
| *Development Workshops* of the Czechoslovak Academy of Sciences, Papírenská, 16610 Prague, Czechoslovakia | C | A | | P | M |
| *Dohrmann, Div. of Envirotechnical Corp.*, 1062 Linda Vista Ave., Mountain View, Calif. 94040, U.S.A. | | A | FC | P | |
| *Durrum Chemical Corp.*, 3950 Fabian Way, Palo Alto, Calif. 94303, U.S.A. | C | A | | | |
| *Fischer-Porter Co.*, Lab Crest Scientific Div., County Line Rd., Warminster, Pa. 18974, U.S.A. | C | | FC | | |
| *Gilson Medical Electronics*, 3000 W. Beltline Hwy., Middleton, Wisc. 53562, U.S.A. | | | FC | | |
| *Hamilton Co.*, 4960 Energy Way, P.O. Box 7500, Reno, Nev. 89502, U.S.A. | | A | | | |
| *Instrumentation Specialities Co.*, 4700 Superior Ave., Lincoln, Nebr. 68504, U.S.A. | C | | FC | P | M |
| *Kontes*, Spruce St., Vineland, N.J. 08360, U.S.A. | C | A | FC | | |
| *Lab Glass*, 1172 Northwest Blvd., Vineland, N.J. 08360, U.S.A. | C | A | FC | | |
| *LaPine Scientific Co.*, 6001 S. Knox Ave., Chicago, Ill. 60629, U.S.A. | | A | | P | |
| *LKB Produkter*, P.O. Box 12220, Stockholm, Sweden | C | A | FC | P | M |
| *Mikrotechna,* U. Průhonu 22, Prague-Holešovice, Czechoslovakia | | | FC | P | |
| *New York Laboratory Supply Co.*, P.O. Box 516, W. Hempstead, N.Y. 11552, U.S.A. | | A | | | |
| *Perkin-Elmer Corp.*, 702-G Main Ave., Norwalk, Conn. 06852, U.S.A. | | A | FC | | M |
| *Pharmacia Fine Chemicals*, Box 175, S-751 04 Uppsala 1, Sweden | C | A | | | M |
| *Phase Separations Ltd.*, Deeside Industrial Estate, Queensferry, Flintshire, Great Britain | | | FC | | |
| *Phy-Lab*, Div. of Physicians and Hospital Supply Co., 1400 Harmon Pl., Minneapolis, Minn. 55407, U.S.A. | C | | FC | | |

TABLE 8.2 *(continued)*

| Company | Equipment available | | | | |
|---|---|---|---|---|---|
| *Pye Unicam Ltd.*, York St., Cambridge, Great Britain | | | FC | P | |
| *H. Reeve Angel Co.*, Ltd., 14 New Bridge Street, London E.C.4, Great Britain | C | | | P | |
| *Scientific Glass Blowing Co.*, P.O. 19353, Houston, Texas 77023, U.S.A. | C | | FC | | |
| *Serva International*, Römerstr. 118, 6900 Heidelberg 1, Postfach 1505, G.F.R. | C | | FC | | M |
| *Shandon Scientific Company Ltd.*, 65 Pound Lane, Willesden, London NW 10, Great Britain | | | FC | P | |
| *Sigmamotor*, 14 Elizabeth St., Middleport, N.Y. 14105, U.S.A. | | | | P | |
| *Technicon Instruments Corp.*, 511 Benedict Ave., Tarrytown, N.Y. 10591, U.S.A. | C | A | FC | P | M |
| *I.W. Towers and Co.*, Widnes, Lancashire, Great Britain | | | FC | | |
| *Utopia Instrument Co.*, Caton Farm Rd., P.O. Box 863, Joliet, Ill. 60434, U.S.A. | C | | | P | M |
| *Varian Aerograph*, 2700 Mitchell Dr., Walnut Creek, Calif. 94598, U.S.A. | C | A | FC | | M |
| *Waters Associates*, 61 Fountain St., Framingham, Mass. 01701, U.S.A. | | A | FC | | |

In commercial products this question has been solved technically in various ways (some producers are listed in Table 8.2). The elimination of dead volumes at both ends of the column is solved for swelling materials by using piston adaptors, for example (Fig. 8.2C). It is evident that when a column is constructed in this manner it must be possible to dismantle it. From the original all-glass construction with ground-glass joints and a spring-clip the development went on to screw closures made from chemically and mechanically resistant plastics. As work at constant temperature is often necessary, sometimes at an elevated temperature and at other times in the cold, the tubes are provided with jackets for temperature control. Large-size preparative columns are discussed in a separate section (p. 120).

Classical materials used for the bed support were glass-wool or a fritted glass disc; today these are gradually being replaced by sintered plastics (polyurethane, PTFE, polyamide).

Although there is an appreciable difference in price between simple and more elaborate columns, it is not wise to economize on the equipment; in comparison with various improvizations, these devices save time and soon become profitable, especially if routine analyses are to be carried out.

### *Accessories*

These include tubes, connectors, fittings to columns and stopcocks. More oomplex devices such as pumps and fraction collectors are not considered here as column accessories.

Connecting tubes should be made of inert material; the most resistant is PTFE, sometimes polyethylene is used, and for work with aqueous solutions poly( vinyl chloride) can

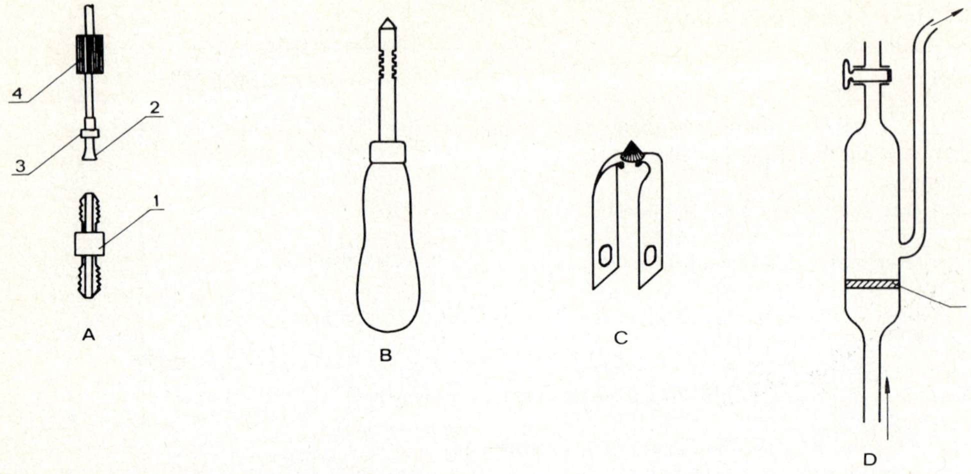


Fig. 8.3. Column accessories: (A) connection; (B, C) device for extending tube diameter; (D) bubble-trap. 1 = Joint; 2 = tube; 3 = mouthpiece; 4 = cap nut.

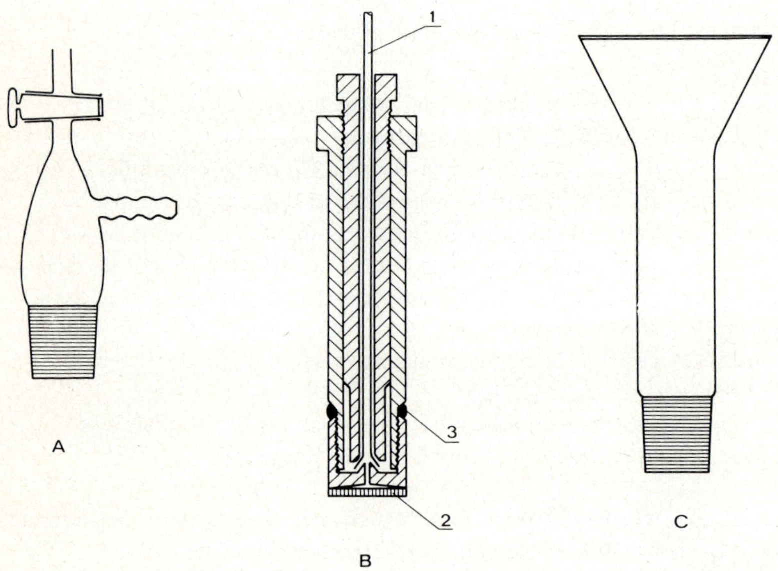


Fig. 8.4. Column adaptors: (A) connecting adaptor; (B) piston flow adaptor; (C) filling adaptor. 1 = Inlet capillary tube; 2 = sintered glass plate; 3 = O-ring.

suffice. The tubes must have a sufficiently small diameter; commonly tubes of 1 mm I.D. are used. For connecting the tubes, connectors with a conical nucleus and a cap nut (Fig. 8.3) are most suitable. A good conical extension of the tube ends can be obtained from a conical piece heated over a burner or electrically with a transformer solderer.

Column fittings of three types exist today: for the connection of the eluent reservoir (Fig. 8.3), for the application of the sample (Fig. 8.21, p. 141), and for the packing of the column (Fig. 8.4).

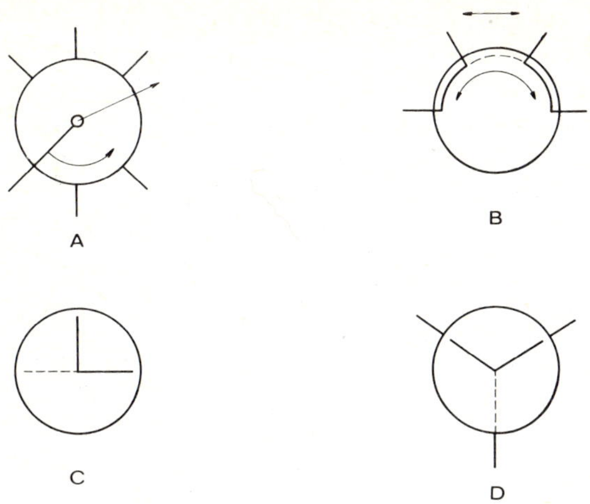


Fig. 8.5. Valves. A = 7-way valve; B = recycling valve; C = L valve; D = Y valve.

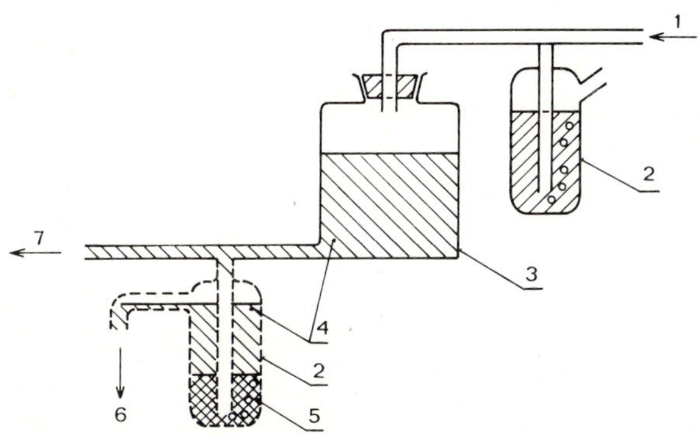


Fig. 8.6. Connection of gas under pressure. 1 = Inlet for gas; 2 = manostat; 3 = reservoir; 4 = mobile phase; 5 = mercury; 6 = waste; 7 = to column.

In Table 8.1 it is stated that for high-efficiency liquid chromatography, high operating pressures are characteristic. Therefore, the construction of the stopcocks must be quite different. In addition to ground-glass stopcocks with glass or PTFE nuclei, capillary stopcocks of various types (Fig. 8.5) are also used in classical column chromatography: common two-way, branched and special recirculation stopcocks (the concept of recirculation chromatography is explained in the section on elution techniques). They are regulated manually or electromagnetically.

If the column packing consists of excessively fine particles, the back-pressure of the column increases so that a forced flow must be applied. The simplest way is to expose the mobile phase level to a pressure of gas (several millimetres of mercury). The pressure is regulated by a simple mercury manostat connected either to the gas inlet into the reservoir or to the mobile phase outlet from the reservoir (see Fig. 8.6).

However, the insertion of pumps is more common, as a faster flow of mobile phase can be achieved even in longer columns. Low-pressure pumps are used (see p. 106). A pressure of several tens of millimetres of mercury up to several atmospheres can be achieved with simple piston or peristaltic pumps, for example, a simple syringe of a 10–20 ml volume with a mechanically driven piston can be used. The advantage of multi-channel peristaltic pumps is that the solvent can be pumped on to several chromatographic columns simultaneously or part of the effluent can be split off for analysis. Piston pumps with a distribution by seat valves are very sensitive to gas bubbles in the mobile phase; a bubble formed in the valve may completely upset its function, and a bubble-trap is connected to the liquid inlet of such a pump (Fig. 8.3). Naturally, the use of pumps makes it necessary to reconstruct the connections of the column to the mobile phase. The techniques used are similar to those described on p. 107.

*Commercially available equipment*

In Table 8.2 a review of some producers of equipment for classical liquid chromatography is given. Basic equipment is usually included in the production programme of all producers of laboratory apparatus, while glass parts are manufactured by all of the larger glass firms. Some producers of sorbents have also extended their production programmes to include columns and other accessories for chromatography.

As we were unable to test in practice and compare the various equipment available, this section cannot serve as a "buyer's guide". The names of the models, technical parameters and prices change rapidly, and the production programmes and the names of firms also change. In the course of an approximately 5-year period, approximately one third of the data will become obsolete (compare Holeyšovský). Therefore, we decided not to give a detailed review here and the reader should refer to up-to-date sources, for example the annual instrument guides in the journals *Science* and *Analytical Chemistry*, exhibitions of apparatus, etc.

It seems that the trend in development is towards ready-for-use packed columns.

## Column preparation and introduction of sample

*Column preparation*

The preparation of the columns is a fundamental operation. All theoretical considerations are valid only for a correctly packed column; only on a perfectly packed column can a good separation be achieved. In this respect no procedure, however perfect, can replace experimental skill. The technique itself may differ for various types of sorbents (*cf.*, Chapters 12 and 13) and for high-efficiency liquid chromatography it is rather different.

The main requirement is that the column should be packed regularly so that it does not contain either air bubbles or channels through which an excessively large part of the mobile phase could flow ineffectively.

Of the three possible methods of column packing – in suspension, dry sorbent into the solvent, and dry packing – the first is probably most often used. Only this method permits an effective degassing of the sorbent which is very important especially in ion-exchange chromatography.

The column should be fastened in a vertical position and the bottom of the column carefully sealed with a porous layer of wool or a similar material. The column is filled to one third of its height with the mobile phase, and a thick suspension of suitably adjusted, degassed sorbent in the mobile phase is poured into it. It is recommended that the sorbent suspension should be poured into the column immediately in order to prevent stratification, *i.e.*, classification directly in the column according to particle size, with coarser particles at the bottom and finer particles in the upper section, which is undesirable.

A certain time is necessary before the material becomes settled and a compact bed is formed (for a column 1 m high, approximately 5 h are required; this time is longer the finer are the sorbent particles and the lower is its density). The column is then stabilized by washing it with the mobile phase.

Care must always be taken to ensure that the level of the mobile phase never drops below the top of the column, so that it does not become dry. Sometimes an easily permeable filter-paper disc is put on the top in order to prevent swirling of the sorbent.

Filling the column with the sorbent poured into the solvent, or by the dry packing technique, is limited to liquid–solid or liquid–liquid chromatography (silica gel, alumina sorbents, etc.). When the column is packed in a wet state, the tube is first filled with a larger amount of the mobile phase (up to 70–90% of the column length) and the outflow is regulated with the stopcock. The prepared classified sorbent is poured into the tube through a funnel, which should not touch the level of the mobile phase. As some materials may agglomerate, it is advisable to pour the adsorbent through a dry sieve with meshes of a slightly larger diameter than that of the particles used. After a sufficient amount of sorbent has been introduced into the column, the material is allowed to settle until a compact column is obtained. When the column is packed in a dry state, regular settling of the sorbent is enhanced by tapping the column walls.

### *Introduction of sample*

For the introduction of samples, four methods can be considered: on the top of the sorbent bed, under the mobile phase, with a doser, and in a layer of sorbent.

The first method is classical. The solvent is allowed just to soak into the sorbent bed. The high column of the sorbent should not become dry, because bubbles might form after refilling with the solvent. Then the sample is introduced on to the column with a syringe or pipette over the whole column section, as homogeneously as possible. Again, care should be taken to ensure that the sorbent is not disturbed and that the solvent level drops just to the top of the sorbent. Immediately after the introduction of the sample, a small amount of solvent is added, which is allowed just to soak into the sorbent bed, and only then a sufficient supply of the mobile phase is poured into the column, or the column is connected with the solvent reservoir.

It should be stressed again that neither during the addition of the solvent nor during the introduction of sample should the surface of the sorbent be swirled or disturbed. If the surface is disturbed, however, it is possible to swirl the whole sorbent surface to a

*References p.168*

depth of 1 cm and to allow the sorbent to settle again. However, this operation may impair the separation.

The introduction of the sample by "underlayering" is more advantageous in some instances, and is useful for materials that are very easily disturbed, for example rare types of Sephadex. In this method, it is not necessary to wait until the mobile phase has soaked into the bed; the sample is simply injected through a thin tube inserted on a syringe under the solvent layer above the top of the bed, *i.e.*, just above the surface of the sorbent. Obviously, the sample solution should be denser than the solvent itself.

The automatic introduction of the sample from a doser or from a dosing loop is a common method in high-efficiency liquid chromatography and therefore it is described elsewhere (p. 143). In contrast, the application of sample in an adsorbed state is used only rarely and only for columns packed with dry sorbent, *i.e.*, in the so-called dry column technique. The mixture to be analyzed is transferred into a small amount of adsorbent, which is placed on the top of a column of dry adsorbent, and is then covered with another layer of the same sorbent.

### Techniques of elution

The techniques of frontal and displacement chromatography are used very little in practice. The first technique was utilized mainly for theoretical studies and for the description of sorption phenomena in chromatography. From the analytical point of view, the frontal analysis technique can be used for qualitative purposes only, especially for the determination of the number of components in the mixture being analyzed on the basis of the number of fronts. The displacement technique was mainly applied in preparations, primarily for the preparation and analysis of hydrocarbon mixtures.

Over the past 20 years, the expression "column chromatography" may be considered as synonymous with elution chromatography. Almost all analytical and preparative procedures described are based on the elution technique. At the top of the column, a mixture of substances to be separated is introduced, which are then washed through the column with an elution agent. When the elution is ended before the investigated substances have left the column, the method is called column elution, a technique used very little nowadays. The main argument for the introduction of this technique was the difficult elution of strongly sorbed components of the sample mixture. Today, these components can be eluted with a stronger eluent, by the use of the gradient technique, and in some instances by the introduction of a temperature gradient.

There are instances when chromatography with one mobile phase in a single elution run does not lead to the required results. Sometimes it is sufficient to repeat the separation under the same conditions, *i.e.*, in recycling chromatography, in which the effluent from the column is led back to the top of the column.

#### *Dry column technique*

This procedure represents the passage from classical TLC to preparative-scale chromatography. According to Loev and Snader, this method has several advantages in comparison with the elution technique, and in particular the resulting separation is virtually identical with that in TLC on the same material. The technique consists in the introduction on a column packed with a dry adsorbent of a mixture of substances; most commonly, a maximally concentrated sample is placed on the top layer of the adsorbent with a pipette Then the solvent is introduced so that it lies above the adsorbent bed to a height of no more than 1–2 cm. The small column of liquid thus formed does not suffice for the generation of a flow of mobile phase by gravity through the column and the transport along the column is due almost exclusively to the capillary movement of the mobile phase, as in thin-layer chromatography. As soon as the mobile phase front (solvent front) approaches a distance of several millimetres above the end of the column, the development is interrupted and the column containing the separated substances is then expelled from the tube and cut into slices. This may be facilitated by using a column cut vertically into two halves, the two halves being held together with tape (cellophane, adhesive plastic tape, etc.) wound around the column in a spiral (Fig. 8.7). When the chromatogram has been developed, the tape is removed, one half of the column is taken off and the exposed column of adsorbent contains further worked-up components of the original mixture. The fractions separated on the column are then detected by procedures used in thin-layer and paper chromatography, for example, inspection of the column under UV light or chemical detection.

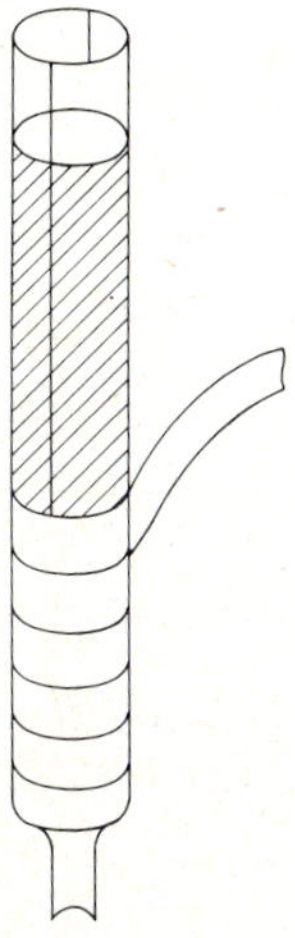

Fig. 8.7. Segment column for elution technique.

*Gradient elution*

Elution with a single solvent, usually used in gel chromatography, is not always sufficient. Of possible gradient techniques in liquid chromatography (temperature gradient, stationary phase gradient, gradient of the mobile phase strength; see p. 252), the technique of gradient elution proved best and has been used longest. In this technique, the composition (strength) of the mobile phase is changed either stepwise or continuously.

*References p.168*

A description of only the most commonly used devices for gradient elution are given below.

Three of many possibilities of producing gradients are shown in Fig. 8.8. The oldest device, with a closed mixer, has been purposely omitted, as it permitted only a single gradient form (exponential) and was little used in practice. The arrangement shown in Fig. 8.8A was found to be suitable. If two vessels (beakers) of equal cross-section and shape are used as the reservoir and mixer, a linear gradient is obtained. The mobile phase should be led on to the column with a pump. A concave or a convex gradient may be

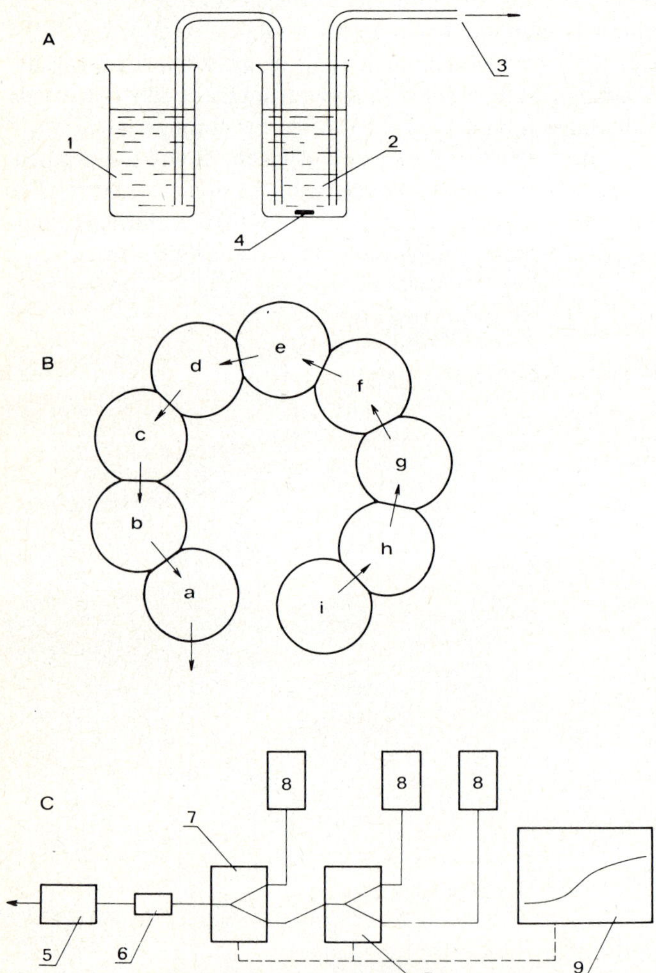


Fig. 8.8. Gradient-forming devices: (A) simple device; (B) Varigrad (Technicon; for description, see text); (C) photoelectric sensor (Ultragrad, LKB). 1 = Reservoir; 2 = mixing chamber; 3 = to pump; 4 = stirring bar; 5 = pump; 6 = small-volume mixing chamber; 7 = valves; 8 = reservoirs of mobile phase; 9 = photoelectric gradient monitor.

obtained if the mixer or reservoir vessels are of different shapes. It should be stressed that these arrangements fail if the densities of the two solutions differ too much.

The apparatus described by Peterson and Sober and available commercially under the name Varigrad (Technicon, Tarrytown, N.Y., U.S.A.) is very popular (Fig. 8.8B). It permits the form of the gradient to be changed at a certain point on the chromatogram in a very simple manner. It consists of nine equal cylindrical vessels provided with stirrers, interconnected in such a manner that the solvent is led from the vessel i to h, from h to g, etc., and from vessel a it is introduced into the column. Each of the vessels may contain a different solution.

Fig. 8.8C shows a rather complicated device, the commercial name of which is Ultragrad (LKB, Stockholm, Sweden). The solution is led by a pump on to the column from three reservoirs via two valves. Their regulation is ensured by a regulating unit which follows the form of a gradient photometrically.

A detailed review of the techniques and the theory of gradient elution has been given by Snyder (1965).

*Flow-rate*

An increase in the flow-rate of the mobile phase through the column leads to a shortening of the time necessary for the separation, so that it is preferable to work with a maximum flow-rate. This, however, has its limitations: an increase in flow-rate leads to a decrease in efficiency, as was demonstrated for gel chromatography by Flodin. For other types of chromatography (for example, ion-exchange chromatography), this effect becomes apparent only when the flow-rate exceeds a certain limit, *i.e.*, as soon as the flow-rate of the mobile phase is excessively large with respect to the rate of mass transfer between the mobile and the stationary phase (see Chapter 4). The dependence of the efficiency of the process on the quality and particle size of the sorbent is described elsewhere. The sorbent characteristics steadily improve in this respect. Earlier, a flow-rate of 5–10 ml/h · cm$^2$ was common, and such a flow-rate corresponds to a linear rate of *ca.* 0.001–0.01 cm/sec, while today flow-rates of 30–120 ml/h · cm$^2$ are more usual. It should be kept in mind that for some materials, the softness of their structure is the main limiting factor, *i.e.*, the critical flow-rate at which the material becomes compressed is reached relatively quickly.

**Analysis of effluent**

Up to the present day, it has been usual to analyze each collected fraction separately for solute content by chemical or physical methods. A general description of the possible methods available for this purpose is useless, because it would need to cover the whole of analytical chemistry. Recommended procedures for actual cases are given elsewhere in this book.

This "classical" approach is unsuitable if the number of fractions increases to hundreds or if several parameters have to be followed in the effluent. Therefore, a continuous analysis, *i.e.*, the monitoring of the effluent, is being used increasingly, even within the area of "classical" liquid chromatography.

*References p.168*

In principle, the detectors used are the same as those described on p. 146, and of these, at least in some fields of biochemistry, the UV monitor is most popular, either in its simplest form with a mercury lamp and recording at 254 nm, or in a more advanced form with the possibility of recording at two or more wavelengths. An apparatus with a perfect spectrophotometer with a prism or grating monochromator is less common today and the application of fluorimetric detectors is also rather scarce. As fluorimetry is substantially more sensitive than UV spectrophotometry, its importance in routine analysis is increasing, and it may be expected that in the near future it will be used more extensively in liquid chromatography.

Particularly if two or more analytical values have to be determined in the eluate (*e.g.*, total protein content and enzymatic activity), it is profitable to consider the automation of the analytical determination. This may be achieved, in principle, by inserting between the column and the flow-through photometer a means of continuous addition of the reagent and a reactor in which the colour reaction takes place. It is not necessary to consume the whole of the effluent in this manner, but it is possible to withdraw only an aliquot of it with a peristaltic pump, or to split the stream into two or more analytical channels. This system was introduced commercially by Technicon and today similar analyzers can be obtained from several sources.

A useful method for the detection of radioactive substances is the continuous measurement of the radioactivity of the effluent. These detectors are also commercially available.

When changing the composition of the mobile phase, the agreement between the true and the calculated gradients is of interest. In such a case, either the pH or the effluent conductivity is measured in the flow-through cell; in work with organic solvents, refractometry is usually more convenient.

### *Fraction collectors*

The eluate from the column is usually collected in a fraction collector. If there is not a more suitable apparatus at hand, a fraction collector is indispensable for most experiments. The size (volume) of the collected fractions is given by the least volume required for the analysis of the effluent, *i.e.*, for the determination of the solute content in the mobile phase. The larger the number of fractions collected and the smaller the effluent content in each fraction, the more accurately can the chromatogram be recorded, giving an idea of the properties of the column, or of the course of the separation. Nowadays, a fraction collector is used mainly for preparative purposes whether it is desired to obtain a pure substance or a fraction of a certain solute for further detailed analysis.

The exchange of the collecting vessel at the effluent outlet is possible in two different ways. Either the collecting vessel is moved together with the leading device, or a capillary is deflected through which the effluent from the column flows out. The second method usually requires an additional lead, and as a consequence it also causes a larger transportation retardation. This fact is of importance especially when the impulse for the exchange of the vessel is derived from the detector signal. Today, fraction collectors based on the movement of the collecting vessels are more often used. The simplest are circular plates in which collecting vessels are fixed in one or several rings. The circular plate is given a rotatory movement by either a mechanical or an electrical drive. A mechanical shift of the circular transporter is due to a weight and a snap, which determines the moment of the disc movement and limits the distance to which the collecting vessel is shifted. In the case of electrical monitoring of the transporter, an electric motor is either in the centre or on the circumference of the plate. The disadvantage of the collectors with a circular plate is the necessity for transferring the tubes into ordinary racks. Therefore, collectors have been developed with segments that can be taken out and manipulated as racks (in this type, the outlet capillary is moved).

Nowadays, a large number of very good fraction collectors are commercially available (about twenty types are listed in Table 8.2). The fundamental advantage of some types is their compactness and the large number of monitored collecting vessels. A miniaturization of the apparatus is possible mainly because the circular type of transporter of collecting vessels is being abandoned. Thus, for example, in accurate metallic models, metal collecting vessel holders are moved and are regulated by chains with ball joints. The collecting vessels are not moved in circles but along straight lines. The chromatographic column outlet is not moved and therefore the connections between the column (detector) and the fraction collector can be shortened and the transport delay of the whole arrangement is decreased. Another solution consists in the use of plastic vessels with catches that permit them to be connected in an endless chain. There is also a progressive solution in which the tubes are placed in rectangular removable racks. Further technical measures ensure that the liquid flow is stopped at the moment when the collecting vessels change, or that the liquid is led into a bigger reservoir when all collecting tubes are full, for example.

The regulation of the movement of the fraction collector is based on several principles shown schematically in Fig. 8.9. The parameters that can be measured as the basis for regulation are time, volume of the effluent and detector response (Snyder, 1967).

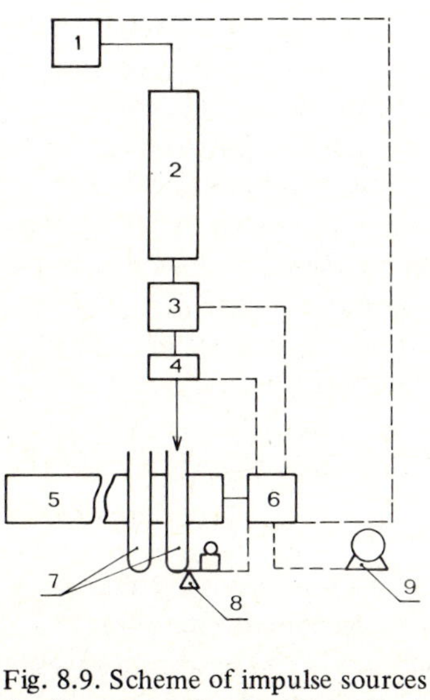


Fig. 8.9. Scheme of impulse sources for a fraction collector. 1 = Pump; 2 = column; 3 = detector; 4 = flow meter; 5 = fraction collector; 6 = control unit of fraction collector; 7 = collecting vessel; 8 = balance; 9 = time switch.

In the simplest case, the impulse for the transporter shift is derived from time. A timer gives impulses at regular intervals (from several seconds to about 1 h), which cause a change of the collecting vessel below the column outlet. Hence, if the solute concentration in single fractions is plotted against time, approximate retention times can be obtained for given solutes.

However, retention volumes are a much more reliable parameter. For this, a device is often inserted between the column and the fraction collector which measures the outgoing effluent, or one for the measurement of the liquid flow. The collecting vessels are then exchanged when filled with a pre-selected volume of liquid. The same effect can be achieved if the passing liquid volume is derived from the number of rotations of the piston pump. The effluent volumes are evidently not measured too accurately. The results obtained by means of impulses from the pump and by means of the volumes passed through are approximately the same. The measurement of volume by weighing is little utilized today. If the weight of the vessel itself is too high, a stable light vessel is used, which is weighed and emptied into the collecting vessel after it has attained a certain weight.

For preparative purposes, it is best if the signal for the fraction collector is derived from the detector signal, which is a more recent and so far little used technique. In this manner, a whole fraction of the chosen solute can be collected in a single collecting vessel. The chosen response level is set in the detector and as soon as this level is exceeded an impulse is emitted into the fraction collector and an empty vessel is placed below the liquid outlet from the detector. This vessel is filled with the effluent until the detector response drops below the chosen level. At this moment, another collecting vessel is shifted. Experience from gas chromatography in which programmes have been developed that permit a relatively successful preparation of even incompletely separated fractions can also be applied in this field.

More expensive fraction collectors sometimes have selectable regulation principles. In comparing them, it is useful to be aware of some of their advantages and disadvantages. Time-regulated collectors are highly reliable, which is a great advantage, and in this respect experience with them has been good. Further, they permit the choice of fraction volumes within a very broad range. A disadvantage is that the flow-rate must be kept constant, which is not necessary in volume-regulated collectors. In working with drop counters, difficulties may sometimes arise with the constancy of the volumes of the drops when the surface activity of the effluent changes during the operation. Weight-regulated collectors have numerous disadvantages: the apparatus is sensitive to heating, the adjustment of balance is exacting, they often require vessels with identical weights and they cannot be used if the density of the effluent changes.

*Apparatus for flow measurement*

Apparatus for measurement of the solvent flow is not only utilizable as a regulating sensor of the fraction collector, but also specially as a source of information for the accurate determination of retention volumes. With few exceptions, all measuring devices are based on the principle of the integral measurement of the liquid passed and as a consequence they are not sufficiently accurate for the measurements of the small volumes that are characteristic of efficient chromatography. Integral flow meters are used instead of the simple measurement of the liquid volume which has flowed through using a calibrated cylinder or a measuring vessel of known volume.

Very often, integral flow meters based on the syphon principle are used (Fig. 8.10). As soon as the liquid level in the measuring space attains the height of the syphon outlet, the whole device is emptied immediately and completely. The flowed-through volume of the mobile phase is determined from the number of volumes emptied per unit time. The time of emptying the volume is mostly detected with a photo-tube, but sometimes with a contact manometer, capacity detector, etc. The main risk of obtaining inaccurate measurements on the liquid results from imperfect emptying of the syphon (which is also the cause of the contamination of subsequent fractions). This occurs especially when the capillary forces cannot keep the remainder of the liquid in the draining tube. The number of volumes emptied is plotted against time. The simplest work-up is brought about by an impulse from the detector (photo-tube, etc.) directly on to the recording millivolt meter, where the moment of the syphon emptying is recorded either with a separate pen or directly with the pen recording the detector response. From the knowledge of one volume of the syphon, the rate of shifting of the recorder chart and the number of volumes emptied, it is possible to calculate the total volume of the mobile phase which has flowed through and the flow-rate.

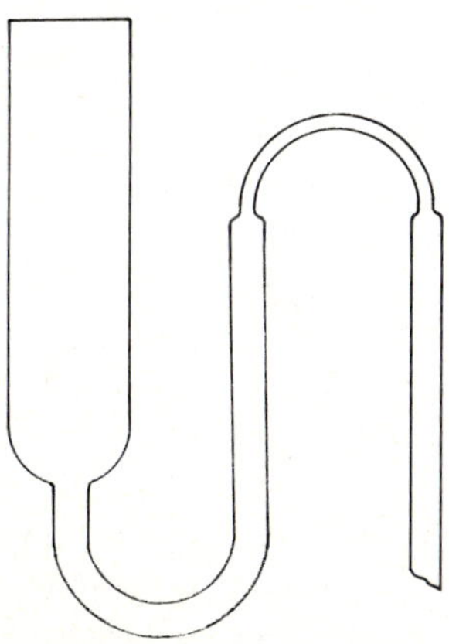

Fig. 8.10. Syphon flow meter.

In many instances, the volume of even a single syphon is too large, and then a device for counting the drops is used. Drop counters are mostly based on the shielding of a light beam by falling drops. The necessary impulse is generated by a photo-cell (or by a photo-resistance). As the measured volumes are small, these units must be provided with a counter that permits the recording of only a certain number of drops. The device must be calibrated in order to permit the calculation of the flowed-through volume from the number of drops.

Very small volumes of liquids can be measured with a bubble flow meter. An air (or nitrogen) bubble is introduced into a stream of liquid, which passes through the calibrated volume (burette). The time necessary for the gas bubble to pass between two marks is the time necessary for a volume of liquid defined by marks to flow out of the column. The devices that operate on this principle are very accurate (for example, for a 250-$\mu$l volume, a ±0.5% accuracy is given) and they do not require calibration. It is only necessary to check the calibration of the burette or of the tube with the marks. A bubble flow meter can be fully automated. Bubbles are introduced into the effluent stream from an air pump and their passage between the marks is recorded by photo-cells. The whole evaluating system and the impulse generation is automatic, while the size of the bubbles is also controlled.

A differential flow meter of considerable sensitivity can be constructed on the principle of a calorimeter. Two thermocouples (thermistors) are located in the flow-through tube between which a thermal source, most commonly a spiral heated with a current from a stabilized source, is connected. The first thermocouple measures the temperature of the ingoing liquid and the second the temperature of the liquid beyond the heat source (spiral). The temperature measured by the second thermocouple, provided that constant conditions (quality of the liquid, temperature of the entering liquid) are maintained, is proportional to the flow-rate of the measured liquid. The device may have a very small volume (several tens of microlitres) and still be sufficiently sensitive. Of course, the temperature stability of the whole device, but especially of the entering liquid, plays an important role.

### Preparative and industrial liquid chromatography

After 1967, liquid chromatography developed mainly in the direction of increasing the efficiency of separation, especially at the cost of a decrease in the amount of sample used. Procedures which up to recently were considered to be analytical only are today considered as semi-preparative or even preparative. On the other hand, preparative chromatography is taking over the basic procedures of analytical chromatography but increasing the column dimensions (capacity). For the production of pure chemicals, columns up to 180 cm in diameter and of a corresponding length, with a mobile phase flow-rate of 10 l/min are already in use.

Instead of the preparative columns the diameter and length of which correspond to the increased dimensions of the analytical columns, the so-called preparative segment columns are often employed. These columns are composed of segments the diameter of which is usually about twice their height (for example, 37 cm diameter and 15 cm height). The segments are connected in series and a separation efficiency can thus be achieved which approaches that of the same chromatographic system (sorbent, mobile phase) used in a classical analytical column. This is usually explained by the fact that the transfer of mobile phase from one segment to the other causes intense mixing and thus the suppression of the radial concentration gradient in the column. The inlet and outlet of the mobile phase into and out of the segments is provided for by several tubes, mostly made of PTFE, which are led to a single junction between the segments, which is provided with a connecting tube with a closing valve. During the packing and emptying of the columns, single segments can be separated from each other hydraulically and mechanically. As a consequence of the large diameters of the segments used, the pressure drop in the column is usually not large.

The process of discontinuous manufacture or preparation of pure chemicals can be automated to an appreciable extent. Automatic samplers are similar to those in gas chromatography, based on piston (linear) dosers of suitable volume, monitored with a timer. The collection of the separated samples is most commonly regulated by means of the signal derived from the detector response. Under these circumstances, it is possible to prepare in a single cycle a maximum amount of the required component. The fraction collectors are set so that at the moment of the repeated out-going of the prepared fraction from the column the same collecting vessel is returned to the collecting position. This technique has been developed to a high standard, especially in the preparation of amino acids and rare earths by ion-exchange chromatography.

In spite of the fact that discontinuous procedures helped to prepare a large number of pure substances, efforts have for a number of years been concentrated on making elution chromatography a continuous procedure. Experiments based on the principle of counter-current movement of the solid and the liquid phase (hypersorption) have not been particularly successful in liquid chromatography. The solid sorbent falls against the rising liquid phase. At a suitable position in the column, the sample is introduced into the system. Those components of the mixture which are sufficiently weakly sorbed move through the column in the direction of flow of the mobile phase, while more strongly sorbed components are carried in the opposite direction together with the sorbent. This system has several disadvantages. In view of the low diffusion coefficient in liquids, the moving sorbent cannot form an efficient column and, further, the regeneration of the sorbent is necessary, which in addition to possible deactivation also undergoes considerable mechanical stress. As a consequence of the mechanical particle destruction, the hydrodynamic conditions in the system must also be corrected systematically.

In 1955, Svenson *et al.* proposed a principle (an analogy of which exists in paper

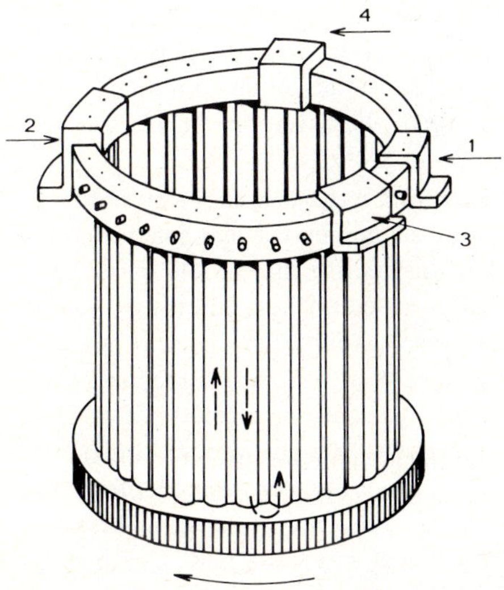


Fig. 8.11. Tube bundle of a continuous working chromatograph. 1 = Mobile phase inlet; 2 = feed part; 3 = product A off-take; 4 = product B off-take.

*References p.168*

chromatography) based on the synchronization of the rate of movement of fractions by a series of vertical columns forming a cylinder and rotating at a convenient rate around the axis of this cylinder (Fig. 8.11). The inlet part of the columns is connected with a ring through which the mobile phase is introduced into all columns at a suitable rate. The sample is introduced at a single position on the ring (*i.e.*, at a given moment on a single column only). For a suitably chosen column packing, mobile phase flow-rate and rate of rotation of the system of columns, it can be arranged that the separated fractions always leave the separating system at a single position below the rotating column system. As a result, the collecting vessels standing below the outlet of a series of columns are always filled with the same part of the solvent from one (in fact discontinuous) chromatographic run. The larger the number of the columns used, the closer is the process to being continuous. Just as this technique can be applied in paper chromatography with a chromatographic paper rolled into a cylinder, a single column can also be made which is formed by the space between two coaxial cylinders. The basic problem in the realization of the above described continuous chromatographic columns is the preparation of a series of identical columns, *i.e.*, their homogeneous packing.

**How to learn the technique**

In order to use this book effectively, it is useful if a procedure for the separation of a mixture of interest found in the "Applications" part of the book can be compared with the procedure recommended in the corresponding chapter and to check in the Chapters 11–15 whether it corresponds to actual ideas. The introduction of the most progressive techniques into general practice is not uniform and it is possible that an empirical procedure may be proposed, especially at the beginning of the development of a technique, that is not the best.

It is never ideal to acquire practical skill on the basis of a written description only, even when it is accompanied by figures. Therefore, it is recommended that some theoretical knowledge should first be acquired and then some instructive film should be viewed, a summer school or course attended or work should be carried out in a laboratory where the method is well known. Regular courses on separation techniques in the field of biopolymers are organized in collaboration with the Federation of European Biochemical Societies (FEBS) and the European Molecular Biology Organization (EMBO) in Sweden as a summer school. Short courses are organized from time to time by various firms, producers of sorbents and chromatographic equipment. Films are also available, for example on ion-exchange chromatography (Czechoslovak Academy of Sciences, Prague, Czechoslovakia), on chromatography on cellulose ion exchangers (Whatman, Colnbrook, Great Britain) and on gel chromatography (Pharmacia, Uppsala, Sweden).

For those who wish to proceed individually, a demonstration experiment with model substances and sorbents is useful.

*A beginner's demonstration experiment*

This section is devoted exclusively to those readers who have no experience in liquid chromatography.

The demonstration experiment is taken from a basic laboratory chemistry course; it is almost identical with the experiment described in the Pharmacia University Demonstration Kit. The experiment demonstrates the separation of a high-molecular-weight and a low-molecular-weight component by gel chromatography.

For the experiments, a tube of dimensions 20 × 1.2 cm or similar with a stopcock (see Fig. 8.2), 4 g of Sephadex G-25 (medium), blue dextran (Pharmacia; it can be replaced by haemoglobin, for example) and potassium chromate are required.

Sephadex G-25 is allowed to swell in a beaker for at least 3 h at room temperature (or 1 h on a boiling water-bath) in an excess of water. The column is fixed in a vertical position on a laboratory stand. Into the bottom part, a glass-wool plug is pushed (or a disc of soft polyurethane) and the tube is filled with distilled water to one third of its height. Excess of water is decanted from the Sephadex so that the suspension is thicker (in about 50 ml), then it is well stirred with a glass rod and poured into the tube. Care must be taken that no air bubbles are present in the suspension. Excess of water can be drained off slowly. The next portion of the suspension is added before the sorbent bed is completely settled. When all of the suspension is in the column, the Sephadex column should be about 15 cm high. The stopcock is opened completely and the column is stabilized by washing with distilled water for about 30 min. During the addition of the solvent, the surface of the sorbent should not be disturbed.

The sample is prepared by dissolving about 5 mg of blue dextran and 5 mg of potassium chromate in 1 ml of distilled water (blue dextran dissolves slowly, best under trituration). Water is then allowed to soak into the column, the stopcock is closed and the green solution is introduced carefully into the column with a pipette (preferably along the tube walls), The stopcock is opened and after the sample has soaked in, the residue is rinsed with one or two 1-ml portions of water, the tube is then filled to the top with water and this is completed during the experiment. From the moment when the sample has soaked into the column, fractions are collected into 3-ml calibrated test-tubes. Uniform zones mean that the column was prepared regularly and correctly. The volume necessary for the elution of blue dextran gives the free volume of the column (*cf.*, Chapter 5).

## TECHNIQUES OF HIGH-EFFICIENCY LIQUID CHROMATOGRAPHY

### Principal differences between classical and high-efficiency liquid chromatography

After some decades of use of liquid chromatography, a revival and increase in the possibilities of successful separations is now taking place. This trend is due primarily to the new theoretical treatment of the problems of liquid chromatography, which permits the optimization of the process. (These problems are discussed in Chapters 4–7.) The requirements for optimization are reflected in the parameters of the apparatus, of which

*References p.168*

greater demands are made. It is important to note that the basis of the new theoretical aspects from which modern liquid chromatography is developing lies in gas chromatography, which in many instances has already attained the limits of its theoretical possibilities. The fundamental directions which the development of the apparatus for liquid chromatography is taking can be demonstrated from simple relationships.

One of the basic advantages of liquid chromatography is the variability of the system selectivity, consisting mainly in the variability of phase equilibria between the solute and the stationary phase and the solute and the mobile phase. Very roughly, we could say that the shift in the equilibrium to the advantage of the stationary phase leads to an increase in retention time, $t_R$, and capacity ratio, $k = (t_R - t_M)/t_M$, where $t_M$ is the dead retention time. The same effect is also produced by a decrease in the solubility of the solute in the mobile phase. Quantitatively, the measure of system selectivity can be expressed as relative retention:

$$\alpha_{1,2} = \frac{t_{R,1} - t_M}{t_{R,2} - t_M} = \frac{k_2}{k_1} \tag{8.1}$$

For quantitative expression of the degree of separation, Purnell proposed a relationship for the resolution, $R_{1,2}$, as a function of relative retention:

$$R_{1,2} = \frac{t_{R,2} - t_{R,1}}{\sigma_2^2} = \frac{1}{4} \cdot \frac{\alpha - 1}{\alpha} \cdot N_{\text{eff}} \tag{8.2}$$

where $\sigma_2^2$ is the variance of the peak of stronger sorbed (subscript 2) solute and the effective number of theoretical plates, $N_{\text{eff}} = \left(\frac{t_R - t_M}{2\sigma}\right)^2$ (it is assumed that $\sigma_1 \approx \sigma_2$). From eqn. 8.2, it can be seen that with a decrease in the selectivity of the system ($\alpha$ converges to unity), the requirements of the number of theoretical plates increase if a constant resolution, $R_{1,2}$, is to be maintained. The attainment of a high separation efficiency of the system (achievement of a large number of theoretical plates, or effective theoretical plates) in a sufficiently short analysis time is the fundamental aim of the development of modern liquid chromatography. The relationship between $\alpha$ and $N_{\text{eff}}$ for a constant $R_{1,2}$

TABLE 8.3
DEPENDENCE OF THE REQUIRED NUMBER OF EFFECTIVE THEORETICAL PLATES, $N_{\text{eff}}$, ON RELATIVE RETENTION, $\alpha$, FOR VARIOUS RESOLUTION VALUES, $R_{1,2}$

| $\alpha$ | $N_{\text{eff}}$ | | |
|---|---|---|---|
| | $R_{1,2} = 1$ | $R_{1,2} = 1.5$ | $R_{1,2} = 2$ |
| 1.001 | $16 \cdot 10^6$ | $36 \cdot 10^6$ | $64 \cdot 10^6$ |
| 1.01 | $16 \cdot 10^4$ | $36 \cdot 10^4$ | $64 \cdot 10^4$ |
| 1.05 | 6,944 | 15,625 | 27,777 |
| 1.10 | 1,932 | 4,347 | 7,728 |
| 1.15 | 932 | 2,098 | 3,729 |
| 1.2 | 574 | 1,291 | 2,294 |
| 1.3 | 299 | 674.65 | 1,199 |

is evident from Table 8.3. The values were calculated using eqn. 8.2.

The increase in separation efficiency of the column, keeping the required time for analysis sufficiently short, is possible on the basis of the optimization of the relationship of the functional dependence of the number of theoretical plates on time, $N = f(t)$, or the optimization of the HETP on the linear flow-rate of the mobile phase in the column, $H = f(u)$. Using the general formulation by Giddings, this relationship for liquid chromatography can be written in the form

$$H = C_S u + C_M u + (1/A + 1/C_M u)^{-1} \tag{8.3}$$

where $C$ is the coefficient of the mass transfer resistance in the stationary phase (subscript $S$) and in the mobile phase (subscript $M$), $A$ is the coefficient of eddy diffusion and $u$ is the linear flow-rate of the mobile phase (in this expression, the longitudinal diffusion coefficient in the mobile phase is not considered because the diffusion coefficient in liquids is approximately $10^4$ times lower than the diffusion coefficient in gases; therefore, in contrast to gas chromatography, the value of this coefficient is negligible). The limiting members of the equation in liquid chromatography are the coefficients of the mass transfer resistance in the stationary and the mobile phases. The coefficient of mass transfer resistance in the mobile phase is proportional to the square of the particle diameter, $d_p$, and indirectly proportional to the diffusion coefficient of the solute in the mobile phase $D_M$, $C_M = \omega d_p^2/D_M$, where $\omega$ is a constant for the given system. The mass transfer resistance coefficient in the mobile phase (and hence the HETP) decreases with a decrease in the particle diameter of the material used.

The mass transfer resistance coefficient in the stationary phase, $C_S$, is formulated in different ways for various chromatographic systems. For example, for the liquid–liquid system it is proportional to the square of the effective film thickness of the stationary phase, $d_f$, and indirectly proportional to the diffusion coefficient of the solute in the stationary phase; for the system liquid–adsorbent the mass transfer resistance coefficient in the stationary phase is a function of the porosity of the material and of the effective coefficient of the radial diffusion in the particle, of the obstructive factor, etc. Most of these parameters also decrease with decrease in the particle size. If we take into account that the value of the eddy diffusion coefficient ($A = \lambda d_p$, where $\lambda$ is a constant decreasing with increasing regularity of the column packing) is dependent on the particle size of the column packing, it is evident that the diminution of the particle of the sorbent in the chromatographic column is the first step necessary for an increase in separation efficiency in liquid chromatography.

If the analysis time also has to be sufficiently short, the work should be carried out at a linear flow-rate of the mobile phase, which converges to the linear flow-rates in gas chromatography, *i.e.*, about 1 cm/sec. From Darcy's law, it follows that the linear flow-rate of the liquid, $u$, is proportional to the permeability constant, $K$, and the pressure drop, $\Delta p$, and indirectly proportional to the viscosity of the liquid, $\eta$, and the column length, $L$:

$$u = \Delta p \cdot \frac{K}{\eta L} \tag{8.4}$$

*References p.168*

As according to the Kozeny–Carman equation

$$K = \frac{d_p^2}{180} \cdot \frac{\epsilon^2}{(1-\epsilon)^2} \tag{8.5}$$

the permeability constant for regularly filled columns is a function of the diameter of filling particles, $d_p$, and porosity, $\epsilon$, a higher pressure drop should be used in order to achieve the required linear rates. Halász *et al.* discussed the possibility of using a simplified expression for the permeability constant, $K = d_p^2/10^3$, under the assumption that the particles of the column filling range from 30 to 200 $\mu$m and porosity is $\epsilon = 0.4$. Taking into account the above value, it can be expected that the necessary pressure drop in the column will amount to several hundred kp/cm$^2$. Simultaneously, this represents the requirement put on the pump for high-efficiency liquid chromatography. The application of capillary chromatographic columns will require even higher pressures (by an order of magnitude), as was demonstrated by Pretorius and Smuts. However, this possibility of analytical separations has not as yet been utilized in practice in liquid chromatography.

A broadening of the concentration peak of the solute takes place not only on the column, but also in the extra-column space. The importance of this fact for liquid chromatography was stressed by Huber (1969a) and Halász *et al.*

The necessity for decreasing the HETP is connected with the necessity for minimizing the dead volumes between the columns, between the injection block and the column and between the column and the detector. Spreading caused in these spaces may exceed by several times the spreading of the concentration pulse in the column. Simultaneously, increased consideration was given to the technique of packing the column in view of achieving maximum homogeneity of the packing. Elsewhere (Chapter 9), attention is paid to new materials which should decrease the mass transfer coefficient between the stationary and the mobile phases, for example, glass beads coated with a thin layer of silica gel.

However, these materials, which in many instances permit an appreciable increase in efficiency and the selectivity of separation (of the chemically bonded phase), mostly have a low sorption capacity. As columns that have a smaller diameter (1–5 mm) are often used simultaneously the detector sensitivity must be increased. Although nowadays detectors of high sensitivity and universal use exist, their development is not yet complete. Detectors do not exist that would combine sufficient sensitivity with sufficient versatility, as is the case with the flame ionization detector in gas chromatography. However, further development may be expected in this direction.

From the point of view of the practice of liquid chromatography, the following basic elements, differing from classical liquid chromatography, may be envisaged:

(1) use of new sorbing materials with a high homogeneity of the size and the shape of particles;

(2) use of materials with small particle diameters (10–80 $\mu$m);

(3) utilization of new and better procedures for column packing;

(4) application of higher inlet pressures (up to 300 kp/cm$^2$);

(5) reduction of the dead volumes in the separating system of the chromatograph to the minimum;

(6) use of highly sensitive detectors with very small measuring cells.

## The function of a liquid chromatograph

A liquid chromatograph consists of three basic functional parts. As the construction of the apparatus is modular, each of its parts is located in a separate box, which increases the possibilities of combinations and also the possibility of exchanging single blocks or modules (exchange of detectors, etc.).

(i) The source of the mobile phase flow consists of a reservoir, a pump and a filter. According to the construction of single elements, this part also includes a gradient-forming device, a degasser and a pulse-damping device, if the type of detector and pump chosen require it.

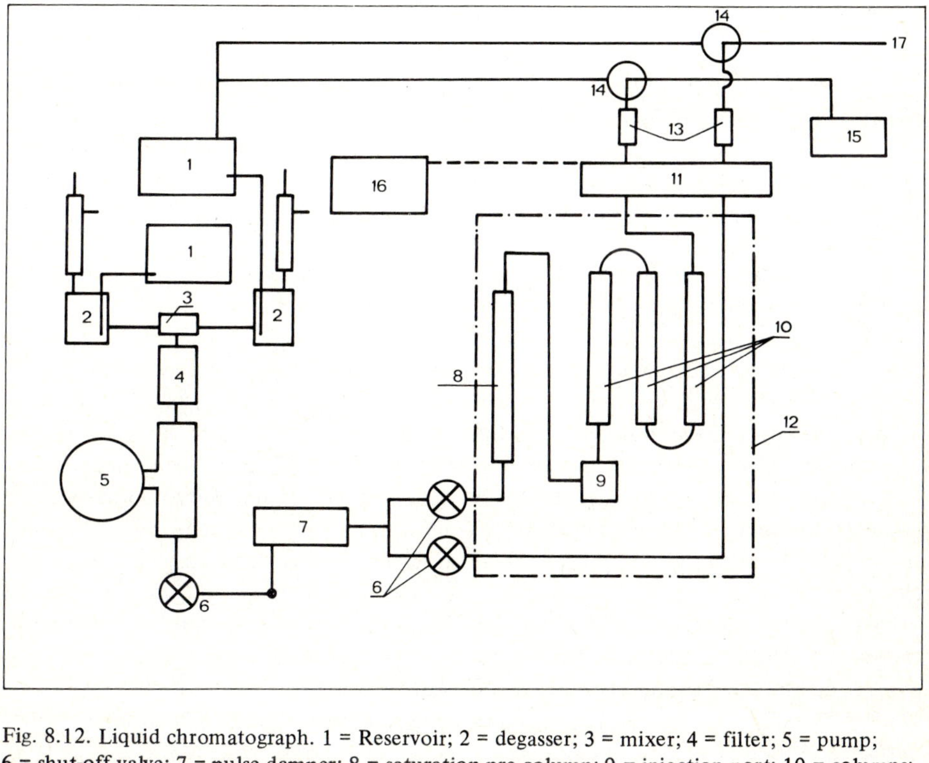


Fig. 8.12. Liquid chromatograph. 1 = Reservoir; 2 = degasser; 3 = mixer; 4 = filter; 5 = pump; 6 = shut-off valve; 7 = pulse damper; 8 = saturation pre-column; 9 = injection port; 10 = columns; 11 = detector; 12 = thermostat; 13 = flow meters; 14 = stopcock; 15 = fraction collector; 16 = recorder; 17 = drain.

*References p.168*

(ii) The separating part of the chromatograph contains a sample introduction device, chromatographic columns and sometimes a saturation pre-column and a thermostat.

(iii) The detection part of the chromatograph is a detector or a suitably connected system of detectors, and sometimes also a fraction collector and a flow meter.

A scheme of a liquid chromatograph is shown in Fig. 8.12. The liquid mobile phase is located in a reservoir and is usually led through the degasser, which decreases the content of dissolved gases in the mobile phase by a temporary temperature increase. When the gradient elution technique is applied, the mobile phase components pass through a mixer and are led through a filter into the pump. When a pulsating pump is used, a pulse-damping device is inserted, which attenuates the pressure pulses (changes in flow-rate) created in the expelling phase of the pump. For detectors fitted with a measuring and a reference cell, the mobile phase flow must be split. The technique involving two parallel mobile phase streams is not yet general. In the thermostatted space, a saturation pre-column is located (used exclusively for liquid–liquid systems), then a sample introduction device and the analytical columns themselves. Most chromatographs are not yet fitted with thermostats, *i.e.*, the columns operate at room temperature. The column outlet is connected with a capillary of a minimum dead volume with the detector. Detectors, if their satisfactory functioning requires it, are fitted with a separate thermostat. The detector response is recorded continuously by the recorder. Behind the detector, a flow meter and a fraction collector, and also valves for mobile phase recycling, may be inserted.

Currently used liquid chromatographs are usually not supplied with all the elements given in this scheme. The construction of a chromatograph depends to a large extent on the analytical aims for which the chromatogram obtained is to be used. A detailed description and the characteristics of single constructional units are given below.

**Mobile phase reservoirs**

The reservoir is a vessel in which the liquid mobile phase is stored. The material of which it is constructed must be chemically inert to the mobile phases used and it should be placed in a separate space, isolated from the heat and sparking sources. Most commonly glass and stainless steel are used, but in some instances stainless steel coated with PTFE or polyethylene foil is used. The volume of the reservoir is dependent on the type of chromatography which is to be performed on the apparatus; it is usually 0.5–5 l. For high-speed liquid chromatography, reservoirs of smaller volumes (about 0.5–1 l) are suitable.

For some types of phases, it is advantageous if the space above the liquid is rinsed with an inert gas. Therefore, an inlet for the rinsing gas is connected to the reservoir. If the rinsing is not necessary, this inlet can be used for the elimination of gases or solvent vapour with a vacuum aspirator. Other types of mobile phases require work at an elevated temperature and therefore provision must be made for heating the reservoir.

Degassing of the mobile phase is advantageous because it decreases the probability of the formation of bubbles, which might disturb detection, especially in measuring cells of small volume. It should be kept in mind that degassing is only one part of the prevention of bubble formation. It is imperative to tighten all connections between the pump, the columns and the detector. Especially in the high-pressure part of the hydraulic system, connections are often found which are apparently tight (no liquid leaks through them), but air may penetrate through these connections and dissolve in the mobile phase; eventually, when the pressure drops in the detector, bubbles may appear. In instances when a heater and stirrer are inserted into the reservoir, mobile phase reservoirs are employed that can be blown through with an inert gas or connected to a vacuum.

However, it is more advantageous if an independent degasser is placed between the reservoir and the pump. Degassers are based on the possibility of sufficient heating of the mobile phase and the separation of the gases released. When a volatile mobile phase is used, a reflux condenser must be inserted and then the temperature of the degasser need not be regulated and it suffices if the input of the heating body is set so that the temperature does not exceed the boiling point of the mobile phase. The reflux condenser is connected either with the outer atmosphere or with a vacuum pump. Stripping with an inert gas is often used to increase degassing efficiency, especially if it is necessary to eliminate oxygen from the mobile phase. Usually, the insertion of a simple heated tube of sufficient diameter (1–5 cm), fitted with a gas separator on the upper part, is adequate.

A filter is inserted in front of the pump in order to capture mechanical impurities, which impair the function of the return valves of the pump and may cause a change in the hydraulic resistance behind the pump. Most often glass, stainless steel or PTFE porous discs are used, positioned either directly behind the reservoir or in front of the liquid inlet of the pump. The filters must have a relatively low hydraulic resistance in order to prevent the formation of a gas pillow in the suction pump, especially if a volatile mobile phase is used, which could impair the regularity of the pumping.

## Gradient-forming devices

At present, there are four possible techniques of programming, *i.e.*, gradient formation, in liquid chromatography: solvent programming, stationary phase programming (coupled columns), temperature programming and flow programming. From the point of view of separation efficiency per unit time, the programming of the concentration of the mobile phase (solvent) is most effective. Snyder (1970) found that other techniques are less efficient, the efficiency decreasing in the order given above.

The programming of the concentration of a more strongly sorbed component in the mobile phase (programming of the concentration of the mobile phase) is the most commonly used procedure for the regulation of sorption properties of the system liquid–adsorbent and liquid–ion exchanger. The use of chemically bonded liquid stationary phases indicates the possibility of using gradient elution even in liquid–liquid systems. The fundamental advantage of gradient chromatography is the shortening of the time of analysis, work in the linear field of the sorption isotherm (symmetrical peaks), narrowing of the width of the chromatographic peak and, as a consequence, a decrease in the minimum detectable amount of solute in the chromatograph, *i.e.*, an increase in the sensitivity of the chromatographic analysis.

The choice of the device for the creation of a gradient is dependent on the requirements put upon it. Simple devices can be used when a single gradient form is required. When

*References p.168*

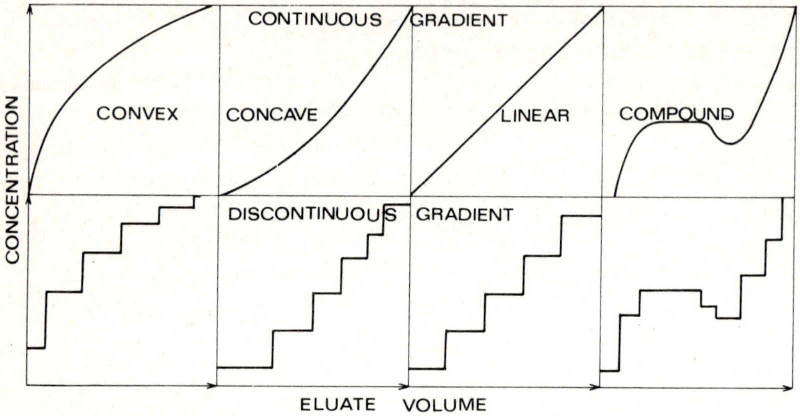


Fig. 8.13. Gradient types and shapes.

various special types of programmes are necessary, more complex devices have to be employed. The types and the shapes of the gradients used are represented in Fig. 8.13. According to their course, the gradients are divided into continuous and discontinuous. In view of the time axis (mobile phase volume) there may be convex, concave, linear or compound gradients of concentration of the stronger sorbed component in the mobile phase. Steep gradients, or stepwise techniques, are used mainly when components that differ strongly in their sorption properties are to be separated. For the separation of components that are poorly separated, the use of flat gradients is advantageous, or the use of a mobile phase of constant composition or of negative gradients (*i.e.*, the concentration of the stronger sorbed component in the mobile phase decreases with time). Although a series of procedures for the calculation of retention characteristics in gradient chromatography has been proposed (reviewed by Snyder, 1965, 1970), for the selection of an optimum gradient experimental work is indispensable.

Devices for creating gradients in high-efficiency liquid chromatography are divided into two groups, low-pressure devices, which are placed in the hydraulic circuit in front of the high-pressure pump, and high-pressure devices, inserted behind the high-pressure pump.

Low-pressure devices are further divided into those which give "exponential" gradients (Snyder, 1965) or proportional gradients. The first consist, in principle, of one reservoir with the strongest solvent and a mixer containing a mobile phase of the initial composition. By its gravity or by using a gas pressure over the liquid surface, the liquid overflows from the reservoir into the mixer, from which the mobile phase of varying composition flows out. These devices are often used because they tend to give a convex gradient without further arrangement; this convex gradient often corresponds to a linear eluent concentration gradient. According to Snyder (1965), the basic advantage of these gradients consists in the formation of peaks with constant width and, consequently, also in the constancy of the amount of the solute detectable in the effluent.

The ratio of the amount of solvent to the amount of the most strongly sorbed solvent component (moderator) can be regulated by the form of the reservoirs connected via the mixer in a manner enabling their simultaneous emptying. These devices are not in great use in view of the difficulty in changing the gradient shape, as this requires a change of shape of one or both of the reservoirs used. Continuous gradients can also be obtained by using devices for discontinuous gradients. The simplest device for the preparation of a discontinuous gradient is based on the gradual emptying of individual reservoirs containing the prepared solvent mixture. The greater the number of the reservoirs used and the smaller the differences in concentrations in the single reservoirs, the more the course of such a gradient approaches a continuous gradient.

The second group of the low-pressure devices for gradient formation in the mobile phase are devices for achieving a volume proportional gradient. These devices permit the introduction, under pressure, of both mobile phase components at a previously programmed rate. For such a purpose, any of the pumps that give an easily regulated volume rate of the expelled liquid can be used. Often piston pumps, also called pulse-less pumps (with a large piston volume), driven by a worm-gear drive, are used for such purposes. The choice of the form of gradient can be made either by directly setting the corresponding values on the programming device, or by using a follower with a photoelectric cell, which sets the corresponding values on the programming device automatically when following the dependence of the moderator concentration in the mobile phase on time, represented graphically on a corresponding scale.

High-pressure devices for making concentration gradients are based on the utilization of high-pressure pumps. In this case also, pulse-less pumps, often pneumatically monitored, are used, which enable the mobile phase flow-rate to be programmed most easily. When the pumps are connected directly into the high-pressure chromatographic circuit, the importance of a high-quality performance of the mixer increases. Perfect mixing of both components forming the gradient in a minimum volume is required in order to keep the time constant of the whole arrangement at a minimum. The use of two pumps connected in parallel also permits the formation of stepwise gradients. A significant advantage of this arrangement is the ease of manipulation which permits the rapid selection of optimum conditions for the analysis, or of optimum gradients.

In gradient chromatography, detectors must be used that do not give a response to the mobile phase on changing composition. UV, fluorescence and transporter detectors are especially suitable. In some instances, it is possible to use even a spectrophotometric detector working in the visible and infrared regions of the spectrum. Although mention is made in the literature of the use of a refractometer for gradient chromatography, this application will be possible only in exceptional cases, especially at lower detection sensitivities.

Regardless of whether a continuous or discontinuous gradient is used, it should always be borne in mind that equilibrium is not attained in the column after elution with an amount of eluent corresponding to the hold-up volume of the column. Here frontal chromatography of the solvent takes place. The more strongly sorbed moderator reaches the end of the column later than the front of the lesser sorbed mobile phase. In view of the fact that the knowledge of the immediate equilibrium state on the column may afford significant qualitative chromatographic information, Scott and Lawrence (1970a) proposed gradient elution under conditions of axial equilibrium. The principle of this method consists in temperature programming on the column, which is in solvent–moderator–adsorbent equilibrium. On increasing the column temperature, part of the

*References p.168*

sorbed moderator is desorbed, thus changing the mobile phase concentration simultaneously with the adsorbent activity. In addition to the factors common in the gradient technique, this technique also introduces two additional important advantages that can be utilized in modern liquid chromatography:

(1) during the whole chromatographic process, the content of the solvent and the moderator is constant in the whole column regardless of the temperature;

(2) on returning the column to the original temperature, the column also returns to its original state, which does not require special activation or an exchange of the filling.

Little attention has so far been paid to programming the temperature in liquid chromatography. The technique of temperature programming is similar to that in gas chromatography and is described on p. 145 in connection with thermostats. It should be noted that possibilities exist of positive and negative temperature programming. Both programmes may lead, as demonstrated by Maggs, to decreased retention volumes. The temperature programme comes into consideration in liquid–adsorbent and liquid–ion exchanger systems.

The importance of programming the stationary phases by selectable combinations of several columns in series, as well as the importance of programming the flow, has not yet penetrated into chromatographic practice. Its theoretical and practical evaluation was discussed by Snyder (1970).

## Manipulation

The main risks in the operation of liquid chromatographs follow from the properties of the mobile phases used. Flammable and toxic liquids used as the mobile phase require suitable measures and great caution.

During work with flammable liquids, the possibility of contact of the vapour or the liquid with hot parts of the equipment, for example the heating cells of the deaerators, thermostats, etc., and also with places where sparking may take place, for example the motors of pumps, electric drives, etc., must be prevented. The parts of the chromatograph in which contact of the flammable mobile phase with the sources of sparking, etc., may take place must be very well aerated, or preferably treated with an inert gas such as nitrogen. Laboratories in which chromatographs are placed should be well ventilated and provided with suitable fire extinguishers.

During work with toxic mobile phases, it is recommended that the chromatograph should be operated under a hood and that the laboratory should be well ventilated.

Neither inflammable nor toxic materials should be stored in laboratories in amounts greater than would correspond to approximately one filling of the reservoir. Such materials should be kept in safety containers so that in the case of accidental breakage of flasks no fire or poisoning could occur.

The high pressures under which modern liquid chromatographs are often operated do not represent a serious risk if the mobile phase is kept at a temperature lower than its critical temperature.

**Pumping systems**

The pump ensures a constant flow of mobile phase through the chromatographic system (column and detector) under given conditions, *i.e.*, temperature, flow-rate and pressure. The pump should fulfil several basic requirements following from the operating conditions of the chromatographic separation. Those parts of the pump which come into contact with the mobile phase should be inert to it, so that they are not damaged by it, and *vice versa*, so that the mobile phase should not be contaminated by dissolution or extraction of some parts (for example, packing) or of the lubricant. With pumps used in high-efficiency liquid chromatography, great accuracy and reproducibility of the flow of mobile phase is required. Usually, a flow stability of better than 2–3% is required for the necessary flow-rate range of about 0.01–20 ml/min. The requirement of the flow stability is usually difficult to fulfil, especially at very low flow-rates, when the functioning of the check valve is not always sufficiently reliable. The possibility of a rapid and simple resetting of the pump piston lift, and thus the possibility of a change in the flow-rate, is fulfilled in almost all modern pumps.

The pumps used in liquid chromatography can be classified according to several criteria. From the point of view of their use in high-efficiency liquid chromatography, they are classified as high-pressure and medium-pressure pumps. High-pressure pumps can be used up to a pressure of about 300 kp/cm$^2$ (or higher) and are suitable for use in high-efficiency chromatography with column fillings with a very small particle size. Medium-pressure pumps usually operate at pressures up to 75–100 kp/cm$^2$ and are suitable for use when the column filling is made of compressible organic material (GPC, IEC), which does not allow the application of higher pressures. The pumps of this type are often installed in single-purpose chromatographs, such as equipment for gel chromatography and amino acid analyzers.

According to their stroke frequency, pumps are classified as pulsating and pulse-free. The former are pumps with a small working volume of the expelled liquid per lift, and in the latter the volume of the expelled liquid is approximately the same as the volume that passes through the chromatograph during several tens of minutes to several hours. The former type of pumps produce a liquid flow that pulses synchronously with the pump lifts, and consequently disturbances in the functioning of the detector may occur. The flow can be compensated with a damping system. The disadvantage of pulse-free pumps is their large volume, because they must pump in the mobile phase in the shortest possible time. Therefore, two pistons are utilized; in the interval when one expels the liquid, the other is filled with it.

The pumps are classified according to their construction as piston, membrane and peristaltic pumps. Peristaltic pumps are usually used only for low-pressure variants of chromatographic experiments. They are based on the expulsion of the pumped liquid from plastic deformable tubes (hoses) by means of rotating cylinders. This type of pump is common in analyzers used in clinical biochemistry laboratories. It is advantageous especially when aqueous liquids are to be pumped. No sufficiently chemically resistant and simultaneously flexible materials exist as yet for hydrocarbon phases.

Piston or plunger pumps are composed of a cylinder, piston or plunger, and check valves. The body and the cylinder of the pump are usually made of stainless steel and the

*References p.168*

cylinder volume is usually 0.1–1 ml. The piston or plunger can be made of stainless steel, but other inert materials are also used, for example agate. Plungers are mostly sealed with plastic O-rings, which represent limiting factors as regards both the durability of the pump used and its inertness towards the liquids being pumped. It should be noted that in most instances the pumps are constructed so that a change of packing causes no serious difficulty. The materials used are chosen so as to resist the pumped liquids as much as possible. The piston construction permits the regulation of its stroke, and thus also of the volume of the liquid expelled, within broad limits. A scheme of a typical construction of a piston pump used in liquid chromatography is given in Fig. 8.14.

Important components of the pumps are the valves, which regulate the direction of flow of the pumped liquid. Check valves are commonly used which, in the phase of expulsion, close the suction tubing, and during the suction lock the expulsion tubing. The construction of the check valves is shown schematically in Fig. 8.15. Ball valves are usually employed and are usually constructed as multi-step valves with various diameters of the seats and balls. In the phase when the valve is to be closed, the balls are pressed into the seat either by the pressure of the liquid only or by springs of a suitable strength. The functioning of the check valves determines the efficiency of the pump. With increasing pressure at the pump outlet, a larger force is necessary in order to close the valve; imperfect closing of the valve may cause a back flow of the liquid. The dependence of the efficiency of the pump on pressure is shown in Fig. 8.16. The efficiency, $\eta$, is the ratio of the liquid volume actually expelled and the theoretical volume that can be expelled by the piston per unit time, *i.e.*, $\eta = V_{act.}/V_{theor.}$. The basic requirement put on the pumps used in chromatography is a good reproducibility of their efficiency, so that at a chosen stroke and working pressure of the pump, its efficiency must be constant. This requirement is fulfilled when the functioning of the check valves is perfect.

At a low piston stroke, insufficient pressure is exerted on the check valves and their functioning is defective, the efficiency of the pump changes and, as a consequence, the flow of the mobile phase through the chromatograph also changes. This situation can be prevented by allowing the pump to operate at a higher piston stroke, which makes its

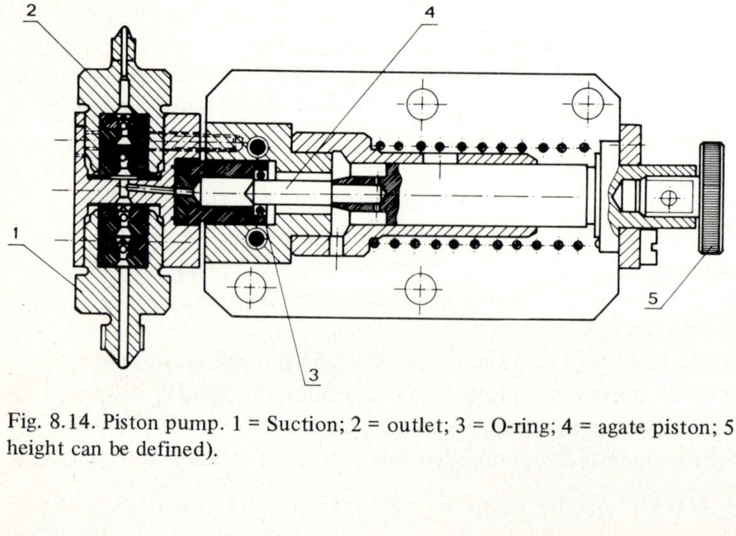


Fig. 8.14. Piston pump. 1 = Suction; 2 = outlet; 3 = O-ring; 4 = agate piston; 5 = driven disc (stroke height can be defined).

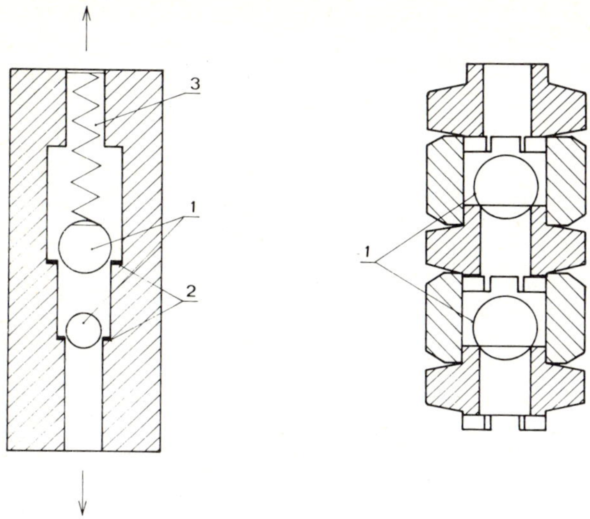


Fig. 8.15. Check valve. 1 = Stainless-steel ball; 2 = seat; 3 = spring.

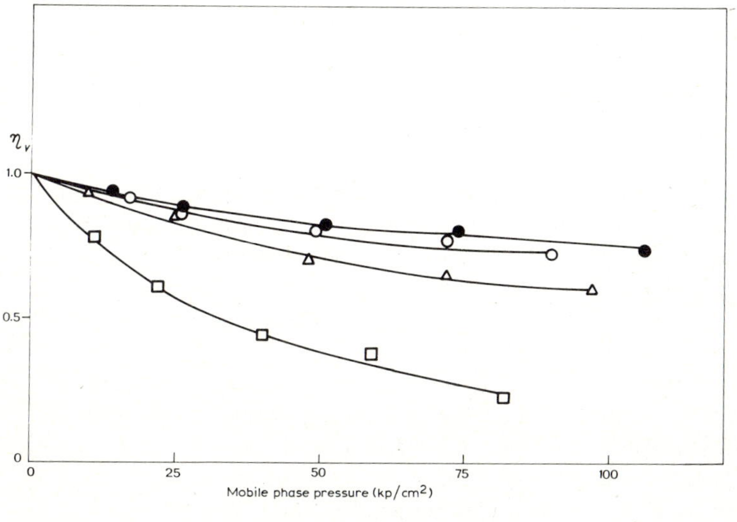


Fig. 8.16. Dependence of the efficiency, $\eta$, on the pump stroke and mobile phase pressure. Stroke: •, 10 mm; ○, 7.5 mm; △, 5 mm; □, 2.5 mm.

efficiency higher and usually constant. The mobile phase flow should then be divided into two streams of a suitable ratio and part of the pumped liquid fed back into the reservoir.

In diaphragm pumps, elastic diaphragms are used to expel liquids. The movement of the diaphragm is monitored by a plunger pump with the compression liquid. The scheme of the diaphragm pump is shown in Fig. 8.17. The advantage of diaphragm pumps is that the liquid pumped comes into contact only with the stainless-steel parts. Therefore, these

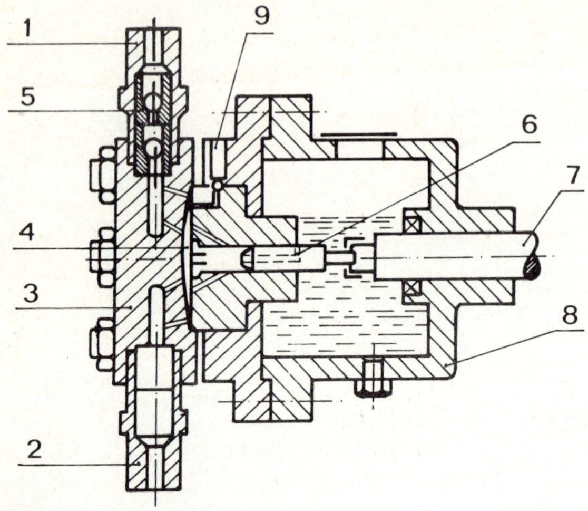


Fig. 8.17. Diaphragm pump (by courtesy of Orlita, Giessen, G.F.R.). 1 = Pressure gauge; 2 = suction piping; 3 = diaphragm head; 4 = diaphragm; 5 = valve; 6 = piston drilling; 7 = piston pull-rod; 8 = reservoir of hydraulic liquid; 9 = shut-off screw.

pumps can be used with advantage for the operation of chromatographs that contain corrosive liquids. The volume delivered by a single piston stroke is usually smaller (about 0.1 ml) than with piston pumps and therefore, higher frequencies of the pump are used, or double or triple pumps with a phase-shifted diaphragm stroke.

According to their type of drive, the pumps are divided into mechanical and pneumatic. Piston and diaphragm reciprocating pumps are driven by a mechanical propulsion directly connected to an electric motor. A pneumatic propulsion is mostly used in pulse-free pumps. Gas pressure acts on a piston of large diameter, which expels the liquid with a piston of smaller diameter. This construction works as a pressure amplifier, so that with lower gas pressures a sufficient pressure for the expulsion of the pumped liquid can be attained.

While reciprocating pumps driven mechanically operate as a pressure-independent source of liquid flow, *i.e.*, the changes in the resistance in the delivery tubes have no effect on the flow of the liquid being pumped, pulse-free pumps with a pneumatic propulsion work at constant pressure on the liquid outlet from the pump, so that the liquid flow is dependent on the resistance in the outlet tubing of the pump.

The fact that most pumps used nowadays in liquid chromatography operate as a pressure-independent source of flow brings with it the risk that the pump would be damaged if the resistance in the hydraulic system behind the pump increased excessively, which often happens when locks made of plastic are used for the introduction of the sample on to the column with a syringe. A safety cap is inserted behind the pump, which allows the solvent to flow back into the reservoir when the pressure exceeds the acceptable limit. The simplest type of such a safety valve is a stainless-steel membrane inserted parallel with the liquid flow. The size of the membrane is chosen for a particular pressure and if this pressure limit is exceeded the membrane is broken and the pressure decreases.

An electric safety device can also be inserted which switches off the pump when the pressure becomes too high. The pressure signal for the safety device can be taken either from a contact located on a suitable diaphragm reacting to pressure or from the recording of a tensiometer placed on the tubing behind the pump.

### Pressure pulse-damping device

Piston and diaphragm pumps operating at a relatively high frequency (120–20 strokes/min) expel the liquid under a pressure that changes depending on the position of the piston. The change of the pressure with time is approximately sinusoidal. A partial suppresion of pressure pulses can be achieved by connecting a double or triple pump, the operation phases of which are mutually shifted by 180° or 120°.

The pulse-damping device suppresses short-time pressure pulses generated by the pump. This arrangement ensures a flow of mobile phase with an approximately constant rate through the column and especially through the detector, where the short-term flow-rate changes of the mobile phase may have an appreciable effect on the noise increase in the detector and in some instances may make the use of the detector impossible.

Pressure pulses are damped in some instances by the volume of gas retained in the apparatus. This is disadvantageous because the mobile phase is saturated with the gas, which may in some instances disturb the functioning of the detector or even the separation process in the column. The stabilization of the mobile phase flow-rate also requires a long time, which is dependent on the volume of gas and the volume rate of the pumped liquid.

The pulse-damping devices should fulfil several requirements in order to ensure disturbance-free operation of the chromatograph. Firstly, the separation of the mobile phase from the gas is required, if the latter is used in the damping device. An easy and rapid exchange of the mobile phase in the damping device is a further requirement, which should permit the pressure pulse damper to be used for various mobile phases, or make its use possible in the gradient technique.

The required efficiency of the device depends predominantly on the detector used. For example, for detectors with an effluent transporter it is not necessary to use pressure pulse dampers. In other instances, damping devices with a 95–98% efficiency suffice, *i.e.*, the changes in the flow-rate vary from 2 to 5%.

The principle of the suppression of pressure pulses is comparable with the principle utilized for smoothing the alternating current (Fig. 8.18). The simplest arrangement consists of a number of pneumatic capacitances and resistances connected in series. The higher the number of these components the better is the smoothing of the pulses that can be achieved. Pneumatic resistances can be constructed simply from capillaries of suitable diameter and length and pneumatic capacitances can be made from vessels filled partly with gas and partly with the pumped liquid. The use of vessels that can change their volume as a function of pressure is more advantageous. Most often such vessels are bellows, Bourdon tubes or vessels of which one wall is made of an elastic diaphragm. The required pressure dependence of these capacitances is usually attained either by the properties of the material and construction used (for example Bourdon tubes, as used by

*References p.168*

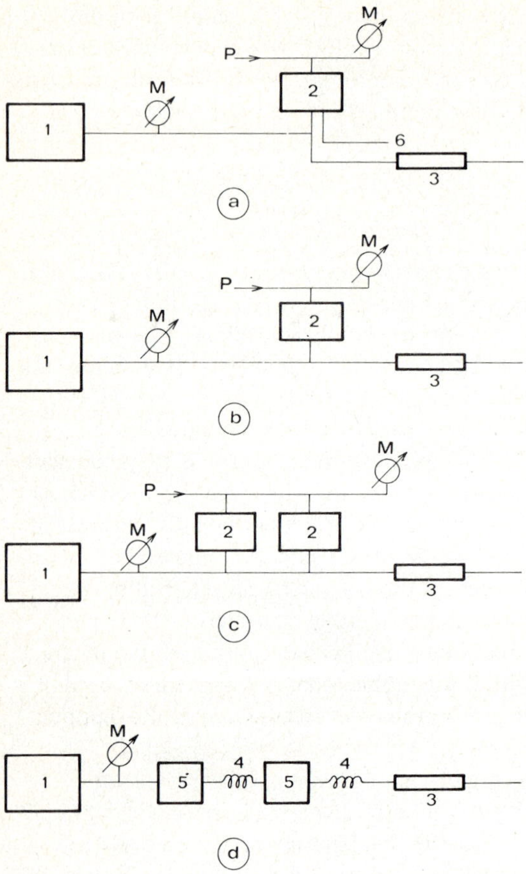


Fig. 8.18. Connection of the pulse damper. (a) Diaphragm-type damper with draining of access liquid; (b) diaphragm-type damper; (c) double diaphragm damper; (d) damper based on series connection of hydraulic capacitance and hydraulic resistor. 1 = Pump; 2 = diaphragm damper; 3 = chromatographic column; 4 = hydraulic resistance; 5 = hydraulic capacitance; 6 = drain; P = compensating pressure; M = manometer.

Locke), or by a suitable gas pressure or springs (for example, diaphragm dampers described by Halász *et al.*).

It is further possible to smooth the pressure pulses by inserting a unit which maintains a constant pressure in front of the capacitance and resistance unit. These units are well known from gas chromatography and are formed by a globe valve which closes the liquid inlet depending on the pressure attained at the outlet side of the device. In some instances, a splitter of the liquid flow is placed behind the pulse-damping device, which returns the excess liquid back into the suction pipes of the pump. This arrangement is advantageous mainly when the pump malfunctions as a result of being set at excessively low values.

The functioning of the membrane damper with a pressure-dependent liquid volume (Fig. 8.19) consists in forcing the liquid into the space limited on one side by the dia-

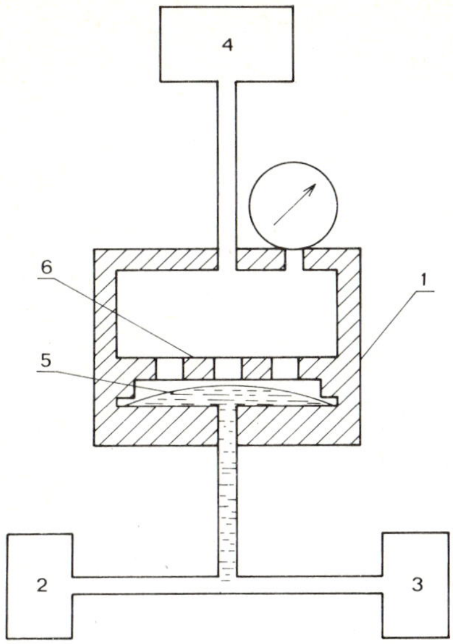


Fig. 8.19. Diaphragm pulse damper. 1 = Pulse damper; 2 = pump; 3 = column; 4 = source of the compensating pressure; 5 = diaphragm; 6 = diaphragm support.

phragm during the expulsion phase. In this space an approximately constant pressure is maintained either by a gas introduced over the membrane, or by a spring exerting an adequate pressure on the membrane. In the suction phase of the pump, the liquid is forced into the piping in the direction of the column.

When a damping device with a liquid splitter is used, the valve opens depending on the pressure in the damper and the excess liquid returns into the reservoir or into the suction pipes. When the pressure in the damper drops to the required value, the valve is again closed and all the liquid is led on to the column. Damping devices of this type are useful when double or triple pumps are used; for a single-action pump, a device of this type cannot be used.

When pulse-damping devices are inserted into a circuit, it should be kept in mind that the following parameters should be coordinated: the volume of the liquid expelled by one piston stroke and the damper volume dependent on pressure, the hydraulic resistance of the column (the required mobile phase flow) and the compensation pressure in the damper. In Fig. 8.20, the dependence of the efficiency of the damper on the compensation pressure used in the device is shown. It is evident that this dependence has a clear optimum. Similarly, when Bourdon manometric tubes are used, the nominal pressure in these tubes should be chosen correctly.

## Sample introduction devices

The techniques used for introducing the sample into the chromatograph in modern liquid chromatography are closer to those used in gas chromatography than to those in

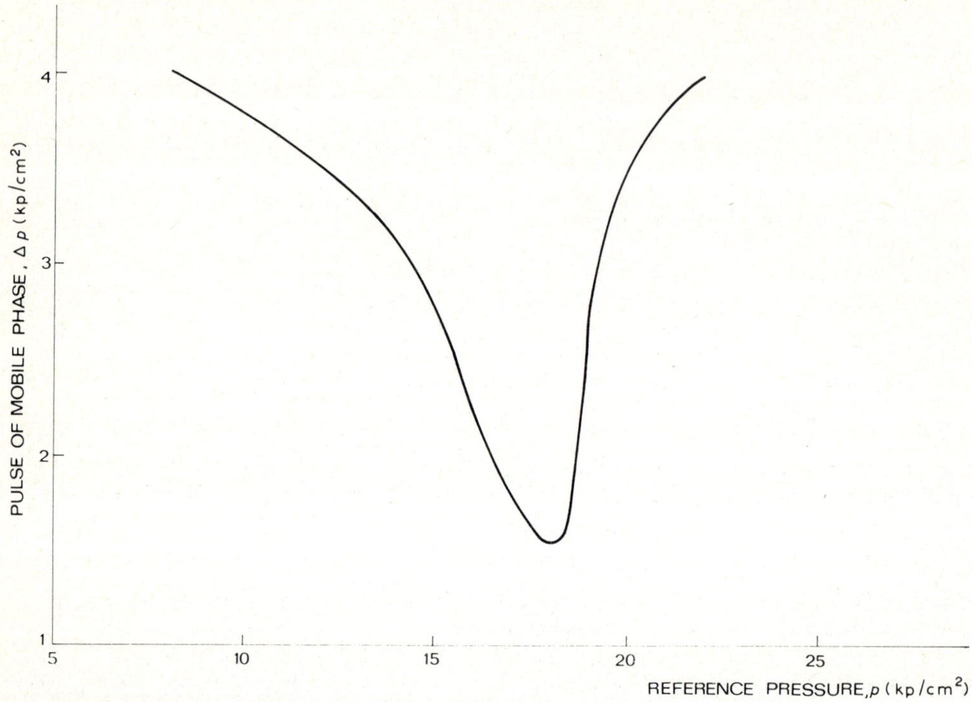


Fir. 8.20. Dependence of the efficiency of the damper on the compensating pressure.

the classical liquid chromatography, which is connected with spreading of the concentration pulse before the column inlet. In spite of the similarity of the devices used for the introduction of samples in liquid and gas chromatography, there are several fundamental differences that in many instances are also reflected in the technical construction of the sample introduction device.

Weak diffusion of the solute in the liquid mobile phase does not permit a sufficiently fast equalization of the radial concentration of the introduced solute before it enters the column. Therefore, the introduction of the solute into the centre of the cross-section of the column filling should be ensured with maximum accuracy. Any disturbance of the profile of the mobile phase flow (for example, as a consequence of erosion of the part of the septum which is washed out into the column inlet) results in a deformation of the concentration pulse, and hence also a decrease in column efficiency. A similar deformation of the flow profile may also be caused by an air bubble entering the mobile phase from the introduction device. This may decrease the separation efficiency and cause difficulties with the detector (detectors with a transporter of the effluent are an exception). On this occasion, it should be remembered that a degassed liquid mobile phase may be contaminated by gas even from minute leakages in the hydraulic system, and hence also in the injection port, without this being evident from escape of the mobile phase from the chromatograph. On the other hand, a low diffusion of the solute in the liquid mobile phase permits the application of the stop-flow method of injection, without a significant adverse effect on efficiency.

High pressures increase the need for the sample introduction device to be gas-tight. In many instances, this device is constructed so that it permits the inlet of the liquid into the injection port to be closed after the injection; thus the mobile phase is led on to the column directly without passing through the sample introduction device.

Another important fact is the relatively high sensitivity of the detectors used, permitting (the low sorption capacity of the modern column fillings even requires it) the introduction of small volumes (*ca.* 0.1–5 $\mu$l) of the samples. For a micro-preparative scale method this volume is 40–100 $\mu$l.

The requirement for the sample to be introduced directly into the centre of the filling at the column inlet is best fulfilled by dosing with syringes through the closing septum. The use of high pressures of the liquid led to the construction of injection ports with leading guides, which ensure the correct injection of the syringe and its introduction into the centre of the column filling, and also permit a tighter compression of the closing septum with nuts and thus an increase in the tightness of the injection port. The scheme of the injection port with the leading guides according to Kroneisen is shown in Fig. 8.21.

Injection at high pressure requires the use of special high-pressure syringes. When the work is carried out at medium pressures, normal syringes with long needles are used, because the pressure gradient in them is suitably distributed and because they work more reliably. In view of its small diameter, a syringe needle always functions as a pressure attenuator.

The choice of a suitable septum remains a problem. Up to now, no universal material exists which resists both aqueous and non-aqueous solvents, and even materials that are sufficiently resistant to a given mobile phase tend to swell after continued use. As a result of their strong compression, the syringe needles can cut out small parts of these materials, which are then washed out together with the sample on to the column, where they deform the profile of the flow and increase the column resistance. In extreme cases they

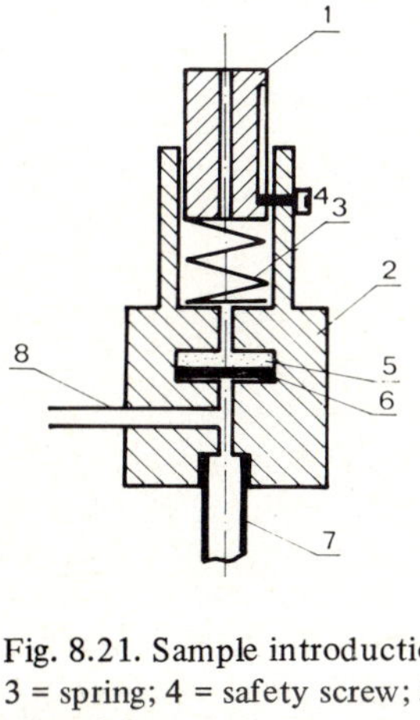


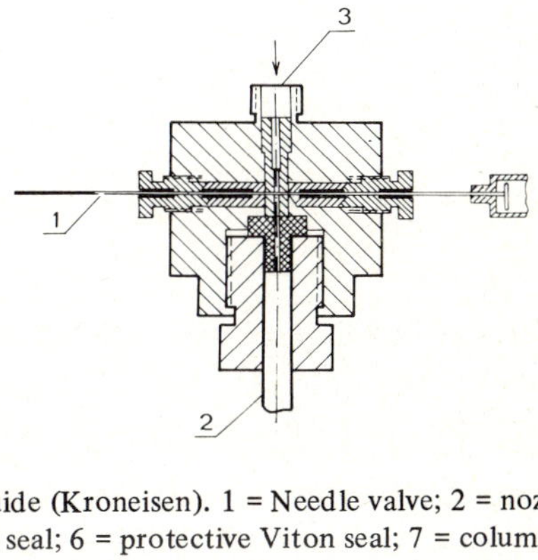


Fig. 8.21. Sample introduction device with a guide (Kroneisen). 1 = Needle valve; 2 = nozzle body; 3 = spring; 4 = safety screw; 5 = silicone rubber seal; 6 = protective Viton seal; 7 = column; 8 = mobile phase inlet.

Fig. 8.22. Sample introduction device with a closed injection port (by courtesy of Siemens AG, Karlsruhe, G.F.R.). 1 = Side opening; 2 = column; 3 = mobile phase.

*References p.168*

may block the column. This disadvantage can be eliminated by using a closed sample introduction system in which the septum need not be pierced (Fig. 8.22). This introduction system is again based on a syringe with a suitable volume (1–10 μl), fitted with a needle with an outlet orifice drilled in the side. The needle is terminated with a full rod, which ensures the correct guiding of the needle into the packing of the dosage block. The packing is made of a material that is not attacked by the mobile phase (for example, PTFE) and it can be tightened with nuts as required. Regulable stops further ensure the placing of the dosage orifice of the needle exactly into the centre of the column, which is connected with the dosage block. A disadvantage of this type of injection port is the requirement for larger samples for analysis than those used when a free syringe is employed. The sample cannot be located on the bottom of a micro-tube, but must first be transferred into a capillary with an open end, through which the rod of the syringe could pass. It can be sucked into the dosage syringe only from this capillary.

In Fig. 8.23, a sample introduction device is shown in which the mobile phase flow is separated from the septum after sampling. The closing septa are then no longer exposed to high pressure and the possibility of leakage decreases. As already stated, the interruption of the flow of mobile phase has no substantial effect on the peak broadening and under such conditions the use of high-pressure syringes is not necessary.

Although in gas chromatography devices are now used that permit automatic introduction of the sample with a hypodermic needle, such devices are not utilized in liquid chromatography and, in view of the small resistance of the closing septum to the mobile phase, their wide use cannot be expected. For the automation of sample introduction, taps are most often used, and also rod introduction devices. The application of these dosers does not usually permit the introduction of the sample directly into the column and it is introduced via a capillary either directly or through a fritted disc, which should provide for a homogeneous distribution of the mobile phase over the cross-section of the

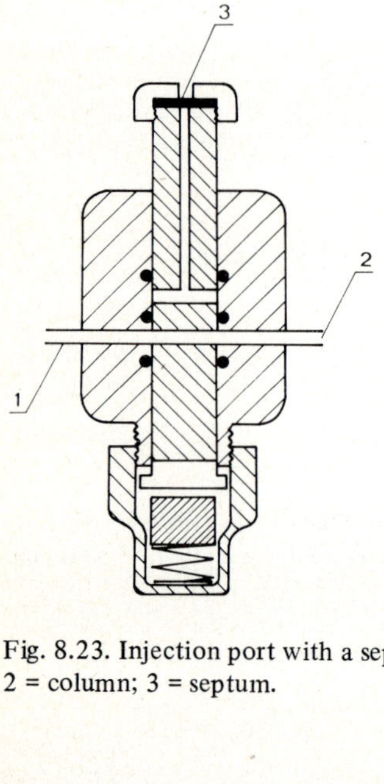


Fig. 8.23. Injection port with a septum separated from the mobile phase. 1 = Mobile phase inlet; 2 = column; 3 = septum.

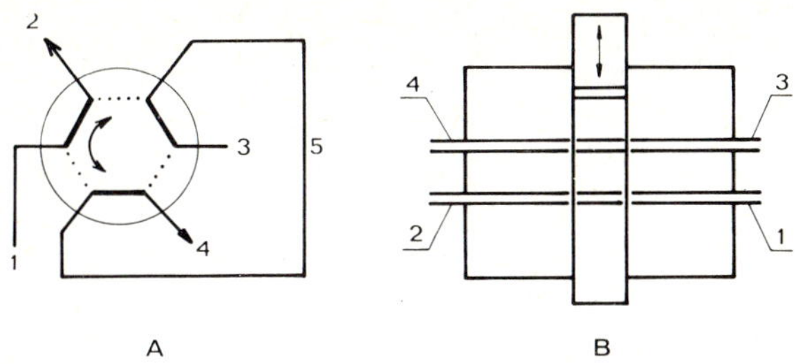


Fig. 8.24. Automatic sample introduction devices. A = Rotation; B = piston. 1 = Mobile phase inlet; 2 = mobile phase outlet; 3 = sample inlet; 4 = sample outlet; 5 = sampling loop.

column. Although the sample introduction devices achieve a good efficiency, the spreading of the inlet concentration pulse is usually larger than with the introduction systems utilizing hypodermic needles.

A scheme for a six-port sample valve and a rod doser is shown in Fig. 8.24. The main technical problems with these sample introduction devices are concerned with their gas-tightness and rather small spaces between the column and the injection port. Injection ports are constructed that are gas-tight up to 350 atm. For automatic monitoring of the introduction devices, either mechanical gears or pneumatic equipment are used.

**Columns**

The most frequently used materials of construction for columns are stainless steel, glass, aluminium, copper and PTFE. In view of the very small diffusion coefficient of the solute in the mobile phase, and with respect to the profile of the mobile phase flow, an appreciable influence of the materials of construction of the columns on their efficiency might be expected. Karger and Barth demonstrated that this effect is smaller than the effects caused by other factors connected with the column preparation (nature of the sorption material, type of column filling, mobile phase viscosity, etc.). They could not find significant differences in column efficiency when stainless steel, aluminium or copper was used. For example, it seems that the constancy of the column cross-section plays an important role in the efficiency of single columns; this parameter is usually worse in glass columns than in metallic tubes.

Nowadays, columns of 2 mm I.D. are most commonly used. However, the column diameter does not affect the column efficiency as significantly as might be supposed. DeStefano and Beachel showed that with columns of 10 mm diameter, virtually the same results for their efficiency can be achieved as with columns of small diameter. This fact is important from the point of view of the efficiencies that can be attained in types of column chromatography where the application of high-pressure gradients, which are indispensable for columns of small diameter, is impossible (for example, when organic material without a solid core is used). The most commonly used column lengths range from several centimetres to several metres. The shortest columns are used for liquid–adsorbent and liquid–liquid chromatography, especially when particles smaller than

40 μm are used. In other cases, longer columns are used, composed of short segments (usually 50 cm) connected with capillaries. This trend is different in principle from that of gas chromatography where long, spiral-form columns are in common use. Barth *et al.* found that the dependence of the column efficiency on the shape of the chromatographic column was significant; for example, the height equivalent to a theoretical plate, $H$, increased with the column diameter and the diameter of the spiral into which it was wound. Thus, for example, a column of 2.4 mm I.D. wound into a spiral of 2.4 cm diameter showed a four-fold increase in the value of $H$. They also noted the dependence of the column efficiency on the direction of the spiral winding (a figure "8" or a letter "S" configuration). This effectively represents an experimental check on Giddings' hypothesis that the molecule present on the inner wall of a spirally wound column precedes the molecule on the outer wall, both in view of its shorter path (which must be covered at the same linear rate in the same time interval) and in view of the larger pressure gradient that exists on the shorter path of the inner side of the spiral column. A deterioration in column efficiency was not observed when the column was wound into a spiral such that in single loops of the spiral the direction of flow of the mobile phase was always reversed. The above effect was then cancelled and the decrease in column efficiency was not observed.

The most important factor affecting column efficiency remains the method of column filling (disregarding for the present purpose the character of the adsorbent). The columns are closed at both ends by stoppers made of strong filter-paper discs, porous PTFE, or porous stainless-steel or fritted glass discs. The pore diameters of these materials should be as large as possible. The main function of these stoppers is the prevention of the sweeping out of fines from the column filling into the detector, where they could cause interference, or into the capillary connections between the columns, which could become clogged or cause undesirable resistance in the hydraulic system.

As already stressed, the suppression of dead volumes between the columns, the injection port and the column, or the column and the detector, is one of the basic conditions for the achievement of high efficiency in liquid chromatography. The connection of the column with the injection port is made, as far as possible, by direct insertion of the

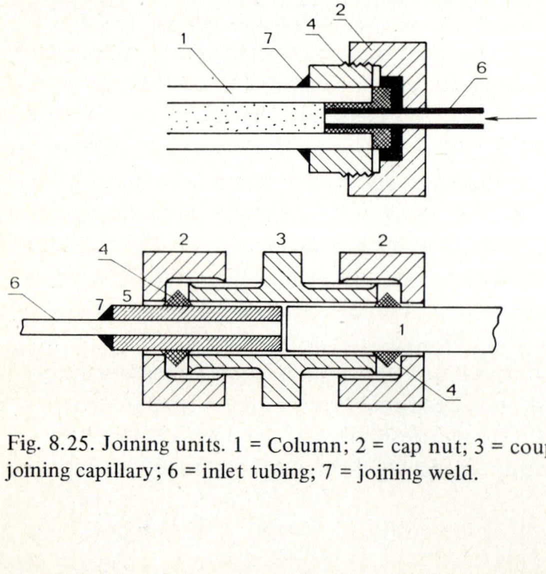


Fig. 8.25. Joining units. 1 = Column; 2 = cap nut; 3 = coupling; 4 = seal; 5 = reinforcement of the joining capillary; 6 = inlet tubing; 7 = joining weld.

column into the injection port in a manner that allows the syringe needle to reach to just over the column filling. If other sampling devices are employed, the injected solute should be introduced into the column through a capillary of minimum volume; usually, capillaries of 0.1–0.05 mm I.D. are used for this purpose. Several systems of connections with minimum dead volumes have been proposed. An example of a connection proposed by Huber (1969a, b), which is used both for the connection of the injection port and the detector to the column and for the connection of the columns to each other, is shown in Fig. 8.25.

### Thermostats

A liquid chromatograph is usually provided with two independent thermostatted circuits. One thermostat serves for the temperature regulation of the columns and the second for the regulation of the detector temperature. Detectors with an effluent transporter are an exception. While a temperature regulation of 0.1°C is satisfactory for the column, a temperature constancy in the range between $10^{-2}$ and $10^{-5}$°C is required for detectors. Commercially available detectors are usually provided with a thermostat. When an external thermostatting circuit is to be connected, liquid thermostats are used. For detectors, an accurate temperature is less often required than a high temperature stability and, owing to its high heat capacity, the liquid thermostatting circuit is therefore very suitable.

In contrast, for thermostatting columns, air thermostats, which have been well developed in gas chromatography, are usually used. They are completely satisfactory from the point of view of the stability of the set temperature. The liquid volume in the columns used in high-efficiency liquid chromatography is usually not large and therefore the heat capacity of the column is only slightly higher than that in gas chromatography. Circulating air thermostats for liquid chromatography are usually fitted with a device for flushing the thermostatted space with nitrogen. The air in the heated space is flushed with nitrogen, especially when inflammable liquids are used as the mobile phase. Thermostats are usually provided with a safety device that prevents the temperature from exceeding the chosen limit (usually 70–150°C). In the thermostat space, a heat-exchanger is usually located, which can be used, after being connected with the external cooling circuit, for cooling the thermostat space to below room temperature.

The use of a circulating air thermostat has several advantages in comparison with manipulation with a liquid thermostat. The volume of the thermostat space is sufficiently large to contain an analytical column or several connected analytical columns completed by a saturation column, and in some instances also for the location of columns in the reference side of the detector. The setting of the required temperature is rapid and easily changeable. The use of a warm-air thermostat also permits work with a temperature programme that may be necessary in high-efficiency chromatography.

When liquid thermostats are used, water jackets for the columns are usually used. These jackets are often constructed so that several columns can be located in them. The main advantage of these thermostats is that it is possible to construct simple equipment in a laboratory that is equipped with a liquid thermostat.

*References p.168*

## Detectors

### *Introduction*

Attempts to record as accurately as possible the concentration profile of the solute at the chromatographic column outlet led to the use of continuously working detectors. The basic requirements of a detector are high sensitivity and a small volume of the cell. The former parameter permits work with low concentrations of the sample being analyzed that will not result in flooding of the column, while the latter decreases the probability of the broadening of the concentration peak in the mobile phase. From these requirements, it further follows that the simple and previously widely used detection techniques in liquid chromatography cannot be employed.

The development of detectors is at present expanding very rapidly. However, the level of perfection attained in gas chromatography has still not been achieved. The fundamental difference is that as yet it has been impossible to apply the detection principle in liquid chromatography where the parameter utilized as an analytical property is sufficiently different for the solute and the mobile phase. Therefore, liquid chromatography lacks such universal detectors as are known in gas chromatography (for example, the flame ionization or thermal conductivity detector). The question remains as to whether it is at all possible to attain such a state, in view of the much greater variability of the mobile phases used compared with those in gas chromatography. It seems that for some time laboratories will have to be equipped with several types of detectors chosen selectively for single analytical applications.

Chromatographic detectors are generally continuously operating analyzers of binary mixtures leaving the column. They utilize the physical or physico-chemical properties of the mobile phase and the solute, which have a precisely defined relationship with the amount and the quality of the analyzed solute. The parameter sensed by the detector is called "the analytical property" of the system, and detectors usually have names that include the analytical property utilized for detection. The detector response, $R_i$, to a substance $i$ is proportional to the difference between the analytical properties of the effluent, $a_{i0}$, and the mobile phase, $a_0$. Generally

$$R_i = K[a_{i0}(c) - a_0] \tag{8.6}$$

where $a_{i0}(c)$ represents the functional dependence of the analytical property of the effluent, $a_{i0}$, on the concentration of the solute in the mobile phase, $c$. In modern liquid chromatography, the analytical properties most often utilized are the absorption of light of a chosen wavelength, refractive index, dielectric constant, electrolytic conductivity, and sometimes also the heat of adsorption and flame ionization. The minimum detectable difference between the analytical properties of the effluent and the mobile phase is called the "absolute detector sensitivity" and is expressed in units of the chosen analytical property.

For the full exploitation of the selectivity of the chromatographic system, extensive groups of solvents should be used that differ considerably in their physico-chemical properties. As a consequence, systems are often met in which the analytical property of the solute differs only negligibly from the analytical property of the mobile phase. From eqn. 8.6, it follows that the response to the same concentration of solute in solvent is weak because $a_{i0}(c) \approx a_0$. The detector sensitivity is small and is usually accompanied by a small range of the linearity.

The detector sensitivity is the detectable response given by the detector for a minimum amount of solute. Depending on whether the time integrals of the responses (area under the peak) are dependent on the flow-rate of the mobile phase, the sensitivities are expressed as response to unit concentration or response per solute weight in unit time. Double the value of the noise is considered as the minimum detectable response. In a similar manner, the minimum detectable concentration or the minimum detectable amount is obtained (see Table 8.4).

For the purpose of quantitative analysis, the response can be utilized only when it is linearly proportional to the concentration. Eqn. 8.6 can then be rewritten in the simple form

$$R = Kc \tag{8.7}$$

It is assumed that the response is directly proportional to the concentration of the substance being analyzed, but this assumption especially if the above properties of the effluent in the detector are considered, is seldom fulfilled. The following equation describes the relationship between the response and the solute concentration more realistically:

$$R = Kc^x \tag{8.8}$$

where $K$ is the proportionality constant and $x$ the exponent that characterizes the deviation from linearity. A detector is considered to be working in the linear range when this exponent lies within the range 0.98–1.02.

In the logarithmic form, eqn. 8.8 is suitable for the graphical determination of the value of $x$:

$$\log R = x \log c + \log K \tag{8.9}$$

The range of linearity is determined as the difference $\log c_{max.} - \log c_{min.}$, or $c_{max.}/c_{min.}$, where $c_{max.}$ is the highest concentration and $c_{min.}$ the lowest concentration in which $x$ lies within the required interval (0.98–1.02).

Modern electronic circuitry permits the electric output of the detector to be increased almost infinitely. The parameters that limit the signal amplification are the short-term noise and the detector drift.

The noise is generated both directly in the detector (the major part of the noise) and in the transfer channel of the detection system (amplifier, etc.). The baseline oscillates about the mean value and the movement mostly follows stochastic laws, but in some instances the baseline movement is systematic. Thus, for example, in detectors the response of which is dependent on the flow-rate of the mobile phase, a regular oscillation of the baseline may be observed, which is caused by the change in the flow-rate of the mobile phase due to the activity of the pump. The damping systems may change this noise substantially. In the same way as it is possible to limit the fluctuation of the flow through the detector by inserting damping devices, the short-term noise can also be limited by electric capacitance–resistance filters. As the introduction of this electric circuit into the detector outlet is also accompanied by an increase in the time constant of

TABLE 8.4
PROPERTIES OF SELECTED DETECTORS

| Type of detector | Analytical property units | Noise (0.5 sensitivity in units of the analytical property | Linear dynamic range | Selectivity of detector | Response to flow | Temperature dependence | Minimum detection for a favourable sample* | Cell volume ($\mu$l) | Suitability for gradient technique |
|---|---|---|---|---|---|---|---|---|---|
| UV absorption | Absorbance ($A$) | $10^{-4}$ | $2.5 \cdot 10^{4}$ | Selective | Independent | Small | $10^{-10}$ g/ml | 10 | Suitable |
| Refractometric | Refractive index ($n$) | $10^{-7}$ | $10^{4}$ | Non-selective | Independent | Appreciable | $10^{-7}$ g/ml | 2–10 | Unsuitable |
| Capacitance | Dielectrical constant (e) | $10^{-6}$ | $10^{4}$ | Non-selective | Independent | Appreciable | $10^{-6}$ g/ml | 2–10 | Unsuitable |
| Conductivity | Electrolytic conductivity ($\mu\Omega^{-1}$) | $10^{-3}$ | $10^{6}$ | Selective | Independent | Yes | $10^{-8}$ g/ml | 0.5–2 | Unsuitable |
| Fluorimetric | Fluorescence | | | Selective | Independent | Small | $10^{-9}$ g/ml | 7 | Suitable |
| IR absorption | Absorbance ($A$) | | | Selective | | | $10^{-6}$ g/ml | 15 | Suitable |
| Heat of sorption | Temperature (°C) | $10^{-5}$ | $10^{4}$ | Non-selective | Dependent | Appreciable | $10^{-9}$ g/sec | 5–10 | Unsuitable |
| Polarographic | Electrolytical current (A) | | | Selective | Dependent | Yes | $10^{-9}$ g/ml | 10 | Unsuitable |
| Wire, with FID | Ionization current (A) | $10^{-13}$ | $10^{5}$ | Non-selective | Dependent | No | $10^{-7}$ g/sec | | Suitable |
| Wire, with alkali FID | Ionization current (A) | | | Selective | Dependent | No | $10^{-7}$ g/sec | | Suitable |
| Disc | Ionization current (A) | $10^{-11}$ | $10^{5}$ | Non-selective | Dependent | No | $10^{-8}$ g/sec | | Suitable |

*For concentration-dependent detectors (peak area depends on mobile phase velocity) in g/ml. For mass-dependent detectors (peak height depends on mobile phase velocity) in g/sec. (See Chapter 15.)

the detector, the limitation of the noise is simultaneously accompanied by deformation of the peaks, mainly their decrease. In order to permit the choice of the optimum value of the filters and yet preserve the quantitative responses of the detector, Dressler and Deml proposed a simple calculation for the filtration circuit.

Sometimes a long-term noise is also encountered that causes an oscillation of the baseline in time intervals approximately similar to the time of outlet of an appreciably sorbed component from the detector. This type of noise decreases the possibilities of identifying single peaks. Sometimes it is caused by an inhomogeneous composition of the mobile phase entering the detector.

The long-term shift of the baseline is called drift and is usually caused by inadequate equalization of the working conditions in the detector (temperature, mobile phase flow-rate and composition), and sometimes by spoiling the column (for example, bleeding of the stationary phase in liquid–liquid systems) or the wrong choice of column.

An important parameter in modern detectors is the volume of the measuring cell. Nowadays, the cell volume does not exceed a few microlitres. The contribution of extra-column spaces plays a much greater role in liquid chromatography than in gas chromatography. Deininger and Halász deduced the relationship between the experimentally determined height equivalent to a theoretical plate (HETP), $H^*$, and the broadening caused by the column and also characterized by the HETP, $H$, and other column parameters:

$$H^* = L \cdot \frac{\sigma_v^2 + Hq^2 L(1+k')(1+\kappa)^2}{[V' + q L(1+k')(1+\kappa)]^2} \tag{8.10}$$

where $\sigma_v$ is standard deviation in volume units, $L$ is the column length, $q$ is the free cross-section of the column, $k'$ is the capacity factor, $\kappa = V/V_p$ is the ratio of the interparticle column volume $V$ and the pore volume $V_p$ of the support accessible for mobile phase, and $V'$ is the volume of the extra-column spaces. For gas chromatography, $\sigma_v^2 \ll h q^2 L$ and $V' \ll q L$, and therefore $H \approx H^*$. Naturally, these conditions do not apply in liquid chromatography.

From eqn. 8.10, it follows that the peak streaking is contributed to not only by the detector cell volume, but also by the connection of the column outlet with the detector. The extent of peak streaking is also in some instances proportional to the shape of the volume detector through which the liquid flows, or to the character of the current in the detector. The slit cells of the detectors usually cause a steep increase in $H^*$ at higher flow-rates of the mobile phase.

*Ultraviolet absorption detector*

The UV photometric detector is the most commonly utilized detector in liquid chromatography. A large number of biochemically important substances absorb ultraviolet light. Constructional simplicity and the possibility of using these detectors in gradient chromatography are the basic reasons for their popularity. For the detector response, the following expression can be used:

$$R_i = K \cdot \frac{\epsilon_i}{V_{Mi}} \cdot v_i \tag{8.11}$$

*References p.168*

where $K$ is the apparatus constant, including the optical path of the detector, $\epsilon_i$ is the molar absorption coefficient, $V_{Mi}$ is the molar volume of the solute and $\nu_i$ is the volume fraction of the solute.

The direct proportionality of the solute concentration to the response can be utilized by recording the signal in a narrow concentration range, up to an absorbance of approximately $10^{-2}$. Above this value, a logarithmic amplifier should be used. A double-beam detector with logarithmic amplifiers is shown schematically in Fig. 8.26. In most detectors, a single source is used in order to eliminate the possible intensity oscillation by a compensatory connection of the reference and measuring detector components. Most commercial detectors operate with a single wavelength and most often a low-pressure mercury lamp is used as a source (254 nm), its radiation sometimes being used for the excitation of suitable materials. Fluorescence of these materials further serves as a secondary radiation source for the detector (for example, phosphorus for emission at 280 nm). In some apparatus, a medium-pressure mercury lamp is used as a source for the 280 nm region.

The principle of the detector permits a relatively easy construction of a cell with a very small volume. The basic requirement that the whole of the working volume of the cell should be swept by the mobile phase is best realized by three fundamental types of flow-through cells with Z, H and U shapes (Fig. 8.27). The optical path in flow-through

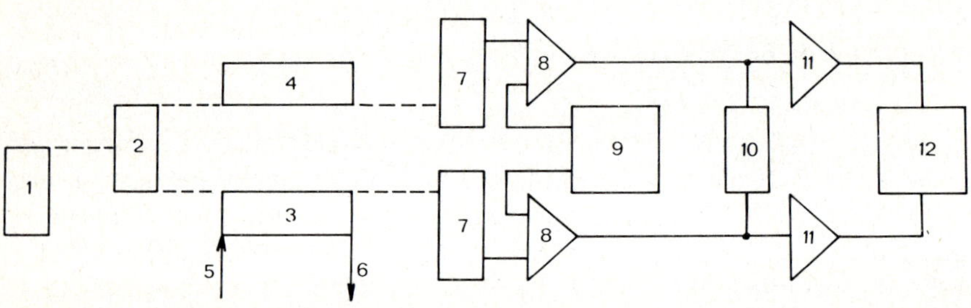


Fig. 8.26. Ultraviolet absorption detector. 1 = Light source; 2 = beam splitter and optics; 3 = sample cell; 4 = reference cell; 5 = column effluent in; 6 = column effluent out; 7 = photo-tubes; 8 = logarithmic amplifier; 9 = zero suppression; 10 = balance control; 11 = linear amplifier; 12 = recorder.

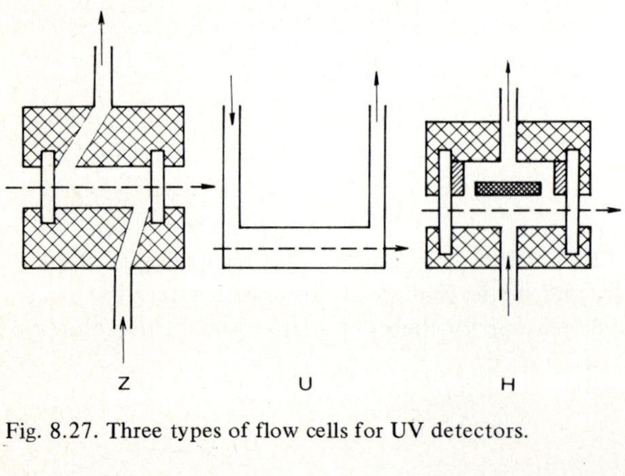


Fig. 8.27. Three types of flow cells for UV detectors.

cells of several microlitres volume is usually 0.5–10.0 mm and the dynamic linear range in good quality detectors is up to $2.5 \cdot 10^4$. The maximum working range of the detectors is in the range 0.0004–5 absorbance units.

The relatively small dependence of the detector response on temperature and flow-rate of the mobile phase and the selectivity of the detector are the basic advantages of these detectors. If mobile phases are used to which the detector does not give a response, then the parameters of single-beam and double-beam systems are virtually identical. In this case, both systems can also be used for gradient chromatography. When a mobile phase must be used that absorbs in the operative frequency range of the detector, it is necessary to use a double-beam system both for the compensation of unimportant changes caused by temperature and the mobile phase flow-rate and also for the compensation of appreciable absorption changes arising during gradient chromatography. The use of a compensatory system in high-efficiency liquid chromatography is, however, connected with much greater difficulties than in the use of analogous systems in gas chromatography.

A very advantageous variant of absorption detectors is the use of spectrophotometers that operate within a sufficiently broad spectrum (for example, 185–2500 nm) equipped with flow-through cells. Another advantage of this type of detector, especially with spectrophotometers with monochromators, is the possibility of increasing their versatility by choosing the most suitable wavelength for the component or group of components used.

*Other photometric detectors*

Spectrophotometers that operate in the visible region of the spectrum can also be utilized for detection. The flow-through cells are of similar construction to those used in UV absorption spectrometry (Fig. 8.27). The cells of commercially available apparatus often have a larger volume than that required in high-efficiency liquid chromatography (0.1 ml or more).

For the selective detection of selected components, fluorimetric detectors are useful. In most instances the fluorimeters do not attain the sensitivity of UV absorption detectors. Cassidy and Frei described a fluorescence detector with a cell volume of 4.5 $\mu$l and a sensitivity of $1.5 \cdot 10^{-11}$ g for fluorigenic substances.

Detectors for the infrared region cause some problems owing to the complexity of the spectra obtained. No mobile phase can be found that would not absorb in the infrared region. Therefore, it is necessary to work in the so-called "solvent transmission windows", *i.e.*, at wavelengths at which the absorption bands of the mobile phase are absent. Flow-through sensors have been constructed, for example, which have volumes of 15 $\mu$l and an optical path of 1 mm, or 50 $\mu$l with an optical path of 3 mm. In simple apparatus, the wavelength can be changed continuously within approximately 1.5–14 $\mu$m limits. The greatest advantage of the use of IR detectors is their high selectivity. Especially advantageous is the combination of an non-selective detector (for example, a refractometer) and a spectrophotometer operating in the IR region.

*Refractometric detectors*

The response, $R_i$, of refractometric detectors is proportional to the difference between the refractive index of the pure mobile phase, $n_0$, and that of the mixture of the mobile

*References p.168*

phase and the solute $i$, $n_{i0}$. A simple relationship can be written:

$$R_i = K(n_{i0} - n_0) \tag{8.12}$$

where $K$ is the apparatus constant. It is evident that within certain concentration limits, the response is directly proportional to the concentration of the solute in the mobile phase, because

$$n_{i0} = n_0 v_0 + n_i v_i$$

where $v$ is the volume fraction of the solute (subscript $i$) and of the mobile phase (subscript 0).

Refractometric detectors are universal detectors that give a response for all substances the refractive index of which is different from that of the mobile phase. The detection sensitivity is, of course, dependent on the difference between the refractive indices of the mobile phase and the solute.

The refractometer cells should be carefully thermostatted in view of the appreciable dependence of the refractive index on temperature. The temperature should be kept constant with a deviation not exceeding approximately $\pm 10^{-6}$ °C. In order to achieve a constant temperature of the reference and the measuring cells, the temperatures of both are usually controlled by a forced flow of thermostatted water. The leading capillaries usually function as heat exchangers, the role of which is to attain the same temperature for the effluent emerging from the column and the liquid already present in the cells. These heat exchangers should be constructed with great care in order to prevent the broadening of the concentration peaks of the solute even before they enter the cells proper. The same principles should be observed as for the connection of columns.

The refractometric detectors used are based on two different principles. The scheme of the principles used is shown in Fig. 8.28 with more details of the detector characteristics.

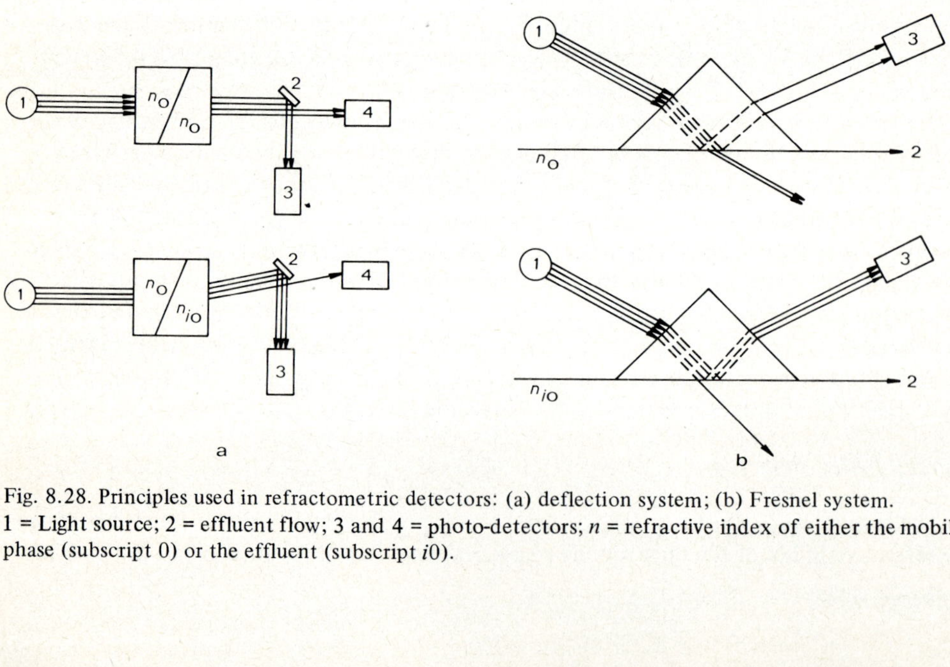


Fig. 8.28. Principles used in refractometric detectors: (a) deflection system; (b) Fresnel system. 1 = Light source; 2 = effluent flow; 3 and 4 = photo-detectors; $n$ = refractive index of either the mobile phase (subscript 0) or the effluent (subscript $i0$).

*Deflection refractometer*

The light beam passes through the reference cell, filled with mobile phase or through which the mobile phase flows, and the measuring cell rinsed with the effluent from the chromatographic column. In the presence of the solute in the measuring cell, the passing beam is deflected and falls on a photo-detector, which is sensitive to the position of the beam. The photo-detector signal is amplified and recorded with a recording millivolt meter (Fig. 8.28a). The cells are formed from prisms through which liquid flows. The volume of the cells used in high-efficiency liquid chromatography is about 10 $\mu$l. The advantage of a detector of this type is the broad range of refractive indices within which the detectors can operate without adjustment of the cells, and partial compensation of small changes in the refractive index by utilization of the reference cell. The detector can be used within a broad range of flow-rates of the mobile phase.

*Fresnel refractometer*

The Fresnel refractometer or critical-angle refractometer is based on the measurement of the intensity of the light transmitted or reflected by a liquid–glass interface (Fig. 8.28b); this intensity is dependent on the refractive index of the liquid and the angle of the incidental light. Transmittance is used more often than reflectance. In order to achieve the maximum possible sensitivity, it is necessary to work close to the critical angle and therefore a suitable glass prism should be used for liquids with a refractive index within a particular interval. When switching over to mobile phases with substantially different refractive indices, these prisms must be changed in the apparatus.

The flow-through cells are formed from slits between the glass prism and the reflection surface. The gaps are limited by PTFE masks with suitable slits. This constructional principle permits work with exceptionally small cell volumes (for example, 3 $\mu$l).

A differential arrangement provided with a measuring and reference cell is used for the compensation of temperature changes, changes in intensity of the light source, etc.

The sensitivity of both systems (deflection and Fresnel) of refractometric detectors is approximately $10^{-7}$ refractive index units, which for a substance being analyzed that has a sufficiently different refractive index from that of the mobile phase (for example, $n_{CH_3COOC_2H_5} = 1.3814$; $n_{C_6H_6} = 1.5011$) represents a sensitivity of about $10^{-8}$ mole/ml.

In Table 8.5, refractive indices and other physical properties of substances that are often used as mobile phases are given. Although all commercially available detectors today are constructed as differential detectors with a measuring and a reference cell, in practice it is very difficult to use refractometric detectors for gradient chromatography. Table 8.5 shows that the refractive index of the mobile phase in gradient chromatography may change so significantly that eqn. 8.12 cannot be utilized for the detector response. For these reasons, a differential connection can be used in gradient chromatography only to a very limited extent, and the detector responses cannot be utilized for quantitative purposes under these conditions. The appreciable temperature dependence of the detector requires work under isothermal conditions. The temperature dependence of the refractometer is also the main source of the noise and the instability of the baseline, depending on the change in flow-rate of the mobile phase. If the flow-rate changes, then the temperature of the liquid entering the detector is not controlled so as to be equal to the temperature of the detector cell.

TABLE 8.5
SELECTED ANALYTICAL PROPERTIES OF SOME COMMON MOBILE PHASES

| Mobile phase | Refractive index, $n$ | Dielectric constant, $\epsilon$ | UV cut-off (nm) | B.p. (°C) | Viscosity at 20°C (cP) |
|---|---|---|---|---|---|
| $n$-Pentane | 1.358 | 1.844 | 210 | 36 | 0.23 |
| $n$-Decane | 1.412 | 1.991 | 210 | 174 | 0.92 |
| Cyclohexane | 1.427 | 2.023 | 210 | 81 | 1.00 |
| Cyclopentane | 1.406 | 1.965 | 210 | 49.5 | 0.47 |
| l-Pentene | 1.371 | 2.100 | | 29.2* | |
| Carbon disulphide | 1.626 | 2.641 | 380 | 45 | 0.37 |
| Carbon tetrachloride | 1.463** | 2.238 | 265 | 76.7 | 0.97 |
| Xylene | 1.50 | 2.270–2.568 | 290 | 136–145 | 0.62–0.81 |
| Diisopropyl ether | 1.368 | 3.88 | 220 | 69 | 0.37 |
| Toluene | 1.496 | 2.379*** | 285 | 110.8 | 0.59 |
| Chlorobenzene | 1.525 | 5.708 | | 132 | 0.80 |
| Benzene | 1.501 | 2.284 | 280 | 80.2 | 0.65 |
| Diethyl ether | 1.354§ | 4.335 | 220 | 34.5 | 0.23 |
| Chloroform | 1.489 | 4.806 | 245 | 61.2 | 0.57 |
| Dichloromethane | 1.424 | 9.08 | 245 | 40 | 0.44 |
| Methyl ethyl ketone | 1.381** | 18.5 | 330 | 79.6 | |
| Acetone | 1.359 | 20.7*** | 330 | 56 | 0.32 |
| Dioxane | 1.422 | 2.209*** | 220 | 101.4 | 1.54 |
| Ethyl acetate | 1.372 | 6.02*** | 260 | 77.2 | 0.45 |
| Methyl acetate | 1.362 | 6.68*** | 260 | 57.5 | 0.37 |
| Amyl alcohol | 1.410 | 13.9*** | 210 | 138.3 | 4.1 |
| Aniline | 1.586 | 6.89 | 330 | 184.4 | 4.4 |
| Pyridine | 1.510 | 12.3*** | 305 | 115.4 | 0.94 |
| Isopropanol | 1.378 | 18.3*** | 210 | 82.2 | 2.86§§ |
| $n$-Propanol | 1.385 | 20.1*** | 210 | 97.2 | 2.25 |
| Ethanol | 1.361 | 24.30*** | 210 | 78.3 | 1.20 |
| Methanol | 1.331** | 32.63*** | 210 | 64.7 | 0.60 |
| Ethylene glycol | 1.432 | 37.7*** | 210 | 197.2 | 19.9 |
| Acetic acid | 1.372 | 6.15 | | 118.1 | 1.26 |
| Water | 1.330 | 80.37 | | 100 | 1.002 |

*At 740 mmHg.
**At 15°C.
***At 25°C; other values for $\epsilon$ at 20°C.
§At 17°C.
§§At 30°C.

Differential refractometers are non-destructive concentration detectors. As cells of very small volume can be used, no significant broadening of the peak takes place in the detector.

*Solute transport detectors*

Part of the effluent emerging from the column, or all of the effluent, is transported by a mechanical transporter into the detection system in which the mobile phase is evapo-

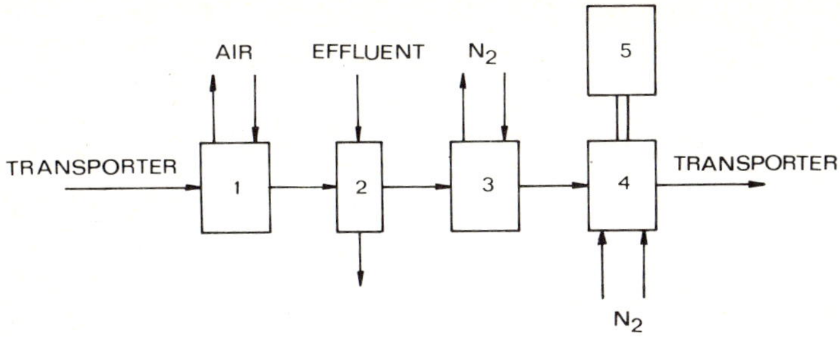


Fig. 8.29. Schematic diagram of the detector with effluent transport. 1 = Cleaner oven; 2 = coating block; 3 = evaporation oven; 4 = pyrolysis oven; 5 = detector.

rated from the transporter, and the remaining solute is usually pyrolyzed and detected with a flame ionization detector, or with an argon ionization detector, an electron capture detector or an alkali flame ionization detector. This type of detector is shown schematically in Fig. 8.29. A sufficiently long wire, tape, spiral, chain or disc is used as the transporter, and must be freed from impurities, which are the main source of the noise. The cleaner oven is usually heated to 700–900°C for this purpose and blown through with an inert gas (nitrogen, argon) or air. The cleaned transporter is led into the coating block located at the outlet of the chromatographic column. Part of the effluent is caught by the transporter in this coating block, which is then carried into the evaporation oven, where the mobile phase is evaporated at a suitable temperature (80–250°C) and then removed with a stream of inert gas. In the pyrolysis oven, pyrolysis or volatilization of the substance being analyzed takes place and the gaseous products formed are introduced by the flowing nitrogen stream into the flame ionization detector.

The analytical property of the detector is the ionization of the solute, which is proportional to the number of effective carbon atoms. The response can be expressed by the equation

$$R_i = K\,\alpha\,(\Sigma C_{\text{eff}})_{ip}(\text{d}N_{i\,\text{eff}}/\text{d}t) \tag{8.13}$$

where $K$ is the apparatus constant, $\alpha$ is the ionizing efficiency of the sensor, $C_{\text{eff}}$ is the number of effective carbon atoms of the substance entering the sensor, $t$ is time and the subscript $ip$ indicates that the solute $i$ was changed in the pyrolysis process to a mixture of chemically different substances. The effective number of moles, $N_{i\,\text{eff}}$, is given by the equation

$$N_{i\,\text{eff}} = \beta(N_{id} - N_{is})$$

where $\beta$ is the efficiency of the pyrolysis oven, $N_{id}$ is the number of moles introduced into the detector by the transporter and $N_{is}$ is the number of moles carried off from the transporter by the nitrogen leaving the drying oven.

Detectors of this type are destructive detectors. Only that part of the solute which is introduced by the transporter into the detection system undergoes destruction, and the remaining part of the solute can be collected and used further. The sensitivity of solute transport detectors depends on a number of factors (for example, the constructional character of the detector type of the transporter used, construction of the coating block), operating conditions of the detector (for example, temperatures of single ovens, rate of transporter movement), or even the solute properties (to what extent pyrolysis or evaporation takes place under given conditions, composition of the volatile products).

Detectors attain sensitivities of up to $10^{-3}$ A or, for example for squalane, $10^{-8}$ mole/sec. Detectors of this type can be used only for solutes that differ sufficiently in their volatility from the mobile phase. Only when selective detection systems are used, which do not give a response for the mobile phase (for example, electron capture detector or alkali flame ionization detector with a hydrocarbon mobile phase) is it possible to omit the drying zone from the system and introduce the whole effluent carried by the transporter into the detector.

Variants of detector construction are known which enable its sensitivity to be increased by approximately an order of magnitude. Scott and Lawrence (1970b) replaced pyrolysis by solute combustion, the carbon dioxide formed being brought into contact with hydrogen and converted into methane, which was then detected in the flame ionization detector. A scheme of this apparatus is shown in Fig. 8.30. Dubský used a disc (wire gauze) as the transporter, by which the whole column effluent is transported into the detection system. The disc has a coating, evaporating and pyrolyzing zone over which a flame ionization detector is located.

All types of solute transport detectors are subject to appreciable noise and electric capacitance–resistance filters are therefore often used, which decrease the short-term noise but simultaneously increase the time constant of the detector. Detectors have a very small effective volume and therefore contribute little to the broadening of the chromatographic zones. A large time constant of the detector, however, distorts the chromatographic zones, especially when the mobile phase flow-rate is higher. For quantitative utilization of the detectors, very careful calibration and maintenance of constant operating conditions, such as stabilization of the transporter rate, stabilization of the gas flow through the ovens and oven temperature, are indispensable.

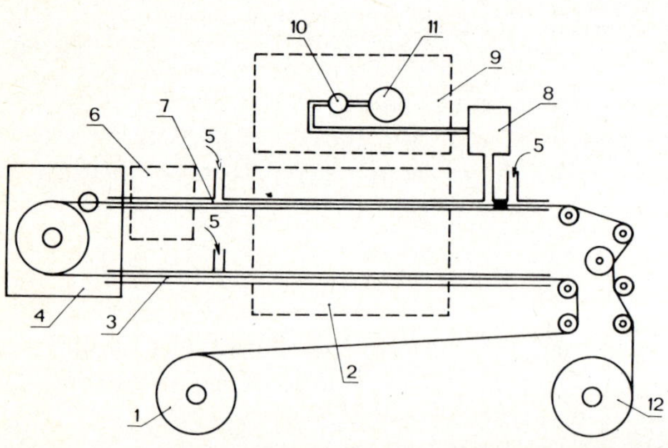


Fig. 8.30. Transport detector with subsequent conversion of carbon dioxide into methane. 1 = Feed spool; 2 = cleaner-oxidizer oven; 3 = cleaner (glassware); 4 = coating block; 5 = air; 6 = evaporator oven; 7 = evaporator-oxidizer (glassware); 8 = molecular entrainer; 9 = reactor oven; 10 = reactor chamber; 11 = flame ionization detector; 12 = collecting spool. (By courtesy of Pye Unicam, Cambridge, Great Britain.)

*Permittivity detector*

The dielectric constant is utilized as an analytical property in permittivity detectors. For the detector response, the following general expression can be used:

$$R_i = K(\epsilon_{i0} - \epsilon_0) \tag{8.14}$$

where $K$ is the proportionality constant, $\epsilon_0$ is the dielectric constant of the mobile phase and

$$\epsilon_{i0} = \epsilon_0 \cdot \nu_0 + \epsilon_i \cdot \nu_i$$

where $\nu$ is the volume fraction of the solute (subscript $i$) and the mobile phase (subscript 0).

The cell of the permittivity detector is a condenser and each cell must contain two electrically isolated electrodes. The condenser dielectric consists of the column effluent. Cells can be constructed with a fairly small volume (down to 10 µl). In addition to the temperature dependence of the dielectric constant, the temperature dependence of the cell dimensions (thermal dilatation) should also be taken into account during construction. For this reason, cells with electrodes made of two coaxial cylinders and a dielectric present in the space limited by the electrodes and the isolants forming the cell fronts seem most advantageous. The use of an identical reference cell permits, in some instances, compensation of the effect of temperature and the mobile phase composition within a not too large interval. These detectors are not suitable for use in the gradient technique for the same reasons as refractometric detectors.

Two possible principles of electric connections of the permittivity detector are utilized: resonance and bridge connections.

Vespalec and Hána described the use of a permittivity detector with the measuring cell connected in the resonance circuit of the oscillator. The change in the dielectric constant of the condenser dielectric, *i.e.*, the change in the effluent in the detector measuring cell, causes a change in the oscillator frequency, which is transferred by the interference method and using an analogous frequency meter converted into a signal suitable for the recorder. A schematic representation of the connections is shown in Fig. 8.31.

Bridge methods are based either on the principle of the Wheatstone bridge fed with an alternating current or on the utilization of the transformer bridge. The change in the

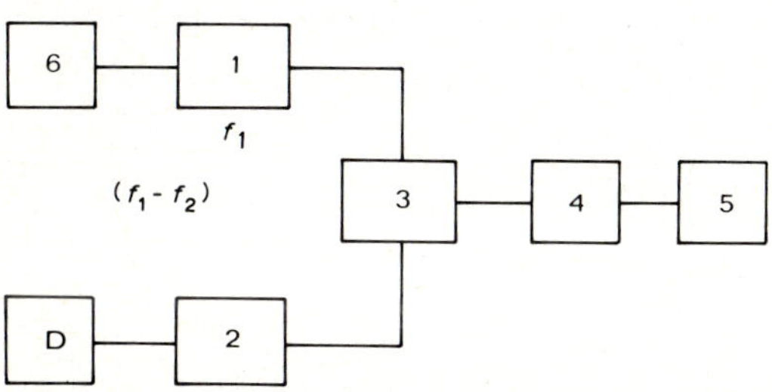


Fig. 8.31. A resonant circuit for the permittivity detector (Vespalec and Hána). D = detection cell; 1 = oscillatory circuit with a constant frequency $f_1$; 2 = measuring oscillatory circuit with a variable frequency $f_2$; 3 = mixer; 4 = frequency convertor; 5 = recorder; 6 = fine tuning device.

*References p.168*

composition of the effluent in the measuring cell brings the bridge out of balance. The recording device is used in the diagonal part of the bridge. In connection with the use of bridge methods, Haderka pointed out the possibility of the simultaneous recording of the conductivity component and the loss of complex permittivity. This possibility would eliminate one of the important limitations to the utilization of permittivity detectors, which is that components with a sufficiently high conductivity exclude the detector, especially detectors in resonance circuits, from use. A cell construction in which the electrodes are covered with an insulator has not yet been used.

The properties of permittivity detectors are similar to those of differential refractometers. The sensitivity of the detectors is $10^{-6}$ $\epsilon$, which represents, for example, approximately $10^{-8}$ mole/ml for benzene ($\epsilon^{20}$ = 2.284) in hexane ($\epsilon^{20}$ = 1.890). The range where the response is a linear function of the concentration is about $10^4$. The permittivity detector is suitable for solutes with a dielectric constant that differs from that of the mobile phase used (see Table 8.5). When circuits are used simply for measuring the capacity of the measuring condenser, the detector cannot be used for substances with a substantial conductivity component.

*Conductivity detector*

This type of detector is based on the measurement of the difference in the electrolytic conductivities of the mobile phase and the solute. A pair of chemically inert metallic electrodes (platinum or stainless steel) is connected into a Wheatstone bridge fed either with a direct or alternating current. Usually an alternating current is used in order to avoid the difficulties that arise from the polarization of electrodes. A scheme of a Wheatstone bridge is shown in Fig. 8.32. Conductivity detectors are used mainly with aqueous solutions and the use of these detectors for mobile phases with a high conductivity is not advantageous. The principle of the detector permits a cell to be constructed with a very small volume (less than 1 $\mu$l), which does not cause appreciable peak broadening. The detector sensitivity is about $10^{-5}$ $\Omega^{-1}$ of the conductivity change of the effluent, which represents, for example for sodium chloride in water, about $10^{-8}$ g/ml.

The thermostatting of the detector is advantageous but not indispensable. This detector is not suitable for gradient chromatography during which the conductivity of the mobile phase changes significantly.

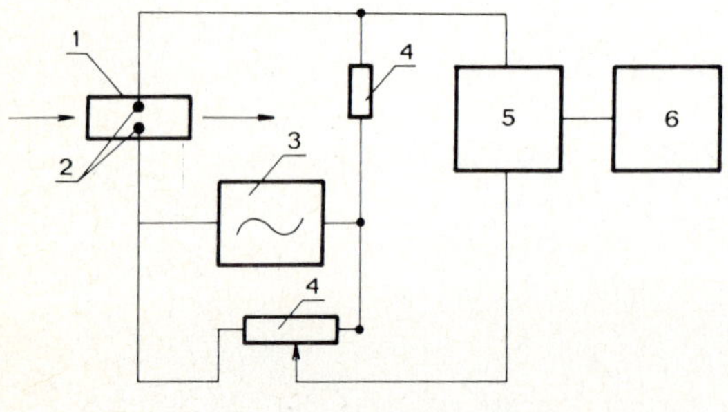


Fig. 8.32. The bridge for the conductivity detector. 1 = Detection cell; 2 = electrodes; 3 = a.c. voltage source; 4 = resistor; 5 = phase detector; 6 = recorder.

*Heat of sorption detector*

This detector makes use of the heat of sorption as an analytical property. The great variability of the sorbent filling of the detector permits its universal utilization. In practice, it gives a response to all substances that are sorbed.

The principle of its function is illustrated in Fig. 8.33. The cell usually consists of two geometrically identical chambers, one of which is filled with the active sorbing material and the other, the compensation chamber, is filled with non-sorbing inactive glass micro-beads. Thermocouples are located in the chambers or, more often, the more sensitive thermistors which measure the temperature of the chamber filling. In some instances, the detector consists of four chambers of which three are filled with inert material. The temperature-dependent elements in the chambers form the arms of a Wheatstone bridge. It is evident from the principle of the detector that thermistors (thermocouples) can also be located directly in the column (in front of the effluent outlet), which makes it possible for the detector volume to be very small.

When the solute enters the chamber filled with the active sorbing material, heat is released and the temperature in the chamber increases. Endothermic desorption, which takes place during the issue of the solute from the chamber, causes the temperature to decrease below the original, equilibrium, value. A typical curve, differing substantially from common chromatographic peaks, is shown in Fig. 8.34. This curve is similar to a certain extent to the derivation of a Gaussian chromatographic peak. The perpendicular distance A–B in Fig. 8.34 between the positive and the negative maxima of the curve corresponds to the quantitative content of the solute. It is evident from the character of the curve that the detector can be utilized only for sufficiently separated curves. Even a small degree of overlapping of two curves leads to a rather complex curve, which is almost unusable for quantitative evaluations, and utilizable for qualitative differentiation only to a very limited extent.

The advantages of the detector are that the broadening of the curve in the detector can be very small, especially if the same filling is used for the measuring cell as was used in the separating column, so that the measuring chamber then represents an additional micro-column. The use of the same filling in the detector and the column has the further significant advantage that the detector sensitivity increases with increasing solute reten-

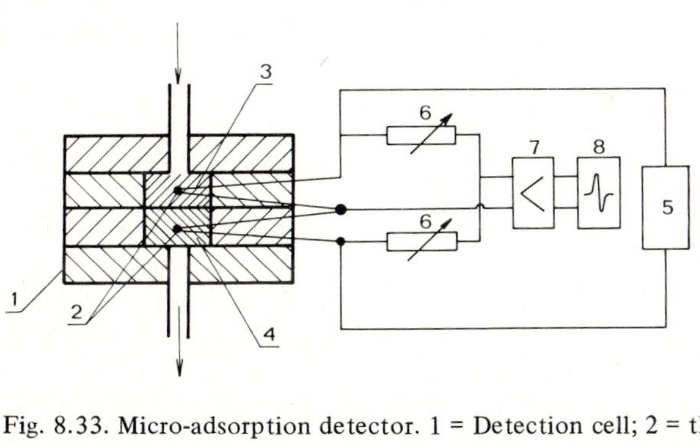


Fig. 8.33. Micro-adsorption detector. 1 = Detection cell; 2 = thermistors; 3 = sorbent; 4 = glass beads; 5 = d.c. voltage source; 6 = resistor; 7 = amplifier; 8 = recorder.

*References p.168*

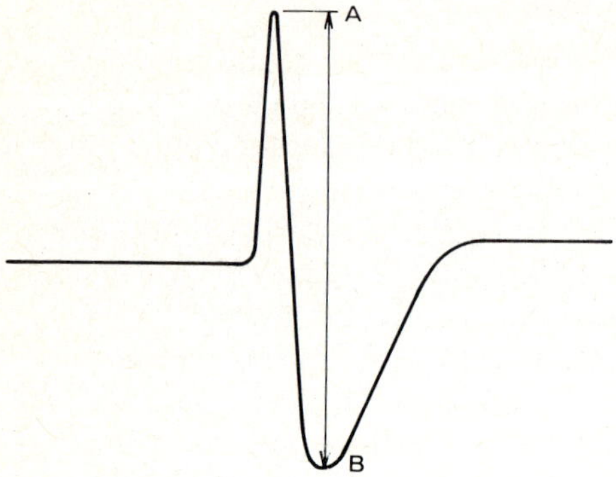


Fig. 8.34. Response of the micro-adsorption detector.

tion (with increase in retention, the heat of sorption also increases). Hence, the filling of the measuring chamber may consist of selected liquid–solid, liquid–liquid, and also ion-exchange or gel permeation systems.

The disadvantage of heat of sorption detectors is their appreciable temperature dependence. Although the function of the measuring element located in the glass beads is to compensate for the temperature changes created in both chambers, it becomes clear that both the cells and the entering effluent should be kept at a constant temperature to within approximately $10^{-5}$ °C.

For work at high sensitivities, it is advantageous to use thermistors; their application leads to an appreciable dependence of the detector response on the mobile phase flow-rate. Thermistors are themselves heat sources and their cooling is dependent on the mobile phase flow-rate.

As already stated, the detector volume may be sufficiently small (*ca.* 5 $\mu$l) and does not contribute substantially to the broadening of the peaks. The sensitivity of detectors is $10^{-4}$–$10^{-5}$ °C and is very dependent on the chosen solute–mobile phase–sorbent system. For a suitably chosen system, a sensitivity of up to about $10^{-9}$ g/sec can be attained. The utilization of the detector in the gradient elution technique is possible only to a very limited extent. Frequent control of the calibration of the detector is recommended because, as a result of the irreversible sorption of some components of the mobile phase or of the mixture being analyzed, the sorption activity of the measuring chamber filling may change.

*Polarographic detectors*

The analytical property of these detectors is the current intensity between the polarizable and the non-polarizable electrode at a chosen, constant, voltage. The circuit scheme is shown in Fig. 8.35. The fundamental problem with these detectors, arising during their use in high-efficiency, high-speed liquid chromatography, is the construction of a cell that would have a sufficiently small effective volume.

Koen *et al.* made use of the classical mercury drop electrode with which the effluent is in contact only within a small space. The effective cell volume is several microlitres. As in

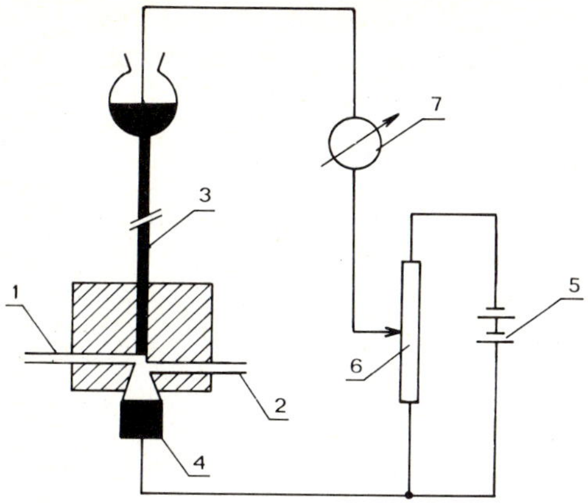


Fig. 8.35. Polarographic detector. 1 = Mobile phase inlet; 2 = mobile phase outlet; 3 = cathode; 4 = anode; 5 = source of d.c. voltage; 6 = potentiometer; 7 = electric current recording.

all polarographic apparatus of this type, the signal recording requires an appreciable electric damping in order to suppress the noise, which results in an increased time constant of the detector. Joynes and Maggs used a polarizable carbon electrode with a non-polarizable platinum electrode. This arrangement eliminated some difficulties connected with mercury electrodes.

The use of polarographic detectors is possible only for electrolytic conducting mobile phases and solutes that are capable of oxidation–reduction reactions. The possibility of choosing the polarization voltage permits work with a selective response for some substances. The detector sensitivity is $10^{-8}$ $\mu$A, which corresponds for a suitable substance to about $10^{-9}$ g/ml.

*Other types of detectors*

In liquid chromatography, a detector sensitive to radioactive radiation may be very useful. This type of detector, giving a response to radioactively labelled compounds, is very sensitive and highly selective. Its relatively rare use may be explained by the fact that technical evolution in this area has not yet reached the stage of commercial availability. Its first disadvantage is that the volume of the cell is large, usually 0.1–1 ml. However, its more extensive utilization may be expected in the future.

Other physico-chemical properties of the effluent can also be made use of for detection, such as density, viscosity, heat conductivity and vapour pressure. All these parameters are of course temperature dependent, as can be also seen in the case of the similar non-selective detectors described earlier. In many instances their application will be subject to the difficulty of finding a suitable measuring principle and maintaining a small measuring cell volume.

Gas chromatography has also been utilized for detection in liquid chromatography. Alishoyev *et al.* employed a gas chromatograph for the analysis of the effluent from a liquid chromatographic column. The conditions of the gas chromatography were set so

*References p.168*

that only two components of the effluent were detected: the mobile phase and the solute. This method permitted an appreciable acceleration of the gas chromatographic analysis and the possibility of subtracting the sample from very small effluent volumes. This technique cannot be considered to have much prospect, however, in view of the fact that it is a discontinuous process and that the indispensable condition for its application is the volatility of the mobile phase and the solute. Under these conditions, the analysis can usually be carried out directly by gas chromatography.

The combination of liquid chromatography with mass spectrometry, which can afford highly selective information on the solute structure, seems more promising. A problem that is still unsolved is the question of the transfer of the non-volatile solute into the ionization source of the mass spectrometer.

### Evaluation of different detectors

From the above description of the various types of detectors, and also from Table 8.4, it is evident that a large range of detectors exists which often differ in principle in their properties and possibilities. No general rules can be given for the choice of detector type. However, it is certain that when several types of liquid chromatographic analyses are to be carried out, a single detector is insufficient. Most often a combination of a non-selective detector (for example, a refractometer) with a selective detector (for example, UV absorption) is used.

In spite of the fact that some types of chromatographs have been developed to a high degree of sophistication, further developments may be expected in the near future.

## COUNTER-CURRENT CHROMATOGRAPHY

This technique is essentially a type of liquid–liquid partition without a solid support. Several different variants of this technique have been described (Ito and Bowman, 1970; for a review, see Ito and Bowman, 1971).

Helix counter-current chromatography is illustrated in Figs. 8.36 and 8.37. In principle, partitioning occurs in a horizontal helical tube filled with one phase of a two-phase liquid system. The other phase is fed into this coil through one end and passes through the first phase owing to the vertical direction of flow. Segments of the two phases along the helical tube are thus formed. Continued flow may cause displacement of the stationary (first) phase. A liquid–liquid partitioning system is therefore established, although a number of practical problems must be overcome if this procedure is to be of analytical use. The coils must comprise several thousand turns. In order to prevent a situation in which injection of the second phase will simply push the other phase through the helix, it is necessary to increase gravity by applying centrifugal force.

The equipment used for this purpose consists of a centrifugal head in which the separation tube is supported at its periphery. The separating coil is fed from a coaxially rotating syringe. Ito and Bowman (1970) used two methods of arranging the tubing into the centrifuge head. In the first method, the tubing is wound tightly on a flexible rod support,

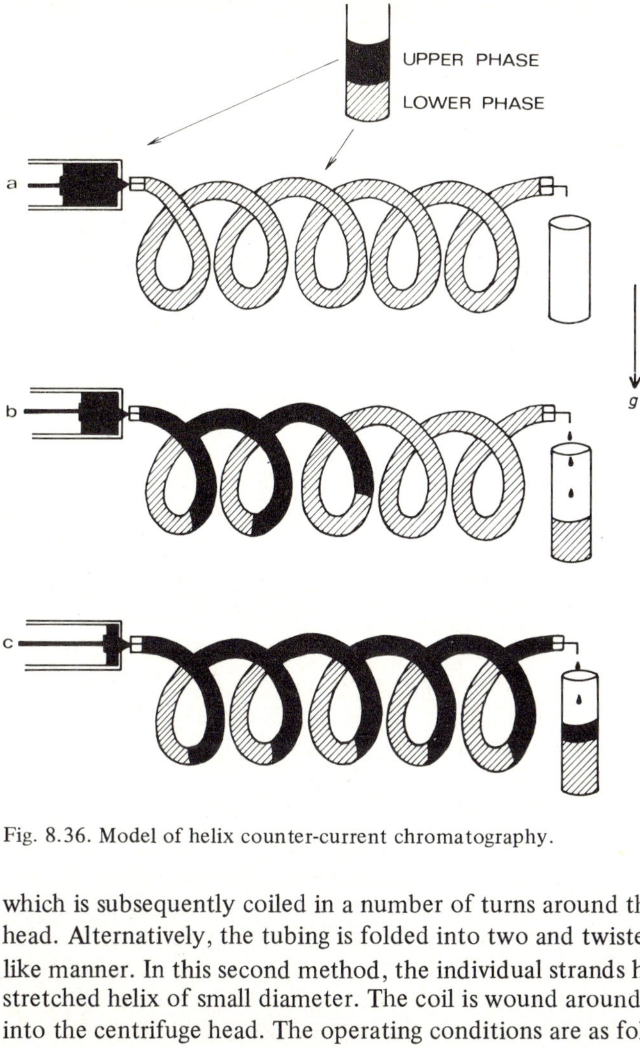


Fig. 8.36. Model of helix counter-current chromatography.

which is subsequently coiled in a number of turns around the inside of the centrifuge head. Alternatively, the tubing is folded into two and twisted along its length in a rope-like manner. In this second method, the individual strands have the appearance of a stretched helix of small diameter. The coil is wound around a drum support which fits into the centrifuge head. The operating conditions are as follows: 900–950 rpm; 16–20 atm overpressure; 125–820 μl/h.

Fig. 8.38 shows the separation of seven DNP-amino acid derivatives on a twisted column of 17,000 turns prepared from an 80-m length of PTFE tubing of 2.0 mm I.D.

The droplet counter-current chromatography technique was introduced by Tanimura *et al.* and was developed for practical use by Ito and Bowman (1970). In principle, the system is based on the observation that a light phase with low surface affinity forms discrete droplets that rise through the more dense phase with a distinct interfacial motion. Ito and Bowman (1970) suggested that under ideal conditions, each bubble of fluid can be considered to be a plate. The procedure is preferably used for preparative purposes. The system consists of long (20–60 cm) columns of narrow-bore silanized glass tubing with fine capillary tubes used to interconnect the system. Discrete droplets at the tips of the finer tube inserted into the bottom of the long glass tube were made to follow one

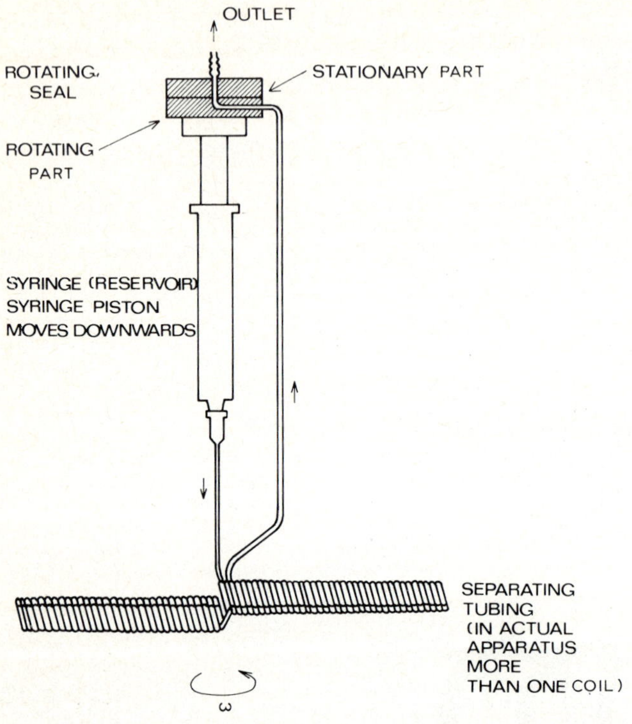


Fig. 8.37. The centrifuge head used in helix counter-current chromatography.

another with a minimal space between and a diameter similar to that of the internal bore of the column. These droplets divide the column into discrete segments that prevent longitudinal diffusion along the length of the columns as they mix locally near equilibrium. The fine PTFE tubing interconnecting the individual columns preserves the integrity of the process with minimum diffusion and helps to form new droplets at the bottom of the next column.

The equipment used by Ito and Bowman (1970) consisted of 300 glass tubes 60 cm long and 1.8 mm I.D. The operational overpressure was 150 p.s.i. at a flow-rate of 16 ml/h.

The apparatus used for locular counter-current chromatography (Ito and Bowman, 1970) is shown schematically in Fig. 8.39. Liquid–liquid partition occurs in a column that is inclined at an angle to the horizontal. The column consists of multiple segments of PTFE tubing; longitudinal diffusion of the solute is prevented by centrally perforated partitions spaced across the tube. Circular stirring is ensured by rotating the column. At the beginning of the separating procedure, the column is filled with the lower phase and the upper phase is fed into the column through the first rotating seal connection while the column is set into motion. The upper phase displaces the lower phase in each segment down to the hole that communicates with the next segment. Further feeding of the lower phase results in displacement of the upper phase only, thus leaving the appropriate amount of the lower phase in each segment of the column. Solute fed into the column is

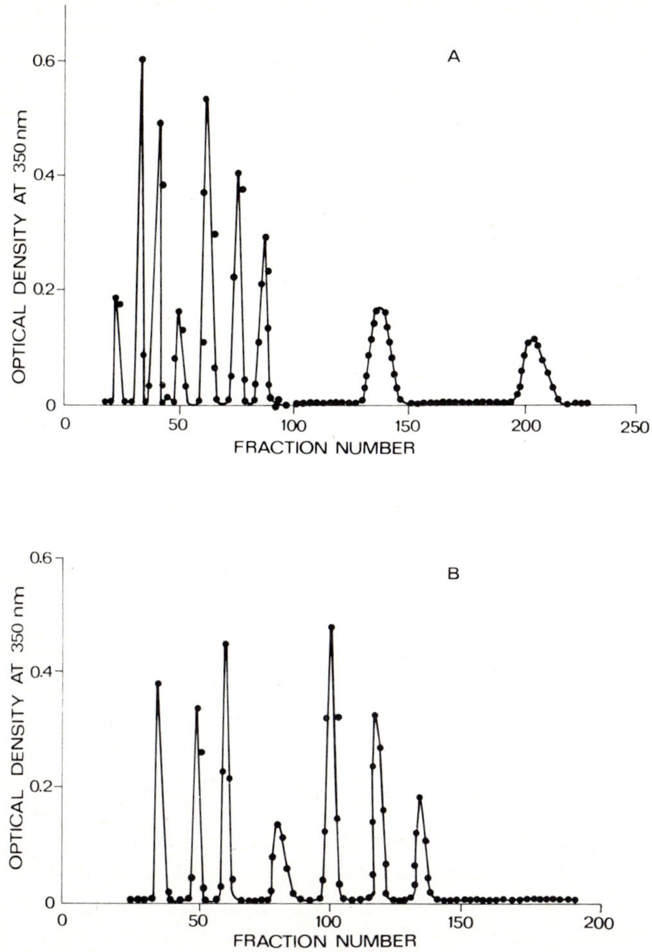


Fig. 8.38. Separation of DNP-amino acid derivatives by helix counter-current chromatography: (A) coiled column; (B) twisted column. Peak identification, from the left: DNP-Orn, DNP-Asp, DNP-Glu, DNP-CysH, DNP-$\beta$-Ala, DNP-Ala, DNP-Pro, DNP-Val, DNP-Leu; valine and leucine derivatives are missing in the (B) version.

therefore subjected to a number of partition steps along the column and is finally collected from the second rotating seal at the top of the device. Essentially, separation occurs due to gravity while rotation of the column ensures adequate mixing.

The column consisted of 5000 locules (2.6 mm I.D., 3 mm long) and a total capacity of 100 ml including the dead space of the column, which was estimated to be 5%. The column was rotated at 180 rpm at an angle of 30° to the horizontal. The upper phase was pumped at a flow-rate of 5 ml/h with a Beckman Accu-Flow pump. The maximum pressure during the operation did not exceed 20 p.s.i. The efficiency was of the order of 3000 theoretical plates, which resulted in a 50% efficiency in each locule. The elution time was 11 h for the first peak and 68 h for the last peak.

*References p.168*

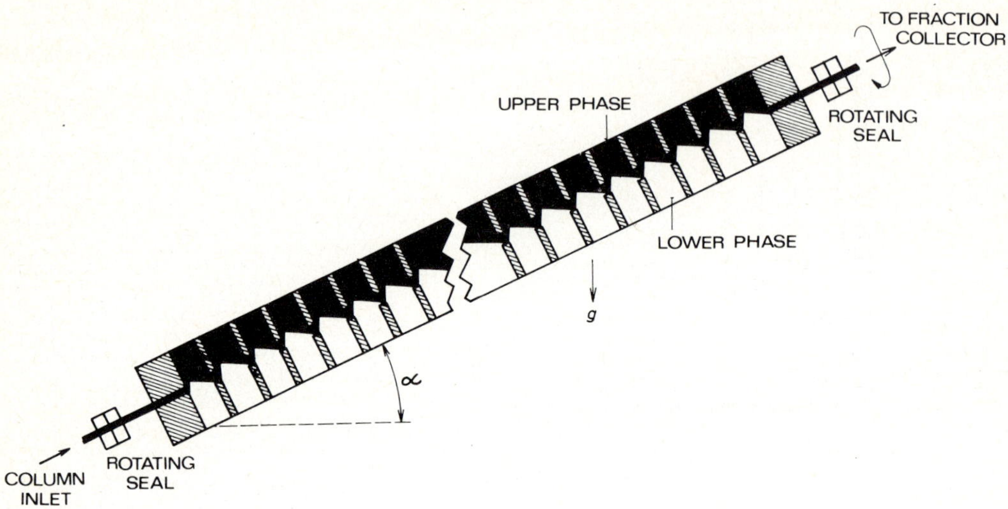


Fig. 8.39. Mechanism of rotation locular counter-current chromatography.

According to the experience of Ito and Bowman (1970), the general efficiency of the procedure increases with increase in column length, decrease in the locular length and decrease in the flow-rate. For a particular column, the efficiency increases with the angular velocity of rotation up to 180 rpm. It has been observed that a higher speed of rotation fails to improve the separation, probably owing to the cancellation of the gravitational separation by centrifugal force.

The main advantage of this procedure lies in the possibility of using virtually any pair of solvents, without any special demands upon their physical properties, which also means that the procedure is applicable not only to the particular problems of counter-current chromatography of DNP-amino acids. Ito and Bowman (1970) also suggested another alternative of this technique, namely gyration locular counter-current chromatography. Separation occurs in a column constructed in a manner similar to the above, but the column, instead of being rotated, is gyrated with a fixed eccentricity (1.24 cm). The rotation of the holding plate is continuously adjustable to 800 rpm. As indicated in Fig. 8.40, the successive positions of one locule as it is gyrated provide a steady change in the direction of the centrifugal force, which prevents the effect of separation from collapsing due to increasing centrifugal force. Hence the overall efficiency of the whole system might be further increased.

The main disadvantage of counter-current chromatography is the use of the rotating seals, which, if not manufactured precisely, cause frequent leaks in the systems. This is, of course, deleterious to the separation procedure. The problem can be overcome, according to Ito *et al.*, by using an elution centrifuge. The principle of this system is shown in Fig. 8.41. A cylindrical holder containing the column rotates both in the horizontal and vertical directions, which is ensured by a pair of toothed pulleys and toothed belt feed and outlet tubes are led through the central hole of the holder and tightly supported to

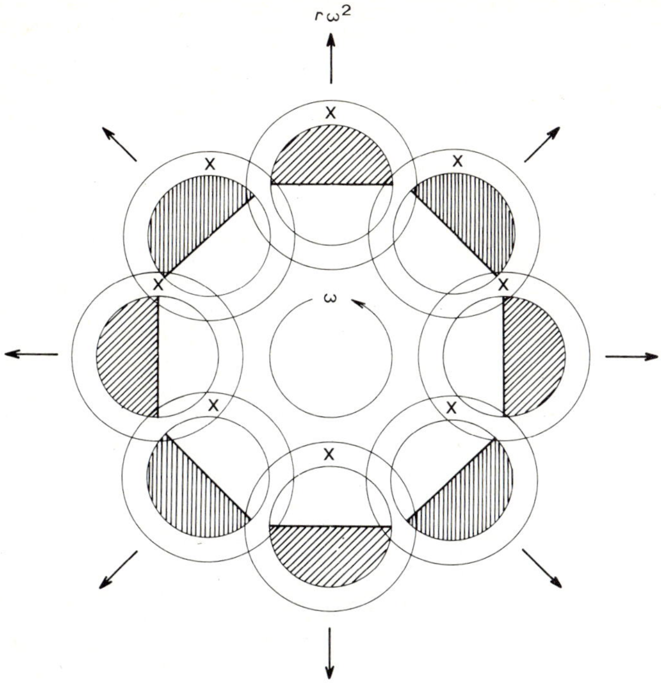


Fig. 8.40. Mechanism of gyration locular counter-current chromatography. Arrows indicate the continuous change of direction of the centrifugal force $r\omega^2$. The figure illustrates the gyration effect upon the liquid interface in the locule on the cross-section of successive positions of the same locule. $r$ = Radius of gyration.

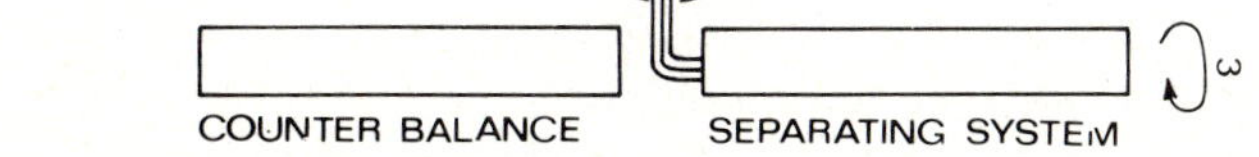


Fig. 8.41. Mechanism of elution centrifuge. The cylindrical column holder revolves around the central axis of the system in a horizontal plane while it rotates about its own axis at the same angular velocity to prevent twisting the lead tubes. Thus the system allows the movement of liquids in and out through the lead tubes without rotating seals.

the centre of the apparatus by a guide tube fixed to the centrifuge frame. The synchronous rotation in two perpendicular planes unwinds the twisting of the tubes caused by their revolution.

## REFERENCES

Alishoev, V. R., Berezkin, V. G. and Tatarinskii, V. S., *Zavod. Lab.*, 34 (1968) 148.
Barth, H., Dallmeier, E. and Karger, B. L., *Anal. Chem.*, 4 (1972) 1726.
Carman, P. C., *Trans. Inst. Chem. Eng.*, 15 (1937) 150.
Cassidy, R. M. and Frei, R. W., *J. Chromatogr.*, 72 (1972) 293.
Deininger, G. and Halász, I., *J. Chromatogr. Sci.*, 9 (1971) 83.
DeStefano, J. J. and Beachel, H. C., *J. Chromatogr. Sci.*, 8 (1970) 434.
Dressler, M. and Deml, M., *J. Chromatogr.*, 56 (1971) 23.
Dubský, H., *J. Chromatogr.*, 71 (1972) 395.
Dunnill, P. and Lilly, M. D., *Biotechnol. Bioeng. Symp.*, No. 3 (1972) 97.
Flodin, P., *J. Chromatogr.*, 5 (1961) 103.
Giddings, J. C., *Dynamics of Chromatography*, Marcel Dekker, New York, 1965.
Haderka, S., *J. Chromatogr.*, 57 (1971) 181.
Halász, I., Kroneisen, A., Gerlach, H. O. and Walkling, P., *Z. Anal. Chem.*, 243 (1968) 81.
Holeyšovský, V., in O. Mikeš (Editor), *Laboratory Handbook of Chromatographic Methods*, Van Nostrand, London, 1966.
Huber, J. F. K., *J. Chromatogr. Sci.*, 7 (1969a) 85.
Huber, J. F. K., *J. Chromatogr. Sci.*, 7 (1969b) 172.
Ito, Y. and Bowman, R. L., *J. Chromatogr. Sci.*, 8 (1970) 315.
Ito, Y. and Bowman, R. L., *Anal. Chem.*, 43 (1971) 69A.
Ito, Y., Bowman, R. L. and Noble, F. W., *Anal. Biochem.*, 49 (1972) 1.
Joynes, P. L. and Maggs, R. S., *J. Chromatogr. Sci.*, 8 (1970) 427.
Karger, B. L. and Barth, H., *Anal. Lett.*, 4 (1971) 592.
Kočent, A., in M. Rybák, Z. Brada and I. M. Hais (Editors), *Saulenchromatographie an Cellulose-Ionenaustauschern*, VEB Gustav Fischer, Jena, 1966, p. 62.
Koen, J. G., Huber, J. F. K., Poppe, H. and Den Boef, G., *J. Chromatogr. Sci.*, 8 (1970) 192.
Kozeny, P. K., *Sitzungsber. Akad. Wiss. Wien, Math.-Naturwiss. Kl., Abt. 2B,* 136 (1927) 271.
Kroneisen, A., *Dissertation*, J. W. Goethe-Universität, Frankfurt am Main, 1969.
Locke, D. C., *J. Gas Chromatogr.*, 5 (1967) 202.
Loev, B. and Snader, K. M., *Chem. Ind. (London)*, (1965) 15.
Maggs, R. S., *J. Chromatogr. Sci.*, 7 (1969) 145.
Peterson, E. A. and Sober, M. A., *Anal. Chem.*, 31 (1959) 857.
Pretorius, V. and Smuts, T. W., *Anal. Chem.*, 38 (1966) 274.
Purnell, J. H., *J. Chem. Soc.*, (1960) 1268.
Reichstein, T. and Shoppee, C. W., *Discuss. Faraday Soc.*, 7 (1949) 305.
Scott, R. P. W. and Lawrence, J. G., *J. Chromatogr. Sci.*, 8 (1970a) 65.
Scott, R. P. W. and Lawrence, J. G., *J. Chromatogr. Sci.*, 8 (1970b) 619.
Snyder, L. R., *Chromatogr. Rev.*, 7 (1965) 1.
Snyder, L. R., in E. Heftmann (Editor), *Chromatography*, Reinhold, New York, 1967, p. 80.
Snyder, L. R., *J. Chromatogr. Sci.*, 8 (1970) 692.
Svenson, H., Agrell, C. E., Dehlén, S. O. and Hagdahl, L., *Sci. Tools*, 2 (1955) 17.
Tanimura, T., Pisano, J. J., Ito, Y. and Bowman, R. L., *Science*, 169 (1970) 54.
Vespalec, R. and Hána, K., *J. Chromatogr.*, 65 (1972) 53.
Zweig, G. and Sherma, J., *Anal. Chem.*, 44 (1972) 42R.

*Chapter 9*

# Sorbents

J. JANÁK, J. ČOUPEK, M. KREJČÍ, O. MIKEŠ and J. TURKOVÁ

## CONTENTS

## RATIONAL CLASSIFICATION

The choice of the mobile and stationary phases is still mostly determined by experience in the appropriate field of application and it is usually limited to several well established systems. Therefore, it will be useful to consider the range of substances that may fulfil the function of the solvent or the sorbent and make an attempt to classify them more rationally on the basis of their physical nature and function during chromatographic sorption. In this respect, the following section should be considered as an introduction to this chapter.

By applying such an approach, it can be seen that sharp boundaries between single types of phases (gas, liquid, adsorbent) are removed, as is well known from chromatographic practice, and a continuous transition from gases up to solid substances does indeed exist (Table 9.1). A sufficiently compressed gas in the supercritical region already has the properties of a liquid, and this applies even more to a completely liquefied gas. A characteristic feature of sorption in liquids is a three-dimensional sorption of solute within the limits of its solubility. When we consider the limitation of this spacially free sorption (random sorption), it can be seen that a completely continuous transition exists from liquids, via liquid crystals, gel solvents with an associated or swollen gel matrix, to gels with a rigid matrix, up to solid porous substances. Even in the last substances there is a continuous change of properties, starting with organic polymers with a certain elasticity of the network and ending with inorganic adsorbents.

A similar continuous transition also occurs in the construction of porous materials, from strictly geometrically regular networks of porous crystals to substances with random porosity. The transition from substances with small to negligible specific surface area (for example, below 1 $m^2/g$), which may be considered as mechanical supports, to substances with a highly developed surface (for example, $10^3$ $m^2/g$), functioning as adsorbents, is also continuous.

It is important to consider all these relationships when choosing systems of chromatographic phases. Even from a first approach it appears that the use of liquefied gases as mobile phases is very promising. The investigation of carriers and adsorbents with a regulated porosity and chemical character can also be considered to have good prospects.

TABLE 9.1
CLASSIFICATION OF SORBENTS WITH PROSPECTIVE UTILIZATION IN LIQUID CHROMATOGRAPHY

| Phase | Behaviour in chromatography | Utilization | Examples | Notes |
|---|---|---|---|---|
| *Liquids* | | | | |
| Liquid gases | Unrestricted three-dimensional sorption; discrimination of solute molecules on the basis of solubility | Not yet; possible | $CO_2$; $SF_6$; light Freons | Solute may be trapped easily by evaporation of liquefied gas behind pressure restriction valve |
| Liquids | Unrestricted three-dimensional sorption; discrimination of solute molecules on the basis of solubility | Frequent | Hydrocarbons; tetrahydrofuran; diethyl ether; aqueous solutions; buffer solutions | Mutual insolubility of phases expected |
| Liquid crystal solvents | Three-dimensional sorption partially restricted by parallel-layered lattice structure of such liquids in the nematic region | Not yet; probably poor | *p*-Decyloxybenzoic acid (122–142°C); 4,4′-azoxyphenetol (139–168°C); bis(phenetidyl)-terephthalaldehyde (200–330°C) | Nematic region frequently exists at higher temperatures |
| *Gels* | | | | |
| Gel solvents | Partial discrimination of solute molecules on the basis of molecular shape due to temporarily existing pores of molecular size (by intermolecular bonding) in liquid phase | Not yet; but probably poor | No commercial equivalents | Gels exist in certain region between swelling and coagulation; sensitive to pressure |
| Gels with inert lattice | Discrimination on the basis of diffusivity of solute molecules of different size and shape into the pores (of molecular size) of a rigid lattice | Frequently | Sephadex; Spheron; Bio-Gel; Styragel; Poragel; Merckogel | There are materials with different exclusion limits |
| Gels with lattice of chemical functionality | Discrimination on the basis of diffusivity of solute molecules of different sizes and shape and/or on the basis of chemically qualified adsorption | Used; promising | Ion exchangers such as Dowex, Amberlite; materials for affinity chromatography; Aquapack; Chelex 100; cellulose ion exchangers | |

*(Continued on p. 172)*

*References p.227*

TABLE 9.1 (*continued*)

| Phase | Behaviour in chromatography | Utilization | Examples | Notes |
|---|---|---|---|---|
| *Solids* | | | | |
| Organic polymers with controlled porosity | Adsorption with certain solution effects; discrimination on the basis of partially restricted sorption by rigid lattice with homogeneously porous structure | Poor | *E.g.*, Porapak; Synachrom; Chromosorb 101 | Particles are deformed at pressures of about 20 atm and more |
| Inorganic solids with chemically bonded liquids | Solution with certain adsorption effects; discrimination of solute molecules with different special oriented groups on the basis of sorption by chemically modified surface of inorganic matter of rather homogeneous porosity | | | |
| | *Montmorillonite-type:* planar developed silicate lattice with hydroxyl groups modified by alkylammonium salts | Not yet; possible | Bentone | Powder must be distributed on inert porous materials; swelling effects |
| | *Porous glass type:* etherified or esterified hydroxyl groups on surface of porous glass originating from partial etching of special silicate melts (controlled forming of porous, rather random network of hard glass) | Used | Durapak; Permaphase | Different stability of chemical bonds in different solvents, stable also in strong solvents |
| Inorganic solids with constructed porosity | Adsorption on inorganic network with rather higher homogeneity of porosity or solution in liquid coating supports; | | | |
| | *Porous glass type:* remainder of lattice material is homogeneously porous in the whole matter or only on the surface of compact beads with controlled surface area and porosity | Frequently used; promising | Spherosil; Merckosorb; Bio-Sil; Sil-X; Porasil; Bio-Glass; Corasil; Zipax; Vydac; Pellosil; Ligna-Chrom; Pesisorb; ion exchangers | Convenient for high-pressure flow |

| | | | | |
|---|---|---|---|---|
| | *Molecular sieve type:* crystal network, three dimensional, physically or chemically treated to obtain strictly homogeneous porous structure of molecular size dimensions | Few applications; probably promising | Molecular sieves; carbon molecular sieve | |
| | *Crystallized surface type:* three-dimensional crystal network | Used; promising | Graphitized carbon black | |
| Inorganic solids with random porosity | Adsorption on surface or solution on liquids deposited on such surfaces | | | |
| | *Small specific surface:* selected natural or synthesized materials with rather high porosity but low specific surface area | Frequently used | *E.g.*, Chromosorb W, P | |
| | *Large specific surface:* selected natural or synthesized materials with large porosity of small pore diameter | Frequently used | Silica gel; alumina; active charcoal | Since pores are homogeneous, sorption isotherms tend to linearity in the region of concentration used in chromatography |

## SORBENTS FOR LIQUID–SOLID CHROMATOGRAPHY

### Physical characteristics of adsorbents

Adsorbents are solid substances on the surface of which molecules of solute present in a solution (or in the gas phase) may agglomerate. The solute concentration thus obtained on the adsorbent surface is in equilibrium with the concentration of the solute in the solvent (gas). The adsorbent consists of micro-particles, which are mostly spherical, but also may have the form of plates, rods, tubes and hexagonal prisms or other crystal forms. The conglomerates of these building units compose the structure of the adsorbent. The most important physical characteristics derived from these building units are the specific surface area of the adsorbent ($m^2/g$) and the volume of adsorbent pores (ml).

The specific surface area of the adsorbent is a function of the particle size and for spherical particles it is given by the relationship

$$S = \frac{6}{\bar{d}_p} \cdot \frac{\Sigma(\gamma_i \lambda_i^2)}{\Sigma(\gamma_i \lambda_i^3)} \tag{9.1}$$

where $\bar{d}_p$ is the most frequently occurring particle diameter and $\lambda$ is the diameter of the particle the fraction of which is $\gamma_i$. If the surface is formed by the connection or agglomeration of the building particles, it is called the internal surface. However, if it is a surface of completely independent particles, it is called the external surface, in most instances adsorbents have both types of surface. The external surface is formed from pores the diameter of which is greater than their depth and the specific internal surface area is therefore usually larger than the specific external surface area. However, this situation does not apply if the particle diameter is close to $10^{-1}$–$10^{-2}$ $\mu$m, where there is no distinct boundary between the external and internal surfaces. In addition, one type of surface may be easily transformed into the other. For example, when small particles conglomerate, the original external surface of these particles forms the internal surface of the conglomerate formed.

In this manner, pores are formed the volume ($V_\pi$) of which bears a simple relationship to the surface area ($S$):

$$d_\pi = 4\ V_\pi / S \tag{9.2}$$

where $d_\pi$ is the most frequently occurring pore diameter. As the pores are formed between contact points, or contact curves and planes of single constructional units of the adsorbent, they can assume various forms. However, in any one type of adsorbent pores that approximate to one form usually predominate, or at most pores of only a few types. Pores with cylindrical and ink-bottle shapes, and pores with defined plates or spheres, are most commonly encountered. The pores may be open (*i.e.*, communicate at both ends with the mobile phase) or closed.

Adsorbents do not have pores of equal diameters. In every adsorbent, a distribution of pore diameters exists which is of Gaussian or Poissonian character. When the pore volume of adsorbents of one type (for example, silica gel) is approximately constant, the most frequently occurring pore diameter is, according to eqn. 9.2, indirectly proportional to the specific surface area of the adsorbent. The larger the specific surface area the smaller is the most frequently occurring pore diameter of the adsorbent. This state exists, for example, in porous glasses or silica gels with a regulated surface area. Adsorbents of the same type may occur, the pore sizes of which range from tens to several thousands of Ångstroms, and with surface areas from several to several hundred square metres per gram. When the pore diameter is greater than *ca.* 20 Å, it has no longer an influence on the strength of adsorption.

The structure and the volume of the pores also affect the rate of attainment of adsorption equilibrium by the solute between the liquid phase and the adsorbent. Diffusion into the pores of the adsorbent particles becomes the controlling process. Therefore, the adsorbent should contain as few pores of small diameter as possible, because solute molecules diffuse into them less easily. Part of the mobile phase which is in the pores does not move and hence the communication of the solute between the bulk liquid and the adsorbent surface is mediated by a layer of stationary liquid. With increase in the thickness of this layer, the resistance to the mass transfer between the phases also increases, which is the main reason why preference is given in high-efficiency liquid chromatography to adsorbents with very small particle diameters: the distance between the external surface of the particle and the centre of the particle is decreased.

Golay has shown theoretically the advantage of using a non-porous support with a thin layer of adsorbent, and Halász and Horváth used this material for the first time in gas chromatography. A thin layer (1–2 μm) of adsorbent on the surface of a carrier substantially diminishes the diffusion distance of the solute and increases the column efficiency. The thinner the adsorption layer, the smaller is the mass transfer coefficient. However, the adsorbent capacity decreases simultaneously because its specific surface area is also decreased. Therefore, a compromise should be made between the mass transfer resistance and the adsorbent capacity. Such a type of adsorbent is called today "porous layer beads".

The adsorption of the solute from the liquid phase is a complex process because the adsorption energies of the solute and the solvent are usually similar. During adsorption, a monomolecular coating of the adsorbent takes place. When the mobile phase (subscript 1) consists of a single substance, the number of moles of the liquid 1 present on the adsorbent surface, $(n_1^s)^0$, is proportional to the specific surface area of the adsorbent, $S$:

$$S = S_1 (n_1^s)^0 \tag{9.3}$$

where $S_1$ is the surface area of the adsorbent covered with 1 mole of sorbate 1, and can be expressed as

$$NA_{M1} = S_1 \tag{9.4}$$

where $N$ is Avogadro's number and $A_{M1}$ is the area (Å$^2$) occupied by one molecule of the sorbate. For the calculation of the specific surface area of the adsorbent and the monolayer capacity, the following equation is very often used:

$$S = \frac{x_1}{M} \cdot N \cdot A_{M1} 10^{-20} \tag{9.5}$$

where $M$ is the molecular weight of the sorbate 1.

*References p.227*

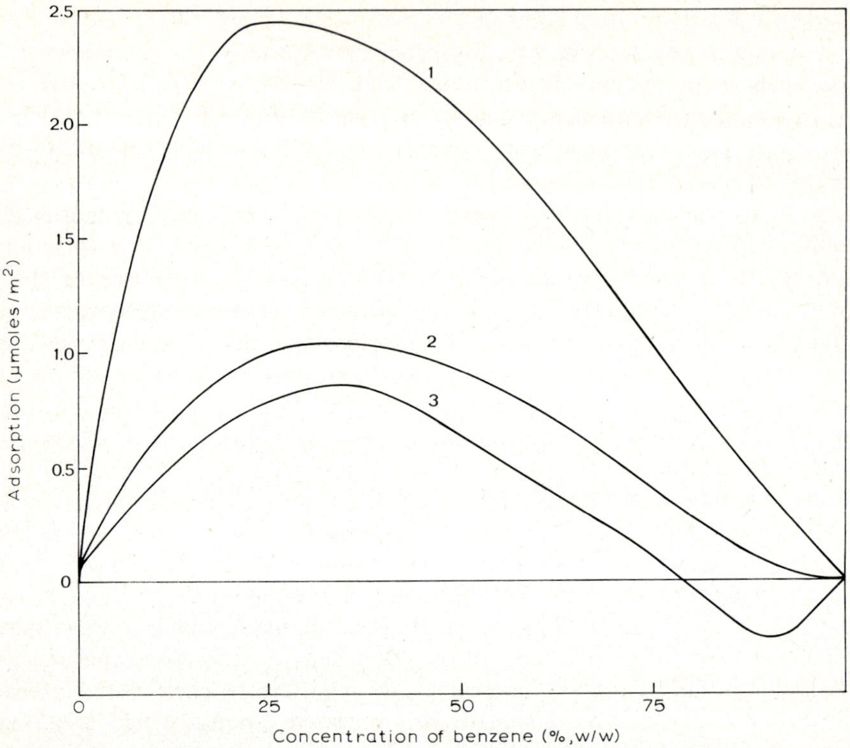


Fig. 9.1. Example of adsorption isotherm. Solvent: *n*-hexane–benzene. 1 = Hydroxylated silica gel; 2 = dehydroxylated silica gel; 3 = graphitized carbon black.

For the solute (subscript 2), an equation analogous to eqn. 9.3 can be used. As

$$S_1 + S_2 = S \tag{9.6}$$

then

$$\frac{n_1^s}{(n_1^s)^0} + \frac{n_2^s}{(n_2^s)^0} = 1 \tag{9.7}$$

where $n_1^s$ and $n_2^s$ are the number of moles of the adsorbed components 1 and 2, respectively. These numbers of moles can be calculated from the experimentally determined differences in the concentrations of solutes in the liquid phase before and after attainment of equilibrium. If the liquids are completely miscible, the result can be described by an isotherm, an example of which is given in Fig. 9.1.

## Chemical character of the adsorbent surface

The specific surface area is a quantitative characteristic of adsorption. The adsorption forces (see Chapter 4) which may arise depend on the chemical properties of the adsorbent surface (and, naturally, on the chemical and physical properties of the sorbate), and they represent the so-called "surface affinity". The surface affinity represents the adsorption potential of a given sorbent–sorbate system and an example is the isotherm in Fig. 9.1. In contrast to conventional procedures, in this case there is a constant sorbate (benzene) and solvent (*n*-hexane) and only the chemical character of the adsorbent surface changes. As the isotherm values are calculated per unit area, the effect of the magnitude of the specific surface area on the amount sorbed (μmoles) vanishes. Maximum sorption per unit area takes place (*i.e.*, the surface has a maximum affinity towards benzene) in the case of a completely hydroxylated surface. In this case, interactions of the following character may take place:

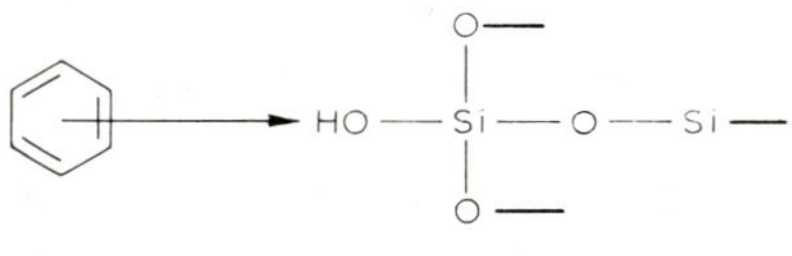


Graphitized carbon black has a smaller affinity for benzene.

It is evident that sites exist on the adsorbent surface (adsorption centres) on which adsorption takes place. The magnitude of the adsorption potential (for a given solute) depends predominantly on the chemical character of these sites. In sorbents that have a sufficiently large specific surface area to be utilized in practice, however, there are not only centres with a single value of the adsorption potential but also, as in the case of pore diameters, a distribution of the adsorption energies exists within certain limits. This distribution function may (but need not) be of a Gaussian type, for example, when the maximum number of centres has a medium value of the adsorption potential. The "adsorbent activity" is estimated according to the number of centres with a maximum adsorption potential. On centres with maximum adsorption energy, the sorbate molecules are adsorbed preferentially and hence, if a suitably strongly sorbed substance is chosen (a moderator), the sites with maximum adsorption potential energy may be saturated and thus the "adsorbent activity" may be decreased. Such circumstances apply to ternary systems in most instances. Scott and Lawrence have demonstrated the effect of the pre-

TABLE 9.2
EFFECT OF THE MODERATOR (ISOPROPANOL) ON RELATIVE RETENTION VOLUMES (RELATIVE TO SQUALANE)

Solvent: heptane.

| Solute | Isopropanol (%, v/v) | | | | | | | | |
|---|---|---|---|---|---|---|---|---|---|
| | 0.05 | 0.10 | 0.20 | 0.30 | 0.40 | 0.50 | 0.75 | 1.00 | 1.25 |
| Squalane | 1.00 | 1.00 | 1.00 | 1.00 | 1.00 | 1.00 | 1.00 | 1.00 | 1.00 |
| Phenanthrene | 1.81 | 1.61 | 1.42 | 1.33 | 1.31 | 1.26 | 1.21 | 1.20 | 1.18 |
| Methyl palmitate | 4.85 | 2.67 | 1.88 | 1.60 | 1.50 | 1.42 | 1.29 | 1.25 | 1.23 |
| Di-*n*-nonyl phthalate | – | 6.42 | 3.33 | 2.48 | 2.00 | 1.72 | 1.48 | 1.38 | 1.33 |
| Di-*n*-butyl phthalate | – | – | 3.75 | 3.08 | 2.60 | 2.29 | 2.08 | 1.92 | 1.79 |
| Tritolyl phosphate | – | – | – | – | – | 3.38 | 2.63 | 2.17 | 1.93 |
| *n*-Decanoic acid | – | – | – | – | – | – | 7.79 | 5.33 | 4.67 |

*References p.227*

adsorption of isopropanol on silica gel (Table 9.2). From Table 9.2, it can be seen that the moderator affects least the components that are sorbed predominantly by dispersion forces. The retention volume of the components that require polar functional groups blocked by the moderator preferentially for their adsorption is affected most. It was stated that the absolute values of the retention volumes are dependent on the magnitude of the specific surface and therefore, when estimating the adsorbent activity, it is more advantageous to make use of relative retention data than of absolute data.

From the above discussion, it is evident that the basic parameters that should be defined for the determination of the adsorbent activity are the type of adsorbent, the standard solutes (sorbates) used, the type of moderator used or the method for regulating the adsorbent activity and the method of evaluating the sorption data. A number of procedures have been proposed for the determination of adsorbent activity of which some, however, do not respect all four basic conditions. Most often the method described by Brockmann and Schodder is used, involving the use of six azo dyes as solutes: azobenzene, *p*-methoxyazobenzene, benzeneazo-β-naphthol, azobenzeneazo-β-naphthol, *p*-aminoazobenzene and *p*-hydroxyazobenzene. Light petroleum and benzene should be used as solvents. The adsorbent activity is divided into five classes, of which activity I represents the most active adsorbents and activity V the least active adsorbents. As a deactivator, especially in the case of silica gel and alumina, water is used and depending on the type of adsorbent, an amount ranging from 0 to 15% (w/w) is necessary in order to obtain the required adsorbent deactivation.

The use of moderators then permits the occupation of the centres with the highest adsorption energy. In the most frequently used sorbent (silica gel), these are usually free hydroxyl groups on the surface. As in gas chromatography, it is also possible to utilize chemical modification of the adsorbent surface in liquid chromatography. Substances may be bound to the hydroxyl groups that can change the character of the surface substantially. A scheme of the chemical modification of the silica gel surface is shown in Fig. 9.2. For example, the use of trimethylchlorosilane as a modifying agent may change the originally polar adsorbent surface of silica gel to a surface that is essentially non-polar. Organic substituents on the adsorbent surface permit the regulation of the adsorbent polarity within broad limits and thus bridge the earlier vacancy that existed between adsorbents of the polar type, such as alumina and silica gel, and of the non-polar type, such as carbon black (charcoal).

Inorganic materials with chemically bound deactivators are further complemented by a

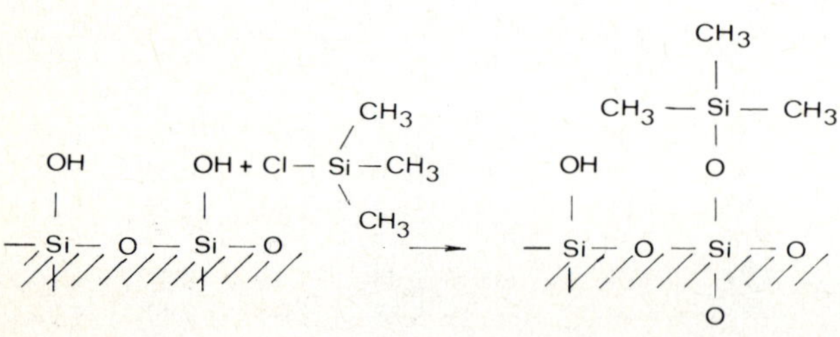


Fig. 9.2. Chemical modification of the adsorbent surface.

wide range of organic porous polymers whose surface polarity can also be selected within broad limits.

The wide range of adsorbents used today permits the programming of the operating properties of chromatographic columns. When columns are connected in parallel, the selection of a suitable basic adsorption activity is possible and a change in the mobile phase property (use of a gradient) regulates these properties to a still finer degree.

## Adsorbents

### *Silica gel*

Silica gel of the general formula $SiO_2 \cdot x\,H_2O$ is one of the most commonly used adsorbents for chromatography. It is classified as a polar adsorbent the polarity of which is mostly caused by the surface hydroxyl groups. The basic units of the adsorbent are tetrahedrons of silicon and oxygen $(SiO_4^{4-})_x$ and polymers of these units form a porous structure which is characteristic through its internal surface. By changing the pH of the solution from which the gel is prepared, the specific surface area of the adsorbent can be regulated approximately in the 200–800 $m^2/g$ range. Larger specific surface areas can be obtained when more acidic solutions are used (pH 4). Much wider use is made of silica gels with substantially lower specific surface areas (1–200 $m^2/g$), the preparation of which was described by Ashinskaya *et al.* The magnitude of the specific surface area is directly proportional to the adsorbent pore diameter. Adsorbents with a specific surface area greater than 500 $m^2/g$ usually have a mean pore diameter below 100 Å. For a specific surface area of 30 $m^2/g$, the mean pore diameter is about 400 Å, and for a silica gel with a specific surface area of 5 $m^2/g$, the mean pore diameter is about 2500 Å.

From the point of view of adsorption, hydroxyl groups bound to silicon, which form the adsorbent surface, are of importance. Snyder (1968a, b) demonstrated that the formation of hydrogen bonds with surface hydroxyl groups is the mechanism of adsorption that occurs most often on silica gel. Kiselev and Snyder (1968b) studied the mechanism of adsorption and the properties of silica gel in detail.

Snyder (1968a) differentiated three basic types of hydroxyl groups on the adsorbent surface. In Fig. 9.3, single types of functional groups are shown schematically. The free type represents an adsorption centre on which adsorption may take place, especially an adsorption directed by association by hydrogen bonding; such saturated centres on the surface are then called bound. If association by hydrogen bonding takes place between the hydroxyl groups bound on the adsorbent surface, these hydroxyl groups are called reac-

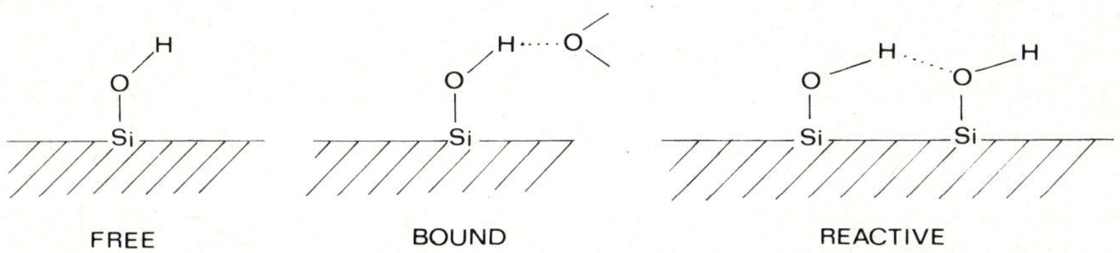


Fig. 9.3. Types of hydroxyl radicals on the adsorbent surface (Snyder, 1967).

*References p.227*

tive. The difference between single types of hydroxyl groups on the surface is evident during the thermal activation of the adsorbent. Up to about 200°C, components sorbed physically on the bound hydroxyl groups are released, and on further increase in the temperature the free hydroxyl groups may be transformed into reactive groups which, up to about 400°C, may liberate a molecule of water to form an oxygen bridge between two atoms of silicon. The adsorption centre thus formed is substantially less active. In agreement with this, an increase in silica gel activity was observed on activation by heating to about 200°C, while activation above this temperature led to a decrease in adsorption of substances that can form hydrogen bonds. On addition of water or water vapour to activated silica gel, the approximately original level of the silica gel activity can be achieved. The process is appreciably exothermic and the adsorbent is strongly heated. Irreversible dehydroxylation of the surface takes place in such instances when the formation of a new $(SiO_4^{4-})_x$ unit takes place as a consequence of the liberation of water. It is evident that this also affects the structure of the adsorbent.

If silica gel is not specially treated, the water which was used to wash it may display an acidic reaction. The acidity of the silica gel surface does not disturb some substances. If a neutral adsorbent is required, the silica gel is washed with distilled water. The pH of the eluate, or the decantation water, is controlled and the process is continued to neutrality.

Although silica gel, as the most commonly used adsorbent, is available from several firms, it may sometimes be advantageous if it is prepared in the laboratory. Therefore, the reliable method of Pitra and Štěrba for the preparation of silica gel with various surface areas and pore diameters will be described here.

Water-glass is diluted with a four-fold volume of distilled or deionized water to a density of 1.070 and 30 ml of concentrated ammonia solution are added per litre of solution. The solution is freed from cations by filtration through Wofatit KPS-200 ($H^+$) cation exchanger. Silica gel is eluted from the column at a pH of 2.5–3.0. The sol is stable for several days. Drying at 120°C gives a narrow-pore silica gel and the pore size can be regulated by adding a suitable amount of 10% (w/w) ammonium hydrogen carbonate to vigorously stirred silicic acid sol. The larger the amount of the ammonium hydrogen carbonate added, the larger will be the pores in the silica gel formed. After drying at 120°C, the silica gel obtained is ground and classified by some of the procedures the detailed conditions of which can be found in the original paper by Pitra and Štěrba.

### *Alumina*

Alumina is the second most often used adsorbent and, of the possible crystalline forms of alumina, the $\gamma$-form is most often used in chromatography. The specific surface area is about 200 $m^2/g$. King and Benson gave an idealized scheme of the adsorbent surface, which is composed of a surface layer of oxygen ions, and cations ($Al^{3+}$) from the lower layer, parallel with the layer of oxygen ions. In view of the valence of aluminium, the sites for cations in the cation layer are only two-thirds occupied and the remaining third of the positions in the lattice forms "vacancies". Under normal conditions, water does not form hydroxyl groups on the adsorbent surface, but is present in a chemisorbed state. An increase in temperature leads to a partial desorption of water from the adsorbent surface, and partly to a chemical reaction connected with the formation of hydroxyl groups on the adsorbent surface. The surface concentration of the hydroxyl groups attains higher values than those found for silica gel. However, at temperatures above 800°C, the surface loses almost all of the hydroxyl functional groups.

Snyder (1968a) pointed out that the hydroxyl groups on the adsorbent surface are not a basic component of the sorption properties of alumina. The surface structure of the oxygen and aluminium ions generates a strong electrostatic field, which forms in principle three basic types of adsorption centres. A strong positive field is formed by the centres with an acidic character, centres of a basic type have a proton acceptor character and are probably formed above the oxygen ions or are caused by the ionizable hydroxyl groups, and the third group consists of centres with electron acceptors. The electrostatic field on the alumina surface causes the main mechanism of adsorption to be surface field induction. The fact that no mechanism of adsorption similar to that in silica gel is involved to a significant extent was demonstrated on the basis of the observation that the adsorbent activity increases with increasing activation temperature, without regard to the fact that the number of hydroxyl groups on the surface diminishes and that above 800°C they are no longer present. In comparison with silica gel, alumina may display a greater catalytic effect and chemisorption, which is the main reason why activating temperatures above 150–200°C are not recommended.

Alkaline, neutral and acidic alumina can be obtained commercially. After preparation (precipitation), alumina has a rather alkaline extract (up to pH 11) and neutral alumina is obtained from this material by washing it with distilled water. In some instances, when acidic alumina is required, the adsorbent is adjusted with an acid in suspension until the extract has a pH of 4–6.

### *Other adsorbents*

#### *Charcoal*

The adsorbent surface consists mainly of carbon, which enables adsorption to take place on the principle of dispersion forces. According to the technique used for the preparation of the adsorbent, three main types of material exist today.

For the preparation of active charcoal, salts are used that can be eliminated from the adsorbent only with difficulty. Active charcoal forms an exceptionally differentiated internal surface of 300–1000 $m^2/g$ with a very inhomogeneous distribution of pore diameters. Although the basic construction unit is again carbon, active charcoal is partly a polar adsorbent because oxide, carbonyl and hydroxyl groups and unwashed salts may occur on its surface.

Graphitized charcoal is prepared at temperatures above 1000°C. Kiselev demonstrated, for example, that virtually all undesirable elements can be eliminated from the adsorbent by suitable treatment. The surface area of such an adsorbent does not attain the usual magnitude of that of active charcoal, but is non-polar and is usually employed as a model material for adsorption on the principle of dispersion forces.

Carbon molecular sieves are prepared from chlorine-containing polymers of the Saran type by thermal degradation. Kaiser found that a pure porous carbon can be obtained even below 180°C. The specific surface area of the material exceeds 1000 $m^2/g$ and the distribution of pores of 12.4 Å diameter is very homogeneous. The homogeneity of the pore diameters and the magnitude of the specific surface area permit the use of this material as molecular sieves in gas chromatography.

Carbon adsorbents are not used extensively for analytical purposes in liquid chromatography nowadays. New materials and the possibilities of obtaining large separation effects on columns seem to be leading to wider applications of such materials. The exceedingly high inertness of their surface permits their use even with strongly reactive substances.

*Magnesium silicates*

The surface of magnesium silicate has similar properties to those of aluminosilicates used as cracking catalysts. Untreated magnesium silicates usually have a very acidic surface. If water is used for the partial deactivation of the adsorbent, a chromatographic separation is achieved which is between those obtained on alumina and silica gel. Acid-washed alumina is closest to magnesium silicate in its properties. Magnesium silicate is supplied commercially as Florisil, Magnesol and magnesium trisilicate. It is difficult to prepare this material as coarser granules and therefore inert porous carriers are sometimes used (see Chapter 11) in order to decrease the hydraulic resistance of the columns used.

*Magnesium oxide*

Magnesium oxide is weakly alkaline and its properties are similar to those of alumina. Its advantage, according to Snyder (1967), consists in a high selectivity for carbon–carbon double bonds. The adsorbent activity increases with increasing activation temperature in the range 100–500°C, then decreases and finally disappears at 1000°C. The sorbed water, used for deactivation, is not bound firmly and the adsorbent activity may therefore change during a chromatographic run, so that the mobile phase should be pre-saturated to a suitable degree. As with magnesium silicate, magnesium oxide can be obtained only in a very fine form and therefore porous adsorbent carriers should be used in this case also.

## SUPPORTS AND STATIONARY PHASES FOR LIQUID–LIQUID CHROMATOGRAPHY

### General aspects

The fundamental work of Martin and Synge on the use of liquid–liquid systems extended the possibilities of chromatography as a result of the wider range of linearity of sorption isotherms and a greater choice of utilizable phases. Only recently the application of theoretical and practical advances in gas chromatography to the conditions of liquid chromatography has renewed attention to liquid–liquid column systems. Such systems are promising in view of the appreciable variability of phases, reproducibility of sorbent preparations, rate of separations, etc. In this section the most important liquid stationary phases are described, and also their fixation on the support, types of support, and the role that supports play in the chromatographic process.

**Basic types and characteristics of sorbents**

First the term "sorbent" should be defined. A sorbent is a liquid with sorption activity fixed on a solid support, which may be either inert or may affect the properties of the fixed liquid. In some instances the co-activity of the solid support may be advantageous – sometimes it is even indispensable for the achievement of a suitable selectivity of the system – but in other instances it may cause undesirable effects (non-linearity of the sorption isotherm, catalytic effects, etc.). Three fundamental types of liquid–liquid systems can be differentiated:

(1) systems in which an amount of the stationary liquid is used which exceeds that necessary for coating the carrier with a monomolecular layer of the liquid;

(2) systems in which the amount of the stationary liquid is less than would correspond to a monomolecular layer;

(3) systems in which the active sorbing phase is bound chemically on the support surface.

It is evident that especially among the systems which belong to the first and the second categories (from the point of view of sorbing properties, *i.e.*, from the point of view of retention characteristics), no strict boundary exists. The support surface may exert its sorbing property even when the amount of stationary phase considerably exceeds the amount necessary for covering the carrier with a monomolecular layer. Halász and Holdinghausen found that a superposition of the sorbing properties of the support and of the stationary liquid can be detected even when the carrier is coated by 40 molecular layers of stationary phase. The extent of this phenomenon is dependent on the properties of the support surface and its area as well as on the properties of the stationary liquid and the solute.

*Systems with large amounts of stationary phase on the support surface*

A large amount of stationary phase on the support surface permits a higher capacity ratio to be obtained on shorter columns, or on columns with a smaller weight of the support. Usually, 15–20% (w/w) of the stationary liquid is fixed on the support (per unit carrier weight). In most instances, the sorbing effect of the carrier is no longer apparent. Supports can be divided in three groups:

(i) supports with a large specific surface area (adsorbents, most commonly silica gel);

(ii) supports (used in gas chromatography) with a small specific surface area, a low sorption activity of the surface, and a large volume of internal pores of the support of relatively large pore diameter (0.1–1 $\mu$m);

(iii) supports of controlled surface porosity with a small specific surface area (usually several $m^2/g$; in view of the fact that most of the weight of the material is due to the non-porous glass beads, the specific surface area of the porous layer on the beads is high, of the order of 100 $m^2/g$, which is a value common for silica gels).

The stationary phase is applied on the support by a technique described in the part Practice of liquid chromatography. The first portions of the applied liquid cover the support surface with a monomolecular layer and further liquid then collects in the carrier pores. Hence a film, the thickness of which would be proportional to the liquid

*References p.227*

applied on the support, is not formed. With an increasing amount of liquid, the interface between the mobile and the stationary phases decreases. An example is shown in Fig. 9.4. The decrease of the surface area is characteristic and is a function of the pore volume and the specific surface area. Relative values listed in Fig. 9.4 are expressed per unit surface area of the original dry carrier. Specific surface areas of some supports are given in Table 9.3. The decrease in the relative specific surface area that Krejčí found for silica gels is the same for samples with a specific surface area of 140, 50, 15, 10 and 5 $m^2/g$.

As stated above, a larger amount of liquid on the support leads to an irregular increase in the stationary phase layer and a decrease in the contact area between the stationary and

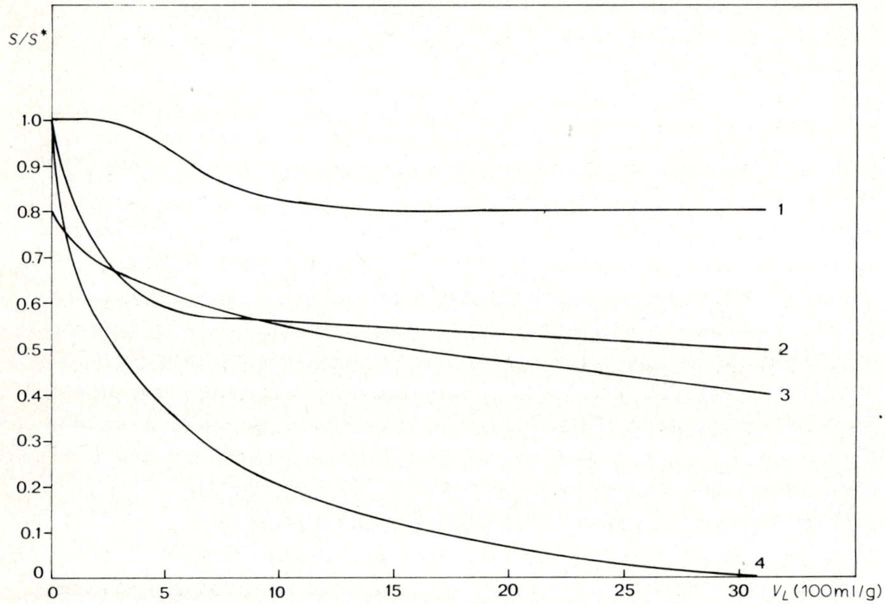


Fig. 9.4. Decrease in the adsorbent specific surface area as a consequence of introducing a liquid. S = Surface area of wet adsorbent; $S^*$ = surface area of dry adsorbent; $V_L$ = volume of coated liquid. 1, Silica gel with a chemically modified surface; 2, silica gel; 3, Chromosorb W; 4, Rysorb.

TABLE 9.3
SPECIFIC SURFACE AREAS OF SOME SUPPORTS

| Material | Specific surface area ($m^2/g$) |
|---|---|
| Charcoal | 1000–400 |
| Silica gel | 400–5 |
| Alumina | 200–100 |
| Diatomaceous earth | 7–1 |
| Corasil | 7 |
| Zipax | 1 |
| Surface-etched beads | 0.4 |

the mobile phases. Both of these effects should lead to a slowing down of the mass transfer between the phases, and hence to an increase in the mass transfer coefficient between the phases and to an increase in the HETP. Halász *et al.* (1970) used heavy loaded columns that contained 50% (w/w) of the stationary phase (1 g of stationary phase per gram of support), and they could not find an appreciable increase in the HETP. They explained this phenomenon by the fact that spaces exist between the particles and in the sorbent pores in which stagnant mobile phase is present. Diffusion coefficients in the mobile and the stationary phases are virtually identical and therefore the rates of mass transfer between the phases are equal, as in columns with a smaller amount of stationary phase. (The boundary between the mobile and the stationary phases is in fact formed between the mobile phase and the stagnant mobile phase.) With an increasing amount of stationary phase applied on to the carrier, the capacity ratio also increases and hence the mass transfer coefficient between the phases decreases for more strongly sorbed components.

The maintenance of the column efficiency as the amount of stationary phase on the surface increases cannot, however, be considered as generally valid. Kirkland (1969, 1971), who used a support of controlled surface porosity (*i.e.*, a support with a low specific surface area and a small volume of internal pores), found an increase in the HETP when the carrier was wetted above a certain limit. He explained this decrease in efficiency by the fact that under these conditions the stationary phase is already on the outer surface of the particles (the internal pore volume is already filled). Consequently, bridges of liquid are formed between single particles, which then stick together. The columns are less easily packed with sticking conglomerates of particles and the flow of the mobile phase between the particles may also be impeded. More extensive secondary spaces are formed with the stagnant mobile phase.

The main advantage of columns that contain a larger amount of stationary phase on the support surface is the increase in the capacity ratio of the column, so that the sample volume that can be applied on to the column can be increased without decreasing its efficiency and hence less sensitive detectors can be used. The absence of a sorption effect of the support is also considered to be an advantage because, in most instances, the variability of the properties and the mobile phase is sufficient for the achievement of the required analytical separation.

### *Systems with small amounts of stationary phase on the support surface*

The fundamental criterion for the estimation of the effect of the support on the retention characteristics in liquid–liquid systems is the dependence of the retention volume (or time) on the volume (or weight) of the stationary liquid. In an ideal case when the sorption activity of the support surface does not come into operation, the dependence is linear, *i.e.*, the retention volume increases proportionally with the amount of the fixed stationary liquid. However, the support is wetted only weakly, especially when the coating of the support is less than would correspond to a monomolecular layer, and this relationship need not be the general rule. Often systems are met in which the retention times decrease with an increase in the amount of stationary phase or, for a certain range of amounts of stationary phase, they remain constant. This result indicates that deactivation of an active sorbing surface of the support (in many instances an adsorbent) has occurred. Although these examples are on the boundary between chromatography in liquid–adsorbent and liquid–liquid systems, they are usually classified as liquid–liquid systems because they are dependent on the sorption properties of the solute such that even in a single chromatographed mixture examples typical of both liquid–liquid systems (increasing linear dependence, $t_R = f(V_L)$, where $t_R$ is the retention time and $V_L$ is the stationary liquid volume) and of adsorbent deactivation (decreasing dependence, $t_R = f(V_L)$) can be found.

Coating the surface with a stationary phase volume which is less than that which would correspond to a monomolecular layer, and covering the surface with a volume which exceeds that of a monomolecular layer, also results in a fundamental experimental difference. In the first instance, the concentration of the fixed liquid in the mobile solvent may be lower than or at most equal to that corresponding to saturation. Therefore, in the second instance, a solvent saturated with the liquid used as the stationary phase must always be used. From the above principle, it is evident that columns with weak wetting can be prepared by washing the carrier with the mobile phase and a corresponding concentration of the stationary phase. Karger *et al.* showed that by using this method, columns can be prepared very reproducibly and with a homogeneous distribution of the stationary phase on the support surface. Of course, the same mobile phase must then be used during the actual chromatographic run. Naturally, the increase in concentration of the fixed liquid in the mobile phase leads to an increase in the stationary phase concentration per unit surface area of the adsorbent. A decrease in this concentration leads to a decrease in the stationary phase concentration of the support surface.

The concentrations of the stationary phase corresponding to saturation in the mobile phase are usually low and the preparation of such low concentrations is usually difficult. The use of pre-columns filled with a support containing the required amount of the stationary phase on its surface has been found to be the most satisfactory. The liquid passing through these pre-columns is saturated to equilibrium with the stationary phase. The chromatographic column proper then retains a constant amount of the stationary phase on its surface.

Supports with a stationary phase of low concentration are usually adsorbents, most often based on silica gel. Supports of controlled surface porosity (Kirkland, 1970) are popular, consisting of small beads on the surface of which a thin layer of silica gel (1–2 $\mu$m) is applied. Good results were achieved with such material. With materials with a low concentration of stationary phase on the surface, it is advantageous to use supports with a large specific surface area.

### *Chemically bonded stationary phases*

Chemically bonded stationary phases suppress the bleeding that often occurs when physically sorbed stationary phases are employed and the sorption properties of the column remain unchanged for a longer time. Simultaneously, they permit the use of mobile phases with a changing eluent concentration (gradient elution technique). Chemically bonded stationary phases can be classified into two different groups: (i) materials obtained by esterification (etherification) of silica gel and (ii) materials with surface-bound silicones.

By chemical modification of the hydroxyl groups on the adsorbent surface, a monomolecular layer of chemically bonded organic functional groups is formed on it. Their polarity can be regulated over a wide range, from non-polar hydrocarbon chains to residues of considerable polarity. Halász and Sebestian, for example, used 3-hydroxypropionitrile. The organic part of the sorbent is oriented into space, perpendicular to the inorganic matrix; these materials are therefore called "brush-type" sorbents.

This form of stationary phase on the support surface has a favourable effect on the increase in mass transfer between the stationary and the mobile phase. It is mainly the internal structure of the matrix that determines the value of the mass transfer resistance coefficient. Therefore, not only silica gel, with the classical internal structure, was used successfully as a support, but also supports of controlled surface porosity. A disadvantage of the materials prepared by esterification is their tendency to split off the introduced group, and therefore proteolytically active liquids cannot be used as mobile phases.

The preparation of materials with silicones bound on the sorbent surface was described by Aue and Hastings. The synthesis of these materials again starts from silica gels, but bifunctional silanization reagents are used. On the basic chlorosilane unit fixed in the monomolecular layer on the adsorbent surface, further monomeric substances are copolymerized, thus forming a stronger layer of polymer bound on the surface. The considerable variability of the properties and higher chemical and thermal stability are among the advantages of these materials. They permit the use of a proteolytically strong mobile phase and the use of the gradient technique.

## COLUMN PACKINGS FOR GEL CHROMATOGRAPHY

### General aspects

As in gas or liquid–liquid chromatography, the choice of packing for gel chromatographic columns greatly affects the success of analysis. Generally, gel packings can be divided into universal and those that have a specific effect or "tailor-made" packings. A universal gel covering a broad molecular-weight distribution can give the first, but often very important, information on the molecular-weight distribution of the components of an unknown sample. Such preliminary assessment allows the subsequent choice of a gel that has the optimum properties for the given separation problem.

In the literature, one often encounters descriptions of gel chromatographic analyses that involve the use of standard commercially available combinations of gel packings. However, a closer inspection reveals that the choice of column packings for the system to be analyzed was made without a thorough analysis of the possibilities offered by the gel material in question, and also that the final results could have been more satisfactory if the specific properties and working range of the available packings had been considered better. Most importance is assigned to the choice of packing for chromatographic analyses, where separation into individual chemical compounds is the ultimate aim, in the low-molecular-weight and oligomeric range and also in the polymeric range when the compounds to be separated are defined high-molecular-weight compounds, such as proteins,

*References p.227*

polysaccharides and nucleotides. When determining the molecular-weight distribution of polydisperse systems, one should always bear in mind departures from linearity of the relationship between the logarithm of the molecular weight and the elution volume, which leads to a deformation of the resulting gel chromatographic distribution curve. If high-efficiency gel chromatographic columns are used, the differences between the real and the measured molecular-weight distributions, which require a numerical correction, become smaller.

The main properties that should be exhibited by gel chromatographic packings are good mechanical, chemical and thermal stabilities, which, if need be, would permit elevated pressures and temperatures to be applied. Good flow properties of the gel are a necessary condition for high-speed, high-resolution chromatography involving a wide range of solvents of different viscosities required by the commonly analyzed systems. These requirements are due to the fact that although gel chromatography is a much faster, more versatile and less labour-consuming analytical method than classical methods used for the determination of the molecular-weight distribution and other characteristics of polymer systems, the time (4 h) needed for a single analysis is still comparatively long compared with gas and liquid chromatography, and it can be further reduced.

The requirement of an easy flow of the eluent, and also that of the highest possible values of the capacity ratio and the separation efficiency of the gel packing, can be met only then if the gel particles are perfectly spherical and have been carefully fractionated according to size. The presence of even a small proportion of particles of an irregular shape considerably lowers the flow-rate through the column. A similar effect is produced by an imperfectly fractionated packing; the separation efficiency is greatly reduced. The producers of packed columns are well aware of these effects and they guarantee a certain separation efficiency for the recommended solvent flow-rate. The choice of the particle size of the packing is a result or a compromise between the requirement of a high column efficiency and the flow-rate of the eluent. When verifying the validity of the relationship of Van Deemter *et al.* and Giddings and Mallik between the flow-rate of the eluent and the separation efficiency, Heitz and Čoupek found that with tetrahydrofuran as eluent at room temperature, the particle size of the styrene–divinylbenzene gel should be 20–25 μm if the required efficiency about 10,000 theoretical plates per metre is to be obtained, assuming that the optimum conditions are used. With respect to technological difficulties involved in the perfect fractionation of such small particles, well fractionated gels in the range 30–50 μm can be recommended, which ensure an efficiency of 4000–8000 theoretical plates per metre for the optimum flow-rate.

Another important requirement in gel chromatographic separations is the absolute inertness of the packing under the conditions of analysis. In principle, a minimum interaction between the gel and the solute is required. The gel must not contain ionogenic groups, which might react with the solutes via exchange reactions or strongly interact with them. For a number of gel types that do not fulfil this requirement owing to their chemical nature, treatments have been developed that lead to an improvement in their inertness. Hydrophilic packings for gel chromatography in aqueous solutions contain only non-ionogenic hydrophilic groups, such as hydroxyl or amide groups. Even a low content of carboxylic or amine groups in the gel packing produces very strong and frequently irreversible adsorption effects, which disturb gel chromatographic analyses.

## Types of gel packings

Gels have been divided into groups on the basis of several criteria. With respect to polarity, they are hydrophilic*, organophilic and universal. Another possibility is a classification according to the chemical nature of the gel matrix. The colloidal structure of the gels makes it possible to divide them into xerogels, the volume of which changes considerably in solvents as a result of swelling, and aerogels with a permanent porosity even in the dry state, the volume of which changes only slightly on contact with solvent. Bombaugh divided gel materials according to the rigidity of their structures into soft, semi-rigid and rigid, as shown in Table 9.4. This classification seems to be sufficiently general, and reflects adequately the gel porosity, which plays a decisive role in separation. It can be seen from Table 9.4 that macroporous inorganic packings are regarded as rigid gels, while organic polymers, which, owing to the wide variations in their properties, belong to the most frequently used materials for gel chromatography, have been classified as soft or semi-rigid gels. This classification is justified by the fact that even the most densely cross-linked organic gel may swell somewhat when it comes into contact with a solvent.

TABLE 9.4
COLUMN PACKING MATERIALS FOR GEL CHROMATOGRAPHY

| Soft | | Semi-rigid | | Rigid | |
|---|---|---|---|---|---|
| Organophilic | Hydrophilic | Organophilic | Hydrophilic | Organophilic | Hydrophilic |
| Sparsely cross-linked polystyrene | Polydextran | Polystyrene | Modified polystyrene | Glass | Glass |
| | Polyacrylamide | Poly(vinyl acetate) | Sulphonated polystyrene | Silica | Silica |
| Chemically modified polydextran | Agarose | Poly(methyl methacrylate) | Merrifield derivatives | Modified silica | Modified silica |
| Poly(vinyl acetate) | Hydroxyalkyl methacrylates | Hydroxyethyl methacrylates | Hydroxyethyl methacrylates | | |
| Hydroxyalkyl methacrylates | | | | | |

Heitz divided organic gel packings according to their preparation into homogeneous, semi-heterogeneous and heterogeneous gels differing in their porosity.

Homogeneous gels are formed by the cross-linking of linear polymers and the network density of these gels is determined by the molecular weight of the initial linear polymer and by the amount of the cross-linking agent. Another possibility for the preparation of homogeneous gels consists in the copolymerization of monovinylic compounds with divinylic or polyvinylic cross-linking agents where the only factor determining the network density of the final product is cross-linking agent concentration. Homogeneous gels are

*In gel chromatography, only water and aqueous solutions are to be considered as hydrophilic solvents.

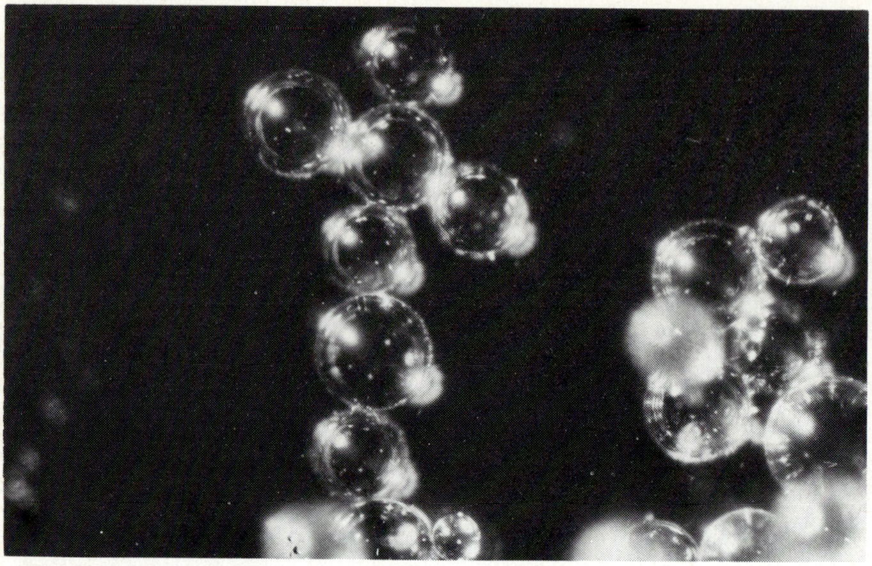

Fig. 9.5. Homogeneous gel: particles of Spheron 1.

transparent (Fig. 9.5) and do not exhibit any measurable porosity in the dry state. They swell in solvents, and the equilibrium degree of swelling attained is inversely proportional to the network density of the gel. The network density determines the working range of the gel; its decrease leads to an increase in the exclusion limit of the gel packing while at the same time reducing the mechanical strength of the particles, which is the limiting factor for the applicability of the packing. It can be said that a homogeneous copolymer is a homogeneously cross-linked gel in the statistical sense of the term. Typical representatives of hydrophilic gels are dextrans cross-linked with epichlorohydrin (Sephadex), copolymers of acrylamide and bisacrylamide (Bio-Gel P), or of 2-hydroxyethyl methacrylate and ethylene dimethylacrylate (Spheron P-1), or organophilic copolymers of styrene and divinylbenzene (Poragel), or of vinyl acetate and divinyl adipate (Merckogel PGM 2000), etc. A special position in this category is held by gels with physical cross-linkages, such as starch or agar or agarose gels (Sepharose).

Semi-heterogeneous gels are formed by the suspension copolymerization of monovinylic and divinylic monomers in the presence of a good polymer solvent. Owing to the limited swelling capacity of the three-dimensional polymer thus formed, separation of phases occurs during the copolymerization, especially at elevated concentrations of the cross-linking agent, and the later stages of polymerization proceed via a heterogeneous mechanism. The degree of conversion at which the separation of phases takes place is obviously a function of the initial concentration of the cross-linking agent in the system and of the solvation capacity of the solvent. As a result, a semi-heterogeneous gel is obtained (Fig. 9.6) the structure of which is characterized by the presence of heterogeneous regions which considerably raise the mechanical stability of particles, by a higher exclusion limit of molecular weight (*ca.* 50,000), by a reduced swelling and by a certain, although not too high, porosity in the dry state.

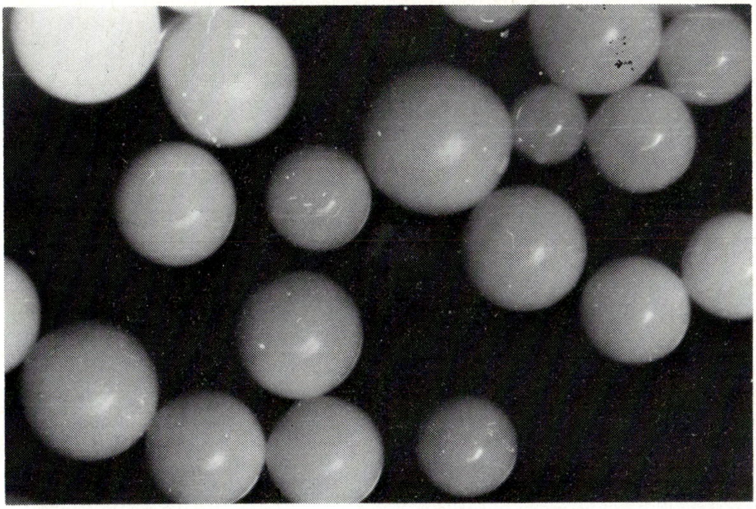

Fig. 9.6. Semi-heterogeneous gel: particles of Spheron 40.

Fig. 9.7. Heterogeneous gel: particles of Spheron 700.

A heterogeneous gel is obtained as the final product if the inert component of the copolymerization system is represented by a thermodynamically poor solvent for the polymer (with the requirement, however, that the solvent should be a good solvation agent for monomers), which already in the early stages of copolymerization causes precipitation of the polymer in suspended particles, so that the reaction can be considered to be heterogeneous over a wide range of conversion. After completion of copolymerization and removal of solvents from the gel, an opaque heterogeneous gel is obtained, which exhibits

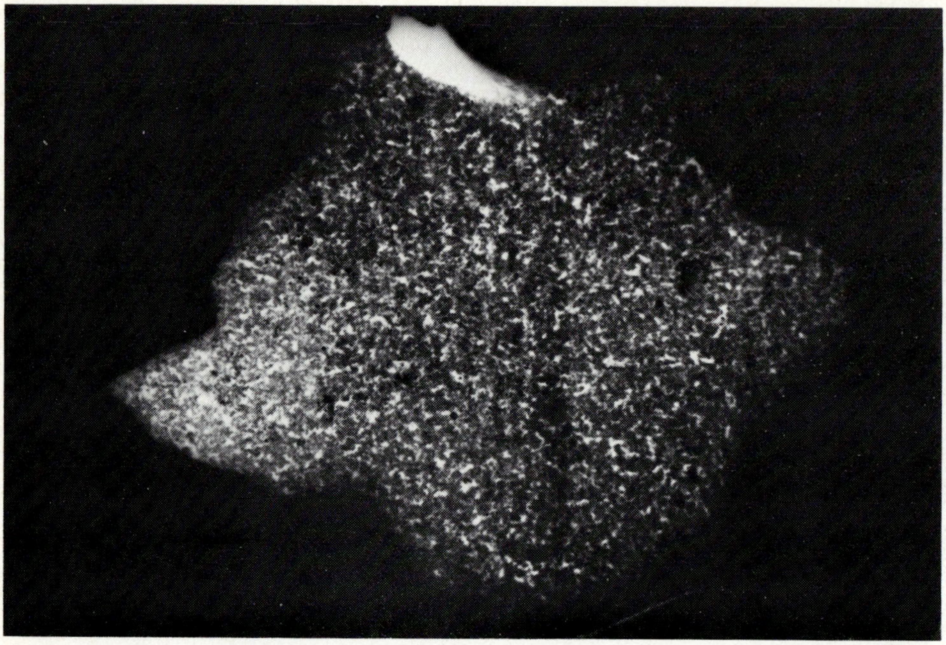

Fig. 9.8. Electron microphotograph of porous structure of Spheron 700 (enlarged 4200 ×).

a lasting porosity even in the dry state and is characterized by a large internal surface area (up to several hundred $m^2/g$). The heterogeneity of the gel can easily be seen under an optical microscope (Fig. 9.7) or electron microscope (Fig. 9.8). An appropriately chosen system of monomers and inert solvent allows gels to be obtained that have exclusion limits from $10^{-7}$ to $10^{-8}$ molecular-weight units with a very good mechanical strength of the packing and excellent flow properties of columns packed with these gels.

*Hydrophilic gels*

Agar gels were the first known hydrophilic gels; Friedman found that they gave different diffusion rates with urea and glycerin. Deuel *et al.* published a survey of the results of chromatographic separations of macromolecular hydrophilic compounds according to the size and shape of their molecules. A cross-linked polysaccharide galactomannate was prepared by Deuel and Neukom, who used it for the desalination of macromolecular solutions. However, their work has not aroused special interest, although the procedure described and known nowadays as gel filtration, is one of the most frequently used techniques in biochemistry. Lindquist and Storgårds fractionated peptides and amino acids on starch. Polson established that the penetration of proteins into agar gels is a function of the gel concentration and of the size and shape of protein molecules.

The first real chromatographic separation of hydrophilic macromolecular mixtures of polysaccharides and proteins was described by Lathe and Ruthven. Further development was characterized by the preparation and commercial application of dextran gels (1959), agar and agarose gels (1964), polyacrylamide gels (1964), various types of hydrophilized synthetic polymers, cellulose gels (1967), and homogeneous and heterogeneous hydroxy-alkyl methacrylate gels (1972).

*Polysaccharide gels*

Porath and Flodin described the preparation of gels by the cross-linking of linear dextrans obtained by the fermentation of saccharose with epichlorohydrin in an alkaline medium. Water-soluble polymeric dextran chains are cross-linked with glycerin ether bonds. A schematic view of the structure of cross-linked dextran is shown in Fig. 9.9. The cross-linking reaction itself may be accompanied by a number of side-reactions, so that irregularities may be expected in the final structure of the gel. A low content of carbonyl groups in the gel causes deviations from the ideal behaviour of the gel chromatographic material, leading to undesirable side-effects, such as adsorption and ion-exclusion, which must be eliminated. The gels are characterized by a high capacity ratio, good flow properties and a satisfactory hydrophilic stability, particularly when physiological conditions are used. Once swollen, gels with a higher molecular-weight exclusion limit must not be subjected to drying. During application, the packings must be protected by bacteriostats because, as they are polysaccharides, they are attacked by microorganisms. Dextran gels are produced by Pharmacia Fine Chemicals, Uppsala, Sweden, under the trade-name Sephadex in the form of spherical particles with defined dimensions, for various molecular-weight ranges (Table 9.5).

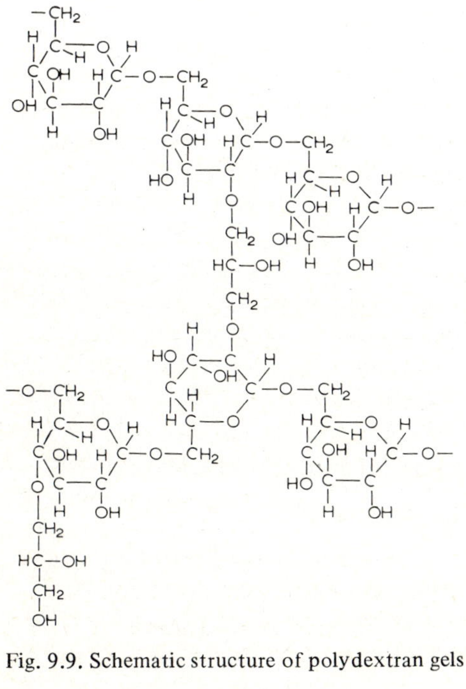

Fig. 9.9. Schematic structure of polydextran gels (Sephadex).

TABLE 9.5
SEPHADEX POLYDEXTRAN GEL CHROMATOGRAPHIC PACKINGS (PRODUCED BY PHARMACIA, UPPSALA, SWEDEN)

| Sephadex | Molecular-weight exclusion limit | | Peptide or protein standards | Recommended solvents |
|---|---|---|---|---|
| | Poly(ethylene glycol) standards | Dextran standards | | |
| G-10 | 700 | – | – | |
| G-15 | 1,500 | – | – | Aqueous solutions, electrolytes |
| G-25 | – | 5,000 | 5,000 | |
| G-50 | – | 10,000 | 10,000 | |
| G-75 | – | 50,000 | 70,000 | |
| G-100 | – | 100,000 | 150,000 | |
| G-150 | – | 150,000 | 400,000 | |
| G-200 | – | 200,000 | 800,000 | |
| LH-20 | – | – | 4,000 | Water, organic solvents (alcohols, tetrahydrofuran, acetone, N,N-dimethylformamide, toluene, dioxane, chloroform, etc.) |

A similar procedure was used to prepare gels containing polysaccharides other than dextran, such as starch or dextrin. However, their properties are not as useful as those of the dextran gels. Starch is difficult to purify, and the gels undergo strong non-specific interactions with a number of solutes. Solms *et al.* and Wiedenhof reported on $\alpha$, $\beta$-cylodextrins cross-linked with 10–25% of epichlorohydrin.

Gels based on cellulose and its derivatives have also proved successful. Determann *et al.* (1968a, b) prepared a rigid and chemically stable gel by precipitating disperse cellulose solutions with organic acids. The pore size in the particles depends on the initial cellulose concentration in solution. Schweitzer's agent was used as the solvent for cellulose. This procedure allowed gels to be obtained with a molecular-weight exclusion limit from $10^4$ to $10^7$. Martin and Rowland found that, by cross-linking cellulose with formaldehyde, the permeability of the former for large particles is reduced and the fractionation efficiency is improved for molecular weights up to $10^3$, and this material is therefore suitable for the separation of oligosaccharides. Another possibility for attaining an internal structure that is suitable for the separation of oligomeric compounds, namely cross-linking of cellulose with epichlorohydrin, was used by Luby *et al.* Cellulose thus modified contains 0.4–0.45 hydroxypropyl bonds per glucose unit. Potato starch was modified in a similar manner.

The investigation of the properties and uses of agar gels proceeded along with the development of the dextran gels. Agar and agarose gels differ considerably from dextran gels. Araki showed that agar consists of two main components, namely neutral agarose and carboxylic and sulpho groups containing agaropectin. Agarose is a linear poly-saccharide of D-galactose and 3,6-anhydro-L-galactose that contains no ionic groups. The concentration of polysaccharide in the gel determines its working range. Laurent established that the gel is composed of a network of long fibres with cross-links formed by hydrogen bonds. Commercial products with 0.5–10% of agarose are listed in Table 9.6. At first, the isolation of agarose from agar seemed to be technologically difficult; however, a more detailed study revealed that methods exist for its preparation on an industrial scale. Agarose gels are mostly supplied in the form of a suspension of particles swollen in water; the suspension should not become dried. Renn and Mueller stated that it is possible to prepare dehydrated agarose, which, on swelling, has the same properties as freshly prepared gels.

TABLE 9.6
AGAROSE GELS

| Type | Molecular-weight exclusion limit | | Recommended solvents | Producer |
|---|---|---|---|---|
| | Dextran standards | Peptide or protein standards | | |
| Sepharose B | 3,000,000 | | Aqueous solutions | Pharmacia, Uppsala, Sweden |
| Sepharose 2B | 20,000,000 | | | |
| Sepharose 4B | 5,000,000 | | | |
| Sepharose 6B | 1,000,000 | | | |
| Sag(Ago-Gel)-10 | | 250,000 | Dilute aqueous solutions of salts; testing before use recommended | Seravac Labs., Great Britain; Mann Labs., New York, N.Y., U.S.A. |
| Sag(Ago-Gel)-8 | | 700,000 | | |
| Sag(Ago-Gel)-6 | | 2,000,000 | | |
| Sag(Ago-Gel)-4 | | 15,000,000 | | |
| Sag(Ago-Gel)-2 | | 150,000,000 | | |
| Bio-Gel A-0.5 | | 500,000 | Dilute aqueous solutions of salts; testing before use recommended | Bio-Rad Labs., Richmond, Calif., U.S.A. |
| Bio-Gel A-1.5 | | 1,500,000 | | |
| Bio-Gel A-5 | | 5,000,000 | | |
| Bio-Gel A-15 | | 15,000,000 | | |
| Bio-Gel A-50 | | 50,000,000 | | |
| Bio-Gel A-150 | | 150,000,000 | | |

Owing to the structure of the agarose gels, which is formed by secondary valence forces (hydrogen bonds), they must be protected against contact with organic solvents or solutions that affect hydrogen bonds, such as urea. Before using agarose in concentrated salt solutions, it is recommended to make a preliminary stability test with a small amount of gel. It is desirable that the eluent used should contain a small amount of bacteriostats in order to prevent the growth of microorganisms which might impair the quality of the gel to a great extent. In spite of these disadvantages, however, agarose gels are widely used both in gel chromatography and in affinity chromatography of macromolecular systems with high molecular weights lying above the exclusion limit of dextran gels.

In recent years, remarkable progress has been made in increasing the thermal and chemical stability of agarose gels by the chemical cross-linking of agarose with epichlorohydrin, epibromohydrin or divinyl sulphone. The material thus obtained is particularly suitable for uses not only in chromatography, but also in analytical and preparative electrophoresis.

*References p.227*

Ghetie and Schell suggested that the carboxymethyl and diethylaminoethyl derivatives of cross-linked agarose could be used for some special separations of biological materials. Uriel *et al.* used mixed agarose–acrylamide gels prepared in a granulated form of various porosity for the separation of proteins.

All of the above procedures may improve some of the adverse properties of the agarose materials. It can be expected, therefore, that after these methods have been put into practice, agarose gels will become even more favoured in the chromatography of molecules that have extremely high molecular weights.

*Polyacrylamide gels*

The first applications of poly(acrylamide) gels were announced by Lea and Sehon and by Hjertén and Mosbach for the fractionation of proteins, and the use of a gel based on the copolymer of N,N′-methylenebisacrylamide and vinylpyrrolidone has been suggested for the same purpose. Ammonium persulphate and a tertiary amine are used as the redox initiator in radical copolymerization. In the presence of riboflavine, the polymerization can also be initiated with light. The porosity of the gel is determined by the concentration of both monomers in the original mixture. Fawcett and Morris prepared polyacrylamide gels with a high content of the cross-linking agent; the structure of these gels differs somewhat from that of the gels prepared in the usual way and their porosity is higher.

Polyacrylamide gels are chemically more stable than polysaccharides; in alkaline medium above pH 9, the amidic bonds may be hydrolyzed with the formation of ionogenic carboxylic groups, the presence of which markedly impairs the gel chromatographic properties of the gel. As polyacrylamide gels are synthetic polymers, they are not attacked by microorganisms.

Polyacrylamide gels for chromatographic purposes are prepared in the form of spherical particles. They are produced by Bio-Rad Laboratories (Richmond, Calif., U.S.A.) under the name Bio-Gel P (Table 9.7); their maximum of molecular-weight exclusion limit is about 400,000.

The investigation of gel materials based on acrylamide is still in progress. A procedure for the preparation of gels suitable for the separation of compounds with extremely high

TABLE 9.7
BIO-GEL POLYACRYLAMIDE GELS (PRODUCED BY BIO-RAD LABS., RICHMOND, CALIF., U.S.A.)

| Bio-Gel | Molecular-weight exclusion limit, tested by peptide or protein standards | Recommended solvents |
|---|---|---|
| P-2 | 2,000 | Aqueous solutions |
| P-4 | 4,000 | |
| P-6 | 5,000 | |
| P-10 | 17,000 | |
| P-30 | 50,000 | |
| P-60 | 70,000 | |
| P-100 | 100,000 | |
| P-150 | 150,000 | |
| P-200 | 300,000 | |
| P-300 | 400,000 | |

molecular weights has been suggested by Acuff. Johansson and Joustra have described a two-stage preparation of methylenebisacrylamide gels. In the first stage, methacrylamide is copolymerized with methylenebisacrylamide in the presence of initiator, emulsifier and an inert solvent (toluene), and the beads formed in this stage are subjected to a reaction with acrylamide and N,N-methylenebisacrylamide. The ternary copolymer of acrylamide, methacrylamide and N,N-methylenebisacrylamide is a good packing for gel filtration. A similar procedure was used to obtain the copolymer of acrylamide, methylenebisacrylamide and 1-vinyl-2-pyrrolidone. The so-called "spray polymerization" described by Frisque and Bernet is very interesting from the preparative point of view.

*Glycol methacrylate gels*

Glycol methacrylate gels were prepared in bulk by the copolymerization of glycol methacrylate and ethylene dimethacrylate by Kubín *et al.* The gel was mechanically disintegrated for chromatographic purposes. When hydrophilic methacrylate monomers are subjected to suspension copolymerization with a hydrophobic cross-linking agent, it is difficult to keep both comonomers in the polymerizing particle. By using suitable suspension stabilizers and in the presence of inert solvents, Čoupek *et al.* (1972a, b) prepared hydrophilic copolymers with perfectly spherical particles capable of separation over a very broad range of molecular weights (Table 9.8). Macroporous types of hydroxyalkyl methacrylate gels are characterized by a definite internal surface area. As the hydroxyl group is the hydrophilic group of the gel, non-specific sorptions on the gel are relatively very small. The hydrolytic stability of gels is excellent owing to its structure, which is analogous to that of pivalic acid esters. Gels resist the action of acidic and alkaline solutions, even at elevated temperatures. Owing to the macroporous structure, the mechanical and flow properties of gels are very good; packings of hydroxyalkyl methacrylate gels can be used even at elevated inlet pressures. They are marketed under the trade-name Spheron by Lachema, Brno, Czechoslovakia, and can be widely used in gel, liquid–liquid and gas chromatography. After modification with biologically active compounds, the gels proved to be excellent carriers in affinity chromatography and in enzyme catalysis.

TABLE 9.8
SPHERON HYDROXYALKYL METHACRYLATE GELS (PRODUCED BY LACHEMA, BRNO, CZECHOSLOVAKIA)

| Spheron | Molecular-weight exclusion limit, tested by dextran standards | Specific surface area ($m^2$/g) | Recommended solvents |
|---|---|---|---|
| P-1 | 1,000 | – | Aqueous solutions |
| P-40 | 40,000 | 142.3 | Aqueous solutions, organic solvents |
| P-100 | 100,000 | 1.29 | |
| P-200 | 200,000 | 3.09 | |
| P-300 | 300,000 | 77.7 | |
| P-500 | 500,000 | 131.5 | |
| P-700 | 700,000 | 15.28 | |
| P-1000 | 1,000,000 | 22.46 | |
| P-10,000 | 10,000,000 | 5.01 | |

*References p.227*

*Special gels*

Some hydrophilic gels have been prepared for special uses. Schott and Greber prepared derivatives by the reaction of guanosine, trimethylchlorosilane and methacryloyl chloride which, on polymerization, yield gels suitable for the separation of nucleic acids. Samsonov and Ponomareva investigated protein complexes using gels based on polyvinylpyrrolidone. Wolf and Lindemann suggested the use of porous ternary copolymers of styrene, divinylbenzene, ethyl acrylate or vinyl acetate for the chromatographic separation of water-soluble dyes. The third component (vinyl acetate or ethyl acrylate) imparts slightly hydrophilic properties to the polymer.

*Organophilic gels*

At first, the use of organophilic gel materials for chromatographic purposes aroused only minor attention. For the separation of hydrocarbon mixtures, Brewer (1960, 1961) used granulated rubber swollen in toluene or hexane. By using the common suspension polymerization of methyl methacrylate in the presence of ethylene dimethacrylate, Determann *et al.* (1964) prepared gels suitable for the separation of oligomers up to 2500 molecular-weight units. Determann *et al.* (1968a, b) also prepared, in the presence of inert components, macroporous copolymers. A systematic investigation of the effects of the type and concentration of inert components during the copolymerization of methyl methacrylate and ethylene dimethacrylate was carried out by Heitz and Winau, who succeeded in the preparation of macroporous packings with a molecular-weight exclusion limit up to $10^{10}$.

A basic contribution to the gel chromatography of organophilic polymer systems was made by Moore (1964), who used his experience with the preparation of macroreticular ion exchangers based on polystyrene. By using suitable precipitants in appropriate proportions and a suitable concentration of the cross-linking agent, he prepared a series of gels with molecular weights ranging from $10^3$ to $10^{10}$. The polystyrene gels exhibit excellent gel chromatographic properties, are very stable, both mechanically and chemically, and their thermal stability is also very satisfactory. In the form of spherical particles they are marketed by Waters Associates (Framingham, Mass., U.S.A.) under the trade-names Poragel and Styragel; Bio-Rad Labs. market them under the trade-name of Bio-Beads S (Table 9.9). Poragels are suitable for the separation of oligomers, while Styragels, depending on their molecular-weight exclusion limit, can be used for the separation of polymer systems. For chromatography in an aqueous medium, partially sulphonated cross-linked polystyrene gels known as Aquapak A-440 and Bio-Beads SM were used successfully. Berger and Mindner described the preparation of styrene–divinylbenzene copolymers by emulsion polymerization; Greber and Hausmann reported a method for preparation of polystyrene and poly-α-methylstyrene gels cross-linked with substituted chlorosilanes.

The gel chromatographic properties of vinyl acetate gels during their application in tetrahydrofuran were studied by Čoupek and Heitz with the aim of comparing them with polystyrene packings of the same porosity. A detailed investigation of the relationship between the properties of the copolymers of vinyl acetate and butanediol divinyl ether and divinyl adipate was carried out by Heitz and Platt. By common suspension copolymerization, they prepared homogeneous gels with a molecular-weight exclusion limit up

TABLE 9.9
POLYSTYRENE GELS

Poragel, Styragel and Aquapak produced by Waters Ass., Framingham, Mass., U.S.A.; Bio-Beads produced by Bio-Rad Labs., Richmond, Calif., U.S.A.

| Type | Molecular-weight exclusion limit (polystyrene standards) | Exclusion limit* (Å) | Recommended solvents |
|---|---|---|---|
| Poragel 27121 | 2,400 | | Organic solvents (tetrahydrofuran, chloroform, benzene, dimethylformamide, chlorobenzene, etc.), which are generally good swelling agents for cross-linked polystyrene |
| Poragel 27123 | 4,000 | | |
| Poragel 27990 | 8,000 | | |
| Poragel 27126 | 20,000 | | |
| Styragel 39720 | | 60 | |
| Styragel 39721 | | 100 | |
| Styragel 39722 | | 350 | |
| Styragel 39723 | | 700 | |
| Styragel 39724 | | 2,000 | |
| Styragel 39725 | | 5,000 | |
| Styragel 39726 | | 15,000 | |
| Styragel 39727 | | 50,000 | |
| Styragel 39728 | | 150,000 | |
| Styragel 39729 | | 700,000 | |
| Styragel 39730 | | 5,000,000 | |
| Styragel 39731 | | 10,000,000 | |
| Aquapak A-440 | 100,000 | | |
| Bio-Beads SX-1 | 3,500 | | Organic solvents |
| Bio-Beads SX-2 | 2,700 | | |
| Bio-Beads SX-3 | 2,100 | | |
| Bio-Beads SX-4 | 1,400 | | |
| Bio-Beads SX-8 | 1,000 | | |
| Bio-Beads SM-1 | Macroporous | | |
| Bio-Beads SM-2 | Macroporous | | |

*Exclusion limit given by the length of standard polystyrene chains in Ångstroms. An approximate molecular-weight exclusion limit is obtained by multiplication of these values by 41.

to 4000. If the polymerization was carried out in the presence of some organic solvents (octane, heptanol), vinyl acetate gels were obtained with molecular-weight exclusion limits up to $10^6$. The gels are produced by E. Merck (Darmstadt, G.F.R.) under the tradename Merckogel OR (Table 9.10).

By using suspension copolymerization of vinyl acetate and glycidyl methacrylate in an aqueous dispersion medium at 60–70°C, Motomato *et al.* prepared gels with molecular-weight exclusion limits between 300 and 80,000; these gels are stable against both acidic and alkaline hydrolysis.

Hydrophilic gels of the Sephadex type can be hydrophobized by appropriate modifying reactions. The free hydroxyl groups of the polysaccharide chains of Sephadex can be acylated or alkylated, thus yielding gel packings suitable for chromatography in non-aqueous systems. The esterification of Sephadex with acetic anhydride in benzene yielded

TABLE 9.10
MERCKOGEL POLY(VINYL ACETATE) GELS (PRODUCED BY E. MERCK, DARMSTADT, G.F.R.)

| Merckogel | Molecular-weight exclusion limit (styrene standards) | Recommended solvents |
|---|---|---|
| OR-PVA 500 | 500 | Organic solvents |
| OR-PVA 2000 | 2,000 | |
| OR-PVA 6000 | 6,000 | |
| OR-PVA 80,000 | 80,000 | |
| OR-PVA 200,000 | 200,000 | |
| OR-PVA 300,000 | 300,000 | |
| OR-PVA 1,000,000 | 1,000,000 | |

a gel in which *ca.* 50% of the hydroxyl groups were transformed. Determann used this gel successfully for the separation of polystyrene oligomers up to a molecular weight of *ca.* 10,000. If the alkylation reaction is carried out in water, where the gel is swollen, and consequently the hydroxyl groups are more accessible to low-molecular-weight reagents, the number of modified hydroxyl groups is even higher. Nyström and Sjövall demonstrated that the modification of Sephadex G-25 with dimethyl sulphate in a solution of sodium hydroxide gives rise to a gel that is eminently suitable for gel chromatography in organic solvents. Also, reactions of Sephadex G-25 with aliphatic isocyanates in dimethyl sulphoxide lead to gel packings that swell well in organic solvents (Heitz *et al.*).

A number of workers have tried to prepare gels that would have a universal character and could be used for chromatography in a wide range of polar and non-polar solvents. However, when an inappropriate solvent is used, such universal gels often exhibit strong interactions with dissolved compounds, which make them useless as gel chromatographic packings. On the other hand, the adsorption effects can be used in separations in liquid–solid chromatography. The best known gel, swelling both in aqueous solutions and in organic solvents, is Sephadex LH-20, produced by Pharmacia, which was found to have excellent qualities in many applications. Increased temperature reduces the swelling of the gel, and Determann and Lambert interpreted this phenomenon by a change in the internal structure of the gel at an elevated temperature by assuming that the polymeric chains approach each other while at the same time the number of small pores is reduced, and the number of the large pores increases. This effect can be observed only for homogeneous gels; as was demonstrated by Moore and Hendrickson, the porosity of macroporous gels does not change much with temperature.

Another type of universal packing is the Spheron hydroxyalkyl methacrylate gels, the polarity of which can be widely varied, either by varying the proportion of the organophilic or hydrophilic cross-linking agent, or by ternary copolymerization with other more or less polar methacrylate, acrylate or styrene co-monomers. In this way, materials are obtained that can be used in virtually all branches of chromatography.

Heitz and Pfitzner prepared universal gels by the suspension polymerization of oligo(ethylene glycol dimethacrylates) and poly(ethylene glycol dimethacrylates) in the presence of inert compounds and obtained gels with separation efficiencies within a broad range of molecular weights in aqueous systems and organic solvents (tetrahydrofuran). Merckogels thus prepared can be used in an aqueous medium for the separation of oligopeptides or poly(ethylene glycol) (Randau *et al.*).

*Rigid gels*

Rigid gels are inorganic aerogels which, owing to their numerous advantages, are regarded by many workers as the greatest achievement to emanate from research on gel chromatographic materials. The gels do not swell and their rigid structure ensures very easy packing of columns and good flow properties. The constant bed volume of the column, regardless of temperature or type of eluent, and the possibility of the application of a high pressure at the column inlet, provide for an easy application of the gradient elution technique even in high-speed, high-resolution chromatography. However, the presence of active sites on the surface of the inorganic aerogel may lead to some undesirable sorption effects; therefore, a number of deactivation procedures have been developed involving esterification or silanization of active hydroxyl groups present on the surface of porous materials. An advantage of rigid inorganic gels is the possibility of their regeneration by heating them or by treating them with hot mineral acids.

The most important inorganic packings used in gel chromatography are porous glasses, which can be obtained from borosilicate glasses by heating at the temperature of nucleation followed by etching with acidic and alkaline agents. The preparation of silicate glasses with strictly controlled porosity was described by Haller, who also developed treatments leading to the formation of a micro-heterogeneous phase with defined properties. Under the trade-name of Bio-Glass, porous glasses were introduced by Bio-Rad Labs.; under the name of CPG (Controlled Pore Glass), they are marketed by Corning Glass

TABLE 9.11
POROUS GLASS PACKING MATERIALS

| Type | Molecular-weight exclusion limit | | Recommended solvents | Producers |
|---|---|---|---|---|
| | Dextran standards | Polystyrene standards | | |
| Bio-Glass 200 | | 30,000 | Polar and non-polar solvents, except HF and strong bases | Bio-Rad Labs., Richmond, Calif., U.S.A. |
| Bio-Glass 500 | | 100,000 | | |
| Bio-Glass 1000 | | 500,000 | | |
| Bio-Glass 1500 | | 2,000,000 | | |
| Bio-Glass 2500 | | 9,000,000 | | |
| CPG 10-75 | 28,000 | | Polar solvents (tetrahydrofuran, dimethylformamide, etc.) | Waters Ass., Framingham, Mass., U.S.A.; Corning Glass Works, Corning, N.Y., U.S.A. |
| CPG 10-125 | 48,000 | | | |
| CPG 10-175 | 68,000 | | | |
| CPG 10-240 | 95,000 | 120,000 | | |
| CPG 10-370 | 150,000 | 400,000 | | |
| CPG 10-700 | 300,000 | 1,200,000 | | |
| CPG 10-1250 | 550,000 | 4,000,000 | | |
| CPG 10-2000 | 1,200,000 | 12,000,000 | | |

TABLE 9.12
POROUS SILICA PACKING MATERIALS

| Type | Molecular-weight exclusion limit (polystyrene standards) | Recommended solvents | Producer |
|---|---|---|---|
| Porasil 60 | 60,000 | Organic solvents | Waters Ass., Framingham, Mass., U.S.A. |
| Porasil 250 | 250,000 | | |
| Porasil 400 | 400,000 | | |
| Porasil 1000 | 1,000,000 | | |
| Porasil 1500 | 1,500,000 | | |
| Porasil 2000 | 2,000,000 | | |
| Merckogel SI-150 | 50,000 | Organic solvents | E. Merck, Darmstadt, G.F.R. |
| Merckogel SI-500 | 400,000 | | |
| Merckogel SI-1000 | 1,000,000 | | |

Works (Corning, N.Y., U.S.A.) and Waters Ass. (Table 9.11). Polar solvents are recommended as eluents for porous glasses in gel chromatography. If non-polar solvents have to be used, it is advisable to add a trace amount of a polar solvent to the eluent; the solvent should be able to block the active hydroxyl groups on the glass surface. The surface can also be treated by silanization, for example (Cooper and Johnson).

The properties and uses of porous silica beads are similar to those of porous glass. They were introduced by De Vries *et al.* and are supplied by Waters Ass. under the trade-name Porasil (Table 9.12). According to Le Page, a porous silica gel with a defined internal structure is formed by treating silica gel with ammonia in an autoclave at about 500°C. Other types commercially available are Spherosil and Merckogel SI. The porous silica gel must be deactivated, in a similar manner to porous glass. Aue and Hastings recommended treatment with octadecyltrichlorosilane, which forms thermostable and non-extractable coatings on the gel surface.

In recent years, porous aluminas have also been used in the chromatography and molecular-weight determinations of starch and proteins (Sato and Otaka).

If a strong alkaline solvent for cellulose is employed, namely Cadoxen, when the use of porous glass is not recommended, Eriksson *et al.* suggested the use of illites. Plachenov *et al.* prepared a microporous graphite with gel chromatographic properties. Alumina and silica modified with poly(vinyl alcohol) were used by Hadeball and Seide for the fractionation of high-molecular-weight systems.

In most instances, gel chromatographic materials were used as basic carriers for a wide range of further applications in gas, liquid and affinity chromatography, and after modification also in ion-exchange chromatography.

## ION-EXCHANGE MATERIALS

### Inorganic ion exchangers

Inorganic ion exchangers have a long history. In the middle of the 19th century, Way (1850, 1852) and Thompson studied the sorption of ions in clays and discovered the fundamental laws of ion exchange. Way (1850, 1852) found that the substances responsible for the exchange of ions in soils were complex aluminium silicates, and it is now known that many of them can exchange bound alkali metals. For example, zeolites (analcite, chabazite, harmoton, heulandite and natrolite) have exchangeable cations in their crystal lattice, in beidellite and montmorillonite the exchangeable ions lie between the layers of the lattice, whereas the compact glauconites exchange only potassium ions on their surface. Most of these minerals, including hydrated mica-like substances, exchange cations, and some of them also have anion-exchange properties (*cf.*, Table 9.13). At the beginning of the 20th century, some of these natural inorganic ion exchangers were used for industrial purposes in water softening and in decalcification of beet sugar juice, but they were gradually replaced by synthetic inorganic ion exchangers.

Synthetic zeolites were first made by fusing a mixture of soda, potash, quartz and kaolin, and later by precipitating a suitable mixture of aluminium sulphate and sodium

TABLE 9.13
SURVEY OF IMPORTANT ION-EXCHANGE MINERALS

| Mineral | Formula* |
|---|---|
| *Minerals exchanging cations* | |
| Analcite | $Na[AlSi_2O_6]\cdot H_2O$ |
| Apophyllite | $K\,Ca_4[F/(Si_4O_{10})_2]\cdot 8\,H_2O$ |
| Beidellite | $Al_2[(OH)_2/Al_{0.5}Si_{3.5}O_{10}]^{0.5}(Ca, Na)_{0.3}(H_2O)_4$ |
| Chabazite | $(Ca, Na)[Al_2Si_4O_{12}]\cdot 6\,H_2O$ |
| Glauconite | $(K, Na, Ca)_{<1}(Al, Fe^{3+}, Fe^{2+}, Mg)_2[(OH)_2/Al_{0.35}Si_{3.65}O_{10}]$ |
| Harmotome | $Ba[Al_2Si_6O_{16}]\cdot 6\,H_2O$ |
| Heulandite | $Ca[Al_2Si_7O_{18}]\cdot 6\,H_2O$ |
| Natrolith | $Na_2[Al_2Si_3O_{10}]\cdot 2\,H_2O$ |
| Montmorillonite | $(Al_{1.67}Mg_{0.33})[(OH)_2/Si_4O_{10}]^{0.33-}Na_{0.33}(H_2O)_4$ |
| Ultramarine, lasurite | $(Na, Ca)_8[(SO_4, S, Cl)_2/(AlSiO_4)_6]$ |
| *Minerals that display anion-exchange properties* | |
| Apatite | $Ca_5[F/(PO_4)_3]$ |
| Hydroxyapatite | $Ca_5[(OH)/(PO_4)_3]$ |
| Cancrinite | $Na_6Ca[(CO_3)/(AlSiO_4)_6]\cdot 2\,H_2O$ |
| Kaolinite | $Al_4[(OH)_8/Si_4O_{10}]$ |
| Montmorillonite | $(Al_{1.67}Mg_{0.33})[(OH)_2/Si_4O_{10}]^{0.33-}Na_{0.33}(H_2O)_4$ |
| Sodalite | $Na_8[Cl_2/(AlSiO_4)_6]$ |
| Ultramarine, lasurite | $(Na, Ca)_8[(SO_4, S, Cl)_2/(AlSiO_4)_6]$ |

*Formulae according to Strunz.

*References p.227*

silicate by sodium hydroxide. The gel was then dried carefully and ground. The porous amorphous Permutits so prepared had an elementary composition that was in agreement with the hypothetical formula with exchangeable sodium ions bound through oxygen atoms to aluminium and not to silicon:

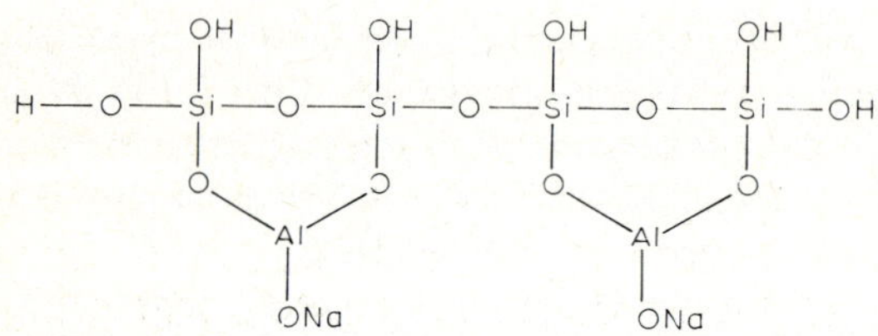

These preparations are of only historical interest.

Zeolites are now prepared by hydrothermal syntheses. These crystalline preparations are more often used as molecular sieves than as ion exchangers. The major disadvantage of both natural and synthetic zeolites is their limited stability, they can operate only in neutral solution and they are decomposed by acids and alkalis.

In addition to hydrated oxides of tetravalent metals (zirconium, thorium, titanium and tin), a special group of modern inorganic ion exchangers are the phosphates, molybdates, tungstates and vanadates of tetravalent metals, especially of zirconium. Zirconium phosphate behaves in a similar manner to cross-linked phosphoric acid and can be regarded as a polymer:

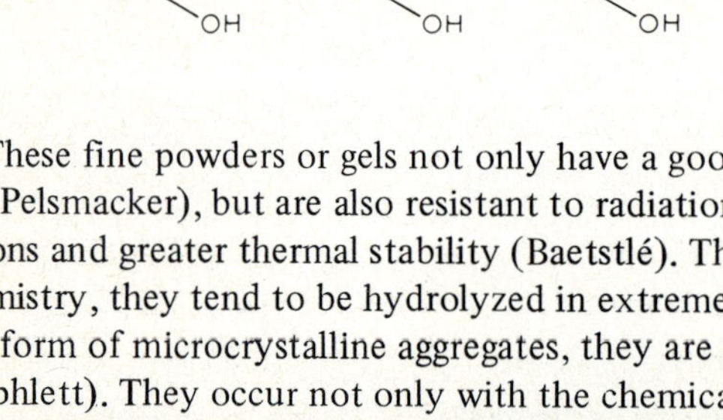

These fine powders or gels not only have a good capacity and rapid exchange (Baetstlé and Pelsmacker), but are also resistant to radiation, have greater selectivity for some types of ions and greater thermal stability (Baetstlé). They are used for separations in inorganic chemistry, they tend to be hydrolyzed in extremely alkaline media. Because they consist of a form of microcrystalline aggregates, they are also called ion exchange crystals (*cf.*, Amphlett). They occur not only with the chemical composition of zirconium phosphate, but also as hydrated zirconium oxide or ammonium molybdophosphate (*cf.*, Table 13.8, p. 351). These preparations have exchange capacities similar to those of conventional resins and have many specialized applications.

**Organic ion exchangers**

Various natural organic substances behave as ion exchangers. Proteins containing both basic and acidic groups on their surfaces behave as amphoteric ion exchangers. Some aliphatic alcohol groups of cellulose are easily oxidized to carboxyl groups and thus cellulose often behaves as a cation exchanger with low capacity. Humic acids in soil act as ion exchangers with good capacity. Certain living organisms exchange ions (*e.g.*, parts of some plants and algae producing alginic acids). Peat and certain types of ground coal exchange cations to a certain degree owing to the presence of a small amount of carboxylic and phenolic groups. Forty years ago it was discovered that the exchange ability of coal is increased substantially if sulphonic groups are introduced into coal by sulphonation. This material was the first that was able to exchange hydrogen ions in addition to other cations, and represented the first step towards the possibility of demineralization by ion exchange.

The classical experiments of Adams and Holmes led to the development of synthetic insoluble organic ion-exchange resins. They showed that resins prepared by condensing formaldehyde with polyhydric phenols (later with phenolsulphonic acids) yield cation exchangers and anion exchangers were prepared by condensation with amines. The latter resins in the free base form were able to bind acids from solutions and thus the demineralization of water was achieved. Resins prepared by the polycondensation of phenols now have little importance in chromatography because they are not homoionic.

From the point of view of the chemical composition of the matrices, the organic ion exchangers used at present for chromatography can be divided into three main classes:

(a) ion-exchange resins prepared by polymerization or polycondensation;
(b) cellulose ion exchangers;
(c) ion-exchange derivatives of polydextran.

*Ion-exchange resins*

The matrices of the most suitable resins for many chromatographic purposes are based on polystyrene or polyacrylic acid. Styrene polymerizes to give a linear polymer:

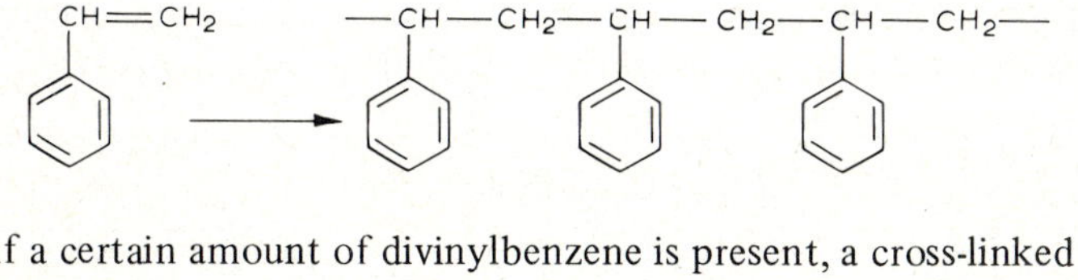

but if a certain amount of divinylbenzene is present, a cross-linked copolymer is obtained:

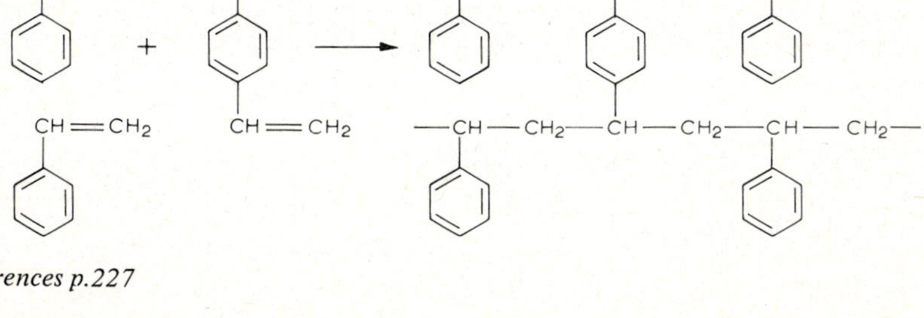

The amount of divinylbenzene controls the degree of cross-linking. The percentage of divinylbenzene in the starting mixture is expressed by the X value of the prepared ion exchanger (*cf.*, Chapter 6). Both components are copolymerized in bead form and the functional groups are introduced by subsequent treatment. Sulphonation produces a strongly acidic ion exchanger (cation exchanger). Strongly basic anion exchangers are prepared by chloromethylation and treatment with tertiary amines. Various functional groups can also be introduced into these copolymers.

In other instances, the functional groups are incorporated in the form of a monomer. A weakly acidic cation exchanger is prepared by copolymerizing methyl methacrylate with a cross-linking agent and hydrolyzing the product, *e.g.*:

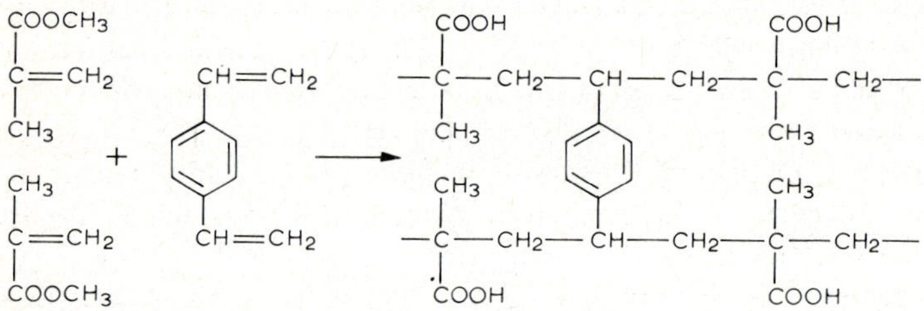

Another carboxylic ion exchanger is manufactured by copolymerizing acrylic acid with N,N′-methylenebisacrylamide:

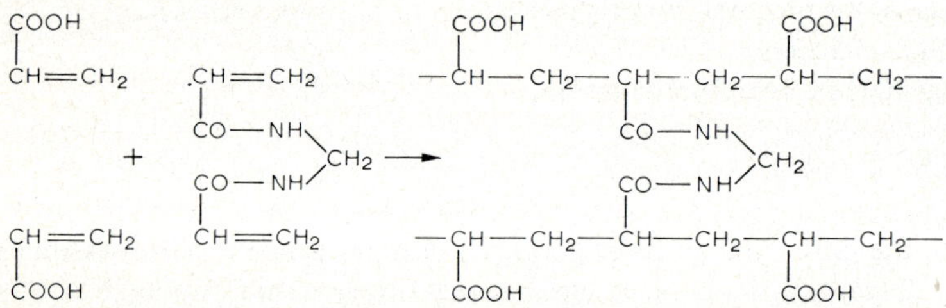

Some ion-exchange resins are prepared by condensation reactions. Heteroionic cation exchangers can be prepared by condensing formaldehyde with phenolsulphonic acid:

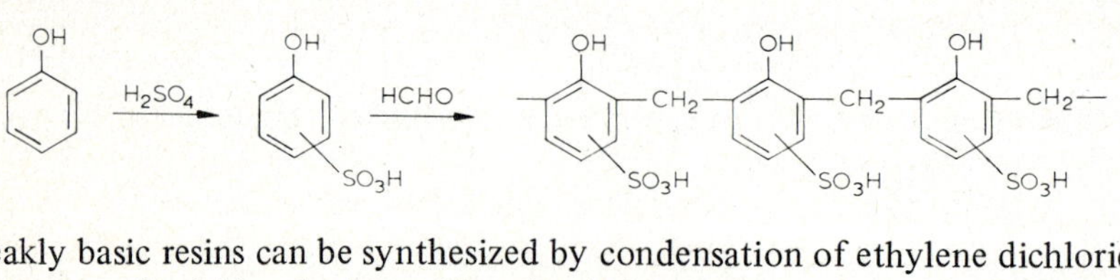

Weakly basic resins can be synthesized by condensation of ethylene dichloride with polyamines or of epichlorohydrin with polyethyleneimine:

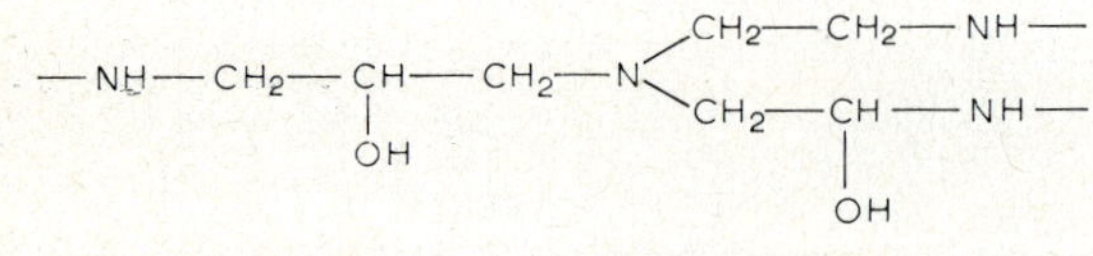

An extensive survey of various methods of preparing synthetic organic ion exchangers was given by Helfferich.

*Cellulose ion exchangers*

Cellulose fibres were found to be excellent carriers for ionogenic functional groups for the ion-exchange chromatography of proteins, nucleic acids and their high-molecular-weight fragments. Their fibrous nature makes the functional groups accessible to macromolecules from the solution and their hydrophilic properties do not cause undesirable denaturing hydrophobic interactions that occur with other ion exchangers containing aromatic matrices.

Native cellulose is a polycondensate of anhydroglucose monomers and is composed of thousands of cellobiose units:

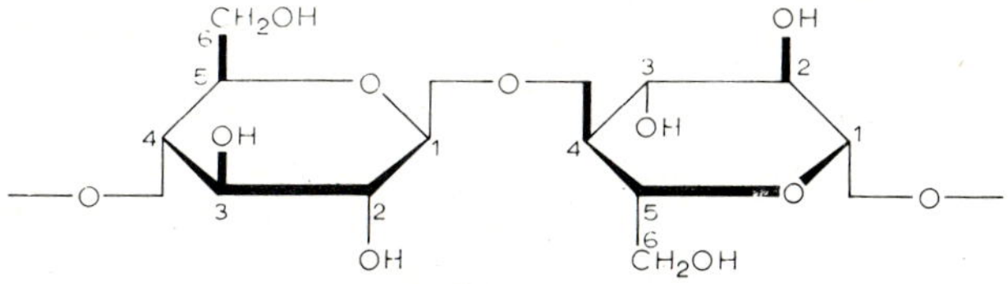

The cellulose chains associate in fibres to form "crystalline" areas (micelles). The free hydroxyl groups in positions 2 and 6 (and sometimes also in position 3) can be easily substituted. Guthrie introduced ionogenic groups into cellulose, but these derivatives were not used in chromatography until Peterson and Sober, Porath, Sober *et al.* and Sober and Peterson described methods by which cellulose ion exchangers can easily be prepared in the laboratory and showed the importance of these materials in the chromatography of biopolymers. Cellulose swollen in alkali reacts with chlorine derivatives, which can be illustrated by the synthesis of carboxymethylcellulose from cellulose and chloroacetic acid:

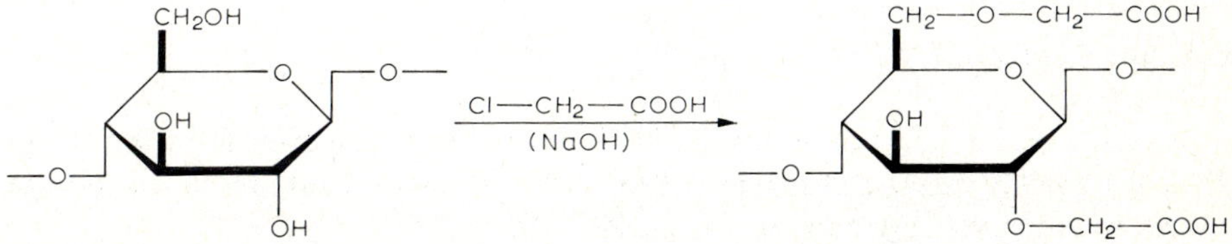

These ion exchangers are now commercially available in various forms (*cf.*, Table 13.6, p. 346/347).

*Ion-exchange derivatives of polydextran*

Ion-exchange derivatives of polydextran have favourable properties for the chromatography of biopolymers similar to those mentioned for cellulose derivatives. Dextran is formed from sugar solutions by *Leuconostoc mesenteroides* and, after copolymerization with epichlorohydrin, cross-links are formed and the product represents a suitable carrier for ionogenic groups (Porath and Lindner). The structure of cross-linked polydextran is

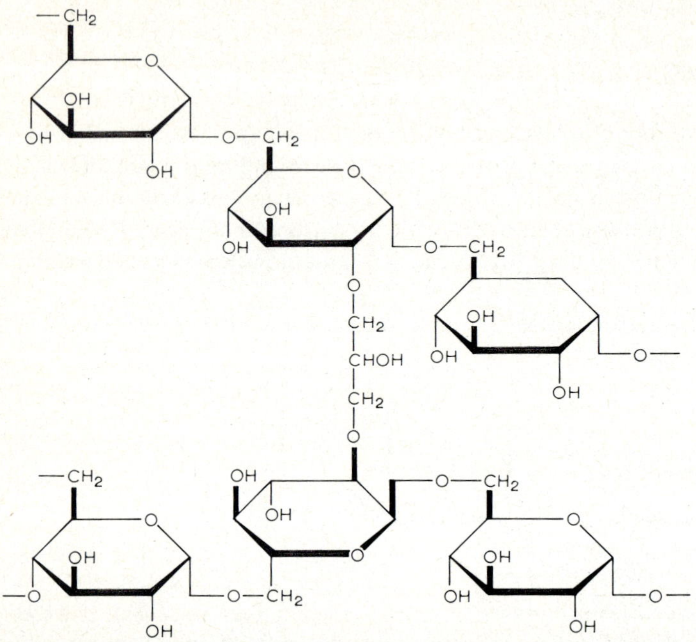

The free hydroxyl groups are available for substitution in the same way as with cellulose. In alkaline media, they react with suitable halogen derivatives to produce ion exchangers.

These ion exchangers are commercially available as Sephadex®. They are produced with two porosities corresponding to Sephadex G-25 and G-50 (*cf.*, Table 13.7, p. 348/349) and contain all necessary ionogenic groups. With respect to the porosity of the matrix, ion-exchange chromatography of biopolymers on these derivatives can be influenced by the molecular sieving process.

## Functional groups of ion exchangers

A survey of the most important functional groups in cation exchangers is given in Table 9.14 and in anion exchangers in Table 9.15. In ion-exchange resins, the ionogenic groups are often attached to aromatic nuclei of the matrix:

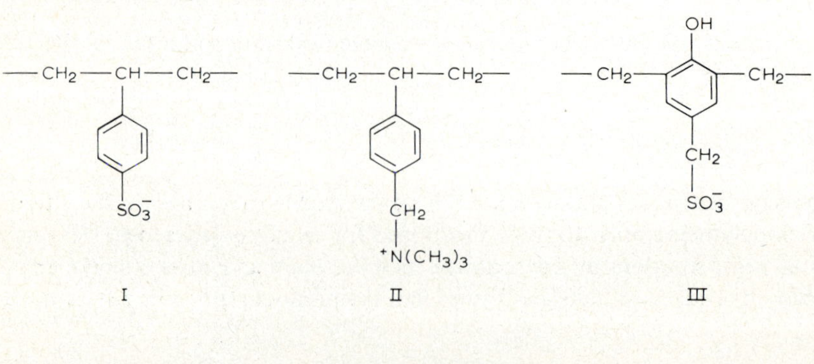

I II III

(I = cation exchanger; sulphonated polystyrene; II = anion exchanger; chloromethylated polystyrene treated with a tertiary amine; III = cation exchanger; sulphonated phenolic resin). The functional groups can also be incorporated into the lattice in the form of aliphatic monomers.

In ion-exchange derivatives of cellulose and polydextran, the functional groups are bound through the oxygen atoms of the hydroxyl groups.

TABLE 9.14
FUNCTIONAL GROUPS IN ORGANIC CATION EXCHANGERS

| Group | Formula | Type | Carriers |
|---|---|---|---|
| Sulphonic | $-SO_3^-$ | Strongly acidic | Aromatic nuclei in polystyrene or phenolic resins |
| Sulphomethyl | $-CH_2-SO_3^-$ | Strongly acidic | Aromatic nuclei in phenolic resins |
| Sulphoethyl (SE) | $-C_2H_4-SO_3^-$ | Strongly acidic | Hydroxyl groups in cellulose or polydextran; aromatic nuclei in a phenolic lattice |
| Sulphopropyl (SP) | $-C_3H_6-SO_3^-$ | Strongly acidic | Hydroxyl groups in polydextran |
| Phosphonic (P) | $-PO_3^{2-}$ | Moderately acidic | Aromatic nuclei in resins; hydroxyl groups in cellulose or polydextran |
| Carboxyl | $-COO^-$ | Weakly acidic | Polyacrylic resins or polyacrylic gels |
| Carboxymethyl (CM) | $-CH_2-COO^-$ | Weakly acidic | Hydroxyl groups in cellulose or polydextran; phenolic groups in resins |

TABLE 9.15
FUNCTIONAL GROUPS IN ORGANIC ANION EXCHANGERS

| Group | Formula | Type | Carriers |
|---|---|---|---|
| Trimethylamino-methyl | $-CH_2\overset{+}{N}(CH_3)_3$ | Strongly basic (type I) | Aromatic nuclei in resins |
| Hydroxyethyl-dimethylamino-methyl | $-CH_2\overset{+}{N}(CH_3)_2C_2H_4OH$ | Strongly basic (type II) | Aromatic nuclei in resins |
| Triethylamino-ethyl (TEAE) | $-C_2H_4\overset{+}{N}(C_2H_5)_3$ | Strongly basic | Hydroxyl groups in cellulose |
| Quaternary aminoethyl (QAE) | $-C_2H_4\overset{+}{N}(C_2H_5)_2CH_2CH(OH)CH_3$ | Strongly basic | Hydroxyl groups in polydextran |
| Guanidinoethyl (GE) | $-C_2H_4NHC(\overset{+}{N}H_3)(=\overset{+}{N}H_2)$ | Strongly basic | Hydroxyl groups in cellulose |
| Methylpyridinium | $-C_5H_5\overset{+}{N}CH_3$ | Strongly basic | Polystyrene lattice |
| Diethylamino-ethyl (DEAE) | $-C_2H_5\overset{+}{N}H(C_2H_5)_2$ | Intermediate basic | Hydroxyl groups in cellulose and polydextran |

*Continued on p. 210*

*References p.227*

TABLE 9.15 (*continued*)

| Group | Formula | Type | Carriers |
|---|---|---|---|
| Mixed amino (ECTEOLA) | Undefined | Intermediate basic (mixture of weakly, moderately and strongly basic groups) | Hydroxyl groups in cellulose |
| Mixed amino | $\left\{ \begin{array}{l} -\overset{+}{N}H(CH_3)_2 \\ -\overset{+}{N}(CH_3)_2C_2H_4OH \end{array} \right.$ | Intermediate basic (mixture of tertiary and quaternary ammonium groups) | Epoxypolyamine lattice |
| Aminoethyl (AE) | $-C_2H_4\overset{+}{N}H_3$ | Intermediate basic | Hydroxyl groups in cellulose |
| Polyethyleneimine (PEI) | $-(C_2H_4\overset{+}{N}H_2)_nC_2H_4\overset{+}{N}H_3$ | Intermediate basic | Hydroxyl groups in cellulose |
| Alkylamino | $-\overset{+}{N}HR_2$ | Weakly basic | Aromatic nuclei in polystyrene or phenolic polyamine resin |
| *p*-Aminobenzyl (PAB) | $-CH_2C_6H_4\overset{+}{N}H_3$ | Weakly basic | Hydroxyl groups in cellulose |

TABLE 9.16
FUNCTIONAL GROUPS IN SPECIAL RESINS

| Group | Formula | Type | Carriers |
|---|---|---|---|
| Trimethylammoniummethyl and carboxyl | $\left\{ \begin{array}{l} -CH_2\overset{+}{N}(CH_3)_3 \\ -COO^- \end{array} \right.$ | Amphoteric resin | Aromatic nuclei in polystyrene and polyacrylic resin |
| β- Alanine | $>C=\overset{+}{N}H-(CH_2)_2-COO^-$ | Dipolar ion exchanger | Hydroxyl groups in polydextran |
| Bis(carboxymethyl)iminomethyl | $-CH_2N\begin{array}{l} \diagup CH_2COOH \\ \diagdown CH_2COOH \end{array}$ | Chelating resin | Aromatic nuclei in resins |
| Hydroquinoyl | $-C_6H_3O_2$ | Electron-exchange resins (redox resins) | Vinyl group of vinylhydroquinone–divinylbenzene copolymer |
| Quinine | $-O(C_{20}H_{23}N_2O)$ | Resin for resolution of enantiomers | Carbonyl group of carboxylic resin |

## Special exchangers

In addition to cation and anion exchangers, special types of ion exchangers have been developed (Table 9.16).

Amphoteric resins are prepared by saturating a strongly basic anion exchanger with acrylic acid and polymerizing the latter (Retardation IIA8 is manufactured in this way from Dowex 1). Another type is prepared by saturating a strongly acidic cation exchanger with vinylpyridine followed by polymerization. In both instances the linear polymer formed from the additive monomers is trapped in the network of the original resin and cannot diffuse out (Hale *et al.*, Hatch *et al.*). Owing to the marked shape of the polyelectrolyte, these types of exchangers are also called "snake cage resins". The introduced ionogenic groups with opposite charge are oriented to the vicinity of the functional groups originally present and are capable of interacting with them. These ion exchangers must be distinguished from mixed bed resins.

The newest form of amphoteric ion exchangers are dipolar ion exchangers (Porath and Fornstedt, Porath and Fryklund). These exchangers are prepared by covalent binding of dipolar monomers (*e. g.*, amino acids) to a suitable carrier (*e. g.*, polydextran). They are used advantageously for the separation of biopolymers.

Mixed bed resins are mixtures of beads of common cation exchangers in the $H^+$ form and anion exchangers in the $OH^-$ form and are used for the demineralization of water and aqueous solutions in a single column (a single bed). After being exhausted by desalting, the individual exchangers must be separated for regeneration, which is carried out separately, then the ion exchangers are mixed again.

Chelating ion exchangers of the iminodiacetic acid type have been prepared (Pepper and Hale) from chloromethylated cross-linked polystyrene:

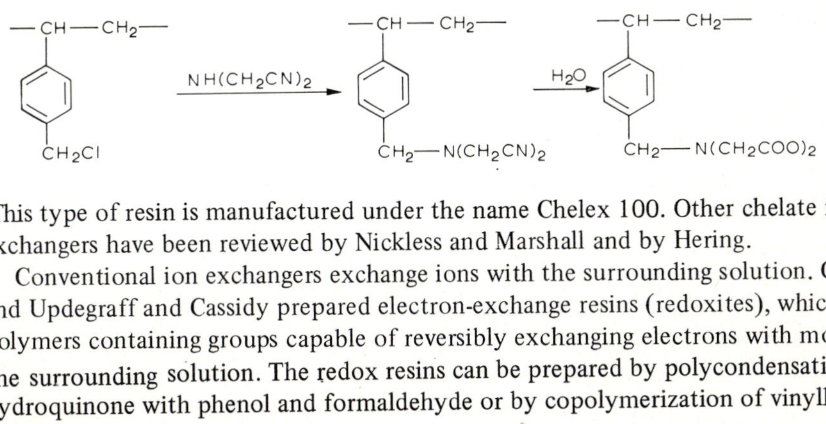


This type of resin is manufactured under the name Chelex 100. Other chelate ion exchangers have been reviewed by Nickless and Marshall and by Hering.

Conventional ion exchangers exchange ions with the surrounding solution. Cassidy, and Updegraff and Cassidy prepared electron-exchange resins (redoxites), which are polymers containing groups capable of reversibly exchanging electrons with molecules in the surrounding solution. The redox resins can be prepared by polycondensation of hydroquinone with phenol and formaldehyde or by copolymerization of vinylhydroquinone with divinylbenzene:

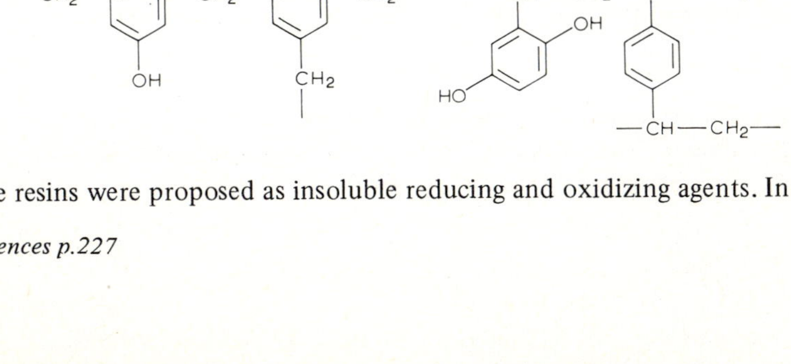


These resins were proposed as insoluble reducing and oxidizing agents. In spite of the large number of possibilities for utilizing them, they have not been used extensively in practice.

Many unsuccessful attempts have been made to copolymerize optically active monomers into various resins in order to prepare materials that are capable of separating optical isomers from racemic mixtures. Grubhofer and Schleith prepared a carbonyl chloride derivative from a monofunctional carboxylic resin by treatment with thionyl chloride. This resin, containing –COCl groups, then underwent a reaction with quinine via esterification of its secondary alcohol group so that the asymmetric carbon atom in the quinine molecule was retained:

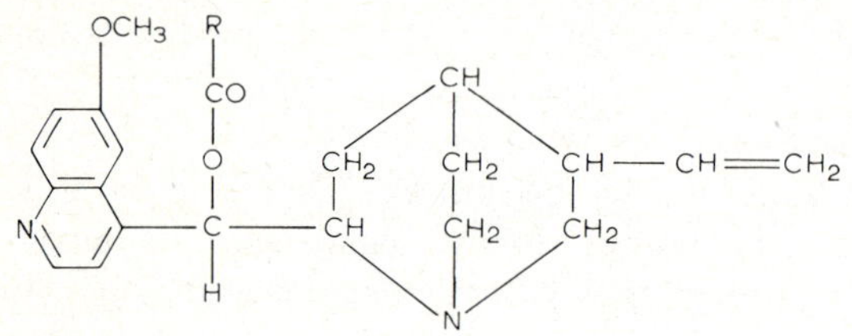


(R = resin). This resin was used successfully for the column separation of racemic mandelic acid.

### Form of ion exchangers

Ion exchangers are usually solid, but liquid ion exchangers have been developed for special purposes.

Solid exchangers can be prepared in the form of membranes, tubes and fibres for various ion-exchange applications, especially for continuous processes and electrophoresis. For column chromatography, the granular form of ion exchangers is the most usual. Ion exchangers are supplied commercially for these purposes as ground grains or, more usually, as spherical beads. The difference between them can be seen in Fig. 9.10.

Grains prepared by grinding the dried gel have an irregular shape. In comparison with beads, they offer greater losses caused by friction, less mechanical resistance and greater flow resistance of the column. The bead form of ion exchangers has a regular spherical shape and is obtained by polymerization in suspension. The liquid monomer mixture is stirred with water in such a way that stable drops of reacting mixture are formed in the form of regular spheroids, which are converted into solid spheres of the same size by polymerization (Groggins). The functional groups are introduced into the spheres by subsequent treatment, which does not destroy their form. The washed products are termed ion-exchange beads. Regular small beads can also be prepared by dropping the reaction mixture from capillary tubes of suitable dimensions. The bead form of ion exchangers, with its regular dimensions, is important in modern column chromatography, because the lower flow resistance of the beads in columns enables the particle size to be diminished and the greater mechanical stability permits the use of higher pressures, which results in a substantial increase in the speed of the chromatographic process.

A special porous form of ion exchanger with large pores has been developed that

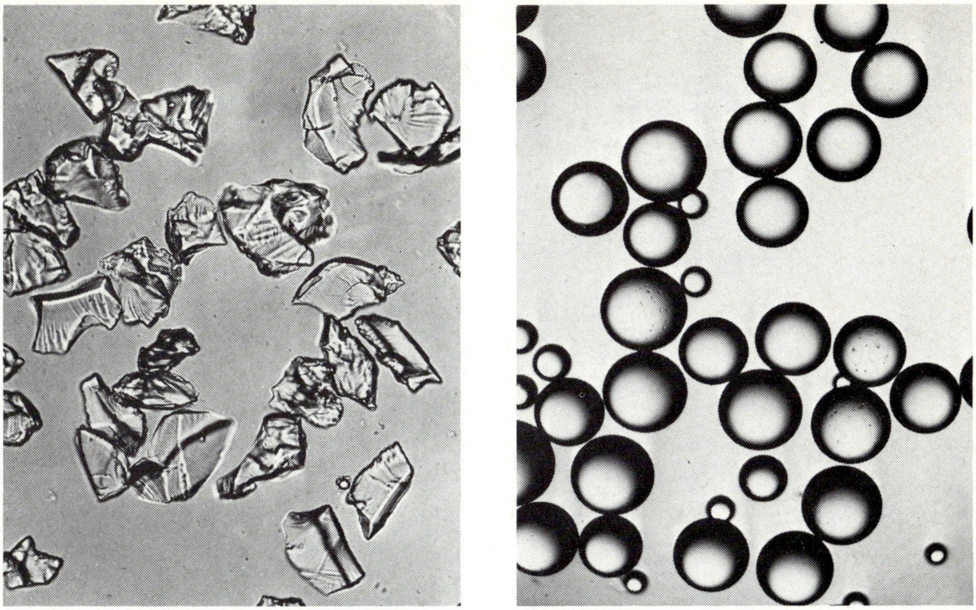

(A) (B)

Fig. 9.10. Microphotographs of granular ion exchangers. (A) Grains prepared by grinding of the gel; (B) beads prepared by submersion polycondesation in emulsion.

contains two types of cross-linkages (*cf.* Samuelson). Initial cross-linking is obtained in the normal way by copolymerization of styrene with small amounts of divinylbenzene in the bead. Additional cross-linking is achieved by Friedel-Crafts catalyzed methylene bridging in the swollen state. Dowex 21K is an example of this type of resin. It has volume changes and a capacity comparable with those of X8 resin but otherwise it has many of the desirable properties of X4 resin, especially a higher permeability and a faster equilibration rate.

A special modern form of ion exchanger are the so-called macroreticular resins (Kun and Kunin; Kunin *et al.*, 1962a, b). The beads (Fig. 9.11) are aggregates consisting of tiny granules (with diameters of a few hundred Ångstroms) with large pores between them (diameter up to 1000 Å). They are prepared by suspension copolymerization in the presence of a good solvent for monomers (styrene and divinylbenzene), which is simultaneously a poor swelling agent for the polymer. Macroreticular resins have a large internal surface area (several tens of $m^2/g$). These ion exchangers allow the penetration of large molecules and are also able to operate in non-aqueous solutions. This type of resin was introduced commercially under the name Amberlyst.

For high-speed analytical chromatography, pellicular ion-exchange resins have been developed (*cf.* Kirkland, 1971). This type of ion-exchange bead consists of an inert solid core that displays no ion-exchange properties, covered with a thin shell of ion-exchange resin (Fig. 9.11). Diffusion into the very thin ion-exchange film lasts only a few seconds and equilibrium is reached very quickly in comparison with the usual beads. Therefore, this type of ion exchanger exhibits a much higher chromatographic efficiency than conven-

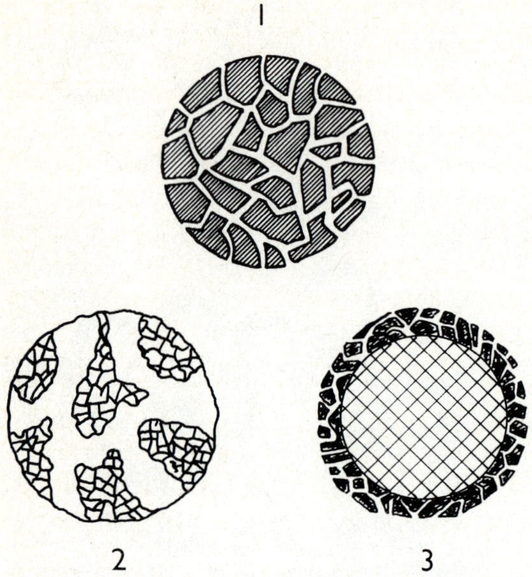


Fig. 9.11. Schematic representation of the inner structure of three types of ion-exchange beads. (1) Microreticular resins are composed of a gel containing micropores defined by the X value. (2) Macroreticular resin beads contain macropores (several hundred Ångstroms wide) with a large inner surface area. The structural part is composed of a highly cross-linked gel with narrow micropores. (3) Pellicular resins (porous layer beads) contain a solid inert core surrounded by a shell (*ca.* 1 $\mu$m thick film) composed of cross-linked microporous ion-exchange gel.

tional ion-exchange resins. Kirkland (1969, 1970) published the following relative data: cation exchange of 5′-uridine monophosphoric acid 3.5 and anion exchange of 2-aminobenzimidazole 8 theoretical plates per second; conventional gel ion exchangers often give results of less than 0.1–0.5 theoretical plates per second. The carrier velocity has high values, *e.g.*, 2–3 cm/sec in columns of 2–3 mm I.D.

The non-porous impervious spherical silica support particles (3–40 $\mu$m) are uniformly coated with a thin (1 $\mu$m) porous crust of adsorbent and these types of beads are generally called controlled surface porosity ion exchangers*. The strongly acidic cation-exchange shell consists of a very stable fluoropolymer containing sulphonic acid groups with a capacity of 3.5 $\mu$equiv./g of $Zn^{2+}$. The strongly basic anion exchanger has a capacity of 12 $\mu$equiv./g and contains tetraalkylamonium groups.

These porous layer beads of ion exchangers are suitable for use in high-speed chromatography. The dry packing technique can be used for preparing the columns, because these ion exchangers undergo no detectable changes due to swelling when the pH and/or ionic strength or buffer composition are varied. The dimensions of the columns used vary, with ratios of diameter to length from 1:10 to very high values, and stainless-steel columns (straight or coiled) are used with lengths up to 3 m and I.D.s of several milli-

*E. I. du Pont de Nemours & Co. (Wilmington, Del., U.S.A.) uses the trade-name Zipax for controlled surface porosity supports.

metres (Horváth and Lipsky). The mechanical stability of the beads allows a column input pressure of up to 5000 p.s.i. to be used. The separation is accelerated by increasing the temperature.

The disadvantage of this type of ion exchanger is the low capacity. Owing to the lower retention of substances, the ionic strength usually has to be reduced in comparison with the resolution of similar samples using conventional ion exchangers.

The possibility of increasing the efficiency of high-speed chromatography when only a thin sorbing porous layer will be active on the surface of the beads has been considered theoretically by various workers (*e.g.*, Bohemen and Purnell, Golay, Knox, Knox and McLaren, Parrish, Weiss). The desirability of using porous layer beads in ion-exchange chromatography has been demonstrated by Horváth *et al.* Other information and examples, in addition to the literature cited, have been given by Burtis *et al.*, Gere, Schmit, Uziel and others.

For some continuous or counter-current processes with ion exchange, especially in industry and in nuclear chemistry, liquid ion exchangers are used (Högfeldt). The ion exchange can occur not only between the crystal or gel and the solution of an electrolyte, but in principle also between two non-miscible phases. One phase is an aqueous electrolyte solution and the other can be a solution of a base or acid in an organic solvent, *e.g.*, long-chain amines (Moore, 1960), or a special liquid compound. Substances used as liquid ion exchangers must contain an ion-exchanging group and a large non-polar part that is insoluble in aqueous phase. Di-(2-ethylhexyl)phosphoric acid is an example of a weakly acidic liquid ion exchanger, and dinonylnaphthalenesulphonic acid is a strongly acidic liquid exchanger. Liquid ion exchangers are also used in analytical chemistry (Coleman *et al.*) and can be applied in the impregnation of inert supports for column processes (Cerrai).

## SORBENTS FOR AFFINITY CHROMATOGRAPHY

In their review article, Cuatrecasas and Anfinsen (1971a) mention the properties which a sorbent for affinity chromatography should possess. Firstly, it should have a minimal non-specific adsorption. When preparing an insoluble affinant, it is important that this should be on the carrier only in the form of covalently bound molecules. The molecules of the affinant that are not bound covalently must be washed out, which is difficult with supports that strongly adsorb the affinant molecules. Similarly, when substances that form a specific and reversible complex with the bound affinant are isolated, it is important that, as far as possible, only their retention should take place on the column of insoluble affinant and only in the form of a specific complex with the bound affinant. This is one of the main reasons why carriers that contain ionogenic groups, such as the copolymer of ethylene with maleic anhydride, which set carboxyl groups free after the affinant had been attached, have never been as widely applied as the neutral agarose in affinity chromatography. For a smooth course of affinity chromatography, good flowing properties are important, *i.e.*, the eluent should penetrate the support column at a sufficient rate even when the affinant is bound on to it. The support must possess a sufficient number of chemical groups which can be activated or modified in such a manner as to be able to bind affinants. The capacity of a specific adsorbent prepared by the attach-

*References p.227*

ment of the affinant to the solid support is dependent on the number of these groups present. The activation or modification should take place under conditions that do not change the structure of the support. No less important are the chemical and mechanical stabilities of the carrier under the conditions of the affinant binding, but also at various pHs, temperatures and ionic strengths, in the presence of denaturating agents, etc., which are necessary for good sorption and elution of the isolated substance. The possibility of repeated use of a specific adsorbent depends on these stabilities. Should the isolated substance be sorbed on the bound affinant, the support must be sufficiently porous to provide for sufficient freedom of formation of the specific complex. An example is the sorption of β-galactosidase on to sorbents prepared by binding the inhibitor of *p*-aminophenyl-β-D-thiogalactopyranoside through a hydrocarbon chain both to the polydextran gel Sepharose and to the polyacrylamide gel Bio-Gel P-300 (Steers *et al.*). The contents of the bound inhibitor were virtually identical in the two instances, but the isolation of the β-galactosidase by affinity chromatography was successful only when Sepharose was used. In spite of a high concentration of inhibitor (50 μmole/ml), the enzyme was not retained on Bio-Gel P-300. This might be explained by an excessively large volume of the tetramer of β-galactosidase (molecular weight 540,000; Craven *et al.*), which could not enter into the Bio-Gel pores. On the other hand, when nuclease of molecular weight 17,000 (Cuatrecasas, 1970a) was isolated from *Staphylococci*, Bio-Gel P-300 appeared to be a suitable support. A high porosity of the solid support is further indispensable for the isolation of substances with a relatively weak affinity for the bound affinant (dissociation constant $\geqslant 10^{-5}$). The concentration of the bound affinant, freely accessible to the isolated substance, should be very high in this instance, in order to achieve a strong interaction which would retain physically the isolated substances migrating down the column. When choosing gels, their specific surface area should also be taken into consideration in addition to the pore size. In Table 9.17 are given the amounts of chymotrypsin bound to 1 ml of hydroxyalkyl methacrylate gels of various pore sizes (given by the value of exclusion molecular weights), and various specific surface areas (Turková *et al.*, 1973). From Table 9.17 it is evident that the amount of bound chymo-

TABLE 9.17
AMOUNTS OF CHYMOTRYPSIN BOUND TO HYDROXYALKYL METHACRYLATE GELS (SPHERON) AS A FUNCTION OF THE MAGNITUDE OF THEIR SPECIFIC SURFACE AREAS

| Spheron | Molecular-weight exclusion limit | Specific surface area ($m^2$/ml) | Amount of bound chymotrypsin (mg/ml) | Relative proteolytic activity (%) |
|---|---|---|---|---|
| $10^5$ | $10^8$ | 0.96 | 0.73 | – |
| $10^3$ | $10^6$ | 5.9 | 7.8 | 44 |
| 700 | 700,000 | 3.6 | 6.7 | 49 |
| 500 | 500,000 | 23 | 17.1 | 37 |
| 300 | 300,000 | 19.5 | 17.7 | 44 |
| 200 | 200,000 | 0.6 | 6.9 | 53 |
| 100 | 100,000 | 0.2 | 2.6 | 38 |

trypsin is directly dependent on the magnitude of the specific surface area, which is at a maximum in Spheron 300 and 500. The relative proteolytic activity is also given in this table.

### Cellulose and its derivatives

The binding of affinants of a predominantly proteinic nature on to cellulose and its derivatives was discussed in a review by Silman and Katchalski, the binding of enzymes by Crook *et al.*, and the binding of nucleotides, polynucleotides and nucleic acids by Gilham. These reviews mention very varied methods of binding. In view of the present limited use of cellulose in affinity chromatography, we shall briefly mention some of them here.

The most commonly used method of binding substances with a free amino group on to cellulose (CEL) is the Curtius azide method, used for the first time by Micheel and Ewers and today applied mostly in the modification of Hornby *et al.*:

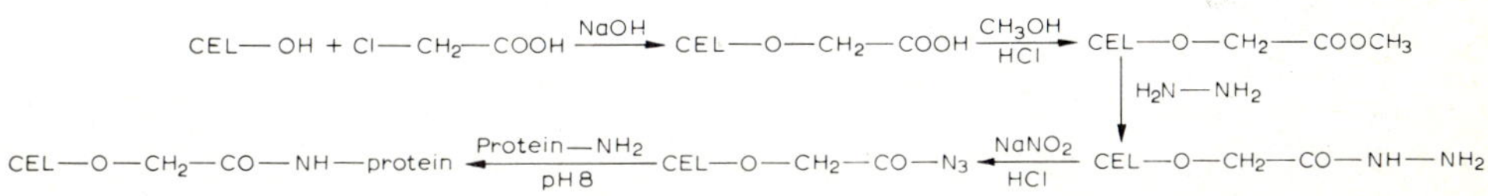


After the preparation of carboxymethylcellulose azide by Curtius rearrangement, an isocyanate is formed on to which the amino group of the affinant is bound.

Affinants with basic amino groups can be further coupled to the carboxyl groups of carboxymethylcellulose in the presence of carbodiimide (Weliky *et al.*):

$$\text{CEL-O-CH}_2\text{-COOH} + \text{R-NH}_2 \xrightarrow[\text{N,N}'\text{-dicyclohexylcarbodiimide}]{} \text{CEL-O-CH}_2\text{-CO-NH-R} + \text{H}_2\text{O}$$

Kay and Lilly worked out the triazine method of protein binding. 2-Amino-4,6-dichloro-*s*-triazine is bound to the hydroxyl group of cellulose and reacts further with the amino group of the protein:

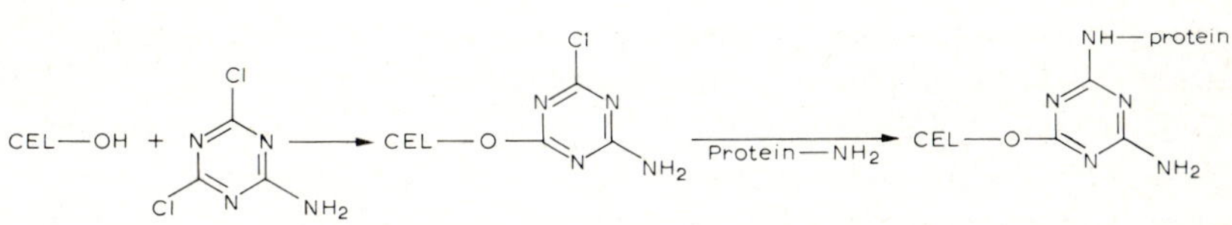


Jagendorf *et al.* developed a method of protein binding based on the acylation of the hydroxyl group in cellulose with bromoacetyl bromide and subsequent alkylation of the amino group of the protein:

$$\text{CEL-OH} + \text{Br-CO-CH}_2\text{-Br} \rightarrow \text{CEL-O-CO-CH}_2\text{-Br} \xrightarrow{\text{Protein-NH}_2} \text{CEL-O-CO-CH}_2\text{-NH} - \text{protein}$$

The first attachment of an affinant on to cellulose was carried out by means of diazonium groups (Campbell *et al.*):

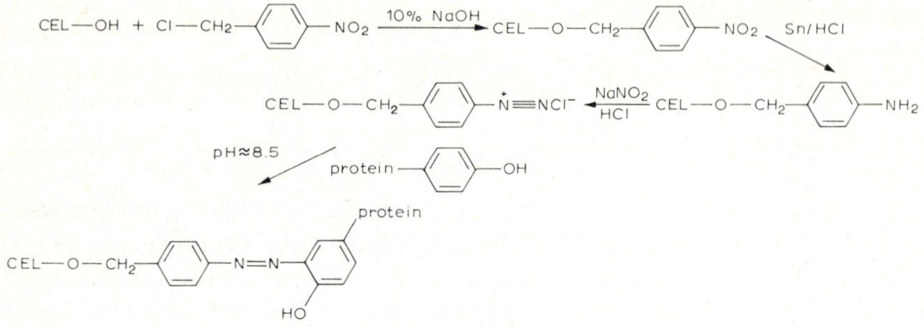


The affinants are bound by their aromatic residues, in the case of proteins mainly by tyrosine and histidine, but also non-specifically and more slowly by their amino groups (Gundlach *et al.*, Tabachnik and Sobotka).

Nucleic acids are bound to aminoethylcellulose mostly by using periodate oxidation (Gilham).

Cellulose and its derivatives are produced by a number of firms*. In addition to Whatman (Maidstone, Great Britain), Serva (Heidelberg, G.F.R.) lists both current cellulose derivatives and bromoacetylcellulose (BA-cellulose) and *p*-aminobenzylcellulose (PAB-cellulose). Bio-Rad Labs. (Richmond, Calif., U.S.A.) supply *p*-aminobenzylcellulose under the trade-name Cellex PAB and aminoethylcellulose under the name Cellex AE. Miles-Seravac and Miles-Yeda (Maidenhead, Great Britain) produce a hydrazide derivative of CM-cellulose (Enzite-CMC-hydrazide), bromoacetylcellulose (BAC) and *m*-aminobenzyloxymethylcellulose (ABMC). In addition to the supports for the binding of affinants, they also supply insoluble affinants, such as CM-cellulose-bound cysteine, and further the enzymes trypsin, chymotrypsin, papain, protease from *Streptomyces griseus*, bromelain, ficin, peroxidase, RNase, amylase and cytochrome *c*, DEAE-cellulose-bound leucineaminopeptidase, alcohol dehydrogenase (YADH), glucoso-oxidase and urease under the names consisting of the name of the enzyme with the prefix Enzite. Cellulose-bound trypsin is also produced by E. Merck (Darmstadt, G.F.R.).

Although cellulose was used as a carrier mainly during the initial period of the development of affinity chromatography, it is still used. Some workers (Dean and Lowe; Lowe and Dean) still consider cellulose to be the most suitable support for nicotinamide–adenine dinucleotide (NAD) and they isolated on it the NAD-binding dehydrogenase. Similarly, Fritz and his co-workers (Fink *et al.*; Fritz *et al.*, 1970; Tschesche *et al.*) still make use of cellulose-bound proteases for the isolation of their inhibitors from most various sources, and Sato *et al.* used halogenoacetylcellulose for the isolation of aminoacylase.

*Only firms known to the authors are mentioned in the text. Therefore, the list is necessarily incomplete and it should in no case be considered as implying recommendation of any particular firm or product.

### Copolymer of ethylene and maleic anhydride

The linkage of affinants to a copolymer of ethylene and maleic anhydride (EMA) was discussed in a review by Goldstein. The method of binding enzymes to this support was worked out by Levin *et al.* in 1964. The protein is bound to anhydro groups of the polymer by its amino groups:

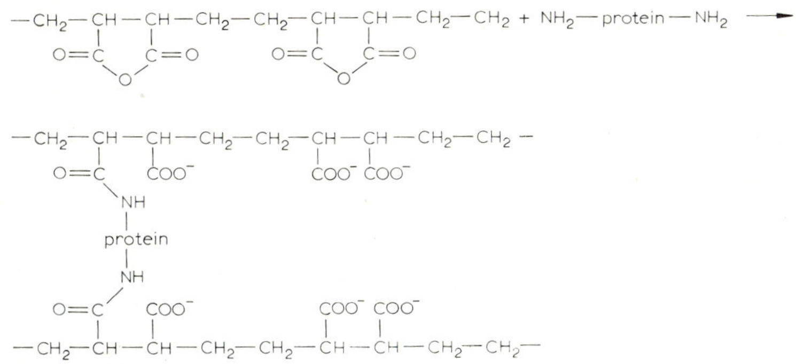

When the affinant is bound, carboxyl groups are set free (either after the binding with proteins or hydrolysis in aqueous medium), which give the support a polyanionic character.

The copolymer of ethylene and maleic anhydride is produced by Monsanto (St. Louis, Mo., U.S.A.). The British firm Miles-Yeda binds the enzyme trypsin, chymotrypsin, papain and subtilopeptidases A and B on to this polymer and supplies them under the names Enzite-EMA-(enzyme name). These preparations are characterized by a high content (about 60%) of the bound enzyme.

The properties of enzymes bound to EMA carriers have been intensively investigated (Goldstein and Katchalski, Silman and Katchalski). This support with bound proteases was utilized mainly by Fritz, Werle and co-workers for the preparation of a series of inhibitors of proteolytic enzymes (Fritz *et al.*, 1967, 1968, 1972; Hochstrasser *et al.*, 1972a, b) and with bound inhibitors for the isolation of proteases (Fritz *et al.*, 1969).

### Agarose and its derivatives

Agarose is a polydextran carrier which is presently most used in affinity chromatography. As demonstrated by Cuatrecasas and Anfinsen (1971a), agarose fulfils virtually all the requirements of an ideal carrier.

The most utilized method of affinant binding on Sepharose activated by cyanogen bromide was developed by Porath *et al.* and Axén *et al.* The affinants are bound by means of primary aliphatic or aromatic amino groups. Axén and Ernback proposed a two-step activation scheme. In the first step, a labile intermediate is formed under the effect of halogenocyanates, which is then converted in the second step into an inert carbamate and a reactive imidocarbonate. The amino group of the affinant is then bound to the latter in a weakly alkaline medium with the formation of a stable covalent bond. On the basis of a study of model reactions with methyl 4,6-O-benzylidene-α-D-glucopyranoside (Ahrgren *et al.*), this course seems improbable and at present the prevailing view is that the main part

*References p.227*

of the affinant binding to the support takes place via isourea derivatives.

The degree of agarose activation, measured on the basis of the capacity of binding of small peptides (Axén and Ernback), is directly proportional to the pH. Activation takes place at pH 11. The whole procedure for the activation of Sepharose with cyanogen bromide (Axén *et al.*, Cuatrecasas, Porath *et al.*) is described in detail below.

*Coupling of affinants on agarose*

Washed and decanted Sepharose is suspended in distilled water (1:1). The suspension is placed in a well ventilated hood, a pH meter electrode pair is immersed in the suspension, and finely divided cyanogen bromide (50–300 mg per millilitre of packed Sepharose) is added gradually with constant stirring. The suspension is maintained at pH 11 by addition of sodium hydroxide solution. The concentration of the sodium hydroxide solution used depends on the amounts of Sepharose and cyanogen bromide present; for 5–10 ml of packed Sepharose and 1–3 g of added cyanogen bromide, Cuatrecasas recommends the use of 2 *M* sodium hydroxide, and for 100–200 ml of packed Sepharose and 20–30 g of cyanogen bromide an 8 *M* solution is suitable. The temperature should not exceed 20°C; if cooling is necessary, ice may be added. The reaction is completed in 8–12 min. After a rapid transfer of the suspension on a büchner funnel under constant suction, the activated Sepharose is washed with cooled buffer solution of the same composition as will be used for the subsequent coupling of the affinant. The buffer volume should be 10–15 times that of the Sepharose to be activated. The washed Sepharose is suspended as rapidly as possible in an equal volume of the affinant solution. According to Cuatrecasas, the washing, addition of affinant solution and mixing should not take more than 90 sec. Even at a low temperature, the activated Sepharose is unstable.

The conditions used for the linking of the affinant, such as pH, buffer composition and temperature, are dependent on the nature of the bound affinant. The binding reaction proceeds most satisfactorily at pH 8–10, but if the nature of the affinant requires a lower pH, the binding of a sufficient amount of affinant can usually be carried out successfully even at that pH if the amount of cyanogen bromide is increased during activation and the amount of affinant during the binding. The active groups that remain after the binding of the affinant are eliminated by blocking them with 1 *M* ethanolamine. A detailed procedure of the binding of affinant to cyanogen bromide-activated Sepharose is given below.

Cyanogen bromide-activated Sepharose 4B is produced commercially by freeze-drying with the addition of dextran and lactose, which must be washed out before use. The manufacturer (Pharmacia) gives the following procedure for coupling the affinant with cyanogen bromide-activated Sepharose. The required amount of gel is allowed to swell in $10^{-3}$ *M* hydrochloric acid and the solution is then used for washing the gel for 15 min. The volume of 1 g of the freeze-dried gel when swollen is approximately 3.5 ml. It is recommended that 200 ml of the solution per gram of dry gel should be used for the washing, in several batches. Immediately after washing, the solution of the affinant to be coupled is added.

The optimum conditions for coupling the affinant, *i.e.*, pH, buffer composition and temperature, are dependent to an appreciable extent on its character. Generally, the coupling reaction takes place most effectively at pH 8–10, but if the nature of the coupled affinant requires it, lower pH values can also be used. The affinant, especially if it is of a proteinic character, is dissolved in a buffer of high ionic strength (about 0.5) in order to prevent non-specific adsorption. The higher ionic strength then facilitates subsequent washing. Carbonate or borate buffers with sodium chloride added can be used. The amount of the affinant coupled depends on the proportion of the affinant in the reaction mixture and the volume of gel, the pH of the reaction, the nature of the coupled affinant (number of reactive groups, etc.), and also on the reaction time and temperature. For example, when chymotrypsin is coupled with 2 ml of cyanogen bromide-activated Sepharose at pH 8, only 5 mg were coupled when 10 mg of protein were present, at 20 mg of protein *ca.* 8 mg were coupled, and at 30 mg the amount coupled was *ca.* 10 mg. At room temperature (20–25°C), the coupling is usually completed after 2 h, while at lower temperatures it is recommended that the mixture should be allowed to stand overnight. During the coupling, the reaction mixture must be stirred. Stirring with a magnetic stirrer is not recommended, as it may cause mechanical destruction of the gel. When the coupling is completed, the gel with the coupled affinant is transferred on to a sintered-glass filter and washed with the buffer used during the coupling. In order to eliminate the remaining active groups, the manufacturer (Pharmacia) recommends blocking them with 1 *M* ethanolamine at pH 8 for 2 h. The final product should then be washed four or five times alternately with buffer solutions of high and low pH. For example, acetate buffer (0.1 *M*, pH 4) and borate buffer (0.1 *M*, pH 8.5) can be used, each being 1 *M* in sodium chloride. As already stated, all non-covalently bound substances should be eliminated during the washing.

In addition to the binding of affinant on to agarose, described above, the triazine method of binding affinants on agarose is also used. It was originally developed by Kay and Lilly for binding of affinants to cellulose (see p. 217), using 2-amino-4,6-dichloro-*s*-triazine.

Affinity chromatography of some substances takes place only if the affinant is sufficiently distant from the surface of the solid support (Steers *et al.*), which can be achieved by inserting between them a sufficiently long bridge. For this purpose, aliphatic diamines are usually used, for example ethylenediamine, which can be bound directly to cyanogen bromide-activated Sepharose in the conventional manner. In order to eliminate the undesirable formation of additional cross-linkages owing to the reaction of both terminal amino groups, a large excess of diamine should be employed (Cuatrecasas, 1970a). A detailed procedure for the preparation of ω-aminoalkyl derivatives of agarose is given below.

Aliphatic diamino compounds, for example ethylenediamine, can be coupled directly on to cyanogen bromide-activated Sepharose. In order to avoid the undesirable formation of additional cross-linkages owing to the reaction of both terminal amino groups, a large excess of the diamine is used. Into a suspension of Sepharose 4B in water (1:1), 250 mg of cyanogen bromide per millilitre of packed Sepharose are added and the reaction is carried out as described in the procedure for the activation of Sepharose with cyanogen bromide. An equal volume of cold distilled water containing 2 mmole of ethylenediamine per millilitre of packed Sepharose is added to the washed and collected activated Sepharose, which has been adjusted with 6 *M* hydrochloric acid to pH 10. After 16 h of reaction at 4°C, the gel is washed with a large volume of distilled water. Thus Sepharose derivatives can be obtained containing about 12 μmole of aminoethyl group per millilitre of packed Sepharose. By use of various diamino compounds of the general formula $NH_2(CH_2)_xNH_2$, various ω-aminoalkyl derivatives can be prepared.

Affinants that contain a free carboxyl group may be bound to aminoethyl-agarose by water-soluble carbodiimides (Cuatrecasas, 1970a), as mentioned in the procedure given below for the preparation of estradiol-Sepharose.

A 300-mg amount of 3-O-succinyl-[$^3$H]estradiol dissolved in 400 ml of dimethylformamide is added to 40 ml of packed aminoethyl-Sepharose 4B. The dimethylformamide is needed in order to solubilize the estradiol, and it is not necessary for affinants that are soluble in water. The suspension is maintained at pH 4.7 with 1 *N* hydrochloric acid. Then 500 mg (2.6 mmole) of 1-ethyl-3-(3-dimethylaminopropyl) carbodiimide, dissolved in 3 ml of water, are added to the suspension over 5 min and the reaction is allowed to proceed at room temperature for 20 h. Substituted Sepharose, after being transferred into the column, is washed with 50% aqueous dimethylformamide until the eluate is no longer radioactive. It is recommended that the derivative should be washed with *ca.* 10 l of the washing liquid over 3–5 days. Using this procedure, about 0.5 μmole of estradiol may be bound per millilitre of packed Sepharose. Of other Sepharose derivatives, Cuatrecasas (1970a) prepared bromoacetamidoethyl-Sepharose by reaction of O-bromoacetyl-N-hydroxysuccinimide with aminoethyl-Sepharose, succinylaminoethyl-Sepharose by reaction with succinic anhydride, diazonium derivatives from *p*-aminobenzamidoethyl-Sepharose, and also tyrosyl-Sepharose and sulfhydryl-Sepharose. He bound affinants with free carboxyl groups to sulfhydryl-Sepharose by thiol–ester bonds using water-soluble carbodiimides. The preparation of various derivatives of Sepharose 4B was reviewed by Cuatrecasas (1970b).

As regards the stability of agarose, it is stable in the pH 4–9 range and temperatures below 0°C or above 40°C are not recommended. Sepharose is resistant to high salt concentrations, urea (7 *M*) and guanidine hydrochloride (6 *M*) (Cuatrecasas, 1970a). It is stable even when exposed at room temperature to 0.1 *M* sodium hydroxide solution and 1 *M* hydrochloric acid for a short time (2–3 h). For the affinity chromatography of weakly water-soluble substances, 50% dimethylformamide or 50% ethylene glycol can also be used. Lyophilization can be carried out only after addition of protective substances, for example dextran, glucose and serum albumin.

The main manufacturers of agarose are Pharmacia (Uppsala, Sweden), who produce agarose under the name Sepharose, and Bio-Rad Labs., who produce agarose under the name Bio-Gel A. Both preparations are produced in several varieties that differ in their pore sizes and the molecular-weight exclusion limits depending on them. For facilitating the binding of affinants, Pharmacia produce lyophilized cyanogen bromide-activated Sepharose. They also produce the ready-made insoluble affinant Con A-Sepharose, representing concanavalin A bound to Sepharose, which is suitable for the isolation of polysaccharides or glucoproteins containing the glycosyl residue as the terminal group (Aspberg and Porath, Edelman *et al.*, Lloyd, Yariv *et al.*). Miles-Seravac produce a series of insoluble affinants based on agarose, for example agarose-bound enzymes such as trypsin, chymotrypsin, papain and protease from *Streptomyces griseus.* Their commercial names are Enzite-agarose-(name of enzyme). For the isolation of chymotrypsin, Miles-Seravac produce agarose-bound ε-aminocaproyl-D-tryptophan methyl ester (Cuatrecasas *et al.*), for the isolation of papain agarose-bound tetrapeptide Gly–Gly–Tyr (O-benzyl)–Arg (Blumberg *et al.*), for the isolation of ribonuclease agarose-5′(4′-aminophenyl)uridine-2′(3′)-phosphate (Wilchek and Gorecki), for the isolation of L-tryptophan-binding proteins agarose–L-tryptophan and agarose–D-tryptophan (Sprossler and Lingens), for L-tyrosine-binding proteins agarose–L-tyrosine (Chan and Takahashi), and for thyroxine-binding proteins agarose–thyroxine (Pensky and Marshall). Concanavalin A bound to agarose is produced under the name Glycosylex A. Among immunoadsorbents, Miles-Seravac produce for the isolation of antibodies agarose-bound antigens, for example bovine serum albumin, human and goat immunoglobulins, and insoluble haptens, for example agarose-bound dinitrophenyl, arsanilic acid, gibberellic acid and 3-indolylacetic acid. For the isolation of antigen, they produce agarose-bound antibodies which are formed against bovine albumin, growth hormone, glucagon, human IgG, dinitrophenol, gibberellic acid and 3-indolylacetic acid.

It is expected that the number of commercially produced affinants bound to agarose and other supports will increase substantially in the future, which is indicated by the present rapid increase in the number of publications that describe the affinity chromatography of a wide range of substances on supports prepared from agarose. The number of such papers is already so large that it is impossible to review them within the scope of this chapter: in 1968, only two papers describing the use of affinity chromatography on agarose appeared, in 1969 the number of such papers was ten, in 1970 twenty, in 1971 forty-six, in 1972 sixty-seven and in 1973 seventy. A review of most of the papers utilizing Sepharose for affinity chromatography is offered by Pharmacia in a leaflet on cyanogen bromide-activated Sepharose 4B used for the immobilization of biopolymers.

### Polyacrylamide supports and their derivatives

Polyacrylamide gels are composed of a hydrocarbon skeleton on to which carboxamide groups are bound:

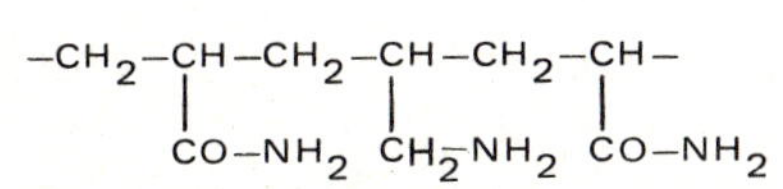

On reaction with a suitable compound, they can be converted into solid carriers suitable for the binding of a series of affinants (Inman and Dintzis). Their aminoethyl derivatives may be prepared by using a large excess of ethylenediamine at 90°C, and hydrazide derivatives by using excess of hydrazine at 50°C. Aminoethyl derivatives of polyacrylamide gels can be converted into their *p*-aminobenzamidoethyl derivatives by reaction with *p*-nitrobenzoylazide in the presence of N,N-dimethylformamide, triethylamine and sodium thiosulphate.

After activation with nitrous acid, the hydrazide derivative can bind affinants with its amino groups:

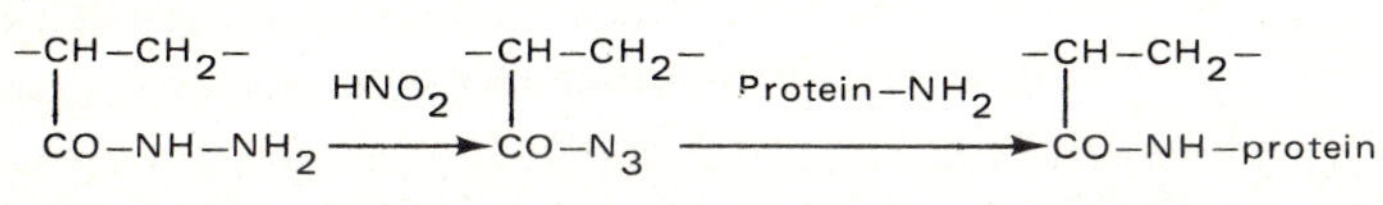

Polyacrylamide gels containing residues of aromatic acids, when diazotized with

*References p.227*

nitrous acid, bind affinants mainly through their aromatic residues:

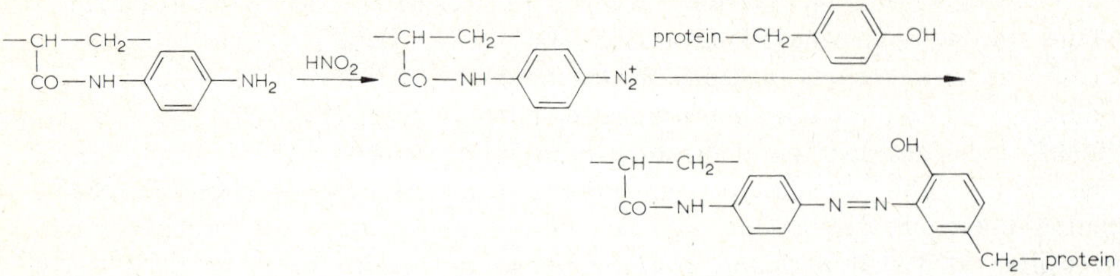

The same gels, when activated with thiophosgene, bind affinants by means of their free amino groups:

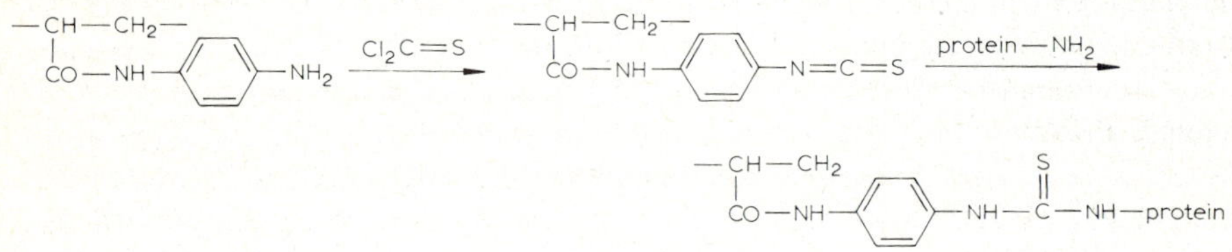

The procedures for the binding of proteins on to all three derivatives of polyacrylamide gels are given below.

*Coupling of proteins with commercially produced polyacrylamide derivatives (Enzacryls)*

*Coupling of affinants on polyacrylamide gels containing aromatic amino acid residues (Enzacryl AA) after activation with nitrous acid*

To a suspension of 100 mg of Enzacryl AA in 5 ml of 2 *M* hydrochloric acid, cooled to 0°C, 4 ml of an ice-cold 2% solution of sodium nitrite are added and the mixture is stirred magnetically for 15 min. The diazo-Enzacryl formed is then washed four times with the buffer in which the affinant will undergo coupling (for example, a phosphate buffer of pH 7.5). After centrifugation and decantation, the affinant is added, for example an enzyme (2.5 mg) in a suitable buffer (0.5 ml). The coupling is allowed to proceed with magnetic stirring for 48 h. The reaction is terminated by addition of an ice-cold solution of phenol (0.01%) in sodium acetate (10%). After a further 15 min, the Enzacryl with the coupled affinant is first washed with a dilute buffer, then with the same buffer made 0.5 *M* in sodium chloride. This washing should be carried out very carefully. The manufacturer (Koch-Light) recommends carrying out the whole experiment first with non-diazotized Enzacryl, in order to determine the best conditions for washing out all of the adsorbed material.

*Coupling of affinants on polyacrylamide gels containing aromatic amino acid residues (Enzacryl AA) after activation with thiophosgene*

To a suspension of 100 mg of Enzacryl AA in 1 ml of phosphate buffer (3.5 *M*, pH 6.8–7.0), well stirred with a magnetic stirrer, 0.2 ml of a 10% thiophosgene solution in chloroform is added. After vigorous stirring for 20 min, a further 0.2 ml of the thiophosgene solution is added, and after additional stirring for 20 min the NCS-Enzacryl is washed once with acetone, twice with 0.5 *M* sodium hydrogen carbonate solution and twice with a buffer suitable for coupling (for example, a borate buffer of pH $>$ 8.5). After centrifugation and decantation, 0.5 ml of an affinant solution (for example, 2.5 mg of enzyme) is added and the coupling is carried out as described in the preceding section.

Activation of the hydrazine derivative of polyacrylamide gel (Enzacryl AH) with nitrous acid and subsequent coupling is carried out in the same manner as described for Enzacryl AA.

*Coupling of proteins on polyacrylamide gels by using glutaraldehyde*

Weston and Avrameas developed a method of direct binding of affinants on to polyacrylamide gels using glutaraldehyde, which, if present in excess, reacts with one of its two aldehydic groups, which is bound to the free amido group present in the polyacrylamide gel. The remaining free active group then reacts with the amino group of the affinant added during the subsequent binding reaction. Thus a firm bond is formed between the support and the affinant.

Bio-Gel P-300 is allowed to swell in water and is washed twice with a four-fold volume of 0.1 *M* phosphate buffer of pH 6.9. Then 19.4 ml of gel (1 g of dry beads per 45 ml) are mixed with glutaraldehyde solution (4.8 ml; 25%, v/v) and incubated at 37°C for 17 h. The gel is washed and centrifuged four times with a four-fold volume of 0.1 *M* phosphate buffer of pH 6.9, then three times with 0.1 *M* phosphate buffer of pH 7.7. The coupling of the protein is carried out after mixing of 3 ml of gel in 13.5 ml of a buffer of pH 7.7 with 0.3 ml of enzyme solution (20 mg/ml) at 4°C for 18 h on a shaker. After the reaction, the gel is centrifuged and washed. Using this method, 70 mg of acid phosphatase could be coupled per gram of dry gel.

Polyacrylamide gels are stable in the pH range 1–10 and they support well all generally employed eluents. They do not contain charged groups, and so ion exchange with the chromatographed substances is minimal. They are biologically inert and, as they are synthetic polymers, they are not attacked by microorganisms. As the gel particles adhere strongly to clean glass surfaces, Inman and Dintzis recommend the use of siliconized glass or polyethylene laboratory vessels.

The main producer of polyacrylamide gels is Bio-Rad Labs., who produce them under the commercial name Bio-Gel P by copolymerization of acrylamide and N,N′-methylenebisacrylamide. Bio-Gel P of various pore sizes range from Bio-Gel P-2 with a molecular-weight exclusion limit of 1800 up to Bio-Gel P-300 with a molecular-weight exclusion limit of 400,000. All brands are of 50–100, 100–200, 200–400 and 400 mesh size. In addition to these gels, Bio-Rad Labs. produce ion-exchanging derivatives of the gels, for example the weakly acidic cation exchanger Bio-Gel CM, and also intermediates for affinity chromatography, such as the aminoethyl and hydrazide derivatives of Bio-Gel P-2 and P-60.

For the linking of affinants, mainly enzymes, Koch-Light (Colnbrook, Great Britain), produce Enzacryls. Enzacryl AH is a hydrazide derivative of polyacrylamide gels, and Enzacryl AA is a polyacrylamide gel containing aromatic acid residues. Enzacryl polythiolactone, polythiol and polyacetate are currently being introduced.

Although Cuatrecasas (1970a) demonstrated the suitability of acrylamide gels for the isolation of staphylococcal nuclease by affinity chromatography and Truffa-Bachi and

*References p.227*

Wofsy isolated specific cells on Bio-Gel P-6 with bound hapten, the utilization of these gels in affinity chromatography is still limited.

### Hydroxyalkyl methacrylate gels

Hydrophilic hydroxyalkyl methacrylate gels are prepared by polymerization of a suspension of hydroxyalkyl esters of methacrylic acid and alkylene dimethylacrylate (Čoupek *et al.*, 1972a, b) by varying the ratio of the concentrations of monomer and inert components. The number of reactive groups, porosity, and the specific surface area of the gel may be changed within broad limits. The gel has the following structure:

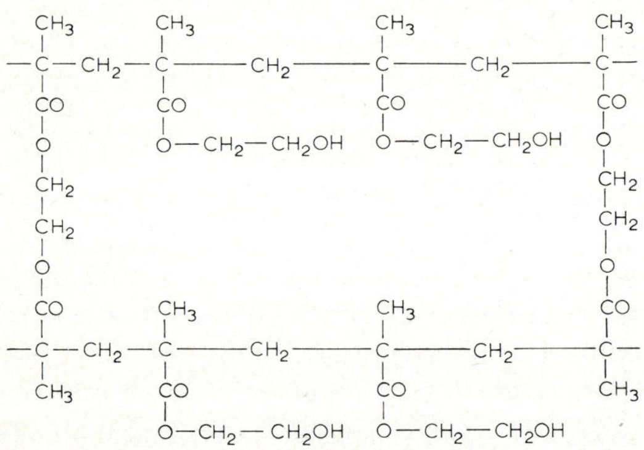

Hydroxyl groups of the gel possess analogous properties to those of agarose. After cyanogen bromide activation, they bind the affinants in the same manner as Sepharose by their amino groups (Turková *et al.*, 1972, 1973). The gel activation and the affinant binding are virtually identical with those described in the procedures for the binding on agarose.

The gels form regular beads with excellent chemical and physical stabilities. They stand chromatography well under pressure, and do not change their structures after heating for 8 h in 1 *N* sodium glycolate solution at 150°C or after boiling for 24 h in 20% hydrochloric acid. They are biologically inert and, as acrylamide gels, are not attacked by microorganisms.

Their production is being started by Lachema (Brno, Czechoslovakia) under the name Spheron. Spherons of various pore sizes and exclusion molecular weights are manufactured from Spheron 100, with a molecular-weight exclusion limit of 100,000, up to Spheron $10^5$, with an exclusion molecular weight of $10^8$. The production of cyanogen bromide-activated dried Spheron is already also under consideration.

Spherons have been used successfully for a series of affinity chromatographic procedures, of which only the isolation of the chymotrypsin inhibitor from potatoes on Spheron 300 (with bound chymotrypsin) and the chromatography of chymotrypsin on Spheron 300 with bound trypsin-inhibitor has been published up to the present time (Turková *et al.*, 1972).

### Glass and its derivatives

Workers at Corning (Corning, N.Y., U.S.A.) demonstrated that glass, when treated with $\gamma$-aminopropyltriethoxysilane, becomes a suitable support for a series of affinants (Line *et al.*; Messing; Weetall, 1969a, b, 1970, 1971; Weetall and Baum; Weetall and Hersh, 1969, 1970; Weibel and Bright; Weibel *et al.*). Affinants can be bound on to the amino groups of the glass derivatives by their carboxyl group using soluble carbodiimides, and by their amino groups after activation with thiophosgene, and by aromatic residues with the azo-bond to arylamine derivatives.

Glass derivatives have outstanding stability and are not attacked by microorganisms. However, in some instances they cause undesirable unspecific adsorption (Cuatrecasas and Anfinsen, 1971b).

Corning produces a series of glass supports of various bead and pore sizes, which are, however, not yet commercially available.

In affinity chromatography, only Weibel *et al.* have used a glass support as a specific adsorbent prepared by diazo-coupling of nicotinamide–adenine dinucleotide, which proved an effective coenzyme for the apoenzyme of alcohol dehydrogenase from yeast.

## REFERENCES

Acuff, R. D., *Ger. Pat., Offen*, 1,900,517 (1968).
Adams, B. A. and Holmes, E. L., *J. Soc. Chem. Ind., London, Trans. Commun.*, 54 (1935) 1T.
Ahrgren, L., Kågedal, L. and Åkerström, S., *Acta Chem. Scand.*, 26 (1972) 285.
Amphlett, C. B., *Inorganic Ion Exchangers*, Elsevier, New York, 1964.
Araki, C., *Bull. Chem. Soc. Jap.*, 29 (1956) 543.
Ashinskaya, N. V., Kiselev, A. V. and Nikitin, Yu. S., *Zh. Fiz. Khim.*, 37 (1963) 927.
Aspberg, K. and Porath, J., *Acta Chem. Scand.*, 24 (1970) 1839.
Aue, W. A. and Hastings, C. R., *J. Chromatogr.*, 42 (1969) 319.
Axén, R. and Ernback, S., *Eur. J. Biochem.*, 18 (1971) 351.
Axén, R., Porath, J. and Ernback, S., *Nature (London)*, 214 (1967) 1302.
Baetstlé, L., *J. Inorg. Nucl. Chem.*, 25 (1963) 271.
Baetstlé, L. and Pelsmacker, J., *J. Inorg. Nucl. Chem.*, 21 (1961) 124.
Berger, R. and Mindner, K., *Ger. Plast.*, 78 (1970) 383.
Blumberg, S., Schechter, I. and Berger, A., *Eur. J. Biochem.*, 15 (1970) 97.
Bohemen, J. and Purnell, J. H., *J. Chem. Soc.*, (1961) 360.
Bombaugh, K. J., in J. J. Kirkland (Editor), *Modern Practice of Liquid Chromatography*, Wiley-Interscience, New York, 1971, p. 248.
Brewer, P. J., *Nature (London)*, 118 (1960) 934.
Brewer, P. J., *Nature (London)*, 190 (1961) 625.
Brockmann, H. and Schodder, H., *Chem. Ber.*, 74 (1941) 73.
Burtis, C. A., Gere, D. R., Gill, J. M. and McDonald, F. M., in E. S. Kováts (Editor), *Column Chromatography, 5th Int. Symp. Separ. Methods, 1969*, Sauerländer, Aarau, 1970.
Campbell, D. H., Luescher, E. L. and Lerman, L. S., *Proc. Nat. Acad. Sci. U.S.*, 37 (1951) 575.
Cassidy, H. G., *J. Amer. Chem. Soc.*, 71 (1949) 402.
Cerrai, E., *Chromatogr. Rev.*, 6 (1964) 129.
Chan, W. W. C. and Takahashi, M., *Biochem. Biophys. Res. Commun.*, 37 (1969) 272.
Coleman, C. F., Blake, C. A. and Brown, K. B., *Talanta*, 9 (1962) 297.
Cooper, A. R. and Johnson, J. F., *J. Appl. Polym. Sci.*, 13 (1969) 1487.
Čoupek, J. and Heitz, W., *Macromol. Chem.*, 112 (1968) 286.

Čoupek, J., Křiváková, M. and Pokorný, S., *IUPAC International Symposium on Macromolecules, Helsinki, 1972a; J. Polym. Sci., Part C*, in press.

Čoupek, J., Turková, J. and Křiváková, M., *2nd Discussion Conference on Macromolecules, Prague, 1972b*, Abstr. No. D9/3 b.

Craven, G. R., Steers, Jr., E. and Anfinsen, C. B., *J. Biol. Chem.*, 240 (1965) 2468.

Crook, E. M., Brocklehurst, K. and Wharton, C. W., *Methods Enzymol.*, 19 (1970) 963.

Cuatrecasas, P., *J. Biol. Chem.*, 245 (1970a) 3059.

Cuatrecasas, P., *Nature (London)*, 228 (1970b) 1327.

Cuatrecasas, P. and Anfinsen, C. B., *Methods Enzymol.*, 22 (1971a) 345.

Cuatrecasas, P. and Anfinsen, C. B., *Annu. Rev. Biochem.*, 40 (1971b) 259.

Cuatrecasas, P., Wilchek, M. and Anfinsen, C. B., *Proc. Nat. Acad. Sci. U.S.*, 61 (1968) 636.

Dean, P. D. G. and Lowe, C. R., *Biochem. J.*, 127 (1972) 11P.

Determann, H., *Angew. Chem.*, 76 (1964) 635.

Determann, H., Kriever, M. and Wieland, T., *Makromol. Chem.*, 114 (1968a) 256.

Determann, H. and Lambert, K., *J. Chromatogr.*, 56 (1971) 140.

Determann, H., Lüben, G. and Wieland, T., *Makromol. Chem.*, 73 (1964) 168.

Determann, H., Rehner, H. and Wieland, T., *Makromol. Chem.*, 114 (1968b) 263.

Deuel, H. and Neukom, H., *Advan. Chem. Ser.*, 11 (1954) 51.

Deuel, H., Solms, J. and Anyas-Weisz, L., *Helv. Chim. Acta*, 33 (1950) 2171.

De Vries, A. J., Le Page, M., Beau, R. and Guillemin, C. L., *3rd International Gel Permeation Chromatography Seminar, Geneva, 1966*, Waters Ass., Framingham, Mass., 1966.

Edelman, G. M., Rutishauser, U. and Millette, C. F., *Proc. Nat. Acad. Sci. U.S.*, 68 (1971) 2153.

Eriksson, K. E., Petterson, B. A. and Steenberg, B., *Sv. Papperstidn.*, 71 (1968) 695.

Fawcett, J. S. and Morris, C. J. O. R., *Separ. Sci.*, 1 (1966) 9.

Fink, E., Jaumann, E., Fritz, H., Ingrisch, A. and Werle, E., *Hoppe-Seyler's Z. Physiol. Chem.*, 352 (1971) 1591.

Friedman, L., *J. Amer. Chem. Soc.*, 52 (1930) 1311.

Frisque, A. J. and Bernet, K., *U.S. Pat.*, 3,644,305.

Fritz, H., Brey, B. and Béress, L., *Hoppe-Seyler's Z. Physiol. Chem.*, 353 (1972) 19.

Fritz, H., Brey, B., Schmal, A. and Werle, E., *Hoppe-Seyler's Z. Physiol. Chem.*, 350 (1969) 617.

Fritz, H., Gebhardt, M., Mester, R., Illchmann, K. and Hochstrasser, K., *Hoppe-Seyler's Z. Physiol. Chem.*, 351 (1970) 571.

Fritz, H., Hochstrasser, K., Werle, E., Brey, E. and Gebhardt, B. M., *Z. Anal. Chem.*, 243 (1968) 452.

Fritz, H., Schult, H., Hutzel, M., Wiederman, M. and Werle, E., *Hoppe-Seyler's Z. Physiol. Chem.*, 348 (1967) 308.

Gere, D. R., in J. J. Kirkland (Editor), *Modern Practice of Liquid Chromatography*, Wiley-Interscience, New York, 1971, p. 417.

Ghetie, V. and Schell, H. D., *Rev. Roum. Biochim.*, 4 (1967) 179.

Giddings, J. C. and Mallik, K. L., *Anal. Chem.*, 38 (1966) 997.

Gilham, P. T., *Methods Enzymol.*, 21 (1971) 191.

Golay, M. J. E., in R. P. W. Scott (Editor), *Gas Chromatography 1960*, Butterworths, London, 1960, p. 139.

Goldstein, L., *Methods Enzymol.*, 19 (1970) 935.

Goldstein, L. and Katchalski, E., *Z. Anal. Chem.*, 243 (1968) 375.

Greber, G. and Hausmann, P., *Angew. Chem., Int. Ed. Engl.*, 7 (1968) 394.

Groggins, P. H., *Unit Processes in Organic Synthesis*, McGraw-Hill, New York, 3rd ed., 1947, p. 803.

Grubhofer, N. and Schleith, L., *Naturwissenschaften*, 40 (1953) 508; *Hoppe-Seyler's Z. Physiol. Chem.*, 296 (1954) 262.

Gundlach, G., Kohne, C. and Turba, F., *Biochem. Z.*, 336 (1962) 215.

Guthrie, J. D., *Ind. Eng. Chem.*, 44 (1952) 2187.

Hadeball, W. and Seide, *Plast. Kaut.*, 16 (1969) 418.

Halász, I., Engelhardt, H., Asshauer, J. and Karger, B. L., *Anal. Chem.*, 42 (1970) 1460.

Halász, I. and Holdinghausen, F., in H. G. Struppe (Editor), *Gas Chromatographie 1968*, Akademie Verlag, Berlin, 1968, p. 311.

Halász, I. and Horváth, C., *Anal. Chem.*, 36 (1964) 1178.
Halász, I. and Sebestian, I., *Angew. Chem., Int. Ed. Engl.*, 8 (1969) 453.
Hale, D. K., Hawdon, A. R., Jones, J. I. and Packham, D. I., *J. Chem. Soc.*, (1952) 3503.
Haller, W., *J. Chem. Phys.*, 42 (1965) 686.
Hatch, M. J., Dillon, J. A. and Smith, H. B., *Ind. Eng. Chem.*, 49 (1957) 1812.
Heitz, W., *J. Chromatogr.*, 53 (1970) 37.
Heitz, W. and Čoupek, J., *Makromol. Chem.*, 105 (1967) 280.
Heitz, W. and Pfitzner, K., *Ger. Pat., Offen.*, 1,907,666 (1969).
Heitz, W. and Platt, K. L., *Makromol. Chem.*, 127 (1969) 113.
Heitz, W., Ullner, H. and Höcker, H., *Makromol. Chem.*, 98 (1966) 42.
Heitz, W. and Winau, H., *Makromol. Chem.*, 131 (1970) 75.
Helfferich, F. G., *Ion Exchange*, McGraw-Hill, New York, 1962.
Hering, R., *Chelatbildende Ioneaustauscher*, Akademie-Verlag, Berlin, 1967.
Hjertén, S. and Mosbach, R., *Anal. Biochem.*, 3 (1962) 109.
Hochstrasser, K., Reichert, R., Matzner, M. and Werle, E., *Z. Klin. Chem. Klin. Biochem.*, 10 (1972a) 104.
Hochstrasser, K., Reichert, R., Schwarz, S. and Werle, E., *Hoppe-Seyler's Z. Physiol. Chem.*, 353 (1972b) 221.
Högfeldt, E., in J. A. Marinsky (Editor), *Ion exchange*, Vol. 1, Marcel Dekker, New York, 1966, p. 139.
Hornby, W. E., Lilly, M. D. and Crook, E. M., *Biochem. J.*, 98 (1966) 420.
Horváth, C. G. and Lipsky, S. R., *Anal. Chem.*, 41 (1969) 1227.
Horváth, C. G., Preiss, B. A. and Lipsky, S. R., *Anal. Chem.*, 39 (1967) 1422.
Inman, J. K. and Dintzis, H. M., *Biochemistry*, 8 (1969) 4074.
Jagendorf, A. T., Patchornik, A. and Sela, M. P., *Biochim. Biophys. Acta*, 78 (1963) 516.
Johansson, J. H. and Joustra, M. K., *Ger. Pat., Offen*, 2,120,493 (1969).
Kaiser, R., *Chromatografia*, 3 (1970) 38.
Karger, B. L., Engelhardt, H., Conroe, K. and Halász, I., in N. Stock and S. G. Perry (Editors), *Gas Chromatography 1970*, Institute of Petroleum, 1970, p. 112.
Kay, G. and Lilly, M. D., *Biochim. Biophys. Acta*, 198 (1970) 276.
King, Jr., J. and Benson, S. W., *J. Chem. Phys.*, 44 (1966) 1007.
Kirkland, J. J., *J. Chromatogr. Sci.*, 7 (1969) 7 and 361.
Kirkland, J. J., *J. Chromatogr. Sci.*, 8 (1970) 72.
Kirkland, J. J., in J. J. Kirkland (Editor), *Modern Practice of Liquid Chromatography*, Wiley-Interscience, New York, 1971, p. 161.
Kiselev, A. V., *Advan. Chromatogr.*, 4 (1967) 113.
Knox, J. H., *Anal. Chem.*, 38 (1966) 253.
Knox, J. H. and McLaren, L., *Anal. Chem.*, 35 (1963) 449.
Koch-Light, *Catalogue KL4*, Koch-Light Labs., Colnbrook, 1973, p. 476.
Krejčí, M., *Collect. Czech. Chem. Commun.*, 32 (1967) 1152.
Kubín, M., Špaček, P. and Chromeček, R., *Collect. Czech. Chem. Commun.*, 32 (1967) 3881.
Kun, K. A. and Kunin, R., *J. Polym. Sci., Part B*, 2 (1964) 587 and 839.
Kunin, R., Meitzner, E. F. and Bortnick, N., *J. Amer. Chem. Soc.*, 84 (1962a) 306.
Kunin, R., Meitzner, E. F., Oline, J. A., Fisher, S. A. and Frisch, N. W., *Ind. Eng. Chem., Prod. Res. Develop.*, 1 (1962b) 140.
Lathe, H. and Ruthven, C. R. J., *Biochem. J.*, 62 (1956) 665.
Laurent, T. C., *Biochim. Biophys. Acta*, 136 (1967) 199.
Lea, D. A. and Sehon, A. H., *Can. J. Chem.*, 40 (1962) 159.
Le Page, M., *Ger. Pat., Offen.*, 2,127,629 (1969).
Levin, Y., Pecht, M., Goldstein, L. and Katchalski, E., *Biochemistry*, 3 (1964) 1905.
Lindquist, B. and Storgårds, *Nature (London)*, 175 (1955) 511.
Line, F. W., Kwong, A. and Weltall, H. H., *Biochim. Biophys. Acta*, 242 (1971) 194.
Lloyd, K. O., *Arch. Biochem. Biophys.*, 137 (1970) 460.
Lowe, C. R. and Dean, P. D. G., *FEBS Lett.*, 14 (1971) 313.

Luby, P., Kuniak, L. and Berek, D., *J. Chromatogr.*, 59 (1971) 79.
Martin, A. J. P. and Synge, R. L. M., *Biochem. J.*, 35 (1941) 91.
Martin, J. and Rowland, H., *J. Chromatogr.*, 22 (1967) 139.
Messing, R. A., *Enzymologia*, 39 (1970) 12.
Micheel, F. and Ewers, J., *Macromol. Chem.*, 3 (1949) 200.
Moore, F. L., *U.S. At. Energy Comm., NAS-NS*, (1960) 3101.
Moore, J. C., *J. Polym. Sci., Part A-2*, (1964) 835.
Moore, J. C. and Hendrickson, J. G., *J. Polym. Sci., Part C*, 8 (1965) 233.
Motomato, Y., Hiraymo, Ch. and Matsumoto, K., *Kogyo Kagaku Zasshi*, 74 (1971) 1904.
Nickless, G. and Marshall, G. R., *Chromatogr. Rev.*, 6 (1964) 154.
Nyström, E. and Sjövall, J., *J. Chromatogr.*, 17 (1965) 574.
Parrish, J. R., *Nature (London)*, 207 (1965) 402.
Pensky, J. and Marshall, J. S., *Arch. Biochem. Biophys.*, 135 (1969) 304.
Pepper, K. W. and Hale, D. K., in *Ion Exchange and its Applications*, Society of Chemical Industry, London, 1955, p. 13.
Peterson, E. A. and Sober, H. A., *J. Amer. Chem. Soc.*, 78 (1956) 751.
Pharmacia, *CNBr-activated Sepharose 4B for immobilization of biopolymers*, Pharmacia, Uppsala, 1972.
Pitra, J. and Štěrba, J., *Chem. Listy*, 57 (1963) 389.
Plachenov, T. G., Musakina, V. P., Sevryugov, L. B. and Falchuk, V. M., *Zh. Prikl. Khim.*, 42 (1969) 2020.
Polson, A., *Biochim. Biophys. Acta*, 19 (1956) 53.
Porath, J., *Ark. Kemi*, 11 (1957) 97.
Porath, J., Axén, R. and Ernback, S., *Nature (London)*, 215 (1967) 1491.
Porath, J. and Flodin, P., *Nature (London)*, 183 (1959) 1657.
Porath, J. and Fornstedt, N., *J. Chromatogr.*, 51 (1970) 479.
Porath, J. and Fryklund, L., *Nature (London)*, 226 (1970) 1169.
Porath, J. and Lindner, E. B., *Nature (London)*, 191 (1961) 69.
Randau, D., Bayer, H. and Schnell, H., *J. Chromatogr.*, 57 (1971) 77.
Renn, D. W. and Mueller, G. P., *J. Org. Tech.*, (1968) 277.
Samsonov, G. V. and Ponomareva, R. B., *Stud. Biophys.*, 24-25 (1970) 399.
Samuelson, O., *Ion Exchange Separations in Analytical Chemistry*, Wiley, New York, London, 1963, p. 40.
Sato, S. and Otaka, Y., *Chem. Econ. Eng. Rev.*, 3 (1971) 40.
Sato, T., Mori, T., Tosa, T. and Chibata, I., *Arch. Biochem. Biophys.*, 147 (1971) 788.
Schmit, J. A., in J. J. Kirkland (Editor), *Modern Practice of Liquid Chromatography*, Wiley-Interscience, New York, 1971, p. 375.
Schott, H. and Greber, B., *Makromol. Chem.*, 114 (1971) 333.
Scott, R. P. W. and Lawrence, J. G., *J. Chromatogr. Sci.*, 7 (1969) 65.
Silman, I. H. and Katchalski, E., *Annu. Rev. Biochem.*, 35 (1966) 873.
Snyder, L. R., *J. Chromatogr.*, 28 (1967) 300.
Snyder, L. R., *Principles of Adsorption Chromatography*, Marcel Dekker, New York, 1968a.
Snyder, L. R., *J. Phys. Chem.*, 72 (1968b) 489.
Sober, H. A., Gutter, F. J., Wyckoff, M. M. and Peterson, E. A., *J. Amer. Chem. Soc.*, 78 (1956) 756.
Sober, H. A. and Peterson, E. A., *J. Amer. Chem. Soc.*, 76 (1954) 1711.
Sprossler, B. and Lingens, F., *FEBS Lett.*, 6 (1970) 232.
Steers, E., Cuatrecasas, P. and Pollard, H., *J. Biol. Chem.*, 246 (1971) 196.
Strunz, H., *Mineralogische Tabellen*, Akademische Verlagsgesellschaft, Geest & Portig K. G., Leipzig, 1970.
Tabachnik, M. and Sobotka, H., *J. Biol. Chem.*, 235 (1960) 1051.
Thompson, H. S., *J. Roy. Agr. Soc. Engl.*, 11 (1850) 68.
Truffa-Bachi, P. and Wofsy, L., *Proc. Nat. Acad. Sci. U.S.*, 66 (1970) 685.
Tschesche, H., Dietl, T., Marx, R. and Fritz, H., *Hoppe-Seyler's Z. Physiol. Chem.*, 353 (1972) 483.

Turková, J., Hubálková, O., Křiváková, M. and Čoupek, J., *Abstr. National Meeting Czech. Biochem. Soc., Nitra, 1972*, Abstr. No. A-18, p. 10.
Turková, J., Hubálková, O., Křiváková, M. and Čoupek, J., *Biochim. Biophys. Acta*, 322 (1973) 1.
Updegraff, I. H. and Cassidy, H. G., *J. Amer. Chem. Soc.*, 71 (1949) 407.
Uriel, J., Berges, J., Boschetti, E. and Tixier, R., *C.R. Acad. Sci., Ser. D*, 273 (1971) 2358.
Uziel, M., Koh, C. K. and Cohn, W. E., *Anal. Biochem.*, 35 (1968) 77.
Van Deemter, J. J., Zuiderweg, F. J. and Klinkenberg, A., *Chem. Eng. Sci.*, 5 (1956) 271.
Way, F. T., *J. Roy. Agr. Soc. Engl.*, 11 (1850) 313.
Way, F. T., *J. Roy. Agr. Soc. Engl.*, 13 (1852) 123.
Weetall, H. H., *Nature (London)*, 232 (1969a) 959.
Weetall, H. H., *Science*, 166 (1969b) 615.
Weetall, H. H., *Biochim. Biophys. Acta*, 212 (1970) 1.
Weetall, H. H., *Nature (London)*, 232 (1971) 473.
Weetall, H. H. and Baum, G., *Biotechnol. Bioeng.*, 12 (1970) 399.
Weetall, H. H. and Hersh, L. S., *Biochim. Biophys. Acta*, 185 (1969) 464.
Weetall, H. H. and Hersh, L. S., *Biochim. Biophys. Acta*, 206 (1970) 54.
Weibel, M. K. and Bright, H. J., *Biochem. J.*, 124 (1971) 801.
Weibel, M. K., Weetall, H. H. and Bright, H. J., *Biochem. Biophys. Res. Commun.*, 44 (1971) 347.
Weiss, P. E., *Aust. J. Appl. Sci.*, 4 (1953) 510.
Weliky, N., Weetall, H. H., Gilden, R. V. and Cambell, D. H., *Immunochemistry*, 1 (1964) 219.
Weston, P. D. and Avrameas, S., *Biochem. Biophys. Res. Commun.*, 45 (1971) 1574.
Wiedenhof, N., *Stärke*, 21 (1969) 163.
Wilchek, M. and Gorecki, M., *Eur. J. Biochem.*, 11 (1969) 491.
Wolf, P. and Lindemann, M., *Plaste Kaut.*, 19 (1972) 6, 573.
Yariv, J., Kalb, A. J. and Levitzki, A., *Biochim. Biophys. Acta*, 165 (1968) 303.

*Chapter 10*

# Mobile phases

O. MIKEŠ and R. VESPALEC

## CONTENTS

## MOBILE PHASES FOR LIQUID–LIQUID CHROMATOGRAPHY

### Equilibrium between the mobile and stationary phases

In liquid–liquid chromatography, the fundamental process of distribution is based on dissolution. Hence, this technique may be looked upon as a repeated distribution of samples between two mutually immiscible liquids. The high efficiency of the distribution process may be achieved by a frequent repetition of the elementary distribution process effected by the movement of at least one liquid phase. From these basic characteristics, several conclusions ensue directly:

(1) The stationary and the mobile liquids should be chosen so that they are mutually sufficiently poorly soluble, but sufficiently good solvents for all the components of the sample.

(2) The dissolution isotherm is usually a straight line or can be at least approximated to a straight line better than and in a broader interval than in adsorption. Therefore, symmetrical elution zones can be easily obtained.

(3) As in gas–liquid chromatography, it is usually desirable to suppress maximally the adsorption of the sample on the carrier used.

(4) If the mobile phase and the sample are poorly soluble in the stationary phase employed, the controlling mechanism may consist in the adsorption of the sample on the surface of the stationary liquid. In this case, the adsorbing surface is energetically substantially more homogeneous than the surface of the commonly used adsorbents.

(5) Two completely immiscible and mutually insoluble liquids do not exist. Therefore, the experimental conditions should be chosen so that the passage of the mobile phase through the column does not result in washing out of the stationary phase and hence a change in the column properties.

The prerequisite for a chromatographically utilizable sample retention in a liquid–liquid system is the achievement of a measurable concentration of solute 2 in the mobile phase $m$ and the stationary phase $s$. The magnitude of the elementary partition step is characterized by the experimental distribution coefficient, $K$, which is equal to the ratio of the solute concentrations in the stationary and the mobile phases:

$$K = \frac{[2]_s}{[2]_m} = \lim_{x_2 \to 0} \left( \frac{n_2^s}{V_s} \cdot \frac{V_m}{n_2^m} \right) \tag{10.1}$$

where $n_2^i$ is the number of moles of solute dissolved under the equilibrium state in phase $i$, the volume of which is $V_i$. For very dilute solutions, which are of interest in chromatography, $n_2^i = x_2^i \cdot n^i$, where $x_2^i$ is the molar fraction of the solute in phase $i$, of which $n^i$ moles are present in the column. Then,

$$K = \frac{x_2^s \cdot n_2^s \cdot V_m}{x_2^m \cdot n_2^m \cdot V_s} = \frac{x_2^s \cdot v_m^0}{x_2^m \cdot v_s^0} \tag{10.2}$$

where $v_i^0$ is the molar volume of the pure phase $i$.

At a given pressure and temperature, the requirement for the equilibrium of the component in the mobile and the stationary phases is the equality of chemical potentials:

$$\mu_2^s(T, P) = \mu_2^m(T, P) \tag{10.3}$$

From this basic condition and from the relationship for the chemical potential of the solution component, Locke and Martire deduced the theory of retention for LLC.

The chemical potential of a solution component at any chosen pressure $P$ and temperature $T$, $\mu_2(T, P)$, can be expressed by using the standard chemical potential of the component at the given temperature and standard pressure $P^*$, $\mu^{*,1}(T, P^*)$, the molar volume of the pure solute $v_A^0$, the molar fraction of the solute $x_2^1$, and the activity coefficient at the given pressure and temperature $\gamma_2(P, T)$:

$$\mu_2(T, P) = \mu_2^{*,1}(T, P^*) + (P - P^*)v_2^0 + RT \ln x_2^1 \gamma_2(T, P) \qquad (10.4)$$

Substituting into eqn. 10.3, it follows that

$$\mu_2^{*,m}(T, P^*) - \mu_2^{*,s}(T, P^*) = RT \ln \frac{x_2^s \cdot \gamma_2^s(T, P)}{x_2^m \cdot \gamma_2^m(T, P)} \qquad (10.5)$$

If the state of the pure liquid is chosen as the standard state of the solute for dissolution in a chosen phase at a chosen pressure and temperature, *i.e.*, if $\lim\limits_{x_2^i \to 1} \gamma_2^i = 1$, then

$$\mu_2^{*,m}(T, P^*) = \mu_2^{*,s}(T, P^*) \qquad (10.6)$$

The activity coefficient of the component at a chosen pressure $P$ can be expressed, when the activity coefficient at standard pressure $P^*$ is used, by the equation

$$RT \ln \gamma_2(T, P) = RT \ln \gamma_2(T, P^*) + (\bar{v}_2 - v_2^0)(P - P^*) \qquad (10.7)$$

where $\bar{v}_2$ is the partial molar volume of the component 2 in solution. Combining eqns. 10.5, 10.6 and 10.7, and taking into consideration eqn. 10.2, a relationship for $K$ is obtained. If $P^* = 1$ atm and $P$ (the pressure at which equilibrium is attained in the column) is taken as being equal to the average pressure $\bar{P}$, then

$$\ln K = \ln \left( \frac{\gamma_2^{m,\infty}(T, 1) \cdot v_m^0}{\gamma_2^{s,\infty}(T, 1) \cdot v_s^0} \right) + \frac{(P - 1)}{RT} \cdot \Delta\bar{v}_2 \qquad (10.8)$$

The solutions of the solute 2 in both the pure stationary and mobile liquids, of molar volumes $v_s^0$ and $v_m^0$, respectively, can be considered to be infinitely diluted during chromatographic conditions. Therefore, in eqn. 10.8, the symbols $\gamma_2^m$ and $\gamma_2^s$ are replaced by $\gamma_2^{m,\infty}$ and $\gamma_2^{s,\infty}$. $\Delta\bar{v}_2$ is the difference of partial molar volumes of the solute in the mobile and the stationary phase. If the specific retention volume, $V_g$, is defined by the relationship

$$V_g = V_N / w_s = K / \rho_s \qquad (10.9)$$

where $V_N$ is the corrected retention volume, $w_s$ is the weight of the stationary phase in the column and $\rho_s$ is its density, then

$$\ln V_g = \ln \left( \frac{\gamma_2^{m,\infty}(T, 1) M_m}{\gamma_2^{s,\infty}(T, 1) M_s \cdot \rho_m(T)} \right) + \frac{(\bar{P} - 1)}{RT} \Delta\bar{v}_2 \qquad (10.10)$$

where $M_i$ is the molecular weight of the phase $i$ and $\rho_m(T)$ is the density of the mobile phase at temperature $T$. As $\Delta\bar{v}_2$ is small and the mean pressure on the column is usually several tens of atmospheres, the magnitude of the last term in eqns. 10.8 and 10.10 can be neglected in current analytical practice. Then, approximately

$$K = \frac{\gamma_2^{m,\infty} v_m^0}{\gamma_2^{s,\infty} v_s^0} \tag{10.11}$$

and

$$V_g = \frac{\gamma_2^{m,\infty} M_m}{\gamma_2^{s,\infty} M_s \rho_m} \tag{10.12}$$

From eqns. 10.8 and 10.10–10.12, distribution constants and retention volumes can be predicted for systems in which the activity coefficients of solutes have been determined either by calculation from the theory of solutions or by independent measurements. In contrast, activity coefficients can be derived from chromatographic measurements.

For the relative retention, $\alpha$, of two solutes A and B, which is a measure of selectivity of their separation, it follows from eqns. 10.10 and 10.12 that

$$\alpha = (V_g)_A/(V_g)_B = \left(\frac{\gamma_2^{m,\infty}}{\gamma_2^{s,\infty}}\right)_A \cdot \left(\frac{\gamma_2^{s,\infty}}{\gamma_2^{m,\infty}}\right)_B \tag{10.13}$$

Hence the resolution of two components in LLC is given only by the difference in their activities (deviations from ideality) in the chosen phases. Therefore, the selectivity of the separation of a chosen pair of substances can be regulated by the choice of both the stationary and the mobile phase. As the activity coefficients are temperature dependent, temperature can also be utilized for the control of selectivity.

Deviations from ideal behaviour of the solute in solutions may be positive or negative and be of different magnitudes. In solutions, where only weak interactions of molecules occur due mainly to dispersion forces, the values of activity coefficients may be expressed in values close to unity. Strong intermolecular interactions lead to negative deviations characterized by small values of $\gamma_2^\infty$, which, during complex formation, may be even of the order of $10^{-3}$. In contrast, if the solute--solvent interactions are weaker than the solute–solute or solvent–solvent interactions, then the deviations are positive and the activity coefficients may be high. In analytical practice, it is usually required that $1 \leqslant K \leqslant 100$. The capacity factor, $k'$, which affects the resolution distinctly (*cf.* eqn. 10.28), is connected with the term $K$ by the relationship

$$k' = K \cdot \frac{V_s}{V_m} \tag{10.14}$$

As usually $V_s/V_m \ll 1$, values of $K < 1$ as a rule lead to a low resolving power even when the relative retention of the separated components is good. It therefore follows that the stationary phase should be chosen so that the activity coefficients of the separated components are lower in that phase than in the mobile phase, so that it is a better solvent for the sample.

If the activity coefficient according to Pierotti *et al.* (1956, 1959) is expressed as a function of the structure of the solute and the solvent, then in LLC the activity coefficients of solutes that form homologous series may be described for both phases by the first two terms only. Then, approximately

$$\ln V_g = k_1 + k_2 n_2 \tag{10.15}$$

where $n_2$ is approximately equal to the number of carbon atoms in the solute molecule and $k_1$ and $k_2$ are the differences in the empirical constants of the first two members of the relationship of Pierotti *et al.* for the stationary and the mobile phase. Hence, for each homologous series, a linear relationship between the number of carbons and the logarithm of the specific retention volume may be expected.

The effect of the molecular weights of the stationary and the mobile phases on retention, following from eqn. 10.12, is demonstrated for the case of the stationary phase. Liquids with as low a molecular weight as possible are usually chosen as the mobile phase. However, any liquid can be used as the stationary phase. Often high-molecular-weight compounds are used, as in gas chromatography, and sometimes polymeric compounds. Therefore, it can be stated that the relationship $M_m/M_s \leqslant 1$ applies generally. Let us consider two columns in which two chemically similar liquids, X and Y, of molecular weights $M_X$ and $M_Y$ and densities $\rho_X$ and $\rho_Y$, are used. From the theory of Flory and Huggins, it follows that

$$\ln \frac{(V_g)_X}{(V_g)_Y} \approx \rho_s v_2^0 \left( \frac{1}{M_X} - \frac{1}{M_Y} \right) \tag{10.16}$$

For the deduction, it is assumed that $\rho_X = \rho_Y = \rho_s$ and that the molecular weight of each of the two stationary phases is higher than the molecular weight of the sample.

Hence, with an increase in the molecular weight of the stationary phase (of the same type), the retention decreases, and the decrease is greatest for $M_X/M_Y < 20$ (Fig. 10.1). Similarly, for the relative retention, $\alpha = (V_g)_A/(V_g)_B$, of the separated substances A and B on chemically similar stationary phases X and Y,

$$\ln \frac{\alpha_X}{\alpha_Y} = (v_A^0 - v_B^0) \cdot \rho_s \cdot \left( \frac{1}{M_X} - \frac{1}{M_Y} \right) \tag{10.17}$$

From eqns. 10.16 and 10.17, it is evident that if the molar volumes of the pure solutes, $v_A^0$ and $v_B^0$, are equal, then by changing the molecular weight of the stationary phase only the retention values can be affected. The relative retentions of substances A and B, which represent a measure of the separation selectivity, do not change. If $v_A^0 > v_B^0$, then the relative retention decreases with increasing molecular weight of the stationary phase. In the opposite case, the selectivity increases with increase in $M_X$. The higher the difference in the molar volumes of substances A and B, the lower are the retention volumes and a better resolution is obtained on increasing the molecular weight of the stationary phase. The effect of the molecular weight of the mobile phase is the reverse. In order to achieve short analysis times, it is therefore advisable to choose a mobile phase with a low molecular weight and a stationary phase with a high molecular weight.

The temperature dependence of the retention volume is expressed by the equation

$$R \cdot \frac{\mathrm{d} \ln V_g}{\mathrm{d}(1/T)} \approx \Delta \bar{h}_1 + RT^2 \alpha_m \tag{10.18}$$

where $\Delta \bar{h}_1$ is the partial molar energy of the transfer of the solute from the mobile to the stationary phase, which is equal to the difference in the partial energies of mixing of the solute with the mobile and the stationary phases, and $\alpha_m$ is the temperature coefficient of

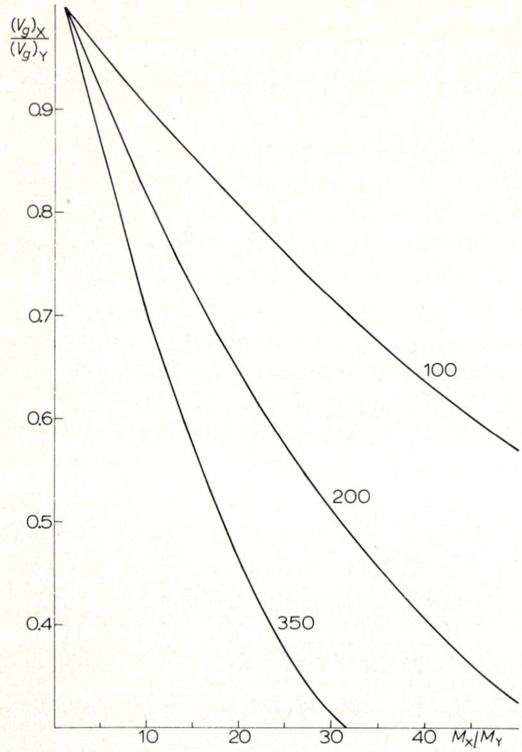


Fig. 10.1. Change in the retention of the solute with change in molecular weight of the stationary phase, calculated according to eqn. 10.16. $M_X = 2 \cdot 10^4$; $\rho_S \cdot v_2^0 = 100$, 200 and 350.

the mobile phase dilatation. In typical cases, $R \cdot \frac{\mathrm{dln}\, V_g}{\mathrm{d}(1/T)}$ usually has a value of about 3 kcal/mole, which is of the same order as those in gas chromatography. It therefore follows that in order to achieve good reproducibility of retention measurements in LLC, the temperature of the column should be thermostatted to *ca.* ±0.1°C. It was found also experimentally that in LLC the elution volume changes distinctly with temperature. Schmit *et al.* found that on increasing the temperature by 30°C, the elution volume decreases by about 50%.

The effect of pressure on the specific retention volume is very small. At medium pressures in the column, *i.e.*, units to tens of atmospheres, the simplified equation 10.12 can be employed. The general relationship of eqn. 10.10 can be used even for very accurate measurements up to mean pressures of $10^2$–$10^3$ atm, because the difference in the partial molar volumes of the solute in the mobile and the stationary phases, $\Delta\bar{v}$, is very small. The effect of pressure on $\Delta\bar{v}$ should be considered only for pressures higher than $10^3$ atm. Hence, it may be said that for current analytical practice, the simple relationship of eqn. 10.12 should suffice in most instances. Only for very accurate measurements, carried out at mean pressures in the column, *i.e.*, about 100 atm and higher, should a more accurate relationship be used.

**Selectivity of the liquid–liquid systems and solubility coefficients**

The most important criterion for the choice of the partition system in liquid–liquid chromatography is its selectivity. Selectivity is a property characteristic of a given system of two phases with respect to the chosen solutes. Therefore, it is possible to discuss only the effect of a single phase on the system selectivity in an actual system. In addition, each component of the chosen liquid pair, especially if both are of low molecular weight, can be used, in principle, as either the stationary or the mobile phase. When selectivity is calculated on the basis of the relative retention of the separated components, it is irrelevant which of the liquids was considered as the stationary or mobile phase.

In practice, the system of phases is most often chosen empirically. The main reasons for this are the inadequacy of the theory of solutions for giving an effective and sufficiently precise prediction of partition coefficients in any given system, a lack of data necessary for the prediction (molar volumes, critical temperatures and solubility parameters) and, to a certain extent, also the difficult calculations encountered. Another route based on the theory of solutions is the determination of partition coefficients from the characteristic parameters of the sample and both liquid phases, which was proposed by Huber *et al.* In this case, a procedure can be used based on the theory of regular solutions, according to Hildebrand and Scott (1962, 1964), or on the method of structural increments (Deal *et al.*; Derr *et al.*; Huggins; Martire; Pierotti *et al.*, 1956, 1959; and others). The sum of the empirical partition coefficients may also be useful when choosing a system of phases.

In general, it can be stated that one of the conditions necessary for the achievement of adequate selectivity in an LLC system are the differences in the properties of the phases that are usually indicated as differences in polarity. Therefore, before discussing selectivity, a description of these properties should be given.

The so-called polarity of a liquid phase is, in fact, the result of intermolecular interactions in the liquid. These intermolecular interactions are divided into two basic types: unspecific, which are caused by dispersion forces, and specific, which are caused by other interactions. The different types and intensities of these interactions are evident not only from the different behaviour of liquids as solvents, but also from their boiling points and latent heats of evaporation. The stronger are the intermolecular interactions, the more effective the given liquid usually is as a solvent for polar solutes and thus it also has a higher heat of evaporation. Its magnitude, relative to unit volume, was used by Hildebrand and Scott (1962, 1964) for the quantitative characterization of the magnitude of intermolecular forces. They introduced as one of the fundamental concepts of the theory of regular solutions the so-called solubility parameter, $\delta$, defined as the square root of the ratio of the molar heat of evaporation, $\Delta E^V$, and the molar volume, $V$:

$$\delta = \left(\frac{\Delta E^V}{V}\right)^{\frac{1}{2}} \tag{10.19}$$

The solubility parameters of liquids used as mobile phases range between 5.00 for perfluoroalkanes and 21 for water. In addition to Hildebrand and Scott's data, the solubility parameters of various compounds are listed in papers by Burrel, Crowley *et al.*, Gardon, Lieberman, Morrison and Freiser, and Polak. The critical properties of liquids, necessary for calculations, were summarized by Kudchadker *et al.*

TABLE 10.1
SELECTED SOLVENT CHARACTERISTICS

| Solvent | Molecular weight | Density (g/ml) | Viscosity (cP, 20°C) | Surface area of the molecule, $S^A$* | $\epsilon^0$ ($Al_2O_3$)** | $\delta$*** | $\delta_d$ | $\delta_o$§ | $\delta_a$§ | $\delta_h$§ |
|---|---|---|---|---|---|---|---|---|---|---|
| Fluoroalkanes§§ | | | | | −0.25 | 6.0 | 6.0 | 0 | 1.0 | 0 |
| $CFCl_2-CF_3$ | 170.92 | 1.455 | | | | 6.2 | 5.9 | 1.5 | 0 | 0 |
| Isooctane§§§ | 114.23 | 0.691 | 0.50 | 7.6 | 0.01 | 7.0 | 7.0 | 0 | 0 | 0 |
| Diisopropyl ether | 102.18 | 0.724 | 0.37 | 5.1 | 0.28 | 7.0 | 6.9 | 0.5 | 0.5 | 0 |
| *n*-Pentane | 72.15 | 0.626 | 0.23 | 5.9 | 0.00 | 7.1 | 7.1 | 0 | 0 | 0 |
| $CCl_3-CF_3$ | 187.38 | 1.579 | | | | 7.1 | 6.8 | 1.5 | 0.5 | 0 |
| *n*-Hexane | 86.18 | 0.660 | 0.33 | | 0.01 | 7.3 | 7.3 | 0 | 0 | 0 |
| *n*-Heptane | 100.21 | 0.684 | 0.41 | | 0.01 | 7.4 | 7.4 | 0 | 0 | 0 |
| Diethyl ether | 74.12 | 0.714 | 0.28 | 4.5 | 0.38 | 7.4 | 6.7 | 2 | 2 | 0 |
| Triethylamine | 101.19 | 0.725 | | | | 7.5 | 7.5 | 0 | 3.5 | 0 |
| *n*-Decane | 142.29 | 0.730 | 0.92 | 10.3 | 0.04 | 7.8 | | | | |
| Cyclopentane | 70.13 | 0.751 | 0.47 | 5.2 | 0.05 | 8.1 | 8.1 | 0 | 0 | 0 |
| Cyclohexane | 84.17 | 0.779 | 1.00 | 6.0 | 0.04 | 8.2 | 8.2 | 0 | 0 | 0 |
| *n*-Propyl chloride | 78.54 | 0.892 | 0.35 | 3.5 | 0.30 | 8.3 | 7.3 | 3 | 0 | 0 |
| Isopropyl chloride | 78.54 | 0.859 | 0.33 | 3.5 | 0.29 | 8.4 | | | | |
| Tetrachloromethane | 153.84 | 1.595 | 0.97 | 5.0 | 0.18 | 8.6 | 8.6 | 0 | 0.5 | 0 |
| Diethyl sulphide | 90.19 | 0.837 | 0.45 | 5.0 | 0.38 | 8.6 | 8.2 | 2 | 0.5 | 0 |
| Ethyl acetate | 88.10 | 0.901 | 0.45 | 5.7 | 0.58 | 8.6 | 7.0 | 3 | 2 | 0 |
| Propylamine | 59.11 | 0.719 | | | | 8.7 | 7.3 | 4 | 6.5 | 0.5 |
| Ethyl bromide | 108.99 | 1.456 | | 3.4 | 0.37 | 8.8 | 7.8 | 3 | 0 | 0 |
| *m*-Xylene | 106.16 | 0.864 | 0.62 | 7.6 | 0.26 | 8.8 | 8.8 | 0 | 0.5 | 0 |
| Toluene | 94.12 | 0.866 | 0.59 | 6.8 | 0.29 | 8.9 | 8.9 | 0 | 0.5 | 0 |
| Chloroform | 119.39 | 1.489 | 0.57 | 5.0 | 0.40 | 9.1 | 8.1 | 3 | 0.5 | 0† |
| Tetrahydrofuran | 72.10 | 0.888 | | 5.0 | 0.45 | 9.1 | 7.6 | 4 | 3 | 0 |
| Methyl acetate | 74.08 | 0.934 | 0.37 | 4.8 | 0.60 | 9.2 | 6.8 | 4.5 | 2 | 0 |
| Benzene | 78.12 | 0.879 | 0.65 | 6.0 | 0.32 | 9.2 | 9.2 | 0 | 0.5 | 0 |
| Perchloroethylene | 236.74 | 2.091 | | | | 9.3 | 9.3 | 0 | 0.5 | 0 |
| Ethyl methyl ketone | 72.10 | 0.805 | | 4.6 | 0.51 | 9.3 | | | | 0 |
| Acetone | 58.08 | 0.792 | 0.32 | 4.2 | 0.56 | 9.4 | 6.8 | 5 | 2.5 | 0 |
| Dichloromethane | 84.94 | 1.326 | 0.44 | 4.1 | 0.42 | 9.6 | 6.4 | 5.5 | 0.5 | 0 |
| Chlorobenzene | 112.56 | 1.107 | 0.80 | 6.8 | 0.30 | 9.6 | 9.2 | 2 | 0.5 | 0 |

| | | | | | | | | | | |
|---|---|---|---|---|---|---|---|---|---|---|
| Anisole | 108.14 | 0.995 | 1.32 | | | 9.7 | 9.1 | 2.5 | 2 | |
| 1,2-Dichloroethane | 98.96 | 1.256 | 0.79 | 4.8 | 0.49 | 9.7 | 8.2 | 4 | 0 | 0 |
| Methyl benzoate | 166.18 | 1.157 | | | | 9.8 | 9.2 | 2.5 | 1 | 0 |
| Dioxane | 88.10 | 1.035 | 1.54 | 6.0 | 0.56 | 9.8 | 7.8 | 4 | 3 | 0 |
| Amyl alcohol | 88.16 | 0.814 | 4.1 | 8.0 | 0.61 | 9.8 | | | | |
| Methyl iodide | 141.94 | 2.279 | 0.50 | | | 9.9 | 9.3 | 2 | 0.5 | 0 |
| Bromobenzene | 157.03 | 1.499 | 1.19 | | | 9.9 | 9.3 | 1.5 | 0.5 | 0 |
| Carbon disulphide | 76.14 | 1.263 | 0.37 | 3.7 | 0.15 | 10.0 | 10.0 | 0 | 0.5 | 0 |
| *n*-Propanol | 60.09 | 0.803 | 2.3 | 8.0 | 0.82 | 10.2 | 7.2 | 2.5 | 4 | 4 |
| Pyridine | 79.10 | 0.982 | 0.93 | 5.8 | 0.71 | 10.4 | 9.0 | 4 | 5 | 0 |
| Benzonitrile | 103.12 | 1.010 | 1.24 | | | 10.7 | 9.2 | 3.5 | 1.5 | 0 |
| Nitromethane | 61.04 | 1.130 | 0.67 | 3.8 | 0.64 | 11.0 | 7.3 | 8 | 1 | 0 |
| Nitrobenzene | 123.12 | 1.200 | 2.03 | | | 11.1 | 9.5 | 4 | 0.5 | 0 |
| Ethanol | 46.07 | 0.789 | 1.20 | 8.0 | 0.88 | 11.2 | 6.8 | 4.0 | | |
| Phenol | 94.12 | 1.071 | 12.7 | | | 11.4 | 9.5 | | | |
| Dimethylformamide | 73.10 | 0.944 | | | | 11.5 | 7.9 | | | |
| Acetonitrile | 41.05 | 0.783 | 0.37 | 10.0 | 0.65 | 11.8 | 6.5 | 8 | 2.5 | 0 |
| Methylene iodide | 267.87 | 3.226 | | | | 11.9 | 11.3 | 1 | 0.5 | 0 |
| Acetic acid | 60.05 | 1.049 | 1.26 | 8.0 | 1.0 | 12.4 | 7.0 | | | |
| Dimethyl sulphoxide | 78.13 | 1.101 | 2.2 | 4.3 | 0.6 | 12.8 | 8.4 | 7.5 | 5 | 0 |
| Methanol | 32.04 | 0.791 | 0.60 | 8.0 | 0.95 | 12.9 | 6.2 | 5 | 7.5 | 7.5 |
| Ethanolamine | 61.08 | 1.018 | | | | 13.5 | 8.3 | | | |
| Ethylene glycol | 67.02 | 1.109 | 19.9 | 8.0 | 1.11 | 14.7 | 8.0 | | | |
| Formamide | 45.04 | 1.134 | | | | 17.9 | 8.3 | | | |
| Water | 18.02 | 0.998 | 1.00 | | | 21 | 6.3 | | | |
| Amyl chloride | 106.60 | 0.883 | 0.43 | 4.2 | 0.26 | | | | | |
| Isobutyl methyl ketone | 100.16 | 0.801 | | 5.3 | 0.43 | | | | | |
| 1-nitropropane | 89.09 | 1.022 | | 4.5 | 0.53 | | | | | |
| Aniline | 93.13 | 1.022 | 4.4 | 6.7 | 0.62 | | | | | |
| Diethylamine | 73.14 | 0.711 | 0.38 | 7.5 | 0.63 | | | | | |

*The surface area of the molecule as derived by Snyder from chromatographic measurements, expressed in relative units. For transformation to Å², the values given should be multiplied by 8.5.

**Elution strength in liquid–solid systems for elution from alumina.

***Solubility parameters calculated from boiling points.

§ Approximate values.

§§ Average values for various compounds.

§§§ 2,2,4-Trimethylpentane.

† Some papers give higher values.

When liquids are ordered according to their increasing solubility parameters, a series is obtained which is very similar to the sequence of some eluotropic series. Hence, the solubility parameter $\delta$ can also be considered to be a measure of the elution power of a solvent in liquid–liquid systems. In spite of this, when this series is compared with the actual elution behaviour, in some instances (for example, according to Macek and Procházka), appreciable discrepancies are found. This is probably due to the effect of specific interactions of various types in these liquids, which, however, need not display their effects to the same extent during the dissolution of samples of various types. It follows that the magnitude of $\delta$, characterizing the magnitude of intermolecular forces, may not always be sufficiently reliable. Keller *et al.* have shown that, assuming that individual contributions to the solubility parameter can be added to give a total, resultant, solubility parameter, $\delta$ may be divided into four contributions: the contribution of dispersion forces, $\delta_d$; the contribution of the orientation of molecules, $\delta_o$; the contribution given by the ability of accepting protons, $\delta_a$; and the contribution due to the ability of donating protons, $\delta_h$. Each of these contributions, with the exception of $\delta_d$, is in fact a measure of another type of "polarity", and the percentage fraction of each is different for different substances. It therefore follows that a single precise and generally applicable eluotropic series cannot exist. When an eluotropic series is composed for an actual case, in addition to the dispersion contribution, those contributions to the total polarity which might apply in the considered case must be taken into account. However, the lack of necessary numerical data is a practical problem. A review of selected characteristics that are important for the selection of mobile phases is given in Table 10.1. This table shows that there is no direct relationship between $\epsilon^0$, characterizing the elution power in liquid–solid systems, and the solubility parameter, $\delta$. The reasons for this were discussed in detail by Keller and Snyder.

From the thermodynamic point of view, the problem of selectivity in LLC for similar (and hence poorly separable) solutes was investigated by Martire and Locke. Assuming that

$$\ln \gamma_2 = \ln \gamma_2^{\text{th}} + \ln \gamma_2^{\text{ath}} \tag{10.20}$$

they demonstrated that even for relative retention the following equation can be written:

$$\ln \alpha = (\ln \alpha)^{\text{th}} + (\ln \alpha)^{\text{ath}} \tag{10.21}$$

The component $(\ln \alpha)^{\text{ath}}$ is given by athermal, configurational contributions of activity coefficients, $\gamma_2^{\text{ath}}$, following from the different sizes of the solute and solvent molecules. It only reflects the effects of the arrangement and therefore it can be considered to be a purely entropic term. The thermal component $(\ln \alpha)^{\text{th}}$ is determined by thermal contributions of activity coefficients, $\gamma_2^{\text{th}}$, following from the intermolecular forces between the solute and solvent molecules. According to Martire, it is given by both enthalpic and entropic contributions to non-ideality.

Using Flory–Huggins' relationships, it can be shown for athermal contributions to the relative retention that

$$(\ln \alpha)^{\text{ath}} = \ln \left(\frac{\gamma_A^m}{\gamma_B^m}\right)^{\text{ath}} - \ln \left(\frac{\gamma_A^s}{\gamma_B^s}\right)^{\text{ath}} = (v_B^0 - v_A^0)\left(\frac{1}{v_m^0} - \frac{1}{v_s^0}\right) \tag{10.22}$$

From eqn. 10.22, it follows that the greater is the difference between the molar volumes of the pure stationary and mobile phases, $\nu_s^0$ and $\nu_m^0$, and those of the two solutes A and B, $\nu_A^0$, $\nu_B^0$, the greater is also the athermal term. The contribution of the athermal term to the relative retention leads to the result that even compounds that undergo absolutely identical energetic interactions with both phases (for example, isomeric solutes) and that differ sufficiently only in their molar volumes (densities), can be separated successfully by LLC. The solute with the lower molecular weight dissolves preferentially in phases with a larger molar volume, because a larger increase in entropy takes place during dissolution.

For the thermal contribution to the relative retention, it can be deduced that

$$(\ln \alpha)^{\text{th}} = \ln \left(\frac{\gamma_A^m}{\gamma_B^m}\right)^{\text{th}} - \ln \left(\frac{\gamma_A^s}{\gamma_B^s}\right)^{\text{th}} = \frac{\nu_A^0}{RT} \cdot (\delta_s^2 - \delta_m^2) \cdot \frac{T_A^* - T_B^*}{T_A^*} \tag{10.23}$$

where $\nu_A^0$ is the molar volume of component A, approximately equal to $\nu_B^0$; $\delta_s$ and $\delta_m$ are Hildebrand and Scott's (1962) solubility parameters for the mobile and the stationary phase; and $T_A^*$ and $T_B^*$ are the critical temperatures of substances A and B. The thermal contribution increases with increase in the difference between the intermolecular forces in the mobile and stationary phases, which always exist if the miscibility of the liquids is limited. For a given pair of solvents, the thermal contribution increases with the magnitude of the product $\nu_A^0 \cdot \frac{T_A^* - T_B^*}{T_A^*}$ and hence a requirement for separation, based on energy terms, is a difference between the critical temperatures of the solutes A and B. A substance with a higher critical temperature is dissolved preferentially in the phase with a higher coefficient of solubility. If this solvent is the stationary phase, the substance with a higher critical temperature has a greater retention, while in the opposite case it is eluted preferentially. In view of the approximations that have not been discussed for the sake of brevity, the relationship for the thermal contribution is not applicable (a) to systems in which the solvent–solvent interactions are stronger than solute–solvent interactions, (b) to the separation of substances that differ appreciably in their behaviour and (c) to systems in which adsorption phenomena in interphases occur.

Every system of solvents can therefore be characterized by the terms $\left(\frac{1}{\nu_m^0} - \frac{1}{\nu_s^0}\right)$ and $(\delta_s^2 - \delta_m^2)$. If the stationary phase is less polar than the eluent (which is common in separations of non-polar or weakly polar solutes), then the elution sequence is in accordance with the increase in the molecular weights. If the eluent molecules are smaller than the molecules of the stationary phase, as is usual, then the contribution of the thermal term to the total retention is negative. In addition, $\delta_m^2 > \delta_s^2$ and, because $T$ decreases with increasing molecular weight, the thermal contribution term is positive and larger than the athermal term. During analyses of polar solutes, the stationary phase must be more polar than the mobile phase. If, in this instance, the stationary phase has a higher molecular weight, both contributions are positive. It therefore follows that for separation in a liquid–liquid system, the value of the thermal term is usually decisive. The athermal (configurational) term, the absolute contribution of which is usually smaller, may either enhance or adversely affect the separation. As the contribution of both terms during the separation of strongly polar solutes is positive, it is evident that LLC has potentially a higher selectivity for the separation of polar substances. The contribution of the thermal

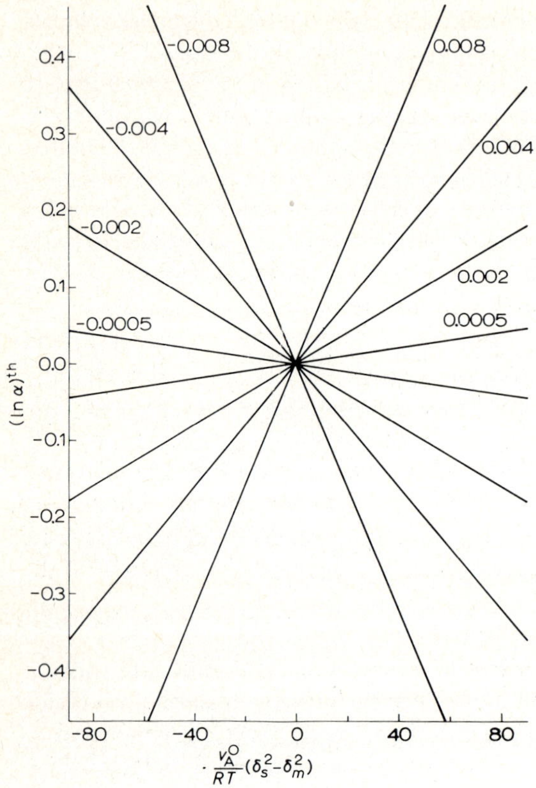


Fig. 10.2. Contribution of the thermal term, $(\ln \alpha)^{th}$, to the relative retention, $\alpha$, for various values of $f(T^*) = (T_A^* - T_B^*)/T_A^*$, calculated according to eqn. 10.22.

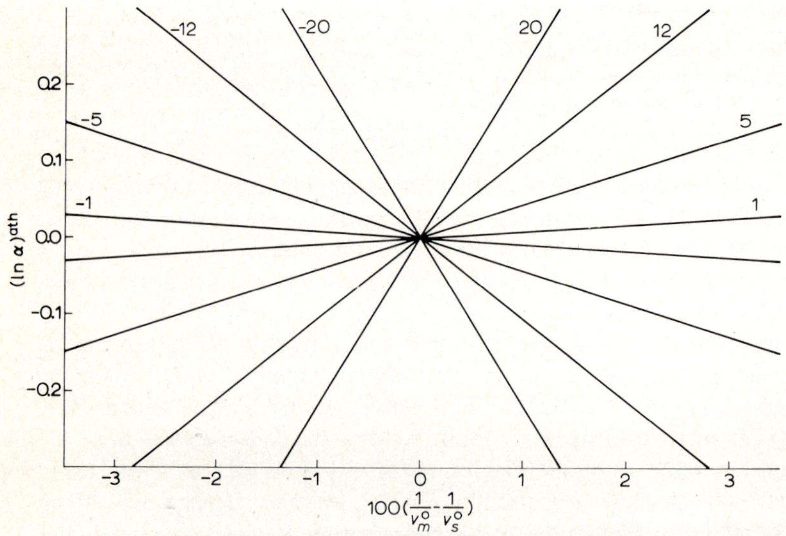


Fig. 10.3. Contribution of the athermal term, $(\ln \alpha)^{ath}$, to the relative retention, $\alpha$, for various values of $(v_A^0 - v_B^0)$, calculated according to eqn. 10.23.

and athermal terms to the relative retentions is illustrated by Figs. 10.2 and 10.3.

As already mentioned, a requirement for selectivity in LLC is the immiscibility of both chromatographic phases. From the theory of solutions, it follows that the higher is the heat of mixing of two liquids, $\Delta H^M$, the less miscible they are. For the mobile and the stationary phases, the corresponding relationship assumes the form:

$$\Delta H^M = (x_m v_m^0 + x_s v_s^0)(\delta_s^2 - \delta_m^2)\,\Phi_s\,\Phi_m \qquad (10.24)$$

where $x_s$ and $x_m$ are the molar fractions of the stationary and mobile phases, respectively, and $\Phi_s$ and $\Phi_m$ are their corresponding volume fractions in the resulting solution. Hence, the miscibility is the lower the larger are the molar volumes and the greater is the difference between their solubility parameters. In view of the magnitude of the solubility parameters, the phase should be chosen so that $|\delta_s - \delta_m| \geqslant 4$ and the solubility parameters of all separated components are between the values $\delta_s$ and $\delta_m$. If, however, $\delta_s$ and $\delta_m$ differ too widely, adsorption could take place in the liquid–liquid interphase. The immiscibility of the liquids $m$ and $s$ may be enhanced by the choice of a stationary liquid with a high molecular weight. On the other hand, it is advisable to avoid the use of mobile phases of high molecular weight, because they have an excessively high viscosity, which is disadvantageous from the point of view of column efficiency (see p.249).

The requirements for the choice of a suitable carrier for a given type of stationary phase are discussed in Chapter 9.

**Programming of the mobile phase**

When analyzing samples containing components of various types, capacity factors of the components very often exceed the optimum range $1 \leqslant k_i \leqslant 10$, so that the choice of constant conditions does not lead to a satisfactory separation. In a system that permits a good resolution of the most easily eluted components the last components are usually eluted over excessively long intervals and their zones are extremely broadened, which impairs detection or makes it impossible. Under the conditions suitable for elution of the slowest components, $k_i'$ values for the first components are much less than unity, so that even at a sufficient column effectivity and suitable values of the relative retentions, the separation in this region is not effective. The so-called effect of small $k'$ applies here. The problem can be solved either by changing the column efficiency or the $k_i'$ value during the analysis; the latter method is more effective and is therefore more often used.

*Programming of the solvent flow*

The column efficiency can be controlled by changing the linear flow-rate of the mobile phase in a continuous or discontinuous manner. On decreasing the flow-rate at the beginning of the analysis, the column efficiency may be increased. An increase in the flow-rate at the end of the analysis, accompanied by a decrease in column efficiency, can be utilized only if the last components are sufficiently well separated. The programming of the flow is suitable only when the lowest $k_i'$ values are not much less than unity, and the highest values do not exceed a value of about 10. The condition for this is that the column efficiency should change sufficiently with the flow-rate, *i.e.*, the constant $C$ in eqn. 10.29

should have a sufficiently high value. For systems characterized by a very small dependence of the efficiency on the mobile phase flow-rate, the technique of flow programming is not very effective.

From the practical point of view, this technique is very advantageous because of its simplicity and low demand on apparatus. Chemical properties and the physical state of the column do not change during the flow programme, which is important from the point of view of rapid repeatability of the analyses. As the value of $k'$ does not change during the flow programming, the elution volumes remain constant.

*Gradient elution*

The most commonly used technique of programming in liquid chromatography is the regulated change of the mobile phase composition, *i.e.*, the so-called gradient elution. Its most important effect is the control of capacity factors, which are regulated so that their values for all eluted components vary within the limits $1 \leqslant k' \leqslant 10$. The effect of changes in the viscosity of the mobile phase on the column efficiency, caused by the change in composition, is usually negligible. Concentration gradients may be either discontinuous, created by one or several stepwise concentration changes of two or more components, or continuous of various types. Snyder and Saunders have shown that the most suitable is a continuous logarithmic concentration gradient, which gives the most effective separations with respect to time. In this method of separation, the value of $k'$ for each eluted substance during its passage through the column is close to the optimum. The composition, and hence also the elution power of the mobile phase can be changed within a broad range, which permits the effective control of the magnitude of the capacity factor over an exceptionally wide range.

However, the successful application of this method assumes the fulfilment of several conditions. Firstly, a suitable stationary phase must be chosen, or the programme should be adapted to a particular stationary phase. In systems with a liquid stationary phase, the danger of increased washing out arises during the changes in the mobile phase polarity. In some types of chemically bound phases, chemical reactions with the components of the mobile phase may take place under certain conditions, even if these are generally characterized by a much higher stability. Adsorbents show the highest resistance towards the effect of the mobile phase and therefore they are the most suitable for use with concentration gradients. Therefore, in classical adsorption liquid chromatography or in thin-layer chromatography, where the preparation of the chromatographic bed is rapid and cheap, the use of simple gradients has found wide application.

In high-efficiency liquid chromatography, complicated devices (see Chapter 8) are necessary for the generation of concentration gradients at high pressures. During gradient elution, the capacity factors change in time as a function of the chosen form of the gradient. If reproducibility of elution volumes and quantitative data is to be achieved, which would be comparable with the results obtained during work at a constant composition of the mobile phase, it is necessary that the form of the concentration gradient and its course with time should be reproducible as precisely as possible. After each analysis, complete regeneration of the column used is essential. This regeneration requires the complete washing out of all components of the mobile phase that passed through the column during the programme and the re-attainment of the equilibrium state that existed in the column before the sample injection.

In liquid–solid systems, regeneration is often time consuming, and it is advantageous to increase the mobile phase flow-rate or to change the column temperature. If the column is not completely regenerated, reproducibility of the measurements cannot be achieved. The gradient elution adversely affects the conditions of the functioning of detectors that are sensitive to the composition of the mobile phase used, especially in binary detectors (refractometers, etc.).

In spite of these disadvantages, the gradient technique remains the most effective means for the analysis of complex mixtures that contain substances of different chemical types. The development of the necessary instrumentation and the investigations devoted to the theory and practice of gradient elution indicate that this technique will find wide utilization in modern liquid column chromatography.

*Temperature programming*

Relatively little attention has so far been paid to the possibility of temperature programming, which in liquid–liquid systems follows from the dependence of activity coefficients on temperature, and in liquid–solid systems from the values of heats of adsorption. During temperature programming, both the capacity factors and the column efficiency change. According to Schmit *et al.*, an increase in temperature of 60°C decreases the HETP by half. The capacity factor usually decreases with increasing temperature; in some instances the increase was observed only in systems with a multicomponent mobile phase.

In liquid–liquid systems, this technique meets with difficulties caused by the increased solubility of the stationary phase in the mobile phase at elevated temperatures. If the temperature increases over the whole analytical column simultaneously, provision cannot be made that the mobile phase present in the column should be in equilibrium with the stationary phase at every moment and at each point in the column, even if the temperature of the pre-column changes at the same rate. A partial solution of this problem consists in heating the pre-column to a temperature slightly higher than that of the column (Kirkland, 1971b). However, the temperature programme of the pre-column should be chosen carefully with respect to the column diameter and the flow-rate.

If the mobile phase is a single compound, the situation is similar to that in temperature programming in gas chromatography. Lebedeva *et al.* have shown that in this way the retention can be regulated in liquid–solid systems even when the heats of adsorption of the solutes are of the order of kilocalories per mole. The higher the heat of adsorption, the more distinct is the effect of temperature.

If a temperature programme is applied to a system with a binary or a more complex phase, then the increase in temperature also leads to changes in equilibrium between the mobile and the stationary phases, which result in changes in the composition of the mobile phase during the programme. This effect was made use of by Scott and Lawrence in the programme of a temperature-regulated gradient elution characterized by a stable, instantaneous equilibrium between the stationary and mobile phases along the whole column. The resulting effect is comparable with the influence of the concentration gradient prepared in the conventional manner, but with the advantage of a better resolution of the eluted components. Its disadvantage is that a relatively lengthy regeneration of the column and pre-column is required, which, however, can be shortened by using exchangeable pre-columns.

## MOBILE PHASES FOR LIQUID–SOLID CHROMATOGRAPHY

### Properties of the mobile phase

When comparing gas and liquid chromatography, it is evident that almost all of the basic differences between them, both theoretical and practical, follow from the different properties and the behaviour of the mobile phase in the chromatographic system. They appear in all basic aspects: in the thermodynamics of the chromatographic process, in its dynamics, in the operating technique and in the instrumentation.

The physical basis of these differences is the different intermolecular interactions in the liquid and in the gas, which is directly evident, for example, from the very great differences in their physical properties (density, compressibility, viscosity, etc.) and in the diffusion coefficients. In a chromatographic system with a liquid mobile phase, and also in other liquid systems, intermolecular forces are the cause of the dissolution of the sample components in the mobile phase. They lead to their molecular dispersion, which is a prerequisite of the chromatographic process. Energetically important interactions of the mobile phase molecules with those of the stationary phase distinctly affect and modify the interaction of the sample with the stationary phase. Together with the mobile phase–solute interactions, they often distinctly regulate the resulting selectivity of the chromatographic system used, so that in liquid chromatography, the mobile phase cannot be regarded as an inert component and its presence cannot be neglected or its effects underestimated.

As the effect of the mobile phase in liquid chromatography is more complex than in gas chromatography, the criteria for its choice must also be more varied. It should also be kept in mind that the choice is not only dependent on other parameters (separation selectivity, equipment, operating technique, etc.) but often also exerts a reverse influence. If the basic requirement for achieving the chromatographic process is fulfilled, *i.e.*, the sample is sufficiently soluble in the chosen mobile phase, it is important that this mobile phase should also fulfil some other requirements, as follows:

(1) In combination with the stationary phase, it should ensure, at least for the most important components of a given sample, a sufficient selectivity of separation, which is essential for good resolution. For the capacity factors of single samples, $k_i'$, $1 \leqslant k_i' \leqslant 10$ should usually apply.

(2) It should enable such experimental conditions or operating procedure to be chosen as would prevent changes in the column properties. If conditions (1) and (2) cannot be fulfilled simultaneously, another mobile or stationary phase should be selected.

(3) It should satisfy the requirements for work with the most suitable or at least a satisfactory type of detector.

(4) It should possess suitable properties that permit the maximum efficiency of the chromatographic bed and the shortest time of analysis to be achieved (low viscosity, high diffusion coefficients of the sample components).

(5) It should permit the recovery of the separated components.

(6) It should be inert to all of the materials of construction of the chromatograph with which it comes into contact during the measurements.

(7) It should be safe and economically acceptable, or be capable of regeneration after use.

Of course, all of these requirements do not always apply to the same extent. The first two requirements, even if only on broad lines and according to general principles, should be considered in relation to the operating techniques used, and the basic types of chromatographic systems or the type of sample serve as criteria.

Points (3), (6) and (7) do not require further discussion. The recovery of the separated components from the eluate is usually effected by the difference between their boiling points and those of the mobile phase. The effect of the physical properties of the mobile phase on the efficiency of the chromatographic bed will differ according to the shape of the bed (thin layer or a column) and, to a certain extent, also according to the method chosen. Point (4) applies equally to both column adsorption chromatography (LSC) and partition chromatography (LLC) because, from the point of view of the dynamics of the chromatographic process, the role of the mobile phase is the same in both instances. Therefore, the effect of the physical properties of the mobile phase on the efficiency of the chromatographic column will be discussed before the analysis of the role of the mobile phase in LSC and LLC.

## Effect of the physical properties of the mobile phase on column efficiency

On the supposition that a system with an optimum, or at least an acceptable, selectivity has been found for a given analysis, mainly two criteria are important for estimating the efficiency of the analysis: the resolution of individual components and the time necessary for the analysis. Both parameters are affected to a considerable extent by the experimental arrangement (column length, pressure drop, linear flow-rate of the mobile phase) and the dynamics of the transport of the sample through the column.

Assuming that the total time of analysis is equal to the elution time of the last detected zone, $t_Z$, then

$$t_Z = t_0(1 + k'_Z) = \frac{L}{u}(1 + k'_Z) \tag{10.25}$$

For an efficient column and a sufficiently large capacity factor, $k'_Z$, and for a linear distribution isotherm, the elution time of the latter part of the last zone is negligible with respect to the total time of analysis. For given experimental conditions, the column length, $L$, and the capacity factor, $k'_Z$, are constants. The elution time of the unretained component, $t_0$, is indirectly proportional to the linear flow-rate of the mobile phase in the column, $u$ (cm/sec). From the equation for the column permeability, $K^0$, the mobile phase flow-rate in the column can be expressed as a function of the pressure drop, $\Delta P$, the mobile phase viscosity, $\eta$, the column length, $L$, and its total porosity, $f$:

$$u = K^0 \cdot \frac{\Delta P}{\eta L f} \tag{10.26}$$

Substituting this expression into eqn. 10.25, it follows that

$$t_Z = \frac{L^2 f \eta}{K^0 \Delta P} \cdot (1 + k'_Z) \tag{10.27}$$

It is evident that $t_Z$ is directly proportional to $\eta$, if the other terms are constant. The viscosity of the mobile phase can in practice be changed either by a change in the composition of the mobile phase or by an increase in temperature. However, an increase in temperature is generally accompanied by a change, usually a decrease, in the capacity factor. The change in the composition can usually be carried out so that $k'_Z$ remains constant if other conditions do not change.

The resolution, $R_s$, of two components A and B can be expressed by the relationship

$$R_s = \frac{2(t_B - t_A)}{w_B + w_A} = \frac{1}{4} \cdot \underset{(a)}{\frac{\alpha - 1}{\alpha}} \cdot \underset{(b)}{\frac{k'_B}{1 + k'_B}} \cdot \underset{(c)}{\sqrt{N}} \tag{10.28}$$

where $t_A$ and $t_B$ are uncorrected elution times, $w_A$ and $w_B$ are the peak widths at the baseline expressed as time, $\alpha$ is the relative retention, $k'_B$ is the capacity factor of the later eluted component B and $N$ is the number of theoretical plates of the column used. Terms (*a*) and (*b*) are unambiguously determined by the chosen chromatographic system at a given temperature and are independent of the physical properties of the mobile phase. The effect of the physical properties of the mobile phase on the resolution becomes evident only in term (*c*), which characterizes the efficiency of the column used. Consideration of the effect of the physical properties of the mobile phase on the resolution can therefore be replaced by their effect on the efficiency.

The effect of the properties of the mobile phase on the column efficiency is complex because it is connected with the dependence of the plate height on the linear flow-rate of the mobile phase. For the sake of clarity, a simple equation for the theoretical plate height may serve as the starting point:

$$H = A + Cu \tag{10.29}$$

This equation is most often used to describe the results in high-speed liquid chromatography. The mass transfer resistance coefficient, $C$, can be divided into the contribution of the resistance to the mass transfer in the stationary phase, $C_s$, in the mobile phase fraction contained in the interparticulate space, $C_m$, and in the immobile fraction of the mobile phase contained within the packing particles, $C'_m$:

$$C = C_s + C_m + C'_m \tag{10.30}$$

The contribution of the mobile phase can be expressed in the form

$$C_m = \omega \cdot \frac{d_p^2}{D_m} \tag{10.31}$$

$$C'_m = \varphi(\Phi', k') \frac{d_p^2}{D_m}$$

where $\omega$ is a constant the magnitude of which is given by the non-uniformity of the mobile phase flow through the bed, $d_p$ is the average particle diameter, $D_m$ is the diffusion coefficient of the solute in the mobile phase, $\varphi(\Phi', k')$ is a term dependent on the mobile phase fraction occupying the space within the particles, $\Phi'$, and $k'$ is the capacity factor of the solute. For our purposes, the contribution of the stationary phase may be considered to be constant. The term $A = 2\lambda d_p$ is also independent of the physical properties of the mobile phase because the constant $\lambda$ characterizes the bed geometry. After substituting in eqn. 10.29, it follows that

$$H = A + \left\{C_s + \frac{d_p^2}{D_m}[\omega + \varphi(\Phi', k'_B)]\right\} u \tag{10.32}$$

$$= A + \left(C_s + \frac{d_p^2}{D_m} \cdot \text{constant}\right) u$$

From eqn. 10.32, it follows that at a given flow-rate, the contribution to the plate height of the mass transfer resistance in the mobile phase is indirectly proportional to the diffusion coefficient, so that with increasing $D_m$, the column efficiency, and hence the resolution, also increase.

Substituting eqn. 10.26 for $u$ in eqn. 10.32, it becomes evident that the column efficiency is also affected by $\eta$:

$$H = A + \left(C_s + \frac{d_p^2}{D_m} \cdot \text{constant}\right) \cdot \frac{\Delta P}{\eta L f} \cdot K^0 \tag{10.33}$$

As the viscosity of the mobile phase also affects the magnitude of the diffusion coefficients (a more viscous liquid lowers the diffusion rate and decreases the diffusion coefficient value), $D_m$ is not constant when $\eta$ is changed in eqn. 10.33.

The effect of the viscosity on the efficiency in liquid–solid systems was investigated experimentally by Snyder (1967, 1969). He found that when the viscosity of the mobile phase is doubled, the analysis time is also doubled when the pressure drop remains constant. For a constant time of analysis (constant $u$), however, the column efficiency decreases. It can be deduced that by increasing the viscosity 2½-fold, the plate height is approximately doubled. The effect of the viscosity on the separation efficiency in liquid–liquid systems was described by Kirkland (1971a). Reversed-phase systems were studied from this point of view by Schmit *et al.*, who found that when the temperature is increased from 20°C to 80°C, the column efficiency approximately doubles. From the point of view of the rate of analysis and the resolution, it is therefore preferable if mobile phases with as low a viscosity as possible are chosen, or if the work is performed at elevated temperatures.

*References p.277*

As a series of liquids of very low viscosity (about 0.2–0.3 cP) exists, the viscosity of the mobile phase can be kept below 0.4–0.6 cP even when more viscous liquids are added. Only when aqueous solutions or phases that contain a large proportion of water are used can viscosities below 1 cP not be achieved.

**Strength of the mobile phase**

In the conventional arrangement of high-efficiency adsorption column chromatography, the adsorbent is wetted with the mobile phase. If the molecules of the transported sample A are to be retained by the surface, then a certain number of solvent molecules must first be expelled from the adsorbed layer. On the other hand, the surface set free by sample desorption is immediately occupied by molecules of the mobile phase M. In the zone moving through the column, a process is taking place that is characterized by the equilibrium equation

$$A_m + n\,M_s \rightleftharpoons A_s + n\,M_m \tag{10.34}$$

The subscripts $m$ and $s$ indicate molecules in the mobile and stationary phase, respectively. For the sake of simplicity and clarity, it is useful to assume that the mobile phase is composed only of a single type of molecules, that the injected sample is dissolved in the mobile phase, and that the interaction of the sample with the mobile phase does not lead to the so-called secondary solvent effects (hydrogen bonds, complex formation, etc.). The coefficient $n$ in eqn. 10.34 is equal to the ratio of the effective adsorption cross-sections of the sample molecule and the mobile phase, and is therefore not a whole number.

The net energy of adsorption during this process, $\Delta E$, can be expressed as

$$\Delta E = E_s^{\mathrm{A}} + n\,E_m^{\mathrm{M}} - E_m^{\mathrm{A}} - n\,E_s^{\mathrm{M}} \tag{10.35}$$

where $E_s^{\mathrm{A}}, E_s^{\mathrm{M}}, E_m^{\mathrm{A}}, E_m^{\mathrm{M}}$ are the molar free energies of the interaction of sample molecules with the mobile phase molecules in the adsorbed layer and in the solution. Under the above conditions, to a first approximation the liquid phase energy terms $E_m^{\mathrm{A}}, nE_m^{\mathrm{M}}$ can usually be ignored. Snyder (1964b) also corroborated experimentally that their contribution in typical adsorption systems is less than 10% of the contribution of the corresponding adsorption terms. This finding can be easily explained. If no specific interactions are involved in the solution with the mobile phase (in the case of non-polar or weakly polar solvents), the intermolecular interactions are caused by non-specific Van der Waals forces. The magnitude of these dispersion forces per molecule is proportional to its cross-section. From eqn. 10.34, it follows that the surface area of the sample molecule is equal to $n$ times the area of the mobile phase molecule. In solution, the sample molecule is surrounded by mobile phase molecules and, at low sample concentrations and the covering of the adsorbent surface usual in chromatography, mutual interaction of the sample molecules does not come into consideration in the adsorbed layer either. Contact of the solvent molecules with the sample molecules is therefore disturbed only at the site of contact of the sample molecule with the adsorbent surface. Therefore, with complete solvation of the adsorbed sample molecules, a maximum of 50% of the solvation energy, represented by the term $E_m^{\mathrm{A}}$, can be released after desorption. Similar conditions also exist during the desorption of the mobile phase molecules, which causes the contribution of the dissolving members to the total energy change to decrease further. In addition, the terms that characterize adsorption interactions are usually much higher, especially the term $E_s^{\mathrm{A}}$. In the opposite case, the value of the distribution coefficient is so low with respect to the relationship

$$\Delta E = RT \ln K \tag{10.36}$$

that a satisfactory separation cannot be achieved, so that the total energy effect of the adsorption process is approximately

$$\Delta E \approx E_s^{\mathrm{A}} - n\,E_s^{\mathrm{M}} \tag{10.37}$$

where the terms $E_s^{\mathrm{A}}$ and $E_s^{\mathrm{M}}$ represent the molar free adsorption energies of the sample and the mobile phase, respectively, on the pure adsorbent surface.

It is evident that the more strongly the mobile phase molecules are bound by the surface, the lower is the value of $\Delta E$ for a given sample during elution. Therefore, more strongly adsorbed mobile phases decrease the retention and act as more effective, stronger elution agents. Hence, the adsorption value of the mobile phase molecules is a measure of their elution strength.

On the basis of the above, with respect to the magnitude of the adsorbing surface area and its activity, $\beta$, and assuming the validity of Langmuir's isotherm for the adsorption of the mobile phase and sample adsorption, Snyder (1968) derived a very useful relationship for the prediction of the adsorption distribution coefficient, $K$, of the component A:

$$\log K = \log V_a + \beta(E^{\mathrm{A}} - S^{\mathrm{A}}\epsilon^0) \tag{10.38}$$

The magnitude of $K$ gives the ratio of the sample concentrations in the stationary and mobile phases. Numerically, it is also equal to the corrected retention volume in millilitres and referred to 1 g of adsorbent. The volume of the adsorbed phase, $V_a$, is equal to the product of the specific adsorbent surface area, $S$ ($m^2/g$), and the thickness of the adsorption layer. To a first approximation, when it can be considered as being equal to the thickness of a monomolecular layer of adsorbed water,

$$V_a = 0.000{,}35\, S \tag{10.39}$$

Under this supposition, $V_a$ is a constant characteristic of a given adsorbent and independent of the nature of the sample. However, in general, the above supposition is not completely valid because the layer thickness depends on the shape and arrangement of the molecules. $S^{\mathrm{A}}$ is the area of the adsorbent surface occupied by the sample molecule (adsorption cross-section). $E^{\mathrm{A}}$ is proportional to the energy of adsorption of the sample molecule from pentane solution*.

The symbol $\epsilon^0$, *i.e.*, the strength of the mobile phase, characterizes the magnitude of the interaction of the mobile phase molecules with the surface. According to Snyder (1968), its numerical value is proportional to the interaction energy of the mobile phase

*Snyder (1968) used a different approach, so that with his original symbols eqn. 10.38 has the following form:

$$\log K^0 = \log V_a + \alpha(S^0 - A_s\epsilon^0)$$

with the pure adsorbent surface less the interaction of the pentane molecule with the adsorbent surface relative to unit area. For pentane as the mobile phase, $\epsilon^0 = 0$.

From the mobile phase elution strength, it follows that the numerical value of $\epsilon^0$ is dependent on the adsorbent used. Experimentally, it was found that the result determined for elution from alumina can be applied directly to other adsorbents of the oxide type. For silica gel, magnesium oxide and Florisil, the following relationships can be used:

$$\epsilon^0(SiO_2) = 0.77\ \epsilon^0(Al_2O_3) \tag{10.40a}$$

$$\epsilon^0(\text{Florisil}) = 0.53\ \epsilon^0(Al_2O_3) \tag{10.40b}$$

$$\epsilon^0(MgO) = 0.58\ \epsilon^0(Al_2O_3) \tag{10.40c}$$

The scattering of the calculated values, ±0.04 unit, differs only slightly from the experimental results and it can therefore be considered that a single eluotropic series can be proposed for all adsorbents of the oxide type.

During chromatography on non-polar adsorbents, polar interactions that are important in adsorbents of the oxide type cannot apply. Therefore, the polarity and polarizability of the molecules virtually does not affect the elution strength, and the magnitude of the non-specific Van der Waals forces is the only factor. The elution strength in these instances increases approximately in proportion with the increase in the molecular size and the sequence of the elution strength is generally reversed in comparison with polar adsorbents, as was shown, for example, by Schorn, Williams *et al.* and others (compare Tables 10.1 and 10.2).

The effect of the elution strength of the mobile phase on the distribution coefficient during the elution of a sample from an adsorbent follows from eqn. 10.38. For the ratio of the distribution coefficients of the given solute during elution with the mobile phases 1 and 2, the equation

$$\log\frac{K_1}{K_2} = \beta S^A(\epsilon_2^0 - \epsilon_1^0) \tag{10.41}$$

applies.

TABLE 10.2
ELUOTROPIC SERIES FOR ELUTION FROM CHARCOAL ACCORDING TO SCHORN

The elution strength increases from the first member (water) to the last (benzene).

| Solvent | Surface area of the molecule (Å²) |
|---|---|
| Water | 12.7 |
| Methanol | 24.6 |
| Ethanol | 32.3 |
| Acetone | 35.7 |
| Propanol | 40.0 |
| Diethyl ether | 47.6 |
| Butanol | 47.6 |
| Ethyl acetate | 48.4 |
| *n*-Hexane | 57.8 |
| Benzene | 51.0 |

Without changing the surface area or the type of adsorbent, the sample retention can be changed merely by changing the mobile phase elution strength. The retention change is the greater the more active is the adsorbent used and the greater the area occupied by the sample. The greater these values, the smaller is the change in the elution strength necessary for the required shift of $K$. On the other hand, a given pair of mobile phases affects the retention of solutes that differ in the cross-section of the molecule in different ways. Eqn. 10.41 represents the theoretical basis for the gradient elution technique in LSC.

*Elution strength of the binary mobile phase*

The description and quantitative evaluation of the function of a multicomponent mobile phase in a chromatographic system is complicated. For the sake of simplicity, only two-component systems will be described here, as they represent a special case of more complex phases. As two-component mobile phases permit a wide control of the selectivity and the elution strength, their use in practice is also more advantageous than more complex phases. The more complicated the mobile phase used, the more difficult is a preliminary estimation of its function in a chromatographic system, its preparation and the maintenance of constant properties during the measurement. The choice of a simpler mobile phase is also preferred from the point of view of the interpretation of the separation processes.

A mixed mobile phase can be considered as a simple mobile phase with the addition of a sample. Therefore, in a similar manner to adsorption of the sample from a simple mobile phase, a dynamic equilibrium takes place when a binary mobile phase is introduced. This is described by the equation

$$\mathrm{M}_m + \mathrm{N}_s \rightleftharpoons \mathrm{M}_s + \mathrm{N}_m \qquad (10.42)$$

where N represents the more strongly sorbed component.

For the formulation of the relationship for the elution strength of a binary mobile phase, Snyder (1968) made the following assumptions:

(1) No secondary effects of the mobile phase are observed in the system.

(2) During the passage of the mobile phase through the adsorbent bed, no change in composition takes place which might cause the formation of a front of the stronger component of the solvent mixture, also indicated by demixing of the mobile phase. This condition is fulfilled only in the elution method after the passage of a volume of mobile phase that permits the front of the more strongly sorbed component to leave the column. If the mobile phase is introduced on to a dry adsorbent during development (dry column technique), or if a continuous concentration gradient is applied, this condition is never fulfilled. The formation of a front can be neglected, as a rule, only if a mobile phase is employed in which the concentrations of components are comparable and the elution strengths not too different, and if the most rapidly eluted sample component has a retention such that it is eluted only after the front of the more strongly sorbed component.

(3) The molecules of the mobile phase components are of equal size ($n = 1$).

(4) Adsorption of both components on the surface of the stationary phase is described by Langmuir's equation.

*References p.277*

(5) In the description of the competing adsorption equilibrium, solvation terms are neglected (see eqn. 10.35). The interaction of molecules in the adsorbed layer is also negligible.

For the elution strength of a binary mobile phase, $\epsilon^0_{MN}$, he obtained the following relationship:

$$\epsilon^0_{MN} = \epsilon^0_M + \frac{\log(x_N 10^{\beta S^N}(\epsilon^0_N - \epsilon^0_M) + 1 - x_N)}{\beta S^N} \qquad (10.43)$$

where $\epsilon^0_M$ and $\epsilon^0_N$ are the elution strengths of the pure components on a given adsorbent of activity $\beta$, $S^N$ is the area occupied by a molecule N of the mobile phase and $x_N$ is the molar fraction of the stronger sorbed component.

The elution strength of the binary mobile phase, and hence of every more complex mobile phase, depends not only on its composition and the elution strengths of the individual components and their molecular size, but also on the activity of the adsorbent used. Therefore, an eluotropic series that consists of simple and binary mobile phases, which would be the same for different adsorbents of the same type and also for the same adsor-

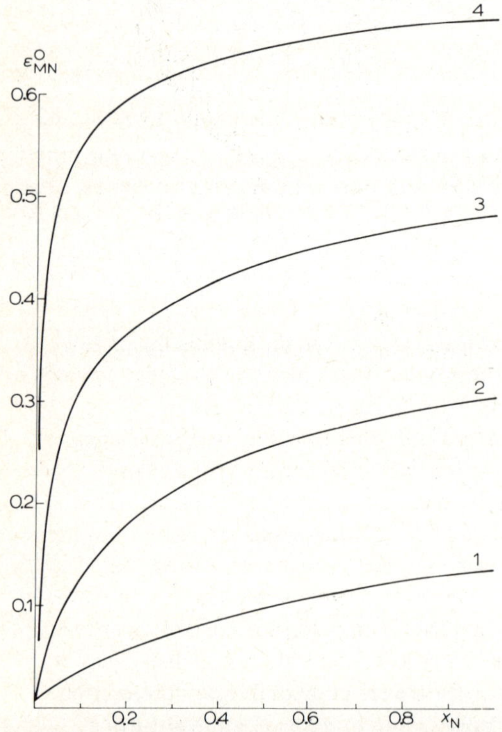


Fig. 10.4. Dependence of the elution strength of a binary mobile phase on its composition, for elution from silica gel. Specific surface area *ca.* 300 m$^2$/g, deactivation with 2% of water; $\beta$ = 0.71. 1 = Hexane–carbon tetrachloride; 2 = hexane–chloroform; 3 = hexane–diethylamine; 4 = hexane–ethanol. The curves were calculated from eqn. 10.43. In the region of the validity of the equation, the calculated values differ from the experimental values by 0 ± 0.02 to 0.03 $\epsilon^0$ units.

bents with different activities, cannot be devised. It is important that the elution strength of a binary mobile phase changes most rapidly with the content of the stronger component in the region of low concentrations (Fig. 10.4). The component present in very low concentrations was termed a moderator by Maggs.

Eqn. 10.43 can also be used for the estimation of the elution strength of a multicomponent system if it is chosen so that the elution strengths of the two strongest components are much higher than the strengths of the remaining components. In such a case, the adsorbent surface is covered only by molecules of the strongest components, and the difference $1 - x_N$ should be replaced in eqn. 10.43 by the molar fraction of the strongest component.

A detailed analysis of the elution strength of the mobile phase and of the conditions of the validity and applicability of the above relationships was given by Snyder (1968). Quantitative data characterizing the mobile phases are summarized in Table 10.1. Other magnitudes (adsorbent activities, $\beta$; areas of the surface occupied by the molecules of various samples, $S^A$; volumes of adsorbed layers, $V_a$) necessary for calculations according to eqns. 10.38, 10.41 and 10.43 were also given by Snyder (1968).

### *Demixing effects*

The basis for the deduction of the sample retention and the relationship for the elution strength of a binary mobile phase consisted in competing adsorption of the molecules on the adsorbing surface, characterized at each point of the column by dynamic equilibrium between the molecules in the mobile and the adsorbed phases. From the point of view of this process, it is irrelevant whether the component added to the mobile phase acts as the sample or another mobile phase component; only the method of its introduction into the column is of importance. If the component is introduced into the column as a discrete pulse, it is considered to be a sample. As a result of its passage through the column, its original shape, usually rectangular, is gradually changed until at the column outlet an elution zone of the conventional shape is obtained. The maximum concentration of the zone, $c_{max.}$, is always lower than the originally introduced concentration, $c_0$.

If the added component of original concentration $c_0$ is introduced into the column continuously, the technique is called continuous or frontal injection, which leads to the formation of a frontal chromatographic zone. The maximum equilibrium concentration in the frontal zone is equal to the injected concentration, $c_0$. Its formation, in relation to the shape of the adsorption isotherm, was described qualitatively for the first time by De Vault. He demonstrated that if the adsorption system is characterized by a convex isotherm (isotherm of type I), then the frontal zone has a sharp front and a diffuse rear. Systems with a convex isotherm (type III) lead to frontal zones with a diffuse front and a steep rear (for types of isotherms, see Brunauer, Brunauer *et al.* or Young and Crowell). As a consequence of the added component being caught on the surface, its concentration decreases to zero in the first fractions of the injected sample. Its elimination by adsorption leads to a situation where the frontal zone formed moves through the column at a lower rate than the pure mobile phase. The stronger is the added component adsorbed and the lower its concentration, $c_0$, the more slowly its front moves through the column. It is evident that the frontal zone is formed in all instances, regardless of whether the component acts as the sample or an additional component of the mobile phase. The position and the shape of the zone are determined by the shape of the adsorption isotherm, concentration $c_0$, amount of adsorbent and the flow-rate. For binary liquid systems, these processes were treated quantitatively, for equilibrium and non-equilibrium courses of the adsorption process, by Glueckauf (1945; 1947a, b, c; 1949).

If several components are injected simultaneously with a simple mobile phase, the number of fronts formed equals the number of components present. However, it can be demonstrated that the zones are also formed during the changes in concentration of the mobile phase components. The amount of the corresponding component in the stationary phase must always be changed so that it is in equilibrium with its content in the mobile phase. If the concentration increases, an adsorption front is formed, and if it decreases, a desorption front appears. As the attainment of equilibrium always takes a certain time, it appears that in this process, during the control of the mobile phase composition at the column outlet, more strongly adsorbed components, *i.e.*, those with a greater elution strength, added at the column inlet were demixed in the column.

The formation of frontal chromatographic zones in the mobile phase, the so-called demixing, is, therefore, a general phenomenon caused by the attainment of adsorption distribution equilibria in the chromatographic system under dynamic conditions, during changes in the composition of the frontally injected mobile phase. It always takes place regardless of the bed shape (column, thin-layer) and whether and how the bed was previously wetted. Therefore, a description and quantitative evaluation of the demixing effect should be undertaken from these points of view. When passing from a single to a binary mobile phase, the shape and the course of the zone formed, and also the volume of the phase necessary for its elution from the column, can generally be described by using known procedures. Certainly, for the interpretation or the prediction of the retention data, the described effects represent an undesirable and complicating factor, especially in work with changing gradients. As is shown for polyzonal thin-layer chromatography, sometimes they can be utilized with advantage. In other instances a volume of mobile phase must be allowed to pass through the column such that all zones formed during the change in composition are eluted. Only then can the column be considered to be conditioned and able to give equilibrium data reproducibly.

## Selectivity of the mobile phase

The basic equation 10.38 for the magnitude of the retention in liquid–solid systems was deduced on the assumption that the contribution of the liquid phase terms to the total energy change of the process (see eqn. 10.35) is negligible. This assumption is usually satisfactory for weak and not too strong mobile phases. When such phases are used, usually samples in which the components are chemically related or very similar in their adsorption behaviour and dissolving properties are analyzed. From the point of view of molecular interactions, this means that the interactions in the mobile phase are weak and non-specific (see the section Strength of the mobile phase). The interactions of the molecules of single sample components with the surface of the stationary phase are substantially of the same type and do not differ in principle from the type and magnitude of interactions of the surface with the mobile phase molecules. Large differences lead to excessively high distribution coefficients, which are unsuitable from the chromatographic point of view.

In strong mobile phases (with respect to polar sorbents of the silica gel type), both assumptions are, however, only rough approximations. From the definition of the concept of elution strength, it follows directly that in the case of strong mobile phases during the interaction with the surface of the adsorbent, interactions of other types (usually indicated as polar or specific) are involved in addition to dispersion intermolecular interactions. The molecules of the sample components must also undergo specific interactions during contact with the adsorbent. However, these interactions are even stronger, and if this were not so, elution of all components would occur in the dead volume. However, if the values of the distribution coefficients are to be maintained within the range $1 \leqslant K \leqslant 100$, it is necessary, with respect to eqn. 10.36, for $\Delta E$ to remain within the same range of values as in the work with weak mobile phases. Only in a few instances can it be assumed, according to eqn. 10.37, that the difference in the adsorption interactions of the sample and mobile phase molecules will fulfil this condition. Therefore, it follows from eqn. 10.35 that the liquid phase terms must play an important role in the total energy balance. Larger differences in interactions in an adsorbed state must be compensated by larger differences in interactions in the mobile phase. This, however, is possible only if specific interactions between solvent molecules or between molecules of solute and solvent will, even in solution (in the mobile phase), occur at least in some instances. Then, of course, the basic conditions from which eqn. 10.38 was deduced are no longer valid and therefore it becomes at most a rough approximation. The magnitude of specific interactions in the mobile phase, indicated in the literature as secondary solvent effects, should be amended by introducing a correction term, $\Delta$. Then

$$\log K = \log V_a + \beta(E^A - S^A \epsilon^0) + \Delta \qquad (10.44)$$

For a given adsorbent, $\Delta$ is a function of the solute and solvent compositions.

It is evident that the secondary solvent effects complicate the quantitative interpretation of the results and the prediction of the retention value according to eqn. 10.38. On the other hand, they permit the improvement of the separation selectivity or the achievement of the separation of components that are inseparable in the usual systems (*i.e.*, in systems without secondary effects). From the secondary effects of the mobile phase, the second important mobile phase characteristic follows, *i.e.*, selectivity.

The magnitude of the relative adsorption and the migration rate of the sample through the bed are regulated by the choice of the elution strength of the solvent and the values of the distribution coefficients are optimized. The elution strength of the mobile phase can be considered to be independent on the sample type. However, the differences in the relative migration rates of the components of a given sample can be changed by controlling the selectivity of the mobile phase. It is evident that the selectivity of a certain mobile phase can be discussed only in connection with the given solutes.

During the practical solution of an analytical problem, first a suitable elution strength of the mobile phase should be found for a given sample and a certain adsorbent, in order to optimize the distribution coefficients. Then, if the resolution of some sample components is insufficient, even after some simple adjustments of experimental conditions (column length, flow-rate, column temperature, etc.), the required improvements can often be achieved by changing the selectivity of the mobile phase by changing its composition. However, the elution strength must remain constant even if the composition is changed. The so called equi-eluotropic series according to Neher can serve as a guide (Fig. 10.5). For each mixture of two components in this nomogram, the elution strength increases from left to right. The composition of the binary mobile phases of identical

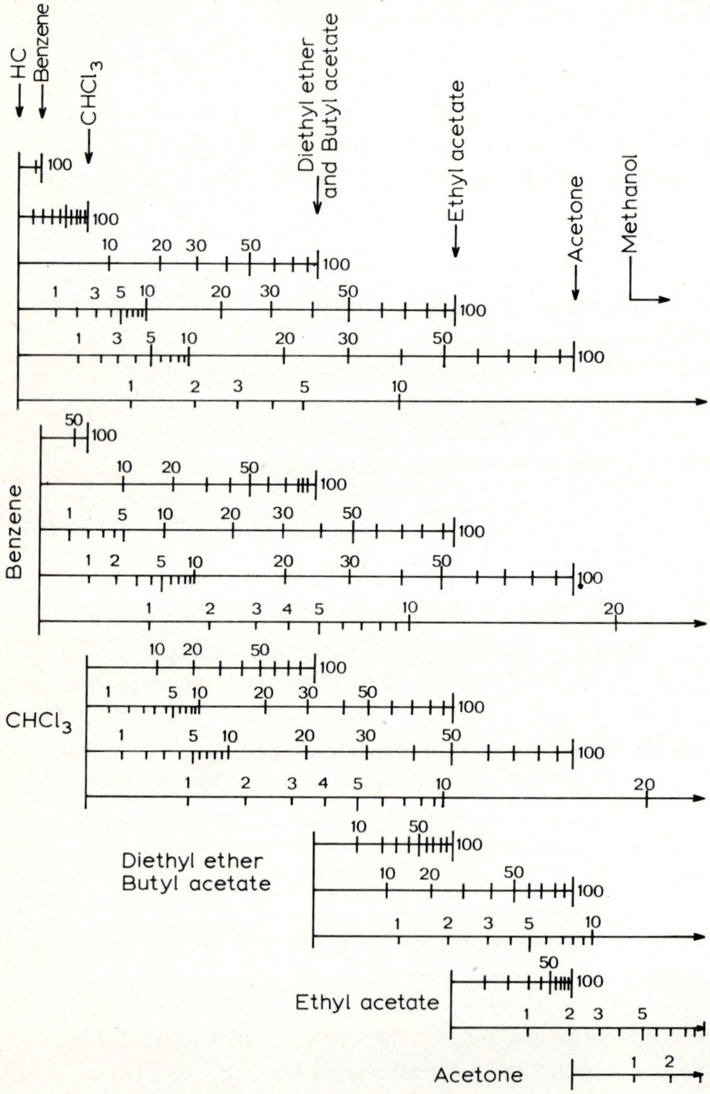


Fig. 10.5. Equi-eluotropic series according to Neher. The numbers correspond to the content (%, v/v) of the stronger component in a binary mobile phase. The composition of mobile phases with equal elution strengths can be read from the intersection points of the lines parallel to the Y (coordinate) axis (in the direction of the arrows). HC = saturated hydrocarbons and cyclic hydrocarbons.

elution strength can be read from the intersection points in a vertical line. The values obtained serve for orientation only.

If preliminary considerations are inconclusive as to which solvent could affect the selectivity of the resulting mobile phase satisfactorily, the consecutive testing of single members of the equi-eluotropic series may prove time consuming. Insufficient attention has been devoted to the study of mobile phase selectivity in liquid–solid systems and therefore no general rules, relationships or numerical data can be given that would permit the choice of a mobile phase of suitable selectivity.

## MOBILE PHASES FOR ION-EXCHANGE CHROMATOGRAPHY

### Aqueous solutions

Most ion-exchange processes and ion-exchange chromatography are carried out in aqueous solutions, which are the best solvents for these purposes. Electrolytic dissociation in water immediately splits the dissolved substance that is to be sorbed into ions and the swelling of ion exchangers in water enables the ions so formed quickly to penetrate to functional groups, which in water are also in a dissociated or partly dissociated form. All these circumstances are in favour of rapid exchange in comparision with non-aqueous or mixed solutions.

#### *Importance of ionic strength and pH of the mobile phase*

There are two main properties that must be considered carefully in ion exchange in aqueous solutions: pH and ionic strength. Very acidic or very basic solutions regenerate cation and anion exchangers to their $H^+$ or $OH^-$ (base) form or they convert the anion and cation exchanger completely into the corresponding salt forms. Any sorption of other ions under these conditions is not possible or is very limited. Therefore, in most instances ion-exchange chromatography is effective only in a certain range of pH values, which can be derived from titration curves of ion exchangers and are summarized in Table 10.3 for

TABLE 10.3
USEFUL RANGE OF pH VALUES OF MOBILE PHASES FOR ION-EXCHANGE CHROMATOGRAPHY

| Type of ion exchanger | pH range for maximal capacity of ion exchanger* |
|---|---|
| Strongly acidic cation exchanger | 2–12 |
| Moderately acidic cation exchanger | 4–12 |
| Weakly acidic cation exchanger | 5–12 |
| Strongly basic anion exchanger | 12–2 |
| Moderately basic anion exchanger | 8–2 |
| Weakly basic anion exchanger | 6–2 |

*This pH range is limited by the stability of ion exchangers. The producer of Sephadex (Pharmacia, Uppsala, Sweden) recommends the following pH ranges: SP-Sephadex, pH 2–10; CM-Sephadex, pH 6–10; QAE-Sephadex, pH 10–2; DEAE-Sephadex, pH 9–2.

various exchangers. Within these limits, the ion exchangers display a good capacity for sorption of corresponding ions.

The ionic strength, $\mu$, is a measure of the intensity of the electrical field due to the ions in the solution. It is defined as half the sum of the terms obtained by multiplying the concentration, $c_i$ (or molality), of each ionic species in the solution by the square of its valence, $z_i$:

$$\mu = \frac{1}{2} \sum c_i z_i^2$$

The ionic strength of mobile phases strongly influences the available capacity of exchangers. A high ionic strength in general decreases the sorption of sorbed ions (or zwitter-ions) also at non-extreme values of pH. Therefore, at the beginning of the chromatography the mobile phase must have a sufficiently low ionic strength. When operating with buffered ion exchangers (see below), the starting value for many applications is $\mu$ = 0.05 or 0.1 and the final value is about $\mu$ = 0.5; it does not usually exceed $\mu$ = 2, at which value regenera-

TABLE 10.4
CHOICE OF THE MOBILE PHASE FOR ION-EXCHANGE CHROMATOGRAPHY USING CATION EXCHANGERS

| Matrix and mobile phase | Form of ion exchanger | | |
|---|---|---|---|
| | $H^+$ | Mixed form ($H^+$ + another cation) | Buffered columns |
| Composition of the matrix | Stable resins | Resins | Resins, cellulose, polydextran |
| Usual aqueous mobile phases | (1) Aqueous solution of a suitable acid<br>(2) Sequence of aqueous solutions of the same acid with increasing concentration<br>(3) Sequence of aqueous solutions of various acids with increasing acidity and concentration | (1) Aqueous solution of a weak or medium base<br>(2) Sequence of solutions of bases with increasing basicity | (1) Anionic buffer with constant pH and low ionic strength<br>(2) Sequence of anionic buffers with constant pH and increasing ionic strength<br>(3) Sequence of buffers with increasing basicity and increasing ionic strength |
| Mixed mobile phases | Solutions of polar organic solvents in water or dilute acids | Addition of organic solvents to mobile phase in instances when the desorbed base is not soluble in the aqueous phase | Admixture of small amount of organic solvent influences the relative sorption of solutes |
| Special mobile phases | Solutions of substances capable of forming complexes or derivatives with sorbed cations (selective elution) or solutions capable of keeping the solute in a complex form during the whole chromatographic process. | | |

TABLE 10.5
CHOICE OF THE MOBILE PHASE FOR ION-EXCHANGE CHROMATOGRAPHY USING ANION EXCHANGERS

| Matrix and mobile phase | Form of ion exchanger | | |
|---|---|---|---|
| | $OH^-$ | Mixed form ($OH^-$ + another anion) | Buffered columns |
| Composition of the matrix | Stable resins | Resins | Resins, cellulose, polydextran |
| Usual aqueous mobile phases | Aqueous solutions of bases or their mixtures with increasing concentration or basicity | Aqueous solutions of acids or their mixtures with increasing concentration or acidity | (1) Cationic buffer with constant pH and low ionic strength (2) Sequence of cationic buffers with constant pH and increasing ionic strength (3) Sequence of buffers with increasing acidity and increasing ionic strength |
| Mixed mobile phases | Solutions of polar organic solvents in water or dilute bases | Addition of organic solvents to mobile phase in instances when the desorbed acid is not soluble in the aqueous phase | Admixture of small amount of organic solvent influences the relative sorption of solutes |
| Special mobile phases | Solutions of substances capable of forming or decomposing complexes or derivates with sorbed anions (selective elution) or solutions capable of keeping the desired solute in a complex form during the whole chromatographic process. | | |

tion proceeds. In some instances, a lower ionic strength (*e.g.*, $\mu = 0.01$) is used when substances are sorbed on ion exchangers or in the first stages of the chromatography. Some types of flexible ion exchangers with a low degree of cross-linking (*e.g.*, polydextran ion exchangers) swell considerably in the range $\mu = 0.1$–$0.05$ and lower, so that column processes are not possible with these ion exchangers in this range.

The ionic strength of a buffer solution can be increased by the admixture of some non-buffering soluble salts, *e.g.*, sodium chloride or potassium chloride. This mixing with a salt is often used for selective elution and can be realized either stepwise or continuously with a gradient device (*cf.*, Chapter 8).

*Choice of the mobile phase*

There are three possibilities for operating with ion-exchange columns as follows.

(1) Columns in the $H^+$ or $OH^-$ form can be used. After the sorption of the sample, the

column is washed with water and then elution is started, gradually increasing the concentration of the acidic or basic solution. This procedure is possible with chemically stable ion-exchange resins only. It is essentially a simple regeneration of the exchanger accompanied by chromatographic separation because the displaced ions moving down the column are continuously sorbed and desorbed.

(2) Buffered columns can be used. After the equilibrium of the column with a buffer with a sufficiently low ionic strength and a suitable pH, the sample is sorbed and elution is begun with a system of buffers starting with the buffer of the lowest ionic strength.

(3) Complexing agents can be used for the selective elution of the sorbed ions.

The choice of a suitable mobile phase depends upon the operating conditions and is summarized in Tables 10.4 and 10.5.

*Increasing concentrations of acids and bases*

This type of chromatography can be used for the separation of some inorganic ions and the separation of simple organic acids and bases. The sorbed cations are gradually eluted from cation-exchange columns in the $H^+$ form with several solutions of increasingly concentrated acid or with a series of different acids with increasing dissociation constants. For example, for orienting experiments, solutions can be used beginning with 0.001 *N* and finishing with 6 *N* hydrochloric acid. A sequence of increasing concentrations of acetic, formic, chloroacetic, hydrochloric and sulphuric acids is also possible. $H^+$ ions compete with other cations in this type of chromatography. Gradient elution can be used in order to change gradually the composition of the mobile phase. In special instances, dilute bases (*e.g.*, solutions of ammonia) can be used for the selective separation of zwitter-ions (*e.g.*,amino acids) or weaker bases from inorganic cations bound on the strongly acidic cation-exchange column.

The sorbed anions (or zwitter-ions) can be eluted from an ion-exchange column in the base form in a similar manner by using a series of increasingly concentrated solutions of a base or a series of several bases with increasing dissociation constants. For example, 0.01–2 *M* solutions of pyridine, collidine, amino alcohols, ammonia or aliphatic amines and their mixtures can be used for orienting experiments. In other instances the anion-exchange column is used in a mixed salt form (*e.g.*, $OH^-$ plus $Cl^-$, $HCOO^-$ or $CH_3COO^-$ form). The sorbed anions (*e.g.*, mixture of organic acids) are gradually eluted with a sequence of eluting solutions containing an increasing concentration of acids (*e.g.*, from 0.01 *N* to 6 *N* hydrochloric acid) or a sequence of solutions of weak and stronger acids of various concentrations.

*Mobile phases for chromatography on buffered columns*

Chromatography on buffered columns is usually used for the separation of zwitter-ions, especially in biochemistry, but also for other purposes when very fine separations are required. The ion exchangers are used in certain mixed forms. They are equilibrated with buffers of a certain pH value and a known, low, ionic strength. For cation exchangers, the starting pH is low (*cf.*, Table 10.3) and the subsequent buffers have higher pH values. The chromatography on anion exchangers begins with a high pH and the following buffers have lower pH values. In both instances, the subsequent buffers have higher ionic strengths.

For the chromatography of proteins, the changes in pH of eluting buffers are of limited use. In many separations, they are omitted because the variation in the dissociation of the amphoteric counter-ions allowing selective desorption are compensated by variations in charge of the functional groups of the exchanger, which increase the sorption, and *vice versa*. Therefore, chromatography at a constant pH and increasing ionic strength only is often used in this instance. The appropriate pH value should be at least 1 pH unit below the isoelectric point of the protein when cation-exchange chromatography is used and at least 1 pH unit above the isoelectric point for anion-exchange chromatography (Boman). The final decision depends upon certain special conditions, and in particular the stability of the protein should be considered.

After the production of fractions containing many inorganic and non-volatile organic buffers, a large amount of salts contaminates the product. From the point of view of further processing of the fractions of the sample, volatile buffers may be important (Holeyšovský *et al.*, Keilová and Keil, Rudloff and Braunitzer, Schroeder *et al.*, Tomášek *et al.*, Vaněček *et al.*, and others). Their components are only weakly mutually bound to form salts, which are decomposed and escape when the water is evaporated in a rotating evaporator or by lyophilization. The pure solid substances of the chromatographic peaks are then obtained. The most usual volatile buffers contain aliphatic amines or amino alcohols, pyridine, α-picoline, 2,4,6-collidine or N-ethylmorpholine and formic or acetic acid. The disadvantage of using trimethylamine buffers is their offensive odour. Other semi-volatile buffers (ammonium carbonate, ammonium acetate) are decomposed in a high vacuum and at elevated temperatures and can be removed by sublimation. Examples of buffers are given in Tables 10.6 and 10.7.

TABLE 10.6

VOLATILE BUFFERS FOR THE CHROMATOGRAPHY OF PEPTIDES FROM TRYPTIC DIGEST OF S-SULPHOTRYPSINOGEN ON A STRONGLY ACIDIC CATION EXCHANGER (HOLEYŠOVSKÝ *et al.*)

The solution of pyridine was adjusted with acid to the given pH and the volume was then made up to the final volume with water.

| No.* | Pyridine (*M*) | Acid | pH |
|---|---|---|---|
| S | 0.10 | Formic | 2.85 |
| 1 | 0.10 | | 3.0 |
| 2 | 0.15 | | 3.0 |
| 3 | 0.20 | Acetic | 4.0 |
| 4 | 0.40 | | 5.0 |
| 5 | 0.80 | | 7.0 |

*S = buffer for dissolving the sample and application on the column, Zerolite 225, 100–200 mesh, length 150 cm, equilibrated with buffer No. 1. 1–5 = buffers used for subsequent gradient elution. The fractions were evaluated by paper chromatography or paper electrophoresis.

*References p.277*

TABLE 10.7

VOLATILE BUFFERS FOR THE CHROMATOGRAPHY OF PROTEINASES FROM ASCITES FLUIDS AND ASCITES CELLS (KEILOVÁ AND KEIL)

The solution of trimethylamine was adjusted with acetic acid to the given pH and then the volume was made up to the final volume with water.

| CM-cellulose column | | | DEAE-cellulose column | | |
|---|---|---|---|---|---|
| No. | Trimethylamine (*M*) | pH | No. | Trimethylamine (*M*) | pH |
| 1 | 0.005 | 5 | 1 | 0.005 | 7 |
| 2 | 0.01 | 5 | 2 | 0.02 | 7 |
| 3 | 0.01 | 6 | 3 | 0.02 | 6 |
| 4 | 0.02 | 6 | 4 | 0.1 | 6 |
| 5 | 0.02 | 7 | 5 | 0.1 | 5 |
| 6 | 0.04 | 8 | 6 | 0.1 (+0.1 *M* NaCl) | 5 |

TABLE 10.8

ANTIMICROBIAL REAGENTS FOR MOBILE PHASES IN ION-EXCHANGE CHROMATOGRAPHY

| Substance | Type of ion exchanger often used | Concentration in the solution (%) | Type of medium where the agent is active |
|---|---|---|---|
| Caprylic acid (octanoic acid) | Cation-exchange resins (amino acid analysis) | 0.01 | Weakly acidic |
| Chloretone (trichlorobutanol) | Ion-exchange derivatives of cellulose and polydextran | 0.05 | Weakly acidic |
| Hibitane (chlorohexidine) | Anion-exchange derivatives of cellulose and polydextran | 0.002 | Weakly acidic, neutral, weakly basic |
| Merthiolate (thimerosal, ethylmercury(II) thiosalicylate) | Cation-exchange derivatives of cellulose and polydextran | 0.005 | Weakly acidic |
| Pentachlorophenol | Cation-exchange resins (amino acid analysis) | 0.0005* | – |
| Phenylmercury(II) salts | Anion-exchange derivatives of cellulose and polydextran | 0.001 | Weakly alkaline |
| Sodium azide | Cation-exchange derivatives of cellulose and polydextran | 0.02 | Weakly acidic, neutral, weakly basic |
| Butanol, carbon tetrachloride, chloroform, toluene | Ion-exchange cellulose | Traces | – |

*Used in the form of a 0.5% (w/v) solution in 95% ethanol.

When choosing buffers for ion-exchange chromatography, the following recommendations can be made. The buffering ions should not react with the functional groups of the exchanger. With cation exchangers, anionic buffers should be used when possible, and with anion exchangers, cationic buffers are recommended. If the buffering-active ions interact with the functional groups of the exchanger, there is a danger of the formation of steps in gradient elution, because the continuity of the gradient is lost. However, there are examples in the literature in which these recommendations were not followed and good separations were nevertheless achieved. Anionic buffers are those with which the buffering activity is caused by the anionic component (*e.g.*, acetate, citrate, phosphate and also glycine). In cationic buffers, the active components are ammonia and amines, aminoethanol, imidazole, pyridine and Tris. Barbital displays anionic properties and, below pH 7.5, also cationic properties.

*Special additives to buffers*

In some instances when ion exchange is used for analytical purposes, certain substances are added to the mobile phase in order to improve the chromatography. A small admixture (up to 5%) of methanol, ethanol, *tert.*-butanol, benzyl alcohol, methyl Cellosolve or phenetol influences the relative positions of amino acids, and several of these compounds have been used by some workers for accelerated amino acid analysis (Dévényi, Ertingshausen *et al.*, Hubbard, Moore and Stein, Nauman, Pobel, Vrátný and Zbrožek, Zuev *et al.*, and others). Detergents, *e.g.*, Brij 25, are sometimes used for improving the wetting of the resins by the mobile phase and for preventing the formation of bubbles when warming the mobile phase before colorimetry. Antioxidants are added in order to protect sensitive solutes against the influence of air (*e.g.*, thiodiglycol to prevent oxidation of methionine). Antimicrobial agents (*cf.*, Table 10.8) keep the mobile phases and ion exchangers free from bacteria and other microbial contamination*. Urea at a concentration up to 8 *M* in the buffers is used to dissolve less soluble modified protein or large peptide fragments and to keep them in solution during chromatography. All of these additives should be chosen so that they are not firmly bound to the exchanger.

*Complex-forming mobile phases and phases that alter the solutes*

Some organic acids (citric acid, ethylenediaminetetraacetic acid, lactic acid, oxalic acid, α-hydroxybutyric acid, nitrilotriacetic acid, uramildiacetic acid or their salts) and thenoylnitrofluoroacetone display special properties for the selective elution of inorganic cations in the complex form from cation exchangers. If some of them are used for the desorption of a very complicated mixture (*e.g.*, fission products from atomic reactors), groups of similar ions are gradually desorbed and these methods can be used for the first fractionation. Complexing agents are also suitable for the selective desorption of individual inorganic ions. One of the first successes of ion-exchange chromatography with inorganic substances was the separation of rare earths with citrate solutions.

*From the point of view of microbial contamination, the most dangerous is the long presence of the substrate in contact with the ion exchanger in phosphate buffers.

Some inorganic acids and their salts (*e.g.*, phosphoric acid, cyanides) also display complex-forming activity when used for the elution of cations of heavy metals. In other instances the complex-forming ability of concentrated hydrochloric acid (or other halogen acids), which forms chloride complexes with transition metals, was used for fractionation. They can then be separated on anion exchangers using a decreasing concentration of hydrochloric acid as the eluting agent, which gradually decomposes the complexes. Other examples of the use of complex-forming mobile phases in inorganic chemistry have been described by Helfferich.

A similar principle of elution is also sometimes used in ion exchange in organic chemistry. The chromatography of saccharides or other non-ionic hydroxy compounds containing vicinal hydroxy groups (*e.g.*, glycols) can be carried out on anion exchangers using buffers containing boric acid. These compounds undergo the following reactions:

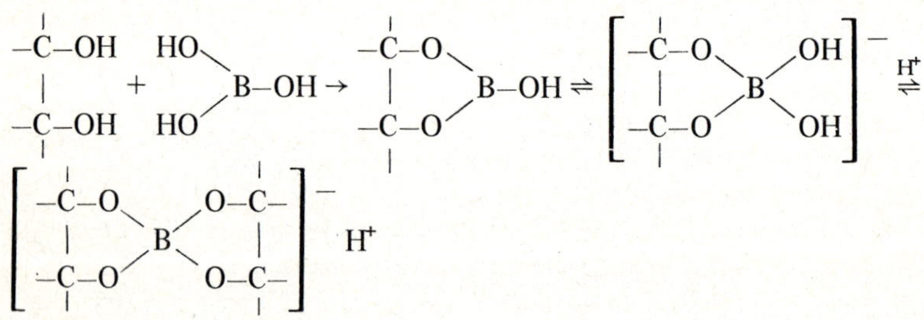

The complex anions formed are much stronger acids than the less dissociated forms of boric acid or of the transient form. Their dissociation constants differ with individual organic compounds, which leads to the possibility of effective ion-exchange chromatography (Khym and Zill, 1951, 1952; Khym *et al.*; Sargent and Rieman; Zager and Doody).

In a similar manner, hydrosulphite buffers can be used for the ion-exchange chromatography of aldehydes and ketones, which in other buffers are not retained on ion exchangers. The following reactions permit the binding:

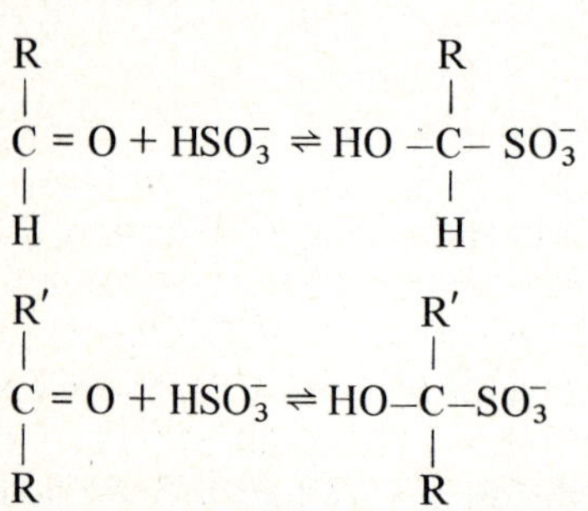

The hydroxysulphonic acids of ketones can be eluted from anion exchangers with hot water, while the hydroxysulphonic acids of aldehydes are eluted with solutions of alkalis or salts. There is also the possibility of the chromatographic separation of individual derivatives (Gabrielson and Samuelson, Ruff, Sherma and Rieman).

The so-called ligand-exchange chromatography is mentioned elsewhere (Chapter 6).

### Mixed and non-aqueous phases

Some considerations concerning non-aqueous phases for ion exchange have already been discussed in Chapter 6. The resins prefer more polar solvents and therefore after equilibrium of ion exchanger particles with the mixed solution the actual concentration of the organic solvent (*e.g.*, alcohol) will be greater in the aqueous phase than in the resin phase. The swelling and the useful exchange capacity of the resin will be decreased. On the other hand, the sorption of polar non-electrolytes (*e.g.*, sugars, polyalcohols) from the solution will be greater in comparison with their uptake from the aqueous solution. The sorption of less polar solutes (ketones, esters, phenols, hydrocarbons), which are sorbed by Van der Waals forces, is influenced in the opposite way.

The presence of organic solvents influences the selectivity coefficients for ions, sometimes very substantially. However, the time required to attain equilibrium is much longer, and this effect is increased with increase in the concentration of the organic solvent. In pure organic solvents, the rate of exchange can be about three orders of magnitude less than in water.

The most commonly used organic solvents for studies of mixed or non-aqueous phases are methanol, ethanol, propanol, acetone, ethyl methyl ketone, dioxane, ethylene glycol and glycerol. The admixture of organic solvents with the mobile phase is also used in instances when the desorbed substance, after having lost its electrical charge, is not soluble in aqueous solution. In other instances, the addition of a suitable organic solvent can lead to the separation of ions, which could not be separated in aqueous solutions only. The altered solvation of ions in mixed solutions can be reflected in differences in the chromatographic separation. Because of the presence of two phases with different solvent compositions (gel and outer solution), the factors of partition chromatography may play a role.

### Temperature of mobile phases

Most chromatographic experiments are carried out at room temperature and many biochemical separations are carried out at 1–4°C. The lability of chromatographed substances does not allow work at normal temperature.

Temperatures of up to 100°C are sometimes used with aqueous solutions, because the higher temperature accelerates the achievement of the equilibrium of ion exchange, diminishes the viscosity of the solvent and makes the whole chromatographic process much quicker. The elevating of temperature is of course possible only when stable substances are treated.

The influence of temperature on the relative positions of the peaks of separated substances is not too important if true ion exchange is considered. The temperature usually does not change the exchange potential substantially, but if adsorption of the sorbed substances due to Van der Waals forces takes place in parallel with ion exchange, an increase in temperature in general diminishes the sorption.

## CALCULATION OF GRADIENTS

### General aspects

Elution chromatography is the most widely used form of liquid column chromatography. In simple elution, the solvent entering the column has a constant composition. If a series of solvents with different compositions is used for the elution of particular components of a complex chromatographed mixture, the procedure is called stepwise elution. Tiselius and his collaborators (Alm, Alm *et al.*, Hagdahl *et al.*, Williams) proposed a chromatographic procedure in which elution is carried out with a solvent the composition of which is changed continuously and they proposed the term gradient elution chromatography for this procedure. The shortened term gradient elution is also used. Other workers have discussed or used a similar principle independently, although they did not consider it to be so (Busch *et al.*; Busch and Potter, 1952, 1953; Donaldson *et al.*; Mitchell *et al.*; Nervik; Strain, 1951, 1960; Synge). Gradient elution suppresses the tailing of zones and hence improves their symmetry. In addition, it improves and refines the separation process, especially for biopolymers.

The first thorough theoretical discussion of concentration gradient elution was published by Drake, and of pH gradient elution by Piez. The method was rapidly accepted and widely used and has been the subject of several reviews (Dorfner, Henry, Lebreton, Mikeš, and others), and an extensive paper by Snyder (1965) gives a detailed discussion. Gradient elution chromatography is described in this book from the point of view of methods and apparatus in Chapter 8. The aim of this chapter is to present equations for the calculation of single types of gradients.

### Classification of gradients

Gradient elution chromatography is divided into concentration gradient elution, the aim of which is to change the polarity or ionic strength of the eluent caused by a continuous change in component concentration in the eluent, and pH gradient elution, aiming at a continuous and defined change of pH of the eluent. In view of various devices used today for the creation of gradients various types and shapes of gradients are also known. They are summarized schematically in Fig. 10.6, where the terminology of Snyder (1965) is used. As regards the suitability of the shapes of gradients for chromatography, linear gradients are generally most suitable for first experiments. For repeated chromatography of the same substances, it is most convenient to find experimentally the optimum gradients (so-called "custom gradients"), which may have a complex course ("compound gradients"). For the formation of a gradient, not only its shape but also its steepness is of importance. A gradient that is too mild often leads to broad zones, while a gradient that is too steep makes the zones narrow but also brings closer together zones which would otherwise be well separated with an optimum gradient.

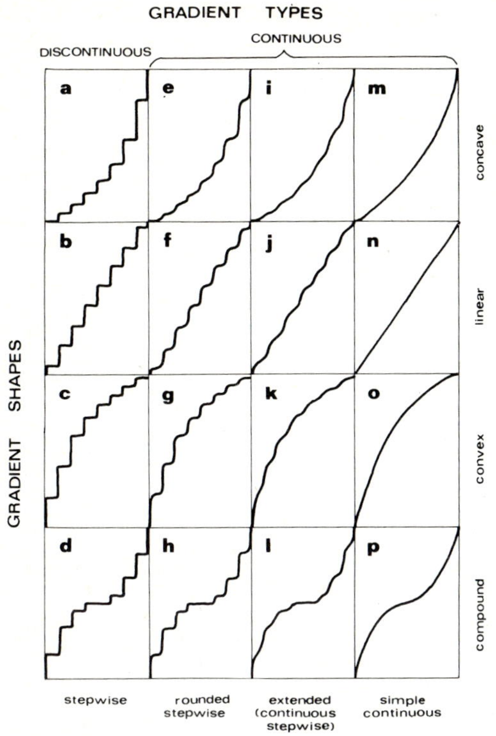


Fig. 10.6. Limit and transitional types and forms of gradients. Vertical axis: concentration of the component or pH in the eluent running through the column. Horizontal axis: eluent volume.

## Gradient formation

Gradient elution (see also Chapter 8, p. 113) can be carried out easily by the continuous mixing of two or more solutions before they enter the column. According to the method of mixing, four different methods of gradient preparation can be distinguished and are represented schematically in Fig. 10.7 and discussed below.

(1) "Exponential" methods of mixing two liquids are illustrated in Fig. 10.7a and b. From the reservoir, A, of an optional volume, the effective "solvent" flows through a mixer, B, of constant volume or through two mixers, $B_1$ and $B_2$, of constant volumes. The mixer B contains an ineffective "diluent" with which the column is also filled at the start of chromatography. When two mixers are used, they may contain the same diluent or an active solvent diluted to various degrees. The corresponding device is called a "mixer of constant volume". The only mixer with a constant volume is the device originally proposed for gradient elution. It is still often used even though it generally does not give an optimum gradient. Its convex course is shown in Fig. 10.6o. The use of two mixers is more suitable because an almost linear gradient can be obtained (Drake).

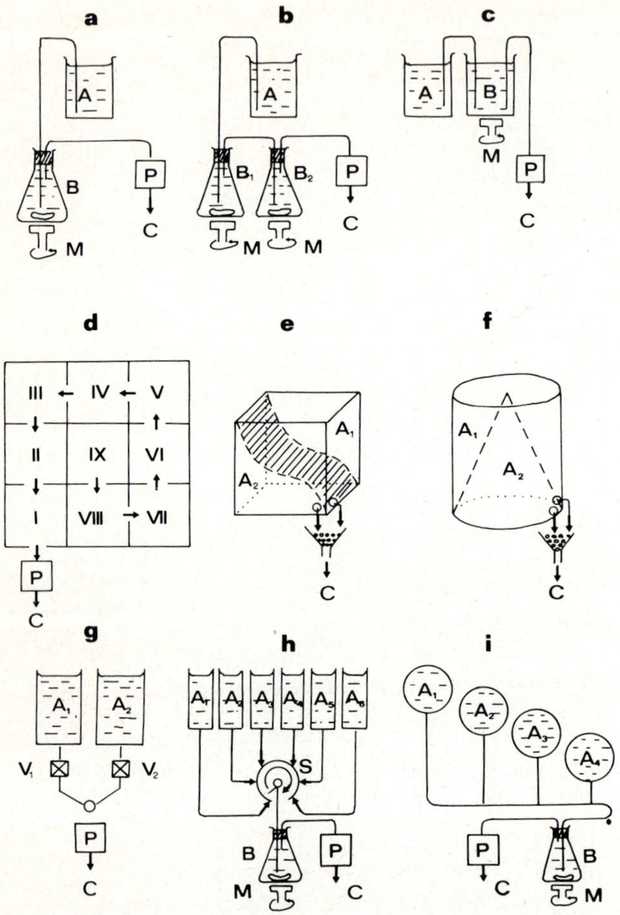


Fig. 10.7. Principles of gradient formation. a, b: device for forming "exponential" gradients; c, d: device for forming "proportional" gradients; e, f, g: device for forming "disproportional" gradients; h, i: device for forming multicomponent gradients. A, reservoirs containing an actively eluting component ("solvent"); $A_1 - A_6$, reservoirs with liquids of increasing elution strength; B, mixers with liquids of minimum elution strength ("diluents") used for the initial column (C) equilibration; M, magnetic stirrer; P, laboratory pump; S, six-way valve; $V_1$ and $V_2$, suitably conjugated valves; I–IX, mixers as connected vessels.

(2) "Proportional" methods of mixing two liquids based on the principle of connected vessels are shown in Fig. 10.7c and d. The device represented in Fig. 10.7c is also called an "open mixer". Reservoir A and mixer B are open connected vessels, and A contains the solvent and B the weaker solvent or inactive diluent. When the solution is pumped on to the column, the levels of the liquids in both vessels decrease at the same rate. Depending on the ratio of the cross-sections of the vessels, the liquids are mixed in different but constant proportions, giving gradients of various shapes. If the cross-sections of the vessels are equal (A = B), a linear gradient is formed (Fig. 10.6n), if $A > B$ a convex gradient is obtained (Fig. 10.6o), while if $A < B$ a continuously increasing concave gradient results

(Fig. 10.6m). In a similar manner, a series of connected vessels (Fig. 10.7d) can be joined in series and stirred with the same motor (for example, the nine-chamber Technicon Varigrad according to Peterson and Sober (1958, 1959) and Peterson and Rowland).

(3) "Disproportional" methods of mixing two liquids (Fig. 10.7e, f and g) lead to various forms of gradients. The principle of the method according to Bock and Ling is based on two vessels, $A_1$ and $A_2$, to which the required profiles were given (Fig. 10.7e). From these vessels, the liquids flow into a small mixer, for example a funnel filled with glass beads. The shape of the vessels enables a ratio of mixing of the liquids to be used that changes with time and thus creates the special gradient required. Other arrangements of the two vessels according to Bock and Ling (a cone within a cylinder) are represented in Fig. 10.7f. The same effect can be achieved, according to Zahn and Stahl, by regulating the effluent from two vessels, $A_1$ and $A_2$, with suitable conjugated valves, $v_1$ and $v_2$ (Fig. 10.7g). Instead of valves $v_1$ and $v_2$, two programmable, reversely proportional pumps can be used, the total output of which is constant, and partial outputs can be regulated with time with an optical device reading the required gradient from a pre-drawn curve. By these methods, gradients of virtually any shape may be obtained.

(4) The last method is the formation of gradients by mixing several liquids. When a series of liquids that differ slightly in composition are introduced on to the column, a rounding of the sharp steps in the composition of solutions typical of stepwise elution takes place as a consequence of mixing in the dead spaces of the leading tubes and under the effect of the pump (Fig. 10.6, "rounded stepwise"). A further rounding of the steps takes place on the particles in the upper part of the column. If a small mixer of constant volume is inserted into the tubing leading the eluent (Fig. 10.7h), a still coarser rounding takes place. If, according to Anderson *et al.* (Fig. 10.7i), the solution is mixed gradually with half of the preceding and then with half of the following solution, an extended gradient (continuous stepwise gradient) is obtained (Fig. 10.6). In ion-exchange chromatography, the buffering function of the ion exchanger causes the total equalization of the rounded stepwise to a continuous gradient (Mikeš *et al.*). By gradual mixing of solutions of similar composition (Fig. 10.6a–d) and by using a suitable arrangement, any of the continuous gradients may be obtained (Fig. 10.6m–p).

## Calculation of concentration gradients

For the sake of simplicity, in the following calculations the liquid with a high elution strength in the reservoir A (Fig. 10.7) is designated as the "solvent" and the liquid with a minimum elution strength in the mixer B as the "diluent". A mixture of solvent and diluent flowing on to the column is indicated as the eluent.

### *"Exponential" gradients*

For the gradient of the solvent in the liquid flowing out of a single closed mixer, Alm *et al.* derived the equation

$$\frac{v}{V} = 2.303 \log \frac{X}{X - x} \qquad (10.45)$$

where $x$ is the concentration of the solvent in the mixer B after a volume $v$ has entered the column, $X$ is the concentration of the solvent in the reservoir A and $V$ is the volume of the mixer. Cherkin *et al.* derived an equation expressing the dependence of the outflowing solvent on its original concentration and the ratio of the volume that has flowed through to the mixer volume. Solving the corresponding differential equation, they obtained the expression

$$\frac{C}{C_0} = \frac{e^K - 1}{e^K} \tag{10.46}$$

where $C$ is the concentration of the solvent flowing out of the mixer, $C_0$ is the concentration of the solvent in the reservoir, and $K$ is the ratio of the total volume $v$ of the consumed eluate to the volume $V$ of the diluent in the mixer ($K = v/V$). Only when $K$ does not exceed unity has the gradient an approximately linear shape. If a linear course is required, the volume of the diluent in the mixer should be at least equal to the total volume of the consumed eluate. In this case, however, the final concentration of the solvent from the reservoir cannot be attained in the column during elution. This must be kept in mind in advance and therefore a higher concentration chosen. If the total eluate volume $v$ exceeds the original diluent volume $V$ in the mixer more than two-fold ($K > 2$), the gradient effect of further elution is small or negligible. Both eqns. 10.45 and 10.46 are more or less identical. A similar equation was proposed by Mader:

$$X = B \ln \frac{A}{A - y} \tag{10.47}$$

where $X$ is the eluate volume that has flowed out of the column, $B$ is the mixer volume, $A$ is the solvent concentration in the reservoir and $y$ is the solvent concentration at the column inlet.

Bock and Ling, and Drake derived for this type of gradient equations formulated in a different way, which, however, are also virtually identical. If $C_{in}$ is the inlet concentration of the solvent introduced into the column, then

$$C_{in} = C_r - (C_r - C_m) \cdot e^{-(v/V_m)} \tag{10.48}$$

where $C_r$ is the concentration of the solvent in the reservoir, $C_m$ the starting concentration of the solvent in the mixer, $v$ the volume of the liquid that has passed through the column and $V_m$ the volume of liquid in the mixer. The solution of the gradient of one closed mixer and of its practical testing was also studied by Donaldson *et al.*

The calculation of the gradient with the use of two closed mixers (Fig. 10.7b) was carried out by Drake. If the mixers have different volumes ($B_1 \neq B_2$), then the equation

$$C_{in} = C_r - \frac{V_1}{V_1 - V_2}(C_r - C_1)\, e^{-(v/V_1)} - \left(\frac{C_r V_2 - C_1 V_1}{V_1 - V_2} - C_2\right) e^{-(v/V_2)} \tag{10.49}$$

applies, where $C_{in}$ and $C_r$ are as in eqn. 10.48, $V_1$ and $V_2$ are the volumes of the two mixers and $C_1$ and $C_2$ are the starting concentrations of the solvent in the two mixers. The gradients with a smaller ratio of the volumes $V_2/V_1$ are less steep. For the special case of eqn. 10.49 for which both mixers have the same volume ($V_1 = V_2$), eqn.

10.50 applies:

$$C_{in} = C_r - (C_r - C_1)\frac{v}{V_2}\,\mathrm{e}^{-(v/V_2)} - (C_r - C_2)\,\mathrm{e}^{-(v/V_2)} \tag{10.50}$$

In this case, in the interval $C_1 = 0$ to $C_1 = C_r/2$, the course of the gradient approaches linearity for $v = 2.5\ V_2$ (maximum).

*"Proportional" gradients*

Lakshmanan and Lieberman investigated the formation of gradients in which the liquid would flow from the reservoir into the mixer at a constant rate, $R_1$, different from the rate at which it would flow out from the mixer into the column ($R_2$). If $V_0$ is the original volume of the diluent in the mixer, $C_0$ the concentration of the solvent in the reservoir and $C$ the concentration of the solvent in the mixer at time $t$, then

$$C = C_0\left[1 - \left(\frac{a}{a + bt}\right)^{\frac{1}{b}}\right] \tag{10.51}$$

where $a = V_0/R_1$ and $b = 1 - (R_2/R_1)$. If $b < -1$ or $R_2 > 2R_1$, the dependence of the concentration on time (or on eluate volume) is concave, if $b = -1$ or $R_2 = 2R_1$, the gradient is linear, and if $b > -1$ or $R_2 < 2R_1$, the curve is convex. At $R_1 = R_2$, the curve is exponential because in fact it is a case of the solutions in eqns. 10.45–10.48. The realization of this method by the procedure mentioned would, of course, be subject to difficulties connected with fluctuations of the level in the mixer.

This principle, however, can be easily carried out according to Fig. 10.7c. By choosing the ratio of the cross-sections of the two cylinders, the ratio of the in-flowing and out-flowing rates can be controlled in the mixer, because the levels must decrease uniformly. In this manner, all of the gradients mentioned can be obtained. Drake, and Bock and Ling deduced for this type a formally identical equation

$$C_{in} = C_r - (C_r - C_m)\left(1 - \frac{v}{V_{tot}}\right)^{\frac{V_r}{V_m}} \tag{10.52}$$

in which the symbols have the same meaning as in eqns. 10.48–10.50, $V_r$ is the starting volume of the liquid in the reservoir, $V_m$ is the starting volume of the liquid in the mixer and $V_{tot} = V_r + V_m$. If the cylinders A and B (Fig. 10.7c) are equally wide, *i.e.*, $V_r = V_m$, eqn. 10.52 becomes linear:

$$C_{in} = C_r - (C_r - C_m)\left(1 - \frac{v}{V_{tot}}\right) = C_m + (C_r - C_m)\frac{v}{V_{tot}} \tag{10.53}$$

and this gradient is therefore called linear. The composition of the buffer entering the column can easily be found on the straight line connecting the ordinates of a simple rectangular graph constructed on the basis of both starting concentrations $C_r$ and $C_m$. The volume which has flowed through $v$ is read on the abscissa of total length $V_{tot}$. This graphical method is most often used.

If the mixer has a larger volume than the reservoir (Fig. 10.7c), *i.e.*, $V_m > V_r$, a continuously increasing concave gradient is formed (Fig. 10.6m). However, if the reservoir is

*References p.277*

larger than the mixer ($V_r > V_m$), a decreasing convex gradient is obtained. Both cases can be expressed by eqn. 10.52; Kočent published a nomogram facilitating the calculations. A similar nomogram was published by Warner and Lands. If a large number of equal-sized vessels are connected in series (Varigrad, Fig. 10.7d), of which any internal one contains the solvent of concentration $L$ and the others contain the diluent, then the concentration gradient $C$ of the out-going liquid can be calculated according to the equation

$$\frac{C}{L} = \frac{(N-1)!}{(N-n)!\,(n-1)!} \cdot \left(1 - \frac{v}{V}\right)^{N-n} \left(\frac{v}{V}\right)^{n-1} \tag{10.54}$$

where $N$ is the total number of vessels in the system, $n$ is the serial number of the vessel containing the solvent of concentration $L$, $v$ is the volume of liquid entering the column and $V$ is the original volume of liquid in the mixing system.

*"Disproportional" gradients*

The gradient of the system illustrated in Fig. 10.7e is pre-determined by its geometry and, in principle, it can be read from the profile of the vessel. For the system according to Fig. 10.7f, the following equation applies:

$$C = C_1 + (C_2 - C_1)\left(\frac{v}{V}\right)^2 \tag{10.55}$$

where $C_1$ and $C_2$ are the concentrations of the solvent in compartments 1 and 2 in Fig. 10.7f, $v$ is the volume of liquid that has flowed through and $V$ is the total volume of the system. If the internal vessel of the system (Fig. 10.7f) has a cone, the square of the radius of which is equal to the product of the height and the jacket radius (*i.e.*, $r^2 = k \cdot h$), then a linear gradient is achieved, characterized by the equation

$$C = C_1 + (C_2 - C_1)\frac{v}{V} \tag{10.56}$$

The calculation of the formation of a gradient obtained by the gradual mixing of several liquids can be simplified by dividing the process into several stages according to some of the above described methods; in this procedure, only two mixing liquids are considered at once and the gradients can therefore be calculated in stages.

### pH gradients

The calculation of pH gradients was described by Piez, who differentiated between three cases: (1) the gradient is prepared by mixing one buffer with the other; (2) it is prepared by mixing a weak base or acid with the buffer; (3) it is obtained by mixing a strong base or acid with the buffer. The calculation of pH gradients is more complex and the range of validity is limited. Buffers that consist of a monovalent acid and its salt effectively buffer only over a range of two pH units and the calculated gradients are also formed in this range only. Buffers of polyvalent acids (*e.g.*, citric acid) would buffer over a range of up to four pH units. We do not consider it necessary to mention the corresponding equations here, and we refer to the original paper by Piez.

In ion-exchange chromatography, the most convenient systems are generally those in which a concentration and a pH gradient take place simultaneously and where the concentration gradient is linear and the pH gradient continuously increasing (concave). This can easily be achieved with two equally broad, connected vessels (according to Fig. 10.7c), when a more concentrated buffer from a reservoir is mixed into the more diluted buffer in the mixer. For the separation of proteins, an ionic strength gradient (*i.e.*, concentration gradient) is often used with a constant pH throughout the whole experiment.

### Theory of gradient elution

The aim of this chapter is to explain the calculation of gradients and not the theory of gradient elution. However, we consider it appropriate to summarize in this last section the theoretical papers concerning this topic. The bases of a general theory were forwarded by Drake, and general equations for the understanding of elution were also derived by Said and by Snyder (1964a, b). A theory of ion-exchange chromatography for discontinuously and continuously changing eluent compositions was published by Freiling (1955, 1957), and a theory of gradient elution in ion-exchange chromatography was developed by Schwab *et al.* The latter authors, as well as Maslova *et al.*, also compared theoretical calculations with experimentally observed peak positions of separated compounds. The theory of ion-exchange chromatography was further developed by Koguchi *et al.* and Ohashi and Koguchi. An extensive discussion of the theory of gradient elution was published by Snyder (1965) (*cf.* also Snyder and Saunders). A series of valuable considerations on gradient chromatography of proteins was summarized by Peterson and Sober (1960), Sober *et al.* and Sober and Peterson; Novotný *et al.* described the linear dependence of the specific electric charge of immunoglobulin fragments and the ionic strength of the buffer at which the fragments are eluted from QAE-Sephadex. Novotný published an equation permitting the optimization of the elution of proteins and their fragments on ion exchangers when an ionic strength gradient is used.

## REFERENCES


Alm, R. S., *Acta Chem. Scand.*, 6 (1952) 1186.

Alm, R. S., Williams, R. J. P. and Tiselius, A., *Acta Chem. Scand.*, 6 (1952) 826.

Anderson, N. G., Bond, H. E. and Canning, R. E., *Anal. Biochem.*, 3 (1962) 472.

Bock, R. M. and Nan Sing Ling, *Anal. Chem.*, 26 (1954) 1543.

Boman, H. G., in K. Paech and M. V. Tracy (Editors), *Modern Methods in Plant Analysis*, Springer, Berlin, 1962.

Brunauer, S., *The Adsorption of Gases and Vapours*, Clarendon Press, Oxford, and Princeton Univ. Press, Princeton, 1945.

Brunauer, S., Deming, L. S., Deming, W. E. and Teller, E., *J. Amer. Chem. Soc.*, 62 (1940) 1723.

Burrel, H., *Interchem. Rev.*, 14 (1953) 3 and 31.

Busch, H., Hurlbert, R. B. and Potter, V. R., *J. Biol. Chem.*. 196 (1952) 717.

Busch, H. and Potter, V. R., *Cancer Res.*, 12 (1952) 660.

Busch, H. and Potter, V. R., *Cancer Res.*, 13 (1953) 168.
Cherkin, A., Martinez, F. E. and Dunn, M. S., *J. Amer. Chem. Soc.*, 75 (1953) 1244.
Crowley, J. D., Teague, Jr., G. S. and Lowe, Jr., J. W., *J. Paint Technol.*, 34 (1966) 269.
Deal, C. H., Derr, E. R. and Papadoupoulos, M. N., *Ind. Eng. Chem., Fundam.*, 1 (1962) 17.
Derr, E. L., Deal, C. H. and Pierotti, G. J., *Amer. Soc. Test. Mater., Spec. Tech. Publ.*, No. 244 (1957) 111.
De Vault, P., *J. Amer. Chem. Soc.*, 65 (1943) 532.
Dévényi, T., *Acta Biochim. Biophys.*, 3 (1968) 429.
Donaldson, K. O., Tulane, V. J. and Marshall, L. M., *Anal. Chem.*, 24 (1952) 185.
Dorfner, K., *Chem. Ztg.*, 87 (1963) 871.
Drake, B., *Ark. Kemi*, 8 (1955) 1.
Ertingshausen, G., Adler, H. J. and Reichler, A. S., *J. Chromatogr.*, 42 (1966) 355.
Flory, P. J., *J. Phys. Chem.*, 10 (1942) 51.
Freiling, E. C., *J. Amer. Chem. Soc.*, 77 (1955) 2067.
Freiling, E. C., *J. Phys. Chem.*, 61 (1957) 543.
Gabrielson, G. and Samuelson, O., *Sv. Kem. Tidskr.*, 62 (1950) 214.
Gardon, J. L., *J. Paint Technol.*, 38 (1966) 43.
Glueckauf, E., *Nature (London)*, 156 (1945) 748.
Glueckauf, E., *Nature (London)*, 301 (1947a) 301.
Glueckauf, E., *J. Chem. Soc.*, (1947b) 1302.
Glueckauf, E., *J. Chem. Soc.*, (1947c) 1321.
Glueckauf, E., *J. Chem. Soc.*, (1949) 3280.
Hagdahl, L., Williams, R. J. P. and Tiselius, A., *Ark. Kemi*, 4 (1952) 193.
Helfferich, F. G., *Ion Exchange*, McGraw-Hill, London, 1962.
Henry, R. A., in J. J. Kirkland (Editor), *Modern Practice of Liquid Chromatography*, Wiley-Interscience, New York, 1971, pp. 80–89.
Hildebrand, J. H. and Scott, R. L., *Regular Solutions*, Prentice-Hall, Englewood Cliffs, N.J., 1962.
Hildebrand, J. H. and Scott, R. L., *Solubility of Non-Electrolytes*, Dover, New York, 1964.
Holeyšovský, V., Alexijev, B., Tomášek, V., Mikeš, O. and Šorm, F., *Collect. Czech. Chem. Commun.*, 27 (1962) 2665.
Hubbard, R. W., *Biochim. Biophys. Res. Commun.*, 19 (1965) 679.
Huber, J. F. K., Meijers, C. A. M. and Hulsman, J. A. R. J., *Anal. Chem.*, 44 (1972) 111.
Huggins, M. L., *Ann. N.Y. Acad. Sci.*, 43 (1942) 1.
Keilová, H. and Keil, B., *Collect. Czech. Chem. Commun.*, 27 (1962) 2193.
Keller, R. A., Karger, B. L. and Snyder, L. R., in N. Stock and S. G. Perry (Editors), *Gas Chromatography 1970*, Institute of Petroleum, London, 1970, p. 125.
Keller, R. A. and Snyder, L. R., *J. Chromatogr. Sci.*, 9 (1971) 346.
Khym, J. X. and Zill, L. P., *J. Amer. Chem. Soc.*, 73 (1951) 2399.
Khym, J. X. and Zill, L. P., *J. Amer. Chem. Soc.*, 74 (1952) 2050.
Khym, J. X., Zill, L. P. and Cohn, W. E., in C. Calmon and T. R. E. Kressman (Editors), *Ion Exchangers in Organic and Biochemistry*, Interscience, New York, 1957.
Kirkland, J. J., *J. Chromatogr. Sci.*, 9 (1971a) 206.
Kirkland, J. J., in J. J. Kirkland (Editor), *Modern Practice of Liquid Chromatography*, Wiley-Interscience, New York, 1971b, p. 161.
Kočent, A., *J. Chromatogr.*, 6 (1961) 324.
Koguchi, K., Waki, H. and Ohashi, S., *J. Chromatogr.*, 25 (1966) 398.
Kudchadker, A. P., Alani, G. H. and Zwolinski, B. J., *Chem. Rev.*, 68 (1968) 659.
Lakshmanan, T. and Lieberman, S., *Arch. Biochem. Biophys.*, 45 (1953) 235.
Lebedeva, N. P., Frolov, I. I. and Jashin, J., *J. Chromatogr.*, 58 (1971) 11.
Lebreton, P., *Bull. Soc. Chim. Fr.*, (1960) 2188.
Lieberman, E. P., *Off. Dig., Fed. Soc. Paint Technol.*, 34 (1962) 30.
Locke, D. C. and Martire, D. E., *Anal. Chem.*, 39 (1967) 921.
Macek, K. and Procházka, Ž., in I. M. Hais and K. Macek (Editors), *Paper Chromatography*, Academic Press, New York, 1963, p. 115.

Mader, C., *Anal. Chem.*, 26 (1954) 566.
Maggs, R. J., *J. Chromatogr. Sci.*, 7 (1969) 145.
Martire, D. E., in A. B. Littlewood (Editor), *Gas Chromatography 1966*, Elsevier, Amsterdam, 1967, p. 21.
Martire, D. E. and Locke, D. C., *Anal. Chem.*, 43 (1971) 68.
Maslova, G. B., Nazarov, P. P. and Khmutov, K. V., *Ionoobmen. Sorbenty Prom.*, (1963) 103.
Mikeš, O., *Chem. Listy*, 54 (1960) 578.
Mikeš, O., Tomášek, V. and Holeyšovský, V., *Chem. Listy*, 53 (1959) 609.
Mitchell, H. K., Gordon, M. and Haskins, F. A., *J. Biol. Chem.*, 180 (1949) 1071.
Moore, S. and Stein, W. H., *J. Biol. Chem.*, 192 (1951) 663.
Morrison, G. H. and Freiser, H., *Solvent Extraction in Analytical Chemistry*, Wiley-Interscience, New York, 1957.
Nauman, L. W., *J. Chromatogr.*, 36 (1968) 398.
Neher, R., in G. B. Marini-Bettòlo (Editor), *Thin-Layer Chromatography*, Elsevier, Amsterdam, 1964, pp. 75–86.
Nervik, W. E., *J. Phys. Chem.*, 59 (1955) 690.
Novotný, J., *FEBS Lett.*, 14 (1971) 7.
Novotný, J., Franěk, F. and Šorm, F., *Eur. J. Biochem.*, 16 (1970) 278.
Ohashi, S. and Koguchi, K., *J. Chromatogr.*, 27 (1967) 214.
Peterson, E. A. and Rowland, J., *J. Chromatogr.*, 5 (1961) 330.
Peterson, E. A. and Sober, H. A., *Fed. Proc., Fed. Amer. Soc. Exp. Biol.*, 17 (1958) 288 and 1116.
Peterson, E. A. and Sober, H. A., *Anal. Chem.*, 31 (1959) 857.
Peterson, E. A. and Sober, H. A., in E. W. Putnam (Editor), *The Plasma Proteins*, Vol. I, Academic Press, New York, 1960, pp. 105–141.
Pierotti, G. J., Deal, C. H. and Derr, E. L., *Ind. Eng. Chem.*, 51 (1959) 95.
Pierotti, G. J., Deal, C. H., Derr, E. L. and Porter, P. E., *J. Amer. Chem. Soc.*, 78 (1956) 2989.
Piez, K. A., *Anal. Chem.*, 28 (1956) 1451.
Pobel, E. S., *Anal. Biochem.*, 18 (1967) 406.
Polak, J., *Collect. Czech. Chem. Commun.*, 31 (1966) 1483.
Rudloff, V. and Braunitzer, G., *Hoppe-Seyler's Z. Physiol. Chem.*, 323 (1961) 129.
Ruff, E., *Anal. Chem.*, 31 (1959) 1626.
Said, A. S., *AIChE J.*, 2 (1956) 477.
Sargent, R. and Rieman, III, W., *Anal. Chim. Acta*, 16 (1957) 144.
Schmit, J. A., Henry, R. A., Williams, R. C. and Dieckman, J. F., *J. Chromatogr. Sci.*, 9 (1971) 645.
Schorn, P. J., *Z. Anal. Chem.*, 205 (1964) 298.
Schwab, H., Rieman, W. and Vaughan, P. A., *Anal. Chem.*, 29 (1957) 1357.
Scott, R. P. W. and Lawrence, G. J., *J. Chromatogr. Sci.*, 8 (1970) 619.
Sherma, J. and Rieman, III, W., *Anal. Chim. Acta*, 19 (1958) 134.
Shroeder, W. A., Jones, T. R., Cornick, J. and McCalla, K., *Anal. Chem.*, 34 (1962) 1570.
Snyder, L. R., *J. Chromatogr.*, 13 (1964a) 415.
Snyder, L. R., *Advan. Anal. Chem. Instrum.*, 3 (1964b) 251.
Snyder, L. R., *Chem. Rev.*, 7 (1965) 1.
Snyder, L. R., *Anal. Chem.*, 39 (1967) 698.
Snyder, L. R., *Principles of Adsorption Chromatography*, Marcel Dekker, New York, 1968.
Snyder, L. R., *J. Chromatogr. Sci.*, 7 (1969) 352.
Snyder, L. R. and Saunders, D. L., *J. Chromatogr. Sci.*, 7 (1969) 195.
Sober, H. A., Gutter, F. J., Wyckoff, M. M. and Peterson, E. A., *J. Amer. Chem. Soc.*, 78 (1956) 756.
Sober, H. A. and Peterson, E. A., in J. T. Edsall (Editor), *Amino Acids, Proteins and Cancer Biochemistry*, Academic Press, New York, 1960, p. 61.
Strain, H. H., *Anal. Chem.*, 23 (1951) 25.
Strain, H. H., *Anal. Chem.*, 32 (1960) 3R.
Synge, R. L. M., *Discuss. Faraday Soc.*, 7 (1949) 167.
Tomášek, V., Holeyšovský, V., Mikeš, O. and Šorm, F., *Biochim. Biophys. Acta*, 38 (1960) 570.
Vaněček, J., Meloun, B., Kostka, V., Keil, B. and Šorm, F., *Biochim. Biophys. Acta*, 37 (1960) 169.

Vrátný, P. and Zbrožek, J., *J. Chromatogr.*, 76 (1973) 482.
Warner, H. R. and Lands, W. E. M., *J. Lipid Res.*, 1 (1960) 248.
Williams, R. J. P., *Analyst (London)*, 77 (1952) 905.
Williams, R. J. P., Hagdahl, C. and Tiselius, A., *Ark. Kemi*, 7 (1954) 1.
Young, D. H. and Crowell, A. D., *Physical Adsorption of Gases*, Butterworths, London, 1962, p. 4.
Zager, S. E. and Doody, T. C., *Ind. Eng. Chem.*, 43 (1951) 1570.
Zahn, P. K. and Stahl, I., *Hoppe-Seyler's Z. Physiol. Chem.*, 302 (1955) 204.
Zuev, S. N., Kozarenko, T. D. and Chernov, A. B., *Zh. Anal. Khim.*, 25 (1970) 2039.

# PRACTICE OF LIQUID CHROMATOGRAPHY

*Chapter 11*

# Operation of a modern liquid chromatograph

R. VESPALEC and M. KREJČÍ

## CONTENTS



## PREPARATION OF THE APPARATUS

For the preparation of an apparatus for chromatographic measurements, the first aspects to be considered are the demands of the problem being examined. Firstly, suitable dimensions for the column should be chosen and it should be filled by using one of the methods described below. Depending on the character of the packing, it must be decided whether or not a pre-column will also be necessary and the type of material that would be suitable for filling it. If it is necessary to change the composition of the mobile phase during the analysis, then a device for the production of a gradient should be chosen and connected with the column, and its shape and time programme selected. In this case, a pre-column is usually not connected. However, if a pre-column is indispensable (for example, for complete drying of some of the mobile phase components), it should be inserted before the gradient device. A suitable detector should be connected to the column outlet, followed by a flow meter. Finally, a fraction collector or a waste-container is connected. Further procedures depend on whether the analysis is carried out with a constant or changing mobile phase composition. First, work with a mobile phase of constant composition will be described (Fig. 11.1).

From the reservoir, 1, and degasser, 2, the old mobile phase is emptied through the stopcock, 3 (in some instances, the reservoir also functions as a degasser) and the reservoir is filled with a new phase. The performance of the pump, 4, is then chosen, and if the construction of the pressure valve, 5, permits it, the maximum permissible pressure in the apparatus is set. This operation is especially important if the mechanical strength of some components of the chromatograph is limited (for example, when working with glass columns). The pump should be protected by an independent break-through membrane so as to prevent its damage. The pressure pulse attenuator, 6, is set so that it operates with

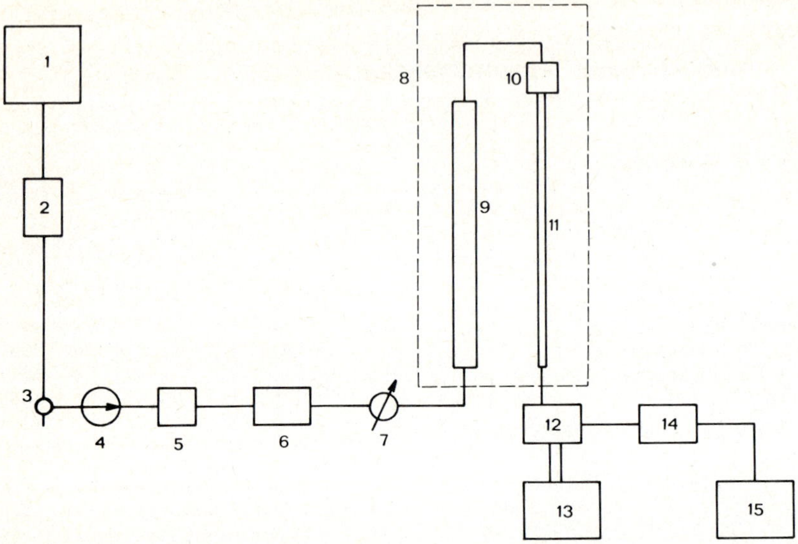


Fig. 11.1. Schematic diagram of a liquid chromatograph during operation with a mobile phase of constant composition. 1 = Reservoir; 2 = degasser; 3 = emptying and switching valve; 4 = pump; 5 = pressure trap; 6 = pressure damping; 7 = manometer; 8 = thermostat; 9 = pre-column; 10 = injection port; 11 = column; 12 = detector; 13 = recorder; 14 = flow meter; 15 = fraction collector or waste container.

maximum efficiency at the pump outlet pressure. In chromatographs with pulse-free pumps, the damper is not used. The inlet pressure on the column is read from the control manometer, 7. The temperature of the thermostat, 8, in which the pre-column, 9, the injection port, 10, and the chromatographic column, 11, are placed, is then set.

The whole chromatograph should be carefully washed with the mobile phase so as to remove the residues of the old mobile phase and air. Therefore, it is useful if all parts are flushed with the mobile phase and if their volumes are as small as possible. Complete de-aeration of the column and pre-column is important, especially if detectors with a flow-through cell are employed. After de-aeration, the detector is switched on. The injection of the sample can be started only when equilibrium is attained in the column, which is often signalled by the disappearance of the baseline drift of the detector. If the detector response is independent of the mobile phase composition, the equilibrium in the column can be ensured by repeated injection of a model mixture. Perfect conditioning of the column is important, especially in adsorption chromatography with a multicomponent mobile phase and when the so-called demixing effects take place (see Chapter 10). Also, when the temperature of the column changes, perfect attainment of the equilibrium state should be achieved. The time necessary for the washing of the chromatograph, degassing and conditioning of the column can be shortened by increasing the flow-rate.

After equilibrium in the column has been attained, the pump performance is regulated, the flow-rate of the mobile phase measured and the detector sensitivity and recorder chart speed are set. Unless the total volume of the mobile phase that has flowed through the column is recorded, the stability of the flow should be controlled even during the analysis, either by measuring the flow-rate or by reading the pressure meter data. During prolonged measurements, periodic refilling of the pre-column is necessary.

When concentration gradients are used, the chromatograph must be provided with several reservoirs. Usually, a different reservoir is used for each component. The preparation of the apparatus for measurement is, in principle, the same as that used in work with a simple mobile phase. Increased attention must be paid to column regeneration after each analysis. It is not sufficient if only the mobile phase in the column is replaced, but equilibrium must be attained between the stationary and mobile phases in the whole column. Inadequate column regeneration substantially impairs the reproducibility of the measurements. If mobile phases with large differences in elution strength are employed for gradient preparation, the regeneration may be time consuming, but the time required can be shortened by increasing the rate of flow of the mobile phase and by changing (usually increasing) the temperature. In work with concentration gradients in the mobile phase, it is important that the volume between the outlet of the gradient-forming device and the column inlet should be as small as possible. Large volumes might affect the shape and the time course of the gradient.

## SORTING OF SORBENTS ACCORDING TO PARTICLE SIZE

In other parts of this book, the importance of particle size (Chapter 3) and the homogeneity of its distribution (Chapter 3) for liquid chromatography has been stressed. The preparation of efficient columns of various types and diameters is connected with the use of sorbents with a particle size of *ca.* 5–500 μm (Table 11.1). Although commercially available sorbents and supports, recommended for certain uses, often have a satisfactory particle-size distribution, in many instances it is necessary to sort out and remove particles that are either too small or too large. Laboratory methods can be divided into basic groups: classification with sieves and fluid methods.

Materials up to *ca.* 25 μm particle size can theoretically be classified by sieving. When sieving in the dry state is carried out, the uppermost sieve has the largest mesh diameter, while the lowest sieve has smallest mesh size. The material passes through the sieves by its own weight. In order that the fine fractions may fall through the sieves, the sorted filling should be stirred continuously. The sieves are agitated either by hand or on a vibrator. An excessively high vibration frequency causes a strong erosion or fragmentation of the material with ensuing redistribution of particles. Vibrations not exceeding 5 cycles/sec are satisfactory. A receiver is placed under the sieves, in which the material from the finest sieve is collected.

When sieving is carried out in the wet state, water flows in the same direction as the classified particles. The receiver must have a sufficiently large diameter that even the finest particles that have passed through all of the sieves may sediment.

During the sieving, especially in the dry state, the particle-size distribution of the sorted material is often changed owing to mutual erosion or fragmentation. The powder and very fine particles formed adhere to larger particles from which they are separated with great difficulty. Some materials, especially organic, may have an electrostatic charge on

*References p.300*

TABLE 11.1
CONVERSION OF U.S. AND BRITISH SIZES OF SIEVES INTO METRIC SIZES
Taken from the catalogue of E. Merck (Darmstadt, G. F. R.) with their kind permission.

| Sieve opening (mm) | U.S. Standard, ASTM E 11–61 | Number of meshes (Tyler mesh/in.) | British Standard, BS 410:1962 (mesh/in.) |
|---|---|---|---|
| 0.037 | 400 | 400 | – |
| 0.044 | 325 | 325 | – |
| 0.045 | – | – | 350 |
| 0.053 | 270 | 270 | 300 |
| 0.063 | 230 | 250 | 240 |
| 0.074 | 200 | 200 | – |
| 0.075 | – | – | 200 |
| 0.088 | 170 | 170 | – |
| 0.090 | – | – | 170 |
| 0.105 | 140 | 150 | 150 |
| 0.125 | 120 | 115 | 120 |
| 0.149 | 100 | 100 | – |
| 0.150 | – | – | 100 |
| 0.177 | 80 | 80 | – |
| 0.180 | – | – | 85 |
| 0.210 | 70 | 65 | 72 |
| 0.250 | 60 | 60 | 60 |
| 0.297 | 50 | 48 | – |
| 0.300 | – | – | 52 |
| 0.354 | 45 | 42 | – |
| 0.355 | – | – | 44 |
| 0.420 | 40 | 35 | 36 |
| 0.500 | 35 | 32 | 30 |
| 0.595 | 30 | 28 | – |
| 0.600 | – | – | 25 |
| 0.707 | 25 | 24 | – |
| 0.710 | – | – | 22 |
| 0.841 | 20 | 20 | – |
| 1.00 | 18 | 16 | 16 |
| 1.19 | 16 | 14 | – |
| 1.20 | – | – | 14 |
| 1.41 | 14 | 12 | – |
| 1.68 | 12 | 10 | 10 |
| 2.00 | 10 | 9 | 8 |

their surface, which causes aggregation of the particles. The aggregates are retained on the sieves, so that the size of the conglomerates and not the true size of the particles corresponds to their mesh size. Agglomeration occurs particularly in the separation of small particles (approximately 10 $\mu$m diameter and less).

Many of these disadvantages are avoided in fluid methods, which are based on Stokes' law, from which the rate of movement of particles in a fluid medium can be deduced. At a critical linear rate of the fluid, $u_k$, moving upwards through the separator, the rate of movement of a spherical particle moving under the effect of gravity in the opposite direction to that of the fluid flow is equal to the rate of fluid flow if

$$u_k = \frac{R^2}{8\eta L} \cdot \frac{P_1^2 - P_2^2}{2P_0} = g \cdot \frac{d_p^2 (s - \rho)}{18\eta} \tag{11.1}$$

where $R$ is the diameter and $L$ the length of the separator, $\eta$ is the dynamic fluid viscosity, $P_1$ and $P_2$ are the inlet and outlet pressures in the separator, $P_0$ is the pressure at which the flow through the compressible fluid (gas) is measured, $\rho$ the fluid density, $d_p$ the particle diameter, $s$ the particle density and $g$ the gravitational acceleration. At a fluid flow-rate of less than $u_k$, equal-sized spherical particles ($s$ and $d_p$ constant) move against the liquid flow, while at rates above $u_k$ they are conveyed in the direction of the flow.

Fluid methods of separation are based on this principle. From eqn. 11.1, it can be deduced that in a moving fluid, spherical particles of diameter $d_p$ will float if

$$d_p^2 = \frac{9R^2}{4(s - \rho)gL} \cdot \frac{P_1^2 - P_2^2}{P_0} \tag{11.2}$$

Particles of larger diameter sediment, while particles of smaller diameter are conveyed from the separator. If the classified particles have irregular shapes, their dimensions can be replaced by an equivalent diameter $d_{p\ \mathrm{equiv.}}$, for which

$$d_{p\ \mathrm{equiv.}} = \frac{8\varphi Re\eta^2 L}{R^2 \rho} \cdot \frac{P_0}{P_1^2 - P_2^2} \tag{11.3}$$

where the Reynolds number $Re = u_k d_p \rho/\eta$. The values of the correction factor, $\varphi$, depend on the particle size (see Table 11.2). From eqns. 11.2 and 11.3, it is evident that the material can be classified, *i.e.*, fractions of a definite $d_p$ can be obtained, by changing the rate of fluid flow (which is achieved by changing $P_1$ and $P_2$) or by changing the radius of the grading device $R$, if other conditions are constant.

TABLE 11.2
DEPENDENCE OF CORRECTION COEFFICIENT, $\varphi$, ON $Re^2$

| $Re^2$ | Particle shape | | | | |
|---|---|---|---|---|---|
| | Globular | Round | Angular | Oblong | Flat |
| 8000 | 1 | 0.805 | 0.68 | 0.61 | 0.45 |
| 10,000 | 1 | 0.80 | 0.678 | 0.595 | 0.441 |
| 20,000 | 1 | 0.79 | 0.672 | 0.59 | 0.43 |
| 50,000 | 1 | 0.755 | 0.65 | 0.564 | 0.42 |
| 100,000 | 1 | 0.753 | 0.647 | 0.562 | 0.408 |
| 200,000 | 1 | 0.74 | 0.635 | 0.560 | 0.392 |

Tesařík and Nečasová described a simple laboratory device with which they achieved good results, in which both a gas (air) and a liquid were used as the carrier fluid. A scheme of the device is shown in Fig. 11.2. The basic separation unit consists of a glass tube of I.D. $R$ = 16 mm and length $L$ = 640 mm. In the bottom part of the tube, a porous glass plate is sealed, which ensures a homogeneous distribution of the introduced carrier

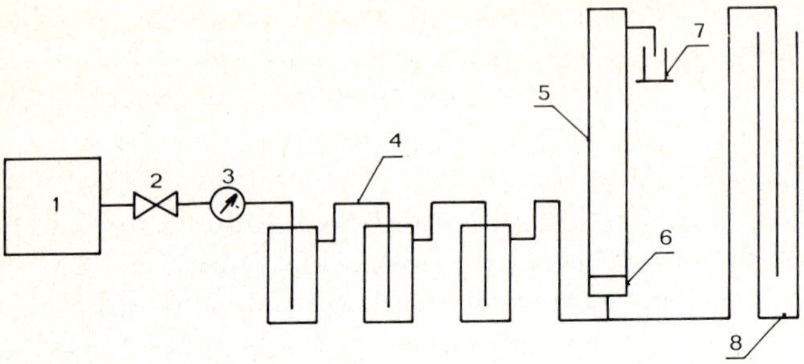


Fig. 11.2. Scheme of the grading device according to Tesařík and Nečasová. 1 = Membrane pump; 2, 3 = flow regulation elements; 4 = pulse dampers; 5 = grading tube; 6 = porous plate; 7 = fraction collector; 8 = water pressure regulator.

medium over the whole cross-section of the tube. When gas is used as the carrier medium, air is led from an efficient membrane pump (outlet pressure 2 kp/cm$^2$) through a conical regulation valve, a flow meter, a drying column and pressure pulse attenuators (capillaries serving as a pneumatic resistance connected in series and bottles of 250 ml volume serving as pneumatic capacities) into the grading tube. Parallel with this tube, a simple water manostat is connected, which permits the fine setting of accurate flow-rates of air through the grading tube. When water is used as the carrier medium, the use of a manostated gas of precisely known constant pressure is recommended for expelling the water from the reservoir. The grading tube remains unchanged.

The classification proper is carried out by introducing the sample of adsorbent (carrier, etc.), up to 3 ml in volume for the given experimental device, on to the porous plate in the grading tube. Then a stream of carrier medium is introduced at such a rate as to convey the smallest particles and the particles with lowest density in the direction of fluid flow. They are collected in the fraction receiver. As soon as only those particles are present in the tube which are not carried off by the carrier medium, the flow of the carrier fluid is increased either by increasing the water level in the manostat connected in parallel with the grading tube (if gas is used), or by increasing the gas pressure above the water in the reservoir (if water is used as the grading medium). Then the sorted out fraction that has a larger particle diameter or a higher specific weight is collected in another fraction receiver. The classification process is repeated if necessary until the whole of the volume has been classified.

If the technique described is employed and the fluctuation in the gas flow-rate is 4%, then the standard deviation, $d_p$, of the size of spherical particles is about 2%, assuming that the density of the material is homogeneous. However, the sizes of the particles may differ by up to 50% in instances when the value of $s - \rho$ varies within the range 0.8–2.1 g/cm$^3$. With particles of irregular shape, the standard deviation of the particle diameters may increase up to 25% on changing the flow-rate (in agreement with Table 11.2).

Today, various types of grading devices can be purchased. The most widely used are fluid graders based on the principle described by Rumpf and Leschonski, which use separators, the shorter part of which, adjacent to the membrane gas inlet, is cylindrical and the longer part is conical with a diameter that decreases in the direction of gas flow. This grading vessel vibrates at a suitable frequency and amplitude. At constant flow-rate of the carrier medium, *i.e.*, gas, the material can be sorted into two fractions. The principle of the classification also follows from eqns. 11.2 and 11.3. In a simplified manner, the grading process may be viewed so that the classified material is mechanically thrown by the vibration into the conical part of the separator (smaller $R$), and, depending on the particle radii and their density, it either falls back on the membrane or is conveyed out from the device into the collector of the fine fraction. The performance of this apparatus is higher than that of the above described laboratory device, but its construction is difficult under laboratory conditions.

The particle-size distribution can be determined by measuring a sufficient number of particles microscopically. The resulting dependence is shown in the form of a histogram, showing the number of particles of a given radius present in the measured mixture. The lowest number of measured particles should be 100–500.

## Classification of silica gel particles according to size by sedimentation

For a single classification of 2–3 kg of crude silica gel, five 10-l broad-necked flasks and one *ca.* 20-l vessel may be used.

In flask No. 1, the added silica gel is stirred with 9–10 l of water and then allowed to sediment for 1 min. The suspension is decanted (in thirds of the volume) into flask No. 2, where silica gel is allowed to sediment for 2 min. The suspension is decanted gradually from the full flask No. 2, again in thirds of the volume, up to flask No. 5, allowing sedimentation in flask No. 3 to take place for 4 min, in flask No. 4 for 8 min and in flask No. 5 for 16 min. From the last flask, the suspension is decanted (in halves of the volume) into the larger vessel, where the fines are allowed to sediment. This procedure is continued until 40–50 l of water have passed through the sedimentation vessel. Single silica gel fractions from flasks Nos. 1–5 are then filtered off under suction and dried at 120°C for 12 h. The fine powder from the large sedimentation vessel can be used for thin-layer chromatography, after drying under the same conditions. The procedure is summarized in Table 11.3. It should be noted that, in order to economize with the use of distilled water, the first three runs can be carried out with tap water and only the last two runs with distilled water. For stirring, a strong rod (preferably made of a hard polyvinyl plastic) is used, which is rinsed with water after each stirring.

TABLE 11.3
CLASSIFICATION BY SEDIMENTATION

| Number of flask | Time of sedimentation (min) | End of stirring (min) | Time of decantation (min) |
|---|---|---|---|
| 5 | 16 | 0 | 16 |
| 4 | 8 | 8.5 | 16.5 |
| 3 | 4 | 13 | 17 |
| 2 | 2 | 15.5 | 17.5 |
| 1 | 1 | 17 | 18 |

*References p.300*

**Regeneration of silica gel**

The silica gel is boiled with a 5–10-fold amount of 1% sodium hydroxide solution for 30 min. After checking that the solution is strongly alkaline to phenolphthalein, the suspension is filtered hot and washed three times with distilled water. Further boiling is carried out with a 3–6-fold volume of 5% acetic acid for 30 min. After filtration, the particles are washed with distilled water until neutral, then with methanol, and twice with distilled water. After filtration under suction, the material is dried. Activation is carried out at 120°C for 12 h.

## DETERMINATION OF THE ACTIVITY OF ALUMINA BY THIN-LAYER CHROMATOGRAPHY

Heřmánek *et al.* spread alumina (*ca.* 10 g) on a glass plate (for example, 20 × 10 cm) and smoothed the surface by rolling it with a glass rod, 0.6 mm of each end being strengthened with insulation tape. The length of rod between these tapes is about 4 cm, which produces on the plate a layer of this width and 0.6 mm thick. Solutions (0.02 ml) of standards (azo dyes: azobenzene (30 mg), *p*-methoxyazobenzene, Sudan yellow, Sudan red and *p*-aminoazobenzene (20 mg each) in 50 ml of dry, distilled tetrachloromethane) are applied at a distance of 3 cm from the edge of the plate, which is then developed in a slightly oblique position in a low tank containing tetrachloromethane. The $R_F$ values of single azo dyes (measured from the centres of the spots) are correlated with the activities as determined by Brockmann and Schodder (see Table 11.4). For other standardization methods, see Engelhardt and Wiedemann.

The activity of commercial alumina is usually less than I. If alumina of this activity is needed, it is heated at 350°C for 6–8 h, or at 120°C in a vacuum (oil pump) for 2–3 h. Lower activities are achieved by adding the corresponding amount of distilled water to this most active alumina (see Table 11.5).

TABLE 11.4
$R_F$ VALUES (±0.04) OF INDIVIDUAL AZO DYES ON ALUMINA OF DIFFERENT GRADES

| Azo dye | Grade of alumina according to Brockmann and Schodder | | | |
|---|---|---|---|---|
| | II | III | IV | V |
| Azobenzene | 0.59 | 0.74 | 0.85 | 0.95 |
| *p*-Methoxyazobenzene | 0.16 | 0.49 | 0.65 | 0.89 |
| Sudan yellow | 0.01 | 0.25 | 0.57 | 0.78 |
| Sudan red | 0.00 | 0.10 | 0.33 | 0.56 |
| *p*-Aminoazobenzene | 0.00 | 0.03 | 0.08 | 0.19 |

TABLE 11.5
ACTIVITY OF ALUMINA, SILICIC ACID AND MAGNESIUM SILICATE DEPENDING ON WATER CONTENT

The values given are averages depending on the type and the manufacturing procedure used for individual adsorbents.

| Activity grade | Water added (%) | | |
|---|---|---|---|
| | To alumina | To silicic acid | To magnesium silicate |
| I | 0 | 0 | 0 |
| II | 3 | 5 | 7 |
| III | 6 | 15 | 15 |
| IV | 10 | 25 | 25 |
| V | 15 | 38 | 35 |

## COLUMN PREPARATION

Today, the factors are known that determine the efficiency of regular filled columns in which the ratio of the column diameter, $d_c$, to the mean particle diameter of the filling, $d_p$, is $d_c/d_p > 10$. Although columns can be prepared reproducibly that are comparable with gas chromatographic columns with respect to their HETP, development in this area is still continuing. The technique of filling the columns can be divided into two basic groups: (1) filling in the wet state and (2) filling in the dry state.

(1) When the column is filled in the wet state, a suspension of the column filling in a suitable liquid is prepared, which is then introduced into the column in such a way that a bed that is as homogeneous as possible is obtained, in which the particles are settled as densely as possible. The prepared suspension should be stable and no agglomeration or fractionation of solid particles during the preparation of the suspension and filling it into the column should take place. Both of these effects lead to inhomogeneity of the bed, which impairs the quality of the prepared column. Therefore, in an optimum case, the particles in the slurry should float (the density of the suspending liquid, preferably a binary mixture, must be equal to the mean density of the support particles) and the column should be filled as rapidly as possible. The suspending liquid must also often induce the necessary swelling of the filling particles. In a glass vessel, the column filling is stirred with an amount of a binary liquid such that the resulting suspension contains 10–25% (w/w) of particles. Perfect mixing and degassing of the slurry is achieved by ultrasonic stirring for 2 min. If the particles sediment or if they aggregate at the surface, the density of the suspending solvent is adjusted by adding the heavier or the lighter component until the entire filling remains suspended for at least 10 min. The balanced slurry is then stirred again and transferred rapidly into the reservoir (Fig. 11.3), to which a column filled with the suspending solvent is connected. In the reservoir, the suspension is carefully overlayered with an immiscible liquid that functions as a piston. As the filling should be carried out as rapidly as possible, a column connection should be chosen with as large a diameter as possible, and a fritted glass septum closing the column outlet with

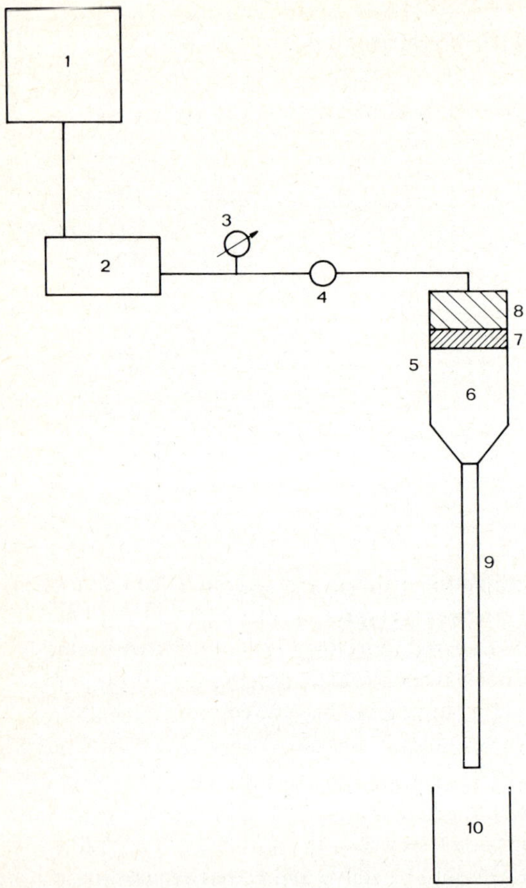


Fig. 11.3. Schematic arrangement for packing the column with equilibrated suspension. 1 = Reservoir containing hexane; 2 = high-pressure pump; 3 = manometer; 4 = high-pressure valve; 5 = reservoir with suspension; 6 = suspension; 7 = water; 8 = hexane; 9 = column; 10 = waste container.

maximum admissible pores (with respect to the grain size of the filling) and the maximum possible working pressure (up to 340 atm) should be used. The reservoir containing the suspension is connected with the pump after the required pressure at the pump outlet has been achieved.

The liquid flow-rate through the column decreases during the filling and therefore the pump speed should be regulated. The column is filled when a calculated volume of the suspending solvent or the first fraction of the liquid serving as a piston has passed through it. After the column has been filled, the introduction of the mobile phase should be stopped until the pressure drops spontaneously. If the pressure is decreased before entering the column, the homogeneity of the bed is seriously disturbed and the column efficiency decreases. This effect applies generally.

Classical methods of filling columns with suspensions (sedimentation, filtration with increasing pressure) are not of much use in practice owing to the low efficiency of the columns so prepared. Later methods involving the use of a balanced density solvent permit the preparation of high-quality, high-efficiency columns even when very fine materials are used, which so far cannot be filled in the dry state. The whole procedure is, however, tedious and time consuming, although the filling can be carried out very rapidly. Fillings are not applicable that would be changed in an undesirable manner by the suspending solvent or the liquid used for the subsequent purification and adjustment of the activity of the filling of the prepared column (changes in the degree of wetting in work with stationary liquids, reactions with some types of chemically bound stationary phases, etc.).

In order to illustrate the results that can be achieved by the described method of column packing (in the literature, it is indicated as the "balanced slurry column packing procedure"), several examples are presented here.

When a material the particles of which adhere to each other in dry state was packed, for example Zipax with a chemically bonded stationary phase, Kirkland (1971b) achieved with particles of size 37–44 $\mu$m an efficiency 30–50% better than when the dry packing procedure was used. With finer particles, the increase in efficiency was even higher. The efficiency of the packed columns was reproducible to ±10%. For the suspension of the packing, he used a mixture of 21 parts of tetrabromomethane and 15 parts of Perclene (DuPont, Wilmington, Del., U.S.A.). The suspending liquids were eluted from the column with methanol, which was then expelled by the mobile phase.

When packing the column with silica gel of 5–10 $\mu$m particle diameter, Majors suspended 1 g of the packing dried at 200°C for 2 h in 10 ml of a solvent consisting of 60.6% (w/w) of tetrabromomethane and 39.4% (w/w) of tetrachloroethylene; undried silica gel agglomerates in this liquid. For the elimination of the remainder of the suspending liquid and for adjusting the activity of the silica gel, he used a procedure of activation by solvent, proposed by Snyder. At a linear flow-rate of the mobile phase of 1.18 cm/sec, the column had an HETP of 0.1 mm.

The best results so far were obtained by Kirkland (1972b). Spherical silica gel particles with a regulated porosity of 5–6 $\mu$m diameter were packed as an aqueous suspension stabilized by the addition of 0.001 *M* ammonia solution. For the component with capacity factor $k' = 12$, at a linear flow-rate of the mobile phase of 0.44 cm/sec, he measured with a 250 × 3.2 mm column an HETP of 0.038 mm. This efficiency is comparable with the efficiency of capillary columns in gas chromatography.

(2) When columns are filled in the dry state, small portions of sorbent are introduced into a tube that is closed at the bottom with a metallic, glass or PTFE filter disc, glass-wool or a filter-paper disc. The homogeneity of the sorbent bed can be improved by providing for the movement of the support particles, which permits their subsequent orientation. The amount and the method of supplying the mechanical energy for moving the added particles (shaking down, vibration, compression) should be chosen so that no fractionation of the particles according to their size or even the disturbance of the formed bed occur. Hence, it is advisable to add the packing slowly, most often discontinuously and in small portions, so that the amount of packing that has to be moved each time is as small as possible. Each portion of the material added should be such that its packed length does not exceed 1–5% of the length of the prepared column. The amount of mechanical energy introduced by shaking, vibration or compression should be regulated according to

particle size, shape and mean density. Therefore, a series of variants of this method have been described for various granulations and types of materials, as illustrated in the examples below.

It is important that the packings should have as narrow a particle-size distribution as possible. The larger the distribution, the easier is the separation of the particles according to their size along the column. The smaller the granules used, the more carefully the column should be packed and the longer is the time necessary for the packing. The basic limitation of all procedures in which dry materials are used is that they cannot be utilized for packings of low specific weight, with increased adhesion of particles, for packings that change their volume under the effect of the mobile phase, and for very fine granules (the lower limit varies from 20 to 50 $\mu$m). In spite of these disadvantages, a number of workers agree that whenever possible it is preferable to use the dry column packing procedure.

A few examples of packing chromatographic columns are presented below, from which differences connected with various materials follow.

Halász and Naefe showed that for silica gel particles, the surfaces of which were esterified with polyethylene glycol 200 and which were larger than 50 $\mu$m, the same method can be used as in gas chromatography. Silica gel was added continuously under simultaneous vibration (60 cycles/sec). Then the packing was shaken down by allowing the tube to fall from a height of 20 cm and the filling was completed. For particles of 10–50 $\mu$m diameter, it was preferable if the packing was added in such amounts as would give a column (zone) about 1 cm long after each addition. After each addition, the column was vibrated and finally filled to completion as in the preceding case, tapping the sides to effect the packing. The efficiency of such columns was reproducible to ±10%.

Kirkland (1971a) stated that relatively light materials (silica gel, Kieselguhr) with non-spherical shapes cannot be packed in the dry state satisfactorily if their diameter is less than 50 $\mu$m. He recommended that the support should be added in such a manner as to increase the bed height after each addition by 0.4 cm, with continuous vertical tapping on the floor (2–3 taps/sec) and also tapping the side of the column at the level of the filling. The filling of the column should be completed after tapping for 5–10 min. The packed column is then run for half an hour at a pressure exceeding the operating pressure during the measurement. If, during this interval, the bed is shortened the column was not packed well and it must be remade. A 1 m × 2 mm column can be packed by this technique in 15–30 min, depending on the particle size and the type of packing. Kirkland (1972b) mentioned that for packing a column with Zipax without a bound stationary phase, the most advantageous procedure is the "modified tap–fill" procedure. The packing is added in aliquots of 100–200 mg and, after each addition, the column is tapped vertically on the floor and also on the side (80–100 times, with a frequency of 2–3 taps/sec). Then the packing is consolidated by gently tapping the column on the floor for 15–20 sec, without tapping the side. The procedure can be also used for Corasil, Durapak and Porasil, and for particles smaller than 37 $\mu$m.

Zipax wetted with the stationary phase was packed by Done and Knox. A column of 2.1 mm I.D. was placed into a mechanical apparatus which rotated the column at 180 rpm and simultaneously allowed it to fall from a height of 1 cm 100 times per minute. During the filling, the column was tapped gently on the side at the level of the packing. The packing was introduced into the column continuously with a stream of dry nitrogen. Karger *et al.* recommended filling the column with surface-etched glass beads by tamping with a glass rod. When packing large diameter columns, Sie and Van den Hoed found that rotating the column was advantageous.

Randau and Schnell compared packing columns by sedimentation with three methods of dry packing: vibration on a vibrational table, mechanical tamping and tapping on the base. The best columns were obtained by simple tapping, the column prepared by vibration was less satisfactory and the column obtained by sedimentation was the worst (Table 11.6).

TABLE 11.6
EFFICIENCY OF COLUMNS PACKED BY VARIOUS METHODS (RANDAU AND SCHNELL)

Silica gel of particle diameter 60 Å (40–63 μm), coated with 40% (w/w) of 1,2,3-tris-(2-cyanoethoxy)-propane; column length 50 cm, flow-rate 0.68 cm/sec.

| Method of packing | HETP (cm) (2-cyanoethoxy) |
|---|---|
| Tapping | 0.80–0.84 |
| Vibration | 0.85–1.27 |
| Tamping | 0.96–1.08 |
| Sedimentation | 1.46–1.94 |

## SAMPLE PREPARATION AND APPLICATION

When samples are prepared, two fundamental criteria must be considered: the detector sensitivity and the capacity of the column packing indicating the maximum amount of sample that can be introduced into the column. In liquid–solid systems, the concept of the linear capacity of the packing is common. It is usually considered to be equal to the amount of solute that would result, during its introduction into the column, a maximal 10% decrease in the specific elution volume of a zero amount of a given solute. This specific elution volume of the zero amount of solute is obtained by extrapolation of a graph of the specific elution volume plotted against the amount introduced; it is usually given in grams of solute per gram of support. In adsorption systems, the capacity of the packing is critical, especially in work with non-deactivated adsorbents with large surface areas or with adsorbents with very low specific surface areas. When the support capacity is exceeded, the non-linearity of the distribution isotherm is reflected in the asymmetry of peaks. The occurrence of peak broadening, either at the front or back of the zone (the latter is more common), leads to a poorer resolution of the components and, from the point of view of separation efficiency, it is therefore undesirable.

In systems with a linear distribution isotherm, when the support capacity is exceeded, an anomalous spreading of the zones occurs, even when their symmetry is preserved. In partition chromatography, the capacity of the support is the limiting factor at low wetting, for example, if glass beads or porous layered beads are used as supports.

Hence, an amount of the separated component should be injected per plate of the chromatographic column such that the capacity of the plate is not exceeded, which differs

*References p.300*

for various materials. However, the total amount injected should be such as would permit the detection of the injected component after its passage through the column. For qualitative analysis, it is required that the response should be equal to at least double the noise, while for quantitative purposes the response should exceed the noise by at least five times. This condition, however, would not be fulfilled in many instances, especially for components with a high retention, if the sample were to be introduced only on to the first plate of the column. In addition to the capacity of the stationary phase, the solubility of the sample in the mobile phase also can be a factor that limits the maximum utilizable sample concentration.

For a maximum volume of the injected sample, $V_S$ (which would cause at most a 5% increase in the zone width), Klinkenberg derived the relationship

$$V_S = 1.1\,(v_m + Kv_s)\sqrt{N} = 1.1 \cdot \frac{V_m + KV_s}{\sqrt{N}} \tag{11.4}$$

where $N$ is the number of theoretical plates of the column used, $v_m$ the volume of the mobile phase in one plate (HETP), $v_s$ the volume of the stationary phase in one plate, $V_m$ the total volume of the mobile phase in the column, $V_s$ the total volume of the stationary phase in the column and $K$ the distribution coefficient of the injected sample. For a component with zero retention, the equation assumes a simpler form:

$$V_S = 1.1\,v_m\sqrt{N} = 1.1 \cdot \frac{V_m}{\sqrt{N}} \tag{11.5}$$

which represents the strictest condition for the introduced sample volume. The term $v_m$ can be made equal to the ratio of the dead volume $V_0$ to the number of plates of the column $N$ only if the volume of the connections is negligible with respect to $V_m$. From eqns. 11.4 and 11.5, it follows that the better the column of given dimensions is prepared, *i.e.*, the more efficient it is, the smaller is the injection volume that can be used.

Scott has shown that the sample can be introduced dissolved in the mobile phase, as a solution in the stationary phase (if the stationary phase is a liquid), or as a solution in another, better solvent. The possibility of using an auxiliary solvent with a better dissolving ability than that of the mobile phase for the dissolution of the solute follows from the fact that in the chromatographic column a decrease in the concentration of the solute in the mobile phase takes place as a consequence of its partitioning between the stationary and mobile phases. In the case of a solute characterized by a capacity factor $k'$, of the total amount injected into the first plate, only a fraction $1/(1 + k')$ remains in the mobile phase. Therefore, for the injection, a concentration of the solute in the auxiliary solvent can be used that exceeds $(1 + k')$ times its solubility in the mobile phase. It is therefore evident that the utilization of an auxiliary solvent is important in practice and is possible only when solutes with a higher retention are injected. For a component with a low retention (small $k'$), a decrease in its solubility in the mobile phase and its separation from the mobile phase can easily take place after the substitution of the mobile phase for the auxiliary solvent. The same phenomenon can be observed with excessively large differences in the solubilities in the mobile phase and the auxiliary solvent, even for solutes with a higher $k'$, if the solubility of the solute in the auxiliary solvent is higher than (1 +

$k'$) times its solubility in the mobile phase. The precipitate is then dissolved slowly in the pure eluting agent and the solution introduced into the column. This phenomenon is equivalent to the injection of an excessive volume of a dilute sample. The result is a very broadened asymmetric zone.

The basic requirements for the injection can be summarized as follows. The accurate measurement of a known volume must be ensured, and this volume must be introduced into the column as rapidly as possible. As diffusion coefficients in liquids are low, it is important that the sample should be introduced accurately into the centre of the column, as an excentric injection decreases the efficiency. High inlet pressures on columns require the use of high-pressure injection ports or special techniques of sample introduction. The choice of organic solvents or of aqueous solutions of acids and bases as mobile phases places appreciable requirements on the construction of the injection ports and the materials used. A detailed description of these devices is given in Chapter 8.

However, it can be stated that no system described so far in the literature is universally satisfactory if the following requirements are placed upon it simultaneously: high accuracy and reproducibility of the injection with low consumption of the sample; the possibility of injecting against pressures of several hundreds of atmospheres; ensuring a rapid transport of sample into the centre of the column without diluting the mobile phase or loss of the sample; universality (with respect to the injected volumes) within a range of three orders of magnitude; operational reliability; and simplicity of maintenance.

## GENERAL COMMENTS

A good result in chromatographic analysis is not dependent only on a suitable choice of the chromatographic system, *i.e.*, by an appropriate combination of stationary and mobile phases, and on the separation conditions, such as temperature, mobile phase flow-rate and the use of gradients of various types. The preparation of a sufficiently effective chromatographic column and, last but not least, the choice of an optimal and sufficiently sensitive detector, are also of particular importance.

The choice of components is determined by the types of separations or chromatographic methods and techniques that most often come into consideration (ion-exchange chromatography, gel chromatography, chromatography on adsorbents or stationary phases, preparative chromatographic procedures), and thus also by whether the apparatus will be used for a single or a few routine analyses or for a more demanding research work. The choice of detectors depends firstly on the type of substances to be separated and the requirements placed on the sensitivity of their detection. Usually, the composition of the mobile phase must also be taken into consideration, and also the changes in its composition during the analysis when gradients are used. The choice of a particular detector is usually a compromise between the requirements placed on the sensitivity and the universality of the detector, but often the possibility of working under different experimental conditions (flow-rate, temperature, mobile phase composition, etc.) must also be envisaged.

It is necessary for all materials with which the mobile phase passing through the apparatus will come into contact to be perfectly inert towards it. This condition is usually

*References p.300*

fulfilled by stainless steel, PTFE and glass, which are used as basic constructional materials. The resistance of rubbers, which are used as packings at some points in the apparatus, depends on the composition of the mobile phase. So far, no type of rubber that would be perfectly suitable in all instances is known.

Efficient degassing of all spaces through which the mobile phase flows is always useful. It can be enhanced by various methods: preliminary evacuation of the chromatograph, its flushing with a gas that is soluble in the mobile phase used, increasing the pressure of the mobile phase in the system, which enhances the dissolution of gases, or degassing the mobile phase before it enters the pump. As the degassing of the mobile phase is also useful during the operation, the degasser is usually an integral part of liquid chromatographs run at higher pressures. In an apparatus that is not provided with a degasser, the formation of bubbles often occurs, which disturbs the operation of detectors with a flow-through measuring cell. A dynamic resistance at the outlet, which increases the pressure in the system and thus suppresses the formation of bubbles, can be connected only if the detector can operate at least at a mildly increased pressure. In some instances, the degassing of the mobile phase may be useful even from the point of view of long-term stability of the column packing.

When the mobile phase is changed or when a concentration gradient is employed, efficient washing of the whole system and bringing the column into the original state are essential. Imperfect exchange of the phases or poor equilibration in the column results in irreproducibility of the elution data. If the detector response is dependent on the mobile phase composition (in the case of binary detectors, for example, refractometers), an imperfect phase exchange may also affect the quantitative data.

The requirement of perfect sealing of the whole system is obvious. Leakages in components inserted before the column cause a pressure decrease in the system and a disturbance in the correlation between the pump speed and the mobile phase flow-rate through the column. A leakage of liquid after the injection site is also reflected in irreproducibility of the detector response. Leakages also allow the penetration of air into the apparatus, thus facilitating the undesirable formation of bubbles in the detector. In work with organic liquids, there is also a risk of ignition of the escaping vapours. In addition to the input pressure, the flow-rate of the mobile phase through the column should also be followed.

The device for the measurement of the flow-rate does not always constitute a part of the chromatograph. Of various methods of measuring flow-rate (syphons, integral measurements, weighing of the liquid passed; see Chapter 8), only the determination of the immediate flow-rate is usually suitable for work with narrow analytical columns; the flow-rate is measured on the basis of the time of passage of an air bubble, introduced into the current of liquid behind the detector, through a calibrated volume of a narrow tube; this method was proposed by Gerding and Hagel. Commercial flow meters also operate on this principle, enabling an integral measurement of the passed-through volume of the mobile phase to be made. When working with the degassed mobile phase the air bubble is significantly dissolved in it. The error in the determination of the flow-rate caused by this effect, which may be up to 30%, can be eliminated by substituting the air with a liquid immiscible with the mobile phase used. For accurate quantitative work, it is important to know, in addition to the immediate flow-rate, the total volume passed as a function of time. The accuracy of the calculation of the passed-through volume from the flow-rate and time is dependent on the constancy of flow-rate, conditioned by the stability of the pump performance and the pressure drop in the column. The pressure drop in the column of a perfectly sealed apparatus usually changes only if the column is obstructed by pieces of the eroded septum during the introduction of the sample with syringes. The resistance of the septum depends not only on the material used but also on the composition of the mobile phase and the diameter of the syringe needle used for injection.

An important condition for accurate work, especially for the measurement of the retention data and during analyses utilizing these data, is to ensure a known, constant temperature of the column to an accuracy of several tenths of 1°C. The effect of temperature on retention, demonstrated by a number of workers, is comparable, in principle, with the effect in gas chromatography if the system consists of a simple mobile phase. Maggs, and Scott and Lawrence found that in multicomponent mobile phase systems, where the conditions for the attainment of partition equilibria are much more complex, an increase in temperature may cause not only changes in relative retention or the reversal of elution sequences, but even an increase in the retention of a certain component. The need for thermostatic control of column temperatures should also be stressed because, by tradition, it is often still not carried out in liquid column chromatography. However, it is desirable that with the development of modern instrumentation, permitting the analysis of complex mixtures and work at temperatures other than room temperature, it should become equally common as in gas chromatography.

In view of the effect of the mobile phase on retention and separation selectivity, and sometimes of the response, increased attention must be paid to the composition and purity of the mobile phase. In analyses carried out without a concentration gradient, it should be ensured that the composition of the mobile phase does not change during the whole series of measurements. Therefore, it is advisable to use as mobile phases either individual chemicals or to prepare large amounts of mixed mobile phases of known composition from components of known or at least defined compositions. In accurate and demanding measurements, it is useful if the purity or the composition of the mobile phase is controlled. The elimination of constituents (impurities) that could function as moderators or substances that appreciably affect the elution strength of the mobile phase is of special importance. In practice, this usually means the elimination of constituents that have a chemical nature different from that of the mobile phase or its components (for example, trace amounts of polar substances from hydrocarbons). If a concentration gradient is used, it is desirable that the changes in concentration with time should be reproduced as accurately as possible during individual analyses. The purity of all of the components must satisfy the requirements mentioned in the preceding case. Only under these conditions can qualitative and quantitative reproducibility of the analysis be achieved that would be comparable with the results obtained with a mobile phase of constant composition.

In conclusion, it should be stressed that the successful utilization of narrow and especially short columns of high efficiency is dependent on the minimum volume of the connections between the injection port and the detector. The detector volume is also important in these instances. The dependence of the column efficiency on extracolumnar parameters is discussed in Chapter 8. Krejčí and Pospíšilová also found that the contribution to the

*References p.300*

zone width, given by zone spreading in the connections, depends not only on the magnitude, but also on the shape, of these volumes. For example, the contribution caused by the spreading of the zone in a 40-μl volume, corresponding to 45 cm of a capillary of 0.25 mm diameter, is virtually equal to the contribution created in 176 μl of free volume of a capillary of 45 cm × 1 mm, packed with glass beads. Scott and Kucera derived equations that permitted the calculation of the dimensions of empty tubes connecting the columns that would prevent an increase in the width of each eluted zone by spreading in these connections above 5%.

## REFERENCES


Brockmann, H. and Schodder, H., *Ber. Deut. Chem. Ges.*, 74 (1941) 73.
Done, J. N. and Knox, J. H., *J. Chromatogr. Sci.*, 10 (1972) 606.
Engelhardt, H. and Wiedemann, H., *Anal. Chem.*, 45 (1973) 1641.
Gerding, J. J. Th. and Hagel, P., *J. Chromatogr.*, 31 (1967) 218.
Halász, I. and Naefe, M., *Anal. Chem.*, 44 (1972) 76.
Heřmánek, S., Schwarz, V. and Čekan, Z., *Collect. Czech. Chem. Commun.*, 26 (1961) 3170.
Karger, B. L., Conroe, K. and Engelhardt, H., *J. Chromatogr. Sci.*, 8 (1970) 242.
Kirkland, J. J., in J. J. Kirkland (Editor), *Modern Practice of Liquid Chromatography*, Wiley-Interscience, New York, 1971a, p. 178.
Kirkland, J. J., *J. Chromatogr. Sci.*, 9 (1971b) 206.
Kirkland, J. J., *J. Chromatogr. Sci.*, 10 (1972a) 129.
Kirkland, J. J., *J. Chromatogr. Sci.*, 10 (1972b) 593.
Klingenberg, A., in R. P. W. Scott (Editor), *Gas Chromatography 1960*, Butterworths, London, 1961, p. 182.
Krejčí, M. and Pospíšilová, N., *J. Chromatogr.*, 73 (1972) 105.
Maggs, R. J., *J. Chromatogr. Sci.*, 7 (1969) 145.
Majors, R. E., *Anal. Chem.*, 44 (1972) 1722.
Randau, D. and Schnell, W., *J. Chromatogr.*, 57 (1971) 83.
Rumpf, H. and Leschonski, K., *Chem.-Ing.-Tech.*, 39 (1967) 1231.
Scott, R. P. W., *J. Chromatogr. Sci.*, 9 (1971) 449.
Scott, R. P. W. and Kucera, P., *J. Chromatogr. Sci.*, 9 (1971) 641.
Scott, R. P. W. and Lawrence, J. G., *J. Chromatogr. Sci.*, 7 (1969) 65.
Sie, S. T. and Van den Hoed, N., *J. Chromatogr. Sci.*, 7 (1969) 257.
Snyder, L. R., *J. Chromatogr.*, 25 (1966) 274.
Tesařík, K. and Nečasová, M., *J. Chromatogr.*, 75 (1973) 1.

*Chapter 12*

# Practice of gel chromatography

J. ČOUPEK, M. KUBÍN and Z. DEYL

## CONTENTS



## CHOICE OF GEL PACKING

The choice of the gel packing, together with an adequate design of the apparatus, plays a decisive role in gel chromatographic separations. Besides fulfilling basic requirements towards the gel, *viz*., its chemical inertness, minimal adsorption of separated compounds, swelling capacity or at least wettability with an eluent, etc., the gel packing must also have an adequate resolution as required by the given problem. Heitz *et al.* discussed the factors affecting the resolution, defined by $R = \Delta V_e/4\sigma$, where $\Delta V_e$ is the difference between the elution volumes and $\sigma$ is the mean width of the chromatographic zone of compounds being separated that do not differ much in their molecular weights. The $\sigma$ values of such compounds can be regarded as being approximately identical and can be calculated from the relationship suggested by Van Deemter or Giddings (Heitz and Čoupek). The dependence of $\sigma$ on particle size, flow-rate of the eluent, diffusion coefficients of the compounds used and column length was investigated. $\Delta V_e$ depends on the physical structure of the gel packing or, more exactly, on the pore-size distribution of the gel in the swollen state. This dependence is shown in Fig. 12.1; it can be seen that the basic information on the application of the gel for a given problem can be read off conveniently from a calibration graph of the logarithm of the molecular weight of standard test compounds plotted against their elution volumes. The differences between the

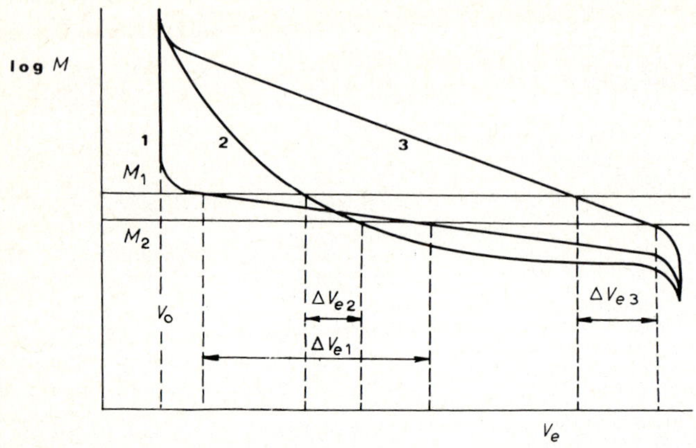


Fig. 12.1. Relationship between log $M$ and the elution volume, $V_e$ ($M$ = molecular weight). 1 = Tailor-made gel with a suitable molecular-weight range; 2 = gel with an unfavourable pore distribution; 3 = universal gel with a molecular-weight range that is too wide for the given separation problem.

elution volumes of two compounds with molecular weights $M_1$ and $M_2$, and thus also their resolution, will be the greater the smaller is the slope of the calibration curve defined by the relationship $d\log M_w/dV_e$. For a gel with good separation properties, the calibration graph must be linear over the largest possible range of elution volumes (curve 1). It may be expected from the non-linear form of curve 2 that the separation properties of a gel characterized by this curve will not be satisfactory. Curve 3 is the calibration graph of a gel with a wide range of applications, which, in spite of its poorer resolution, may be used successfully for preliminary analyses of samples that have an unknown molecular-weight distribution. The emphasis that used to be laid on the use of spherical packing particles for gel chromatographic elution is nowadays somewhat obsolete, because virtually all gel producers meet this requirement. However, another problem remains, namely, that of the particle size and particle-size distribution of gel packings. The column separation efficiency increases with decreasing particle diameter according to known relationships; on the other hand, analyses become too time consuming, and also the techniques used for the fractionation of the gel particles become more difficult, thus raising the price of the packing material. If the particle-size distribution of the packing used for separation becomes broader, the separation efficiency may be considerably impaired.

The apparatus used for separation is another important factor, and its effect on the choice of the gel must not be neglected. The simple technique involving the use of a glass column with elution by hydrostatic pressure, which is still frequently used for preparative separations in aqueous systems, has much lower requirements towards the geometrical and mechanical properties of a gel packing than modern high-speed, high-resolution chromatographic procedures with columns that have separation efficiencies of thousands of theoretical plates per metre of the column length.

It is therefore desirable that all of the factors outlined above should be taken into account when choosing the packing in order to achieve a compromise between the separation efficiency of the packing and the time of analysis, and between the use of a universal gel and a "tailor-made" gel; it is also useful to consider the solvation of the gel with an eluent and the possible adsorption effects that may arise here and affect considerably the elution volume of the compound, thus causing undesirable deformations of the chromatographic zone (tailing). No lesser importance should be assigned to the chemical and mechanical stabilities of the gel, which play a decisive role in the lifetime of the packing and the stability of the flow-rates through the column. For exacting separations, ready made columns are commercially available (Waters Ass., Framingham, Mass., U.S.A.; Knauer, Berlin, G.F.R., etc.) with a comparatively broad (virtually universal) range of applications. Gel packings with a rigid structure, such as porous glass or porous silica gel, are less exacting than organic gels with respect to the precise experimental conditions that must be used and are therefore simpler to use.

## CHOICE OF SOLVENT AND OPERATING TEMPERATURE

When choosing the solvent, it must be borne in mind that the preparation of a high-quality gel chromatographic column is time consuming and requires experimental dexterity. It is therefore desirable that the lifetime and application range of columns should be as high as possible. The solvent should be of a universal character, dissolve the greatest possible number of compounds, be readily available and easy to be freed from impurities. During analysis or storage, it should not undergo any chemical changes, should not be aggressive towards the materials of construction of the chromatographic apparatus and must not react chemically with the gel packing or with the compounds being separated. The requirement that the eluent must be perfectly inert should be observed as strictly as in other chromatographic methods.

The eluent must also meet some other requirements as to its physico-chemical and physical properties. Firstly, it must be a good swelling agent for the gel packing, or at least perfectly wet its surface (macroporous inorganic packings). The viscosity of the solvent should be as low as possible under the conditions of analysis and it must not be too volatile.

Special requirements for the eluent ensue from the detection system used. Spectrophotometric detectors require the use of solvents that have a minimal absorbance at the wavelengths used; differential refractometers call for solvents with the lowest possible refractive index or, on the contrary, a very high refractive index – in general, it can be said that the refractive index must differ from that of the sample as much as possible (the detector response should have the same polarity and a high value for all components of the mixture being separated); finally, wire or chain liquid chromatographic detectors with flame ionization units require a certain volatility of the eluent as well as certain chemical properties.

The eluent used for gel chromatographic analyses of polymers must be a thermodynamically good solvent. Some polymer systems necessitate the use of elevated temperatures, as the polymer is insoluble in any solvent at room temperature. With increasing temperature of analysis, the requirements for accurate thermostatting increase correspondingly, and the apparatus becomes more complicated.

It can be seen from the requirements outlined above that there is no ideal solvent that meets all of them, and that there will be only a few solvents that meet most of them satisfactorily. For hydrophilic systems, suitable solvents are water, various buffer solutions and polar organic solvents such as methanol, ethanol, acetone, dimethyl sulphoxide and dimethylformamide. For organophilic compounds, the most universal eluent (which, however, forms explosive peroxides in contact with air) is tetrahydrofuran, which has a low viscosity and refractive index and no absorption at the UV wavelengths commonly used for detection (250 and 280 nm). Benzene, toluene, chlorinated aromatic hydrocarbons, chloroform, tetrachloroethane, *m*-cresol and trifluoroethanol are also frequently used.

Apart from its effects on the solubility of the sample, the temperature also affects other operating variables. The elution volume, depending on the hydrodynamic volume of the molecule, is a function of temperature. The dependence of the elution volume of polystyrene fractions on temperature has been measured by Cantow *et al.* The decrease in the viscosity of the eluent with increasing column temperature results in a lower pressure drop for a given flow-rate. The equilibrium degree of swelling of homogeneous gels varies with temperature; the gels dilate with increasing temperature, which in turn leads to a reduction in the interstitial volume and to a possible change in the distribution of the packing. However, these effects probably play only a minor role compared with the changes in the hydrodynamic volume of the molecule. Little is known about the effect of changes in the diffusion coefficients of compounds with temperature on the separation parameters. Theta-temperatures are recommended for use in theoretical studies (Moore and Arrington).

## APPARATUS FOR GEL CHROMATOGRAPHY

A reliable apparatus is necessary for successful gel chromatographic separations, in addition to high-quality gel packings. Descriptions of equipment that operates without pressure can be found in a number of papers devoted mainly to the preparative aspects of the problem. In this section we discuss in more detail the design of a gel chromatograph that meets the requirements for an apparatus operating with elevated overpressures of the eluent, *i.e.*, high-speed systems.

Even if high-resolution columns are used, the result of the analysis may be impaired by the use of an inadequate experimental apparatus (zone spreading in dead spaces of the apparatus, strong pulses of the pump and irregularities in the flow, imperfect thermostatting of the temperature-sensitive detection system, inadequate injection valve, etc.). We shall describe here an apparatus that consists of the necessary parts, and also the principles that should be observed if the apparatus is not to have an adverse effect on the separation process. For general considerations regarding this theme, see Chapter 8.

The basic scheme of a gel chromatograph is shown in Fig. 12.2: The gel chromatograph consists of three main parts, *viz.*, a pumping and injection system, columns, and a recording detector system.

The eluent is stored in solvent tanks under an inert atmosphere (1), which prevents reaction with oxygen and the access of air moisture. The solvent then passes into a degasser (2), where it is freed from dissolved gases. The degasser consists of a boiler (2a),

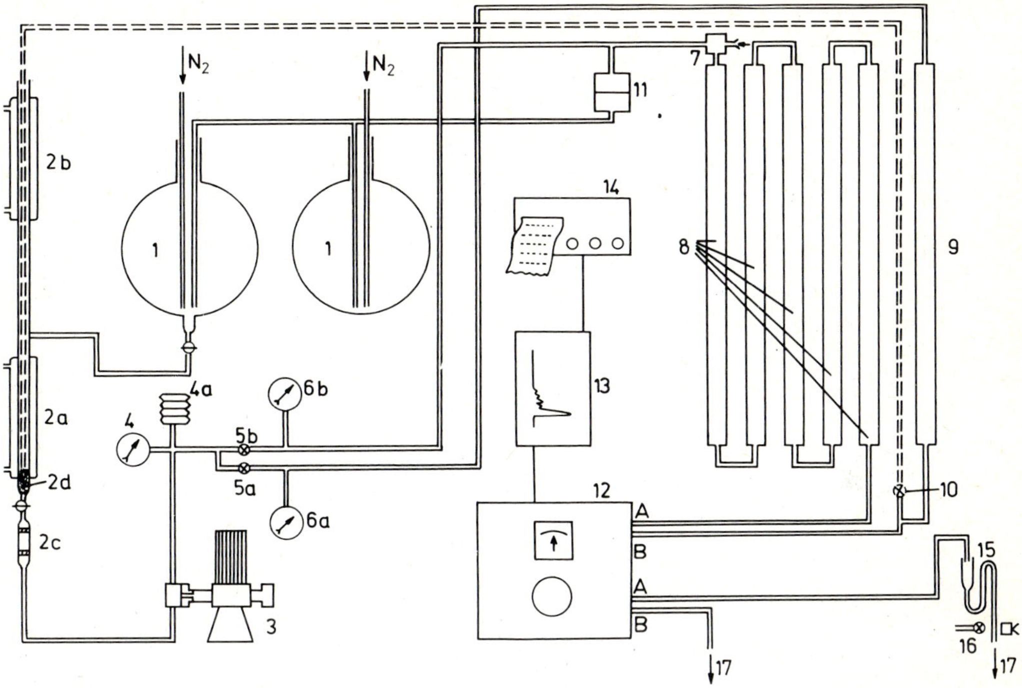


Fig. 12.2. Scheme of a gel chromatograph. For a description of the components, see the text.

heated either with a liquid or electrically, a water or air cooler (2b) and a filter (2c). The boiling of the solvent, and thus its degassing, is facilitated by the presence of a boiling centre (a glass-wool or sintered filter) inserted in the lower part of the heater (2d). The filter may consist of a glass tube provided with fused-in plates of sintered glass, or is made of metal with asbestos padding. From the degasser, the solvent enters the pump (3), which must be very reliable, with minimal pulses, a minimal dead space and a controllable amount of delivered liquid. The dosed amount in rhythmically operating pumps is controlled by the length of stroke of the plunger or piston, or by the revolutions of the driving motor.

The pressure of the liquid leaving the pump is read off on a manometer (4); the pump pulses are balanced by a pulse damper, beyond which the flow of the eluent splits in two streams and passes through regulation throttles (5a, 5b); manometers (6a, 6b) are built-in in the individual branches. One branch is led through the injection valve (7) into a system of columns (8), while the other leads to a reference column (9). If the arrangement is to be further simplified, the reference column can be omitted, and the solvent may pass through the reference cell after throttling with the regulation valve (10) directly from the lower part of the heater of the degasser; or the reference cell may be filled with the eluent and then closed (care should be taken that the solvent does not evaporate if the analysis takes a long time). A safety membrane or globe valve (11) prevents damage to the apparatus if the pressure becomes too high.

In the columns (8) the components of the mixture are separated into chromatographic zones. From the last column, the effluent is led into the detector or the detector system (12), from where the electrical signal reaches the recorder (13). An integrator (14) used for the quantitative evaluation of the chromatogram may serve as an additional device. From the measuring cell of the detector, the solvent passes into an integral volume meter (15), which is used for following the volume of mobile phase passing through the columns and whose pulses are recorded on the chromatogram (marking). As a rule, systems are employed that record the overflowing of the measuring siphon or, more simply, the drop counters.

If the individual fractions of the sample to be analyzed must be preserved, a fraction collector controlled by an integral volume meter can be added. In other instances the solution is led into a waste solvent tank (17).

The first commercial gel chromatograph produced by Waters Ass. (Framingham, Mass., U.S.A.) was described by Maley. Other firms that produce gel chromatographs are Varian Aerograph (Palo Alto, Calif., U.S.A.), DuPont (Wilmington, Del., U.S.A.), Siemens (Karlsruhe, G.F.R.) and recently also Knauer (Berlin, G.F.R.) and others.

**Pumping systems and injection**

The importance of the pumping system becomes more pronounced with increasing requirements for a constant flow-rate and increasing inlet pressures on the column. The flow-rate of the eluent should be adjustable over the range *ca.* 0.1–5.0 ml/min at a back-pressure of up to 30–50 atm. For these purposes, there are a large number of commercially available devices in the form of plunger pumps, membrane pumps or pumps based on linear dosers which are able to pump the eluent without undesirable pulses. In common pumps with a rhythmic operation, the pulses are damped in most instances by dampers consisting of stainless-steel bellows connected in parallel with the capillary and pressurized to the required value with gas or saturated vapour in equilibrium with the liquid at a suitable temperature. Liquid (mercury) pulse dampers operate in a similar manner (Mulder and Buytenhuys).

Good prospects seem to be offered by a pulse-free double-acting rhythmic pump with a programmed plunger movement (Waters Ass., Knauer). Two glass plungers are driven by an electronically controlled step motor in such a manner that the sum of the amount dosed is constant at each moment. Digital adjustment of the electronic control allows the eluent to be dosed within the range from 0.1 ml to 9.9 ml/min with high-reproducibility at an overpressure of up to 400 atm. The operation of the pulse-free double-acting pump is shown in Fig. 12.3.

The sample may be introduced into a pressure-free column by using the procedure described by Determann. The top of the gel column is covered with sample solution, the sample solution is allowed to soak into the gel bed while the column outlet is left half opened, the gel is then covered with the pure eluent and the sample is eluted.

Two types of injection valves are used in devices that operate under an overpressure of the eluent, *viz.*, multi-port valves and injection with a hypodermic syringe. In the first instance, the sample is inserted into a sample loop and, at the moment of injection, the loop volume is placed in a stream of eluent by turning the valve. In the second instance,

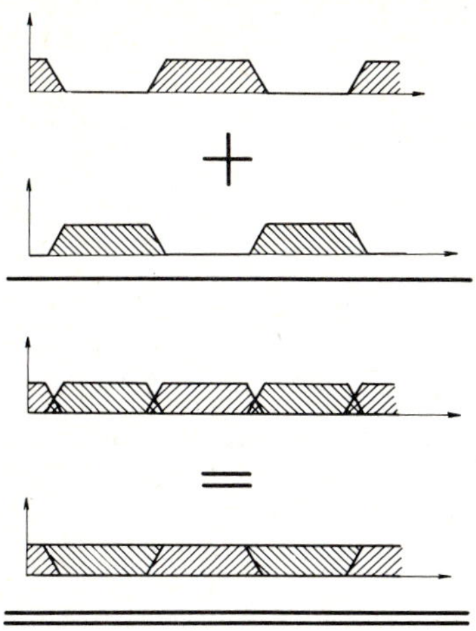


Fig. 12.3. Operation principle of a Knauer pulse-free double-acting high-pressure pump.

the sample is placed in the stream of eluent by perforating a rubber septum with the syringe. The advantages of a multi-port valve are the perfect reproducibility of the injected volume and the feasibility of automation of the injection within exactly defined time intervals in routine continuous analyses. An obvious disadvantage of the septum system is that the exactness of the amount injected is insufficiently guaranteed for quantitative analyses; on the other hand, a much smaller amount of sample is usually needed for the procedure, because the losses caused by filling and washing the loop are avoided.

**Columns**

The column design depends on the manner in which gel chromatographic separation is performed. The simplest procedure can be carried out by using a straight glass tube, 15–25 mm in diameter and 300–500 mm long (Fig. 12.4a), tapered at the lower end. The tapering is such that it enables the minimal dead space of the column below the gel bed to be filled with sand, small glass beads or glass-wool, and also enables the column to be connected with the detector through a plastic tube 1 mm in diameter. A detailed description of such a column has been given by Determann, together with a survey of types and producers. In order to eliminate wall effects, it is recommended that the column should be silylated when working in aqueous systems.

More efficient columns (Fig. 12.4b) for low-pressure gel chromatography have mobile plungers that allow contractions or dilations of the gel bed without the risk of damaging the gel bed or the column.

High-speed gel chromatography has led to the wide implementation of columns made of stainless steel, 4–25 mm in diameter. By analogy with other chromatographic methods, it would be expected that the column efficiency will continue to increase with increase in the ratio of the column diameter to the particle diameter. The columns most frequently used are 1000–1200 mm long and are connected in series by means of small-diameter

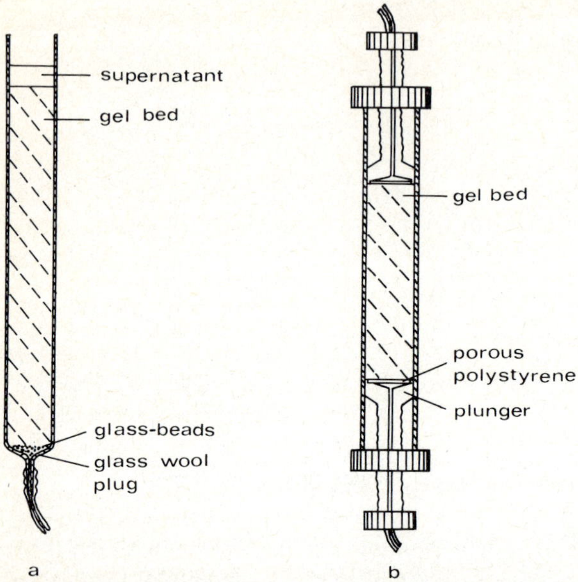


Fig. 12.4. Laboratory made and commercial gel chromatographic columns. (a) Commercial column; (b) laboratory made column (proposed by Koch-Light, Colnbrook, Great Britain).

capillaries so as to achieve the required length. This arrangement is particularly suitable for soft gels, which, if subjected to pressure, easily plug the column. For preparative purposes, sectional columns combining the advantages of good resolution and high output have been developed. Carnegie, Horton and Chernoff and Stouffer *et al.* reported the use of micro-columns 1–3 mm in diameter; Wasteson used polyethylene tubes and glass capillaries for analytical studies. However, with decreasing column diameter, the difficulties connected with a reproducible packing increase, and the columns usually have a lower separation efficiency than the theoretical value. The ultimate in micro-scale arrangements was probably attained by Boguth and Repges, who used individual gel particles. A detailed survey of special designs of gel chromatographic columns has been given by Fischer.

### Detectors

The fractions leaving the column can be detected by means of a chemical or instrumental analysis of the collected fractions. However, in routine analyses, continuous detection is obviously desirable, which at the same time may become a source of a signal for an automatic fraction collector.

Through-flow ultraviolet spectrophotometers operating with one or more well defined wavelengths or provided with a monochromator have a wide field of application and have become popular. The spectrophotometer is a highly sensitive detector that is particularly suitable for working in aqueous solutions when analyzing natural products. Spectrophotometers are produced as both single-beam and differential analyzers.

A through-flow differential refractometer suitable for work in organic solvents and characterized by a linear concentration response can be used for a variety of purposes. If the work is to be carried out at higher sensitivities, this detector requires accurate thermostatting and a good design of the input–output and geometry of the measuring cell. These problems have been solved satisfactorily by Waters Ass. and Knauer.

In some instances, an infrared spectrophotometer (Ross and Castro) is recommended; if used for the detection of polyethylene in perchloroethylene, it exhibits a higher sensitivity and a smaller temperature drift than a differential refractometer. Another advantage is the possibility of the quantitative determination of the functional groups in copolymers.

Saunders and Pecsok described the design of a simple and very sensitive conductivity detector for inorganic electrolytes; Kondo *et al.* recommended that colorimetric detection should be used for the analysis of polyether polyols. Jackson measured the difference between the dielectric constants of analyzed compounds as a means of detection; following the concentration of the analyzed compounds by radioactive labelling and polarographically also proved successful.

Very sensitive detection is given by a procedure in which the effluent leaving the column is deposited on a moving wire or conveyor belt, the solvent is evaporated and the non-volatile residue is led into a pyrolysis cell. The products of pyrolysis are detected by a flame ionization detector. Instead of an infinite conveyor or wire, Janák used a rotating stainless-steel grid, thus achieving a quantitative response of the FID.

It can be seen from this brief survey of detectors used for GPC (a detailed description is given in Chapter 8) that there is a wide choice. The most suitable method of detection will depend on the properties of the system to be analyzed and on the information required. A combination of a universal detector, *e.g.*, a differential refractometer, with a selective detector seems to be very useful; it has many advantages and is frequently employed. Both principles of detection have been combined in a single measuring cell by Knauer (Fig. 12.5). The scheme shows the passage of UV and visible light through a single measuring cell.

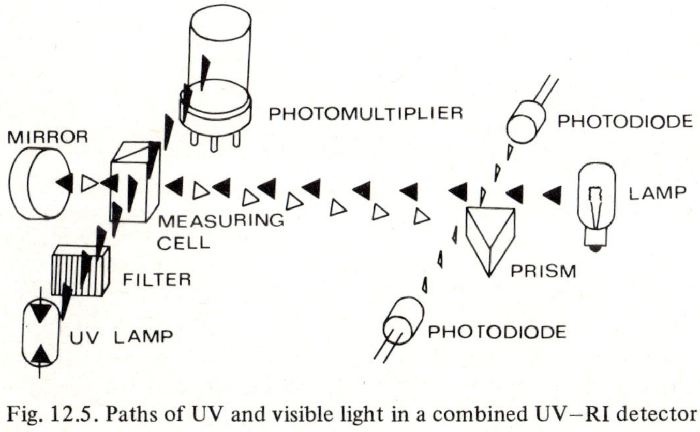


Fig. 12.5. Paths of UV and visible light in a combined UV–RI detector made by Knauer.

*References p.321*

**Column packing**

According to Van Deemter's equation, which defines the dependence of the reduced theoretical plate height ($h = H/d_p$, where $H$ is the height equivalent to a theoretical plate and $d_p$ is the diameter of the particle in the swollen state) on the linear reduced flow-rate ($\nu = vd_p/D$, where $v$ is the linear rate of elution and $D$ is the diffusion coefficient of the solute in the mobile phase):

$$h = a + b/\nu + c\nu \tag{12.1}$$

The constant $a$ depends on the regularity of the column packing, while $b$ reflects the broadening of the zone due to diffusion and $c$ has its origin in non-equilibrium conditions. From this equation it follows that a perfect and reproducible column packing may play an important role in attempts to attain a high separation efficiency.

The methods of column packing have been described by Altgelt (1965), Determann, Flodin, Moore, Peaker and Tweedale and by Sie and Van den Hoed. A more detailed investigation of the relationship between the separation efficiency and the experimental conditions of gel chromatography, and also of the problems of reproducible column packing, was carried out by Heitz and Čoupek. After comparing various methods of packing columns with gels, they recommend that the column should be packed with dilute gel suspension at a constant flow-rate of the solvent with mechanical vibration of the column. Neither the particle size nor the chemical character of the gel had any effect on the dependence of the separation efficiency on the flow-rate, which, in the case of the optimum packing procedure, is linear over a wide range of flow-rates. The packing equipment shown schematically in Fig. 12.6 consists of a solvent tank (1), degasser (2), pump (3), pressure vessel for the gel suspension with a volume approximately 10 times the column volume (4), with a column (5) attached to it by an extension tube and vibrated mechanically (6). The contents of the pressure container are stirred with a magnetical stirrer (7). The solvent is recirculated, and the flow-rate during packing is approximately

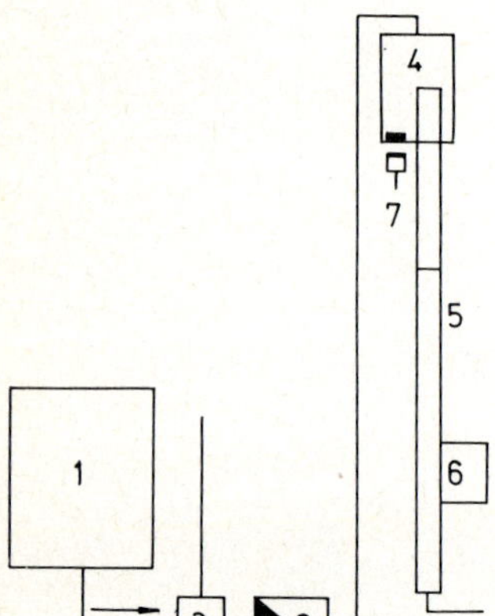


Fig. 12.6. Schematic representation of a column packing apparatus. 1, solvent tank; 2, degasser; 3, pump; 4, stainless-steel pressure vessel; 5, column; 6, vibrator; 7, magnetic stirrer.

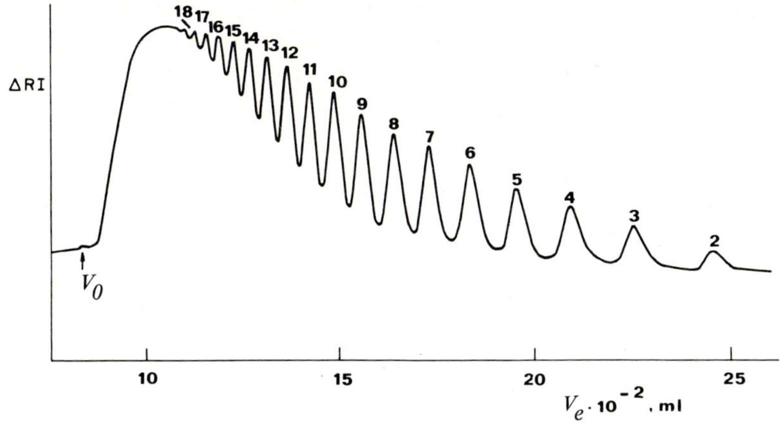


Fig. 12.7. Preparative-scale separation of styrene oligomers. Elution with tetrahydrofuran; column 200 × 5 cm; polystyrene gel containing 2% of divinylbenzene. Detection by refraction.

twice the highest flow-rate actually used for analysis. After packing, the tube used for extension is dismantled and the column is closed with an end-fitting.

An ingenious method of packing preparative columns (up to a diameter of 50 mm) has been described by Heitz and Ullner. The container is designed in a similar way, but the column rotates about its own axis instead of being mechanically vibrated. In order to prevent the centrifugation of particles in the column during sedimentation, both the direction and angle of each revolution of the column are random. Such columns exhibited an excellent separation efficiency, and their operation at the optimum elution rate was as good as those of the best analytical columns (Fig. 12.7).

## SPECIAL GEL CHROMATOGRAPHIC TECHNIQUES

In order to increase the resolution of gel chromatographic columns, Porath and Bennich used the recycling technique known from gas chromatography. They used a peristaltic pump with a low overpressure and a low flow-rate. Bombaugh *et al.* and Bombaugh and Levangie applied a small-volume reciprocating pump and obtained optimum conditions with a minimum broadening of the chromatographic zone. By using the recycling technique connected with a concentration of the eluate in a film evaporator without mixing the chromatographic zones, Heitz *et al.* separated oligomers into the individual species on a preparative scale. Smith *et al.* described a technique that eliminates the risk of "overtaking".

The details of the design, mode of operation and some applications of continuous gel chromatography were reported by Fox *et al.* and by Nicholas and Fox. Barker *et al.* also described continuous automatic gel chromatography.

In order to increase the flow-rate through the columns with soft homogeneous gels, the upwards-flow technique has been recommended, because it reduces the risk of plugging the columns. Another possibility is the application of short column sections, which can easily be connected in series and used in large-scale separations on the softest gel types (Type KS 370, Pharmacia, Uppsala, Sweden). Little *et al.* optimized the operating conditions of a chromatographic apparatus with the aim of increasing the flow-rate, which allowed them to reduce eight-fold the time needed for analysis without any substantial decrease in the separation efficiency. In order to increase the molecular-weight exclusion limit, Hellsing used a solution of a suitable neutral polymer as an eluent.

The high loads applied to the column frequently impair the efficiency owing to viscosity effects. Altgelt (1970) found that good separations can be achieved even with overloaded columns on the assumption that solutions that have relative viscosities higher than 2 are avoided and the effects of the difference between the densities of the eluent and solution are reduced by decreasing the diameter of the connecting capillaries.

Excellent results can be obtained by an appropriate combination of gel chromatography with another separation or analytical technique such as, for example, a combination of thin-layer gel chromatography with electrophoresis or immunoelectrophoresis, of column gel chromatography with mass, IR or NMR spectrometry, preparative gel chromatographic fractionation followed by analysis of the fractions by GLC, etc. In all combinations, specific features of each of the methods used and of the systems analyzed must be borne in mind. If this requirement is met, gel chromatography will be a valuable addition to other chromatographic methods, thus extending the possibilities offered by other analytical and preparative separations.

## EVALUATION OF GEL PERMEATION CHROMATOGRAPHIC DATA

### Determination of molecular weight and molecular-weight averages

The simplest method for obtaining the average molecular weight from GPC data, *viz.*, reading directly off a calibration curve the molecular weight corresponding to the retention volume of the maximum on the elution curve, is not recommended for polydisperse samples. It is obviously applicable only when the elution curve has a single maximum and, moreover, this value ($\overline{M}_{GPC}$) is an average of an unknown type (it has been shown by Berger and Schultz that the inequality $\overline{M}_n \leqslant \overline{M}_{GPC} \leqslant \overline{M}_w$ holds, where $\overline{M}_n$ and $\overline{M}_w$ are the number- and weight-average molecular weights, respectively). The simplest procedure is first to determine the molecular-weight distribution curve (a problem discussed below) and to use it for calculating the molecular-weight averages directly from their definitions.

Several workers have studied the accuracy of GPC for determining molecular weights of macromolecular samples by comparing the results of GPC with those of osmometry and light scattering. Typical results are given in Table 12.1. The accuracy of $\overline{M}_n$ and $\overline{M}_w$ values determined by GPC and the effect of the correction for zone broadening were discussed by Duerksen and Hamielec (1968a).

Data typical for the precision of GPC were given by Hill, who analyzed the same polymer repeatedly and found a standard deviation for $\overline{M}_n$ of about 6.8% and for $\overline{M}_w$ of about 1.03%.

Useful information on the possibilities and limitations of GPC as a method for deter-

TABLE 12.1
COMPARISON OF MOLECULAR-WEIGHT AVERAGES DETERMINED BY GPC AND CONVENTIONAL METHODS

| Polymer | GPC | | Osmometry and light scattering | |
|---|---|---|---|---|
| | $\bar{M}_n \cdot 10^{-3}$ | $\bar{M}_w \cdot 10^{-3}$ | $\bar{M}_n \cdot 10^{-3}$ | $\bar{M}_w \cdot 10^{-3}$ |
| Polyglycols* | 0.37 | 0.40 | 0.42 | 0.49 |
| | 0.69 | 0.74 | 0.70 | 0.76 |
| | 0.88 | 0.98 | 0.92 | 1.01 |
| | 1.30 | 1.43 | 1.13 | 1.27 |
| | 2.13 | 2.20 | 1.95 | 2.10 |
| Narrow | 4.87 | 5.58 | 3.52 | 3.6 |
| polystyrene | 9.20 | 10.9 | 9.7 | 10.3 |
| samples** | 47.0 | 55.9 | 49.0 | 51.0 |
| | 107.0 | 138.0 | 115.0 | 122.0 |
| | 357.0 | 495.0 | 392.0 | 411.0 |
| | 653.0 | 1017.0 | 773.0 | 860.0 |
| Broad | 16.8 | 34.0 | 18.2 | 33.0 |
| polystyrene | 35.4 | 99.7 | 38.6 | 94.0 |
| samples** | 43.2 | 130.0 | 51.5 | 117.0 |
| | 126.7 | 280.0 | 136.0 | 257.0 |
| Polybutadienes*** | 137 | 254 | 207 | 230 |
| | 143 | 320 | 140 | 310 |

*Baijal and Blanchard.
**Alliet.
***Adams *et al.*

mining molecular weights and its comparison with other techniques can be gained from a paper by Strazielle and Benoit, giving results of simultaneous measurements of molecular weights and distributions on a series of commercial polydisperse polymers in a number of laboratories throughout the world.

### Calculation of molecular-weight distribution from gel permeation chromatographic data

The simplest procedure for calculating the molecular-weight distribution (MWD) of a polydisperse sample from the GPC results is based on the assumption that no overlapping of neighbouring peaks takes place. Under this assumption of infinitely high resolution, the area of the chromatogram between two vertical and sufficiently close lines is proportional to the weight fraction of the polymer, whose molecular weight can be read off on the calibration curve for the corresponding retention volume. The procedure is exemplified by calculations summarized in Table 12.2.

The first column gives the elution volumes (in counts) and the second column the corresponding heights of the peaks in the experimental chromatogram, $h_i$. Molecular weights corresponding to the given elution volumes (counts), read off on the appropriate calibration curve, are listed in the third column.

*References p.321*

TABLE 12.2
EVALUATION OF GPC DATA

| Counts | $h_i$ (mm) | $M_i \cdot 10^{-3}$ | $w_i = h_i/\Sigma h_i$ | $I_i = w_i/2 + \sum_{j=1}^{i-1} w_j$ |
|---|---|---|---|---|
| 45 | 1.5 | 2186 | 0.0016 | 0.9996 |
| 46 | 5.5 | 1550 | 0.0058 | 0.9959 |
| 47 | 14.0 | 1123 | 0.0148 | 0.9856 |
| 48 | 30.0 | 861 | 0.0319 | 0.9622 |
| 49 | 49.5 | 601 | 0.0525 | 0.92 |
| 50 | 72.0 | 436 | 0.0763 | 0.8557 |
| 51 | 89.5 | 309 | 0.0948 | 0.7701 |
| 52 | 97.0 | 224 | 0.1028 | 0.6713 |
| 53 | 99.5 | 167 | 0.1054 | 0.5670 |
| 54 | 98.5 | 120 | 0.1043 | 0.4621 |
| 55 | 91.0 | 85 | 0.0964 | 0.3617 |
| 56 | 79.0 | 63 | 0.0837 | 0.2717 |
| 57 | 64.0 | 44 | 0.0678 | 0.1960 |
| 58 | 50.0 | 33 | 0.0530 | 0.1356 |
| 59 | 38.0 | 29 | 0.0403 | 0.089 |
| 60 | 28.0 | 17 | 0.0297 | 0.0539 |
| 61 | 19.5 | 12 | 0.0207 | 0.0288 |
| 62 | 12.0 | 9 | 0.0127 | 0.0122 |
| 63 | 5.5 | 6 | 0.0058 | 0.0029 |
| | $\Sigma h_i = 944$ | $\bar{M}_w = \Sigma w_i M_i = 224 \cdot 10^3$ | | |
| | | $\bar{M}_n = (\Sigma w_i/M_i)^{-1} = 65 \cdot 10^3$ | | |

The weight fraction of the polymer eluted between the two vertical lines corresponding to counts $i$ and $i + 1$ is approximated by the ratio $h_i/\Sigma h_i$ (for higher precision, the area of the chromatogram between the vertical lines corresponding to counts $i$ and $i + 1$ should be calculated by some more appropriate method, *e.g.*, the trapezoidal rule or Simpson rule, but for the present purpose this was not deemed necessary). The partial sums in the fourth column then correspond to the cumulative distribution function; the differential weight distribution can be derived therefrom by graphical or numerical differentiation. $\bar{M}_n$ and $\bar{M}_w$ can now be calculated from their definitions, as indicated at the foot of the table. A computer program written in Fortran, which can be used for these calculations, was published by Pickett *et al.* (1966).

It was pointed out by several workers and confirmed experimentally, (see, for example, Tung, 1966a) that this procedure gives satisfactory results for polymers with a broad MWD, provided that long columns with a sufficiently high resolving power are used. Yau and Fleming recommended a suitable calculation method that is also valid for non-linear calibration curves. Schrager developed a simple graphical method for determining the average molecular weights from GPC, based on the assumption that the distribution can be adequately approximated by the Schultz distribution function.

For low-resolution columns and/or samples with a narrow MWD, a correction for zone broadening must be applied. A number of methods for solving numerically the integral equation 5.6 were developed.

The problem can be simplified by replacing the integral equation by a set of linear algebraic equations, which can then be solved either by direct matrix inversion or by some minimization procedure; Hess and Kratz, Pickett *et al.* (1968), Smith, and Tung (1966b) all used this approach. In these methods, the kernel $W(v, y)$ can have any form; however, owing to the ill-posed nature of the problem, the low relative precision in reading the ordinates on the chromatogram near its ends often leads to artificial oscillations in the corrected function $w$. In order to avoid this difficulty, Smit *et al.* suggested a rapid iteration procedure and Rosen and Provder (1971) made use of the singular value decomposition method.

Other investigators, assuming that the kernel has the convolution form $W(v - y)$, solved the problem by Fourier transform (Pierce and Armonas; Rosen and Provder, 1970; Tung, 1969a; Vladimiroff). Tung (1966b, 1969a) used a polynomial approximation for calculating the corrected chromatograms. Chang and Huang (1969) reformulated the problem into an equivalent variational problem, which was then solved by a method of steepest descent in the function space. In a later paper (Chang and Huang, 1972), they generalized their treatment to cope with even the most general forms of the kernel $W(v, y)$.

Hamielec and Ray showed that, if the calibration curve is linear and the spreading function $W$ in eqn. 5.6 has a Gaussian form, the corrected values of the average molecular weights, $\bar{M}_n$, $\bar{M}_w$, $\bar{M}_z$ and $\bar{M}_{z+1}$, can be calculated directly by means of an analytical solution of Tung's integral equation. Their expressions for the corrected number-($\bar{M}_n^*$) and weight-($\bar{M}_w^*$) average molecular weights are

$$\bar{M}_n^* = \bar{M}_n \exp (D^2/4h) \tag{12.2}$$

$$\bar{M}_w^* = \bar{M}_w \exp (-D^2/4h) \tag{12.3}$$

where the molecular weights without asterisks are calculated directly from the uncorrected MWD; $h$ (assumed to be constant) has the same meaning as in eqn. 5.7 and $D = 2.303\ B$, where $B$ was defined by eqn. 5.9.

Later, Balke and Hamielec, and Hamielec generalized this method to include cases of unsymmetrical kernels, using the empirical general form of the spreading function proposed by Provder and Rosen. Two new promising iterative methods, applicable to very general shapes of the spreading function, were published by Ishige *et al.*

All these methods require the use of digital computers; a different approach, attractive for its simplicity, was developed by Frank *et al.*, which enables one to perform the necessary calculations on a desk calculator.

Some of the older methods of zone spreading correction were critically evaluated by Duerksen and Hamielec (1968a, 1968b).

The accuracy of MWDs determined by GPC has been tested by numerous workers by comparing the results with those of conventional techniques. Results typical of the excellent agreement usually observed are shown in Fig. 12.8 for the difficult system polypropylene–1,2,4-trichlorobenzene.

*References p.321*

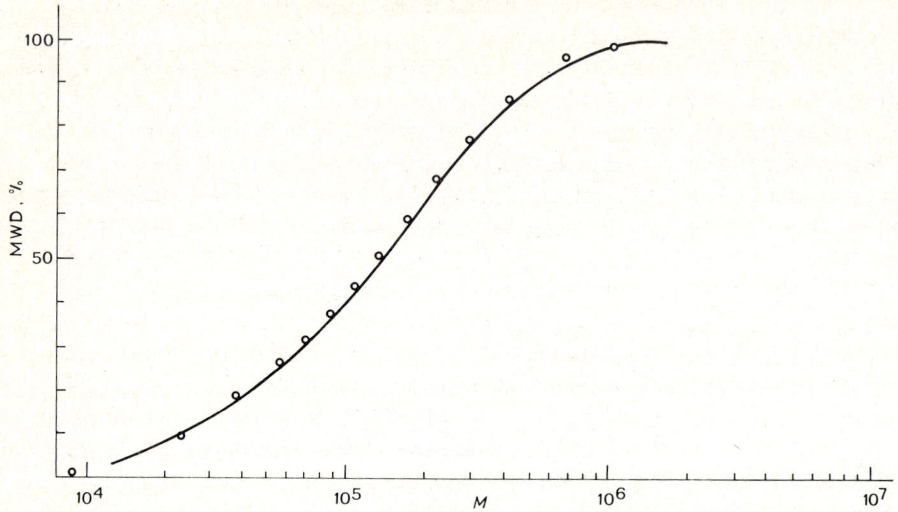


Fig. 12.8. Cumulative MWD curve of a commercial polypropylene sample as determined by GPC (solid line) and gradient elution fractionation (circles) (Crouzet *et al.*). GPC operating conditions: Spherosil packing; 1,2,4-trichlorobenzene (1 ml/min); 135°C.

**Simultaneous determination of polydispersity in molecular weight and the chemical heterogeneity of copolymers**

GPC has been successfully employed in studies on the inhomogeneity in the chemical composition of copolymers. Thus, Crammond *et al.* analyzed several block and random copolymers of isoprene and styrene by GPC. Pavelich and Livigni studied chemical composition and its correlation with molecular weight in butadiene–styrene block copolymers. Terry and Rodriguez monitored the effluent of a GPC column by means of a continuous infrared detector at different wavelengths and were able to follow the chemical heterogeneity of a styrene–methyl methacrylate copolymer. Runyon *et al.* discussed the advantages of a GPC unit equipped with two suitably selected detectors connected in series for the concurrent analysis of MWD and the heterogeneity in the chemical composition of copolymers.

**Attempts to determine the degree of branching**

The simultaneous analysis of MWD and the distribution of branches in polydisperse branched polymers is of great importance, but the amount of experimental work required is considerable. It is therefore not surprising that several investigators explored the possibility of using GPC as a simple automated method for these studies.

Most of these attempts are based on the now classical papers of Benoit *et al.* and Grubisic *et al.*, who showed that, regardless of the degree of branching, the separation in GPC is controlled by the hydrodynamic volume of the chain polymer in solution; as branching reduces the hydrodynamic volume, branched molecules are eluted later than linear molecules of the same molecular weight.

By 1967 some workers had studied the effect of chain branching on the elution behaviour of polymers in GPC experiments (Salovey and Hellman, Wild and Guliana, Yamada *et al.*). A combination of at least two methods is required in order to obtain information about branching and polydispersity. Drott and Mendelson (1970a, 1970b) discussed and tested experimentally the possibility of determining the branching index of polydisperse polyethylene samples from concurrent GPC and viscosity measurements. Their conclusions were confirmed by Červenka and Bates. Schultz analyzed theoretically the elution behaviour of polymers with randomly distributed tri- or tetrafunctional branches.

Tung (1969b) proposed the combination of GPC and sedimentation velocity and applied this method to branched styrene–divinylbenzene copolymers. In a later paper (Tung, 1971), he proposed a simplified calculation scheme valid for a certain class of branched polymers (in which the degree of branching increases monotonically with molecular weight), based on the results of GPC, sedimentation and viscosity measurements.

However, Pannell questioned whether the hydrodynamic volume as measured by the quantity $[\eta] \cdot M$ is an adequate parameter to use for describing unequivocally the elution behaviour of linear and branched polystyrenes and it was shown that, even if the correct size separation parameter were known, GPC could not be used to characterize branched polymers unambiguously. It therefore seems that more work is required before the question of the applicability of GPC to these difficult measurements can be definitely settled.

## DETERMINATION OF MOLECULAR WEIGHTS OF NATURALLY OCCURRING MACROMOLECULAR COMPOUNDS BY MOLECULAR SIEVE CHROMATOGRAPHY

Investigations that led to a widening of the applicability of chromatography were carried out by Lathe and Ruthven; the observation that the elution volumes of polysaccharides and polypeptides on starch beds are related to their molecular weights and further work carried out by Porath and Flodin with cross-linked dextran gels, known nowadays as Sephadex, were the beginning of molecular-weight determinations by techniques other than classical ultracentrifugation. An even more powerful method is perhaps gel electrophoresis, but column chromatography nevertheless remains important (Hjertén). The great advantage of determining molecular weights by molecular sieve chromatography is that the sample assayed need not necessarily be in a pure form and the only requirement is the availability of an adequate detection method. On the other hand, it must be emphasized that molecular weights determined by this technique must be considered as tentative, as the differences between actual and determined values are sometimes high, especially when the solute has a very asymmetric molecular shape. Most of the theoretical considerations in this respect refer to strictly globular molecules, which is not necessarily always the case. An extreme example is the behaviour of polyethylene glycol of molecular weight 15,000 on Sephadex G-200; here the particular compound is eluted at the void volume (Tyle), which leads to an erroneous determination of the molecular weight as about 100,000.

*References p.321*

TABLE 12.3

SYSTEMATIC INVESTIGATIONS ON THE CORRELATION BETWEEN MOLECULAR WEIGHT AND ELUTION VOLUME (DETERMANN AND MICHEL)

| Gel | Type of substance | Solvent | Relationship* | References |
|---|---|---|---|---|
| Sephadex | Polysaccharides | Buffer | $V_e/V_t \approx \log M$ | Granath and Flodin |
| Agar | Proteins | Buffer | $V_e = f(\log M)$ | Andrews (1962) |
| Sephadex | Polysaccharides | Buffer | $\sqrt[3]{K_d} \approx \sqrt{M}$ | Porath |
| Sephadex | Proteins | Buffer | $\sqrt[3]{K_d} \approx \sqrt{M}$ | Auricchio and Bruni, Wieland *et al.* |
| Sephadex | Proteins | Buffer | $V_e/V_0 \approx \log M$ | Whitaker |
| Sephadex | Proteins | Buffer | $V_e \approx \log M$ | Andrews (1964) |
| Sephadex | Oligothymidylic acid | Buffer | $\ln K_d \approx M$ | Hohn and Pollmann |
| Poly(methyl-methacrylate) | Oligostyrenes | Chloroform | $\sqrt[3]{K_d} \approx \sqrt{M}$ | Determann *et al.* |
| Agarose | Proteins | Buffer | $V_e \approx \log M$ | Largier and Polson |
| Sephadex | Proteins | Buffer | $K_{av} = f(r_s, r_r)$ | Laurent and Killander |
| Sephadex | Proteins | Buffer | $K_d = f(a/r)$ | Ackers |
| Sephadex | Proteins | Buffer | $\sqrt[3]{(V_e/V_0)} \approx \sqrt[3]{M}$ | Squire |
| Sephadex | Proteins | Buffer | $\log V_e \approx \log M$ | Van Thoai *et al.* |
| Sephadex | Peptides | Phenol–acetic acid–water | $\sqrt[3]{K_d} \approx \sqrt{M}$ | Carnegie |
| Polyacrylamide | Proteins | Buffer | $K_d \approx \log M$ | Sun and Sehon |
| Sephadex | Proteins | Buffer | $\{\sqrt{K_d} \approx \sqrt{M}; K_d = f(r_s, r_r); K_d = f(a/r)$ | Siegel and Monty |

* $V_e$ = Elution volume; $V_t$ = total volume of the column; $V_0$ = void volume; $K_d$ = distribution constant (coefficient) $\left(K_d = \frac{V_e - V_0}{V_t - V_g - V_e}; K_{av} = \frac{V_e - V_0}{V_t - V_0} = \frac{V_t - V_g - V_0}{V_t - V_0} \cdot K_d\right)$; $V_g$ = volume fraction not accessible to the solvent; $M$ = molecular weight; $r$ = diameter of a cylindrical channel in the column bed; $a$ = Stokes radius of a macromolecule; $a/r$ = radius quotient; $\eta$ = viscosity of the medium; $r_r$ = radius of the rod-like particle constituting the gel network; $r_s$ = radius of spherical particles.

References p.321

TABLE 12.4

MOLECULAR WEIGHTS, STOKES RADII AND ELUTION CONSTANTS OF PROTEINS ON SEPHADEX G-100, G-75 AND G-50 GELS

| No. | Protein | $r$ (nm) | $M \cdot 10^{-3}$ | G-200 | | | G-100 | | G-75 | | G-50 | |
|---|---|---|---|---|---|---|---|---|---|---|---|---|
| | | | | $K_d$ | $V_e/V_o$ | $R_{cyt.}$[m] | $K_d$ | $V_e/V_o$ | $K_d$ | $V_e/V_o$ | $K_d$ | $V_e/V_o$ |
| 1 | Cytochrome *c* | 1.74[a] | 13 | 0.76[b] | 2.61[b] | 1.00 | 0.70[b] | 2.14[c] | – | 1.99[c] | – | – |
| 2 | Ribonuclease | 1.92[g] | 13.6 | 0.76[d] | 2.61[d] | 1.04 | – | – | 0.35[h] | – | 0.26 | 1.45 |
| 3 | Methaemoglobin | – | 17 | 0.68[d] | 2.48[d] | – | – | – | – | – | – | – |
| 4 | Soyabean trypsin inhibitor | 2.26[a] | 21.5 | – | – | 1.20 | – | 1.93[b] | – | 1.65[k] | – | – |
| 5 | α-Chymotrypsin | 2.28[g] | 22.5 | – | – | 1.18 | 0.54[b] | 1.88[b] | 0.28[b] | 1.66[b] | – | – |
| 6 | Trypsin | 2.41[a] | 24 | 0.62[e] | 2.30[e] | 0.83 | 0.54[b] | 1.84[b] | 0.28[h] | 1.71[c] | – | – |
| 7 | Pepsin | – | 35.5 | 0.59[d] | 2.26[d] | 1.40 | – | 1.62[b] | 0.25[h] | 1.40[c] | – | – |
| 8 | α-Hydroxysteroid dehydrogenase | – | 47 | – | – | – | – | 1.56[b] | – | – | – | – |
| 9 | Peroxidase I | 3.02[g] | 40 | – | – | 1.50 | – | 1.57[b] | – | – | – | – |
| 10 | Ovalbumin | 2.80[a] | 45 | – | 2.08[l] | 1.45 | 0.29[i] | – | 0.20[n] | 1.35[n] | – | – |
| 11 | Phosphoglycerate mutase | – | 64 | – | – | – | – | – | – | 1.21[b] | – | – |
| 12 | Serum albumin (bovine) | 3.61[a] | 67 | 0.43[d] | 1.90[d] | 1.69 | 0.16[h] | 1.32[b] | – | 1.10[c] | – | – |
| 13 | Malate dehydrogenase | – | 79 | 0.38[e] | 1.80[e] | – | – | – | – | – | – | – |
| 14 | Enolase | – | 80 | – | – | – | – | 1.29[b] | – | – | – | – |
| 15 | Creatine phosphate kinase | – | 81 | – | – | 1.66 | – | 1.25[b] | – | 1.09[b] | – | – |
| 16 | Transferin | 4.00[a] | 88 | 0.32[h] | – | – | 0.11[h] | – | – | – | – | – |
| 17 | Glyceraldehyde phosphate dehydrogenase | 4.30[a] | 117 | 0.31[d] | 1.66[d] | 1.81 | – | – | – | – | – | – |
| 18 | Serum albumin (dimer) | – | 134 | 0.26[d] | 1.54[d] | – | – | – | – | – | – | – |
| 19 | Aldolase (yeast) | – | 147 | 0.27[d] | 1.59[d] | 1.89 | – | – | – | – | – | – |
| 20 | Alcohol dehydrogenase | 4.55 | 150 | – | 1.54[l] | 2.00 | – | – | – | – | – | – |
| 21 | γ-Globulin (human) | 5.22[a] | 140 | 0.26[d] | 1.61[d] | 2.02[i] | – | – | – | – | – | – |
| 22 | Catalase | – | 225 | – | 1.38[l] | 2.19 | – | – | – | – | – | – |
| 23 | γ-Globulin | – | 200 | – | 1.44[l] | – | – | – | – | – | – | – |
| 24 | Chymotrypsinogen A | – | 25 | – | – | – | – | – | – | – | 0.12 | 1.20 |
| 25 | Lima bean trypsin inhibitor | – | 8.4 | – | – | – | – | – | – | – | 0.42 | 1.72 |
| 26 | Kallikrein inhibitor | – | 6.5 | – | – | – | – | – | – | – | 0.49 | 1.85 |

[a]Determann and Michel. [b]Determann and Michel (data obtained on columns of 300–150 ml and 2 cm width. The proteins were applied in 1–2 ml of buffer and eluted (10–20 ml/h) with phosphate buffer of pH 7.2 ($I$ = 0.075) + 0.5 $M$ NaCl). [c]Whitaker. [d]Wieland *et al.* [e]Auricchio and Bruni. [f]Squire. [g]Laurent and Killander. [h]Determann and Gelotte. [i]Calculated from Morris. [k]Andrews. [l]Leach and O'Shea. [m]Determann and Michel (data obtained by thin-layer chromatography on Sephadex G-200 superfine; bed volume, 0.05 × 20 × 40 cm; phosphate, pH 6.6; ionic strength, 0.2; $R_{cyt.}$ = relative mobility to cytochrome *c*. [n]Fritz *et al.*

TABLE 12.5
SEPARATION RANGES OF DIFFERENT GELS

| Trade Name | | Separation range for globular materials (mol. wt.) |
|---|---|---|
| Sephadex | G-10 | up to 700 |
| | G-15 | up to 1500 |
| | G-25 | 1000–5000 |
| | G-50 | 1500–30,000 |
| | G-75 | 3000–70,000 |
| | G-100 | 4000–150,000 |
| | G-150 | 5000–400,000 |
| | G-200 | 5000–800,000 |
| Bio-Gel | P-2 | 200–2600 |
| | P-4 | 500–4000 |
| | P-6 | 1000–5000 |
| | P-10 | 5000–17,000 |
| | P-30 | 20,000–50,000 |
| | P-60 | 30,000–70,000 |
| | P-100 | 40,000–100,000 |
| | P-150 | 50,000–150,000 |
| | P-200 | 80,000–300,000 |
| | P-300 | 100,000–400,000 |
| | A-0.5 m | $< 10{,}000$–500,000 |
| | A-1.5 m | $< 10{,}000$–$1.5 \cdot 10^6$ |
| | A-150 m | $1 \cdot 10^6$– $> 150 \cdot 10^6$ |
| Sepharose | 6B | up to $2 \cdot 10^6$ |
| | 4B | 300,000–$20 \cdot 10^6$ |
| | 2B | $2 \cdot 10^6$–$25 \cdot 10^6$ |
| Sagavac | 10 | 10,000–250,000 |
| | 9 | 25,000–500,000 |
| | 8 | 25,000–700,000 |
| | 7 | 50,000–$1.5 \cdot 10^6$ |
| | 6 | 50,000–$2 \cdot 10^6$ |
| | 5 | 50,000–$7 \cdot 10^6$ |
| | 4 | 200,000–$15 \cdot 10^6$ |
| | 3 | 500,000–$50 \cdot 10^6$ |
| | 2 | 500,000–$150 \cdot 10^6$ |
| Gelarose | 2%, 4%, 6%, 8%, 10% | Depends on gel concentration |

On the other hand, the proportionality of the elution volume and the logarithm of the molecular weight has been repeatedly verified and is now definitely established (Andrews, 1962, 1964, 1965; Andrews *et al.*; Bess and Hnilica; Bought *et al.*; Burges *et al.*; Burke and Ross; Dellacha *et al.*; Dimigen *et al.*; Downey and Andrews; Fritz *et al.*; Jungwirth and Bodo; Lanchantin *et al.*; Largier and Polson; Leach and O'Shea; Lisowski; Nakayama and Miyake; Nieschlag and Otto; Ostrowski and Rybarska; Přistoupil; Schane; Selby and Maitland; Whitaker). Granath and Flodin were the first to derive this relationship based on polysaccharide data. Determann and Michel also collected a large series of data related to the correlation between the elution volume and molecular weight, and the results are presented in Table 12.3.

In practice, however, when looking for the molecular weight of an unknown protein, a calibration series is necessary. The most useful in this respect is the list of proteins summarized in Table 12.4. There are, as one would expect from the nature of Sephadex gels, distinct limitations according to the type of gel used. The range within which every gel type is capable of yielding a reliable value of molecular weight can be judged best from Fig. 12.1.

In addition to cross-linked dextrans, cross-linked polyacrylamide and agarose are also available today for this type of analysis. The separation ranges and commercial names of these materials are summarized in Table 12.5. However, these data should serve only as a rough guide as exceptions are frequent.

Data obtained from different laboratories were summarized by Determann and Michel and fitted the general equations calculated from the actual values remarkably well.

Similar linear relationships were published by Leach and O'Shea, but in contrast to the series of equations summarized by Determann and Michel, the former are of only limited validity.

## REFERENCES

Ackers, G. K., *Biochemistry*, 3 (1964) 723.
Adams, H. E., Farhat, K. and Johnson, B. L., *Ind. Eng. Chem., Prod. Res. Develop.*, 5 (1966) 126.
Alliet, D. F., *Appl. Polym. Symp.*, 8 (1969) 39.
Altgelt, K. H., *Makromol. Chem.*, 88 (1965) 75.
Altgelt, K. H., *ACS Symposium on Gel Permeation Chromatography, Houston, Texas, February 1970*, Waters Ass., Framingham, Mass., 1970.
Andrews, P., *Nature (London)*, 196 (1962) 36.
Andrews, P., *Biochem. J.*, 91 (1964) 222.
Andrews, P., *Biochem. J.*, 16 (1965) 595.
Andrews, P., Bray, R. C., Edwards, P. and Shooter, K. V., *Biochem. J.*, 93 (1964) 627.
Auricchio, F. and Bruni, C. B., *Biochem. Z.*, 340 (1964) 321.
Baijal, M. D. and Blanchard, L. P., *J. Appl. Polym. Sci.*, 12 (1968) 169.
Balke, S. T. and Hamielec, A. E., *J. Appl. Polym. Sci.*, 13 (1969) 1381.
Barker, S. A., Hatt, B. W. and Somers, P. J., *Carbohyd. Res.*, 11 (1969) 355.
Benoit, H., Grubisic, Z., Rempp, P., Decker, D. and Zilliox, J., *J. Chim. Phys. Physicochim. Biol.*, 63 (1966) 1507.
Berger, H. L. and Schultz, A. R., *J. Polym. Sci., Part A*, 3 (1965) 3643.
Bess, L. G. and Hnilica, L. S., *Anal. Biochem.*, 12 (1965) 421.
Boguth, W. and Repges, R., *Z. Wiss. Mikrosk.*, 68 (1967) 241.
Bombaugh, K. J., Dark, W. A. and Levangie, R. F., *J. Chromatogr. Sci.*, 7 (1969) 42.
Bombaugh, K. J. and Levangie, R. F., *Separ. Sci.*, 5 (1970) 751.
Bought, W., Kirsch, K. and Niemann, H., *Biochem. Z.*, 341 (1965) 149.
Burges, R. A., Brammer, K. W. and Coombes, L. D., *Nature (London)*, 208 (1965) 894.
Burke, D. C. and Ross, J., *Nature (London)*, 208 (1965) 1297.
Cantow, M. J. R., Porter, R. S. and Johnson, J. F., *J. Polym. Sci., Part A-1*, 5 (1967) 987.
Carnegie, P. R., *Nature (London)*, 206 (1965) 1128.
Červenka, A. and Bates, T. W., *J. Chromatogr.*, 53 (1970) 85.
Chang, K. S. and Huang, R. Y. M., *J. Appl. Polym. Sci.*, 13 (1969) 1459.

Chang, K. S. and Huang, R. Y. M., *J. Appl. Polym. Sci.*, 16 (1972) 329.
Crammond, D. N., Hammond, J. M. and Urwin, J. P., *Eur. Polym. J.*, 4 (1968) 451.
Crouzet, P., Fine, F. and Mangin, P., *J. Appl. Polym. Sci.*, 13 (1969) 205.
Dellacha, J. M., Enero, M. A. and Faiferman, I., *Experientia*, 22 (1966) 16.
Determann, H., *Gel Chromatography*, Springer Verlag, New York, 1968.
Determann, H. and Gelotte, B., in H. M. Rauen (Editor), *Biochemisches Taschenbuch*, Vol. 2, Springer, Berlin, Göttingen, Heidelberg, 1964, p. 905.
Determann, H., Lüben, G. and Wieland, Th., *Makromol. Chem.*, 73 (1964) 168.
Determann, H. and Michel, W., *J. Chromatogr.*, 25 (1966) 303.
Dimigen, J., Klink, F. and Richter, D., *Z. Naturforsch. B*, 20 (1965) 924.
Downey, W. K. and Andrews, P., *Biochem. J.*, 94 (1965) 642.
Drott, E. E. and Mendelson, R. A., *J. Polym. Sci., Part A-2*, 8 (1970a) 1361.
Drott, E. E. and Mendelson, R. A., *J. Polym. Sci., Part A-2*, 8 (1970b) 1373.
Duerksen, J. H. and Hamielec, A. E., *J. Polym. Sci., Part C*, 21 (1968a) 83.
Duerksen, J. H. and Hamielec, A. E., *J. Appl. Polym. Sci.*, 12 (1968b) 2225.
Fischer, L., *An Introduction to Gel Chromatography*, North-Holland, Amsterdam, 1969.
Flodin, P., *J. Chromatogr.*, 5 (1961) 103.
Fox, Jr., J. B., Calhoun, R. C. and Eglinton, W. J., *J. Chromatogr.*, 43 (1969) 48.
Frank, F. C., Ward, I. M. and Williams, T., *J. Polym. Sci., Part A-2*, 6 (1968) 1357.
Fritz, H., Trautschold, I. and Werle, E., *Hoppe-Seyler's Z. Physiol. Chem.*, 342 (1965) 253.
Granath, K. A. and Flodin, P., *Makromol. Chem.*, 48 (1961) 160.
Grubisic, Z., Rempp, P. and Benoit, H., *J. Polym. Sci., Part B*, 5 (1967) 753.
Hamielec, A. E., *J. Appl. Polym. Sci.*, 14 (1970) 1519.
Hamielec, A. E. and Ray, W. H., *J. Appl. Polym. Sci.*, 13 (1969) 1319.
Heitz, W., Bömer, B. and Ullner, H., *Makromol. Chem.*, 121 (1969) 102.
Heitz, W. and Čoupek, J., *J. Chromatogr.*, 36 (1968) 290.
Heitz, W. and Ullner, H., *Makromol. Chem.*, 120 (1968) 58.
Hellsing, K., *J. Chromatogr.*, 36 (1968) 170.
Hess, M. and Kratz, R. F., *J. Polym. Sci., Part A-2*, 4 (1966) 731.
Hill, J. A., *International Gel Permeation Chromatography Seminar, Boston, 1965*, Waters Ass., Framingham, Mass.
Hjertén, S., in A. Niederwieser and G. Pataki (Editors), *New Techniques in Amino Acid, Peptide and Protein Analysis*, Ann Arbor Sci. Publ., Ann Arbor, Mich., 1971, p. 227.
Hohn, Th. and Pollmann, W., *Z. Naturforsch. B*, 18 (1963) 919.
Horton, B. F. and Chernoff, A. I., *J. Chromatogr.*, 47 (1970) 493.
Ishige, T., Lee, S.-I. and Hamielec, A. E., *J. Appl. Polym. Sci.*, 15 (1971) 1607.
Jackson, A., *J. Chem. Educ.*, 42 (1965) 447.
Janák, J., private communication.
Jungwirth, C. and Bodo, G., *Biochem. Z.*, 339 (1964) 382.
Kondo, K., Mori, M. and Hattori, M., *Bunseki Kagaku (Jap. Anal.)*, 16 (1967) 414.
Lanchantin, G. F., Friedmann, J. A. and Hart, D. W., *J. Biol. Chem.*, 260 (1965) 3276.
Largier, J. F. and Polson, A., *Biochim. Biophys. Acta*, 79 (1964) 626.
Lathe, G. H. and Ruthven, C. R. J., *Biochem. J.*, 62 (1956) 665.
Laurent, T. C. and Killander, J., *J. Chromatogr.*, 14 (1964) 312.
Leach, A. A. and O'Shea, P. C., *J. Chromatogr.*, 17 (1965) 245.
Lisowski, J., *Biochim. Biophys. Acta*, 113 (1966) 321.
Little, J. N., Waters, J. L., Bombaugh, K. J. and Pauplis, W. J., *J. Polym. Sci., Part A-2*, 7 (1969) 1775.
Maley, L. E., *J. Polym. Sci., Part C*, 8 (1965) 253.
Moore, J. C., *J. Polym. Sci., Part A*, 2 (1964) 835.
Moore, J. C. and Arrington, M. C., *3rd International Symposium on Gel Permeation Chromatography, Geneva, May 1966*, Waters Ass., Framingham, Mass., 1966.
Morris, C. J. O. R., *J. Chromatogr.*, 16 (1964) 167.
Mulder, J. L. and Buytenhuys, F. A., *J. Chromatogr.*, 51 (1970) 459.
Nakayama, F. and Miyake, H., *J. Lab. Clin. Med.*, 65 (1965) 638.

Nicholas, R. A. and Fox, Jr., J. B., *J. Chromatogr.*, 43 (1969) 61.
Nieschlag, E. and Otto, K., *Hoppe-Seyler's Z. Physiol. Chem.*, 340 (1965) 46.
Ostrowski, W. and Rybarska, J., *Biochim. Biophys. Acta*, 105 (1965) 196.
Pannell, J., *Polymer*, 13 (1972) 277.
Pavelich, W. A. and Livigni, R. A., *J. Polym. Sci., Part C*, 21 (1968) 215.
Peaker, F. W. and Tweedale, C. R., *Nature (London)*, 216 (1967) 75.
Pickett, H. E., Cantow, M. J. R. and Johnson, J. F., *J. Appl. Polym. Sci.*, 10 (1966) 917.
Pickett, H. E., Cantow, M. J. R. and Johnson, J. F., *J. Polym. Sci., Part C*, 21 (1968) 67.
Pierce, P. E. and Armonas, J. E., *J. Polym. Sci., Part C*, 21 (1968) 23.
Porath, J., *Pure Appl. Chem.*, 6 (1963) 233.
Porath, J. and Bennich, H., *Arch. Biochem. Biophys., Suppl.*, No. 1 (1962) 152.
Porath, J. and Flodin, P., *Nature (London)*, 183 (1959) 1657.
Přistoupil, T. I., *J. Chromatogr.*, 19 (1965) 64.
Provder, T. and Rosen, E. M., *Separ. Sci.*, 5 (1970) 437.
Rosen, E. M. and Provder, T., *Separ. Sci.*, 5 (1970) 485.
Rosen, E. M. and Provder, T., *J. Appl. Polym. Sci.*, 15 (1971) 1687.
Ross, J. H. and Castro, M. E., *J. Polym. Sci., Part C*, 21 (1968) 143.
Runyon, J. R., Barnes, D. A., Rudd, J. F. and Tung, L. H., *J. Appl. Polym. Sci.*, 13 (1969) 2359.
Salovey, R. and Hellman, M. Y., *J. Polym. Sci., Part A-2*, 5 (1967) 333.
Saunders, D. and Pecsok, R. L., *Anal. Chem.*, 40 (1968) 44.
Schane, H. P., *Anal. Biochem.*, 11 (1965) 371.
Schrager, M., *J. Appl. Polym. Sci.*, 15 (1971) 83.
Schultz, A. R., *Eur. Polym. J.*, 6 (1970) 69.
Selby, K. and Maitland, C. C., *Biochem. J.*, 94 (1965) 578.
Sie, S. T. and Van den Hoed, N., *J. Chromatogr. Sci.*, 7 (1969) 257.
Siegel, L. M. and Monty, K. V., *Biochim. Biophys. Acta*, 112 (1966) 346.
Smit, J. A. M., Hoogervorst, C. J. P. and Staverman, A. J., *J. Appl. Polym. Sci.*, 15 (1971) 1479.
Smith, J. K., Eaton, R. H., Whitby, L. G. and Moss, D. W., *Anal. Biochem.*, 23 (1968) 84.
Smith, W. N., *J. Appl. Polym. Sci.*, 11 (1967) 639.
Squire, P. G., *Arch. Biochem. Biophys.*, 107 (1964) 471.
Stouffer, J. E., Oakes, P. C. and Schlatter, J. E., *J. Gas Chromatogr.*, 1 (1966) 89.
Strazielle, C. and Benoit, H., *Pure Appl. Chem.*, 26 (1971) 451.
Sun, K. and Sehon, A. H., *Can. J. Chem.*, 43 (1965) 969.
Terry, S. L. and Rodriguez, F., *J. Polym. Sci., Part C*, 21 (1968) 191.
Tung, L. H., *J. Appl. Polym. Sci.*, 10 (1966a) 1271.
Tung, L. H., *J. Appl. Polym. Sci.*, 10 (1966b) 375.
Tung, L. H., *J. Appl. Polym. Sci.*, 13 (1969a) 775.
Tung, L. H., *J. Polym. Sci., Part A-2*, 7 (1969b) 47.
Tung, L. H., *J. Polym. Sci., Part A-2*, 9 (1971) 759.
Tyle, A. P., *Nature (London)*, 206 (1965) 1256.
Van Deemter, J. J., Zuiderweg, F. J. and Klinkenberg, A., *Chem. Eng. Sci.*, 5 (1956) 271.
Van Thoai, N., Kassab, R. and Pradel, L. A., *Biochim. Biophys. Acta*, 110 (1965) 532.
Vladimiroff, T., *J. Appl. Polym. Sci.*, 14 (1970) 1397.
Wasteson, A., *Biochim. Biophys. Acta*, 177 (1969) 152.
Whitaker, R., *Anal. Chem.*, 35 (1963) 1950.
Wieland, Th., Duesberg, P. and Determann, H., *Biochem. Z.*, 337 (1963) 303.
Wild, L. and Guliana, R., *J. Polym. Sci., Part A-2*, 5 (1967) 1087.
Yamada, S., Imai, Sh. and Kitahara, S., *Chem. High Polym. (Tokyo)*, 24 (1967) 12.
Yau, W. W. and Fleming, S. W., *J. Appl. Polym. Sci.*, 12 (1968) 2111.

*Chapter 13*

# Practice of ion-exchange chromatography

O. MIKEŠ

## CONTENTS



## INTRODUCTION

The technique of ion-exchange chromatography has rapidly developed from classical procedures to the modern high-speed methods, and has been described in specialized chapters in several books (*e.g.*: Borman; Dorfner; Hale; Inczédy; Kirkland, 1971; Mikeš; Morris and Morris; Samuelson; Scott, 1971; Walton) and also in commercial booklets (*e.g.*: Thompson, Pharmacia). The procedures considered in this chapter are based on published work correlated with personal experience. The author's intention was to describe briefly the fundamental techniques for the most important types of ion exchangers used in liquid column chromatography (resins, cellulose and polydextran derivatives).

## CHOICE OF SUITABLE ION EXCHANGERS

When selecting an ion exchanger for chromatographic purposes, the following factors should be considered: (1) the type of process and (2) the extent of the operation.

Ion exchangers can be used in both batch and column processes. The former is essentially static, the exchanger being mixed with the solution in a beaker and, after equilibrium has been reached, it is separated mechanically. The latter process is essentially dynamic; the solution flows through the column, equilibrium being repeatedly renewed, and it can be divided into ion exchange and chromatography. For non-chromatographic processes (deionization of water, separations of cations from anions, exchange of one type of ion for another, etc.), there is a wide choice of ion-exchange resins (Tables 13.1–13.6), and

TABLE 13.1
CATION EXCHANGERS

Lists of producers of the resins are given in Tables 13.5 and 13.6.

| Type | Trade name | Cross-linking | Active group* | Supplied form | Lattice and particles** | |
|---|---|---|---|---|---|---|
| Strongly acidic | AG 50 or AG 50W (Bio-Rad) | X1 | $C_6H_5{-}SO_3^-$ | $H^+$ | PS | s |
| | | X2 | | $H^+$ | | s |
| | | X4 | | $H^+$ | | s |
| | | X8 | | $H^+$ | | s |
| | | X10 | | $H^+$ | | s |
| | | X12 | | $H^+$ | | s |
| | | X16 | | $H^+$ | | s |
| | | X16 | | $H^+$ | | g |
| | AG MP-50 (Bio-Rad) | Macro-porous | $C_6H_5{-}SO_3^-$ | $H^+$ | PS | s<br>g |
| | Amberlite 200 | | $C_6H_5{-}SO_3^-$ | | PS | g |
| | Amberlite CG-120 | X8 | $C_6H_5{-}SO_3^-$ | $Na^+$ | PS | s |
| | Amberlite IR-120 | X8 | $C_6H_5{-}SO_3^-$ | $Na^+$ | PS | s |
| | Amberlite IR-122 | X10 | $C_6H_5{-}SO_3^-$ | $Na^+$ | PS | s |
| | Amberlite XE-69 | X8 | $C_6H_5{-}SO_3^-$ | $Na^+$ | PS | g |
| | Bio-Rex 40 | Medium pore size | $C_6H_4(OH){-}CH_2{-}SO_3^-$ | $H^+$ | PH | g |
| | Bio-Rex RG 50W | X8 | $C_6H_5{-}SO_3^-$ | $H^+$, $NH_4^+$ or $Li^+$ | | s |
| | Dowex 50 or Dowex 50W | X1 | $C_6H_5{-}SO_3^-$ | $H^+$ | PS | s |

| Particle size (U.S. mesh dry grading) | Capacity | | Thermal stability, °C (form) | Chemical stability*** | | | Notes |
|---|---|---|---|---|---|---|---|
| | Dry resin (mequiv./g) | Resin swollen in water (mequiv./l) | | O | R | S | |
| 50–100 | 5.0 | 0.4 | 150 ($H^+$, $Na^+$) | + | ++ | ++ | Analytical grade, processed from Dowex 50 or Dowex 50W |
| 50–100, 100–200, 200–400 | 5.2 | 0.7 | | | | | |
| 20–50, 50–100, 100–200, 200–400, <400 | 5.2 | 1.2 | | | | | |
| 20–50, 50–100, 100–200, 200–400, <400 | 5.1 | 1.7 | | | | | |
| 20–50 | 5.0 | 1.9 | | | | | |
| 20–50, 50–100, 100–200, 200–400, <400 | 5.0 | 2.3 | | | | | |
| 20–50 | 4.9 | 2.6 | | | | | |
| 50–100, 100–200, 200–400 | 4.9 | 2.6 | | | | | |
| 20–50<br>50–100, 100–200, 200–400 | | 1.6–1.8 | | | | | |
| | | | | ++ | + | + | |
| 100–200, <200 | 4.3 | | 120 | | | | Chromatographic grade processed from IR-120 |
| 0.45–0.60 mm | 4.25 | 1.9 | 120 | | | | Commercial grade |
| 100–325 | 4.3 | | 120 | | | | Fine grain IR-120 |
| 20–50, 50–100, 100–200, 200–400 | 2.9 | 1.2 | 40 | – | + | + | Analytical grade, processed from Duolite C-3 |
| 20–50 | 5.1 | 1.7 | 150 | + | ++ | ++ | Reactor grade, processed from Dowex 50W |
| 50–100 | 5.0 | 0.4 | 150 | + | ++ | ++ | Commercial grade |



(Continued on p. 328/329)

TABLE 13.1 (*continued*)

| Type | Trade name | Cross-linking | Active group* | Supplied form | Lattice and particles** | |
|---|---|---|---|---|---|---|
| Strongly acidic | Dowex 50 or Dowex 50W | X2 | | $H^+$ | | s |
| | | X4 | | $H^+$ | | s |
| | | X8 | | $H^+$ | | s |
| | | X12 | | $H^+$ | | s |
| | | X16 | | $H^+$ | | s |
| | Duolite C-3 | | $C_6H_4(OH)-CH_2-SO_3^-$ | $Na^+$ | | |
| | Merck Lewatit S-1080 | | $C_6H_5-SO_3^-$ | $Na^+$ | | |
| | Merck Lewatit SP-1080 | Macro-porous | $C_6H_5-SO_3^-$ | $Na^+$ | | |
| | Zeo-Karb 225 | X1 | $C_6H_5-SO_3^-$ | $Na^+$ | PS | s |
| | | X2 | | $Na^+$ | | s |
| | | X4.5 | | $Na^+$ | | s |
| | | X8 | | $Na^+$ | | s |
| | | X12 | | $Na^+$ | | s |
| | | X20 | | $Na^+$ | | s |
| Intermediate acidic | Bio-Rex 63 | Large pore size | $C_6H_5-PO_3^{2-}$ | $Na^+$ | PS | s |
| | | | | $Na^+$ | | g |
| | Duolite C-63 | Large pore size | $C_6H_5-PO_3^{2-}$ | | PS | s |

| Particle size (U.S. mesh dry grading) | Capacity | | Thermal stability, °C (form) | Chemical stability*** | | | Notes |
|---|---|---|---|---|---|---|---|
| | Dry resin (mequiv./g) | Resin swollen in water (mequiv./l) | | O | R | S | |
| 50–100, 100–200, 200–400 | 5.2 | 0.7 | | | | | |
| 20–50, 50–100, 100–200, 200–400, <400 | 5.2 | 1.2 | | | | | |
| 20–50, 50–100, 100–200, 200–400, <400 | 5.1 | 1.7 | | | | | |
| 50–100, 100–200, 200–400 | 5.0 | 2.3 | | | | | |
| 200–400 | 4.9 | 2.6 | | | | | |
| 16–50 | 2.9 | 1.2 | | | | | Commercial grade |
| 70–150 | 4.0 | | | | | | Analytical grade |
| 70–150 | 4.0 | | | | | | Analytical grade |
| 14–52, 52–100, 100–200, < 200§ | 4.5–5.0 | 2.1 | 120 | + | + | + | Chromatographic grade |
| 14–52, 52–100, 100–200, < 200§ | | | | | | | |
| 14–52, 52–100, 100–200, < 200§ | | | | | | | |
| 14–52, 52–100, 100–200, < 200§ | 4.8 | 2.1 | | | | | |
| 14–52, 52–100, 100–200, < 200§ | | | | | | | |
| 14–52, 52–100, 100–200, < 200§ | | | | | | | |
| 20–50 | 6.6 | 3.1 | 100 | ++ | ++ | ++ | Analytical grade, processed from Duolite C-63 |
| 50–100, 100–200, 200–400 | 6.6 | 3.1 | | | | | |
| 16–50 | 6.6 | 3.1 | 100 | ++ | ++ | ++ | Commercial grade |

*References p.368*

(*Continued on p. 330/331*)

TABLE 13.1 (*continued*)

| Type | Trade name | Cross-linking | Active group* | Supplied form | Lattice and particles** | |
|---|---|---|---|---|---|---|
| Weakly acidic | Amberlite CG-50 | X2–3 | R–COO$^-$ | $H^+$ | PM | |
| | Amberlite IRC-50 | X2–3 | R–COO$^-$ | $H^+$ | PM | s |
| | Amberlite XE-64 | X2–3 | R–COO$^-$ | $H^+$ | PM | g |
| | Amberlite XE-64 (micro-powder) | X2–3 | R–COO$^-$ | $H^+$ | PM | g |
| | Bio-Rex 70 | Large pore size (macro-reticular resin) | R–COO$^-$ | $Na^+$ | PA | s |
| | | | | $Na^+$ | | g |
| | Duolite CS-101 | Large pores | R–COO$^-$ | $H^+$ | PA | s |
| | Merck Lewatit CP-3050 | Macro-porous | R–COO$^-$ | $H^+$ | | |
| | Zeo-Karb 226 | X2.5 | R–COO$^-$ | $H^+$ | PA | s |
| | | X4.5 | | $H^+$ | | s |

*R = Aliphatic chain; $C_6H_5-$ or $C_6H_4(OH)-$ = bond to aromatic matrix.
**Type of lattice: EP = epoxypolyamine; PA = polyacrylic; PM = polymethacrylic; PH = phenolic; PS = polystyrene. Type of particles: s = spheres (beads); g = granular form.
***Stability: O = oxidation; R = reduction; S = organic solvents. – = Unstable; + = stable; ++ = very stable.
§ BBS mesh wet grading.

| Particle size (U.S. mesh dry grading) | Capacity | | Thermal stability, °C (form) | Chemical stability*** | | | Notes |
|---|---|---|---|---|---|---|---|
| | Dry resin (mequiv./g) | Resin swollen in water (mequiv./l) | | O | R | S | |
| 100–200, <200 | 10 | | 120 | | | | Chromatographic grade IRC-50 |
| 0.33–0.5 mm | 10 | 3.5 | 120 | | | | Commercial grade |
| 100–325 | 10 | | 120 | | | | Fine grain IRC-50 |
| <325 | 10 | | 120 | | | | Fine grain IRC-50 |
| 20–50 | 10.2 | 3.3 | 100 | + | + | + | Analytical grade, processed from Duolite CS-101 |
| 50–100, 100–200, 200–400, <400 | 10.2 | 3.3 | | | | | |
| 16–50 | 10.2 | 3.2 | 100 | + | + | + | |
| 70–150 | 12 | | | | | | Analytical grade |
| 14–52, 52–100, 100–200, < 200§ | 9–10 | 3.5 | 100 | + | + | + | Chromatographic grade |
| 14–52, 52–100, 100–200, < 200 | | | | | | | |

TABLE 13.2
ANION EXCHANGERS

Lists of producers of the resins are given in Tables 13.5 and 13.6.

| Type | Trade name | Cross-linking | Active group* | Supplied form | Lattice and particles** | |
|---|---|---|---|---|---|---|
| Strongly basic | AG 1 (Bio-Rad) | X1 | $C_6H_5-CH_2-\overset{+}{N}(CH_3)_3$ | $Cl^-$ | PS | s |
| | | X2 | | $Cl^-$ | | s |
| | | X4 | | $Cl^-$ | | s |
| | | X8 | | $Cl^-$ | | s |
| | | X10 | | $Cl^-$ | | s |
| | AG 2 (Bio-Rad) | X4 | $C_6H_5-CH_2-\overset{+}{N}(CH_3)_2C_2H_5OH$ | $Cl^-$ | PS | s |
| | | X8 | | $Cl^-$ | | s |
| | | X10 | | $Cl^-$ | | s |
| | AG 21K (Bio-Rad) | Novel type similar to X4 | $C_6H_5-CH_2-\overset{+}{N}(CH_3)_3$ | $Cl^-$ | PS | s |
| | AG MP-1 | Macro-porous | $C_6H_5-CH_2-\overset{+}{N}(CH_3)_3$ | $Cl^-$ | PS | s<br>g |
| | Amberlite CG-400 I, II (XE-67) | X8 | $C_6H_5-CH_2-\overset{+}{N}(CH_3)_3$ | | PS | |
| | Amberlite IRA-400 | | $C_6H_5-CH_2-\overset{+}{N}(CH_3)_3$ | $Cl^-$ | PS | s |
| | Amberlite IRA-401 (XE-75) | Highly porous | $C_6H_5-CH_2-\overset{+}{N}(CH_3)_3$ | $Cl^-$ | PS | |
| | Amberlite IRA-410 | | $C_6H_5-CH_2-\overset{+}{N}(CH_3)_2C_2H_4OH$ | $Cl^-$ | PS | s |
| | Amberlite IRA-411 (XE-98) | Porous | $C_6H_5-CH_2-\overset{+}{N}(CH_3)_2C_2H_4OH$ | $Cl^-$ | PS | s |
| | Amberlite XE-67 | | $C_6H_5-CH_2-\overset{+}{N}(CH_3)_3$ | $Cl^-$ | PS | |

| Particle size (U.S. mesh dry grading) | Capacity | | Thermal stability, °C (form) | Chemical stability*** | | | Notes |
|---|---|---|---|---|---|---|---|
| | Dry resin (mequiv./g) | Resin swollen in water (mequiv./l) | | O | R | S | |
| 50–100 | 3.2 | 0.4 | 50 ($OH^-$)<br>150 ($Cl^-$) | – | ++ | ++ | Analytical grade, processed from Dowex 1 |
| 50–100, 100–200, 200–400, <400 | 3.5 | 0.8 | | | | | |
| 20–50, 50–100, 100–200, 200–400, <400 | 3.5 | 1.2 | | | | | |
| 20–50, 50–100, 100–200, 200–400, <400 | 3.2 | 1.4 | | | | | |
| 50–100, 100–200, 200–400, <400 | 3.0 | 1.5 | | | | | |
| 20–50 | 3.2 | 1.2 | 30 ($OH^-$)<br>150 ($Cl^-$) | – | + | ++ | Analytical grade, processed from Dowex 2 |
| 20–50, 50–100, 100–200, 200–400 | 3.2 | 1.4 | | | | | |
| 50–100, 100–200, 200–400 | 3.0 | 1.5 | | | | | |
| 16–20, 20–50, 50–100 | 4.5 | 1.3 | 50 ($OH^-$)<br>150 ($Cl^-$) | – | + | ++ | Analytical grade, processed from Dowex 21K |
| 20–50<br>50–100, 100–200, 200–400 | | 1.0 | | | | | |
| 100–200, 200–400 | | 1.2 | 60 ($OH^-$)<br>75 ($Cl^-$) | | | | Analytical grade, processed from Amberlite IRA-400 |
| 0.35–0.45 mm | 3.3 | 1.2 | 60 ($OH^-$) | | | | Commercial grade |
| 0.40–0.55 mm | 3.3 | 1.2 | 60 ($OH^-$) | | | | Porous form of IRA-400 |
| 0.35–0.45 mm | 3.0 | 1.2 | 40 ($OH^-$) | | | | Commercial grade |
| 0.35–0.50 mm | 3.0 | 0.7 | 40 ($OH^-$) | | | | Porous form of IRA-410 |
| 100–325 | 3.3 | 1.2 | 60 ($OH^-$) | | | | Fine grain IRA-400 |

*References p.368*

*(Continued on p. 334/335)*

TABLE 13.2 (*continued*)

| Type | Trade name | Cross-linking | Active group* | Supplied form | Lattice and particles** | |
|---|---|---|---|---|---|---|
| Strongly basic | Bio-Rex 9 | | 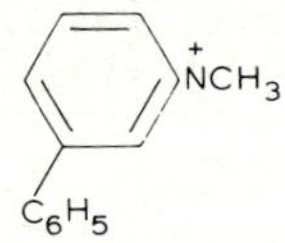 | $Cl^-$ | PS | s<br>g |
| | Bio-Rex RG 1 | X8 | $C_6H_5-CH_2-\overset{+}{N}(CH_3)_3$ | $OH^-$ | PS | s |
| | Bio-Rex RG 2 | X8 | $C_6H_5-CH_2-\overset{+}{N}(CH_3)_2C_2H_5OH$ | $OH^-$ | PS | s |
| | De Acidite FF | X2–3 | $C_6H_5-CH_2-\overset{+}{N}(CH_3)_3$ | $Cl^-$ | PS | s |
| | | X3–5 | | $Cl^-$ | | s |
| | | X7–9 | | $Cl^-$ | | s |
| | Dowex 1 | X1 | $C_6H_5-CH_2-\overset{+}{N}(CH_3)_3$ | $Cl^-$ | PS | s |
| | | X2 | | $Cl^-$ | | s |
| | | X4 | | $Cl^-$ | | s |
| | | X8 | | $Cl^-$ | | s |
| | Dowex 2 | X4 | $C_6H_5-CH_2-\overset{+}{N}(CH_3)_2C_2H_5OH$ | $Cl^-$ | PS | s |
| | | X8 | | $Cl^-$ | | s |
| | Dowex 21K | Novel type similar to X4 (porous) | $C_6H_5-CH_2-\overset{+}{N}(CH_3)_3$ | $Cl^-$ | | s |
| | Merck Lewatit M-5080 | | $C_6H_5-CH_2-\overset{+}{N}(CH_3)_3$ | $Cl^-$ | | |
| | Merck Lewatit MP-5008 | Macro-porous | $C_6H_5-CH_2-\overset{+}{N}(CH_3)_3$ | $Cl^-$ | | |
| Intermediate basic | Bio-Rex 5 | | $R-\overset{+}{N}(CH_3)_2-C_2H_5OH$ and $R-\overset{+}{N}H(CH_3)_2$ | $Cl^-$ | EP | s |
| | | | | $Cl^-$ | | g |

| Particle size (U.S. mesh dry grading) | Capacity | | Thermal stability, °C (form) | Chemical stability*** | | | Notes |
|---|---|---|---|---|---|---|---|
| | Dry resin (mequiv./g) | Resin swollen in water (mequiv./l) | | O | R· | S | |
| 20–50 | 3.7 | 1.3 | 38 | ++ | | | Analytical grade |
| 50–100, 100–200, 200–400 | 3.7 | 1.3 | | | | | |
| 20–50 | 3.2 | 0.4 | 50 ($OH^-$) 150 ($Cl^-$) | – | + | ++ | Reactor grade, processed from Dowex 1 |
| 20–50 | 3.2 | 1.2 | 30 ($OH^-$) 150 ($Cl^-$) | – | + | ++ | Reactor grade, processed from Dowex 2 |
| 14–52, 52–100, 100–200, <200§ | 4.2 | 1.5 | 60 ($OH^-$) | – | + | ++ | Chromatographic grade |
| 14–52, 52–100, 100–200, <200 | | | | | | | |
| 14–52, 52–100, 100–200, <200 | | | | | | | |
| 50–100 | 3.6 | 0.4 | 50 ($OH^-$) | – | + | ++ | Analytical grade |
| 50–100, 100–200, 200–400, <400 | 3.6 | 0.9 | 150 ($Cl^-$) | | | | |
| 20–50, 50–100, 100–200, 200–400, <400 | 3.5 | 1.2 | | | | | |
| 20–50, 50–100, 100–200, 200–400 | 3.5 | 1.4 | | | | | |
| 20–50 | 3.7 | 1.2 | 30 ($OH^-$) | – | + | ++ | Analytical grade |
| 20–50, 50–100, 100–200, 200–400 | 3.6 | 1.4 | 150 ($Cl^-$) | | | | |
| 16–20, 20–50, 50–100 | 4.5 | 1.3 | 50 ($OH^-$) 150 ($Cl^-$) | – | + | ++ | Analytical grade |
| 70–150 | 3.5 | | | | | | Analytical grade |
| 70–150 | 3.0 | | | | | | Analytical grade |
| 20–50 | 8.8 | 2.8 | 60 | + | ++ | + | Analytical grade, processed from Duolite A-30 |
| 50–100, 100–200, 200–400 | 8.8 | 2.8 | | | | | |



*(Continued on p. 336/337)*

TABLE 13.2 (*continued*)

| Type | Trade name | Cross-linking | Active group* | Supplied form | Lattice and particles** | |
|---|---|---|---|---|---|---|
| Intermediate basic | De Acidite H | X2–3 | $C_6H_5-CH_2-\overset{+}{N}H(CH_3)_2$ and $(C_6H_5-CH_2)_2-\overset{+}{N}(CH_3)_2$ | $Cl^-$ | PS | |
| | | X3–5 | | | | |
| | | X7–9 | | | | |
| | Duolite A-30 | | $R-\overset{+}{N}(CH_3)_2-C_2H_4OH$ and $R-\overset{+}{N}H(CH_3)_2$ | $Cl^-$ | EP | s |
| Weakly basic | AG 3 (Bio-Rad) | X4 | $C_6H_5-\overset{+}{N}HR_2$ | $Cl^-$ | PS | s |
| | | | | $Cl^-$ | PS | g |
| | Amberlite CG-45 | | Polyamine | $OH^-$ | PS | s |
| | Amberlite CG-4B I, II | | Amine | | PH | |
| | Amberlite IR-4B | | Amine | $OH^-$ | PH | g |
| | Amberlite IR-45 | | Polyamine | $OH^-$ | PS | s |
| | Amberlite XE-58 | | Amine | $OH^-$ | PH | g |
| | Amberlite XE-58 (micro-powder) | | Amine | $OH^-$ | PH | g |
| | Amberlyst A-21 | Macro-reticular | *tert.*-Amine | | | |
| | De Acidite G | X2–3 | $-CH_2-\overset{+}{N}H(C_2H_5)_2$ | $Cl^-$ | PS | s |
| | | X3–5 | | | | s |
| | | X7–9 | | | | s |
| | De Acidite M | X3–5 | Polyamine | $Cl^-$ | | |
| | Dowex 3 | X4 | $C_6H_5-\overset{+}{N}HR_2$ | $Cl^-$ | PS | s |
| | Merck Lewatit MP-7080 | Macro-porous | | $Cl^-$ | | |

*R = Aliphatic chain; $C_6H_5-$ or $C_6H_4(OH)-$ = bond to aromatic matrix.

**Type of lattice: EP = epoxypolyamine; PH = phenolic; PS = polystyrene. Type of particles: s = spheres (beads): g = granular form.

***Stability: O = oxidation; R = reduction; S = organic solvents. – = Unstable; + = stable; ++ = very stable.

§ BBS mesh wet grading.

| Particle size (U.S. mesh dry grading) | Capacity | | Thermal stability, °C (form) | Chemical stability*** | | | Notes |
|---|---|---|---|---|---|---|---|
| | Dry resin (mequiv./g) | Resin swollen in water (mequiv./l) | | O | R | S | |
| 14–52, 52–100, 100–200, < 200 § | 3.8 | | | | | | Chromatographic grade |
| 14–52, 52–100, 100–200, < 200 § | | | | | | | |
| 14–52, 52–100, 100–200, < 200 § | | | | | | | |
| 16–50 | 8.8 | 2.8 | 60 | + | ++ | + | Commercial grade |
| 20–50 | 2.8 | 1.9 | 65 | + | | ++ | Analytical grade, processed from Dowex 3 |
| 100–200, 200–400 | 2.8 | 1.9 | | | | | |
| 100–200, <200 | 5 | 2 | 100 | | | | Chromatographic grade |
| 100–200, 200–400 | | 2.5 | | | | | Analytical grade, processed from Amberlite IR-4B |
| 0.4–0.6 mm | 10 | 3.0 | 40 | | | | Commercial grade |
| 0.4–0.55 mm | 5 | 2 | 100 | | | | Commercial grade |
| 100–325 | 10 | 2.5 | 40 | | | | Fine grain IR-4B |
| <325 | 10 | 2.5 | 40 | | | | Fine grain IR-4B |
| 50–100 μm | | 1.7 | | | | | Chromatographic grade |
| 14–52, 52–100, 100–200, < 200 § | 3.5 | 1.3 | | | | | Chromatographic grade |
| 14–52, 52–100, 100–200, < 200 § | | | | | | | |
| 14–52, 52–100, 100–200, < 200 § | | | | | | | |
| 14–52, 52–100, 100–200, < 200 § | 6 | | | | | | Chromatographic grade |
| 20–50 | 2.8 | 1.9 | 65 | + | | ++ | Commercial grade |
| 70–150 | 4.0 | | | | | | Analytical grade |

*References p.368*

TABLE 13.3
MIXED BED RESINS

| Trade name | Producer or distributor | Components | Supplied form | Maximum temperature (°C) |
|---|---|---|---|---|
| Amberlite MB-1 | Rohm & Haas (Philadelphia, Pa., U.S.A.) | Amberlite IR-120 | $H^+$ | 60 |
| | | Amberlite IRA-400 | $OH^-$ | |
| Amberlite MB-2 | | Amberlite IR-120 | $H^+$ | 40 |
| | | Amberlite IR-410 | $OH^-$ | |
| Amberlite MB-3 | | Amberlite MB-2 + indicator | | 40 |
| Amberlite MB-4 | | Amberlite IR-120 | $H^+$ | |
| | | Amberlite IR-45 | $OH^-$ | |
| Amberlite MB-5 | | Amberlite IRC-50 | | |
| | | Amberlite IRA-410 | | |
| Amberlite MB-6 | | Amberlite IRC-50 | | |
| | | Amberlite IR-45 | | |
| Bio Deminrolit | Permutit (London, Great Britain) | Zeo-Karb 225 | $H^+$ | 60 |
| | | De Acidite FF | $OH^-$ | |
| Bio-Rad AG 501-X8 | Bio-Rad Labs. (Richmond, Calif., U.S.A.) | AG 50W-X8 | $H^+$ | |
| | | AG 1-X8 | $OH^-$ | |
| Bio-Rad AG 501-X8 (D) | | Bio-Rad AG 501-X8 + indicator | | |
| Bio-Rex RG 501-X8 | | Bio-Rex RG 50W-X8 | $H^+$ or $Li^+$ | |
| | | Bio-Rex RG 1-X8 | $OH^-$ | |
| Bio-Rex RG 502-X8 | | Bio-Rex RG 50W-X8 | $H^+$ or $Li^+$ | |
| | | Bio-Rex RG 2-X8 | $OH^-$ | |
| Duolite GPM-331 | Diamond Shamroc (Redwood City, Calif., U.S.A.) | Duolite C-20 | $H^+$ | |
| | | Duolite A-101D | $OH^-$ | |
| Duolite GPM-331A | | Duolite C-20 | $H^+$ | |
| | | Duolite A-102D | $OH^-$ | |
| Indicator-Bio Deminrolit | Permutit (London, Great Britain) | Bio-Deminrolit + indicator | | 60 |
| Permutit M-100 | Permutit (New York, N.Y., U.S.A.) | Permutit Q | $H^+$ | |
| | | Permutit S-100 | $OH^-$ | |
| Permutit M-103 | Permutit (New York, N.Y., U.S.A.) | Permutit Q | $H^+$ | |
| | | Permutit S-200 | $OH^-$ | |

uniform grain size is not necessary. For chromatographic separations, the exchangers must be carefully chosen. The important factors to be considered are as follows.

(*a*) For the separation of biopolymers (proteins, nucleic acids and their large fragments), ion-exchange derivatives of cellulose (Table 13.7) or polydextran (Table 13.8) are recommended. In some instances, microgranular resins with an acrylic type of lattice (Table 13.1) can also be used.

(*b*) For other purposes, ion-exchange resins are suitable.

(*c*) Ion-exchange crystals (Table 13.9) are used for selective separations of some inorganic ions in addition to resins.

(*d*) Homoionic exchangers are to be preferred (very often with a polystyrene lattice).

(*e*) A bead form (spheres) is generally better than ground grains.

(*f*) The grain size should be as uniform as possible. The grain sizes usually used are given in Table 13.10.

(*g*) Cross-linking is an important factor. For the separation of low-molecular-weight ions, X8 is usually used with cation exchangers and X8 or X4 with anion exchangers. The relationship between X and WR values is discussed in Chapter 6. For larger ions (*e.g.*, simple peptides consisting of only a few amino acids), X4 cation exchangers are recommended. Polypeptides from enzymic digests of proteins are often separated using X2 resins. For separations in non-aqueous or mixed solutions, extremely porous macroreticular resins can be recommended.

(*h*) For many conventional laboratory or commercial applications, the commercial grade of ion exchangers is adequate. Resins for chromatographic purposes often require purification (repeated cycling and extraction with organic solvents) and this pre-treatment can be carried out in the laboratory. As it is also carried out commercially in addition to efficient grain sizing (AG resins, Bio-Rex resins, etc.), it is possible to choose ion exchangers that are ready for use.

(*i*) For most separations, strongly acidic or strongly basic resins are to be preferred. Weakly acidic or weakly basic resins are usually used only for special purposes, *e.g.*, in biochemistry.

(*j*) The chemical and thermal stability of the resin under the conditions of the separation should be considered.

(*k*) Strongly acidic cation exchangers in the $H^+$ form and strongly basic anion exchangers in the $OH^-$ form catalyze some side-reactions (hydrolysis of esters and of peptides, Cannizzaro reactions, etc.), which can interfere with the chromatographic process.

(*l*) For the high-speed analysis of low- and medium-molecular-weight substances, the pellicular form of resins (porous layer beads) can be recommended. The diffusion into and out of the thin layer proceeds in a short period of time and this permits the use of a very rapid carrier velocity, thus allowing the shole analysis to be completed within several minutes. A high pressure of the influent and the use of special equipment are necessary.

(*m*) Consideration of related experiments described in the literature is very useful for the selection of a suitable ion exchanger. The Applications part of this book gives many such examples. The major producers will send references on the application of their products on request.

TABLE 13.4
SPECIAL ION EXCHANGERS

| Type | Trade name | Producer or distributor* | Active groups** |
|---|---|---|---|
| Ion retardation resins | Bio-Rad AG 11-A8 | B-R | $C_6H_5-CH_2-\overset{+}{N}(CH_3)_3$, $R-COO^-$ |
| | Retardion 11-A8 | D Ch | $C_6H_5-CH_2-\overset{+}{N}(CH_3)_3$, $R-COO^-$ |
| Chelating resins | Chelex 100 | B-R | $C_6H_5-CH_2-\overset{+}{N}H(CH_2COO^-)_2$ |
| | Dowex A-1 | D Ch | $C_6H_5-CH_2-\overset{+}{N}H(CH_2COO^-)_2$ |
| Resins for biochemical analysis | Amberlite IR-120/2835 AS | R H | $C_6H_5-SO_3^-$ |
| | Amberlite IR-120/3545 AS | R H | $C_6H_5-SO_3^-$ |
| | Amberlite IR-120/4560 AS | R H | $C_6H_5-SO_3^-$ |
| | Amberlite IRC-50/80100 AS | R H | $R-COO^-$ |
| | Aminex 50W | B-R | $C_6H_5-SO_3^-$ |
| | Aminex 50W | B-R | $C_6H_5-SO_3^-$ |
| | Aminex 50W | B-R | $C_6H_5-SO_3^-$ |
| | Aminex A-4 | B-R | $C_6H_5-SO_3^-$ |
| | Aminex A-5 | B-R | $C_6H_5-SO_3^-$ |
| | Aminex A-6 | B-R | $C_6H_5-SO_3^-$ |
| | Aminex A-7 | B-R | $C_6H_5-SO_3$ |
| | Aminex A-14 | B-R | $C_6H_5-CH_2-\overset{+}{N}(CH_3)_3$ |
| | Aminex A-25 | B-R | $C_6H_5-CH_2-\overset{+}{N}(CH_3)_3$ |

| Resin lattice | Cross-linking | Particle size | Uses | Notes |
|---|---|---|---|---|
| Polystyrene + acrylic acid | | 50–100 mesh | Desalting | Analytical grade, processed from Retardion 11-A8 |
| Polystyrene + acrylic acid | | 50–100 mesh | Desalting | Prepared by polymerizing acrylic acid inside Dowex 1 |
| Polystyrene | | 50–100, 100–200, 200–400 mesh | Specific sorption of transition metals, amino acids, etc. | Analytical grade, processed from Dowex A-1 |
| Polystyrene | | 50–100, 100–200, 200–400 mesh | Specific sorption of transition metals, amino acids, etc. | |
| Polystyrene | X8 | 28–35 $\mu$m | Accelerated amino acid analysis | Ground particles |
| Polystyrene | X8 | 35–45 $\mu$m | Standard amino acid chromatography | Ground particles |
| Polystyrene | X8 | >45 $\mu$m | Preparative amino acid chromatography | Ground particles |
| Polyacrylic acid | | 80–100 $\mu$m | Chromatography of basic amino acids | Ground particles |
| Polystyrene | X2 | 200–325 mesh | Analytical separation of peptides | |
| Polystyrene | X12 | 21–29 $\mu$m | Amino acid analysis; buffer gradient system | |
| Polystyrene | X4 | 20–30 $\mu$m, 30–35 $\mu$m | Analytical separation of peptides and iodo-amino acids | |
| Polystyrene | X8 | 16–24 $\mu$m | Amino acid analysis; low-pressure 2-column system | |
| Polystyrene | X8 | 13 ± 2 $\mu$m | Amino acid analysis; low- and high-pressure 2-column system; also in physiological fluids | |
| Polystyrene | X8 | 17.5 ± 2 $\mu$m | Nucleoside and amino acid analysis; high-pressure 2-column system | |
| Polystyrene | X8 | 7–11 $\mu$m | Amino acid analysis of physiological fluids | |
| Polystyrene | X4 | 17–23 $\mu$m | Carbohydrate analysis; stepwise buffer system | |
| Polystyrene | X8 | 17.5 ± 2 $\mu$m | Carbohydrate analysis; gradient buffer system | |

*References p.368*

(*Continued on p. 342/343*)

TABLE 13.4 (*continued*)

| Type | Trade name | Producer or distributor* | Active groups |
|---|---|---|---|
| Resins for biochemical analysis | Aminex A-27 | B-R | $C_6H_5-CH_2-\overset{+}{N}(CH_3)_3$ |
| | Aminex A-28 | B-R | $C_6H_5-CH_2-\overset{+}{N}(CH_3)_3$ |
| | Aminex MS, fraction B | B-R | $C_6H_5-SO_3^-$ |
| | Aminex MS, fraction C | B-R | $C_6H_5-SO_3^-$ |
| | Aminex MS, fraction D | B-R | $C_6H_5-SO_3^-$ |
| | Aminex Q-150S | B-R | $C_6H_5-SO_3^-$ |
| | Aminex Q-15S | B-R | $C_6H_5-SO_3^-$ |
| | Beckman M-72 | B-S | $C_6H_5-SO_3^-$ |
| | Beckman PA-28 | B-S | $C_6H_5-SO_3^-$ |
| | Beckman PA-35 | B-S | $C_6H_5-SO_3^-$ |
| | Beckman UR-30 | B-S | $C_6H_5-SO_3^-$ |
| | Bio-Rad AG 1 | B-R | $C_6H_5-CH_2-\overset{+}{N}(CH_3)_3$ |
| | Chelex 100 | B-R | $C_6H_5-CH_2-\overset{+}{N}H(CH_2COO^-)_2$ |
| | Durrum DC-4A | DU | $C_6H_5-SO_3^-$ |
| Other special ion exchangers | Bio-Gel HT | B-R | Modified calcium phosphate |
| | Bio-Gel HTP | B-R | Powdered Bio-Gel HT |
| | DNA-grade Bio-Gel HTP | B-R | Powdered Bio-Gel HT |
| | Brushite | B-R | $CaHPO_4 \cdot 2\,H_2O$ |

*Producers or distributors: B-R = Bio-Rad Labs., Richmond, Calif., U.S.A.; B-S = Beckman-Spinco, Palo Alto, Calif., U.S.A.; D Ch = Dow Chemical, Midland, Mich., U.S.A.; DU = Durrum, Palo Alto, Calif., U.S.A.; R H = Rohm & Haas, Philadelphia, Pa., U.S.A.

**R = Aliphatic chain; $C_6H_5-$ = bond to aromatic matrix.

| Resin lattice | Cross-linking | Particle size | Uses | Notes |
|---|---|---|---|---|
| Polystyrene | X8 | 12–15 $\mu$m | Analysis of UV-absorbing components in physiological fluids | |
| Polystyrene | X8 | 7–11 $\mu$m | | |
| Polystyrene | X8 | 25–30 $\mu$m | Amino acid analysis, fraction collector system | |
| Polystyrene | X8 | 33–47 $\mu$m | | |
| Polystyrene | X8 | 47–65 $\mu$m | | |
| Polystyrene | X8 | 20–35 $\mu$m | Preparative chromatography of amino acids | |
| Polystyrene | X8 | 19–25 $\mu$m | | |
| Polystyrene | | | Amino acid analysis (one-column system) | |
| Polystyrene | X7.5 | 16 ± 6 $\mu$m | Amino acid analysis (acidic and neutral amino acids) | |
| Polystyrene | X7.5 | 13 ± 6 $\mu$m | Amino acid analysis (basic amino acids) | |
| Polystyrene | | | Amino acid analysis (acidic and neutral amino acids) | |
| Polystyrene | X2 | 200–400 mesh | Separation of iodo-amino acids | |
| Polystyrene | | 20–30 $\mu$m | Base analysis of nucleic acid components | |
| Polystyrene | | 8 ± 1 $\mu$m | Analysis of amino acids and physiological fluids | |
| Hydroxy-apatite | | Large granules | Chromatography of polypeptides and proteins | |
| | | Powder | Chromatography of lipids | |
| | | Powder | Chromatography of DNA | |
| | | | Chromatography of proteins | |

TABLE 13.5
INTERCHANGEABLE CATION-EXCHANGE RESINS (TRADE NAME INDEX)

Resins with the same active groups and type of lattice produced by different manufacturers can often be used interchangeably. However, they need not be identical in composition and may differ in their physical properties.

| Resin | Manufacturer | Strongly acidic ion exchangers | | Intermediate acidic ion exchangers | | Weakly acidic ion exchangers | |
|---|---|---|---|---|---|---|---|
| | | Polystyrene lattice ($C_6H_5-SO_3^-$ active groups) | Phenolic lattice ($R-CH_2-SO_3^-$ active groups*) | Polystyrene lattice ($C_6H_5-PO_3^{2-}$ active groups) | Polystyrene lattice ($C_6H_5-PO_2^-$ active groups) | Acrylic lattice ($R-COO^-$ active groups*) | Phenolic lattice ($-COO^-$ active groups) |
| Amberlite | Rohm & Haas, Philadelphia, Pa., U.S.A. | IR-112 (X4)** IR-120 or CG-120 (X8) IR-122 (X10) IR-124 (X12) | IR-1*** IR-100*** IR-105*** | | | IRC-50 (CG-50) XE-64 XE-89 | |
| Bio-Rad (analytical grade) | Bio-Rad Labs., Richmond, Calif., U.S.A. | AG 50W (X1–X16) | | | | | |
| Bio-Rex (analytical grade) | Bio-Rad Labs. | | 40 | | | 70 | |
| Dowex | Dow Chemical, Midland, Mich., U.S.A. | 50 or 50W (X1–X16) | | | | | |
| Duolite | Diamond Shamroc, Redwood City, Calif., U.S.A. | C-20 (X8) C-20 (X10) C-20 (X12) | C-3 C-10 | ES-63 | C-62 | CS-101 | CS-100 |

| | | | | | | |
|---|---|---|---|---|---|---|
| Imac | Industrieele Mij. Activit, Amsterdam, The Netherlands | C-22<br>C-12 | | | | C-19 |
| Lewatit | Bayer, Leverkusen, G.F.R. | S-100 | KSN<br>PN | | | CNO |
| Nalcite | Nalco Chemical, Chicago, Ill., U.S.A. | HCR (X8)<br>HGR (X10)<br>HDR (X12) | | X219 | | |
| Permutit | Permutit, New York, N.Y., U.S.A. | Q (X8)<br>Q-100 (X10)<br>Q-110 (X12)<br>Q-130 (X16) | | | Q-210 | H |
| Permutite | Phillips and Pain Vermorel, Paris, France | C-50D | | | | |
| Resex | J. Crosfield and Sons, Warrington, Great Britain | Resex P | Resex | | | |
| Wofatit | VEB Farbenfabrik, Wolfen, G.F.R. | KPS-200 | F<br>P | | CP-300 | CN |
| Zeo-Karb (Zerolit)§ | Permutit, London, Great Britain | 225 (X4)<br>225 (X8) | 215<br>315*** | | 226 | 216 |

*R = Aliphatic chain; $C_6H_5-$ = bond to aromatic matrix.
**X followed by a number is the corresponding cross-linking in cases when it does not form part of the trade name.
***Resins no longer in production (these indications are not complete).
§Marketed under the name Zerolit (with the same numerical designation) by United Water Softeners, London, Great Britain.

TABLE 13.6
INTERCHANGEABLE ANION-EXCHANGE RESINS (TRADE NAME INDEX)

Resins with the same active groups and type of lattice produced by different manufacturers can often be used interchangeably. However, they need not be identical in composition and may differ in their physical properties.

| Resin | Manufacturer | Strongly basic ion exchangers | |
|---|---|---|---|
| | | Polystyrene lattice ($C_6H_5-\overset{+}{N}(CH_3)_3$ active groups) type I | Polystyrene lattice ($C_6H_5-\overset{+}{N}(CH_3)_2C_2H_4OH$ active groups) type II |
| Amberlite | Rohm & Haas, Philadelphia, Pa., U.S.A. | IRA-401 (X4)**<br>IRA-400 or<br>CG-400 (X8) | IRA-411 (X4)<br>IRA-410 (X8) |
| Bio-Rad (analytical grade) | Bio-Rad Labs., Richmond, Calif., U.S.A. | AG 1 (X1–X10)<br>AG 21K | AG 2-X4 |
| Bio-Rex (analytical grade) | Bio-Rad Labs. | | |
| De Acidite (Zerolit) § | Permutit; London, Great Britain | FF (lightly cross-linked)<br>FF | |
| Dowex | Dow Chemical, Midland, Mich., U.S.A. | 1 (X1–X10)<br>21K | 2-X4<br>2-X8 |
| Duolite | Diamond Shamroc, Redwood City, Calif., U.S.A. | A-101 (X4)<br>A-42 (X8) | A-102 (X4)<br>A-40 (X8) |
| Imac | Industrieele Mij. Activit, Amsterdam, The Netherlands | S-3 | |
| Lewatit | Bayer, Leverkusen, G.F.R. | MN | |
| Nalcite | Nalco Chemical, Chicago, Ill., U.S.A. | SBR<br>SBR-P | SAR |
| Permutit | Permutit, New York, N.Y., U.S.A. | S1 (lightly cross-linked)<br>S1<br>S-100 | S-200 |
| Permutite | Phillips and Pain Vermorel, Paris, France | | A-300D |
| Resex | J. Crosfield and Sons, Warrington, Great Britain | Resanex HB | |
| Wofatit | VEB Farbenfabrik, Wolfen, G.F.R. | | |

*R = Aliphatic chain; $C_6H_5-$ = bond to aromatic matrix.

**X followed by a number is the corresponding cross-linking in cases when it does not form part of the trade name.

***Resin no longer in production (these indications are not complete).

§Marketed under the name Zerolit (with the same numerical designation) by United Water Softeners, London, Great Britain.

| | Intermediate basic ion exchangers | Weakly basic ion exchangers | |
|---|---|---|---|
| Polystyrene lattice ( 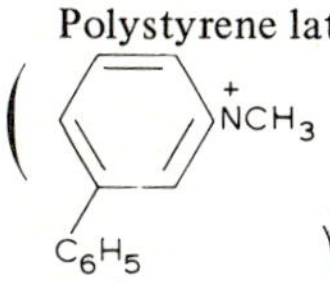 active groups) | Epoxypolyamine lattice (R–$\overset{+}{N}H(CH_3)_2$ and R–$\overset{+}{N}(CH_3)_2C_2H_4OH$ active groups*) | Polystyrene lattice (R–$\overset{+}{N}HR_2$ active groups*) | Phenolic polyamine lattice (R–$\overset{+}{N}HR_2$ active groups*) |
| | | IR-45 | IR-4B |
| | | | IRA-68 |
| | | AG 3-X4 | |
| 9 | 5 | | |
| | F*** | G | E |
| | | 3-X4 | |
| | A-30B | A-2 | A-6 |
| | | | A-7 |
| | | | A-4F |
| | A-17 | | A-21 |
| | A-19 | | |
| | | | MIH |
| | | WBR | |
| A 580 | S-310 | S-300 | S-350 |
| | S-380 | | |
| | | A-240A | A-230A |
| | | | Resanex |
| | L-150 | | N |
| | L-160 | | MD |

*References p.368*

TABLE 13.7
CELLULOSE ION EXCHANGERS

| Type | Trade name | Producer* | Nature** | Active groups | Ionic form | Particle size (μm)*** | Capacity for small ions (mequiv./g) | Notes |
|---|---|---|---|---|---|---|---|---|
| Strongly acidic cation exchangers | Cellex SE | Bio-Rad | SE-c. | $-O-C_2H_4-SO_3^-$ | $Na^+$ | 20–300 | 0.2 ± 0.06 | |
| | SE-cellulose | Serva | SE-c. | $-O-C_2H_4-SO_3^-$ | | 50–200 | 0.2–0.3 | |
| Intermediate acidic cation exchangers | Cellex P | Bio-Rad | Phosphonic acid c. | $-O-P(O)(O^-)_2$ | $Na^+$ | 20–300 | 0.85 ± 0.1 | |
| | P-cellulose | Serva | c. phosphate | $-O-P(O)(O^-)_2$ | | 50–100 | 0.8–0.9 | |
| | P-1-cellulose | Whatman | c. phosphate | $-O-P(O)(O^-)_2$ | $NH_4^+$ | 1000 | 7.4 | Chromatography of peptides, nucleotides up to mol. wt. 8000 |
| | P-11-cellulose | Whatman | c. phosphate | $-O-P(O)(O^-)_2$ | $NH_4^+$ | 50–250 | 7.4 | |
| | No. 79 cellulose | Sch-Sch | c. phosphate | $-O-P(O)(O^-)_2$ | | 100–300 | 0.9 | |
| Weakly acidic cation exchangers | Cellex CM | Bio-Rad | CM-c. | $-O-CH_2-COO^-$ | $Na^+$ | 20–300 | 0.7 ± 0.1 | |
| | CM-cellulose | Serva | CM-c. | $-O-CH_2-COO^-$ | | | | |
| | Carboxymethyl-Neo-Cel | Serva | CM-c. | $-O-CH_2-COO^-$ | | | | |
| | CM-1 cellulose | Whatman | CM-c. | $-O-CH_2-COO^-$ | $Na^+/H^+$ | 1000 | 0.6 | |
| | CM-11 cellulose | Whatman | CM-c. | $-O-CH_2-COO^-$ | $Na^+/H^+$ | 50–250 | 0.6 | |
| | CM-22 cellulose | Whatman | CM-c. | $-O-CH_2-COO^-$ | $Na^+$ | | 0.6 | Lysozyme 600 mg/g, γ-globulin 150 mg/g (CM-22 and CM-23) |
| | CM-23 cellulose | Whatman | CM-c. | $-O-CH_2-COO^-$ | $Na^+$ | | 0.6 | |

| | | | | | | | | |
|---|---|---|---|---|---|---|---|---|
| | CM-32 cellulose | Whatman | CM-c. | $-O-CH_2-COO^-$ | $Na^+$ | | 1.0 | Lysozyme 1260 mg/g, γ-globulin 400 mg/g |
| | CM-52 cellulose | Whatman | CM-c. | $-O-CH_2-COO^-$ | $Na^+$ | | 1.0 | |
| | No. 76 cellulose | Sch-Sch | CM-c. | $-O-CH_2-COO^-$ | | 100–300 | 0.7 | |
| | No. 77 cellulose | Sch-Sch | CM-c. | $-O-CH_2-COO^-$ | | 80–240 | 0.7 | |
| | No. 78 cellulose | Sch-Sch | CM-c. | $-O-CH_2-COO^-$ | | 60–180 | 0.7 | |
| Strongly basic anion exchangers | Cellex GE | Bio-Rad | Guanidoethyl-c. | $-O-C_2H_4-NH-C(=\overset{+}{N}H)-NH_2$ | $Cl^-$ | 20–300 | 0.4 ± 0.1 | |
| | GE-cellulose | Serva | Guanidoethyl-c. | $-O-C_2H_4-NH-C(=\overset{+}{N}H)-NH_2$ | | | 0.2–0.3 | |
| | Cellex T | Bio-Rad | TEAE-c. | $-O-C_2H_4-\overset{+}{N}(C_2H_5)_3$ | $Cl^-$ | 20–300 | 0.5 ± 0.1 | |
| | TEAE-cellulose | Serva | TEAE-c. | $-O-C_2H_4-\overset{+}{N}(C_2H_5)_3$ | | 50–200 | 0.55–0.75 | |
| Intermediate basic anion exchangers | Cellex D (low capacity) | Bio-Rad | DEAE-c. | $-O-C_2H_4-\overset{+}{N}H(C_2H_5)_2$ | $Cl^-$ | 20–300 | 0.4 ± 0.1 | |
| | Cellex D (standard capacity) | Bio-Rad | DEAE-c. | $-O-C_2H_4-\overset{+}{N}H(C_2H_5)_2$ | $Cl^-$ | 20–300 | 0.7 ± 0.1 | |
| | Cellex D (high capacity) | Bio-Rad | DEAE-c. | $-O-C_2H_4-\overset{+}{N}H(C_2H_5)_2$ | $Cl^-$ | 20–300 | 0.9 ± 0.1 | |
| | Cellex PEI | Bio-Rad | Polyethylene-imine-c. | $-(C_2H_4NH)_n-C_2H_4-NH_2$ | | 20–300 | 0.2 ± 0.1 | |
| | Dimethylaminoethyl-cellulose SN | Serva | DEAE-c. | $-O-C_2H_4-\overset{+}{N}H(C_2H_5)_2$ | | 50–100 | 0.40–0.55 | |
| | Dimethylaminoethyl-cellulose SS | Serva | DEAE-c. | $-O-C_2H_4-\overset{+}{N}H(C_2H_5)_2$ | | 50–100 | 0.55–0.75 | |
| | Dimethylaminoethyl-cellulose SH | Serva | DEAE-c. | $-O-C_2H_4-\overset{+}{N}H(C_2H_5)_2$ | | 50–200 | 0.75–0.90 | |
| | Dimethylaminoethyl-cellulose GS | Serva | DEAE-c. | $-O-C_2H_4-\overset{+}{N}H(C_2H_5)_2$ | | 100–1000 | 0.30–0.55 | |
| | Dimethylaminoethyl-Neo-Cel | Serva | DEAE-c. | $-O-C_2H_4-\overset{+}{N}H(C_2H_5)_2$ | | 100–200 | 0.9–1.0 | |
| | Cellex E | Bio-Rad | ECTEOLA-c. | Mixed amines | $Cl^-$ | 20–300 | 0.3 ± 0.05 | |
| | DE-1 cellulose | Whatman | DEAE-c. | $-O-C_2H_4-\overset{+}{N}H(C_2H_5)_2$ | $OH^-$ | 1000 | 1.0 | |
| | DE-11 cellulose | Whatman | DEAE-c. | $-O-C_2H_4-\overset{+}{N}H(C_2H_5)_2$ | $OH^-$ | 50–250 | 1.0 | |
| | DE-22 cellulose | Whatman | DEAE-c. | $-O-C_2H_4-\overset{+}{N}H(C_2H_5)_2$ | $OH^-$ | | 1.0 | Insulin 750 mg/g |
| | DE-23 cellulose | Whatman | DEAE-c. | $-O-C_2H_4-\overset{+}{N}H(C_2H_5)_2$ | $OH^-$ | | 1.0 | |

(*Continued on p. 350*)

TABLE 13.7 (*continued*)

| Type | Trade name | Producer* | Nature** | Active groups | Ionic form | Particle size ($\mu$m)*** | Capacity for small ions (mequiv./g) | Notes |
|---|---|---|---|---|---|---|---|---|
| Intermediate basic anion exchangers | DE-32 cellulose | Whatman | DEAE-c. | $-O-C_2H_4-\overset{+}{N}H(C_2H_5)_2$ | $OH^-$ | | 1.0 | |
| | DE-52 cellulose | Whatman | DEAE-c. | $-O-C_2H_4-\overset{+}{N}H(C_2H_5)_2$ | $OH^-$ | | 1.0 | Insulin 850 mg/g |
| | ECTEOLA-cellulose | Serva | ECTEOLA-c. | Mixed amines | | 50–200 | 0.3–0.4 | |
| | ET-11 cellulose | Whatman | ECTEOLA-c. | Mixed amines | $OH^-$ | 50–250 | 0.5 | |
| | PEI cellulose | Serva | Polyethylene-imine-c. | $-(C_2H_4NH)_n-C_2H_4-NH_2$ | | | | |
| | No. 70 cellulose | Sch-Sch | DEAE-c. | $-O-C_2H_4-\overset{+}{N}(C_2H_5)_2$ | | 100–300 | 0.9 | |
| | No. 71 cellulose | Sch-Sch | DEAE-c. | $-O-C_2H_4-\overset{+}{N}(C_2H_5)_2$ | | 80–240 | 0.9 | |
| | No. 72 cellulose | Sch-Sch | DEAE-c. | $-O-C_2H_4-\overset{+}{N}(C_2H_5)_2$ | | 60–180 | 0.9 | |
| | No. 73 cellulose | Sch-Sch | ECTEOLA-c. | Mixed amines | | 100–300 | 0.3 | |
| | No. 74 cellulose | Sch-Sch | ECTEOLA-c. | Mixed amines | | 80–240 | 0.3 | |
| | No. 75 cellulose | Sch-Sch | ECTEOLA-c. | Mixed amines | | 60–180 | 0.3 | |
| Weakly basic anion exchangers | AE-cellulose | Serva | AE-c. | $-O-C_2H_4-NH_2$ | | | 0.3–0.4 | *cf.*, Semenza |
| | Cellex AE | Bio-Rad | AE-c. | $-O-C_2H_4-NH_2$ | | 20–300 | 0.4 ± 0.1 | |
| | Cellex PAB | Bio-Rad | *p*-Amino-benzoyl-c. | $-O-CH_2-C_6H_4-NH_2$-*p* | | 20–300 | 0.2 ± 0.1 | Fixation of proteins and nucleic acids |
| | PAB-cellulose | Serva | *p*-Amino-benzoyl-c. | $-O-CH_2-C_6H_4-NH_2$-*p* | | | 0.15–0.20 | Fixation of proteins and nucleic acids |
| Special ion exchangers | BD-cellulose | Serva | Benzoyl-DEAE-c. | | | 50–300 | | Chromatography of nucleic acids, *cf.*, Gillam *et al.* |
| | BND-cellulose | Serva | Benzoyl-naphthoyl-DEAE-c. | | | | | Chromatography of nucleic acids, *cf.*, Gillam *et al.* |

*Producers: Bio-Rad = Bio-Rad Labs., Richmond, Calif., U.S.A.; Sch-Sch = Schleicher & Schüll, Zürich, Switzerland; Serva = Serva Feinbiochemica, Heidelberg, G.F.R.; Whatman = Whatman Biochemicals, W. & R. Balston, Maidstone, Great Britain.

**c. = cellulose.

***In this column, the length of the particle is given; the average diameter of the rods is 18 $\mu$m.

TABLE 13.8
POLYDEXTRAN ION EXCHANGERS

Producer: Pharmacia Fine Chemicals, Uppsala, Sweden. Sephadex is the trade name. All derivatives are supplied with particle size 40–120 μm.

| Trade name | Nature | Processed from Sephadex** | Type of exchanger | Functional groups | Ionic form (counter ions) | Capacity for small ions (mequiv./g) | Haemoglobin capacity (g/g) |
|---|---|---|---|---|---|---|---|
| CM-Sephadex C-25 | Carboxymethyl | G-25 | Weakly acidic cation exchangers | $-CH_2-COO^-$ | $Na^+$ | 4.5 ± 0.5 | 0.4 |
| CM-Sephadex C-50 | Carboxymethyl | G-50 | | | | | 9 |
| DEAE-Sephadex A-25 | Diethylaminoethyl | G-25 | Weakly basic anion exchangers | $-(CH)_2-\overset{+}{N}H(C_2H_5)_2$ | $Cl^-$ | 3.5 ± 0.5 | 0.5 |
| DEAE-Sephadex A-50 | Diethylaminoethyl | G-50 | | | | | 5 |
| QAE-Sephadex A-25 | Quaternary aminoethyl | G-25 | Strongly basic anion exchangers | $-(CH_2)_2-\overset{+}{N}(C_2H_5)_2-CH_2CH_2(OH)CH_3$ | $Cl^-$ | 3.0 ± 0.4 | 0.3 |
| QAE-Sephadex A-50 | Quaternary aminoethyl | G-50 | | | | | 6 |
| SE-Sephadex C-25* | Sulphoethyl | G-25 | Strongly acidic cation exchangers | $-(CH_2)_2-SO_3^-$ | $Na^+$ | 2.3 ± 0.3 | 0.2 |
| SE-Sephadex C-50* | Sulphoethyl | G-50 | | | | | 3 |
| SP-Sephadex C-25 | Sulphopropyl | G-25 | Strongly acidic cation exchangers | $-(CH_2)_3-SO_3^-$ | $Na^+$ | 2.3 ± 0.3 | 0.2 |
| SP-Sephadex C-50 | Sulphopropyl | G-50 | | | | | 7 |

*From 1970, SP-Sephadex replaced the earlier SE-Sephadex, which had similar properties.
***Cf.*, Table 9.5.

*References p.368*

TABLE 13.9
ION-EXCHANGE CRYSTALS

These ion-exchange crystals are produced by Bio-Rad Labs., Richmond, Calif., U.S.A.

| Type | Trade name | Composition | Particle size (mesh) | Capacity | | | Chemical stability |
|---|---|---|---|---|---|---|---|
| | | | | mequiv./g | mequiv./ml | Notes | |
| Cation exchangers | Bio-Rad ZP-1 | Zirconium phosphate | 20–50<br>50–100<br>100–200 | 1.5 | 1.5 | $Cs^+$ uptake at pH 4 (the capacity varies with pH from 1 to 4.5) | Strong acid to pH 13 |
| | Bio-Rad ZT-1* | Zirconium tungstate | 50–100<br>100–200 | 0.6 | 0.4 | $Cs^+$ uptake at pH 4 | pH 1–6 |
| | Bio-Rad ZT-2* | Zirconium molybdate | 50–100<br>100–200 | 0.6 | 0.5 | $Cs^+$ uptake at pH 4 | pH 1–5 |
| | Bio-Rad AMP-1 | Ammonium molybdo-phosphate | Micro-crystalline | 1.2 | 0.4** | See footnote** | Strong acid to pH 6 |
| Anion and cation exchanger | Bio-Rad HZO-1 | Hydrated zirconium oxide | 20–50<br>50–100<br>100–200 | 1.5 | 1.4 | Anion-exchange capacity: $Cr_2O_7^{2-}$ uptake at pH 1 | pH 1 to 5 *N* base |

*Both crystals are not listed in catalogues issued in 1971 and 1972, but they were listed in older leaflets.
**$Cs^+$ uptake at pH 4 of AMP–asbestos (1:1, w/w).

TABLE 13.10
TYPICAL GRAIN SIZES OF ION EXCHANGERS USED FOR VARIOUS EXPERIMENTS

| Purpose | Grain size | |
|---|---|---|
| | μm | Mesh |
| Pilot plant experiments | 850–2000 | 10–20 |
| Non-chromatographic laboratory preparation experiments | 300–850 | 20–50 |
| Ion sieve process | >500 | <30 |
| Laboratory chromatographic separation in inorganic chemistry | 140–300 | 50–100 |
| High-sensitivity separations | 75–140 | 100–200 |
| Separations of biochemical materials | 40–80 | 200–400 |
| Ion-exchange cellulose rods | (18–20) × (20–300) | |
| Examples of special resins for high-speed analysis | 20–30<br>7–11 | |
| Lowest limit | 2–3 | |

## METHODS FOR THE FRACTIONATION OF ION EXCHANGERS

Homogeneity of the grain size is best ensured by choosing resins that have been sized carefully by the producer. Refined and carefully sized ion exchangers for chromatographic purposes are obtainable from J. T. Baker (Phillipsburgh, N.J., U.S.A.), Bio-Rad Labs. (Richmond, Calif., U.S.A.), Durrum (Palo Alto, Calif., U.S.A.) and others. If laboratory sizing is necessary, either sieving or a hydraulic method is used.

Sieving can be carried out in either the dry or the wet state. For dry sieving, the resin should be air-dried and then separated by means of a set of standard sieves. Dry sieving is simple, but wet sieving is to be preferred because of the better uniformity of the grains obtained. *Caution:* cation exchangers are never sieved in the $H^+$ form owing to extensive corrosion of the sieves and contamination of the resin by heavy metals. The swollen resin is placed on the coarsest sieve in a large funnel and rinsed with a stream of circulating distilled water into a large cylinder.

Hamilton's hydraulic method is suitable for the separation of very fine particles (<40 μm) and requires special apparatus based on a large separating funnel (33 cm long, 16 cm maximal diameter). A continuous flow of bubble-free distilled water (measured precisely by a flow meter) rises from the bottom into a funnel in which a 400-ml portion (or less) of resin is suspended. Equilibrium between the sedimentation of particles and the counter-flow of water depends on the diameter of the particles, which enables an effective fractionation to be achieved by stepwise changes in the flow-rate. An example for Amberlite IR-120 ($Na^+$) fractionated at 25°C is given in Table 13.11. If the fractions obtained are refractionated, an almost homogeneous grain size can be achieved. Other modifications of the water elutriation method have been published by Scott (1968) and Simonson (*cf.*, also Samuelson, Zmrhal).

Improved methods have been developed that involve the use of long tubes with a laminar flow instead of separation funnels. The resin is fed into a special chamber at the

TABLE 13.11
FRACTIONATION OF AMBERLITE IR-120 ($Na^+$) BY HAMILTON'S HYDRAULIC METHOD AT 25°C

| Flow-rate (ml/min) | Particles flowing off (diameter, μm) |
|---|---|
| 600 | Filling |
| 50 | $\lessdot$25 |
| 110 | 25–30 |
| 280 | 40 ± 7 |
| 580 | 56 ± 9 |
| Remaining in separating funnel | >70 |

bottom and the whole length of the column is used for the separation process. In some instances, hydraulic methods have been replaced by pneumatic methods for the fractionation of grains. It is also possible to prepare homogeneous beads when polymerizing drops of the starting material falling from capillary tubes, and the expensive ion exchangers obtained do not require subsequent sizing.

## DECANTATION AND CYCLING OF ION EXCHANGERS

Ion exchangers must be swollen before use, and decantation can be carried out at the same time. The almost colloidal particles are removed by suspending the resin in ten times its volume of water in a measuring cylinder and then allowing it to stand for about half an hour. The cloudy suspension above the resin is decanted off, a further portion of water is added and the process is repeated until the supernatant remains clear. At the same time, the resin takes up water and becomes swollen to capacity. A similar procedure is followed when decanting cellulose ion exchangers. The settling time depends upon the degree of removal of fines required. With Sephadex ion exchangers, which normally do not require decantation, buffers used in the experiments can be employed instead of water for swelling. Complete swelling of Sephadex derivatives is achieved after 1–2 days at room temperature, but it can be accelerated by warming on a boiling water-bath (at neutral pH in order to prevent hydrolysis). With commercially pre-cycled ion-exchange cellulose, buffers can be used for decantation.

The cycling of resins is important not only for purification but also for suitable stereochemical rearrangement of the matrix. Extraction with alcohol or acetone is also recommended. When using AG ion exchangers refined by the producer, this treatment can be omitted. After a reaction time of 10 min, the solution is filtered off and a new portion is added. The treatment should be repeated until the effluent is colourless. These operations are best carried out in a sintered-glass funnel filled with the resin to half its height or in a wide, low chromatographic column. When cycling an anion exchanger, contact with carbon dioxide from the laboratory atmosphere should be prevented. Cation-exchange resins are usually washed successively with several portions of 1 *N* sodium hydroxide solution, water, several portions of 2–6 *N* hydrochloric acid and water and the resin remains in the $H^+$ form. Anion-exchange resins are cycled with 1–2 *N* hydrochloric acid, water, 0.5–1 *N* sodium hydroxide and water. For a short time, they can be kept in the $OH^-$ form, while for prolonged storage they are converted into the $Cl^-$ form with 1 *N* hydrochloric acid and washed with water. Every washing with water should be prolonged until the resin is free from the electrolyte added. The conversion of weakly acidic cation exchangers and weakly basic anion exchangers from the $H^+$ or $OH^-$ form into a salt form requires a greater time interval, depending on the grain size (sometimes overnight).

The cycling of ion-exchange cellulose is recommended with advanced types of preparations where the separation efficiency is increased (Thompson). CM-cellulose is stirred in a beaker with 15 volumes of 0.5 *N* sodium hydroxide solution, and after 30 min the supernatant is poured off and the cellulose is washed on a büchner funnel (acid-hardened paper) until a pH of 8 in the effluent is reached. The exchanger is suspended twice in 15 volumes of 0.5 *N* hydrochloric acid and treated for 30 min in each instance. After the second decantation, the exchanger is transferred into a büchner funnel and washed thoroughly until the effluent is about neutral. The CM-cellulose is then in the $H^+$ form. DEAE-cellulose is treated in a similar manner, the first solution being 0.5 *N* hydrochloric acid, the intermediate pH is 4, and the second solution is 0.5 *N* sodium hydroxide. The DEAE-cellulose is in the $OH^-$ form after this treatment.

There are commercially available preparations of ion-exchange cellulose that have been pre-swollen and pre-cycled in the factory, and this procedure can be omitted in the laboratory. Polydextran ion-exchange derivatives do not require pre-cycling.

## BUFFERING OF ION EXCHANGERS

Ion exchangers for elution chromatography must be buffered before use. They must first be converted into the ionic form identical with the type of ions present in the buffer, which can be achieved for the $H^+$ and $OH^-$ forms and also for the salt forms of all types of exchangers by suspending them in an excess of a 1–2 *M* solution of the salt of the new counter-ion. (If the new counter-ion forms a weaker base or weaker acid than the originally present counter-ion, *e.g.*, ammonium *versus* sodium or acetate *versus* chloride, a more concentrated solution should be used. However, conversion via the $H^+$ or $OH^-$ form is the best method in such a case.) After equilibrium has been reached, the supernatant is decanted off and a new portion of the same salt solution is poured on the ion exchanger with occasional stirring.

After 15 min, this solution is poured off and the exchanger is suspended in the first buffer that is to be used in the chromatographic process. After several decantations, the pH is checked and corrected by titration using acids or bases that constitute the buffer used. Decantations or repeated washing on a büchner funnel or a sintered-glass funnel follow until equilibrium is reached, *i.e.*, the pHs and electrical conductivities of the effluent and of the buffer are identical.

When the first buffer used for chromatography is very dilute, the use of a cascade of several progressively less concentrated buffers of the same pH and composition will help in achieving equilibrium more rapidly than prolonged washing with the first buffer.

*References p.368*

## DEAERATION OF ION EXCHANGERS AND FILLING OF CHROMATOGRAPHIC COLUMNS

Bubbles in the slurry of the ion exchangers disturb the packing of the column and affect the chromatography. The air adhering to the surface of the grains should be removed if a perfect filling of the column is desired. A round-bottomed flask is half-filled with the resin and buffer (1:1, v/v). The flask is then placed in a water-bath (at a temperature about 5–10°C higher than that of the column operation) and evacuated with a water pump. The bubbles begin to escape and, after about 5 min, no more bubbles escape and the slurry begins to "bump". The evacuation is then stopped and the slurry is transferred into the column. (For the rapid preparation of a conventional ion-exchange bed, deaeration can be omitted.)

The same procedure is used for the removal of carbon dioxide bound to anion exchangers. In this instance, the slurry is prepared by using a dilute solution of an acid corresponding to the anionic component of the buffer. After the escape of carbon dioxide, the ion exchanger is titrated with a solution of a base to the required pH and washed with portions of the eluting buffer. The buffers should be prepared from carbon dioxide-free water (water boiled and cooled before use) and kept free from carbon dioxide.

For many ion-exchange chromatographic applications, conventional simple glass columns are sufficient. Fig. 13.1 gives several illustrations. The ion exchanger can be supported on a cotton-wool or glass-wool plug, covered with a layer of glass beads, on a porous PTFE plate or on a porous sintered-glass plate connected with the column. The top can be closed simply with a rubber stopper, but ball-joints are much better. The surface of the exchanger is protected from disturbance by the inflowing solutions with a plastic or paper disc. For work at elevated temperatures, columns with a jacket allowing the circulation of heated water are used. The columns must not be filled to the top, as the exchanger should be provided with space for the volume changes following the changes in the composition of eluting solutions. In the bottom of the column, a minimum dead space is desirable.

Many workers therefore prefer tubes made of plastics or precision-bore glass with special screw units at both ends containing a porous PTFE plate and capillary tubes for the influent and effluent. Precision chromatographic columns are now commercially available with various fittings, connectors, rotary valves, etc., suitable for connecting the column with tubes and other auxiliary chromatographic devices. Adjustable column ends are available that permit various bed volumes to be selected using the same column. Modern columns, connectors and PTFE tubes are able to operate at elevated pressures and are made of solvent-resistant materials. Many columns are jacketed and thus enable work to be carried out at constant temperatures. For high-speed analysis, the ion-exchange materials are packed into precision-bore stainless-steel columns using metal tubing connected with compression fittings, etc. Many firms will send leaflets on request with a description of various types of modern chromatographic columns and accessories, *e.g.*, Bio-Rad Labs. (Richmond, Calif., U.S.A.), Chromatronix (Berkeley, Calif., U.S.A.), Jobling Laboratory Division (Stone, Great Britain), Metaloglass (Boston, Mass., U.S.A.), Pharmacia (Uppsala, Sweden), Serva (Heidelberg, G.F.R.), Whatman Biochemicals (W. &

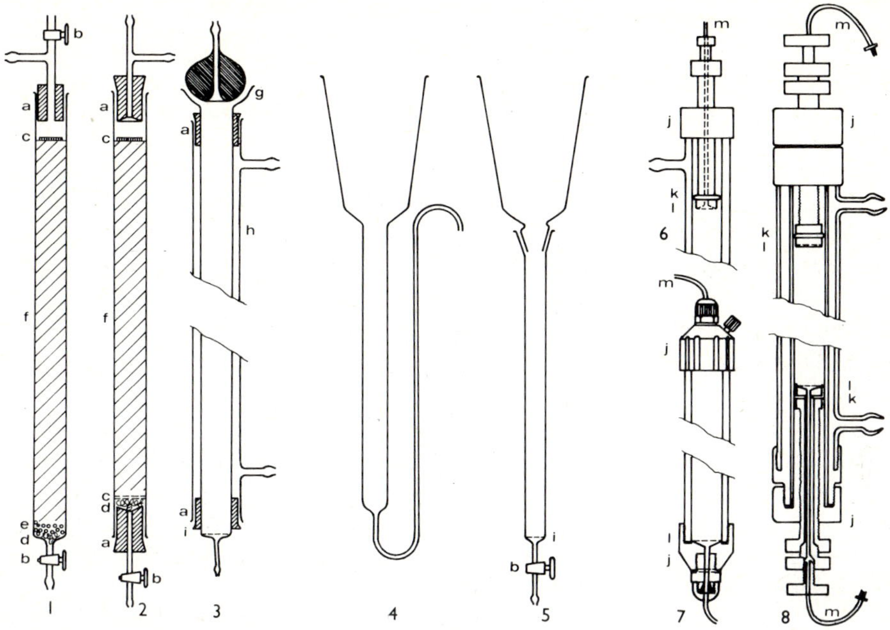


Fig. 13.1. Columns for ion-exchange chromatography. 1–3, Simple glass columns that can easily be made in the laboratory; 4, wide-mouthed column often used for ion exchange with a bent tube to prevent it from running dry; 5, column with tapered joint funnel extension allowing back-washing; 6–8, types of modern commercially available columns for chromatography. a, Rubber stopper; b, stop-cock; c, plastic porous disc (in columns 1 and 2, paper discs can be used); d, glass-wool or cotton-wool plug; e, glass beads; f, ion-exchanger bed; g, ball-joint; h, constant-temperature jacket; i, sintered-glass plate; j, screw end unit; k, adjustable end with tightening rings; 1, porous PTFE plate; m, capillary tubes.

R. Balston, Maidstone, Great Britain), and others.

The relative dimensions of the column (diameter to height) are very important for chromatography. The demands for ion-exchange chromatography of low-molecular-weight substances carried out mostly on ion-exchange resins differ substantially from those for the separation of biopolymers on cellulose ion exchangers or on ion-exchange polydextran. Typical data for chromatography on ion-exchange resins are given in Table 13.12. For most analytical work, a 20 × 1 cm column is convenient. There is normally no great advantage in using very long columns.

TABLE 13.12
DIMENSIONS OF COLUMNS FOR DIFFERENT PURPOSES

| Operation | Diameter:height |
|---|---|
| Regeneration, sorption, deionization | 1:5 to 1:10 |
| Ion exchange and other non-chromatographic operations | 1:10 to 1:20 |
| Chromatographic operations | 1:20 to 1:200 |

*References p.368*

With separations of proteins, true chromatography seldom occurs. The desorption of sorbed proteins is mostly an "all or nothing" operation. At certain values of pH and ionic strength, the macromolecules are released completely from the exchanger and repeated ion-exchange equilibria in the lower part of the column, typical of conventional ion-exchange chromatography, are not realized. Therefore, long columns have little positive effect and, when using gradient or stepwise elution, they only give rise to complications. For ion-exchange polydextran, the standard optimal height of the exchanger is 20 cm. Only exceptionally are higher columns used, when very complex mixtures are to be separated. The diameter of the column depends upon the amount of exchanger required for the separation.

For ion-exchange cellulose, the usual height of the column is 5–15 cm. Given the recommended column packing, the diameter may increase according to the required capacity of the column.

The filling of the warm, deaerated slurry of ion-exchange resins into the column is illustrated by Fig. 13.2. The column must be mounted vertically and is then filled about three quarters full with the solution (buffer) used for the equilibration of the resin; this solution should be of about the same temperature as the slurry. Ion exchangers in the $H^+$ or $OH^-$ form are packed in columns containing distilled (deionized) water. Care must be taken to prevent the entry of air bubbles. The thick slurry is added in several portions down a glass rod and the inner wall of the column without waiting for the preceding portion to settle completely, otherwise separate layers (zones) are formed. The liquid is

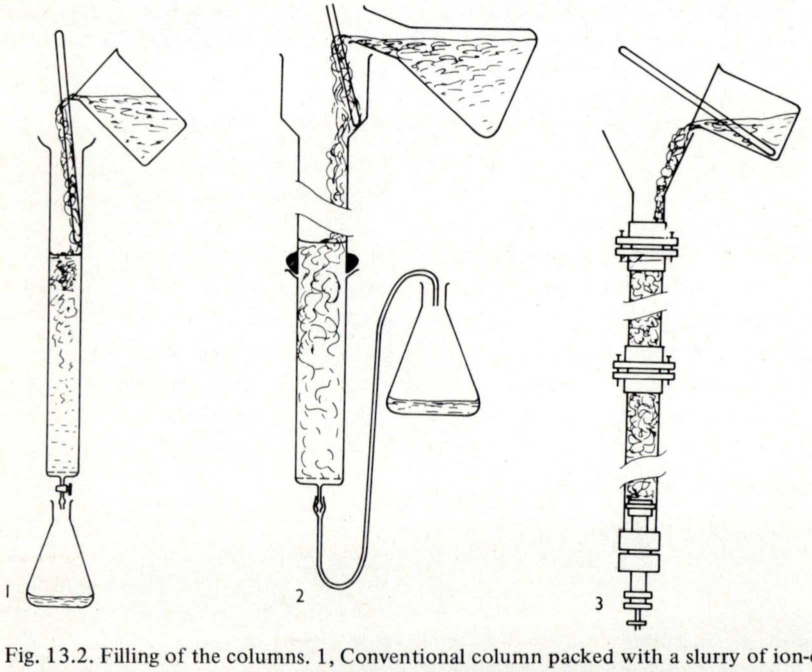


Fig. 13.2. Filling of the columns. 1, Conventional column packed with a slurry of ion-exchange resin; 2, column with extension funnel permitting simple filling with ion-exchange polydextran; 3, filling of the column with ion-exchange cellulose using an extension tube and funnel joined by connectors.

withdrawn at the bottom during filling. When the column has been packed to the desired height, it is washed with the same solution (buffer) until the bed volume is constant. Possible slow changes in the desired height can be corrected by adding further small portions of the resin. The absorbance, pH and conductivity of the effluent must be checked, and the column is ready for chromatography if these values are identical with those of the influent.

For packing very small particles into columns for high-pressure liquid chromatography, a different procedure must be used, as the conventional packing technique is not possible because the settling time of very small particles is very long. Pellicular ion-exchange resin beads can be filled into columns by the dry packing technique (Kirkland, 1969; for more details, see Chapter 9) and all types of small particles can be packed by dynamic packing (Scott, 1971; Scott and Lee). The first method is possible because there is no swelling with the pellicular type of exchangers. After the column has been filled completely with dry material, the air in the bed is displaced by the pumped liquid influent. Very long columns (*e.g.*, 3 m) can be made by connecting several straight columns in series with U-shaped connectors. A significant loss in chromatographic efficiency was observed when long, straight columns were bent after being packed (Kirkland, 1969). Similar effects have been reported with the coiling of columns for other types of liquid chromatography.

The second, dynamic, method uses a slurry and the very small particles are forced into the bed at a velocity that is much greater than their settling velocity. Two principal procedures are possible: (a) the thick slurry is displaced from a special chamber (mounted above the bed) by the pumped buffer, and (b) the ion exchanger prepared in the form of a fixed bed in a special stainless-steel cartridge with a diameter larger than that of the column is thrust into the column by the pressure of the influent. This method can be used both for straight and coiled stainless-steel columns.

The packing of columns with polydextran ion exchangers is simple. The swollen ion exchangers are mixed carefully (in order to prevent the formation of air bubbles) with starting buffer until it can be easily poured, but is still not too liquid. The column is filled with buffer to a height of about 1 cm above the bed support and the outlet is closed. The slurry is then poured into the column down a glass rod in such a way that no air bubbles are trapped in it and transferred into the column. The use of a funnel extension enables all of the suspension to be poured at once. After settling for 10 min, the space above the exchanger is filled with the buffer and the column is connected to the reservoir. The bed is then washed with the buffer using an operating pressure of about 1 cm of water per centimetre height of the bed. When the packing is completed, the surface of the bed is protected by the addition of a 0.5 cm layer of Sephadex G-25 pre-swollen in the same buffer. The comparison of the effluent with the influent is carried out in the manner mentioned above.

The recommended packing of ion-exchange cellulose columns is a little more complicated. It is very important to prevent the separation of particular types of particles (differing in length) from the slurry during the packing (Thompson). Therefore, the filling procedure must be rapid, which is achieved best by using a rapid flow of buffer through the column in order to increase the settling rate. The extension tube is joined with the column by using a suitable connector. The buffer is filled into the column to a height of

*References p.368*

about 1 cm above the bed support. The settled volume of decanted ion exchanger in an auxiliary cylinder is enlarged by the addition of half the volume of buffer and the optimal slurry is formed by means of gentle stirring. The slurry is then poured carefully through a wide-necked funnel into the extension tube, and the end unit is immediately inserted at the top and the buffer is pumped through the extension tube and the column at a flow-rate of 45 ml/h $\cdot$ cm$^2$. After 1 h, the bed will have reached a constant volume and the extension tube is then removed and the column is ready for chromatography.

Whenever filling columns, care must be taken not to remove the excess of solution above the ion-exchanger bed, so that the column cannot run dry. Only immediately before the application of the sample may this excess be removed.

## APPLICATION OF SAMPLES

We must distinguish between ion exchange in columns (which will not be dealt with in this book) and ion-exchange chromatography. In spite of the fact that even within the ion-exchange operation (*e.g.*, demineralization or, in general, the exchange of one type of ion for another) a certain type of chromatographic separation occurs, the main aim is as complete an ion exchange as possible. The column capacity is fully exhausted. However, within the chromatographic process, the main aim of which is the separation of ions, the capacity may be only partly exhausted. Therefore, the total capacity of the column must be calculated and the exhaustible part should be estimated, and the data in Tables 13.1–13.9 (mequiv./ml or mequiv./g) are important for this purpose. We must also calculate the equivalence of sample in milliequivalents (mequiv.). The usual relationships between sample and column capacity for ion-exchange resins are given in Table 13.13.

The relationships of the capacities for the chromatography of biopolymers on polydextran and cellulose ion exchangers depend on the available capacity for the sample substances. For precise procedures, this value should be estimated in a preliminary experiment by using the micro-batch process (a *ca.* 1% suspension of the exchanger in a solution of an excess of sorbed substance in several buffers is shaken for some time; the difference between the concentration of the sample in the supernatant or filtrate and that of the standard solution determines the available capacity under the conditions examined; *cf.* also Pharmacia). For a quick and approximate estimation, the protein capacity (Tables 13.7 and 13.8) should be considered. The capacity estimated by the static micro-batch experiments or found from the tables usually indicates a higher capacity than that which

TABLE 13.13
LOADING OF SAMPLE ON ION-EXCHANGE CHROMATOGRAPHIC COLUMNS

| Type of chromatography | Maximum load (% of the column capacity) |
|---|---|
| Frontal analysis | > 100 |
| Displacement | 25 (occasionally 50) |
| Elution in preparative work | 5–10 |
| Elution in analytical work | 1 |

can be attained in the dynamic chromatographic process. The separation of macromolecules will hardly allow equilibrium to be reached and therefore a much larger amount of exchanger should be used, in spite of the fact that the separation process is much more a selective sorption and desorption than "true chromatography".

Not more than 10–20% of the column capacity should be exhausted. Various workers used sample to exchanger ratios of 1:10 to 1:2 (w/w). In the rare "true chromatography" (*i.e.*, separation during elution by one starting buffer only), the lower part of the column capacity should be exhausted and longer columns are to be used (*e.g.*, with a ratio of diameter to length of 1:60).

The chromatography usually consists of sorption (application of the sample) and desorption (elution), and the sorption process will be discussed here. The effective sorption of the substances to be separated to the top of the column of ion-exchange resins is best realized from a dilute solution with a non-extreme pH value. If the sorption of cations or anions is carried out on the column of ion exchanger in the $H^+$ or $OH^-$ form, the salts introduced are converted into solutions of acids or bases. If the concentration of the original salt solution is too high, there is also a high concentration of acid or base in the effluent and this solution begins to regenerate the ion exchanger to the original $H^+$ or $OH^-$ form. It is clear that this process decreases the sorption and makes the zones of the sorbed substances very diffuse. Therefore, the concentration of salts designated for the sorption at the top of the resin column in the $H^+$ or $OH^-$ form should not exceed the equivalent of 0.1–0.2 *N*. If it is lower, sharper zones of sorbed ions at the top of the column are obtained.

For elution chromatography, if the sample application is carried out under conditions at which the sorption is not effective (*i.e.*, if there is only a small or no difference between the solution in which the sample was dissolved and the first eluting buffer), more concentrated solutions (up to 10%) are used in order to eliminate wide zones. For the application of low-molecular-weight zwitter-ions (amino acids and peptides) on buffered resin columns, acidification of the sample is used for cation-exchange chromatography (the sample solution should be 0.2 pH unit lower than the pH of the buffer used for equilibration of the resin). For anion-exchange chromatography, the sample is made 0.2 pH unit more alkaline.

The sorption of biopolymers on ion-exchange cellulose or ion-exchange polydextran is carried out from solutions with the lowest possible ionic strength, where the sorption process is most effective (*cf.*, Chapter 10). Therefore, the salt-free sample is usually dissolved in the starting buffer, the pH value is corrected by the acid or base component of the buffer and the solution is dialyzed against the starting buffer. Gel permeation chromatography also can be used instead of dialysis, and the sample solution is then applied on the column.

There are several techniques for leading the sample on to the column. The most simple procedure (overlayering), which is often used, is as follows. The level of liquid is lowered just to the surface of the resin without air being allowed to enter the bed. The sample solution is slowly added from a bent pipette down the wall just above the surface of the resin (Fig. 13.3). A disc of thin filter-paper will prevent the disturbance of the resin bed surface if a straight pipette is used. From this moment, the effluent from the column should be taken as the first fraction. The solution is allowed to soak into the bed at the

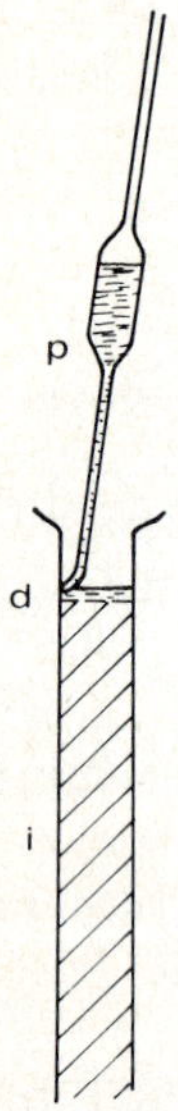


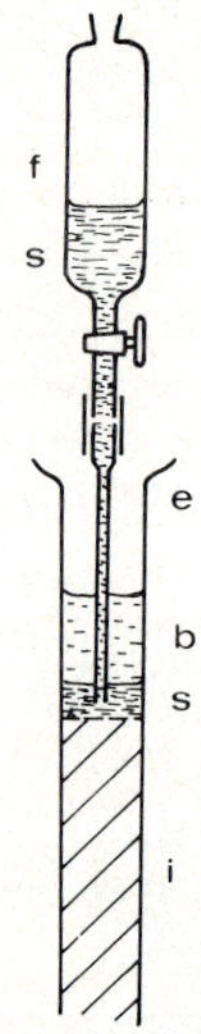


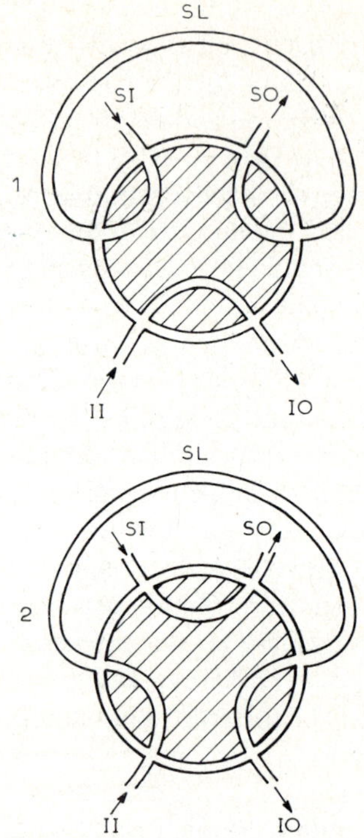


Fig. 13.3. Overlayering method of sample loading. p, Bent pipette; d, disc of porous paper; i, ion exchanger.

Fig. 13.4. Underlayering method of sample loading. f, Separating funnel; e, extension capillary tubing; s, sample solution; b, buffer; i, ion exchanger.

Fig. 13.5. Application of the sample by means of a six-way valve using the sample loop method. 1, Position of the filling of the sample loop (SL); 2, position of injecting the sample into the column; SI, sample in; SO, sample out; II, influent in; IO, influent out (into the column).

same rate that has been calculated for the chromatographic process. After the sample has soaked in, the inner surface of the wall is washed two or three times with small portions of the first eluting buffer, each portion being allowed to soak in separately. With this procedure also, care must be taken to prevent the column from running dry. The column is then filled with the eluting solution, the end unit is joined on the top and the chromatography can be started.

Another technique is the so-called underlayering, which is often used for applications of biopolymers on to columns of polydextran derivatives. The important condition is that the sample solution should have a higher density than that of the eluting buffer. The former can be made denser by the addition of glucose or sucrose. The surface of the ion exchanger should be protected with a porous disc of thin filter-paper. The buffer solution above the exchanger is lowered only partly, a sufficient layer being left above the bed surface. A syringe is filled with the sample solution and this is layered just above the surface of the bed by a capillary tubing. The same procedure can be carried out on a larger scale, as follows. Both a separating funnel and a joined capillary tube are filled with sample solution, and this equipment is fixed above the column, allowing the end of the tubing almost to touch the protecting disc (*cf.*, Fig. 13.4). No air bubbles must be present in the tubing. The solution is slowly layered on the protecting disc until the level in the funnel is lowered to just above the valve. Then the separating funnel is carefully filled with a sufficient portion of elution buffer and layering continues in order to wash the tubing until the first drop of the buffer appears at the end of the tubing. If necessary, the rising level of buffer in the column can be aspirated off. It is possible to start the chromatography when the first thin layer of the sample solution is formed.

The third technique is recommended by the producer of advanced ion-exchange cellulose (Thompson) and consists in pumping the solution with the sample directly to the top of the column at the flow-rate chosen for the elution. Sliding piston-type column end units enable this operation to be carried out very precisely because a porous diffusion disc spreads the solution evenly over the surface of the ion-exchanger.

In the fourth technique, a three-way valve is used, which facilitates the injection of the sample to the feed line. This procedure forms a basis for automation of successive applications of samples and allows operation at high pressures.

The fifth method for the introduction of the sample is effective for high-speed analysis. The sample solution is injected into the influent with a hypodermic syringe through a suitable septum just above the top of the bed. This method is applicable only up to a pressure of 1000 p.s.i.

The sixth method, involving the use of a sample loop, is generally applicable even at high pressures. It requires a special valve (the most usual is a six-port valve), illustrated in Fig. 13.5. For automation of analysis, special systems have been developed, allowing the stepwise application of large series of samples on the same column at suitable time intervals (see Chapter 32; for instrumentation details, see Chapter 8).

## METHODS OF ELUTION

There are three principal methods of elution. (1) Simple elution only with starting buffer is sometimes sufficient to resolve the desired components; it is also called the "starting condition procedure". This method is often used for high-speed chromatography on pellicular ion-exchange resins, because it does not require regeneration. The column is sometimes very long, *e.g.*, 1 m × 2.1 mm, using a flow-rate of 2 ml/min (Kirkland, 1969). Horwath and Lipsky used a column filled with pellicular ion-exchange beads, of a total length of 3 m. (2) Stepwise elution is a procedure by which several buffers are used with subsequent increase in the pH (for cation-exchange chromatography) or decrease in the pH (for anion-exchange chromatography). Usually the ionic strength increases in both instances. Stepwise elution can also be achieved using buffers that have the same pH and differ only in ionic strength. (3) In gradient elution, at least one additional buffer flowing from the reservoir is continuously mixed into the starting buffer in the mixing chamber and the mixed solution is pumped on to the column. This method permits the generation of an elution solution with a continuously changing composition. The change refers to pH or ionic strength (concentration) or to both pH and ionic strength. The devices for gradient elution are described in Chapter 8, the calculation of the gradients is discussed in Chapter 10 and suitable buffers are mentioned in Chapter 10.

Of these methods, linear gradient elution is generally the best. Sometimes (for example, for the separation on ion-exchange resins of complex enzymic digests of proteins that contain many peptides) it is not sufficient to mix only two buffers and a sequence of several buffers is necessary, each following one being mixed in the same constant-volume mixing chamber after the preceding one is exhausted (*cf.*, Table 10.6, p. 265). The ratio of the volumes of two buffers used for the creation of the gradient to the bed volume of the ion exchanger determines the steepness of the gradient. The steeper the gradient, the closer to each other do the peaks emerge in the effluent, which can cause a poorer resolution in some instances. However, a too slow gradient leads to flat peaks. For the separation of biopolymers on polydextran or cellulose ion exchangers, it is recommended that the total volume of buffer for the gradient should be about five times the bed volume.

## CALCULATION OF FLOW-RATES

The flow-rate is one of the decisive factors for the establishment of equilibrium in ion-exchange chromatography. The other factors (grain size, temperature, viscosity of the solution, rate of diffusion into grains, etc.) are also important and therefore the flow-rate must be chosen in correlation with them in order to be able to work as near to the equilibrium conditions as possible. The finer the particles, the higher is the flow-rate that can be used, but this is limited by the flow resistance of the column and a higher pressure may be necessary. The use of high pressures is suitable only with ion-exchange resins, especially in bead form. For example, some types of ion-exchange polydextrans decrease the flow-rate with an increase in pressure drop of more than 2 cm of water per centimetre of the bed. An increase in temperature favours the establishment of equilibrium, lowers the viscosity and higher flow-rates are possible, but too high pressures should not be used. Higher temperatures usually cannot be used in the chromatography of labile biochemical substances.

The flow-rate can be expressed in several ways:

| *Type of flow* | *Units* |
|---|---|
| Linear | $ml/h \cdot cm^2 = cm/h$;<br>$ml/min \cdot cm^2 = cm/min$ |
| Volumetric | $ml/h \cdot cm^3$; $ml/min \cdot cm^3$ |

The definition of linear flow is the volume (millilitres) of the effluent flowing in unit time through 1 $cm^2$ of cross-section (this is equivalent to the difference in height of the level of the solution if it were present above the bed in a column of the same diameter). This value should be identical in all columns for a particular chromatographic experiment, independent of their cross-section. Therefore, if the operational data are to be transferred from one column to another of different dimensions, the same height of the bed should be used and the flow-rate should be proportional to the cross-sectional area of the new column. The capacity of the column and the size of the fractions are changed in the same proportions. A convenient linear flow-rate for many inorganic applications is 2 $ml/min \cdot cm^2$ using a 100–200 mesh resin.

The volumetric flow-rate defines the volume (millilitres) of effluent flowing in unit time through a 1 $cm^3$ volume of the bed. If the relative dimensions of the column are not known, the flow-rate cannot be transferred exactly to columns of different dimensions and therefore the linear flow-rate should be used. In spite of this volumetric data were often used in the literature when the usual ratio of the bed diameter to the bed height (*ca.* 1:20) was assumed. For ion-exchange resins, typical volumetric flow-rates used for the chromatography of various small ions are given in Table 13.14.

TABLE 13.14
TYPICAL VOLUMETRIC FLOW-RATES FOR ION-EXCHANGE RESINS

| Type of use | Diameter of particles (mm) | Volumetric flow-rate ($ml/min \cdot cm^3$) |
|---|---|---|
| Preparative chromatography | 0.5–0.15 | 0.005–0.05 |
| Non-equilibrium exchange | 0.5–0.15 | 0.05–0.1 |
| Outer limit | 0.5–0.15 | 0.5 |
| Inorganic applications | 0.15–0.075 | 0.1–0.5 |
| High-resolution chromatography | 0.075–0.038 | 0.003–0.02 |
| High-speed chromatography (porous layer beads; 1-μm layer) | 0.003–0.020 | 0.1–1.0 |

Modern ion-exchange resins with perfectly fractionated micro-grains enable excellent resolutions to be achieved using relatively high flow-rates. Higher pressures must be used for feeding the bed. Examples of analytical applications are given in Table 13.15 (*cf.*, Gere; Green; Kesler; Ohms *et al.*; Uziel *et al.*).

The recommended flow-rates for ion exchange cellulose columns (Bio-Rad Cellex) are 4–30 $ml/h \cdot cm^2$ but may reach 50 $ml/h \cdot cm^2$ or more. The flow-rate $F$ (ml/h) can also

TABLE 13.15
SEPARATIONS USING ION-EXCHANGE RESIN MICRO-GRAINS

| Separation | Resin | Particle size (μm) | Column dimensions (cm) | Operating pressure (p.s.i.) | Linear flow-rate ($ml/min \cdot cm^2$) |
|---|---|---|---|---|---|
| Amino acids | Aminex A-6 | 17.5 ± 2 | 0.9 × 50 | 260–300 | 3.2 |
| Amino acids | Aminex A-5 | 13 ± 2 | 0.9 × 7 | 30–40 | 3.2 |
| Peptides | Aminex A-5 | 13 ± 2 | 0.9 × 18 | 50 | 0.8 |
| Nucleosides | Aminex A-6 | 17.5 ± 2 | 0.6 × 19 | 6–16 | 1 |
| Carbohydrates | Beckman 1-S | 17 ± 6 | 0.9 × 55 | <300 | 1.8 |
| Nucleosides | Aminex A-7 | 8.5 ± 1.5 | 0.24 × 25 | 1000 | 9.30 |
| Nucleosides | Pellicular type | 5 ± 2 | 0.24 × 25 | 4800 | 32.5 |

be calculated from the equation

$$F = \frac{k \cdot \Delta P \cdot C}{L}$$

where $\Delta P$ is the pressure drop across the column (g/cm$^2$ = p.s.i. × 70.3), $C$ is the cross-section of the column (cm$^2$), $L$ is the length of the bed (cm) and $k$ is a constant characterizing particular types of exchangers excluding separation conditions. According to the data of Thompson, the usual values of $k$ for Whatman DEAE-celluloses are 5 for DE-52 and 20 for DE-23, and for Whatman CM-celluloses 2 for CM-52 and 40 for CM-23.

The flow-rates used for polydextran ion-exchange columns are about 2–70 ml/h · cm$^2$ and depend upon the type of exchanger used and the substances being separated. A-50 or C-50 types of ion-exchange derivatives of Sephadex used for the chromatography of protein macromolecules with molecular weights from 30,000 up to 200,000 do not allow the use of high flow-rates because a higher pressure drop (above 2 cm of water per centimetre) has a reverse effect on the flow-rate. The A-25 and C-25 types do not show this effect. The smaller the ions separated, the higher are the flow-rates that can be used. The high-molecular-weight biopolymers usually require relatively lower flow-rates.

## EVALUATION OF FRACTIONS

Continuous monitoring of fractions is carried out by measuring the absorbance of the effluent at 280 nm (proteins), 220 nm (peptides), 254–260 nm (nucleic acid constituents) or at a wavelength corresponding to maximal absorbance of other substances. If a gradient of ionic strength is used for the elution, determination of the conductivity of every fifth or tenth fraction is used. This is accompanied by the measurement of pH if a pH gradient is used. These data are important for the characterization of points of emergence of particular peaks, which are usually very well reproducible.

When substances labelled with radioactive isotopes are separated, the continuous measurement of radioactivity is used. The separation of enzymes is checked by determination of the activity of aliquots in addition to absorbance data. Other analytical methods are used in special instances; for example, amino acids are detected by ninhydrin colorimetry, peptides by auxiliary paper or thin-layer chromatography or electrophoresis and biologically active substances by bioassays. The most universal method is the measurement of refractive index. A comparative continuous refractometer is very useful in this respect, but not for gradient elution.

The evaluation of fractions when drawn on paper in the form of elution curves is the key to their collection. They are then usually evaporated in a rotary evaporator or freeze-dried before further processing.

## REGENERATION AND STORAGE OF ION EXCHANGERS

After use, all ion exchangers are converted into the form in which they are required for use again. They should never be stored for long periods in columns or in aqueous suspension, because there is a risk of microbial infection. Regeneration purifies the exchangers, and decantation and complete cycling is the best means of regenerating resins. Cation exchangers can be stored in the $H^+$ form or salt form (*e.g.*, $Na^+$), while anion exchangers should be kept only for short periods in the base form, the salt form (*e.g.*, $Cl^-$) being preferred for longer storage. In some repeated applications, shorter processes can be used for regeneration, consisting in washing the column with a concentrated (*e.g.*, 2 *M*) solution of salt containing the desired ion. Then the resin is thoroughly washed with water, the surface water is dried off briefly on a sintered-glass funnel by suction and the slightly wet resin is stored in hermetically closed flasks or in sealed polyamide or other plastic bags. The resin should never be allowed to dry out completely. When it is to be used after a year or a longer period, it is advisable to regenerate it again.

Ion-exchange cellulose can be regenerated in the following manner. The column is checked for the presence of some pigment or other undesired material (*e.g.*, denaturated protein) bound on the top of the bed, and in such a case the top 1–2 cm of the cellulose is discarded. The remaining part of the bed is then extruded, washed with concentrated buffer solution and equilibrated again. When this procedure is not sufficient, the exchanger should be cycled as described earlier (p. 354). If virtually no organic material remains associated with the column after the elution, the regeneration between successive repeated operations of the same type may be carried out in a simple manner: the column is washed with strong salt solution (*e.g.*, 1 *M* sodium chloride solution) or strong buffer solution and equilibrated again.

Regeneration of ion-exchange polydextran is carried out as follows. The insoluble impurities at the top of the bed, together with the layer of Sephadex G-25, are discarded. The ion exchanger is then washed with salt solution containing a counter-ion which should be bound after equilibration. An increasing concentration should be used up to an ionic strength of 2. The exchanger is then removed from the column and thoroughly washed with water. If other contaminants are present, the producer of Sephadex recommends that 0.1 *N* sodium hydroxide solution should be used for washing out proteins, and an alcoholic solution or non-ionic detergent (Pluronic F-68) for removal of lipids. The ion exchanger is then washed with water and salt solution and equilibrated with the starting buffer. The ion exchanger is then ready for use and can be packed into the column. If long periods of standing will occur, antimicrobial agents should be added (Table 10.8, p. 266).

If the polydextran ion exchanger is to be kept out of use for very long periods (more than 6 months), it should be regenerated, washed with a very dilute neutral buffer solution and shrunk slowly on a büchner funnel with alcoholic solutions of increasing concentrations up to 96%. For CM-, DEAE- and QAE-Sephadex, ethanol should be used, while for SP-Sephadex methanol is recommended. The alcohol is then aspirated off and the exchanger is washed with diethyl ether and allowed to dry. Because this material is hygroscopic, it should be kept in well sealed bottles, and it is then stable for several years.

For storage of all types of ion exchangers, the container should be labelled with all the necessary information characterizing the exchanger: trade name of the ion exchanger; its ionic form; cross-linking; grain size; date and process of regeneration; and reference to the last experiment in which the exchanger was used, and name of the worker.

When the ion exchanger is kept for some time in bed form under solution, this information should be labelled on the column and also the type of the antimicrobial agent should be noted. The exchanger should never be kept for long periods containing non-eluted zones of organic substances, especially in phosphate buffers.

## REFERENCES

Borman, H. G., in K. Paech and M. V. Tracy (Editors), *Modern Methods of Plant Analysis*, Vol. V, Springer, Berlin, 1962.

Dorfner, K., *Ionenaustausch-Chromatographie*, Akademie-Verlag, Berlin, 1963.

Gere, D. R., in J. J. Kirkland (Editor), *Modern Practice of Liquid Chromatography*, Wiley-Interscience, New York, 1971, p. 417.

Gillam, I., Millwards, S., Blew, D., Tigerstrom, M., Wimmer, E. and Teuer, G. M., *Biochemistry*, 6 (1967) 3043.

Green, J. G., *Nat. Cancer Inst., Monogr.*, No. 21 (1966) 447.

Hale, D. K., in C. Calmon and T. R. E. Kressman (Editors), *Ion Exchangers in Organic and Biochemistry*, Interscience, New York, 1957, p. 157.

Hamilton, P. B., *Anal. Chem.*, 30 (1958) 914.

Horwath, C. and Lipsky, S. R., *Anal. Chem.*, 41 (1969) 1227.

Inczédy, J., *Analytical Applications of Ion Exchangers*, Pergamon Press, New York, 1966.

Kesler, R. B., *Anal. Chem.*, 39 (1967) 1416.

Kirkland, J. J., *J. Chromatogr. Sci.*, 7 (1969) 361.

Kirkland, J. J. (Editor), *Modern Practice of Liquid Chromatography*, Wiley-Interscience, New York, 1971.

Mikeš, O., in O. Mikeš (Editor), *Laboratory Handbook of Chromatographic Methods*, Van Nostrand, London, 1966, p. 247.

Morris, C. J. O. R. and Morris, P., *Separation Methods in Biochemistry*, Interscience, New York, 1963.

Ohms, J. I., Zec, J., Benson, J. V. and Patterson, J. A., *Anal. Biochem.*, 20 (1967) 51.

Pharmacia, *Sephadex Ion Exchangers: A Guide to Ion Exchange Chromatography*, Pharmacia, Uppsala, 1971.

Samuelson, O., *Ion Exchange Separations in Analytical Chemistry*, Wiley, New York, London, 1963, p. 158.

Scott, C. D., *Anal. Biochem.*, 24 (1968) 292.

Scott, C. D., in J. J. Kirkland (Editor), *Modern Practice of Liquid Chromatography*, Wiley-Interscience, New York, 1971, p. 287.

Scott, C. D. and Lee, N. E., *J. Chromatogr.*, 42 (1969) 263.

Semenza, G., *Helv. Chim. Acta*, 43 (1960) 1057.

Simonson, R., *Sv. Kem. Tidskr.*, 73 (1961) 531.

Thompson, C. M., *Whatman Advanced Ion Exchange Celluloses: Laboratory Manual*, W. & R. Balston, Maidstone, 1972.

Uziel, M., Koh, G. K. and Cohn, W. E., *Anal. Biochem.*, 25 (1968) 77.

Walton, H. F., in E. Heftman (Editor), *Chromatography*, Reinhold, New York, 2nd ed., 1967, p. 325.

Zmrhal, Z., unpublished work, cited in O. Mikeš (Editor), *Laboratory Handbook of Chromatographic Methods*, Van Nostrand, London, 1966, p. 271.

*Chapter 14*

# Practice of affinity chromatography

J. TURKOVÁ

## CONTENTS



## PREPARATION OF THE SOLID SUPPORT WITH A BOUND AFFINANT

As the optimum conditions for binding (*i.e.*, affinant and gel concentration, pH and composition of the buffer, temperature and duration of the binding reaction) depend on the nature of the solid carrier and the affinant, a single optimum procedure cannot be prescribed. Nonetheless, several general principles can be laid down which, in most instances, increase the effectiveness of affinity chromatography.

When choosing the pH of the binding reaction, the stabilities of both the solid carrier and the affinant should be taken into consideration. During the binding, thorough stirring is necessary. A magnetic stirrer should not be used as it may cause the destruction of the gel, and reciprocal or rotational shaking of the vessel is the most suitable technique. For an affinant of proteinic character, the use of buffers of higher ionic strength (*ca.* 0.5 *M*) during the binding is recommended, in order to minimize adsorption of the protein on the coupled protein as a result of the polyelectrolytic nature of proteins and thus facilitate the subsequent washing. The washing out of all components present in the material in addition to the covalently bound affinant should be performed with great care and the efficiency of the washing procedure should be checked. The washing out facilitates the change of buffers of high and low pH values, for example 0.1 *M* acetate buffer of pH 4.1 and 0.1 *M* borate buffer of pH 8.5, each 1 *M* in sodium chloride (Axén and Ernback). The elution of non-covalently bound affinants is usually achieved by four or five washing cycles. With bonded aromatic compounds, washing with organic solvents is advantageous, while with proteins washing with denaturation agents is preferred under conditions such that neither the carrier nor the binding capacity of the affinant is affected (Cuatrecasas and Anfinsen). The thoroughness of the washing can be checked by measuring the effect of changing the concentration of the insoluble derivative or by incubation with various buffers, etc.

The determination of the amount of affinant bound to the solid carrier can be determined by a method depending on the nature of the affinant, usually after its release by acid or alkaline hydrolysis. If the affinant is of a peptidic nature, its amount

*References p.376*

can be determined on the basis of the quantitative determination of amino acids after acid hydrolysis of the adsorbent (Axén and Ernback). The measurement of radioactivity is also advantageous with radioactive affinants. In a review, Cuatrecasas and Anfinsen recommended that the concentration of the bound affinant should be expressed in micromoles or milligrams of affinant per millilitre of the packed gel, rather than indication per unit dry weight of the support as used formerly.

## SORPTION CONDITIONS

The original conditions for sorption should be chosen so that adsorption of the isolated substance on it is the strongest. The choice of the starting buffer is completely dependent on the optimum conditions for complex formation between the affinant and the isolated substance and is related, in addition to pH and ionic strength, to the content of metal ions and other specific factors. The high ionic strength of the starting buffer decreases the non-specific adsorption of polyelectrolytes on possible charged groups of the bound affinant. Therefore, the recommended content of sodium chloride is about 0.5 *M*.

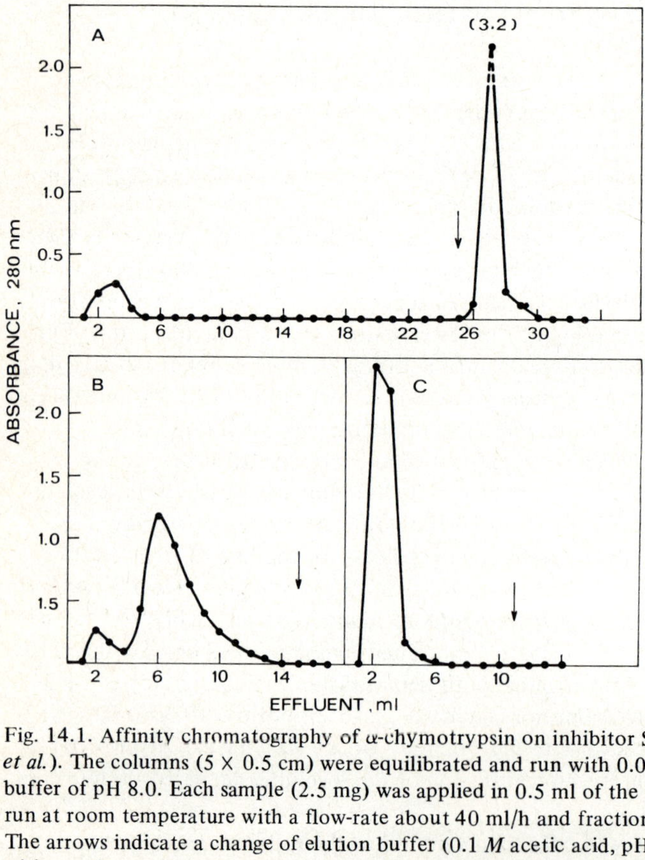


Fig. 14.1. Affinity chromatography of α-chymotrypsin on inhibitor Sepharose columns (Cuatrecasas *et al.*). The columns (5 × 0.5 cm) were equilibrated and run with 0.05 *M* Tris–hydrochloric acid buffer of pH 8.0. Each sample (2.5 mg) was applied in 0.5 ml of the same buffer. The columns were run at room temperature with a flow-rate about 40 ml/h and fractions containing 1 ml were collected. The arrows indicate a change of elution buffer (0.1 *M* acetic acid, pH 3.0). (A), Sepharose coupled with ε-aminocaproyl-D-tryptophan methyl ester; (B), Sepharose coupled with D-tryptophan methyl ester; (C), unsubstituted Sepharose. The first peaks in A and B were devoid of enzyme activity.

A sample of the substance to be isolated is preferably dissolved in the starting buffer and, if necessary, the change of the composition of the salts in the sample should be carried out by dialysis or gel filtration. If a substance is isolated which, under these conditions, forms a strong complex with the affinant, the volume of the sample introduced on to the column is irrelevant. However, if affinity chromatography is employed for the isolation of substances with a low affinity for the bound affinant, the volume of sample applied should not exceed 5% of the hold-up volume, so as to prevent the elution of the isolated substance together with non-adsorbed material. If a substance with a low affinity for the bound affinant is isolated, its elution from the column often occurs even without a change of buffer. In such a case, the isolated substance is obtained in a dilute form. As an example, the affinity chromatography of α-chymotrypsin is shown in Fig. 14.1 both on Sepharose coupled with ε-aminocaproyl-D-tryptophan methyl ester (A) and on Sepharose coupled with D-tryptophan methyl ester (B), in comparison with the chromatography on unsubstituted Sepharose (C) (Cuatrecasas *et al.*). In the first case (A), the bonded inhibitor has a high affinity for α-chymotrypsin and the enzyme can be released from the complex only by decreasing the pH of the eluting buffer. By using 0.1 *M* acetic acid of pH 3.0, the chymotrypsin fraction is eluted as a sharp peak and the volume of the eluted chymotrypsin does not depend on the volume of the sample applied on to the column. In the second case (B), the inhibitor coupled directly on Sepharose has a much lower affinity for the isolated α-chymotrypsin, owing to steric hindrance. In this instance, a change of buffer is not necessary for enzyme elution and, as can be seen from the graph, the enzyme is eluted in a much larger volume closely after the inactive material. In order to verify that non-specific adsorption on the carrier did not take place under the given experimental conditions, it was also necessary to carry out the chromatography of α-chymotrypsin on an unsubstituted carrier (C).

If a small amount of substance is isolated from the crude mixture by using an affinant of high affinity, a batch process combined, if necessary, with the elution after the transfer on to a column can be applied.

The effects of pH and the ionic strength on the affinity adsorption of α-chymotrypsin on Sepharose columns coupled with ε-aminocaproyl-D-tryptophan are illustrated in Fig. 14.2. The first peak (fractions 2–4) contained, in all instances, a material that was devoid of chymotryptic activity. The specific activity of chymotrypsin eluted in subsequent fractions was constant. Hence, it is evident that the decrease in the pH of the first buffer leads to a decrease in the stability of the complex of α-chymotrypsin with the bound inhibitor, which results in the gradual elution of α-chymotrypsin with the first buffer. The elution of α-chymotrypsin from the insoluble affinant is greater at higher ionic strengths.

For the isolation of high-molecular-weight substances, it is important to keep the flow-rate of the adsorbent through the column slow. The attainment of adsorption equilibrium depends not only on the number of collisions between the molecules of the isolated substances and the coupled affinant, but also requires a certain mutual orientation of the binding sites. In Fig. 14.3, the isolation of β-galactosidase from an extract from *Escherichia coli* (Steers *et al.*) is shown. Fig. 14.3A represents the chromatography of an extract from *E. coli* on unsubstituted Sepharose 4B, and Fig. 14.3B chromatography on Sepharose 4B with bonded *p*-aminophenyl-β-D-galactopyranoside. The experimental conditions are specified in the caption.

*References p.376*

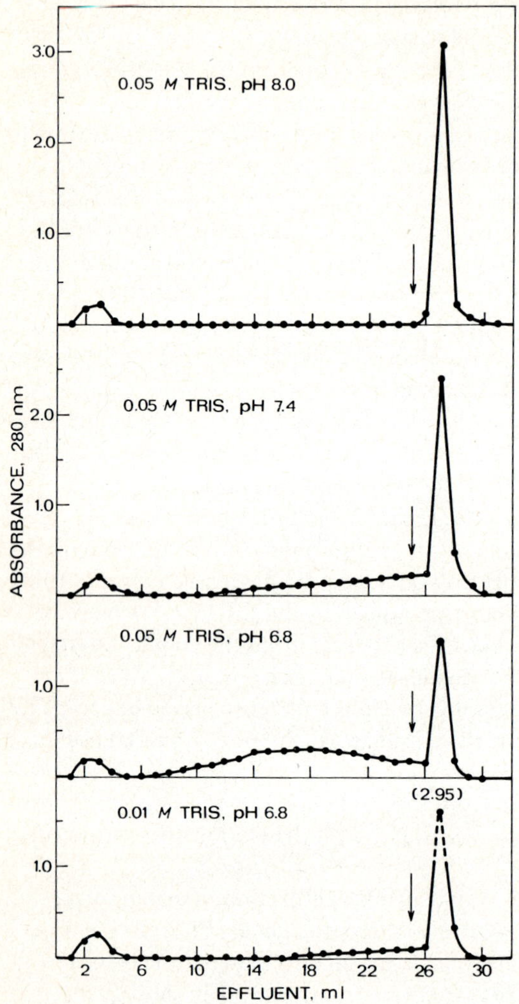


Fig. 14.2. Effects of pH and ionic strength on the affinity adsorption of α-chymotrypsin on a column of Sepharose coupled with ε-aminocaproyl-D-tryptophan methyl ester (Cuatrecasas *et al.*). The columns (5 × 0.5 cm) were equilibrated and run with Tris buffers as shown in the figure. The other conditions were identical with those specified in Fig. 14.1.

## CONDITIONS FOR ELUTION

While substances without affinity for the bound affinant are usually eluted with the hold-up volume, the adsorbed material mostly requires a change of the buffer system. In many instances, a change in pH (Cuatrecasas *et al.*; Feinstein, 1970, 1971; Wilchek and Gorecki) or an increase in ionic strength (Baggio *et al.*, Baker and Siebeneick, Kalderon *et al.*, Lowe and Dean) suffice for the dissociation of the complex and subsequent elution.

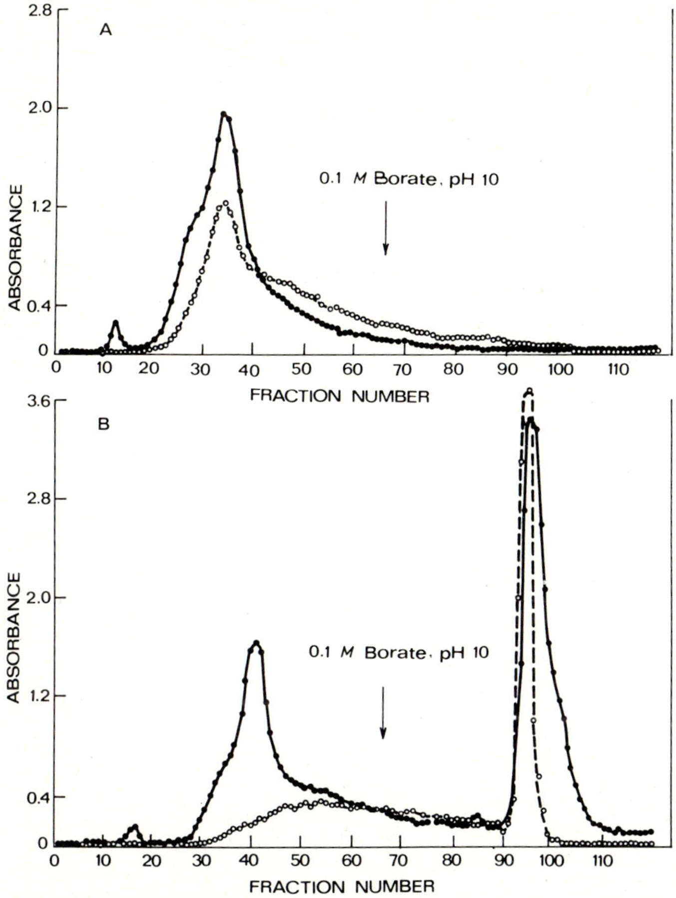


Fig. 14.3. Chromatography of extracts from *Escherichia coli* on unsubstituted Sepharose 4B (A) and on the Sepharose 4B coupled with *p*-aminophenyl-β-D-galactopyranoside (B) (Steers *et al.*). The columns were equilibrated and chromatographed with 0.05 *M* Tris–hydrochloric acid buffer of pH 7.5. Elution of adsorbed protein was carried out with 0.1 *M* sodium borate solution of pH 10.05. The extract (20 ml) was applied to a 22 × 1.5 cm column and 0.8-ml fractions were collected. The flow-rate was 80 ml/h and the experiments were performed at room temperature (23°C). •–•, Absorbance (280 nm); ○- -○, enzymatic activity (the substrate was *o*-nitrophenyl-β-D-galactopyranoside and the rate of release of *o*-nitrophenol was measured spectrophotometrically at 420 nm).

As an example of elution resulting from a change in pH, the isolation of trypsin by affinity chromatography on Sepharose with bound ovomucoid is illustrated in Fig. 14.4 (Feinstein, 1970). In addition to affinity chromatography on insoluble affinant proper (B), it was necessary (as always) to check by chromatography on an unmodified solid support (A) whether only the elution of the specific material takes place under the conditions of affinity chromatography. As an example of the elution of adsorbed material from an insoluble affinant by means of an increase in ionic strength, the affinity chromatography of the RNA polymerase from *Escherichia coli* on DNA–agarose is illustrated in

Fig. 14.5 (Nüsslein and Heyden). From 95 ml of the RNA polymerase solution obtained by preceding chromatography on Bio-Gel A, inactive material was separated in the first peak by affinity chromatography, while when a linear gradient of 0.25–1.25 *M* potassium chloride solution was applied two different RNA polymerases were obtained. The experimental conditions are specified in the caption.

In the affinity chromatography of enzymes, elution with buffers containing a low-molecular-weight soluble affinant is sometimes employed, for example, elution with the substrate or the inhibitor used for the preparation of the insoluble affinant at a higher concentration than that bound to the carrier, or with a stronger competitive inhibitor (Baker and Siebeneick, Porath, Steers *et al.*). However, sometimes it is necessary to use a

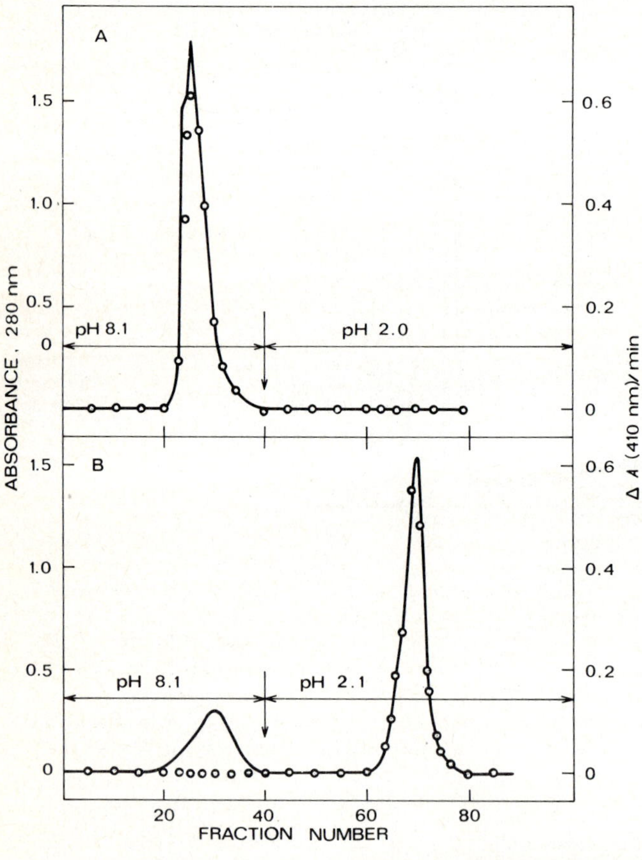


Fig. 14.4. Affinity chromatography of trypsin on Sepharose (A) and ovomucoid–Sepharose (B) (Feinstein, 1970). The columns (35 × 1.7 cm) were equilibrated and run with 0.10 *M* triethanolamine buffer solution of pH 8.1, containing 0.02 *M* calcium chloride. Each sample of crystalline bovine trypsin (20 mg) was applied in 2.0 ml of the same buffer. Fractions of 2.8 ml were collected. The vertical arrows indicate a change of elution buffer (0.20 *M* potassium chloride–hydrochloric acid, pH 2.0). The curves represent the absorbance at 280 nm; o, tryptic activity [the substrate was N-benzoyl-dl-arginine-*p*-nitroanilide and its rate of hydrolysis was followed by the increase in absorbance at 410 nm(———)].

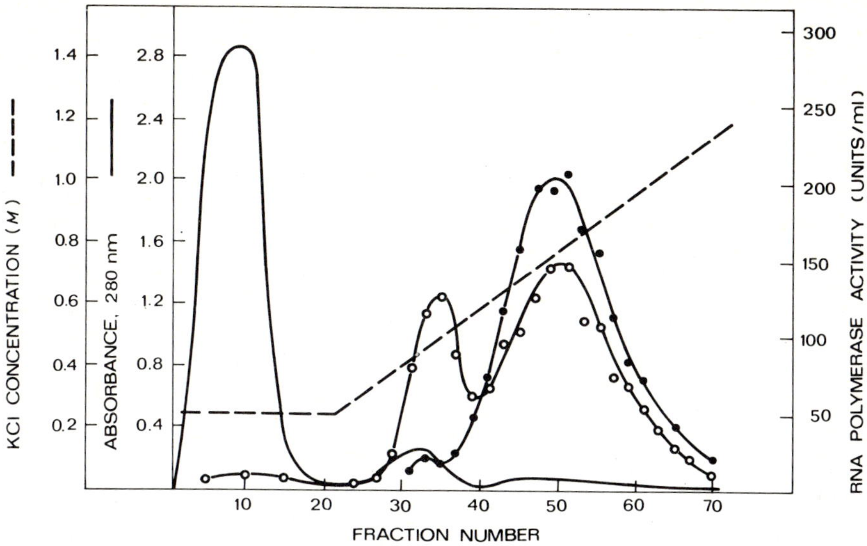


Fig. 14.5. Affinity chromatography of RNA polymerase on DNA–agarose (Nüsslein and Heyden). The column (15 × 1 cm) was equilibrated and chromatographed with standard buffer solution (0.01 *M* Tris, pH 8.0; $10^{-3}$ *M* EDTA, $10^{-4}$ *M* dithioerythritol, 5% glycerol) and 0.25 *M* potassium chloride solution. Elution of adsorbed protein was carried out with a 600 ml linear gradient of 0.25–1.25 *M* potassium chloride in standard buffer solution. The fraction of RNA polymerase (95 ml) was applied on to the column. A constant flow-rate during loading and elution was maintained by use of a peristaltic pump. Fractions of 10 ml were collected and assayed for RNA polymerase activity using calf thymus DNA (○) and $T_4$ DNA (●) as template. The total recovery of the RNA polymerase activity from the column was 80%. Solid line, absorbance; broken line, potassium chloride concentration.

buffer for elution, with a high concentration of the dissociating agent and a low pH (Bodanszky and Bodanszky, Cuatrecasas and Wilchek; see Fig. 14.6).

If the affinant is attached to its matrix by an azo bond or by thiol or alcohol–ester bonds, the complex of the affinant with the isolated substance can be split from the solid matrix and the affinant then separated by dialysis or gel filtration. Of course, this prevents the repeated use of the affinant matrix (Cuatrecasas).

## PRESERVATION OF SOLID SORBENTS WITH A BOUND AFFINANT

Most affinants used at present are biologically active substances characterized by a much lower stability than that of the solid carriers on to which they are bound. Therefore, it is necessary to choose storage conditions that would not cause a loss in activity of the bound affinant. It is of great advantage that a whole series of affinants, especially of proteinic character, acquires a higher stability by their binding to a solid support (with respect to temperature, pH, etc.). In many instances, the storage of the packed specific adsorbent at low temperatures in the presente of a suitable bacteriostatic agent is most suitable. The choice of the storage buffer depends on the properties of the bound affinant (Axén and Ernback, Line *et al.*).

*References p.376*

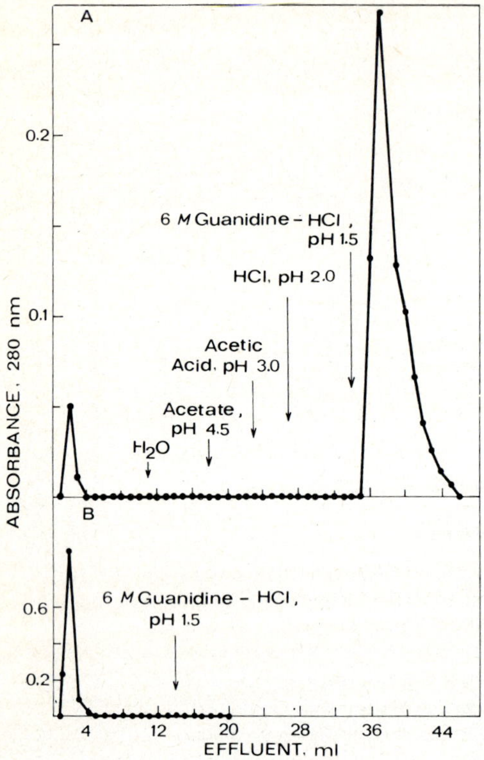


Fig. 14.6. Affinity chromatography of avidin on biocytin–Sepharose (A) and unsubstituted Sepharose (B) columns (Cuatrecasas and Wilchek). The columns (5 × 0.5 cm) were equilibrated and run with 0.2 *M* sodium hydrogen carbonate solution of pH 8.7. Each sample of avidin (0.75 mg) was applied in 0.5 ml of the same buffer. The columns were run at room temperature with a flow-rate of about 30 ml/h and 1-ml fractions were collected. Elution was attempted by varying the conditions as indicated by arrows. The small protein peak that emerges early in A represents an impurity.

## REFERENCES

Axén, R. and Ernback, S., *Eur. J. Biochem.*, 18 (1971) 351.
Baggio, B., Pinna, L. A., Morel, V. and Siliprandi, N., *Biochim. Biophys. Acta*, 212 (1970) 515.
Baker, B. R. and Siebeneick, H. U., *J. Med. Chem.*, 14 (1971) 799.
Bodanszky, A. and Bodanszky, M., *Experientia*, 26 (1970) 327.
Cuatrecasas, P., *J. Biol. Chem.*, 245 (1970) 3059.
Cuatrecasas, P. and Anfinsen, C. B., *Methods Enzymol.*, 22 (1971) 345.
Cuatrecasas, P. and Wilchek, M., *Biochem. Biophys. Res. Commun.*, 33 (1968) 235.
Cuatrecasas, P., Wilchek, M. and Anfinsen, C. B., *Proc. Nat. Acad. Sci. U.S.*, 61 (1968) 636.
Feinstein, G., *FEBS Lett.*, 7 (1970) 353.
Feinstein, G., *Biochim. Biophys. Acta*, 236 (1971) 73.
Kalderon, N., Silman, I., Blumberg, S. and Dudai, Y., *Biochim. Biophys. Acta*, 207 (1970) 560.
Line, W. F., Kwong, A. and Weetall, H. H., *Biochim. Biophys. Acta*, 242 (1971) 194.
Lowe, C. R. and Dean, P. D. G., *FEBS Lett.*, 14 (1971) 313.
Nüsslein, C. and Heyden, B., *Biochem. Biophys. Res. Commun.*, 47 (1972) 282.
Porath, J., *Biotechnol. Bioeng. Symp.*, No. 3 (1972) 145.
Steers, E., Cuatrecasas, P. and Pollard, H., *J. Biol. Chem.*, 246 (1971) 196.
Wilchek, M. and Gorecki, M., *Eur. J. Biochem.*, 11 (1969) 491.

*Chapter 15*

# Analytical utilization of chromatograms

J. NOVÁK, J. JANÁK and S. WIČAR

## CONTENTS



## IDENTIFICATION

Chromatography is primarily a separation method and, owing to its extremely high separation efficiency, its use as an ancillary technique can substantially facilitate the problems involved in the analysis of complex mixtures by other methods. A combination

*References p.401*

of either a physical or a chemical analytical method with a suitable chromatographic technique makes it possible to undertake analyses the solution of which is practically impossible without an efficient pre-analytical separation of the mixture. The possibilities of accomplishing such combinations are very wide; a number of physical analytical methods, such as mass, infrared, ultraviolet, and nuclear magnetic resonance spectrometry, can be considered in connection with liquid chromatography. A simpler yet very efficient approach is the use of special (selective) identification detectors.

In addition to the above combinations, correlations of retention data with various physical properties or structural features of the solute components, particularly if data obtained by chromatography in different systems are considered, can give valuable information on the compounds. The identification postulated as a result of these correlations is then confirmed by chromatographing standard substances. This simple approach can be very helpful in identifying the individual components of a mixture of compounds of a particular chemical type, but it often fails if the determination of the functional group is the aim. In this respect, the procedures of classical chemical analysis can be more efficient but, on the other hand, they are less satisfactory for the identification of the components of complicated mixtures. Hence, the combination of chromatography and chemical tests leads to a very powerful and inexpensive analytical technique.

There are numerous procedures based on the chemical alteration of particular groups of compounds, either in the sample or in the column effluent, which leads either to the subtraction of some group of compounds or to characteristic shifts of the respective peaks in the chromatogram. When their results are combined with the information gained from chromatographic retention data, the chemical procedures afford a very detailed qualitative analysis of column eluates. In many instances, the efficacy of the above methods can be amplified by combining different chromatographic techniques, particularly column techniques with planar techniques.

## Principles of identification by means of retention data

Treatments involving retention data are usually introduced with the statement that the retention data are characteristic of the substance chromatographed. We shall try to be more precise by saying that retention data can generally be characteristic of any of the substances that constitute the chromatographic system, *i.e.*, the solute component as well as the substances that compose the mobile and stationary phases.

It follows from the theory of chromatography in liquid–liquid systems that the retention time, $t_R$, is given (at a given temperature) by

$$t_R = \frac{L}{u}(1 + k) = \frac{L}{u}\left(1 + \frac{\gamma_{im}^0 M_m \rho_s \varphi_s}{\gamma_{is}^0 M_s \rho_m \varphi_m}\right) \tag{15.1}$$

where $L$ and $u$ are the column length and forward velocity of the mobile phase, $\gamma_{im}^0$ and $\gamma_{is}^0$ are the Raoult's law activity coefficients of solute in the mobile and stationary phases, $M_m$, $\rho_m$ and $\varphi_m$ are the molecular weight, density and cross-sectional area of the mobile phase, and $M_s$, $\rho_s$ and $\varphi_s$ are the same quantities referring to the stationary phase. $L$, $u$, $\varphi_s$ and $\varphi_m$ are adjustable (within certain limits) instrumental parameters, $M_m$, $\rho_m$ and $M_s$, $\rho_s$

are characteristics merely of the mobile and stationary phases, and only $\gamma^0_{im}$ and $\gamma^0_{is}$ have any reference to the solute compound. However, as the value of $\gamma^0$ is a function of the composition of the given mixture, $\gamma^0_{im}$ and $\gamma^0_{is}$ are characteristics of the solute $i$ and the mobile and stationary phases, respectively. Hence, the retention data reflect the properties of all the components involved in the system. In other words, the retention time of a given solute component can be used to characterize either the solute or the phases, provided that two of the three have been specified.

Gas chromatography leads to a much simpler situation in this respect. As $t_R$ is given by

$$t_R = \frac{L}{u}\left(1 + \frac{RT\rho_s\varphi_s}{\gamma^0_{is}f^0_i M_s\varphi_m}\right) \tag{15.2}$$

where $R$ is the universal gas constant, $T$ is the absolute temperature of the column and $f^0_i$ is the fugacity of the pure solute (at temperature $T$), the GLC retention data reflect the properties of only the solute compound and the stationary phase (provided that there are no appreciable interactions between the solute vapour and the carrier gas).

*Relative retention data*

It is expedient that retention data should be expressed in a form independent of the non-specific instrumental parameters ($L$, $u$, $\varphi_s$ and $\varphi_m$). This can be achieved by carrying out the appropriate corrections, but sometimes it may be difficult to specify some of the parameters, especially the phase cross-sections. It is therefore very convenient to work with relative retention data. As the ratio $L/u$ is equal to the dead retention time, $t_m$, eqn. 15.1 can be rewritten to read

$$t_R - t_m = t_m k = t_m \cdot \frac{\gamma^0_{im}M_m\rho_s\varphi_s}{\gamma^0_{is}M_s\rho_m\varphi_m} \tag{15.3}$$

where $t_R - t_m$ is the net retention time. Now, if the net retention time of the substance being analyzed ($i$) is divided by that of an auxiliary reference substance ($a$) chromatographed in the same system under the same conditions, the relative retention time, $t_{ir}$, is given by

$$t_{ir} = (t_R - t_m)_i/(t_R - t_m)_a = \gamma^0_{im}\gamma^0_{as}/\gamma^0_{is}\gamma^0_{am} = K_i/K_a \tag{15.4}$$

where $K_i$ and $K_a$ are the respective distribution constants. Eqn. 15.4 shows that $t_{ir}$ determined in a system of a given composition is a function of only temperature and pressure, the effect of the latter being negligible. Further, it is evident that the ratio of the net retention times is equal to the ratio of the corresponding net retention volumes. Hence, relative retention data are very suitable for use if results obtained in the same system with different arrangements are to be compared.

The simplest means of achieving a tentative identification by means of retention data is a direct comparison of the retention data of the unknown with the same type of retention data for standard compounds. It is obvious that the retention data of the unknown and of the standard must be measured in the same systems under identical conditions in order to obtain a meaningful comparison. Either data obtained by direct measurement or compiled from the literature can be employed for this comparison. The coincidence of the retention data of the unknown with that of a standard substance indicates that the compounds may be identical. However, coincidence of the results obtained from measurements in a single chromatographic system does not indicate the identity of the unknown with certainty. A reasonable degree of certainty can be achieved only by carrying out measurements in several different systems and positive proof of identity is obtained if the retention data of the unknown and of the standard coincide in each of the systems employed.

The proposed identity can further be confirmed by modifying chemically both the substance being analyzed and the standard by using reactions that are typical for the given species of compound. This modification results in characteristic shifts or elimination of some peaks in the chromatogram, and if the same shifts are obtained with both the analyzed and the standard compounds, the identity is confirmed. Various chemical reactions can be applied while employing several different chromatographic systems, so that it is possible to confirm the identity positively in this way.

### Correlations between the constitution of the solute and chromatographic behaviour

#### *Group contributions to the logarithm of the distribution constant*

Empirical correlations of chromatographic retention data with the number and chemical nature of the structural units of the solute molecules reveal that there is some additivity of the contributions to retention due to the individual units of the molecules of the substance. This concept, originally proposed by Martin, is the basis of numerous empirically observed regularities between the structure and retention data of solute components. which are applicable, under certain circumstances, to all types of chromatography. Correlations based on this concept represent another significant means of identification of chromatographic eluates. In this section, the above concept is discussed in terms of the thermodynamics of chromatography presented in the theoretical part of this book.

The following equation was derived for liquid–liquid systems:

$$\frac{\Delta G_s^*}{RT} = \frac{\Delta G_{is}^E - \Delta G_{im}^E}{RT} = \ln K + \ln\left(\frac{M_m \rho_s}{M_s \rho_m}\right) \tag{15.5}$$

where $\Delta G_s^*$ is the standard molar Gibbs free energy of sorption of the solute at infinite dilution and $\Delta G_{is}^E$ and $\Delta G_{im}^E$ are the partial molar excess Gibbs free energies of the solute in the stationary and the mobile phase, respectively.

Let us consider a series of monofunctional straight-chain compounds of the type $i \equiv CH_3(CH_2)_nX$, where X is a functional group, and assume that the Gibbs free energy of sorption of substance $i$ is additively composed of partial Gibbs free energies of sorption of the individual units that constitute the molecule of the substance. Thus, we can write

$$\Delta G_s^*(i) = n\Delta G_s^*(CH_2) + \Delta G_s^*(CH_3) + \Delta G_s^*(X) \tag{15.6}$$

and, consequently

$$\Delta G_{is}^E = n\Delta G_{is}^E(CH_2) + \Delta G_{is}^E(CH_3) + \Delta G_{is}^E(X) \tag{15.7}$$

$$\Delta G_{im}^E = n\Delta G_{im}^E(CH_2) + \Delta G_{im}^E(CH_3) + \Delta G_{im}^E(X) \tag{15.8}$$

The combination of eqns. 15.5, 15.7, and 15.8 gives

$$\ln K = n \cdot \frac{\Delta G_{is}^E(\mathrm{CH_2}) - \Delta G_{im}^E(\mathrm{CH_2})}{RT} + \frac{\Delta G_{is}^E(\mathrm{CH_3}) - \Delta G_{im}^E(\mathrm{CH_3})}{RT} + $$

$$+ \frac{\Delta G_{is}^E(\mathrm{X}) - \Delta G_{im}^E(\mathrm{X})}{RT} - \ln\left(\frac{M_m \rho_s}{M_s \rho_m}\right) \tag{15.9}$$

The individual terms of the right-hand side of eqn. 15.9 can be looked upon as additive contributions to $\ln K$, and the equation can be rewritten to read

$$\ln K = n \ln K(\mathrm{CH_2}) + \ln K(\mathrm{CH_3}) + \ln K(\mathrm{X}) - \ln\left(\frac{M_m \rho_s}{M_s \rho_m}\right) \tag{15.10}$$

Hence, when plotting $\log K$ against the methylene or carbon number for a homologous series of compounds chromatographed in a given system, we obtain a straight line with a slope

$$\frac{\mathrm{d}\log K}{\mathrm{d}n} = \log\left(\frac{K_{i(n+1)}}{K_{i(n)}}\right) = \frac{\Delta G_s^E(\mathrm{CH_2}) - \Delta G_m^E(\mathrm{CH_2})}{2.303\,RT} = \log K(\mathrm{CH_2}) \tag{15.11}$$

the intercept being given by $\log K(\mathrm{CH_3}) + \log K(\mathrm{X}) - \log (M_m \rho_s / M_s \rho_m)$.

This phenomenon is common in gas chromatography. However, in contrast to GC systems, where graphs of $\log K$ *versus* carbon number almost exclusively have positive slopes, the slopes of the same graphs plotted using retention data measured in LC systems may have values ranging from positive to negative values, passing through zero. The reason for this distinction is as follows. Whereas the gas chromatographic $\Delta G_s^*(\mathrm{CH_2})$, being given by $\Delta G_{is}^E(\mathrm{CH_2}) + \Delta G_c(\mathrm{CH_2})$ (the Gibbs free energy of condensation), is mostly positive and large, the liquid chromatographic $\Delta G_s^*(\mathrm{CH_2})$ can easily become either positive or negative, depending on the values of $\Delta G_{is}^E(\mathrm{CH_2})$ and $\Delta G_{im}^E(\mathrm{CH_2})$. A practical implication of this situation is that there may occur either normal $[\Delta G_s^*(\mathrm{CH_2}) > 0]$, reverse $[\Delta G_s^*(\mathrm{CH_2}) < 0]$ or no $[\Delta G_s^*(\mathrm{CH_2}) = 0]$ separation of the individual homologues of a given series in liquid chromatography.

Experience has shown that, when plotting $\log K$ against methylene number for several homologous series in a given system, one obtains a family of nearly parallel lines. This result indicates two important facts: the linearity is proof that the $\log K(\mathrm{CH_2})$ values for a given homologue are independent of the number and positions of the $\mathrm{CH_2}$ groups in the molecule (except for the first homologues), and the parallelism shows that the $\log K(\mathrm{CH_2})$ value is independent of the type of functional group present. Actually, this implies that the partial $\log K$ values involved are virtually independent of each other and are therefore additive. Thus, having constructed such graphs for various homologous series in several chromatographic systems, it is possible to estimate both the functionality and hydrocarbon chain-length by fitting the retention data of the unknown to lines with which the data correspond to integral carbon numbers. If the data fitted in this manner fall upon a line of the same homologous series with all the systems employed, the corresponding carbon number being always the same, the estimation of identity is confirmed.

Other correlation procedures, such as fitting the retention data of the unknown to plots of retention data of various homologous series in two different chromatographic systems, are merely variations of the above-described method. It is a matter of simple arithmetic to show that all the correlations can be performed with relative retention data.

The concept of the additivity of the partial log $K$ values can further be generalized; for a substance $i \equiv (CH_3)_m(CH_2)_n(CH)_oC_pX_xY_yZ_z \ldots$, $\log K = m \log K(CH_3) + n \log K(CH_2) + o \log K(CH) + \ldots$. Unfortunately, large deviations from the additivity principle are encountered with complex compounds. It can be assumed that all the regularities discussed and the correlations derived from them also apply to retention data measured in liquid–solid chromatography systems.

*Retention index in liquid chromatography*

The conventional Kováts retention index of a given substance ($i$), as employed in gas chromatography, is defined as 100 times the carbon number of an $n$-alkane (generally hypothetical) that displays the same retention in the given system as the substance under test. Hence it follows that the retention data of two homologous $n$-alkanes as reference compounds are necessary in order to express the retention index. The calculation is carried out by using the equation

$$I_i = 100\,n + 100 \cdot \frac{\log K_i - \log K_{p(n)}}{\log K_{p(n+1)} - \log K_{p(n)}} \tag{15.12}$$

where $I_i$ is the Kováts retention index of substance $i$, $K_i$ is the distribution constant of substance $i$ and $K_{p(n+1)}$ and $K_{p(n)}$ are the distribution constants of the reference alkanes with carbon numbers $n + 1$ and $n$, respectively. In place of $K$, the net retention time, volume or relative retention data can be employed.

It is not difficult to infer from the definition of the Kováts retention index that the suitability of employing a retention index system is qualified by the adequacy, with respect to the chromatographic system, of the reference compounds used to determine the retention index scale. Further, the advantages of the retention index system largely rest with the fact that the graphs of the logarithm of the retention data *versus* carbon number are parallel for different homologous series. It is evident that in many instances $n$-alkanes may be very unsuitable as reference compounds in liquid chromatography, but this shortcoming can be avoided by employing another series of compounds instead of $n$-alkanes, in order to define the retention scale. As the graphs of the logarithm of the retention *versus* carbon number were shown to be linear and parallel in liquid chromatography systems also, there appears to be no serious argument against employing retention indices in liquid chromatography. The above extension of the definition of the retention index does not alter the validity of eqn. 15.12, except for the fact that $K_{p(n+1)}$ and $K_{p(n)}$ refer to homologues of the new series of compounds (other than $n$-paraffins), the subscript $n$ being the carbon number of the straight-chain part of the hydrocarbon moiety. However, the concept of retention index will definitely become meaningless in instances when $\Delta G_s^*(CH_2) = 0$.

*Relationship between retention data measured in column and flat-bed systems*

At a given temperature and pressure, the distribution constant of a solute component in a two-phase system is determined by the composition of the phases, regardless of their geometrical arrangement. Hence, in a column and a flat-bed system containing the same combination of phases and maintained at the same temperature and pressure, the zone migration is controlled by the same distribution constant. This fact has some important practical implications and it may be very useful to employ either paper or thin-layer chromatography in order to find by trial and error an optimum pair of phases to be used in a column arrangement. In addition, the literature contains a large amount of retention data measured by flat-bed techniques, especially by paper chromatography (*cf.*, Hais and Macek; Macek *et al.*, 1968, 1972, 1976). It is therefore expedient to discuss the possibilities of correlating the retention data measured in a given system of phases by column and flat-bed techniques. Actually, this correlation refers to only formal aspects of the problem, *viz.*, different ways of expressing retention data in column and planar arrangements, the physical bases of retention being the same in both instances.

The fundamental retention equations give

$$t_R/t_m = V_R/V_m = 1 + k = 1 + K(\varphi_s/\varphi_m) = 1/\boldsymbol{R} \tag{15.13}$$

where $\boldsymbol{R}$ is the retardation factor, equal (with some reservations) to the $R_F$ value employed in planar chromatographic techniques. Thus, denoting the terms referring to column and planar arrangements by subscripts $c$ and $p$, respectively, eqn. 15.13 and the condition $K_c = K_p = K$ lead to the relationship

$$k_c = [(1/R_F) - 1]\,(\varphi_m/\varphi_s)_p(\varphi_s/\varphi_m)_c \tag{15.14}$$

or

$$K = [(1/R_F) - 1]\,(\varphi_m/\varphi_s)_p \tag{15.15}$$

Eqn. 15.15 shows that the value of $K$, applicable to any arrangement of the given system of phases, can be obtained from measurements carried out completely in a planar arrangement. Unfortunately, it is sometimes not feasible to specify the ratio $(\varphi_m/\varphi_s)_p$. Bate-Smith and Westall introduced the quantity $R_M$, equal to log $[(1/R_F) - 1]$. Hence, eqn. 15.15 can be rewritten as

$$\log K = R_M + \log\,(\varphi_m/\varphi_s)_p \tag{15.16}$$

Eqn. 15.16 implies that the additivity principle postulated for log $K$ applies in the same manner for the quantity $R_M$. In context with the previous discussion (eqns. 15.5–15.10), eqn. 15.16 is a proof of Martin's theory of the additivity of $\Delta R_M$ values (Martin and Synge, Consden *et al.*).

The possibilities of correlations between column and paper chromatographic data were demonstrated for a group of steroids by Kabasakalian and Talmage. A computational method for the determination of the number of functional groups and structural properties of a molecule by processing the $R_M$ and $\Delta R_M$ values of the substance in various chromatographic systems was described by Schauer and Bulirsch. However, when determining the distribution constant from data obtained by a planar technique, one must be aware of the fact that the ratio $(\varphi_m/\varphi_s)_p$ may vary with respect to both time and position in the chromatographic bed, and that the distribution constant in this arrangement may be erroneous as a result of spurious sorption effects. In this respect, paper chromatography is more satisfactory than thin-layer chromatography.

**Use of selective detection and coupling with analytical methods**

Combinations of the use of selective detectors and various analytical methods are widely practised in gas chromatography. However, the translation of this methodology into liquid chromatographic work is not straightforward in all instances. Although solutions of samples can be subjected to many of the above analytical methods (UV, IR, NMR), the solvent that has to be used as eluent in the chromatographic stage may be totally unsuitable for the analytical method. With mass spectrometry, the solute compound should be free from all solvent. Hence, the eluent must be eliminated before the analysis proper in many instances.

In contrast to the situation in gas chromatography, it is rather difficult to remove the liquid carrier from the solute component in the column effluent, especially under continuous operating conditions, so that the procedures developed for gas chromatography need special modifications to make them applicable in liquid chromatography.

If two different detectors responding to different properties of a solute component are connected in series or in parallel to the column outlet, two records are obtained that give two complementary types of information on the component. The use of a parallel arrangement of a flame ionization and an electron capture detector in gas chromatography can serve as a typical example. In liquid chromatography, similar possibilities are afforded by the combination of a refractometer and an ultraviolet detector, the first detector providing a response related mostly to the size of the molecule while the second responds selectively to certain functional groups (aromatic rings, conjugated multiple bonds, –CONH–, etc.). According to suggestions by Haderka, conductivity–permittivity detectors can provide two independent responses simultaneously, corresponding to the real and the imaginary components of the complex permittivity of the eluate.

As the zone is subject to excessive broadening on passage through extra-column spaces, parallel arrangements of detectors are to be preferred although this results in a decrease in sensitivity owing to the column effluent being split into two streams.

Of physical analytical methods for the identification of eluates, the coupling of liquid chromatography with mass spectrometry has great potential. This combination is the subject of intensive development at present, the main problem being the simplification of a rich mass spectrum due to residues of the eluent. In this respect, it may be very advantageous to use liquefied gases, for example carbon dioxide, as liquid carriers.

*Combinations of column and flat-bed techniques*

Planar chromatograms can provide two types of information on the substances chromatographed, one based on the $R_F$ values and the other on the investigation of chemical or physical properties of the individual solute components located in the chromatographic zones. Flat-bed chromatography enables virtually selective detection to be achieved with the use of suitable chemical reagents. Hence, if the column eluates are deposited on a planar chromatographic bed and subsequently chromatographed and detected in the usual manner for laminar techniques, while monitoring the column effluent with a column chromatographic detector, three types of information can be obtained, or even four types if the column chromatographic detector is selective to particular species of compounds. It is also possible to analyze by column chromatography a spot isolated from the planar chromatogram.

There are a number of possible alternatives to the above combination, as shown in Fig. 15.1. Generally, with respect to the possibilities of quantitative processing of chromatograms, combinations terminated with a column technique are advantageous if quantitative analysis is a main aim. However, because paper and thin-layer chromatograms are particularly amenable to chemical testing, procedures with planar chromatography as the last step are preferable in identification problems. In this case, it is possible either to collect column chromatographic fractions and then treat them by flat-bed techniques, or to apply the column effluent directly on to the starting line of the chromatographic paper or thin layer. In the latter version, the flat bed can be moved either continuously, at a constant or programmed velocity, or intermittently. This is exemplified schematically in Fig. 15.2.

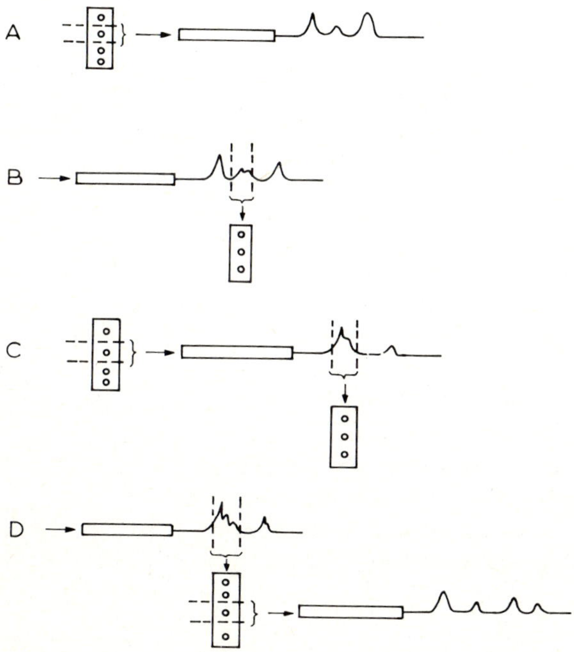


Fig. 15.1. Examples of the combination of column and flat-bed techniques. A, column–flat bed; B, flat bed–column; C, column–flat bed–column; D, flat bed–column–flat bed.

The combination of column and flat-bed techniques was originally developed for the direct coupling of gas and thin-layer chromatography (Janák, 1964, 1971), but results have already appeared that were obtained with a direct coupling of classical liquid column chromatography and thin-layer chromatography (Van Dijk). Owing to the limited volatility of liquid eluents, certain difficulties can be encountered as a result of slow evaporation of the solvent during the deposition of the eluates on the flat bed. In this respect, the use of liquefied gases as the column eluent may be advantageous.

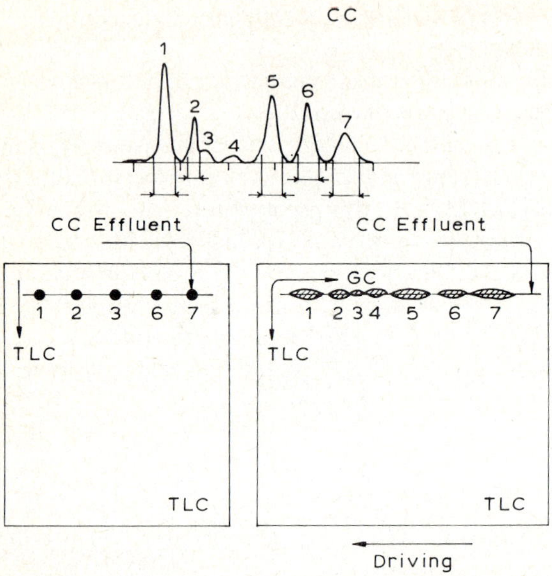


Fig. 15.2. Intermittent and continuous application of the column effluent on to the starting line of a flat bed.

## QUANTITATION

### Basic concepts

Quantitative analysis by means of any version of modern column elution chromatography is basically a combination of the processes of chromatographic separation of an $n$-component mixture into $n$ binary solute–mobile phase mixtures and the on-line determination of the solute constituents in the binary mixtures by using a special analyzer. This analyzer consists of a chromatographic detector and a recording and data processing system. Hence, chromatographic quantitative analysis involves the combined problems of

(*i*) chromatographic separation;

(*ii*) detection;

(*iii*) evaluation of the chromatogram.

A method of chromatographic quantitative analysis is a combination of systems and procedures relevant to the above three problems and the procedures of sample preparation and calibration.

The chromatogram of a given substance is actually a record of the changes with time of the concentration or the amount of the substance present in the detector; in some instances, the rate at which the substance is supplied to the detector is the main factor. These changes with time follow approximately a Gaussian curve. While the position of the curve in the chromatogram (retention time, retention volume, etc.) is characteristic of the substance in the given chromatographic system, the area under the peak, bound by the curve and the baseline, is proportional to the amount of the solute component that has passed through the detector. If the detector response is linearly proportional to the amount of the solute component fed into the sensor, the area under the peak is independent of all other parameters of the zone, such as its velocity of migration, dispersion and shape (symmetry). It is sometimes convenient to characterize the size of a peak by its maximum (height). With symmetrical peaks, the height is proportional to the area, but depends on both the velocity of migration and the dispersion of the zone, so that the analytical significance of the peak height is rather limited. With asymmetric peaks, the analytical utility of the peak height is doubtful, as there is even a lack of proportionality between the peak height and peak area in this instance.

## Relationship between the detector response and the amount of solute components in the chromatographic zone

### *Concentration of the solute components in the column effluent*

Provided that there is no accumulation or loss of solute during the migration of a zone down the column, the total amount of the solute component that has passed through the detector is given by the amount injected into the column. Let us consider this amount in units based on the number of moles; in order to avoid ambiguity, all symbols referring to the number of moles of solute and to the solute mole fraction in either phase of the sorption system will be designated by the subscript *i*. In the chromatographic system, the given (injected) amount, $n_i$, is distributed between the stationary and the mobile phases, *i.e.*

$$n_i = n_i'' + n_i' \tag{15.17}$$

where $n_i''$ and $n_i'$ are the numbers of moles of substance *i* in the stationary and the mobile phase, respectively; for symbols quoted without definition, see Chapter 1. As

$$n_i'' = K\, n_i'\, V_s/V_m \tag{15.18}$$

then

$$n_i' = n_i/[1 + K(V_s/V_m)] \tag{15.19}$$

The mean concentration of substance *i* in the mobile part of the chromatographic zone migrating down the column, $\bar{c}_z'$, can be expressed by

$$\bar{c}_z' = n_i'/\Delta V_{m,z} \tag{15.20}$$

where $\Delta V_{m,z}$ is the volume of the mobile phase bound by the zone in the column. This volume can be expressed as $\varphi_m\, \Delta z_z$, where $\varphi_m$ is the column void cross-section and $\Delta z_z$ is the width of the zone as measured in the column. Hence, the combination of eqns. 15.19 and 15.20 gives

$$\bar{c}_z' = n_i/\varphi_m \Delta z_z\,[1 + K(V_s/V_m)] \tag{15.21}$$

The subscript *z* indicates that the symbol refers to a quantity dependent on the axial position in the column. This implies that the respective equations apply for any position of the zone within the column, including the outlet of the column. For symbols that refer only to the column outlet, the subscript $z$ is omitted. Thus, the zone width as measured at the column outlet can be expressed by the equation

$$\Delta z = u_i \Delta t = u\Delta t / [1 + K(V_s/V_m)]$$

Further, as $u\varphi_m = v$, then the mean concentration in the column effluent is given by

$$\bar{c}' = n_i / v\Delta t \tag{15.22}$$

where $\Delta t$ is the time interval between the beginning and the end of elution of the zone. The product $v\Delta t$ actually denotes the volume of the solute–mobile phase mixture in which the solute component is contained after separation. The right-hand side of eqn. 15.21 can be expressed even more explicitly. It was shown in Chapter 3 (pp. 34 and 35) that $\Delta z = 4\,\sigma_L$ and $\sigma_L^2/L = H$, so that $\Delta z = 4(HL)^{½}$. Further, as $\varphi_m = \pi\epsilon d^2/4$, where $\epsilon$ is the total porosity of the column packing and $d$ is the column diameter, $\bar{c}'$ is given by

$$\bar{c}' = n_i / \pi\epsilon d^2 (HL)^{1/2} [1 + K(V_s/V_m)] \tag{15.23}$$

Eqn. 15.23 shows readily the influence of the basic column parameters on the solute concentration in the column effluent. If the porosity is expressed by $\epsilon = V_m/V = 4V_m/\pi d^2 L$, where $V$ is the total volume of the empty column, and the relationships $L/H = N$ and $V_m + KV_s = V_R$ are employed, we obtain finally

$$\bar{c}' = n_i N^{1/2} / 4V_R \tag{15.24}$$

With respect to the properties of Gaussian profiles, the solute concentration corresponding to the concentration maximum of the zone, $c'_{\text{max.}}$ is given approximately by

$$c'_{\text{max.}} = (2\pi)^{½}\,\bar{c}' \tag{15.25}$$

The mean concentration of the solute component in the column effluent can be formally expressed by

$$\bar{c}' = (1/\Delta t) \int_{t_1}^{t_2} c' \mathrm{d}t \tag{15.26}$$

where $c'$ is the instantaneous solute concentration in the column effluent and $t_1$ and $t_2$ are the time of the beginning and the end of elution of the zone, respectively ($\Delta t = t_2 - t_1$). Hence, on combining eqns. 15.26 and 15.22, we obtain

$$n_i = v \int_{t_1}^{t_2} c' \mathrm{d}t \tag{15.27}$$

It should be noted that $c'$ is defined as the number of moles per unit volume and $v$ is the volume flow-rate of the mobile phase.

### *Role of the detector*

Before deriving the relationships between the detector response and the solute concentration in the column effluent, it is necessary to discuss briefly some properties of detectors. The nature of the detector can substantially modify the above relationships, no matter whether peak height *versus* peak-maximum solute concentration or peak area *versus* total amount of solute in the zone relationships are considered. As the problems of detectors are dealt with in greater detail in a special chapter (Chapter 8), only those properties which determine directly the above relationships will be discussed here. In this context, three different classifications, each having a certain position with respect to quantitative analysis, are applicable, and are discussed below.

*Differential versus integral detectors*

This classification is based on the course of the time dependence of the detector response. Differential detectors react to the instantaneous amount of a substance present in the sensor whereas integral detectors yield data on the total amount of the substance that has passed through the detector from the entry of the substance into the sensor up to a given moment. With differential detectors, the chromatogram reflects directly the course of the time dependence of the solute concentration in the column effluent, modified to a certain extent by the specificity of detection towards the solute substance. Chromatograms obtained with differential and integral detectors are related to each other as differential and the corresponding integrated chromatographic records.

*Non-destructive versus destructive detectors*

With non-destructive detectors, the source of the signal is the detected substance itself. The signal is given merely by the presence of the substance in the sensor and is produced as long as the substance is present there, independent of the rate of supply of the substance into the sensor. With destructive detectors, the signal is given by the intensity of a quantity that accompanies some change in the substance to be detected. The intensity is determined by the rate of feeding the substance into the sensor, no matter whether this rate is controlled by the concentration of the substance in the mixture being introduced or by the rate of introduction of the mixture. The reaction products do not produce a signal, so that when the introduction of the substance into the sensor ceases, the signal disappears.

*Concentration-sensitive versus mass-sensitive detectors*

Concentration-sensitive detectors respond to the relative amounts of the solute and mobile phases, independent of the total amount of solute in the sensor, whereas mass-sensitive detectors respond either to the total amount of the solute in the sensor or to the rate of introduction of solute into the sensor, independent of the solute concentration in the mixture supplied.

As far as a particular detector is concerned, it always represents a combination of the above three types, *i.e.*, there are differential–non-destructive–concentration-sensitive detectors (refractometers), differential–destructive–mass-sensitive detectors (FID), etc. The properties of a detector, determined by its correspondence to one of the above classes, are readily apparent from the mathematical expression of the relationships between the detector response or peak area and the amount of solute in the chromatographic zone. In this respect, only the classifications non-destructive–destructive and concentration-sensitive–mass-sensitive will be considered further; the analytical implications of the classification differential–integral are obvious.

*References p.401*

*Relationships between the peak sizes and the amount of solute components in the column effluent*

If the detector response to a substance *i*, $R_i$, is expressed directly in the units of the recorder pen deflection, the peak area, $A_i$, is given by

$$A_i = b \int_{t_1}^{t_2} R_i \, \mathrm{d}t \tag{15.28}$$

where $b$ is the chart speed and $R_i$ is the instantaneous pen deflection. Eqns. 15.27 and 15.28 are generally valid and their combination with the appropriate definitions of the individual types of detector yield the respective relationships between $A_i$ and $n_i$; in all the cases discussed below, it will be assumed that the instantaneous detector response is linearly proportional to $c'$.

*Non-destructive mass-sensitive (NM) detectors*

With this type of detector, the instantaneous response should be proportional only to the number of moles of solute, $n_i^*$, present in the sensor space, $V^*$, *i.e.*,

$$R_i^{\mathrm{NM}} \approx n_i^* \tag{15.29}$$

As $n_i^* = c' V^*$, from eqns. 15.27 and 15.28 we obtain

$$A_i^{\mathrm{NM}} \approx (bV^*/v)n_i \tag{15.30}$$

For the peak height, $h_i$, with regard to eqn. 15.29, the equation

$$h_i \approx V^* c'_{\mathrm{max.}} \tag{15.31}$$

applies, which can be further modified according to eqns. 15.24 and 15.25 to read

$$h_i^{\mathrm{NM}} \approx (V^* N^{\frac{1}{2}}/2V_R)\, n_i \tag{15.32}$$

*Non-destructive concentration-sensitive (NC) detectors*

This type of detector can be characterized, for instance, by

$$R_i^{\mathrm{NC}} \approx x_i' \tag{15.33}$$

where $x_i'$ is the instantaneous solute mole fraction in the column effluent. At low solute concentrations, the mole fraction can be replaced by the mole ratio of solute to mobile phase, $n_i/n_m$, which can further be expressed by $n_i/n_m = c' M_m/\rho_m$, where $M_m$ and $\rho_m$ are the molecular weight and density, respectively, of the mobile phase. Hence, employing eqns. 15.27 and 15.28, we have

$$A_i^{\mathrm{NC}} \approx (bM_m/v\rho_m)\, n_i \tag{15.34}$$

and $h_i$ is given, with regard to eqns. 15.24 and 15.25, by

$$h_i^{\mathrm{NC}} \approx (M_m N^{\frac{1}{2}}/2V_R\rho_m)\, n_i \tag{15.35}$$

*Destructive mass-sensitive (DM) detectors*

This type of detector can be characterized by

$$R_i^{\mathrm{DM}} \approx \mathrm{d}n_i/\mathrm{d}t \tag{15.36}$$

As $\mathrm{d}n_i/\mathrm{d}t = c'v$, we can write

$$A_i^{\mathrm{DM}} \approx bn_i \tag{15.37}$$

and

$$h_i^{\mathrm{DM}} \approx (vN^{\frac{1}{2}}/2V_R)n_i \tag{15.38}$$

The relationships of $A_i$ *versus* $n_i$ and $h_i$ *versus* $n_i$ for individual types of detectors enable the following conclusions to be drawn. With NM detectors, both the peak area and the peak height are proportional to the effective volume of the detector sensor. Further, the peak area is inversely proportional to the flow-rate of the mobile phase while the peak height is independent of the flow-rate. With NC detectors, both $A_i$ and $h_i$ are independent of the sensor volume, the dependence on $v$ being the same as with NM detectors. Finally, with DM detectors, $A_i$ is independent of all of the above parameters and $h_i$ is proportional to $v$.

Closer inspection reveals that the dependence of $A_i$ and $h_i$ on $V^*$ is due to the fact that the detector is mass-sensitive, whereas the proportionality of $A_i$ to $1/v$ and the independence of $h_i$ of $v$ stem from the non-destructive character of the detector. The independence of $A_i$ of all the parameters discussed and the proportionality of $h_i$ to $v$ are typical of destructive detectors. Thus, the dependence of $A_i$ and $h_i$ on $V^*$ can be considered as a criterion of whether the detector is mass-sensitive or concentration-sensitive, whereas the dependence of $A_i$ and $h_i$ on $v$ is a criterion of whether the detector is destructive or non-destructive. It should be noted, however, that these criteria have been derived from idealized models of detectors. Therefore, any real detector can be expected to show certain deviations from the behaviour that could be assumed by virtue of the above characteristics.

## Specificity of detection and response factors

All of the relationships between $A_i$ and $n_i$ derived in the preceding section are based on the assumption of a linear proportionality between $R_i$ and $c'$. This proportionality can be written as

$$R_i = Bc' \tag{15.39}$$

where $B$ is a constant, comprising an apparatus (system of processing and recording the detector response) constant, $B_a$, a constant characterizing the type of detector as discussed above, $B_d$, and a constant reflecting the properties of the column effluent, $B_{im}$. Hence, the relation between $A_i$ and $n_i$ can be expressed by

$$A_i = B_a B_d B_{im} n_i \tag{15.40}$$

which should hold generally, regardless of the type of detection system employed.

*References p.401*

Eqn. 15.40 shows that the peak area corresponds to the amount of the solute component eluted out of the column. Therefore, it is necessary to know the proportionality constant $B$ in order to convert the peak area into the corresponding amount of solute. This constant can be determined either by calibrating the instrument, *i.e.*, by determining the peak area corresponding to a known amount of the given substance, or by theoretical analysis. With the given detector and under constant conditions (detector temperature, flow-rate of mobile phase, electrometer and recorder settings), the constants $B_a$ and $B_d$ are the same for any solute substance and mobile phase, but the constant $B_{im}$ is generally different for different solutes and mobile phases. This variability in the sensitivity of a given detector to different substances can be utilized for selective detection. However, in quantitative chromatography, this selectivity raises the problems of detector response factors and of the appropriate concentration units to be used for expressing the analytical results. In the discussion below, it will be assumed that the chromatographic zone does not produce any discontinuity in the physical state of the fluid flowing through the column after migration of the zone for some time and that a steady-state hydrodynamic regime is established well before the zone enters the detector. These assumptions imply that some volume of the mobile phase is replaced by a volume of the solute component in the zone.

*Analytical properties and molar and relative molar response*

When discussing the problem of the specificity of detection, it is very helpful to employ the concept "analytical property", which is a property of the substance to be analyzed that has a defined relationship with the amount and nature of the substance. Thus, the character of the analytical property is determined by the method of detection while the magnitude of this property is given by the type of the substance. Let us denote the instantaneous analytical property of the solute–mobile phase mixture by $a_{i0}$ and the analytical property of the pure mobile phase by $a_0$. The detector responds to both the solute–mobile phase mixture ($R_{i0}$) and the mobile phase alone ($R_0$), the net response to the solute compound, $R_i$, being given by the difference

$$R_i = R_{i0} - R_0 \tag{15.41}$$

The quantity $a_{i0}$ can be considered as an implicit function of the solute concentration. Hence, one can write for detectors of the NM, NC and DM type, respectively:

$$R_i^{\mathrm{NM}} \approx V^*(a_{i0} - a_0) \tag{15.42}$$

$$R_i^{\mathrm{NC}} \approx (a_{i0} - a_0) \tag{15.43}$$

$$R_i^{\mathrm{DM}} \approx v\,(a_{i0} - a_0) \tag{15.44}$$

Let us assume that $a_{i0}$ is a function of the solute mole fraction in the column effluent, $x_i'$ (*cf.*, Fig. 15.3) as only the solute concentration in the mobile phase will be dealt with in the further discussion, the prime on the symbol $x_i'$ will be omitted. The expansion of $a_{i0}$ into a Maclaurin series yields

$$a_{i0} = a_0 + \frac{1}{1!} \cdot \frac{\mathrm{d}a_{i0}}{\mathrm{d}x_i} \cdot x_i + \frac{1}{2!} \cdot \frac{\mathrm{d}^2 a_{i0}}{\mathrm{d}x_i^2} \cdot x_i^2 + \ldots \tag{15.45}$$

In chromatography, the concentration of the solute in the column effluent is usually small, so that it is possible to take into account only the two first terms of the series and write

$$a_{i0} - a_0 = (\mathrm{d}a_{i0}/\mathrm{d}x_i)\,x_i \tag{15.46}$$

At low solute concentrations, $x_i$ can be expressed by $x_i = c'M_m/\rho_m$ (*cf.*, eqns. 15.33 and 15.34), so that eqns. 15.42–15.44 can be rewritten to read

$$R_i^{\mathrm{NM}} \approx (V^*M_m/\rho_m)(\mathrm{d}a_{i0}/\mathrm{d}x_i)\,c' \tag{15.47}$$

$$R_i^{\mathrm{NC}} \approx (M_m/\rho_m)(\mathrm{d}a_{i0}/\mathrm{d}x_i)\,c' \tag{15.48}$$

$$R_i^{\mathrm{DM}} \approx (vM_m/\rho_m)(\mathrm{d}a_{i0}/\mathrm{d}x_i)\,c' \tag{15.49}$$

Provided that $a_i$ and $a_0$ are additive, *i.e.*, if $a_{i0} = a_ix_i + a_0x_m$, where $x_m$ is the mole fraction of the mobile phase in the mixture, then, as $x_m = 1 - x_i$, we can write

$$a_{i0} - a_0 = (a_i - a_0)\,x_i \tag{15.50}$$

In some instances, it might be more convenient to consider $a_{i0}$ to be a function of the weight or volume fraction of the solute. A procedure analogous to that shown above would then lead to some modifications in the $R_i$ *versus* $c'$ relationships. These alternatives are obvious and will not be discussed in detail here. The dependence of $a_{i0}$ on $x_i$ is shown in Fig. 15.3.

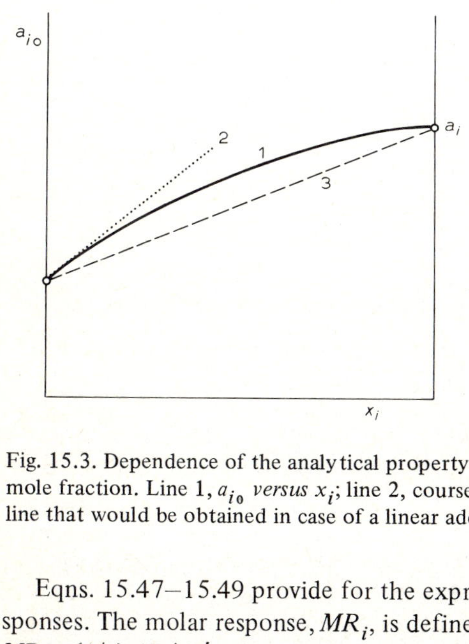


Fig. 15.3. Dependence of the analytical property of the solute–mobile phase mixture on the solute mole fraction. Line 1, $a_{i0}$ *versus* $x_i$; line 2, course of the approximately linear section of line 1; line 3, line that would be obtained in case of a linear additivity of $a_i$ and $a_0$.

Eqns. 15.47–15.49 provide for the expression of the molar and relative molar responses. The molar response, $MR_i$, is defined by $MR_i = \mathrm{d}R_i/\mathrm{d}(\mathrm{d}n_i'/\mathrm{d}t)$. As $\mathrm{d}n_i'/\mathrm{d}t = c'v$, then $MR_i = (1/v)\,\mathrm{d}R_i/\mathrm{d}c'$. The relative molar response, $RMR_{ir}$, is defined by the ratio $RMR_{ir} = MR_i/MR_r$, where $MR_r$ is the molar response of a reference substance $r$, provided that both $MR_i$ and $MR_r$ have been determined under identical conditions. Hence, for any type of detector, the equation

$$RMR_{ir} = (\mathrm{d}a_{i0}/\mathrm{d}x_i)/(\mathrm{d}a_{r0}/\mathrm{d}x_r) \tag{15.51}$$

holds, where the subscript $r$ refers to the above reference substance. The substitution for $\mathrm{d}a_{i0}/\mathrm{d}x_i$ in eqns. 15.47–15.49 from eqn. 13.51 and integration while employing eqns. 15.49 and 15.28 results in

$$A_i^{\mathrm{NM}} \approx (bV^*M_m/\rho_m v)(\mathrm{d}a_{r0}/\mathrm{d}x_r)RMR_{ir}n_i \tag{15.52}$$

$$A_i^{\mathrm{NC}} \approx (bM_m/\rho_m v)(\mathrm{d}a_{r0}/\mathrm{d}x_r)RMR_{ir}n_i \tag{15.53}$$

$$A_i^{\mathrm{DM}} \approx (bM_m/\rho_m)(\mathrm{d}a_{r0}/\mathrm{d}x_r)RMR_{ir}n_i \tag{15.54}$$

Hence, for any kind of detector, the equation

$$A_i = C \cdot RMR_{ir}n_i \tag{15.55}$$

holds, where $C$ represents $B_aB_d$ from eqn. 15.40 and $RMR_{ir}$ represents $B_{im}$. According to eqn. 15.55, $A_r = Cn_r$ ($RMR_{ir} = 1$ for $i = r$), so that

$$RMR_{ir} = (A_i/n_i)/(A_r/n_r) \tag{15.56}$$

Eqn. 15.56 is equivalent to eqn. 15.51 and can be readily employed in experimental determinations of $RMR$ values.

Eqn. 15.55 reveals that if the individual peak areas in the chromatogram are divided by the corresponding $RMR_{ir}$ values, the resulting $A_i/RMR_{ir}$ values are uniformly proportional to the numbers of moles of the respective components in the chromatographic zones. Thus, we can formulate the molar response factor as

$$f_i^n = 1/RMR_{ir} \tag{15.57}$$

where the superscript $n$ indicates that the peak areas multiplied by this factor are proportional to molar units of the solute compounds. As $n_i = w_i/M_i$, where $w_i$ is the weight of substance $i$, one can readily formulate the weight response factor:

$$f_i^w = M_i/RMR_{ir} \tag{15.58}$$

The peak areas multiplied by the corresponding $f_i^w$ values are directly proportional to weights of the solute components.

## Methods of quantitative chromatographic analysis

### *Working techniques*

All of the techniques devised for and employed in quantitative gas chromatography are essentially applicable in liquid chromatography also. As these techniques have been described in detail elsewhere (Novák), only a brief survey of the principles and final equations will be presented here.

In this section, all data on concentration will refer to the material being analyzed rather than to the column effluent. The subscripts $i$ and $s$ refer to the compound being determined and to a standard compound, respectively. Symbols with the subscripts without parentheses refer to pure compounds $i$ or $s$, whereas symbols with the subscripts in parentheses refer to a mixture analyzed for the content of the compound specified by the subscript. The symbols $W$, $V$ and $N$ represent the weight, volume and number of moles of the material handled in sample preparation prior to injection into the chromatograph, whereas the symbols $w$, $v$, and $n$ denote the weight, volume and number of moles of the charge introduced into the chromatograph. $M$ and $\rho$ are molecular weight and density. Symbols with a bar over them represent mean values.

We shall consider the following concentration units in the present survey:

(*a*) number of moles of a component per unit volume of the mixture:

$$m_i = \frac{N_i}{V_{(i)}} \tag{15.59}$$

(*b*) weight of a component per unit volume of the mixture:

$$q_i = \frac{W_i}{V_{(i)}} \tag{15.60}$$

(*c*) mole fraction of a component:

$$y_i = N_i/\Sigma N_i \tag{15.61}$$

(*d*) weight fraction of a component:

$$g_i = W_i/\Sigma W_i \tag{15.62}$$

The symbol $\Sigma$ represents a summation over all the components present in the mixture.

*Absolute calibration technique*

*Calculation procedure.* Separate injections of defined amounts of the material under analysis and of a calibration solution containing a standard substance, $s$, of known concentration are carried out under the same conditions, thus giving two chromatograms with peak areas $A_i$ and $A_s$. The concentration of the substance being determined, $i$, can be calculated by using the following equations:

$$m_i = m_s A_i f_i^n v_{(s)}/A_s f_s^n v_{(i)} \tag{15.63}$$

$$q_i = q_s A_i f_i^w v_{(s)}/A_s f_s^w v_{(i)} \tag{15.64}$$

$$g_i = g_s A_i f_i^w w_{(s)}/A_s f_s^w w_{(i)} \tag{15.65}$$

$$y_i = y_s A_i f_i^n n_{(s)}/A_s f_s^n n_{(i)} \tag{15.66}$$

*Graphical procedure.* A series of different defined amounts of a standard substance is introduced into the column and a calibration graph is constructed by plotting either $m_s v_{(s)}$ or $y_s n_{(s)}$ against $A_s f_s^n$, or $q_s v_{(s)}$ or $g_s w_{(s)}$ against $A_s f_s^w$. The amount of the substance being determined in the amount of the material injected is then found by fitting the corrected peak area ($A_i f_i^n$ or $A_i f_i^w$) to the appropriate calibration graph. With both the calculation and the graphical procedures, the response factors can be omitted if the standard component is identical with that being determined.

*References p.401*

*Internal standard technique*

*Calculation procedure.* A defined amount of the material to be analyzed is mixed with a defined amount of a solution of a standard substance of known concentration, and a certain amount of this mixture is introduced into the chromatograph. The following equations are used for calculating the results:

$$m_i = m_s A_i f_i^n V_{(s)} / A_s f_s^n V_{(i)} \tag{15.67}$$

$$q_i = q_s A_i f_i^w V_{(s)} / A_s f_s^w V_{(i)} \tag{15.68}$$

$$g_i = g_s A_i f_i^w W_{(s)} / A_s f_s^w W_{(i)} \tag{15.69}$$

$$y_i = y_s A_i f_i^n N_{(s)} / A_s f_s^n N_{(i)} \tag{15.70}$$

*Graphical procedure.* Several solutions of different defined concentrations of the compound to be determined are prepared, and each solution is mixed in defined proportions with a defined solution of a standard substance. Samples of the mixtures are chromatographed and the numerical values of the right-hand side of one of eqns. 15.67–15.70 relevant to the concentration units chosen are calculated, without taking into account the response factors. The calibration graph is obtained by plotting the concentrations of the compound being determined in the initial solutions (prior to mixing them with the standard substance solution) against the corresponding numerical values determined as mentioned above. The calibration graph of $g_i$ against $\dfrac{A_i W_{(s)} g_s}{A_s W_{(i)}}$ should be a straight line passing through the origin. In the analytical run proper, the material to be analyzed is handled as described in the calculation procedure, but the data obtained are processed in the manner specified for constructing the calibration graph. The analytical results are obtained by fitting the processed data to the appropriate calibration graph. The use of the graphical procedure eliminates the necessity of knowing the response factors.

*Standard addition technique*

Although this technique can involve a graphical procedure, it will not be discussed here as it is relatively unimportant. The calculation procedure exists in essentially two variants, described below.

*Direct measurement of the sample charges.*

(*a*) A defined amount of the material being analyzed is introduced into the chromatograph and a chromatogram of the original material is obtained. The amount of material introduced in this run and the corresponding peak area will be denoted by subscripts $(i)$ and $i$, respectively.

(*b*) A defined amount of the original material is mixed with a defined amount of a standard solution of known concentration of the substance being determined, and a defined amount of the mixture is introduced into the chromatograph and run under the same conditions as those used for the original material. The amounts of the original material and the standard solution will be denoted by subscripts $(i)$ and $(s)$, respectively, while subscripts $(is)$ and $is$ will be used for the charge of the mixture and the corresponding peak area, respectively. The calculation of the results can be carried out by means of the following equations:

$$m_i = \frac{m_s V_{(s)}/V_{(i)}}{(A_{is}v_{(i)}/A_i v_{(is)})[1 + (V_{(s)}/V_{(i)})] - 1} \tag{15.71}$$

$$q_i = \frac{q_s V_{(s)}/V_{(i)}}{(A_{is}v_{(i)}/A_i v_{(is)})[1 + (V_{(s)}/V_{(i)})] - 1} \tag{15.72}$$

$$g_i = \frac{g_s W_{(s)}/W_{(i)}}{(A_{is}w_{(i)}/A_i w_{(is)})[1 + (W_{(s)}/W_{(i)})] - 1} \tag{15.73}$$

$$y_i = \frac{y_s N_{(s)}/N_{(i)}}{(A_{is}n_{(i)}/A_i n_{(is)})[1 + (N_{(s)}/N_{(i)})] - 1} \tag{15.74}$$

*Use of an auxiliary reference substance.* This version makes it possible to avoid the necessity for defining the amounts charged into the chromatograph. Instead, peak areas of an auxiliary reference substance either present in the original material or added to it are measured in both chromatograms together with the peak areas of the substance to be determined. The type and amount of the reference substance added need not be defined. The analytical procedure is otherwise exactly the same as described in the preceding version except for the requirement of defining the amounts of the sample charges. In the present version, it is the peak area of the reference substance that actually serves as a measure of the charge. Denoting the peak areas of the reference substance in the chromatograms of the original material and of the mixture with the standard solution by $A_a$ and $A'_a$, respectively, we can write the following equations:

$$m_i = \frac{m_s V_{(s)}/V_{(i)}}{(A_{is}A_a/A_i A'_a) - 1} \tag{15.75}$$

$$q_i = \frac{q_s V_{(s)}/V_{(i)}}{(A_{is}A_a/A_i A'_a) - 1} \tag{15.76}$$

$$g_i = \frac{g_s W_{(s)}/W_{(i)}}{(A_{is}A_a/A_i A'_a) - 1} \tag{15.77}$$

$$y_i = \frac{y_s N_{(s)}/N_{(i)}}{(A_{is}A_a/A_i A'_a) - 1} \tag{15.78}$$

Provided that $V_{(i)}$, $W_{(i)}$ and $N_{(i)}$ always refer to the amount of the original material being analyzed, the same equations hold for both the alternative with the reference substance already present in the material and that involving addition of the reference substance. It should be noted that all of the above techniques are unsuitable for calculating the results in mole fractions, as it would be necessary to know the molecular weights of both the material being analyzed and the standard substance, which is usually impractical. Hence, the equations are quoted only in order to make the survey complete. However, the technique of normalization, described below, is well suited for calculating mole or weight fractions.

*Normalization technique*

With this technique, it is necessary for all the components of the material being analyzed to be identified and to produce a measurable peak in the chromatogram. A certain amount of the material is introduced into the chromatograph and a chromatogram with the peaks of all the components present is obtained. The area of each peak is multiplied by the corresponding response factors and the corrected areas are summed. The mole or weight fraction of any component is then obtained by dividing the respective corrected area by the above sum. The following equations are applicable with this technique:

$$y_i = A_i f_i^n / \Sigma A_i f_i^n \tag{15.79}$$

$$g_i = A_i f_i^w / \Sigma A_i f_i^w \tag{15.80}$$

The calculation of the results in the units of $m_i$ and $q_i$ can be carried out by means of the equations

$$m_i = (\rho_{(i)}/M_{(i)})(A_i f_i^n / \Sigma A_i f_i^n) \tag{15.81}$$

$$q_i = \rho_{(i)} A_i f_i^w / \Sigma A_i f_i^w \tag{15.82}$$

which are obviously unsuitable because it is necessary to know the density of the material being analyzed.

*Manual evaluation of chromatograms*

The area of a peak of any shape can be determined planimetrically or by cutting the peak out of the chromatogram and weighing it. With symmetrical peaks, the peak area can also be calculated from linear parameters of the peak. It follows from the analysis of a Gaussian curve that the area of a symmetrical peak is determined by the following equations:

$$A = (\pi/4 \ln 2)^{\frac{1}{2}} h w_{\frac{1}{2}} = 1.06\, h w_{\frac{1}{2}} \tag{15.83}$$

$$A = [(2\pi \mathrm{e})^{\frac{1}{2}}/4]\, h' w' \tag{15.84}$$

where $h$ is the peak height, $w_{\frac{1}{2}}$ is the width at the half-height of the peak, $h'$ is the height of the triangle bound by the baseline and the tangent lines drawn at the points of inflection of the curve, $w'$ is the width of the base of the triangle and e is the base of natural logarithms. The application of the plate theory yields the following relationship:

$$A = (2\pi/N)^{\frac{1}{2}}\, h\, b\, t_R \tag{15.85}$$

where $N$ is the number of plates, $b$ is the recorder chart speed and $t_R$ is the retention time ($b\,t_R$ is the distance of the peak maximum from the starting point). All of the above methods are rather laborious and have therefore mostly been replaced by modern techniques of automatic integration and data processing.

*Automatic evaluation of chromatograms*

The detectors employed in modern liquid chromatography usually provide an electrical signal that is proportional to the concentration of solute in the detector. Such a signal can be processed automatically by methods that have been developed in the last decade in connection with gas chromatography.

There are a number of levels on which it is possible to handle automatically the information provided by the chromatograph, ranging from the simplest mechanical analogue integrators to sophisticated and expensive single-purpose computer systems. A reasonable choice between them represents a compromise between the price of the instrumentation and the required effect. The following general levels are available:

(*i*) mechanical analogue integrators;
(*ii*) digital integrators;
(*iii*) small single-purpose computers;
(*iv*) large computer systems.

Probably the most common type of analogue integrator is the mechanical integrator of the disc and ball type. The basic components of this integrator are a disc of radius $r$, revolving at a constant angular velocity $\omega$ given by the chart speed, and a small ball of radius $\rho$, pressed towards the disc. The ball can be moved by means of a coupling rod along the radius of the disc within the centre and the outside, *i.e.*, $0 \leqslant r(t) \leqslant r$; this movement is usually controlled by the potentiometer of the recorder. The speed of rotation of the ball being rolled by the revolving disc is dependent on the angular velocity of the disc, the time-dependent deflection of the ball and the radius of the ball proper. The total number of revolutions of the ball, $N$, within the interval $<0, t>$ is given by the relationship

$$N = \frac{\omega}{2\pi\rho} \int_0^t r(t)\,\mathrm{d}t \tag{15.86}$$

and is scanned mechanically and recorded by an auxiliary pen.

The integrator, the operation of which is closely associated with the recorder, works linearly over the whole range of the recorder scale. Even in less favourable instances, a remarkable precision of 0.9–1.3% is attained, provided that suitable sensitivity settings are used by switching over the attenuator. However, a pre-condition of the proper operation of this integrator is a precise zeroing of the recorder and a stable zero line. The manufacturers provide facilities for the elimination of the systematic errors incidental to the baseline drift, but their use lengthens the time of processing the record and decreases the precision of the data measured.

The rapid development of computer-based techniques led to the construction of modern digital integrators. These instruments permit qualitatively different working possibilities compared with the above analogue device. Typical features of this modern type of instrumentation are an extremely wide linearity range, a precision better than 0.1% and the ability of both providing the operator with the results in the form of printed retention times and peak areas and referring the data to a computer by means of a punched or magnetic tape.

The basic components of instruments of the above type are an operational amplifier of the detector signal, a device for converting the voltage into a sequence of pulses with a

*References p.401*

frequency proportional to the voltage, and a counter of the generated pulses, together with appropriate memory facilities and output devices. In addition, there are extensive analogue facilities providing for automatic peak detection, baseline tracking and elimination of the baseline drift, and matching the signal to a control line-recorder.

A simplified scheme of a digital integrator is shown in Fig. 15.4. The signal from the detector first enters the operational amplifier, from which it is led, after some modifications, into the recorder, converter and detector of peaks. The converter transforms the input voltage into pulses of frequency in a ratio of about 1 kHz/mV. The operation of the instrument is controlled by the information from both the converter and peak detector by means of control logic. This unit controls the work of the automatic baseline corrector and the data counter (peak areas and retention times). When the integration is completed, the data from the counter are transferred upon the command of the control unit into memory registers, the counters are zeroed and the data stored in the memory are printed out or recorded on a punched or magnetic tape.

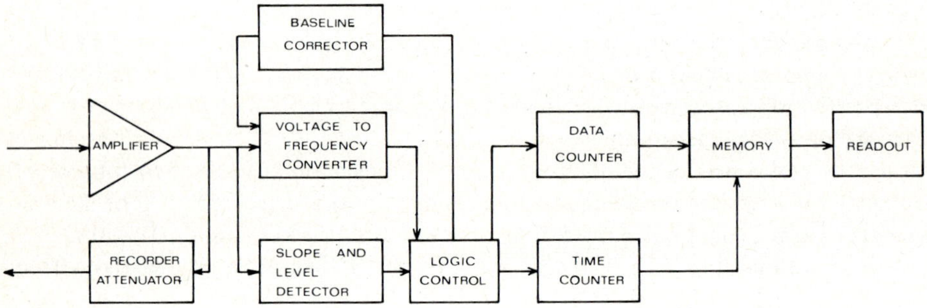


Fig. 15.4. Schematic representation of a digital integrator.

The peak detector can usually operate in two optional modes. In one of these modes, a peak is detected by virtue of the logical sequence of the changes in the slope resulting from the elution of the zone. The beginning of the peak is recognized by the instrument and the command to start the integration is given if the value and time of duration of the positive slope exceed a level chosen by the operator. The end of the peak is determined by a zero slope that follows after a negative one. In the other working mode, the instrument starts integration after the signal exceeds a certain adjustable voltage level and continues until it drops below this level again, regardless of the slope and its changes. Hence, the precision depends appreciably on the operator, as an excessively high suppression of the noise in the line before the peak detector (controlled manually by setting the amplitude and frequency filtering) or too high a speed of tracking the positive baseline drift can decrease the sensitivity of peak detection significantly and cause a systematic negative error in the results of automatic integration.

The combination of an integrator and punched or magnetic tape provides for a computer to perform quantitative and qualitative processing of the data or even to complete the analytical report. In addition to this so-called off-line system of data handling, there are also small special calculators for simultaneous automatic processing of data from a single integrator. Finally, this category of instrumentation involves the so-called computing integrators, which can, in a certain limited way, process simultaneously the output information from four chromatographs.

The last two levels of the automatic handling of chromatographic data provide for processing data from several tens of chromatographs simultaneously. Analogue items of information provided simultaneously by all of the chromatograms enter a multiplexer from which they are introduced periodically into the computer memory via a digital converter. In this unit, the data are processed exclusively on a numerical basis in such a manner that the result represents a complete analytical report. As a more detailed description of these techniques and procedures is beyond the scope of this book, only a brief survey of the characteristics of the above systems is given below.

A small single-purpose computer (8–11 K) can handle data from ten to forty chromatographs simultaneously. In addition to functions analogous to those of digital integrators, it can also perform all the necessary calculations required by the chosen technique of quantitative analysis (normalization, internal standardization, etc.). With respect to the limited capacity of the memory, it is not able to store a library of retention data and thus perform identifications.

A large single-purpose computer system (>30 K) can simultaneously process data from more than forty chromatographs, mass spectrometers, infrared spectrometers, etc. From the chromatographic viewpoint, it can perform the same service as the smaller computer but, in addition, the large capacity of its quick external memory enables a large library of relative retention data and detector response factors to be stored. Therefore, this system makes it possible to perform effective qualitative analyses and provides for a reduction in the number of calibrations that are necessary for the systems of the lower category.

## REFERENCES

Bate-Smith, E. C. and Westall, R. G., *Biochim. Biophys. Acta*, 4 (1950) 427.

Consden, R., Gordon, A. H. and Martin, A. J. P., *Biochem. J.*, 38 (1944) 224.

Haderka, S., *J. Chromatogr.*, 91 (1974) 167.

Hais, I. M. and Macek, K. (Editors), *Paper Chromatography*, Academic Press, London, 1963.

Janák, J., *J. Chromatogr.*, 15 (1964) 15.

Janák, J., in A. Niederwieser and G. Pataki (Editors), *Progress in Thin-Layer Chromatography and Related Methods*, Vol. II, Ann Arbor Sci. Publ., Ann Arbor, Mich., 1971, p. 63.

Kabasakalian, P. and Talmage, J. M., *Anal. Chem.*, 34 (1962) 273.

Kováts, E., *Helv. Chim. Acta*, 41 (1958) 1915.

Macek, K., Hais, I. M., Kopecký, J. and Gasparič, J. (Editors), *Bibliography of Paper and Thin-Layer Chromatography 1961–1965*, Elsevier, Amsterdam, London, New York, 1968.

Macek, K., Hais, I. M., Kopecký, J., Gasparič, J., Rábek, V. and Churáček, J. (Editors), *Bibliography of Paper and Thin-Layer Chromatography 1966–69*, Elsevier, Amsterdam, London, New York, 1972.

Macek, K., Hais, I. M., Kopecký, J., Schwarz, V., Gasparič, J. and Churáček, J. (Editors), *Bibliography of Paper and Thin-Layer Chromatography 1970–1973*, Elsevier, Amsterdam, London, New York, 1976, in press.

Martin, A. J. P. and Synge, R. L. M., *Biochem. J.*, 35 (1941) 1358.

Novák, J., *Advan. Chromatogr.*, 11 (1974) 1.

Schauer, H. K. and Bulirsch, R., *Z. Naturforsch.*, 13b (1958) 327.

Van Dijk, J. H., in E. Kováts (Editor), *Column Chromatography, Lausanne, 1969*, Sauerländer, Aarau, 1970, p. 234.

*Chapter 16*

# Radiochromatographic techniques

I. M. HAIS and J. DRŠATA

## CONTENTS



## INTRODUCTION

The chromatography of radioactive substances is encountered in many fields, but especially in biochemistry and the biological sciences. In radiochemistry, chromatography is used for the isolation and analysis of the products of nuclear and chemical reactions. Isotope dilution and isotope derivative methods are often combined with chromatography.

Column, capillary and flat-bed radiochromatography using a liquid mobile phase generally require a less complicated detector or fraction collector than does the gas–liquid chromatography of labelled compounds (*cf.*, Schutte and Koenders). The speed of modern liquid chromatography may prove to be of special value in the separation of short-lived isotopes and labile compounds.

The chromatographic technique as such, involving sample introduction, columns, sorbents, developers and a pumping system, used for radioactive substances, does not differ in any essential respect from the techniques used for stable nuclides. It should be borne in mind, however, that contamination may present a problem that can be serious owing to the inherently high sensitivity (in terms of the ratio between the signal and the number of molecules of the substance being analyzed) of the technique and the wide range of radioactivity levels that may occur simultaneously in different compounds. In

*References p.413*

this respect, one should ensure the radiochemical cleanliness of the laboratory, glassware (inexpensive items are preferably discarded after use) and the column after completion of the chromatogram. The possibility that radioactive reaction products such as tritiated water from tritiated compounds and $^{14}CO_2$ from $^{14}$C-labelled compounds may be generated should also be considered and atmospheric contamination avoided.

When low concentrations of radionuclides of high specific activity are being used with no displacing molecules present and if the sorption isotherm initially rises steeply owing to the presence of a small number of highly active sorption sites, this may lead to the retention of residual radioactivity until a strong displacer is passed through the column. This problem can be avoided if a cold carrier is added to the sample before chromatography begins.

## DETECTORS

Radiation detectors used during or after liquid column chromatography differ according to the type of radiation involved, its energy and sometimes its level.

### $\gamma$-Radiation detectors

The usual method for detecting $\gamma$-radiation involves the use of an aluminium-covered crystal of thallium-activated sodium iodide coupled to a photomultiplier. The counts may be printed digitally at pre-determined time intervals or converted to give an output on a continuously recording rate-meter. The photomultiplier may be connected to either a simple pulse-height analyzer for rapid monitoring or to a multi-channel $\gamma$-spectrometer if it is intended to use the radiation characteristics of the components of a mixture for the discrimination and possibly even determination of individual nuclides in the mixture.

If Geiger-Müller (GM) counters are used to detect $\gamma$-radiation, they are provided with a metal radiator to improve their low efficiency for this type of radiation. Semiconductor devices have been used for high-resolution $\gamma$-spectroscopy in the form of lithium-drifted germanium detectors (Table 16.1).

### Geiger-Müller detectors

GM counting has been described for liquid samples, either after fraction collection (Bernhardt) or for continuous effluent monitoring using a cell covered with thin Mylar foil (Jordan) (Fig. 16.1). This method is, of course, suitable only for sufficiently hard $\beta$-emitters such as $^{32}$P. However, $^{14}$C- or $^{35}$S-labelled compounds have also been GM counted in paper or aluminium cups after evaporation of the solvent. For tritium, windowless proportional gas-flow counters have been used, but the efficiency is not very high.

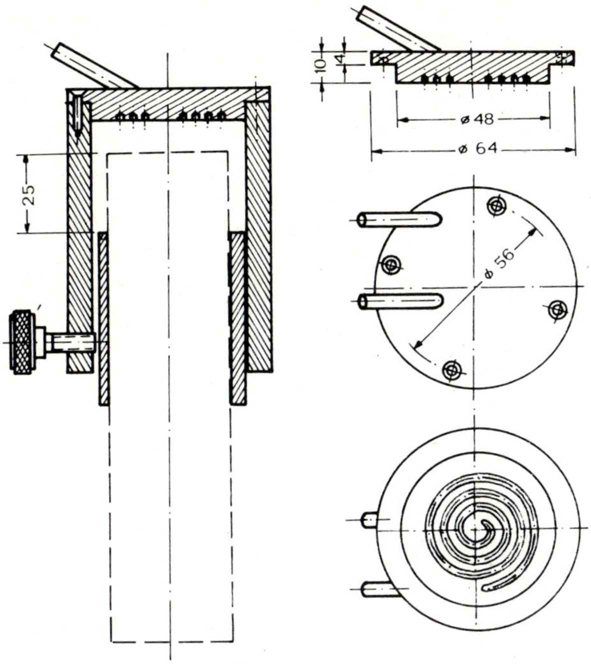


Fig. 16.1. Flow cell for GM counting (Jordan). Dimensions in millimetres. The flow cell proper (in which the spiral groove is machined) and the inlet–outlet tubes are made of PTFE.

### Solid-phase scintillation detectors for β-particles

The scintillator itself may line the walls of a coil-shaped or U-shaped capillary cell. Coils made of NE 102A plastic scintillator (Hertel *et al.*) are available as Nuclear Enterprises NE 801 (0.7 mm I.D. tubing, total standard cell volume 0.35 ml). The material is suitable for aqueous solutions and lower alcohols, but is attacked by most other solvents and exhibits about 5% efficiency for $^{14}C$ in liquids.

The efficiency for less energetic radiation is depressed owing to absorption of particles originating near the axis of the tubular cell. If the phosphor is finely divided, the effluent is forced through the porous mass and shorter path-lengths are thereby achieved. The most important variables include the solubility and the adsorption activity of both the cell material and the phosphor. One of the most frequently used phosphors is anthracene, which has been applied even for tritium (Hunt, Schram and Lombaert). NE 806 and NE 806A are available, 1.2 ml standard spiral cell volume. Anthracene is of course unsuitable for use with organic solvents in which it is soluble or for strongly oxidizing acids (nitric, chromic, etc.). Water or aqueous alcoholic solutions may be used. Many compounds are adsorbed from aqueous solutions, leading to tailing in the "effluent monitoring" mode.

Schutte has compared the properties of various materials in this respect (Table 16.1). He found that counting efficiencies for organic phosphors in U-shaped tubes were good for $^{14}C$ but less satisfactory for $^{3}H$. All of the organic materials tabulated are soluble in organic solvents and highly adsorbent. Calcium fluoride (used in NE 808, 0.5 ml standard spiral cell volume) is dissolved in solutions of ammonium salts and adsorbs nucleotides. The only universally applicable material is therefore glass powder, as a result of its insolubility and lack of adsorption. It is marketed as NE 808-modified (0.5 ml standard spiral cell volume). Its principal disadvantage is the high light-induced phosphorescence, which persists for several days to give high background counts. In Schutte's hands the efficiency was better for the U-tube than for the coils, even when smaller cell volumes were involved.

For aqueous effluents, Sieswerda and Polak prefer POPOP crystals for their suitable flow properties. A detergent in the mobile phase, such as polyethylene lauryl ether, may reduce adsorption contamination of the scintillator.

TABLE 16.1
COUNTING EFFICIENCIES FOR VARIOUS FLOW CELLS (SCHUTTE)

| Shape of cell | Effective volume (μl) | Scintillator | Counting efficiency (%) | |
|---|---|---|---|---|
| | | | $^{14}C$ | $^{3}H$ |
| U | 350 | Anthracene | 37 | 1.0 |
| U | 160 | Anthracene | 31 | 1.0 |
| Coil | 500 | Anthracene | 20 | 0.6 |
| U | 350 | PPO* | 43 | 1.8 |
| U | 500 | Butyl-PBD** | 40 | 1.7 |
| U | 480 | Glass*** | 17 | 0.2 |
| Coil | 430 | Glass*** | 5 | $<0.1$ |
| U | 190 | $CaF_2$ (Eu-activated) | 38 | 0.5 |

*2,5-Diphenyloxazole.
**2-(4′-*tert.*-Butylphenyl)-5-(4″-biphenylyl)-1,3,4-oxadiazole.
***Cerium-activated lithium glass NE 901 (250–1000 μm), from Nuclear Enterprises Ltd., Edinburgh, Great Britain.

## Liquid scintillation detectors

Liquid scintillation counting seems to be currently the most generally applicable method, especially after fractionation. Liquid scintillation is also used in the effluent monitoring mode, as will be shown below (Hunt). Commercial scintillation equipment with automatic quench correction (external standard) is available for this purpose.

The chromatographic solvent may cause certain problems if it is not removed before counting. Highly quenching solvents should be avoided. Where aqueous solvents are concerned, either toluene or dioxane scintillation mixtures can be used (Hunt). The addition of polar organic solvents allows one-phase mixtures to be formed with certain proportions of water.

Toluene–ethoxyethanol scintillation mixture (Gaitonde and Nixey) consists of 0.1% of PPO and 0.04% of POPOP in a mixture of toluene and ethoxyethanol (7:3, v/v). A 10-ml volume of this mixture will dissolve 0.2 ml of water; 0.5 and 1.0 ml of water require the addition of 1.5 and 3 ml of ethoxyethanol, respectively.

Dioxane scintillation mixture (Minard and Mushahwar) consists of 4.2328 g PPO and 63.429 g naphthalene per litre of dioxane. According to Gaitonde and Nixey, 10 ml of the mixture should dissolve up to 4.5 ml of water.

Bray's solution contains 4 g of PPO, 0.2 g of POPOP, 60 g of naphthalene, 20 ml of ethylene glycol, 100 ml of methanol and dioxane up to 1 litre. A 10-ml volume dissolves up to 2.5 ml of water at –5°C.

Gaitonde and Nixey have observed a striking loss of efficiency in the scintillation counting of amino acids in the presence of citrate buffers in non-acidic solutions. This efficiency loss, unlike that due to quenching, could not be corrected by using the external standardization facility provided with the counter. One possible cause considered is the precipitation or adsorption of the radioactive solute on the walls of the vial. These observations should be taken as indicating the necessity for periodically checking, by internal standardization, whether quenching correction is adequate for correct results to be obtained.

Toluene-based scintillation mixtures are sensitive to the presence of salts. Emulsions can be counted after stabilization with Triton X-100. Quenching which occurs in strongly acid solutions is suppressed by Triton X-100.

Mixtures of the base Hyamine 10-X (Evered, Whyman) are used for solutions that contain protein.

### Semiconductor detectors for β-particles

Partially-depleted surface-barrier silicon semiconductor detectors have been used for the assay of β-emitters, especially if the latter occur in sheet form (Tykva, 1971). Main advantages are their long lifetime, very low background and high energy-resolution of simultaneously present isotopes. As the noise depends on the overall area of the detectors, the area of the detector and hence the counted sample area are usually small. The set-up consists, in addition to the detector, of amplifiers, amplitude analyzer and counter. The counts may be recorded in various ways.

Standard surface-barrier silicon detectors employed up to now (such as Ortec Model A-018-007-100 or Princeton Model PD-25-18-1000) had to be used in vacuo and preferably at low temperatures. This caused complications in the design of the set-up. Efficient detectors have now been described (Tykva, 1973; Tykva and Votruba, 1974) which detect tritium at atmospheric pressure and room temperature. In practice, the depletion-layer thickness may be about 100 μm for $^{14}C$ (Tykva *et al.*) and about 600 μm for $^{32}P$ (Tykva and Pánek). For improved amplitude resolution, if $^{32}P$ is measured simultaneously with $^{14}C$, deeper layers are preferable. In the example given by Tykva (1971), pulse amplitudes were standardized with conversion electrons emitted from $^{133}Ba$, $^{109}Cd$ and $^{57}Co$. The window used for counting of $^{35}S$ in the presence of $^{14}C$ corresponded to 156–171 keV (Tykva and Pánek); the counting rate in this window was about 0.1% of that in the "overall" $^{35}S$ window (17.2–171 keV).

### α-Radiation detectors

Weinländer and Höhlein sited an α-detector directly next to the end of the effluent tube above the fraction collector (*cf.*, Fig. 16.2). It consists of a coated surface barrier counter furnished with a 0.1-mm barrier and a 5-mm diameter window. A threshold discriminator is used to suppress the $\beta$-count background. The device as reported is not sufficiently accurate for use in quantitative evaluations.

Liquid scintillation is well suited for the detection of $\alpha$-emitters.

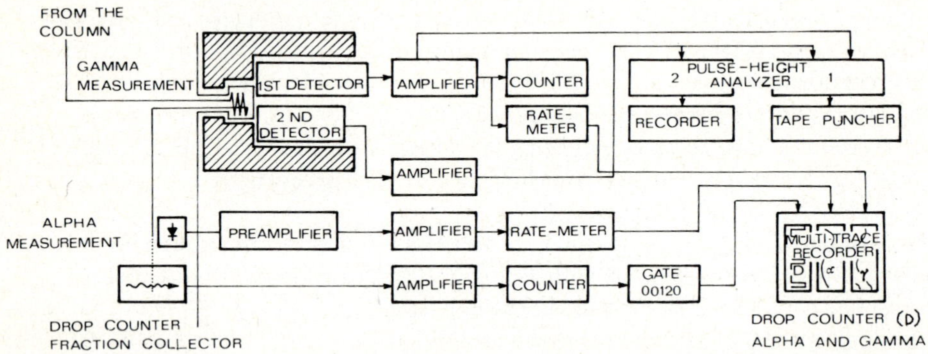


Fig. 16.2. Schematic diagram of the electronic measuring and recording instruments (according to Weinländer and Höhlein).

## DETECTION MODES

The principal detection modes are continuous effluent monitoring and recording, and discontinuous counting of collected fractions. In the latter instance, radiometry can be carried out over longer periods of time than the actual duration of the chromatography. In addition, the effluent can, of course, be also counted discontinuously and individual effluent drops or a streak which have been deposited on a paper band can be counted continuously using a rate-meter. The "sweeping" principle, which is a compromise between the two basic modes, is described below.

Flow-through measurements of certain (physical) quantities can be combined with the measurements of other quantities on the fractions collected; with radioactivity, approximate measurements in flow-through cells can be combined with the results of more precise (although more lengthy) measurements on the fractions (Gaitonde and Nixey). If the chromatography is sufficiently rapid for the operator to observe the process, the fractions can be selected manually and their number minimized, thus reducing the workload on the liquid scintillation counter.

Continuous flow-through measurements and measurements of fractions can be combined either in parallel, using a stream splitter, or in series. The former method is the method of choice if the technique involves mixing with a reagent that interferes with subsequent measurement, *e.g.*, by excessive dilution, quenching, addition of salts to reduce miscibility with the scintillator mixture, precipitation, etc.

The lag-time between measurement of effluent in the flow cell and discharge to the fraction tubes must be determined in order to interpret the respective elution curves correctly. One method is to calculate the inner volume of the tubing, and another is to add a radioactive marker for mutual positioning of the chromatographic curves. The lag-time may change during a series of experiments (or even during a single run) owing to gradual wear and widening of the tubing. It is therefore advisable to check this factor routinely by the marker technique. Correct mutual positioning is, of course, critical when determining specific radioactivities with two flow cells or before and after fractionation.

It is also possible to detect the radioactivity while it is still on the column, *e.g.*, by autoradiography.

**The problem of low-activity samples**

The specific activity (*i.e.*, radioactivity in curies divided by the weight in grams) of pure nuclides is a function of their half-life. It is therefore physically impossible, in the case of long-lived isotopes, to obtain low-weight samples of very high radioactivity*.

In biological and biochemical experiments, substances undergoing detection (especially if labelled with $^{14}C$) are mostly present at low radioactivity levels. This low level is connected with the cost of or difficulties in the preparation of the material, with the radiation hazards of higher radioactivities in biological systems (especially in experiments on human subjects) and the conversion of the original compounds with relatively low yields into a number of products. It is therefore necessary to consider the implications of working with low radioactivities. Firstly, it is necessary to count for sufficiently long periods of time. Both radioactive disintegration and most background counts are random processes and it is therefore impossible to achieve high precision if a low overall number of counts is obtained. The standard deviation of the measurement corresponds to the square root of the number of counts. Let us consider a simple example of 10 cpm, disregarding any background. It is clear that if we count for 6 sec, the result is indistinguishable from zero. If one counts for 1 min (10 ± 3 counts), the coefficient of variation is about 30%, which is unsatisfactory. For a 10% coefficient of variation, it is necessary to count for 10 min, and for 3% to count for *ca.* 100 min. If background counts are considered, low total counts become even less reliable. As only those disintegrations which are counted are relevant, a low efficiency obviously increases the time necessary for a given accuracy.

It is therefore clear that where low activities are concerned, the measurement cannot keep pace with the speed with which the zones of substances emerge from a modern high-efficiency liquid chromatograph. Several modifications of continuous monitoring systems have been described in the literature and commercial equipment is available. It must be kept in mind, however, that in most instances the chromatographic flow-rate used for these methods was rather low. Effective counting times can be increased if higher cell volumes are used or if the time constant of the rate-meter is increased. Both of these procedures may reduce the chromatographic resolution and cause distortion (tailing) of

---

**E.g.*, 1 ng of pure cholesterol labelled on one of its carbon atoms with $^{14}C$ ($T_{\frac{1}{2}}$ = 5650 years) corresponds to 164 pCi (361 dpm).

*References p.413*

the peaks. Thus Weinländer and Höhlein ensured that the volume of the cell did not exceed 1–3% of the individual band volume, but this may be too exacting a requirement in the case of very sharp peaks. Schutte indicated that the practical minimum residence time of the measured effluent volume is 0.5 min and the practical detection limit was 2 nCi for $^{14}C$ and 5 nCi for $^{3}H$. This time may be too long with high-speed chromatographs and these limits too high when smaller samples are being analyzed.

Sufficiently high total counts must, of course, be achieved in any of the channels of a pulse-height analyzer. This requirement makes a multi-channel analyzer too insensitive for immediate quantitative effluent monitoring, but spectral data obtained during sufficiently long periods can be used for the characterization of nuclides and the determination of the purity of the fractions.

The use of fraction collectors permits an increase in counting time by any factor compatible with the available scintillation capacity. (Bands of paper or other materials that are used to absorb the drops of effluent can be also included under the general heading of fraction collectors.)

The counting time could be reduced and thus adjusted to the duration of the chromatographic experiment if subsequent portions of the effluent were counted simultaneously by a number of detectors. The intention is that the effluent would flow through a circular loop along which a number of detectors are situated. The detectors would either move at the same speed as the effluent liquid in the loop or, alternatively, they would be stationary but would be phased in such a way as to obtain integrated values for all counts corresponding to a particular portion of the effluent on its passage through the loop. This "sweeping" principle has not yet been put into practice so far as we know, probably because any advantage in detection speed would be outweighed by the complexity and cost of the equipment required. Each detector unit could be a combination of two or more detectors connected in an appropriate manner. Increasing the number of such units may lead to considerable complications. Fractionation, on the other hand, makes use of standard collectors and the prolongation of the counting time, compared with effluent monitoring, can be up to several orders of magnitude. One basic technical problem of the "sweeping" principle which would have to be solved is that of ensuring synchronization of the detector phasing with the flow of liquid in the loop.

### Effluent monitoring and recording

This technique has the advantage that data are obtained soon after the effluent has left the column and can be directly compared with the results of other measurements carried out on the same effluent (UV absorption, pH etc.; Hertel *et al.*). If the results of two or more detection methods have to be evaluated, the time interval between measuring the same effluent portion with different detectors has to be taken into account, as already discussed.

If the relative positions of the recorder pens can be varied at will, they can be exploited for ensuring correct relative positioning of the traces. Large time differences occur particularly if long reaction coils are inserted between the relevant detectors. The disadvantage of effluent monitoring is its low sensitivity when running fast chromatograms, owing to the short time available for counting the portions of a chromatogram, as discussed above. When higher radioactivities are involved, with low flow-rates and broad peaks, this principle and commercial equipment based on it may be useful.

*Geiger-Müller counting*

Jordan described a cell consisting of a spiral groove machined in polytetrafluoroethylene (Teflon) and covered with a Mylar window glued to it (Fig. 16.1). He reported rapid decontamination as checked by $^{32}$P-labelled nucleic acids (90% decontamination was reached after flushing the cell with 0.5 cell volume of effluent).

Self-absorption is reduced if samples from which the solvent has been evaporated are counted (the label must, of course, reside in non-volatile compounds). As, in principle, there is no basic difference whether the GM counting proceeds at the same or a slower speed as the emergence of the individual portions from the column, the reader is referred to the section on the radiometry of collected fractions.

*Solid-phase scintillation counting*

Cells for $\gamma$-counting can be accommodated within a well-type NaI(Tl) crystal. In the combination of instruments used by Weinländer and Höhlein for lanthanides, actinides and fission products, the effluent is forced through a helical flow cell (5 ml volume) made of polythene which is shielded with lead against the remaining tubing system. Two detectors with their respective multipliers are located to face the coil (Fig. 16.2). One of them produces (in addition to the counter) a signal for the immediate record of overall $\gamma$-radioactivity, which is traced in parallel with the volume (drop number) and $\alpha$-radiation record. The other detector traces a record after passing a supplementary (simple) pulse-height analyzer (2) for provisional identification. The output from both detectors is combined to feed a pulse-height analyzer (1), which consists of two 100-channel sections (for low- and high-energy counts). The respective spectra are recorded simultaneously and punched on to tapes after the storage capacity of the instrument has been reached. These tapes can be evaluated after the separation and assist in the identification of individual fractions. Whereas the first detector operates with constant amplification throughout the chromatography, the amplification can be changed to suit the pulse-height analyzer.

For $\beta$-counting, coils made of plastic scintillators as well as porous scintillator materials have been described on p. 405. According to Sieswerda and Polak, the contribution of the porous solid detector to the peak broadening expressed in $\mu l^2$ is proportional to the flow-rate (ml/h).

*Liquid scintillation counting*

The effluent or aliquot is mixed in a fixed ratio with the scintillation mixture, and a single-phase mixture or an emulsion may result, according to circumstances. In order to ensure good mixing and to prevent distortion and tailing of the peaks due to different laminar flow-rates, the interspacing of bubbles (*i.e.*, the Technicon principle, see Chapter 32) may be advantageous. The technicalities of effluent splitting and proportionate

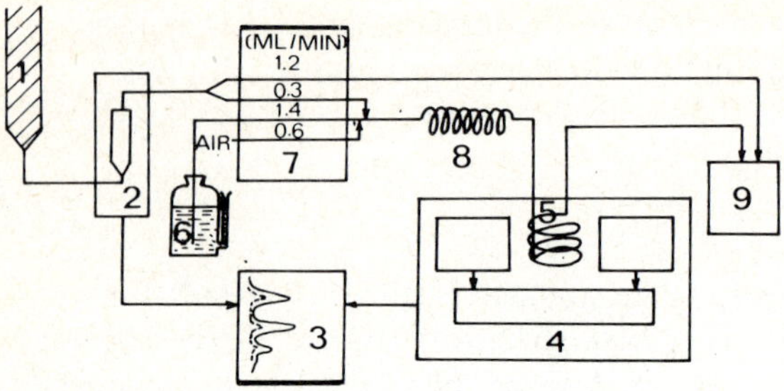


Fig. 16.3. Schematic representation of the homogeneous counting system (Schutte). 1 = Column; 2 = ultraviolet detector, 254 nm; 3 = two-pen recorder; 4 = liquid scintillation spectrometer; 5 = helical flow cell; 6 = scintillator solution reservoir; 7 = proportioning pump; 8 = mixing spiral; 9 = fraction collector.

mixing are also dealt with in Chapter 32. The diagram of Schutte's equipment is shown in Fig. 16.3. The cell is represented by a spiral (2 mm I.D.) of total volume 1.4 ml; the flow-rate through the cell was 2.3 ml/min in the example given.

*Semiconductor counting of β-particles*

As in Geiger-Müller counting, the effluent would have to be counted in a thin layer to limit self-absorption. For $^{14}C$ or $^{35}S$ the wall facing the detector would have to be made of very thin material, such as 0.9 mg/cm$^2$ Mylar foil (Tykva, private communication). Tritium cannot thus be counted. Relatively low efficiencies and small sample areas would usually require higher radioactivity.

**Radiometry of collected fractions**

The whole liquid fraction or an aliquot of it can be subjected to any radiometric procedure; liquid scintillation counting is the method of choice for most β-emitters. The problem of aqueous effluents has been mentioned above. If, in order to ensure sufficiently high count numbers, very high sample volumes would have to be measured, concentration by evaporation might be necessary before the sample is mixed with the scintillator; this would complicate the procedure.

Fraction collectors are dealt with in Chapter 8. One method of fraction collection is to allow the effluent (either the whole stream or an aliquot of it) to drip on to a suitable advancing medium which is left to dry*. This can then be separated in portions of suitable length and subjected to liquid scintillation or any other suitable radiometric procedure. Alternatively, the strip need not be separated beforehand but can be counted continuously. Commercial radioactivity scanning apparatus for PC or TLC can be used, equipped with GM or windowless proportional gas-flow counters, which feed a recording rate-meter. The rate of advancement of the strip, the time constant of the rate-meter and the recording range can mostly be varied at will. GM counting is, of course, unsuitable for use with $^3H$ and of low efficiency for $^{14}C$ and $^{35}S$.

*It is assumed that the solvent is volatile and the radioactive components are not.

*Semiconductography*

Among the techniques used for counting of samples deposited on paper or other sheet materials, "semiconductography" (Tykva, 1971) with its low background may be recommended. The scanning operation may be programmed and the output recorded as a curve (via a rate-meter), numerically or, like in scintigraphy, in the form of (coloured) dots. The set-up published for scanning of paper chromatograms (Tykva, 1971, Tykva and Pánek), gel electrophoreograms (Tykva and Votruba, 1972) or thin-layer chromatograms (Tykva and Votruba, 1974) may be used with minor modifications. The distance between the sample and the detector window is kept to a minimum (less than 0.1 mm; Tykva and Votruba, 1972).

The design is simplified if recently introduced detectors (Tykva, 1973; Tykva and Votruba, 1974) which need not be refrigerated or evacuated are used. Counting efficiencies of 62% for $^{32}P$, 25% for $^{14}C$ and 0.5% for $^{3}H$ have been assessed by Tykva (1974), the background without any screening usually is less than 0.2 cpm (Tykva and Votruba, 1974).

Detectors of very small area would require several scanning operations for each drop, thus increasing total scanning time.

Alimarin *et al.* and Dybczyński used paper and Kyrš and Kadlecová used aluminium as support media. If scintillation counting follows, complications may arise owing to the adsorption or limited solubility of compounds, as is known from PC or TLC (Hais). The low adsorptivity of glass-fibre paper might be of special advantage from this point of view.

With substances that contain polar groups, the paper cutting should preferably be first soaked in a small amount of water and a liquid scintillation mixture suitable for aqueous samples should be added afterwards.

## REFERENCES


Alimarin, I. P., Miklishanskii, A. Z. and Yakovlev, Yu. V., *J. Radioanal. Chem.*, 4 (1970) 45.

Bernhardt, C., *Isotopenpraxis*, 4 (1968) 143.

Bray, G. A., *Anal. Biochem.*, 1 (1960) 279.

Dybczyński, R., *J. Chromatogr.*, 71 (1972) 507.

Evered, D. C., *Int. J. Appl. Radiat. Isotop.*, 20 (1969) 608.

Gaitonde, M. K. and Nixey, R. W. K., *Anal. Biochem.*, 50 (1972) 416.

Hais, I. M., in K. Macek (Editor), *Pharmaceutical Applications of Thin-layer and Paper Chromatography*, Elsevier, Amsterdam, London, New York, 1972, p. 79.

Hertel, W., Sacher, V. and Rohrlich, M., *Z. Anal. Chem.*, 252 (1970) 147.

Hunt, J. A., *Anal. Biochem.*, 23 (1968) 289.

Jordan, B. R., *Anal. Biochem.*, 35 (1970) 244.

Kyrš, M. and Kadlecová, L., *J. Radioanal. Chem.*, 1 (1968) 103.

Minard, F. N. and Mushahwar, I. K., *J. Neurochem.*, 13 (1966) 1.

Schram, E. and Lombaert, R., *Anal. Biochem.*, 3 (1962) 68.

Schutte, L., *J. Chromatogr.*, 72 (1972) 303.

Schutte, L. and Koenders, E. B., *J. Chromatogr.*, 76 (1973) 13.

Sieswerda, G. B. and Polak, H. L., in M. A. Crook, P. Johnson and B. Scales (Editors), *Liquid Scintillation Counting*, Vol. 2, Heyden & Son, London, 1972, p. 49.

Tykva, R., in *Advances in Physical and Biological Radiation Detectors*, Int. At. Energy Agency, Vienna, 1971, p. 211.
Tykva, R., *Excerpta Med. Int. Congr. Ser.*, No. 301 (1973) 455.
Tykva, R., *Czech. Biochem. Soc. Meet., Prague, February 5, 1974.*
Tykva, R., Kokta, L. and Pánek, V., *Radiochem. Radioanal. Lett.*, 10 (1972) 71.
Tykva, R. and Pánek, V., *Radiochem. Radioanal. Lett.*, 14 (1973) 109.
Tykva, R. and Votruba, I., *Anal. Biochem.*, 50 (1972) 18.
Tykva, R. and Votruba, I., *J. Chromatogr.*, 93 (1974) 399.
Weinländer, W. and Höhlein, G., *Kerntechnik*, 10 (1968) 563.
Whyman, A. E., *Int. J. Appl. Radiat. Isotop.*, 21 (1970) 81.

# APPLICATIONS

*Chapter 17*

# Hydrocarbons

J. CHURÁČEK

## CONTENTS



## INTRODUCTION AND GENERAL TECHNIQUES

Liquid chromatography is not a very suitable method for the separation of hydrocarbon mixtures. Lower and medium-sized aliphatic hydrocarbons are either gaseous or volatile compounds and therefore the method of choice for their analysis is gas chromatography. Nevertheless, liquid chromatography can be applied successfully to the separation of higher paraffinic hydrocarbons, especially polycyclic aromatic hydrocarbons. This method is mainly used for the pre-separation of such compounds (Hofman *et al.*, Jaworski and Szewczyk, Johnstone and Entwistle).

The sorption of aliphatic and cyclic hydrocarbons on polar adsorbents is relatively weak and not very selective. However, a large number of investigators have used these adsorbents for separation purposes. It was shown that the properties of silica gel can be utilized with advantage in combination with temperature programming of the column (Lebedeva *et al.*). However, it was found that the use of various gels was more advantageous; on such gels, the separation of hydrocarbons, especially polycyclic hydrocarbons, takes place by the molecular sieving effect.

The analysis of polycyclic hydrocarbons is important when considering the problem of their carcinogenic properties. This problem was described in detail by Schaad in an extensive review. Harmful and carcinogenic substances in cigarette smoke have been investigated by several workers (Johnstone and Entwistle, Neurath *et al.*, Stedman *et al.*, Swain *et al.*).

## CHROMATOGRAPHY ON ADSORBENTS

A good mutual separation of single types of hydrocarbons on an alumina column was achieved by Falk and Steiner. Acid-free alumina was washed with diethyl ether, dried and heated at 130°C for 30 h. Alumina treated in this manner contained about 12% of water and another 1.7% of water was added to it. After complete mixing, the material was allowed to stand in a closed container for 12 h. About 50–100 mg of the benzene-soluble fraction of dust from the air were dissolved in a small amount of chloroform and *ca.* 1 g of alumina was added. The chloroform was then evaporated off and the mixture introduced on to a column (12 X 400 mm), the lower section of which was packed with 9 g of alumina and the upper section with 0.5 g of silica gel. Elution was started with a 100-fold amount of *n*-pentane containing a small amount of diethyl ether. The concentration of ether in the pentane increased stepwise at levels of 0, 3, 6, 9 and 12%. During this 2–3 h operation, the column was protected from light. Gradually, the following hydrocarbons were eluted: aliphatic hydrocarbons, olefins, benzene derivatives, naphthalene derivatives, dibenzofuran, anthracene, pyrene, benzofluorene, chrysene, benz(α)pyrene, benzoperylene and coronene.

Column chromatography is fairly often used in the analysis of petroleum products, especially oils and asphalts, which contain paraffinic and aromatic hydrocarbons. Gradient elution has been used with success for this purpose. In this application this technique has the same importance and utilization as temperature programming in gas chromatography. The work of Middleton serves as an example. He separated a mixture of higher aliphatic and aromatic hydrocarbons on a column filled with alumina containing 2% of water (the course of the gradient is represented in Fig. 17.1, while Fig. 17.2 shows the resulting chromatogram of a six-component mixture). A mixture of polynuclear aromatic hydrocarbons was also separated on the same adsorbent, and the course of the gradient and the separation can be seen in Fig. 17.3 (Popl *et al.*, 1970). Liebisch and Eckardt separated aromatic hydrocarbons present in oil on a column filled with alumina and picric acid.

Asphalt components were well separated by Corbett. Fig. 17.4 represents their separation schematically. The separation of "petrolenes" (components of a heptane extract) and the changes in the elution gradient are shown in Fig. 17.5.

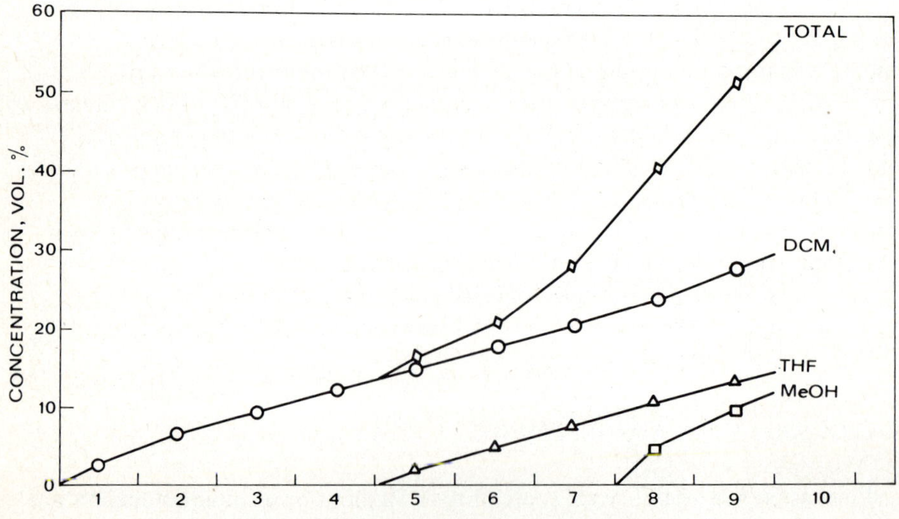


Fig. 17.1. Composition of eluent gradient (Middleton). Polar solvent added to *n*-hexane base. DCM = dichloromethane; THF = tetrahydrofuran; MeOH = anhydrous methanol; TOTAL = sum of polar solvents.

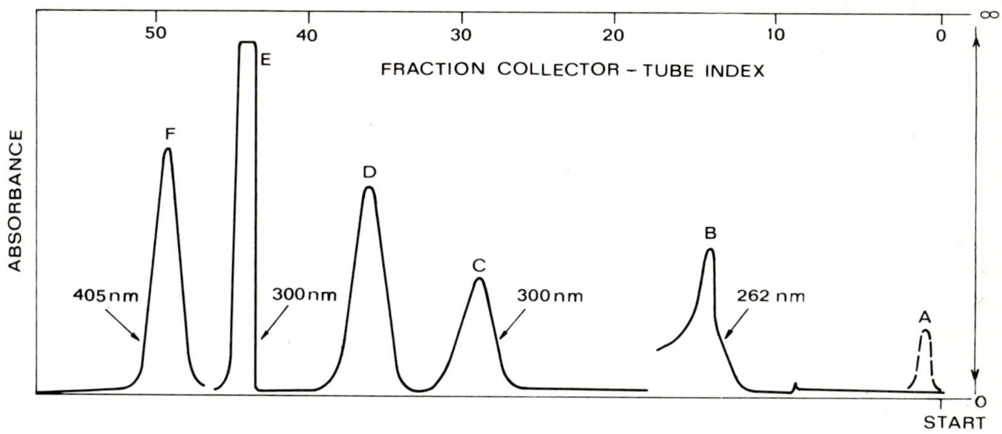


Fig. 17.2. Separation of a six-component broad-range mixture (Middleton). Column: 20 × 1.8 cm. Sorbent: alumina + 2% of water. Eluent: *n*-hexane, diethyl ether, chloroform, benzene, methanol. Operating conditions: gradient elution, see Fig.17.1. Detection: spectrophotometric. A = 1,3-dicyclopentyl-2-dodecylcyclopentane; B = 1-phenylpentadecene; C = 2,6-dioctylnaphthalene; D = 1,2,3,-triphenylethane; E = 9,10-dimethyl-1,2-benzanthracene; F = benz(α)anthracene-7,12-dione.

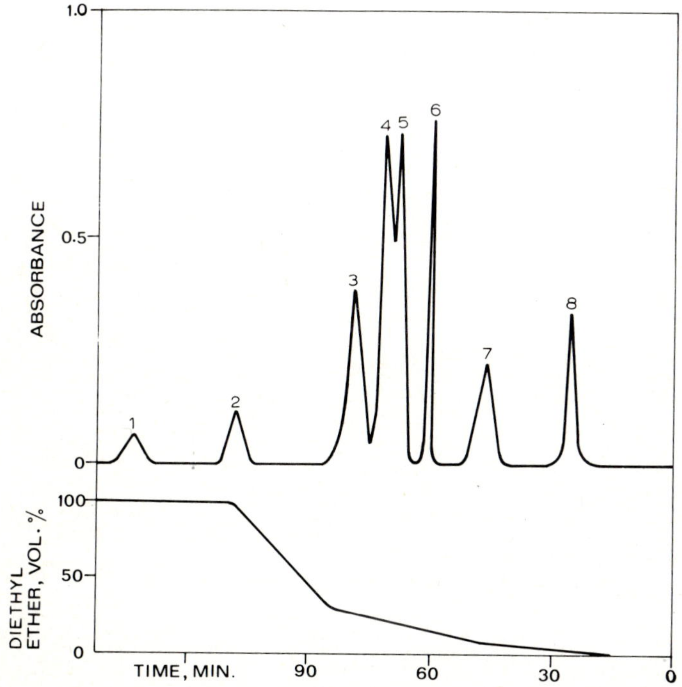


Fig. 17.3. Separation of polycyclic hydrocarbons (Popl *et al.*, 1970). Column: 100 × 0.4 cm. Sorbent: alumina + 2% of water. Eluent: *n*-pentane–diethyl ether. Operating conditions: gradient elution. Detection: spectrophotometric. 1 = carbazole; 2 = chrysene; 3 = pyrene; 4 = phenanthrene; 5 = fluorene; 6 = acenaphthene; 7 = naphthalene; 8 = indane.

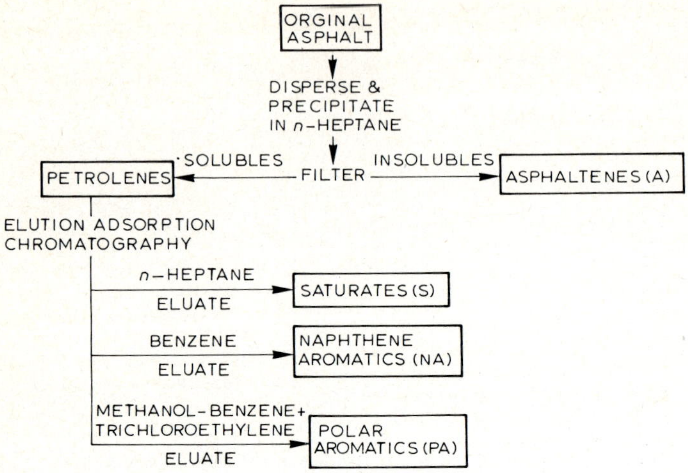


Fig. 17.4. Scheme for the separation of asphalt into four generic components (Corbett).

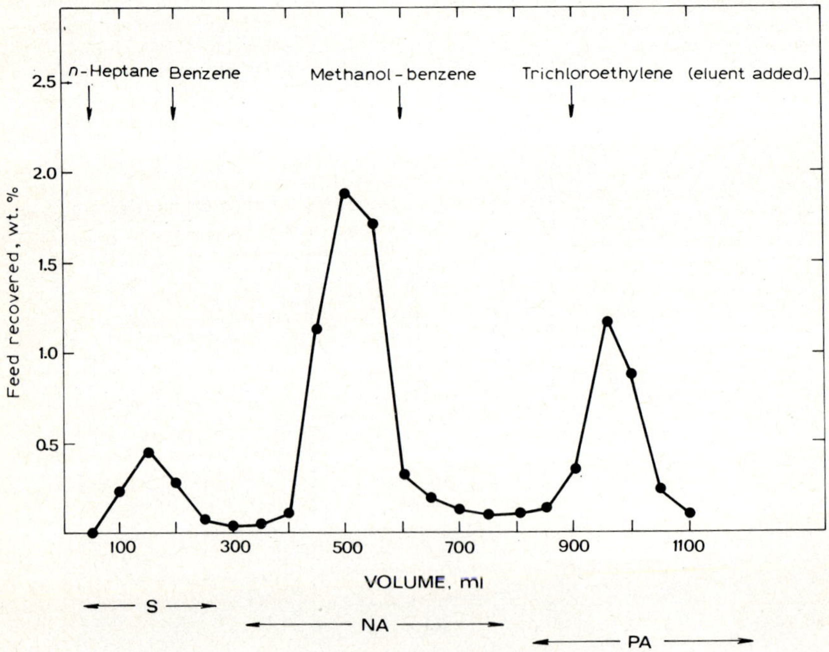


Fig. 17.5. Separation of petrolenes (Corbett). Column: 100 × 3.1 cm. Sorbent: alumina F-20. Mobile phase: *n*-heptane, benzene, methanol–benzene, trichloroethylene. Detection: eluates evaporated to dryness and weighed. S = saturated; NA = naphthene aromatics; PA = polar aromatics.

## CHROMATOGRAPHY ON GELS

Higher hydrocarbon fractions, especially oil fractions, cannot be separated successfully by gas chromatography, and gel chromatography can be used with advantage instead.

Klimisch and Reese used gel chromatography for the preparative fractionation of hydrocarbons in cigarette smoke condensate on a 24.5 × 1090 mm column packed with Bio-Beads SX-8 (200–400 mesh). Tetrahydrofuran was used as the mobile phase. A molecular sieving effect was the predominant factor in the separation, and in addition, the separation was influenced by interactions with the gel matrix or reactions with the solvent, depending on the structure of the compounds. A fraction of cigarette smoke condensate was separated into three sub-fractions, one of which contained polycyclic aromatic hydrocarbons composed of up to six rings.

For the separation of the fractions of coal tar, gel chromatography on Sephadex LH-20 was found to be a convenient method (Hsieh *et al.*). This sorbent can also be used for the separation of paraffins from cycloparaffins and alkylbenzenes from cyclic benzene derivatives on the basis of selective sorption (Mair *et al.*). Higher hydrocarbon fractions also containing polynuclear hydrocarbons were separated on Merckogel (Oelert, Oelert and Weber), and a correlation between the elution volume and the logarithm of the molecular weight was also found. Elution volumes of the separated hydrocarbons are given in Table 17.1. For the determination of the distribution of hydrocarbons in mixtures according to their molecular weights, Styragel (Coleman *et al.*) or Poragel A-1 (Weber and Oelert) can be used. Mate and Lundstrom used gel chromatography for the determination of the molecular mass distribution in fractions of aromatic, naphthenic and paraffinic oils.

TABLE 17.1
GEL CHROMATOGRAPHY OF HYDROCARBONS (OELERT)

Values are elution volumes (ml) obtained using the following flow-rates of the mobile phases: cyclohexane 0.2 mm/sec; methylene chloride 0.3 mm/sec; isopropanol 0.1 mm/sec. Column: 60 × 1.0 cm. Gel: Merckogel OR-500 vinyl acetate gel. Temperature: 22°C. Detector: RI.

| Compound | Mobile phase | | |
|---|---|---|---|
| | Cyclohexane | Methylene chloride | Isopropanol |
| *n*-Hexatriacontane | 17.5 | 7.7 | 16.8 |
| *n*-Hexadecane | 19.3 | 16.2 | 22.0 |
| *n*-Dodecane | 19.9 | 17.6 | 22.9 |
| *n*-Heptane | 20.8 | 21.6 | 25.0 |
| Squalene | – | 10.9 | 19.7 |
| Diphenyl | 20.9 | 21.4 | 24.6 |
| *p*-Terphenyl | 20.5 | 19.0 | 22.0 |
| Benzene | 22.0 | 26.3 | 27.8 |
| Picene | 20.8 | 20.9 | 36.3 |
| Chrysene | 21.1 | 21.8 | 33.4 |
| Phenanthrene | 21.6 | 22.2 | 31.6 |
| Naphthalene | 22.8 | 23.3 | 29.6 |

*(Continued on p.422)*

*References p.429*

TABLE 17.1 (*continued*)

| Compound | Mobile phase | | |
|---|---|---|---|
| | Cyclohexane | Methylene chloride | Isopropanol |
| Rubrene | 19.0 | 15.8 | – |
| Truxene | – | 19.1 | – |
| Pyrene | 19.0 | 23.6 | 47.8 |
| Coronene | 29.9 | 32.8 | 51.8 |
| Decacyclene | – | 45.5 | – |
| Cyclohexane | – | – | 21.3 |
| Decalin | 20.7 | 21.1 | 22.8 |
| Indane | 21.4 | – | – |
| Tetralin | 21.1 | 22.9 | 22.0 |
| Fluoranthene | – | 22.1 | – |
| Dihydrophenanthrene | – | 21.3 | 25.6 |
| Octahydrophenanthrene | – | 21.4 | 26.0 |
| Dihydrotetracene | – | 21.0 | – |
| Dioctylbenzene | 18.5 | 14.5 | 19.3 |
| Tridecylbenzene | 19.3 | 15.8 | 20.0 |
| Diisopropylbenzene | 20.6 | 19.3 | – |
| Ethylbenzene | – | 22.7 | – |
| Dimethylnaphthalene | – | 22.3 | – |
| Trimethylnaphthalene | – | 21.8 | – |
| Methylanthracene | – | 22.0 | – |
| Cholesteryl stearate | – | 12.2 | – |
| Cholesteryl palmitate | – | 12.6 | – |
| Cholesteryl laurate | – | 13.2 | – |
| Cholesterol | – | 18.3 | – |

TABLE 17.2
GEL CHROMATOGRAPHY OF ALIPHATIC HYDROCARBONS (CAZES AND GASKILL)

Column: diameter 3/8 in.; length, four 4-ft. columns in series having exclusion limits of $3 \times 10^3$, 250, 60 and 60 Å, respectively. Gel: rigid, cross-linked polystyrene gel (Waters Ass., Framingham, Mass., U.S.A.); Mobile phase: *o*-dichlorobenzene. Flow-rate: 1 ml/min. Temperature: 130°C. Detection: RI.

| Compound* | Elution volume (ml) | Compound* | Elution volume (ml) |
|---|---|---|---|
| *n*-Pentane | 176.0 | *n*-Octadecane | 141.3 |
| *n*-Hexane | 172.2 | *n*-Eicosane | 138.3 |
| *n*-Heptane | 168.0 | *n*-Octacosane | 130.3 |
| *n*-Octane | 164.4 | *n*-Pentatriacontane | 125.4 |
| *n*-Nonane | 161.2 | *n*-Hextriacontane | 124.5 |
| *n*-Decane | 158.4 | 2,2,4-Trimethylpentane | 166.4 |
| *n*-Undecane | 155.7 | 3-Methylpentane | 173.5 |
| *n*-Dodecane | 153.1 | 2-Methylpentene-1 | 174.5 |
| *n*-Tridecane | 150.7 | 4-Methylpentene-1 | 174.1 |
| *n*-Tetradecane | 148.2 | | |

*Samples were 2-ml aliquots of 0.25–1.2% solutions.

TABLE 17.3
GEL CHROMATOGRAPHY OF HYDROCARBONS AND SOME ETHERS (HENDRICKSON)

Column: 12 ft., except for figures in parentheses, where it was 8 ft. × 3/8 in. Gel: 40 Å styrene–divinylbenzene gel, permeable to alkanes with mol. wt. < 450. Mobile phase: benzene. Flow-rate: 1 ml/min. Temperature: 24°C. Detection: RI.

| Compound* | Elution volume (ml) | Peak width** (ml) | Compound* | Elution volume (ml) | Peak width** (ml) |
|---|---|---|---|---|---|
| $(n\text{-}C_{18}H_{37})_2O$ | 62.8 | 3.60 | *p*-Xylene | (76.2) | (–) |
| $(n\text{-}C_{12}H_{25})_2O$ | 67.4 | 3.70 | Cumene | (71.0) | (–) |
| Dioctyl ether*** | 76.0 | 3.81 | Naphthalene | (81.5) | (6.20) |
| *n*-Dodecane*** | 84.2 | 3.96 | Biphenyl | (75.2) | (5.92) |
| *n*-Decane | 88.9 | 3.82 | Diphenylmethane | 102.7 | 5.08 |
| *n*-Heptane*** | 99.2 | 4.57 | Diphenylethane | 96.0 | 4.94 |
| *n*-Hexane | 101.5 | 3.89 | Dicumyl peroxide | 84.8 | 6.02 |
| *n*-Pentane*** | 105.4 | 3.97 | Anisole | 114.6 | 4.56 |
| Diethyl ether*** | 107.5 | 4.16 | Phenetole | 107.6 | 4.43 |
| Dioctyl ether*** | (53.3) | (4.24) | Cyclohexane*** | 110.7 | 4.59 |
| *n*-Dodecane*** | (58.3) | (4.25) | Cyclohexane*** | (76.2) | (5.44) |
| *n*-Nonane | (63.0) | (4.20) | Styrene | 114.2 | 4.82 |
| *n*-Heptane*** | (67.3) | (4.41) | 2-Methylpentene-1 | 91.8 | 4.85 |
| *n*-Pentane*** | (72.4) | (4.48) | 4-Methylpentene-1 | 90.6 | 4.21 |
| Diethyl ether*** | (73.6) | (4.47) | Heptene-1 | (67.3) | (4.41) |
| Toluene | (81.0) | (–) | Decene-1 | 82.7 | 4.56 |
| Ethylbenzene | (75.2) | (–) | Isooctane | (65.6) | (4.28) |
| | | | Methyl methacrylate | 98.2 | 4.95 |

*Sample sizes were typically 0.1 ml of a 4% solution in benzene.
**Peak widths were obtained by drawing tangents to each side of the curves and are expressed as the number of millilitres of eluent at the base of the triangle.
***Compound run more than once.

Using a cross-linked polystyrene gel and *o*-dichlorobenzene as the mobile phase, higher aliphatic hydrocarbons can be separated by gel chromatography at an elevated temperature (130°C). Elution volumes of the separated substances are presented in Table 17.2 (Cazes and Gaskill).

Medium-sized aliphatic and volatile aromatic hydrocarbons can be separated on styrene-divinylbenzene gels using benzene as the mobile phase (Hendrickson). Table 17.3 lists the elution volumes of the separated hydrocarbons.

## OTHER METHODS OF CHROMATOGRAPHY OF HYDROCARBONS

The property of olefins to form complexes with silver ions was exploited by Janák *et al.* for their chromatography in a liquid–solid system. Porapak G was used as the sorbent and $AgNO_3$ (0.08 g/ml) was added to the mobile phase, which consisted of *n*-propanol and water (2:1). A similar sorbent, Porapak T, was used by Martinů and Janák for the separation of polycyclic hydrocarbons using high-pressure LC. The results of the measurement of relative retention times are given in Table 17.4. A very good separation of phenanthrene derivatives is shown in Fig. 17.6.

*References p.429*

TABLE 17.4

RELATIVE RETENTION TIMES, $t'_R$, OF SOME AROMATIC HYDROCARBONS AND THEIR PARTIALLY HYDROGENATED DERIVATIVES ON PORAPAK T IN A COLUMNAR ARRANGEMENT (LC) (MARTINŮ AND JANÁK)

Column: 50 × 0.2 cm. Sorbent: Porapak T (0.853 g). Mobile phase: *n*-hexane. Flow-rate: 0.3 ml/min. Detection: UV (254 nm). Results are expressed relative to naphthalene = 1.00.

| Substance | $t'_R$ |
|---|---|
| Indene | 0.59 |
| Hydrindene | 0.25 |
| Naphthalene | 1.00 |
| Tetralin | 0.86 |
| Acenaphthylene | 1.91 |
| Acenaphthene | 1.10 |
| Anthracene | 2.33 |
| 9,10-Dihydroanthracene | 1.18 |
| 1,2,3,4-Tetrahydroanthracene | 0.82 |
| Octahydroanthracene | 0.21 |
| Phenanthrene | 2.68 |
| 9,10-Dihydrophenanthrene | 1.44 |
| Octahydrophenanthrene | 0.87 |
| Fluoranthene | 3.69 |
| 1,2,3,4-Tetrahydrofluoranthene | 1.47 |
| Tetracene | 5.33 |
| 5,12-Dihydrotetracene | 3.67 |
| Pyrene | 3.31 |
| Dihydropyrene | 2.24 |
| *sym.*-Hexahydropyrene | 0.83 |

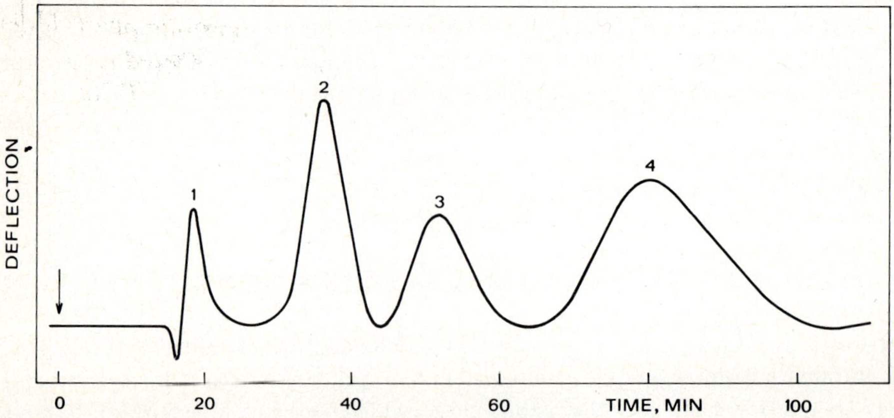


Fig. 17.6. Separation of a mixture of phenanthrenes (Janák *et al.*). Column: 500 × 2 mm. Sorbent: Porapak T (0.853 g). Mobile phase: *n*-hexane. Operating conditions: flow-rate 0.15 ml/min. Detection: spectrophotometric. 1 = benzene; 2 = octahydrophenanthrene; 3 = 9,10-dihydrophenanthrene; 4 = phenanthrene.

Polynuclear aromatic hydrocarbons were also separated successfully by high-speed LC using Permaphase ODS and a linear gradient from methanol–water (1:1) to pure methanol. A very good separation is represented in Fig. 17.7 (Schmit *et al.*).

Reversed-phase chromatography was used for the separation of a series of medium-sized aliphatic and volatile monocyclic aromatic hydrocarbons (Locke). For example, a 3-m column packed with 25% squalene on silanized Chromosorb P was used with acetonitrile as eluent. This chromatographic procedure can be performed at various temperatures, which should not, however, exceed 32°C. The results of the separations are shown in Table 17.5. For the separation of hydrocarbon mixtures (Sergienko *et al.*), the formation

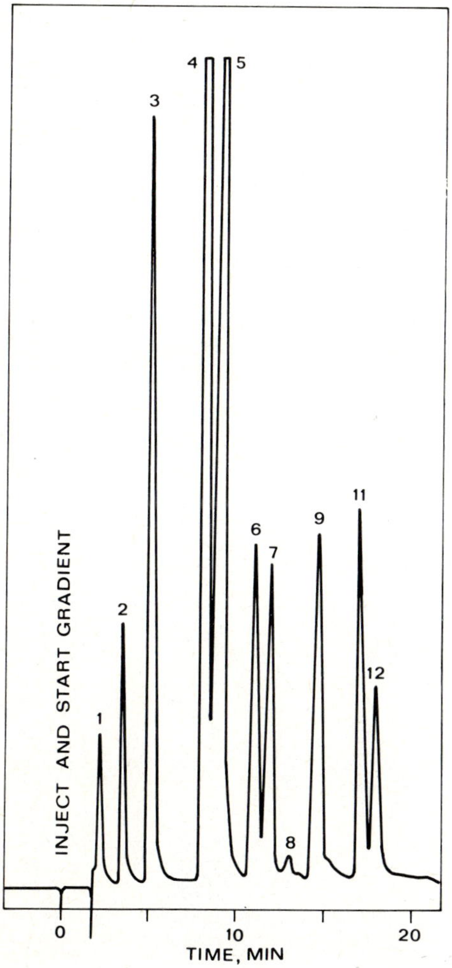


Fig. 17.7. Separation of fused-ring aromatics (Schmit *et al.*). Column: 1 m × 2.1 mm I.D. precision-bore, stainless steel. Sorbent: Permaphase ODS. Mobile phase: linear gradient from methanol–water (1:1) to methanol. Operating conditions: temperature 50°C; flow-rate 1 ml/min; pressure 100 p.s.i. Detection: UV photometer. 1 = Benzene; 2 = naphthalene; 3 = biphenyl; 4 = phenanthrene; 5 = anthracene; 6 = fluoranthrene; 7 = pyrene; 8 = unknown; 9 = chrysene; 11 = benz(*e*)pyrene; 12 = benz(α)pyrene.

*References p.429*

TABLE 17.5
PARTITION CHROMATOGRAPHY OF HYDROCARBONS (LOCKE)

Values are specific retention volumes (ml per gram of squalene). Column: 3 m × ¼ in. Sorbent: 25% (w/w) squalene coated on silanized Chromosorb P (100–200 mesh). Mobile phase: de-aerated acetonitrile. Detection: RI.

| Compound* | Temperature (°C) | | | |
|---|---|---|---|---|
| | 15.05 | 20.0 | 25.0 | 35.1 |
| *n*-Butane | 3.33 | 3.24 | 3.16 | 3.01 |
| 2-Methylbutane | 4.62 | 4.38 | 4.12 | 3.70 |
| *n*-Pentane | 4.90 | 4.60 | 4.32 | 3.84 |
| 2,2-Dimethylbutane | 5.24 | 4.84 | 4.48 | 3.86 |
| 2,3-Dimethylbutane | 5.89 | 5.50 | 5.15 | 4.58 |
| 3-Methylpentane | 6.12 | 5.79 | 5.44 | 4.88 |
| 2-Methylpentane | 6.23 | 5.89 | 5.54 | 4.99 |
| *n*-Hexane | 6.87 | 6.36 | 5.90 | 5.11 |
| 2,2-Dimethylpentane | 7.30 | 6.70 | 6.18 | 5.29 |
| 2,3-Dimethylpentane | – | – | 6.82 | – |
| 2,4-Dimethylpentane | 7.40 | 6.83 | 6.32 | 5.48 |
| 3-Ethylpentane | 8.49 | 7.73 | 7.08 | 5.99 |
| 2-Methylhexane | 8.76 | 7.95 | 7.21 | 6.00 |
| *n*-Heptane | 9.50 | 8.61 | 7.82 | 6.51 |
| 2,2,4-Trimethylpentane | 8.65 | 8.11 | 7.62 | 6.75 |
| 2,4-Dimethylhexane | 10.15 | 9.24 | 8.43 | 7.09 |
| *n*-Octane | 13.20 | 11.80 | 10.65 | 8.75 |
| *n*-Nonane | 17.75 | 15.90 | 14.35 | 11.65 |
| 2-Methylbutene-1 | – | – | 2.41 | – |
| 3-Methylpentene-1 | 2.98 | 2.82 | 2.67 | 2.40 |
| 2-Methylpentene-1 | – | – | 2.78 | – |
| 2,3-Dimethylbutene-1 | – | – | 2.74 | – |
| Pentene-1 | 2.10 | 2.00 | 1.91 | 1.75 |
| Hexene-1 | 3.10 | 2.94 | 2.79 | 2.53 |
| 1,5-Hexadiene | 1.47 | 1.38 | 1.31 | 1.17 |
| Heptene-1 | 4.23 | 3.98 | 3.73 | 3.32 |
| Octene-1 | 5.78 | 5.45 | 5.13 | 4.61 |
| 2,4,4-Trimethylpentene-1 | 4.83 | 4.58 | 4.35 | 3.94 |
| Cyclopentane | 4.55 | 4.31 | 4.09 | 3.70 |
| Methylcyclopentane | 5.82 | 5.50 | 5.19 | 4.66 |
| Cyclohexane | 6.60 | 6.11 | 5.68 | 4.93 |
| Methylcyclohexane | 8.16 | 7.71 | 7.29 | 6.52 |
| Ethylcyclohexane | 11.30 | 10.49 | 9.78 | 8.50 |
| Cyclopentene | 2.45 | 2.36 | 2.28 | 2.14 |
| Cyclohexene | 3.26 | 3.08 | 2.92 | 2.64 |
| Benzene | 0.554 | 0.550 | 0.548 | 0.539 |
| Toluene | 0.804 | 0.786 | 0.770 | 0.738 |
| Ethylbenzene | 1.04 | 1.01 | 0.989 | 0.935 |
| *o*-Xylene | 1.06 | 1.03 | 1.00 | 0.950 |
| *m*-Xylene | 1.16 | 1.12 | 1.08 | 1.02 |
| *p*-Xylene | 1.17 | 1.14 | 1.10 | 1.03 |
| *n*-Propylbenzene | 1.38 | 1.33 | 1.26 | 1.13 |

TABLE 17.5 *(continued)*

| Compound* | Temperature (°C) | | | |
|---|---|---|---|---|
| | 15.05 | 20.0 | 25.0 | 35.1 |
| *n*-Butylbenzene | 1.98 | 1.90 | 1.84 | 1.71 |
| *n*-Pentylbenzene | 2.53 | 2.48 | 2.38 | 2.12 |
| *n*-Hexylbenzene | 3.89 | 3.58 | 3.30 | 2.83 |
| *n*-Octylbenzene | 6.84 | 6.37 | 5.92 | 5.18 |
| *n*-Nonylbenzene | 9.41 | 8.56 | 7.80 | 6.52 |
| Isopropylbenzene | – | – | 1.26 | – |
| 1,2,4-Trimethylbenzene | – | – | 1.54 | – |
| 1,3,5-Trimethylbenzene | – | – | 1.57 | – |
| *tert.*-Butylbenzene | – | – | 1.45 | – |
| *sec.*-Butylbenzene | – | – | 1.63 | – |
| *o*-Diethylbenzene | – | – | 1.71 | – |
| *m*-Diethylbenzene | – | – | 1.81 | – |
| *p*-Diethylbenzene | – | – | 1.78 | – |
| 1,3,5-Triethylbenzene | – | – | 2.88 | – |
| 1-Methyl-4-isopropylbenzene | – | – | 1.84 | – |
| Tetralin | – | – | 1.83 | – |
| Naphthalene | 0.643 | 0.620 | 0.593 | 0.550 |
| $CH_2Cl_2$ | – | – | 0.107 | – |
| $CHCl_3$ | – | – | 0.232 | – |
| $CCl_4$ | 1.52 | 1.49 | 1.45 | 1.39 |

*Samples were 0.01–1.0 μl of pure solute.

TABLE 17.6
SURVEY OF DIFFERENT PROCEDURES APPLICABLE TO THE SEPARATION OF HYDROCARBONS AND HALOHYDROCARBONS

| Chromatographic technique | Compounds | Sorbent | Mobile phase | Reference |
|---|---|---|---|---|
| LC | Hydrocarbons (air pollutants) | Alumina | Cyclohexane | Kotin |
| LC | Hydrocarbons (air pollutants) | Alumina | Light petroleum –benzene (10:1); Light petroleum –methanol (100:1) | Héros |
| LC | Polynuclear hydrocarbons | Alumina; silica gel | Methanol–ethanol –water (1:1:1) | Weigert and Mottram |
| LC | Hydrocarbons | Silica gel | *n*-Hexane; *n*-hexane–benzene (4:1) | Hoffman and Wynder |
| LC | Hydrocarbons | Acetylcellulose | Ethanol–toluene–water (17:4:1) | Spotswood |

*(Continued on p.428)*

TABLE 17.6 *(continued)*

| Chromatographic technique | Compounds | Sorbent | Mobile phase | Reference |
|---|---|---|---|---|
| LC | Aromatic hydrocarbons | MgO–Celite (2:1) | *n*-Hexane–benzene–acetone (3:1:1) | Lijinsky, Lijinsky *et al.* |
| GPC | Hydrocarbons | Sephadex LH-20 | Isopropanol; chloroform–cyclohexane (4:1) | Wilk *et al.* |
| GPC | Alkanes (application to a rock extract) | Sephadex LH-20 | Acetone–chloroform (1:1) | Cooper |
| GPC | Halogenated hydrocarbons (products of chloroprene production) | Styrene–divinylbenzene copolymer | Tetrahydrofuran | Čoupek and Bouchal |
| LC | Alkyl chlorides | Silica gel with fluorescent indicator | Isopropanol | Jaworski *et al.* |
| LC | *n*-Halogenated stilbenes (purification and preparation, *cis–trans* isomers) | Alumina | *n*-Hexane | Krueger and Lippert |
| LC | Fluorinated hydrocarbons* | Dowex 50W-X8 | Ethanol–water (1:1) | Miller and Keyworth |

*Determination of F after transforming the bound fluorine into fluoride ion.

of clathrates can be made use of. Oil fractions of higher paraffins and cycloparaffins (boiling range *ca.* 250–450°C) were separated in the form of clathrates by Stejaru and Popescu. Another method of utilization of clathrates consists in packing the column with a mixture of thiourea and diatomaceous earth and eluting with benzene and methanol. A 120-fold excess of thiourea (relative to the weight of hydrocarbons) was used (Lieberman and Furman).

Ion-exchange resins have also been used for the separation of hydrocarbons. Yamada *et al.* discussed the relationship between the distribution coefficients of benzene derivatives on the hydrogen and sodium forms of styrene-based strong cation-exchange resins (Bio-Rad AG 50 W-X8) and their chemical structures.

A survey of other papers concerning the separation of hydrocarbons and halogenated hydrocarbons both from the point of view of the analyzed substances and that of the sorbents used is given in Table 17.6.

## REFERENCES


Cazes, J. and Gaskill, D. R., *Separ. Sci.*, 4 (1969) 15.

Coleman, H. J., Hirsch, D. E. and Dooley, J. E., *Anal. Chem.*, 41 (1969) 800.

Cooper, B. S., *J. Chromatogr.*, 46 (1970) 112.

Corbett, L. W., *Anal. Chem.*, 41 (1969) 576.

Čoupek, J. and Bouchal, K., *Macromol. Chem.*, 135 (1970) 69; *C.A.* 73 (1970) 45865j.

Falk, H. L. and Steiner, P. E., *Cancer Res.*, 12 (1952) 30.

Hendrickson, J., *J. Chromatogr.*, 32 (1968) 543.

Héros, M., *Rev. Pollut. Atm.*, 5 (1963) 205.

Hoffmann, D. and Wynder, E. L., *Anal. Chem.*, 32 (1960) 295.

Hofman, J., Tománek, O., Vodička, L. and Landa, S., *Collect. Czech. Chem. Commun.*, 34 (1969) 1042.

Hsieh, B. C. B., Wood, R. E., Anderson, L. L. and Hill, G. R., *Anal. Chem.*, 41 (1969) 1066.

Janák, J., Jagarič, Z. and Dressler, M., *J. Chromatogr.*, 53 (1970) 525.

Jaworski, M. and Szewczyk, H., *Chem. Anal. (Warsaw)*, 15 (1970) 53.

Jaworski, M., Zielasko, A. and Szewczyk, H., *Chem. Anal. (Warsaw)*, 14 (1969) 1225; *C.A.*, 72 (1970) 106850d.

Johnstone, R. A. W. and Entwistle, I. D., *J. Chem. Soc. C.*, (1968) 1818.

Klimisch, H. J. and Reese, D., *J. Chromatogr.*, 67 (1972) 299.

Kotin, P., *Cancer Res.*, 16 (1956) 16.

Krueger, K. and Lippert, E., *Chem. Ber.*, 102 (1969) 3233.

Lebedeva, N. P., Frolov, I. I. and Yashin, Ya. I., *J. Chromatogr.*, 58 (1971) 11.

Liberman, A. L. and Furman, D. B., *Neftekhimiya*, 8 (1968) 811; *C.A.*, 72 (1970) 121058v.

Liebisch, G. and Eckardt, H., *Z. Chem.*, 6 (1966) 377; *Anal. Abstr.*, 15 (1968) 838.

Lijinsky, W., *Anal. Chem.*, 32 (1960) 684.

Lijinsky, W., Saffiotti, U. and Shubik, P., *J. Natl. Cancer Inst.*, 18 (1957) 867.

Locke, D. C., *J. Chromatogr.*, 35 (1968) 24.

Mair, B. J., Hwang, P. T. R. and Ruberto, R. G., *Anal. Chem.*, 39 (1967) 838.

Martinů, V. and Janák, J., *J. Chromatogr.*, 65 (1972) 477.

Mate, R. D. and Lundstrom, H. S., *J. Polym. Sci., Part C*, No. 21 (1967) 317; *C.A.*, 68 (1968) 96614f.

Middleton, W. R., *Anal. Chem.*, 39 (1967) 1839.

Miller, M. and Keyworth, D. A., *Talanta*, 14 (1967) 1287; *Anal. Abstr.*, 16 (1969) 761.

Neurath, G., Gewe, J. and Wichern, H., *Beitr. Tabakforsch.*, 4 (1968) 250; *Anal. Abstr.*, 18 (1970) 1863.

Oelert, H. H., *J. Chromatogr.*, 53 (1970) 241.

Oelert, H. H. and Weber, J. H., *Erdöl Kohle, Erdgas, Petrochem.*, 23 (1970) 484; *C.A.*, 73 (1970) 79167j.

Popl, M., Mostecký, J. and Havel, Z., *J. Chromatogr.*, 53 (1970) 233.

Schaad, R. A., *Chromatogr. Rev.*, 13 (1970) 61.

Schmit, J. A., Henry, R. A., Williams, R. C. and Dieckman, J. F., *J. Chromatogr. Sci.*, 9 (1971) 645.

Sergienko, S. R., Chelpanova, M. P., Aidogdyev, A. and Kozyreva, A. S., *Gazokondens. Neft., Mater. Sredneaziat. Nauch. Soveshch. Neftekhim. Khim. Perezab. Uglevodorov, 2nd, 1967*, 1968, p. 129; *C.A.*, 71 (1969) 126840h.

Spotswood, T. M., *J. Chromatogr.*, 3 (1960) 101.

Stedman, R. L., Miller, R. L., Lakritz, L. and Chamberlain, W. J., *Chem. Ind. (London)*, 12 (1968) 394.

Stejaru, D. and Popescu, R., *Rev. Chim. (Bucharest)*, 20 (1969) 629; *C.A.*, 73 (1970) 17035v.

Swain, A. P., Cooper, J. E., Stedman, R. L. and Bock, F. G., *Beitr. Tabakforsch.*, 5 (1969) 109; *C.A.*, 73 (1970) 63321h.

Weber, J. H. and Oelert, H. H., *Separ. Sci.*, 5 (1970) 669.

Weigert, F. and Mottram, J. C., *Cancer Res.*, 6 (1947) 97.

Wilk, M., Rochlitz, J. and Bende, H., *J. Chromatogr.*, 24 (1966) 414.

Yamada, M., Nomura, N. and Shiho, D., *J. Chromatogr.*, 64 (1972) 253.

*Chapter 18*

# Alcohols and polyols

J. CHURÁČEK

## CONTENTS



## INTRODUCTION AND GENERAL TECHNIQUES

This chapter includes aliphatic, cycloaliphatic and aromatic compounds with one or more hydroxy groups in the molecule. Their chromatographic properties are mainly determined by the number of hydroxy groups present. Sugar alcohols are dealt with in the chapter on saccharides (Chapter 22).

Lower monohydric alcohols are best analyzed by gas chromatography, while liquid column chromatography is best used for the separation and determination of higher free alcohols and their derivatives. For the separation of alcohols and glycols by this method, common adsorbents and also ion exchangers and gels are used. New sorbents and stationary phase supports have recently been developed, for example porous silica microspheres, which enable a substantial increase in the separation rate to be achieved (Kirkland). When diols, mainly polymeric compounds of the polyethylene glycol type, are analyzed, the determination of the molecular-weight distribution is the most important task. Gel permeation chromatography is most suitable for this purpose, but silica gel can also be used.

## HIGH-SPEED LIQUID AND GEL PERMEATION CHROMATOGRAPHY OF FREE ALCOHOLS

High-speed liquid chromatography is now used for the separation of some aromatic and aliphatic alcohols. Highly effective phases on 2 m × 2 mm columns are used as sorbents. For example, using Vydac adsorbent as the packing and elution with 1% amyl alcohol in isooctane at a working pressure of 2800 p.s.i., derivatives of benzyl alcohol and cinnamyl alcohol can be separated, the latter having the longest elution time (Chromatronix) (Fig. 18.1.). When a column packed with Permaphase ETH was used at room temperature and a pressure of 250 p.s.i., these substances were also well separated

*References p.439*

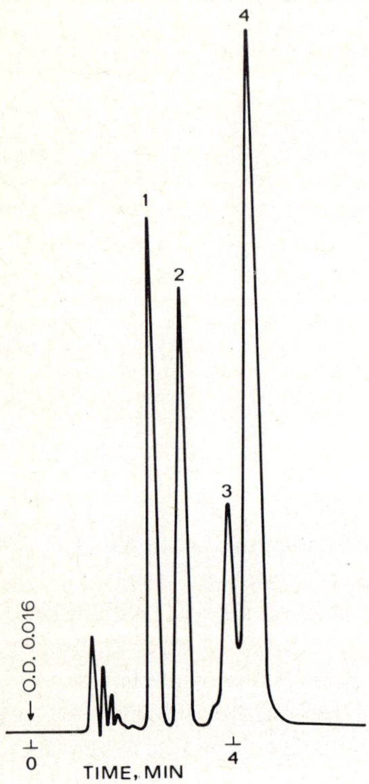


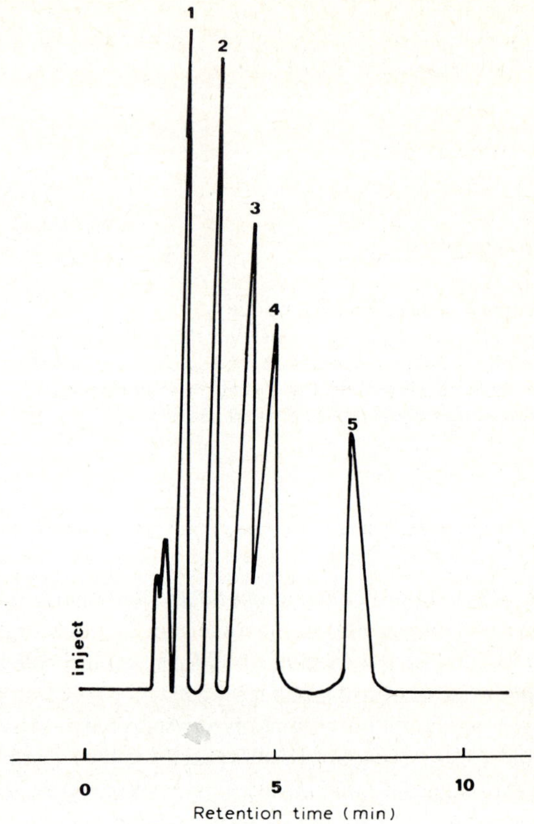


Fig. 18.1. Separation of alcohols (Chromatronix). Column: 2000 × 2 mm. Sorbent: Vydac adsorbent (Chromatronix). Mobile phase: 1% amyl alcohol in isooctane. Operating conditions: flow-rate 3 ml/min; pressure 3000 p.s.i.; temperature ambient. Detection: UV at 254 nm. Instrument: Chromatronix 3100 UV. Sample: 20 $\mu$l. Peaks: 1 = 2-phenyl-2-propanol; 2 = $\alpha$-methylbenzyl alcohol; 3 = benzyl alcohol; 4 = cinnamyl alcohol.

Fig. 18.2. Separation of a mixture of alcohols (DuPont Instruments). Column: 1 m U-shape; DuPont Model 830 chromatograph. Sorbent: Permaphase ETH. Mobile phase: *n*-hexane. Operating conditions: column pressure 250 p.s.i.g.; flow-rate 1 ml/min. Detection: UV photometer. Peaks: 1 = $\alpha$-methylbenzyl alcohol; 2 = $\alpha$, $\alpha'$-dimethylbenzyl alcohol; 3 = 2-phenylethanol; 4 = cinnamyl alcohol; 5 = benzyl alcohol.

but their elution sequence was different. When *n*-hexane was used for elution, cinnamyl alcohol had a lower elution time than benzyl alcohol (DuPont Instruments) (Fig. 18.2). In both instances, the time of analysis was less than 8 min. Kirkland separated similar compounds in a liquid–liquid system using porous silica microspheres as the carrier, impregnated with up to 30% of $\beta,\beta'$-oxydipropionitrile as the stationary phase. *n*-Hexane saturated with the stationary phase was used as the eluent (Fig. 18.3). Under the conditions of liquid–solid chromatography, the same substances were also separated on plain porous silica microspheres as adsorbent. Fig. 18.4 shows how rapid (60 sec) the separation of five aromatic alcohols at 2000 p.s.i. was. A series of homologous alcohols, branched alcohols and some glycols were separated chromatographically on a styrene–divinylbenzene gel using benzene as the mobile phase (Hendrickson), and the results are given in Table 18.1. A styrene gel was also used for the separation of a mixture of alcohols (methanol to *n*-heptanol) and diols with *o*-dichlorobenzene or tetrahydrofuran as eluent (Cazes and Gaskill) (Table 18.2).

TABLE 18.1
GEL CHROMATOGRAPHY OF ALCOHOLS (HENDRICKSON)

Column: 12 ft. (8 ft.) × 3/8 in. Column packing: 40 Å styrene-divinylbenzene gel, permeable to alkanes of mol. wt. 450. Mobile phase: benzene. Flow-rate: 1 ml/min. Detection: RI.

| Compound* | Elution volume (ml) | Peak width** (ml) |
|---|---|---|
| Methanol | (92.0) | (5.47) |
| Ethanol | 127.3 | 4.53 |
| Isopropanol | 117.6 | 4.52 |
| *n*-Propanol | (81.3) | (5.47) |
| Benzyl alcohol | (82.7) | (5.22) |
| *tert.*-Butanol | 110.7 | 4.60 |
| *n*-Butanol | 112.7 | 4.82 |
| 3-Heptanol | (68.7) | 4.65 |
| *n*-Heptanol*** | (69.8) | (4.77) |
| *n*-Heptanol*** | 101.0 | 4.40 |
| Diethylene glycol monomethyl ether | 97.3 | 4.88 |
| 2,2,4-Trimethylpentanol | 98.7 | 4.70 |
| *n*-Decanol*** | 93.5 | 4.70 |
| *n*-Decanol*** | 82.2 | 4.51 |
| Ionol | (60.4) | (5.79) |
| Triphenylcarbinol | (65.8) | (–) |
| Dipropylene glycol | (68.1) | (7.16) |
| Tripropylene glycol | (61.8) | (5.36) |
| Tetraethylene glycol | (86.7) | (4.77) |
| Allyl alcohol | 110.0 | 5.11 |
| 2-Butene-l-ol | 103.8 | 5.30 |

*Sample sizes were typically 0.1 ml of a 4% solution in benzene.
**Peak widths were obtained by drawing tangents to each side of the curves and are expressed as the number of millilitres of eluent at the base of the triangle.
***Compound run more than once.

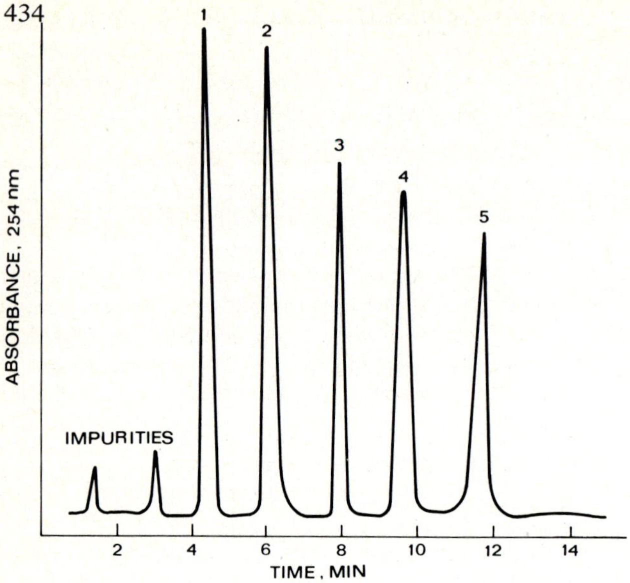


Fig. 18.3. Separation of hydroxylated aromatics (Kirkland). Column: 250 mm × 3.2 mm. Sorbent: 5–6 $\mu$ porous silica microspheres (*ca*. 350 Å) impregnated with 30% of $\beta$, $\beta'$-oxydipropionitrile. Mobile phase: *n*-hexane (saturated with stationary phase). Operating conditions: flow-rate 1 ml/min; pressure 600 p.s.i.g.; temperature 26°C. Detection: absorbance at 254 nm. Peaks: 1 = $\alpha$, $\alpha$-dimethylbenzyl alcohol; 2 = 2,5-dimethylphenol; 3 = $\alpha$-methylbenzyl alcohol; 4 = 3-phenylpropanol; 5 = $\beta$-phenylethanol.

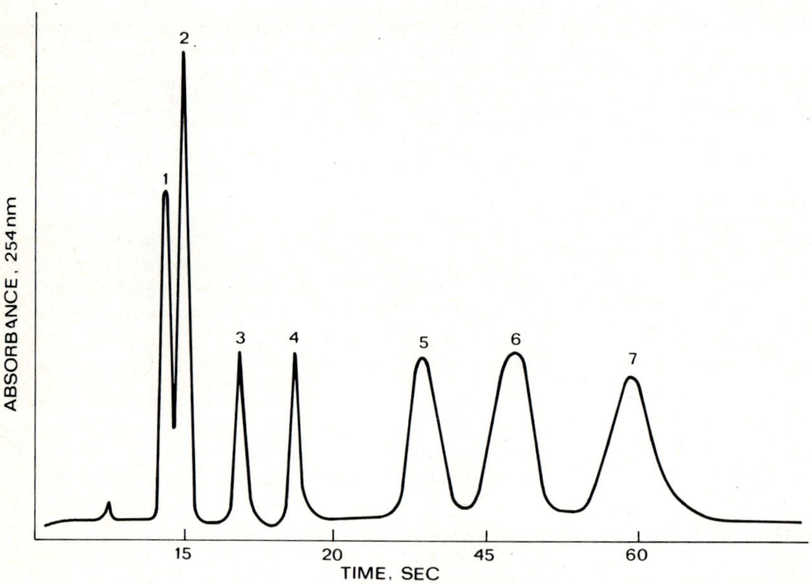


Fig. 18.4. Separation of hydroxylated aromatics (Hendrickson). Column: 250 mm × 3.2 mm. Sorbent: 8–9 $\mu$ porous silica microspheres (*ca.* 75 Å). Mobile phase: dichloromethane, half saturated with water. Operating conditions: flow-rate 10.5 ml/min; pressure 2000 p.s.i.; temperature 27°C. Detection: absorbance at 254 nm. Peaks: 1 = 2,5-dimethylphenol; 2 and 3 = impurities in standards; 4 = $\alpha$-phenylbenzyl alcohol; 5 = $\alpha$, $\alpha$-dimethylbenzyl alcohol; 6 = benzyl alcohol; 7 = 3-phenylpropanol.

TABLE 18.2
CHROMATOGRAPHY OF ALCOHOLS AND GLYCOLS (CAZES AND GASKILL)

Values are elution volumes (ml) obtained under the following conditions. A, column 1, mobile phase *o*-dichlorobenzene, 130°C; B, column 2, mobile phase tetrahydrofuran, 25°C; C, column 3, mobile phase *o*-dichlorobenzene, 130°C. Columns: (1) four 4-ft. columns in series having exclusion limits of $3 \times 10^3$, 250, 60 and 60 Å, respectively; (2) four 4-ft. columns in series having exclusion limits of about 40 Å; (3) four 4-ft. columns in series having exclusion limits of $3 \times 10^3$, 45, 45 and 45 Å, respectively. Diameter of columns 3/8 in. Gel: a rigid, cross-linked polystyrene gel characterized by Waters Ass., Framingham, Mass., U.S.A.

| Compound* | Conditions | | |
|---|---|---|---|
| | A | B | C |
| Methanol | 208.8 | 132.5 | – |
| Ethanol | 196.4 | 131.4 | – |
| *n*-Propanol | 189.7 | 127.3 | – |
| Isopropanol | 186.8 | 125.8 | – |
| *n*-Butanol | 183.6 | – | – |
| Isobutanol | 184.4 | – | – |
| *sec.*-Butanol | 182.5 | – | – |
| *tert.*-Butanol | 181.2 | – | – |
| Allyl alcohol | 194.4 | – | – |
| *n*-Pentanol | 179.3 | – | – |
| *n*-Hexanol | 174.6 | – | – |
| *n*-Heptanol | 170.6 | 113.1 | – |
| 1,2-Ethanediol | – | – | 202.5 |
| 1,2-Propanediol | – | – | 186.7 |
| 1,4-Butanediol | – | – | 182.9 |
| 1,3-Butanediol | – | – | 179.7 |
| 2,2-Dimethyl-1,3-propanediol | – | – | 175.8 |
| 2,2,4-Trimethyl-1,3-pentanediol | – | – | 180 |

*Samples were 2-ml aliquots of 0.25–1.0% solutions.

## CHROMATOGRAPHY OF ALCOHOLS ON ION EXCHANGERS

Ion-exchange chromatography is much less important for the separation of free alcohols than other chromatographic techniques. Salting-out chromatography on Dowex 1-X8 ($SO_4^{2-}$) columns was used for the separation of less than milligram amounts of diacetyl, acetoin and 2,3-butylene glycol (Speckman and Collins). Sodium or ammonium sulphate solution (0.5 *M*) was used for elution. The substances are eluted in order of decreasing polarity, *i.e.*, 2,3-butylene glycol, acetoin and diacetyl.

Higher alcohols were separated on a Dowex 50-X8 ($H^+$) (200–400 mesh) column by gradual elution with acetic acid of increasing concentration (1–3 *N*) (Sherma and Rieman) (Table 18.3). Sherma *et al.* also studied the possibility of the separation of higher alcohols on Dowex 1 ($CH_3COO^-$) by elution with aqueous–organic salt solutions. In these separations, both the salting out and the solubility play a role and therefore the dependence

TABLE 18.3
SOLUBILIZATION CHROMATOGRAPHY OF ALCOHOLS (SHERMA AND RIEMAN)

Column: 20 cm × 2 cm. Column packing: Dowex 50-X8 ($H^+$) (200–400 mesh). Flow-rate: 0.4–0.5 cm/min. Detection: 6-ml fractions were mixed with 0.22 *M* sodium dichromate in concentrated $H_2SO_4$, diluted with 25 ml of water and the resulting Cr(III) was measured spectrophotometrically.

| Alcohol | Water | Acetic acid | | | |
|---|---|---|---|---|---|
| | | 1 *M* | 2 *M* | 3 *M* | 4 *M* |
| *tert.*-Amyl alcohol | 2.19 | 1.86 | – | – | – |
| *n*-Amyl alcohol | 4.61 | 3.85 | 3.30 | 2.21 | 1.43 |
| *n*-Hexanol | 7.97 | 6.12 | 4.88 | 2.88 | 1.46 |
| *n*-Heptanol | 14.9 | 10.8 | 8.28 | 3.81 | 1.65 |
| *n*-Octanol | 27.5 | 20.7 | 13.8 | 5.39 | 1.85 |
| *n*-Nonanol | 55.1 | 36.7 | 21.6 | 7.47 | 1.98 |
| *n*-Decanol | – | – | – | 9.60 | 2.69 |
| *n*-Undecanol | – | – | – | – | 3.55 |
| *n*-Dodecanol | – | – | – | – | 4.77 |
| Benzyl alcohol | 5.71 | 4.30 | 3.76 | 2.53 | 1.55 |
| Cyclohexanol | 4.80 | 4.04 | 3.58 | 2.55 | 1.67 |

of the logarithm of the distribution coefficient of the separated substance on the electrolyte concentration is not always linear. These workers have shown that the partial separation of higher alcohols (*n*-hexanol, *n*-heptanol and *n*-octanol) with 4 *M* LiCl in methanol as eluent is possible. However, an appreciable broadening of the elution curves was observed. Sherma and Lowry studied solubilization chromatography on a macroreticular ion-exchange resin. The separation of a homologous series of aliphatic alcohols by solubilization chromatography on the macroreticular ion-exchange resin Amberlyst 15 was not as successful as that on conventional gel resins. A method of determination of alcohols in aqueous solution by thermodetection liquid chromatography has also been proposed (Suzuki *et al.*). A series of sorbents was tested but the best results were achieved on Dowex 50W-X8.

The possibility of the separation of organic compounds that are only poorly soluble or are completely insoluble in water can be improved by increasing their sorption affinity to the resin. An anion-exchange resin in the form of amphophilic anions of higher organic acids was used by Small and Bremer for this purpose. On such resins, in addition to the interaction forces between the organic compounds and the resin skeleton, these substances also interact with the amphophilic anion, which increases their sorption affinity to the resin. On such modified resins, some water-insoluble substances of low polarity can be separated more satisfactorily. Amphophilic resins have a greater tendency to swell in organic solvents than in water, and the separation of substances can be considerably affected by temperature. On a Dowex 1-X1 column in the form of fatty acid anions [or anions of other acids, for example di-(2-ethylhexylphosphoric acid)], an almost complete separation of a mixture of propylene glycol and *tert.*-butanol can be achieved if the elution is carried out with water at 70°C. Satisfactory separations of a mixture of ethanol and *n*-propanol, and of ethylene glycol and *n*-propanol, can be achieved by the same method. Only a weak overlapping of the elution curves took place, while these substances were not separated at all when comparative attempts were made on a chloride form of this resin. A mixture of alcohols was separated on a cation-exchange column of Dowex 50W-X2 ($K^+$) with water as the mobile phase (Wu and McCready). Alcohol mixtures (10 μl each of methanol, ethanol, 1-propanol, 2-propanol, 1-butanol, 2-butanol, 2-methyl-1-propanol and 2-methyl-2-propanol in 1 ml of water) were applied to the column and eluted with water at an elution rate of 0.5 ml/min and with fractions of 2.5 ml. The alcohol content in single fractions was determined in 20-μl aliquots with a Beckman Carbon Analyzer.

## CHROMATOGRAPHY OF DERIVATIVES OF ALCOHOLS AND GLYCOLS

Lower alcohols were separated by liquid chromatography in the form of their *p*-(N,N-dimethylamino)benzene-*p*′-azobenzoates on a 240 × 2.7 mm column packed with Dowex 50W-X2 (200–400 mesh) cation exchanger (Churáček and Jandera). Distribution coefficients for optimum amounts of hydrochloric acid in the eluent were found by elution with aqueous alcoholic hydrochloric acid solution (Table 18.4). A good resolution of these coloured derivatives can also be achieved by adsorption chromatography on a silica gel CH column (5–40 μm) using a cyclohexane–ethyl acetate mixture as the mobile phase.

The quantitative micro-determination of alcohols as esters of pyruvic acid 2,4-dinitrophenylhydrazone was carried out on an alumina column (Schwartz). The liquid chromatographic determination of trace amounts of glycols in the form of their 3,5-dinitrobenzoates was carried out on a 2 m × 1/8 in. column filled with Corasil II, with ethyl acetate–*n*-heptane (1:3, v/v) as the mobile phase, at 1200 p.s.i. (Carey and Persinger).

TABLE 18.4
VOLUME DISTRIBUTION COEFFICIENTS, $D_v$, OF SOME ESTERS OF N,N-DIMETHYL-*p*-AMINOBENZENEAZOBENZOIC ACID ON THE CATION EXCHANGER DOWEX 50W-X2 IN A 0.925 *M* SOLUTION OF HYDROCHLORIC ACID IN 80.5% ETHANOL (CHURÁČEK AND JANDERA)

$D_v$ is defined as the ratio of the amount of compound in a unit volume of the ion-exchanger phase to the same volume of external solution.

| Ester | $D_v$ |
|---|---|
| Methyl | 6.3 |
| Ethyl | 5.2 |
| *n*-Propyl | 4.5 |
| *n*-Butyl | 3.9 |
| *n*-Amyl | 3.5 |
| *n*-Hexyl | 3.0 |
| *n*-Octyl | 2.3 |
| *n*-Nonyl | 2.1 |
| *n*-Decyl | 1.9 |
| Isopropyl | 4.3 |
| Isobutyl | 3.6 |

*References p.439*

## SEPARATION OF POLYOLS AND POLYMERIC DIOLS

Silicone-treated Celite 545 was used for the analysis of polyethylene glycol in polyoxyethylene-type non-ionic surfactants. A mixture of *n*-butanol and water, or *n*-butanol in 10% aqueous sodium chloride solution, was used for elution (Konishi and Yamaguchi). For the determination of the terminal hydroxy group, the possibility of separating the 3,5-dinitrobenzoates of glycols from the excess of 3,5-dinitrobenzoyl chloride and 3,5-dinitrobenzoic acid on a silica gel column was made use of by Han. The column was eluted with chloroform and the absorption of the eluate measured at 528 nm.

Linear elution adsorption chromatography was also used for the fractionation of polyethylene glycol derivatives of the Ph–S–$(CH_2-CH_2-O)_n$–$CH_2-CH_2$–S–Ph type (*n* indicates the degree of polymerization) (Calzolari *et al.*). The effect and the method of activation or deactivation of silica gel with trimethylchlorosilane on the separation was investigated. In column chromatography, the degree of activation of silica gel has a marked effect on the separation of ethylene glycol oligomers. At maximum activation, the wide distribution of the adsorption energy of the active silanol centres that form hydrogen bonds with the ether groups of the oligomers favours tailing. The elimination of the more active centres by partial deactivation with water was used earlier in order to reduce tailing and the retention volume of the peaks. Polyols can also be separated by gel chromatography using a cross-linked polystyrene and 1,2-dichloroethane as eluent. The content of polyether polyols in the effluent was determined colorimetrically, as Co–SCN complexes, at 620 nm. Good linearity was observed between the effluent volume and the logarithm of the molecular weights of the compounds (Kondo *et al.*). The chromatographic separation of glycols on hydrophobic gels was described by Cazes and Gaskill (Table 18.2). A column packed with Merckogel PGM-2000 was used for the separation of polyethylene glycols in combination with water or tetrahydrofuran as the mobile phase (Randau *et al.*). A 100 × 1.4 cm or 100 × 0.4 cm column was used and the applied pressures were 0, 10 and 15 atm. At 10 atm pressure and a column diameter of 0.4 cm, the time necessary for the separation of four polyethylene glycols of various molecular weights was 20 min, at an effluent flow-rate of 34 ml/h, while at laboratory temperature (20–22°C), the flow-rate, which is given by the resistance of the column packing, was 8 ml/h and the total time of analysis was about 13 h. A method for the chromatographic separation of glycols on a 20 × 2.28 cm column of Dowex 1-X8 (200–300 mesh) in the borate form, using 0.925 *M* sodium metaborate or 0.02 *M* sodium tetraborate solution for elution, was also proposed (Sargent and Rieman), and permitted the quantitative separation of diethylene glycol, ethylene glycol, 1,2-propylene glycol, glycerol, *meso*-2,3-butanediol and D,L-2,3-butanediol. Propylene glycol is eluted with metaborate together with glycerol, while on elution with tetraborate it is eluted in admixture with *meso*-2,3-butanediol. For the complete separation and determination of all of these glycols, both chromatographic procedures have to be combined. The determination requires 10 min.

Chromatography on a cation-exchange column (KU-2) was used for the separation of polyols (xylitols, glycerol and ethylene glycol) and water was used for elution (Dabagov and Balandin). On a Dowex 50-X12 (200–400 mesh) column or a column of KU-2 cross-linked to the extent of 12%, a mixture of polyols was separated with water as the mobile phase at 60°C. The components were eluted in the following order: xylitol, erythritol, glycerol, ethylene glycol and 1,2-propanediol. The separation required 24–26 h. The fractions were analyzed using an Abbe refractometer or colorimetrically after oxidation with dichromate. The best separation was achieved when the exchanger was in the $H^+$ form. On an exchanger in the $Ca^{2+}$ form, a change in the elution sequence was observed. Xylitol is sorbed selectively on this exchanger and is therefore eluted last. The separation of a small amount of xylitol from ethylene glycol on the free $H^+$ form of the cation exchanger was also proposed. This method can be used for the analytical control of the large-scale hydrolysis of sugars and polyols.

## REFERENCES


Calzolari, C., Favretto, L. and Stancher, B., *J. Chromatogr.*, 47 (1970) 209.
Carey, M. A. and Persinger, H. E., *J. Chromatogr. Sci.*, 10 (1972) 537.
Cazes, J. and Gaskill, D. R., *Separ. Sci.*, 2 (1967) 426; 4 (1969) 15.
Chromatronix, *Liquid Chromatography Application, No. 14,* Chromatronix, Berkeley, Calif.
Churáček, J. and Jandera, P., *J. Chromatogr.*, 53 (1970) 69.
Dabagov, N. S. and Balandin, A. A., *Izv. Akad. Nauk SSSR, Ser. Khim.*, (1966) 1308 and 1315.
DuPont Instruments, *Product Bulletin No. 830 PB1*, DuPont, Wilmington, Del., September 1971.
Han, K. W., *Analyst (London)*, 92 (1967) 316.
Hendrickson, J. G., *J. Chromatogr.*, 32 (1968) 543.
Kirkland, J. J., *J. Chromatogr. Sci.*, 10 (1972) 593.
Kondo, K., Hori, M. and Hattori, M., *Bunseki Kagaku (Jap. Anal.)*, 16 (1967) 414; *C.A.*, 68 (1968) 35636t.
Konishi, K. and Yamaguchi, S., *Anal. Chem.*, 40 (1968) 1720.
Randau, D., Bayer, H. and Schnell, W., *J. Chromatogr.*, 57 (1971) 77.
Sargent, R. and Rieman, W., *Anal. Chim. Acta*, 16 (1957) 144.
Schwartz, D., *Anal. Biochem.*, 38 (1970) 148.
Sherma, J., Locke, D. and Bassett, D., *J. Chromatogr.*, 7 (1962) 273.
Sherma, J. and Lowry, J. D., *Anal. Lett.*, 1 (1968) 707; *C.A.*, 69 (1968) 80571z.
Sherma, J. and Rieman, W., *Anal. Chim. Acta*, 18 (1958) 214.
Small, H. and Bremer, D. N., *Ind. Eng. Chem., Fundam.*, 3 (1964) 361.
Speckman, R. A. and Collins, E. B., *Anal. Biochem.*, 22 (1968) 154.
Suzuki, Y., Ishii, D. and Takeuchi, T., *Bunseki Kagaku (Jap. Anal.)*, 18 (1969) 858.
Wu, C.-M. and McCready, R. M., *J. Chromatogr.*, 57 (1971) 424.

*Chapter 19*

# Phenols

J. CHURÁČEK and J. ČOUPEK

## CONTENTS



## INTRODUCTION

The sorption energy of phenols toward aluminium oxide and silica gel is greater than that of alcohols owing to the more acidic character of the phenolic hydroxyl group compared with alcoholic hydroxyl groups. Important factors in chromatographic separations are the number of hydroxyl groups in the phenol molecule, their mutual positions and the presence of other functional groups. These properties can be used not only in the gel chromatography of phenols, but also in liquid and ion-exchange chromatography.

## GEL CHROMATOGRAPHY

In recent years, an important role in the separation of phenols and their derivatives has been played by gel chromatography on hydrophilic and organophilic gels, and in relatively polar solvents, in addition to the effects of the shape and size of the molecule, the effect of solvation of the phenolic hydroxyl group also strongly influences the separation.

When investigating separations of low-molecular-weight compounds on styrene–divinylbenzene gels in tetrahydrofuran, Hendrickson and Moore observed that molecules that contain certain functional groups appear to be much larger than would be expected from their molecular dimensions. This phenomenon was ascribed to the formation of hydrogen bonds between the compound being analyzed and the eluent. Hendrickson (1967) found that chloroform behaved as a hydrogen donor, in contrast with tetrahydrofuran. The elution volumes in benzene, determined by Hendrickson (1968), differed from those measured in tetrahydrofuran. Edwards and Ng, who measured the elution volumes of more than 100 model compounds in tetrahydrofuran, also observed anomalous behaviour with some types of compounds; this behaviour was attributed to the solvation of compounds with solvent and to sorption on the gel.

The effect of the formation of the hydrogen bonds between the phenols under investigation and the eluent was shown by Yoshikawa *et al.* By comparing the theoretical values

*References p.448*

of the elution volumes of a series of substituted phenols with the measured values, using tetrahydrofuran as the eluent, they concluded that the phenol molecule is not solvated by tetrahydrofuran if the structure of the phenol enables an intramolecular hydrogen bond to be formed. A bulky substituent in the *ortho*-position limits the solvation with tetrahydrofuran; 2,6-substituted derivatives are not solvated at all. The acidity of the phenolic hydroxyl group enhances the solvation of the molecule. These conclusions were confirmed by Čoupek *et al.* (1972a), who used a series of 42 monohydric phenols and concentrated their attention on the effects of substitution in order to find a relationship between the contribution of the elution volume and the type and position of the substituent. The elution volumes give information on the steric requirements of the alkyl group on the aromatic ring. By determining the effects of substitution and the character of the connecting bond in the bisphenol series on their chromatographic behaviour, Čoupek *et al.* (1973) again demonstrated the important part played by solvation in the gel chromatographic analysis of polynuclear phenolic and extensively substituted compounds. The results obtained were used by Čoupek *et al.* (1971, 1972b) for gel chromatographic determinations of the content of phenolic antioxidants and light stabilizers in polymers. Additive systems in several commercial polymers were analyzed by Howard in connection with the determination of the effect of the treatment and ageing of polymers on the change in the content of the individual components. Gel chromatography on organophilic packings has also been used extensively in the study of molecular-weight distributions of phenolic resins (Drum, Gardikes and Konrad, Quinn *et al.*, Varishth *et al.*, Hagen and Schröder).

## ADSORPTION CHROMATOGRAPHY

The separation of phenols can also be based on the fact that the phenolic hydroxyl group forms bonds with the amide group present in, for example, polyamide gels (Endres and Hormann, Grassmann *et al.*). The bond strength depends on the number and positions of the hydroxyl groups in the phenol molecule. The capacity of polyamides is usually relatively high, thus making these gels suitable for the separation of isomers. The eluotropic series of solvents used for the elution of phenols from a polyamide column is represented by water, ethanol, methanol, acetone, formamide, dimethylformamide (Hörhammer and Wagner). Ligroin (60–70°C) removed mainly *ortho,para*-substituted phenols, while benzene removed mainly *ortho,para*-substituted phenols containing a methyl group in the *para*-position only. Benzene containing 5% of methanol, and methanol itself, removed mainly phenol and *para*-substituted phenols. Martin found that methanol removed catechol and its *para*-substituted derivatives.

The sorption properties of phenols on a cross-linked polyamide gel (Bio-Gel P-2 and P-6) were studied by Streuli, who found very good correlations between the molecular structures and the adsorption properties of phenols. The column was calibrated by using a 0.5% aqueous solution of blue dextran to determine the interstitial volume ($V_i$), and acetone and tetrahydrofuran to determine the void ($V_0$) plus pore volume. The values of $V_0$ and $V_i$ measured using the above solvents were smaller than those for methanol and larger than those for dimethyl sulphoxide. Acetone and tetrahydrofuran were considered to be molecules with the lowest probability of interaction with gels. A dilute solution of

TABLE 19.1
$K_d$ VALUES FOR BENZENE DERIVATIVES ON CROSS-LINKED GELS (STREULI)

| Compound | Bio-Gel P-2 (100–200 mesh) | | | Bio-Gel P-6 | Sephadex G-25 |
|---|---|---|---|---|---|
| | Exptl. | Calc. | Difference | Exptl. | Exptl. |
| Benzene | 1.69 | – | | | |
| Phenol | 2.20 | 2.24 | 0.04 | 1.43 | 2.0 |
| Benzoic acid | 1.04 | 0.95 | −0.09 | | |
| 2-Hydroxyphenol | 2.81 | 2.79 | −0.02 | | 1.9 |
| 3-Hydroxyphenol | 2.90 | 2.79 | −0.11 | | 2.65 |
| 4-Hydroxyphenol | 2.58 | 2.79 | 0.21 | 1.60 | 2.4 |
| 2,3-Dihydroxyphenol | 3.41 | 3.34 | −0.07 | | 2.4 |
| 3,5-Dihydroxyphenol | 3.50 | 3.34 | −0.16 | | 3.15 |
| 2-Methylphenol | 2.42 | 2.39 | −0.03 | | 2.55 |
| 3-Methylphenol | 2.44 | 2.39 | −0.05 | | |
| 2,4-Dimethylphenol | 2.48 | 2.54 | 0.06 | | |
| 2,6-Dimethylphenol | 2.44 | 2.54 | 0.10 | | |
| 2-Carboxyphenol | 1.42 | 1.50 | 0.08 | | 2.1 |
| 3-Carboxyphenol | 1.40 | 1.50 | 0.10 | | 1.8 |
| 4-Carboxyphenol | 1.45 | 1.50 | 0.05 | | 1.45 |
| 2-Chlorophenol | 3.04 | 3.26 | 0.22 | | 3.4 |
| 3-Chlorophenol | 3.40 | 3.26 | −0.14 | | |
| 4-Chlorophenol | 3.18 | 3.26 | 0.08 | 1.77 | |
| 2,4-Dichlorophenol | 4.48 | 4.28 | −0.20 | | |
| 2,4,6-Trichlorophenol | 4.98 | 5.30 | 0.32 | | |
| 2-Nitrophenol | 2.94 | 3.78 | 0.84 | | 2.45 |
| 3-Nitrophenol | 3.74 | 3.78 | 0.04 | | |
| 4-Nitrophenol | 3.75 | 3.78 | 0.03 | 1.94 | |
| 2,4-Dinitrophenol | 2.37 | 5.32 | 2.95 | 1.76 | |

sodium chloride was used as eluent. Distribution coefficients ($K_d$) for all of the phenols studied, calculated from the standard equation $K_d = (V_e - V_0)/V_i$, (where $V_e$ *is* the elution volume), are listed in Table 19.1. The pooled standard deviation for all results was 0.04.

Cross-linked dextran gel was used for separation by Gelotte. Woof and Pierce used Sephadex G-25 for the separation of polyhydroxyphenols. This procedure proved to be very effective and can therefore be recommended.

With water, elution occurred according to the number of hydroxyl groups contained in the phenol. Adsorption occurred in all instances, and fractionation was therefore achieved not on the basis of molecular size and shape but as a result of differing degrees of bonding between the phenol and the residual carboxyl groups present in the gel matrix. There was little or no difference between *ortho*- and *meta*-substituted compounds (Table 19.2).

It will be noted from inspection of the $K_d$ values that the separation of mixtures of these phenols should be possible and water or acid is probably the most suitable medium. Fig. 19.1 shows two traces (a and b) of the elution pattern obtained automatically with an AutoAnalyzer. Fig. 19.1a was obtained with water as eluent, and Fig. 19.1b with dilute acetic acid. Fig. 19.1c shows the separation of various phenol derivatives using acetic acid with NaCl as eluent. Acid was used in this case because of the anomalous behaviour of hydroxy acids in water.

*References p.448*

TABLE 19.2

$K_d$ VALUES FOR PHENOLS IN AQUEOUS SOLUTIONS FROM SEPHADEX G-25 COLUMNS (WOOF AND PIERCE)

| Compound | Eluting medium | | | |
|---|---|---|---|---|
| | Water | 0.5 *M* NaCl | 0.1 *M* $NH_4OH$ | 0.1 *M* $CH_3COOH$ |
| Phenol | 1.8 | 2.0 | 1.1 | 1.95 |
| Hydroquinone | 1.05 | 2.4 | 1.7 | 2.4 |
| Resorcinol | 2.1 | 2.05 | 1.6 | 2.5 |
| Catechol | 2.05 | 1.9 | 1.95 | 2.5 |
| Phenolglucinol | 2.3 | 3.15 | 1.1 | 3.25 |
| Pyrogallol | 2.4 | 2.4 | 1.3 | 3.25 |
| *o*-Hydroxybenzoic acid | 1.9 | 2.1 | 1.6 | 3.4 |
| Arbutin | 1.45 | 1.55 | 0.85 | |
| *o*-Nitrophenol | 1.45 | 2.45 | 1.8 | 2.4 |
| Saligenin | 1.85 | 2.1 | 1.55 | 1.9 |
| Guaiacol | 2.2 | 1.8 | 1.45 | 2.1 |
| *o*-Hydroxybenzaldehyde | 2.3 | 2.3 | 1.55 | 2.3 |
| *o*-Cresol | 2.6 | 2.55 | 1.85 | 3.0 |
| Orcinol | 3.0 | 2.6 | 1.5 | – |
| *o*-Chlorophenol | 3.15 | 3.4 | 1.55 | 3.0 |
| *p*-Hydroxydiphenyl | 4.8 | 5.4 | 2.9 | – |

The behaviour even of simple phenols on Sephadex columns is complex and it is difficult to predict $K_d$ values from the structures. The factors that determine elution volume include both the number and positions of hydroxyl groups, substituent groups that affect strength of bonding, and the medium. In addition, steric factors may govern penetration into gel grains. Neither the $K_d$ values nor the change in $K_d$ values on passing from neutral to alkaline medium seemed to be related directly to the p*K*s of the phenols. Only nitro and carboxylic acid groups decreased bonding.

In many instances, silica gel (Hanson and Zucker) was also used for preparative separation of a number of phenols. Van Dijk and Mijs separated phenols on silica gel presaturated with solvent vapour, using a mixture of non-polar and polar solvents in a 1:1 ratio as the eluent. Zakupra *et al.* separated high-molecular-weight alkylphenols on silica gel KSK and aluminium silicate, and achieved a good separation of 2,4- and 2,6-dialkylphenols from 2- and 4-monoalkylphenols. However, the procedure is time consuming, a good separation requiring *ca.* 8 h. The identification of alkyl-substituted phenols present in cashew nut-shell liquid was performed by Murthy *et al.* using a 150 × 1.5 cm column packed with silica gel and a column packed with silica gel impregnated with silver nitrate. Pure cardanol was also isolated in this manner (De Vries). Mixtures containing 5, 10 and 20% (v/v) of ethyl acetate in benzene were used as eluents. Berrera separated 21 tar-phenols on a column packed with Celite 535 or with a cellulose powder impregnated with formamide. The eluent was cyclohexane and diethyl ether (98:2 or 95:5), saturated with formamide prior to use.

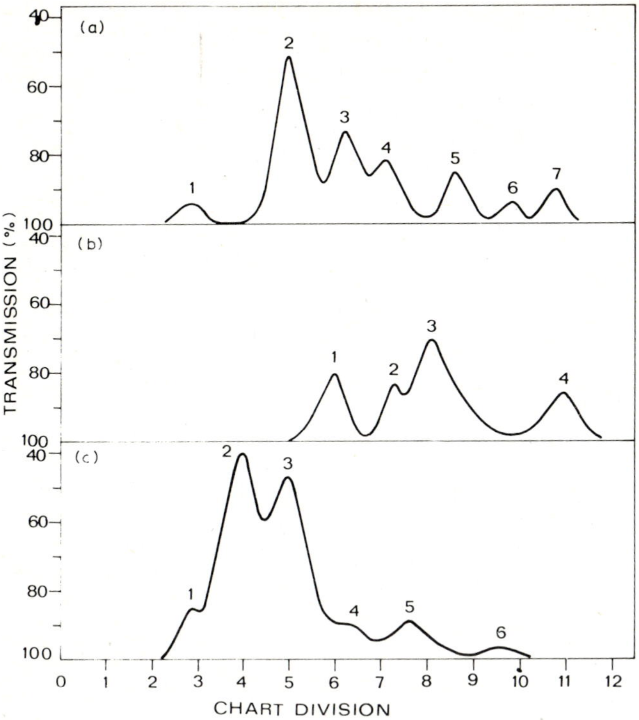


Fig. 19.1. Separation of phenols (Woof and Pierce). Column: (a and b) 35 × 2.5 cm; (c) 0.9 × 25 cm; Sephadex G-25. Mobile phase: (a) water; (b) 0.1 *M* acetic acid; (c) 0.1 *M* acetic acid containing 0.05 *M* sodium chloride. Detection: UV photometric. Traces: (a) 1 = hydroxyquinone; 2 = phenol; 3 = catechol + resorcinol; 4 = guaiacol; 5 = pyrogallol + phloroglucinol; 6 = *o*-cresol; 7 = orcinol. (b) 1 = phenol; 2 = hydroquinone; 3 = resorcinol + catechol; 4 = pyrogallol + phloroglucinol. (c) 1 = saligenin; 2 = nitrophenol + guaiacol; 3 = salicylaldehyde; 4 = catechol; 5 = *o*-cresol + chlorophenol; 6 = salicylic acid.

## ION-EXCHANGE CHROMATOGRAPHY

Phenols, which have an acidic character, can be sorbed on anion-exchangers via the mechanism of anion exchange. Their aromatic nuclei exhibit a great affinity for the matrix of styrene–divinylbenzene ion exchangers, and both ionic and molecular sorption are involved. It was found that phenols were also retained by non-ionic sorption on the functional groups of ion exchangers (Anderson and Hansen), which is why both anion- and cation-exchange resins can be used for the separation of phenols.

A method has been described for the chromatographic separation of mixtures of phenol and *m*-cresol on a column with the cation-exchanger KU-2 ($H^+$) using water as eluent (Stankevich and Skorochod). Elution with sodium citrate solution of pH 3.42 was used in order to separate 10–15 mg of a mixture of phenol derivatives on a Dowex 50

*References p.448*

column in the following order: phenol, cresols and finally xylenols (Krampitz and Albersmeyer).

The quantitative separation of N-acetyl-*p*-aminophenol from *p*-aminophenol on a column packed with the cation-exchanger Amberlite IR-120 ($H^+$) can be achieved by elution of the N-acetyl-*p*-aminophenol with water and of the *p*-aminophenol with a 5% hydrochloric acid. The method was used for the determination of both of these constituents in drugs and urine (Koswy and Lach).

Sherma and Rieman used chromatography on a cation-exchange column (Dowex 50) to separate phenols. Pyrocatechol and phenol were eluted with 1.0 *N* acetic acid, while *o*-cresol and *o*-nitrophenol were eluted with 2.0 *N* acetic acid.

Logie described the chromatographic separation and determination of chlorophenols in commercial 2,4-dichlorophenol on a column of the anion-exchanger De Acidite FF ($CH_3COO^-$). Chlorophenols were eluted gradually with solutions of glacial acetic acid in methanol or with buffers consisting of mixtures of acetic acid and triethylamine.

The same ion exchanger was used by Thomas and Thomas, who first concentrated on the separation of the *ortho*-, *meta*- and *para*-isomers. The separation of mixtures containing 1 mg of each isomeric nitrophenol is shown in Fig. 19.2 (Thomas and Thomas). A similar separation of isomers of tetrachlorophenol on an anion-exchange column by means of gradient elution with acetic acid solutions was described by Skelly.

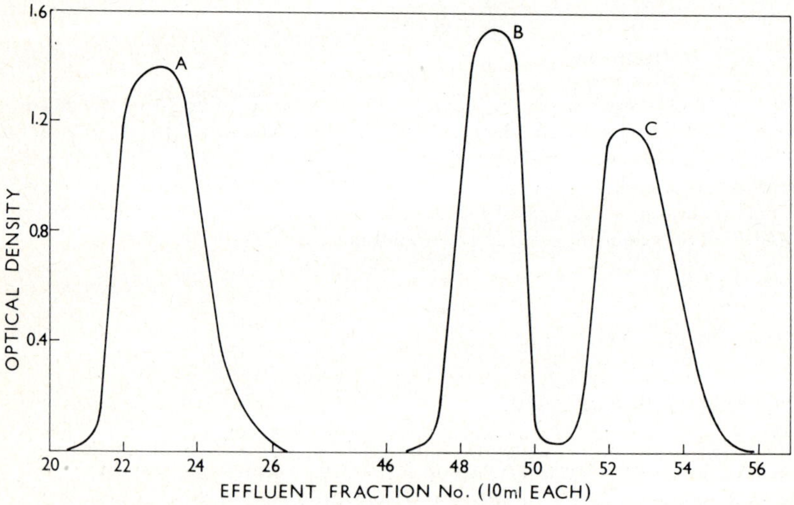


Fig. 19.2. Separation of phenols (Thomas and Thomas). Ion exchanger: De Acidite FF ($CH_3COO^-$; 3–5% divinylbenzene; > 200 mesh). Mobile phase: 4% triethylamine in methanol. Flow-rate: 1 ml/min. Detection: Optical density at 275 nm. (A) *m*-Nitrophenol; (B) *o*-nitrophenol; (C) *p*-nitrophenol.

Seki (1954) used elution with a 1:4 mixture of *n*-propanol and 0.3 *N* hydrochloric acid for the separation of phenol and substituted phenols on columns of the cation exchangers Amberlite IR-120 and IRC-50 ($H^+$ and $Na^+$). In this way, he separated a

TABLE 19.3
SURVEY OF DIFFERENT PROCEDURES APPLICABLE TO THE SEPARATION OF PHENOLS

| Chromatographic technique | Compounds analyzed chromatographically | Sorbent | Eluent | Reference |
|---|---|---|---|---|
| GPC | Phenols | Sephadex | | Somers |
| LC | Hydroxylated alkylphenols (fractionation) | Silica gel Sh, SM | Chloroform, acetone, methanol and their mixtures | Pozdnyshev and Petrov |
| LC | Alkylphenols and alkylanisoles, Preparation, purification, concentration | Silica gel KSK (140–180 μm) | Light petroleum (b.p. 40–60°C)–benzene (10:1, 4:1 and 1:1) | Ivanovskaya and Gorfinkel |
| GPC | Phenolic compounds. Fractionation, isolation | Sephadex G-25 | Distilled water | Rastorgueva |
| LC | Dimethylphenols and polynuclear phenols. Preparation and purification | Silica gel (0.2–0.5 mm) | Methylene chloride, tetrachloromethane–methanol (1:1), light petroleum (b.p. 40–60°C)–benzene (4:1) | Mijs *et al.* |
| LC | Smoke phenols | Celite impregnated with dimethylformamide<br>Alumina<br>Cellulose | Cyclohexane–ethyl acetate saturated with dimethylformamide | Kurko *et al.* |
| LC | Alkylphenol–ethylene oxide adducts in soil | Alumina<br>Silica gel<br>Soda lime | Methanol<br>Chloroform<br>Tetrachloromethane | Weibull and Thorsell |
| LC | Thymol derivative (rough pre-separation). Preparation | Alumina act. II | Diethyl ether–light petroleum (b.p. 40–60°C) (1:10) | Bohlmann *et al.* |
| LC | Tropolones (1-hydroxy-6,7,8-trimethoxynaphthalene) | Neutral alumina act. III, IV | Tetrachloromethane, benzene–diethyl ether (95:5) | Forbes *et al.* |
| LC | 3,3′-Dihydroxydiphenoquinone. Purification and preparation after synthesis | Alumina<br>Silica gel | Benzene<br>Benzene, then chloroform | Musso and Pietsch |
| IEC | Phenols, isomeric *o*-, *m*- and *p*-nitrophenols, halogenated phenols, cresols, isopropylphenols | Bio-Rad AG 50W-X8 ($Na^+$)<br>Dowex 50-X8 ($H^+$) | Aqueous ammonia (pH 9.6)<br>50% Ethanolic ammonia solution | Nomura *et al.* |

(Continued on p.448)

References p.448

TABLE 19.3 (*continued*)

| Chromatographic technique | Compounds analyzed chromatographically | Sorbent | Eluent | Reference |
|---|---|---|---|---|
| IEC | Halogenated phenols | Bio-Rad AG 1-X2 ($CH_3COO^-$) | Acetic acid–methanol (5:95), continuous gradient at 280 nm | Skelly and Crummett |
| High-speed IEC | Phenols and their derivatives (biological applications | Pellicular anion-exchange resin (type LSF) according to *Anal. Chem.*,, 39 (1967) 1422) | 10 m*M* formic acid (pH 3) containing 1 *M* KCl. Temperature 80°C, pressure 800–1000 p.s.i.g. | Andres and Latorre |

mixture of phenol, *p*-cresol and 4-*tert.*-butylphenol; he also separated a mixture of nitrophenols obtained by the nitration of phenol. A mixture of trinitrophenol, 2,6-dinitrophenol, 2,4-dinitrophenol and *o*-, *m*- and *p*-nitrophenol was also separated by Seki (1954) by elution chromatography with an acetate buffer of pH 4.5 on a column of the cation-exchanger Amberlite IR-112. For the separation of the isomers of phenol, cresol, dihydroxybenzenes and various nitrophenols, he also used elution with alcoholic solutions of citrate or acetate buffers on columns of finely granulated cation exchangers. Seki (1960a, b) analyzed the individual fractions photometrically after a prior colour reaction.

A mixture of acetophenone and β-naphthol and a mixture of nitrobenzene and β-naphthol were successfully separated on a column of the cation exchanger Dowex 50-X4 ($H^+$) (200–400 mesh) using an aqueous solution of ethanol for elution (Spitz *et al.*). The eluate fractions were analyzed photometrically in the UV region. Acetophenone or nitrobenzene appear in the eluate earlier than β-naphthol.

Other important papers on the liquid chromatography of phenols, and particularly papers of an applied character published in recent years, are listed in Table 19.3.

## REFERENCES

Anderson, R. E. and Hansen, R. D., *Ind. Eng. Chem.*, 47 (1955) 71.
Andres, M. W. and Latorre, J. P., *J. Chromatogr.*, 55 (1971) 409.
Berrera, J. B., *Rev. Cienc. Apl.*, 21 (1967) 426; *C.A.*, 68 (1968) 70939r.
Bohlmann, F., Niedballa, U. and Schulz, J., *Chem. Ber.*, 102 (1969) 864.
Čoupek, J., Kahovec, J., Křiváková, M. and Pospíšil, J., *Angew. Makromol. Chem.*, 15 (1971) 137.
Čoupek, J., Pokorný, S. Jiráčková, L. and Pospíšil, J., *J. Chromatogr.*, 75 (1973) 87.
Čoupek, J., Pokorný, S. and Pospíšil, J., *11th IUPAC Microsymposium on Macromolecules, Prepr. E-2*, (1972a); *J. Chromatogr.*, 95 (1974) 103.
Čoupek, J., Pokorný, S., Protivová, J., Holčík, J., Karvaš, M. and Pospíšil, J., *J. Chromatogr.*, 65 (1972b) 279.
De Vries, B., *J. Amer. Oil Chem. Soc.*, 40 (1963) 184.
Drum, M. F., *Amer. Chem. Soc., Div. Org. Coatings Plast. Chem., Prepr.*, (1966) 85.
Edwards, D. G. and Ng, Q. Y., *J. Polym. Sci., Part C*, 21 (1968) 105.
Endres, H. and Hormann, H., *Angew. Chem.*, 15 (1963) 288.
Forbes, E. J., Griffiths, J. and Ripley, R. A., *J. Chem. Soc., C*, (1968) 1149.

Gardikes, J. J. and Konrad, F. M., *Amer. Chem. Soc., Div. Org. Coatings Plast. Chem., Prepr.*, (1966) 131.
Gelotte, B., *J. Chromatogr.*, 3 (1960) 330.
Grassmann, W., Enders, H., Paukner, W. and Mathies, H., *Chem. Ber.*, 90 (1957) 1125.
Hagen, E. and Schröder, E., *Plaste Kaut.*, 16 (1969) 335.
Hanson, K. R. and Zucker, M., *J. Biol. Chem.*, 238 (1963) 1105.
Hendrickson, J. G., *4th International Gel Permeation Chromatography Seminar, Miami Beach, 1967*, Waters Ass., Framingham, Mass., 1967.
Hendrickson, J. G., *J. Chromatogr.*, 32 (1968) 543.
Hendrickson, J. G. and Moore, J. C., *J. Polymer Sci., Part A-1*, 4 (1966) 167.
Hörhammer, L. and Wagner, H., *Pharm. Ztg.*, 104 (1959) 783.
Howard, III, J. M., *J. Chromatogr.*, 55 (1971) 15.
Ivanovskaya, L. Y. and Gorfinkel, M. I., *Zh. Org. Khim.*, 4 (1968) 1227.
Koswy, K. T. and Lach, J. L., *Drug Stand.*, 28 (1960) 53.
Krampitz, G. and Albersmeyer, W., *Experientia*, 15 (1959) 375.
Kurko, V. I., Kelman, L. F. and Kuznetsova, A. A., *Sb. Dokl. Uch. Speta. Myas. Prom. SSSR*, (1968) 54; *C.A.*, 71 (1969) 79830t.
Logie, D., *Analyst (London)*, 82 (1957) 563.
Martin, W. N., *Wood Sci.*, 1 (1968) 102; *C.A.*, 70 (1969) 103573p.
Mijs, W. J., Van Dijk, J. H., Huysmans, W. G. B. and Westra, J. G., *Tetrahedron*, 25 (1969) 4233.
Murthy, B. G. K., Siva Samban, M. A. and Aggarwal, J. S., *J. Chromatogr.*, 32 (1968) 519.
Musso, H. and Pietsch, H., *Chem. Ber.*, 100 (1967) 2854.
Nomura, N., Hiraki, S., Yamada, M. and Shiho, D. I., *J. Chromatogr.*, 59 (1971) 373.
Pozdnyshev, G. N. and Petrov, A. A., *Primen. Poverkh. Aktiv. Veschestv. Neft. Prom., Tr. Vses. Soveshch., 3rd, 1965*, (1966) 21: *C.A.*, 69 (1968) 24352s.
Quinn, E. J., Ostergouldt, H. W., Heckles, J. S. and Ziegler, D. C., *Anal. Chem.*, 40 (1968) 547.
Rastorgueva, L. I., *Prikl. Biokhim. Mikrobiol.*, 5 (1969) 591; *C.A.*, 72 (1970) 39640v.
Seki, T., *J. Chem. Soc. Jap., Pure Chem. Sect.*, 75 (1954) 1297.
Seki, T., *J. Chromatogr.*, 4 (1960a) 6.
Seki, T., *J. Chromatogr.*, 3 (1960b) 376.
Sherma, J. and Rieman, W., *Anal. Chim. Acta*, 18 (1958) 214.
Skelly, N. E., *Anal. Chem.*, 33 (1961) 271.
Skelly, N. and Crummett, W. B., *J. Chromatogr.*, 55 (1971) 309.
Somers, T. C., *Nature (London)*, 195 (1962) 184.
Spitz, H. D., Rothbart, H. L. and Rieman, W., *Talanta*, 12 (1965) 395.
Stankevich, I. V. and Skorochod, O. R., *Vestsi Akad. Navuk Belarus. SSR*, (1966) 116; *C.A.*, 66 (1967) 98834k.
Streuli, C. A., *J. Chromatogr.*, 47 (1970) 355.
Thomas, D. E. and Thomas, J. D. R., *Analyst (London)*, 94 (1969) 1099.
Van Dijk, J. H. and Mijs, W. J., *Z. Anal. Chem.*, 236 (1968) 419; *C.A.*, 69 (1968) 24359z.
Varishth, R. C., Schwarz, F. E. and Leong, S. Y., *Spectrovision*, 20 (1968) 4.
Weibull, B. and Thorsell, L., in Asinger, F., *Proc. Int. Congr. Chem. and Phys. Applied Surface-Active Substances, 4th, 1964*, Gordon and Breach, London, 1967, p. 523; *C.A.*, 70 (1969) 114178q.
Woof, J. B. and Pierce, J. S., *J. Chromatogr.*, 28 (1967) 94.
Yoshikawa, T., Kimura, K. and Fujimura, S., *J. Appl. Polym. Sci.*, 15 (1971) 2513.
Zakupra, W. A., Dobrov, V. S., Lebedev, E. V., Blinova, E. K. and Pliev, T. N., *Khim. Tekhnol. Topl. Masel.*, 15 (1970) 50; *C.A.*, 73 (1970) 94418z.

*Chapter 20*

# Ethers and peroxides

J. CHURÁČEK

The liquid column chromatography of ethers and peroxides is used predominantly for their purification and isolation; it is used less for analytical purposes. Volatile ethers are analyzed mainly by gas chromatography because they are thermostable. However, other conditions for analysis have to be chosen for thermolabile substances, such as some types of peroxo compounds. The isolation of ethers and peroxides, their separation and their determination in the reaction products or in commercial products is frequently necessary.

Six and seven-component mixtures of aliphatic ethers ($C_2$–$C_8$) were separated by salting-out chromatography on Dowex 50-X4 (200–400 mesh) using gradient elution with ammoniumsulphate of decreasing concentration (Sargent and Rieman). The same compounds were separated, using the same principle, with water and acetic acid solutions of various concentrations (1–8 *M*). The results are summarized in Table 20.1 (Sherma and Rieman).

TABLE 20.1
SOLUBILIZATION CHROMATOGRAPHY OF ETHERS (SHERMA AND RIEMAN)

Ion exchanger: Dowex 50-X4 ($H^+$) (200–400 mesh). Column: 15 × 2 cm. Flow-rate: 0.4–0.5 cm/min. Detection: fractions of 5 ml were mixed with 5 ml of 0.02 *M* sodium dichromate in concentrated $H_2SO_4$, the mixture was diluted with 25 ml of water, and the resultant Cr(III) measured spectrophotometrically.

| Compound | Mobile phase | | | | |
|---|---|---|---|---|---|
| | Water | Acetic acid | | | |
| | | 1 *M* | 2 *M* | 4 *M* | 6 *M* |
| Diisopropyl ether | 3.51 | 3.11 | 2.88 | 2.13 | 1.30 |
| Di-*n*-propyl ether | 4.94 | 4.16 | 3.37 | 2.08 | 1.65 |
| Ethyl-*n*-butyl ether | 5.25 | 4.19 | 3.60 | 2.48 | 1.69 |
| Di-*n*-butyl ether | 12.5 | 9.36 | 6.81 | 3.90 | 1.83 |
| Anisole | 15.2 | 11.6 | 9.09 | 5.36 | 3.08 |
| Diphenyl ether | – | – | 54.6 | 19.2 | 6.65 |
| Diisoamyl ether | – | – | – | – | 2.28 |
| Di-*n*-amyl ether | – | – | – | – | 3.03 |

Benzoyl peroxide can be adsorbed from non-aqueous media on zirconium oxide (Sokolova and Boichinova). Various non-polar or weakly polar solvents serve as the mobile phase, for example, acetone, benzene, toluene, tetrachloromethane, ethyl acetate, diethyl ether and dichloroethane. The optimum temperature for drying the adsorbent for chromatography is 20–40°C. The adsorption of peroxides is dependent on the amount of water present in the adsorbent and in the solvents which comprise the mobile phase.

*References p.453*

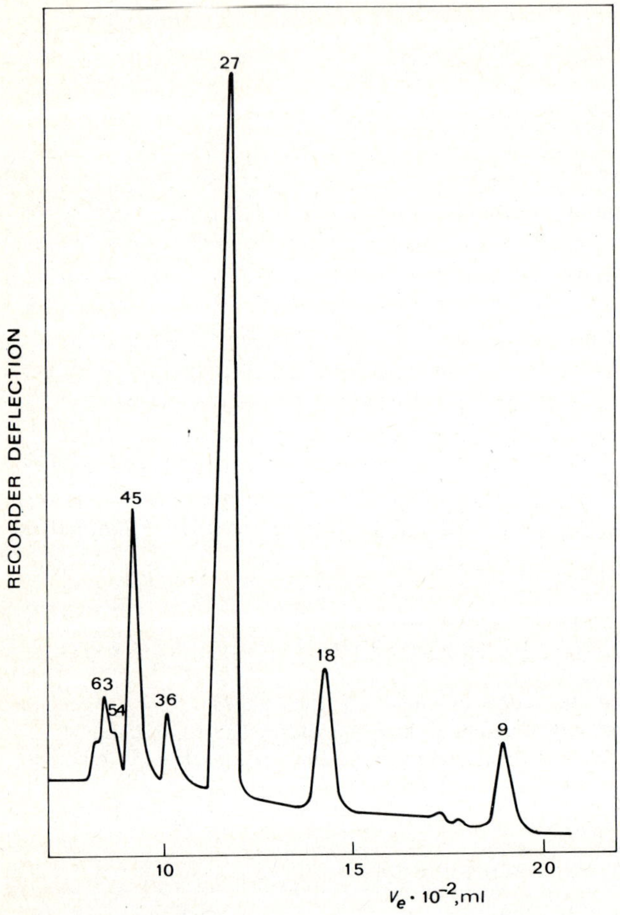


Fig. 20.1. Separation of ethers (Bömer *et al.*). Column: 200 × 5 cm. Sorbent: polystyrene gel, cross-linked with 2% divinylbenzene. Eluent: tetrahydrofuran. Flow-rate: 20 ml/h. Oligomeric ethylene glycols of the $HO-[CH_2-CH_2-O]_n$-H type, where $n = 9x$ and $x = 1, 2, 3, \ldots$. A discontinuous series of homologous polymers; numbers above peaks designate the degree of polymerization, $n$. $V_e$ = elution volume.

For the separation and the preparation of monodisperse polyethylene oxides, gel permeation chromatography (Bömer *et al.*) can be used. Pure oligomers were isolated by this method on a preparative scale using a polystyrene gel cross-linked with 2% divinylbenzene. A chromatogram of homologous series of oligomeric ethylene glycols is shown in Fig. 20.1, and demonstrates a good separation of a whole series of oligomers.

Newer papers on applications, concerning the separation and preparation of ethers and peroxides by liquid chromatography, are listed in Table 20.2.

TABLE 20.2
SURVEY OF PUBLISHED SEPARATIONS OF ETHERS AND PEROXIDES

| Compounds chromatographed | Sorbent | Mobile phase | Reference |
|---|---|---|---|
| Methyl decenyl ether methyl pentenyl ether (purification) | Neutral alumina | Pentane | Damico |
| Diglycidyl ether in epoxy resins (isolation, determination) | Silica gel | Chloroform | Ghosh *et al.* |
| α- and β-Isopropylidene-cyclotriveratrylenes α-(8) and β-(8) | Neutral alumina | Benzene–chloroform (1:1) | Stevens *et al.* |
| Polyethylene glycol monoalkyl phenyl ethers (non-ionic surfactants) | Kieselguhr | Tetrachloromethane–isooctane (1:1) | Huber *et al.* |
| Peroxo derivatives of benzanthracene | Alumina | Dichloromethane | Bailey *et al.* |
| Compounds resulting from autoxidation of acetylenic hydrocarbons | Silica gel KSK | Light petroleum–diethyl ether–methanol | Chirko *et al.* |
| 3,3-Di-*tert.*-butyldiperoxy-phthalide | Silica gel | Light petroleum–diethyl ether (3:1) | Milas and Klein |
| Tetradecalylphenyl peracetate | Alumina | Diethyl ether | Rüchardt and Quadbeck-Seeger |

## REFERENCES


Bailey, P. S., Batterbee, J. E. and Lane, A. G., *J. Amer. Chem. Soc.*, 90 (1968) 1027.
Bömer, B., Heitz, W. and Kern, W., *J. Chromatogr.*, 53 (1970) 51.
Chirko, A. I., Efimova, T. A. and Ivanov, K. I., *Khimiya*, 4 (1968) 44.
Damico, R., *J. Org. Chem.*, 33 (1968) 1550.
Ghosh, P. K., Bandyopadhyay, C. and Saha, A. N., *J. Polym. Sci., Part A-1*, 6 (1968) 3418; *C.A.*, 70 (1969) 48081z.
Huber, J. F. K., Kolder, F. F. M. and Miller, J. M., *Anal. Chem.*, 44 (1972) 105.
Milas, N. A. and Klein, R. J., *J. Org. Chem.*, 33 (1968) 848.
Rüchardt, Ch. and Quadbeck-Seeger, H. J., *Chem. Ber.*, 102 (1969) 3525.
Sargent, R. and Rieman, W., *Anal. Chim. Acta*, 18 (1958) 197.
Sherma, J. and Rieman, W., *Anal. Chim. Acta*, 20 (1959) 357.
Sokolova, I. A. and Boichinova, E. S., *Zh. Prikl. Khim. (Leningrad)*, 43 (1970) 798; *C.A.*, 73 (1970) 38937f.
Stevens, I. D. R., Cookson, R. C. and Halton, B., *J. Chem. Soc., B*, (1968) 767.

*Chapter 21*

# Oxo compounds

J. CHURÁČEK

## CONTENTS



## INTRODUCTION

The use of liquid chromatography for the systematic separation of free aldehydes and ketones is not extensive. Most papers deal with applications and mainly concern the isolation and purification of synthetic products. When aldehydic substances are chromatographed on alumina, it should be borne in mind that they may undergo some catalyzed reactions in alkaline media and form various intermediates. These reactions restrict the general utilization of the liquid chromatography of aldehydes, especially if alumina is used as a sorbent. Carbonyl compounds are most often separated in their hydrogen sulphite form by ion-exchange chromatography on basic resins. The reactivity of the carbonyl compound with hydrogen sulphite ions is made use of, which leads to the formation of α-hydroxysulphonic acids. In the liquid chromatography of oxo compounds, their derivatization is also made use of; mainly oximes and 2,4-dinitrophenylhydrazones are prepared.

The section on quinones includes the liquid chromatography of simple quinones, anthraquinones, ubiquinones and plastoquinones. Liquid chromatography is used mainly for the purification and preparation of synthetic compounds, or for the isolation of active components from natural material. In future, extensive use of the high-speed liquid chromatography of these substances on modern sorbents is to be expected because often they cannot be separated by gas chromatography owing to the low thermal stability of some quinones.

The application of liquid chromatography in lignin chemistry is topical at present. Chromatographic papers in this field can be divided into two main groups. The first group deals with the fractionation of polydisperse lignin derivatives and the determination of the molecular-weight distribution on the dextran gel Sephadex. The second group, dealing with the separation of lignin derivatives (mainly ligninsulphonic acids), uses ion exchangers in which molecular sorption plays a role in addition to ion exchange. This chapter also covers papers that concern the identification of lignin derivatives resulting from degradation processes.

## ALIPHATIC AND CYCLIC ALDEHYDES AND KETONES

Aldehydes and ketones can be separated chromatographically whether they are in the free state or in the form of derivatives. Although carbonyl compounds represent non-ionogenic compounds, most investigators use ion-exchange chromatography for their separation. Very often the problem of the group separation of aldehydes and ketones from organic acids and other substances can also be solved. The acids are bound quantitatively on to a column of the strongly basic anion exchanger Amberlite IRA-400 ($HCO_3^-$). Aldehydes and ketones pass into the filtrate and the acids are then desorbed with sodium carbonate solution (Gabrielson and Samuelson, 1952a). For the separation of ketones from alcohols, quantitative sorption of ketones on anion-exchange resins in the hydrogen sulphite form can be used, while alcohols pass into the eluate. Ketones can be eluted with hot water or with a solution containing a mixture of carbonate and hydrogen carbonate (Gabrielson and Samuelson, 1952b).

Samuelson described a complete analysis of a mixture of acetic acid, ethanol, furfural, and acetone, based on the following principle. Acetic acid is bound on a column of anion exchanger in the hydrogen carbonate form, and acetone and furfural are retained on a column containing the anion exchanger Amberlite IRA-400 ($HSO_3^-$) (0.12–0.30 mm). They are then eluted selectively with water and 1 *N* sodium chloride solution. Acetic acid is eluted with alkali, filtered through a cation exchanger in the $H^+$ form and titrated with alkali; acetone and furfural are determined photometrically by reaction with salicylaldehyde and orcinol, and ethanol is determined by measuring the density of the effluent from the hydrogen sulphite column. A column of anion exchanger in the cyanide form on to which aldehydes and ketones are bound via the formation of addition compounds was not found to be advantageous for their mutual separation, but it can be used for their separation as a group from other types of compounds (Gabrielson).

### Separation of aldehydes and ketones in the form of their derivatives

Liquid chromatography can be used successfully for the determination of carbonyl compounds in the form of derivatives with hydroxylamine, and the oximes formed can be determined by ion-exchange chromatography. A strongly basic anion exchanger was used with water as eluent (Ebel). Using high-speed liquid chromatography, 2,4-dinitrophenylhydrazones of aliphatic aldehydes and ketones can be separated. The derivatives are orange coloured and absorb strongly in the UV region, producing a strong response during detection. Using a 3 m × 2.1 mm column, packed with Corasil II activated by heating *in vacuo* for 3 h at 110°C, and a mixture of *n*-heptane and 3% of ethyl acetate, separations were achieved within *ca.* 35 min (Carey and Persinger) (Fig. 21.1).

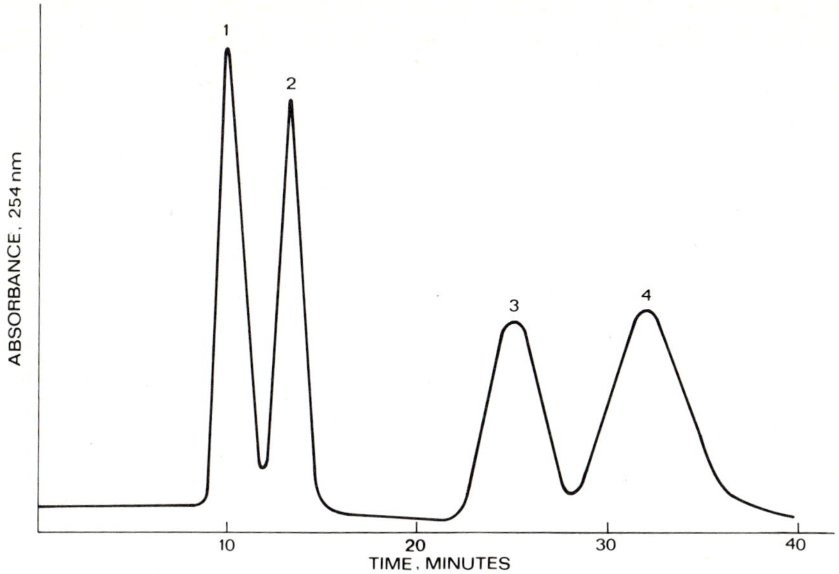


Fig. 21.1. Separation of 2,4-dinitrophenylhydrazones of aliphatic aldehydes (Carey and Persinger). Column: 3 m × 2.1 mm, stainless steel. Sorbent: Corasil II activated for 3 h at 110°C. Eluent: 97:3 (v/v) *n*-heptane–ethylacetate. Operating conditions: flow-rate, 1.7 ml/min; pressure, 1175 p.s.i. Detection: absorption at 254 nm. Peaks: 1 = butyraldehyde; 2 = propionaldehyde; 3 = acetaldehyde; 4 = formaldehyde (all 2,4-dinitrophenylhydrazones).

### Ion-exchange chromatography of free aldehydes and ketones

For the chromatographic separation of aldehydes and ketones, strongly basic anion exchangers in the hydrogen sulphite form were used. On such exchangers, carbonyl compounds are sorbed via the formation of complex α-hydroxysulphonic acids. On the basis of the differences in stability of these complexes, their chromatographic separation can be achieved by using water or carbonate buffers as eluents (Gabrielson and Samuelson, 1950; 1952c). Complexes of aldehydes are more stable than complexes of ketones. The most stable compound is given by formaldehyde, and acetaldehyde, furfural, benzaldehyde, salicylaldehyde, vanillin, glyoxal, acetone and methyl ethyl ketone also form stable, strongly retained complexes. Some ketones can be eluted selectively with hot water and so separated from aldehydes, which remain on the column and which can be eluted with salt solutions (for example, 1 *M* sodium chloride solution). Thus, for example, acetaldehyde and furfural were separated from acetone and methyl ethyl ketone.

Huff separated a mixture of lactic aldehyde, acetone, pyruvic aldehyde and a mixture of formaldehyde and acetaldehyde on a column of Dowex ($HSO_3^-$) by using gradual elution with hydrogen sulphite solutions of increasing concentration. A mixture of carbonyl compounds was separated chromatographically on a 41 × 1.1 cm Dowex-1 ($HSO_3^-$) (150–300 mesh) column using the same procedure as above (Christofferson, 1965). After the removal of hydrogen sulphite from the eluate fractions with iodine, the concentration of the carbonyl compounds could be determined photometrically in UV

TABLE 21.1.
SOLUBILIZATION CHROMATOGRAPHY OF KETONES (SHERMA AND RIEMAN)

Column: 20 × 2 cm. Cation exchanger: Dowex 50-X8 ($H^+$) (200–400 mesh). Flow-rate: 0.4 cm/min. Temperature: room temperature. Detection: fractions of 5 ml were mixed with 5 ml of 0.1 *M* hydroxylamine hydrochloride and the pH determined after 5 min. The difference between this pH and that of a fraction containing no ketone is proportional to the amount of ketone present. Samples consisted of 0.2 mmole of compound dissolved in 1.0 ml of at least 50% of eluent.

| Ketone | Mobile phase | | | |
|---|---|---|---|---|
| | Aqueous methanol | | Aqueous ethanol | |
| | 2.0 *M* | 4.0 *M* | 1.0 *M* | 2.0 *M* |
| Methyl *n*-butyl ketone | 4.14 | 3.04 | 4.33 | 3.58 |
| Methyl *n*-amyl ketone | 5.95 | 4.00 | 6.13 | 4.96 |
| Methyl *n*-hexyl ketone | 10.1 | 6.94 | 11.3 | 8.16 |
| Methyl isobutyl ketone | 3.21 | 2.76 | – | – |
| Methyl *n*-heptyl ketone | 17.1 | 11.6 | – | – |
| Methyl *n*-octyl ketone | 33.6 | 21.3 | – | – |
| Methyl *n*-nonyl ketone | 63.1 | 39.6 | – | – |
| Acetophenone | 12.1 | 9.56 | – | – |

light (Christofferson, 1964). Mixtures containing acetaldehyde, formaldehyde, 5-hydroxymethylfurfural, furfural and vanillin in amounts of 0.05–0.1 mmole could be separated sharply and the relative error was less than 10%.

Ketones were separated on cation-exchange columns of Dowex 50-X8 ($H^+$) (200–400 mesh) by gradual elution with aqueous methanol or ethanol of increasing concentration (Sherma and Rieman). From Table 21.1, it is evident that a successful separation of various methyl ketones was achieved. The separation of aldehydes and ketones by salting-out chromatography on anion-exchange resins can also be recommended (Breyer and Rieman, 1958; 1960). A good separation of a series of carbonyl compounds is shown in Table 21.2.

## Other chromatographic methods of separation of carbonyl compounds

Bell *et al.* separated aromatic ketones from cigarette smoke. By extracting the condensate, acetonitrile fractions were obtained with derivatives of fluoren-9-one, and these fractions were then fractionated on an alumina column by gradient elution. The polarity of the system was gradually increased. The eluent was *n*-hexane to which benzene, diethyl ether and methanol were added gradually. The final identification of the components was carried out by paper or gas chromatography. In Table 21.3, elution data are given for some carbonyl compounds (and esters), which were separated by gel chromatography on styrene–divinylbenzene gel (Hendrickson).

TABLE 21.2
DISTRIBUTION RATIOS OF CARBONYL COMPOUNDS IN SALTING-OUT CHROMATOGRAPHY (BREYER AND RIEMAN, 1958)

Column: 25 × 2 cm. Cation exchanger: Dowex 1-X8 ($SO_4^{2-}$) (200–400 mesh). Flow-rate: 0.2–0.8 cm/min. Temperature: room temperature. Detection: by the differential pH method of Roe and Mitchell in which 5 ml of 0.1 *M* hydroxylamine hydrochloride is added to each fraction and the pH is determined after 5 min; the amount of carbonyl compound is related to the pH. Alternatively, fractions are mixed with 5 ml of 0.1 *N* sodium dichromate in conc. $H_2SO_4$, diluting with 25 ml of water and measuring the absorbance of the resulting Cr(III). Samples consisted of 0.050–0.100 mmole.

| Compound | Mobile phase | | | | | |
|---|---|---|---|---|---|---|
| | Water | Ammonium sulphate solution | | | | |
| | | 0.5 *M* | 1.0 *M* | 2.0 *M* | 3.0 *M* | 4.0 *M* |
| Formaldehyde | 1.05 | 0.93 | 1.00 | 0.98 | 0.97 | 0.88 |
| Acetaldehyde | 0.80 | 0.98 | 1.22 | 1.52 | 2.18 | 2.92 |
| Acetone | 0.70 | 0.93 | 1.53 | 3.00 | 6.12 | 10.9 |
| Acetoin | 0.76 | 0.96 | 1.38 | 2.91 | 6.13 | 10.8 |
| Diacetyl | 1.25 | 2.06 | 2.20 | 4.60 | 8.50 | 13.9 |
| 2,5-Hexanedione | 1.06 | 1.47 | 2.56 | 6.87 | 22.4 | – |
| Diacetone alcohol | 0.78 | 1.36 | 2.31 | 7.31 | 22.5 | – |
| Propionaldehyde | 1.37 | 1.86 | 2.50 | 4.58 | 7.70 | – |
| Methyl ethyl ketone | 1.28 | 2.00 | 3.02 | 8.21 | 21.1 | – |
| Cyclopentanone | 1.70 | 2.59 | 4.01 | 8.83 | – | – |
| 2,3-Pentanedione | 1.92 | 2.59 | 4.49 | 12.8 | – | – |
| 2,4-Pentanedione | 1.90 | 2.73 | 4.34 | 12.9 | 21.8 | – |
| Methyl isopropyl ketone | 2.12 | 3.59 | 5.76 | 16.5 | – | – |
| Butyraldehyde | 2.82 | 4.42 | 5.81 | – | – | – |
| Methyl *n*-propyl ketone | 2.51 | 4.56 | 7.52 | 19.6 | – | – |
| Diethyl ketone | 2.52 | 4.42 | 7.16 | 19.6 | – | – |
| Cyclohexanone | 3.09 | 5.17 | 8.61 | – | – | – |

## QUINONES

Naturally occurring quinones can be isolated from most natural materials by first extracting them with a neutral or alkaline extractant and then fractionating the extracts by column chromatography. Alkali-soluble fractions can be further fractionated on a column packed with deactivated silica gel by elution with benzene (Thomson and Burnett, 1967). Some synthetic quinones of the anthraquinone type were separated by high-speed liquid chromatography on modern sorbents, such as Corasil/$C_{18}$ or Permaphase ODS. In the first instance, methanol–water (1:1) was used as the polar mobile phase at a pressure of 1200 p.s.i.g. (Fig. 21.2) (Waters Ass.). In the second instance, the separation was carried out at an elevated temperature (60°C) using a methanol–water mixture (45:55) at a pressure of 450 p.s.i.g. (DuPont) (Fig. 21.3).

Table 21.4 gives a review of some recent applications in which liquid chromatography is used primarily for purification and preparative purposes.

TABLE 21.3
GEL CHROMATOGRAPHY OF CARBONYL COMPOUNDS (HENDRICKSON)

Column: 12 ft. × 3/8 in. Gel: 40 Å styrene–divinylbenzene gel, permeable to alkanes of mol. wt. < 450. Mobile phase: benzene. Flow-rate: 1.0 ml/min. Temperature: 24°C. Detection: RI.

| Compound* | Elution volume ($V_e$, ml) | Peak width (ml)** |
|---|---|---|
| Acetone*** | (78.6) | (4.66) |
| Acetone*** | 115.3 | 4.34 |
| Acetone*** | 101.5 | 4.52 |
| *n*-Butyraldehyde | 110.4 | 4.46 |
| Methyl ethyl ketone | 110.3 | 4.10 |
| Ethyl acetate | 108.6 | 4.34 |
| Methyl isobutyl ketone | 102.2 | 4.40 |
| Dimethyl terephthalate | 100.6 | 5.30 |
| *n*-Heptaldehyde | 98.1 | 6.40 |
| Dimethyl adipate | 94.2 | 4.17 |
| Dimethyl sebacate | 73.8 | 4.62 |

*Sample sizes were typically 0.1 ml of 4% solute in benzene.
**Peak widths were obtained by drawing tangents to each side of the curves and reporting the number of millilitres of eluent at the base of the triangle.
***Compound run more than once.

TABLE 21.4
SURVEY OF CHROMATOGRAPHIC PROCEDURES APPLICABLE TO THE SEPARATION OF ALIPHATIC AND AROMATIC OXO COMPOUNDS

| Substances chromatographed | Sorbent | Mobile phase | Reference |
|---|---|---|---|
| Cyclobutanone derivatives | Alumina, act. I | Dichloromethane and methanol | Huisger and Feiler |
| Bis-diazoketones | Florisil (60–100 mesh) | Benzene–*n*-hexane (1:1) | Bien and Ovadia |
| Dichlorobenzophenone | Florisil | Light petroleum (b.p. 60–80°C)–diethyl ether (9:1) | Morgan |
| 6,6-Diarylbicyclo[3.1.0] hexan-2-one and its derivatives | Silica gel, Celite, silicic acid | *n*-Hexane–1% diethyl ether | Zimmerman *et al.* |
| Trihydroxycyclo-2,6-dien-5-one (purification) | Silicic acid, Dowex AB 1-X2 | Diethyl ether–benzene (9:1) | Shaw and Smith |

TABLE 21.4 *(continued)*

| Substances chromatographed | Sorbent | Mobile phase | Reference |
|---|---|---|---|
| Tropolones (purification) | Silica gel | Chloroform–benzene (1:1) | Forbes and Griffiths |
| Adamantanone (preparation) | Silica gel | Light petroleum (b.p. 60–80°C)–acetone (5:2) and others | Snatzke and Eckhardt |
| Anthraquinones in natural materials | Silica gel | Benzene | Thomson and Burnett (1968b) |
| | Florisil | Benzene | |
| | Silica gel G | Benzene–light petroleum (b.p. 60–80°C) (1:4) | Thomson and Burnett (1968c) |
| | Florisil | Light petroleum (b.p. 60–80°C) | Thomson and Burnett (1968a) |
| | Magnesium oxide, silica gel | Chloroform | Thomson and Brown |
| Anthraquinone derivatives (purification) | Alumina | Skelly-Solve B | Cristol and Caspar |
| | Acid alumina | Benzene | Bredereck *et al.* |
| Benzoquinone derivatives (purification) | Silica gel | Ethanol (extraction from the column with water) | Teuber and Dietrich |
| | | Chloroform | Cuntze and Musso |
| (α-, β-, γ-) Isomers of rubromycin derivatives of napthoquinone (isolation) | Silica gel G and oxalic acid | Chloroform–acetone (97:3)<br>Chloroform–benzene (1:1) | Brockmann and Zeeck |

## APPLICATIONS IN LIGNIN CHEMISTRY

In lignin chemistry, liquid chromatography is often used for the fractionation of polydisperse lignin derivatives, mostly on Sephadex gels (Kirk *et al.*). For determining molecular-weight distributions, Sephadex G-100 was found to be the most suitable resin and, when the formamide system was applied, enabled a good separation of single fractions to be achieved. Aqueous extracts of lignin derivatives were fractionated successfully on a

*References p.463*

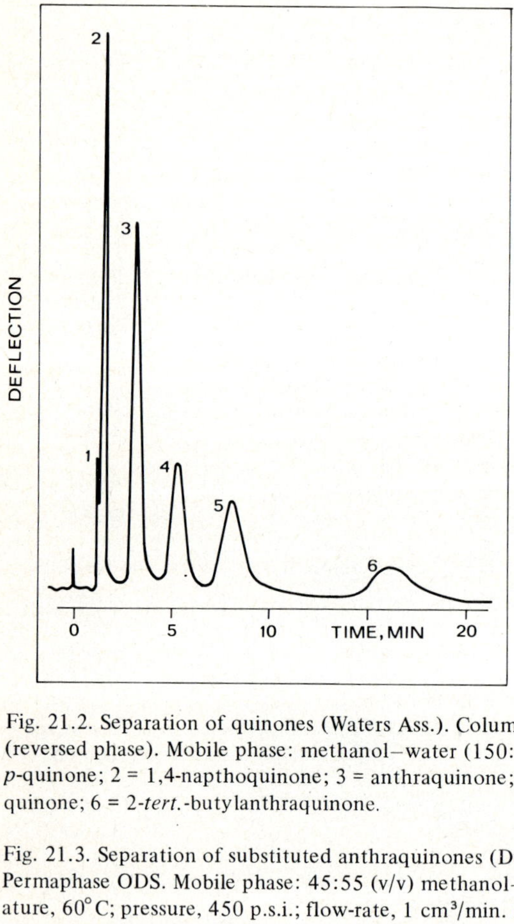


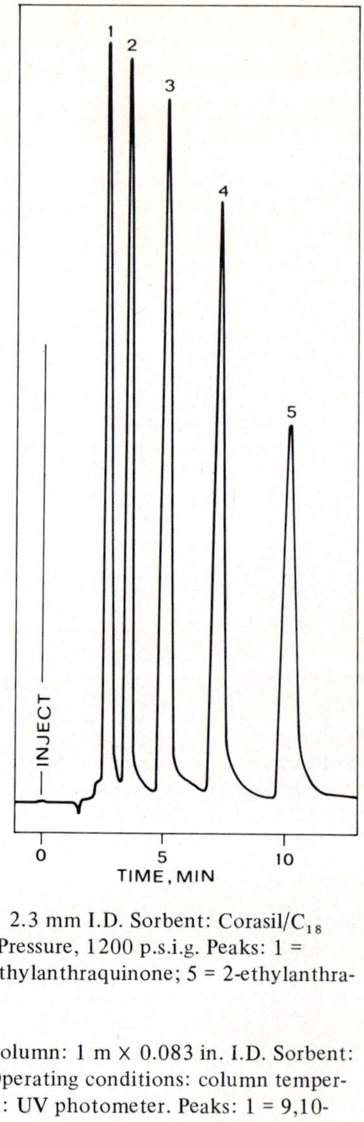


Fig. 21.2. Separation of quinones (Waters Ass.). Column: 4 ft. × 2.3 mm I.D. Sorbent: Corasil/$C_{18}$ (reversed phase). Mobile phase: methanol–water (150:50, v/v). Pressure, 1200 p.s.i.g. Peaks: 1 = *p*-quinone; 2 = 1,4-napthoquinone; 3 = anthraquinone; 4 = 2-methylanthraquinone; 5 = 2-ethylanthraquinone; 6 = 2-*tert.*-butylanthraquinone.

Fig. 21.3. Separation of substituted anthraquinones (DuPont). Column: 1 m × 0.083 in. I.D. Sorbent: Permaphase ODS. Mobile phase: 45:55 (v/v) methanol–water. Operating conditions: column temperature, 60°C; pressure, 450 p.s.i.; flow-rate, 1 $cm^3$/min. Detection: UV photometer. Peaks: 1 = 9,10-anthraquinone; 2 = 2-methyl-9,10-anthraquinone; 3 = 2-ethyl-9,10-anthraquinone; 4 = 1,4-dimethyl-9,10-anthraquinone; 5 = 2-*tert.*-butyl-9,10-anthraquinone.

polyamide column using aqueous methanolic solutions for elution (Hostettler and Seikel, Seikel *et al.*). It was observed that the gel permeation chromatography of wood components, such as hemicelluloses and lignins, is more effective than column electrophoresis. A very positive effect on the separation of these substances in buffered systems is exerted by the presence of carboxyl groups in the gel, and for this reason the polyacrylamide gel Bio-Gel P and similar gels can be recommended for the fractionation (Simonson).

Very often, tasks connected with the fractionation of lignin sulphonates have to be performed. These substances can be isolated from sulphite liquors (wastes) by means of hexamminocobalt trichloride and conversion into their barium salts on an ion-exchange column (Alekseev *et al.*). Using Sephadex G-75 and G-100, fractions of molecular weight up to 100,000 can be obtained. The mechanism of the fractionation of calcium and lithium lignosulphonates by gel chromatography was investigated on a column packed with Sephadex G-25 and G-50 using water, dioxane–water and aqueous solutions of calcium and lithium chlorides as eluents (Stenlund).

On Sephadex G-50, lignosulphonates up to a molecular weight of 40,000 can be well separated, and on Sephadex G-75 up to a molecular weight of 80,000 (Forss and Stenlund).

When ligninsulphonic acids are sorbed on ion exchangers, molecular sorption occurs in addition to ion exchange (Seidl). For sorption, the weakly basic anion exchanger Lewatit MP-60 was found to be the most suitable. The sorption of these acids is partly irreversible. The most efficient desorption agent was a solution of 2 *M* sodium chloride plus 1.5 *M* sodium hydroxide. Anion exchangers with a microporous or visibly porous structure were equally efficient.

A very common task consists in the separation and isolation of degradation products of lignin derivatives obtained either by degradation with thioacetic acid (Nimz, 1969b) or alkali (Johansson and Miksche), or by new degradation procedures (Nimz, 1969a). In the last instance, Sephadex LH-20 is used as sorbent and dimethylformamide as eluent. In other instances, columns packed with silica gel and eluted with acetone and *n*-hexane (or cyclohexane) are used.

## REFERENCES


Alekseev, A. D., Ashina, I. V., Reznikov, V. M. and Sukhaya, T. V., *Obshch. Prikl. Khim.*, (1969) 226; *C.A.*, 73 (1970) 132105e.

Bell, J. H., Ireland, S. and Spears, A. W., *Anal. Chem.*, 41 (1969) 310.

Bien, A. and Ovadia, D., *J. Org. Chem.*, 35 (1970) 1028.

Bredereck, K., Sommermann, F. and Diamantoglon, M., *Chem. Ber.*, 102 (1969) 1053.

Breyer, A. and Rieman, W., *Anal. Chim. Acta*, 18 (1958) 207.

Breyer, A. and Rieman, W., *Talanta*, 4 (1960) 67.

Brockmann, H. and Zeeck, A., *Chem. Ber.*, 103 (1970) 1709.

Carey, M. A. and Persinger, H. E., *J. Chromatogr. Sci.*, 10 (1972) 537.

Christofferson, K., *Anal. Chim. Acta*, 31 (1964) 233.

Christofferson, K., *Anal. Chim. Acta*, 33 (1965) 303.

Cristol, S. J. and Caspar, M. L., *J. Org. Chem.*, 33 (1968) 2020.

Cuntze, U. and Musso, H., *Chem. Ber.*, 103 (1970) 62.

DuPont, *Product Bulletin, Liquid Chromatographs*, No. 820 PB4, DuPont, Wilmington, Del., 1971.

Ebel, S., *Arch. Pharm., Berl.*, 300 (1967) 472.

Forbes, E. J. and Griffiths, J., *J. Chem. Soc., C*, (1968) 575.

Forss, K. and Stenlund, B., *Pap. Puu*, 51 (1969) 93 and 97; *C.A.*, 70 (1969) 88965x.

Gabrielson, G., *J. Appl. Chem.*, 7 (1957) 533.

Gabrielson, G. and Samuelson, O., *Sv. Kem. Tidskr.*, 62 (1950) 214.

Gabrielson, G. and Samuelson, O., *Acta Chem. Scand.*, 6 (1952a) 729.

Gabrielson, G. and Samuelson, O., *Acta Chem. Scand.*, 6 (1952b) 738.

Gabrielson, G. and Samuelson, O., *Sv. Kem. Tidskr.*, 64 (1952c) 150.
Hendrickson, I. G., *J. Chromatogr.*, 32 (1968) 543.
Hostettler, F. G. and Seikel, H. K., *Tetrahedron*, 25 (1969) 2325.
Huff, E., *Anal. Chem.*, 31 (1959) 1626.
Huisger, R. and Feiler, L. A., *Chem. Ber.*, 102 (1969) 3391.
Johansson, B. and Miksche, G. E., *Acta Chem. Scand.*, 23 (1969) 924.
Kirk, T. K., Brown, W. and Cowling, E. B., *Biopolymers*, 7 (1969) 135; *C.A.*, 71 (1969) 14348x.
Morgan, N. L., *Bull. Environ. Contam. Toxicol.*, 3 (1968) 254; *C.A.*, 69 (1968) 85459d.
Nimz, H., *Chem. Ber.*, 102 (1969a) 799.
Nimz, H., *Chem. Ber.*, 102 (1969b) 3803.
Roe, H. and Mitchell, J., *Anal. Chem.*, 23 (1951) 1758.
Samuelson, O., *Z. Elektrochem.*, 57 (1953) 207.
Seidl, J., *Chem. Prům.*, 16 (1966) 273.
Seikel, M. K., Hostettler, F. D. and Johnson, D. B., *Tetrahedron*, 24 (1968) 1475.
Shaw, S. J. and Smith, P. J., *J. Chem. Soc., C*, (1968) 1882.
Sherma, J. and Rieman, W., *Anal. Chim. Acta*, 19 (1958) 134.
Simonson, R., *Sv. Papperstidn.*, 70 (1967) 711; *C.A.*, 68 (1968) 51090r.
Snatzke, G. and Eckhardt, G., *Chem. Ber.*, 101 (1968) 2010.
Stenlund, B., *Pap. Puu*, 52 (1970) 197; *C.A.*, 73 (1970) 26793t.
Teuber, H. J. and Dietrich, M., *Chem. Ber.*, 100 (1967) 2908.
Thomson, R. H. and Brown, P. N., *J. Chem. Soc., C*, (1969) 1184.
Thomson, R. H. and Burnett, A. R., *J. Chem. Soc., C*, (1967) 2100.
Thomson, R. H. and Burnett, A. R., *J. Chem. Soc., C*, (1968a) 850.
Thomson, R. H. and Burnett, A. R., *J. Chem. Soc., C*, (1968b) 854.
Thomson, R. H. and Burnett, A. R., *J. Chem. Soc., C*, (1968c) 2437.
Waters Ass. Inc., *Firm Prospects–Corasil/$C_{18}$*, Waters Ass., Framingham, Mass.
Zimmerman, H. E., Crumrine, D. S., Döpp, D. and Huyffer, P. S., *J. Amer. Chem. Soc.*, 91 (1969) 434.

*Chapter 22*

# Carbohydrates

K. ČAPEK and J. STANĚK, Jr.

## CONTENTS

## INTRODUCTION

It should be said initially that from the viewpoint of carbohydrate chemistry the review given in this chapter is incomplete because a compromise had to be made between the number of papers published and the space available for this type of compound. As the applicability of column chromatography to carbohydrates has been reviewed several times in the past (Binkley; Lederer; Staněk *et al.*, 1963), we focused our attention mainly on modern methods and on the classical methods in which recent improvements in operating conditions have been responsible for their continuing applicability. Moreover, we have omitted, with a few exceptions, all papers in which separations of different types of saccharides (amino sugars from neutral sugars, sugar acids from their lactones, and similar examples) are described, and all papers in which the chromatography was performed in order to purify the synthesized saccharide from trace amounts of unreacted reagents and from some inorganic compounds. These chromatographic separations are simple and in many instances resemble filtration.

The names of compounds mentioned in the separations in this chapter are as given in the original communications; we have changed them only when current nomenclature recommendations dictate that the original name should not be used now.

In descriptions of chromatographic separations, saccharides are always listed in the order in which they are eluted from the columns.

## GENERAL TECHNIQUES

### Liquid–solid chromatography

Many adsorbents have been used for the adsorption chromatography of sugars and their derivatives in the last three decades: charcoal, silica gel, alumina, Fuller's earth clays, calcium acid silicate, hydrated magnesium acid silicate, freshly precipitated calcium carbonate, etc. (Binkley). Nowadays, chromatography on ion-exchange resins and gel permeation chromatography are of major interest and only the first three of the above adsorbents have retained their importance.

*Chromatography on charcoal*

As adsorption on charcoal is governed largely by the molecular weight of the sample (see Chapter 9), this sorbent is used mainly in the fractionation of monosaccharide–oligosaccharide mixtures and in the isolation of sugars and their polar derivatives (alditols, uronic acids, N-acetylated amino sugars, etc.) from natural sources. The advantages of this technique are its simplicity, its ability to handle large amounts of material and the low cost of the adsorbent.

Most of the commercially available charcoals or carbons (Darco G-60, Charcoal BDH, Norit P, Carboraffin, Carbon-Nuchar, etc.) are in the form of fine powders, which require the addition of a filter aid if used in column chromatography. Celite has been found to be the most suitable for this purpose, but cellulose (Jermyn) or alumina (Stefanović) is preferred by some workers. The properties of charcoals can sometimes be improved by treating them with borate (Barker *et al.*, 1955), stearic acid (Alm) or hydrochloric acid (French *et al.*).

Recently, Lammers pointed out that the results obtained with common commercial charcoals are not satisfactorily reproducible when the separation is repeated with the same column. This may be caused by the development of channelling when small particles of charcoal are carried along with the mobile phase. This difficulty can be overcome by flotation of the charcoal in water until it is free from fines.

The most suitable mobile phases for carbon column chromatography are water and aqueous alcohols, such as methanol, ethanol, *n*-butanol and 2-butanol. Taylor and Whelan reported that the pH of the eluting medium has a considerable effect on the manner in which sugars emerge from the column. Non-polar mono- and oligosaccharides appear as sharper bands when the mobile phase is acidic than when it is neutral.

*Chromatography on silica gel*

As the surface of silica gel is covered with polar oxygen functions, the adsorptive forces between the adsorbent and free sugars or polar sugar derivatives are too high for their successful separation under the conditions of LSC. The wide applicability of silica gel chromatography consists, therefore, in the separation of less polar sugar derivatives, such as sugar esters, ethers and halogen-containing derivatives. Owing to the simplicity and low costs of silica gel chromatography, it is often worthwhile first to convert the

*References p.519*

polar sugar or sugar derivatives into less polar (*e.g.*, peracetyl) derivatives, then to separate these less polar derivatives on a silica gel column, and finally to regenerate the parent compounds.

The main advantage of this method lies in the possibility of making an appropriate choice of both the mobile phase and the sorbent activity in advance, by means of TLC. As a guide, the following procedure can be followed.

(a) The sample is examined on micro-plates coated with silica gel G in order to find the most suitable mobile phase; combinations of one polar and one non-polar solvent are particularly recommended.

(b) Next, the sample is tested in the chosen mobile phase on 12 × 20 cm plates coated with the silica gel to be used in the column chromatography. Successful, complete separation of the sample on the column can be expected only when the spots of individual compounds are at least partly resolved. If the opposite is true, it is recommended that another type of silica gel or solvent should be tried.

(c) The chosen silica gel is suspended in light petroleum (b.p. 60–70°C) and loaded into the tube, which is partly filled with the same solvent. The height of the silica gel bed should not be greater than 10 times its diameter. A 30–100-fold excess (related to the weight of the sample) is the most frequently used amount.

(d) The sample, highly soluble in a non-polar solvent, is applied on to the column as a solution. If the syrupy sample is insoluble in the desired solvent, the following procedure is recommended. A few millilitres of non-polar solvent are poured into the flask containing the sample; the wet syrupy sample is then absorbed with a piece of cotton-wool rolled on a spatula until all sample is present in the cotton-wool. The silica gel bed is then covered with the cotton-wool containing the sample, and the remaining solvent, if any, is poured on to the column and allowed to drain through. A solid sample can be applied on to the column in the solid state.

(e) If the mobile phase found to be suitable by means of TLC consists of non-polar and polar solvents, the elution should start with the non-polar solvent alone; the concentration of the polar solvent should then increase slowly.

(f) For monitoring the sugars in the effluent, any appropriate method of automated analysis can be applied; however, TLC is the most useful in common laboratory practice. In complicated situations, in which the compounds overlap to such an extent that virtually no resolution is observed on TLC, the graph relating the weight of substance in a particular fraction to the fraction number provides invaluable information.

Of commercial products, silica gels with particle sizes ranging from 40 to 200 μm have been used most frequently. For the mobile phases, light petroleum, dichloromethane, chloroform and benzene as less polar components, and diethyl ether, ethyl acetate, acetone, methanol and ethanol as more polar components have been used the most.

In some exceptional cases, silica gel catalyses the chemical reaction between the solute and the solvent. Thus, methyl 3,4,6-tri-O-benzyl-α-D-mannopyranoside upon column chromatography on silica gel affords methyl 2-O-acetyl-3,4,6-tri-O-benzyl-α-D-mannopyranoside, using ethyl acetate as the mobile phase (Franks and Montgomery). Detritylation of some trityl sugar derivatives on a silica gel column was reported by Trimnell *et al.* and by Lehrfeld. Kiely and Fletcher observed the decomposition of 3-O-benzyl-1,2-O-isopropylidene-α-D-*xylo*-hexofuranos-5-ulose; this decomposition is avoided when silica gel previously washed with acetic acid–benzene (4:96) is used.

*Chromatography on alumina*

Chromatography on alumina is, in practice, not very different from the procedures described for silica gel. The preliminary examination of an unknown sample can be performed in a manner similar to that already described (p.468).

In comparison with silica gel, chemical reactions of solutes on alumina columns occur more frequently. Jarý *et al.* observed that during the chromatography of methyl 3-acetamido-2,4-di-O-acetyl-3,6-dideoxy-α-D-altropyranoside on an alkaline alumina column using a benzene–ethanol mixture as the mobile phase, partial de-O-acetylation occurred. This phenomenon was later used in the preparation of several partially acetylated methyl 3-acetamido-3,6-dideoxyhexopyranosides (Čapek *et al.*, 1966, 1967, 1968a, b, 1970b) and methyl 3,6-dideoxyhexopyranosides (Čapek *et al.*, 1970a). Ennor *et al.* reported that methyl 4,6-O-benzylidene-2-O-tosyl-α-D-altropyranoside was completely transformed into the corresponding epoxide during chromatography on alumina, even if the alumina was acid washed.

## Liquid–liquid chromatography

There are a vast number of applications of LLC that have been used to solve analytical problems in carbohydrate chemistry over the past 25 years; however, only recently have the exact analytical liquid–liquid method and detailed column efficiency and selectivity studies been carried out. As in GLC, precise retention data and distribution constants form the basis for tabulating carbohydrates and identifying unknown compounds.

*Chromatography on cellulose*

Cellulose has been the most frequently used sorbent in the LLC of carbohydrates, and all theories are equally applicable to column and paper chromatography. This analogy is very useful, mainly in the choice of the appropriate solvent system (Hais and Macek). In addition to liquid–liquid partition, the retention of the solute on cellulose is also frequently influenced by adsorption and ion-exchange, and all of these effects contribute to the distribution constant.

Whatman No. 1 cellulose (and occasionally also Whatman CF-11, Yoshida *et al.* (1969b), or CF-12 cellulose powder, Otter *et al.*) is usually used for the separation of mono- and oligosaccharides and of some of their polar derivatives on a preparative scale. Microgranular cellulose does not give any improvement in separation, tending to pack too tightly, resulting in a reduced flow-rate. The method of packing the column is important for good separation characteristics (Otter *et al.*).

A 45 × 1.5 cm column, used for the quantitative chromatography of oligosaccharides (see p.486), is packed to a depth of 43 cm with CF-12 cellulose powder. Prior to use, the fines are removed from the cellulose by stirring 50 g of the powder with 1 l of distilled water and allowing the mixture to settle. Then, the supernatant liquid together with the fines is decanted off and this procedure is repeated five times. The column is filled with a slurry of 50 g (less fines) of cellulose in 800 ml of water at 90°C and allowed to settle

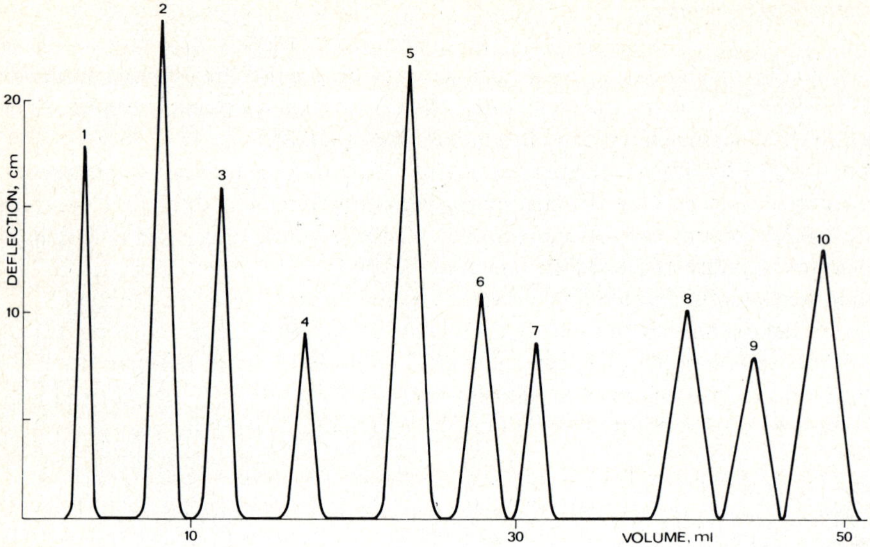


Fig. 22.1. Partition chromatography of alditols on Dowex 50W-X8 ($Li^+$, 14–17 μm). Resin bed 131 × 0.26 cm; mobile phase 85% ethanol; flow-rate 3 ml/min·cm²; temperature 75°C. 1 = formaldehyde; 2 = ethylene glycol; 3 = glycerol; 4 = erythritol; 5 = ribitol; 6 = arabinitol; 7 = xylitol; 8 = mannitol; 9 = glucitol; 10 = galactitol. (Samuelson and Strömberg, 1968).

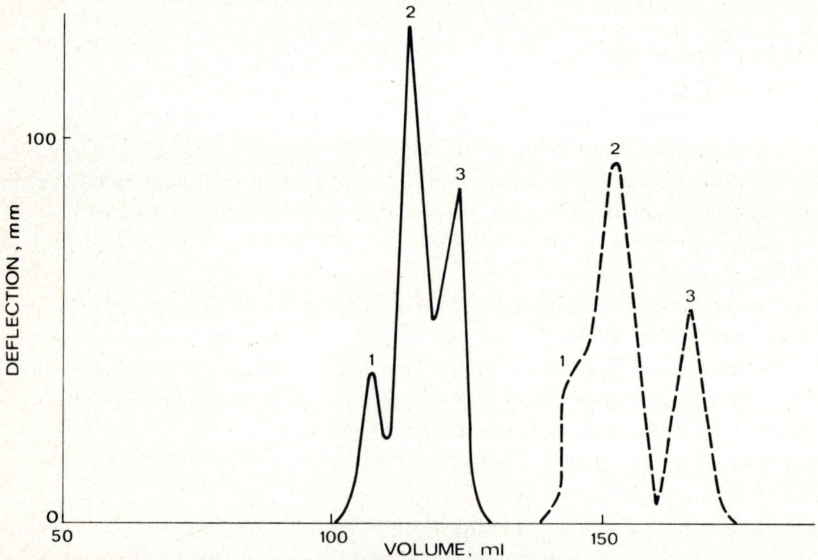


Fig. 22.2. Influence of temperature on the separation of xylose (1), mannose (2) and sorbose (3) on a strongly basic anion exchanger ($SO_4^{2-}$) (Jonsson and Samuelson, 1967b). Resin bed 89 × 0.43 cm; mobile phase 95% ethanol; flow-rate 2.3 ml/min · cm². Solid line 95°C; dashed line 75°C.

under gravity with no flow. The packing is consolidated by aspirating the surplus water through the outlet, and this procedure is repeated until the cellulose is packed to the required depth. When correctly packed, the column should be capable of maintaining a flow-rate of 30–40 ml/h. Water is then displaced by percolating at least 500 ml of water–ethanol–*n*-butanol (24.5:23:52.5) through the column (Otter *et al.*).

*Chromatography on ion-exchange resins*

Recently, chromatography on cellulose has been replaced by partition chromatography on ion-exchange resins with different counter-ions, which offers a great variety of experimental conditions. The separation is achieved as a result of the uneven distribution of the solvent components between the resin and the external solution. With aqueous ethanol, for example, the relative amount of water present is greater in the resin phase than in external solution; this explains why polar solutes are preferentially held by the resin (Samuelson).

Cation-exchange resins have the advantage over anion-exchange resins that they are readily available commercially as smaller, more uniform resin particles (owing to their use in amino acid analyzers). Dowex 50W-X8, Amberlite IR-120, Aminex A-6 and Technicon $T_5C$, Technicon $T_4$ or Dowex 1-X8 with a narrow size in the range 5–20 $\mu$m are the most frequently used. Lithium, sodium, potassium, barium, calcium, sulphate and chloride are usually used as counter-ions. Of the resin forms studied, the most favourable separations of polyols were obtained with the lithium form (Fig. 22.1) (for the procedure for this separation, see p. 504). Recently, organic base counter-ions, such as piperidinium and methyl-, dimethyl- and trimethylammonium have been used (Hobbs and Lawrence, 1972b), the trimethylammonium ion giving the optimum compromise between elution volumes, peak widths and analysis times.

In general, the volume distribution constant, $K_d$, in water–ethanol mixtures increases with increase in the number of hydroxyl groups and decreases with the presence of non-polar groups in the solute; it may also be influenced by the position of the substituent (Samuelson). The $K_d$ values increase markedly with increase in the ethanol concentration, but the order of the elution is (within the concentration range of practical interest) independent of this concentration (Samuelson and Strömberg, 1968). The peak elution volume and the width of the elution curves usually decrease with increase in temperature (see Fig. 22.2); in some instances, however, the separation may be jeopardized at very high temperatures because of the decrease in the ratio of the $K_d$ values of the separated compounds (Jonsson and Samuelson, 1967b).

A factor which limits the applicability of this technique is the extremely low rate of diffusion inside the resin particles. Porous resins and small particles are, therefore, preferable. With such particles, a high-pressure technique must be used in order to force the solution through the column. A striking example, which shows the influence of the particle size, is given in Fig. 22.3 (Dahlberg and Samuelson). In order to reduce the pressure drop in the column, ion-exchange resins in admixture with Celite 545 were used (Arwidi and Samuelson, 1964).

Of course, the solvent–resin interactions and specific interaction forces between polyols and the counter-ion in the resin phase also have a great influence on the equilib-

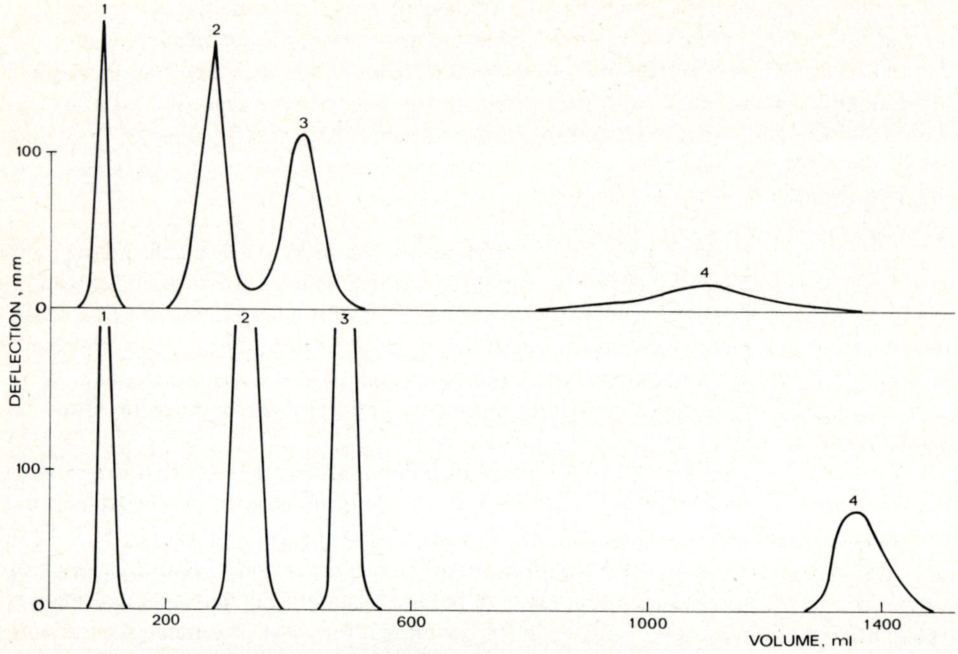


Fig. 22.3. Influence of particle size on the separation of various sugars on Dowex 21K ($SO_4^{2-}$) (Dahlberg and Samuelson). Resin bed 75 × 0.8 cm; mobile phase 74% ethanol; flow-rate 0.8 ml/min · $cm^2$. Upper chromatogram, 45–75 μm; lower chromatogram, 15–40 μm. 1 = 2-deoxy-D-glucose; 2 = D-glucose; 3 = sucrose; 4 = raffinose.

rium on the ion-exchange resin; sieve effects also occur, which in some instances result in a reversed order of elution of some sugars when the counter-ion is changed (Samuelson and Strömberg, 1968).

*Chromatography on silica gel and alumina*

In LLC, silica gel is also sometimes used as a carrier. In this event, ethyl acetate containing dimethyl sulphoxide or dimethylformamide (Otake), methanol–water (7:3) (Heyns *et al.*) or water-saturated ethyl acetate (Zorbach *et al.*) are used as mobile phases. When a solvent system containing water, such as water-saturated butanone, is to be used, a 1:1 mixture of silica gel and alumina is advantageous owing to a better flow-rate (*e.g.*, Micheel and Staněk). It is for this reason that the bottom layers of 2-propanol–chloroform–ammonia–water mixtures in various ratios (Cooper *et al.*; Čapek *et al.*, 1974) are superior. Alumina alone has been used with various water–ethanol mixtures (Bonner *et al.*).

**Gel chromatography**

So far, gel chromatography (for a recent review, see Churms) has been almost exclusively used for the separation of sugars that differ in their molecular size, especially for the fractionation of oligo- and polysaccharide mixtures. However, some papers (Brown, 1970b; Marsden) dealing with the examination of the chromatographic properties of monosaccharides and their derivatives, showed that their $K_d$ values did not decrease monotopically with increase in molecular weight and that they varied even within the aldopentose or aldohexose series. This phenomenon, which is attributed to the differences in the degree of steric hindrance to the entry of the sugar molecule into the gel phase, indicates further possibilities for the use of GPC in the carbohydrate field.

Cross-linked dextran gels (Sephadex) and polyacrylamide gels (Bio-Gel P series) have been the most frequently used so far. Porous glass (Bio-glass) and porous silica beads (Porasil) are now being increasingly applied owing to their resistance to thermal, chemical and bacteriological degradation. Other types of gels, such as agarose gels (Sepharose, Sagarose), cross-linked polystyrene containing hydrophylic groups (Aquapak) and ion-exchange resins have been used on a few occasions. Recently, an improvement in the chromatographic properties of Sephadex G-15 has been reported, produced by treating the gel with 1 *M* hydrochloric acid; this pre-treatment results in increases in the number of theoretical plates, the internal volume and the effective pore size (Goodson and DiStefano).

Concerning the effect of operating conditions on the efficiency of the GPC of saccharides, there are no serious exceptions to the general procedure (see Chapter 6); gels with particles of a well defined pore size and small diameter, long columns with a small I.D. and very slow flow-rates are recommended. The most suitable mobile phases for the GPC of sugars are water, aqueous solutions of sodium chloride and various buffers.

### Ion-exchange chromatography

With ion exchangers, it is usually not easy to decide whether different retention volumes are due merely to the ion exchange. In practice, there is always some partition effect owing to the partial solubility or adsorption of the compound in the resin matrix, and also molecular sieve effects cannot be neglected. Separations based predominantly on a change of the counter-ion or in its chemical identity (complexing) are considered in this chapter.

#### *Chromatography on ion-exchange resins in an $H^+$ or $OH^-$ cycle*

In the separation of acidic or basic sugar derivatives on ion-exchange resins, important factors include not only the equilibrium exchange conditions (solute p*K*), but also the molarity of the displacing ion, the pH of the developing buffer and temperature (Brendel *et al.*, 1967a). The differences in peak widths that do not fit the common pattern that the later a given compound emerges from the column the broader is the peak due to diffusion, are assumed to be due to the presence of various species, anomers and conformers and their relative velocities of rearrangement (Brendel *et al.*, 1967a).

Sometimes, the opposite effect, so-called ion-exclusion, may also take place, for instance, with sugar acids, which are excluded from the matrix of Bio-Rad AG 50W-X2 ($Li^+$) containing ionizable groups, thus decreasing the retention volumes (Barker *et al.*, 1969a).

*References p.519*

With neutral sugar derivatives, such as glycosides, the separation on strongly basic anion-exchange resins is assumed to be due to an ion-exchange process, involving the loss of a proton from one of the hydroxyl groups. According to this assumption, the retention volume is proportional to the p$K$ value of the hydroxyl group which is ionized first. This acidity is determined by the number and the proximity of other polar substituents in the molecule, by the relative angles of various dipoles to each other and to the hydroxyl group which ionizes (Neuberger and Wilson). Owing to the weak acidity of the sugar hydroxyl groups, carbon dioxide must not be present in water used as the mobile phase.

*Chromatography on ion-exchange resins in a borate cycle*

Anion-exchange resins in the borate form are the most common resins amongst ion exchangers with reactive counter-ions. In this method, an anion exchanger in the chloride form is packed in the column and converted several times alternately between the chloride and borate forms until no further settling occurs. All columns prepared in this manner produced a better resolution of sugar mixtures than a column packed once in the chloride form followed by single conversion into the borate form (Kesler). It is also recommended that a large volume of the mobile phase is forced through the column at high speed so as to ensure that the resin bed is closely packed before use (Alfredsson *et al.*).

Kesler compared three anion-exchange resins in the borate form and found Technicon 3/28/VI to be far superior to Dowex 1-X8 (200–400 mesh) and Bio-Rad AG 1-X8 (30–40 μm) for the separation of carbohydrates. The use of a resin with a lower degree of cross-linking, Dowex 1-X4, allows equilibrium to be attained at increased rates and greatly improves the chromatographic behaviour of di- and trisaccharides (Walborg and Lantz); for the procedure for this separation, see p. 492. The resolving power is also enhanced by the use of smaller resin particles and a narrower size range (Green), but probably the most critical parameter that affects the resolving power is the ionic strength of the developing buffer (Ohms *et al.*).

For the optimum over-all separation of mono- and oligosaccharides (see p.491), the following linear buffer gradient at a flow-rate of 1.15 ml/min at 50°C was established. It is formed in a two-chambered device which contains 325 ml of limiting buffer (38.9 g of $Na_2B_4O_7 \cdot H_2O$ and 7 g of $H_3BO_3$ per litre, pH 8.9) in a reservoir and 325 ml of starting buffer (one-fifth dilution of the limiting buffer, 0.042 *M*) in a mixing chamber. All water used for buffer formulation should be deionized on a mixed-bed resin (Amberlite IRA-400 and Dowex 50-X8, both 16–50 mesh, 2:1) so as to produce an initial specific conductance of $0.5 \cdot 10^{-6}\ \Omega^{-1}\ cm^{-1}$. As the slope of the linear buffer gradient was increased or decreased by varying the ionic strength of the starting buffer from 0.035 to 0.046 *M*, the resolution of closely emerging doublet peaks was seriously affected. However, the results changed only slightly with a reduction in the pH of the mobile phase to 8.45 (Ohms *et al.*).

It was found that the effect of increasing the column temperature was to shift the eluted bands to longer elution times; simultaneously, the resolution was enhanced (Green; Kesler). However, virtually no further advantage was gained between 55 and 70°C.

As the band widths of various sugars do not change under constant elution conditions, the amount of the saccharide eluted bears a linear relationship to the peak height expressed as absorbance (Kesler). Variations in reproducibility, *i.e.*, in the resolving power of the system, have been explained as being due to contamination by metals and silicate. Incomplete conversion of the resin into the borate form also may result in a reduced resolution after re-packing a column (Ohms *et al.*).

The use of non-alkaline borate buffers, such as boric acid–glycerol (Walborg *et al.*, 1965; Walborg and Lantz) or boric acid–2,3-butanediol (Walborg *et al.,* 1969), minimizes the rearrangement processes on the resin, which makes it possible to achieve absolute saccharide recoveries in the range of 85–100% (see p. 492).

**Automated detection methods**

The rapid increase in the number of chromatographic separations of small amounts of multicomponent (⩾ 10) mixtures obtained from natural material regularly necessitated the introduction of automated procedures for monitoring the column effluent. Generally, there are two different types of automated analysis of effluents that contain saccharides. The first type, which could be termed "non-destructive", is based on the examination of the effluent by physical methods. In the second, "destructive" type, saccharides are detected by means of a chemical reaction. Both types of method are used mainly for analytical separations. For preparative purposes, either all of the effluent is led to the flow cell, where changes in some physical property are followed ("non-destructive" type, as no changes in the chemical composition of the solute are caused), or an aliquot of the effluent is continuously analyzed for the sugar content by a "destructive" chemical reaction.

It should be pointed out that for preparative separations of sugar mixtures that are carried out only occasionally, manual methods are preferably used in order to avoid the waste of time resulting from preliminary tests and calibrations.

*Non-destructive methods*

Although less sensitive than colour reactions, these methods have the advantages of being easy to run with minimal costs of reagents and being highly reproducible. The separated compounds may afterwards be recovered from the eluate, and for preparative purposes, where the highest quantitative accuracy is not demanded, they find wide application.

*Optical rotation*

For qualitative purposes, the optical rotation of the column effluent is monitored. An automatic polarimeter is a convenient and sensitive instrument for following the elution, *e.g.*, of glycosides from resin columns (Evans *et al.*), where the anomeric nature of the eluted compound is indicated by the sign of the rotation. The same procedure was also used by Neuberger and Wilson, Yoshida *et al.* (1969b), and others. With more complicated mixtures, where some overlapping occurs, a disadvantage of this method appears, due to a broad spectrum of specific rotation values of the compounds in question. The quantitative evaluation of the sugars present is possible only with well separated solutes, for which the specific rotation is known, a constant flow-rate being another necessary condition.

*Ultraviolet spectrophotometry*

With sugar derivatives that are UV-active, the quality of the separation and the amounts of substances involved can be followed easily by measuring the absorbance at an appropriate wavelength. Unfortunately, there are only a few types of compounds (benzyl and trityl ethers, esters of aromatic acids, aromatic glycosides, arylidene derivatives and some others), which can be considered, and, on the other hand, there are solvents such as benzene and acetone, which are very frequently used in LSC but which cannot be used in this case. For applications of the method, in which UV photometers are usually used according to the manufacturers' instructions, see, for instance, Micheel and Pick, Micheel and Staněk and Neuberger and Wilson.

*Refractometry*

Refractometry is widely used in the sugar industry to measure the concentration of sucrose solutions (Charles and Meads); recently, numerous applications of a differential, automatically recording refractometer in GPC have been described (*e.g.*, Brown, 1970b). A differential refractometer was also used for monitoring the IEC separations of neutral sugars as borate complexes (Liljamaa and Hallén). The sensitivity to variations in the buffer concentration together with the presence of false peaks, however, causes some difficulties.

*Destructive methods*

When using destructive methods, guidance in the choice in working conditions is usually obtained from the manual methods that are commonly used in sugar analysis (Pigman and Horton, Staněk *et al.*, 1963). Principally, saccharide contained in the effluent is converted into a coloured or UV-active product, which is then determined colorimetrically.

*Orcinol–sulphuric acid*

The orcinol–sulphuric acid method was used in the chromatography of free carbohydrates on anion-exchange resins in the borate form (Kesler). It was also used for the hydrolyzates from wood and wood pulp separated on an anion-exchange resin in the sulphate form with 92% aqueous ethanol as the mobile phase (Arwidi and Samuelson, 1965). The orcinol–sulphuric acid reagent also proved to be useful in the automated analysis of the effluents from the GPC of polysaccharides (Bathgate). The schematic representation of this system using the Technicon AutoAnalyzer is given in Fig. 22.4. The column effluent is mixed with 1% aqueous orcinol before being mixed with 72% sulphuric acid. The reaction mixture is then heated to 95°C, cooled, and its absorbance is monitored at 420 nm (Bathgate). A double glass coil (24 m × 2 mm I.D.) was used for colour development (John *et al.*). For other applications of this method, which is probably the most frequently used, see, for instance, Larsson and Samuelson (1967), Martinsson and Samuelson, and Vrátny.

*Anthrone–sulphuric acid*

The partition chromatography of monosaccharides on cross-linked dextran containing quaternary ammonium ions in the sulphate form (ethanol–water) was followed auto-

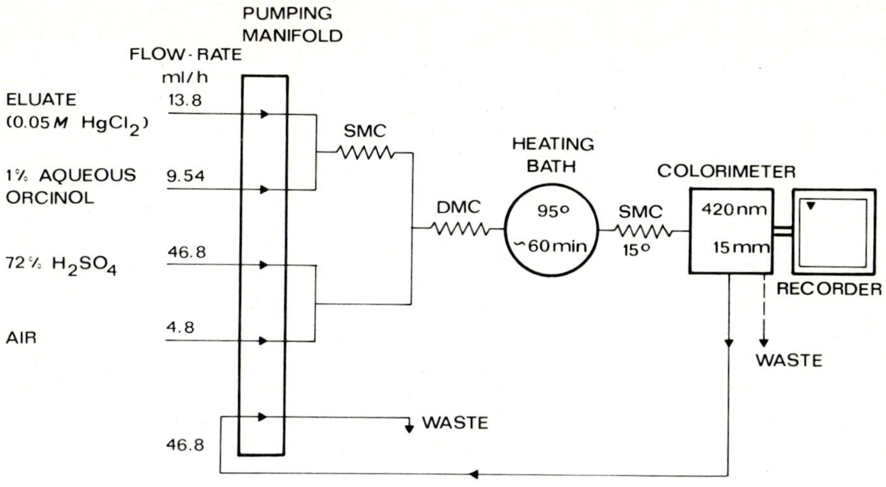


Fig. 22.4. Flow scheme for the orcinol–sulphuric acid method (Bathgate). SMC = single mixing coil; DMC = double mixing coil. Tubing that contains sulphuric acid is made from Acidflex, the remainder from polyethylene.

matically using the anthrone–sulphuric acid reagent with an accuracy of 5% (Jonsson and Samuelson, 1967b; see also Jonsson and Samuelson, 1966; 1967a). The effluent was mixed with the reagent (2 g of anthrone per 1 l of sulphuric acid) in the ratio 1:2. The piston used to feed the reagent into the analyzing system was covered with PTFE tubing in order to prevent corrosion. After the eluate and the reagent had been mixed, the colour was developed by passing the solution through PTFE tubing (I.D. 1.2 mm) submerged in a heating bath at 100°C. The time of the reaction was about 1 min. The transmittance was measured at 625 nm in a 2-mm flow cell and recorded automatically.

A modification of this method was developed for cellulose column separations (Otter *et al.*) because the *n*-butanol-containing mobile phase (water–*n*-butanol–ethanol gradient mixtures) interferes with the reaction between a saccharide and anthrone. The most reproducible results were obtained when the mixture was heated for 20 min at 80°C instead of the more usual temperature of 100°C. The reagent used for colour development was a 0.1% solution of anthrone in 85% sulphuric acid, and was freshly prepared at the beginning of each run. The stock bottle containing the anthrone reagent was kept below 4°C before being mixed, at a flow-rate of 14 ml/h, with the effluent which was cooled to the same temperature. The ratio of the reagent to the column effluent is important and should be maintained at 2:1. Even small changes in this ratio result in a serious decrease in the absorbance measured at 640 nm (Otter *et al.*). For details of this procedure, see the analyzer flow diagram (Fig. 22.5).

*Phenol–sulphuric acid*

The phenol–sulphuric acid method was used in an automated sugar analyzer by Green, Ohms *et al.* and others. Neutral sugars eluted from the anion-exchange resin with a borate buffer were treated continuously with 5% aqueous phenol and concentrated sulphuric

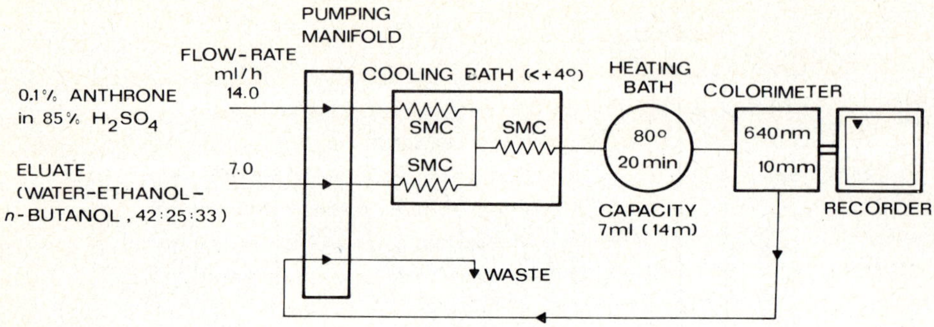


Fig. 22.5. Flow scheme for the anthrone–sulphuric acid method; a modification for an *n*-butanol-containing mobile phase (Otter *et al.*). SMC = single mixing coil (2 m). All tubing is made from PTFE, I.D. 1.0 mm.

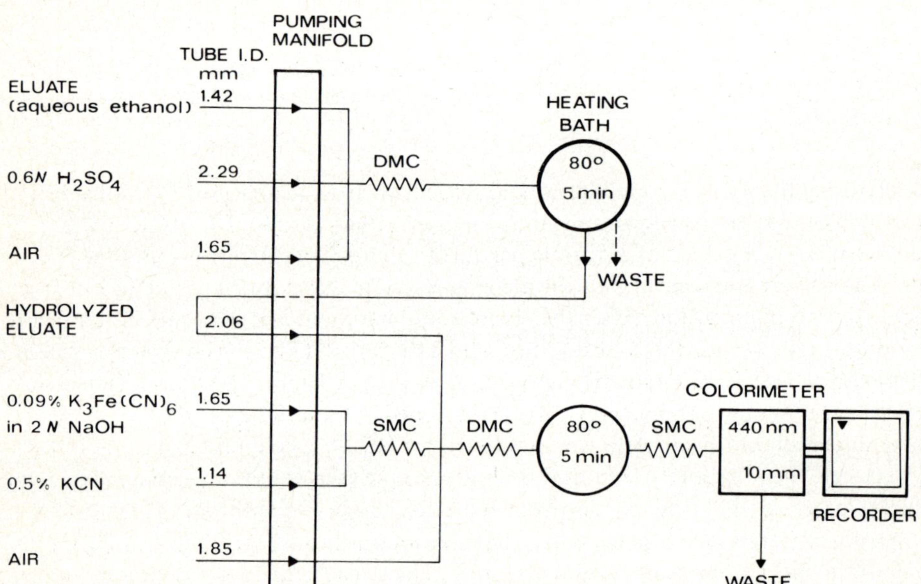


Fig. 22.6. Flow scheme for the potassium hexacyanoferrate(III) assay, together with the hydrolysis of oligosaccharides (Samuelson and Swenson). SMC = single mixing coil; DMC = double mixing coil. All tubing is made from Tygon.

acid, the absorbance of the resulting stream being monitored by a colorimeter with filters at 480, 486 and 490 nm. The flow-rates of phenol and sulphuric acid added to the eluate stream were 0.6 and 3.05 ml/min, respectively (Ohms *et al.*). Based on this method of detection, a Mark II prototype carbohydrate analyzer in which high-resolution ion-exchange chromatography was used for the separation of sugars, with detailed operating conditions, was described by Jolley *et al.* (1969) and was used in the determination of carbohydrates in physiological fluids (Jolley and Freeman; Jolley *et al.*, 1970).

*Potassium hexacyanoferrate(III) assay*

In 1963, Samuelson and Swenson used potassium hexacyanoferrate(III) for detection in the Technicon AutoAnalyzer system. As oligosaccharides were involved in the separation on ion-exchange resins with ethanol–water as the mobile phase, a hydrolysis step was also used in the analyzer (see Fig. 22.6). The hydrolysis to monosaccharides was not complete, but the results were reproducible. Part of the solution obtained after the hydrolysis (the remainder of the solution was discarded) was mixed with a 0.09% solution of potassium hexacyanoferrate(III) in 2 *N* sodium hydroxide, 0.5% potassium cyanide solution and air, then heated at 80°C for about 5 min and, after cooling, the light absorption at 440 nm was measured. Differences in flow-rates were produced by using tubing of various internal diameter.

*Cysteine–sulphuric acid*

In the fractionation of dextran on porous silica beads, an automatic, continuous

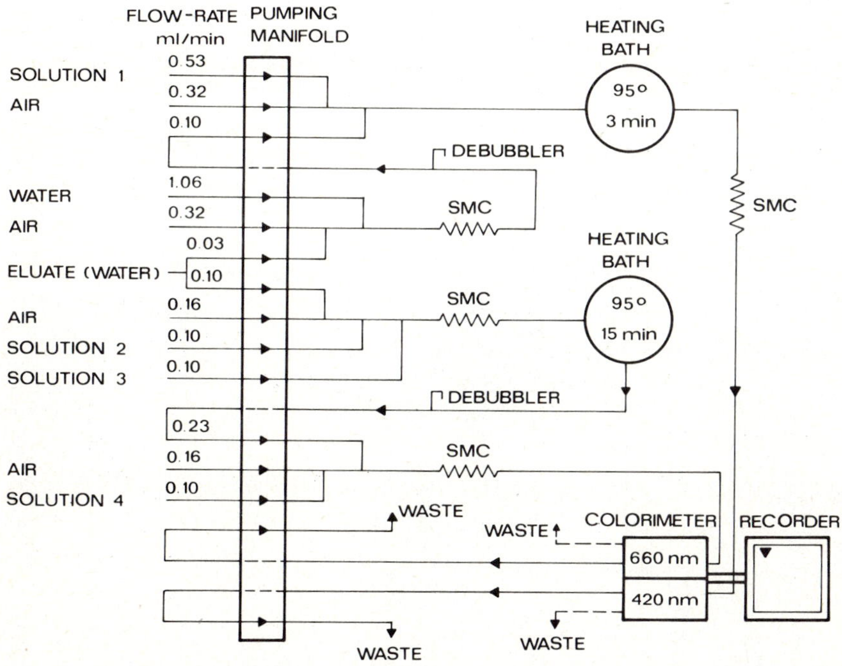


Fig. 22.7. Flow scheme for the cysteine–sulphuric acid and potassium hexacyanoferrate(III) assays (Barker *et al.*, 1969b). SMC = single mixing coil. Technicon AutoAnalyzer modular equipment was used throughout. Solution 1 is a 0.07% (w/v) solution of L-cysteine hydrochloride in 86% (v/v) sulphuric acid. Solution 2 is sodium carbonate (0.53%) and potassium carbonate (0.065%) in water. Solution 3 is potassium hexacyanoferrate(III) (0.05%) in water. Solution 4 is ammonium iron(III) sulphate (0.75%) and sodium lauryl sulphate (0.5%) in 0.05 *N* sulphuric acid.

cysteine–sulphuric acid assay for the total hexose concentration, in addition to a reducing end-group assay (using the potassium hexacyanoferrate(III) reagent mentioned above) was used (Barker *et al.*, 1969b). The column eluate was mixed with water and added at a flow-rate of 0.1 ml/min to a 0.07% (w/v) solution of L-cysteine hydrochloride in 86% (v/v) sulphuric acid, which had a flow-rate of 0.53 ml/min. The reaction stream was heated for 3 min at 95°C, cooled, and the absorbance measured at 420 nm. The complete flow scheme of this analyzer is given in Fig. 22.7. The combination of these two assays obviates the need for any previous calibration of the column by eluting samples of known molecular weight, as the two assays would give the molecular weight of the species eluted at any point. Care must be taken to keep the waste from the two assays separate, in order to avoid the production of hydrogen cyanide gas.

*Periodate oxidation*

The colour reaction of formaldehyde, resulting from periodate oxidation, with 2,4-pentanedione was used in the application of ion-exchange resin chromatography to mixtures of alditols and aldoses (Samuelson and Strömberg, 1966). The flow-scheme for this system is given in Fig. 22.8 (compare also Fig. 22.9). The periodate oxidation of the effluent (flow-rate 0.4 ml/min) is carried out at pH 7.5 or 1.0 (flow-rate 0.6 ml/min) for about 3.5 min. At pH 1.0, most sugars give rise to negligible amounts of formaldehyde,

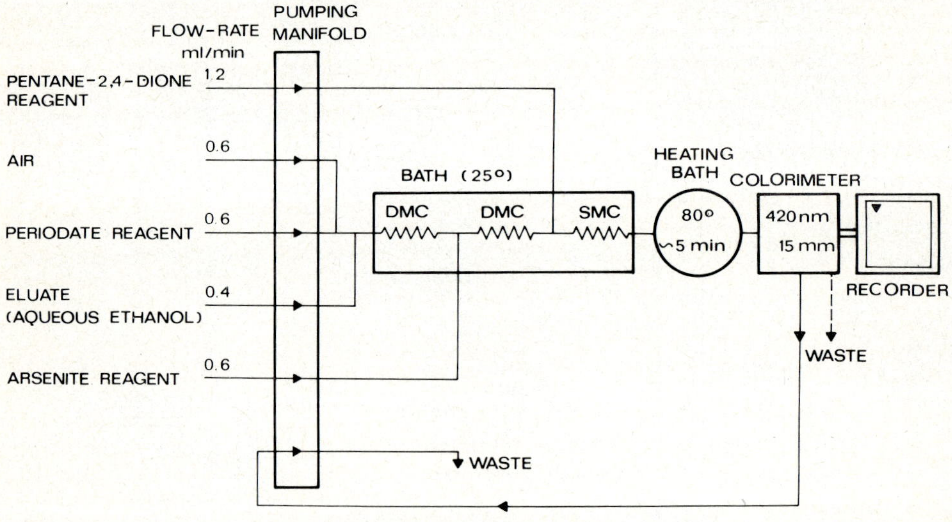


Fig. 22.8. Flow scheme for the assay involving periodate oxidation and formation of formaldehyde (Samuelson and Strömberg, 1966). SMC = single mixing coil; DMC = double mixing coil. Pentane-2,4-dione reagent: 2 *M* ammonium acetate + 0.05 *M* acetic acid + 0.02 *M* pentane-2,4-dione. Arsenite reagent: 0.5 *M* sodium arsenite neutralized with hydrochloric acid to pH 7. Periodate reagent: (a) oxidation at pH 7.5, 0.015 *M* periodic acid neutralized with ammonia and buffered to pH 7.5 with a phosphate buffer (100 ml/l); (b) oxidation at pH 1.0, 0.015 *M* sodium periodate in 0.12 *M* hydrochloric acid.

while the alditols react without difficulty. At pH 7.5, formaldehyde is formed in high yield from both alditols and aldoses. Before the colorimetric determination, the unreacted periodate is reduced to iodate or iodide (depending on the pH) by the addition, at a flow-rate of 0.6 ml/min, of 0.5 *M* sodium arsenite neutralized with hydrochloric acid to pH 7. This precaution is necessary because periodate destroys the colour obtained with the 2,4-pentanedione reagent (flow-rate 1.2 ml/min). The absorbance at 420 nm is measured in a 15-mm flow cell.

For complicate mixtures, a two-channel analyzer, involving the combination of this method with the orcinol–sulphuric acid method (no response with alditols), was used (Samuelson and Strömberg, 1966). This principle of using several channels, each of which specifically detects only one type of sugar derivative, proved to be very useful also in other situations, as described below.

*Chromic acid and carbazole assays*

Samuelson and Thede used a two-channel analyzer in the chromatography of sugar acids on Dowex 1-X8 ($CH_3COO^-$) (see p. 513). Chromic acid oxidation gives a response with all eluted hydroxy acids, while the carbazole method gives a strong response only with uronic acids and some keto acids, no response with aldonic acids and a weak reaction with aldobionic and lactic acids (Johnson and Samuelson). With complicated mixtures

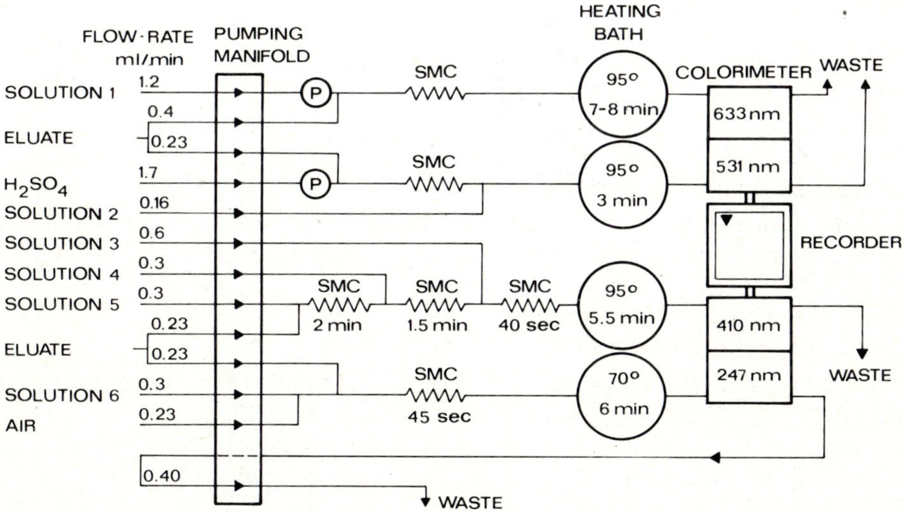


Fig. 22.9. Flow scheme for a four-channel manifold, using chromic acid, carbazole and two periodate assays for the analysis of ion-exchange chromatograms of sugar acids, with acetic acid or aqueous sodium acetate as the mobile phase (Carlsson and Samuelson, 1970). SMC = single mixing coil (25°C); P = pulse suppressor. PTFE tubing of I.D. 0.8 or 1.2 mm is used throughout. Solution 1: 5 volumes of conc. sulphuric acid + 2 volumes of a 2.45 g/l aqueous solution of potassium dichromate. Solution 2: 0.15% (by weight) solution of carbazole in 70% (w/w) ethanol. Solution 3: 0.02 *M* pentane-2,4-dione in 2 *M* ammonium acetate solution, 0.05 *M* with respect to acetic acid. Solution 4: 0.2 *M* sodium arsenite neutralized with hydrochloric acid to pH 7. Solution 5: 0.015 *M* sodium metaperiodate buffered at pH 7.0.

that contain other types of hydroxy acids, a third channel was introduced in the pumping manifold (Carlsson *et al.*, 1968; Carlsson and Samuelson, 1969), in which the periodate oxidation of the eluted compounds and the subsequent determination of the formaldehyde formed was recorded. As there were still some types of acids that could not be distinguished from one another, Carlsson and Samuelson (1970) introduced a fourth channel, in which the consumption of periodate is determined automatically by UV spectrophotometry. The complete scheme with all flow-rates, reagent compositions, temperatures and delays in the reaction coils is given in Fig. 22.9.

*Ninhydrin*

The behaviour of mono-, oligo- and deoxy saccharides and related compounds towards ninhydrin with respect to their ability to produce coloured products (see Table 22.5, p. 492), and the use of an amino acid analyzer in their separation on an Amberlite IR-120 column with 0.2 *N* citrate buffer (pH 2.2), was described by Zacharius and Porter. The results from this and other studies emphasized earlier warnings on the hazards of interpreting all ninhydrin peaks as products of nitrogen compounds.

The amino acid analyzer was also used in the separation of hexosamines (*e.g.*, Donald, Stern), for the quantitative determination of amino sugar antibiotics and their degradation products (Inouye and Ogawa), for the determination of O-methylated derivatives of amino sugars (Adams *et al.*), etc.

*Acetic acid–aniline–orthophosphoric acid*

In this method (Walborg and Kondo), the reagent (acetic acid–aniline–85% orthophosphoric acid, 100:3:50, v/v) developed for the quantitation of neutral mono- and oligosaccharides in buffers containing glycerol (Walborg and Christenson) or butane-2,3-diol was used. The eluate and the reagent at flow-rates of 20 and 60 ml/h, respectively, were forced (Fig. 22.10) into the mixing manifold, which was maintained at 60°C so as to prevent the formation of the precipitate there. For the same reason, the connection between the single mixing coil and the heating bath was insulated so as to reduce the

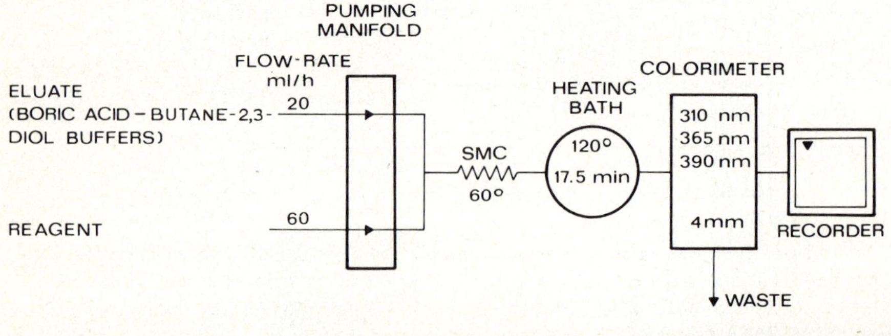


Fig. 22.10. Flow scheme for acetic acid–aniline–orthophosphoric acid assay (Walborg and Kondo). SMC = single mixing coil. PTFE tubing was used throughout; the reaction coil consists of a 13-m length of PTFE tubing, I.D. 1.5 mm, with an internal volume of 23 cm³. The reagent was prepared by adding 200 ml of glacial acetic acid to 6 ml of redistilled aniline, then adding 100 ml of 85% orthophosphoric acid to the mixture.

cooling of the mixture. Colour development was then carried out at 120°C for 17.5 min. A back pressure of 4 kg/cm$^2$ was maintained on the reaction coil so as to prevent the reaction mixture from boiling. The absorbance was measured at 310, 365 and 390 nm; the ratio of the adsorptions at 356 and 369 nm was also utilized (Walborg and Kondo) as a qualitative method for distinguishing between classes of saccharides. Aldopentoses yield a ratio of 1.78, aldohexoses and ketohexoses 1.29, 6-deoxyaldohexoses 1.10. The precision of the method is *ca.* ±5%, the reagent being stable for at least 2 months at room temperature.

*Dyed polysaccharides*

The use of dyed polysaccharides (for instance, dyed with Procion Brilliant Red $M_2B$) leads to a simple method for monitoring the performance of GPC. It is believed that with these reactive dyes, a low degree of substitution introduces sufficient colour to make the product visible, but does not affect the properties upon which the separation depends (Dudman and Bishop). A solution of 2.5 mg of dye in 0.25 ml of water was added to a solution of 2.5 mg of polysaccharide in 0.25 ml of water; after 5 min, sodium chloride was added to give a final concentration of 2 *M*. The dyed sample (1–3 mg) in 2 *M* sodium chloride (0.5 ml) was applied with care to the top of the column (35 × 1.5 cm) and eluted with 1 *M* sodium chloride at a flow-rate of 0.5–1.0 ml/min. The column effluent was passed directly into the cell of a colorimeter (Anderson *et al.*).

*Flame ionization detector for liquid chromatography*

For the quantitative analysis of sugar mixtures on a strong cation-exchange resin in the $Li^+$ form with 85% aqueous ethanol as the mobile phase, a modified Pye moving-wire detector was connected to the column. This detector provides a direct trace of the column effluent on the potentiometric recorder without the necessity for forming a coloured derivative. The detector responses were found to be linear for carbohydrates with a pentoses:hexoses:disaccharides:deoxy sugars ratio of 1.00:1.00:1.05:1.10 (Hobbs and Lawrence, 1972a, b). Direct combustion in liquid chromatography with a hydrogen flame ionization detector was also described (Foster and Weiss), but this system was not recommended for long, continuous operations.

## MONO-, OLIGO- AND DEOXY SACCHARIDES

The separation of free saccharides is carried out by adsorption, partition, ion-exchange and gel permeation chromatography, the last three methods being of increasing importance in the past decade. The rapid development in this field is connected with the exploitation of the automated analysis of effluents.

### Chromatography on charcoal–Celite

Carbon column chromatography is especially suitable (see p. 467) for the fractionation of mixtures that contain sugars of different chain lengths. This technique is now

TABLE 22.1
SEPARATION OF MONO-, OLIGO- AND DEOXY SACCHARIDES

| Compounds separated | Sorbent | Mobile phase | Reference | Notes |
|---|---|---|---|---|
| Products of invertase action on sucrose | BDH activated charcoal–Celite 535 | Linear ethanol gradient | Bacon | |
| D-Glucose<br>1,6-Anhydro-β-D-glucopyranose<br>Gentiobiose<br>Gentiotriose<br>Gentiotetraose | Charcoal–Celite (1:1) (75 × 5 cm) | Water<br>Water<br>5% Ethanol<br>15% Ethanol<br>25% Ethanol | McGrath *et al.*, 1969a | With gentiotetraose, an additional filter sheet chromatographic step was necessary in order to obtain the pure compound |
| D-Glucose<br>Laminaribiose<br>Laminaritriose<br>Higher oligosaccharides | Charcoal–Celite (1:1) (20 × 5 cm) | Water<br>10% Ethanol<br>15% Ethanol<br>50% Ethanol | McGrath *et al.* 1969b | Higher oligosaccharides have a degree of polymerisation of 4–7 |
| Mono- and oligosaccharides from *Araucaria bidwilii* | Charcoal–Celite (1:1) (52 × 5.5 cm) | Water–ethanol gradients (0–15% ethanol) | Aspinall and McKenna | |
| Linear maltodextrins | Darco G-60–Celite 560 (4:3) (40 × 3.6 cm) | *tert.*-Butanol–water linear gradient (0–10% *tert.*-butanol) | French *et al.* | Charcoal was pre-treated with a 2.5% solution of stearic acid in ethanol |
| α- and β-Cyclodextrin | Darco G-60–Celite 560 (4:3) (45 × 3 cm) | *n*-Butanol–water gradient (3–7% *n*-butanol) | Lammers | |
| D-Glucose–cellohexaose series | Charcoal–Celite (800 cm × 113 $cm^2$) | Water–ethanol gradient (0–50% ethanol) | Brown, 1970a | Charcoal treated with stearic acid |

TABLE 22.1 – *continued*

| | | | | |
|---|---|---|---|---|
| α,α-Trehalose<br>6-O-Mesyl-α,α-trehalose<br>6,6′-Di-O-mesyl-α,α-trehalose | Charcoal–Celite (1:1) (54 × 3 cm) | Water<br>Water<br>Water–methanol (1:1) | Birch and Richardson | *ca.* 23 g of the mixture were separated |
| D-Xylose<br>D-Glucose<br>Primeverulose<br>Primeverose | Cellulose, Whatman No. 1 (92 × 4 cm) | Satd. aq. *n*-butanol–*n*-butanol (1:3)<br>(1:1)<br>Satd. aq. *n*-butanol | Rutherford and Richtmyer | 2.95 g were separated (product of partial rearrangement of primeverose) |
| L-Rhamnose<br>L-Arabinose<br>D-Galactose | Cellulose, Whatman No. 1 (60 × 4 cm) | *n*-Butanol–water–pyridine–benzene (5:3:3:1), upper layer | Anderson and Cree | 5-g sample, hydrolyzate of *Accacia nubica* |
| 3-O-α-D-Glucopyranosyl-D-ribose<br>3-O-α-D-Glucopyranosyl-D-arabinose | Cellulose (85 × 7 cm) | Acetone–water (7:1) | Furda and Perry | 13 g of the mixture were applied |
| L-Rhamnose<br>L-Arabinose<br>Arabinobiose<br>3-O-β-L-Arabinopyranosyl-L-arabinose | Cellulose | Satd. aq. *n*-butanol–*n*-butanol (1:1)<br>Satd. aq. *n*-butanol<br>*n*-Butanol–ethanol–water (40:11:19) | Chalk *et al.* | |
| Tetruloses and pentuloses | Dowex 50W-X8 ($Li^+$, 14–17 μm) (815 × 4.5 mm) or Technicon $T_5C$ ($SO_4^{2-}$, 14–17 μm) (740 × 4.3 mm) | 85% ethanol at 75°C | Havlicek *et al.* | 2-Pentuloses and 3-pentuloses give a strong response in the periodate–formaldehyde channel in contrast to aldoses, see p. 479 |

applied to complex mixtures mainly as a method of pre-treatment, followed by other types of chromatography to resolve the particular sugar group.

Of the various modifications of the charcoal chromatography of free sugars developed, the method of Whistler and Durso gives the most consistent results. A mixture of equal amounts of Darco G-60 and Celite 535 is packed into a 23 × 3.4 cm tube and washed with water. The ratio of the size of the sample (applied as a 10% aqueous solution) to the amount of charcoal is 1:100; monosaccharides are eluted with water, disaccharides with 5% ethanol and trisaccharides with 15% ethanol. For example, Mansford used this method with BDH activated charcoal-Celite 545 (2:1) for the analysis of commercial starch-conversion products; D-glucose, maltose, maltotriose and maltotetraose were eluted by increasing the ethanol concentration stepwise (0–12% ethanol).

By means of gradient elution, a successful separation can be achieved even in instances (Bacon) in which the stepwise increase in ethanol concentration is not capable of separating the mixture. Some typical examples are summarized in Table 22.1.

The application of silica gel for the separation of mono- and oligosaccharides was described by Otake. The capacity of silica gel for sugars was about 10 mg/g, with a dimethyl sulphoxide–ethyl acetate mixture as the mobile phase.

**Chromatography on cellulose**

Since the pioneering work of Hough *et al.* (1949), who used a cellulose column to separate a mixture of L-rhamnose, D-ribose, L-arabinose and D-galactose, with water-saturated *n*-butanol as solvent, a large number of applications of cellulose for preparative purposes appeared (Table 22.1). This is due to the analogy of this technique with paper chromatography, which is frequently used for the analysis of sugar mixtures. Flat-bed chromatography is also a very popular method for tracing effluents. Oligosaccharides are eluted from the cellulose powder column according to their migration on paper (Bacon).

The heated cellulose column (45 × 1.5 cm; 60°C), packed to a height of 43 cm with Whatman CF-12 cellulose powder (see p. 469), was used for the separation and quantitative analysis of oligosaccharides from four to fifteen glucose units (Otter *et al.*). The mobile phase was pumped from a gradient elution device consisting of a 250-ml conical flask filled to a volume of 200 ml with water–ethanol–*n*-butanol (24.5:23:52.5) mixture, connected to an empty 50-ml flask, which was connected in turn to a 1-l flask containing 800 ml of water–ethanol–*n*-butanol (42:25:33). All the solvents should be deaerated and kept free from air at 60°C. The mobile phase is pumped to the top of the column at a fixed flow-rate of 28 ml/h; 0.1–0.2 g of the oligosaccharide mixture is dissolved in the minimum amount of degassed water, and the viscous solution is layered on to the column. Anthrone-sulphuric acid was used for detection (see p. 477). It is necessary to wash the column after each run by running hot deaerated water through it overnight, finally displacing the water by 500 ml of eluting solvent (water–ethanol–*n*-butanol, 24.5:23:52.5).

Attention is called (Huffman *et al.*, Whistler) to the presence of extraneous carbohydrates (other than those placed on the column) in the eluates. These sugars, of a hemicellulose nature, are extracted from the cellulose, particularly if water is passed through the column.

**Chromatography on ion-exchange resins**

*Liquid–liquid chromatography*

In 1965, Samuelson *et al.* used the anion-exchange resin Technicon $T_4$ or $T_5B(SO_4^{2-})$ (3–17 μm) for the separation of the following sugars: 2,6-dideoxy-D-*ribo*-hexose, 2-deoxy-D-*erythro*-pentose, 2-deoxy-D-*lyxo*-hexose, L-rhamnose, L-fucose, D-ribose, D-lyxose, arabinose, D-xylose and hexuloses. Aqueous ethanol (94%) was used for the elution at a flow-rate of 3.5 ml/cm$^2$·min, the effluent being assayed by means of the automated orcinol–sulphuric acid method. This type of chromatography is described in detail for alditols (see p. 504).

A comparison of the sulphate and chloride forms of strongly basic anion-exchange resins with respect to their resolving power for monosaccharides and deoxy sugars was reported by Jonsson and Samuelson (1967b), while Jonsson and Samuelson (1967a) dealt with the application of sulphonated styrene–divinylbenzene resins in the lithium, sodium and potassium forms for the separation of carbohydrates. Table 22.2 serves as a guide for the choice of the appropriate form of a particular resin. Various ethanol concentrations were used in extensions of this work, and alditols and some simple aliphatic carbonyl compounds were also included (Samuelson and Strömberg, 1968). The latter compounds were included into the mixture as the potential application of this technique is in structural studies of polysaccharides where in total hydrolyzates all these compounds may occur side by side. For these results see p. 502.

TABLE 22.2
VOLUME DISTRIBUTION CONSTANTS IN 92.4% ETHANOL AT 75°C AND 100°C DETERMINED ON 460 × 6 mm COLUMNS OF AMBERLITE IR-120 ($Li^+$, $Na^+$ AND $K^+$) (8% CROSS-LINKAGE) (JONSSON AND SAMUELSON, 1967a)

| Compound | $Li^+$ | | $Na^+$ | | $K^+$ | |
|---|---|---|---|---|---|---|
| | 75°C | 100°C | 75°C | 100°C | 75°C | 100°C |
| Digitoxose | 0.8 | 0.6 | 0.8 | 0.6 | 0.8 | 0.6 |
| 2-Deoxy-D-ribose | 1.5 | 1.3 | 2.0 | 1.4 | 2.0 | 1.4 |
| 2-Deoxy-D-glucose | 2.1 | 1.7 | 2.9 | 2.2 | 3.2 | 2.4 |
| 2-Deoxy-D-galactose | 2.9 | 2.3 | 4.3 | 3.1 | 4.3 | 2.9 |
| Rhamnose | 1.4 | 1.2 | 3.3 | 2.3 | 4.1 | 2.7 |
| Fucose | 2.4 | 1.8 | 7.1 | 4.2 | 8.9 | 5.1 |
| Lyxose | 2.9 | 2.4 | 6.5 | 4.4 | 8.1 | 5.5 |
| Xylose | 3.0 | 2.4 | 7.3 | 4.5 | 8.6 | 5.7 |
| Arabinose | 3.8 | 3.0 | 11.2 | 6.2 | 14.2 | 7.9 |
| Ribose | 4.0 | 3.1 | 12.3 | 6.8 | 9.2 | 5.6 |
| Tagatose | 4.5 | 3.7 | 11.3 | 7.6 | 10.9 | 7.4 |
| Sorbose | 4.6 | 3.7 | 13.0 | 8.1 | 13.4 | 8.5 |
| Fructose | 5.6 | 4.3 | 18.7 | 10.0 | 20.1 | 11.3 |
| Glucose | 5.4 | 4.4 | 16.8 | 10.3 | 22.5 | 13.6 |
| Mannose | 5.3 | 4.4 | 17.4 | 10.7 | 25.0 | 14.6 |
| Galactose | 6.8 | 5.3 | 22.0 | 12.6 | 27.5 | 15.2 |

Rapid separations of multicomponent mixtures were achieved on Aminex A-6 ($(CH_3)_3N^+$; 17.5 $\mu$m) (Hobbs and Lawrence, 1972b). With this resin, as well as other organic-base counter-ions tested, even the separation of D-glucose from D-mannose was successful (column, 100 × 0.4 cm; temperature, 75°C; solvent, 85% aqueous ethanol; flow-rate, 0.45 ml/min; inlet pressure, 30 atm).

A comparison of the applicability of Dowex 50W-X8 ($Li^+$) and the strong anion-exchange resin Technicon $T_5C$ ($SO_4^{2-}$) at 75°C using aqueous ethanol as the mobile phase was presented by Martinsson and Samuelson. It is obvious from Table 22.3 that allose could not be separated from altrose on this anion exchanger, but these compounds were well resolved on Dowex 50W-X8 ($Li^+$). Sometimes even a reversed order of elution occurs, *e.g.*, with tetroses or 1,6-anhydro-β-D-glucoses. In general, a combination of both methods is recommended for the complete resolution of complex mixtures.

Martinson and Samuelson also reported the complete separation of oligosaccharides in the glucose–cellohexaose series; the volume distribution constants of these compounds are given in Table 22.4. Fig. 22.11 shows that the plot of the logarithm of $K_d$ for particular oligosaccharides against the number of hexose units is a straight line (Martin's rule).

TABLE 22.3
VOLUME DISTRIBUTION CONSTANTS OF SOME MONO- AND DISACCHARIDES AT 75°C (MARTINSSON AND SAMUELSON)

| Saccharide | Technicon $T_5C$ ($SO_4^{2-}$) (mobile phase, 88% ethanol) | Dowex 50W-X8 ($Li^+$) (mobile phase, 92.4% ethanol) | Dowex 50W-X8 ($Li^+$) (mobile phase, 96% ethanol) |
|---|---|---|---|
| Erythrose | 3.1 | 1.9 | 2.3 |
| Threose | 3.8 | 1.4 | 1.8 |
| Arabinose | 10.1 | 3.5 | 4.2 |
| Lyxose | 8.9 | 2.7 | 3.2 |
| Ribose | 6.6 | | 4.6 |
| Xylose | 12.5 | 2.7 | 3.2 |
| Allose | 16.6 | 6.0 | 8.3 |
| Altrose | 16.6 | 4.2 | 5.6 |
| Galactose | 23.4 | | 7.6 |
| Glucose | 28.1 | 4.8 | 6.0 |
| Gulose | 20.4 | 4.9 | 6.3 |
| Mannose | 16.4 | | 5.9 |
| Talose | 10.7 | 5.8 | 8.3 |
| Fructose | 13.5 | | 6.8 |
| 3-Ribohexulose | 9.8 | 4.7 | 6.6 |
| 6-Deoxy-D-glucose | 7.3 | 0.9 | 1.1 |
| Rhamnose | 4.8 | 1.4 | 1.6 |
| 1,6-Anhydro-β-D-glucofuranose | 6.3 | 1.4 | 1.6 |
| 1,6-Anhydro-β-D-glucopyranose | 3.8 | 2.3 | 2.9 |
| 2,3-Di-O-methyl-6-deoxy-D-allose (mycinose) | 0.53 | 0.4 | 0.4 |
| Mannobiose | 41.6 | | |
| Xylobiose | 35.8 | 5.4 | 7.3 |

TABLE 22.4
VOLUME DISTRIBUTION CONSTANTS OF OLIGOSACCHARIDES IN THE CELLOBIOSE–CELLOHEXAOSE SERIES AT 75°C (MARTINSSON AND SAMUELSON)

| Oligosaccharide | Technicon $T_5C$ ($SO_4^{2-}$) (mobile phase, 70% ethanol) | Dowex 50W-X8 ($Li^+$) (mobile phase, 85% ethanol) |
|---|---|---|
| D-Glucose | 3.1 | 3.0 |
| Cellobiose | 4.4 | 4.9 |
| Cellotriose | 5.8 | 8.7 |
| Cellotetraose | 7.9 | 15.4 |
| Cellopentaose | 10.9 | 27.4 |
| Cellohexaose | 15.2 | 44.4 |

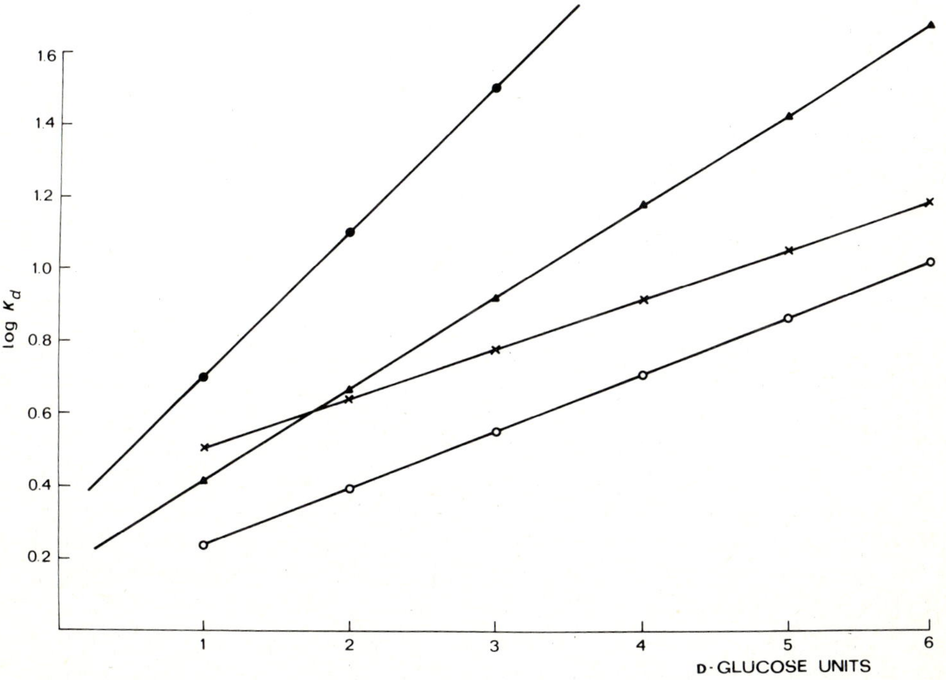


Fig. 22.11. Relationship between log $K_d$ and the number of D-glucose units in the oligosaccharide at 75°C (Martinsson and Samuelson). x = Technicon $T_5$ C ($SO_4^{2-}$); mobile phase, 70% ethanol. o = Dowex 50W-X8 ($Li^+$); mobile phase, 80% ethanol. ▲ = Dowex 50W-X8 ($Li^+$); mobile phase, 85% ethanol. • = Dowex 50W-X8 ($Li^+$); mobile phase, 92.4% ethanol.

However, in the chromatography of xylan oligosaccharides Havlicek and Samuelson found that the slope of this line varies with the ethanol concentration. At a high ethanol concentration, the saccharides are eluted in the order of increasing molecular size, whereas at a low concentration the elution order is reversed (Fig. 22.12). The consequence of this

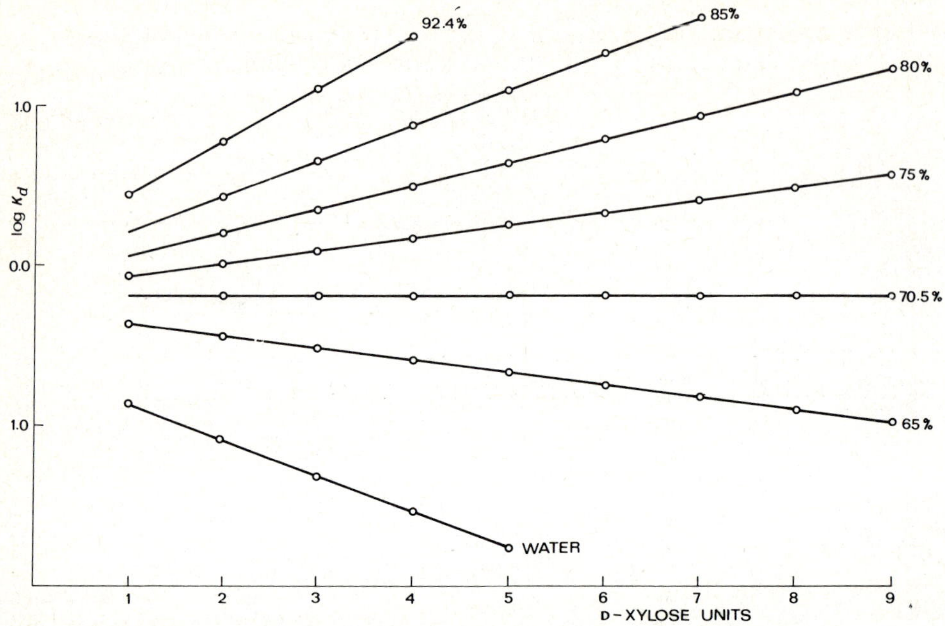


Fig. 22.12. Relationship between log $K_d$ and the number of D-xylose units in the oligosaccharides at various concentrations of ethanol (Havlicek and Samuelson). Dowex 50W-X8 ($Li^+$); temperature 75°C.

effect is that a critical mobile phase composition exists (70.5% ethanol for the lithium resin, 60% ethanol for the sulphate resin) at which all of the oligomers exhibit the same distribution constant, *i.e.*, no separation occurs.

*Chromatography on ion-exchange resins in the borate form*

An anion-exchange resin in the borate form was first used by Khym and Zill. In 1967, the rapid quantitative anion-exchange chromatography of carbohydrates on this basis was described by Kesler, who used Dowex 1-X8 ($BO_3^{3-}$, 200–400 mesh), Bio-Rad AG 1-X8 ($BO_3^{3-}$) or Technicon 3/28/VI ($BO_3^{3-}$) to separate a 17-component mixture (see Fig. 22.13) in 4–6 h. It has also been established that under these conditions sorbose is eluted together with xylose, lyxose with ribose, raffinose with cellobiose and trehalose with sucrose. Increasing or decreasing the flow-rate by 50% made little difference in the resolution of sugars. This method can be used for analytical (0.3 cm I.D. column) as well as for preparative (2.4 cm I.D. column) purposes, the amounts of individual sugars varying from less than a microgram to several hundred milligrams.

The anion-exchange resin Technicon 3/28/VI proved to be of high value for separating a mixture of D-fructose, D-mannose and D-glucose, using a gradient elution with borate buffers (pH 8.5–9.3) at 30°C (MacLaurin and Green, 1969a). Orcinol–sulphuric acid was used for the quantitative determination and gave an accuracy ±2–3%. Under the same conditions, a mixture of the above three compounds plus cellobiose, cellobiulose and

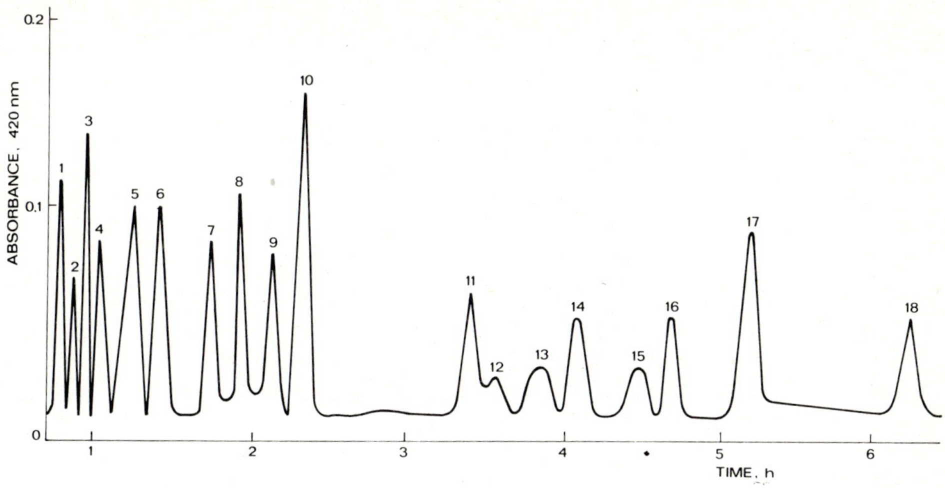


Fig. 22.13. Chromatogram of a solution containing 17 known components (Kesler) on a 75 × 0.3 cm column. Gradient: chamber 2 (70 ml of 0.60 *M* boric acid, pH 10.0) to chamber 1 (70 ml of 0.125 *M* boric acid, pH 7.0) to column. Flow-rate, 0.30 ml/min; temperature, 53°C. 1 = Furfural (1.0 $\mu$g); 2 = hydroxymethylfurfural (2.0 $\mu$g); 3 = sucrose (2.4 $\mu$g); 4 = cellotetraose (2.2 $\mu$g); 5 = cellotriose (2.3 $\mu$g); 6 = cellobiose (2.4 $\mu$g); 7 = maltose · $H_2O$ (2.2 $\mu$g); 8 = rhamnose · $H_2O$ (2.0 $\mu$g); 9 = lactose · $H_2O$ (2.2 $\mu$g); 10 = ribose (1.6 $\mu$g); 11 = mannose (2.0 $\mu$g); 12 = unknown; 13 = fructose (2.4 $\mu$g); 14 = arabinose (1.1 $\mu$g); 15 = galactose (1.3 $\mu$g); 16 = xylose (0.8 $\mu$g); 17 = glucose (2.5 $\mu$g); 18 = gentiobiose (1.6 $\mu$g).

4-O-$\beta$-D-glucopyranosyl-D-mannose has been resolved (MacLaurin and Green, 1969b). The strong basic anion-exchange resin Beckman 1-S (11–23 $\mu$m), specifically designed for the automated analysis of neutral sugars, was used (Ohms *et al.*) with a linear gradient of borate buffer to separate a synthetic mixture of sucrose, raffinose, cellobiose, maltose, lactose, ribose, rhamnose, mannose, fructose, arabinose, galactose, xylose and glucose.

Prior to column packing, the resin is converted sequentially into the chloride form with 1 *N* hydrochloric acid, the hydroxide form with 1 *N* sodium hydroxide and the borate form with 0.5 *M* boric acid. The resin is then equilibrated on a büchner funnel with 10 volumes of a borate buffer (pH 8.9, 0.215 *M*) containing 0.01% of Brij-35, and packed as an approximate 1:1 slurry in a 0.9 cm I.D. column to a bed height of 55 cm with a pump flow-rate of 1.15 ml/min at 50°C. The use of a thick slurry and the addition of Brij-35 improves the reproducibility of column packing. The packed column is equilibrated for 8 h with 0.042 *M* borate buffer without Brij-35.

A synthetic mixture of saccharides (0.5–1.0 $\mu$mole each) in a volume of 110 $\mu$l is mixed with 200 $\mu$l of the limiting buffer (see p. 474) and maintained at 50°C for 15 min so as to form a sugar–borate complex; the sample is then forced into the resin bed, followed by two washings with starting buffer. In order to maintain the initial gradient, the top of the column is filled with an adaptor so as to eliminate all but 2 mm of liquid headspace above the resin bed. One run with a linear buffer gradient (see p. 474) requires about 7.5 h; when D-glucose is eluted, the column should be re-equilibrated for 7 h with the starting buffer in preparation for the next run (Ohms *et al.*).

Walborg and coworkers used boric acid–glycerol buffers (Walborg *et al.*, 1965; Walborg and Lantz) at pH 6.8 in order to minimize any alkaline rearrangement of saccharides on the resin. The acetic acid–aniline–orthophosphoric acid reagent was used for the detection.

A further improvement was obtained (Walborg *et al.*, 1969) when 2,3-butanediol, which complexes less strongly with borate, was used instead of glycerol; a stepwise, two-buffer elution system was used for the separation of mono-, di- and trisaccharides. A Dowex 1-X4 (10–50 μm) column (100 × 0.6 cm), equilibrated with buffer A (0.8 *M* boric acid, 1.0 *M* 2,3-butanediol, 0.1% of Brij-35 and 0.5 ml of toluene per litre; pH 7), was washed with 500 ml of buffer B (0.15 *M* boric acid, 0.5 *M* 2,3-butanediol, 0.1% of Brij-35 and 0.5 ml of toluene per litre). Samples (about 1 μmol of each sugar) were applied in 1-ml aliquots of buffer B. Elution was begun with buffer B at 40°C at a flow-rate of 20 ml/h, and after the consumption of 230 ml, the temperature was increased to 60°C and the elution was continued with buffer A.

### *Other ion-exchange separations*

A number of sugars was chromatographed on the amino acid analyzer, using Amberlite IR-120 at 30°C with 0.2 *M* citrate buffer (pH 2.2) (Zacharius and Porter). The ability of these compounds to produce coloured products with ninhydrin, the 440:570nm absorption ratio of the products, elution volumes and approximate colour factors were ascertained (see Table 22.5).

TABLE 22.5
ELUTION VOLUMES AND NINHYDRIN PEAKS FOR SOME SUGARS CHROMATOGRAPHED ON AN AMINO ACID ANALYZER (ZACHARIUS AND PORTER)

Amberlite IR-120, column 150 × 0.9 cm; mobile phase, 0.2 *N* citrate buffer (pH 2.2); temperature, 30°C.

| Compound | Elution volume (ml) | 440:570 nm absorption ratio | Colour factor |
|---|---|---|---|
| Turanose | 45 | 2.0 | 0.00081 |
| Lactose | 45 | 2.0 | 0.00063 |
| Glucose | 51–52 | 3.3 | 0.00072 |
| Galactose | 54 | 3.3 | 0.00093 |
| Mannose | 54 | 3.3 | 0.0010 |
| Sorbose | 54–55 | 5.69 | 0.026 |
| Rhamnose | 57–58 | 3.3 | 0.00099 |
| Fructose | 57–58 | 5.23 | 0.0132 |
| Ascorbic acid | 65–66 | 2.57–3.94 | 0.384–0.440 |
| Ribose | 69 | 4.71 | 0.0084 |
| Laevulinic acid | 79–80 | 1.34 | 0.250 |
| Kojic acid | 137–139 | 4.28 | 0.63 |
| Xylose | 58–59 | 4.0 | 0.0027 |
| Arabinose | 62–63 | 5.1 | 0.0020 |
| Lyxose | 58–59 | 2.72 | 0.014 |

Free sugars are retained in the column when a strong anion-exchange resin in the hydroxide form (such as Amberlite IRA-400) is used (Roseman *et al.*). On the contrary, the weakly basic Amberlite IR-45 ($OH^-$) was shown to be able to deionize aqueous, sugar-containing eluates without retaining any saccharide (Mizelle *et al.*). Murphy *et al.* described a reversible reaction between D-xylose and this resin, which produced a Schiff's base or a glycosylamine.

Hough *et al.* (1960) examined the possibility of carbonate complexing on De Acidite SRA-68 ($CO_3^{2-}$ or $HCO_3^-$) resin (3.5% cross-linked, < 200 mesh) for the separation of sugars using water as the mobile phase. They observed no such effect, but they found that the carbohydrates were fractionated in order of decreasing molecular size; with a 100 × 2 cm column and a flow-rate of 6–12 ml/h, 200 mg each of raffinose, sucrose and D-glucose were separated. This resin probably functions like a molecular sieve.

## Chromatography on molecular sieves

As mentioned earlier, GPC is predominantly used for the separation of mixtures that contain sugars of different molecular weights. John *et al.* reported the successful use of the highly cross-linked polyacrylamide gel Bio-Gel P-2 in the separation of a series of α-D-(1→4)-linked oligosaccharides obtained by the action of *Escherichia coli* ML 30 on maltose. Fig. 22.14 illustrates the complete resolution of a D-glucose–oligosaccharide mixture containing up to 11 D-glucose–units. The separation of D-glucose from D-ribose and the fractionation of a D-glucose–maltose–isomaltose–maltotriose–isomaltotriose mixture were described in the same paper.

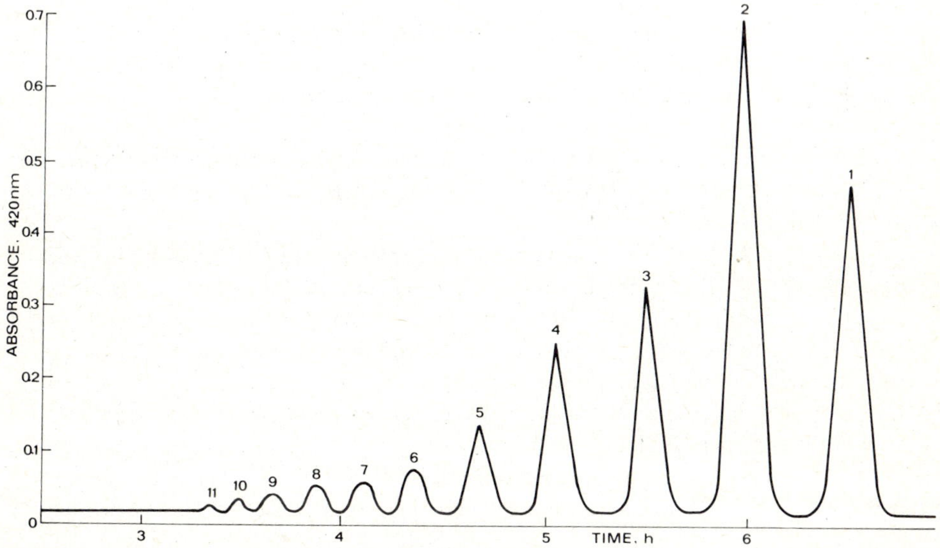


Fig. 22.14. Separation of D-glucose and oligosaccharides synthesized by the action of *E. coli* ML 30 on maltose (John *et al.*). Bio-Gel P-2 (–400 mesh); column, 127 × 1.5 cm; mobile phase, water; flow-rate, 28 ml/h; temperature, 65°C. Peaks 1–11: glucose (1) to maltoundecaose (11).

*References p.519*

A water-jacketed column (100 × 2.5 cm) was filled with water and air bubbles were carefully removed. A glass tube was then attached to the top of the column by means of a screw joint. Sufficient Bio-Gel P-2 (–400 mesh) suspension, previously degassed in a vacuum flask, was poured into the lengthened column and the gel was allowed to settle overnight. Tight and close packings were obtained by pumping water at 30°C through the column at a flow-rate of 28 ml/h until settling of the gel was complete. An exactly fitting, punched-plastic cylinder was then inserted on the surface of the gel in order to minimize the dead volume. A 1-g amount of the oligosaccharide mixture was dissolved in 3 ml of water and applied to a column through the injection equipment. The separation was carried out at a column temperature of 65°C and a pressure of 6 kp/cm$^2$, and a Technicon AutoAnalyzer with orcinol–sulphuric acid was used for the detection. For analytical purposes, a Duran-glass tube (127 × 1.5 cm) with not more than 20 $\mu$l of the sample was used. The columns were coated by rinsing with a 1% solution of dichloromethylsilane in benzene so as to prevent wall effects (John *et al.*).

The same column, with distilled water as the mobile phase, was used for the separation of the lower members of the fructosane series, which has members from sucrose to inulin (mol. wt. *ca.* 5000) with one fructosyl unit difference between the subsequent members (Pontis).

The separation of D-glucose and the cellodextrin series up to cellohexaose, using water for elution, was performed on both Bio-Gel P-2 and Sephadex G-15 (Brown, 1970a). An interaction between the gel and the solute, which increased with molecular weight, was observed, especially when Sephadex G-15 was used. Sephadex G-15, pre-treated with 1 *M* hydrochloric acid, proved to be valuable in the separation of a blue dextran–stachyose–maltose–D-glucose mixture, when water was used as the mobile phase (Goodson and DiStefano).

Gels can also function as the support for partition chromatography. Zeleznick resolved a mixture of L-rhamnose, 2-acetamido-2-deoxy-D-glucose, D-glucose and 2-amino-2-deoxy-D-glucose on the column packed with highly cross-linked Sephadex G-25, using *n*-butanol–acetic acid–water (62:15:25, v/v) as the mobile phase.

In some instances, an ion-exchange resin can function like a molecular sieve. Thus, for instance, Saunders used Dowex 50W-X4 ($K^+$; 200–400 mesh) for the separation of the saccharides listed in Table 22.6.

The resin (120 × 4.5 cm) was used without regeneration between experiments. Sugars (150 mg each) were applied in about 5 ml of water and chromatographed with water as the mobile phase, and fractions of 5 ml at a flow-rate of 0.7–0.8 ml/min were collected and tested by the phenol–sulphuric acid method. In all experiments, the yield of a recovered sugar was not less than 95% (Saunders).

From Table 22.6, it is obvious that the fractionation of oligosaccharides, and the separation of oligosaccharides from the hexoses group and the pentoses group, is determined by the molecular size; however, the differences in the $K_d$ values of particular sugars within an individual group give some evidence that also other factors affect the elution pattern.

In similar work by Barker *et al.* (1969a), the chromatographic behaviour of various sugars and sugar derivatives on the cation-exchange resin Bio-Rad AG 50W-X2, -X4 and -X8 ($Li^+$, $Ba^{2+}$ and $Ca^{2+}$) was studied. It can be seen from Table 22.7 that when Bio-Rad

TABLE 22.6
SEPARATION OF VARIOUS SUGARS ON DOWEX 50W-X4 ($K^+$; 200–400 MESH) (SAUNDERS)

Column, 120 × 4.5 cm; mobile phase, water; flow-rate, 0.7–0.8 ml/min.

| Sugar | Distribution constant | Sugar | Distribution constant |
|---|---|---|---|
| Stachyose | 0.27 | D-Ribose | 0.85 |
| Raffinose | 0.33 | D-Xylose | 0.67 |
| Sucrose | 0.39 | 2-Deoxy-D-*erythro*-pentose | 0.68 |
| D-Glucose | 0.58 | Methyl-α-D-glucoside | 0.51 |
| D-Galactose | 0.66 | Methyl-β-D-glucoside | 0.48 |
| D-Mannose | 0.68 | Methyl-α-D-mannoside | 0.67 |
| D-Fructose | 0.69 | 1,2-O-Isopropylidene-α-D-glucofuranose | 0.66 |
| L-Rhamnose | 0.55 | 1,2:5,6-Di-O-isopropylidene-α-D-glucofuranose | 0.85 |
| L-Arabinose | 0.78 | 2,3:4,5-Di-O-isopropylidene-β-D-fructopyranose | 0.89 |
| D-Lyxose | 0.71 | | |

TABLE 22.7
FRACTIONATION OF VARIOUS CARBOHYDRATES ON BIO-RAD AG 50W-X2 ($Li^+$; 200–400 MESH) (BARKER *et al.*, 1969a)

Column, 150 × 0.6 cm; mobile phase, water; flow-rate, 0.23 ml/min; temperature, 25°C.

| Compound | Retention factor | Compound | Retention factor |
|---|---|---|---|
| Amylopectin | 0.37 | Laminaritriose | 0.61 |
| Dextran 80 | 0.37 | Laminaribiose | 0.70 |
| $Li^+$ | 0.37 | Maltose | 0.70 |
| N-Acetylneuraminic acid | 0.37 | D-Glucose | 0.78 |
| D-Gluconic acid | 0.37 | D-Mannose | 0.78 |
| Glyceric acid | 0.37 | D-Galactose | 0.78 |
| Maltohexaose | 0.55 | L-Fucose | 0.78 |
| Maltohexaitol | 0.55 | Galactitol | 0.78 |
| Laminaripentaose | 0.59 | 2-Acetamido-2-deoxy-D-glucose | 0.78 |
| Raffinose | 0.61 | 2-Acetamido-2-deoxy-D-mannose | 0.78 |

AG 50W-X2 ($Li^+$) is eluted with water, neutral saccharides are separated according to molecular size only; polysaccharides emerge from the column earlier than oligo- and monosaccharides. No resolution within a group of sugars of similar molecular weights was observed, even for aldose–alditol pairs. It is of interest that acidic carbohydrates such as N-acetyl-neuraminic acid and D-glucuronic acid were eluted in the same position as compounds of higher molecular weight, apparently due to the ion-exclusion principle.

A comparison of several methods for the chromatography of the oligomeric sugars of the β-(1→4)-linked D-xylose series (Havlicek and Samuelson) revealed that for preparative purposes permeation chromatography on polyacrylamide gel is inferior to the charcoal–Celite method and to partition chromatography on ion-exchange resins. In the separation of lower homologues (up to five residues), however, gel permeation chromatography on ion-exchange resins in water was found to give the best results. The charcoal–Celite method was the least satisfactory for analytical purposes; on the other hand, this method was useful for preparative treatment in combination with LLC on ion-exchange resins, because a large amount can be handled and stepwise elution can be applied without disturbances. A disadvantage of partition chromatography is the use of high temperatures for the volatile ethanol–water solvent system; in addition, the reversed relationship between molecular weight of the polysaccharide and its solubility in ethanol limits the use of this solvent system to lower saccharides (Kesler). Disturbances due to the formation of ethyl glycosides occur above 75°C with 2-deoxyaldoses, and above 90°C with some others (Samuelson). Decomposition of ketoses and sucrose was observed when organic-base counter-ions were used (Hobbs and Lawrence, 1972b). The borate method may be more difficult for preparative treatment. Sometimes, transformations of sugars during the borate ion-exchange chromatography were observed (Carubelli, Kesler).

## AMINO SUGARS

Virtually the only means of separating mixtures of amino sugars and/or complex mixtures of amino sugars and amino acids is ion-exchange chromatography. Elution from the ion-exchange resin column is performed with dilute hydrochloric acid or with various buffers. For this purpose, amino acid analyzers have been often used, the column effluents being assayed by the reaction with ninhydrin (see p. 482).

### Free amino sugars

*Ion-exchange chromatography with dilute hydrochloric acid as the mobile phase*

Hydrochloric acid was used as the mobile phase first by Gardell and later by other workers (Crumpton, Smith). Crumpton examined the effect of the normality of hydrochloric acid on the separation of D-glucosamine and D-galactosamine on Zeo-Karb 225 resin. He found that, within the range 0.2–0.9 *M*, the best resolution of the mixture was obtained with 0.33 *M* hydrochloric acid. The $R_{\text{glucosamine}}$ values of several amino sugars and their derivatives are listed in Table 22.8.

The same mobile phase proved to be effective even for the isolation of 3-amino-3,6-dideoxy-D-glucose from the hydrolyzate of the phenol-soluble lipopolysaccharide *C. freundii* 8090 (Raff and Wheat). 3-Amino-3,6-dideoxy-D-glucose was separated from D-glucosamine on a Dowex 50 ($H^+$) resin column. The chromatographic properties of the above amino sugar compared with those of other known 3-amino-3,6-dideoxyhexoses (see Table 22.9) supported the proposed configuration.

*Ion-exchange chromatography in buffered systems*

An exhaustive study of the cation-exchange chromatography of amino sugars was made by Brendel *et al.* (1967a). They examined the behaviour of 29 amino sugars by means of

TABLE 22.8

$R_{glucosamine}$ VALUES OF AMINO SUGARS AND THEIR DERIVATIVES (CRUMPTON)

Zeo-Karb 225 ($H^+$; 8% cross-linkage, particle size > 200 mesh); column, 43 × 0.8 cm; mobile phase, 0.33 *M* hydrochloric acid; flow-rate 2 ml/h.

| Compound | $R_{glucosamine}$ |
|---|---|
| D-Glucosaminuronic acid | 0.71 |
| Isomuramic acid | 0.87 |
| D-Glucosamine | 1.00 |
| D-Mannosamine | 1.07 |
| Muramic acid | 1.10 |
| D-Gulosamine | 1.20 |
| D-Galactosamine | 1.20 |
| D-Allosamine | 1.23 |
| D-Xylosamine | 1.43 |
| D-Fructosamine | 1.50 |
| D-Talosamine | 1.60 |
| D-Fucosamine | 1.94 |

TABLE 22.9

$R_{glucosamine}$ VALUES OF SOME 3-AMINO-3,6-DIDEOXY HEXOSES (RAFF AND WHEAT)

Dowex 50 ($H^+$); column, 50 × 1 cm; mobile phase, 0.33 *M* hydrochloric acid.

| Compound | $R_{glucosamine}$ |
|---|---|
| 3-Amino-3,6-dideoxy-L-glucose | 1.37 |
| 3-Amino-3,6-dideoxy-D-mannose | 1.28 |
| 3-Amino-3,6-dideoxy-D-idose | 1.44 |
| 3-Amino-3,6-dideoxy-D-galactose | 1.58 |
| 3-Amino-3,6-dideoxy-L-talose | 2.37 |

a Technicon amino acid analyzer, using Chromobeads B resin (125 cm × 0.35 $cm^2$) and pyridine–acetate buffers at 35°C. It was found that the amino sugars examined emerged from the column in the following order: muramic acids, hexosaminuronic acids, hexosamines, deoxyhexosamines and methyl hexosaminides, diaminohexoses. Satisfactory resolution of amino sugars was achieved only for some of the amino saccharides; a single, complete run for all of the amino sugars and their derivatives examined failed. An almost linear gradient between 0.1 *M* pyridine–acetate buffer of pH 2.8 (5.0 *M* acetic acid) and 0.133 *M* pyridine–acetate buffer of pH 3.85 (0.82 *M* acetic acid) proved to be successful for separating some hexosamines and their acid derivatives. Straight pyridine–acetate buffers of 0.1–0.2 *M*, with a pH between 3.85 and 4.15, separated four out of six 2-amino-2-deoxyhexoses examined and were useful even for the separation of 2-amino-2,6-dideoxyhexoses or 3-amino-3,6-dideoxyhexoses. 2,6-Diamino-2,6-dideoxyhexoses were partially resolved on a shorter column (20 cm), using 3.1 *M* pyridine–acetate buffer of pH 4.5.

Yaguchi and Perry described the separation of seven 2-amino-2-deoxy-D-hexoses by means of a Technicon NC-1 amino acid analyzer, using a column of Chromobeads B resin (133 × 0.6 cm) jacketted at 60°C. Borate–citrate buffer of pH 7.24 was used for the elution at a flow-rate of 0.5 ml/min and was prepared by adjusting a mixture of 77.9 g of boric acid, 11.8 g of sodium citrate dihydrate, 100 ml of 2 *M* sodium hydroxide, 40 ml of Brij-35 solution (100 g in 300 ml of water) and 200 μl of octanoic acid (preservative) to pH 7.24 at 25°C with 6 *M* hydrochloric acid, the final volume being 4 l.

Under similar conditions (Chromobeads A), Adams *et al.* separated all of the possible methyl ethers of 2-amino-2-deoxy-D-glucopyranose (see Fig. 22.15). Sodium citrate buffer of pH 4.40 was used for the elution; between the runs, the column was washed with 0.2 *M* sodium hydroxide for 30 min, and then regenerated with sodium citrate buffer of pH 3.10 (made in a similar manner to the pH 4.4 buffer except that the pH was adjusted to 3.1 with 6 *M* hydrochloric acid). This method could be of particular interest in the analysis of polysaccharides containing D-glucosamine.

Bella and Kim studied the separation of amino alditols and amino sugars, and overcame the difficulties that had occurred in previous work on this topic (Donald, Weber and Winzler). The chromatography was performed at 65°C in less than 4 h by means of a Beckman-Spinco Model 120 C amino acid analyzer equipped with a 56 × 0.9 cm column of Beckman UR-30 resin. A citrate–borate buffer of pH 5.06 (flow-rate 40 ml/h) was used for the elution and was prepared from 0.35 *M* sodium citrate buffer (pH 5.28) by adding 18.55 g/l of boric acid. The effluent was analyzed automatically (ninhydrin, flow-rate

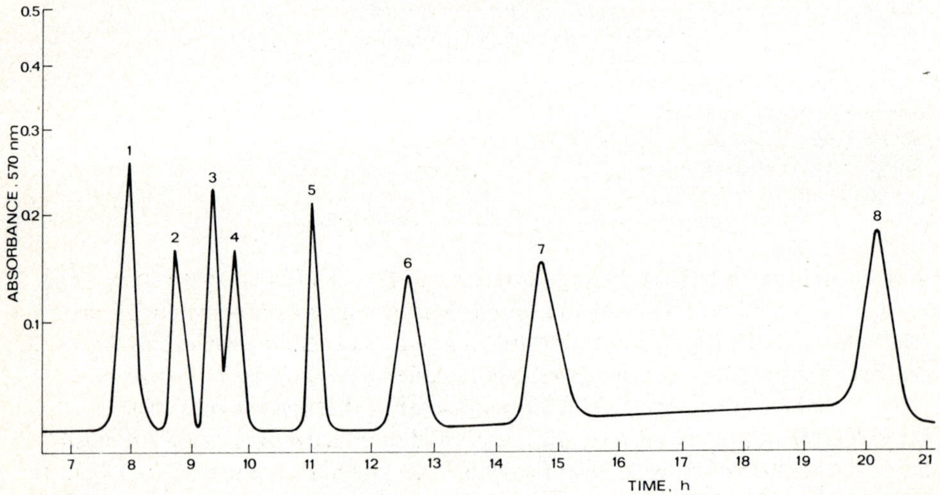


Fig. 22.15. Separation of methyl ethers of 2-amino-2-deoxy-D-glucopyranose (Adams *et al.*). Chromobeads A; column, 133 × 0.6 cm; mobile phase, sodium citrate buffer (pH 4.40); flow-rate, 0.5 ml/min; temperature, 60°C. 1 = 2-amino-2-deoxy-3-O-methyl-D-glucose; 2 = 2-amino-2-deoxy-4-O-methyl-D-glucose; 3 = 2-amino-2-deoxy-6-O-methyl-D-glucose; 4 = 2-amino-2-deoxy-D-galactose; 5 = 2-amino-2-deoxy-4,6-di-O-methyl-D-glucose; 6 = 2-amino-2-deoxy-3,4-di-O-methyl-D-glucose; 7 = 2-amino-2-deoxy-3,4,6-tri-O-methyl-D-glucose; 8 = 2-amino-2-deoxy-3,6-di-O-methyl-D-glucose.

20 ml/h). The following retention times were estimated (relative to D-glucosamine): D-galactosaminitol 0.72, D-glucosaminitol 0.76 and D-galactosamine 1.12.

**Mutual separation of amino sugars and amino acids**

Brendel's method (see p. 496) was extended to the separation of mixtures of amino sugars and of amino acids (Brendel *et al.*, 1967b). The effluent was assayed for amino groups (ninhydrin method) as well as for reducing properties (potassium hexacyanoferrate(III) method). It was concluded that, using pyridine–acetic acid buffers of decreasing acid content and of increasing pyridine molarity and pH, amino sugars and amino acids, $R-CH{<}^{COO^-}_{NH_3^+}$, emerged from the column in the following simplified order:

*R containing* $-SO_3H$, $-OSO_3H$, $-OPO_3H_2$ *groups*

secondary amino acids

muramic acids

*R containing –COOH groups*

*R containing alcoholic –OH groups*

*R = alkyl*

*R containing both* $-NH_2$ *and –COOH groups*

hexosaminuronic acids

*R containing aryl groups*

*R containing phenolic –OH groups*

monoamino sugars

*R containing* $-NH_2$ *groups*

ammonia

*R containing strongly basic groups*

diamino sugars

Efficient mobile phases were found only for the resolution of small groups of the substances examined. Thus, 0.1 *M* pyridine–acetic acid buffer of pH 2.8 (5.0 *M* acetic acid) proved to be particularly useful for separating the acidic derivatives of amino acids and amino sugars. A gradient of the pyridine–acetic acid buffers of pH 2.8 and 3.85, respectively, allowed the separation of acidic and neutral amino acids from muramic acid and hexosaminuronic acids, as well as glucosamine, galactosamine, quinovosamine and fucosamine. A single buffer consisting of 0.133 *M* pyridine–acetic acid (0.82 *M* acetic acid), with a pH of 3.85, gave a satisfactory separation of 2-amino-2-deoxy- and 3-amino-3-deoxyhexoses and their 6-deoxy derivatives without interference from any amino acids.

Fig. 22.16 illustrates the separation of a mixture containing D-glucosamine, D-galactosamine and various amino acids (Monsigny), carried out at 60°C on a Technicon amino acid analyzer equipped with nine compartments (Autograd) and a 65 × 0.6 cm column of Chromobeads C-2. Each of the buffers used contained 17.85 g/l of sodium citrate pentahydrate and 5 g/l of Brij-35 and was adjusted to the desired pH with 5.6 *M* hydrochloric acid. The buffer of pH 2.75 contained 10% of methanol and 0.5% of thiodiglycol, buffers of pH 2.875 and 3.80 contained 0.05% of thiodiglycol, and the buffer of pH 6.10 was 1 *M* in sodium chloride. The gradient used is shown in Table 22.10.

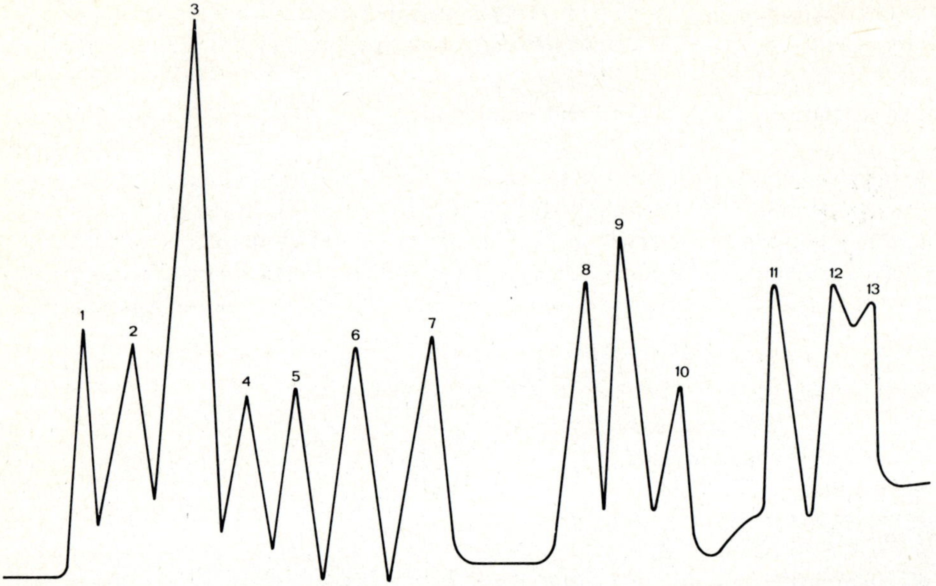


Fig. 22.16. Separation of a mixture of D-glucosamine, D-galactosamine and various amino acids (Monsigny). Chromobeads C-2; column, 65 × 0.6 cm; mobile phase, see Table 22.10; temperature, 60°C. 1 = Glycine; 2 = alanine; 3 = D-glucosamine; 4 = D-galactosamine; 5 = valine; 6 = cysteine; 7 = methionine; 8 = isoleucine; 9 = leucine; 10 = norleucine; 11 = tyrosine; 12 = phenylalanine; 13 = ammonia.

**Derivatives of amino sugars and chromatographic methods used in the synthesis of amino sugars**

The separation of amino sugars and their isolation from natural or synthetic mixtures can also be achieved in an indirect manner as their derivatives. The chromatographic behaviour of amino sugar derivatives is determined by the number and the nature of the substituents. It is generally true that if the amino group of an amino sugar is blocked, the chromatographic behaviour of the derivative formed no longer resembles that of the parent aminosaccharide. For instance, the chromatographic properties of N-acetylated amino sugars resemble those of free neutral saccharides rather than those of amino sugars. This phenomenon affects the choice of the chromatographic technique to be used for a particular separation. Whereas for free amino sugars or derivatives that still carry an amino group, cation-exchange chromatography is the method almost exclusively used, carbon column and cellulose column chromatography are of importance in the resolution of N-acetylated sugars (see, for instance, Gibbs *et al.*, Kuhn *et al.*, Perry and Webb). Less polar amino sugar derivatives were most frequently separated on silica gel (Albano and Horton, Jeanloz *et al.*) or on alumina (Čapek and Jarý).

In the synthesis of amino sugars, it is sometimes advantageous to separate the isomers already in form of precursors of amino sugars. For example, the nitromethane condensation of "dialdehyde" formed in the periodate oxidation of methyl α-L-rhamnopyranoside

TABLE 22.10
COMPOSITION OF THE BUFFER GRADIENT ENABLING THE SEPARATION OF D-GLUCOSAMINE, D-GALACTOSAMINE AND VARIOUS AMINO ACIDS (MONSIGNY)

See Fig. 22.16.

| Compartment No. | Volume (ml) | | | | | |
|---|---|---|---|---|---|---|
| | pH 2.75 | pH 2.875 | pH 3.80 | pH 6.10 | 2.5 *M* NaCl | Methanol |
| 1 | 38 | – | – | – | – | 2 |
| 2 | 12 | 28 | – | – | – | – |
| 3 | – | 32 | 5 | – | 3 | – |
| 4 | – | 25 | 10 | – | 5 | – |
| 5 | – | 10 | 25 | – | 5 | – |
| 6 | – | – | – | 40 | – | – |
| 7 | – | – | – | 40 | – | – |
| 8 | – | – | – | 40 | – | – |
| 9 | – | – | – | 40 | – | – |

produces a mixture of four methyl 3,6-dideoxy-3-nitro-α-L-hexopyranosides, 2.5 g of which is than applied on a 40 × 2 cm silica gel column; 100:1 benzene–ethanol elutes the isomer with the *talo*-configuration, 100:1.5 benzene–ethanol elutes the *manno*- and *gluco*-isomers (only partially resolved) and 100:3 benzene–ethanol elutes the *galacto*-isomer (Baer and Čapek). The alumina column chromatography of azido sugars has been reported, *e.g.*, by Guthrie and Murphy.

## SUGAR DERIVATIVES

### Alditols

Owing to the number of hydroxyl groups present in the molecule, the methods used for the separation of alditols resemble those for free sugars (see p. 483). In these methods, the most frequently used sorbents are carbon, cellulose and ion-exchange resins.

*Chromatography on charcoal*

A carbon column (50% Celite) proved to be satisfactory for the fractionation of the carbohydrate mixture obtained from extracts of *Umbilicaria pustulata* (L.) Hoffm., when stepwise elution with aqueous ethanol (1–15%) was used (Lindberg and Wickberg). Monitored by optical rotation measurements, fractions containing arabinitol, mannitol, α,α-trahalose, umbilicin and sucrose were collected.

*Chromatography on cellulose*

Richtmyer (1970a) reported the separation of a mixture of polyhydric alcohols, obtained from avocado seeds, on cellulose columns. For other applications, see Begbie and Richtmyer, Richtmyer (1970b, c), etc.

TABLE 22.11

VOLUME DISTRIBUTION CONSTANTS OF VARIOUS ALDITOLS, ALDOSES AND SIMPLE ALIPHATIC CARBONYL COMPOUNDS ON DOWEX 50W-X8 ($Li^+$, $Na^+$ AND $K^+$) AT 75°C AND VARIOUS ETHANOL CONCENTRATIONS (SAMUELSON AND STRÖMBERG, 1968)

| Compound | Volume distribution constants | | | | | | | | | | | | | | |
|---|---|---|---|---|---|---|---|---|---|---|---|---|---|---|---|
| | 80% ethanol | | | 85% ethanol | | | 90% ethanol | | | 95% ethanol | | | 97% ethanol | | |
| | $Li^+$ | $Na^+$ | $K^+$ | $Li^+$ | $Na^+$ | $K^+$ | $Li^+$ | $Na^+$ | $K^+$ | $Li^+$ | $Na^+$ | $K^+$ | $Li^+$ | $Na^+$ | $K^+$ |
| Ethylene glycol | 0.81 | 0.71 | 0.61 | 0.84 | 0.68 | 0.68 | 0.95 | 0.72 | 0.66 | 1.1 | 0.93 | 1.1 | 1.2 | 1.2 | 1.3 |
| Formaldehyde | 0.19 | 0.21 | 0.18 | 0.17 | 0.16 | | 0.16 | 0.13 | 0.16 | 0.17 | 0.10 | 0.25 | 0.16 | 0.10 | 0.25 |
| Glycolaldehyde | 0.37 | 0.45 | 0.45 | 0.37 | 0.41 | 0.48 | 0.37 | 0.40 | 0.48 | 0.39 | 0.48 | 0.74 | 0.40 | 0.67 | 0.95 |
| Glyceraldehyde | 0.52 | 0.71 | 0.82 | 0.59 | 0.70 | 0.92 | 0.67 | 0.75 | 1.1 | 0.80 | 1.3 | 2.1 | 0.87 | 1.8 | 2.9 |
| Glycerol | 1.2 | 1.0 | 1.0 | 1.4 | 1.2 | 1.1 | 1.7 | 1.4 | 1.5 | 2.2 | 2.5 | 3.0 | 2.7 | 3.6 | 4.3 |
| Erythrose | 1.0 | 1.4 | 1.6 | 1.1 | 1.7 | | 1.4 | 2.3 | | 1.8 | 4.2 | | 2.1 | 6.3 | |
| Erythritol | 1.6 | 1.4 | 1.6 | 2.1 | 1.8 | 1.8 | 2.9 | 2.7 | 2.7 | 4.3 | 5.4 | 6.5 | 5.5 | 9.0 | 10.5 |
| Ribitol | 2.1 | | | 3.0 | 2.6 | | 4.4 | 4.1 | 4.4 | | | | | | |
| Arabinitol | 2.5 | 2.5 | | 3.7 | 3.4 | 3.1 | 5.7 | 5.7 | 5.4 | | | | | | |
| Xylitol | 2.8 | 3.2 | | 4.1 | 4.5 | 4.1 | 6.4 | 7.9 | 7.6 | | | | | | |
| Mannitol | 3.5 | 3.5 | | 5.5 | 5.1 | 4.7 | 9.5 | 9.6 | 9.1 | | | | | | |
| Galactitol | 4.1 | | | 6.7 | 6.7 | | 11.8 | 13.1 | 11.6 | | | | | | |
| Glucitol | 3.8 | 4.5 | | 6.1 | 6.9 | 6.0 | 10.4 | 13.4 | 12.2 | | | | | | |
| Xylose | | 2.2 | | 1.7 | 2.9 | | 2.1 | 4.6 | | | | | | | |
| Arabinose | 1.7 | 3.0 | | 2.2 | 4.2 | | 2.8 | 6.9 | | | | | | | |
| Glucose | 2.0 | 3.6 | | 2.8 | 5.3 | | 3.8 | 9.7 | | | | | | | |
| Mannose | | 3.6 | | 2.8 | 5.3 | | 3.8 | 9.7 | | | | | | | |
| Galactose | | 4.5 | | 3.4 | 6.7 | | 4.7 | 12.5 | | | | | | | |

A mixture of polyhydric alcohols (8.4 g) was mixed with powdered cellulose and transferred to the top of a column (34 × 3.7 cm) containing washed cellulose powder (Whatman); 14-ml fractions were collected as soon as a sample of the eluate left a residue on evaporation. With water-saturated *n*-butanol–pure *n*-butanol (1:3) as the mobile phase, glycerol (fractions 1–50), D-arabinitol (126–175), galactitol (301–625), volemitol and a "perseitol–octitol" fraction were eluted. The last-mentioned fraction (5.9 g) could be re-chromatographed on a larger cellulose column (83 × 5 cm) by elution with water-saturated *n*-butanol–pure *n*-butanol (1:1), giving perseitol, D-*erythro*-D-*galacto*-octitol and *myo*-inositol (Richtmyer, 1970a).

### *Chromatography on ion-exchange resins*

The separation of alditols by LLC on ion-exchange resins has been studied in detail by Samuelson and Strömberg (1966). The successful resolution of alditols on the anion-exchange resin Technicon $T_5B$ ($SO_4^{2-}$) is illustrated in Fig. 22.17. The efficiency of resolution is dependent on the temperature and the exchange capacity of the resin. The great influence of the form of the cation-exchange resin on the distribution constant was described later by Samuelson and Strömberg (1968). It follows from Table 22.11 that the most favourable separations are achieved with the lithium form of the resin.

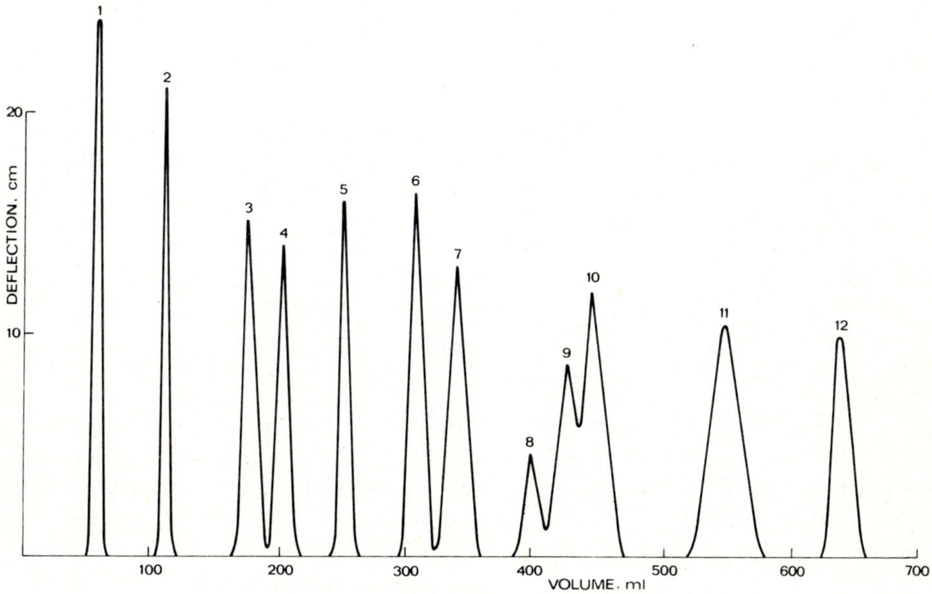


Fig. 22.17. Separation of various alditols and monosaccharides (Samuelson and Strömberg, 1966). Technicon $T_5B$ ($SO_4^{2-}$); resin bed, 85.2 × 0.6 cm; mobile phase, 86% ethanol; flow-rate, 2.51 ml/min · cm²; temperature, 75.5°C. 1 = glycerol; 2 = erythritol; 3 = xylitol; 4 = arabinitol; 5 = arabinose; 6 = xylose; 7 = glucitol; 8 = mannose; 9 = galactitol; 10 = mannitol; 11 = galactose; 12 = glucose.

Dowex 50W-X8 ($Li^+$) with a particle size of 14–17 $\mu$m and an exchange capacity of 5.1 mequiv./g of dry resin ($H^+$) in a water-jacketted glass column (100–150 cm length, 1.2–2.6 cm I.D.) at a flow-rate of 1–5 ml/min $\cdot$ cm$^2$ is used for separations such as that shown in Fig. 22.1. The chromatogram in Fig. 22.1 was obtained from a run at 75°C in 85% ethanol, which took *ca.* 5 h. Before this type of column is packed, the resin is slurried with boiled aqueous ethanol of the same concentration as that used as the mobile phase, and is kept in this solution so that all air bubbles disappear. After sedimentation, a concentrated slurry (1:2, v/v) is poured in the column and the mobile phase is pumped through until a uniform resin bed is formed. After the column has been packed, mobile phase should be circulated through it for at least 16 h before a chromatographic run is started. With the use of a column which is about 25 cm longer than the resin bed, the mobile phase can be pre-heated inside the column even when high flow-rates are used, so that a separate pre-heater can be omitted. The sample is applied on to the column as a solution in ethanol of the same concentration as that used as the mobile phase, in order to prevent swelling changes. If the temperature in the column is above the boiling-point of the mobile phase at atmospheric pressure, it is important to allow the temperature to decrease before removing the top fitting. A stainless-steel piston pump is used to feed the mobile phase into the column. The pressure is followed on a manometer equipped with a circuit-breaker that stops the pump and heating baths if the desired pressure (80 atm) is exceeded or if the pressure decreases because of leakage. Periodate oxidation is used for detection.

A frequently used method for the separation of alditols is based on the resolution of their borate complexes on strongly basic anion-exchange resins in the borate form. Thus, Spencer succeeded in separating a mixture of glycerol, threitol, erythritol, xylitol, arabinitol, ribitol, glucitol, galactitol and mannitol, using De Acidite FF ($BO_3^{3-}$), and 0.18 and 0.36 *M* boric acid (adjusted to pH 9 with triethylamine) at 35°C as the mobile phase. For micromolar amounts, a 60 × 0.8 cm column with a flow-rate of 25 ml/h was used. Rhamnitol and fucitol overlapped with arabinitol and glucitol, respectively. A substantial temperature dependence of the resolving power was observed.

## Glycosides

In the synthesis of glycosides, the most common task is the separation of the resulting mixture of anomers, and/or the removal of the unreacted starting sugar or its derivative. This problem is also very frequent in the Koenigs–Knorr and Fischer syntheses of glycosides and oligosaccharides.

### *Glycosides with simple aglycones*

The chromatography of these unsubstituted glycosides is similar to that of free monosaccharides, *i.e.*, LLC and IEC are now preferably used.

For partition chromatography, cellulose (for a review see Mowery) is still of great interest. Thus, for example, Yoshida *et al.* (1969b) resolved a mixture (*ca.* 10 g) of ethyl β-D-galactofuranoside, ethyl α-D-galactopyranoside and ethyl β-D-galactopyranoside (α-furanoside was not isolated) on a 80 × 4.5 cm column of Whatman CF-11 cellulose powder, with ethyl acetate–*n*-propanol–water (5:3:2) as the mobile phase. For detection, the optical rotation of the eluate was measured.

The most promising preparative method seems to involve the utilization of a strongly basic ion-exchange resin in the hydroxide form, such as Dowex 1-X2 ($OH^-$; 200–400 mesh); any free sugar present in the mixture is retained by the resin (Roseman *et al.*). Water is used as the mobile phase in such a separation, which is rapid and gives high recoveries; in general, furanosides are adsorbed more strongly than pyranosides. The efficiency of Dowex 2, which has a smaller particle size but a higher degree of cross-linkage (8%) than the above Dowex 1 resin, was not sufficient for the complete separation of methyl D-glucopyranosides (Austin *et al.*).

Dowex 1 resin (450 g; 2% cross-linked; 200–400 mesh) was recycled twice between the hydroxide and chloride forms with 2 *M* sodium hydroxide and 2 *M* hydrochloric acid. The resin was finally washed in an acrylic column (60 × 3 cm) with deionized, distilled, carbon dioxide-free water. The sample (1–3 g) was dissolved in 3 ml of water and washed on to the column with 1 ml of water. The column was then eluted with deionized, distilled, carbon dioxide-free water from a reservoir fitted with a soda-lime trap. After 400 ml of the mobile phase had been consumed (flow-rate 27–30 ml/h; peristaltic pump), 5-ml fractions were collected and monitored by optical rotation (Neuberger and Wilson).

The differences between the volume distribution constants of methyl α- and β-D-glucopyranoside, methyl α- and β-D-mannopyranoside and methyl α- and β-D-galactopyranoside were discussed in terms of the anomeric effect and other forms of dipole interactions by Neuberger and Wilson (see also p. 474). However, Evans *et al.* assumed from the influence of the $C_5$ substituent of the pyranose ring on the $K_d$ value that factors other than the sugar hydroxyl acidity can also be important in separations on resins in the hydroxide form.

It should be pointed out that the presence of an amino group in the molecule may cause the free aldose to be eluted from the resin, see, for example, the separation of methyl 3,6-diamino-3,6-dideoxy-α- and -β-D-glucopyranoside from 3,6-diamino-3,6-dideoxy-D-glucose on Dowex 1-X2 ($OH^-$) with water as the mobile phase (Kovář and Jarý).

With the chloride form of the resin and water as the mobile phase, the sorption probably involves Van der Waals and polar interactions (Evans *et al.*). For the separation of the α- and β-anomers of methyl D-glucopyranoside, methyl 6-deoxy-D-glucopyranoside and methyl 6-chloro-6-deoxy-D-glucopyranoside, 75% aqueous *n*-propanol was also used instead of water. The $K_d$ values decrease with changes in the substituent on the $C_5$ atom of the pyranose ring: $-CH_2OH > -CH_2Cl > -CH_3$. As the order of elution is the same with the chloride and the hydroxide forms of the resin, Evans *et al.* concluded that the separation is mainly due to partition in both instances.

These methods can also be used for analytical purposes by combining them with an automated analytical procedure. However, GLC of the trimethylsilyl derivatives seems to be more efficient (see for instance Evans *et al.*; Smirnyagin and Bishop; Yoshida *et al.*, 1969a).

*Complex glycosides*

With complex glycosides, a bulky aglycone becomes one of the dominating factors influencing the chromatography, so that besides LLC even GPC can be used (Table 22.12). By this means, the separation of sucrose, D-glucose and their phenolic glycosides from the bark and the leaves of *Populus tremula* was achieved by gel filtration on Sephadex G-25 and LH-20, using water as the mobile phase (Repaš and Nikolin; Repaš *et al.*). However, in synthetic applications (Koenigs–Knorr synthesis, etc.) the separation is usually carried out before the removal of the screening groups, by means of LSC; the solvent system chosen usually follows from TLC, which is used also for the detection. For several examples, see Table 22.12.

The same is also true for internal glycosides and all anhydro-sugars. The use of cellulose or ion-exchange resins for free anhydro derivatives, and silica gel for substituted derivatives, is advisable (Table 22.12).

## Ethers and acetals

For the separation of di-, tri- and tetra-O-methyl derivatives of aldoses, ketoses and their glycosides, respectively, silica gel, alumina, cellulose and Celite were used (see Table 22.13). A large number of applications of these methods in the analysis of natural compounds were listed in a book by Lederer; recently, the GLC–MS method, applied after reduction of the compounds to the corresponding alditols and acetylation or trifluoroacetylation, has become the most widely used technique for analytical purposes.

Kefurt *et al.* described the separation of a methylation mixture (1.6 g) of methyl 4,6-dideoxy-α-D-*xylo*-hexopyranoside on a silica gel column (90 g, 40 × 2.8 cm, 70–200 μm) with benzene–ethanol mixtures as the mobile phase. After a volume of 1800 ml (in 100-ml fractions) of the effluent (0.5% ethanol) had been discarded, 50-ml fractions were collected and the di-O-methyl derivative (0.5% ethanol, fractions 19–22), 3-O-methyl ether (0.5%; fractions 30–35), 2-O-methyl ether (1.0%; fractions 45–50) and starting diol (2.0%; fractions 61–70) were eluted. For the separation of an analogous mixture with the *lyxo* configuration, the use of an alumina column resulted in better resolution; with silica gel, the mono-O-methyl ethers overlapped to a serious extent.

For the separation of benzyl ethers, LSC on silica gel can always be used (Table 22.13). Good resolution can be obtained even when the solutes are very similar in structure, such as per-O-benzyl α,α-, α,β- and β,β-trehalose, followed by 2,3,4,6-tetra-O-benzyl-D-glucopyranose (Micheel and Pick). For analytical purposes, *ca.* 20 mg of the mixture were eluted from a 15 x 1.2 cm column of Merck silica gel (0.08 mm) with water-saturated methylene chloride–ethyl acetate–methanol (100:1.3:0.75) at a flow-rate of 20 ml/h. The effluent, containing the above compounds in the order given, was monitored continuously for benzyl ethers with a UV photometer at a wavelength of 255 nm; one run required *ca.* 2 h.

By blocking the hydroxyl groups of carbohydrates with acetal groups, *e.g.*, with isopropylidene, benzylidene or ethylidene groups, one can achieve a significant increase in the hydrophobicity of the molecule; thus, LSC is most frequently used for the separation of such compounds (see Table 22.13). In some instances, the resolution of diastereoisomeric -ylidene derivatives can also be achieved, *e.g.*, from the diastereoisomeric benzylidene derivatives of methyl 4,6-O-benzylidene-2,3-di-O-methyl-α-D-glucopyranoside, the "unusual" (S) isomer being eluted first (silica gel; 4:1 benzene–diethyl ether; Bagget *et al.*).

Sometimes, alumina is used with an aqueous solvent: see, for instance, Bonner *et al.*, who separated 2,3-O-butylidene-D-glucitol from D-glucitol with 93–100% aqueous ethanol as the mobile phase.

### Esters

In the synthesis of sugar esters, the starting compound and the product usually differ substantially in their chromatographic mobilities, so that their separation by LSC is not difficult. However, the separation of mixtures arising from the partial acylation of sugars or their derivatives requires more care. Here, the problem of the quantitative separation of substances, starting with per-O-acylated derivatives, followed by different incompletely substituted derivatives and ending with unreacted starting material, is usually solved by utilizing gradient elution.

In the chromatography of a partial acetylation mixture of methyl 3-acetamido-3,6-dideoxy-α-D-mannopyranoside on silica gel (70–200 μm), 100:3 benzene–ethanol first elutes the 2,4-di-O-acetyl derivative. Then the 4-O-acetate, followed by the 2-O-acetate, are obtained with 20:1 benzene–ethanol, and for the elution of the starting compound, 10:1 benzene–ethanol should be used (Čapek *et al.*, 1968b). These solvent systems were successfully used in the separation of analogous mixtures of other configurations (Čapek *et al.*, 1967, 1968a, 1970b). Some other applications are given in Table 22.13.

As the $K_d$ values of positional isomers differ only slightly, the choice of an appropriate solvent system and sorbent is very important because these systems affect the $K_d$ ratios of the different isomers. Thus, 2,3-di-O-tosyl-, 2-O-tosyl-, 3-O-tosyl- and methyl 4,6-O-benzylidene-β-D-glucopyranoside, the first two being virtually unresolved in chloroform and 20:1 chloroform–methanol on alumina (Guthrie *et al.*), can be separated without difficulty with 10:1 benzene–diethyl ether on silica gel (Staněk *et al.*, 1974). The same is true for the corresponding benzoates.

The fact that alkaline alumina may sometimes cause deacetylation (see p. 469) was used in the partial deacetylation of various per-O-acetyl derivatives; alumina served simultaneously as the reagent and as the sorbent (Jarý *et al.*).

Recently, GLC (particularly when combined with MS) of volatile derivatives has become more frequently used in analytical studies of partially acylated derivatives of saccharides (see, *e.g.*, De Belder and Norrman).

### Sugar acids

The method most widely used for the analysis and separation of uronic, aldobiuronic, aldonic and other sugar acids is based on anion-exchange chromatography. This method, in connection with carbon–Celite or cellulose chromatography, is also used for the isolation of these acids from natural sources.

TABLE 22.12
SEPARATION OF GLYCOSIDES AND THEIR DERIVATIVES

| Compounds separated | Sorbent | Mobile phase | Reference | Note |
|---|---|---|---|---|
| Anomeric butyl 2-acetamido-2-deoxy-3,4,6-tri-O-methyl-D-glucopyranosides | Silica gel, 0.05–0.2 mm (Merck) | Dichloromethane–diethyl ether–methanol (20:10:1) | Salo and Fletcher | The α-anomer was eluted first |
| Methyl [methyl 3,4-O-isopropylidene-2-O-(2,3,4,6-tetra-O-acetyl-β-D-glucopyranosyl)-α-D-galactopyranosid] uronate<br>Corresponding α-D-(1→2)-linked disaccharide | Silica gel, 75–250 mesh (Merck) | Benzene–acetone (8:2) | Šipoš and Bauer | Dry column method was used |
| Benzyl 2-O-(2,3,5-tri-O-benzoyl-α-L-arabinofuranosyl)-3,4-O-isopropylidene-β-L-arabinopyranoside<br>1,3,5-Tri-O-benzoyl-β-L-arabinofuranose<br>2,3,5-Tri-O-benzoyl-α-L-arabinofuranosyl bromide<br>Benzyl 3,4-O-isopropylidene-β-L-arabinopyranoside<br>Benzyl 2-O-(2,3,5-tri-O-benzoyl-α-L-arabinofuranosyl)-β-L-arabinopyranoside | Silica gel | Benzene–ethyl acetate (1:1) | Chalk *et al.* | |
| 2-O-Acetyl-1,6-anhydro-3-O-(3,4,6-tri-O-benzoyl-2-deoxy-2-dichloroacetamido-β-D-glucopyranosyl)-β-D-galactopyranose<br>Corresponding 1→4-linked disaccharide | Silica gel, Davison grade 950, 60–200 mesh | Dichloromethane–diethyl ether (20:3) | Shapiro *et al.*, 1967 | |

| | | | | |
|---|---|---|---|---|
| 1,5-Anhydro-2,3-O-isopropylidene-β-D-lyxofuranose<br>Methyl 2,3-O-isopropylidene-β-L-ribofuranoside<br>Methyl 2,3-O-isopropylidene-D-lyxoside<br>Methyl 2,3-O-isopropylidene-α-L-ribofuranoside | Silica gel (15 × 2.5 cm) | Toluene–acetone (17:3) | Brimacombe *et al.*, 1968b | |
| α- and β- anomers of 3β-(2-deoxy-D-*lyxo*-hexopyranosyloxy)-14β-hydroxy-5β-card-20(22)-enolide | Silica gel (50–200 μm), 220 g pre-mixed with 100 ml of water | Satd. aq. ethyl acetate | Zorbach *et al.* | |
| 1,2-O-Ethylene-α-D-glucofuranose<br>1,2-O-Ethylene-β-D-glucopyranose<br>1,2-O-Ethylene-α-D-glucopyranose<br>2-O-(2-Hydroxyethyl)-D-glucose | Whatman No. 1 cellulose (69 × 7 cm) | Satd. aq. *n*-butanol | Srivastava *et al.* | |
| Sucrose, D-glucose, salicin, tremuloidin | Sephadex G-25 (106 × 1 cm) | Water | Repaš and Nikolin | |
| 1,4-Anhydro-D-altritol and 1,4-anhydro-D-mannitol<br>3,6-Anhydro-D-altritol<br>1,5-Anhydro-L-glucitol | Whatman (standard grade) cellulose (50 × 4 cm) | Acetone–water (9:1) | Buchanan and Edgar | 1 g of the sample was applied |
| 1,5-Anhydro-L-glucitol<br>1,5-Anhydro-D-altritol | Dowex 1-X2 ($OH^-$) (50 ml) | Water | Buchanan and Edgar | 0.4 g of the sample was applied |
| 1,6-Anhydro-4-deoxy-β-D-*ribo*-hexopyranose<br>4-Deoxy-D-*ribo*-hexose | Cellulose (150 g) | Satd. aq. *n*-butanol | Černý *et al.* | 1 g of sample |

*References p.519*

TABLE 22.13
SEPARATION OF ETHERS, ACETALS AND ESTERS

| Compounds separated | Mobile phase | Sorbent | Reference |
|---|---|---|---|
| Methyl 3,6-anhydro-2-O-methyl-β-D-galactopyranoside<br>Methyl 2,3,6-tri-O-methyl-β-D-galactopyranoside<br>Methyl 2,3-di-O-methyl-β-D-galactopyranoside | Ethyl acetate<br>Ethyl acetate<br>Methanol | Silica gel | Brimacombe and Ching |
| Mixture of methyl terminal-4-O-methylmalto-oligosaccharides | Water–ethanol (1–15% ethanol) | Deactivated carbon–Celite (1:1) | BeMiller and Wing |
| Various methyl ethers of D-galactose and L-arabinose | Light petroleum–*n*-butanol (7:3) satd. with water | Cellulose | Anderson and Cree |
| Benzyl alcohol<br>Benzyl 2,3,4-tri-O-benzyl-β-D-ribopyranoside<br>2,3,4-Tri-O-benzyl-D-ribose | Benzene–diethyl ether (9:1) | Silica gel | Tejima *et al.* |
| Benzyl ethers of 1,6-anhydro-β-D-glucopyranose | Chloroform–ethyl acetate (4:1) | Silica gel | Seib |
| 1,2:3,5-Di-O-isopropylidene-α-D-apio-L-furanose<br>1,2:3,5-Di-O-isopropylidene-α-D-apio-D-furanose | Diethyl ether–*n*-hexane (1:1) | Silica gel | Ball *et al.* |
| 3-Deoxy-1,2:5,6-di-O-isopropylidene-D-*xylo*-hexofuranose<br>3-Deoxy-3-fluoro-1,2:5,6-di-O-isopropylidene-α-D-galactofuranose | Benzene–diethyl ether mixtures | Silica gel | Brimacombe *et al.*, 1968a |
| 1,2:3,4-Di-O-isopropylidene-6-O-(methylthio)methyl-α-D-galactopyranose<br>6-O-Acetyl-1,2:3,4-di-O-isopropylidene-α-*D*-galactopyranose | Benzene–diethyl ether (9:1) | Silica gel | Godman and Horton |

| | | | |
|---|---|---|---|
| 1-C-Cyclohexyl-2,3:4,5-di-O-isopropylidene-D-*gluco*-pentitol | Diethyl ether–light petroleum (3:17) | Silica gel | Inch *et al.* |
| 1-C-Cyclohexyl-2,3:4,5-di-O-isopropylidene-D-*manno*-pentitol | | | |
| 3-O-Benzyl-6,7-dideoxy-1,2-O-isopropylidene-α-D-*gluco*-hept-6-ynofuranose | Dichloromethane–diethyl ether (3:1) | Silica gel | Horton and Swanson |
| 3-O-Benzyl-6,7-dideoxy-1,2-O-isopropylidene-β-L-*ido*-hept-6-ynofuranose | | | |
| 2,4-Di-O-acetyl-1,6-anhydro-β-D-glucopyranose | Methylene chloride–ethyl acetate (1:1) | Silica gel | Shapiro *et al.*, 1970 |
| 3,4-Di-O-acetyl-1,6-anhydro-β-D-glucopyranose | | | |
| 2,3-Di-O-acetyl-1,6-anhydro-β-D-glucopyranose | | | |
| Octa-O-acetyl-β-maltose | Benzene–ethyl acetate (1:1) | Silica gel | Dick *et al.* |
| 1,2,6,2′,3′,4′,6′-Hepta-O-acetyl-β-maltose | | | |
| Methyl 2,3-di-O-acetyl-4,6-O-benzylidene-β-D-glucopyranoside | *n*-Hexane–chloroform (1:1) | Silicic acid | Tulloch and Hill |
| Methyl 2-O-acetyl-4,6-O-benzylidene-β-D-glucopyranoside | (1:1) | | |
| Methyl 3-O-acetyl-4,6-O-benzylidene-β-D-glucopyranoside | Chloroform | | |
| Methyl 4,6-O-benzylidene-β-D-glucopyranoside | Chloroform | | |

*References p.519*

*Uronic acids*

The first successful separation of uronic acids on anion exchangers was reported by Khym and Doherty, who separated galacturonic acid and glucuronic acid on Dowex 1 ($CH_3COO^-$) with 0.15 *M* acetic acid as the mobile phase. Free saccharides (arabinose and galactose), which were not adsorbed, were collected in the first fraction. Larsen and Haug described the separation of glucuronic acid and mannuronic acid from each other and from a partially resolved mixture of guluronic and galacturonic acids on a Dowex 1-X8 ($CH_3COO^-$) column (45 × 2 cm), with a linear gradient of acetic acid (from 0.5 to 2.0 *M*) at a flow-rate of 0.3–0.5 ml/min.

Johnson and Samuelson successfully separated a mixture of 4-O-methyl-D-glucuronic, D-galacturonic, L-guluronic, D-glucuronic and D-mannuronic acids at 30°C on a Dowex 1-X8 column (88 × 0.6 cm), using as the mobile phase 0.05 *M* sodium acetate solution buffered with acetic acid at pH 5.9 (flow-rate 1.06 ml/min). The effluent was analyzed with a two-channel detector (carbazole and chromic acid methods). In another paper, Carlsson and Samuelson (1970) established the distribution constants of various uronic acids (Table 22.14). A four-channel detector (see p. 481) was used for the analysis of the effluent.

The separation of oligogalacturonic acids (products of the enzymatic hydrolysis of pectic acid) was reported by several workers. Derungs and Deuel separated a mixture of mono-, di-, tri- and tetragalacturonic acids on Dowex 3 ($HCOO^-$) with a formic acid gradient (0.1–2.5 *M*). Reid separated galacturonic acid and di- and trigalacturonic acids on De Acidite FF resin ($HCOO^-$) with a linear formic acid gradient (0.2–0.5 *M*).

Nagel and Wilson resolved a series of oligogalacturonic acids, including di-, tri-, tetra-, penta-, hexa-, hepta- and octagalacturonic acids, on a Dowex 1-X8 ($HCOO^-$) column (100 × 3 cm), using 8 l of 0.2–0.9 *M* sodium formate (linear gradient). The carbazole method was used to analyze the effluent. Nagel and Wilson also described the separation of unsaturated di-, tri-, tetra- and pentagalacturonic acids, which was performed under similar conditions. However, stepwise elution was preferred in this separation. For instance, for the resolution of this mixture of unsaturated acids, the sodium formate elution scheme was as follows: 0.06 *M*, 1.5 l; 0.08 *M*, 1.5 l; 0.1 *M*, 1.5 l; 0.2 *M*, 0.9 l; 0.3 *M*, 1.5 l; 0.4 *M*, 3 l; 0.5 *M*, 2.8 l; 0.6 *M*, 2 l; and 0.7 *M*, 2 l. The unsaturated acids were detected by measuring the absorbance at 232 nm.

The formation of sugar–borate complexes was used by Hallén for the separation of galacturonic and glucuronic acids on Dowex 2-X8 resin. Mannose, fucose, galactose and glucose were also present; these neutral sugars were eluted at room temperature with 0.01 *M* borax in 0.2 *N* sodium hydrogen carbonate, and, after the appearance of the last neutral saccharide peak, the uronic acids were eluted with 0.03 *M* borax in 0.6 *M* sodium hydrogen carbonate (flow-rate 1.5 ml/min).

The analytical methods mentioned above were applied by Carlsson *et al.* (1969) to the separation of the acids obtained by the isomerization of D-glucuronic acid in neutral aqueous solution; D-*lyxo*-5-hexulosonic acid, D-alluronic acid, the unresolved mixture of D-altruronic and D-mannuronic acids, D-glucuronic acid and L-*ribo*-5-hexulosonic acid were eluted with 1 *M* acetic acid from Dowex 1-X8 resin. The unresolved mixture of D-altruronic and D-mannuronic acids was re-chromatographed with 0.08 *M* sodium acetate.

TABLE 22.14
VOLUME DISTRIBUTION CONSTANTS OF ALDOBIURONIC, HEXURONIC AND HEXULOSONIC ACIDS (CARLSSON AND SAMUELSON, 1970)

Dowex 1-X8 (24–27 μm); column, 76 × 0.6 cm; mobile phase, 0.5 *M* acetic acid, 1 *M* acetic acid or 0.08 *M* sodium acetate adjusted with acetic acid to pH 5.9; flow-rate 4.4 ml/min · cm²; temperature, 30° C.

| Acid | Volume distribution constant | | |
|---|---|---|---|
| | Acetic acid (0.5 *M*) | Acetic acid (1 *M*) | Sodium acetate (0.08 *M*) |
| 2-O-(α-D-Galactopyranosyluronic acid)-L-rhamnose | 8.4 | | 4.16 |
| 4-O-(α-D-Galactopyranosyluronic acid)-D-xylose | 12.0 | | |
| 6-O-(β-D-Glucopyranosyluronic acid)-D-galactose | 15.5 | | |
| 2-O-(4-O-Methyl-α-D-glucopyranosyluronic acid)-D-xylose | 13.2 | | |
| Cellobiuronic acid | 26.0 | | 6.0 |
| Alluronic acid | | 17.1 | |
| Galacturonic acid | | 11.0 | 9.4 |
| Glucuronic acid | | 22.4 | 12.8 |
| Guluronic acid | | 12.3 | |
| Mannuronic acid | | 18.6 | 14.2 |
| Taluronic acid | | 14.2 | |
| 4-O-Methyl-D-glucuronic acid | | 18.0 | 8.0 |
| *arabino*-5-Hexulosonic acid | | 13.2 | |
| *lyxo*-5-Hexulosonic acid | | 15.4 | |
| *ribo*-5-Hexulosonic acid | | 28.2 | 17.4 |
| *xylo*-4-Hexulosonic acid | | 22.3 | |

*Other sugar acids*

Samuelson and Thede reported the distribution constants of 16 aldonic acids and some other sugar acids, separated at 30°C on Dowex 1-X8 (Table 22.15). Acetic acid (1 *M*) or sodium acetate (0.08 *M*) buffered with acetic acid to pH 5.9 were used for elution. It can be seen that, within the aldonic acid series, acids that contain a greater number of hydroxyl groups appear in the sodium acetate effluent earlier than those with a smaller number of hydroxyl groups. This elution pattern, in which mannonic and D-*glycero*-L-*manno*-heptonic acids are exceptions, could be explained by the assumption that the hydrated ionic volumes have a predominating influence on the uptake of the anions, those with a smaller hydrated volume being held more firmly by the resin. The dissociation constants of particular aldonic acids probably determine their elution pattern in acetic acid. This mobile phase gives more favourable separation factors and is therefore more suitable for the fractionation of isomers.

A mixture of the non-volatile monoprotic acids isolated from a hydrolyzate of unbleached cotton (Larsson and Samuelson, 1969) was analyzed by the above method; 2-O-(4-O-methyl-

TABLE 22.15
VOLUME DISTRIBUTION CONSTANTS OF VARIOUS SUGAR ACIDS (SAMUELSON AND THEDE)

Dowex 1-X8 (26–32 μm); column, 135 × 0.6 cm; mobile phase, 0.5 *M* acetic acid or 0.08 *M* sodium acetate adjusted with acetic acid to pH 5.9; flow-rate, 2.8–7.7 ml/min · cm$^2$; temperature, 30°C.

| Acid | Volume distribution constant | |
|---|---|---|
| | Acetic acid | Sodium acetate |
| *Aldonic acids* | | |
| Glycolic | 18.6 | 14.8 |
| Glyceric | 20.2 | 11.9 |
| D-Erythronic | 18.8 | 10.4 |
| D-Threonic | 19.5 | 10.6 |
| D-Arabinonic | 14.2 | 8.89 |
| D-Lyxonic | 19.5 | 10.3 |
| D-Ribonic | 9.17 | 9.24 |
| D-Xylonic | 15.7 | 8.19 |
| D-Galactonic | 11.3 | 7.51 |
| D-Gluconic | 12.5 | 7.21 |
| D-Gulonic | 13.5 | 7.70 |
| D-Mannonic | 17.5 | 9.49 |
| D-Talonic | 6.25 | 7.37 |
| D-*glycero*-L-*manno*-Heptonic | 14.2 | 7.70 |
| D-*glycero*-D-*gulo*-Heptonic | 10.8 | 6.63 |
| 6-Deoxy-D-mannonic | 19.3 | 10.9 |
| *Aldobionic acids* | | |
| Cellobionic | 5.86 | 3.71 |
| Lactobionic | 5.09 | 3.15 |
| Maltobionic | 7.35 | 3.69 |
| Melibionic | 4.62 | 2.64 |
| *Methylated aldonic acids* | | |
| 2,3,5-Tri-O-methyl-D-galactonic | 5.38 | 2.64 |
| 3,5,6-Tri-O-methyl-D-gluconic | 7.12 | 3.90 |
| 2,3,4,6-Tetra-O-methyl-D-gluconic | 4.76 | 2.18 |
| *Saccharinic acids* | | |
| 2-Hydroxypropionic (lactic) | 15.1 | 13.8 |
| D-*threo*-2,3-Dihydroxybutyric | 16.9 | 12.0 |
| 2,4-Dihydroxybutyric | 14.6 | 11.7 |
| 3,4-Dihydroxybutyric | 3.39 | 9.31 |
| 2-Methyl-2,3-dihydroxypropionic | 14.4 | 11.1 |
| D-*threo*-2,4,5-Trihydroxyvaleric | 11.8 | 9.17 |
| α-D-Glucoisosaccharinic | 6.09 | 6.06 |
| β-D-Glucoisosaccharinic | 14.8 | 6.46 |
| α-D-Glucosaccharinic | 5.41 | 6.72 |
| α-D-Glucometasaccharinic | 6.84 | 7.16 |
| β-D-Glucometasaccharinic | 9.56 | 7.59 |

TABLE 22.15 (*continued*)

| Acid | Volume distribution constant | |
|---|---|---|
| | Acetic acid | Sodium acetate |
| *Uronic acids* | | |
| D-Galacturonic | 21.4 | 8.40 |
| D-Glucuronic | 44.1 | 11.7 |
| L-Guluronic | 24.0 | 10.7 |
| L-Iduronic | 29.9 | 12.7 |
| D-Mannuronic | 36.5 | 12.9 |
| *Aldehydo and keto acids* | | |
| Glyoxylic | 65.6 | 20.8 |
| Laevulinic | 3.64 | 13.2 |
| 2-Keto-D-gluconic | 95.0 | 13.7 |
| 5-Keto-D-gluconic | 38.8 | 13.9 |

α-D-glucopyranosyluronic acid)-D-xylose, cellobiuronic acid, 4-O-methylglucuronic acid, gluconic acid, galacturonic acid, anhydrosaccharinic acid, glucuronic acid and laevulinic acid were detected in the hydrolyzate.

An example of a synthetic application is the separation of epimeric pairs of 3-deoxy-D-hexulosonic acids resulting (together with pyruvate) from the condensation of oxalo-acetic acid with D-glyceraldehyde. The chromatography was performed on Dowex 1 ($HCOO^-$; 200–400 mesh, 75 × 3.8 cm) with a linear gradient of formic acid, 0.23–0.46 *M* (Portsmouth).

## Sugar phosphates

As with other sugar derivatives of an ionic character, ion-exchange chromatography is the most valuable method for the analysis of sugar phosphate mixtures and for their isolation from natural sources.

### *Ion-exchange chromatography of borate complexes*

Khym and Cohn first used the complexing of sugar phosphates with borate for their chromatographic separation; glucose-1-phosphate, glucose-6-phosphate, fructose-6-phosphate and ribose-5-phosphate were separated by this procedure. Fractionation was performed on Dowex 1 ($Cl^-$), using ammonia solution ($10^{-3}$ or $2.5 \cdot 10^{-3}$ *M*)–ammonium chloride ($2.5 \cdot 10^{-2}$ *M*) buffers with varying stepwise concentrations of borate ($10^{-2}$–$10^{-5}$ *M* potassium tetraborate) as the mobile phase.

Later, Lefebvre *et al.* separated a mixture of glucose-1-phosphate, galactose-1-phosphate, fructose-6-phosphate, fructose-1-phosphate, glucose-6-phosphate and fructose-1,6-diphosphate. This was first adjusted to pH 8 with ammonia solution and then applied to a 45 × 0.5 cm column of Dowex 1-X4 ($BO_3^{3-}$; 200–400 mesh). After washing the column with water,

*References p.519*

the sugar phosphates were eluted with a linear gradient (0.1–0.4 $M$) of triethyl ammonium tetraborate solution at a flow-rate of 1.0–1.5 ml/min). This solution was prepared by mixing a freshly made boric acid solution with triethylamine. For instance, 0.4 $M$ triethylammonium tetraborate was prepared by dissolving 99.2 g of boric acid and 112 ml of triethylamine in water and making the volume up to 1 l. The same mobile phase (linear gradient from 0.1 to 0.14 $M$) proved to be useful for the resolution of a mixture of N-acetylglucosamine-1-phosphate and N-acetylgalactosamine-1-phosphate.

Bedetti *et al.* used a concave gradient of ammonium chloride and potassium tetraborate for the separation of labelled compounds, listed in Fig. 22.18, on the anion-exchange resin Bio-Rad AG 1-X4 ($Cl^-$; 200–400 mesh). After applying the mixture of sugar phosphates, the resin bed (30 × 1 cm) was eluted with 100 ml of $10^{-3}$ $M$ ammonium hydroxide in order to remove free sugars that may be present. The concave gradient used for the elution of sugar phosphates consisted of $2.5 \cdot 10^{-3}$ $M$ ammonia solution + $2.5 \cdot 10^{-1}$ $M$ ammonium chloride in the reservoir (1.95 l in a 2.5-l erlenmeyer flask) and of $2.5 \cdot 10^{-3}$ $M$ ammonia solution + $5 \cdot 10^{-3}$ $M$ potassium tetraborate + $3 \cdot 10^{-2}$ $M$ ammonium chloride in the mixing chamber (2 l in a 4-l Mariotte bottle). The radioactivity in the effluent was monitored (flow-rate 0.5 ml/min). For preparative purposes, borate was removed from the eluted compounds by means of three or four evaporations with methanol, and ammonia by passing the solution down a 10 × 1 cm column of the resin Bio-Rad AG 50W-X4 ($H^+$). The substantial differences observed in the chromatographic behaviour of

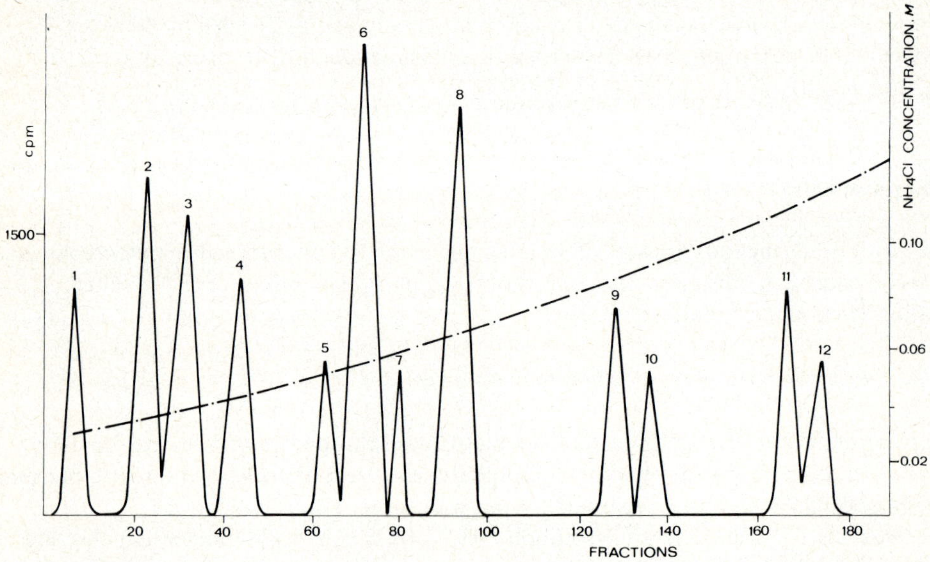


Fig. 22.18. Separation of some sugar phosphates on Bio-Rad AG 1-X4 ($Cl^-$; 200–400 mesh) (Bedetti *et al.*). Resin bed, 30 × 1 cm; mobile phase, concave gradient of ammonium chloride and potassium tetraborate; flow-rate, 0.5 ml/min; room temperature. 1 = glucose; 2 = lactate; 3 = pyruvate; 4 = glucose-1-phosphate; 5 = dihydroxyacetone phosphate; 6 = glucose-6-phosphate; 7 = glyceraldehyde-3-phosphate; 8 = fructose-6-phosphate; 9 = 3-phosphoglyceric acid; 10 = phosphoenolpyruvate; 11 = fructose-1,6-diphosphate; 12 = 2,3-diphosphoglyceric acid.

members of the same series (*e.g.*, glucose-1-phosphate, glucose-6-phosphate and glucose-1,6-diphosphate) are attributed to the resultant effect of p*K* values of the phosphate groups and the stability constants of the borate complexes. Moreover, the steric hindrance of the phosphate group can diminish the stability of the borate complex and thereby lower the binding of the sugar phosphate to the anion exchanger. This effect, together with the participation of hydroxyl groups arising from the formation of a furanose ring in reactions with boric acid, could be responsible for the higher mobility of aldose phosphates compared with ketose phosphates. The above method does not permit the chromatographic separation of 1,3-diphosphoglycerate because of its decomposition under the conditions used.

*Other ion-exchange separations*

Another very useful method for the separation of sugar phosphates was developed by Bartlett (1968a), who used Dowex 1-X8 ($HCOO^-$) and formate buffers as the mobile phase. The linear gradient of 0–5 *M* ammonium formate proved to be the most successful of the various formate buffers of different concentrations and pH values tested. This buffer was prepared by mixing four parts of 5 *M* formic acid with one part of 5 *M* ammonium formate (pH *ca.* 3 at 1 *M*). Table 22.16 lists some of the compounds examined together with their elution positions.

TABLE 22.16
ELUTION PATTERN OF SOME SUGAR PHOSPHATES (BARTLETT, 1968a)

Approx. 10 μmole each of known compounds were chromatographed on Dowex 1-X8 ($HCOO^-$; 100–325 wet mesh); column, 20 × 1 cm; mobile phase, 4 l of a linear gradient of 0–5 *M* ammonium formate (formic acid–ammonium formate, 4:1 v/v); flow-rate, 0.5–0.8 ml/min.

| Compound | Elution position* |
|---|---|
| Octulose-8-phosphate | 6 |
| Sedoheptulose-7-phosphate | 7 |
| Glucose-1-phosphate | 7.5 |
| Glucose-6-phosphate | 8 |
| Fructose-1-phosphate | 8 |
| Fructose-6-phosphate | 8.5 |
| Deoxyribose-5-phosphate | 8.7 |
| Ribose-5-phosphate | 9 |
| Ribulose-5-phosphate | 9.5 |
| Xylulose-5-phosphate | 9.7 |
| 6-Phosphogluconate | 18 |
| Octulose-1,8-diphosphate | 26 |
| 5-Deoxyoctulose-1,8-diphosphate | 26 |
| Glucose-1,6-diphosphate | 26 |
| Mannose-1,6-diphosphate | 27.3 |
| Sedoheptulose-1,7-diphosphate | 27.5 |
| Fructose-1,6-diphosphate | 29 |
| Ribose-1,5-diphosphate | 30 |
| Inositol-hexaphosphate | 100 |

*The figures give the centre of the elution position as a percentage of the total elution volume.

*References p.519*

This method has been found advantageous in the elucidation of biochemical problems. For example, it has been found that the metabolism of deoxyinosine and deoxyadenosine by fresh or stored erythrocytes caused the accumulation of large amounts of fructose diphosphate, triose phosphate and deoxyribose-5-phosphate; in addition, two unexpected intermediates, 5-deoxyxylulose-1-phosphate and 2-keto-5-deoxyoctulose-1,8-diphosphate, were determined (Bartlett, 1968b). For similar examples, see Bartlett (1968c), Bartlett and Bucolo, Itasaka, etc.

Some workers used dilute hydrochloric acid for the elution of sugar phosphates from anion-exchange resins. For example, Hashimoto and Yoshikawa separated a mixture of D-glucose-1-phosphate and D-glucose-1,6-diphosphate using Dowex 1 ($Cl^-$). The former compound was eluted with $1.5 \cdot 10^{-2}$ *M* hydrochloric acid, whereas the diphosphate was eluted with $10^{-1}$ *M* hydrochloric acid. When further purification of the latter compound was required, a linear gradient elution with ammonium formate by the method of Bartlett (1959) was applied.

Cosgrove used this method for the separation and identification of some inositol tetra- and pentaphosphates, formed during the hydrolysis of the four naturally occuring inositol hexaphosphates. For instance, when *myo*-inositol hexaphosphate was hydrolyzed at 110°C and pH 4 for 100 min, the resulting *myo*-inositol phosphate mixture was first resolved on Dowex AG 1-X2 ($Cl^-$), by gradient elution with hydrochloric acid, in four fractions: tetraphosphate, two peaks of pentaphosphate and hexaphosphate. Re-chromatography of each of the pentaphosphate fractions, using the same ion exchanger and 0.48 *M* hydrochloric acid, led to the isolation of all four possible pentaphosphates. One of these was shown to be identical with "bird-blood phytate", *i.e.*, with 1,3,4,5,6-*myo*-inositol pentaphosphate.

TABLE 22.17

DISTRIBUTION CONSTANTS OF *myo*-INOSITOL POLYPHOSPHATES ON SEPHADEX G-50 AND G-25 (100–300 μm) (STEWARD AND TATE)

Mobile phase, aqueous solutions of lithium chloride; room temperature.

| Compound | Molarity of mobile phase (*M*) | | | |
|---|---|---|---|---|
| | 0.01 | 0.20 | 0.01 | 0.10 |
| | Sephadex G-25 | | Sephadex G-50 | |
| *myo*-Inositol hexaphosphate | 0.06 | 0.19 | 0.45 | 0.61 |
| *myo*-Inositol tripyrophosphate* | 0.06 | 0.19 | 0.48 | 0.61 |
| *myo*-Inositol pentaphosphate* | 0.06 | 0.19 | 0.54 | 0.64 |
| *myo*-Inositol tetraphosphate** | 0.12 | 0.25 | 0.57 | 0.68 |
| *myo*-Inositol triphosphate** | 0.19 | 0.33 | 0.64 | 0.75 |
| *myo*-Inositol diphosphate** | 0.31 | 0.44 | 0.70 | 0.79 |
| *myo*-Inositol monophosphate | 0.44 | 0.59 | 0.86 | 0.86 |
| *myo*-Inositol | 0.75 | 0.75 | 0.83 | 0.86 |

*Johnson and Tate.

**Tomlinson and Ballou.

Recently, the successful use of GPC in the examination of inositol phosphates was reported by Steward and Tate. They established the distribution constants of several *myo*-inositol phosphates, listed in Table 22.17. It follows from Table 22.17 that, in general, the distribution constants decrease as the degree of phosphorylation of the inositol increases. However, the anion-exclusion effect contributes to the observed $K_d$ values of charged compounds to a substantial extent; this follows from the change in $K_d$ values when mobile phases of different molarity are used. The anion-exclusion effect can be diminished by increasing the concentration of the eluent to 2.0 *M*.

## REFERENCES


Adams, G. A., Yaguchi, M. and Perry, M. B., *Carbohyd. Res.*, 12 (1970) 267.
Albano, E. L. and Horton, D., *Carbohyd. Res.*, 11 (1969) 485.
Alfredsson, B., Gedda, L. and Samuelson, O., *Anal. Chim. Acta.*, 27 (1962) 63.
Alm, R. S., *Acta Chem. Scand.*, 6 (1952) 1186.
Anderson, D. M. W. and Cree, G. M., *Carbohyd. Res.*, 6 (1968) 385.
Anderson, D. M. W., Hendrie, A. and Munro, A. C., *J. Chromatogr.*, 44 (1969) 178.
Arwidi, B. and Samuelson, O., *Anal. Chim. Acta*, 31 (1964) 462.
Arwidi, B. and Samuelson, O., *Svensk. Papperstidn.*, 68 (1965) 330.
Aspinall, G. O. and McKenna, J. P., *Carbohyd. Res.*, 7 (1968) 244.
Austin, P. W., Hardy, F. E., Buchanan, J. G. and Baddiley, J., *J. Chem. Soc.*, (1963) 5350.
Bacon, J. S. D., *Biochem. J.*, 57 (1954) 320.
Baer, H. H. and Čapek, K., *Can. J. Chem.*, 47 (1969) 99.
Baggett, N., Mosihuzzaman, M. and Webber, J. M., *Carbohyd. Res.*, 11 (1969) 263.
Ball, D. H., Carey, F. A., Klundt, I. L. and Long, L., Jr., *Carbohyd. Res.*, 10 (1969) 121.
Barker, S. A., Bourne, E. J. and Theander, O., *J. Chem. Soc.*, (1955) 4276.
Barker, S. A., Hatt, B. W., Kennedy, J. F. and Somers, P. J., *Carbohyd. Res.*, 9 (1969a) 327.
Barker, S. A., Hatt, B. W. and Somers, P. J., *Carbohyd. Res.*, 11 (1969b) 355.
Bartlett, G. R., *J. Biol. Chem.*, 234 (1959) 449.
Bartlett, G. R., *Biochim. Biophys. Acta*, 156 (1968a) 221.
Bartlett, G. R., *Biochim. Biophys. Acta*, 156 (1968b) 254.
Bartlett, G. R., *Biochim. Biophys. Acta*, 156 (1968c) 231.
Bartlett, G. R. and Bucolo, G., *Biochim. Biophys. Acta*, 156 (1968) 240.
Bathgate, G. N., *J. Chromatogr.*, 47 (1970) 92.
Bedetti, G., D'Agnolo, G. and Pocchiari, F., *J. Chromatogr.*, 49 (1970) 53.
Begbie, R. and Richtmyer, N. K., *Carbohyd. Res.*, 2 (1966) 272.
Bella, A. M., Jr. and Kim, Y. S., *J. Chromatogr.*, 51 (1970) 314.
BeMiller, J. N. and Wing, R. E., *Carbohyd. Res.*, 6(1968) 197.
Binkley, W. W., *Advan. Carbohyd. Chem.*, 10 (1955) 55.
Birch, G. and Richardson, A. C., *Carbohyd. Res.*, 8 (1968) 411.
Bonner, T. G., Bourne, E. J., Cleare, P. J. V. and Lewis, D., *J. Chem. Soc., B,* (1968) 822.
Brendel, K., Roszel, N. O., Wheat, R. W. and Davidson, E. A., *Anal. Biochem.*, 18 (1967a) 147.
Brendel, K., Steele, R. S., Wheat, R. W. and Davidson, E. A., *Anal. Biochem.*, 18 (1967b) 161.
Brimacombe, J. S. and Ching, O. A., *Carbohyd. Res.*, 9 (1969) 287.
Brimacombe, J. S., Foster, A. B., Hems, R. and Hall, L. D., *Carbohyd. Res.*, 8 (1968a) 249.
Brimacombe, J. S., Hunedy, F. and Tucker, L. C. N., *J. Chem. Soc., C,* (1968b) 1381.
Brown, W., *J. Chromatogr.*, 52 (1970a) 273.
Brown, W., *J. Chromatogr.*, 53 (1970b) 572.
Buchanan, J. G. and Edgar, A. R., *Carbohyd. Res.*, 10 (1969) 295.
Čapek, K., Čapková-Šteffková, J. and Jarý, J., *Collect. Czech. Chem. Commun.*, 35 (1970a) 321.

Čapek, K. and Jarý, J., *Collect. Czech. Chem. Commun.*, 31 (1966) 2558.
Čapek, K., Staněk, J., Jr. and Jarý, J., *Collect. Czech. Chem. Commun.*, 39 (1974) 1462.
Čapek, K., Šteffková, J. and Jarý, J., *Collect. Czech. Chem. Commun.*, 31 (1966) 1854.
Čapek, K., Šteffková, J. and Jarý, J., *Collect. Czech. Chem. Commun.*, 32 (1967) 2491.
Čapek, K., Šteffková, J. and Jarý, J., *Collect. Czech. Chem. Commun.*, 33 (1968a) 781.
Čapek, K., Šteffková, J. and Jarý, J., *Collect. Czech. Chem. Commun.*, 33 (1968b) 1750.
Čapek, K., Šteffková, J. and Jarý, J., *Collect. Czech. Chem. Commun.*, 35 (1970b) 107.
Carlsson, B., Isaksson, T. and Samuelson, O., *Anal. Chim. Acta*, 43 (1968) 47.
Carlsson, B. and Samuelson, O., *Carbohyd. Res.*, 11 (1969) 347.
Carlsson, B. and Samuelson, O., *Anal. Chim. Acta*, 49 (1970) 247.
Carlsson, B., Samuelson, O., Popoff, T. and Theander, O., *Acta Chem. Scand.*, 23 (1969) 261.
Carubelli, R., *Carbohyd. Res.*, 2 (1966) 480.
Černý, M., Staněk, J., Jr and Pacák, J., *Collect. Czech. Chem. Commun.*, 34 (1969) 1750.
Chalk, R. C., Stoddart, J. F., Szarek, W. A. and Jones, J. K. N., *Can. J. Chem.*, 46 (1968) 2311.
Charles, D. F. and Meads, P. F., *Methods Carbohyd. Chem.*, 1 (1962) 520.
Churms, S. C., *Advan. Carbohyd. Chem. Biochem.*, 25 (1970) 13.
Cooper, D. J., Yudis, M. D., Marigliano, H. M. and Traubel, T., *J. Chem. Soc., C*, (1971) 2876.
Cosgrove, D. J., *Ann. N. Y. Acad. Sci.*, 165 (1969) 677.
Crumpton, M. J., *Biochem. J.*, 72 (1959) 479.
Dahlberg, J. and Samuelson, O., *Acta Chem. Scand.*, 17 (1963) 2136.
De Belder, A. N. and Norrman, B., *Carbohyd. Res.*, 8 (1968) 1.
Derungs, R. and Deuel, H., *Helv. Chim. Acta*, 37 (1954) 657.
Dick, W. E., Jr., Baker, B. G. and Hodge, J. E., *Carbohyd. Res.*, 6 (1968) 52.
Donald, A. S. R., *J. Chromatogr.*, 35 (1968) 106.
Dudman, W. F. and Bishop, C. T., *Can. J. Chem.*, 46 (1968) 3079.
Ennor, K. S., Honeyman, J., Shaw, C. J. G. and Stening, T. C., *J. Chem. Soc.*, (1958) 2921.
Evans, M. E., Long, L., Jr. and Parrish, F. W., *J. Chromatogr.*, 32 (1968) 602.
Foster, E. P. and Weiss, A. H., *J. Chromatogr. Sci.*, 9 (1971) 266.
Franks, N. E. and Montgomery, R., *Carbohyd. Res.*, 6 (1968) 286.
French, D., Robyt, J. F., Weintraub, M. and Knock, P., *J. Chromatogr.*, 24 (1966) 68.
Furda, I. and Perry, M. B., *Can. J. Chem.*, 47 (1969) 2891.
Gardell, S., *Acta Chem. Scand.*, 7 (1953) 207.
Gibbs, C. F., Williams, D. T. and Perry, M. B., *Can. J. Chem.*, 47 (1969) 1479.
Godman, J. L. and Horton, D., *Carbohyd. Res.*, 6 (1968) 229.
Goodson, J. M. and DiStefano, V., *J. Chromatogr.*, 45 (1969) 139.
Green, J. G., *National Cancer Institute Monogr.*, No. 21 (1966) 447.
Guthrie, R. D. and Murphy, D., *J. Chem. Soc.*, (1963) 5288.
Guthrie, R. D., Prior, A. M. and Creasey, S. E., *J. Chem. Soc., C*, (1970) 1961.
Hais, I. M. and Macek, K. (Editors), *Paper Chromatography: A Comprehensive Treatise*, Academic Press, London, 1963.
Hallén, A., *Acta Chem. Scand.*, 14 (1960) 2249.
Hashimoto, T. and Yoshikawa, H., *J. Biochem. (Tokyo)*, 61 (1967) 260.
Havlicek, J., Petersson, G. and Samuelson, O., *Acta Chem. Scand.*, 26 (1972) 2205.
Havlicek, J. and Samuelson, O., *Carbohyd. Res.*, 22 (1972) 307.
Heyns, K., Pflughaupt, K.-W. and Paulsen, H., *Chem. Ber.*, 101 (1968) 2800.
Hobbs, J. S. and Lawrence, J. G., *J. Sci. Food Agr.*, 23 (1972a) 45.
Hobbs, J. S. and Lawrence, J. G., *J. Chromatogr.*, 72 (1972b) 311.
Horton, D. and Swanson, F. O., *Carbohyd. Res.*, 14 (1970) 159.
Hough, L., Jones, J. K. N. and Wadman, W. H., *J. Chem. Soc.*, (1949) 2511.
Hough, L., Priddle, J. E. and Theobald, R. S., *Chem. Ind. (London)*, (1960) 900.
Huffman, G. W., Rebers, P. A., Spriestersbach, D. R. and Smith, F., *Nature (London)*, 175 (1955) 990.
Inch, T. D., Ley, R. V. and Rich, P., *J. Chem. Soc., C.*, (1968) 1693.
Inouye, S. and Ogawa, H., *J. Chromatogr.*, 13 (1964) 536.
Itasaka, O., *J. Biochem. (Tokyo)*, 63 (1968) 347.

Jarý, J., Kovář, J. and Čapek, K., *Collect. Czech. Chem. Commun.*, 30 (1965) 1144.
Jeanloz, R. W., Walker, E. and Sinaÿ, P., *Carbohyd. Res.*, 6 (1968) 184.
Jermyn, M. A., *Aust. J. Chem.*, 10 (1957) 55.
John, M., Trénel, G. and Dellweg, H., *J. Chromatogr.*, 42 (1969) 476.
Johnson, L. F. and Tate, M. E., *Can. J. Chem.*, 47 (1969) 63.
Johnson, S. and Samuelson, O., *Anal. Chim. Acta*, 36 (1966) 1.
Jolley, R. L. and Freeman, M. L., *Clin. Chem.*, 14 (1968) 538.
Jolley, R. L., Johnson, W. F. and Scott, C. D., *U.S. Clearinghouse for Federal Scientific and Technical Information PB Report*, PB-183556 (1969); *C.A.*, 71 (1969) 88280r.
Jolley, R. L., Warren, K. S., Scott, C. D., Jainchill, J. L. and Freeman, M. L., *Amer. J. Clin. Pathol.*, 53 (1970) 793.
Jonsson, P. and Samuelson, O., *Sci. Tools*, 13 (1966) 17.
Jonsson, P. and Samuelson, O., *Anal. Chem.*, 39 (1967a) 1156.
Jonsson, P. and Samuelson, O., *J. Chromatogr.*, 26 (1967b) 194.
Kefurt, K., Kefurtová, Z. and Jarý, J., *Collect. Czech. Chem. Commun.*, 38 (1973) 2627.
Kesler, R. B., *Anal. Chem.*, 39 (1967) 1416.
Khym, J. X. and Cohn, W. E., *J. Amer. Chem. Soc.*, 75 (1953) 1153.
Khym, J. X. and Doherty, D. G., *J. Amer. Chem. Soc.*, 74 (1952) 3199.
Khym, J. X. and Zill, L. P., *J. Amer. Chem. Soc.*, 73 (1951) 2399.
Kiely, D. E. and Fletcher, H. G., Jr., *J. Org. Chem.*, 33 (1968) 3723.
Kovář, J. and Jarý, J., *Collect. Czech. Chem. Commun.*, 34 (1969) 2619.
Kuhn, R., Baer, H. H. and Gauhe, A., *Chem. Ber.*, 89 (1956) 2513.
Lammers, J. N. J. J., *J. Chromatogr.*, 41 (1969) 462.
Larsen, B. and Haug, A., *Acta Chem. Scand.*, 15 (1961) 1397.
Larsson, K. and Samuelson, O., *Carbohyd. Res.*, 11 (1969) 144.
Larsson, L.-I. and Samuelson, O., *Mikrochim. Acta*, (1967) 328.
Lederer, E., *Chromatographie en Chimie Organique et Biologique*, Vol. 2, Masson, Paris, 1960.
Lefebvre, M. J., Gonzalez, N. S. and Pontis, H. G., *J. Chromatogr.*, 15 (1964) 495.
Lehrfeld, J., *J. Org. Chem.*, 32 (1967) 2544.
Liljamaa, J. J. and Hallén, A. A., *J. Chromatogr.*, 57 (1971) 153.
Lindberg, B. and Wickberg, B., *Acta Chem. Scand.*, 7 (1953) 140.
McGrath, D., Lee, E. E. and O'Colla, P. S., *Carbohyd. Res.*, 11 (1969a) 453.
McGrath, D., Lee, E. E. and O'Colla, P. S., *Carbohyd. Res.*, 11 (1969b) 461.
MacLaurin, D. J. and Green, J. W., *Can. J. Chem.*, 47 (1969a) 3947.
MacLaurin, D. J. and Green, J. W., *Can. J. Chem.*, 47 (1969b) 3957.
Mansford, K. R. L., *Analyst (London)*, 88 (1963) 649.
Marsden, N. V. B., *Ann. N.Y. Acad. Sci.*, 125 (1965) 428.
Martinsson, E. and Samuelson, O., *J. Chromatogr.*, 50 (1970) 429.
Micheel, F. and Pick, E.-D., *Tetrahedron Lett.*, (1969) 1695; for details, see Pick, E. D., *Dissertation*, University of Münster, G.F.R., 1972.
Micheel, F. and Staněk, J., Jr., *Justus Liebigs Ann. Chem.*, 759 (1972) 37.
Mizelle, J. W., Dunlap, W. J. and Wender, S. H., *J. Chromatogr.*, 28 (1967) 427.
Monsigny, M., *Bull. Soc. Chim. Biol.*, 50 (1968) 2188.
Mowery, D. F., Jr., *Methods Carbohyd. Chem.*, 2 (1963) 328.
Murphy, P. T., Richards, G. N. and Senogles, E., *Carbohyd. Res.*, 7 (1968) 460.
Nagel, C. W. and Wilson, T. M., *J. Chromatogr.*, 41 (1969) 410.
Neuberger, A. and Wilson, B. M., *Carbohyd. Res.*, 17 (1971) 89.
Ohms, J. I., Zec, J., Benson, J. V., Jr. and Patterson, J. A., *Anal. Biochem.*, 20 (1967) 51.
Otake, T., *Seito Gijutsu Kenkyukaishi*, 20 (1968) 60; *C.A.*, 70 (1969) 79344s.
Otter, G. E., Popplewell, J. A. and Taylor, L., *J. Chromatogr.*, 49 (1970) 462.
Perry, M. B. and Webb, A. C., *Can. J. Chem.*, 47 (1969) 1245.
Pigman, W. and Horton, D., *The Carbohydrates: Chemistry and Biochemistry*, Vols. IA and IB, Academic Press, New York, London, 1972.
Pontis, H. G., *Anal. Biochem.*, 23 (1968) 331.

Portsmouth, D., *Carbohyd. Res.*, 8 (1969) 193.
Raff, R. A. and Wheat, R. W., *J. Biol. Chem.*, 242 (1967) 4610.
Reid, W. W., *Methods Carbohyd. Chem.*, 1 (1962) 309.
Repaš, A. and Nikolin, B., *J. Chromatogr.*, 35 (1968) 99.
Repaš, A., Nikolin, B. and Dursun, K., *J. Chromatogr.*, 44 (1969) 184.
Richtmyer, N. K., *Carbohyd. Res.*, 12 (1970a) 135.
Richtmyer, N. K., *Carbohyd. Res.*, 12 (1970b) 139.
Richtmyer, N. K., *Carbohyd. Res.*, 12 (1970c) 233.
Roseman, S., Abeles, R. M. and Dorfman, A., *Arch. Biochem. Biophys.*, 36 (1952) 232.
Rutherford, D. and Richtmyer, N. K., *Carbohyd. Res.*, 11 (1969) 341.
Salo, W. L. and Fletcher, H. G., Jr., *J. Org. Chem.*, 34 (1969) 3026.
Samuelson, O., *Methods Carbohyd. Chem.*, 6 (1972) 65.
Samuelson, O., Larsson, L.-I. and Ramnas, O., *2nd Technicon Symposium, London, 1965*, p. 168; *C.A.*, 68 (1968) 65533z.
Samuelson, O. and Strömberg, H., *Carbohyd. Res.*, 3 (1966) 89.
Samuelson, O. and Strömberg, H., *Acta Chem. Scand.*, 22 (1968) 1252.
Samuelson, O. and Swenson, B., *Anal. Chim. Acta*, 28 (1963) 426.
Samuelson, O. and Thede L., *J. Chromatogr.*, 30 (1967) 556.
Saunders, R. M., *Carbohyd. Res.*, 7 (1968) 76.
Seib, P. A., *Carbohyd. Res.*, 8 (1968) 101.
Shapiro, D., Acher, A. J. and Rachaman, E. S., *J. Org. Chem.*, 32 (1967) 3767.
Shapiro, D., Rabinsohn, Y., Acher, A. J. and Diver-Haber, A., *J. Org. Chem.*, 35 (1970) 1464.
Šipoš, P. and Bauer, Š., *Carbohyd. Res.*, 6 (1968) 494.
Smirnyagin, V. and Bishop, C. T., *Can. J. Chem.*, 46 (1968) 3085.
Smith, E. J., *Biochem. Biophys. Res. Commun.*, 15 (1964) 593.
Spencer, N., *J. Chromatogr.*, 30 (1967) 566.
Srivastava, H. C., Ramalingam, K. V., Doshi, N. M. and Chaudhari, A. S., *Carbohyd. Res.*, 12 (1970) 23.
Staněk, J., Černý, M., Kocourek, J. and Pacák, J., *The Monosaccharides*, Academic Press, New York, London, 1963.
Staněk, J., Jr., Čapek, K. and Jarý, J., *Collect. Czech. Chem. Commun.*, (1974) in press.
Stefanović, V. Dj., *J. Chromatogr.*, 5 (1961) 453.
Stern, E. L., *Advan. Automat. Anal., Technicon Int. Congr., 1969*, 1970, Vol. 1, p. 343; *C.A.*, 73 (1970) 21993y.
Steward, J. H. and Tate, M. E., *J. Chromatogr.*, 45 (1969) 400.
Taylor, P. M. and Whelan, W. J., *Chem. Ind. (London)*, (1962) 44.
Tejima, S., Ness, R. K., Kaufman, R. L. and Fletcher, H. G., Jr., *Carbohyd. Res.*, 7 (1968) 485.
Tomlinson, R. V. and Ballou, C. E., *Biochemistry*, 1 (1962) 166.
Trimnell, D., Doane, W. M., Russell, C. R. and Rist, C. E., *Carbohyd. Res.*, 13 (1970) 301.
Tulloch, A. P. and Hill, A., *Can. J. Chem.*, 46 (1968) 2485.
Vrátny, P., *Listy Cukrov.*, 85 (1969) 233; *C.A.*, 72 (1970) 102064c.
Walborg, E. F., Jr. and Christensson, L., *Anal. Biochem.*, 13 (1965) 186.
Walborg, E. F., Jr., Christensson, L. and Gardell, S., *Anal. Biochem.*, 13 (1965) 177.
Walborg, E. F., Jr. and Kondo, L. E., *Anal. Biochem.*, 37 (1970) 320.
Walborg, E. F., Jr. and Lantz, R. S., *Anal. Biochem.*, 22 (1968) 123.
Walborg, E. F., Jr., Ray, D. B. and Öhrberg, L. E., *Anal. Biochem.*, 29 (1969) 433.
Weber, P. and Winzler, R. J., *Arch. Biochem. Biophys.*, 129 (1969) 534.
Whistler, R. L., *Science*, 120 (1954) 899.
Whistler, R. L. and Durso, D. F., *J. Amer. Chem. Soc.*, 72 (1950) 677.
Yaguchi, M. and Perry, M. B., *Can. J. Biochem.*, 48 (1970) 386.
Yoshida, K., Honda, N., Iino, N. and Kato, K., *Carbohyd. Res.*, 10 (1969a) 333.
Yoshida, K., Iino, N., Kamata, T. and Kato, K., *Chem. Pharm. Bull.*, 17 (1969b) 1123.
Zacharius, R. M. and Porter, W. L., *J. Chromatogr.*, 30 (1967) 190.
Zeleznick, L. D., *J. Chromatogr.*, 14 (1964) 139.
Zorbach, W. W., DeBernardo, S. L. and Bhat, K. V., *Carbohyd. Res.*, 11 (1969) 567.

*Chapter 23*

# Polysaccharides

K. ČAPEK and J. STANĚK, Jr.

## CONTENTS



## INTRODUCTION

The chemistry and biochemisty of polysaccharides is an area in which chromatographic separation methods are mentioned in almost every communication, and its recent rapid development is indebted mainly to the application of chromatography. However, the main applicability of chromatography in this area is in isolations from natural material, in which polysaccharides are admixed with various compounds, such as proteins, nucleic acids and substances of lower molecular weight. The last compounds are very different in nature from polysaccharides and these separations are therefore not considered here.

There is no general method for the purification of all polysaccharides, and the procedures used in each instance depend on the nature of the particular polysaccharide and on possible contaminants. In a book by Pigman and Horton, the whole of Volume II is devoted to the chemistry and biochemistry of polysaccharides and their isolation, fractionation and purification, so that we decided in this review to consider only the mutual chromatographic separation of polysaccharides.

The general techniques are similar to those discussed in Chapter 22 for carbohydrates; they include mainly ion-exchange and gel permeation chromatography, and automated detection methods. Moreover, there are many examples of the fractionation of various dextrans, mannodextrins, cellodextrins, etc., which have been described in Chapter 22, in which these compounds were separated together with mono- and oligosaccharides in order to establish the relationship between $K_d$ and molecular size.

After the first attempts to separate polysaccharides by using common sorbents such as cellulose, alumina, silica gel and calcium carbonate (see Lederer), two types of column chromatography are now in general use: ion-exchange chromatography for the separation of acidic and neutral polysaccharides and for the fractionation of their borate complexes, and gel permeation chromatography for separations according to molecular size.

Only recently, an application of affinity chromatography of branched-chain polysaccharides was described by Kennedy and Rosevear. The branched-chain polysaccharides were fractionated by elution from a concavalin A column, immobilised on Sepharose 4B, with aqueous phosphate buffer at near neutral pH values as eluent. Even the mixtures of saccharides which cannot be separated by the usual complex formation with a concavalin A solution, were separated.

## ION-EXCHANGE CHROMATOGRAPHY

Some difficulties may arise when separating mixtures that contain several acidic polysaccharides, in which the strongly acidic sulphate group masks other differences in the ionic properties of the polymer and greatly decreases the efficiency of IEC based on these differences. For this reason, the fractionation of mucopolysaccharides is discussed in a separate chapter (Chapter 24).

In the pioneering work of Neukom *et al.*, it was shown that basic cellulose ion exchangers, such as DEAE-cellulose, are more suitable for the fractionation of polysaccharides than are the anion-exchange resins that were formerly used (Steiner *et al.*), because of the larger active surface of DEAE-cellulose. Acidic carbohydrates are adsorbed at neutral pH, and neutral carbohydrates only under alkaline conditions. For a similar type of separation, ECTEOLA-cellulose was used (Ringertz and Reichard).

Fractionation of a pectin from lemon-peel on DEAE-cellulose ($PO_4^{3-}$) gave two main fractions of acidic polysaccharides together with a small fraction of a neutral polysaccharide. Preparatively, this neutral polysaccharide was removed on DEAE-Sephadex A-50 ($HCOO^-$) (Aspinall *et al.*).

Fractionation of the polysaccharide exudate from *Araucaria bidwillii* on a DEAE-cellulose ($PO_4^{3-}$) column (30 × 4 cm) with 0.025, 0.05, 0.1, 0.25 and 0.5 *M* sodium dihydrogen phosphate buffer (500 ml each) at pH 6 gave a different elution pattern than when it was chromatographed on a 60 × 4.5 cm column of DEAE-Sephadex A-50 ($HCOO^-$) with water containing 1% (3 l), 5% (4 l) and 10% (3 l) of formic acid as the mobile phase; phenol–sulphuric acid and carbazole assays were used for detection (Aspinall and McKenna).

Neutral and acidic polysaccharides formed by *Polyporus fomentarius* and *P. igniarius* were separated by treatment with DEAE-cellulose ($CH_3COO^-$), on which the latter were adsorbed from aqueous solution and subsequently desorbed by treatment with 2 *M* potassium acetate solution. Sephadex G-150 and Sepharose 4B columns (50 × 0.8 cm) were used for further separation attempts (Björndal and Lindberg).

A heteropolysaccharide composed of D-mannose, D-galactose, D-glucose and D-glucuronic acid, from the cell wall of *Aureobasidium pullulans,* was purified from contaminating β-glucan on DEAE-Sephadex A-25 by Brown and Lindberg, using a linear gradient of potassium acetate (0–2 *M*); β-glucan was eluted first.

Neukom *et al.* found that by using the borate form of a cellulose ion exchanger (elution with sodium borate), the fractionation on DEAE-cellulose was increased. However, with Dowex 1-X8 (sodium chloride–boric acid mobile phase), no separation of xylan and mannan was achieved as the xylan "smeared" through the whole run (Kesler).

Pierce and Liao described the contamination of the effluent with D-xylose artefacts in DEAE-cellulose chromatography. Thus, with increased improvements in analytical methods, care must be taken to ensure that these contaminants from chromatographic media are not to be mistaken for minor constituents of the substance being studied.

## GEL PERMEATION CHROMATOGRAPHY

GPC is probably the most frequently used method (for a review, see Churms) because it is the simplest method for the fractionation of polysaccharides that have a broad molecular-weight distribution; simultaneously, it provides a means of determining the molecular weights of polysaccharides. As mild conditions are used, the technique is particularly useful for labile biological materials. A selection of several hundred papers that deal with this subject is listed in Table 23.1.

GPC on Sephadex G-25 (28 × 2.5 cm) with water as the mobile phase was used by Lornitzo and Goldman for the purification of a soluble, low-molecular-weight (4400) polysaccharide containing 60% D-glucose, 40% 6-O-methyl-D-glucose and one acid group. Filtration through a column of Dowex 50 in water removed small amounts of biuret- and ninhydrin-reacting materials; the polysaccharide was not retained in this step.

TABLE 23.1
GEL PERMEATION CHROMATOGRAPHY OF POLYSACCHARIDES

| Packing | Mobile phase | Sample | Reference |
|---|---|---|---|
| Sephadex G-25 | Water | Fractionation of dextrans, $\bar{M}_n$ 5000 | Granath and Flodin |
| | Water | Purification of synthetic arabinan | Kochetkov *et al.* |
| Sephadex G-50 | Water | Fractionation of dextrans, $\bar{M}_n$ 1000–7000 | Granath and Flodin |
| | 0.2 *M* Ammonium hydrogen carbonate | Biosynthesis of mannans | Kozak and Bretthauer |
| Sephadex G-75 | Water | Fractionation of starch dextrins | Nordin |
| Sephadex G-100 | 0.1 *M* Ammonium hydrogen carbonate | Fractionation of arabino-galactans from larch wood | Ettling and Adams |
| Sephadex G-200 | 0.2 *M* Sodium chloride | Fractionation of polymolecular dextran | Laurent and Granath |
| Bio-Gel A | Water | Fractionation of dextrans | Bathgate |
| Bio-Gel P-2 | Water | Homologous glucose oligomers | John *et al.* |
| Bio-Gel P-10 | Water | Fractionation of degradation products of *A. elata* gum polysaccharide | Churms and Stephen |
| Bio-Gel P-300 | 1 *M* Sodium chloride | Determination of mol. wt. distribution of arabinogalactans from plant gums | Anderson *et al.* (1968) |
| Bio-Gel A-50m | Water | Cuprammonium rayon, CM-cellulose | Petterson |
| Sepharose 4B | 1 *M* Sodium chloride | Fractionation of arabino-galactans of high mol. wt. | Anderson *et al.* (1969) |
| Bio-Glass 500 | 0.25 *M* Ammonium formate | Average mol. wt. of *Citrus limonia* gum | Stoddart and Jones |
| Porasil C and D | Water | Fractionation of dextrans | Barker *et al.* |
| Aquapak | Water | Fractionation of dextrans | Bombaugh *et al.* |

*References p.527*

Hummel and Smith investigated the fractionation of dextrans in the molecular-weight range 5000 to 300,000 on Sephadex, agar, agarose and polyacrylamide gels, and found that the best fractionation of dextrans of high molecular weight was achieved on 6.7% agar gel at 4°C on a 145 cm × 1.2 cm$^2$ column with 0.2 *M* Tris hydrochloride buffer (pH 8.0) as the mobile phase. A 4% agarose gel, eluted with water, also proved to be effective. The remaining two gels were found to be unsuitable for the fractionation of dextrans with molecular weights above 100,000. In general, the samples of dextran (10–50 mg) were dissolved in 1.0–1.5 ml of mobile phase and applied to the column; the fractions were detected with phenol–sulphuric acid reagent.

Granath and Kvist described a method of determining the molecular-weight distribution of dextrans in the range 10,000–150,000. A 75 × 1.4 cm column packed with a mixture of two Sephadex gels, G-200 and G-100, in the dry-weight ratio of 1:2 (so that the two gels occupied equal volumes when swollen) was used. The mobile phase was percolated through the bed at 20°C several days before use so as to ensure complete equilibration. After the column had been calibrated with 17 dextran fractions, the dextrans under examination were chromatographed under the same conditions, *i.e.*, using 0.3% sodium chloride solution, containing chlorobutanol as a preservative, at a flow-rate of 6–7 ml/h as the mobile phase. With the automated anthrone–sulphuric acid procedure, amounts of polymer as small as 2 mg (dissolved in 2 ml of water) are detected.

The prior calibration of the column is not necessary if the continuous, reducing end-group assay, in which the response is proportional to the number of molecules, in addition to a cysteine–sulphuric acid assay for total hexose concentration, in which the response is proportional to the weight of the sample, is used. Barker *et al.* developed this technique for the fractionation of dextran on porous silica beads (Porasil D, 75–100 mesh; column, 82.8 × 1.08 cm). Water was used as the mobile phase and was pumped through the column at a flow-rate of 10 ml/h·cm$^2$.

In 1966, Anderson and Stoddart described the use of the polyacrylamide gel Bio-Gel P-300 in the determination of molecular weights of fractions of arabinogalactan gum. The column (50 × 6 cm), pre-treated with 1% dichlorodimethylsilane in benzene at 60°C, was packed with Bio-Gel P-300 that had been allowed to swell in 1 *M* sodium chloride for 2 days. In order to stabilise the soft top-surface, 1-cm layers of Bio-Gel P-200 and Bio-Gel P-10 were applied successively to the column. Sodium chloride (1 *M*) was allowed to flow for 2 days before the column was calibrated with dextran fractions of known number-average molecular weight. Polysaccharide (*ca.* 10 mg), dissolved in 1 ml of 1.5 *M* sodium chloride, was then applied to the column and eluted with 1 *M* sodium chloride. Fractions (2 ml) were screened by the phenol–sulphuric acid method. Elution volumes were estimated to the nearest millilitre from peak maxima.

On calibration of the gel column with dextran fractions of known number-average molecular weight, $\bar{M}_n$, the correlation between the elution volume and log $\bar{M}_n$ was found to be linear over the $\bar{M}_n$ range from 5000 to 125,000. Values of $\bar{M}_n$ within this range could therefore be estimated from elution volumes by reference to this calibration graph. This method was successfully applied in structural studies on various *Araucaria* (Anderson and Munro) and *Acacia* (Anderson *et al.*, 1968) gums.

Bathgate pointed out the disadvantage of Bio-Gel P-300 that considerable compression of this gel occurred when a proportioning pump was used. An agarose gel (Bio-Gel A,

0.5 *M*, 200–400 mesh) did not compress to the same extent and, despite its being a carbohydrate, proved to be eminently suitable for an automated system for the determination of molecular weights. A 30 × 0.9 cm column, pre-treated with dichlorodimethylsilane, was packed with Bio-Gel A (0.5 *M*, 200–400 mesh) according to the manufacturer's instructions. Before use, the column was eluted for several days with a 0.05 *M* solution of mercury(II) chloride in order to replace the sodium azide preservative on the agarose gel, because azide interferes strongly with the orcinol used for detection. The mercury(II) chloride fulfills the double role of preservative and mobile phase. The void volume of the column was determined by eluting Blue Dextran 2000 (Pharmacia, Uppsala, Sweden) and the column was then calibrated by using standard polysaccharides. Although requiring careful handling in order to prevent microbial contamination, agarose gel gives high resolutions for a wide range of molecular weights (10,000 to *ca.* 100,000) on short columns and at a relatively high flow-rate (13.8 ml/h); a complete determination could be carried out in less than 3 h (Bathgate).

It is obvious that the determination of polysaccharide molecular-weight distribution depends on the measurement of elution volumes; they must therefore be completely reproducible. An important study was undertaken by Churms *et al.* to investigate the effect of sample concentration on elution volume. They found that, within the concentration range 2–20 mg/ml, the elution volumes, $V_e$, of dextran of $\bar{M}_w$ 500,000, 70,000 and 10,000 and that of D-glucose on Bio-Gel P-300 were independent of concentration, within the degree of uncertainty. However, a marked dependence of $V_e$ on concentration for D-glucose and dextran of $\bar{M}_w$ 10,000 was observed on Bio-Gel P-10. With the dextran, the observed 10% increase in $V_e$ would correspond to a decrease of *ca.* 20% in an estimated molecular-weight value. In contrast, the elution volume of dextran of $\bar{M}_w$ 500,000 was independent of concentration on this gel. Hence molecular weight obtained on Bio-Gel P-10 will be meaningful only if the concentrations of both the sample and the calibration solutes are the same.

## REFERENCES


Anderson, D. M. W., Dea, I. C. M. and Hirst, E., *Carbohyd. Res.*, 8 (1968) 460.
Anderson, D. M. W., Dea, I. C. M. and Munro, A. C., *Carbohyd. Res.*, 9 (1969) 363.
Anderson, D. M. W. and Munro, A. C., *Carbohyd. Res.*, 11 (1969) 43.
Anderson, D. M. W. and Stoddart, J. F., *Carbohyd. Res.*, 2 (1966) 104.
Aspinall, G. O., Craig, J. W. T. and Whyte, J. L., *Carbohyd. Res.*, 7 (1968) 442.
Aspinall, G. O. and McKenna, J. P., *Carbohyd. Res.*, 7 (1968) 244.
Barker, S. A., Hatt, B. W. and Somers, P. J., *Carbohyd. Res.*, 11 (1969) 355.
Bathgate, G. N., *J. Chromatogr.*, 47 (1970) 92.
Björndal, H. and Lindberg, B., *Carbohyd. Res.*, 10 (1969) 79.
Bombaugh, K. J., Dark, W. A. and King, R. N., *J. Polym. Sci, Part C*, 21 (1968) 131.
Brown, R. G. and Lindberg, B., *Acta Chem. Scand.*, 21 (1967) 2383.
Churms, S. C., *Advan. Carbohyd. Chem. Biochem.*, 25 (1970) 13.
Churms, S. C. and Stephen, A. M., unpublished work.
Churms, S. C., Stephen, A. M. and Van der Bijl, P., *J. Chromatogr.*, 47 (1970) 97.
Ettling, B. V. and Adams, M. F., *Tappi*, 51 (1968) 116.
Granath, K. A. and Flodin, P., *Makromol. Chem.*, 48 (1961) 160.
Granath, K. A. and Kvist, B. E., *J. Chromatogr.*, 28 (1967) 69.

Hummel, B. C. W. and Smith, D. C., *J. Chromatogr.*, 8 (1962) 491.
John, M., Trénel, G. and Dellweg, H., *J. Chromatogr.*, 42 (1969) 476.
Kennedy, J. F. and Rosevear, A., *J. Chem. Soc., Perkin Trans. I.*, (1973) 2041.
Kesler, R. B., *Anal. Chem.*, 39 (1967) 1416.
Kochetkov, N. K., Bochkov, A. F. and Yazlovetsky, I. G., *Carbohyd. Res.*, 9 (1969) 49.
Kozak, L. P. and Bretthauer, R. K., *Biochemistry*, 9 (1970) 1115.
Laurent, T. C. and Granath, K. A., *Biochim. Biophys. Acta*, 136 (1967) 191.
Lederer, E., *Chromatographie en Chimie Organique et Biologique*, Vol. 2, Masson, Paris, 1960.
Lornitzo, F. A. and Goldman, D. S., *Biochim. Biophys. Acta*, 158 (1968) 329.
Neukom, H., Deuel, H., Heri, W. J. and Kündig, W., *Helv. Chim. Acta*, 43 (1960) 64.
Nordin, P., *Arch. Biochem. Biophys.*, 99 (1962) 101.
Petterson, B., *Svensk Papperstidn.*, 72 (1969) 14.
Pierce, J. G. and Liao, T.-H., *Anal. Biochem.*, 24 (1968) 448.
Pigman, W. and Horton, D., *The Carbohydrates: Chemistry and Biochemistry*, Vols. IIA and IIB, Academic Press, New York, London, 1970.
Ringertz, N. R. and Reichard, P., *Acta Chem. Scand.*, 13 (1959) 1467.
Steiner, K., Neukom, H. and Deuel, H., *Chimia*, 12 (1958) 150.
Stoddart, J. F. and Jones, J. K. N., *Carbohyd. Res.*, 8 (1968) 29.

*Chapter 24*

# Polysaccharide–protein complexes

M. JUŘICOVÁ and Z. DEYL

## CONTENTS



## GLYCOSAMINOGLYCANS (MUCOPOLYSACCHARIDES)

### Introduction

Basically, there are two different approaches for the separation of anionic glycosaminoglycans (acid mucopolysaccharides). The most common technique in clinical analysis is based on the selective dissociation of cetylpyridinium complexes of these compounds in solutions of different salt concentrations. This procedure has been developed by Antonopoulos *et al.* (1967), Schiller *et al.* and Scott. Selective elution from an ion-exchange column has recently been used by several investigators and a number of ion exchangers have been used for this purpose, such as Dowex 1, ECTEOLA-cellulose, DEAE-Sephadex and calcium phosphate gels. In model systems, the separations usually dealt with are those of hyaluronic acid, chondroitin-4-sulphate (chondroitin sulphate A), chondroitin-6-sulphate (chondroitin sulphate C), dermatan sulphate (chondroitin sulphate B), keratan sulphate and heparin.

Anionic glycosaminoglycans are liberated from the tissue by non-specific proteolysis of the contaminating proteins with papain (E.C. 3.4.4.10). Digestion is usually carried out at 65°C for a 3-h period (Antonopoulos *et al.*, 1967).

It is surprising that little was known about the optimum conditions for the separation of these compounds until the appearance of recent papers by Antonopoulos *et al.* (1964, 1967), Braselmann and Ramm and Pearce *et al.*

### The cetylpyridinium chloride procedure

The separation of glycosaminoglycans on cellulose columns on a micro-scale has been described by Antonopoulos and Gardell. To the column (60 × 3 mm, provided with a pear-shaped extension at the top with a volume of about 4 ml), 20 μg of each polysaccharide were added and a stepwise gradient consisting of the following mobile phases was used:

(1) 1% cetylpyridinium chloride in water;
(2) 0.3 *M* sodium chloride in water containing 0.05% of cetylpyridinium chloride;
(3) *n*-propanol–methanol–acetic acid–water (40:20:1.5:38.5) containing 0.4% of cetylpyridinium chloride;
(4) 0.75 *M* magnesium chloride in 0.1 *M* acetic acid;
(5) 0.75 *M* sodium chloride containing 0.05% of cetylpyridinium chloride.

Between the application of solvents 2 and 3, 3 and 4 and 4 and 5, the column was eluted with a 0.05% aqueous cetylpyridinium chloride solution in order to remove any solvent that remained from the preceding step.

The above procedure gives a very good separation of chondroitin-4-sulphate, chondroitin-6-sulphate and dermatan sulphate. As cetylpyridinium chloride tends to crystallize (depending on the batch) at about 20–22°C, the eluting solvents that contain cetylpyridinium chloride should be maintained above this temperature and the elution should also be carried out above this temperature. This is particularly important for the alternative version of the cetylpyridinium chloride procedure described by Antonopoulos *et al.* (1964), in which elution is carried out with an increasing concentration of neutral salts without introducing an organic solvent. The general experimental procedure is similar; micro-columns are eluted stepwise with 1-ml portions of the following mobile phases:

(1) 1% cetylpyridinium chloride;
(2) 0.5 *M* sodium chloride in 0.05% cetylpyridinium chloride;
(3) 0.7 *M* magnesium chloride in 0.05% cetylpyridinium chloride;
(4) 1.25 *M* magnesium chloride in 0.05% cetylpyridinium chloride;
(5) 6 *N* hydrochloric acid.

During the separation, hyaluronic acid is eluted with 0.5 *M* sodium chloride (solvent No. 2), chondroitin sulphates with 0.7 *M* magnesium chloride (solvent No. 3) and heparin with 1.25 *M* magnesium chloride (solvent No. 4). The recovery is 80–100% regardless of the composition of the mixture analyzed. As this procedure is not capable of separating chrondroitin-4-sulphate and chondroitin-6-sulphate, the cetylpyridinium chloride procedure with organic solvents is usually preferred.

### Separation on Dowex 1-X2 ($Cl^-$)

On the macro-scale, the separation is carried out on a 0.9 × 44 cm column packed with Dowex 1-X2 ($Cl^-$; 200–400 mesh) (Schiller *et al.*). Anionic glycosaminoglycan (5–10 mg) is applied as an aqueous solution, the loaded column is washed with distilled water and a stepwise gradient of increasing salt concentration is introduced (see Fig. 24.1). This

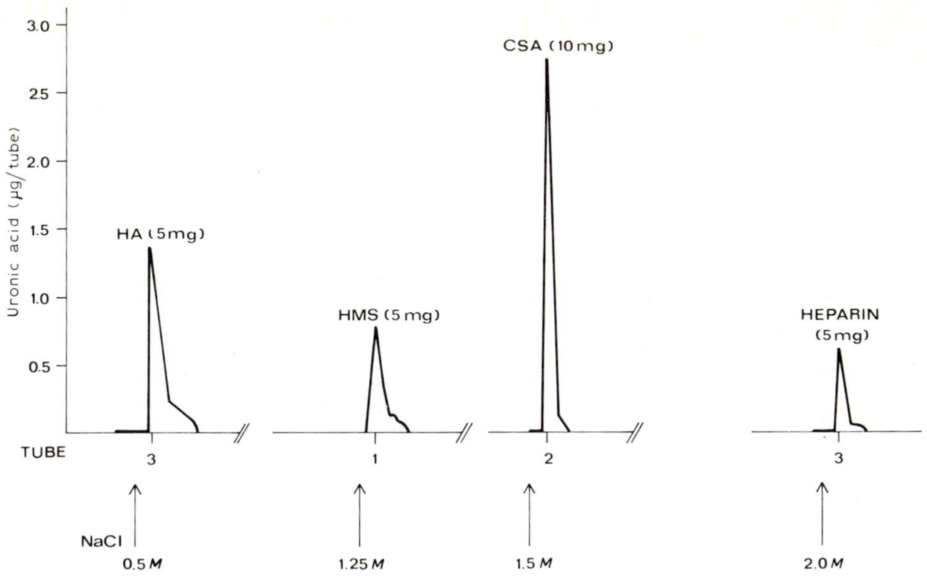


Fig. 24.1. Elution diagrams for hyaluronic acid (HA), heparin monosulphuric acid (HMS), chondroitin sulphuric acid A (CSA) and heparin chromatographed individually on Dowex 1-X2 ($Cl^-$) columns (0.9 × 44 cm) (Schiller *et al.*). Stepwise elution was used with sodium chloride solutions of increasing concentration (0.5–2.0 *M*). The concentration of sodium chloride at which each of the substances was eluted is indicated as well as the number of the tube in which each substance appeared in the effluent.

procedure gives a good separation of hyaluronic acid, heparin monosulphuric acid, chondroitin-4-sulphate and heparin. Keratan sulphate is eluted with 3 *M* sodium chloride solution and can also be completely separated.

In the more developed form of this technique described by Pearce *et al.*, micro-scale operation is possible, and a micro-scale column is used as shown in Fig. 24.2. The amount of each glycosaminoglycan to be separated is within the range 3–5 μmole, the amount of Dowex 1-X2 (100–200 mesh) used is 400 mg per column and elution is carried out with a linear salt gradient in 8 *M* urea. The addition of urea minimizes the non-electrostatic binding of proteins to cellulose-based ion exchangers. The presence of 8 *M* urea in the salt gradient shifts the peaks of glycosaminoglycans to lower elution volumes, indicating that in the absence of urea hydrophobic contacts play a considerable role in the separation process. The peaks are also better resolved in the presence of urea, which suggests that non-electrostatic binding opposes the separation effect. The micro-scale separation of glycosaminoglycans with Dowex 1-X2 is, according to Pearce *et al.*, superior to the classical version of the cetylpyridinum chloride procedure (without organic solvents).

TABLE 24.1
SEPARATION OF AN ARTIFICIAL MIXTURE OF GLYCOSAMINOGLYCANS (HYALURONIC ACID, CHONDROITIN-4-SULPHATE, CHONDROITIN-6-SULPHATE, DERMATAN SULPHATE AND KERATAN SULPHATE)

A 10-mg amount of each polysaccharide was applied to a 2 × 20 cm ECTEOLA-cellulose column; formate solutions contained 1.34 *M* ammonia (Antonopoulos *et al.*, 1967).

| Fraction | Recovery | | Uronic acid* | Hexo-samine* | Hexose* | Gluco-samine** | Galacto-samine** | Predominating glycosaminoglycans in fraction |
|---|---|---|---|---|---|---|---|---|
| | mg | % | | | | | | |
| 0.02 *M* hydrochloric acid | – | – | | | | | | |
| 1.0 *M* ammonium formate | 9.3 | 93.0 | 31.0 | 33.6 | – | 100 | – | Hyalorunic acid |
| 1.5 *M* ammonium formate | 2.0 | 6.7*** | | 24.5 | – | – | 100 | Chondroitin-4-sulphate, chondroitin-6-sulphate |
| 2.0 *M* ammonium formate | 17.1 | 57.0*** | 29.1 | 26.4 | – | – | 100 | |
| 2.75 *M* ammonium formate | 9.2 | 30.7*** | 23.3 | 26.3 | – | – | 100 | Chondroitin-4-sulphate |
| 2.0 *M* sodium chloride | 9.8 | 98.0 | 2.8 | 25.0 | 26.6 | 90 | 10 | Keratan sulphate |

*Expressed as a percentage of air-dried material.
**Expressed as a percentage of total hexosamines.
***Calculated on the sum of all galactosaminoglycans added to the column.

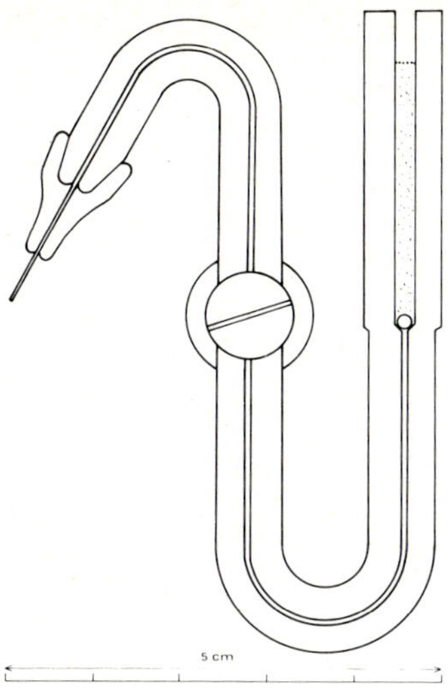


Fig. 24.2. Micro ion-exchange column. A 7-cm length of glass capillary tubing, 3 mm I.D., was sealed to a capillary stop-cock, 1 mm I.D., which was beaten into a goose-neck shape (Pearce *et al.*). A 21-gauge disposable needle (Becton-Dickinson) was cut at right-angles 10 mm from the end and cemented to the glass with expoxy cement.

## Separation on ECTEOLA-cellulose

This procedure was devised by Ringertz and Reichard (1959, 1960) and developed further by Anseth and Laurent; the microgram version of this procedure was described by Trudle and Mann. By using an ECTEOLA-cellulose column and eluting with an increasing concentration of sodium chloride in dilute hydrochloric acid, it is possible to separate hyaluronic acid, chondroitin sulphates and heparin. Antonopoulos *et al.* (1967) used a stepwise gradient for the separation of hyaluronic acid, chondroitin-4-sulphate, chondroitin-6-sulphate, dermatan sulphate and keratan sulphate. In general, galactosaminoglycans could be eluted from the column with a lower concentration of ammonium formate in ammonia than that needed for keratan sulphate (Table 24.1).

## Separation on DEAE-cellulose

Although Pearce *et al.* reported that in DEAE-cellulose chromatography the peaks of individual glycosaminoglycans are eluted close together and that this ion exchanger cannot be recommended for such a separation, a method that gives fairly good results has been developed by Braselmann and Ramm. As in the cetylpyridinium chloride procedure, elution was carried out with an increasing concentration of magnesium chloride solution. The whole procedure requires not more than 50 mg of material and small columns (50 × 4.5 mm) can be used. A typical run is shown in Fig. 24.3. It is obvious that in this instance the separation of chondroitin-6-sulphate, chondroitin-4-sulphate and dermatan sulphate is incomplete, but heparan sulphate and hyaluronic acid are clearly separated. This procedure, as with many others in glycosaminoglycan chromatography, uses a stepwise gradient with magnesium chloride concentrations as indicated in Fig. 24.3. The first elution is carried out with 0.2 *M* magnesium chloride in 0.002 *M* acetic acid, which helps to elute hyaluronic acid. This step has been used in several procedures, such as those of Antonopoulos *et al.* (1967) and Thunell. In summary, the advantages and disadvantages of DEAE-cellulose chromatography are as follows. The capacity of the column is rather high, which means that the procedure would be more suitable for preparative than for analytical purposes. It also allows the accumulation of glycosaminoglycans from very dilute solutions, which occasionally occurs when handling natural material. In some instances, the saccharidic material adheres strongly to the ion exchanger, which causes tailing and disturbances in separation; according to Braselmann and Ramm, this problem can be overcome by using mixed beds of DEAE-cellulose and some other ion exchangers. Braselmann and Ramm believe that these disturbances are not the result of hydrophobic binding, as described by Antonopoulos *et al.* (1967) for Dowex 1-X2, as the application of urea proved ineffective in DEAE-cellulose chromatography. The nature of the DEAE-cellulose separation does not, of course, permit the application of organic solvents as used in the cetylpyridinium chloride separation by Antonopoulos and Gardell.

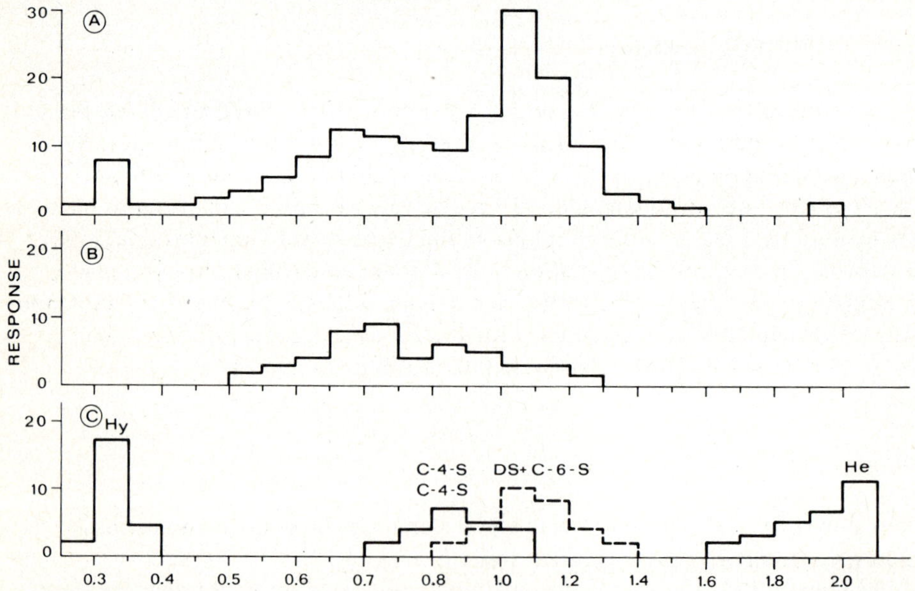


Fig. 24.3. Separation of glycosaminoglycans on DEAE-cellulose: (A) glycosaminoglycans from rabbit aorta (43.4 mg); (B) the same glycosaminoglycans, treated with hyaluronidase; (C) standards, Hy = hyaluronic acid, C-4-S = chondroitin-4-sulphate, DS = dermatan sulphate, C-6-S = chondroitin-6-sulphate, He = heparin (Braselmann and Ramm).

## Separation on DEAE-Sephadex

For this purpose, DEAE-Sephadex A-25 ($Cl^-$) in columns of dimensions 2 × 20 cm was used by Schmidt. After pre-treatment with 0.5 *M* sodium hydroxide, 0.5 *M* hydrochloric acid and 0.1 *M* sodium chloride, a suspension of the ion exchanger in 0.1 *M* sodium chloride was packed into the column. A stepwise gradient was then introduced, as follows:

(1) 0.50 *M* sodium chloride;
(2) 1.25 *M* sodium chloride in 0.01 *M* hydrochloric acid;
(3) 1.50 *M* sodium chloride in 0.01 *M* hydrochloric acid;
(4) 2.0 *M* sodium chloride in 0.01 *M* hydrochloric acid.

Before starting the gradient, the column was washed with 5 ml of distilled water; 5 ml fractions were collected during the separation procedure, the result of which is shown in Fig. 24.4. The isomeric chondroitin sulphates cannot be separated by this system and are eluted as a single peak.

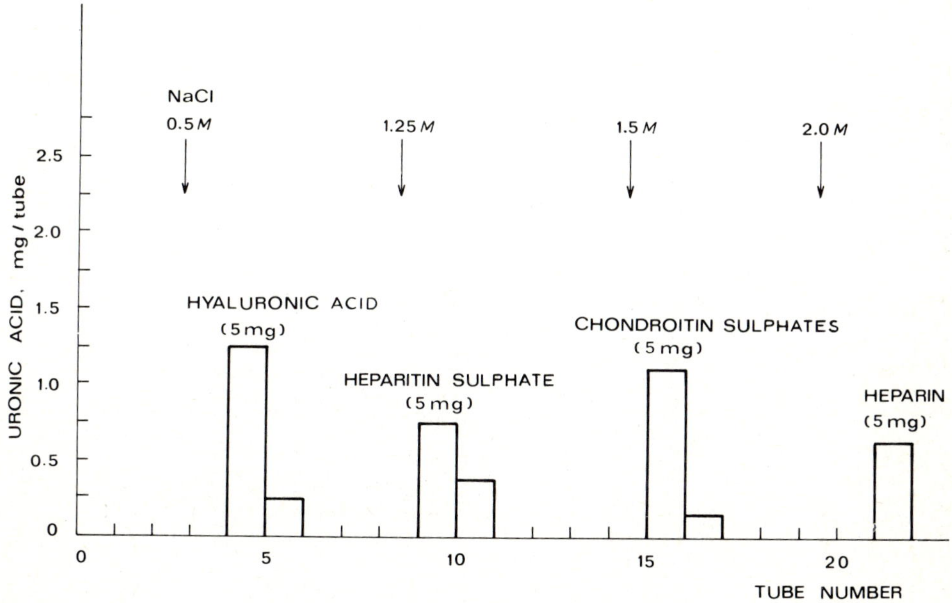


Fig. 24.4. Stepwise elution of a mixture of acid mucopolysaccharides (Schmidt).

## Sephadex gel chromatography and molecular-weight distribution

There are a number of diseases in which the metabolism of anionic glycosaminoglycans is deranged and specific anionic glycosaminoglycans may accumulate in organs and tissues or may be excreted in the urine. There is some evidence that these glycosaminoglycans may differ from their counterparts in healthy individuals in their degree of polymerization.

*References p. 541*

TABLE 24.2
EVALUATION OF DIFFERENT CHROMATOGRAPHIC PROCEDURES USED IN GLYCOSAMINOGLYCAN SEPARATIONS

| Chromatographic procedure | Glycosaminoglycans separated | Pairs of glycosaminoglycans not separated | Reference |
|---|---|---|---|
| Cetylpyridinium chloride procedure (cellulose as sorbent: with organic solvents) | Chondroitin-4-sulphate, chondroitin-6-sulphate, dermatan sulphate | Information not available | Antonopoulos and Gardell |
| As above: without organic solvents | Heparin, chondroitin sulphates, hyaluronic acid | Isomeric chondroitin sulphates | Antonopoulos *et al.* (1964) |
| Dowex 1-X2 | Hyaluronic acid, chondroitin sulphates, heparin | Isomeric chondroitin sulphates, keratan sulphate acid, dermatan sulphates A and B | Pearce *et al.* |
| DEAE-cellulose | Hyaluronic acid, heparin; chondroitin-4-sulphate is partially separated from the mixed peak of dermatan sulphate and chondroitin-6-sulphate | Dermatan sulphate and chondroitin-6-sulphate | Braselmann and Ramm |
| ECTEOLA-cellulose | Hyaluronic acid, chondroitin sulphates, keratan sulphate | The separation of isomeric chondroitin sulphates is only partial | Antonopoulos *et al.* (1967) |
| DEAE-Sephadex A-25 | Hyaluronic acid, heparitin sulphate, chondroitin sulphates, heparin | Isomeric chondroitin sulphates | Schmidt |

Most of the methods other than chromatography are much too time consuming to be of practical value for these purposes. We describe here briefly the method of Constantopoulos *et al.* A sample of chondroitin-4-sulphate is chromatographed on a column of Sephadex G-200, giving a broad peak corresponding to the variety of polymers present in the sample. From this separated fraction, narrow cuts are made and the molecular weight of each is determined by a suitable method such as the diffusion sedimentation coefficient procedure (surprisingly, the results indicate that these cuts are not too polydisperse) and the retention volumes are plotted against the estimated molecular weights. The whole procedure is carried out because the calibration graph of the logarithm of the molecular weight *versus* retention time obtained in this manner is not identical with the calibration graph obtained with any standard protein calibration series. The advantage of this procedure, however, is that dermatan sulphate, chondroitin-6-sulphate and heparitin sulphate also follow the same relationship.

### Additional procedures

A chromatographic procedure involving the use of a mixed bed of Celite and calcium phosphate and gradient elution with phosphate buffers of pH 6.5 and ionic strength up to 0.2 *M* enables chondroitin sulphates to be separated from hyaluronic acid or a sample of hyaluronic acid to be separated into fractions with different molecular weights (Bowness, 1960).

### Evaluation of different chromatographic procedures

It is obvious that none of the separation procedures can be generally recommended for an unknown mixture of glycosaminoglycans and that one has to specify what particular type of glycosaminoglycan is being sought before starting the analysis. Some idea of the applicability of particular separations can be gained from Table 24.2. Molecular sieving alone is not suitable for the separation of these compounds. A complex procedure involving a combination of Sephadex G-50 and DEAE-cellulose has been described by Hallén. In addition to their use for purely analytical purposes, chromatographic techniques have also been applied in glycosaminoglycan chemistry to demonstrate their interactions with proteins, namely with collagen (Wasteson and Öbrink).

As far as detection is concerned, the column effluent can be scanned first by a simple colour reaction, such as the carbazole reaction for uronic acids (Dische; Bowness, 1957). A special modification of this procedure suitable for different types of automated analysis has been described by Balazs *et al.* The Elson–Morgan reaction for hexosamines is equally applicable (Boas). Further tests on those fractions which contain glycosaminoglycans are desirable. Also, purified glycosaminoglycans can be obtained from these fractions after dialysis followed by precipitation with ethanol in calcium acetate buffer.

The applicability of chromatography analysis in the field of glycosaminoglycan analysis coincide with those generally applicable for macromolecular substances of a hydrophilic nature, mainly polysaccharides.

*References p. 541*

## GLYCOPROTEINS AND GLYCOPEPTIDES

Chromatographic separations of glycoproteins are virtually indistinguishable from procedures used in protein and peptide chemistry. Also, there are a number of proteins which, in the usual sense, are not considered to be glycoproteins though they carry one or several glycosidic residues. Of course, there is a wide choice of different applications and many different glycoproteins have been separated; a detailed description of these procedures is not very appropriate from the chromatographer's point of view. On the other hand, there is a chromatographic procedure that is specific for the structural analysis of glycoproteins and glycopeptides; the specificity is based on the suitable arrangement of the detection system rather than on the separation procedure itself. The method has been described by Brummel *et al.* and will be briefly discussed here.

Carbohydrates and their acyl- and alkyloxy-derivatives react with phenol and sulphuric acid to give a yellow-orange chromogen that exhibits an absorption maximum at 485 nm for pentoses and at 489 nm for hexoses. Furthermore, it has been reported by Montgomery that amino acids and proteins do not interfere in this reaction. As this method is both simple and sensitive and does not require special treatment after the reactants necessary for generating the yellow chromogen have been mixed together, it was the method chosen for analyzing large series of samples generated from proteolytic digests of glycopeptides. Brummel *et al.* developed an automated version of this procedure that is suitable for the continuous analysis of column effluents. The segmented technique with a Technicon AutoAnalyzer has been used for this purpose. This technique was applied previously to carbohydrate analyses using orcinol (Judd *et al.*, Kesler), anthrone (Garza and Weissler, Syamanada *et al.*), Molish (Johnson *et al.*) or phenol (Robyt and Bemis) as colorimetric reagents.

### Simple carbohydrate unhydrolyzed peptide analysis

The procedure is, perhaps, best understood from the flow-sheet shown in Fig. 24.5. The effluent from the chromatographic column (B) is split into two streams, which feed the carbohydrate and peptide lines. The overflow is collected in a fraction collector through line S. The stream entering the carbohydrate line is mixed with a 2% aqueous phenol solution (line C) containing 0.2% of ARW-7 detergent (Technicon, Ardsley, N.Y., U.S.A.). The carbohydrate line is segmented with air (line D), passed through a mixing coil (F) and then to the time-delay coil (G). The length of the delay coil is adjusted such that corresponding peaks are read in the colorimeters from the peptide manifolds. The detergent is used to maintain an even segmentation in the delay coils. This system can also be used for carbohydrate analysis; the delay coil is then by-passed and the detergent is not needed.

The phenol–carbohydrate segments are mixed with concentrated sulphuric acid at a glass joint (H). An additional device that helps to smooth the base-line is the pulse suppressor (a Technicon PC 1 can be used for this purpose, but any other type will suffice) located in the sulphuric acid line immediately after the pumping manifold (E).

The complete reaction mixture is passed through another mixing coil (I) and enters a heating bath in which the temperature of the mixture is increased to 96°C. The minimum holding time required is 15 min, but the delay is frequently longer at the lower pumping speeds used for the peptide analyzer. The resulting solution is cooled (K) and recorded (L) at 490 nm in a 15-mm tubular flow-through cell. An additional pull-through ensures a steady flow through the cell.

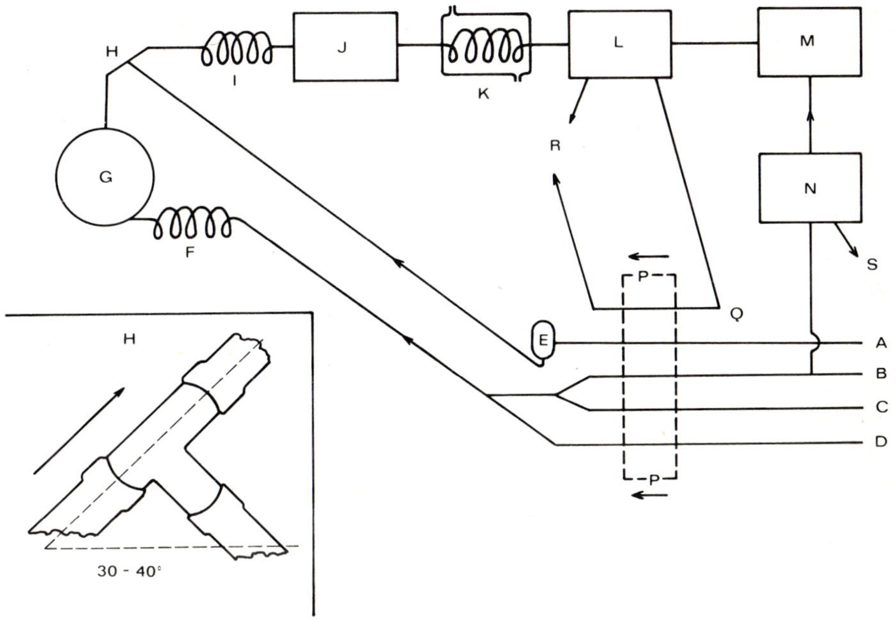


Fig. 24.5. Automatic continuous carbohydrate analyzer (Brummel *et al.*). Components (those which do not appear in Fig. 24.5 occur in Fig. 24.6): A, 0.056-in. Acidflex tubing for concentrated sulphuric acid; B, 0.020 or 0.0075-in. clear standard tubing for column effluent; C, 0.020 or 0.025-in. Solvaflex tubing for 2% (w/v) phenol and 0.2% (v/v) ARW-7 or 1.2% (w/v) phenol and 0.1% ARW-7, respectively; D, 0.035-in. clear standard for air segmentation; E, pulse supressor; F, glass mixing coil; G, time delay coil consisting of about 100 ft. of Intramedic PE 240 polyethylene tubing; H, T-junction for mixing concentrated sulphuric acid with the segmented stream; I, Glass mixing coil; J, glass coil heating bath (96°C); K, water-jacketed cooling coil; L, 490-nm colorimeter with 15-mm tubular flow cell; M, three-point recorder; N, standard peptide manifold and apparatus for ninhydrin colour production on base-hydrolyzed and unhydrolyzed column effluent; P, single-speed proportioning pump and roller assembly; Q, 0.056-in. Acidflex pull-through; R, drain; S, to fraction collector; T, 0.025 or 0.0075-in. clear standard tubing for column effluent; U, 0.045 or 0.056-in. clear standard for 13% (w/v) sodium hydroxide and 0.4 (v/v) ARW-7 or 10.3% sodium hydroxide and 0.3% ARW-7, respectively; V, 0.025-in. clear standard for nitrogen segmentation; W, 0.045-in. Solvaflex tubing for 0.3% (w/v) ninhydrin in methyl Cellosolve, water and acetate buffer (pH 5.5); X, 0.040-in. clear standard for nitrogen segmentation; Y, 0.049 in. clear standard for base-hydrolyzed desegmented stream; Z, 0.040-in. Acidflex tubing for acetic acid to neutralize the base; DB, de-bubbler; F′, glass mixing coil with side-tap for mixing ninhydrin with the nitrogen segmented stream; J′, glass coil heating bath (96°C); J″, PTFE coil heating bath (96°C): K′, water-jacketed cooling coil; L′, 570-nm colorimeter with 15-mm tubular flow cell; P′, variable-speed proportioning pump and roller assembly; Q′, 0.056-in. Acidflex pull-through.

In the schematic representation of this apparatus shown in Fig. 24.5, the peptide line in the pumping manifold is omitted for the sake of clarity as it does not differ from the usual arrangement.

### Carbohydrate–peptide-hydrolyzed peptide analysis

This variation involves a hydrolysis step; the result of the analysis is then three lines, two ninhydrin detections corresponding to the hydrolyzed and unhydrolyzed sample, respectively, and the third which results from the carbohydrate analysis. The flow-sheet of the manifold arrangement is shown in Fig. 24.6. As the system is rather complex and the use of tubing of appropriate dimensions is of decisive importance, the values used by Brummel *et al.* are summarized in Table 24.3.

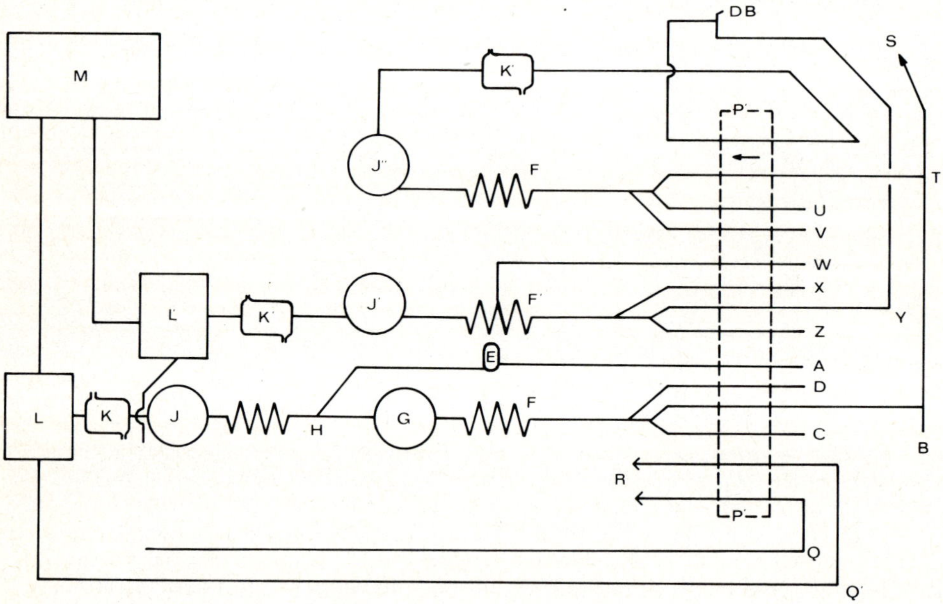


Fig. 24.6. Manifold arrangement for peptide analysis involving hydrolysis (Brummel *et al.*). The components are identified in Fig. 24.5.

Column effluent B is split into three streams. One stream enters the carbohydrate analysis system, which is identical with that shown in Fig. 24.5, the second enters the normal peptide analysis system (not included in Fig. 24.6) and the last stream is subjected to hydrolysis, as follows. After being mixed with 13% (w/v) sodium hydroxide solution and segmented with nitrogen, it passes through a heating coil in which it is heated to 96°C, cooled in a cooling coil (K′), de-bubbled in a de-bubbler (DB) and returned to the manifold at point Y. Here it is neutralized (line Z), mixed with ninhydrin in methyl Cellosolve, segmented with nitrogen and subjected to normal treatment for amino acid analysis. Effluents from the colorimeters (L and L′) are again drawn through the manifold so as to ensure a steady flow through the colorimetric cells.

TABLE 24.3
NOMINAL FLOW-RATE AS FUNCTION OF I.D. OF DELIVERY TUBES IN BOTH NORMAL AND REDUCED SYSTEMS

| System* | I.D. (in.) | | | | | Nominal flow-rate (ml/min) | | | | |
|---|---|---|---|---|---|---|---|---|---|---|
| | Tube** | | | | | Tube** | | | | |
| | B | D | A | Q | C | B | D | A | Q | |
| Normal | 0.020 | 0.035 | 0.056 | 0.056 | 0.020 | 0.16 | 0.42 | 0.92 | 0.92 | 0.16 |
| Reduced | 0.0075 | 0.035 | 0.056 | 0.056 | 0.025 | 0.03 | 0.42 | 0.92 | 0.92 | 0.23 |

*Two different sets of tubes can be used which are specified here as "normal" and "reduced".
**Letter designations refer to the following pump tubes: A, concentrated sulphuric acid; B, takeoff from column effluent; C, phenol; D, air; Q, pull-through.

All of the devices described above have to be run until equilibration of all pumps, as shown by a smooth line on the recorder, is achieved. The sample line is then dipped into solutions that contain known amounts of carbohydrate, amino acid or peptide for 2 min. The areas (peak height multiplied by the width at half-height) of the resulting carbohydrate peaks can be expressed in terms of equivalent amounts of standard sugars, but those which result from the alkaline hydrolysis of the peptide residues cannot easily be quantified as the hydrolysis occurs to different extents with the different peptide bonds.

The column effluents can be analyzed for 70–100 h, after which the Acidiflex tubing in the sulphuric acid line must be replaced. The remaining lines are replaced after a week of more or less continuous operation.

## REFERENCES


Anseth, A. and Laurent, T. C., *Exp. Eye Res.*, 1 (1961) 25.
Antonopoulos, C. A., Fransson, L.-Å., Heinegård, D. and Gardell, S., *Biochim. Biophys. Acta*, 148 (1967) 158.
Antonopoulos, C. A. and Gardell, S., *Acta Chem. Scand.*, 17 (1963) 1474.
Antonopoulos, C. A., Gardell, S., Szirmai, J. A. and de Tyssonsk, E. R., *Biochim. Biophys. Acta*, 83 (1964) 1.
Balazs, E. A., Berntsen, K. O., Karossa, J. and Swann, D. A., *Anal. Biochem.*, 12 (1965) 547.
Boas, N. F., *J. Biol. Chem.*, 204 (1953) 553.
Bowness, J. M., *Biochem. J.*, 67 (1957) 295.
Bowness, J. M., *Arch. Biochem.*, 91 (1960) 86.
Braselmann, H. and Ramm, Ch., *Acta Biol. Med. Ger.*, 24 (1970) 409.
Brummel, M. C., Mayer, H. E. and Montgomery, R., *Anal. Biochem.*, 33 (1970) 16.
Constantopoulos, G., Debakan, A. S. and Carroll, W. R., *Anal. Biochem.*, 31 (1969) 59.
Dische, Z., *J. Biol. Chem.*, 167 (1947) 189.
Garza, A. C. and Weissler, H. E., *Abstracts, Technicon Symposium 1967, New York.*
Hallén, A., *J. Chromatogr.*, 71 (1972) 83.
Johnson, E. A., Rigas, D. A. and Jones, R. T., *Abstracts, Technicon Symposium 1966, New York.*
Judd, J., Clouse, W., Ford, J., van Eys, J. and Cunningham, L. W., *Anal. Biochem.*, 4 (1962) 512.
Kesler, R. B., *Anal. Chem.*, 39 (1967) 1416.
Montgomery, R., *Biochim. Biophys. Acta*, 48 (1961) 378.
Pearce, R. H., Mathieson, J. M. and Grimmer, B. J., *Anal. Biochem.*, 24 (1968) 141.
Ringertz, N. and Reichard, P., *Acta Chem. Scand.*, 13 (1959) 1467.

Ringertz, N. and Reichard, P., *Acta Chem. Scand.*, 14 (1960) 303.
Robyt, J. F. and Bemis, S., *Anal. Biochem.*, 19 (1967) 56.
Schiller, S., Slover, A. and Dorfman, A., *J. Biol. Chem.*, 236 (1961) 983.
Schmidt, M., *Biochim. Biophys. Acta,* 63 (1962) 346.
Scott, J. E., *Biochim. Biophys. Acta,* 18 (1955) 428.
Syamanada, R., Staples, R. C. and Bloch, R. J., *Contrib. Boyce Johnson Inst.*, 21 (1962) 363.
Thunell, S., *Acta Univ. Lund,* 9 (1967).
Trudle, I. S. and Mann, G. U., *Biochim. Biophys. Acta,* 101 (1965) 127.
Wasteson, Å. and Öbrink, B., *Biochim. Biophys. Acta,* 170 (1968) 201.

*Chapter 25*

# Lower carboxylic acids

J. CHURÁČEK and P. JANDERA

## CONTENTS



## INTRODUCTION

In this chapter, the separation of lower carboxylic acids and medium-sized fatty acids (higher fatty acids are dealt with in Chapter 27), as well as of di- and tricarboxylic and keto acids (aliphatic, aromatic and cyclic), is described. However, the mechanisms of the separations do not permit a more detailed classification into sections according to these different types of compounds and hence a classification based on the different separation mechanisms is convenient (Churáček and Jandera). In the work reported in most papers, virtually all groups of acids, when they occur in mixtures, have been separated by one of the methods described below. The separation of sugar acids is discussed in detail in the chapter on carbohydrates (Chapter 22). In this chapter, only those papers are discussed which describe the chromatographic separation of sugar acids when they are present in mixtures with other types of acids.

## GENERAL TECHNIQUES

The most important technique used for the chromatography of acids is ion-exchange chromatography. Such separations are due mainly to the presence of the carboxyl group, the ionogenic properties of which permit the separation of acids on the basis of various ion-exchange mechanisms. The anions of organic acids are retained on anion exchangers by a force that depends on their acidity. This fact was made use of in chromatographic separation of organic acids on anion exchangers in the hydroxyl, carbonate, formate, acetate and chloride forms. Water, dilute formic, acetic and hydrochloric acids or buffer solutions are used for elution. The separation of two acids is the more effective the larger is the difference between their dissociation constants. According to Davies, the separation of two weak acids is sharpest when the pH of the elution liquid is one to two units lower than the value of $\frac{1}{2}(pK_1 + pK_2)$, where $K_1$ and $K_2$ are the dissociation constants of the separated acids. The separation efficiency can be increased by the use of mixtures of solvents. Organic acids are eluted from anion-exchange columns more rapidly by a methanol–water mixture than by pure water (Carroll). A non-polar solvent (*e.g.*, dioxane) decreases the degree of dissociation of weak acids, thus increasing the difference in dissociation and hence improving the separation. The use of organic solvents results in the suppression of the sorption of undissociated molecules, which decreases the sharpness of the separation.

The type of ion-exchange chromatography to be used in a particular instance depends on the nature of the acids to be separated and on the form in which the sample is supplied. If only acids are present in the sample, the situation is simple and a suitable method can be chosen directly. If, however, certain amounts of other substances are also present in the sample, it is advisable to remove them before analysis. If these constituents are of a non-ionogenic nature, the isolation should be carried out by binding the acids on a column filled with a basic anion exchanger. Non-ionogenic substances pass through the column and then the acids are displaced from the column, affording a sample which can be separated further by selected methods. This preliminary separation is necessary in almost all instances when the isolation of acids from natural material or from complex mixtures resulting from an industrial preparation of acids is needed. Owing to molecular sorption, non-ionogenic substances may also be retained on the column and interfere in the separation of the acids themselves.

Samuelson and Simonson (1962a) described a method for the separation of aldonic acids from strong acids and from substances that are not sorbed. For this separation, they made use of the sorption of aldonic acids from aqueous ethanolic solutions. It was found that it is more advantageous to use an anion exchanger in the sulphate form than a cation exchanger in the hydrogen form, which sorbed the acids weakly, or in the potassium form, when the acids were retained too strongly.

The separation of carboxylic acids on ion exchangers may take place on the basis of various mechanisms. In addition to molecular sorption caused by the interaction of the remaining part of the molecule with the ionic resin skeleton, ion exchange is often involved as well as partition based on solubility and salting-out chromatography. The molecular sorption of fatty acids on anion exchangers increases with an increase in the length of their aliphatic chain; on cation exchangers in the hydrogen form it is higher than on other forms, owing to the suppression of dissociation. If a suitable mobile phase is chosen, molecular sorption can be suppressed to a minimum and the mechanism of ion exchange enhanced.

Chromatography of acids on anion exchangers in acetate medium or in the presence of acetates of complexing cations that form complexes with acids has also been used extensively. When a borate buffer was employed, use was made of the formation of complex ions with polyhydroxy compounds. Using this principle, polyhydroxy acids can be well separated from monohydroxy acids or similar compounds.

All mechanisms of separation of carboxylic acids on ion exchangers are very efficient and their choice is therefore based only on the type and quality of the acids and on the purpose of the separation.

## SEPARATION OF CARBOXYLIC ACIDS ON THE BASIS OF MOLECULAR SORPTION, USING AQUEOUS AND NON-AQUEOUS ORGANIC SOLVENTS

Although the elution behaviour of acids is dependent on their acidity, the elution volumes of some acids differ appreciably from the values expected on the basis of their dissociation constants. This effect occurs particularly with aromatic acids, for which a strong interaction between the aromatic nuclei of the aromatic acids and the aromatic skeleton of the resinous anion exchangers (based on styrene–divinylbenzene copolymers) takes place. Purely physical adsorption of weak organic acids, which was observed on cation exchangers, increases with an increase in the molecular weight of the acid; monocarboxylic acids are more strongly bound than dicarboxylic acids. Oxalic acid and mineral acids are not sorbed (Erler). Adsorption increases with an increase in the size of the particles of the ion exchanger and with an increase in the concentration of the sorbed acid. Quantitative desorption may be achieved by elution with water. It was found that the affinity for molecular sorption of the acids on the cation exchanger increases with a decrease in the degree of cross-linking in the exchanger (Patel and Bafna, 1965). Amino acids are sorbed particularly strongly as a consequence of the interaction between the amino group and the sulphonic group of the resin (Skorokhod and Tabulo). The sorption of aliphatic acids on styrene anion exchangers is higher than on cation exchangers, and ion exchange and molecular sorption take place simultaneously; the latter can even predominate (Skorokhod and Sembur). Molecular sorption on cation exchangers in the hydrogen form is greater than on other forms as a consequence of the suppression of dissociation owing to the high concentration of $H^+$ in the ion exchanger phase (Patel and Bafna, 1968). The formation of a covalent bond between the amino groups of the anion exchanger and the carboxylic group of acids may also take place (Robinson and Mills). The molecular sorption of aliphatic fatty acids on strongly basic anion-exchange resins (Dowex 1) increases with an increase in the fatty acid chain-length. According to Starobinets and Gleim, the extent of anion exchange decreases from formic acid to butyric acid; from valeric acid onwards, both ion exchange and molecular sorption increase, which can be explained on the basis of the conformational isomerization of the hydrocarbon chains of acids.

In higher acids, a large number of rotational isomers is possible, which probably increases the dielectric constant of the acid close to the carboxyl groups and thus causes a higher degree of ionization in the resinous phase and a greater selectivity of sorption of acids with longer hydrocarbon chains (Starobinets and Gleim, Starobinets *et al.*). The selectivity depends on the size of the pores of the ion exchanger. In the narrow pores of

*References p.572*

highly cross-linked ion exchangers, rotation is hindered and therefore changes in selectivity as a consequence of conformational isomerization have not been observed (Alenitskaya and Starobinets). The molecular sorption of branched-chain carboxylic acids on cation exchangers depends on the distance of the side-chain from the carboxyl group (Patel and Bafna, 1968).

In chromatography on anion-exchange resins in aqueous organic solvents, the mechanisms of ion exchange, molecular sorption and partition (according to solubility) function interdependently. If a suitable solvent mixture is chosen, the complete suppression of molecular sorption and predominance of the ion-exchange mechanism may be achieved. In some instances the separation is improved, in comparison with the situation in purely aqueous solutions, on the basis of acidity differences. Davies and Owen used 35% aqueous dioxane and, by using displacement chromatography with hydrochloric acid solutions in this solvent they separated quantitatively a mixture of formic and acetic acids on a column filled with Dowex 1. In the same manner, they also separated mixtures of formic and phenylacetic acids, butyric and phenylacetic acids, propionic, acetic and phenylacetic acids, and acetic, phenylacetic and benzoic acids.

Harlow and Morman developed a method for the automatic separation and determination of more complex mixtures of water-soluble acids on cation-exchange resins in the hydrogen form using water as the eluent. Dowex 50W-X12, with a high degree of cross-linking, provides the best resolution of weakly ionized organic acids, and is to be preferred for general use. However, Dowex 50W-X2, with a low degree of cross-linking, provides a better separation of stronger acids, for example, a mixture of formic, acetic and

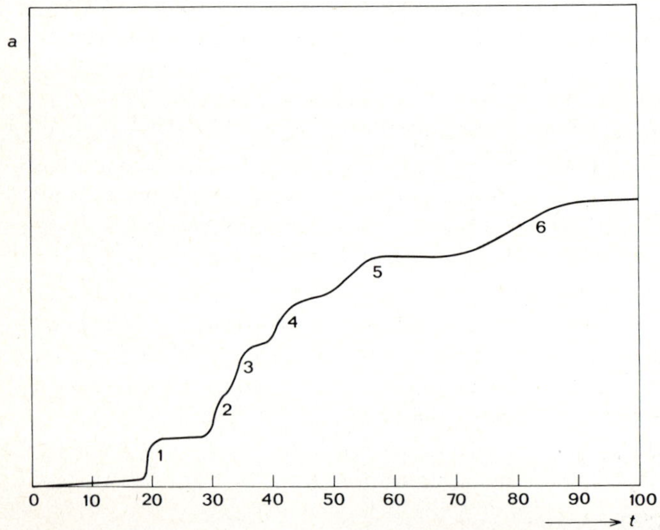


Fig. 25.1. Automatic ion-exclusion–partition chromatogram of a mixture of common acids (Harlow and Morman). Column: 50 cm × 8 mm O.D. Ion exchanger: Dowex 50W-X12 ($H^+$; 200–230 mesh). Mobile phase: water. Flow-rate: 0.2 ml/min. Detection: titrimetric. $a$ = Volume of titrant; $t$ = emergence time (min). 1 = Hydrochloric acid; 2 = formic acid; 3 = acetic acid; 4 = propionic acid; 5 = $n$-butyric acid; 6 = valeric acid.

chloroacetic acids. The effluent emerging from the column enters the titration cell, and the presence of an acid decreases the pH in the cell. The automatic titrator, which is set to maintain the pH of the solution at 8.5, senses the change through a combination glass-reference electrode. The imbalance created activates a relay in the automatic titrator, which supplies power to the motor of the titrant syringe unit, which delivers titrant (0.1 *N* sodium-hydroxide) to the cell. The pH of the solution is thus restored to 8.5, at which point the titrator stops the syringe drive. A potentiometer geared to the syringe drive provides a voltage output that is proportional to the syringe displacement and is, therefore, proportional to the amount of acid eluted. This voltage is recorded as a function of time, resulting in an integral curve as shown in Figs. 25.1 and 25.2. The amount of acid is calculated from the height of each step between lines drawn parallel to the baseline. Results are readily obtained for each acid as the number of milliequivalents per millilitre or as a percentage of the total acidity. No calibration for individual acids is required. As the detector is specific for acids, other materials generally do not interfere.

Fig. 25.1 shows an example of a separation obtained with a 50 cm × 8 mm O.D. column of Dowex 50W-X12 (200–230 mesh) with water as eluent at a flow-rate of 0.2 ml/min. Formic acid is not completely separated from acetic acid under these conditions, but this pair of compounds can be completely resolved by using a smaller amount of sample and by increasing the column length, as shown in Fig. 25.2.

Binary mixtures of naphthalene-2-sulphonic acid with acetic or *N*-caproic acid can be separated on a column of Dowex 50W-X4. The sulphonic acid was eluted with water as the eluent and then the carboxylic acid was eluted with water or aqueous acetone. The advantage of using aqueous acetone was that the longer-chain carboxylic acid could be eluted faster than when water was used as the eluent (Mehta *et al.*).

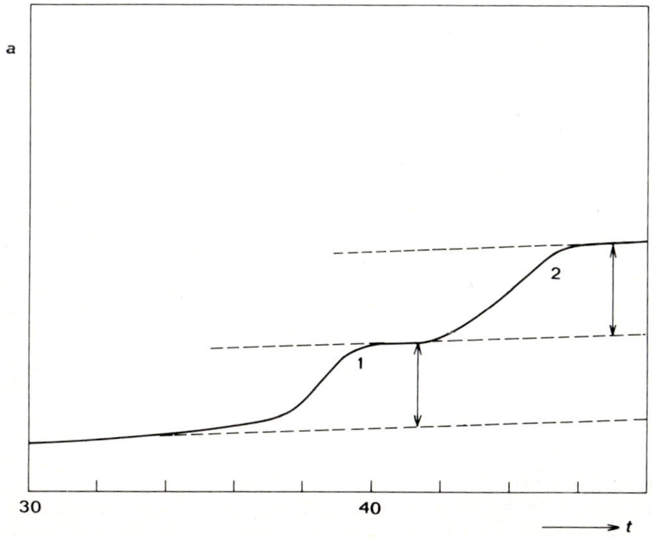


Fig. 25.2. Separation of formic acid and acetic acid (Harlow and Morman). Column: 90 cm × 8 mm O.D. Ion exchanger: Dowex 50W-X12 ($H^+$; 200–230 mesh). Mobile phase: water. Flow-rate: 0.3 ml/min. $a$ = Volume of titrant; $t$ = emergence time (min). 1 = Formic acid; 2 = acetic acid.

A column of KU-2 ($Na^+$) sulphonic acid cation-exchange resin (8 mm I.D., length of bed 230 mm) has been found to be suitable for the quantitative separation of two-component mixtures of caprylic acid with formic, acetic, butyric or *n*-caproic acid (Kresakov and Kolosova). The lower acid was quantitatively eluted in the first 50–75 ml of effluent with water, while 80% ethanol or isopropanol had to be used as the eluent for the complete desorption of caprylic acid, which was then eluted in a further 25–35 ml. The acids have been separated at a ratio from 10:1 to 1:10 with an error of less than 1%. The acids have been determined in the eluate by potentiometric titration.

Thomas separated a mixture of micromolar amounts of some phenolcarboxylic acids on a column packed with Amberlite CG-50 II ($H^+$; 200–400 mesh) using a mixture of methyl ethyl ketone–acetone–0.2 *N* hydrochloric acid (2:1:9) as eluent. The hydrochloric acid suppresses the ionization of the carboxylic groups of the ion exchanger, so that ion exchange does not take place. The substances are separated according to their polarity, polar substances being eluted faster than the less polar ones. In this manner, the following mixtures were separated successfully: 3,4-dihydroxymandelic, 3,4-dihydroxyphenylglyoxylic and protocatechuic acids and protocatechuic aldehyde; 3-methoxy-4-hydroxymandelic acid, vanilloylglycine, 3-methoxy-4-hydroxyphenylglyoxylic acid, vanillic acid and vanillin; 3,4-dihydroxymandelic acid, 3-methoxy-4-hydroxymandelic acid, vanilloylglycine, protocatechuic acid, protocatechuic aldehyde, vanillic acid and vanillin. The fractions were analyzed photometrically in UV light at 260 or 280 nm. The acids were identified by paper chromatography and UV spectroscopy. Seki (1960) described a method for separating two isomeric butyric acids and four isomeric valeric acids (pentanoic acids) as their 2,4-dinitrophenylhydrazides. For this separation he used Amberlite IRC-50 ($H^+$; 200–300 mesh) and a mixture of methyl ethyl ketone–acetone–water (2:1:9) as eluent (Fig. 25.3). The amounts of the dinitrophenylhydrazides were determined by measuring their absorption in the UV region at 340 nm.

Practical examples of the utilization of molecular sorption and elution with aqueous organic solvents for the separation of carboxylic acids are presented in Table 25.1.

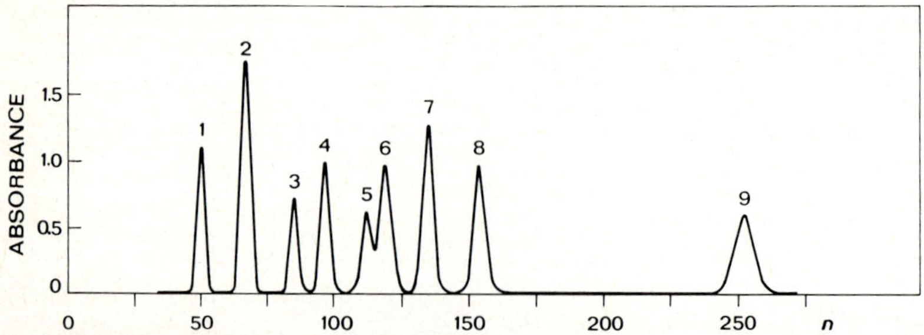


Fig. 25.3. Separation of the 2,4-dinitrophenylhydrazides of lower fatty acids (Seki, 1960). Ion exchanger: Amberlite IRC-50 (200–300 mesh). Mobile phase: methyl ethyl ketone–acetone–water (2:1:9). Detection: spectrophotometric. *n* = Fraction number. The compounds in order of their elution from the column are the 2,4-dinitrophenylhydrazides of: 1, acetic acid; 2, propionic acid; 3, isobutyric acid; 4, *n*-butyric acid; 5, trimethylacetic acid; 6, α-methylbutyric acid; 7, isovaleric acid; 8, *n*-valeric acid; 9, *n*-caproic acid. The peak of the formic acid derivative, if present, overlaps that of acetic acid.

TABLE 25.1
SEPARATION OF CARBOXYLIC ACIDS BASED ON THE PRINCIPLE OF MOLECULAR SORPTION USING WATER AND AQUEOUS ORGANIC SOLUTIONS

| Substances separated | Ion exchanger | Eluent | References |
|---|---|---|---|
| *o*-, *m*- and *p*-isomers of chlorobenzoic, nitrobenzoic and aminobenzoic acids | Styrene cation exchanger KU-2 with 8% divinylbenzene | Elution of *o*-isomers with water, *m*- and *p*-isomers with 85% ethanol | Skorokhod and Tabulo |
| Phenolcarboxylic acids and phenols | Amberlite CG-50 ($H^+$) | Methyl ethyl ketone–acetone–0.2 *N* HCl (2:1:9) | Thomas |
| Lower fatty acids (formic–lauric) | Amberlite IRC-50 ($H^+$) | Acetone–methyl ethyl ketone–water (2:1:9) and (3:1:4) | Seki (1958) |
| Phenolcarboxylic acids | Amberlite IRC-50 ($H^+$) | Methyl ethyl ketone–acetone–0.2 *N* HCl (various ratios) | Seki *et al.* |
| Lower fatty acids and phenylacetic acid | Dowex 50W-X4 ($H^+$; 100–200 mesh) | Water | Patel and Bafna (1965, 1968) |
| Citric, malic and tartaric acids | Amberlite CG-120 ($H^+$; 200–300 mesh) | Acetone–dichloromethane–water (160:100:9) | Seki (1966) |
| Fumaric, glutaric, succinic, citric, malic, tartaric acids | Amberlite CG-120 ($H^+$; 200–300 mesh) | Acetone–dichloromethane–water (20:15:1) | Seki (1966) |

**Salting-out chromatography**

A chromatographic method was suggested for the determination of impurities in terephthalic acid using salting-out chromatography on the cation exchanger Amberlite CG-45 (0.075–0.15 mm) (Calmanovici 1966, 1969). Sodium chloride solutions (0.1–0.5 *N*) in 30–70% methanol were used for elution. The acids were eluted as follows: 4-hydroxymethylbenzoic, *p*-toluic, terephthalic and trimesic acids. Coulometric titration was used for the analysis of the acids in the fractions or, if potassium salt solutions were analyzed, potassium was determined by flame photometry. The same method can also be used for semi-quantitative analyses.

*References p. 572*

Salting-out chromatography on a cation exchanger was also used by Funasaka for the separation of chlorobenzoic acids. The analysis of potassium terephthalate–benzoate mixtures by column chromatography was reported by Scoggins. Amberlite XAD-2 was used and the acid mixtures were separated by stepwise elution with saturated sodium chloride solutions and water. Glass columns, 1.2 × 25 cm and 1.8 × 35 cm, with a solvent reservoir at the top were used for analyzing salt mixtures and high-purity terephthalic acid, respectively.

A 5-ml aliquot of sample containing *ca.* 5 mg of potassium benzoate–terephthalate mixture and saturated with sodium chloride was transferred to a resin column, which had been pre-treated with saturated sodium chloride solution. The sample was adsorbed on the resin and eluted with saturated sodium chloride solution at a flow-rate of 2–3 ml/min. Air should not be allowed to enter the resin bed. The initial 100 ml of eluate were collected and reserved for the determination of potassium terephthalate. The adsorbed potassium benzoate was then eluted with water and the initial 150 ml were collected in a flask containing 2 or 3 drops of concentrated hydrochloric acid. This solution was reserved for the determination of potassium benzoate.

The column was reconditioned by washing it with several bed volumes of methanol and then with water. If air bubbles entered the column during the regeneration cycle, the methanol wash was repeated, as it was difficult to remove air bubbles during the water wash cycle.

The absorbances of the two collected solutions were measured in 1-cm cells *versus* a reagent blank, which was taken from the column prior to the addition of the sample, by scanning the 360–225 nm region. Potassium terephthalate and benzoic acid have absorption maxima at 240 and 230 nm, respectively. The concentrations of potassium terephthalate and benzoate were calculated by using previously prepared calibration graphs (calibrating solutions are not passed through the column). Both compounds obeyed Lambert-Beer's law up to concentrations of 4 and 3 mg per 100 ml of solution, respectively.

Benzoic acid in high-purity terephthalic acid was determined in a similar manner. Approximately 0.5 g of sample, dissolved in dilute potassium hydroxide solution and saturated with sodium chloride, was adsorbed on the resin and the terephthalate was eluted with 500–600 ml of saturated sodium chloride solution (complete elution of terephthalate was confirmed by scanning a portion of the eluate in the 240-nm region). Water (20 ml) was added to the column and, when the water had just entered the resin bed, elution with methanol was started. The benzoate was collected in the initial 100 ml of aqueous methanol eluate in a calibrated flask containing a few drops of concentrated hydrochloric acid. A calibration graph was prepared by eluting known amounts of benzoic acid (0.05–3.0 mg) from the column in the same manner as for the samples.

## ION-EXCHANGE CHROMATOGRAPHY OF CARBOXYLIC ACIDS IN VARIOUS AQUEOUS ACIDS OR BUFFERED SOLVENT SYSTEMS

### Chromatography of acids on anion-exchange resins in acetate medium

Ion-exchange chromatography of carboxylic acids with sodium acetate and acetic acid solutions has been widely utilized. With this method, it was possible to separate even very complex mixtures of hydroxy acids, which is important in sugar chemistry. Sodium acetate solution is a suitable eluent for the separation of ions of various monocarboxylic acids. Aldonic and uronic acids are eluted in order of decreasing molecular weight. A comparison of the elution behaviour of acids with an equal number of carbon atoms but with a different number of hydroxyl groups indicates that the forces which interact with the resin skeleton increase with a decrease in the number of these groups. It is possible to separate a number of stereoisomers, which may be ascribed to the differences in hydration and in interionic forces. Some isomeric acids, however, cannot be separated by elution with sodium acetate, and in such instances the use of acetic acid is more advantageous. During elution with acetic acid, the most important factor is the acidity of the separated acids. Weak acids are eluted more easily than strong acids. When acids are eluted with buffers composed of mixtures of acetic acid and sodium acetate, the effect of the composition of the eluting mixture on the elution volumes can be easily calculated from the law of mass action.

Elution chromatography of organic acids on anion exchangers in acetate medium has also been found advantageous for the analysis of some acids in fruit juices. Goudie and Rieman separated quantitatively a mixture of 4–9 mg of malic, tartaric and citric acids in fruit juices and separated them from sugars with a 2.0 *M* acetic acid and 0.4 *M* sodium acetate (pH = 4) on a 10 cm × 0.95 cm$^2$ column of Dowex 1-X8 ($CH_3COO^-$) at a flow-rate of 0.5 cm/min. The acids in the eluate fractions were determined by oxidation by heating them with potassium dichromate in sulphuric acid and subsequently measuring the extinction at 591 nm against the reference sample. Sugars were eluted first, then gradually malic, tartaric and citric acids. According to Goudie and Rieman, the determination took about 7 h and the standard deviation was 1% (Fig. 25.4).

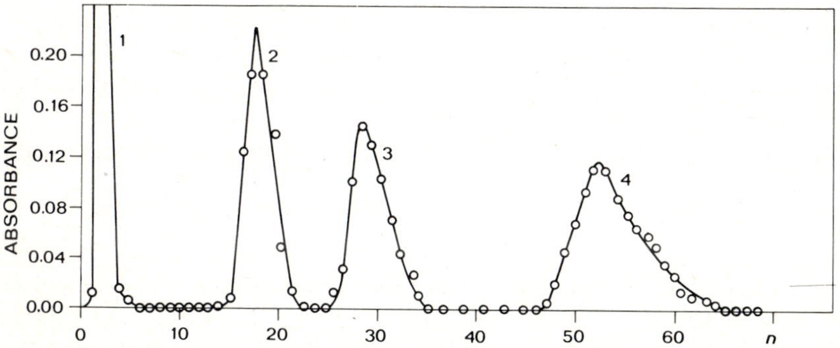


Fig. 25.4. Separation of sugar and fruit acids (Goudie and Rieman). Column: 10 cm × 0.95 cm$^2$. Ion exchanger: Dowex 1-X8 (200–400 mesh). Mobile phase: 2 *M* acetic acid + 0.4 *M* sodium acetate. Operating conditions: flow-rate, 0.50 cm/min; fractions, 2.92 ml. Detection: spectrophotometric. *n* = fraction number. 1 = sugar; 2 = malic acid; 3 = tartaric acid; 4 = citric acid.

By using gradient elution with acetic acid of increasing concentration, it was possible to separate a mixture of lactic, malic and tartaric acids (Courtoisier and Ribereau-Gayon). From the column containing Dowex 2-X8 ($CH_3COO^-$), lactic acid alone was eluted first, followed by a mixture of lactic, lactyl-lactic and malic acids, and finally tartaric acid emerged completely separated from the other acids. However, for the complete elution of tartaric acid, it was necessary to apply a large amount of concentrated acetic acid. Therefore, a procedure was used in which the elution of less strongly bound acids with acetic acid was combined with subsequent displacement of tartaric acid with 3 *N* formic acid. During this procedure, tartaric acid was separated completely from other acids and eluted with 150 ml of eluate. When 0.05 *M* sodium carbonate was used for displacement chromatography, the elution curves of lactic and malic acids overlapped to a certain extent, but the separation of lactic acid from malic and tartaric acids was satisfactory. In this instance, no formation of lactyl-lactic acid from lactic acid took place. The acids in the eluate were determined by oxidation with chromic acid at 100°C; malic acid was determined by oxidation with 0.1 *N* cerium sulphate in sulphuric acid with chromium(III) sulphate as catalyst.

Dowex 1-X8 is a very suitable ion exchanger for the separation of substituted aromatic acids. Elution volumes of various derivatives of benzoic acid are summarized in Table 25.2. The relationships between the chemical structure and chromatographic behaviour were discussed by Katz and Burtis.

Benzoic acid, which is eluted at 1073 ml, can be considered to be the base. The phenyl ring, through conjugation and induction, contributes to the stability of the benzoate ion; for example, benzoic acid was observed to be eluted later than aliphatic acids, with the exception of aconitic acid. Compounds with functional groups that decrease the stability of the benzoate ion would be expected to be eluted earlier than benzoic acid; conversely, compounds with functional groups that increase the stability should be eluted later. *p*-Aminobenzoic acid, which is eluted at 863 ml, illustrates the reduction of stability through resonance of the amino group in the *para*-position, while *o*-aminobenzoic acid, which is eluted at 1064 ml, shows increased stability through hydrogen bonding of the amino group in the *ortho*-position. The inductive power of a *meta*-hydroxyl group for stabilizing the anion is evident from the later elution of 3-hydroxyanthranilic acid at 1287 ml.

Skelley and Crumett described the separation of a mixture of 0.05–0.1 g of benzoic acid and three isomeric hydroxybenzoic acids by chromatography on a 13 × 330 mm column of the strongly basic anion exchanger Dowex 2-X8 (300–400 mesh), using a gradient of acetic acid in methanol for elution. They used an elution technique and method of detection described earlier by Skelly; 15% acetic acid eluted first benzoic acid, then hydroxy acids in order of increasing acidity, *i.e.*, *p*-hydroxybenzoic acid before *m*-hydroxybenzoic acid. Salicylic acid, which is much stronger, had to be eluted with glacial acetic acid. A good separation of benzoic or salicylic acid from *m*- and *p*-hydroxybenzoic acids was achieved. The mutual separation of these two acids, which have similar acidities, was much more difficult, but satisfactory results were still achieved. The authors indicated the possibility of using this technique also for the separation of other acids with p*K* values above 3.

Scheffer *et al.* separated mixtures of 1-cyclohexene-3,4,5-trihydroxy-l-carboxylic, tartaric, quinic, malonic, citric and maleic acids by column chromatography on the weakly basic dextran anion exchanger Sephadex A-25 at 25°C. A gradient of acetic and formic acids was used for elution. Volatile acids were eliminated from the eluate fractions

TABLE 25.2
ANION-EXCHANGE CHROMATOGRAPHY OF BENZOIC ACID DERIVATIVES AND RELATED COMPOUNDS (KATZ AND BURTIS)

Column: 316 × 0.45 cm. Ion exchanger: Dowex 1-X8 (5–10 μm). Mobile phase: acetate buffer, 0.015 *M* (pH 4.4) at the start, up to 6.0 *M* (pH 4.4) at the end. Flow-rate: 28 ml/h. Temperature: 25°C at the start, changed to 60°C after 15 h.

| Compound | Elution volume, $V_e$ (ml) |
|---|---|
| Phenol | 636 |
| Hippuric acid | 817 |
| *p*-Cresol | 834 |
| *p*-Aminobenzoic acid | 863 |
| 3-Methoxy-4-hydroxymandelic acid | 871 |
| Homovanillic acid | 916 |
| Syringic acid | 975 |
| *p*-Hydroxyphenyl-lactic acid | 1010 |
| *p*-Hydroxymandelic acid | 1019 |
| *p*-Hydroxyphenylacetic acid | 1020 |
| Anthranilic acid | 1064 |
| Benzoic acid | 1073 |
| *m*-Hydroxyphenylacetic acid | 1125 |
| Vanillic acid | 1156 |
| Phloretic acid | 1194 |
| Folic acid | 1212 |
| *o*-Hydroxyhippuric acid | 1253 |
| Salicylacetic acid | 1272 |
| 3-Hydroxyanthranilic acid | 1287 |
| *o*-Hydroxyphenylacetic acid | 1318 |
| *m*-Hydroxybenzoic acid | 1343 |
| *p*-Hydroxybenzoic acid | 1349 |
| *o*-Hydroxybenzoic acid | 1350 |
| 3-Methoxy-4-hydroxycinnamic acid | 1390 |
| Homogentisic acid | 1417 |
| α-Resorcylic acid | 1430 |
| *p*-Hydroxycinnamic acid | 1443 |

by evaporation in a vacuum, over a mixture of calcium chloride and sodium hydroxide (1:1). The solid residue was extracted with 50% methanol and titrated with 0.1 *N* sodium hydroxide solution. The method was used for the separation, isolation and determination of acids in the extracts of the roots of cereals.

Carlsson *et al.* and Samuelson and Thede developed a method for the separation of hydroxy acids in acetate medium, which is sometimes more effective than the separation in borate buffer and is complementary to it. The undesirable formation of lactones in acidic medium was prevented by prior saponification of the mixture and elution with a solution of sodium acetate instead of acetic acid. By this method, they achieved an excellent separation over a broad concentration range of the eluting solvent. The logarithms of the elution volumes are a linear function of the logarithm of the concentration of acetate ions, and this may be utilized to shorten the elution times of substances with sufficiently large separation factors when a more concentrated eluting agent is applied. On the other hand, if the concentration is decreased, the separation may be improved (the HETP decreases). The elution curves of most hydroxy acids are symmetrical and narrow. For solutions that contain acids that are strongly retained (for example, formic acid), gradual elution with sodium acetate solutions of increasing concentration is advantageous. In this way, a sharp separation can be achieved without it being necessary to use an excessive amount of the eluent for the elution of acids that are most strongly retained. On Dowex 1-X8 (40–80 μm) columns, the epimers of some hydroxy acids were separated by using 0.08 *M* sodium acetate as the eluent. Elution with 0.1 *M* sodium acetate was used by Alfredsson *et al.* (1963) for the separation of a mixture of seven hydroxy acids.

Gradient elution with sodium acetate at 67°C was successful in the separation of a mixture of pyruvic, glutaric, citric, 2-ketoglutaric and *trans*-aconitic acids when a column of Dowex 1-X8 (200–400 mesh) was used. The eluate was analyzed automatically by measuring the discoloration of a dichromate solution after oxidation of the acids. The relative error was not greater than 5%. At temperatures below 67°C, severe tailing took place and the separation was poor (Zerfing and Veening). A very good separation of a mixture of acids is represented in Fig. 25.5.

An increase in temperature usually causes a decrease in separation selectivity. The separation of some acids is, however, better at elevated temperatures (for example, glyceric from lactic acid). If an ion exchanger with a fine particle size is used, an important narrowing of the elution curves may be achieved by increasing the temperature. The acceleration of the diffusion within the ion exchanger particles decreases the HETP.

With aldonic acids, partial epimerization may take place and at elevated temperatures it may cause destruction. Therefore, during chromatography the mixtures should not contain excess of alkali and the solutions must not be heated too strongly (Larsson *et al.*, 1966b), as in the saponification of lactones.

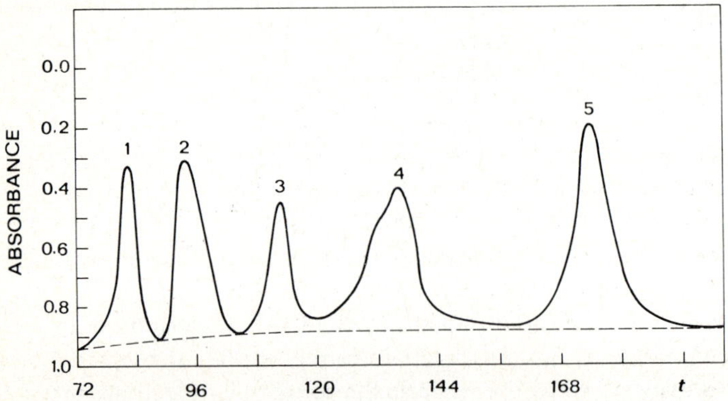


Fig. 25.5. Separation of aliphatic acids (Zerfing and Veening). Column: 23.5 × 0.9 cm. Ion exchanger: Dowex 1-X8 ($CH_3COO^-$; 200–400 mesh). Mobile phase: Sodium acetate gradient, 0.0–1.2 *M*. Operating conditions: flow-rate, 2 ml/min; temperature, 67°C. Detection: spectrophotometric at 424 nm. *t* = Elapsed time (min). 1 = Pyruvic acid, 1.3 mg; 2 = glutaric acid, 35 mg; 3 = citric acid, 1.3 mg; 4 = 2-ketoglutaric acid, 1.5 mg; 5 = *trans*-aconitic acid, 3.2 mg.

The studies of chromatographic separations in acetate medium were later extended to 44 organic acids, mainly hydroxy acids, in order to find suitable conditions for practical analyses in various branches of sugar chemistry (Samuelson and Thede). Some aldonic, aldobionic, methylated aldonic, uronic and biuronic, aldehydo and keto acids, and acids derived from some heterocycles, were investigated by using a column of the strongly basic anion exchanger Dowex 1-X8 (26–32 $\mu$m) at 30°C, after prior saponification of lactones with an automatic titrator, which maintained the mixtures at pH 8 for 5 h prior to analysis. The most convenient separation factors can be obtained in either pure sodium acetate solution or pure acetic acid. By using mixtures of these two eluents, a good separation of some substances with different acidities can be achieved. Such substances are not separated well if sodium acetate alone is used for elution, and the elution with the mixture is much faster than when acetic acid alone is used. For example, 2-ketogluconic and 5-ketogluconic acids, which cannot be separated with sodium acetate as eluent and the elution of which with acetic acid is very slow, are separated well with a mixture of acetic acid and sodium acetate.

The dissociation constants of acids are affected by various factors, such as hydrogen bonds, resonance, inductive and steric effects, and therefore the elution behaviour may be derived from the structure in only the simplest instances. Aliphatic hydroxy acids with the hydroxyl group in the $C_2$ position are stronger than those substituted in the $C_3$ position, and, therefore, 2,4-dihydroxybutyric acid is eluted later than 3,4-dihydroxy-butyric acid, and the stronger hexuronic acids appear in the eluate only after hexonic acids.

With acyclic substances, for example aldonic acids, hydroxyl groups on carbon atoms close to the carboxyl group (*i.e.*, 2-, 3- and 4-) exert a greater effect on the elution pattern than hydroxyl groups on more remote carbon atoms. Pentonic and hexonic acids are eluted in the following order of configurations: *ribo*$<$*arabino*$<$*xylo*$<$*lyxo*. Hence, from the positions on the chromatogram, the structures of some acids may be tentatively deduced, which can be of help in the identification of acids when authentic samples are not available and the dissociation constants are unknown. This rule is also valid for the $\alpha$-, $\beta$- and $\gamma$-positions in the Fischer projection of cyclic uronic acids.

The separation factors of most isomeric acids in acetic acid are more favourable than those in sodium acetate medium, and work with acetic acid is therefore more convenient. For example, by using acetic acid for elution, a mixture of 12 hydroxy acids could be completely separated in 7 h. The elution behaviour in both media is affected by different structural factors and it is therefore possible to separate even relatively complex mixtures by two-step chromatography, using the two reagents in sequence.

In their earlier work on the anion-exchange separation of carboxylic acids, Samuelson and his colleagues transferred fractions from the fraction collector into the sampling device of a Technicon AutoAnalyzer. The acids were oxidized with dichromate and the unreacted dichromate was determined automatically by its reaction with diphenylcarbazide (Samuelson and Simonson, 1962b). Later, a method was developed in which the eluate was passed directly into a monitor, which consisted of peristaltic pump, reaction coils and multichannel photometer with recorder (LKB, Stockholm, Sweden). By means of the peristaltic pump, the eluate was divided into three streams (Johnson and Samuelson) and later into four (Carlsson *et al.*) and five streams (Carlsson and Samuelson). One of these

*References p.572*

streams was passed into an automatic fraction collector, where fractions were taken for additional identification. One of the other streams was used for the automatic determination of uronic acids by the carbazole method. The second stream of the eluate was oxidized with chromic acid in order to determine all oxidizable organic acids. Later, a third channel was added, where the formaldehyde formed by the cleavage of hydroxy acids after the oxidation with periodate was determined with pentane-2,4-dione. Finally, a fourth analytical channel was introduced, where the consumption of periodate was determined automatically by UV spectrophotometry with the aim of obtaining additional information concerning the identity of various hydroxy acids.

After the eluate streams had been mixed with the reagent solutions in T-fittings and mixing coils, the mixtures were passed through PTFE reaction coils of I.D. 1.2 mm maintained at 100°C. The principle of the analysis system is shown in Fig. 25.6.

Martinsson and Samuelson used the same apparatus for the separation of about 40 hydroxy acids. The distribution coefficients of the acids in 0.5 *M* acetic acid and 0.08 *M* sodium acetate (pH 5.9) are listed in Table 25.3.

Martinsson and Samuelson also investigated the effect of different amounts of the acids applied on to the column on their elution parameters. They showed that a characteristic feature of anions that show large contributions from non-polar interaction forces to their ion-exchange affinities is that, independent of the flow-rate in the column, the elution curves tail, whereas the other species exhibit largely symmetrical curves under the operating conditions used. The chromatograms reproduced in Fig. 25.7 show that the strongly polar anions corresponding to xylonic and 2,4-dihydroxybutyric acids give almost symmetrical curves, whereas the peaks that represent three less polar acids tail.

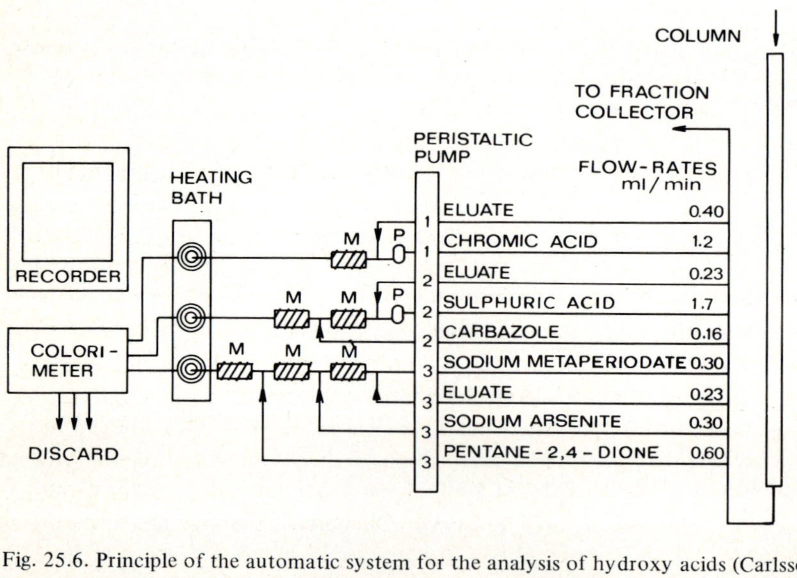


Fig. 25.6. Principle of the automatic system for the analysis of hydroxy acids (Carlsson *et al.*). 1 = chromic acid channel; 2 = carbazole channel; 3 = periodate–formaldehyde channel; M = mixer; P = pulse suppressor.

TABLE 25.3
VOLUME DISTRIBUTION COEFFICIENTS ($D_v$) IN 0.5 *M* ACETIC ACID AND IN 0.08 *M* SODIUM ACETATE (pH 5.9) (MARTINSSON AND SAMUELSON)

$K_{HB}$ is the dissociation constant of the acid.

| Acid | $D_v$ | | $K_{HB} \cdot 10^4$ |
|---|---|---|---|
| | In acetic acid | In sodium acetate | |
| 2-Hydroxyvaleric | 30.3 | 27.7 | 1.4 |
| 4-Hydroxyvaleric | 2.93 | 10.6 | 0.3 |
| 2-Hydroxybutyric | 20.3 | 17.6 | 1.4 |
| 3-Hydroxybutyric | 3.98 | 11.2 | 0.4 |
| 4-Hydroxybutyric | 2.94 | 10.6 | 0.3 |
| 3-Hydroxypropionic | 3.60 | 11.6 | 0.3 |
| 2-Hydroxy-3-methylvaleric | 43.7 | 44.7 | 1.2 |
| 2-Hydroxyisovaleric | 27.8 | 26.1 | 1.3 |
| 2-Hydroxy-2-methylbutyric | 17.2 | 17.8 | 1.2 |
| 2-Hydroxyisobutyric | 11.3 | 13.6 | 1.0 |
| 2-Hydroxymethylisobutyric | 3.29 | 15.1 | 0.2 |
| D-Allonic | 9.20 | 8.32 | 1.5 |
| D-Altronic | 11.8 | 7.50 | 1.9 |
| D-Gluconic | 12.5 | 7.21 | 2.2 |
| D-Idonic | 12.6 | 6.82 | 2.3 |
| 2-Deoxy-D-*arabino*-hexonic | 2.06 | 6.44 | 0.3 |
| 2-Deoxy-D-*lyxo*-hexonic | 1.97 | 6.74 | 0.2 |
| 3-Deoxy-D-*lyxo*-hexonic | 7.07 | 6.61 | 1.2 |
| 3-Deoxy-D-*xylo*-hexonic | 8.82 | 6.85 | 1.6 |
| 6-Deoxy-D-galactonic | 11.5 | 8.31 | 1.7 |
| 2,6-Dideoxy-D-*ribo*-hexonic | 2.32 | 8.45 | 0.2 |
| 2-Deoxy-DL-*erythro*-pentonic | 2.65 | 8.44 | 0.3 |
| 3-Deoxy-D-*erythro*-pentonic | 9.5 | 8.3 | 1.4 |
| 2,5-Dihydroxyvaleric | 13.6 | 10.8 | 1.5 |
| DL-3,5-Dihydroxy-3-methylvaleric | 4.21 | 11.8 | 0.4 |
| DL-*erythro*-2,3-Dihydroxybutyric | 22.2 | 11.5 | 2.5 |
| 3-Deoxy-2-C-hydroxymethyltetronic | 15.8 | 8.5 | 2.4 |
| D-2,4-Dihydroxy-3,3-dimethylbutyric | 19.8 | 22.0 | 1.1 |
| 2-Tetrahydrofuroic | 18.7 | 10.9 | 2.2 |
| 2,5-Anhydro-D-gluconic (chitaric) | 33.2 | 12.9 | 3.4 |
| 2,5-Anhydro-D-mannonic (chitonic) | 25.5 | 10.6 | 3.2 |
| 2,5-Anhydro-D-talonic | 21.7 | 8.55 | 3.4 |
| Arabinuronic | 37.4 | 15.2 | 3.3 |
| Lyxuronic | 29.4 | 11.6 | 3.4 |
| Riburonic | 20.9 | 8.4 | 3.3 |
| Xyluronic | 29.3 | 14.0 | 2.7 |
| 6-Deoxy-*gluco*-hepturonic | 3.16 | 7.52 | 0.4 |
| Quinic | 14.5 | 7.50 | 2.5 |
| Shikimic | 6.50 | 11.1 | 0.7 |
| 3-Hydroxy-2-methyl-1,4-pyrone (maltol) | 1.14 | 1.40 | |

*References p. 572*

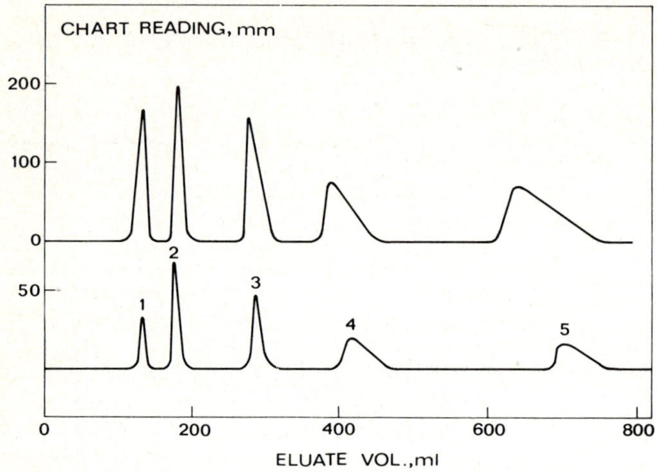


Fig. 25.7. Influence of the amount of acid applied on the elution volume (Martinsson and Samuelson). Column: 1065 × 4 mm. Ion exchanger: Dowex 1-X8. Mobile phase: 0.08 *M* sodium acetate (pH 5.9). Flow-rate: 5.2 ml/min·cm². Detection: spectrophotometric. 1 = Xylonic acid, 0.45 mg (1.78 mg); 2 = 2,4-dihydroxybutyric acid, 1.09 mg (4.3 mg); 3 = 2-methylbutyric acid, 1.07 mg (4.26 mg); 4 = 2-hydroxyisovaleric acid, 1.44 mg (5.74 mg); 5 = 2-hydroxy-3-methylvaleric acid, 2.05 mg (8.2 mg). Amounts in parentheses refer to the upper trace, and the other amounts to the lower trace.

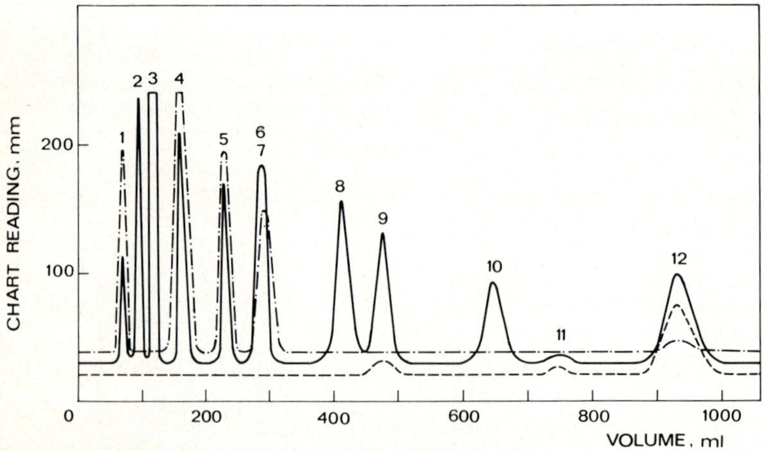


Fig. 25.8. Separation of 2-deoxygalactonic acid, 1.8 mg (1); 4-hydroxyvaleric acid, 7 mg (2); 3-hydroxybutyric acid, 3.5 mg (3); talonic acid, 1.5 mg (4); allonic acid, 1.4 mg (5); 6-deoxygalactonic acid, 1.5 mg (6); altronic acid, 1.2 mg (7); 2-hydroxy-2-methylbutyric acid; 2.2 mg (8); 2-hydroxybutyric acid, 6.5 mg (9); chitonic acid, 2.4 mg (10); xyluronic acid, 2.0 mg (11); and arabinuronic acid, 3.0 mg (12) (Martinsson and Samuelson).——, chromic acid method; – – –, carbazole method; –·–·–, periodate–formaldehyde method. Eluent: 0.5 *M* acetic acid. Flow-rate: 4.3 ml/min·cm². Resin bed: 6 × 750 mm, Dowex 1-X8 ($CH_3COO^-$).

Another observation of practical importance is that the peak elution volumes of the latter acids decrease markedly with an increase in the loading on the column. These results show that the non-polar species exhibit non-linear (convex) exchange isotherms, which means that with these species the observed $D_v$ values do not represent the true volume distribution coefficients. The later the position of these compounds on the chromatogram, the more pronounced is the tailing and the shift in position as a result of changes in the amount applied to the column.

In the same paper, the effect of the detection sensitivity on the detection selectivity of single acids was also reported. A chromatogram from a run with a mixture of 12 monocarboxylic acids is reproduced in Fig. 25.8. It can be seen that 11 discrete peaks were recorded in the chromic acid channel. One of the peaks (denoted 6,7) contained both 6-deoxygalactonic and altronic acids. Of these two acids, only altronic acid contains a primary hydroxyl group with a vicinal hydroxyl group and therefore only this acid is recorded in the periodate channel. The results demonstrate the usefulness of multiple-channel analyzers, not only for identification purposes, but also for the resolution of overlapping peaks that contain more than one compound.

### Chromatography of acids using acetates of complexing cations

Experiments aimed at speeding up the separation of acids by using eluting agents that contain a cation which can form a complex with the separated acids have been to some extent successful. Samuelson derived theoretical relationships that express the effect of the complexing constant on the elution behaviour of acids. In actual chromatography, the central ion is present in an appreciable concentration, while ligands occur only in trace amounts, and therefore the use of the deduced relationships is not absolutely justified; hence only crude qualitative predictions can be made with their use. Substances which form non-sorbable complexes, as for example aldonic acids, appear rapidly in the eluate. These acids can then be easily separated from other acids, for example uronic acids, which do not form complexes. Appreciable differences in elution behaviour exist among the various acids.

For the elution of aldonic and uronic acids, 0.05 *M* copper(II) acetate was originally used (see also Chapter 22). Aldonic acids formed strong complexes, which would not be sorbed and which were, therefore, eluted rapidly. Uronic acids were eluted much later. Hence, the conditions for the group separation of aldonic acids from uronic acids and the subsequent separation of some uronic acids are favourable. In spite of this, however, satisfactory separations could not be achieved because uronic acids were oxidized, with the simultaneous formation of copper(I) oxide (Johnard and Samuelson; Samuelson). For these reasons, 0.05 *M* zinc acetate was used as a complexing agent by Larsson *et al.* (1966a). On Dowex 1 of particle diameter 40–60 μm, the separation of galactonic, lactic, galacturonic, glucuronic, formic and pyruvic acids was achieved; on an anion exchanger with even finer particles (13–18 μm), a mixture of galactonic, arabinonic, glycolic, levulinic, glucuronic, glyoxylic and formic acids was well separated. As most acids form non-sorbable complexes with $Zn^{2+}$ ions, the distribution coefficients were appreciably lower than in sodium acetate solutions. The order of elution is given by the stability constants of the complexes and the selectivity coefficients of anions that do not form complexes. The separation factors of some acids differed to a certain extent on both columns with ion exchangers of different particle size. With decreasing concentration of the eluting agent, the separation improved, similarly as in the elution with sodium acetate. The elution curves were broadened, but the distances between them increased. For example, the separation of galactonic and lactic acids was more satisfactory at lower concentrations of the eluting agent, but the order of elution of the acids was independent of the concentration of the eluting agent. In order to prevent the formation of complexes being influenced by changes in the acidity of the solutions, it is necessary to maintain a constant pH. The optimum recommended pH is 4.6; however, an increase to pH 6 does not affect the separation substantially.

The group separation of aldonic acids from galacturonic and glucuronic acids serves as a practical application, as it is quantitative even when the amount of the separated substances is large. If zinc acetate is used, the separation is sometimes appreciably improved in comparison with the separation when sodium acetate is used as the eluent (for example, the separation of lactic and glucuronic acids, or erythronic and levulinic acids, which with sodium acetate are eluted in the same elution zone). In other instances, separation with sodium acetate may give better results, or the methods may complement each other.

A solution of zinc acetate cannot be used for the separation of mixtures that contain oxalic acid, because zinc oxalate is only slightly soluble. For the chromatography of these mixtures, elution with 0.2 *M* magnesium acetate has been used (Lee and Samuelson). A number of acids form non-sorbable complexes with magnesium ions. Their stability may be appreciably affected by the choice of pH. At low pH values, the dissociation of dicarboxylic acids is suppressed and their anions behave as monovalent ions, and the distribution coefficients of those dicarboxylic acids which do not form complexes with magnesium ions therefore also decrease. Maleic acid is an exception because it is strongly retained by the resin, as the forces of interaction of the double bond with the resin skeleton evidently predominate. Oxalic acid forms non-sorbable complexes that are stable within a wide pH range and it is eluted first with the magnesium acetate solutions, while in sodium acetate medium it is strongly retained. Tartaric acid is eluted much later and its elution can be accelerated by increasing the magnesium acetate concentration. However, maleic acid is strongly retained even in 1 *M* magnesium acetate.

Lactic and glucuronic acids could not be separated by elution with sodium acetate, but when zinc acetate of pH 4.6 was used non-sorbable zinc lactate was formed, which could be separated from glucuronic acid. In 0.05 *M* magnesium acetate, these acids do not separate at pH 4.6 but they can be separated at pH 3.3. At the latter pH, a complex is not formed to any appreciable extent. The improved separation at the lower pH may be explained on the basis of differences in the acidities of the two acids. From this result, it is evident that magnesium acetate is sometimes less effective than zinc acetate.

An increase in temperature causes a decrease in elution volume and hence also a narrowing of the elution curves, which is reflected in a general improvement and acceleration of the separation.

Elution with magnesium acetate was used for the separation of a mixture of milligram amounts of acids at 83°C (Lee and Samuelson). By a gradual elution with 0.2, 0.5 and 0.1 *M* magnesium acetate of pH 4.8, a mixture of oxalic, tartronic, tartaric and maleic acids was completely separated.

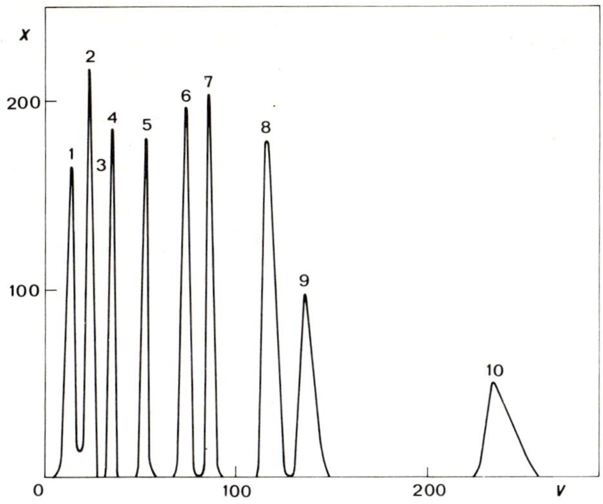


Fig. 25.9. Separation of di- and tri-carboxylic acids (Bengtsson and Samuelson). Column: 135 cm × 3.2 mm. Ion exchanger: Dowex 1-X8 (0.017–0.02 mm). Mobile phase: 0.3 *M* magnesium acetate, pH 7.0. Operating conditions: flow-rate, 4.5 ml/min·cm$^2$ ; temperature, 70°C. Detection: spectrophotometric. *X* = chart reading (mm); *V* = eluate volume (ml). 1 = Oxalic acid (5 mg); 2 = malonic acid (1.5 mg); 3 = dihydroxytartaric acid (1.0 mg); 4 = malic acid (0.5 mg); 5 = galactaric acid (0.5 mg); 6 = citraconic acid (0.8 mg); 7 = maleic acid (0.5 mg); 8 = itaconic acid (1.0 mg); 9 = *cis*-aconitic acid (1.0 mg); 10 = suberic acid (2.0 mg).

Glucaric and galactaric acids appeared in the same elution band in the run at 30°C, but at 83°C the separation was good enough to permit a quantitative evaluation. For this reason, the columns were maintained at 70°C during the chromatography (Bengtsson and Samuelson). The application of magnesium acetate solution at pH 7 for practical analytical separation of acid mixtures is illustrated in Fig. 25.9. A column 3.2 mm in diameter and 135 cm long, filled with Dowex 1-X8 with a particle diameter of 0.017–0.020 mm, was used. The temperature was maintained at 70°C and the flow-rate was 4.5 ml/min·cm$^2$.

Two of the acids involved, malonic and dihydroxytartaric, exhibited only slightly different distribution coefficients, and their elution curves overlapped almost completely. At pH 3.9, the distribution coefficients of these two acids differ markedly and an excellent separation was achieved.

Sharper bands of the last eluted compounds in the separation shown in Fig. 25.9 can be obtained by increasing the concentration of the magnesium acetate, and the speed of the separation can be increased by the application of gradient elution.

## Chromatography of acids on anion-exchange resins in borate medium

For the separation of sugars on anion-exchange resins, Khym and Zill made use of the well-known fact that polyhydroxy compounds form complex ions with borates. This method was also used for the separation of hydroxy acids. The mechanism of their sorption in borate buffer is a combination of the anion exchange of carboxylic acids and the formation of a complex between borate ions and the hydroxy groups of the acids. Better separation factors and a more effective separation can therefore be achieved in a borate-containing medium than in media in which only the ion-exchange mechanism operates, as for example in separations in sodium acetate medium. This is especially true for anions with several hydroxy groups, which acquire a larger charge owing to the formation of complexes with borates. Therefore, many polyhydroxy acids are strongly retained on anion exchangers in the borate form.

Schenker and Rieman employed this method, *i.e.*, elution of a Dowex column with borate buffers, for the separation of malic, tartaric and citric acids in fruits. The separation took about 8 h and the acids were determined with an accuracy of ± 0.1 mg. A 25 cm × 3.8 cm$^2$ column of Dowex 1 (100–200 mesh) was used, with a flow-rate of 0.8 ml/min · cm$^2$. The eluents were: A, 0.08 *M* sodium nitrate, 0.0013 *M* sodium tetraborate and 0.3 *M* boric acid; and B, 0.16 *M* sodium nitrite, 0.0013 *M* sodium tetraborate and 0.3 *M* boric acid. The fruit juice was filtered and 1 ml of the filtrate titrated with 0.1 *N* sodium hydroxide. The total concentration of the fruit acids was then calculated and solid boric acid added until a 0.30 *M* solution was obtained. A sample containing not more than 24 mg of acids was introduced on to the column and eluted with 627 ml of eluent A, followed by 500 ml of eluent B. A good separation was obtained, with the following sequence of elution of the acids: malic, tartaric, citric. The determination of the acids was carried out by total oxidation with permanganate to carbon dioxide and water.

This work was later repeated and extended; for the analysis of the eluate, a flow-through differential refractometer was used (Shimomura and Walton). In the 1–10 mg range, the height of the peaks was directly proportional to the content of acids in the sample. In this study, borate buffer of pH 7.1 was used, which had the best buffering properties. It was observed that dicarboxylic acids were retained more strongly than monocarboxylic acids, and that a higher number of hydroxy groups and double bonds caused stronger sorption, which is in agreement with theoretical considerations. The affinity for the resin decreased in the following order: oxalic>malonic>tartaric acid. Fumaric acid was bound more strongly than maleic acid.

The dependence of the logarithms of the volume distribution coefficients of hydroxy acids on the logarithm of the concentration of the eluting agent is linear. At lower concentrations of borate ions, the separation factors of many acids and their separations are improved. By elution with 0.04 *M* sodium tetraborate, α,β- and β,γ-dihydroxybutyric acids were separated quantitatively, while the separation of lactic and glycolic acids was successful when 0.015 *M* sodium tetraborate was used. For the complete separation and elution of more complex acid mixtures, for example lactic, glycolic and the two dihydroxybutyric acids, as well as acids derived from sugars with five and six carbon atoms, a very large volume of eluting solution is necessary, which prevents practical application. In this event gradual elution with solutions of increasing borate concentration should be applied. In the first step, lower acids are eluted with 0.04 *M* borate; after elution of β,γ-dihydroxybutyric acid, the borate concentration is increased to 0.07 *M* in order to increase the rate of elution of higher acids (Alfredsson *et al.*, 1962). The successful separation of lower acids (for example, glycolic and lactic acid) is simpler and more effective in acetate medium.

Elution curves in borate medium are usually broader than those in acetate medium. Excessive broadening of the elution curves is often caused by too slow an equilibration due to slow diffusion within the resin particles or to the slow attainment of complex-formation equilibrium. At an elevated temperature, equilibration is accelerated in both instances, which often leads to narrowing of the elution curves and to an improved separation.

In tetraborate solution, equilibria exist between borate anions with different charges. An increase in temperature causes an equilibrium shift in a 0.1 *M* solution towards the formation of anions with one negative charge, which are weaker eluents than ions with higher charges. Therefore, the elution volumes of aldonic acids and formic acid increase with an increase in temperature, which contrasts with the behaviour in metaborate solutions.

An anion-exchange chromatographic method was developed for the automatic determination of idonic and gluconic acids on a 200 × 9 mm column of Dowex 1-X8 ($Cl^-$; 400 mesh), in which the elution was performed with a 0.4 *M* borate buffer of pH 7.35 containing 0.05 *M* sodium chloride at 30°C (Aoki *et al.*). The acids were determined by the periodate consumption method. The addition of sodium chloride and the use of an elevated column temperature favoured the separation of the elution bands. This method was also useful for the separation and determination of 5-oxogluconic, arabonic and 2-oxogluconic acids in reaction mixtures.

**Chromatography of acids on anion-exchange resins in the formate, nitrate and chloride forms**

The separation of a mixture of carboxylic acids by displacement chromatography on a column of Duolite A-40 was described by Lesquibe and Lesquibe and Rumpf. The acids were displaced with an acid that was stronger than all of those present in the mixture (0.05 *N* nitric acid), and they appeared in the eluate in order of increasing acidity constants. The pH and the concentration of the acids were measured in the eluted fractions by titration with 0.01 *N* sodium hydroxide. A weak auxiliary acid was added to the mixture, which was eluted first and reacted with trace amounts of alkalis that remained on the column after incomplete regeneration and which otherwise caused low results in the determination of the weakest acid in the mixture. Before separation, an acid slightly stronger than the strongest acid in the mixture was added in order to prevent the penetration of a small amount of nitric acid into the strongest acid and thus avoid higher results. Using this method, it was possible to separate lactic, tartaric and oxalic acids, even when present in a sample at a concentration of $6 \cdot 10^{-4}$ mequiv./g, with an error of less than 10%.

Lawson and Purdie studied the conditions for the chromatography of organic acids on anion exchangers using formic acid as eluent. They found that the degree of cross-linking of the strongly basic anion exchanger Dowex 1 does not affect the order of elution of carboxylic acids. The molarity of formic acid necessary for the elution of 94 acids from a Dowex 1-X10 column was determined by Davies *et al.* The elution behaviour depends on the p*K* of the separated acids, and the solubility of the acids in formic acid is also important, because it affects tailing. By choosing a suitable concentration gradient of

*References p.572*

formic acid, a complete or partial separation of some acids was achieved (malic from mesotartaric, succinic from adipic and tartaric from quinolinic acid) and the tailing was suppressed. The eluate was collected in fractions the composition of which was analysed by paper chromatography after the prior elimination of formic acid by vacuum evaporation to dryness over silica gel.

The method of Lawson and Purdie was used with a 200 × 10 mm column containing Dowex 1-X10 (200–400 mesh) for the determination of the content of non-volatile organic acids in apples by Salkova and Nikiforova. Using gradient elution with formic acid after the removal of sugars, it was possible to determine the contents of malic acid, citric acid and succinic acid, which are the main acidic components, and the content of some other acids, including chlorogenic, shikimic and quinic acids.

On a 12 × 1 cm column of Dowex 1-X8 (200–400 mesh), micromole amounts of quinolinic acid were separated from other pyridine derivatives by gradient elution with 0–4 *M* formic acid. The eluted acid was determined photometrically at 254 nm after decarboxylation to nicotinic acid (Pallini). By elution with formic acid, some uronic and aldobiuronic acids were successfully separated on modified Dowex 1-X4 and Dowex 1-X8 resins by Fransson *et al.*

Egashira (1961) investigated the separation of organic acids on a column of the strongly basic anion exchanger Dowex 1-X8 ($Cl^-$). He calculated theoretical elution volumes of the acids from their characteristics and found a linear relationship between the elution volumes and $[Cl^-]^z$, where $[Cl^-]$ is the concentration of chloride ions in the eluent and $z$ is the number of acidic groups in the completely dissociated acid. The elution volume was considerably affected by temperature and the peak width was directly proportional to the square-root of the flow-rate of the eluent. The sodium chloride solutions used for elution were buffered, then the buffer was eliminated from the eluate by means of a cation-exchange column (Amberlite XE-64 and Dowex 50W) and the acids were determined by titration with 0.01 *N* sodium hydroxide or by measuring the coloration produced with bromophenol blue. In this manner, acetic, succinic, maleic, fumaric and citric acids were separated, using 0.01 *M* sodium chloride solution at pH 2, 0.1 *M* sodium chloride at pH 4, or 0.15 and 0.2 *M* sodium chloride at pH 12 for elution. In view of the linearity of the development of coloration and the instability of the indicator solution, the determination was not quantitative (Egashira, 1966, 1968).

A column of a weakly basic anion exchanger in the chloride form was used for the separation of mixtures of mono- and dichloroacetic acids (Anderson). Sulphurous acid or some other acid with an ionization constant between that of mono- and dichloroacetic acid was added to the sample, which was then passed through the column. The acids being separated were then displaced with 1 *N* hydrochloric acid. Monochloroacetic acid was eluted first, followed by sulphurous acid and then dichloroacetic acid. The acids were partially separated. Multiple cycles can be used to improve the resolution.

Mixtures of mono-, di- and trichloroacetates can be quantitatively separated by stepwise elution with 0.05, 0.1 and 2 *N* sodium chloride, respectively; mixtures of monochloroacetate, 2,4-dichlorophenoxyacetate and trichloroacetate can also be separated with the same sequence of sodium chloride eluting solutions (Tsitovich and Kuzmenko).

An interesting possibility for the chromatographic separation of some aromatic acids was reported by Lee *et al.* These compounds are usually strongly retained by anion-exchange resins and therefore a high concentration and large volume of eluent and a long elution time are required for such acids. A method was suggested involving the use of iron(III) chloride–organic solvent solution as eluent, which reacts with some aromatic organic acids such as salicylic acid, aromatic hydroxamic acids and phenols to form coloured, stable, non-adsorbable complexes. The formation of such a complex leads to reduced sorption on the anion-exchange resin.

## HIGH-SPEED ION-EXCHANGE CHROMATOGRAPHY OF CARBOXYLIC ACIDS WITH ANION EXCHANGERS OF CONTROLLED SURFACE POROSITY

The modern high-speed chromatography of carboxylic acids permits rapid separations, but in fact this method is in principle a highly developed form of ion-exchange chromatography involving the use of modern highly effective carriers and phases.

High-speed ion-exchange chromatographic separations can be carried out by using ion-exchange column packings of controlled surface porosity introduced by Kirkland. These anion exchangers consist of hard, spherical siliceous particles with a solid, impervious core that is surrounded by a thin superficially porous shell, about 2 μm in thickness. The ion-exchange medium is a methacrylate polymer containing strongly basic tetraalkylammonium groups.

These Zipax-supported strong anion exchangers are of low capacity (about 12 μequiv./g) and are intended for use with small samples in analytical-scale applications with equipment with a low dead volume and with high-sensitivity detectors. The most useful mobile phase for such a column is distilled water, with variations in the pH and ionic strength used to vary the resolution and retention times. Sodium sulphate, sodium nitrate and sodium acetate are usually used to change the ionic strength. The retention times are greatly influenced by very small changes in ionic strength. For instance, terephthalic acid is strongly retained in 0.004 *M* sodium nitrate, but by increasing the salt concentration to 0.012 *M*, the elution is easy to perform.

A mobile phase can usually be found such that the sample will be resolved in a few minutes without changing the mobile phase. In this respect, Zipax ion-exchange columns of controlled surface porosity do not behave like a classical ion-exchange column, but rather like adsorptive columns with special affinities for charged solutes.

An example of the use of high-speed, high-pressure anion-exchange chromatography for the rapid separation of a binary mixture containing maleic and fumaric acids is shown in Fig. 25.10. The isomeric acids were separated in about 90 sec, which is considerably faster than by conventional gel ion-exchange chromatography.

The same column (1000 $\times$ 2.1 mm, packed with controlled surface porosity Zipax support coated with anion-exchange resin) was used for the separation of three aromatic carboxylic acids, which were eluted in the order benzoic, toluic, terephthalic acid. Using distilled water buffered to pH 9.2 with the ionic strength adjusted to 0.02 *M* by the addition of ammonium nitrate, it was possible to achieve baseline resolution of the three acids in less than 10 min (Henry and Schmit).

Fig. 25.11 demonstrates another example of the anion-exchange chromatographic separation of carboxylic acids with an anion exchanger of controlled surface porosity

*References p.572*

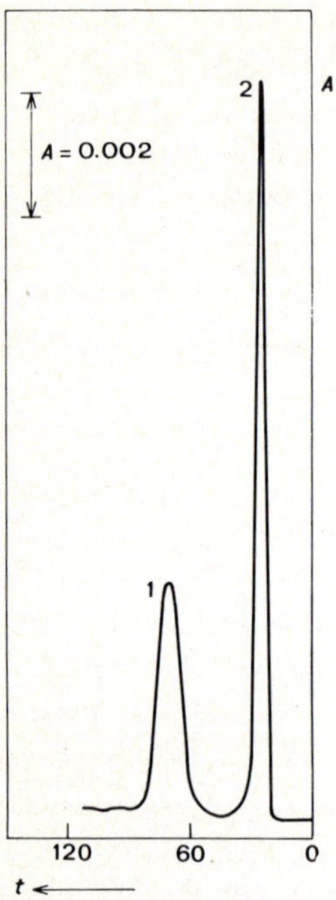


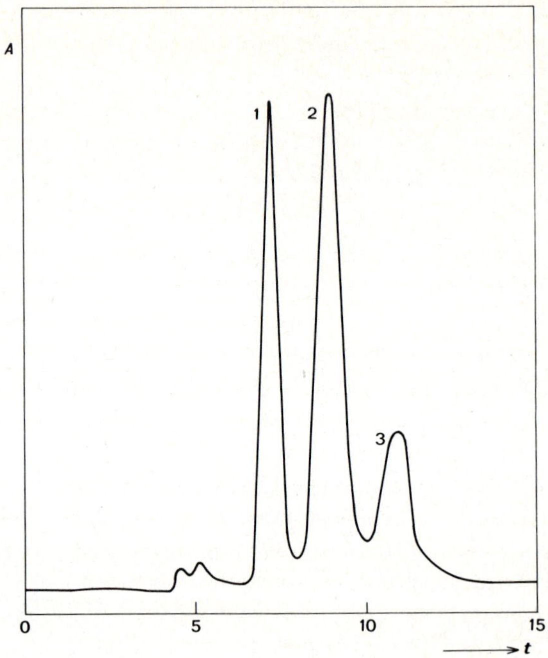


Fig. 25.10. Separation of isomeric acids by controlled surface porosity anion-exchange chromatography (Kirkland). Column: 1000 × 2.1 mm. Ion exchanger: anion exchanger of controlled surface porosity. Mobile phase: 0.01 *N* nitric acid. Operating conditions: carrier flow-rate, 2.73 ml/min; input pressure, 1900 p.s.i.; temperature, 60°C. Detection: spectrophotometric, $t$ = time (sec); $A$ = absorbance. 1 = Maleic acid; 2 = fumaric acid. Sample, 3 $\mu$l of 0.5 mg/ml each in 0.01 *N* nitric acid.

Fig. 25.11. Separation of phthalic acid isomers (Henry and Schmit). Column: 1000 × 2.1 mm. Anion exchanger: controlled surface porosity Zipax. Mobile phase: borate buffer, pH 9.2, containing 0.02 *M* sodium nitrate. $t$ = Retention time (min); $A$ = absorbance. 1 = Phthalic acid; 2 = terephthalic acid; 3 = isophthalic acid.

(Henry and Schmit). Phthalic acid isomers are separated within 15 min by elution with 0.02*M* sodium nitrate at pH 9.2. These acids all decompose or rearrange when heated and therefore cannot be vaporised to allow gas chromatographic analysis.

pH has a great effect on retention and resolution, and this effect may change the elution order of the sample components. For instance, by adjusting the pH to 2.75, the order of elution of phthalic acid isomers becomes terephthalic, isophthalic, phthalic acid.

Longbottom determined nitrilotriacetic acid by high-speed ion-exchange chromatography. A column packed with Zipax support coated with a strong anion exchanger was used, the mobile phase being 0.02 *M* $Na_2P_4O_7$. The flow-rate was 0.5 ml/min and the column inlet pressure 1000 p.s.i.

## OTHER SEPARATION TECHNIQUES FOR CARBOXYLIC ACIDS

### Separation of carboxylic acids on silica gel columns

Ion exchangers are not the only packings for chromatographic columns used in the separation of carboxylic acids. In some papers, for example, the use of silica gel is described (Moehler and Pires, Nakajima and Tanenbaum, Stamley and Moseley. Markova and Smirnov separated chloroacetic, acetic and succinic acids in this way. An example of a qualitative separation of organic acids from plant material on silica gel is shown in Table 25.4 (Freeman).

TABLE 25.4
SEQUENCE OF ELUTION OF ACIDS FROM A SILICA GEL COLUMN (FREEMAN)

The acids were eluted with chloroform containing a progressively increasing proportion of *n*-butanol. The stationary phase was 0.5 *N* sulphuric acid. Fraction volume: 2.6 ml.

| Acid | Elution range (fraction numbers) | Peak maximum (fraction number) | Acid | Elution range (fraction numbers) | Peak maximum (fraction number) |
|---|---|---|---|---|---|
| *n*-Butyric | 1–10 | 2 | 5-Pyrrolidone-2-carboxylic | 65–80 | 69 |
| *n*-Valeric | 1–9 | 2 | Glyoxylic | 53–81 | 70 |
| Isobutyric | 1–12 | 3 | Diglycollic | 64–76 | 70 |
| Propionic | 5–19 | 8 | Oxalacetic | 67–79 | 73 |
| Acetic | 22–29 | 25 | Oxalic | 76–95 | 81 |
| Mesaconic | 24–30 | 27 | Tricarballylic | 80–93 | 88 |
| Pyruvic | 28–38 | 32 | Glycollic | 84–98 | 92 |
| Adipic | 28–38 | 32 | Nitric | 87–147 | 93 |
| Formic | 31–38 | 34 | *cis*-Aconitic | 97–108 | 103 |
| Glutaric | 32–37 | 34 | DL-Malic | 103–120 | 111 |
| Citraconic | 34–45 | 39 | Citric | 134–153 | 141 |
| Itaconic | 39–47 | 42 | DL-Glyceric | 152–170 | 160 |
| Maleic | 38–50 | 45 | DL-Isocitric | 176–195 | 183 |
| Fumaric | 38–52 | 45 | Sulphuric | 178–192 | 181 |
| Thymol blue indicator | 47–53 | 50 | | 193–230 | 198 |
| Succinic | 57–63 | 60 | Shikimic | 175–206 | 198 |
| Lactic | 57–63 | 60 | D(+)-Tartaric | 200–233 | 211 |
| α-Ketoglutaric | 54–67 | 62 | Phosphoric, L-Glutamic | >239 | – |
| *trans*-Aconitic | 66–72 | 69 | | | |
| Malonic | 64–75 | 69 | Quinic, L-Aspartic | >252 | – |

Kesner and Muntwyler developed an automatic method for the analysis of organic acids. The technique consisted in chromatography on silica gel columns with chloroform–*tert.*-amyl alcohol mixtures as eluent. The concentration of *tert.*-amyl alcohol in the eluent was continually increased in a Varigrad gradient apparatus and the mixture was pumped on to the column. The individual separated acids in the eluate reacted with an indicator (*o*-nitrophenol in absolute methanol), which was continually fed into the effluent stream and the coloration developed was recorded with a flow-photometric detector operating at 350 nm. This method was successfully applied to the separation of a number of physiologically important acids, such as the Krebs cycle intermediates. A routine separation can be performed with a sensitivity about 40 times higher than that in the conventional manual method. The accuracy is greater than ± 3%. Furthermore, no preliminary deproteinization and extraction (with the possible loss of volatiles and formation of artifacts) was required prior to introduction of the sample.

### Separation of carboxylic acids on Sephadex columns

In this case, the separation is based on the sieving effect and molecular size. In most separations weakly cross-linked Sephadex G-10 was used (Brock, Brock and Housley, Schiller and Chung). Sephadex G-25 was also used by Woof and Pierce for the separation of phenolic acids.

Monocarboxylic acids were not separated on Sephadex when eluted with water. As reported by Gelotte, the carboxylic acid group has a "negative sorption effect" and in most instances elution occurred much earlier than would be expected from the parent phenol (Table 25.5). Intramolecular hydrogen bonding is not always responsible as there is no difference between *o*- and *p*-hydroxybenzoic acid in water and the 2,4-dihydroxy acid was eluted early while the 2,3-dihydroxy acid was not. In electrolyte solutions, this

TABLE 25.5
GEL FILTRATION OF AROMATIC ACIDS AND OTHER AROMATIC COMPOUNDS (GELOTTE)

Column: 35 × 3.5 cm. Gel: Sephadex G-25 with a water regain of 2.9 g of water per gram of dry substance and a wet density of 1.099 (50–100 mesh). The Sephadex was swelled in 0.05 *M* sodium chloride for 30 min and the fine particles were removed by decantation before packing; the column was equilibrated with the solvent before addition of the sample. Mobile phases: (a) distilled water; (b) 0.05 *M* sodium chloride; (c) phosphate solution, $\mu$ = 0.05, pH = 7; (d) 0.01 *M* ammonia solution, pH = 10.6. Flow-rate: 2 ml/min with a hydrostatic pressure of 60 cm. Temperature: ambient.

| Compound* | Mobile phase | | | |
|---|---|---|---|---|
| | a | b | c | d |
| | $K_d$ values | | | |
| Benzoic acid | 0.5 | – | – | – |
| Anthranilic acid | 0.6 | – | – | – |
| Sulfanilic acid | 0.3 | 1.1 | – | – |
| Picric acid | 0.4 | 2.5 | 2.7 | 0.4 |
| Cinnamic acid | 0.3 | – | – | – |
| Phthalic acid | 1.1 | – | – | – |
| Phenol | 0.7 | 1.7 | – | 1.0 |
| Aniline | 1.5 | – | – | – |
| Benzyl alcohol | 1.3 | – | – | 1.3 |
| Salicyl alcohol | 1.4 | – | – | – |

*0.5–142 mg samples tested.

"negative sorption" effect disappeared in all acids. The rate of elution seemed to depend on the number and orientation of free hydroxyl groups (*i.e.*, those not involved in internal hydrogen bonding) and the extent to which these groups are accessible to the carboxyl groups of the gel. Most of the acids were eluted very early with ammonia solution as they also carry a negative charge and will be repelled by similar acid groups in the gel. The formation of a complex with molybdate resulted in earlier elution, as would be expected, except for the 2,6-dihydroxy acid, which, although both groups can complex, was very strongly adsorbed. Methylation, as in vanillic and syringic acids, resulted in behaviour very like that of a monohydroxy acid. They were not differentiated on Sephadex columns.

From Table 25.6, it can be seen that an electrolyte is required if the separation of mixtures of phenolic acids is to be achieved.

Fig. 25.12 shows the separation of a mixture of acids already reported to be present in barley. Resolution in the first stage is not complete but fractions can be collected as shown and completely resolved by re-running the samples using water as solvent.

Downey *et al.* used a 54 × 2.4 cm column filled with Sephadex LH-20 with a flow-rate of 1 ml/min for the separation of fatty acids in the presence of phospholipids and chloroplast pigments. Using chloroform only as the column eluent, tristearin, tributyrin and stearic, capric, butyric and acetic acids were separated (Fig. 25.13) into well defined peaks (the elution volumes, $V_e$, were 65, 85, 225, 320, 450 and 575 ml, respectively). When linolenic acid was included in this mixture, it was not separated from stearic acid. The capric acid appears to have contained a fatty acid contaminant, as indicated by the inflection in its elution curve (Fig. 25.13), which was also observed on chromatography

TABLE 25.6

$K_d$ VALUES OF PHENOLIC ACIDS IN AQUEOUS ELUTING MEDIA (WOOF AND PIERCE)

Column: 35 × 2.5 cm Sephadex G-25 (medium).

| Phenolic acid | Eluting medium | | | |
|---|---|---|---|---|
| | Water | NaCl | Ammonia solution | $Na_2MoO_6$ |
| *o*-Hydroxybenzoic acid | 1.0 | 2.1 | 1.6 | 1.2 |
| *m*-Hydroxybenzoic acid | 1.0 | 1.8 | 0.85 | – |
| *p*-Hydroxybenzoic acid | 1.0 | 1.45 | 0.8 | – |
| 2,3-Dihydroxybenzoic acid | 2.6 | 1.6 | 0.4 | 2.0 |
| 2,4-Dihydroxybenzoic acid | 0.85 | 2.7 | 1.1 | 1.4 |
| 2,5-Dihydroxybenzoic acid | 1.0 | 3.0 | 1.7 | 1.4 |
| 2,6-Dihydroxybenzoic acid | 0.7 | 1.45 | 0.4 | 2.95 |
| 3,4-Dihydroxybenzoic acid | 1.7 | 2.1 | 0.7 | – |
| 3,5-Dihydroxybenzoic acid | 2.2 | 1.9 | 0.7 | – |
| 3,4,5-Trihydroxybenzoic acid | 1.05 | 2.5 | 1.05 | 1.9 |
| 2,3,4-Trihydroxybenzoic acid | 2.05 | 2.65 | 1.1 | 2.0 |
| Vanillic acid | 0.85 | 1.95 | 0.85 | – |
| Syringic acid | 0.8 | 1.95 | 0.85 | – |
| Chlorogenic acid | 1.6 | 3.9 | 1.35 | – |

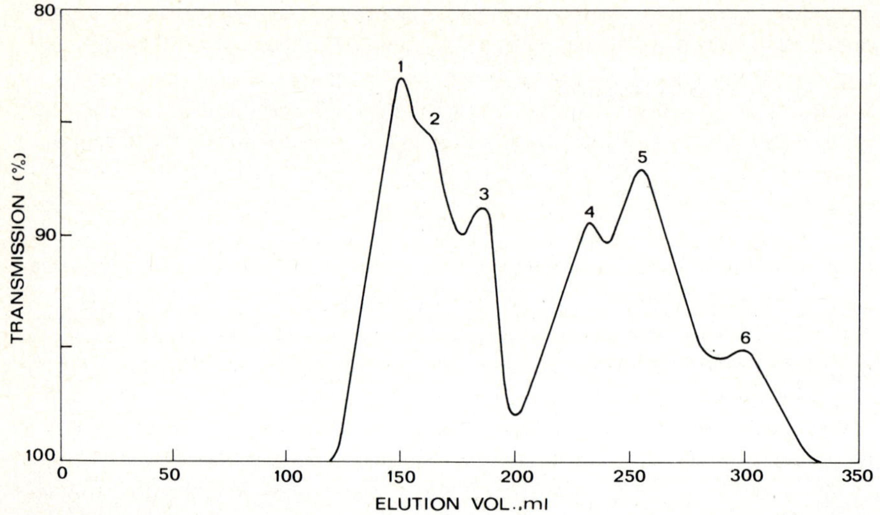


Fig. 25.12. Separation of a phenolic acid mixture (Gelotte). Column: 35 × 2.5 cm. Sorbent: Sephadex G-25. Mobile phase: 0.1 *M* sodium chloride. Detection: spectrophotometric. 1 = vanillic + syringic acids; 2 = 3,4-dihydroxybenzoic acid; 3 = gallic acid; 4 = ferulic acid; 5 = sinapic acid; 6 = chlorogenic + caffeic acids.

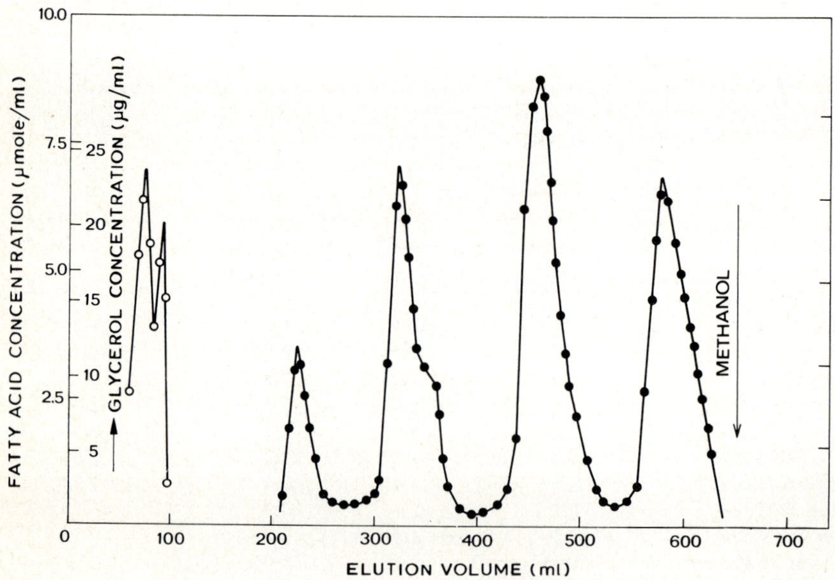


Fig. 25.13. Fractionation and separation of triglycerides and fatty acids (Downey *et al.*). Column: 54 × 2.4 cm. Sorbent: Sephadex LH-20; Mobile phase: chloroform. Flow-rate: 1 ml/min. Detection: electrometric titration. ○, triglycerides; ●, fatty acids. Elution volumes: tristearin 65 ml; tributyrin 85 ml; stearic acid 225 ml; capric acid 320 ml; butyric acid 450 ml; acetic acid 575 ml.

TABLE 25.7
GEL CHROMATOGRAPHY OF ALIPHATIC CARBOXYLIC ACIDS ON POLYACRYLAMIDE GEL (STREULI)

Column: 97 × 0.5 cm. Gels: (a) Bio-Gel P-2 (100–200 mesh) polyamide gel, exclusion limit 200–2600; (b) Bio-Gel P-6 (100–200 mesh) polyamide gel, exclusion limit 1000–5000. Mobile phase: 0.01 $M$ sodium chloride.

| Compound* | Sorbent | |
|---|---|---|
| | a | b |
| | $K_d$ value | |
| Hydrochloric acid | 1.29 | – |
| Trichloroacetic acid | 1.17 | – |
| Chloroacetic acid | 0.96 | – |
| Acetic acid | 1.08 | – |
| Lactic acid | 0.79 | 0.98 |
| Acrylic acid | 0.97 | – |
| Crotonic acid | 1.28 | 1.00 |
| Oxalic acid | 1.04 | – |
| Succinic acid | 1.22 | – |
| Malic acid | 1.00 | 1.03 |
| Tartaric acid | 1.08 | – |
| Maleic acid | 1.26 | 1.11 |
| Fumaric acid | 0.93 | 0.96 |
| Citric acid | 1.19 | 1.04 |
| Glycine | 0.93 | – |
| 4-Aminobenzoic acid | 2.75 | – |

*Samples of 50 μl.

TABLE 25.8
GEL CHROMATOGRAPHY OF ALIPHATIC CARBOXYLIC ACIDS (CAZES AND GASKILL)

Columns: four 4 ft. × 3/8 in. columns in series. Gel: rigid, cross-linked polystyrene gel. Mobile phase: *o*-dichlorobenzene. Flow-rate: 1 ml/min. Temperature: 130°C.

| Acid* | $V_e$ (ml) |
|---|---|
| Acetic | 194.4 |
| Propionic | 184.6 |
| *n*-Butyric | 178.4 |
| *n*-Valeric | 174.7 |
| *n*-Hexanoic | 168.9 |
| *n*-Heptanoic | 165.8 |
| *n*-Octanoic | 160.2 |
| *n*-Nonanic | 157.1 |
| *n*-Decanoic | 154.0 |
| *n*-Undecanoic | 150.8 |
| *n*-Dodecanoic | 148.0 |
| Myristic | 143.1 |
| Palmitic | 139.6 |
| Stearic | 132.8 |

*Samples are 2 ml aliquots of 0.25 to 1.0% solutions.

*References p. 572*

of capric acid alone. The recoveries of the individual fatty acids and triglycerides were approximately 85%.

Monocarboxylic and dicarboxylic aliphatic acids were separated by Streuli on a Bio-Gel P-2 column, as indicated in Table 25.7.

The homologous series of $C_2-C_{18}$ aliphatic acids was successfully separated on cross-linked polystyrene gel using *o*-dichlorobenzene as the mobile phase at 130°C. The results obtained by Cazes and Gaskill are summarized in Table 25.8.

## REFERENCES

Alenitskaya, S. R. and Starobinets, G. L., *Vestn. Akad. Nauk Belorus. SSR, Ser. Khim. Nauk*, (1967) 28; *C.A.*, 67 (1967) 57528r.
Alfredsson, B., Bergdahl, S. and Samuelson, O., *Anal. Chim. Acta*, 28 (1963) 371.
Alfredsson, B., Gedda, L. and Samuelson, O., *Anal. Chim. Acta*, 27 (1962) 63.
Anderson, R. E., *U.S. pat.*, 3,409,667 (1968).
Aoki, I., Hori, M. and Matsumaru, H., *Bunseki Kagaku (Jap. Anal.)*, 18 (1969) 346.
Bengtsson, L. and Samuelson, O., *Anal. Chim. Acta*, 44 (1969) 217.
Brock, A. J. W., *J. Chromatogr.*, 39 (1969) 328.
Brock, A. J. W. and Housley, S., *J. Chromatogr.*, 42 (1969) 112.
Calmanovici, B., *Rev. Chim. (Bucuresti)*, 17 (1966) 170.
Calmanovici, B., *Rev. Chim. (Bucuresti)*, 17 (1966) 374.
Carlsson, B., Isaksson, T. and Samuelson, O., *Anal. Chim. Acta*, 43 (1968) 47.
Carlsson, B. and Samuelson, O., *Anal. Chim. Acta*, 49 (1970) 247.
Carroll, K. K., *Nature (London)*, 176 (1955) 398.
Cazes, J. and Gaskill, D. R., *Separ. Sci.*, 2 (1967) 426 and 4 (1969) 15.
Churáček, J. and Jandera, P., *Chem. Listy*, 64 (1970) 756.
Courtoisier, A. J. and Ribereau-Gayon, J., *Bull. Soc. Chim. Fr.*, (1963) 350.
Davies, C. W., *Biochem. J.*, 45 (1949) 38.
Davies, C. W., Hartley, R. D. and Lawson, G. J., *J. Chromatogr.*, 18 (1965) 47.
Davies, C. W. and Owen, B. D. R., *J. Chem. Soc.*, (1956) 1681.
Downey, W. K., Murphy, R. F. and Keogh, M. K., *J. Chromatogr.*, 46 (1970) 120.
Egashira, S., *Bunseki Kagaku (Jap. Anal.)*, 10 (1961) 1225.
Egashira, S., *Bunseki Kagaku (Jap. Anal.)*, 15 (1966) 1356,
Egashira, S., *Bunseki Kagaku (Jap. Anal.)*, 17 (1968) 958.
Erler, K., *Z. Anal. Chem.*, 131 (1950) 106.
Fransson, L. A., Roden, L. and Spach, M. L., *Anal. Biochem.*, 23 (1968) 317; *C.A.*, 67 (1968) 112034z.
Freeman, G. G., *J. Chromatogr.*, 28 (1967) 338.
Funasaka, W., *Bunseki Kagaku (Jap. Anal.)*, 15 (1966) 835.
Gellotte, B., *J. Chromatogr.*, 3 (1960) 330.
Goudie, A. J. and Rieman, W., *Anal. Chim. Acta*, 26 (1962) 419.
Harlow, G. A. and Morman, D. H., *Anal. Chem.*, 36 (1964) 2438.
Henry, R. A. and Schmit, J. A., *Chromatographia*, 3 (1970) 116.
Johnard, B. and Samuelson, O., *Sv. Kem. Tidskr.*, 73 (1961) 586.
Johnson, S. and Samuelson, O., *Anal. Chim. Acta*, 36 (1961) 1.
Katz, S. and Burtis, C. A., *J. Chromatogr.*, 40 (1969) 270.
Kesner, L. and Muntwyler, E., *ACS Winter Meeting, Phoenix, Ariz., January 17–21, 1966.*
Khym, J. X. and Zill, L. P., *J. Amer. Chem. Soc.*, 73 (1951) 2399.
Kirkland, J. J., *J. Chromatogr. Sci.*, 7 (1969) 361.

Kreshkov, A. P. and Kolosova, I. F., *Zh. Anal. Khim.*, 25 (1970) 1234.
Larsson, U. B., Isakson, T. and Samuelson, O., *Acta Chem. Scand.*, 20 (1966a) 1965.
Larsson, U. B., Norstedt, I. and Samuelson, O., *J. Chromatogr.*, 22 (1966b) 102.
Lawson, G. L. and Purdie, J. W., *Mikrochim. Acta*, (1961) 415.
Lee, K. S., Lee, D. W. and Lee, E. K. Y., *Anal. Chem.*, 42 (1970) 554.
Lee, K. S. and Samuelson, O., *Anal. Chim. Acta*, 37 (1967) 359.
Lesquibe, F., *C.R. Acad. Sci., Paris,* 251 (1960) 2690.
Lesquibe, F. and Rumpf, P., *C.R. Acad. Sci., Paris,* 260 (1965) 5006.
Longbottom, J. E., *Anal. Chem.*, 44 (1972) 418.
Markova, A. V. and Smirnov, V. A., *Zh. Anal. Khim.*, 24 (1969) 1271.
Martinsson, E. and Samuelson, O., *Chromatographia*, 3 (1970) 405.
Mehta, M. J., Bhatt, R. A., Hegde, R. S., Patel, D. J. and Bafna, S. L., *J. Indian Chem. Soc.*, 130.
Moehler, K. and Pires, R., *Z. Lebensm.-Unters.-Forsch.*, 139 (1969) 337; *C.A.*, 71 (1969) 89992m.
Nakajima, S. and Tanenbaum, S. W., *J. Chromatogr.*, 43 (1969) 444.
Pallini, V., *Boll. Soc. Ital. Sper.*, 41 (1965) 676; *C.A.*, 64 (1966) 86.
Patel, D. J. and Bafna, S. L., *Ind. Eng. Chem., Prod. Res. Develop.*, 4 (1965) 1.
Patel, D. J. and Bafna, S. L., *Indian J. Chem.*, 6 (1968) 199.
Robinson, P. A. and Mills, G. F., *Ind. Eng. Chem.,* 41 (1949) 2221.
Salkova, E. G. and Nikiforova, T. A., *Dokl. Akad. Nauk SSSR*, 179 (1968) 218.
Samuelson, O., *Sv. Kem. Tidskr.*, 76 (1964) 635.
Samuelson, O. and Simonson, R., *Anal. Chim. Acta*, 26 (1962a) 110.
Samuelson, O. and Simonson, R., *Sv. Papperstidn.*, 65 (1962b) 363.
Samuelson, O. and Thede, L., *J. Chromatogr.*, 30 (1967) 556.
Scheffer, F., Kickuth, R. and Lorenz, H., *Qual. Plant. Mater. Veg.*, 12 (1965) 342.
Schenker, H. M. and Rieman, W., *Anal. Chem.*, 25 (1953) 1637.
Schiller, J. G. and Chung, A. E., *J. Biol. Chem.*, 245 (1970) 5857.
Scoggins, M. W., *Anal. Chem.*, 44 (1972) 1285.
Seki, T., *J. Biochem. (Tokyo)*, 45 (1958) 855.
Seki, T., *J. Chromatogr.*, 3 (1960) 376.
Seki, T., *J. Chromatogr.*, 22 (1966) 498.
Seki, T., Inamori, K. and Sano, K., *J. Biochem. (Tokyo)*, 49 (1959) 1653.
Shimomura, K. and Walton, H. F., *Anal. Chem.*, 37 (1965) 1012.
Skelly, N. E., *Anal. Chem.*, 33 (1961) 271.
Skelly, N. E. and Crummett, W. B., *Anal. Chem.*, 35 (1963) 1680.
Skorokhod, O. R. and Sembur, M. E., *Ionity, Ionnii Obmen. Akad. Nauk SSSR, Sb. Statei*, (1966) 152; *C.A.*, 67 (1967) 94319n.
Skorokhod, O. R., and Tabulo, M. L., *Ionoobmen. Technol.*, (1965) 186; *C.A.*, 63 (1965) 10648a.
Stamley, J. B. and Moseley, P. B., *J. Amer. Oil. Chem. Soc.*, 46 (1969) 241; *Index Chem.*, (1969) 117104.
Starobinets, G. L. and Gleim, I. F., *Zh. Fiz. Khim.*, 39 (1965) 2188.
Starobinets, G. L., Gleim, I. F., Alenitskaya, S. R. and Chizhevskaya, A. B., *Vestn. Akad. Nauk Belorus. SSR, Ser. Khim. Nauk*, (1965) 5; *C.A.*, 64 (1966) 54e.
Streuli, C. A., *J. Chromatogr.*, 47 (1970) 355.
Thomas, H., *J. Chromatogr.*, 34 (1968) 106.
Tsitovich, I. K. and Kuzmenko, E. A., *Zh. Prikl. Khim.*, 42 (1969) 2066.
Woof, J. B. and Pierce, J. S., *J. Chromatogr.*, 28 (1967) 94.
Zerfing, R. C. and Veening, H., *Anal. Chem.*, 38 (1966) 1312.

*Chapter 26*

# Higher carboxylic acids

J. POKORNÝ

## CONTENTS



## INTRODUCTION AND GENERAL REMARKS

Higher carboxylic acids, usually termed fatty acids, are generally analyzed by gas–liquid chromatography nowadays. Liquid column chromatography is used only for non-volatile or thermolabile derivatives, such as oxidation and polymerization products or highly unsaturated or polar derivatives. The use of column chromatography for the analysis of common fatty acids was recently reviewed (Kaufmann, Stein and Slawson).

## SEPARATION AS FATTY ACID DERIVATIVES

Fatty acids are often converted into methyl or ethyl esters before their analysis in order to improve the separation. Treatment with diazomethane or with alcohol and boron trifluoride or sulphuric acid as catalysts are the most commonly used esterification procedures. The following procedure of Schlenk and Gellerman is suitable.

The apparatus consists of three test-tubes with side-arms. A stream of nitrogen is saturated with diethyl ether in the first tube (150 × 16 mm) and carries diazomethane generated in the second tube (85 × 15 mm) into the third tube (85 × 15 mm) where the esterification takes place. The side-arms (7 mm O.D.) are bent downwards and reach close to the bottom of the next tube. They are drawn out at the ends to an O.D. of *ca.* 1 mm. Rubber stoppers are used for connections.

The flow of nitrogen through the diethyl ether in tube 1 is adjusted to *ca.* 6 ml/min. Tube 2 contains 0.7 ml of 2-($\beta$-ethoxyethoxy)ethanol (purified by heating it to 110°C for 1 h with 5% potassium hydroxide solution followed by distillation at *ca.* 90°C and 12 mm Hg pressure), 0.7 ml of peroxide-free diethyl ether and 1 ml of a solution of 6 g of potassium hydroxide in 10 ml of water. Between 5 and 30 mg of fatty acids dissolved in 2–3 ml of diethyl ether which contains 10% of methanol are placed in tube 3. About 2 mmole of N-methyl-N-nitroso-*p*-toluenesulphonamide per milliequivalent of fatty acids are dissolved in 1 ml of diethyl ether and the mixture is added to tube 2. Connection is made immediately with tube 1 while the yellow colour of diazomethane is already appearing in tube 2. As soon as a yellow tinge becomes visible against a white background in tube 3, this tube is disconnected and the slight excess of diazomethane is consumed by adding acetic acid diluted with diethyl ether or is removed by a stream of nitrogen. The whole procedure requires 10–12 min.

Chlorophenacyl esters of saturated and unsaturated fatty acids can be separated on polyethylene-coated Celite columns (Kibrick and Skupp). Fatty acid esters can be converted into hydroxamic acids and separated on cellulose powder (Davenport).

The separation between saturated and unsaturated fatty acids is improved by preliminary conversion of the latter into mercury adducts by boiling them with mercury(II) acetate (Jantzen and Andreas), *e.g.*, by the following procedure as modified by Kuemmel. Free fatty acids are converted into methyl esters, 0.7 g of the esters are dissolved in 15 ml of methanol and the mixture is refluxed with an appropriate amount of mercury(II) acetate depending on their iodine value (1.8, 1.7, 1.6 and 1.5 g of mercury(II) acetate are used for methyl esters with iodine values of 120–135, 96–119, 83–95 and 70–82, respectively) for 70 min on a magnetic stirrer–hot plate. After cooling, 160 ml of water are added to the methanolic solution and the mixture is extracted with five 35-ml portions of diethyl ether followed by two 30-ml portions of chloroform. The combined diethyl ether extracts are washed with three 50-ml portions of water, and the combined water washings are then back-extracted with the combined chloroform extracts. The combined diethyl ether and chloroform extracts are dried and filtered and the solvent is removed. The sample should then be refrigerated until it is subjected to chromatographic separation, which should be carried out within 24 h of the preparation of the mercury derivatives. The yield of mercury derivatives ranges from 97 to 100% of theory.

Kuemmel obtained relatively pure saturated, monoenoic and polyenoic fractions by the following method. A 390 × 20 mm column of alumina was used for the separation of adducts of a 700-mg sample. The flow-rate was adjusted to 15–18 ml/min. The saturated acids were eluted with 350–400 ml of light petroleum (b.p. 30–45°C) and the monoenoic acid derivatives with 350 ml of a 5% solution of methanol in diethyl ether. If the polyunsaturated acid content is greater than 40% of the total fatty acids, it is more advantageous to use a 1% solution of methanol in diethyl ether and to continue the elution until a 50-ml fraction of eluate contains not more than 2 mg of the substance. The polyunsaturated acid derivatives retained in the column are then decomposed by leaving the column overnight containing 200 ml of hydrochloric acid–methanol (1:10), and recovered from the solution. Bromomercurimethoxy adducts are suitable for the isolation of minor polyenoic fatty acids in mixtures (Craske and Edwards).

## CHROMATOGRAPHY ON ADSORBENTS IN GENERAL USE

Liquid–liquid chromatographic procedures are still important for the preparation of pure polyunsaturated acids, *e.g.*, esters of arachidonic acid (Privett *et al.*), and for the separation of oxygenated fatty acids. Gunstone and Sykes separated various mono-, di-, tri- and tetrahydroxy, mono- and diacetoxy, and epoxystearic acids by reversed-phase chromatography on siliconized Hyflo Supercel coated with neutralized liquid paraffin or acetylated castor oil. *Erythro-* and *threo*-isomers of polyhydroxylic acids were not separated. Radin reviewed various procedures for the separation of hydroxylic fatty acids, and discussed the advantages and disadvantages of reversed-phase chromatography. He recommended the use of polystyrene beads as the support. Naturally occurring epoxy acids were separated by partition chromatography on Celite using acetonitrile as the stationary phase and *n*-hexane as the mobile phase (Morris *et al.*). Piretti *et al.* isolated pure hydroperoxides from oxidized methyl oleate on 100-mesh silica gel (Mallinckrodt, St. Louis, Mo., U.S.A.) containing 5% of water by eluting with *n*-hexane–diethyl ether–methanol (94:5:1) at 18°C (300 × 30 mm column; 1 g of sample; flow-rate 2.6 ml/min). The present author also obtained an excellent separation on Florisil under similar conditions. The silicic acid chromatography was applied with success to the separation of the products of the interaction of linoleic acid hydroperoxide with tocopherols (Gardner *et al.*) and to the separation of some acidic cleavage products of oxidized methyl linoleate (Esterbauer *et al.*). Mixtures of *n*-hexane with diethyl ether were used as eluents in both instances.

Fatty acid oxidative and thermal dimers were satisfactorily fractionated on silica gel impregnated with a 16% solution of methanol in benzene and eluted with a 2% solution of methanol in benzene (Evans *et al.*). Figge used elution with cyclohexane–benzene (6:4) and diethyl ether for the separation of oleic acid thermal dimers. Mounts *et al.* described a method for the separation of oxidative dimers on silica gel.

Cason *et al.* used a mixture of Darco G-60 charcoal and Celite 521 (1:2) and elution with 95% ethanol, anhydrous ethanol and mixtures of ethanol and benzene for the separation of fatty acids of tubercle bacillus lipids.

## CHROMATOGRAPHY ON SPECIFIC ADSORBENTS

A specific adsorbent for the separation of unsaturated fatty acids is silica gel impregnated with silver nitrate, as introduced by De Vries. Silver ions form adducts with double bonds so that the unsaturated acids are retained in the column to an extent that depends on their degree of unsaturation, *trans*-acids being eluted before the corresponding *cis*-isomers. For the preparation of the silver nitrate-impregnated adsorbent, the following procedure reported by Dolev and Olcott is suitable.

A 50-g amount of silicic acid (silica gel G, according to Stahl, 100–200 mesh) is suspended in 100 ml of 40% silver nitrate solution, and the suspension is brought to boiling with stirring, allowed to cool, filtered and dried overnight in an oven at 130°C. The cooled powder is ground in a ball-mill and stored in a desiccator. All operations should be protected from light. Florisil can also be impregnated with silver nitrate (Willner). Artman and Alexander combined a chromatographic separation on silica gel using a stepwise gradient elution with *n*-hexane–benzene, re-absorption of the fractions, and a further separation with the use of columns of silica gel impregnated with silver nitrate in order to separate fatty acids from heated fats.

Chromatography on urea columns is a selective method of separation of straight-chain fatty acids from their branched and cyclic isomers (Cason *et al.*) because urea forms adducts only with normal fatty acids. As Cason *et al.* used a 1% solution of methanol in isooctane as eluent, only saturated acids remained in the column, while both monounsaturated and branched fatty acids passed through into the eluate. According to the present author's experience (Pokorný and El-Tarras) all monounsaturated acids remain in the column if methanol saturated with urea is used as the eluent. Urea chromatography is particularly suitable for the preparation of cyclic or dimeric thermal or oxidation products of fatty acids (Sagredos, 1967, 1969). Molecular sieves of the zeolite type (5A, 10X, 13X) are suitable for separations according to molecular size (Martínez Moreno and López Ruiz).

## GEL AND ION-EXCHANGE CHROMATOGRAPHY

Gel chromatography has found many applications in the analysis of fatty acids in the last few years. Chang succeeded in separating tall oil acids into normal fatty acids and monomeric, dimeric and trimeric resin acids by using Bio-Beads SX-2 and SX-8. The

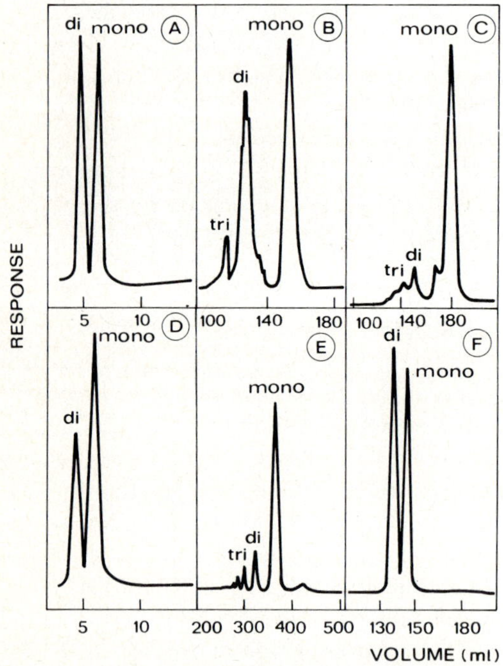


Fig. 26.1. Separation of lipidic oligomers by gel chromatography. A, methyl esters of monomeric and dimeric oleic acid: column, 750 × 25 mm; Sephadex LH-20; eluent, ethanol (after Aitzetmüller, 1972a). B, methyl esters of fatty acid oligomers: column size, 331 ml; sorbent, Sephadex LH-20; weight of adsorbent, 67.1 g; eluent, chloroform–methanol (7:3); elution-rate, 39 ml/h; detection, gravimetric (after Hase and Harva). C, methyl esters of polymerized soyabean oil: column, 6000 × 8 mm; adsorbent, Spheron P; eluent, tetrahydrofuran; flow-rate, 35 ml/h; detection, differential refractometer (after Pokorný *et al.*). D, heat-polymerized groundnut oil; conditions as in A. E, methyl esters of fatty acid oligomers: column, 2300 × 23 mm; temperature, 48–51°C; adsorbent, Sephadex LH-20; eluent, dimethylformamide; flow-rate, 20 ml/h; detection, differential refractometer (after Inoue *et al.*). F, heat-polymerized soyabean oil; conditions as in C.

trimethylsilyl derivatives of Sephadex G-25 and other resins were very satisfactory for the separation of mixtures containing fatty acid esters of fatty alcohols (Tanaka and Konishi).

Zinkel and Zank were able to separate methyl esters of fatty acids on 2000 × 10 mm Styragel columns using diethyl ether as the solvent, but methyl linoleate and methyl linolenate formed critical pairs with methyl myristate and methyl laurate, respectively. Resin acids were separated from one another and from fatty acids.

Aitzetmüller (1972c) used Merckogel SI-50 columns for the analysis of fatty acid oxidation products, *viz.*, methyl 9,10-epoxystearate in the presence of methyl oleate or methyl 12-keto-9-octadecanoate, and ricinoleate in the presence of oleate. Methyl linoleate hydroperoxide was isolated by chromatography on Sephadex LH-20 as the hydroperoxidic fraction was retarded on the column during elution with chloroform (Rubach *et al.*).

The most extensive use of the gel chromatography of fatty acids has been in the analysis of polymerized oils. Inoue *et al.* separated from monomers up to pentamers of unsaturated fatty acids, methyl esters and the corresponding fatty alcohols (Fig. 26.1). They used 2300 × 20 mm or 1000 × 15 mm columns packed with Sephadex LH-20, which was left to swell for 24 h in dimethylformamide before the filling; a differential refractometer was used as the detector; 300 or 50 mg of sample were eluted with dimethylformamide at a controlled temperature, *e.g.*, 50°C.

Aitzetmüller (1972a) tested Bio-Beads SX-1 in addition to Sephadex LH-20. His results and also those of some other workers are shown in Fig. 26.1. For preparative chromatography, Aitzetmüller (1972b) used Sephadex SR-25/100 columns in conjunction with an LKB fraction collector. A 100-μl aliquot of each collector fraction was injected

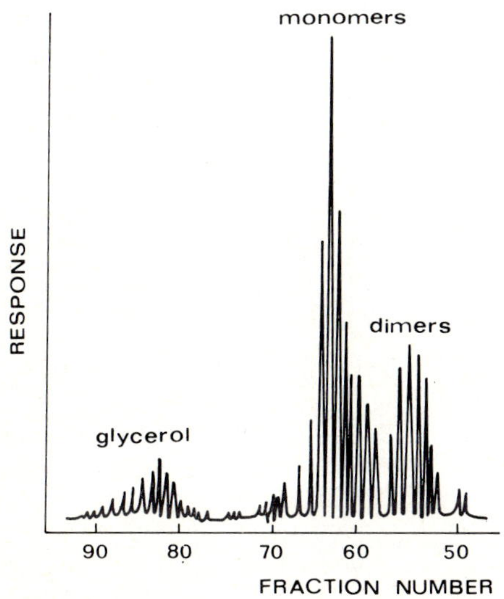


Fig. 26.2. Preparative chromatography of fatty acid methyl esters from frying fats (after Aitzetmüller, 1972a). Column: 800 × 25 mm (Sephadex SR-25/100). Sorbent: Sephadex LH-20. Eluent: ethanol. Operating conditions: see text. Detection: Pye LC detector.

*References p. 580*

with an Eppendorf pipette into a constant flow of ethanol rinsing the wall of a narrow tube which had an opening to accept the tip of the pipette. The ethanol flow transported the sample and rinsed the tubing and coating block between subsequent injections. The frequency of injection was 2–4 $min^{-1}$ at a flow-rate of 30–60 drops/min. An example of the chromatogram is shown in Fig. 26.2.

Only limited use has been made of ion-exchange chromatography for the separation of fatty acids. The procedure of Emken *et al.* for the separation of *cis*- and *trans*-monoenoic acids on Amberlyst XN-1005 treated with silver nitrate was recommended by Applewhite.

## REFERENCES


Aitzetmüller, K., *Fette, Seifen, Anstrichm.*, 74 (1972a) 598.
Aitzetmüller, K., *J. Chromatogr.*, 72 (1972b) 355.
Aitzetmüller, K., *J. Chromatogr.*, 73 (1972c) 248.
Applewhite, T. H., *J. Amer. Oil Chem. Soc.*, 42 (1965) 321.
Artman, N. R. and Alexander, J. C., *J. Amer. Oil Chem. Soc.*, 45 (1968) 643.
Cason, J., Sumrell, G., Allen, C. F., Gillies, G. A. and Elbert, G. S., *J. Biol. Chem.*, 205 (1953) 435.
Chang, T.-L., *Anal. Chem.*, 40 (1968) 989.
Craske, J. D. and Edwards, R. A., *J. Chromatogr.*, 53 (1970) 253.
Davenport, J. B., *Chem. Ind. (London)*, (1955) 705.
De Vries, B., *J. Amer. Oil Chem. Soc.*, 40 (1963) 184.
Dolev, A. and Olcott, H. S., *J. Amer. Oil Chem. Soc.*, 42 (1965) 624.
Emken, E. A., Scholfield, C. R. and Dutton, H. J., *J. Amer. Oil Chem. Soc.*, 41 (1964) 388.
Esterbauer, H., Just, W. and Sterk, H., *Fette, Seifen, Anstrichm.*, 74 (1972) 13.
Evans, C. D., McConnell, D. G., Frankel, E. N. and Cowan, J. C., *J. Amer. Oil Chem. Soc.*, 42 (1965) 764.
Figge, K., *Chem. Phys. Lipids*, 6 (1971) 178.
Gardner, H. W., Eskins, K., Grams, G. W. and Inglett, C. E., *Lipids*, 7 (1972) 324.
Gunstone, F. D. and Sykes, P. J., *J. Chem. Soc. (London)*, (1960) 5050.
Hase, A. and Harva, O., *Kem. Teollisuus*, 25 (1968) 134.
Inoue, H., Konishi, K. and Taniguchi, N., *J. Chromatogr.*, 47 (1970) 348.
Jantzen, E. and Andreas, H., *Chem. Ber.*, 94 (1961) 628.
Kaufmann, H. P., *Fette, Seifen, Anstrichm.*, 72 (1970) 505.
Kibrick, A. C. and Skupp, S. J., *Anal. Chem.*, 31 (1959) 2057.
Kuemmel, D. F., *Anal. Chem.*, 34 (1962) 1003.
Martínez Moreno, J. M. and López Ruiz, J. L., *Grasas Aceites*, 22 (1971) 351.
Morris, L. J., Hayes, H. and Holman, R. T., *J. Amer. Oil Chem. Soc.*, 38 (1961) 316.
Mounts, T. L., McWeeny, D. J., Evans, C. D. and Dutton, H. J., *Chem. Phys. Lipids*, 4 (1970) 197.
Piretti, M., Capella, P. and Taddia, M., *Riv. Ital. Sostanze Grasse*, 46 (1969) 324.
Pokorný, J. and El-Tarras, M. F., unpublished results.
Pokorný, J., Kundu, M. K., Pařízková, H., Luan, N.-T., Čoupek, J., Pokorný, S. and Janíček, G., *Fette, Seifen, Anstrichm.*, 74 (1972) 625.
Privett, O. S., Weber, R. P. and Nickell, E. C., *J. Amer. Oil Chem. Soc.*, 36 (1959) 443.
Radin, N. S., *J. Amer. Oil Chem. Soc.*, 42 (1965) 569.
Rubach, K., Schormüller, J. and Melchert, H.-U., *Lebensm. Ernähr.*, (1971) 166.
Sagredos, A. N., *Fette, Seifen, Anstrichm.*, 69 (1967) 707.
Sagredos, A. N., *Fette, Seifen, Anstrichm.*, 71 (1969) 863.
Schlenk, H. and Gellerman, J. L., *Anal. Chem.*, 32 (1960) 1412.
Stein, R. A. and Slawson, V., *Progr. Chem. Fats Other Lipids*, 8 (1966) 375.
Tanaka, H. and Konishi, K., *J. Chromatogr.*, 64 (1972) 61.
Willner, D., *Chem. Ind. (London)*, (1965) 1839.
Zinkel, D. F. and Zank, L. C., *Anal. Chem.*, 40 (1968) 1145.

*Chapter 27*

# Lipids

J. POKORNÝ

## CONTENTS



## INTRODUCTION AND GENERAL REMARKS

Liquid column chromatography and thin-layer chromatography are the most important techniques for the purification and fractionation of lipids, the former being mainly used for preparative and the latter for analytical purposes. The immense amount of literature that exists on this subject has been reviewed from various standpoints (Carroll, 1963; Creech; Kaufmann; Rouser *et al.*, 1967a; Stein and Slawson).

The sample often has to be purified prior to the fractionation by removing water and non-lipidic contaminants, *e.g.*, by chromatography on a Sephadex G-25 column (Williams and Merrilees), especially if it has been obtained by extraction with chloroform and methanol. Lipid–protein complexes may be present, which can be resolved by chromatography on cross-linked polystyrene gels (Fisher). The separation of highly unsaturated lipids is preferably carried out under nitrogen, as otherwise oxidation could take place in the column (Marinetti). Silica gel can protect polyunsaturated fatty acids and their derivatives against autoxidation (Slawson and Mead). Phospholipids may be hydrolyzed to lysophospholipids during chromatography on silica gel (Renkonen, 1962; Camejo). Passage through Amberlite IRA-400 resulted in losses of some components of natural lipids and incomplete recovery of free fatty acids from the ion-exchange resin (Bottcher *et al.*). Lipids, even triacylglycerols, can be partially methanolyzed and transesterified (Schlenk and Holman).

## SEPARATION OF LIPIDS INTO CLASSES

The complex mixture of lipids is usually first separated into classes. Among numerous modifications, three methods are the most widely used. The procedure of Hirsch and Ahrens (Table 27.1), designed especially for the fractionation of neutral lipids and the separation of phospholipids, was reviewed by Creech, and in more detail by Sweeley, who compared the procedure with those of other workers who used silica gel columns.

TABLE 27.1
SEPARATION OF LIPIDS INTO CLASSES (HIRSCH AND AHRENS)

Column 25.0 × 1.3 cm; 10 g of silica gel; 200 mg of sample.

| Amount (ml) | Solvent | Lipids separated |
|---|---|---|
| 200 | 1% Diethyl ether in light petroleum (b.p. 40–60°C) | Cholesterol esters |
| 250 | 4% Diethyl ether in light petroleum (b.p. 40–60°C) | Triacylglycerols and free fatty acids |
| 150 | 20% Diethyl ether in light petroleum (b.p. 40–60°C) | Free cholesterol |
| 70 | Diethyl ether | Diacylglycerols and carotenoids |
| 70 | Chloroform | Monoacylglycerols |
| 100 | Chloroform–methanol (4:1) | Phosphatidylethanolamine |
| 100 | Chloroform–methanol (3:2) | Phosphatidylcholine |
| 200 | Chloroform–methanol (1:4) | Sphingomyelins, lysophosphatidylcholine |
| 100 | Methanol | Remaining phospholipids |

Carroll (1961 and 1963) proposed the use of acid-washed Florisil as an adsorbent (Table 27.2). His procedure is suitable for neutral lipids and various applications have been discussed by Carroll and Serdarevich. Radin, in his review paper, pointed out the tenacious adsorption of phospholipids and carboxylic acids on this column, contrary to the situation on silica gel columns. Sulpholipids were eluted readily with chloroform–methanol (3:1) but free fatty acids could be eluted only with an acidic solvent, *e.g.*, diethyl ether–acetic acid (96:4).

The procedure of Rouser *et al.* (1967a, b and 1969) is designed for heterolipids (Table 27.3). Rouser *et al.* (1961) used chromatography on DEAE-cellulose and on silicic

TABLE 27.2
SEPARATION OF LIPIDS INTO CLASSES (CARROLL, 1961)

Column A: 17.0 × 2.0 cm; 30 g of Florisil; 40 mg of lipids. Column B: 15.0 × 1.2 cm; 12 g of Florisil; 30 mg of lipids.

| Amount of eluent (ml) | | Eluent | Lipids eluted |
|---|---|---|---|
| Column A | Column B | | |
| 50 | 20 | *n*-Hexane | Hydrocarbons |
| 120 | 50 | 5% Diethyl ether in *n*-hexane | Cholesterol esters |
| 150 | 75 | 15% Diethyl ether in *n*-hexane | Triacylglycerols |
| 150 | 60 | 25% Diethyl ether in *n*-hexane | Free cholesterol |
| 150 | 60 | 50% Diethyl ether in *n*-hexane | Diacylglycerols |
| 150 | 75 | 2% Methanol in diethyl ether | Monoacylglycerols |
| 150 | 75 | 4% Acetic acid in diethyl ether | Free fatty acids |

acid columns for the fractionation of brain lipids. Various modifications for the separation and determination of phospholipids and glycolipids were discussed by Rouser *et al.* (1967a). Wells and Dittmer succeeded in separating brain lipids into 24 classes by a combination of several column-chromatographic techniques. Leeder and Clark reported a micro-procedure for the analysis of serum lipids on silicic acid columns. Hardy *et al.* used silica gel G columns for the chromatography of neutral lipids from fish.

Several papers have been published on the separation of lipids into classes by gel chromatography. Sephadex LH-20 was the adsorbent most frequently used, and its use has been reviewed by De Lange. Nyström and Sjövall used methylated Sephadex G-25 and G-50. Davenport reported a satisfactory separation of phospholipids from neutral lipids, contrary to previous results obtained by Berry and Kaye. The separation is probably due to the formation of larger micelles of phospholipids in non-polar solvents (Tipton *et al.*). The fractionation of neutral lipids on Sephadex LH-20 using chloroform (Calderon and

TABLE 27.3
SEPARATION OF LIPIDS INTO CLASSES (ROUSER *et al.*, 1969)

Column, 20.0 × 2.5 cm; adsorbent, DEAE-cellulose; 100 mg of sample.

| Volume (number of bed volumes) | Solvent | Classes of lipids separated |
|---|---|---|
| 8–10 | Chloroform | Neutral lipids |
| 9 | Chloroform–methanol (9:1) | Cerebrosides, monoglycosyldiglycerides, diglycosyldiglycerides, phosphatidylcholine, lysophosphatidylcholine, sphingomyelins |
| 9 | Chloroform–methanol (7:3) | Ceramide aminoethylphosphonate, ceramide dihexosides, ceramide polyhexosides, phosphatidylethanolamine |
| 9 | Chloroform–methanol (1:1) | Lysophosphatidylethanolamine, oxidation products of phosphatidylethanolamine, traces of phosphatidylethanolamine. |
| 10 | Methanol | Oxidation products of phosphatidylethanolamine, salts, phosphatidylethanolamine |
| 10 | Chloroform–acetic acid (3:1) | Free fatty acids, glycin-conjugated bile acids, unconjugated bile acids, pigments, non-lipidic substances soluble in organic solvents |
| 10 | Acetic acid | Phosphatidylserine, proteins |
| 4 | Methanol | No eluate, to remove acetic acid from the column only |
| 10 | Chloroform–methanol–ammonia–salt | Phosphatidic acid, phosphatidylinositol, phosphatidylglycerol, diphosphatidylglycerol, sulpholipid, cerebroside sulphate, unknown compounds, salts |
| 10 | Methanol | Salts, traces of lipids |

Baumann, 1970a) and ethanol (Calderon and Baumann, 1970b) as eluents has been described. The first procedure was found to be useful for the separation of glycerol esters from diol esters, ether–esters and other neutral lipids (Fig. 27.1). A combination of gel chromatography and silicic acid thin-layer chromatography was necessary for complete resolution. Calderon and Baumann (1971) analyzed mixtures of glycol lipids, glycerol lipids, hydrocarbons and waxes. The best results were obtained by a combination of gel chromatography on Sephadex LH-20 with partition chromatography (Bende).

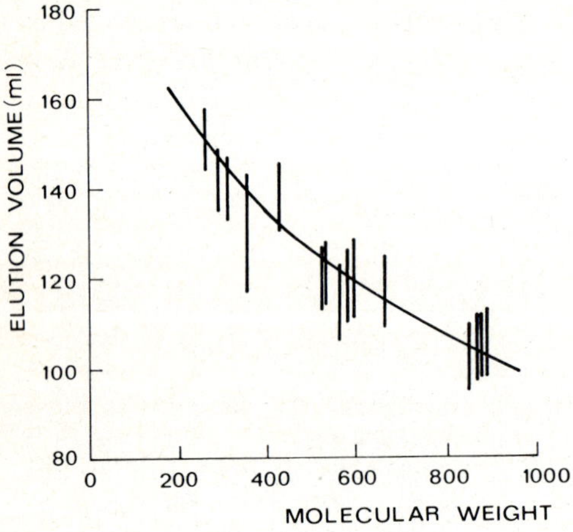


Fig. 27.1. Separation of neutral lipids (Calderon and Baumann, 1970a). Column: 700 × 25 mm. Sorbent: Sephadex LH-20. Eluent: chloroform. Sample: 100 mg. Flow-rate: 0.25 ml/min. Detection: gravimetric detection of 1-ml fractions. The following compounds were separated (molecular weights in parentheses): trioctadecylglycerol (849.5); 1,2-dioctadecyl glycerol octadecenoate (863.5); 1-octadecylglycerol dioctadecenoate (877.5); glycerol trioctadecenoate (891.5); dioctadecyl ethanediol (567.0); cholesteryl octadecenoate (653.1); octadecyl ethanediol octadecanoate (581.0); ethanediol dioctadecanoate (595.0); dioctadecyl ether (523.0); octadecyl octadecanoate (537.0); octadecyl ethanediol acetate (356.6); cholesteryl acetate (428.7); octadecyl acetate (312.5); methyl octadecanoate (298.5); octadecanal (268.5).

The use of gradient elution adsorption chromatography for the analysis of lipid classes has been suggested by Stolyhwo and Privett. The chromatographic system consisted of a pressurized apparatus (inlet pressure 20 p.s.i.) and involved the use of a continuous series of gradient changes of *n*-pentane, diethyl ether, chloroform, and methanol containing 8% ammonia. The 1000 × 2.8 mm column was packed with Corasil II (37–55 $\mu$m), modified by treatment with 28% ammonia. The following set of standard lipids was separated: methyl oleate, trioleyl glycerol, cholesterol, 1,3-dioleyl and 1,2-dioleyl glycerol, 1-monooleyl glycerol, beef brain cerebrosides, phosphatidyl ethanolamine, phosphatidyl choline, and beef brain sphingomyelins. The procedure has been applied for the separation of lipids of rat red blood cells.

## SEPARATION OF GLYCEROL ESTERS AND OTHER NEUTRAL LIPIDS

Triacylglycerols are the most difficult lipids to subfractionate within the neutral lipid classes because of the very similar physical properties of the various compounds. Silylated Celite was a satisfactory adsorbent for the separation of trisaturated glycerol esters and for the fractionation of cocoa butter if acetone–*n*-heptane–water systems were used as eluents (Black and Hammond). The procedure could be useful in combination with modern continuous detectors. Evans *et al.* used their method of column chromatography on silica gel and mixtures of methanol and benzene as eluents, which they had applied previously to the fractionation of saturated glycerol esters, for the separation of triacyl glycerols of oils that contain hydroxy and keto acids, such as isano, oiticica, castor and kamala seed oils.

De Vries utilized the formation of adducts between silver nitrate and the double bonds of unsaturated fatty acids, and succeeded in resolving triacylglycerols that differed by one double bond only (Fig.27.2) using a column of silica gel impregnated with silver nitrate. Dolev and Olcott found the procedure to be efficient, even for the separation of polyunsaturated acylglycerols. Another approach involves the transformation of unsaturated triacylglycerols into the corresponding mercury adducts. The chromatography of the adducts is possible on either silicic acid (Hirayama) or deactivated Florisil (Kerkhoven and Deman). The latter procedure was especially suitable for the determination of trisaturated acylglycerols. Bombaugh *et al.* fractionated saturated triacylglycerols on Styragel (cross-linked polystyrene) using 48,770 × 9.5 mm columns with a flow-rate of 0.4 ml/min and tetrahydrofuran as the eluent. In spite of the extreme efficiency of the column, the peak width was of the same order as the difference between the elution times of triacylglycerols that differed by six carbon atoms. From this example, it can be seen how difficult it is to separate triacylglycerols of natural fats. Gel chromatography seems to be promising for the analysis of mixed triacylglycerols (Mulder and Buytenhuys).

The dimerization of glycerides during heating could be followed by chromatography on either Sephadex LH-20 (Aitzetmüller) or Spheron S-232 (Pokorný *et al.*, 1972) gel (Fig. 26.1). Scharmann and Unbehend described the fractionation of heated fats into monomers, dimers and higher oligomers of triacylglycerols using gel chromatography. Liquid chromatography on silica gel using a Pye LC detector was used for the analysis of oxidized triacylglycerols. The oxidation and polymerization of glycerol esters during heating, especially frying, has often been studied after prior conversion into fatty acids (Kajimoto and Katsunori), as their solubility is better and the differences due to oxidation are more pronounced.

Perkins *et al.* compared the separation of fats heated under the conditions of deep fat frying using Sephadex LH-20 and Bio-Beads SX-1 gels. Chloroform, acetone, chloroform–methanol mixtures, and tetrahydrofuran were used as eluents with a Waters 401 differential refractometer as detector. The efficiencies of the gels S-832 and S-232 for the separation of heated oils and of methyl esters prepared from the heated oils were compared by Pokorný *et al.* (1974) who used tetrahydrofuran as eluent, five 1200 × 5 mm columns and a combination of a differential refractometer and a UV analyzer as detectors.

Aitzetmüller (1973b) reported recently on the estimation of total polar products in frying oils by liquid chromatography. He used a modified Pye LC solute transport

*References p.590*

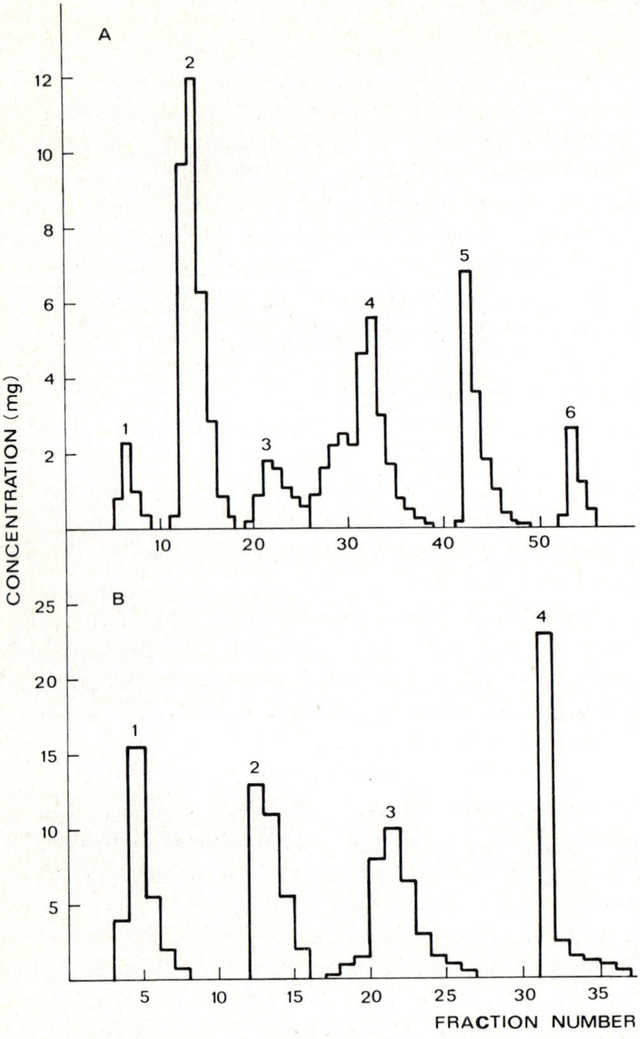


Fig. 27.2. Separation of triglycerides (De Vries). Column: 40 × 1.1 cm. Sorbent: silica gel impregnated with silver nitrate. Eluent: fractions 1–40, benzene; 40–60, diethyl ether. Operating conditions: flow-rate 30 ml/h; 10-ml fractions collected and weighed; temperature 15°C. Detection: gravimetrically; the composition of the whole fraction was determined by GLC. A = palm oil: 1, tripalmitoylglycerol; 2, dipalmitoyloleylglycerol; 3, dipalmitoyllinoleylglycerol; 4, palmitoyldioleylglycerol; 5, mixture of palmitoyloleyllinoleylglycerol and trioleylglycerol; 6, glycerides containing linoleic, oleic and palmitic acids. B = synthetic mixture: 1, tristearoylglycerol; 2, dipalmitoyloleylglycerol; 3, stearoyldioleylglycerol; 4, trioleylglycerol.

detector, 200 × 4 mm columns filled with Merckogel SI-50 or Porasil A (both 36–75 μm). Usually, 10–30 μl of a 20–30% solution of the sample in *n*-heptane as injected. The Ultrograd gradient mixer was used to switch the solvents. Combinations of *n*-heptane, diisopropyl ether, ethanol and water were used as solvents. Similarly, samples of heated oils may be separated, on 300 × 8 mm columns of silicic acid by suitable elution with various concentrations of diisopropyl ether in *n*-heptane followed by ethanol in diisopropyl ether, into several fractions (Aitzetmüller, 1973a) which are subfractionated by gel chromatography.

By analyzing model substances, such as dimeric trioleyl glycerol or monoepoxytrioleyl glycerol Aitzetmüller (1973c) showed that both non-polar and polar artefacts in frying oils are eluted in one peak by frontal elution liquid chromatography after a procedure similar to that given before (Aitzetmüller, 1973b). A combination of a UV detector and a moving wire detector showed that most UV-active substances in heated oils belong to the polar artefacts (Aitzetmüller, 1973d). The amounts of polar and non-polar oligomers could be determined by the combination of liquid chromatography and gel chromatography.

Mono- and diacyglycerols are generally determined on silicic acid columns by the method of Quinlin and Weiser. The method was recently adopted by IUPAC as an international standard method. Silica gel, of particle size 0.07–0.15 mm, should be free from ether-soluble substances and free from iron, and should contain 5% of water. A slurry prepared from 30 g of silica gel and 50 ml of light petroleum is transferred to a 19 × 300 mm column. A solution of 1 g of sample in 15 ml of chloroform (the temperature of the solution should be kept below 40°C) is added to the column, and an additional 5 ml of chloroform are used for rinsing out the beaker containing the sample solution. The flow-rate is adjusted to 2 ml/min and the triacylglycerol fraction is eluted with 200 ml of benzene. Benzene–diethyl ether (90:10) is then used to elute diacylglycerols and free fatty acids. The monoacylglycerol fraction is then eluted by passing 200 ml of diethyl ether through the column. The amount of free fatty acids in the second fraction is determined by titration, by means of the known medium molecular weight of fatty acids, the weight of free fatty acids is calculated, and the weight of the diacylglycerol fraction is thus corrected. The method gives erroneous results if the free fatty acid content is higher than 5%. If the free fatty acid content exceeds 2%, all of the fractions should be tested and corrected for the presence of free fatty acids. The interfering effect of short-chain and hydroxylic acids on the separation was studied by Franzke *et al.*

Because of their different structures, the isomeric acylglycerols can be resolved by gel chromatography. Joustra *et al.* studied the separation of 1,3- and 1,2-dipalmitoylglycerols on Sephadex LH-20. Pokorný *et al.* (1973) studied the separation of 1- and 2-monostearoylglycerol on Spheron S gel using 8 × 6000 mm columns, a flow-rate of 35 ml/h, tetrahydrofuran as the eluent and a Waters Model R-4 differential refractometer and a UV flow analyzer as detectors.

Sucrose esters of fatty acids were satisfactorily resolved into mono-, di-, tri- and higher esters on Sephadex LH-20 columns using dimethylformamide as eluent (Konishi *et al.*), sorbitans on silica gel columns (Cedras *et al.*) and polyethylene glycol esters on silica gel pre-wetted with benzene (Papariello *et al.*). Wickbold separated polyethylene glycol esters of fatty acids into free glycol, free fatty acids, monoester and diester by chromatography on silylated silica gel columns, eluting with isopropanol–water mixtures after a modified procedure of C.I.D.*

*C.I.D. = Comité International des Dérivés Tensioactifs, Paris, France.

*References p. 590*

Natural waxes are fractionated into classes either on alumina of various degrees of activation (Wiedenhof) or on silica gel (Netting). A complicated procedure for the systematic analysis of waxes by chromatography on silica gel, an ion-exchange resin and silica gel impregnated with non-polar solvents was described by Scholz. Waxes of *Vernix caseosa* and skin lipids were fractionated into sterol ester and wax ester fractions by chromatography on magnesium oxide MX-66 columns with use of *n*-hexane and a 1% solution of acetone in *n*-hexane as eluents (Nicolaides *et al.*).

## SEPARATION OF PHOSPHOLIPIDS AND OTHER POLAR LIPIDS

The phospholipidic and glycolipidic fractions, isolated by the above methods of separation into lipid classes, can be further sub-fractionated by various column chromatographic procedures. For the major components, combinations of column and thin-layer chromatography can be used, while for the minor components the use of preparative column chromatography is necessary, even for the sub-fractionation. The most suitable technique for separating the sample under study will depend on the composition of the substrate. Various methods based on chromatography on DEAE-cellulose (Table 27.3) or TEAE-cellulose (Table 27.4) columns were reviewed by Rouser *et al.* (1969). The present author considers his system to be the most suitable and adaptable to various substrates. A combination of column chromatography on DEAE-cellulose followed by re-chromatography on silica gel was successful for the separation of phospholipids of rapeseed gum (Weenink and Tulloch).

Cerebrosides were separated after preliminary removal of phospholipids by chromatography on Florisil deactivated with 10% of water using mixtures of chloroform–methanol–water as solvents (Mehl and Jatzkewitz). A similar procedure on silica gel was recommended by Klenk and Schorsch for the analysis of brain cerebrosides. Earlier methods for the analysis of gangliosides were based on DEAE-cellulose chromatography (Trams and Lauter, Wolfe and Lowden). Extracts containing gangliosides were purified by chromatography on Sephadex G-100 columns (Raveglia and Ghittoni). Kwiterovich *et al.* described the separation of liver glycolipids and phospholipids into classes. Gel chromatography on Sepharose gave good results for the purification of lipopolysaccharides in comparison with ultracentrifugation, and the chromatographic procedure was simple and rapid (Romanowska).

Prostaglandins are best isolated from natural biological material by extraction and purification of the extract by silica gel column chromatography combined with preparative thin-layer chromatography (Clausen and Srivastava).

The procedure of Rouser *et al.* (1967a, b) is still the most widely used for the fractionation of phospholipids. Slight modifications were suggested by Hladík and Pokorný for the separation of oxidized phospholipids. Shimojo *et al.* (1962, 1971) separated phospholipids into 13 fractions by chromatography on cellulose equilibrated with chloroform and eluted with chloroform containing increasing amounts of methanol. Renkonen (1963) separated serum lipids into neutral lipids, cephalins, lecithins, sphingomyelins and lysolecithin by chromatography on silica gel columns and subsequent sub-fractionation

TABLE 27.4
SEPARATION OF TOTAL LIPIDS ON TEAE-CELLULOSE (SUITABLE FOR ACIDIC LIPIDS) (ROUSER *et al.*, 1969)

Column: 25 × 200 mm. Adsorbent: Selectacel, regular grade (Brown, Berlin, N.H., U.S.A.), or equivalent TEAE-cellulose; 15 g of adsorbent. Sample load: 100–300 mg in 5–10 ml of chloroform. Flow-rate: 3 ml/min. Bed volume: *ca.* 75 ml. Preparation of column: washing with 4 column volumes of 0.01 *N* potassium hydroxide in methanol, followed by 6–8 column volumes of methanol, 4 column volumes of methanol–chloroform (1:1) and 4 column volumes of chloroform.

| Volume (number of bed volumes) | Solvent | Classes of lipids separated |
|---|---|---|
| 5 | Chloroform | Free sterols, sterol esters, glycerides, hydrocarbons, carotenes, xanthophylls, phytins, chlorophylls |
| 8 | Chloroform–methanol (9:1) | Cerebrosides, glycosyl diglycerides, phosphatidylcholine, sphingomyelin, more polar xanthophylls |
| 8 | Chloroform–methanol (2:1) | Ceramidepolyhexosides |
| 8 | Methanol | Inorganic substances formed by ion exchange with acidic lipids |
| 6 | Chloroform–methanol (2:1) + 1% glacial acetic acid | Phosphatidylethanolamine, lysophosphatidylethanolamine, ceramideaminoethylphosphonate, free fatty acids, free bile acids, glycine-conjugated bile acids, phorbides, xanthophylls containing carboxyl groups |
| 6 | Glacial acetic acid | Phosphatidylserine, residual protein |
| 3 | Methanol | Only wash for removal of excess of acetic acid |
| 8 | Chloroform–methanol (4:1) made 0.01–0.1 *M* in ammonium or potassium acetate, to which 20 ml of 22% aqueous ammonia is added | Final acidic lipid fractions, such as phosphatidic acid, diphosphatidylglycerol, phosphatidylglycerol, cerebroside sulphate, plant sulpholipid, phosphatidylinositol |
| 6 | Methanol | Remaining lipids |

into individual components by further column chromatography, *e.g.*, on DEAE-cellulose or neutral alumina. The chromatography on silica gel columns can be combined with another technique, *e.g.*, ultracentrifugation in the investigation of lipoproteins (Nelson and Freeman). A similar technique on silica gel columns was applied to the analysis of milk phospholipids (Smith and Freeman).

A suitable ion exchanger for the fractionation of phospholipids was prepared by treating chlorohydroxypropylated Sephadex and cellulose with ammonia or amines, *e.g.*, the dibutylaminohydroxypropyl derivative (in the acetate form) of Sephadex LH-20 was used for the separation of egg phospholipids (Almé and Nyström).

Sphingomyelins are isolated by pre-fractionation on silica gel, inter-esterification of ester phospholipids with sodium methylate catalyst and a second fractionation on alumina

*References p.590*

(Hausheer *et al.*). Plasmalogens are determined after treatment of the column chromatographic fractions with 2,4-dinitrophenylhydrazine and phosphoric acid (Rhee *et al.*). The 2,4-dinitrophenylated and methylated phospholipids can be fractionated on cellulose columns (Collins and Shotlander). Robles and Roels recommended the preliminary deacylation of phosphoglycerides by alkaline hydrolysis followed by silica gel chromatography.

Tipton *et al.* studied the separation of phospholipids by gel chromatography on polystyrene cross-linked with 2% of divinylbenzene. Egg lecithin was separated reasonably well from neutral lipids. Kisselev fractionated lipids from brain tissue on Sephadex LH-20. In chloroform–methanol (2:1), Sephadex behaved as a weakly basic anion exchanger so that phospholipids were separated according to their basic properties. In chloroform–methanol–water (65:35:8), however, it acted as a molecular sieve so that the substances were separated according to their molecular weights. The use of methylated Sephadex was reviewed by Ellingboe *et al.*; Sephadex G-25 was methylated so as to contain 40% of methoxyl groups, and lipids were eluted with chloroform–methanol–water (85:85:30). Under these conditions, phospholipids were eluted in the first fraction together with lipopeptides and glycolipids. DEAE-Sephadex LH-20 was found to be advantageous for the fractionation of acidic lipids of *Escherichia coli* (Dittmer). Downey *et al.* and Shimojo *et al.* (1971) also successfully isolated phospholipids by gel chromatography on Sephadex LH-20.

## REFERENCES


Aitzetmüller, K., *Fette, Seifen, Anstrichm.*, 74 (1972) 598.
Aitzetmüller, K., *Fette, Seifen, Anstrichm.*, 75 (1973a) 14.
Aitzetmüller, K., *Fette, Seifen, Anstrichm.*, 75 (1973b) 256.
Aitzetmüller, K., *J. Chromatogr.*, 79 (1973c) 329.
Aitzetmüller, K., *J. Chromatogr.*, 83 (1973d) 461.
Almé, B. and Nyström, E., *J. Chromatogr.*, 59 (1971) 45.
Bende, H., *Fette, Seifen, Anstrichm.*, 70 (1968) 937.
Berry, J. F. and Kaye, B., *Lipids*, 3 (1968) 386.
Black, B. C. and Hammond, E. G., *J. Amer. Oil Chem. Soc.*, 40 (1963) 575.
Bombaugh, K. J., Dark, W. A. and Levangie, R. F., *J. Chromatogr. Sci.*, 7 (1969) 42.
Bottcher, C. J. F., Woodford, F. B., Boelsma-van Houte, E. and Van Gent, C. M., *Rec. Trav. Chim. Pays-Bas*, 78 (1959) 794.
Calderon, M. and Baumann, W. J., *J. Lipid Res.*, 11 (1970a) 167.
Calderon, M. and Baumann, W. J., *Boichim. Biophys. Acta*, 210 (1970b) 7.
Calderon, M. and Baumann, W. J., *Biochim. Biophys. Acta*, 231 (1971) 52.
Camejo, G., *J. Chromatogr.*, 21 (1966) 6.
Carroll, K. K., *J. Lipid. Res.*, 2 (1961) 135.
Carroll, K. K., *J. Amer. Oil Chem. Soc.*, 40 (1963) 413.
Carroll, K. K. and Serdarevich, B., *Lipid Chromatogr. Anal., 1967–1969*, 1 (1967) 205.
Cedras, J., Carlier, A., Puisieux, F. and Lettir, A., *Ann. Pharm. Fr.*, 25 (1967) 553.
Clausen, J. and Srivastava, K. C., *Lipids*, 7 (1972) 415.
Collins, F. D. and Shotlander, V. L., *J. Lipid Res.*, 1 (1960) 352.
Creech, B. G., *J. Amer. Oil Chem. Soc.*, 38 (1961) 540.
Davenport, J. B., *Lipids*, 4 (1969) 308.
De Lange, J. H., *Chem. Tech. (Amsterdam)*, 26 (1971) 276.

De Vries, B., *J. Amer. Oil Chem. Soc.*, 41 (1964) 403.
Dittmer, J. C., *J. Chromatogr.*, 43 (1969) 512.
Dolev, A. and Olcott, H. S., *J. Amer. Oil Chem. Soc.*, 42 (1965) 624, 1046.
Downey, W. K., Murphy, R. F. and Keogh, M. K., *J. Chromatogr.*, 46 (1970) 120.
Ellingboe, J., Nyström, E. and Sjövall, J., *Methods Enzymol.*, 14 (1969) 317.
Evans, C. D., McConnell, D. G., Hoffmann, R. L. and Peters, H., *J. Amer. Oil Chem. Soc.*, 44 (1967) 281.
Fisher, N., *J. Chromatogr.*, 47 (1970) 501.
Franzke, C., Kretzschmann, F., Rüstow, B. and Rugenstein, H., *Pharmazie*, 22 (1967) 487.
Hardy, R., Smith J. and Mackie P. R., *J. Chromatogr.*, 57 (1971) 142.
Hausheer, L., Pedersen, W. and Bernhard, K., *Helv. Chim. Acta*, 46 (1963) 601.
Hirayama, O., *Nippon Nogei Kagaku Kaishi*, 35 (1961) 437.
Hirsch, J. and Ahrens, Jr., E. A., *J. Biol. Chem.*, 233 (1958) 311.
Hladík, J., Jiroušova, J. and Pokorný, J., *Zeszyty Probl. Postepow Nauk Roln.*, 136 (1973) 87.
IUPAC, Fat and Oil Section, *Standard Methods, Method II.C 7*, Butterworths, London, 1972.
Joustra, M., Söderqvist, B. and Fischer, L., *J. Chromatogr.*, 28 (1967) 21.
Kajimoto, G. and Katsunori, M., *Eiyo-to-Shokuryo*, 17 (1965) 319.
Kaufmann, H. P., *Fette, Seifen, Anstrichm.*, 72 (1970) 505.
Kerkhoven, E. and Deman, J. M., *J. Chromatogr.*, 24 (1966) 56.
Kisselev, G. V., *Biokhimiya*, 34 (1969) 483.
Klenk, E. and Schorsch, E. U., *Hoppe-Seyler's Z. Physiol. Chem.*, 348 (1967) 1061.
Konishi, K., Inoue, H. and Taniguchi, N., *J. Chromatogr.*, 54 (1971) 367.
Kwiterovich, Jr., P. O., Sloan, H. R. and Fredrickson, D. S., *J. Lipid Res.*, 11 (1970) 322.
Leeder, L. G. and Clark, D. A., *Michrochem. J.*, 12 (1967) 396.
Marinetti, G. V., *J. Lipid Res.*, 3 (1962) 1.
Mehl, E. and Jatzkewitz, H., *Naturwissenschaften*, 50 (1963) 227.
Mulder, J. L. and Buytenhuys, F. A., *J. Chromatogr.*, 51 (1970) 459.
Nelson, G. J. and Freeman, N. K., *J. Biol. Chem.*, 235 (1960) 578.
Netting, A. G., *J. Chromatogr.*, 53 (1970) 507.
Nicolaides, N., Fu, H. C., Ansari, M. N. A. and Rice, G. R., *Lipids*, 7 (1972) 506.
Nyström, E. and Sjövall, J., *Anal. Biochem.*, 12 (1965) 235.
Papariello, G. J., Chulkaratana, S., Higuchi, T., Martin, J. E. and Kuceski, V. P., *J. Amer. Oil Chem. Soc.*, 37 (1960) 396.
Perkins, E. G., Taubold, R. and Hsieh, A., *J. Amer. Oil Chem. Soc.*, 50 (1973) 223.
Pokorný, S., Čoupek, J., Luân, N.-T. and Pokorný, J., *J. Chromatogr.*, 84 (1973) 319.
Pokorný, J., Kundu, M. K., Pařizková, H., Luân, N.-T., Čoupek, J., Pokorný, S. and Janiček, G., *Fette, Seifen, Anstrichm.*, 74 (1972) 625.
Pokorný, J., Kundu, M. K., Pokorný, S., Bleha, J. and Čoupek, J., *J. Chromatogr.*, (1974) in press.
Quinlin, P. and Weiser, Jr., H. J., *J. Amer. Oil Chem. Soc.*, 35 (1958) 325.
Radin, N. S., *Methods Enzymol.*, 14 (1969) 268.
Raveglia, I. F. and Ghittoni, N. E., *J. Chromatogr.*, 58 (1971) 288.
Renkonen, O., *J. Lipid Res.*, 3 (1962) 181.
Renkonen, O., *Acta Chem. Scand.*, 17 (1963) 1925.
Rhee, K. S., Del Rosario, R. R. and Dugan, L. R. Jr., *Lipids*, 2 (1967) 334.
Robles, E. C. and Roels, G. F. M., *Chem. Phys. Lipids*, 6 (1971) 31.
Romanowska, E., *Anal. Biochem.*, 33 (1970) 383.
Rouser, G., Bauman, A. J., Kritchevsky, G., Heller, D. and O'Brien, J. S., *J. Amer. Oil. Chem. Soc.*, 38 (1961) 544.
Rouser, G., Kritchevsky, G., Simon, G. and Nelson, G. J., *Lipids*, 2 (1967a) 37.
Rouser, G., Kritchevsky, G. and Yamamoto, A., *Lipid Chromatogr. Anal.*, 1 (1967b) 99.
Rouser, G., Kritchevsky, G., Yamamoto, A., Simon, G., Galli, C. and Bauman, A. J., *Methods Enzymol.*, 14 (1969) 272.
Scharmann, H. and Unbehend, M., *Hauptversammlung, Ges. Deut. Chem.; Angew. Chem.*, 83 (1971) 929.

Schlenk, H. and Holman, R. T., *J. Amer. Oil Chem. Soc.*, 30 (1953) 103.
Scholz, G. H., *Fette, Seifen, Anstrichm.*, 69 (1967) 565 and 651.
Shimojo, T., Kanon, H. and Ohno, K., *J. Biochem. (Tokyo)*, 69 (1971) 255.
Shimojo, T., Yokoyama, A. and Ohno, K., *J. Biochem. (Tokyo)*, 51 (1962) 293.
Slawson, V. and Mead, J. F., *J. Lipid Res.*, 13 (1972) 143.
Smith, L. M. and Freeman, N. K., *J. Dairy Sci.*, 42 (1959) 1450.
Stein, R. A. and Slawson, V., *Progr. Chem. Fats Other Lipids*, 8 (1966) 375.
Stolyhwo, A. and Privett, O. S., *J. Chromatogr. Sci.*, 11 (1973) 20.
Sweely, C. C., *Methods Enzymol.*, 14 (1969) 254.
Tipton, C. L., Paulis, J. W. and Pierson, M. D., *J. Chromatogr.*, 14 (1964) 486.
Trams, E. G. and Lauter, C. J., *Biochim. Biophys. Acta*, 60 (1962) 350.
Weenink, R. D. and Tulloch, A. P., *J. Amer. Oil Chem. Soc.*, 43 (1966) 327.
Wells, M. A. and Dittmer, J. C., *Biochemistry*, 5 (1966) 3405.
Wickbold, R., *Fette, Seifen, Anstrichm.*, 74 (1972) 578.
Wiedenhof, N., *J. Amer. Oil Chem. Soc.*, 36 (1959) 297.
Williams, J. P. and Merrilees, P. A., *Lipids*, 5 (1970) 367.
Wolfe, L. S. and Lowden, J. A., *Can. J. Biochem.*, 42 (1964) 1041.

*Chapter 28*

# Steroids

Ž. PROCHÁZKA

## CONTENTS



## INTRODUCTION

The number of papers that describe or mention the use of column chromatography in the separation of steroid derivatives is enormous. It would be interesting to know the percentage of papers on steroids that do not mention chromatography. Taking into consideration that the number and variety of known steroids are also tremendous, it is impossible to review here all, or even most, of the applications of column chromatography in the steroid field. All that can be done is to restrict the survey to more recent papers, to indicate various types of chromatographic methods, to compare them (if they are comparable), and to try to recommend *some* types of chromatographic procedures for certain particular types of steroids or separations.

Some steroid mixtures can be, and actually have been, separated satisfactorily by several chromatographic methods that differ in principle. As an illustration, a quotation by Cavina *et al.* (1971) may suffice, indicating that similar results can be achieved by two chromatographic methods. They used silicic acid columns (*cf.*, Figs. 28.2 and 28.3) and a gradient of diethyl ether in light petroleum for elution. However, they state: "A recent paper by Fernandez and Noceda describes separations of progestins and estrogens by means of column chromatography on Sephadex LH-20 with methanol–water (85:15) as eluent; our results compare favourably with those reported by these authors, especially regarding better separations between different progestins".

Experience shows that a synthetically or pharmaceutically oriented chemist will probably prefer the silicic acid method, while some biologically or clinically oriented biochemists will choose the Sephadex LH-20 method.

However, the laboratory worker should always keep in mind the purpose the separation serves. For example, a chemist isolating a small amount of a new or unknown steroid from a natural material will probably have a double task: first the elimination of ballast in a preliminary run (on a large amount of sorbent, but a shorter column), and then the required separation and purification on a small diameter, but very long, column. Finally, a clinical biochemist, often working with minute amounts of labelled steroids present in biological liquids, will have the problem of achieving accurate and economical results with expensive material and sophisticated equipment, especially when his work is routine. All these problems make the recommendation of a general procedure for sample preparation, application, chromatography and detection in the steroid field very difficult, and in the space devoted to steroids in this book impossible.

## GENERAL TECHNIQUES

### Introductory and theoretical considerations

For the choice of the most suitable method for a particular separation problem in the steroid field, several criteria must be kept in mind: (a) scale, *i.e.*, the amount of the mixture to be separated; (b) proportions of the required or analyzed steroid or steroids in the mixture, *i.e.*, the number and amounts of components; (c) the physico-chemical character of the steroids to be separated, *i.e.*, polarity, solubility, etc.; and (d) the structures of the steroids to be separated. The differences in polarity among steroids are enormous, ranging from sterols esterified by fatty acids, the lipophilic character of which is similar to those of fats and paraffins, to steroid glycosides or bile acid conjugates, which are appreciably soluble in water. Nonetheless, most steroids are of medium polarity, with a tendency towards lipophilicity owing to their large hydrocarbon skeleton. This is the reason why adsorption chromatography with solvents of low polarity largely prevails in this field, while such methods as gel permeation or affinity chromatography are used much less often. The last criterion, structure, may also become decisive when choosing the most suitable method to use. For example, the use of ion-exchange resins seems to be suitable for the separation of ionisable steroids, such as bile acids or some ionisable steroid conjugates. It is also known that homologous series are poorly separated on adsorbents, but are well separated in liquid–liquid systems. In the latter case, the steroids that are poorly separated in a system of partly miscible phases may be converted into more lipophilic derivatives (for example, by acylation) and separated in a pair of very immiscible solvents, or on very active adsorbents that seem adequate for substances of very weak polarity. The formation of steroid derivatives is also currently used for making the steroids tractable, for example volatile and thus suitable for gas chromatography.

### Sample preparation and application

Because of the variety of steroids that exist, no general procedure for sample preparation can be given, but a few hints may be useful. As most steroids are neutral, partition of an extract from natural material between an organic solvent (most commonly toluene, benzene, chloroform, dichloromethane, diethyl ether and ethyl acetate) and an aqueous alkali solution is recommended for the elimination of organic acids or other acidic material. Washing an organic extract with dilute hydrochloric acid is useful in cases when alkaloids or other basic impurities are present. However, highly polar neutral steroids can behave like estrogens, which are weakly acidic owing to the presence of a phenolic group, or bile acids when partitioned between a non-polar solvent, as for example toluene or chloroform and stronger aqueous alkali (Eberlein). In this case, filtration through an ion-exchange resin column may also lead to a high enrichment of the sample (Hobkirk and Nilsen, 1969a, b). Except for sterol esters and some very unpolar steroids, crude organic extracts from plant or, especially, animal material containing steroids can be pre-purified before application by partition between light petroleum (or *n*-hexane, *n*-heptane or other hydrocarbons) and 90–95% methanol. Common steroids remain in the polar phase, while paraffins, fats and the above exceptions pass into the paraffinic solvent phase. The effect of enrichment by such partitioning can be considerably increased by applying the counter-current distribution technique.

Application of the chromatographed mixture on to a classical preparative column can be carried out in several ways (see Chapter 8). Introduction of the sample into the head of an analytical, high-efficiency, micro-column is usually achieved by injection of a solution of the sample in the mobile phase. The heads of such columns are of a special construction (see Chapter 8).

### Liquid–solid chromatography

The best and most commonly used sorbents for the liquid–solid chromatography of steroids are alumina, silica gel and Florisil, with the use of silica gel largely prevailing today. Silver nitrate-impregnated adsorbents have been used successfully for difficult separations of steroids that differ only in the position or the number of double bonds. This method, although used extensively nowadays, is relatively new and therefore deserves the following more detailed description.

*Examples of the preparation of silver nitrate-impregnated or silvered adsorbents*

*Silicic acid*

A 25-g amount of silver nitrate was dissolved in 500 ml of distilled water; 100 g of silicic acid (Unisil; Clarkson Chemical Co., Williamsport, Pa., U.S.A.) were added to this solution, the water was removed from the slurry with a rotary evaporator and the adsorbent was dried overnight in an oven at 110°C. If the adsorbent is not protected from light, it steadily becomes darker, but this effect does not affect its activity to any appreciable extent. (Vroman and Cohen.)

*Silica gel G and Celite*

Silver nitrate (16.8 g), silica gel G (41.6 g) and Celite (41.6 g) were mixed as a suspension in water. (Canonica *et al.*, Galli and Grossi-Paoletti.)

*Alumina and Hyflo Supercel*

Neutral alumina AG-7 (30 g) was mixed with Hyflo Supercel (Johns-Manville, Denver, Colo., U.S.A.) (15 g), a solution of 9 g of silver nitrate in 75 ml of water was added and the mixture swirled. The suspension was frozen and lyophilized and stored *in vacuo* overnight. Columns were prepared from a suspension of the powder in a solvent (usually 98:2 chloroform–methanol). (Paliokas *et al.*)

*Silvered Florisil*

Florisil (10 g, 60–100 mesh) was suspended in a solution of 1.5 g of glucose in 150 ml of water. A solution of 160 mg of silver nitrate in 20 ml of water containing 1 ml of concentrated ammonia solution was then added with stirring at room temperature, and, after heating the mixture at 50°C in a water-bath for 10 min with constant stirring, the silica gel was filtered and washed thoroughly with water. The product was dried at 80°C for 40 min. (Ercoli *et al.*)

In the classical preparative elution chromatography (the so-called Reichstein chromatography) on alumina, a 1:30 ratio of sample to adsorbent is most commonly used, the elution being started with a non-polar solvent (usually light petroleum), the polarity of the solvent being increased by the gradual addition of a more polar solvent or the solvent being changed to a more polar solvent when solid material can no longer be eluted with the first solvent. With silica gel or Florisil, the ratio of sample to adsorbent is usually between 1:50 and 1:100. For difficult separations or analytical purposes, this ratio ranges from 1:100 to 1:1000.

Because of some advantages (rapidity and ease of preparation, and the similarity between the results and those obtained by TLC), dry column chromatography on silica gel has become the method of choice in more difficult preparative separations.

As regards the choice of solvents and the volume of the fractions, the method of Joska (Chmel *et al.*) is recommended. Employing non-adhering thin layers of silica gel, a solvent system is first found in which the least polar component of the separated mixture has an $R_F$ value of *ca.* 0.5. The same solvent is then used for dry column chromatography, but the volume of the fractions collected (in millilitres) should be about one twentieth of the weight (in grams) of the adsorbent in the column. As a rule, if a good separation is achieved on common thin-layer plates with a particular solvent, then a less polar solvent should be used for column chromatography. For optimization of the composition of the solvent system in TLC, Turina proposed a method in which the best separation can be achieved while carrying out the least number of experiments. This method is based on general principles and should also be applicable to column chromatography.

Examples for analytical liquid–solid high-resolution (high-, medium-, and low-speed) chromatography are given in the Applications section of this chapter. At this point the multicolumn high-resolution liquid–solid chromatography systems developed by Vestergaard and his co-workers (*cf.*, Vestergaard, 1973) should be mentioned. His system is of medium to low speed, but the high resolution and the multiplicity and automation of accessories allow the use of his method for large numbers of routine serial analyses of urinary steroids per day, *i.e.*, up to 100 analyses or more per day. The main features of his method are: (a) automated dry-filling of 6 ft. × 1.5 mm PTFE capillary columns; 20 columns are filled at a time in 5–20 min. (Vestergaard, 1967); (b) gradient elution on neutral alumina (100–200 mesh) deactivated with 6% water (for 17-ketosteroids), or on silica gel (100–200 mesh) containing 20% water (for corticosteroids), and the use of a battery of 12 or more chromatographic pumps (Vestergaard *et al.*, Vestergaard and Sayegh) or a nitrogen pressure system (Vestergaard and Jacobsen) for chromatography on 25 columns; (c) a battery of needle valves connected with a battery of simple injection ports and PTFE columns (see under a) immersed in a water-bath; (d) a multi-collector; (e) a computer; (f) running times from about 6 h up to 24 h and more.

Centrifugal liquid–solid chromatography of steroids on microcolumns of microparticulate silica was described by Ribi *et al.* The method seems very rapid (separations in about 7 min) and rather efficient.

### Liquid–liquid chromatography

Liquid–liquid chromatography is much less frequently used in synthetic organic chemistry. The commonest support for the polar stationary phase is diatomaceous earth (Celite or Hyflo Supercel), but cellulose powder and other carriers have also been used. Various polar organic solvents, often containing water, can be used as the stationary phase. Aqueous methanol or acetic acid (for bile acids) and formamide are among the most commonly used, while various hydrocarbons or their mixtures with other non-polar solvents (benzene, toluene, etc.) serve as mobile phases. For column preparation (Engel *et al.*), Celite or Hyflo Supercel is washed with 6 *N* hydrochloric acid in order to remove metals, then washed by elutriation until free from chloride and fines. Finally, it is washed with absolute methanol and light petroleum and dried. The components of the partition system are agitated in a funnel and allowed to equilibrate overnight. Ten parts by weight of the washed Kieselguhr are then treated with 5–6 parts by volume of the lower phase (for example, 90% methanol) and the mixture is stirred until the solvent is uniformly distributed. The upper phase is then added (for example, trimethylpentane) so as to make a thin slurry. After inserting a small plug of glass-wool into the bottom of the chromatographic column, a small amount of the slurry is added and packed in with a stainless-steel perforated packing disc which fits the tube snugly. The plunger is moved up and down slowly in order to remove air bubbles, and finally pressed down firmly. Successive portions of slurry are added and packed in. Care should be taken to exert an even pressure during the packing and to maintain a head of mobile phase above the column at all times. After the column has been packed, the pre-equilibrated upper phase is allowed to flow through

*References p.620*

the column overnight by gravity in order to condition it. The sample for chromatography is dissolved in a small amount of the stationary phase and a small amount of dry Kieselguhr is added until a powdery material is obtained, and this material is also packed on the top of the column with a small amount of the mobile phase above the packing. According to some workers (Butte and Noble, Dixon, Mickan *et al.*, Siiteri), the columns can be prepared by the so-called "dry-pack" procedure. Celite is mixed with the stationary phase (2:1 to 1:1 ratio) and then added to the column in portions and packed in with a glass rod or plunger. The column can be topped with a small amount (*ca.* 8%) of dry Celite and a glass-wool plug. In order to minimize the amount of air trapped in the column, the mobile phase is aspirated through the bottom of the column by applying a vacuum to the top.

With small automated analytical columns, Meijers found that these conventional methods of loading the stationary phase can conveniently be replaced by the so-called "*in situ*" loading, which consists in the injection of the equilibrated stationary phase directly into the column containing the dry support.

As regards the choice of the phases, the data presented by Hulsman may be useful. He measured and calculated "selectivity factors" (the ratio of the partition coefficients of two components) for many combinations of almost 100 steroids in several common solvent systems (light petroleum–toluene–methanol–water, light petroleum–benzene–methanol–water, benzene–methanol–water, light petroleum–methanol–water, formamide–benzene–chloroform, formamide–benzene, propylene glycol–toluene, formamide–cyclohexane–benzene, propylene glycol–ligroin and propylene glycol–cyclohexane–benzene).

Retention volumes of some steroids (androstane and pregnane derivatives and bile acids) were tabulated for Sephadex derivatives and several solvent systems in a review by Ellingboe *et al.* (1969). The ratio of the diameter to height for Kieselguhr columns was 1:20 to 1:30.

For reversed-phase chromatography, *i.e.*, the fixation of the less polar stationary phase, Celite has also been used (Burstein *et al.*, Burstein and Zamoscianyk), but hydrophobized supports or carriers seem to be more advantageous, such as siliconized Kieselguhr (Bergström and Sjövall), Hostalen (polyethylene powder) (Hoshita *et al.*) or methylated Sephadex (Nyström and Sjövall). These techniques have been described thoroughly in a review by Eneroth and Sjövall. Other hydrophobized Sephadexes were also found suitable, such as hydroxypropyl Sephadex, Sephadex LH-20 and hydroxyalkoxypropyl Sephadex (Ellingboe *et al.*, 1968, 1969). The latter is hydrophobized by the reaction of the former with an aliphatic olefin oxide.

In the following sections, a description is given of the methods of preparation of the above and similar supports.

### *Hydrophobization of Kieselguhr (Celite)*

Hyflo Supercel is washed with 6 *N* hydrochloric acid until a colourless supernatant is obtained, then with distilled water until neutral and finally with acetone. The dried material is placed in dishes in a desiccator together with a beaker containing 50–100 ml of dimethyldichlorosilane or a low-boiling mixture of chlorosilanes (British Thomson-Houston Export Co., Rugby, Great Britain), and is left for 2–3 weeks. The support is then washed with ethanol or methanol until the filtrate is free from hydrochloric acid, dried at 100°C and stored in a closed vessel. The hydrophobic support thus obtained has been found to be superior to supports silanized with solutions of dimethyldichlorosilane. (Bergström and Sjövall, Eneroth and Sjövall.)

*Purification of Hostalen*

Hostalen is purified by continuous Soxhlet extraction with ethanol until the extract is colourless. The support is dried below 75°C. (Eneroth and Sjövall.)

*Hydrophobization of Sephadex*

*Methylated Sephadex*

Sephadex G-25 (10 g, fine, bead form) and 65 ml of 34% (w/v) sodium hydroxide solution are stirred under a nitrogen atmosphere for 4 h at room temperature. Dimethyl sulphate (90 ml) is added slowly while cooling the reaction vessel in ice, taking care that the reaction temperature does not exceed 25°C. After stirring the mixture for 4 h, the excess of dimethyl sulphate is hydrolyzed by adding a large amount of water. The reaction mixture is left overnight and then neutralized with ammonia solution. The methylated Sephadex is collected on a sintered glass funnel and washed with water and ethanol. The $-OCH_3$ content is 34.6%; a higher degree of methylation ($-OCH_3$ = 43.6%) is achieved, however, when Sephadex methylated with dimethyl sulphate in aqueous sodium hydroxide solution ($-OCH_3$ = 34.6%) is used as the starting material. This two-stage reaction has been used on a 150-g scale and is preferred to the direct Hakomori reaction, which requires large amounts of the methylsulphinyl carbanion. (Nyström, Nyström and Sjövall.)

*Hydroxypropyl Sephadex*

Sephadex G-50 (20.0 g, fine, bead form) is soaked for 2 h in 4% (w/v) sodium hydroxide solution, after which the excess of the aqueous phase is removed by filtration with suction. The wet Sephadex (190 g) is then suspended in 700 ml of propylene oxide (commercial grade) and refluxed with stirring for 2 h. The product is filtered free from solvent and soluble reaction products, then washed consecutively with acetone, water (until the filtrate is no longer alkaline), acetone and finally light petroleum (b.p. 40–60°C). The product is first dried under suction on the filter, then at 80°C until a constant weight of 41.3 g is obtained. (Ellingboe *et al.*, 1968.)

*Hydroxyalkoxypropyl Sephadex*

Sephadex LH-20 (100 g, superfine, bead form) is soaked in 150 ml of dry methylene chloride. Boron trifluoride ethyl etherate (2 ml, 48% $BF_3$) is added and the constituents are mixed thoroughly. Stirring is continued at room temperature while an aliphatic olefin oxide (Needox 1114, 150 ml) is added slowly. The rate of addition is adjusted so that the resulting exothermic reaction does not cause uncontrolled boiling. After the addition of the olefin oxide, *ca.* 100 ml of dry methylene oxide are added in order to facilitate stirring and the mixture is stirred for a further 20 min at room temperature. The product is filtered free from solvent and washed consecutively with chloroform, ethanol and acetone. The solvents are removed by suction on the filter and the product is brought to a constant weight at room temperature *in vacuo*. The gain in weight is about 125%. The derivative is no longer wetted by water and swells in solvents such as chloroform, benzene and heptane. Acetone and the lower alcohols cause little swelling. (Ellingboe *et al.*, 1969.)

*Carboxymethoxypropyl Sephadex*

Hydroxypropyl Sephadex G-25 or LH-20 (50 g) is suspended in 750 ml of isopropanol. Sodium hydroxide (60 g) is then added, the suspension is stirred for 15 min and then 60 g of sodium chloroacetate are added in small portions. After continuous stirring for 12 h, the Sephadex derivative is collected on a filter and washed consecutively with water, ethanol, chloroform, ethanol, water, 0.1 *N* hydrochloric acid, 0.001 *N* hydrochloric acid, ethanol and diethyl ether. The degree of substitution is estimated by suspending the Sephadex derivative (about 1.0 g) in 10 ml of 0.2 *M* potassium chloride solution and titrating with 0.2 *N* potassium hydroxide solution while measuring the pH electrometrically. (Joustra *et al.*)

*Diethylaminoethoxypropyl Sephadex*

Hydroxypropyl Sephadex G-25 or LH-20 (50 g) is carefully mixed with aqueous sodium hydroxide solution (33.2 g in 142 ml of water) in an ice-cooled flask. After 30 min, 2-chlorotriethylamine hydrochloride (29.2 g in 37.5 ml of water) is added slowly while stirring, and the temperature is then raised to 80–85°C for 45 min. The product is cooled, then filtered free from the aqueous phase and washed with water, 10% ammonia solution, ethanol, 0.1 *N* hydrochloric acid, 0.001 *N* hydrochloric acid, ethanol and diethyl ether, and dried at 60°C. The degree of substitution, as determined by the titration procedure described above for carboxymethoxypropyl Sephadex, is *ca.* 0.8 mequiv./g. (Peterson and Sober.)

Among the advantages of all these lipophilic derivatives are their chemical stability, ease of preparation and handling, and flexibility in application. Variations in solvent systems and in the degree of substitution of the gel can be used to effect different kinds of separations. Columns can be kept in continuous use over long periods. For most purposes, molecular sieving effects are negligible in reversed-phase chromatography on these derivatives, but may be significant in "straight-phase" chromatography. For the above hydrophobized supports, *n*-heptane alone or in admixture with chloroform is commonly used. Sometimes chloroform–isooctanol or the less polar phase of *n*-heptane–acetone–water, *n*-heptane–chloroform–isopropanol–water and other mixtures is used (Ellingboe *et al.*, 1968). Aqueous methanol or the more polar phase of the above mixtures are used as mobile phases. For the chromatography of bile acids, the addition of acetic acid in order to suppress dissociation is advisable.

According to Eneroth and Sjövall, after equilibration of the solvent mixture in a separating funnel, a volume of the stationary phase (4 ml/4.5 g of hydrophobic Hyflo Supercel; 3 ml/4.5 g of Hostalen; 6 ml/4.5 g of methylated Sephadex) is added to the support and the mixture is thoroughly homogenized with a spatula for a short time (about half a minute so as to avoid evaporation of solvent). The mobile phase is added, the slurry is homogenized and poured into a chromatographic tube with a diameter that gives columns with a ratio of height to diameter of 10:1 to 20:1. Air bubbles are removed with a perforated plunger and the column is allowed to settle by gravitation under free solvent flow. Final packing at the top is achieved by light pressure with the plunger. The sample is applied in a small volume of the mobile phase; a few drops of the stationary phase may be added. According to Ellingboe *et al.* (1968), the columns are packed with a slurry of the lipophilic support under gentle pressure. Most important for easy packing and good column performance is that the lipophilic material should be thoroughly equilibrated with the solvent, preferably by agitation in an ultrasonic bath. The solvent flow-rate with these materials is mentioned in the section Gel chromatography.

For automated high-resolution liquid chromatography with reversed phases, Siggia and Dishman studied the use of Amberlite LA-1 [*n*-dodecanal(trialkylmethyl)amine] as the stationary phase fixed to several supports [Anakrom AB diatomaceous earth, a terpolymer consisting mainly of trifluoroethylene (Plaskon CTFE-2300) and Zipax spherical silica beads] and water and water–methanol mixtures as the mobile phase. The hydrophobic CTFE support exhibited lipophilic adsorption even after coating with the stationary phase, which, however, could be utilized to the advantage of separations. For similar purposes (analytical reversed-phase chromatography), Waters Ass. recommend a novel reversed-phase affinity packing, Poragel PN, which is extremely rigid.

In a paper on the analysis of various types of steroids and their dinitrophenylhydrazones by high-speed liquid chromatography, Henry *et al.* described useful separations in liquid–liquid partition systems in which Zipax was used as the support and the following "straight" and reversed phases were used for the partition of various steroids: *(a) "straight" phases:* 1% of β,β-oxydipropionitrile on Zipax as stationary phase and the following solvent mixtures as the mobile phases: *n*-heptane, tetrahydrofuran–*n*-heptane mixtures (2:8, 1:9 and 5:95); 1% ethylene glycol on Zipax as the stationary phase with a moving chloroform–*n*-heptane mixture (3:97); *(b) reversed phases:* stationary 1% hydrocarbon polymer on Zipax with methanol–water (15:85) as the mobile phase; 1% cyanoethyl-silicone on Zipax as stationary phase and methanol–water (2.5:97.5) as eluent; Permaphase ODS (DuPont, Wilmington, Del., U.S.A.; permanently bonded chromatographic support), *i.e.*, octadecylsilane bonded on Zipax, as the stationary phase and the linear gradient water to 50% aqueous methanol (at 5%/min) as the mobile phase. The authors also consider other stationary phases to be useful, such as tris(2-cyanoethoxy)propane, trimethylene glycol, triethylene glycol, cyanoethylsilicones and Carbowaxes.

### Gel chromatography

Various types of Sephadex gel are also commonly used for steroid separations. On lipophilic Sephadex gels, separations are governed by two main mechanisms, liquid–gel partition and molecular sieving. However, Gelotte has shown that aromatic and heterocyclic compounds are adsorbed more strongly to the gel matrix than other types of substances. This effect is important in the gel chromatography of some conjugates of aromatic steroids (estrogens).

*References p. 620*

Gel chromatography is complementary to established techniques for steroid chromatography. The low chemical reactivity of gels and, with lipophilic gels, also the possibility of using non-polar solvents, make these gels highly convenient for the partition or sieving of readily hydrolysable derivatives and also for quantitative analysis. Furthermore, with lipophilic Sephadex gels as a medium for partition chromatography (*cf.*, preceding section), no solvent mixtures have to be pre-equilibrated and no stationary phase has to be applied to the support.

Among hydrophilic gels, DEAE-Sephadex A-25 in combination with a sodium chloride gradient elution and Sephadex G-25 with distilled water as eluent were mainly used for the separation of estrogen conjugates (see below) or protein-bound testosterone (Horton *et al.*, Kato and Horton).

Lipophilic Sephadex LH-20 with various mixtures of organic solvents used for swelling and elution is suitable for the separations of various types of steroids: benzene–methanol for estrogens and some other steroids (Mikhail *et al.*); chloroform–methanol (1:1) for various steroids from bile (Laatikainen and Vihko); 99% *n*-butanol for corticoids and 17-ketosteroids (Seki and Sugase); and dichloromethane for sterols (Van Lier and Smith).

With hydrophilic gels, various ratios of height to diameter of the columns are used, from 6:1 to about 60:1. With lipophilic gels, a ratio of height to diameter of the column of 30:1 is common. The ratio of the volume of the solution of the steroid mixture to the volume of gel in the column varies from about 1:5 to 1:40; a ratio of 1:20 seems reasonable for most separations.

According to Ellingboe *et al.* (1969), the optimum solvent flow-rate for Sephadex derivatives is about 0.1–0.5 ml/min per square centimetre of column cross-sectional area. Pre-washing of the columns is important, as is the uniformity of particles achieved by the sedimentation method. When thoroughly washed after each run, the columns can be used repeatedly, even for analytical purposes.

## Ion-exchange chromatography

This type of chromatography is much less used than other types in the steroid field, because most steroids are non-ionizable. However, ion-exchange resins are commonly used for the enrichment of the sample, *i.e.*, its pre-purification.

A very good and exhaustive paper (Seki, 1969) has been published on the separation of various types of steroids by means of ion-exchange resins.

Henry *et al.* recently described the analysis of Dexamethasone disodium phosphate on a Zipax support coated with 1% of a weak anion exchanger, *viz.* an amine-substituted polyamide, and ethanol–water (1:9) with 0.1 *M* orthophosphoric acid–0.1 *M* sodium orthophosphate as the mobile phase, and the analysis of 17β-estradiol-17β-glucosiduronic acid on Zipax coated with a strong anion-exchange polymer, *viz.* a quaternary amine-substituted lauryl methacrylate polymer, and a linear gradient of pH 9.2 buffer to pH 9.2 buffer + 0.8 *M* sodium perchlorate (at 5%/min) as the mobile phase.

## Detection

Steroids are detected in the eluate by submitting single fractions or aliquots to UV absorptiometry or colorimetry after addition of an appropriate reagent. In automated analytical chromatography, spectrophotometry and refractometry are the most commonly used methods. In classical preparative chromatography, single fractions are evaporated to dryness and the weight of the residue is determined. Today, the fractions are usually analyzed by TLC, applying a few microlitres of each fraction close to each other on the start of a broad thin-layer plate, then developing and detecting. For this purpose spraying with concentrated sulphuric acid, or a 30% solution in methanol or diethyl ether, and heating is a universal detection method. For analytical purposes, especially in advanced laboratories, the effluent may also be monitored continuously by means of a hydrogen flame ionization detector, part of the effluent being continuously drawn off. The operating conditions for the Barber-Colman (Rockford, Ill., U.S.A.) Model 5400 liquid chromatographic detector were reported by Cavina *et al.* (1969). However, for many investigations, labelled steroids are used and are detected and measured in the eluate on the basis of their radioactivity.

Estrogens can be determined by measuring their UV absorption at 280 nm (in alcoholic solution), or by fluorimetry (Bates and Cohen, Eechaute *et al.*). The use of Kober reagents (phenols in sulphuric acid) (Osawa and Slaunwhite) and heating is a suitable method for the colorimetry of estrogen conjugates. The authors gave the following procedure: 1 g of hydroquinone was dissolved in 14 ml of distilled water to which 35 ml of 98% sulphuric acid were added. The reagent should be prepared freshly every day.

All $\Delta^4$-3-ketosteroids or other steroids that contain an $\alpha,\beta$-unsaturated carbonyl group in the molecule can be detected and measured on the basis of their UV absorption at about 254 nm. A great number of important steroids and their derivatives, such as testosterone, androstenedione, progesterone and corticoids, belong to this group.

17-Hydroxysteroids are detected with the Porter–Silber reagent (phenylhydrazine sulphate and sulphuric acid) (Seki, 1967, 1969), while 17-ketosteroids are detected with the Zimmermann reagent (*m*-dinitrobenzene in aqueous potassium hydroxide solution) and determined with the same reagent by the Epstein method (Epstein; Seki, 1969).

### *Porter–Silber reagent*

Phenylhydrazine sulphate (50 mg) is dissolved in a mixture of 21 ml of ethanol (99.5%) and 39 ml of sulphuric acid (158 ml of sulphuric acid to 37 ml of deionized water). The reagent is added (for example, 1 ml) to evaporated fractions dissolved in 60% ethanol (for example, 0.5 ml). The mixture is allowed to stand at room temperature for 15 h and the optical density finally measured at 410 nm. (Silber and Porter.)

### *Zimmermann reagent*

The residue of the evaporated fraction is dissolved in 0.05 ml of methanol; 0.2 ml of a saturated solution of *m*-dinitrobenzene in 5% Hyamine 1622 solution is added, followed by 0.1 ml of 8 *N* aqueous potassium hydroxide solution with mixing. After the mixture

*References p.620*

has stood for 30 min, 2.0 ml of 5% aqueous Hyamine 1622 solution are added and the optical density is measured at 510 nm against deionized water. (Epstein, Seki, 1969.)

For high-resolution analytical column chromatography, some methods of detection and quantitation in the eluate are already in use (UV absorptiometry and refractive index measurement, see above; radioactivity measurement; flame ionization after evaporation of the solvent), but others have still to be devised.

## APPLICATIONS

### Sterols

Sterols and their esters are the most lipophilic steroids and are generally easily separated by various types of chromatographic procedures. Solid–liquid adsorption chromatography is the most commonly used technique. In instances when the differences in structures of the components of the mixture are too small (for example, pairs of sterols that differ only in the number and the position of double bonds), separation can be achieved by the so-called argentation chromatography (silver nitrate-impregnated adsorbents) (Canonica *et al.*, Paliokas and Schroepfer, Vroman and Cohen, Lee *et al.*, Ziller *et al.*). Table 28.1 gives a short survey of some recent separations by various methods.

### Androgens

In the synthetic preparative field, Florisil as adsorbent and benzene as eluent were used recently by Janot *et al.* for the purification of weakly polar androstane derivatives. The same adsorbent and light petroleum (b.p. 30–40°C) as eluent were used by Swann and Turnbull for some 17-mercaptoandrostanes. Grimwalde and Lester separated formylated testosterone derivatives on dry columns of alumina deactivated by the addition of 7% of water, using the mixtures dichloromethane–cyclohexane (1:1) and dichloromethane–cyclohexane–ethyl acetate (20:20:1 and 5:5:1). The classical method of Dingemanse *et al.* for the separation of urinary 17-ketosteroids on alumina is still used and recommended (Forriol). The same is true of the classical Reichstein elution chromatography on alumina, which was used by Nagasawa *et al.* for the separation of the transformation products of sterols by microorganisms using successively light petroleum, benzene and diethyl ether for elution. Drosdowsky *et al.* used alumina and *n*-hexane–chloroform mixtures for the separation of labelled testosterone and epitestosterone. Silica gel was used by McCurdy and Garrett for the separation of 19-nortestololactone from other components. An adsorbent to sample ratio of 100:1 and ethyl acetate as eluent were used. Ambrus and Wix isolated microbial transformation products of 4-androstene-3,17-dione on silica gel (sample to sorbent ratio *ca.* 1:70) by gradient elution with dichloromethane with additions of ethanol. For analytical purposes, where labelled steroids are often used, other types of chromatography seem to predominate. Thus Eckstein *et al.* identified 5α-androstane-3α,

17β-diol as a metabolite of pregnenolone using 90% methanol fixed on Celite as the stationary phase and trimethylpentane as the mobile phase. Huang separated the metabolites of [7-$^3$H]4-androstene-3,17-dione by reversed-phase chromatography on the same support as above, using toluene–70% methanol, and for straight-phase chromatography a solvent system consisting of *n*-heptane–toluene–80% methanol (4:6:10) (*cf.*, Schneider *et al.*).

Hydrophilic conjugates of androstane derivatives were detected and analyzed in natural material predominantly after their cleavage. However, Dray *et al.* separated free steroids from their sulphates on Celite with plasma or distilled water as the stationary phase and isooctane–ethyl acetate–*n*-butanol–methanol–1 *M* ammonia solution (2:4:1:2:3) as the mobile phase. They separated sulphates from glucuronides on alumina by gradient elution according to Crépy *et al.* For the separation of steroids released from their sulphates by solvolysis, they also used Celite with formamide as the stationary phase and *n*-hexane–benzene (7:3 and 3:7) for elution.

In their paper on the analysis of steroid hormones by high-resolution liquid chromatography, Siggia and Dishman tested several types of supports for the fixing of Amberlite LA-1 [*n*-dodecanal(trialkylmethyl)amine] as the stationary phase. They also presented useful data for the separation of androgens (see Fig. 28.1) using Plaskon CTFE-2300, a terpolymer consisting mainly of trifluoroethylene, as the support.

Applying the technique of Henry *et al.* (see p. 601) Fitzpatrick *et al.* developed the high-speed liquid chromatography of derivatized urinary 17-ketosteroids (2,4-dinitrophenylhydrazones). For straight-phase chromatography, Zipax coated with 0.75 or 1.5% of β,β′-oxydipropionitrile was used as the stationary phase and isooctane as the mobile phase; for reversed-phase chromatography, Corasil with permanently bonded $C_{18}$-silane was used as the stationary phase and aqueous ethanol (1:1) as the mobile liquid phase. Good separations were achieved of all four epimeric forms of androsterone from each other and from other androstane derivatives and impurities (from urine and blood plasma). The time required for the separations was about half an hour and the precision was equivalent to a relative standard deviation of about 10% at the microgram and lower levels.

Useful information on the determination of androgens, mainly testosterone, in human plasma and urine etc. can also be found in papers by Henry *et al.*, Saez *et al.*, Peng and Munson and Vermeulen and Verdonck.

Serial analyses of urinary 17-ketosteroids were carried out successfully by Vestergaard and Jacobsen using the multicolumn chromatographic system described in the section on liquid–solid chromatography of steroids.

## Estrogens

Except for their aromatic A ring and the phenolic character, estrogens do not differ much from androgens or gestagens from the chromatographic point of view, and therefore only a few analytical methods will be described here.

Cavina *et al.* (1971) developed an analytical method for mixtures of various steroids, including estrogens, in pharmaceutical preparations. As the adsorbent they used Bio-Rad

TABLE 28.1
SEPARATIONS OF SOME STEROLS

| Technique | Compounds separated* | Sorbent or ion exchanger | Eluent | Note | Reference |
|---|---|---|---|---|---|
| Prep. solid–liquid | 5α-C-1α,2α-diol | Alumina + 10% of 10% AcOH | Benzene–diethyl ether (9:1) | 200-fold excess of sorbent | Davey *et al.* |
| | 5α-C-1β,2β-diol | Alumina + 10% of 10% AcOH | Benzene–diethyl ether (1:1) | 200-fold excess of sorbent | Davey *et al.* |
| Prep. solid–liquid | 3β-OH-C-5-ene, 3β-OH-C-5,7-diene and other pairs | Silicic acid + $AgNO_3$ | *n*-Hexane–benzene (stepwise increase in benzene concentration) | 200-fold excess of sorbent | Vroman and Cohen |
| Prep. solid–liquid | 3β-AcO-C-5-ene, 7-ene, and 8(14)-ene | Alumina + Hyflo Supercel + $AgNO_3$ | *n*-Hexane–benzene (9:1) | 100 × 1 cm column per 6–7 mg of mixture; 2.9-ml fractions | Lee *et al.* |
| Prep. solid–liquid | Coprostanol, C-3β-ol-4-$^{14}$C, cholesterol | 200 g Silicic acid + 136 g $AgNO_3$ (with 20% of water) | Benzene in hexane (12% and 15%) | 200- to 500-fold excess of sorbent | Ziller *et al.* |
| Gel | Cholesteryl sulphate, neutral sterol derivatives | Methylated Sephadex G-25, Sephadex LH-20 | Chloroform | Solute to gel ratio *ca.* 1:130 to 1:260; above 1:1000; 44 × 0.84 cm column | Eneroth and Nyström |
| Gel | β-Sitosterol, campesterol | Hydrophobic hydroxyalkyl Sephadex LH-20 | Methanol–hexane (95:5) | 25 mg on columns of 182- and 432-ml volume | Hyde and Elliot** |

| | | | | | |
|---|---|---|---|---|---|
| Gel | 3β-OH-C-5-en-23-one | Sephadex LH-20 | Dichloromethane | 250 mg of mixture on a 2.5 × 60 cm column, fraction volume 13.5 ml | Lier and Smith |
| Liquid–liquid | 17α,20α-Dihydroxycholesterol, 17α,20β-dihydroxy-20-isocholesterol, cholesterol, 3β,17α-dihydroxypregn-5-en-20-one | Celite 545 as support | Heptane or Skelly-Solve C–methanol–water (10:8:2); heptane–benzene–methanol–water (33:17:40:10); toluene–heptane–methanol–water (8:4:4:1); toluene–propylene glycol; hexane–formamide; methanol–*n*-propanol–water–toluene–isooctane (4:1:1.3:2:2) | The last system was used as a reversed-phase system, *i.e.*, 0.3 ml of the upper phase for wetting 1 g of Celite | Burstein *et al.**** |
| Ion exchange | Cholesteryl sulphate | Amberlyst 15 ($H^+$) (batch No. 625) converted into $NH_4^+$ form and washed with chloroform–methanol | Chloroform–methanol (4:1) | *ca.* 1000-fold excess of the resin | Eneroth and Nyström |
| | | CM-Sephadex LH-20 | Chloroform–methanol (4:1) | Large excess of the ion exchanger, 1.46 × 46.5 cm column | |

*C = cholestan or cholest-.

**Sephadex LH-20 was treated with Needox 1115 (a mixture of $C_{11}$–$C_{14}$ olefin oxides) according to Ellingboe *et al.* (1968, 1969). The preparation had a weight increase corresponding to a hydroxyalkyl content of 49% (w/w), columns 2.5 × 45 and 2.5 × 100 cm.

***Burstein and Zamoscianyk described further mixtures for liquid–liquid chromatography on Celite.

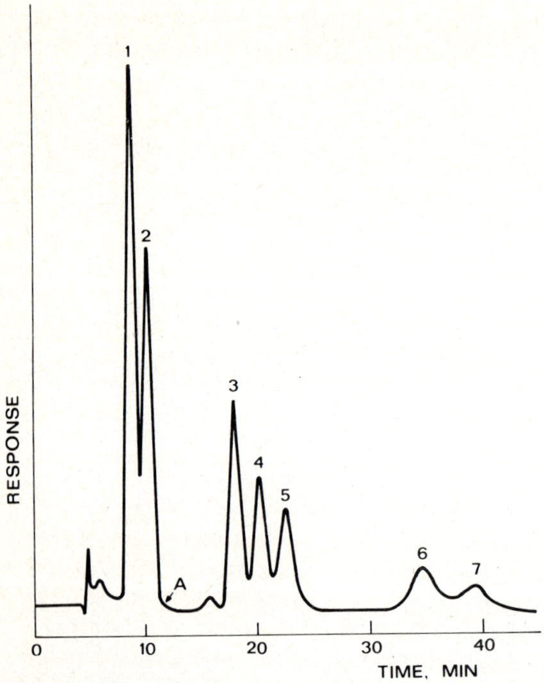


Fig. 28.1. Separation of some androgens (Siggia and Dishman). Column: 0.2 × 48.5 cm I.D. Packing: 23% Amberlite LA-1 on Plaskon CTFE-2300 (see p. 601). Eluent: water. Initial flow-rate: 0.17 ml/min, increased at A to 0.49 ml/min. Detection: battery-operated Beckman DU spectrophotometer equipped with a deuterium source lamp (Beckman Part No. 96280). A Varian G-2000 strip-chart recorder (Varian, Palo Alto, Calif., U.S.A.) equipped with 1–1000 mV variable span was wired across the null meter. Peaks: 1 = 4-androstene-3,11,17-trione; 2 = 4-androstene-11β-ol-3,17-dione; 3= 1,4-androstadiene-17β-ol-3-one; 4 = 19-nor-4-androstene-3-17-dione; 5 = 19-nor-4-androstene-17β-ol-3-one (19-nor-testosterone); 6 = 4-androstene-3,17-dione; 7 = 4-androstene-17β-ol-3-one (testosterone).

silicic acid, 325 mesh, for lipid chromatography, according to Hirsch and Ahrens. The water content was about 9% (9.38 ml of water per 100 g of silicic acid dried at 125°C for 8 h and cooled in a desiccator). Glass columns were packed with the adsorbent, and when 10 g of silicic acid were used, a filling of 56 cm height was obtained. Silicic acid suspended in light petroleum (b.p. 65–75°C) was gradually poured into the column, which was half-filled with solvent, the solvent being allowed to flow slowly out of the column. Samples of steroids (see Table 28.2) were introduced as a solution in oil or in *n*-heptane (1–2 mg/ml). The sample, dissolved in as small a volume as possible (1–2 ml), was transferred on to the column by pipette and washed with light petroleum (3 × 1 ml), allowing a slow efflux by gravity. The solvent was pumped into the column by a metering pump with an adjustable flow-rate. The gradient of diethyl ether in light petroleum was obtained by pumping diethyl ether from the reservoir into the mixing chamber containing light petroleum. In order to monitor the various fractions eluted from the column as concentration peaks, part of the eluate from the column was transferred on to a chain conveyor, which, after the solvent had been evaporated to dryness, introduced the solid residue into a hydrogen flame ionization detector with a moving chain. The splitting ratio was adjusted to deliver 6–7% of the total effluent volume to the detector when the flow-rate was 1 ml/min. The fraction collector was regulated to fractions of 10 min.

The spectrophotometric determination of testosterone propionate or of other steroids that absorb in a similar manner in the UV region, was carried out by collecting all the fractions corresponding to the peak in a tared 50-ml vessel and adjusting the volume to the mark with light petroleum. Aliquots were removed, evaporated to dryness under nitrogen, the residue was dissolved in methanol and the UV spectrum in the range 225–250 nm measured on a Beckman DU-2 spectrophotometer (1-cm cells).

Figs. 28.2 and 28.3 illustrate the separations of some steroids from each other and from carrier oil or impurities.

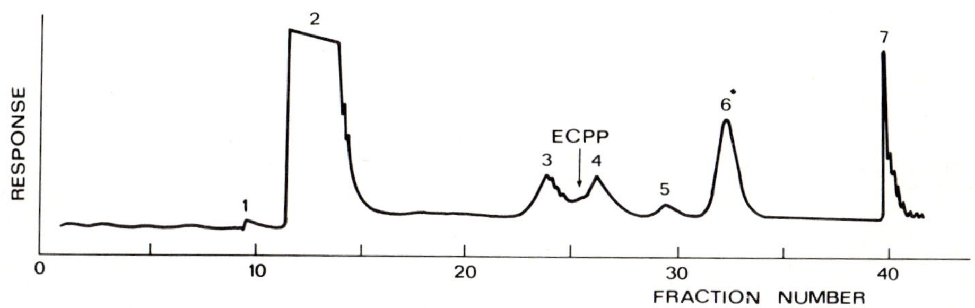


Fig. 28.2. Chromatogram of an oil solution containing 50 mg/ml of testosterone cyclopentylpropionate (TCPP) and 2.5 mg/ml of estradiol cyclopentylpropionate (ECPP) (Cavina *et al.*, 1971). Sample diluted 1 to 25 (v/v) with *n*-heptane; volume analyzed, 1 ml (2 mg of TCPP and 0.1 mg of ECPP). Column: 0.6 × 56 cm I.D. Sorbent: 10 g of silicic acid. Eluent: diethyl ether in light petroleum (b.p. 65–75°C) (gradient elution). Operating conditions: see text. Detection: hydrogen flame ionization detector with moving chain (Barber Colman Model 5400 liquid chromatographic detector). Peaks: 1 = sterol esters; 2 = triglycerides; 3 = sterols; 4 = 1,3-diglycerides + ECPP; 5 = 1,2-diglycerides; 6 = TCPP; 7 = monoglycerides.

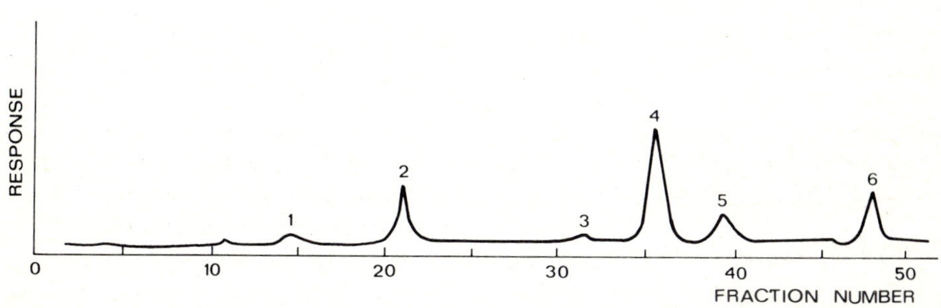


Fig. 28.3. Chromatogram of a mixture of estradiol dipropionate, progesterone, testosterone cyclopentylpropionate and testosterone propionate (Cavina *et al.*, 1971). Sample: 1 ml of an *n*-heptane solution containing 1 mg, 1 mg, 2 mg and 1 mg, respectively, of the listed steroids. Operating conditions: As in Fig. 28.2 except for the detector sensitivity, which in this case was $3 \times 10^{-10}$. Peaks: 1 = triglycerides (small amount as impurities); 2 = estradiol dipropionate; 3 = estradiol monopropionate (small amount as side product of the dipropionate); 4 = testosterone cyclopentylpropionate; 5 = testosterone propionate; 6 = progesterone.

*References p. 620*

TABLE 28.2
ELUTION VOLUMES OF SOME STEROIDS AND RELATED COMPOUNDS (CAVINA *et al.*, 1971)

| Compound | Elution volume (ml) |
|---|---|
| 19-Nortestosterone 17β-decanoate | 310 |
| Testosterone 17β-cyclopentylpropionate | 355 |
| 19-Nortestosterone 17β-propionate | 360 |
| 19-Nortestosterone 17β-phenylpropionate | 360 |
| Testosterone 17β-propionate | 395 |
| Testosterone | >440 |
| Methyltestosterone | >440 |
| 19-Nortestosterone | >440 |
| Sterol esters from oil | 110 |
| Tocopherol acetate | 110 |
| Tocopherol | 125 |
| Triglycerides from oil | 130 |
| Sterols | 250 |
| 1,3-Diglycerides | 280 |
| 1,2-Diglycerides | 315 |
| Monoglycerides | >440 |
| Estradiol dipropionate | 215 |
| Estradiol 17β-cyclopentylpropionate | 270 |
| Mestranol | 275 |
| Estradiol monopropionate | 320 |
| Estradiol 3-benzoate | 360 |
| Ethynylestradiol | 385 |
| Allylestrenol | 145 |
| Lynestrenol | 165 |
| Ethylestrenol | 165 |
| Ethynodiol diacetate | 260 |
| Vinyl estrenolone | 360 |
| Norethinodrel | 395 |
| Chlormadinone acetate | >440 |
| Norethindrone | >440 |
| Progesterone | >440 |

Hulsman (*cf.*, Huber *et al.*) investigated the determination of estriol in pregnancy urine using liquid chromatography for controlling single steps of the conventional procedure. He concluded that hydrolysis of conjugates and extraction does not cause serious losses of estriol. After preparing the sample as described below, quantitative information can be obtained within 15 min by chromatography. However, the simultaneous quantitative determination of estrone and estradiol has not been possible owing to inadequate pre-purification of the complex sample. For pre-purification, Hulsman recommended heating a 50-ml sample of urine to boiling, addition of 7.5 ml of concentrated hydrochloric acid and refluxing for half an hour. After cooling, the mixture is extracted with three 50-ml portions of diethyl ether. The total ethereal phase is treated with 20 ml of concentrated sodium carbonate solution (pH 10.5) and the aqueous layer is discarded. A 4-ml volume of an aqueous solution of 8% (w/w) sodium hydroxide is thoroughly shaken with the ethereal fraction. The alkaline layer is not discarded, but its pH is reduced to 10 and its ionic concentration increased by adding 20 ml of 8% sodium hydrogen carbonate solution, and the aqueous layer is then discarded. The ethereal layer is washed with 4 ml of 8% sodium hydrogen carbonate solution and with 3 ml of water. The aqueous layer is discarded and the ethereal fraction is evaporated to dryness and submitted to chromatography under the conditions given in Fig. 28.4. Before use, the eluent must be deaerated and saturated with the stationary liquid. To ensure equilibrium between the eluent and the stationary liquid, a pre-column is necessary. The detector unit, which was directly connected to the outlet of the column, was a Zeiss PMQ II spectrophotometer with a 7.5-μl flow cell. The signal from the detector was fed to a 2-mV linear recorder (Servogor, Type RE-511).

Celite partition columns and a number of various solvent mixtures were used by Abdel-Aziz and Williams, Smith and Kellie, and Williams and Layne in metabolic studies.

In their study on the aromatic ring hydroxylation of estradiol in man, Fishman *et al.* carried out chromatographic separations of extracts by gradient elution partition chromatography on an acid-washed Hyflo Supercel column with 90% aqueous methanol as the stationary phase and 2,2,4-trimethylpentane with a gradient of 1,2-dichloroethane as the mobile phase. The system was proposed originally by Engel *et al.*

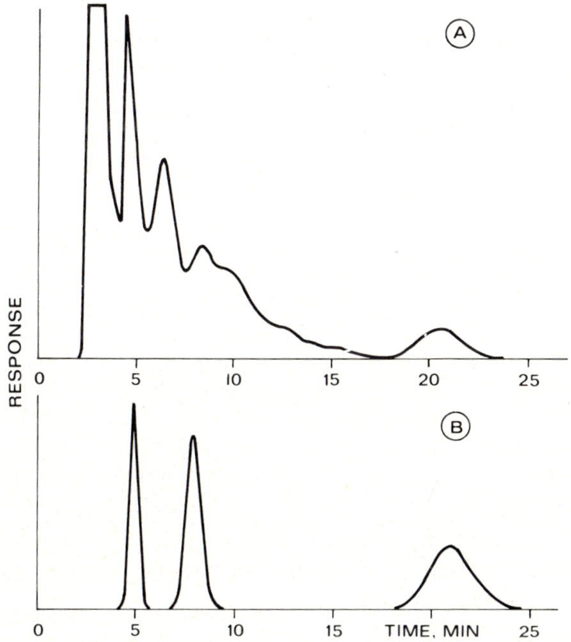


Fig. 28.4. Chromatogram of (A) pre-treated pregnancy urine and (B) a test mixture of estrogens (Hulsman). Column: 0.27 × 50 cm. Support: diatomaceous earth, particle size 28–32 μm. Stationary phase: water-rich phase of a mixture of water–ethanol–2,2,4-trimethylpentane (composition in molar fractions: 0.229:0.680:0.091). Eluent: water-poor phase of the same solvent mixture (composition in molar fractions: 0.019:0.177:0.805). Operating temperature: 22°C. Detection: Zeiss PMQ II spectrophotometer with a 7.5-μl flow cell. The signal of the detector is fed to a 2-mV linear recorder (Servogor, Type RE-511). Peaks: Estrone, estradiol and estriol (from left to right).

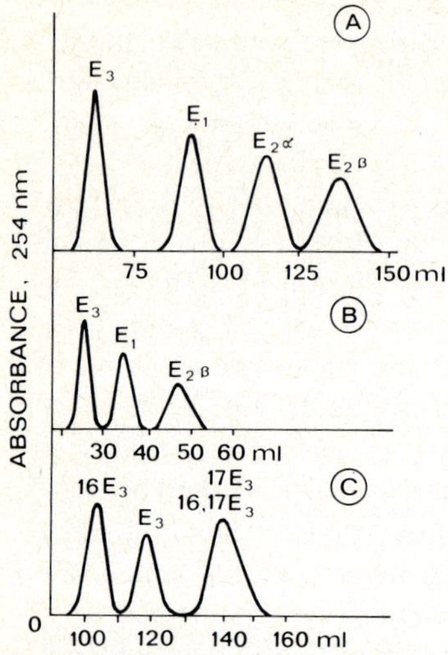


Fig. 28.5. Separation of estradiol and estriol epimers on Sephadex LH-20 (Horst *et al.*). Column: (A) and (C) 1 × 60 cm; (B) 0.9 × 25 cm. Packing: Sephadex LH-20. Eluent: 0.02 *N* NaOH. Operating conditions: Temperature, (A) and (B) 25°C, (C) 55°C; sample application in 0.15 *M* acetate buffer (pH 4.2). Detection: optical density at 254 nm, Uvicord II. For the signal amplification, a square cell with a 10-mm light path was used (Hellma GmbH, Müllheim-Baden, G.F.R.) Peaks: $E_1$ = estrone; $E_2\alpha$ = estradiol-17α; $E_2\beta$ = estradiol-17β; $E_3$ = estriol; $16E_3$ = 16-epiestriol; $16,17E_3$ = 16,17-epiestriol; $17E_3$ = 17-epiestriol.

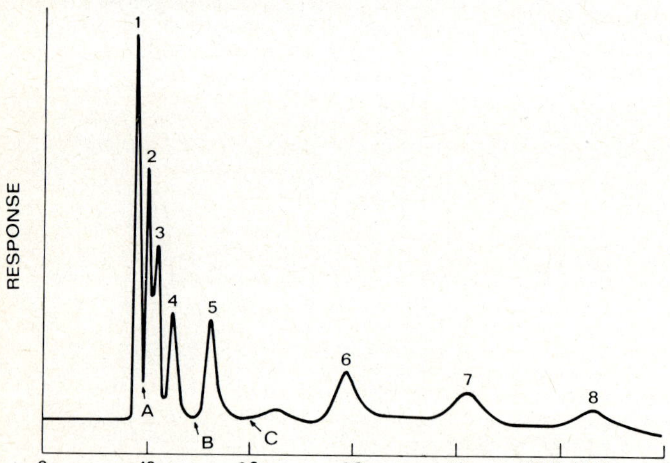


Fig. 28.6. Separation of estrogens (Siggia and Dishman). Column: 0.2 × 48.5 cm I.D. Packing: 28% Amberlite LA-1 on Plaskon CTFE-2300 (*cf.*, p. 601). Eluent: water adjusted to pH 11.5 with sodium hydroxide. Flow-rates: initial, 0.12 ml/min; (A) increased to 0.145 ml/min; (B) increased to 0.19 ml/min; (C) increased to 0.49 ml/min. Detection: see Fig. 28.1. Peaks: 1 = 1,3,5(10)-estratriene-3,17β-diol-17α-glucosiduronic acid (estradiol-17α-glucosiduronic acid); 2 = 1,3,5(10)-estratriene-3,16α,17β-triol (estriol); 3 = 1,3,5(10)-estratrien-3-ol-16,17-dione (16-ketoestrone); 4 = 1,3,5(10)-estratriene-3,17β-diol-16-one (16-ketoestradiol); 5 = 1,3,5(10)-estratriene-3,16β,17β-triol (16-epiestriol); 6 = 1,3,5(10), 6,8-estrapentaen-3-ol-17-one (equilenin); 7 = 1,3,5(10)-estratriene-3,17β-diol (estradiol); 8 = 1,3,5(10)-estratrien-3-ol-17-one (estrone).

A liquid–liquid system for checking the purity of estradiol was proposed by Henry *et al.* (see p. 601).

Estrogens from female urine were chromatographed by Hsu *et al.* on a nylon powder (60–80 mesh) column (1 × 25 cm) using benzene and benzene–ethanol mixtures as the mobile phase. Detection and colorimetry were carried out after reaction with hydroquinone and sulphuric acid.

In their radioimmunoassay of plasma estrone and estradiol, Mikhail *et al.* used repeatedly, for a period of 3–6 months, Sephadex LH-20 (1 × 30 cm, 3 g) and benzene–methanol (85:5).

Jellinck and Fletcher and Jellinck *et al.* applied the gel filtration of water-soluble estrogen conjugates on a Sephadex G-25 column (1.8 × 38 cm) with distilled water as eluent (3-ml fractions).

Horst described an assay for the separation of urinary estrogens on a 1.2 × 60 cm Sephadex G-10 column which could be loaded with samples up to 200 ml. Estrogens and a few other urinary ingredients were reversibly adsorbed by the gel matrix if the test sample was saturated with sodium sulphate and the pH adjusted to 4.6 ± 0.2. Almost all non-estrogenic material was eluted before the three estrogens, which left the column in highly purified fractions. A continuous non-linear gradient elution was used. The gradient was produced by mixing a solution of 16% sodium sulphate with 0.1 *N* sodium hydroxide solution. An automatic estrogen determination could be achieved by fluorimetry in a flow cell.

Useful separations of estrogens were also described by Henry *et al.* (Zipax coated with a strong anion-exchange polymer, pH 9.2 buffer to pH 9.2 buffer + 0.8 *M* sodium perchlorate gradient for elution), Hobkirk *et al.*, Hobkirk and Nilsen (1969a, 1970; prepurification according to Bradlow) (DEAE-Sephadex A-25, 0–0.8 *M* and 0–0.4 *M* sodium chloride gradient for elution), Horst *et al.* (Fig. 28.5), Osawa and Slaunwhite (Amberlite XAD-2, 30% ethanol as eluent) and Van Baelen *et al.* (Sephadex G-25 and LH-20, large amounts of water as eluent).

Siggia and Dishman used a column packed with Plaskon CTFE-2300 (Allied Chemical Co., Morristown, N.J., U.S.A.) on which Amberlite LA-1 (see p. 601) was fixed. The particles of the support were 325 mesh. A dilute alkaline solution was used as the mobile phase. Detection was performed by density recording. The separation achieved is illustrated in Fig. 28.6.

**Gestagens (progestins)**

In the two preceding sections, a large number of applications of various types of chromatography have been mentioned, and in this section we shall limit ourselves predominantly to examples of analytical high-resolution liquid chromatography. As regards preparative chromatography (especially the adsorption and liquid–liquid chromatography of progestins), this does not differ essentially from the chromatography of androgens and estrogens (see Fig. 28.3). In fact, in terms of chromatographic character (*i.e.*, polarity and adsorptivity) progesterone is a homologue and isomer of androstenedione, pregnenolone is very similar to dehydroepiandosterone and a little less to estrone, pregnanediol can be compared to estradiol and androstenediol, and pregnanetriol does not behave very differently from estriol and some androstenetriols. In general, the most frequently used adsorbents have been alumina and silica gel, followed by Florisil and Chromosorb (Gardner *et al.*). For liquid–liquid chromatography, Celite has often been used, with aqueous methanol as the stationary phase and mixtures of benzene with other hydrocarbons as the mobile one. Silver-impregnated Florisil was used for the separation of 17α-ethynyl steroids (Kulkarni and Goldzieher). Sephadex LH-20 was also successfully applied with the solvent system chloroform–methanol (1:1) containing sodium chloride (Laatikainen and Vihko).

An example of high-resolution liquid chromatography of gestagens was proposed by Siggia and Dishman. The operating conditions are much the same as those described for androgens and estrogens. Table 28.3 gives retention data and operating parameters for some progestins. Some retention data for gestagens are also given in Table 28.2.

Another example of analytical high-speed partition chromatographic analysis may be found in the paper by Henry *et al.* (see p. 601).

TABLE 28.3
RETENTION DATA FOR SOME PROGESTINS (SIGGIA AND DISHMAN)

Column, 485 mm × 2 mm I.D.; eluent, 33% (v/v) methanol in water; flow-rate 0.56 ml/min; packing, 28% Amberlite LA-1 on Plaskon CTFE-2300; (p. 601); detection, optical density recording.

| Steroid | $t_R$* (min) | $k$** |
|---|---|---|
| Testosterone | 4.43 | 1.74 |
| 17α-Hydroxyprogesterone | 6.15 | 2.80 |
| Progesterone | 30.6 | 17.5 |
| 4-Pregnene-20β-ol-3-one | 30.6 | 17.9 |

*$t_R$ = retention time.
**$k$ = capacity ratio, defined as the ratio of the weight of a compound in the stationary to that in the mobile phase.

## Corticosteroids

Corticosteroids, even when not conjugated with hydrophilic components such as glucuronic acid, are generally more hydrophilic than other hormonal steroids, owing to their greater oxygen content per molecule. In other respects, they are similar to more hydrophilic progestins and androgens and therefore we shall treat them very briefly here.

For preparative adsorption chromatography, silica gel is probably most often used, followed by alumina and Florisil. Celite with aqueous methanol as the stationary phase and ligroin–benzene (Kelly *et al.*) or a gradient of 1,2-dichloroethane in 2,2,4-trimethylpentane (Dufau and Villee) is used for biosynthetic studies. Reversed-phase chromatography on Sephadex LH-20 is often used for metabolic and general biochemical studies, often with labelled steroids (Seki, 1967; Seki and Sugase; Shapiro and Péron).

Corticosteroids are usually detected on the basis of their UV absorption or by the colour reaction with blue tetrazolium.

The detection of some corticosteroids is sometimes a problem because of their low concentration in various body tissues or liquids; high-resolution chromatography therefore seems to be the method of choice in this field.

Siggia and Dishman separated corticosteroids as shown in Fig. 28.7.

Using a DuPont 820 liquid chromatograph, a UV photometer as detector, Zipax as the stationary phase support, $\beta,\beta'$-oxydipropionitrile as the stationary phase and *n*-heptane–tetrahydrofuran (8:2) as the mobile phase, Henry *et al.* achieved good analytical separations of corticosteroids and similar steroids. They also described other phase systems, normal and reversed, *viz.*, ethylene glycol on Zipax and *n*-heptane–chloroform, and a hydrocarbon polymer (HCP) on Zipax and methanol–water (see p. 601).

Touchstone and Wortmann used a Perkin-Elmer Model 1240 analytical liquid chromatograph, equipped with a UV detector with a 0.3 × 50 cm column packed with a treated silica gel, either SIL-X from Perkin-Elmer (for liquid adsorption chromatography) or SIL-X (RP) (for liquid–liquid chromatography). In Fig. 28.8, the separation of three corticosteroids and androstenedione by high-pressure liquid adsorption chromatography on SIL-X is shown, while Table 28.4 shows the retention times of some corticosteroids and progesterone measured during reversed-phase high-pressure liquid chromatography on SIL-X (RP).

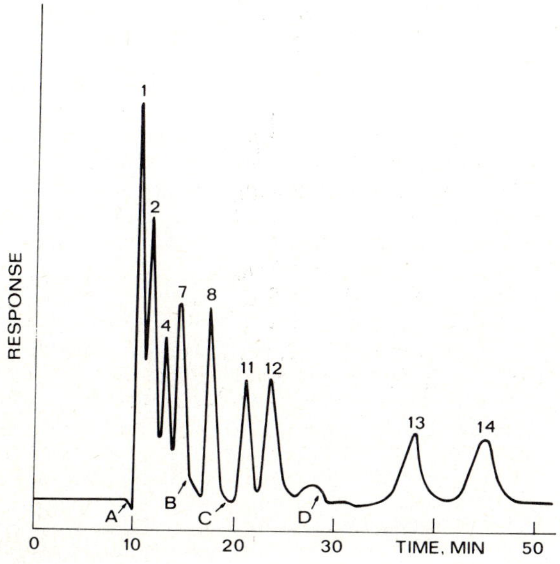


Fig. 28.7. Separation of some adrenal corticosteroids (Siggia and Dishman). Column: 0.2 × 485 cm I.D. Packing: 23% Amberlite LA-1 on Plaskon CTFE-2300 (see p. 601). Eluent: water. Flow-rate: initial, 0.10 ml/min; (A) increased to 0.13 ml/min; (B) increased to 0.21 ml/min; (C) increased to 0.26 ml/min; (D) increased to 0.44 ml/min. Detection: see Fig. 28.1. Peaks: 1 = 4-pregnene-6$\beta$,17$\alpha$-21-triol-3,11,20-trione (6$\beta$-hydroxycortisone); 2 = 4-pregnene-11$\beta$,21-diol-3,20-dione-18-al (aldosterone); 4 = 4-pregnene-17$\alpha$,21-diol-3,11-dione (cortisone); 7 = 4-pregnene-21-o1-3,11,20-trione (11-dehydrocorticosterone); 8 = 4-pregnene-11$\beta$,21-diol-3,20-dione (corticosterone); 11 = 4-pregnene-17$\alpha$,21-diol-3,20-dione (11-deoxycortisol); 12 = 4-pregnene-17$\alpha$,21-diol-3,11,20-trione-21-acetate (cortisone-21-acetate); 13 = 4-pregnene-20$\beta$,21-diol-3-one; 14 = 4-pregnene-21-ol-3,20-dione (deoxycorticosterone).

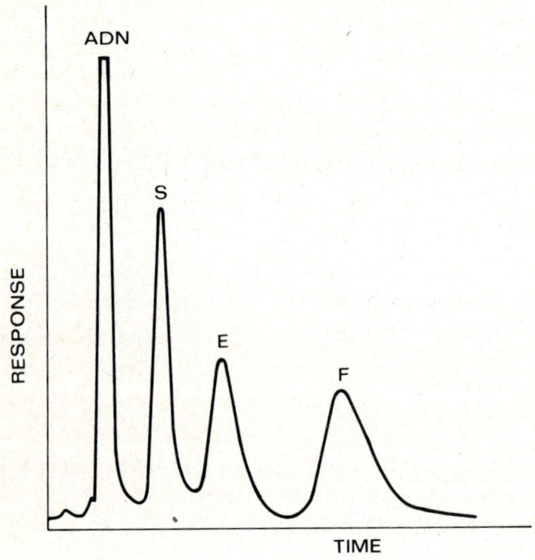


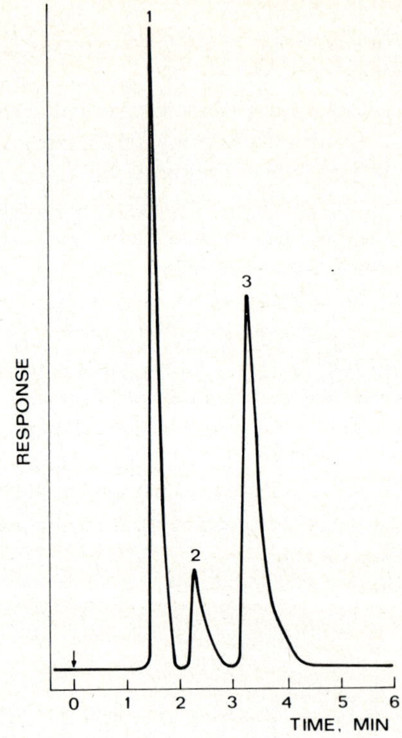


Fig. 28.8. Separation of cortisol (F), cortisone (E), substance S (S) and androstenedione (ADN) by high-pressure liquid adsorption chromatography (Touchstone and Wortmann). Column: 0.3 × 50 cm I.D. Packing: SIL-X. Eluent: chloroform–dioxan (100:4). Operating conditions: flow-rate, 1 ml per 55 sec; pressure, 350–450 p.s.i.; amount injected, 4 $\mu$g each. Detection: UV attenuation, 0.05 O.D.; speed of recorder, 30 cm/h.

Fig. 28.9. Chromatogram of a standard mixture of corticosterone and cortisol (Meijers). Column: 0.27 × 10 cm. Packing: diatomaceous earth (dry), particle size 5–10 $\mu$m. Stationary phase: water-rich phase of water–ethanol–2,2,4-trimethylpentane (18.5:75.0:6.5). Eluent: Water-poor phase of the same solvent mixture (0.1:5.5:94.4). Operating conditions: phase ratio, $28 \cdot 10^{-4}$; flow-rate, 20 ml/h. Detection: Zeiss PMQ II UV spectrophotometer. Peaks: 1 = solvent front; 2 = corticosterone; 3 = cortisol.

Meijers described a system for the separation of corticosterone and cortisol using the phase system water–ethanol–2,2,4-trimethylpentane. The support was diatomaceous earth, which was packed into the column (2.7 mm I.D.) in a dry state. Shaking the column and packing under gentle pressure from a PTFE plunger was necessary. The air was expelled from the column by the mobile phase and the stationary phase was added to the column by injection. Conventional loading of the stationary phase was also satisfactory. The detector was a Zeiss PMQ II UV spectrophotometer. An example of the separation of the two steroids is given in Fig. 28.9.

TABLE 28.4
SEPARATION OF ADRENAL CORTICOSTEROIDS AND PROGESTERONE BY REVERSED-PHASE LIQUID–LIQUID CHROMATOGRAPHY ON SIL-X (RP) (TOUCHSTONE AND WORTMANN)

Chart-speed: 0.26 cm/min.

| Steroid | Retention (cm) | Band width (cm) |
|---|---|---|
| Cortisone | 1.8 | 0.9 |
| Cortisol | 2.0 | 1.0 |
| Aldosterone | 1.7 | 1.0 |
| Corticosterone | 3.1 | 1.6 |
| 11-Deoxycortisol, Reichstein's subst. S | 3.5 | 0.9 |
| Progesterone | 15.0 | 5.8 |

For the determination of methylprednisolone residue in milk, Kreminski *et al.* used high-pressure reversed-phase liquid chromatography with a DuPont 820 liquid chromatograph (3-ft. column) equipped with a precision photometer. A hydrocarbon polymer (HCP, DuPont No. 820950008) fixed on Zipax served as the stationary phase and water–methanol (3:1) as eluent. The sensitivity of their method was in the parts per billion range and the accuracy was 94 ± 4%.

Beyer and Morozowich developed a semi-automated procedure for the quantitation and characterization of steroid phosphates, such as methylprednisolone 21-phosphate, using ion-exchange column chromatography on DEAE-cellulose.

Vestergaard and Sayegh described an excellent separation of 17 corticosteroids on silica gel using their medium-speed automated multicolumn chromatography mentioned on p. 595

## Bile acids and other steroid acids

The ability of bile acids to dissociate, *i.e.*, ionize, leads to new possibilities for their chromatography, but also gives rise to new problems. Except for acid steroid conjugates, the bile acids (and their conjugates) are the only steroids that have also been chromatographed on ion exchangers. On the other hand, their dissociation makes them unsuitable for chromatography unless this dissociation is either suppressed or made complete, as otherwise they will streak. This effect is usually avoided by using acidic systems for their chromatography (*cf.*, Matschiner *et al.*, who used 70% aqueous acetic acid as the stationary phase fixed on Celite and *n*-hexane–benzene as the eluent), in which the dissociation of bile acids is supressed. Another solution to this problem consists in the esterification of the bile acids, usually with diazomethane, which renders them neutral and able to be chromatographed in any common system suitable for other neutral steroids.

In 1969, an excellent article by Eneroth and Sjövall summarized all methods of analysis in the biochemistry of bile acids. Two tables in that paper described separations on silicic acid and aluminium oxide and also the structure–mobility relationships in the bile acids group. It seems that the use of hydrophobized diatomaceous earth (siliconized) or polyethylene powder (Hostalen) as the support for the stationary phase of a reversed-phase system are the most common (see p. 599) for the separation of free acids. Methylated Sephadex has also been used. *n*-Heptane, mixtures of *n*-heptane and chloroform and of chloroform and iso-octanol, served as the stationary phase and water–methanol mixtures as the mobile phase. In analytical applications, labelled bile acids were used so that the chromatograms could be recorded on the basis of their radioactivity.

Other papers include the separation of two new bile acids from rat bile, using liquid–liquid partition on Celite, carried out by Matschiner *et al.* (see above), synthetic work by Mitra and Elliot using low-activity alumina, a metabolic study by Norii *et al.* making use of hydrophobic Celite, a structural study of steroid acids from *Cephalosporium acremonium* by Chou *et al.* and a study on enzymatic saponification of steroid acids by Nguyen Gia Chan and Procházka. In the last two cases, silica gel was used.

**Steroidal glycosides**

Glycosides of various steroids (cardenolides, buffadienolides and spirostanes) are most commonly purified and separated by chromatography on silica gel with chloroform–methanol–water mixtures of various compositions, with chloroform prevailing. However, following the work of Haack *et al.*, many workers used cellulose powder impregnated with formamide as the stationary phase and various mixtures of *n*-heptane with xylene and ethyl methyl ketone as the mobile phase (*cf.*, Nover; Nover *et al.*, 1969a, b, who investigated the relationship between the chemical structure and the chromatographic behaviour of cardiac glycosides). For filling the columns, Haack *et al.* applied the "dry pack" procedure described on p. 599. Before filling the column, cellulose was first suspended in formamide–diethyl ether (1:1), filtered off under suction, spread on a sheet of paper and allowed to stand until the diethyl ether had completely evaporated. However, Angeliker *et al.* found that it is better for some purposes if the cellulose impregnated with formamide and suspended in the mobile phase is not packed into the column, but allowed to settle by gravity. They also used silica gel impregnated with formamide (6 parts of formamide per 40 parts of silica gel) and formamide-saturated chloroform plus 1% of *n*-butanol for the separation of lanatosides. Recently, Singh and Rastogi used formamide fixed on Celite 535 and benzene saturated with formamide as the eluent for the separation of glycosides from *Asclepias curassavica* L. Mackie and Turner found DEAE-cellulose (Whatman DE-32) to be suitable as the stationary phase and phosphate buffer as the eluent for the separation of biologically active glycoside from starfish (*Marthasterias glacialis*). These examples should suffice for glycosides. The separation of aglycones is no problem because they behave like other steroids on alumina or silica gel, Florisil, etc. The introduction of high-resolution micro-column chromatography into this field will lead to great progress.

So far, only Henry *et al.* have described an analytical high-speed micro-separation of digitoxin from digoxin in a reversed-phase system, in which 1% cyanoethylsilicone on Zipax was used as the stationary phase and aqueous 2.5% methanol as the mobile phase (column pressure 600 p.s.i.g., flow-rate 0.5 ml/min, temperature 40°C, UV photometric detection).

**Steroidal insect hormones**

Steroidal insect hormones represent a relatively new group of physiologically active steroids, isolated for the first time from insects, but later discovered also in crustaceans and various plants. They could be defined as polyhydroxylated steroids, mainly characterized as 5β-cholestanols with a conjugated 6-keto group and a 2β,3β-diol grouping. Little remains to be said about their classical chromatography, *i.e.*, on the common adsorbents which were extensively used for their separation and purification. In addition to adsorbents such as alumina, silicic acid and Florisil, active carbon and aqueous methanol and pure methanol were also used for their pre-purification by Imai *et al.* They also used automatic liquid chromatography, utilizing an Amberlite XAD-A column and elution with a linear gradient of 20–70% of aqueous ethanol. The eluate was monitored by the absorption of these substances at 254 nm. Gel chromatography (Sephadex G-25, polyacrylamide gel Bio-Gel P-10) was used for the isolation of crustecdysone from crayfish by Horn *et al.*, and hydrophobic Celite impregnated with *n*-butanol–cyclohexane (9:1 or 7:1; 15 ml per 20 g) and water saturated with the stationary phase as the mobile phase for synthetic work on crustecdysones by Galbraith and Horn and Galbraith *et al.* High-resolution liquid column chromatography was introduced into this field by Waters Associates (Framingham, Mass., U.S.A.), who proposed a novel reversed-phase affinity

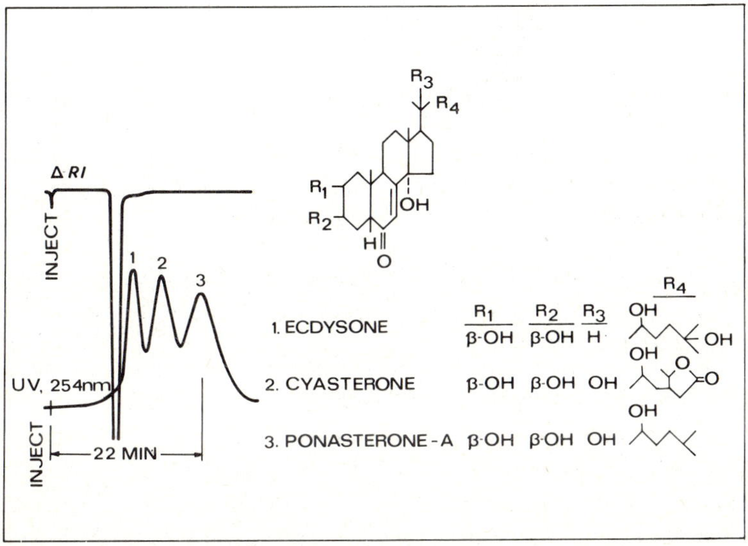


Fig. 28.10. Separation of three insect moulting steroids (Waters Ass.). Column: 0.785 × 91.5 cm. Packing: Poragel PN. Eluent: methanol–water (7:3). Operating conditions: instrument, ALC-100; flow-rate, 3.1 ml/min; sample load, 50 μg total. Detection: UV absorption at 254 nm.

*References p.620*

packing, Poragel PN, which is claimed to be extremely rigid and thus suitable for a scale-up even to large diameter columns. The details of this method are represented in Fig. 28.10 (see also Henry *et al.*).

Less common types of steroids, such as aza-steroids (*cf.*, Mechoulam), various thiosteroids (*cf.*, Djerassi *et al.*), etc., present no special difficulties in their preparative chromatography in various chromatographic systems. It seems that so far no need has arisen to devise analytical high-resolution liquid column chromatographic methods for them. It is probable that with the increasing use of this method in the future, it will be used in all fields of steroid research.

## REFERENCES


Abdel-Aziz, M. T. and Williams, K. I. H., *Steroids*, 13 (1969) 809.
Ambrus, G. and Wix, Gy., *Acta Chim. Acad. Sci. Hung.*, 55 (1968) 99.
Angeliker, E., Barfuss, F. and Renz, J., *Helv. Chim. Acta*, 41 (1958) 479.
Bates, R. W. and Cohen, H., *Fed. Proc., Fed. Amer. Soc. Exp. Biol.*, 6 (1947) 236.
Bergström, S. and Sjövall, J., *Acta Chem. Scand.*, 5 (1951) 1267.
Beyer, W. and Morozowich, W., *Ann. N.Y. Acad. Sci.*, 153 (1968) 393.
Bradlow, H. L., *Steroids*, 11 (1968) 265.
Burstein, S., Kimball, H. L., Chaudhuri, N. K. and Gut, M., *J. Biol. Chem.*, 243 (1968) 4417.
Burstein, S. and Zamoscianyk, H., *Steroids*, 15 (1970) 13.
Butte, J. C. and Noble, E. P., *Acta Endocrinol.*, 61 (1969) 678.
Canonica, L., Fiecchi, A., Kienle, M. G., Scala, A., Galli, G., Paoletti, E. G. and Paoletti, R., *J. Amer. Chem. Soc.*, 90 (1968) 6532.
Cavina, G., Moretti, G., Mollica, A. and Antonini, R., *Int. Symp. VI, Chromatographie, Électrophorèse, Bruxelles, 14–16 September 1970*, Presses Académiques Européennes, Brussels (1971).
Cavina, G., Moretti, G., Mollica, A. and Siniscalchi, P., *J. Chromatogr.*, 44 (1969) 493.
Chmel, K., Pihera, P. and Schwarz, V., *Chem. Listy*, 67 (1973) 649.
Chou, T. S., Eisenbraun, E. J. and Rapala, R. T., *Tetrahedron*, 25 (1969) 3341.
Crépy, O., Jayle, M. F. and Meslin, F., *Acta Endocrinol.*, 24 (1957) 233.
Davey, C. W., McGinnis, E. L., McKeown, J. M., Meakins, G. D., Pemberton, M. W. and Young, R. N., *J. Chem. Soc., C*, (1968) 2674.
Dingemanse, E., Huis in't Veld, L. G. and Hartogh-Katz, S. L., *J. Clin. Endocrinol. Metab.*, 12 (1952) 66.
Dixon, R., *Steroids*, 14 (1969) 718.
Djerassi, C., Lightner, D. A., Schooley, D. A., Takeda, K., Komeno, T. and Huriyama, K., *Tetrahedron*, 24 (1968) 6913.
Dray, F., Mowszowicz, I. and Ledru, M.-J., *Steroids*, 10 (1967) 501.
Drosdowsky, M. A., Nguyen, T. T., Populu, J. and Jayle, M. F., *Bull. Soc. Chim. Biol.*, 50 (1968) 1723.
Dufau, M. L. and Villee, D. B., *Biochim. Biophys. Acta*, 176 (1969) 637.
Eberlein, W. R., *Steroids*, 14 (1969) 553.
Eckstein, B., Mechoulam, R. and Burstein, S. H., *Nature (London)*, 228 (1970) 866.
Eechaute, W., Demeester, G. and Leusen, I., *Steroids*, 13 (1969) 101.
Ellingboe, J., Nyström, E. and Sjövall, J., *Biochim. Biophys. Acta*, 152 (1968) 803.
Ellingboe, J., Nyström, E. and Sjövall, J., *Methods Enzymol.*, 14 (1969) 317.
Eneroth, P. and Nyström, E., *Steroids*, 11 (1968) 187.
Eneroth, P. and Sjövall, J., *Methods Enzymol.*, 15 (1969) 237.
Engel, L. L., Cameron, C. B., Stoffyn, A., Alexander, J. A., Klein, O. and Trofimov, N. D., *Anal. Biochem.*, 2 (1961) 114.
Epstein, E., *Clin. Chim. Acta*, 7 (1962) 735.
Ercoli, A., Vitali, R. and Gardi, R., *Steroids*, 3 (1964) 479.

Fernandez, A. A. and Noceda, V. T., *J. Pharm. Sci.*, 58 (1969) 740.
Fishman, J., Guzik, H. and Hellman, L., *Biochemistry*, 9 (1970) 1593.
Fitzpatrick, F. A., Siggia, S. and Dingman, J., Sr., *Anal. Chem.*, 44 (1972) 2211.
Forriol, E. F., *5th Int. Symp. on Chromatography and Electrophoresis, 1968*, Ann Arbor–Humphrey Sci. Publ., Ann Arbor, Mich., 1969, p. 379; *C.A.*, 72 (1970) 39338c.
Galbraith, M. N. and Horn, D. H. S., *Aust. J. Chem.*, 22 (1969) 1045.
Galbraith, M. N., Horn, D. H. S., Middleton, E. J. and Hackney, R. J., *Aust. J. Chem.*, 22 (1969) 1059.
Galli, G. and Grossi-Paoletti, E., *Lipids*, 2 (1967) 72.
Gardner, J. N., Carlon, F. E. and Gnoj, O., *J. Org. Chem.*, 33 (1968) 1566.
Gelotte, B. J., *J. Chromatogr.*, 3 (1965) 330.
Grimwalde, M. J. and Lester, M. G., *Tetrahedron*, 25 (1969) 4535.
Haack, E., Kaiser, F. and Spingler, H., *Chem. Ber.*, 89 (1956) 1353.
Hakomori, S., *J. Biochem. (Tokyo)*, 55 (1964) 205.
Henry, R. A., Schmit, J. A. and Dieckman, J. F., *J. Chromatogr. Sci.*, 9 (1971) 513.
Hirsch, J. and Ahrens, E. H., *J. Biol. Chem.*, 233 (1958) 311.
Hobkirk, R., Musey, P. and Nilsen, M., *Steroids*, 14 (1969) 191.
Hobkirk, R. and Nilsen, M., *Steroids*, 14 (1969a) 533.
Hobkirk, R. and Nilsen, M., *Steroids*, 15 (1969b) 649.
Hobkirk, R. and Nilsen, M., *Anal. Biochem.*, 37 (1970) 341.
Horn, D. H. S., Fabbri, S., Hampshire, F. and Lowe, M. E., *Biochem. J.*, 109 (1968) 399.
Horst, H.-J., *J. Chromatogr.*, 58 (1971) 227.
Horst, H.-J., Grunert, E. and Stoye, M., *J. Chromatogr.*, 69 (1972) 395.
Horton, R., Kato, T. and Sheridan, R., *Steroids*, 10 (1967) 245.
Hoshita, T., Hirofuji, Sh., Sasaki, T. and Kazuno, T., *J. Biochem. (Tokyo)*, 61 (1967) 136.
Hsu C.-T., Tung Yih-Chih, Lee Hung-Tu, Hanie-Feng, Lo Chi-Ngo, Wu Tsu-Ye and Lee Fang-Zu, *Proc. Asia Oceania Congr. of Endocrinol., 3rd, Manila, 1967; C.A.*, 70 (1969) 74717f.
Huang, W. Y., *Steroids*, 9 (1967) 485.
Huber, J. F. K., Hulsman, J. A. R. J. and Meijers, C. A. M., *J. Chromatogr.*, 62 (1971) 79.
Hulsman, J. A. R. J., *Thesis*, University of Amsterdam, Amsterdam, 1969.
Hyde, P. M. and Elliot, W. H., *J. Chromatogr.*, 67 (1972) 170.
Imai, S., Hori, M., Fujioka, S., Murata, E., Goto, M. and Nakanishi, K., *Tetrahedron Lett.*, (1968) 3883.
Janot, M. M., Milliet, P., Lusinchi, X. and Goutarel, R., *Bull. Soc. Chim. Fr.*, 1967, 4310.
Jellinck, P. H. and Fletcher, R., *Can. J. Biochem.*, 48 (1970) 1192.
Jellinck, P. H., Lewis, J. and Boston, F., *Steroids*, 10 (1967) 329.
Joustra, M., Söderqvist, B. and Fischer, L., *J. Chromatogr.*, 28 (1967) 21.
Kato, R. and Horton, R., *Steroids*, 12 (1968) 631.
Kelly, W. G., Ranucci, S. R. and Shaver, J. C., *Steroids*, 11 (1968) 429.
Kreminski, L. F., Cox, B. L., Perrel, P. N. and Schiltz, R. A., *J. Agr. Food Chem.*, 20 (1972) 970.
Kulkarni, B. D. and Goldzieher, J. W., *Steroids*, 13 (1969) 467.
Laatikainen, T. and Vihko, R., *Eur. J. Biochem.*, 10 (1969) 165.
Lee, W.-H., Lutsky, B. N. and Schroepfer, G. J., *J. Biol. Chem.*, 244 (1969) 5440.
Mackie, A. M. and Turner, A. B., *Biochem. J.*, 117 (1970) 543.
McCurdy, J. T. and Garrett, R. D., *J. Org. Chem.*, 33 (1968) 660.
Matschiner, J. T., Mahowald, T. A., Elliot, W. H., Doisy, E. A., Jr., Hsia, S. L. and Doisy, E. A., *J. Biol. Chem.*, 225 (1957) 771.
Mechoulam, R., *Israel J. Chem.*, 6 (1968) 909.
Meijers, C. A. M., *Thesis*, University of Amsterdam, Amsterdam, 1971.
Mickan, H., Dixon, R. and Hochberg, B., *Steroids*, 13 (1969) 477.
Mikhail, G., Wu, C. H. and Ferin, M., *Steroids*, 15 (1970) 333.
Mitra, M. N. and Elliot, W. H., *J. Org. Chem.*, 33 (1968) 2814.
Nagasawa, M., Bae, M., Tamura, G. and Arima, K., *Agr. Biol. Chem.*, 33 (1969) 1644.
Nguyen Gia Chan and Procházka, Z., *Collect. Czech. Chem. Commun.*, 38 (1973) 2288.

Norii, T., Yamaga, N. and Yamasaki, K., *Steroids*, 15 (1970) 303.
Nover, L., *Arch. Pharm. (Weinheim)*, 302 (1969) 321.
Nover, L., Baumgarten, G. and Luckner, M., *J. Chromatogr.*, 39 (1969a) 450.
Nover, L., Jüttner, G., Noack, S., Baumgarten, G. and Luckner, M., *J. Chromatogr.*, 39 (1969b) 419.
Nyström, E., *Ark. Kemi*, 29 (1969) 99.
Nyström, E. and Sjövall, J., *Anal. Biochem.*, 12 (1965) 235.
Osawa, J. and Slaunwhite, W. R., Jr., *Steroids*, 15 (1970) 73.
Paliokas, A. M., Lee, W. and Schroepfer, G. J., *J. Lipid Res.*, 9 (1968) 143.
Paliokas, A. M. and Schroepfer, G. J., *J. Biol. Chem.*, 243 (1968) 453.
Peng, T.-C. and Munson, P. L., *Steroids*, 11 (1968) 105.
Peterson, E. A. and Sober, H. A., *Biochem. Prep.*, 8 (1961) 39.
Ribi, E., Filz, C. J., Goode, G., Strain, S. M., Yamamoto, K., Harris, S. C. and Simmons, H., *J. Chromatogr. Sci.*, 8 (1970) 577.
Saez, J. M., Saez, S. and Migeon, C. J., *Steroids*, 9 (1967) 1.
Schneider, J. J., Crabbé, P. and Bhacca, N. S., *J. Org. Chem*, 33 (1968) 3118.
Seki, T., *J. Chromatogr.*, 29 (1967) 246.
Seki, T., *Methods Enzymol.*, 15 (1969) 219.
Seki, T. and Sugase, T., *J. Chromatogr.*, 42 (1969) 503.
Shapiro, B. H. and Péron, F. G., *J. Chromatogr.*, 65 (1972) 568.
Siggia, S. and Dishman, R. A., *Anal. Chem.*, 42 (1970) 1229.
Siiteri, P. K., *Steroids*, 2 (1963) 687.
Silber, R. H. and Porter, C. C., *Methods Biochem. Anal.*, 4 (1967) 139.
Singh, B. and Rastogi, R. P., *Indian J. Chem.*, 7 (1969) 1105.
Smith, E. R. and Kellie, A. E., *Biochem. J.*, 104 (1967) 83.
Swann, D. A. and Turnbull, J. H., *Tetrahedron*, 24 (1968) 1441.
Touchstone, J. C. and Wortmann, W., *J. Chromatogr.*, 76 (1973) 244.
Turina, S., private communication.
Van Baelen, H., Heyne, W. and De Moor, P., *J. Chromatogr.*, 30 (1967) 226.
Van Lier, J. E. and Smith, L. L., *J. Pharm. Sci.*, 59 (1970) 719.
Vermeulen, A. and Verdonck, L., *Steroids*, 11 (1968) 609.
Vestergaard, P., *J. Chromatogr.*, 31 (1967) 213.
Vestergaard, P., in E. Heftmann (Editor), *Modern Methods of Steroid Analysis,* Academic Press, New York, London, 1973, p.1.
Vestergaard, P., Hemmingsen, L. and Hansen, P. W., *J. Chromatogr.*, 40 (1969) 16.
Vestergaard, P. and Jacobsen, E., *J. Chromatogr.*, 50 (1970) 239.
Vestergaard, P. and Sayegh, J. F., *Advan. Autom. Anal., Technicon Int. Symp., 1969*, Mediad, New York, 1970, p. 327.
Vroman, H. E. and Cohen, C. F., *J. Lipid Res.*, 8 (1967) 150.
Waters Ass., *Application Highlights, Leaflet 3, Steroids (Insect Moulting)*, Waters Ass., Framingham, Mass., U.S.A.
Williams, K. I. H. and Layne, D. S., *Steroids*, 9 (1967) 275.
Ziller, S. A., Doisy, E. A. Jr. and Elliot, W. H., *J. Biol. Chem.*, 243 (1968) 5280.

*Chapter 29*

# Terpenes

O. MOTL

## CONTENTS



## INTRODUCTION

Terpenic compounds occur mainly as plant components and they are often obtained by steam distillation in the form of essential oils, in addition to conventional extraction procedures, or by collection of the exudates of some shrubs and trees (some structurally less common terpenes, or those which contain less common elements, for example bromine and chlorine, have also been found in animal and insect tissues and also as metabolites of moulds). These products are usually complex mixtures of isomeric terpenic substances, in addition to other types of natural substances, comprising monoterpenic ($C_{10}$), sesquiterpenic ($C_{15}$), diterpenic ($C_{20}$), sesterterpenic ($C_{25}$), triterpenic ($C_{30}$), and tetraterpenic ($C_{40}$) substances*. The complexity of the mixtures, especially of essential oils, which sometimes contain 100–200 components, usually requires a preliminary separation (counter-current distribution, distillation, group separation – see Lawrence, 1971) before the proper chromatographic separation of single isomeric terpenic substances is carried out.

For the analytical separation of lower terpenes (mainly monoterpenic hydrocarbons and oxygen-containing substances – see Guenther *et al.*), GLC is mainly used although in many instances TLC is a very useful complementary method (see, for example, Baines and Jones or Lawrence, 1968a). For preparative purposes, liquid column chromatography is the predominant separation method, and for unstable substances it is the only method other than TLC for their separation. For column chromatography, classical adsorbents are used such as silica gel, alumina and silicic acid, more recently also silica gel and alumina impregnated with silver nitrate, and less often Florisil and charcoal. In the study of terpenic metabolites and very polar terpenic substances, various types of ion exchangers,

*This chapter does not include terpenic alkaloids and tetraterpenes.

Sephadexes and modified celluloses have also been used. In spite of the fact that up to the present time terpenic substances have been separated predominantly by the methods mentioned above, high-speed liquid column chromatography will evidently soon serve for the separation of higher terpenoids.

## HYDROCARBONS

For the separation of most terpenic hydrocarbons, silica gel or neutral alumina impregnated with silver nitrate, and also activated alkaline alumina (activity I–II), can be used; commonly used eluents are *n*-pentane, light petroleum fractions or benzene, in some instances with the gradual addition of small amounts of diethyl ether (gradient elution).

Monoterpenic hydrocarbons usually represent the low-boiling fraction of essential oils or some balsams and resins. The distillation fractions, even when obtained with efficient distillation columns, often contain oxygen-containing substances as impurities. These impurities can be removed by chromatography on alkaline alumina (activity II–III according to Brockmann and Schodder; activity determination by a modified TLC method according to Heřmánek *et al.*) or deactivated silica gel (11% of water), or by displacement chromatography on silica gel with a modified surface (0.7% of Emulphore – Kugler and Kováts). Sometimes the high activity of alumina (I–II) may cause the isomerisation of unstable hydrocarbons (for example, sabinene). If the hydrocarbons are present in the extract, they can be isolated by partitioning between light petroleum (b.p. 40–60°C) and aqueous methanol and subsequent further purification of the light petroleum fraction by the above procedures. In the first chromatographic fractions, paraffins are also present in addition to terpenic hydrocarbons if the mixtures obtained by extraction are submitted to separation. The separation of monoterpenic hydrocarbons from paraffins or higher boiling terpenic hydrocarbons (sesquiterpenic and diterpenic) is usually carried out by simple distillation in a Hickmann flask or in a closed system in a high vacuum at room temperature; condensation takes place in a condenser cooled with dry-ice or liquid nitrogen (Bambagiotti *et al.*). The paraffins can also be separated from terpenic hydrocarbons by chromatography on silver nitrate-impregnated silica gel (a 20–50-fold excess of adsorbent is usually used).

The silver nitrate-impregnated silica gel is prepared as follows. Silver nitrate (45 g) is dissolved in 350 ml of distilled water and the solution is added to 300 g of silica gel. The suspension is evaporated on a rotary evaporator in the vacuum of a water pump. The free-flowing adsorbent is dried to constant weight either in a flask *in vacuo* (bath at 130°C) or in a drying oven at 130°C. Andersen and Syrdal evaporated the suspension on a rotary evaporator at 70°C for 12 h, then activated it by heating at 80°C (0.13 mm Hg) for 6 h, eventually in a drying oven at 120°C for 8 h.

It is recommended that the columns should be protected from light during chromatography in order to prevent the complete darkening of the adsorbent. The solvents used for chromatography on such adsorbents should be freed from traces of sulphur compounds, preferably by filtration through a small amount of the same adsorbent.

When less polar substances are to be separated, the sorbent can be regenerated by washing it with anhydrous diethyl ether which has previously been freed from peroxides, for example by filtration through alkaline alumina of activity I (*ca.* 250 ml of diethyl ether can be used per 25 g of adsorbent). The remaining ether is then displaced with light petroleum (b.p. 40–60°C) and the column is ready for further use.

In view of the physical and chemical properties of monoterpenic hydrocarbons, the most commonly used method for their separation is GLC (analytical and preparative), but from a study of the behaviour of 14 monoterpenic hydrocarbons (Lawrence, 1968b) on thin layers of silica gel, containing various concentrations of silver nitrate (6.25–25%), it follows that they can be separated on this type of adsorbent. In agreement with this, Andersen and Syrdal, who separated monoterpenic hydrocarbons from the essential oil of *Chamaecyparis nootkatensis* leaves, eluted α-pinene, $\Delta_3$-carene, β-pinene, two unidentified hydrocarbons, limonene and myrcene consecutively from a column of silica gel containing 15% of silver nitrate. The first two compounds were eluted with cyclohexane, the others with a 0–50% gradient of benzene in cyclohexane.

Silver nitrate-impregnated neutral alumina is prepared as follows. Silver nitrate (125 g) is dissolved in 380 ml of distilled water and the solution is added, with stirring, to 500 g of neutral alumina. The suspension is evaporated to dryness on a rotary evaporator at 110–130°C bath temperature and under reduced pressure (water pump), thus permitting the smooth distillation of water. When the mixture has been dried, it is activated at 130°C (bath temperature) for 15 min under a full vacuum (water pump). The suitability of this argentised neutral alumina follows from the isolation of the hydrocarbon 1-vinyl-5,5-dimethyl[2.1.1]bicyclohexane (Hogg and Lawrence) from the essential oil of *Mentha cardiaca* (Scotch spearmint). The lowest boiling fractions of the essential oil were separated into 10 fractions by fractional distillation on a column of 35 theoretical plates. The ninth fraction (3 g) was chromatographed on a 40-fold excess of neutral alumina (activity I) by gradient elution (light petroleum–diethyl ether–methanol) into 60 fractions. The combined fractions 1–10 (1.5 g), containing α-pinene, α-thujene and other hydrocarbons, were further chromatographed on a 40-fold excess of neutral alumina containing 15% of silver nitrate, applying gradient elution (light petroleum–diethyl ether–methanol) into 30 fractions of 1 ml. The composition of the fractions was: 9–12, α-pinene; 13 and 14, α-pinene and α-thujene; 17–24, solvents only; 25–28, 1-vinyl-5,5-dimethyl[2.1.1] bicyclohexane, camphene, β-pinene and limonene; and 30, myrcene. It was possible to isolate the required hydrocarbon from fraction 27 by preparative GLC, although it cannot be isolated easily by this method if α-pinene is present. Owing to its supposed lower isomerisation tendency, silica gel impregnated with silver borate was employed by Vokáč *et al.* for the separation of monoterpenic hydrocarbons from albene ($C_{12}H_{18}$), a hydrocarbon which is probably biogenetically similar to terpenes.

Silver borate-impregnated silica gel is prepared as follows. In a 1-l flask with a ground-glass joint, 100 g of silica gel are mixed with 200 ml of a saturated solution of silver borate, which is freed from the substances to be separated before use. The flask with the suspension is heated in a water-bath at 30°C and in a vacuum (water pump) until the swirling of the adsorbent has ceased. Drying is continued in the vacuum of an oil pump (*ca.* 0.05 mm Hg) at a bath temperature of 40°C.

Silver borate is prepared by precipitation of an aqueous silver nitrate solution with a 20% borax solution. The precipitate formed is thoroughly washed with distilled water on a fritted-glass filter.

In contrast to monoterpenic hydrocarbons, sesquiterpenic hydrocarbons represent a very rich class of substances of which some are very similar to each other in their physical

*References p.634*

and chemical properties (positional and spacial isomers of one basic skeleton). These properties impair their separation, and another difficulty consists in their tendency to isomerise. Their separation is most commonly carried out by fractional distillation of the neutral fraction of the essential oil using an efficient column. The distillation process is controlled by GLC and TLC. Corresponding fractions are combined and single components are separated by combined chromatography on alkaline alumina, silica gel or neutral alumina impregnated with silver nitrate (*cf.*, Fig. 29.1), usually under GLC control. When three columns are connected in series, with decreasing diameter and increasing length, up to 15 g of a mixture of sesquiterpenic hydrocarbons can be separated on 250 g of alumina containing 25% of silver nitrate (Lawrence *et al.*) if gradient elution is used.

In view of the fact that separations on strongly activated alumina are dependent on the number of the double bonds (*cf.*, Table 29.1) while on argentised adsorbents the character of the double bonds is more important than their number, a combination of both procedures is a very useful method of separation. Andersen and Syrdal partially separated a complex distillation fraction by this method, eluting first tricyclic hydrocarbons (α-copaene, α-ylangene and longifolene) with light petroleum from a column of alkaline alumina, and then separating all of the remaining hydrocarbons by elution with a more polar system by chromatography on silica gel containing 15% of silver nitrate. The

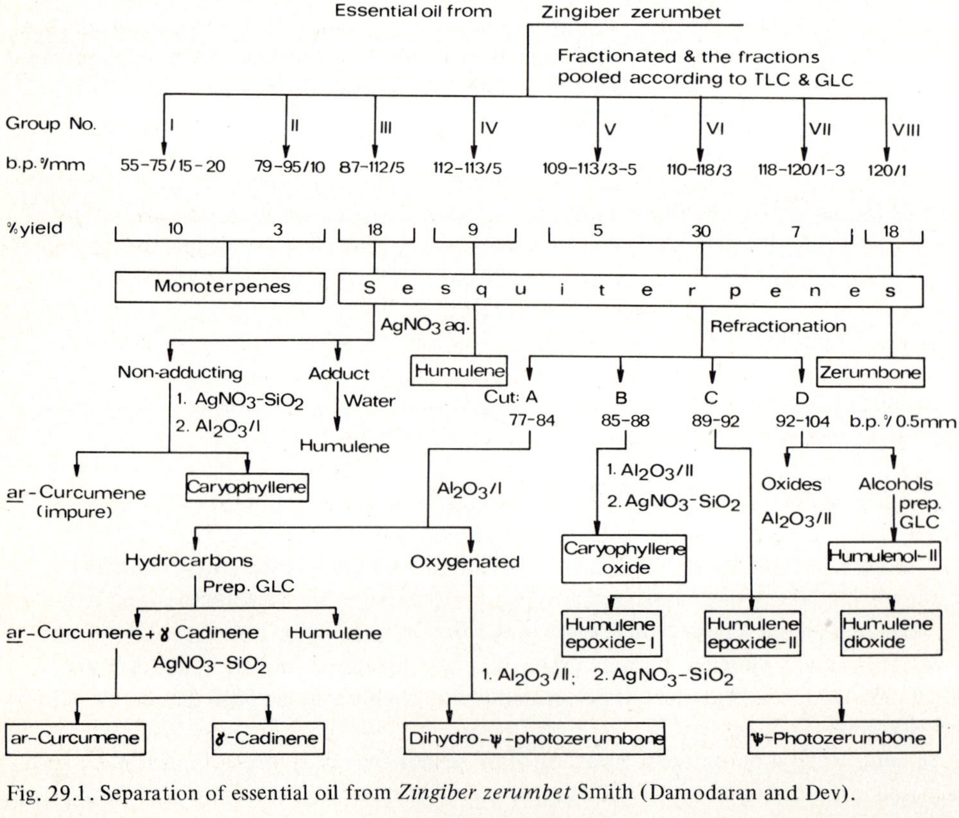


Fig. 29.1. Separation of essential oil from *Zingiber zerumbet* Smith (Damodaran and Dev).

TABLE 29.1
CHROMATOGRAPHY OF SESQUITERPENIC HYDROCARBONS (8.4 g) FROM ATRACTYLIS OIL (CHOW *et al.*)

Sorbent: alkaline alumina (activity I), 1100 g. Eluent: light petroleum (b.p. 40–60°C).

| Fraction | Volume (ml) | Weight (g) | $[\alpha]_D^{20}$ | Main constituent | No. of double bonds |
|---|---|---|---|---|---|
| 1 | 100 | 0.20 | −57.6 | Tricyclic | 1 |
| 2 | 110 | 0.85 | −59.5 | Sesquiterpene | |
| 3 | 210 | 0.65 | +34.7 | Mixture of two | – |
| 4 | 100 | 0.40 | +63.2 | sesquiterpenes | |
| 5 | 200 | 0.55 | +17.1 | | |
| 6 | 200 | 0.60 | −40.6 | α-Isovetivene | 2 |
| 7 | 200 | 0.95 | −45.9 | | |
| 8 | 200 | 0.45 | +23.8 | | |
| 9 | 200 | 0.50 | +39.9 | β-Selinene | 2 |
| 10 | 200 | 0.30 | +40.2 | | |
| 11 | 200 | 0.20 | +28.9 | | |
| 12 | 400 | 0.45 | −2.52 | *ar*-Curcumene | Aromatic |

sequence of the eluted components was the remaining longifolene, "calamenenes", α-alaskene, β-alaskene, δ-cadinene, α-curcumene, β-curcumene, γ-curcumene, γ-cadinene, β-bisabolene and β-farnesene.

Some hydrocarbons are unstable on the adsorbents mentioned and therefore a very mild procedure was used during their isolation. Weinheimer *et al.*, on extraction of gorgonians with hexane and subsequent distillation (up to 35°C), obtained a mixture of sesquiterpenes composed of β-elemene, β-selinene and germacrene A. The mixture (4 g) was separated on 540 g of adsorbent composed of 70% of Florisil, 30% of powdered saccharose and 3% of maize starch. The mixture was homogenised in a blender in benzene. The column temperature was 5°C, elution was carried out with 2% solution of benzene in hexane, the flow-rate was 5 ml/min, the pressure 2.5 p.s.i. under nitrogen and the fractions were each of 25 ml. The authors also pointed out that β-elemene is isomerised to β-selinene when Florisil impregnated with silver nitrate is used.

Less stable azulenes can also be separated on argentised alumina or on neutral deactivated alumina of activity II–III alone. During their preliminary isolation, the formation of the azulenium salts may be utilised, by extracting them with 85% phosphoric acid or dilute hydrochloric acid. The azulenes can be liberated from their colourless salts by dilution with water and extraction with light petroleum (b.p. 40–60°C) or diethyl ether. The separation of the dehydrogenation products of reduced prochamazulenogen was carried out by Herout and Šorm by chromatography on a 100-fold excess of alumina (activity II–III). Elution with light petroleum gave guaiazulene, while with benzene it gave artemazulene. When using alumina impregnated with 5% of silver nitrate (610 g), Bertelli and Crabtree separated a mixture of dihydrochamazulenes (5 g) from absinth; elution with pentane gave 3,6-dihydrochamazulene, while a 5% solution of benzene in pentane eluted chamazulene and pure benzene gave 5,6-dihydrochamazulene. Azulenes can be regenerated easily from their addition compounds (which are used for their

*References p.634*

characterisation) with trinitrobenzene by chromatography on alkaline alumina (activity III–IV) with cyclohexane.

In most instances, diterpenic hydrocarbons can be separated successfully on argentised silica gel by gradient elution, depending on the nature of the original mixture. In one of the first papers describing this type of separation, Norin and Westfelt separated the neutral fraction of the ethereal extract of *Pinus silvestris* wood on an alumina column into three fractions: elution with light petroleum (b.p. 40–60°C) gave hydrocarbons, while benzene gave aldehydes and ethanol gave alcohols. The mixture of diterpenic hydrocarbons thus obtained was separated on a 30-fold excess of silica gel containing 22% of silver nitrate. Elution was carried out with light petroleum (b.p. 40–60°C) and a 0–1% linear gradient of diethyl ether in the same solvent. In the first fractions pimaradiene was eluted and subsequent fractions contained isopimaradiene. If a smaller percentage of silver nitrate was used (*ca.* 9%), light petroleum (b.p. 40–60°C) alone sufficed for the elution, as for example in the isolation of diterpenic hydrocarbons from the leaves of *Cryptomeria japonica*, carried out by Appleton *et al.* The light petroleum extract was chromatographed with light petroleum (b.p. 40–60°C)–diethyl ether (10:1) on neutral alumina of activity

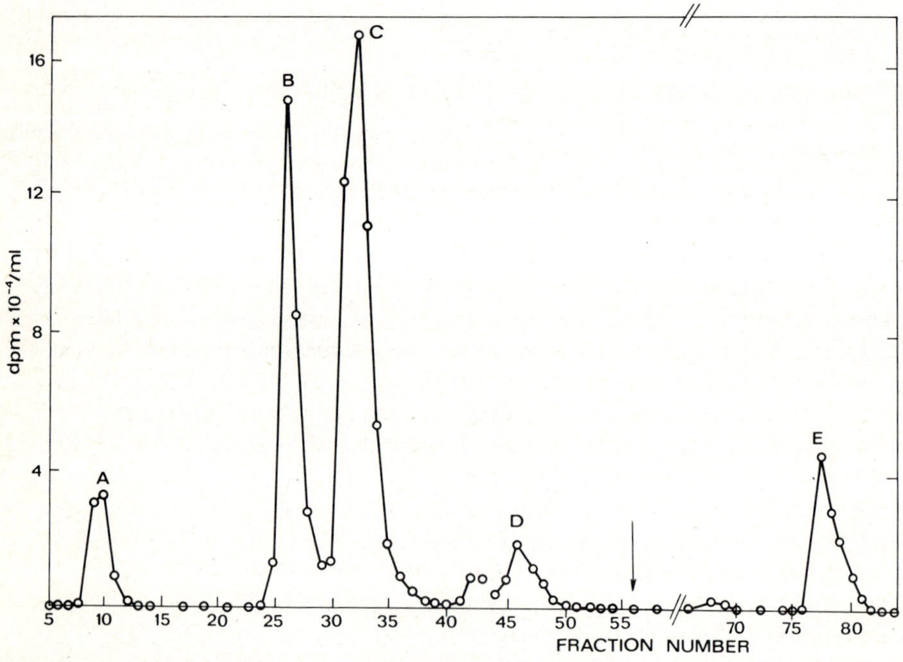


Fig. 29.2. Separation of diterpene hydrocarbons formed from 2-[$^{14}$C]mevalonate by enzyme preparation of castor bean seedlings. Column: 1.7 × 27 cm. Sorbent: 30 g of 5% Bio-Sil HA silver nitrate-impregnated silicic acid (–325 mesh) (Calbiochem, Los Angeles, Calif., U.S.A.). Operating conditions: non-linear gradient of increasing concentrations of benzene in *n*-hexane; starting from fraction 56 (arrow), benzene was replaced with ethyl acetate; fraction volume 5–10 ml; flow-rate 1ml/min. Detection: A 0.1 ml aliquot of each fraction was assayed for radioactivity and the amount of radioactivity per millilitre was plotted against the fraction number. A = trachylobane; B = kaurene; C = sandaracopimaradiene; D = beyerene; E = casbene.

I. The hydrocarbon fraction obtained was re-chromatographed on silica gel impregnated with silver nitrate. Elution with light petroleum (b.p. 40–60°C) gave, consecutively, kaurene, phyllocladene and sclarene. A still lower concentration of silver nitrate (5%) was used by Robinson and West during the isolation of labelled diterpenes obtained on biosynthesis from 2-[$^{14}$C] mevalonate. Their separation is shown in Fig. 29.2.

A mixture of 10 triterpenic hydrocarbons from the leaves of the fern *Adiantum monochlamys* was separated by Ageta *et al.* by a combination of repeated chromatography on alkaline alumina (Wako, Osaka, Japan), elution with *n*-hexane, and silica gel containing 20% of silver nitrate, elution with *n*-hexane or *n*-hexane–diethyl ether (9:1). The course was followed by GLC. The authors were able to identify fern-8-ene, fern-9(11)-ene, ferna-7,9(11)-diene, fern-7-ene, adian-5-ene, neohop-13(18)-ene, neohop-12-ene, filic-3-ene, neohopa-11,13(18)-diene and hop-22(29)-ene. If neutral or acid-washed alumina was used, isomerisation of the double bonds was observed.

## ETHERS, EPOXIDES AND FURANS

The separation of these substances is usually carried out on deactivated neutral alumina of activity II–III or deactivated silica gel (seldom impregnated with silver nitrate) or Florisil. In view of the low polarity of the substances and the use of deactivated adsorbents, light petroleum (b.p. 40–60°C) plus a small amount of benzene, diethyl ether or ethyl acetate is most commonly used for elution. Highly activated alumina cannot be used for the separation because isomerisation might take place (see the detailed study by Joshi *et al.*).

Coates and Melvin purified the reaction product of the dehydration of *cis*-2,2-dimethyl-3-hydroxy-6-methylenecyclohexanemethanol on silica gel, using light petroleum (b.p. 30–60°C) for elution. The separation of very similar sesquiterpenic oxides (differing in the axial and the equatorial positions of the methyl group) obtained on dehydration of 4-hydroxyguaioxide and subsequent hydrogenation was carried out by Ishii *et al.* (1970) on a 200-fold excess of alumina of activity II. The first fractions, obtained on elution with light petroleum, contained guaioxide, and the last fraction was liguloxide.

A more complex example of the isolation of diterpenic isoincensoloxide and incensol-oxide was successfully solved by Forcellese *et al.* in the following manner. The neutral fraction of the resin *Boswellia carteri* was chromatographed on a 30-fold excess of alumina (activity II–III); benzene first eluted incensol and then incensoloxide and isoincensoloxide. The oily fraction containing both oxides was further chromatographed on a 50-fold excess of silica gel; elution with benzene containing 5% of diethyl ether gave fractions from which incensoloxide crystallised out. The mother liquors also contained, according to TLC on silver nitrate-impregnated silica gel, oxides that could not be separated well on silica gel alone (TLC on silica gel gave a single spot). The mixture was converted into benzoates (1.25 g) and separated on silica gel (75 g) with benzene containing 3% of diethyl ether, affording isoincensol benzoate, and with benzene containing 15% of diethyl ether, affording incensoloxide benzoate.

The reaction mixture obtained on oxidation of 11$\beta$-hydroxylanostan-3$\beta$-yl acetate with lead tetraacetate and iodine was chromatographed on deactivated silica gel (15% of water),

*References p. 634*

affording in the first fractions the corresponding iodoether acetate (Roller and Djerassi). On neutral alumina (activity II), the unsaponifiable part of the light petroleum extract of *Polypodium vulgare* rhizomes was separated using light petroleum as eluent. Two triterpenic hydrocarbons were eluted first, followed by 17,21-epoxyhopane (Berti *et al.*).

The effect of the epoxy group on the chromatographic behaviour of sesquiterpenic furan derivatives is apparent from a comparison of the conditions of the isolation of 8,8*a*-epoxyfuranoligularane from the essential oil of *Senecio silvaticus* (described by Schild), and furoligularane obtained by degradation. The native substance was isolated by chromatography of the essential oil on a 100-fold excess of neutral alumina (activity III) with light petroleum (b.p. 40–60°C) containing 2% of ethyl acetate, while furanoligularane was eluted under virtually the same conditions with light petroleum (b.p. 40–60°C) alone. Cimino *et al.* (1972b) separated another mixture of simple sesquiterpenic furans (pleraplysillin and dehydrodendrolasin) on silica gel impregnated with silver nitrate (150 mg of the crude extract on a mixture of 15 g of silica gel and 2.5 g of silver nitrate) in the system light petroleum (b.p. 40–70°C)–benzene (beginning with a 9:1 mixture and then increasing the content of benzene). In connection with the same theme, *viz.*, the study of the components of sea sponges, Cimino *et al.* (1972a) separated two $C_{21}$ furanoterpenes (degraded sesterterpenes) by multiple chromatography on silica gel; using benzene–light petroleum (b.p. 40–70°C) (7:3) they eluted first tetrahydrofurospongin-2 and then dihydrofurospongin-2.

## ESTERS

The most commonly used adsorbents for the separation of simple esters are silica gel and neutral alumina (activity II–III), as well as silica gel impregnated with silver nitrate. Mixtures of light petroleum (b.p. 40–60°C) and benzene (or a small amount of diethyl ether and benzene) were employed as eluents. With substances that contain several double bonds, which are chromatographed on silver nitrate-impregnated silica gel, more polar elution systems should be used, such as benzene containing a small amount of ethanol. The amount of adsorbent is dependent, as in other instances, on the composition and the number of components of the mixture being chromatographed; it is usually from 30 to 50 times the weight of the mixture for alumina, while for silica gel alone or impregnated with silver nitrate it is usually from 15 to 30 times the weight of the mixture.

The oxidation product of dimethyl shellolate was purified on neutral alumina by elution with benzene by Yates and Field. Other esters of sesquiterpenic acids obtained on hydrolysis of *Palas seedlac* and subsequent esterification were also separated by Singh *et al.* on alumina (activity II), using a mixture of benzene and light petroleum (b.p. 40–60°C) (3:1) for the elution of the ester of laccishellolic acid, and benzene for the ester of epilaccishellolic acid. The separation of four stereoisomeric synthetic geranyl esters of farnesylacetic acid, which was poor by fractional distillation or preparative GLC, was carried out by Pala *et al.* on a 25-fold excess of very fine silica gel impregnated with silver nitrate. The first fraction eluted with benzene–ethanol (98:2) indicated a separation of isomer I (*trans,cis*-(4,5) and *cis*-(8,9)) from isomer II (*cis*-(4,5) and *cis*-(8,9)) in only a 5.5:4.5 ratio. However, by triple re-chromatography, pure isomer II could be obtained.

The last fraction eluted with benzene–ethanol (95:5) was a mixture of I and II in a 9.2:0.8 ratio, from which a single additional chromatography gave pure isomer I. In a similar manner, the isomers *trans*-(4,5), *trans*-(8,9) and *cis*-(4,5), *trans*-(8,9) were also separated. The separation of the methyl esters of diterpenic acids obtained on oxidation of a mixture of six diterpenic aldehydes and esterification with diazomethane, which displayed two main peaks and several by-products on GLC analysis, was carried out by Bruns on a 15-fold excess of silica gel containing 28% of silver nitrate; with light petroleum (b.p. 40–80°C) containing 2% of diethyl ether an ester of dehydroabietic acid was obtained, followed by an ester of isopimaric acid, which was purified by crystallisation and re-chromatography on alumina.

Triterpenic acetates (β-amyrenyl acetate and hop-17,(21)-en-3β-yl acetate) were separated by Arthur *et al.* on a 75-fold excess of alumina by elution with light petroleum (b.p. 60–80°C) (in the above sequence). Wahlberg *et al.* separated the following four triterpenic esters on silica gel and then on silver nitrate-impregnated silica gel with light petroleum–diisopropyl ether: α-amyrin palmitate, lupenyl palmitate, β-amyrin palmitate and cycloartenyl palmitate, present in the neutral fraction of the acetone extract of the wood of *Carphephorus odoratissimus.*

## ALDEHYDES AND KETONES

In view of the chemical properties of these compounds it is possible to use either neutral or acidic alumina, as well as silicic acid, silica gel and silver nitrate-impregnated silica gel for their chromatographic separation. Depending on the number of functional groups and the activity of the adsorbents used, the following elution systems can be employed: light petroleum (b.p. 40–60°C), mixtures of light petroleum and diethyl ether, benzene, mixtures of benzene and diethyl ether, and chloroform.

The separation of two sesquiterpenic aldehydes, nuciferal and torreyal, was achieved by Sakai *et al.* on a 15-fold excess of silicic acid by elution with a 3% solution of ether in hexane; nuciferal was eluted first. A mixture of three ketones of the elemane type from the essential oil of the *Acorus calamus* rhizomes was obtained by Yamamura *et al.* by chromatography of the whole essential oil on a 15-fold excess of silica gel with light petroleum–diethyl ether (4:1). This mixture was then separated into single components by repeated chromatography on an 80-fold excess of silica gel, using benzene for elution. The substances were eluted in the following sequence: shyobunone, epishyobunone and isoshyobunone. Canonica *et al.* used a 100-fold excess of silica gel G (Merck)–Celite (1:1) for the chromatographic separation of unstable substances in the acetone extract from the bark of *Cinnamosma fragrans*; elution with benzene and benzene–diethyl ether (9:1) gave fractions containing cinnamodial (a $C_{15}$ dialdehyde containing a carbonyl and a hydroxyl group).

Baker *et al.* isolated from the reaction mixture of 15α,16-epoxyphyllocladane with boron trifluoride–diethyl ether complex 16-epiphyllocladan-15-one using alumina deactivated with 5% (v/v) of aqueous 10% acetic acid; elution was carried out with light petroleum. In a study of the unsaponifiable fraction of the ethereal extract of *Pinus strobus* bark, Zinkel and Evans combined chromatography on silicic acid and silicic acid

*References p. 634*

containing 40% of silver nitrate with gradient elution with light petroleum–diethyl ether in order to isolate, after manoyl oxide, the diterpenic aldehyde strobal; the aldehydes abietal, dehydroabietal, neoabietal, communal and isopimaral followed.

Triterpenic ketones from the light petroleum extract of *Quercus glauca* leaves were separated by Tachi *et al.* first on silica gel (elution with an *n*-hexane–ethyl acetate mixture) and then on silver nitrate–impregnated silica gel (elution with 99:1 *n*-hexane–ethyl acetate). The first fraction contained cyclobalanone, which was followed by 24-methylenecycloartanone. In addition to silica gel, alumina washed with acids is also often used. Djerassi and McCrindle even used alumina of activity I (ratio adsorbent: extract = 40:1) for the separation of the neutral fraction from the methanolic extract of *Tillandsia usneoides*. Elution with light petroleum (boiling range 40–60°C)–diethyl ether (7:3) gave first cycloartenone and then frideline.

## LACTONES

Adsorbents used for the separation of lactones are of neutral character: silica gel, alumina, or silicic acid and Florisil. As most lactones contain additional oxygen-containing functional groups, the elution systems are usually of a more polar character than those for the groups of terpenic substances discussed above. Often mixtures of benzene and chloroform, light petroleum (b.p. 40–60°C) and ethyl acetate, and benzene and diethyl ether are used in which the more polar components account for 10–50%.

Sesquiterpenic lactones, the number of which has increased rapidly in recent years (at present more than 400 of these compounds are known – *cf.*, Devon and Scott) are separated predominantly on silica gel, silicic acid and neutral alumina. Herz *et al.* separated a mixture of six sesquiterpenic dilactones, present in the chloroform extract from the working up of *Mikania scandens*, on silicic acid, using benzene–chloroform (3:2) for elution. Herz and Srinivasan chromatographed a chloroform extract from *Gaillardia amblyodon* on a 10-fold excess of alumina (Alcoa F-20); on elution with benzene–chloroform (3:1), gaillardipinnatin was obtained, while elution with chloroform gave amblyonin.

Diterpenic lactones from the filtrate of *Gibberella fujikuroi* were isolated by Cross *et al.* on a 100-fold excess of a mixture of Celite and silica gel (2:1). Elution with light petroleum (b.p. 60–80°C) containing 15–17% of ethyl acetate gave 7-hydroxykaurenolide, the same mixture, but containing 30–32.5% of ethyl acetate, eluted 7,18-dihydroxykaurenolide, while ethyl acetate–methanol (9:1) gave 7,16,18-trihydroxykaurenolide. In a similar manner, substances of the same type were obtained by Serebryakov *et al.* from the metabolites of *Fusarium moniliforme* when a 6-fold excess of silica gel was used and some fractions were re-chromatographed on neutral alumina (40-fold excess).

The separation of triterpenic lactones was also studied by Chanley *et al.* during the analysis of the components of the toxic principle of holothurin present in sea cucumber (*Actinopygga agassizi*). This principle is a mixture of about six glycosides; aglycones obtained by acid hydrolysis were chromatographed on a 15-fold excess of deactivated alumina (4% of 10% acetic acid solution). Elution with benzene afforded, in the second fraction, a mixture of 22,25-oxidoholothurinogenin and its 17-deoxy derivative. From several chromatographic runs, a mixture of both substances was obtained, which was acetylated and separated chromatographically on alumina deactivated with 3% acetic acid solution. Elution with a 1:4 mixture of benzene and Skelly B (a hydrocarbon mixture containing about 50% of *n*-hexane; Skelly Oil Co., Kansas City, Mo., U.S.A.) gave 17-deoxy-22,25-oxidoholothurinogenin acetate in the first fraction, while a mixture of benzene and Skelly B (1:3–1:1) gave 22,25-oxidoholothurinogenin acetate. Djerassi *et al.* also used neutral alumina (a 60-fold excess) for the separation of acetylated aglycones prepared by hydrolysis and acetylation of the neutral fraction from the ethanolic extract of the cactus *Lemaireocereus stellatus.* On elution with benzene–diethyl ether (1:1) they obtained thurberogenin acetate, and with diethyl ether–chloroform (3:2) they obtained stellatogenin acetate.

## ALCOHOLS

For the separation of terpenic alcohols, neutral alumina and silica gel are mainly used, and silicic acid very rarely. In view of the degree of adsorbent activity and the number of hydroxyl or other functional groups in the chromatographed substances, more polar systems serve as eluents, such as benzene–ethyl acetate, benzene–diethyl ether and chloroform–ethanol.

The separation of 6β- and 7α-hydroxyguaioxides from the microbial transformation of guaioxide was performed by Ishii *et al.* (1971) on a 30-fold excess of alumina (activity V); elution was carried out with light petroleum–diethyl ether (98:2 and 95:5), affording the 6β-hydroxy derivative. Medium fractions, eluted with 9:1 light petroleum–diethyl ether, contained the 7α-hydroxy derivative. Iguchi *et al.* separated calamendiol from isocalamendiol by silica gel chromatography using in the first instance light petroleum–diethyl ether (5:1) for elution, while the second substance was eluted with the same mixture in a 3:1 ratio.

The diterpenic alcohols manool and isopimaradienol from the neutral fraction of the acetone extract of *Dacrydium bidwillii* wood were separated by Grant *et al.* on a 30-fold excess of alumina, activity II, by elution with light petroleum (b.p. 60–80°C)–diethyl ether mixtures. A 4:6 mixture eluted manool, while a 3:7 mixture eluted a mixture of manool and isopimaradienol, which was further separated by TLC. Fujita and Taoka chromatographed an ethereal extract of the leaves of *Isodon lasiocarpus* on a 40-fold excess of silicic acid; elution with dichloromethane–acetone (8:2) gave lasiokaurin (kaurenetriol) and lasiodonin (kaurenetetraol).

The separation of the triterpenic alcohols zeorin and leucotylin from the neutral fraction of the ethereal extract of the lichen *Parmelia leucotyliza* was carried out by Yosioka *et al.* by multiple chromatography on alumina and gradual elution with chloroform, chloroform–methanol and pure methanol.

## ACIDS

The separation of organic acids is usually carried out on the acids either in a free state or in the form of esters, mainly methyl esters (after esterification with diazomethane).

*References p. 634*

The separation of free acids is carried out by adsorption chromatography on alumina, silica gel on a Celite–charcoal mixture, using a variety of systems of different polarity, ranging from light petroleum (b.p. 40–60°C)–benzene, or benzene–diethyl ether, to chloroform–methanol. When partition chromatography was applied, alumina impregnated with dimethyl sulphoxide and diisopropyl ether–acetone as eluent, or silica gel with anchored formic acid and a gradient of ethyl acetate in *n*-hexane were used. The separation of esters was generally carried out under the conditions mentioned in the section on esters.

Two sesquiterpenic hydroxy acids were separated by Runeberg on a 100-fold excess of dimethyl sulphoxide-impregnated alumina. Elution with diisopropyl ether gave hydroxy acid I, while with acetone, hydroxy acid II was eluted. The separation of 33 gibberellins (diterpenic acids, 100 mg) was investigated by Durley *et al.*, using for partition chromatography silica gel (Woelm, Eschwege, G.F.R.) (20 g) on to which 0.5 *M* formic acid saturated with a mixture of ethyl acetate and *n*-hexane (1:9) was fixed as the stationary phase. A gradient of ethyl acetate in *n*-hexane was used as the mobile phase. This mixture was saturated with 0.5 *M* formic acid in a Varigrad system with four mixing chambers (1, ethyl acetate: *n*-hexane 65:35; 2, 20:80; 3, 100:0; and 4, 100:0). The separation achieved was relatively good: groups of 3–6 substances, in some instances even 1–2 substances.

Nagai *et al.* separated two triterpenic acids by chromatography on silica gel using benzene–diethyl ether (9:1) for elution. The early fractions contained urs-12-en-3β-ol-27-oic acid, and were followed by fractions containing olean-12-en-3β-ol-27-oic acid.

## REFERENCES

Ageta, H., Shiojima, K. and Arai, Y., *Chem. Commun.*, (1968) 1105; private communication, 1973.
Andersen, N. H. and Syrdal, D. D., *Phytochemistry*, 9 (1970) 1325.
Appleton, R. A., McCrindle, R. and Overton, K. H., *Phytochemistry*, 9 (1970) 581.
Arthur, H. R., Hui, W. H., Lam, C. N. and Szeto, S. K., *Aust. J. Chem.*, 17 (1964) 697.
Baines, D. A. and Jones, R. A., *J. Chromatogr.*, 47 (1970) 130.
Baker, K. M., Briggs, L. H., Buchanan, J. G. St. C., Cambie, R. C., Davis, B. R., Hayward, R. C., Long, G. A. S. and Rutledge, P. S., *J. Chem. Soc., Perkin Trans. I*, (1972) 190.
Bambagiotti, M. A., Vincieri, F. F. and Cosi, G., *Phytochemistry*, 11 (1972) 1455.
Bertelli, D. J. and Crabtree, J. H., *Tetrahedron*, 24 (1968) 2079.
Berti, G., Bottari, F., Marsili, A. and Morelli, I., *Tetrahedron Lett.*, (1966) 979.
Brockmann, H. and Schodder, H., *Chem. Ber.*, 74 (1941) 73.
Bruns, K., *Tetrahedron*, 25 (1969) 1771.
Canonica, L., Corbella, A., Gariboldi, P., Jommi, G., Křepinský, J., Ferrari, G. and Casagrande, C., *Tetrahedron*, 25 (1969) 3895.
Chanley, J. D., Mezzetti, T. and Sobotka, H., *Tetrahedron*, 22 (1966) 1857.
Chow, W. Z., Motl, O. and Šorm, F., *Collect. Czech. Chem. Commun.*, 27 (1962) 1914.
Cimino, G., De Stefano, S., Minale, L. and Fattorusso, E., *Tetrahedron*, 28 (1972a) 267.
Cimino, G., De Stefano, S., Minale, L. and Trivellone, E., *Tetrahedron*, 28 (1972b) 4761.
Coates, R. M. and Melvin, L. S., *J. Org. Chem.*, 35 (1970) 865.
Cross, B. E., Galt, R. H. B., Hanson, J. R., Curtis, P. J., Grove, J. F. and Morisson, A., *J. Chem. Soc.*, (1963) 2937.
Damodaran, N. P. and Dev, S., *Tetrahedron*, 24 (1968) 4113.
Devon, T. K. and Scott, A. I., *Handbook of Naturally Occurring Compounds, Vol. II, Terpenes*, Academic Press, New York, 1972.

Djerassi, C., Lin, L. H., Farkas, E., Lippman, A. E., Lemin, A. J., Geller, L. E., McDonald, R. N. and Taylor, B. J., *J. Amer. Chem. Soc.*, 77 (1955) 1200.
Djerassi, C. and McCrindle, R., *J. Chem. Soc.*, (1962) 4034.
Durley, R. C., Crozier, A., Pharis, R. P. and McLaughin, G. E., *Phytochemistry*, 11 (1972) 3029.
Forcellese, M. L., Nicoletti, R. and Petrossi, U., *Tetrahedron*, 28 (1972) 325.
Fujita, E. and Taoka, M., *Chem. Pharm. Bull. (Tokyo)*, 20 (1972) 1752.
Grant, P. K., Huntrakul, C. and Sheppard, D. R. J., *Aust. J. Chem.*, 20 (1967) 969.
Guenther, E., Gilbertson, G. and Koenig, R. T., *Anal. Chem.*, 43 (1971) 45R.
Heřmánek, S., Schwarz, V. and Čekan, Z., *Collect. Czech. Chem. Commun.*, 26 (1961) 3170.
Herout, V. and Šorm, F., *Chem. Listy*, 48 (1954) 706.
Herz, W. and Srinivasan, A., *Phytochemistry*, 11 (1972) 2093.
Herz, W., Subramanian, P. S., Santhanan, P. S. and Hall, A. L., *J. Org. Chem.*, 35 (1970) 1453.
Hogg, J. W. and Lawrence, B. M., *Flavour Ind.*, 3 (1972) 321.
Iguchi, M., Nishiyama, A., Koyama, H., Yamamura, S. and Hirata, Y., *Tetrahedron Lett.*, (1969) 3729.
Ishii, H., Tozyo, T. and Nakamura, M., *Chem. Pharm. Bull. (Tokyo)*, 19 (1971) 842.
Ishii, H., Tozyo, T., Nakamura, M. and Minato, H., *Tetrahedron*, 26 (1970) 2911.
Joshi, V. S., Damodaran, N. P. and Sukh, D., *Tetrahedron*, 24 (1968) 5817; 27 (1971) 459 and 475.
Kugler, E. and Kováts, E., *Helv. Chim. Acta*, 46 (1963) 1480.
Lawrence, B. M., *Perfum. Essent. Oil Rec.*, 59 (1968a) 421.
Lawrence, B. M., *J. Chromatogr.*, 38 (1968b) 535.
Lawrence, B. M., *Can. Inst. Food Technol. J.*, 4 (1971) A44.
Lawrence, B. M., Hogg, J. W. and Terhune, S. J., *Perfum. Essent. Oil Rec.*, 60 (1969) 88.
Nagai, M., Izawa, K. and Inouve, T., *Chem. Pharm. Bull. (Tokyo)*, 17 (1969) 1438.
Norin, T. and Westfelt, L., *Acta Chem. Scand.*, 17 (1963) 1826, 1828.
Pala, G., Mantegani, A., Bruzzese, T. and Sekules, G., *Helv. Chim. Acta*, 53 (1970) 1827.
Robinson, D. R. and West, Ch. A., *Biochemistry*, 9 (1970) 70.
Roller, P. and Djerassi, C., *J. Chem. Soc., C*, (1970) 1089.
Runeberg, J., *Acta Chem. Scand.*, 15 (1961) 721.
Sakai, T., Nishimura, K. and Hirose, Y., *Bull. Soc. Chim. Jap.*, 38 (1965) 381.
Schild, W., *Tetrahedron*, 27 (1971) 5735.
Serebryakov, E. P., Simolin, A. V., Kucherov, V. F. and Rosynov, B. V., *Tetrahedron*, 26 (1970) 5215.
Singh, A. N., Upadhye, A. B., Wadia, M. S., Mhaskar, V. V. and Sukh, D., *Tetrahedron*, 25 (1969) 3855.
Tachi, Y., Taga, S., Kamano, Y. and Komatsu, M., *Chem. Pharm. Bull. (Tokyo)*, 19 (1971) 2193.
Vokáč, K., Samek, Z., Herout, V. and Šorm, F., *Tetrahedron Lett.*, (1972) 1665.
Wahlberg, I., Karlsson, K. and Enzell, C. R., *Acta Chem. Scand.*, 26 (1972) 1383.
Weinheimer, A. J., Youngblood, W. W., Washecheck, P. H., Karns, T. K. B. and Ciereszko, L. S., *Tetrahedron Lett.*, (1970) 497.
Yamamura, S., Iguchi, M., Nishiyama, A. and Niwa, M., *Tetrahedron*, 27 (1971) 5419.
Yates, P. and Field, G. F., *Tetrahedron*, 26 (1970) 3135.
Yosioka, I., Nakanishi, T. and Kitagawa, I., *Chem. Pharm. Bull. (Tokyo)*, 17 (1969) 279.
Zinkel, D. F. and Evans, B. B., *Phytochemistry*, 11 (1972) 3387.

*Chapter 30*

# Amines

Z. DEYL

## CONTENTS



## INTRODUCTION

In the separation of amines, several types of compounds have been studied very intensively, particularly catecholamines and tryptophan metabolites, which have been dealt with in many publications using a variety of separation procedures. For other amines, such as aliphatic amines, polyamines and aromatic amines, the situation is less complex, although the distribution into the above types is sometimes difficult to achieve as different types of amines are frequently chromatographed side by side. Also, some types of amines, such as tryptamine and serotonin, are chromatographed together with amino acids. These types of separations are not listed here or in the chapter on amino acid chromatography. However, it is possible to obtain some idea of these separations from the ion-exchange procedures described in this chapter. Ion-exchange chromatography in nearly all its variations has been widely exploited for liquid column chromatographic separations of amines. High-speed techniques and gel permeation chromatography are not popular at present and it is likely that classical ion-exchange techniques will dominate this area as they give good and rapid separations. Another important factor is the obvious applicability of automated amino acid analyzers for this purpose.

## ALIPHATIC MONO-, DI- AND POLYAMINES

The first type of ion-exchange chromatography of aliphatic monoamines that proved suitable for automation was that involving the ion-exchange resin Amberlite CG-120 (Perry and Schroeder).

A summary of the elution volumes for a series of aliphatic monoamines on this ion exchanger is presented in Table 30.1.

TABLE 30.1
ELUTION VOLUMES OF AMINES CHROMATOGRAPHED ON AMBERLITE CG-120 (PERRY AND SCHROEDER)

Authentic compounds were chromatographed in mixtures on Amberlite CG-120 columns 30 cm in length and 0.9–1.0 cm in diameter, at a flow-rate of 30 ml/h and a temperature of 50°C. Chromatograms were developed with 0.2 *N* pyridine acetate buffer (pH 3.50) for the first 600 ml, and thereafter with 0.8 *N* pyridine acetate buffer (pH 5.50). The breakthrough of the second developer occurred at 636 ml.

| Compound | Range of elution (ml) | Elution peak (ml) | Compound | Range of elution (ml) | Elution peak (ml) |
|---|---|---|---|---|---|
| Glucosamine | 115–132 | 121 | γ-Methylmercaptopropylamine | | |
| Galactosamine | 115–132 | 121 | sulphoxide | 290–300 | 295 |
| N-Acetylethylenediamine | 145–160 | 152 | Ammonia | 296–345 | 308 |
| β-Methoxyethylamine | 150–165 | 157 | Propylamine | 312–332 | 322 |
| N-Methylethanolamine | 156–171 | 163 | Isobutylamine | 316–345 | 328 |
| 3-Amino-1-propanol | 175–191 | 183 | Hydroxylamine | 334–371 | 347 |
| Serinol | 175–191 | 183 | 1-Methylhistidine | 340–375 | 361 |
| Dimethylamine | 187–200 | 193 | Cyclopropylamine | 350–382 | 364 |
| β-Hydroxypropylamine | 195–209 | 201 | 3-Methylhistidine | 353–388 | 370 |
| N-Methylethylamine | 198–215 | 206 | *n*-Butylamine | 390–420 | 405 |
| Diethylamine | 212–224 | 218 | Isoamylamine | 430–465 | 445 |
| Ethanolamine | 210–230 | 219 | γ-Methylmercaptopropylamine | 447–490 | 467 |
| 2-Aminobutanol | 220–236 | 228 | *n*-Amylamine | 504–542 | 520 |
| Pyrrolidine | 250–270 | 260 | Histidine | 591–629 | 600 |
| Piperidine | 250–272 | 261 | Ornithine | 646–656 | 650 |
| Methylamine | 262–280 | 273 | Lysine | 653–662 | 656 |
| Ethylamine | 272–294 | 282 | Carnosine | 657–666 | 662 |
| | | | Arginine | 820–855 | 832 |

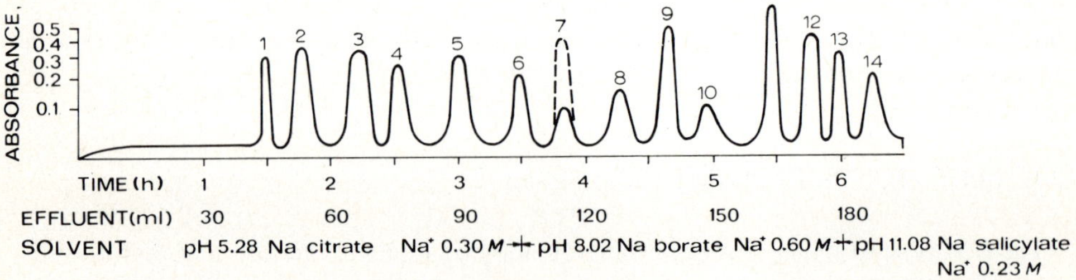


Fig. 30.1. Chromatogram of primary monoamines separated at 50°C on an Aminex A-5 column (10 × 0.6 cm) (Miyagi and Ando). Peaks: 1, methylamine; 2, ethylamine; 3, allylamine; 4, *n*-propylamine; 5, isobutylamine; 6, *n*-butylamine; 7, dopamine; 8, isoamylamine; 9, histamine; 10, *n*-amylamine; 11, tyramine; 12, phenylethylamine; 13, serotonin; 14, *n*-hexylamine.

The separation of primary amines, diamines and polyamines on ion-exchange columns of the Aminex type, using small variations of the equipment normally used for automated amino acid analysis, appears to be the most advanced liquid column chromatographic procedure used for these compounds nowadays. Aminex A-4 and A-5 and Bio-Rex 70 were used for this purpose by several workers (Hatano *et al.*, Miyagi and Ando, Perry and Schroeder, Rosenthal and Tabor, Yoshioka *et al.*). Evaluation is carried out by the conventional ninhydrin procedure. In practice, the individual systems differ in the Varigrad buffer composition, and in the timing of the sudden buffer changes, which, of course, depends greatly upon the mixture being separated and the necessity to spread a particular region or the chromatogram. The problems met here are similar to those which occur in amino acid analysis, as both the nature of the separation process and the equipment used have much in common. Typical runs on primary monamines and diamines are presented in Fig. 30.1. With Aminex A-4 (Hatano *et al.*), elution is carried out with a complex buffer system (Table 30.2). The column size is the same as that with Aminex A-5 (Fig. 30.1); before use, it is packed under 1 atm overpressure and operated thereafter at 50°C. Before application of the sample, the column is conditioned with 0.116 *M* sodium citrate buffer (starting buffer). Buffers I and II (Table 30.2) are used for 70 and 150 min, respectively. After the appropriate amount of buffer II has passed through the column, the system is connected automatically to a three-chamber Technicon Varigrad system containing 120 ml each of the remaining buffers listed under III in Table 30.2. The gradient is allowed to run for an additional 200 min. Flow-rates are maintained at 30 ml/h in both the ninhydrin and sample lines. The retention time of the reaction coil is 11 min 15 sec (exactly) and the coil is maintained thermostatically at 100°C. Absorbance and height:width ratios of eluted peaks for a number of amines were published by Hatano *et al.* A typical example of a separation can be seen in Fig. 30.2.

Bio-Rex 70 (−400 mesh) appears to be the most suitable ion exchanger if polyamines have to be separated. The development of this technique was described in a series of

TABLE 30.2
COMPOSITION AND CONDITIONS OF ELUTING BUFFERS (HATANO *et al.*)

Benzyl alcohol was used for the elution system in order to prevent a tailing effect and to obtain well resolved chromatograms. When benzyl alcohol was added to the eluent, no difficulty was experienced with ninhydrin colour development.

| Buffer system | Concn. (*N*) | Benzyl alcohol concn. (%) | pH* | Ionic strength of sodium ion (*M*)** |
|---|---|---|---|---|
| (I) Sodium citrate | 0.116 | – | 5.28 | 0.35 |
| (II) Sodium borate | 0.025 | 0.5 | 8.02 | 0.60 |
| (III) Gradient | | | | |
| (1) Sodium borate | 0.05 | 0.4 | 10.00 | 0.60 |
| (2) Sodium salicylate | 0.20 | – | 11.50 | 0.65 |
| (3) Sodium salicylate | 0.20 | – | 12.50 | 0.70 |

*Adjusted with 6 *N* sodium hydroxide or hydrochloric acid.
**Adjusted with sodium chloride solution.

*References p. 655*

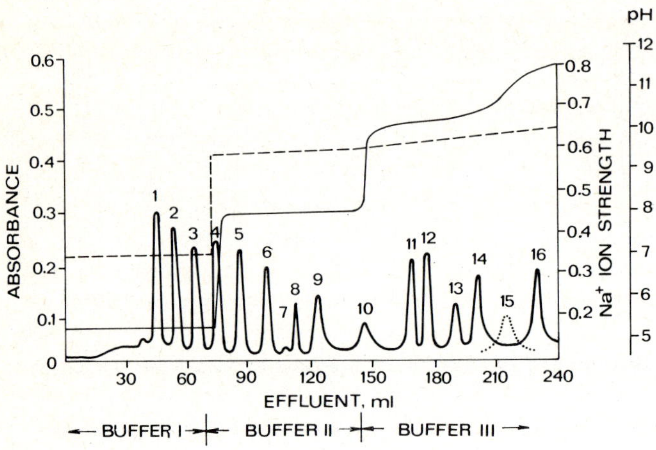


Fig. 30.2. Chromatogram of an authentic mixture of 16 amines and eluting conditions of the buffer system (Hatano *et al.*). Peaks: 1, methylamine; 2, ethylamine; 3, allylamine; 4, *n*-propylamine; 5, isobutylamine; 6, *n*-butylamine; 7, 1,2-propanediamine; 8, histamine; 9, isoamylamine; 10, *n*-amylamine; 11, tyramine; 12, putrescine; 13, phenethylamine; 14, cadaverine; 15, serotonin; 16, hexamethylenediamine. The amount of each amine was 0.4 μmole except for serotonin (1.0 μmole). All curves represent the peaks at 570 nm except for serotonin (440 nm). Monitored pH by a flow pH meter was recorded automatically.

papers by Morris *et al.*, Rosenthal and Tabor, and Tabor *et al.* Columns (7 × 0.9 cm) of the above sorbent are attached to the usual equipment for automated amino acid analysis. Elution is carried out at 2 ml/min with the following series of buffers. The initial buffer is 0.438 *M* pyridinium acetate of pH 7.5. After 100 ml of the initial buffer have passed through the column, this buffer is immediately replaced with a second buffer, consisting of 0.5 *M* pyridinium acetate of pH 4.4. Elution times for polyamines in the above system were published by Dubin and Rosenthal. In the procedure reported recently by Morris, some slight changes in the elution buffer system were introduced. After application of the sample, the elution with the initial buffer was begun, which in this instance was 0.33 *M* pyridinium acetate of pH 5.7. After 100 ml of the initial buffer had passed through the column, the buffer was changed to 0.38 *M* pyridinium acetate buffer of pH 4.4 and elution continued at the same rate (*ca.* 15–30 ml/h). Various polyamines

TABLE 30.3
ELUTION TIMES OF POLYAMINES AND RELATED COMPOUNDS ON BIO-REX 70 (MORRIS *et al.*)

| Compound | Time (min) | Compound | Time (min) |
|---|---|---|---|
| Arginine | 4 | Agmatine | 59 |
| Putrescine | 25 | Spermidine | 64 |
| 1,3-Diaminopropane | 27 | Iminobispropylamine | 64 |
| Cadaverine | 32 | Acetylspermine | 68 |
| Acetylspermidine B | 34 | Spermine | 78 |
| Acetylspermidine A | 42 | | |

and their acetyl derivatives were tested, and their elution volumes are summarized in Table 30.3. All of the commonly occurring polyamines can be quantitatively separated, with the exception of the pairs 1,3-diaminopropane and putrescine, and cadaverine and acetylspemidine, if they are present in a single sample.

High salt concentrations disturb the above separations, and these techniques are therefore not directly suitable for analyzing such materials as tissue culture media, unless the sample is properly desalted by passing it through an ion-exchange desalting column prior to analysis or by a batch process with *n*-butanol.

The choice of different ion exchangers suitable for the separation of polyamines was further extended in the work of Holder and Bremer, who used Amberlite IRP-64 and Dowex 50-X8 for this purpose.

Elution of the Amberlite IRP-64 column is carried out first with a convex salt gradient obtained by continuous mixing of a potassium phosphate buffer of pH 7.1 (0.1 *M* with respect to phosphate) which is simultaneously 3.2 *M* with respect to potassium chloride with distilled water (column 0.9 × 30 cm; flow-rate 30 ml/h). The volume of distilled water used at the beginning of the separation is 200 ml. After 500 ml of the effluent have been collected, elution is continued with saturated potassium chloride solution in order to elute spermine (Fig. 30.3).

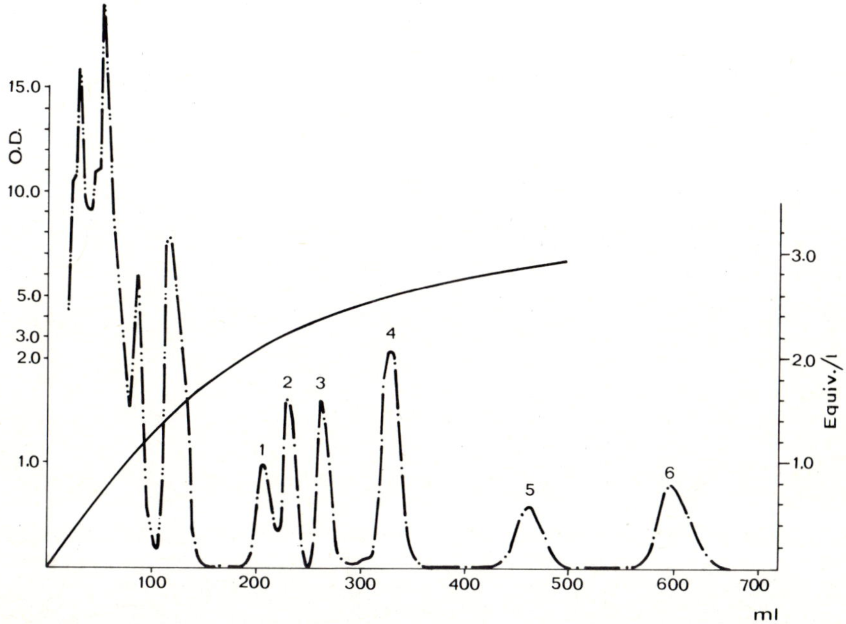


Fig. 30.3. Separation of a mixture of diamines on Amberlite IRP-64 column with potassium chloride buffer (Holder and Bremer). Gradient elution is indicated in the figure; after 500 ml of the mobile phase had passed through the column, gradient elution was replaced with saturated potassium chloride solution in order to eluate spermine. Peaks: 1, 2,2′-dithiobis(ethylamine); 2, cadaverine; 3, putrescine; 4, 1,3-diaminopropane; 5, spermidine; 6, spermine.

*References p. 655*

TABLE 30.4
ELUTION VOLUMES OF SOME AMINES (BLAU)

| Amine | Elution volume (ml) | Shape of peak | Recovery |
|---|---|---|---|
| Trimethylamine N-oxide | 12 | Very sharp | Quantitative |
| Creatine | 20 | Very sharp | Quantitative |
| Tetramethylammonium | 79 | Sharp | Quantitative |
| Diethylamine | 81 | Sharp | Quantitative |
| *n*-Amylamine | 109 | Sharp | Quantitative |
| Isoamylamine | 111 | Sharp | Quantitative |
| Trimethylamine | 118 | Sharp | Quantitative |
| Piperidine | 128 | Sharp | Quantitative |
| *n*-Butylamine | 130 | Sharp | Quantitative |
| Pyrrolidine | 148 | Sharp | Quantitative |
| *n*-Propylamine | 150 | Sharp | Quantitative |
| Dimethylamine | 162 | Sharp | Quantitative |
| Ethylamine | 184 | Sharp | Quantitative |
| Tyramine | 260 | Broad | 85% |
| Glucosamine | 270 | Broad | 85% |
| Methylamine | 280 | Broad | Quantitative |
| Canavanine | 290 | Very broad | 80% |
| Ethanolamine | 300 | Broad | 80% |
| Adrenaline | 310 | Very broad | Not determined |
| Arginine | 330 | Broad | 70% |
| 3-Hydroxytyramine | 430 | Very broad | 90% |
| Noradrenaline | 460 | Very broad | 70% |
| 5-Hydroxytryptamine | 465 | Very broad | 80% |
| Ammonia | 490 | Broad | Quantitative |
| pH 5.0 breakthrough | 600 | – | – |

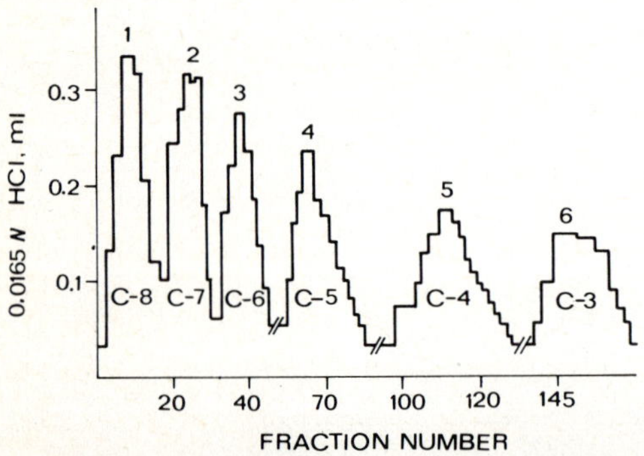


Fig. 30.4. Chromatographic separation of homologous series of volatile primary amines (Clayton and Strong). Solution containing *ca.* 0.05 mequiv. of each amine ($C_3$ to $C_8$) was introduced on to a Celite 545 column, 38 × 100 mm, operated at a flow-rate of 0.5–1.0 ml/min. Fractions of 2-ml volume were extracted into 5 ml of water and titrated with dilute hydrochloric acid. The stationary phase consisted of methanol, ethanol and water; light petroleum, equilibrated with stationary phase, was used as the mobile phase.

While the chromatographic separation on Amberlite IRP-64 is generally applicable, the separation of polyamines on Dowex 50-X8 is limited and is recommended preferably for the separation of simple mixtures such as 1,3-diaminopropane, putrescine and cadaverine mixtures.

Aliphatic diamines and some other amino compounds with free amino groups, including histamine, are easy to separate on Zeo-Karb 226 (Blau). Elution volumes found with this sorbent on a 0.8 × 112.5 cm column are given in Table 30.4. As with chromatography on Dowex 1-X2 (Holder and Bremer), Zeo-Karb 226 is also applicable mainly for concentrating dilute samples of mono- and polyamines.

Partition chromatography on Celite 545 (Clayton and Strong) is one of the rare procedures used in the separation of amines in which the principle of ion exchange is not applied. The result of such a separation is shown in Fig. 30.4.

## AROMATIC AMINES

The chromatographic separation of primary aromatic amines has been studied using a wide variety of sorbents such as Celite, silica gel and Teflon 6. The ion-exchange chromatography of this type of compound is related to both purifications and automated analytical separations, as described later (Tompsett). Recently, Lepri *et al.* described a highly efficient method involving the use of alginic acid and CM-cellulose. Separations on alginic acid were carried out in columns filled with alginic acid of 50–150 mesh. Both columns had a cross-section of 0.94 $cm^2$ and were filled with 4 g of ion exchanger. The columns were eluted with 0.1 *M* acetic acid or with 0.1 *M* hydrochloric acid (flow-rate 2 ml/min). Typical runs are shown in Figs. 30.5 and 30.6. With acetic acid as eluent, the expected differences were observed when comparing column chromatographic separations with thin-layer chromatography. The resolving power of the column separation is less than that of TLC for compounds with high $R_F$ values (above 0.5), while the column is more

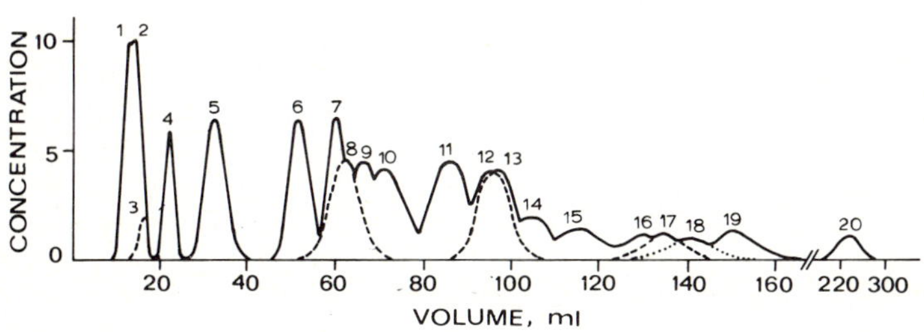


Fig. 30.5. Elution curves for aromatic amines on an alginic acid column with 1 *M* acetic acid as eluent (Lepri *et al.*). The aromatic amines in the effluent were detected with *p*-dimethylaminobenzaldehyde. The total quantity of amine eluted was determined spectrophotometrically after diazotization with nitrous acid followed by coupling with N-(1-naphthyl)ethylenediamine. The concentration is given in arbitrary units. Peaks: 1, sulphanilic, methanilic and orthanilic acids; 2, *o*-arsanilic acid; 3, *o*-nitroaniline, 4, *p*-nitroaniline; 5, *p*-arsanilic acid; 6, 4-aminosalicylic acid; 7, *o*-aminobenzoic acid; 8, sulphanilamide; 9, *p*-aminoacetophenone; 10, 5-aminosalicylic acid; 11, *p*-aminobenzoic acid; 12, *o*-chloroaniline; 13, *p*-aminohippuric acid; 14, *m*-nitroaniline; 15, aniline; 16, *o*- and *p*-toluidine; 17, *m*-aminobenzoic acid; 18, *o*-anisidine and *o*-aminophenol; 19, *m*- and *p*-aminophenol; 20, α- and β-naphthylamine.

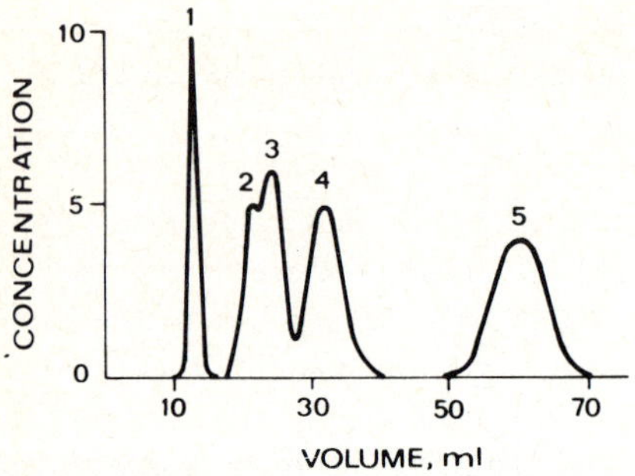


Fig. 30.6. Elution curves for aromatic amines on an alginic acid column with 0.1 *M* hydrochloric acid as eluent (Lepri *et al.*). The aromatic amines in the effluent were detected with *p*-dimethylaminobenzaldehyde. The total quantity of amine eluted was determined spectrophotometrically after diazotization with nitrous acid followed by coupling with N-(1-naphthyl)ethylenediamine. The concentration is given in arbitrary units. Peaks: 1, aniline; 2, *p*-aminodimethylaniline; 3, α- and β-napthylamide, *o*-phenylenediamine and N-phenyl-*p*-phenylenediamine; 4, *m*- and *p*-phenylenediamine; 5, benzidine.

effective than the flat-bed arrangement for compounds with low $R_F$ values. In this case, separations not forseeable from the $R_F$ values are sometimes possible, such as the separation of *o*-chloroaniline and *p*-aminobenzoic acid (corresponding $R_F$ values 0.13 and 0.12) from isomeric toluidines, aminophenols, *o*-anisidine and *p*-aminobenzoic acid, which move with $R_F$ 0.10. On the other hand, the separation of amines that contain a sulphonic group from *o*-arsanilic acid and from *o*-nitroaniline is not possible.

The possibilities of separating isomeric aromatic amines by these techniques were surveyed by Lepri *et al.* The great affinity of diamines can be reduced by using hydrochloric acid as the mobile phase. In this system, *p*-aminodimethylaniline is eluted together with *o*-phenylenediamine. The latter is also incompletely separated from its *m*- and *p*-isomers. The separation with hydrochloric acid as the mobile phase is particularly

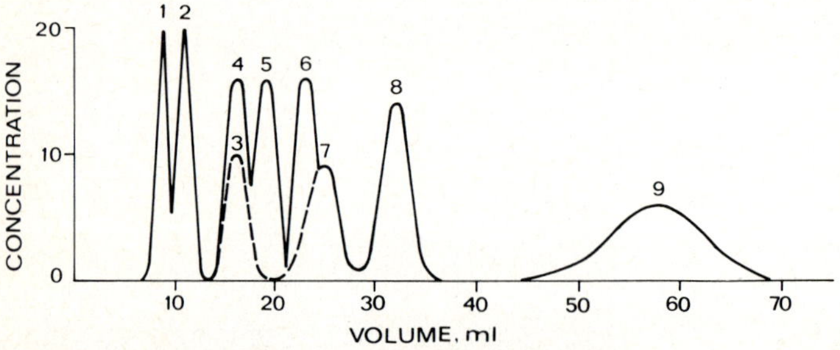


Fig. 30.7. Elution curves for aromatic amines on a carboxylmethylcellulose column with water as eluent (Lepri *et al.*). The aromatic amines in the effluent were detected with *p*-dimethylaminobenzaldehyde. The total quantity of amine eluted was determined spectrophotometrically after diazotization with nitrous acid followed by coupling with N-(1-napthyl)ethylenediamine. The concentration is given in arbitrary units. Peaks: 1, sulphanilic, methanilic and orthanilic acids; 2, *o*- and *p*-arsanilic acids; 3, *o*-nitroaniline and *o*-aminobenzoic acid; 4, 4-aminosalicylic acid and *p*-nitroaniline; 5, sulphonilamide and *p*-aminohippuric acid; 6, *p*-aminoacetophenone; 7, *p*-aminobenzoic acid; 8, *m*-nitroaniline; 9, *m*-aminobenzoic acid.

suitable for the separation of naphthylamines. Instead of the above solvents, 1 *M* monochloroacetic acid in water and in 50% isopropanol were used, but the separating power of these mobile phases is very much lower.

Results achieved in the separation of aromatic amines on a CM-cellulose column are presented in Fig. 30.7 (identical column size as with alginic acid; CM-cellulose in the acid form obtained by treatment of the sodium salt of the ion exchanger with 1 *M* hydrochloric acid and subsequent washing with distilled water until the chloride ions disappeared; flow-rate 1 ml/min). As the affinity of CM-cellulose for aromatic amines is lower with respect to alginic acid, water was used as the mobile phase. This system is capable of separating *m*-aminobenzoic acid from other amines and in particular from *p*-aminobenzoic acid and *m*-nitroaniline. In a similar manner to alginic acid, it is also possible to separate all isomers of aminobenzoic acid. Amines that are strongly bound to CM-cellulose can be eluted with 1 *M* acetic acid. In this solvent, benzidine is separated from *m*- and *p*-phenylenediamine. Reversed-phase chromatography of aromatic amines can be carried out on cyclohexane-loaded Teflon 6 (Hedrick).

## AROMATIC AMINES AND ALIPHATIC POLYAMINES IN MIXTURES

Amberlite CG-50, Type 2, can be recommended for the separation of complex mixtures of aliphatic polyamines and aromatic amines, since both of these types of compounds are eluted from this ion-exchange resin in the same pH range. The recommended procedure was described by Perry and Schroeder. Hirs' purification procedure has been recommended for the resin used. Before packing the column, the resin bed is suspended in a pyridine–acetate buffer of pH 6.32 (0.1 *N*) and rinsed several times before use in order to remove fines and to condition the resin. Most of the amines tested were detected by the conventional ninhydrin procedure. Some of the amines, however, do not give a positive reaction with ninhydrin and must be assayed by another suitable procedure after the effluent from the column has been split into two separate streams, one feeding the ninhydrin line and the other supplying a separate fraction collector (Table 30.5).

## TRYPTOPHAN METABOLITES

Historically, the first liquid column separation of compounds related to tryptophan metabolism was carried out on a silica gel column eluted with a series of mobile phases based on *n*-heptane (Powell).

Also, the separation of these compounds on a molecular sieve such as Sephadex G-25 or G-10 gives only a partial resolution, as reported by Iskric and Keglevic and Schlossberger *et al.* Nowadays, ion-exchange separations clearly predominate. Presumably the use of ion-exchange celluloses in this field is now the fashion, but equally good results are obtained with QAE-Sephadex A-25 and Amberlite IR-120. Dowex chromatography is currently less developed.

The method in which QAE-Sephadex A-25 is used has been developed by Bakri and Carlson; the sodium chloride concentrations at the sudden buffer changes are indicated in

*References p.655*

TABLE 30.5
ELUTION VOLUMES OF AMINES CHROMATOGRAPHED ON AMBERLITE CG-50 (PERRY AND SCHROEDER)

Authentic compounds were chromatographed in mixtures on Amberlite CG-50 columns 45 cm in length and 0.9–1.0 cm in diameter at a flow-rate of 10 ml/h and a temperature of 40°C. Chromatograms were developed with 0.1 *N* pyridine acetate buffer (pH 6.32) for the first 250 ml, and thereafter with 0.2 *N* pyridine acetate buffer (pH 6.12).

| Compound | Range of elution (ml) | Elution peak* (ml) | Compound | Range of elution (ml) | Elution peak* (ml) |
|---|---|---|---|---|---|
| Histidine | 25–31 | 28 | 3-Hydroxy-4-methoxyphenylethylamine | 145–163 | |
| 1-Methylhistidine | 25–31 | 28 | Putrescine | 140–171 | 159 |
| Arginine | 31–42 | 35 | *p*-Hydroxybenzylamine | 151–171 | 162 |
| Ethanolamine | 32–40 | 36 | Cadaverine | 163–183 | |
| Ammonia | 33–42 | 38 | Dopamine | 163–190 | |
| Ethylamine | 38–46 | 42 | Benzylamine | 169–189 | 178 |
| Pyrrolidine | 49–58 | 54 | *p*-Tyramine | 170–196 | 186 |
| N-Acetylhistamine | 49–65 | | 3-Ethoxy-4-hydroxybenzylamine | 177–193 | |
| Pyridoxamine | 68–79 | 74 | *m*-Tyramine | 183–202 | |
| N-Methylmetanephrine | 78–89 | | *p*-Methoxybenzylamine | 192–215 | 203 |
| Metanephrine | 102–115 | 109 | Histamine | 202–225 | 214 |
| Epinephrine | 101–124 | | Bufotenin | 205–235 | |
| Normetanephrine | 112–130 | 121 | Phenylethylamine | 228–254 | 240 |
| 1-Methylhistamine | 112–130 | 121 | *o*-Tyramine | 218–262 | 241 |
| Norepinephrine | 116–132 | | *p*-Methoxyphenylethylamine | 228–261 | 244 |
| Synephrine | 118–135 | | Kynuramine | 229–261 | 244 |
| Isoamylamine | 119–136 | 128 | 2,2′-Dithiobis(ethylamine) | 277–297 | 289 |
| Mescaline | 124–142 | | Serotonin | 282–303 | |
| 3,4-Dimethoxybenzylamine | 127–142 | | Agmatine | 365–321 | 312 |
| Octopamine | 132–147 | | N,N-Dimethyltryptamine | 310–335 | |
| 3-Methoxy-4-hydroxybenzylamine | 134–150 | | 5-Methoxytryptamine | 328–352 | 339 |
| Epinine | 135–151 | | Tryptamine | 380–415 | 397 |
| 3,4-Dimethoxyphenylethylamine | 140–154 | | Spermidine | 408–466 | 438 |
| 3-Methoxytyramine | 141–158 | | 5-Methyltryptamine | 465–505 | 483 |

*Elution peaks were not obtained for a number of amines giving no colour or weak colours with ninhydrin.

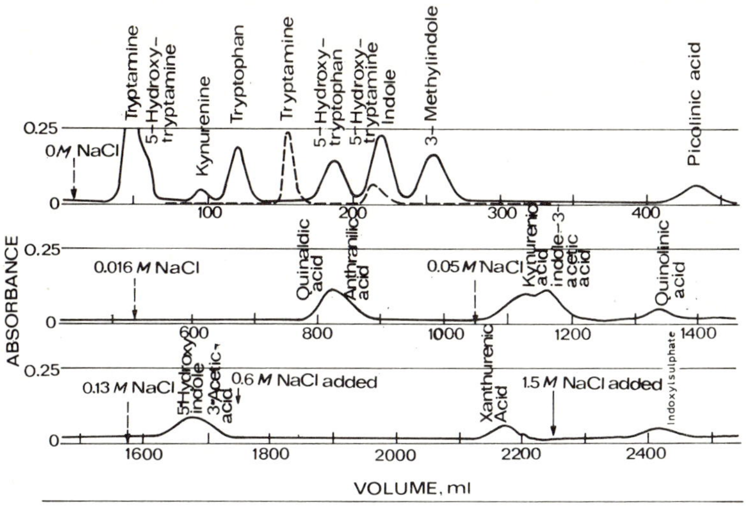


Fig. 30.8. Elution pattern obtained for tryptophan metabolites from a 95 × 0.94 cm QAE-Sephadex A-25 ($Cl^-$) column (solid line) and a 45 × 0.94 cm CM-cellulose ($Na^+$) column (broken line) (Bakri and Carlson). Buffer: 0.05 *M* Tris–hydrochloric acid, pH 7.9. Complex gradient for the QAE-Sephadex A-25 column: 300 ml each of 0, 0.01, 0, 0.076, 0, 0.36, 0 and 0.60 *M* sodium chloride. Temperature: 4°C. The absorbance at 280 nm is plotted against elution volume.

Fig. 30.8, in which the overall pattern of separation can also be seen. The use of an additional CM-cellulose column allows the identification of tryptamine.

According to Arend *et al.*, a mixture of tryptophan metabolites, including anthranilic acid glucuronide, *o*-aminohippuric acid, acetylkynurenine, kynurenine and indoxyl sulphate, can be analyzed in the following way. The metabolites are eluted from a Dowex 50 ion-exchange column with successive washes with hydrochloric acid of increasing concentration. Five acidic fractions are obtained as follows: 0.1 *N*, indoxyl sulphuric acid; 0.5 *N*, anthranilic acid glucuronide; 1.0 *N*, *o*-aminohippuric acid; 2.4 *N*, acetylkynurenine; and 5 *N*, kynurenine. A procedure for separating this type of compounds was described recently by Chen and Gholson, in which DEAE-cellulose was used as an ion exchanger in either the amine or formate form. A mixture of tryptophan metabolites (300 μg each) was loaded on to the column ($HCOO^-$; 1.2 × 30 cm) and the column was then eluted with 80 ml of 0.001 *M* triethylamine–formate (TEA-F) buffer of pH 4.0. Then elution was continued with a gradient formed by placing 250 ml of 0.001 *M* TEA-F of pH 4.0 into a vessel into which 0.1 *M* TEA-F buffer of pH 4.0 was directed from an equal-sized (250 ml) beaker. The first two peaks escaping from the column (Fig. 30.9) were lyophilized and subjected to the second separation step. The second column was packed with DEAE-cellulose (amine form) and eluted with 140 ml of 0.001 *M* TEA-F buffer of pH 8; then elution was continued with a gradient obtained from equal volumes (200 ml) of 0.001 *M* TEA-F buffer of pH 8.0 and 0.05 *M* TEA-F buffer of pH 8.0. During the non-gradient elution 2-ml fractions were collected, while after the gradient has been introduced 4-ml fractions were taken. A typical run on the alkaline column (amine form) is presented in Fig. 30.10.

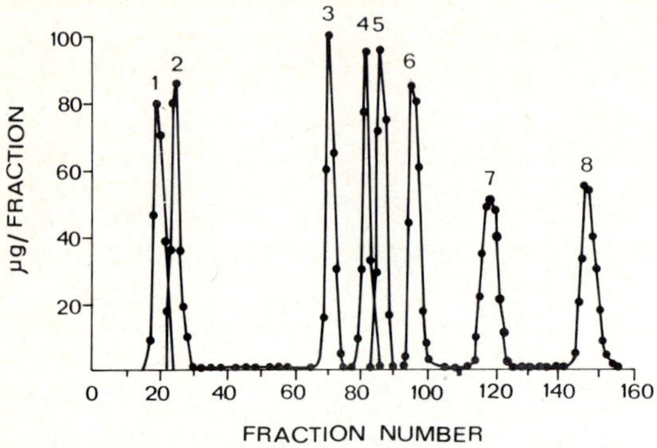


Fig. 30.9. Separation of tryptophan metabolites on the first DEAE-cellulose column ($HCOO^-$ form) (Chen and Gholson). Peaks: 1, tryptophan, tryptamine, 5-hydroxytryptamine, 5-hydroxytryptophan, kynurenine, indole-3-acetonitrile and urea; 2, indole-3-carboxaldehyde; 3, 3-hydroxykynurenine; 4, indole-3-acetic acid; 5, indole-3-aceturic acid; 6, 5-hydroxyindoleacetic acid; 7, kynurenic acid; 8, xanthurenic acid. Some peaks overlap slightly, as shown, beacuse different reagents were used to determine the various compounds.

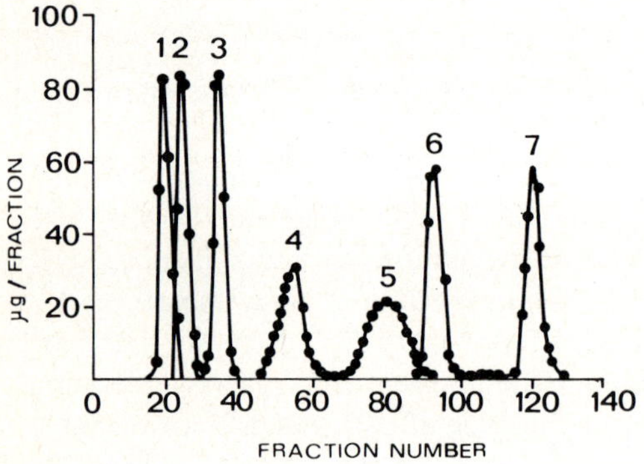


Fig. 30.10. Separation of tryptophan metabolites on the second DEAE-cellulose column (amine form) (Chen and Gholson). Peaks 1 and 2 from the first column were combined and applied to the second column. Peaks: 1, tryptamine and urea; 2, indole-3-acetonitrile; 3, indole-3-carboxaldehyde; 4, kynurenine; 5, 5-hydroxytryptamine; 6, tryptophan; 7, 5-hydroxytryptophan. Some peaks overlap slightly, as shown, because different reagents were used to determine the various compounds.

5-Hydroxytryptophan, 5-hydroxytryptamine and 5-hydroxyindoleacetic acid were separated on Dowex 1 ($CH_3COO^-$) by Nishino *et al.* The column used was 7 × 1 cm in size; elution was carried out step-wise with 100 ml of water, 100 ml of 0.01 *N* acetic acid and 100 ml of 6 *N* acetic acid.

Chromatography on Amberlite IR-120 is the oldest of the various ion-exchange techniques, and it still offers very good separations (Benassi *et al.*) (Fig. 30.11).

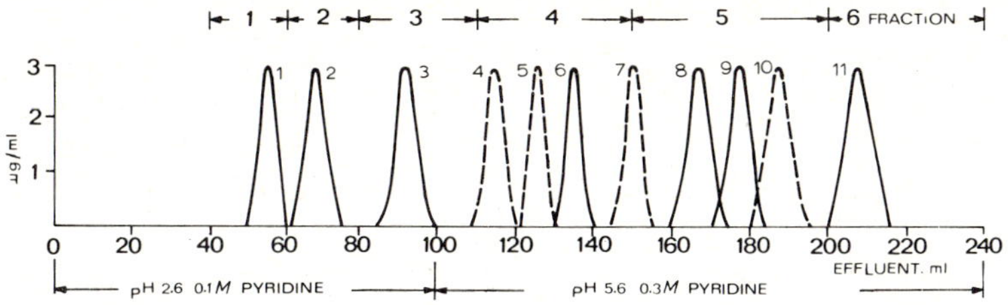

Fig. 30.11. Chromatographic fractionation pattern of tryptophan metabolites on a column of Amberlite IR-120 (pyridine), 28 × 0.9 cm (Benassi *et al.*). The temperature was maintained at 33 ±0.1°C and the flow-rate was adjusted to 12 ml/h with volatile formic acid–pyridine buffer. The concentration values, (μg/ml) were calculated from fluorimetric or colorimetric readings on the paper chromatographic spots eluted after column fractionation. The broken lines show the elution peaks of metabolites usually absent in the urine of normal subjects. Peaks: 1, xanthurenic acid 8-methyl ether; 2, kynurenic acid; 3, xanthurenic acid; 4, N-acetylkynurenine; 5, N-acetyl-3-hydroxykynurenine; 6, *o*-aminohippuric acid; 7, anthranilic acid; 8, kynurenine; 9, 3-hydroxyanthranilic acid; 10, 3-methoxy anthranilic acid; 11, 3-hydroxykynurenine.

## QUATERNARY AMMONIUM COMPOUNDS AND AMINO ALCOHOLS

Since ion-exchange chromatography clearly predominates in the separation of different types of compounds that contain an amino group, it is not surprising that also betaine and quaternary ammonium compounds, which occur in the natural material associated with the oxidation of choline, have been subjected to separations of this type. Important contributions were made by Christianson *et al.*, Gourley *et al.* and Niemann. In the procedure reported by Speed and Richardson, Amberlite CG-50 is used. The flow-rate used ranged from 60 to 80 ml/h and the column was operated at room temperature. A typical run is shown in Fig. 30.12.

The pH 5.3 phosphate–citrate buffer can be also used for separation on Zeo-Karb 226 as published by Cromwell and Richardson.

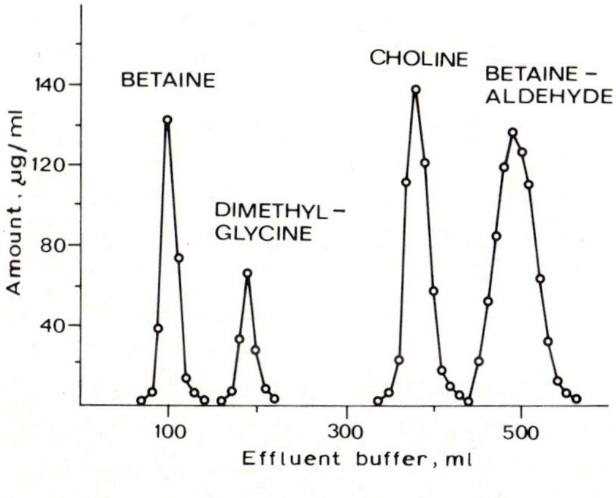


Fig. 30.12. Separation of betaine, dimethylglycine, choline and betainealdehyde on a 125 × 1.2 cm column of Amberlite CG-50 buffered at pH 7.3 and eluted with phosphate–citrate buffer of pH 5.3 (Speed and Richardson).

*References p.655*

For the separation of fatty alcoholamides, elution of silica gel with a stepwise gradient of ethanol–chloroform and methanol–chloroform (with an increasing concentration of the alcoholic component) was used by Hejna and Daly. In this manner, monoethanolamine and diethanolamine were separated. The separation of sphingosine, dihydrosphingosine and phytosphingosine was achieved on 30 × 2.2 cm column containing 50 g of silica gel (0.05–0.2 mm) which had been suspended in chloroform–methanol (1:1) packed by gravity and then washed with chloroform until transparent (Barenholz and Gatt). A gradient of increasing ammoniacal methanol in chloroform was used as the mobile phase. The lower mixing chamber of the gradient system contained 300 ml of chloroform–methanol–2 *M* ammonia solution (90:10:1) and the upper reservoir contained 600 ml of chloroform–methanol–2 *M* ammonia solution (50:50:5). When all of the 600 ml had been delivered from the upper reservoir, the separation was continued with 500 ml of chloroform–methanol–2 *M* ammonia solution (30:70:7). If fractions with a volume of 10 ml were collected, sphingosine appeared in fractions 43–47, fractions 48–58 contained dihydrosphingosine and fractions 63–67 were pure phytosphingosine. Lipids related to sphingosine (sphingosine esters) are dealt with in the chapter on lipids.

## BIOGENIC AMINES

In principle, the isolation of catecholamines from biological material is usually carried out by ion-exchange chromatography in a large series of columns suitable for many parallel samples. Quantitation after chromatography is a very complex operation which has been automated in two different ways. Both methods of evaluation (the ethylenediamine dihydrochloride method and trihydroxyindole method) have been subjected to a number of re-investigations and criticisms, the description of which is beyond the scope of this book. In this section we have limited ourselves to those methods in which individual catecholamines are chromatographically separated. Methods in which liquid column separations were used only for purification purposes have been omitted.

There are not many methods that are suitable for the quantitative separation of the biogenic amines, noradrenaline, adrenaline, dopamine, 5-hydroxytryptamine, histamine and their metabolites, in a single small sample of tissue after a single extraction and purification procedure. Recently, a procedure was reported in which *n*-butanol is used in the organic extraction of catecholamines, 5-hydroxytryptamine and histamine from a small sample of tissue (Sadavongvidad). Atack and Magnusson developed a column chromatographic procedure which permits the total amount of each amine to be concentrated into small fractions. Noradrenaline and adrenaline have been separated from dopamine on a strong cation-exchange column 50 × 4.2 mm in size. The resin used was Dowex 50W-X4 ($Na^+$), 200–400 mesh (Bertler *et al.*, Carlsson and Lindqvist). Noradrenaline, together with adrenaline, is eluted in the first 8 ml and dopamine in the following 12 ml of 1 *N* hydrochloric acid. Adopting the procedure of Green and Erickson and Kahlson *et al.*, Atack and Magnusson were able to elute histamine with 5 ml of dilute hydrochloric acid after eluting both catecholamines. By using a large volume of the mobile phase (up to 20 ml), 5-hydroxytryptamine could be eluted with additional portions of 4–6 *N* aqueous hydrochloric acid or 0.01 *N* sodium hydroxide. The application of an alkaline mobile phase was introduced by Wiegand and Scherfling. The volume of all the fractions could be minimized by using organic solvents; thus elution with 3 *N* ethanolic hydrochloric acid decreases the volumes of the fractions to 4 ml (Schildkraut *et al.*).

Except adrenaline and noradrenaline, all of the other amines are separated by the above methods. According to Atack and Magnusson, the elution of noradrenaline, adrenaline and dopamine is also greatly facilitated by the use of other organic solvents, while the elution of 5-hydroxytryptamine is virtually unaffected.

The procedure described by Atack and Magnusson can be briefly summarized as follows. The sample is loaded into the column at pH 2.5 and the corresponding amino acids are adsorbed together with catecholamines (Bertler *et al.*, Kahlson *et al.*, Wiegand and Scherfling). Elution is carried out with an organic mobile phase as indicated in Fig. 30.13. 5-Hydroxyindoleacetic acid may interfere in the assay of 5-hydroxytryptophan. Aqueous methanol (60%) elutes 5-hydroxyindoleacetic acid and 0.1 *M* phosphate buffer of pH 6.5 can be used for the elution of a mixed band of 5-hydroxytryptamine and dihydroxyphenylalanine. Histidine is usually eluted together with the other amino acids and the subsequent elution of amines is not affected by the use of organic solvents and buffer mobile phases (Atack and Magnusson). The volume of the 5-hydroxytryptamine fraction can be reduced to 3.5 ml by eluting with 1.8 *N* hydrochloric acid–ethylene glycol monoethyl ether (ethyl Cellosolve) (50%), while still permitting the subsequent separation of histamine.

Amberlite IRC-50 gives a good separation of adrenaline, noradrenaline, hydroxytryptamine and dopa (Fig. 30.14, Kirshner and Goodall).

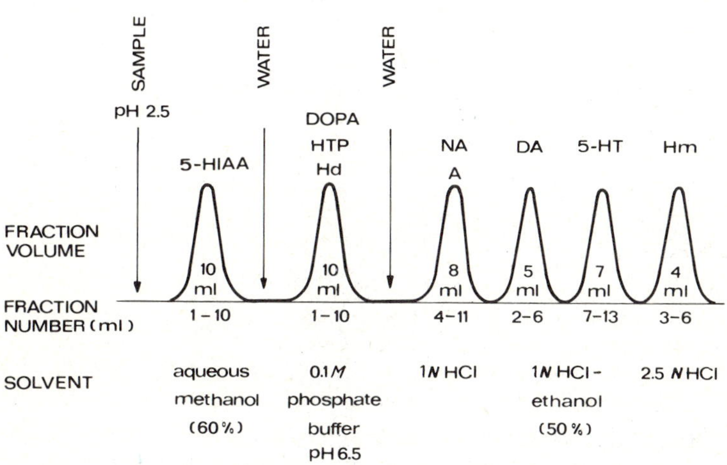


Fig. 30.13. Order of elution of noradrenaline (NA) (together with adrenaline, A), dopamine (DA), 5-hydroxytryptamine (5-HT) and histamine (Hm) and their respective precursors (dihydroxyphenylalanine, DOPA; 5-hydroxytryptophan, HTP; and histidine, Hd) and 5-hydroxyindoleacetic acid (5-HIAA), from a strong cation-exchange resin column (dimensions 72 mm in buffer by 4.0 mm I.D. Dowex 50W-X4 ($Na^+$), 200–400 mesh) (Atack and Magnusson). The volume of 5-HT eluate can be reduced to 3.5 ml by eluting with 1.8 *N* hydrochloric acid–ethylene glycol monoethyl ether (ethyl Cellosolve) (50%), while still permitting the subsequent, separate elution of Hm.

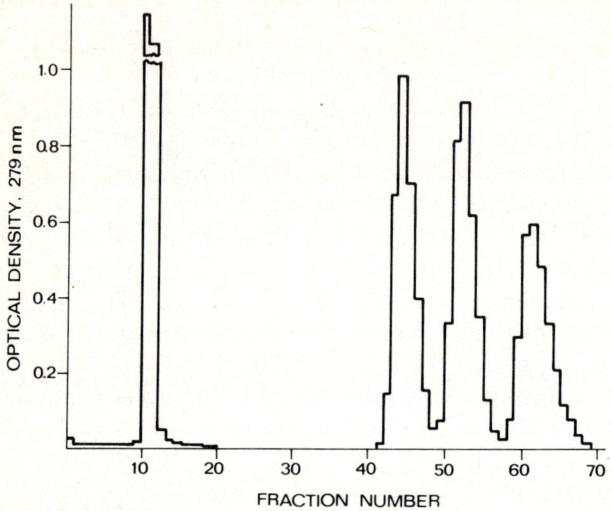


Fig. 30.14. Separation of adrenaline, noradrenaline, hydroxytyramine and dopa on Amberlite IRC-50 (Krishner and Goodall). A 1-ml volume of a solution containing 400 μg of each of the compounds was chromatographed. Column 30 × 0.9 cm, equilibrated with 0.2 *M* ammonium acetate buffer at pH 6.1. Elution with 0.4 *M* ammonium acetate buffer at pH 5.0, starting at fraction 1. Fraction size 1.5 ml. The compounds in order to emergence are dopa, adrenaline, noradrenaline and hydroxytyramine.

Gradient elution chromatography on ion-exchange cellulose (Whatman P-11, cellulose phosphate) also proved to be a very satisfactory method for the separation of sympatomimetic amines. According to the procedure described by Merrills and Farrier, columns 30 × 0.5 cm in size are used. The column is equilibrated with 0.05 *M* ammonium acetate

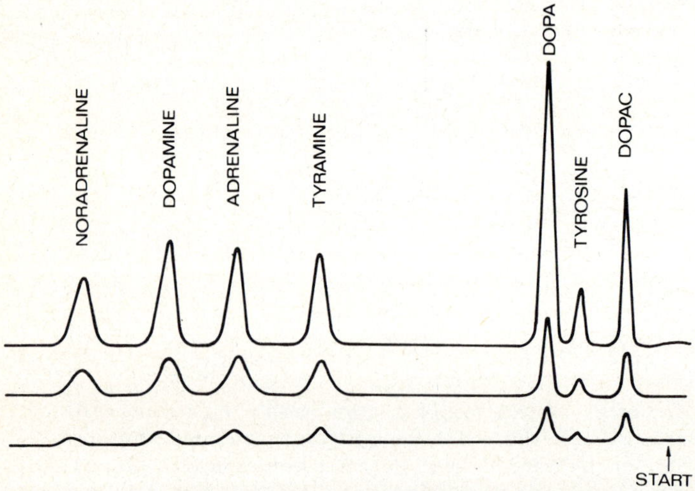


Fig. 30.15. Recorder tracings obtained during chromatographic separation of various amounts of dopac (3,4-dihydroxyphenylacetic acid), L-tyrosine, dopa (3,4-dihydroxyphenylalanine), tyramine, adrenaline, dopamine and noradrenaline (1, 2.5 or 5 μg of each) (Merrills and Farrier).

of pH 6.0 and an exponential salt gradient, formed by pumping a solution of 0.5 *M* ammonium acetate in 20% propanol adjusted to pH 6.0 into a mixing vessel containing 25 ml of 0.05 *M* ammonium acetate at the rate of 0.23 ml/min, is introduced. The mixture is pumped into the column at the same rate. The main advantage of this type of chromatography is that the effluent is suitable for direct scintillation counting, a procedure which, with low concentrations of catecholamines originating in tissues from labelled precursors, can hardly be avoided. In order to identify the substances eluted, it is always necessary to add appropriate carriers, which are detected either by their optical density at 280 nm or by a suitable colorimetric reaction (Merrills). In a later paper Merrills and Offerman reported another detection procedure based on the possibility of recording anodic decomposition potentials of catechol or phenolic substances. Fig. 30.15 gives an example of a separation recorded by this procedure.

In a later modification of this procedure, Deyl *et al.* succeeded in separating adrenaline, noradrenaline, isopropylnoradrenaline and its O-methyl derivative.

For the separation the synthetic catecholamine and its O-methyl derivative from epinephrine and norepinephrine, two columns (40 × 0.6 cm) combined in series were used. These columns were filled up to 35 cm with Whatman CP-11 cellulose phosphate (column No. I) and Whatman CM-cellulose CM-32 (column No. II). The columns were eluted with a linear gradient of ammonium acetate buffer with a concentration change from 0.05 to 0.25 *M*. The pH of the buffer was adjusted previously to 6.1; the flow-rate in both columns was less than 0.8 ml/min. Fractions of 2 ml were collected. The columns were loaded with 0.5 ml of a sample obtained after either purification procedure. The elution of the column was stopped after 100 ml of the eluent had passed through the system. Under these conditions, O-methylisopropylnoradrenaline is eluted in fractions 3–7 and isopropylnoradrenaline in fractions 10–15 (see Fig. 30.16).

Some catechol derivatives can be effectively separated by adsorption on alumina (Drell).

Recently, Routh *et al.* introduced a complex separation technique which combines alumina and two ion exchangers and allows the separation of not only epinephrine and norepinephrine, but also a number of compounds related to catecholamine metabolism such as dopamine, dihydroxyphenylalanine, 4-hydroxy-3-methoxyphenylacetic acid (homovanillic acid), 3,4-dihydroxyphenylacetic acid, 4-hydroxy-3-methoxymandelic acid (vanilmandelic acid) and 3,4-dihydroxymandelic acid. The separation is carried out on a series of three columns, of which the first is packed with Bio-Rad test column packing (catecholamine test kit), the second with Amberlite CG-400 ($Cl^-$), 100–200 mesh, and the third with neutral alumina, Brockman activity I, 80–200 mesh. The dimensions of each column is 5.0 × 0.7 cm.

Flow-rates in this separation are not important and flow is effected by gravity. The procedure is as follows. A 5-ml volume of urine is mixed with 14 ml of $Na_2$EDTA of concentration 1 g/100 ml, adjusted to pH 6.5 with 1 *M* sodium hydroxide solution, and chromatographed on the first column. The column is eluted with distilled water (5 ml) and the combined effluents are used for the following separation step. The first column is further eluted with 10 ml of boric acid of concentration 4 g/100 ml, which brings metanephrine, epinephrine and norepinephrine into solution. The combined effluents from the first column before the elution with boric acid are passed through the second column of

*References p. 655*

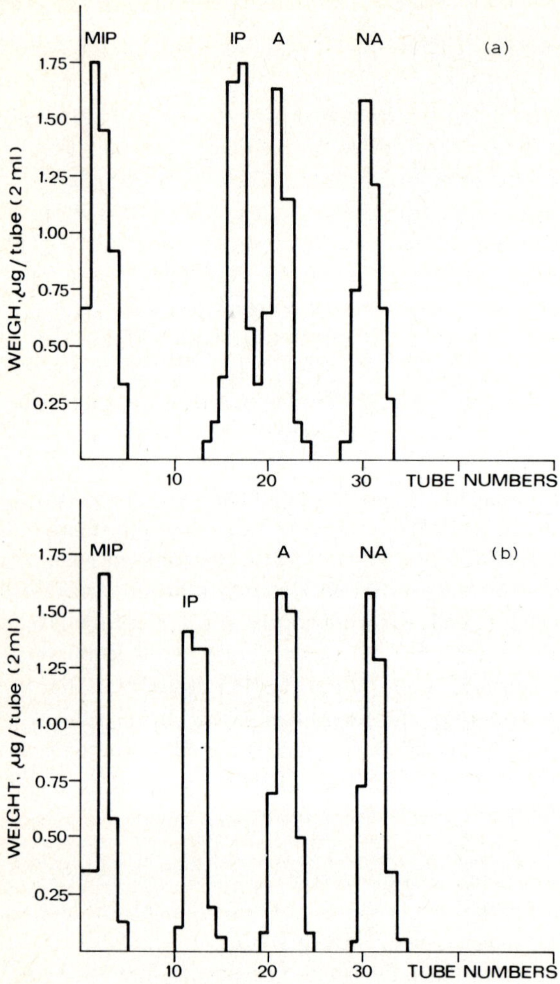


Fig. 30.16. Chromatographic profile of the mixture of O-methylisopropylnoradrenaline (MIP), isopropylnoradrenaline (IP), adrenaline (A) and noradrenaline (NA) on cellulose phosphate (a) and on the confined system of cellulose phosphate and CM-cellulose (b) (Deyl *et al.*).

Amberlite CG-400, which, after draining, is eluted with 5 ml of distilled water. All of the effluent is combined and used for dihydroxyphenylalanine quantitation. Then the second column is eluted with 25 ml of 1 *M* sodium chloride solution, the first 5 ml of the eluate being discarded and the remainder saved for further fractionation. A 5-ml aliquot of the eluate from the second column is mixed with 0.5 ml of 1 *M* sodium acetate of pH 6.5 and 1 ml 2.0 *M* sodium acetate and applied to the third column. This column is eluted with 15 ml of distilled water and the effluent used for homovanillic acid quantitation.

## REFERENCES

Arend, R. A., Leklem, J. E. and Brown, R. R., *Advan. Autom. Anal., Technicon Int. Congr., 1969*, 2 (1970) 195; *C.A.*, 73 (1970) 32189s.
Atack, C. V. and Magnusson, T., *J. Pharm. Pharmacol.*, 22 (1970) 625.
Bakri, M. and Carlson, J. R., *Anal. Biochem.*, 34 (1970) 46.
Barenholz, Y. and Gatt, S., *Biochim. Biophys. Acta*, 152 (1968) 790.
Benassi, C. A., Veronese, F. M. and De Antoni, A., *Clin. Chim. Acta*, 17 (1967) 383.
Bertler, Å., Carlsson, A. and Rosengren, E., *Clin. Chim. Acta*, 44 (1958) 273.
Blau, K., *Biochem. J.*, 80 (1961) 193.
Carlsson, A. and Lindqvist, M., *Biochem. J.*, 54 (1962) 87.
Chen, N. C. and Gholson, R. K., *Anal. Biochem.*, 47 (1972) 139.
Christianson, D. D., Wall, J. S., Cavins, J. F. and Dimler, R. J., *J. Chromatogr.*, 10 (1963) 43.
Clayton, R. A. and Strong, F. F., *Anal. Chem.*, 26 (1954) 579.
Cromwell, B. T. and Richardson, M., *Phytochemistry*, 5 (1966) 735.
Deyl, Z., Pilný, J. and Rosmus, J., *J. Chromatogr.*, 53 (1970) 575.
Drell, W., *Anal. Biochem.*, 34 (1970) 142.
Dubin, D. T. and Rosenthal, S. M., *J. Biol. Chem.*, 235 (1960) 776.
Gourley, W. K., Haas, C. D. and Bakerman, A., *Anal. Biochem.*, 19 (1967) 197.
Green, H. and Erickson, R. W., *Int. J. Neuropharmacol.*, 3 (1964) 315.
Hatano, H., Sumizu, K., Rokushika, S. and Murakami, F., *Anal. Biochem.*, 35 (1970) 377.
Hedrick, C. E., *Anal. Chem.*, 37 (1965) 1044.
Hejna, J. J. and Daly, D., *J. Soc. Cosmet. Chem.*, 21 (1970) 107.
Holder, S. and Bremer, H. J., *J. Chromatogr.*, 25 (1966) 48.
Iskric, S. and Keglevic, D., *Anal. Biochem.*, 7 (1964) 297.
Kahlson, C., Rosengren, E. and Thunberg, R., *J. Physiol. (London)*, 169 (1963) 467.
Kirshner, N. and Goodall, McC., *J. Biol. Chem.*, 226 (1957) 207.
Lepri, L., Desideri, P. G., Coas, V. and Cozzi, D., *J. Chromatogr.*, 49 (1970) 239.
Merrills, R. J., *Autom. Anal. Chem., Technicon Symp., 1965*, Technicon Instruments, Chertsey, Great Britain, 1965.
Merrills, R. J. and Farrier, J. P., *Anal. Biochem.*, 21 (1967) 475.
Merrills, R. J. and Offerman, J., *Biochem. J.*, 99 (1966) 538.
Miyagi, T. and Ando, S., *Annu. Rep. Inst. Food Microbiol., Chiba Univ.*, 6 (1953) 93.
Morris, D. R., *Methods Enzymol.*, 17 (1971) 850.
Morris, D. R., Koffron, K. A. and Okstein, Ch., *Anal. Biochem.*, 30 (1969) 449.
Niemann, A., *J. Chromatogr.*, 9 (1962) 117.
Nishino, M., Moguchi, T. and Kido, R., *Anal. Biochem.*, 45 (1972) 314.
Perry, T. L. and Schroeder, W. A., *J. Chromatogr.*, 12 (1963) 358.
Powell, L. E., *Nature (London)*, 200 (1963) 79.
Rosenthal, S. M. and Tabor, C. W., *J. Pharmacol. Exp. Ther.*, 116 (1956) 131.
Routh, J. I., Bannov, R. E., Fincham, R. W. and Stoll, J. L., *Clin. Chem.*, 17 (1971) 867.
Sadavongvidad, C., *Brit. J. Pharmacol.*, 38 (1970) 353.
Schildkraut, J., Schanberg, S. M., Breese, G. R. and Kopin, I. J., *Biochem. Pharmacol.*, 18 (1969) 1971.
Schlossberger, H. G., Kuch, H. and Buhrow, I., *Hoppe-Seyler's Z. Physiol. Chem.*, 333 (1963) 152.
Speed, D. and Richardson, M., *J. Chromatogr.*, 35 (1968) 497.
Tabor, H., Rosenthal, S. M. and Tabor, C. W., *J. Biol. Chem.*, 233 (1958) 907.
Tompsett, S. L., *Anal. Chim. Acta*, 21 (1959) 535.
Wiegand, R. G. and Scherfling, E., *J. Neurochem.*, 9 (1962) 113.
Yoshioka, M., Ohara, A., Kondo, H. and Kanazawa, H., *Chem. Pharm. Bull.*, 17 (1969) 1276.

*Chapter 31*

# Other non-heterocyclic nitrogen compounds

J. CHURÁČEK

## CONTENTS



## INTRODUCTION

In this chapter, the liquid chromatography of nitro compounds, urea and its derivatives, guanidines, nitrosamines, amides, azo compounds and aromatic hydrazo compounds is described. The chromatography of amides is mentioned only briefly because from the practical point of view the chromatographic separation of peptides and their mixtures is most important and therefore a special chapter is devoted to this aspect (Chapter 34). For the other compounds mentioned, the main aim of liquid column chromatography is their separation from other types of compounds. All modern chromatographic techniques have good prospects of being applied successfully in this field, but little attention has been devoted to them so far. An interesting technique was used for nitro compounds; some of them can be synthesized directly on the column on which they are then separated chromatographically from other components of the reaction mixture.

## NITRO COMPOUNDS

A typical example of the isolation of nitro compounds from industrial mixtures was described by Landram *et al.* Chromatographic separation was used instead of the usual extraction procedures, which represents an important advance in work with explosives. This separation procedure was used satisfactorily for a number of double base propellants containing nitroglycerine, triacetin, 2-nitrodiphenylamine, resorcinol, ammonium perchlorate, aluminium, 2,4,6,8-cyclotetramethylenetriamine and nitrocellulose. The procedure is generally applicable to different types of propellants and it is not necessarily restricted to those which bear a large number of nitro groups. Both Chromosorb T (PTFE) and silica gel can be used for column packing in these instances.

The principle of dry column chromatographic extraction should prove useful also with other non-propellant polymers. The main advantage of this procedure is the decrease in the operating time: in the example described above, the Soxhlet extraction took 1–4 days while the chromatographic procedure took 1 h. In large-scale operations, the reduced consumption of solvents in chromatographic procedures is also important.

In the analytical field, ion-exchange chromatography is the prevailing method for the separation of nitro compounds. This method permits the separation of a series of nitroalkanes and nitro-aromatic compounds. Polarography is mainly used for detection (Kemula); a special chromatopolarographic apparatus was developed by Kemula and Brzozowski for this purpose. The limiting diffusion current produced by organic substances is measured on a mercury drop electrode at a constant input potential (−1 V).

Kemula and Brzozowski also described the separation of some nitroalcohols and nitrobenzoic acids by salting-out chromatography on a thermostatted cation-exchange column at 56, 82, 25 and 71°C. On Wofatit KPS-200, a mixture of four nitroalcohols was separated by elution with 1 *M* ammonium sulphate solution on a 37 × 0.7 cm column. The flow-rate was not an important factor in the separation. The sequence of the eluted compounds was 2-methyl-2-hydroxy-1,3-propanediol, 2-methyl-2-nitropropanol, 2-nitrobutanol and 2-hydroxymethyl-2-nitropentanol.

Kemula and Brzozowski also separated a three-component mixture of nitroalkanes on a Dowex 50 column using ammonium sulphate solution as the mobile phase. The elution sequence observed was nitromethane, 2-nitropropane and 1-nitrobutane. The column dimensions were 11.5 × 0.7 cm and the flow-rate of the mobile phase did not exceed 9 ml/h. On the same ion exchanger, separations at various temperatures can also be effected: at 25°C with 1 *M* ammonium sulphate solution as eluent, nitromethane, 2-nitropropane and 1,3-dinitropropane are eluted first. After increasing the temperature in the column jacket to 71°C and using 0.5 *M* ammonium sulphate solution for elution, 1-nitrobutane and 1-nitropropane appear in the effluent.

Under the same conditions, all three isomeric nitrobenzoic acids can also be separated, using 0.1 *M* ammonium sulphate solution in 0.019 *M* hydrochloric acid as the mobile phase. *o*-Nitrobenzoic acid is eluted at room temperature with a minimum retention time, *p*-nitrobenzoic acid is eluted at 56°C and *m*-nitrobenzoic acid at 82°C. The time required for the separation is about 5 h.

During the analysis of the radiolytic products in the aqueous nitrate–ethylene system, ion-exchange chromatography on Dowex 50W-X8 was also employed. Ammonium sulphate solution of various concentrations served as the mobile phase. Nitromethane, hydrogen peroxide and sodium nitrite, present in the mixture, were detected polarographically. The curves were recorded at a constant potential corresponding to the complete reduction of the dissolved oxygen and lower than the reduction potential of nitrate (Broszkiewicz and Przybylowicz).

Kemula and Sybilska described a method for the separation and determination of 0.05–0.3 mg of a mixture of *o*-, *m*- and *p*-nitroethylbenzene on a column packed with the clathrate nickel γ-picoline thiocyanate. A solution of ammonium thiocyanate and γ-picoline in acetone were used for elution. This separation, based on the formation of clathrates, may be considered to be a special case of ligand exchange.

The specificity of the chromatography of nitro compounds consists in the possibility of preparing alkyl nitrates from the corresponding alkyl bromides on a silica gel column impregnated with silver nitrate, while the chromatographic separation of the mixture takes place simultaneously (Kuemmel). The reaction is unusual in that the breaking and formation of the covalent bonds occur on the adsorbent surface on reaction with the anion from the adsorbent, becoming covalently bound to the product. In addition, olefins are also produced from secondary bromides. The nitrates are readily separated from the olefins, which are strongly adsorbed on the column owing to the formation of silver complexes. A high ratio of adsorbent to sample is required so as to ensure the complete reaction of all the alkyl bromides.

## AMIDES

The possibility of applying liquid chromatography to the separation of coloured homologous N,N-dimethyl-*p*-aminobenzeneazobenzamides was demonstrated by Churáček and Jandera using an apparatus consisting of a pulse-free plunger feeding pump, a narrow bore column, a spectrophotometer with a flow-through measuring cell of their own design and a recorder. A strongly sulphonated styrene–divinylbenzene cation-exchange resin, Dowex 50W-X2 ($H^+$), was used for the separation of the coloured compounds. Fig. 31.1 illustrates the chromatographic separation of some lower secondary amides. Their protonated forms are distributed between the external solution (mobile phase) and the solution in the resin particles in accordance with the basicities of the non-ionic forms. The equilibrium depends on the activity of $H^+$ in the external solution and in the resin particles.

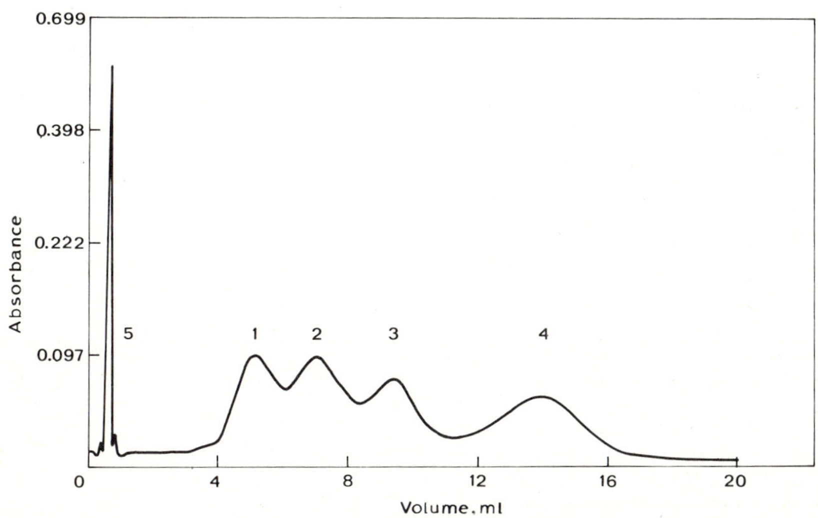


Fig. 31.1. Separation of secondary aliphatic amides of N,N-dimethyl-*p*-aminobenzeneazobenzoic acid (Churáček and Jandera). Column: 240 × 2.7 mm. Ion exchanger: Dowex 50W-X2 ($H^+$) (200–400 mesh). Eluent: 0.925 *M* hydrochloric acid in 80.5% ethanol. Flow-rate: 0.132 ml/min. Detection: optical density at 520 nm. Peaks: 1 = 1.5 μg of di-*n*-butylamide; 2 = 1.5 μg of di-*n*-propylamide; 3 = 1.5 μg of diethylamide; 4 = 1.5 μg of dimethylamide; 5 = inert compound (Ponceau 6R).

Because of the negligible solubility of these compounds in aqueous acid solutions, it is necessary to use mixed aqueous–organic media. The amount of the organic solvent present obviously affects the distribution equilibrium by its solubility and solvation effects, and also by the dielectric constant effect.

*References p.664*

A cation-exchange resin with a low degree of cross-linking (Dowex 50W-X2) was used for the separation in order to improve the accessibility of the ion-exchange phase to the large molecules of the derivatives and to accelerate the diffusion rate in the resin. Coloured amides are sorbed quantitatively on Dowex 50W-X2 ($H^+$), 200–400 mesh, from mixed aqueous–organic solutions (80% ethanol; 80% methanol). The sorbed compounds can be eluted with an aqueous ethanolic or aqueous methanolic solution of hydrochloric acid. The effect of the eluent composition on the chromatographic behaviour of some homologous amides was studied by Churáček and Jandera. Homologous amides are eluted in order of increasing basicities, *i.e.*, in order of decreasing molecular weights. Amides with a higher basicity have a higher distribution coefficient than esters, the basicity of which is lower. Secondary amides are sorbed more strongly than the less basic primary amides.

TABLE 31.1

VOLUME DISTRIBUTION COEFFICIENTS ($D_v$) OF SOME AMIDES OF N,N-DIMETHYL-*p*-AMINOBENZENEAZOBENZOIC ACID ON THE CATION EXCHANGER DOWEX 50W-X2 IN 0.925 *M* HYDROCHLORIC ACID SOLUTION IN 80.5% ETHANOL (CHURÁČEK AND JANDERA)

$D_v$ is defined as the ratio of the amount of compound in unit volume of the ion-exchanger phase to the same volume of external solution.

| Amide derivative | $D_v$ |
|---|---|
| Methylamide | 8.6 |
| Ethylamide | 7.4 |
| *n*-Propylamide | 6.5 |
| *n*-Butylamide | 5.6 |
| *n*-Hexylamide | 4.7 |
| Allylamide | 6.4 |
| Dimethylamide | 9.9 |
| Diethylamide | 6.6 |
| Di-(*n*-propyl)amide | 4.7 |
| Di-(*n*-butyl)amide | 3.5 |

TABLE 31.2

GEL CHROMATOGRAPHY OF AMIDES ON POLYACRYLAMIDES (STREULI)

Column: 97 × 0.5 cm. Gels: 1 = Bio-Gel P-2, 100–200 mesh (polyamide gel), exclusion limit 200–2600; 2 = Bio-Gel P-6, 100–200 mesh, exclusion limit 1000–5000. Mobile phase: 0.01 *M* sodium chloride solution. Flow-rate: 1 ml/min. Detection: RI.

| Compound* | $K_d$ | |
|---|---|---|
| | Gel 1 | Gel 2 |
| Urea | 1.37 | 1.17 |
| Biuret | 1.88 | 1.29 |
| Acetamide | 1.06 | – |
| N-Methylacetamide | 0.94 | – |
| N,N-Dimethylacetamide | 0.86 | 0.89 |
| Acrylamide | 1.17 | – |
| N-*tert.*-Butylacrylamide | 0.97 | – |
| N-Vinyl-2-pyrrolidone | 1.23 | – |

*Samples of 50 µl.

The contribution of the $CH_2$ group to the logarithm of the distribution coefficient proved to be fairly constant for homologous primary aliphatic amides, increasing to some extent with decreasing hydrocarbon chain length. Secondary amides showed greater changes corresponding to the same molecular-weight contribution. Volume distribution coefficients ($D_v$) of iso-derivatives are slightly lower in comparison with those of normal derivatives. The multiple bond contribution to $D_v$ does not seem to be significant (Table 31.1). The hydrochloric acid concentration in the mobile phase influences the equilibrium between the protonated and non-ionic forms of these compounds.

Gel chromatography was also used for the separation of substituted amides. Bio-Gel P-2 and P-6 were used as the stationary phase and 0.01 *M* sodium chloride solution as the mobile phase. Distribution coefficients and other separation conditions are given in Table 31.2 (Streuli).

## GUANIDINE AND UREA DERIVATIVES

Kirkland separated urea derivatives, applied as antidiabetic agents, by means of high-speed liquid chromatography, using a 1 m × 2.1 mm column packed with Permaphase ETH and a 1% solution of dioxane in *n*-hexane as the mobile phase. Fig. 31.2 shows the

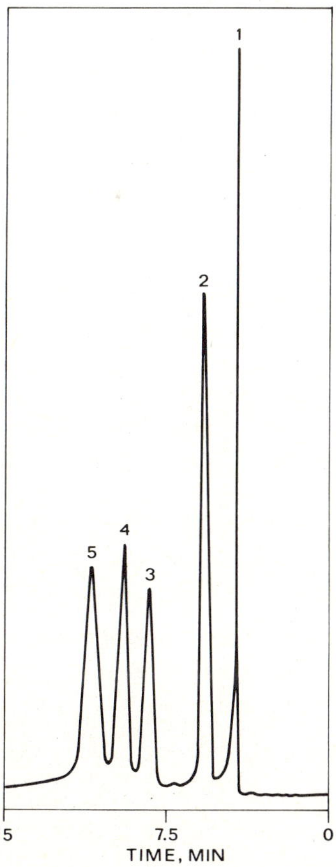


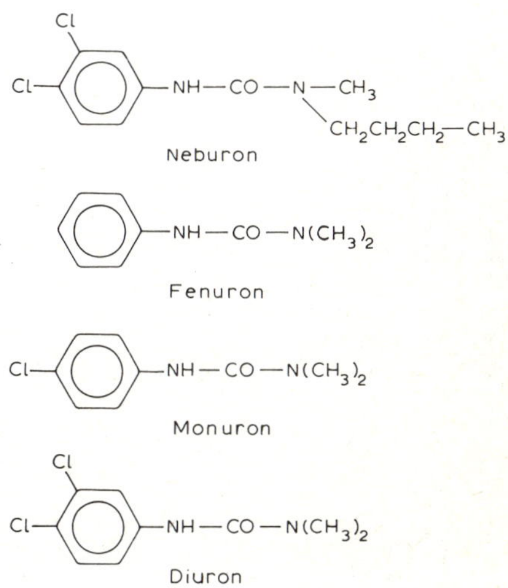


Fig. 31.2. Separation of substituted ureas (Kirkland). Column: 1000 × 2.1 mm. Sorbent: Permaphase ETH. Eluent: 1% solution of dioxane in *n*-hexane. Operating conditions: flow-rate, 1 ml/min; temperature, 27°C; column pressure, 340 p.s.i.g. Detection: UV detector at 254 nm. Peaks: 1 = solvent; 2 = Neburon; 3 = Fenuron; 4 = Monuron; 5 = Diuron. Sample: 1.5 µl of a 0.25 mg/ml solution of each compound in methanol.

separation of a synthetic mixture of substituted aromatic ureas using a relatively non-polar organic carrier. A similar mixture can also be resolved with the same packing using the reversed-phase technique with alcohol-containing water as the mobile phase. The order of elution of these substituted ureas is different in the two systems.

Liquid chromatography was also used for the analysis of gaseous fluorinated organic nitrogen compounds of the guanidine type (Rebertus *et al.*). Silica gel is a satisfactory adsorbent for this type of compound. Alumina and molecular sieves tend to decompose the fluorinated unsaturated compounds.

The separation of tris(difluoroamino)fluoromethane and pentafluoroguanidine is illustrated in Fig. 31.3. Either an inert fluoro-compound or a hydrocarbon such as *n*-heptane can be used as the solvent and mobile phase; however, the former is preferred because of its much greater stability toward oxidation. Tris(difluoroamino)fluoromethane is readily eluted with either of these solvents, but pentafluoroguanidine is removed completely only if large bed volumes are used.

The separation of bis(difluoroamino)difluoromethane–tetrafluoroformamidine mixtures was also effected. The relative affinities of the compounds for silica gel increase in the order $(F_2N)_3CF < (F_2N)_2CF_2 \ll (F_2N)_2C{=}NF < F_2NCF{=}NF$.

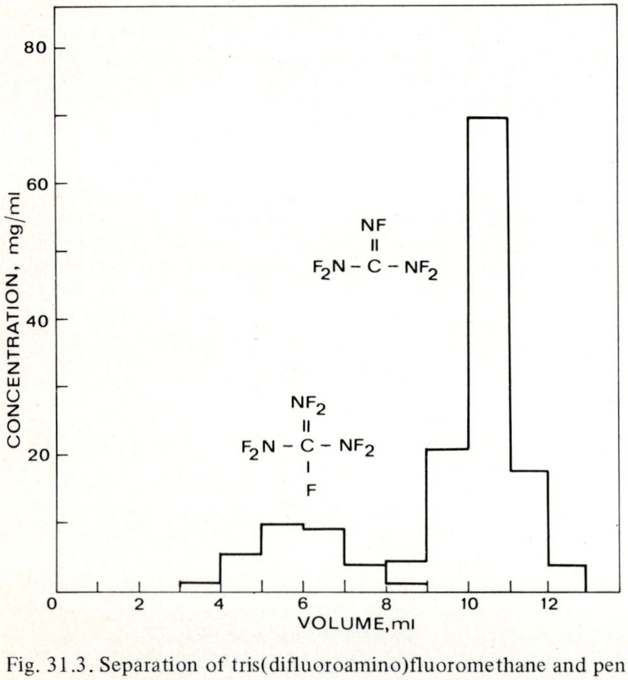


Fig. 31.3. Separation of tris(difluoroamino)fluoromethane and pentafluoroguanidine (Rebertus *et al.*). Column: 0.8 × 8 cm. Sorbent: silica gel (100–200 mesh). Eluent: *n*-heptane. Flow-rate: 1 ml/min. Detection: fractions were collected and elution with the solvent was continued until breakthrough of the trifluoroamidines was indicated by a positive test with 1 *M* potassium iodide solution. Sample: 3 ml of the fluoro-compound containing 1.5% (w/w) of tris(difluoroamino)fluoromethane and 2% (w/w) of pentafluoroguanidine.

TABLE 31.3
ION-EXCHANGE CHROMATOGRAPHY OF COMPOUNDS RELATED TO UREA (NOMURA *et al.*)

Values listed are elution volumes (ml). Column: 15 × 0.8 cm. Flow-rate: 0.46 ml/min. Detection: Automatic thermal reaction energy detection. Temperature: 35°C.

| Compound | Mobile phase | | | | | |
|---|---|---|---|---|---|---|
| | Water | 0.01 *N* HCl | 0.1 *N* HCl | 1.0 *N* HCl | Water | 0.01 *N* HCl |
| | Amberlite CG-120 ($H^+$) (200–400 mesh) | Dowex 50-X8 ($H^+$) (200–400 mesh) | | | Dowex 1-X8 ($Cl^-$) (200–400 mesh) | |
| Biuret | 15.9 | 22.2 | 21.0 | – | 23.5 | 23.7 |
| Thiourea | 21.5 | 27.1 | 26.9 | 26.9 | 34.5 | 34.7 |
| Dicyanodiamide | 27.3 | 33.8 | 34.3 | 34.2 | 34.7 | 34.5 |
| Nitroguanidine | – | 40.7 | 37.5 | – | – | – |
| Urea | 106.7 | 113.1 | 108 | 58 | 18.4 | 18.2 |
| Methylurea | 136.6 | 154.1 | 145.1 | – | 18.9 | 18.4 |
| Ethylurea | – | 226 | 203 | – | 20.2 | 19.6 |
| *tert.*-Butylurea | – | – | – | – | 30.4 | 29.9 |

TABLE 31.4
SURVEY OF DIFFERENT PROCEDURES APPLICABLE TO THE SEPARATION OF NITROGEN COMPOUNDS

| Compounds separated | Sorbent | Mobile phase | Reference |
|---|---|---|---|
| N-Halocyanamides and sulphonamides | Florisil | Dichloromethane, diethyl ether, *n*-hexane and other organic solvents | Neale and Marcus |
| Diphenylcarbazide and phenylsemicarbazide | Polyamide | Water–methanol–acetic acid (1:3:0.04) | Willems *et al.* |
| Automatic quantitative analysis of guanidines (biochemical application) | PQ-28 resin; Dowex 50-X2 | Sodium citrate buffers of various compositions and pH (amino acid analyzer) | Durzan, Carles and Abravanel |
| Substituted acetanilides and similar substances (drugs, analgesics) | LFS pellicular anion-exchange resin | Tris(hydroxymethyl) aminoethane (121 g in 1000 ml aqueous solution) (pH = 9.0, adjusted with dilute HCl). Pressure 1000 p.s.i. at 60°C | Stevenson and Burtis |

*References p.664*

Ion-exchange chromatography of compounds related to urea was carried out on a 15 × 0.8 cm column packed with the cation-exchange resin Amberlite CG-120 or Dowex 50-X8 ($H^+$) or the anion-exchange resin Dowex 1-X8 ($Cl^-$). Solutions of hydrochloric acid of various concentrations (0.01–1.0 *M*) and water were used as the mobile phase (Nomura *et al.*). Elution volumes found under these conditions on various ion exchangers and using hydrochloric acid solutions of various concentrations are given in Table 31.3.

Other, less important, applications of the liquid chromatography of some nitrogen-containing compounds are listed in Table 31.4.

## REFERENCES

Broszkiewicz, R. K. and Przybylowicz, Z., *Anal. Chem.*, 41 (1969) 1121.
Carles, J. and Abravanel, G., *Bull. Soc. Chim. Biol.*, 52 (1970) 453; *C.A.*, 73 (1970) 84454u.
Churáček, J. and Jandera, P., *J. Chromatogr.*, 53 (1970) 69.
Durzan, D. J., *Can. J. Biochem.*, 47 (1969) 657.
Kemula, W., *Rocz. Chem.*, 29 (1955) 1153.
Kemula, W. and Brzozowski, S., *Rocz. Chem.*, 35 (1961) 711.
Kemula, W. and Sybilska, D., *Anal. Chim. Acta*, 38 (1967) 97.
Kirkland, J. J., *Anal. Chem.*, 43 (1971) 43A.
Kuemmel, D. F., *Chem. Ind. (London)*, (1966) 1882; *Anal. Abstr.*, 15 (1968) 815.
Landram, G. K., Wickham, A. A. and DuBois, R. J., *Anal. Chem.*, 42 (1970) 107.
Neale, R. S. and Marcus, N. L., *J. Org. Chem.*, 34 (1969) 1808.
Nomura, N., Shiho, D. I., Ohsuga, K. and Yamada, M., *J. Chromatogr.*, 42 (1969) 226.
Rebertus, L. R., Fiedler, K. R. and Kottong, G. W., *Anal. Chem.*, 39 (1967) 1867.
Stevenson, R. L. and Burtis, C. A., *J. Chromatogr.*, 61 (1971) 253;
Streuli, C. A., *J. Chromatogr.*, 47 (1970) 355.
Willems, G. J., Lontie, R. A. and Seth-Paul, W. A., *Anal. Chim. Acta*, 51 (1970) 544.

*Chapter 32*

# Amino acids

Z. J. ZMRHAL, J. G. HEATHCOTE and R. J. WASHINGTON

## CONTENTS

## ANALYTICAL CHROMATOGRAPHY

The development of the modern analytical chromatography of amino acids was based on two historical events. The first was the advent of methods for the preparation of chemically homogeneous proteins, developed in the 1930s by the school of Northrop (Northrop *et al.*), and the second was the production of ion exchangers and the subsequent development of ion-exchange chromatography in the 1950s. In the meantime, methods based on paper and column chromatography, adsorption and partition, were developed, but were not successful, column techniques probably because of natural materials being utilized as packings the chromatographic properties of which could not be preselected. In spite of this, a successful separation of amino acids was achieved (Stein and Moore, 1948) as the result of a perfectly developed procedure and unusually precise work, typical of the authors, who were awarded the Nobel prize in 1972. However, the method was not suitable for practical use. Disadvantages of this method were that it was tedious and that the standardization of starch was difficult, as its separation properties depended on the method of its isolation and its source. Therefore the authors looked for other chromatographic materials.

At that time, the manufacture of ion exchangers was being started. It was expected that the separation of substances on ion-exchange columns would take place exclusively according to the differences in dissociation constants of ionized substances. The ion exchangers produced, however, also possessed adsorptive properties, although for the quantitative separation of amino acids they were still the most suitable material. Organic solvents used for the elution of amino acids from starch columns could be replaced with aqueous solutions, which thus substantially facilitated and simplified the detection procedure for amino acids (Moore and Stein, 1951). The separation of single amino acids was also better than on starch, the peaks being sharper and the baseline more even. A further advantage of this method was that salts in the chromatographic sample did not interfere.

In order to choose the most suitable conditions for the elution of all amino acids present in protein hydrolyzates, Moore and Stein (1951) determined the rate of elution of single amino acids under different elution conditions. For the elution of basic amino acids in a reasonable time, much more concentrated buffers had to be used. However, with large changes in concentration and pH, the volume of the swelled ion exchanger and the back-pressure of the column also changed. At the same time, the baseline value also increased. The ion exchanger could not be regenerated in the column and had to be removed from the column after each chromatographic run and then refilled. Therefore, it was more advantageous to carry out the elution in two stages, separating two parallel aliquots of sample on two columns. From one column, all amino acids up to the basic acids were eluted, the latter then being displaced with sodium hydroxide solution. From the second, shorter, column, acidic and neutral amino acids were first eluted at once, and then the basic acids were resolved. Thus the so-called two-column system of amino acid chromatography was established. After the quality of ion exchangers and the detection technique had been perfected, the single-column system of perfect quality was also developed.

The new method of amino acid analysis developed rapidly and made possible the solution of a series of formerly insoluble problems, primarily the study of the primary structure of proteins. However, it very soon became evident that the efficiency of the method was inadequate for this purpose and that it was too complicated. A series of papers describing mechanizational and organizational adjustments to the process was intended to make the work easier and to increase the efficiency of the method of amino acid analysis. A fundamental solution was again contributed by the school of Moore and Stein (Spackman *et al.*). The laboratory procedure for the working up of fractions from chromatographic columns was transformed into a continual process by carrying out the reaction of amino acids with ninhydrin in a flow-through capillary reactor and measuring the intensity of colour with a recording flow-through photometer. Thus, from the analytical method an apparatus was developed – the amino acid analyzer – and its production and further development were taken over by industrial producers and their research departments. They pursued two aims, an increase in efficiency and an increase in sensitivity. The efficiency was first limited by the long time required for chromatographic elution. The theoretical study of Hamilton (1958) became the basis for the development of better conditions for chromatography and demonstrated that a further increase in the efficiency of ion-exchange chromatographic columns could be achieved by increasing the flow-rate of the eluent through the column and by decreasing the particle size of the ion exchanger. By means of various modifications to ion exchangers, which still continue to be introduced by various manufacturers, a more than 10-fold decrease in the elution time has been achieved, with no reduction in the quality of resolution of amino acids. The decrease in the time required for one run to less than 8 h made it possible to use an overnight run in order to increase the output of the analyzer. This, however, necessitated the complete automation of the apparatus, *i.e.*, automation of sampling on to the column, switching over the columns to the detection system and the regeneration and stabilization of the columns. Some apparatus at present produced is of this type. There is no suitable differentiation in name to distinguish truly automatic apparatus from the classical, manually monitored apparatus, because the term "automatic analyzer" was incorrectly included in the name of amino acid analyzers from the very beginning of their production. In this chapter, the term "amino acid analyzer" will be used consistently for non-automated apparatus and the term "automatic amino acid analyzer" only for apparatus with the above automated operations.

From its origin, the method of quantitative amino acid chromatography has also been used for the analysis of ninhydrin-positive substances, which occur in natural materials. In this instance, much higher demands were placed on the efficiency of the chromatographic separation technique than in the analysis of amino acids in protein hydrolyzates, because much more complex mixtures are involved. A large proportion of the substances present in these mixtures consists not only of amino acids, but also of substances of different type, their common property being their reaction with ninhydrin reagent with the formation of coloured products. For these purposes, an increase in chromatographic efficiency was achieved by increasing the length of the chromatographic column. The analyses required a much longer time than the analysis of amino acids from hydrolyzates, and it was impossible to develop a universal elution system. The sources of free amino acids differ considerably both quantitatively and qualitatively and therefore these substances should be separated under different elution conditions. The technical equipment used in all elution systems is similar and is an integral part of every analyzer.

*References p. 710*

**Ion-exchange chromatography**

For the column chromatography of amino acids on starch, representing a quantitative analytical method, new technical laboratory equipment had to be developed. This equipment was further improved by substituting ion exchangers for starch and today it is common in every biochemical laboratory, so that it is possible to use the classical method of amino acid analysis in every laboratory. The conditions of chromatography used at present are those of amino acid analyzers because the conditions developed for them were the most simple, so that the colour reaction could be transformed into a continuous process. The chromatography of amino acids with the collection of fractions differs today from the work with the analyzer only in the collection of the eluate in the form of fractions and their manual working up. If the mechanization of all operations involved in working up the fractions is perfect, there is no difference between the time of analysis of amino acids with an analyzer and that of chromatography with the collection of fractions. Chromatography is more tedious than the operation of the amino acid analyzer, but it has advantages. The whole process is standardized by leucine photometry and it is therefore much less dependent on the maintenance of the stability of many technical devices than in the standardization of the amino acid analyzer. Last, but not least, for small series of determinations, the cost of a single analysis is much lower.

*Equipment for amino acid analysis*

*Chromatographic columns*

The dimensions of the chromatographic columns depend on the type of ion exchanger used. Common ion exchangers are roughly classified according to sieve size, for example Amberlite IR-120, 200–400 mesh, and have a weak separation efficiency. Therefore larger amounts of these exchangers, and hence also larger columns, are necessary in order to achieve a separation effect. Classical columns are 165, 65 or 28 cm long, with an I.D. of 0.9 cm. Thermostatting jackets are 155, 55 or 10 cm long, with I.D. 2.5 cm, and are closed at each end with a waterproof fitting through which the body of the column passes. The waterproof fittings may be replaced with rubber stoppers with holes for the column. For the gradient elution of amino acids from Dowex 50-X12 (Piez and Morris), columns of the same construction are used that differ from other columns only in their length (142 cm). The columns are fitted in the bottom section with a sintered-glass disc of G3 porosity, or with a firmly tightened disc made of porous PTFE, polyethylene or poly(vinyl chloride). The space below the sintered disc should be as small as possible and narrowing to a thick-walled capillary of 1 mm I.D. and ending with a ground-glass 8/1 mm semi-ball joint. The upper part of the column ends with an 18/9 mm semi-ball joint (see Fig. 32.1).

With an ion exchanger classified by the method of Hamilton (1958), it is possible to use shorter columns, with a decreased resolving effect (Spackman), for example with columns 60, 20 and 8 cm long and 0.9 cm I.D. Tubes for this purpose are 75, 28 and 20 cm long, with I.D. 0.9 cm, but instead of the 20 cm long columns, a 28 cm long column can be used.

For special ion exchangers, it is preferable to use columns recommended by the manu-

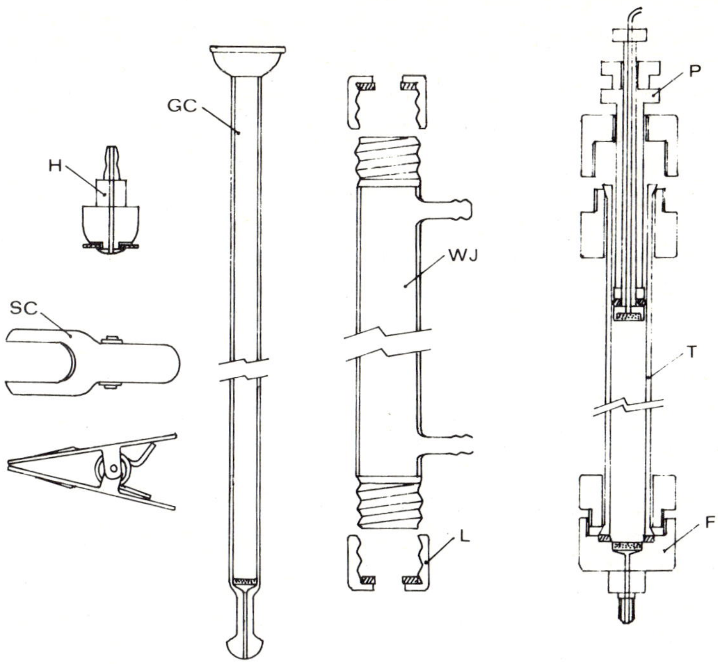


Fig. 32.1. Chromatographic columns. GC = all-glass column for non-automated chromatography; H = plastic semi-ball joint with a silicone rubber seal; SC = stainless-steel spring-clip (used for all sizes of semi-ball joints); T = thick-walled glass column for automated chromatography; P = stainless-steel or plastic plunger closing (may be used at both ends of the tube); F = fixed end; WJ = water jacket (the same for all types of columns); L = plastic lid with a female screw.

facturers (Serva, Heidelberg, G.F.R.; Jobling, Stone, Great Britain; Pharmacia, Uppsala, Sweden; and all manufacturers of amino acid analyzers). With these types of ion exchangers, the optimum separation efficiency depends on the complete elimination of dead spaces below and above the column, and on the accurate introduction of the sample, the volume of which should be small. These conditions can be fulfilled only if the so-called plunger-type column end-fittings are used, which are produced from stainless steel or plastic, and are provided with a fixed disc of porous plastic and tightened into the column with a rubber packing. The column end-fittings may be embedded in the column to an appreciable depth, on to the surface of the ion exchanger column, so eliminating completely the dead space above the column. The lower column end-fitting is also made on the same principle. The column is closed with the lower end-fitting, then filled with the ion exchanger and the prepared column is closed with the upper end-fitting. A column closed in this manner can then be eluted in any direction without damage, and in addition, it is ready for the automation of the whole procedure (see Fig. 32.1). For heating the columns, water-circulating ultrathermostats are commonly used, provided with two contact thermometers that can be switched over interchangeably.

*Pumps*

The eluent flows through the ion-exchange column rather slowly under hydrostatic pressure and therefore it must be pumped with an overpressure. Coarse-grained ion exchangers do not resist the flow of eluent through the column too much, and in order to achieve a sufficient flow-rate an overpressure of a gas (air), which is introduced above the eluent into the reservoir, is used.

The columns of fine-grained ion exchangers require higher overpressures in order to obtain a suitable eluent flow-rate, in the range 1–20 atm, and such pressures are achieved with pumps. Pumps for such purposes are available from most producers of scientific equipment (Milton Roy, Philadelphia, Pa., U.S.A.; Beckman-Spinco, Palo Alto, Calif., U.S.A.; Technicon, Tarrytown, N.Y., U.S.A.; LKB, Stockholm, Sweden; Hitachi, Tokyo, Japan; Bender-Holbein, Munich, G.F.R.; Jeol, Tokyo, Japan; Mikrotechna, Prague, Czechoslovakia). They differ in their construction and when selecting the most suitable pump the following properties should be kept in mind: (a) range of flow-rates (10–300 ml/h); (b) range of achievable pressures (0–30 atm); (c) long-term stability of the flow (better than 0.3%); (d) flow stability on back-pressure change; and (e) resistance to corrosion.

Pumps should always be connected with gauges, and unless they are corrosion-resistant, they must be adjusted so that they do not release metal ions into buffers passing through them (Hamilton and Anderson). Usually elution is carried out by a stepwise pH gradient which is started at a certain time after the start of the chromatogram. For changing eluents, automatic valves are employed in order to avoid wasting time, or when the apparatus is left to run overnight. Automatic switch cocks can be obtained from LKB, Mikrotechna, Technicon, etc., or a simple adapter operated by a solenoid can be made in the laboratory (see Fig. 32.2). Continuous gradient elution is carried out with a pH gradient or a concentration gradient, or both at the same time, which are produced in a mixer (Piez and Morris), and the eluent is then pumped from the mixer. Mixers consist of two vessels, one of which is the mixer proper and the second is a reservoir. If the mixer is closed, the gradient has the shape of a logarithmic curve, which is unsuitable for elution. With open mixers, the gradient depends on the ratio of the cross-sectional areas of both cylindrical

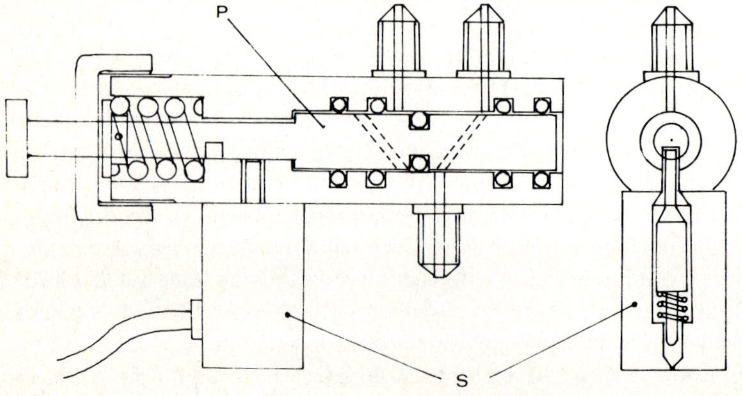


Fig. 32.2. Solenoid-operated double-way valve. S = solenoid; P = spring-driven plunger.

vessels. If both vessels are equal, the gradient is linear. If the reservoir has a smaller cross-sectional area, than the mixer forms a concave gradient; if the ratio is the reverse, the gradient is convex. Multi-chamber mixers are also available under different names, for example the open nine-chamber Technicon Varigrad mixer (Peterson and Sober) (see Fig. 32.3). Some manufacturers (LKB, Development Laboratories of the Czechoslovak Academy of Sciences, Technicon, etc.) have developed pumps that form gradients from storage solutions by changing the ratio of their pumping according to the program inserted for the gradient curve. In this case, the gradient mixer is not used.

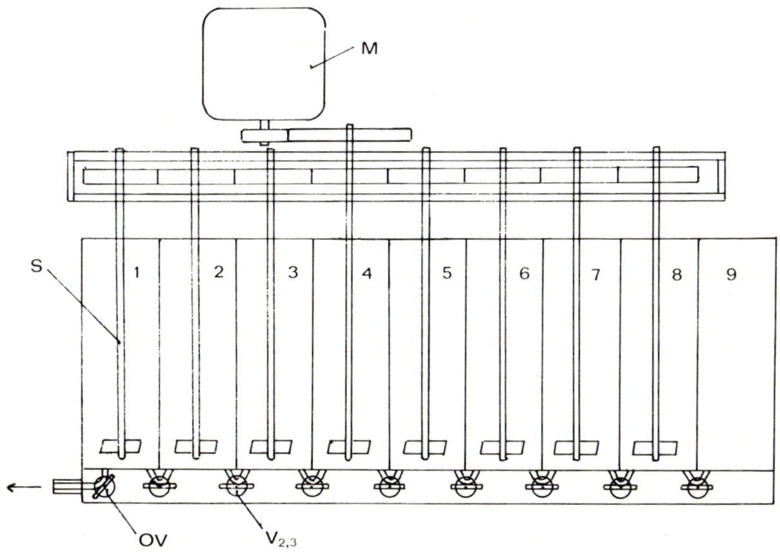


Fig. 32.3. Schematic diagram of Technicon Varigrad gradient mixer. 1–9 = 100-ml Perspex chambers; S = stirrers; OV = outlet valve; $V_{2,3}$ = valve on-line between chambers 2 and 3; M = stirrer drive.

*Fraction collectors*

For the collection of fractions from chromatographic columns, a large number of fraction collectors of various types have been produced. For quantitative analysis, the conditions of fraction collection are very demanding and therefore only a few types of collectors are suitable. The fractions must have equal volumes adjustable accurately to the required value. The collector must be reliable and disturbance free, and it is advantageous if it can be closed, so that the access of ammonia vapour from the air is prevented and evaporation limited. These conditions are best fulfilled by collectors with drop counters. The size of the drops is constant under constant conditions of elution, the size can be measured and the number of drops can be chosen so that the required volume is obtained. The volume of a drop is affected by the density and surface tension of the liquid, which must be kept in mind during elution with a concentration gradient and during the changes in detergent concentration in eluents, so that a correction can be introduced. Some types of collectors are provided with a micro-switch that emits an electric impulse when the collector has been shifted for the chosen number of tubes. This impulse may be utilized

*References p. 710*

for automatic change of eluents and for changes of temperatures in thermostatted columns. If the collector is not provided with such a device, one can easily be fitted. The micro-switch is fixed on to the collector stand below the photocell of the drop counter and then switched with a pin fastened to the chosen tube by a suitable clamp.

*Equipment for colorimetry*

The manual working up of a large number of fractions from each chromatographic run is very tedious and inaccurate. Colorimetry consists in the addition of the reagents or diluents, and heating, followed by measurement of the absorbance. These operations have been mechanized, as follows. For the accurate addition of solutions, automatic pipettes of various types are available, generally based on the principle of a piston pump with a double valve (see Fig. 32.4). It is easy to make such a device in the laboratory from a well tightened all-glass syringe and a double glass valve produced for laboratory pumps. The advantage of these pipetting devices, in addition to rapid and accurate work, is that there is the possibility of connecting the reservoir with the reagent, closed under an inert atmosphere. This type of reservoir is indispensable for ninhydrin reagent.

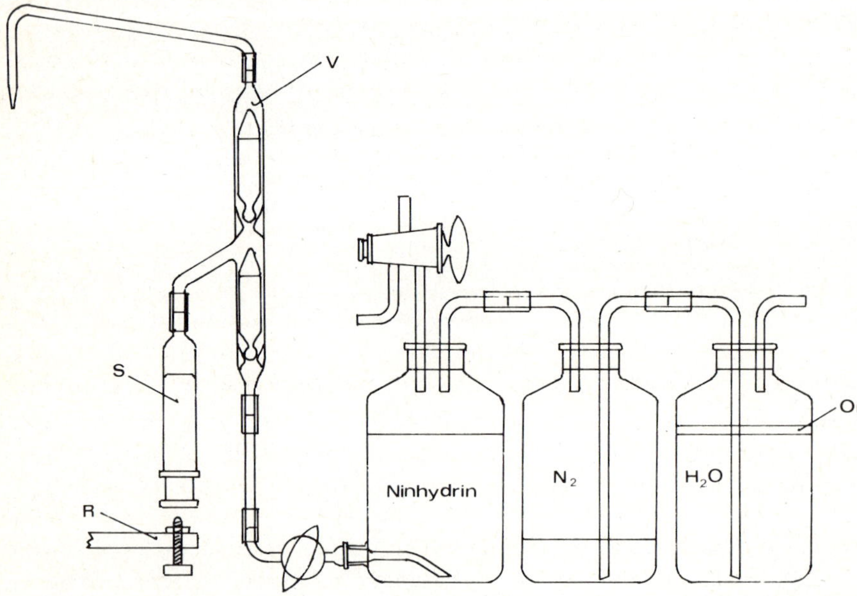


Fig. 32.4. Pipetting apparatus with the ninhydrin reagent reservoir. V = glass double valve; S = all-glass syringe; R = control of the length of piston stroke.

Heating-baths are constructed for a large number of test-tubes, which are dipped in with their racks. The heating is set so that after the rack has been dipped in, the bath temperature should again attain the necessary temperature within 2 min. The bath is closed in order to prevent the solutions in the test-tubes from evaporating and the lid has an oblique shape so that condensed vapour does not drop into the test-tubes. Baths can be obtained from Shandon (Camberley, Great Britain) and other manufacturers or made from a copper or stainless-steel sheet, and provided with electric heating. (A bath for 50 test-tubes, of dimensions 20 × 40 cm, and 5 l of water requires a heater of 2 kW output.) On the side of the bath, the escaping steam is condensed in a reflux condenser. Appreciable saving of time and effort during the measurement of colour intensity can be achieved by providing the fraction collector with test-tubes standardized as cells for colorimeters. The whole analysis of the fractions, including photometry, takes place in such test-tubes without the necessity for pouring the solutions from the test-tubes into photometer cells (Moore and Stein, 1948). For this purpose, strong-walled test-tubes made of hard glass are suitable and can be calibrated with a methyl red solution in 0.03 *M* hydrochloric acid. The dye concentration should be such as would give an absorbance of 0.6–0.7 at 525 nm with water as blank. The outer diameter of the test tubes is measured accurately and tubes of equal diameter are calibrated with a methyl red solution at 525 nm with the absorbance set to 0.010. According to the absorbance values, test-tubes are selected that do not differ by more than ±0.005 absorbance unit. From the remaining test-tubes additional groups that are homogeneous at other absorbance values can be selected. For the measurement of absorbance, the direction of the measurement is indicated with an engraved mark on the front side of the test-tube, and in each subsequent measurement the tubes are oriented in the colorimeter as indicated by this line. As mentioned above, colorimeters are suitable for quantitative analysis of amino acids with cylindrical cells of 18–20 mm diameter, a band width of wavelengths up to ±10 nm, and a good and long-term stability of blank value. If calibrated test-tubes are not available, it is advantageous to use a photometer cell provided with a bottom outlet with a stopcock connected to a water pump. The measured solution is poured into the cell through a firmly fixed funnel, and after the reading of the absorbance it is aspirated through the opened outlet stopcock.

### *Packing of columns*

Perfectly clean columns are packed with an ion exchanger by first suspending it in a three-fold amount of elution buffer and pouring the suspension into a heated vertical column. The exchanger is allowed to settle under a flowing elution buffer pumped at the rate of 30 ml/h. The first portion is poured into a column containing a layer of buffer 5 cm high and, before the suspension of the ion exchanger in the column has settled, the supernatant above the exchanger is removed and a fresh portion of suspension is added. If the column is already almost full, it can be extended by using an extension tube with a ground-glass joint and the filling completed. When the last portion of ion exchanger has settled, the column of the exchanger should be about 5 cm higher than the necessary height, without air bubbles and inhomogeneities, and without visible boundaries between single portions of the packing. Before the first analysis, the columns are regenerated and equilibrated with the first elution buffer (see the procedure for amino acid analyzers, p. 689). During this operation, the column of exchanger is perfectly settled so that it no longer changes its height and it can be adjusted accurately. Short columns are packed with the aid of an extension tube as described above.

*References p. 710*

*Elution*

The procedure for the chromatography of amino acids using a two-column (Moore *et al.*) or a single-column system (Piez and Morris) is the same as that used with amino acid analyzers (see p. 678). The only difference is that the effluent from the columns is collected with a fraction collector provided with a drop counter. The equipment of the chromatographic columns is also the same as that in amino acid analyzers. The column jackets are connected in series in two pairs and heated with water from a thermostat. The columns are situated close to the collector in order to be able to operate them easily, and the column which is eluted is easily connected with the drop counter.

*Colorimetry*

The working up of fractions from the chromatographic columns is carried out in racks containing 50 test-tubes. Solutions of standards and a blank are also present in each rack. Ninhydrin reagent (1 ml) is added with an automatic pipette to each test-tube (containing a 2-ml fraction) and the contents are stirred, then 2 ml of the reagent are added to fractions from gradient elution per 2 ml of fraction. The reagent is modified as follows. Immediately after the addition of the reagent, the rack with the test-tubes is immersed in a boiling water-bath (100°C) for 20 min, then transferred to a bath containing cold tap-water, in which it is cooled for 5 min. The contents of the tubes should be red and are diluted with 5 ml of 60% ethanol using an automatic pipette, taking care that the contents are well mixed. The test-tubes are allowed to stand at room temperature for 1 h, during which time the red colour of hydrindantine disappears as it is oxidised by the oxygen in the air, and only a violet colour persists in test-tubes that contain amino acids. The optical density of the colour is then measured at 570 nm, and fractions that contain proline and hydroxyproline are measured at 440 nm. Blanks are chosen from fractions that do not contain amino acids, according to the average value of their absorbance. Fractions that contain a peak of an amino acid are always measured against a blank selected from fractions eluted with the same buffer. In gradient elution, a gradual increase in effluent absorption takes place as a consequence of the increasing concentration of impurities, and in this case a blank must be chosen from closely preceding fractions. Fractions that have an absorbance value higher than the limit of the linear part of the standard curve are diluted with an additional 5 ml of 60% ethanol and then measured against equally diluted blanks. The absorbance values are converted into the content of amino acids in single fractions by using a standard curve for leucine.

A standard curve for leucine (0.05–0.250 mmole/ml) constructed for three dilutions (5, 10 and 15 ml) with 60% ethanol is used for all amino acids. The yields of other amino acids are then calculated from the so-called leucine units by means of factors (see the calculation of results, p. 705). For facilitating the work, the diagram may be transformed conveniently into a table from which readings can then be taken. The content of ninhydrin-positive substances determined in fractions is plotted against the total effluent volume (from the start) and an elution curve is thus obtained, called a chromatogram. The elution volume (retardation) of each compound, *i.e.*, the volume of the effluent from the start to the peak maximum, is reproducible under constant conditions, is characteristic of each substance and is used for qualitative analysis. The area under the peak is proportional to the amount of the amino acid present.

**Amino acid analyzers**

The development of amino acid analyzers was based on the advances in methods for amino acid analysis. The chromatography of amino acids on ion exchangers did not require extensive modification, only the ensuring of a constant flow through the column. However, the ninhydrin detection technique had to be adapted for a continuous process, which was achieved by modifying two known methods. First, the continuous method of carrying out reactions under flow, known from industry, was adapted to the analytical scale and was performed in capillaries (Spackman *et al.*). Later, the automat for serial colorimetric analyses, developed by Technicon, was used and set the basis for two main constructions of amino acid analyzers, which were then developed further. Both constructions were developed for two-column and single-column systems, taken from classical chromatographic methods. In this manner, four types of amino acid analyzers originated:

(a) basic analyzers with a two-column chromatographic system;
(b) basic analyzers with a single-column chromatographic system;
(c) Technicon analyzers with a two-column chromatographic system;
(d) Technicon analyzers with a single-column chromatographic system.

With the change-over to continuous detection, the method of standardization of the procedure also changed. Instead of standardizing the colorimetry on leucine, the whole procedure using the analyzer was standardized on the basis of the analysis of a standard mixture of amino acids.

Further development took place at first only through random technical and methodological improvements. Specifically directed development started after the publication of theoretical studies concerning the conditions of ion-exchange chromatography (Bogue; Giddings; Hamilton, 1960; Hamilton *et al.*). Ever better ion exchangers for columns were prepared. The dimensions of the columns, their construction and the elution conditions were adjusted for optimum values and a higher resolving power of chromatographic columns was achieved, the rate of elution was improved and the elution procedures were simplified.

The rate and the efficiency of amino acid chromatography began to exceed the levels possible in classical detection systems and further development of the analyzers therefore continued on the basis of deeper studies of the kinetics of the reaction of ninhydrin with amino acids and the study of the construction of the reactor and measuring device (Ertinghausen *et al.*; Spackman; Zmrhal, 1965). This resulted in an increase in resolving power and the sensitivity of the whole apparatus. The time of a single run was shortened to less than a working day and thus the possibility arose of increasing the efficiency of the analyzer by utilizing overnight runs, which required the construction of a fully automated apparatus. At that time, equipment was already available in which most of the operations were automated, but for complete automation an automatic sampler was lacking. The first one-loop sampler, for manual operation, was already known (Crestfield) and it was not difficult to adapt it to an automatic device with a larger number of loops. Later, better

*References p. 710*

cartridges for ion exchangers were developed (Thomson and Eveleigh). In the automatic programmer, the programme was extended to the sampler and a fully automatic amino acid analyzer was thus obtained. The great efficiency of this apparatus makes the evaluation of the results difficult and therefore it is equipped with an integrator or connected with a computer. Non-automated analyzers of the basic type and the Technicon type, and also their operation, are described below. The types of analyzers being produced at present are also briefly surveyed.

*Basic-type analyzer*

The first amino acid analyzers were manufactured by Beckman (Fullerton, Calif., U.S.A.) in 1958 (Model 120 A). The apparatus consisted of two parts – chromatographic and detecting. The conditions for chromatography were those of the conventional fractional method and they were simplified to include as few manual operations as possible. The detection of amino acids was converted into a continuous process by means of a capillary reactor in which the reaction of amino acids with ninhydrin reagent took place continually, under flow, and by a recording photometer with a tubular flow-through cell. In order to keep a correct mixing ratio of the reagent and the column effluent in the detector part, constant conditions for the flow of the reagent and the eluent from the column had to be ensured. The required accuracy of flow was achieved by means of pumps selected from the commercial types available from Milton Roy, and later special pumps were developed for this purpose. The setting of a suitable flow-rate had to be measured accurately and two types of flow meter were developed. The volumetric type of flow meter was inserted in the inlet line before the pump, while the bubble type was connected with the outlet tubing, behind the photometer. Later, it was found that bubble flow meters give a sufficiently accurate measurement and the use of volumetric flow meters was discontinued. An amino acid analyzer is shown schematically in Fig. 32.5.

The starting points for the construction of a reactor were the conditions of the reaction of amino acids with the ninhydrin reagent and the rate of the chromatographic elution. The flow-rate of effluent through the column (30 ml/h) and the mixing ratio of the volumes of the effluent and the reagent (2:1) determined the flow-rate of the reagent, *i.e.*, 15 ml/h. The sum of the two flow-rates (45 ml/h) and the time necessary for heating at 100°C (15 min) determined the reaction volume (11.5 ml). Such a large reactor volume would, however, result in mixing of the zones of substances separated on the column if the tube had a large internal diameter. It was found experimentally that mixing of the liquid by rolling in the tube took place only with internal diameters of the capillary tubing greater than 1 mm, so a capillary of I.D. 0.7 mm was used. The water-repelling and mechanical properties of PTFE made it the material of choice for the capillary, the length of which was 30 m. The electrical heating-bath of the thermostat was set so that water would boil very gently and the vapour would condense in the adjoining air condenser.

Two demands are placed on the flow-through recording photometer for the amino acid analyzer: it must keep the baseline perfectly and for a long period of time, and the cell must be capable of being washed rapidly. The stability of the baseline can be ensured by using an efficient power supply stabilizer. Washable cells cannot have a large volume and they must therefore be constructed of smooth, thin tubes with gradual transitions between

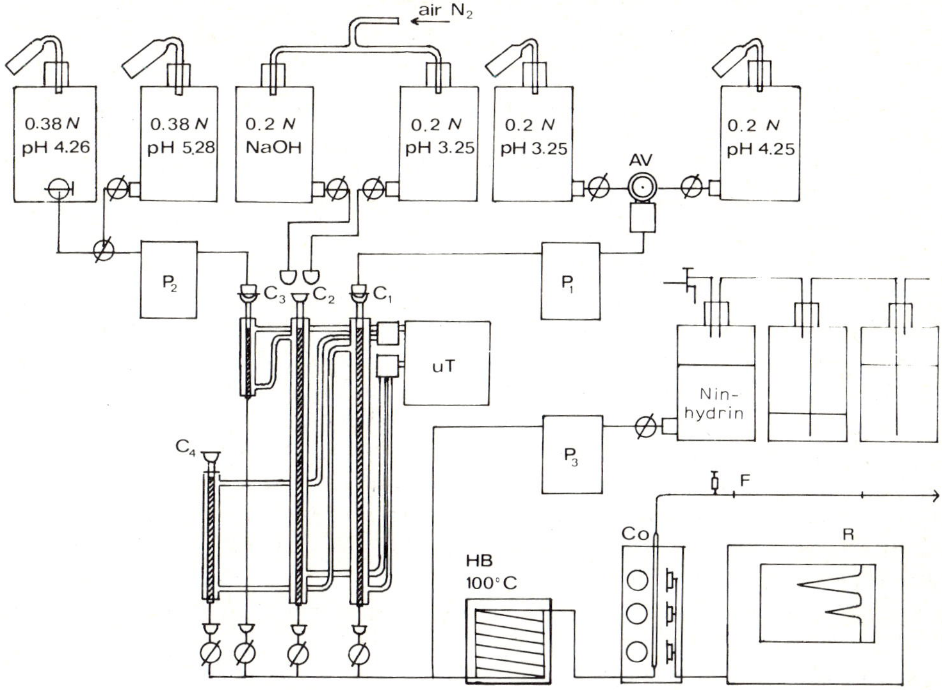


Fig. 32.5. Basic amino acid analyzer with the two-column system. $P_1$, $P_2$ = pumps for buffers; $P_3$ = pump for ninhydrin reagent; $C_1$, $C_2$ = long columns; $C_3$, $C_4$ = short columns; AV = automatic buffer-change valve; HB = reactor (30 m long PTFE capillary tubing, 0.7 mm I.D., in a boiling water-bath); Co = colorimeter, first unit 570 nm, second unit 440 nm and third unit 570 nm, reduced sensitivity; R = recorder (three dotted lines); F = flow meter; uT = ultrathermostat.

the changing inner diameter. However, the sensitivity of the photometry also depends on the thickness of the measured solution layer. In the first types of analyzers, a 2.7 mm layer thickness sufficed, and later tubular cells were developed that can be made with virtually any layer thickness. The coloured products formed during the reaction of amino acids with the ninhydrin reagent are not uniform. Most amino acids give a violet product, the absorption maximum of which is at 570 nm, but proline and its derivatives give a yellow colour with an absorption maximum at 440 nm. Parallel measurements of absorbance of these substances required the incorporation of another photometer, and a third photometer was then necessary for measurements on solutions with excessively high absorbances (570 nm), which in manual procedures must be diluted. The three photometers were placed one above the other on a common long cell into which a glass rod was sealed at the point where the measurement with the decreased sensitivity took place. This rod decreased the thickness of the measured solution by two thirds. The recorder was common to all three photometers and was a recorder printing three dotted lines with a 3-in. shift of chart, divided into tenths, and with a dotting rate of 4 sec, *i.e.*, 12 sec for each of the three recorded lines.

*References p. 710*

The flow meter consisted of an approximately 20 cm long glass capillary tube calibrated to 0.5 ml between two marks, and a bubble injector with a rubber valve. The injector plunger was drilled longitudinally with an approximately 1 mm diameter hole, which could be closed and opened with a finger and which functioned as a suction valve.

Even in the earliest amino acid analyzers, two operations had to be automated in order to permit the exploitation of their full capacity by overnight working: changing the eluent and switching off the apparatus. The eluent was switched over by an electromagnetic valve controlled with a timer. In analyzers with a single-column system, the switching over to the Varigrad gradient mixer remained manual because it coincided with the start of the run. The Varigrad mixer was described in the preceding chapter. The switching off of the apparatus was controlled with another timer. When the programmed time of the end of the analysis was reached, the timer first switched off the ninhydrin reagent pump and the recorder. After 30 min, when all of the ninhydrin reagent had been washed out of the whole system, the timer also switched off the eluent pump.

*Procedure*

For a two-column system of analysis, the amino acid analyzers are provided with two 150-cm columns, one 15-cm column and one 50-cm column. Two 150-cm columns are necessary in order to permit the regeneration of the first when the second is in operation (for the filling of columns, see the section on chromatography with fraction collection, p. 673).

The daily programme for the amino acid analysis of protein hydrolyzates is as follows. In the morning, the analyzer is switched on, the column heater and reactor baths are allowed to reach the necessary temperatures and elution solvents and the reagents are prepared. Then one of the two 2-ml aliquots of the sample is applied on to the 15 cm column. The sample is diluted with 0.2 *N* sodium citrate buffer solution of pH 2.2. The column head is opened and the buffer aspirated off with a pipette down to the bed of ion exchanger. Using a pipette with a bent tip, the sample solution is introduced carefully down the walls from close to the ion exchanger on to its surface; the ion exchanger should not be disturbed. The sample is allowed to soak into the column with a 0.3 atm overpressure of air or nitrogen. In chromatography with a fraction collector, the column outlet is opened before the air pressure is applied so that the eluent can be collected into the first fraction (start). Air should not enter the column. The sample is then washed down into the column with three 0.5-ml volumes of 0.2 *N* sodium citrate buffer solution of pH 2.2. The space in the tube above the column is filled with the first elution buffer, preferably that which was removed from the column earlier, and closed with a ball joint which is fastened by two spring clamps (of 2.5 atm pressure each). After the buffer line has been attached to the column, the pump of the eluent and the run timer are switched on and, when the equilibrium pressure in the column has been attained, the flow-rate is measured. One hour after the start, the pump for the ninhydrin reagent is started; by that time, all the acidic and neutral amino acids have already been eluted from the column. When the pressure becomes constant, the flow-rate is measured and the baseline is switched on. The analysis of basic amino acids lasts 5 h (from the start). When the sample has been introduced into the 15-cm column, the second 2-ml aliquot of the sample is applied on to the equilibrated 150-cm column in the same way. After the end of the elution from the 15-cm column, the eluent pump for the 150-cm column is switched on as well as the run timer (from zero). Simultaneously, the timer of the buffer change (8 h 20 min) and the timer of the end of the analysis (16.5 h) are set to the required time. When the working pressure has been achieved, the 150-cm column is switched over to the reactor instead of the 15-cm column and, at the same time, the latter is switched over from the reactor to the drain. When the equilibrium pressures have been reached, the combined flow of the eluent and the reagent is measured and the baseline is switched on.

The next day, in the afternoon, when the sample is introduced into the 15-cm column and the second 150-cm column, the first 150-cm column which was used for analysis on the previous day is regenerated. The column head is opened and the solution above the ion exchanger is aspirated off with a pipette and discarded. The volume under the column is filled with 0.2 *N* sodium hydroxide solution, then closed and washed with an alkaline solution under air or nitrogen pressure (0.3 atm) or by using a pump (30 ml/h) until the front reaches the middle of the column. A change in colour may be observed. The alkaline solution is then switched over to 0.2 *N* sodium citrate buffer solution of pH 3.25 and the column is washed under air pressure overnight, or for 5 h if a pump is used. The column is then ready for the application of the sample. The 15-cm column is not regenerated.

For the analysis of ninhydrin-positive substances in physiological liquids, 150-cm and 50-cm columns are run overnight. The sample is applied on to the 150-cm column in the morning (as described above) at a column temperature of 30°C. The change from the 0.2 *N* sodium citrate buffer solution of pH 3.25 to the 0.2 *N* sodium citrate buffer solution of pH 4.25 is simultaneous with the temperature change from 30°C to 50°C after 11 h and the whole analysis is terminated after the elution of β-aminobutyric acid, after 22 h. Basic amino acids and related compounds are determined on the 50-cm column, on to which the sample is applied the next day, in the morning. Elution is carried out with a 0.38 *N* buffer of pH 4.26 at 30°C. The temperature change from 30°C to 50°C takes place after the elution of carnosine, after 13 h. The analysis is ended after the elution of arginine, after 21 h. The procedure of starting the analysis is identical with that of the analysis of amino acids in hydrolyzates. The regeneration of the 150-cm columns is carried out during the time when the analysis of basic amino acids takes place on the 50-cm column. The 50-cm column is not regenerated.

The analyzers with the single-column system are provided with one 133 cm long column of 0.9 cm I.D. (for the packing of the columns, see the section on chromatography with the collection of fractions, p. 673). The analysis of amino acids in hydrolyzates and in physiological fluids is carried out by the same procedure at 60°C.

On to the column, equilibrated with the 0.25 *N* sodium citrate buffer solution of pH 2.91, 2 ml (the total amount) of sample solution of pH 2.0 are applied (see the two-column procedure, p. 676). The pump is switched on and the eluent flow-rate is set to 30 ml/h, then 0.25 *N* sodium citrate buffer solution of pH 2.91 is pumped until the pressure above the column reaches equilibrium. The recorder is then switched on and the eluent flow-rate measured, adjusted as necessary, and the pump for the ninhydrin reagent is switched on. Its flow-rate (30 ml/h) is twice as high as in the analysis with the two-column system, because for gradient elution the composition of the ninhydrin reagent is modified in order to increase its buffering capacity. Eventually, the pumping of the 0.25 *N* sodium citrate buffer solution of pH 2.91 is switched over to the Varigrad gradient mixer with the stirring switched on and all chambers filled according to the scheme shown in Table 32.1 are interconnected. About 20 min after the ninhydrin reagent pump is switched on, the baseline is set to 0.020 unit at 570 nm, to 0.000 unit at 570 nm for reduced sensitivity, and to 0.050 optical density unit at 440 nm.

TABLE 32.1
PREPARATION OF THE VARIGRAD GRADIENT MIXER FOR THE CHROMATOGRAPHY OF AMINO ACIDS

| Chamber | Buffer* (ml) | | Water (ml) |
|---|---|---|---|
| | 0.25 *M*, pH 2.91 | 0.8 *M*, pH 6.5 | |
| 1 | 75 | 0 | 0 |
| 2 | 75 | 0 | 0 |
| 3 | 75 | 0 | 0 |
| 4 | 70 | 5 | 0 |
| 5 | 50 | 24 | 0 |
| 6 | 40 | 19 | 15 |
| 7 | 30 | 10 | 34 |
| 8 | 5 | 52 | 15 |
| 9 | 0 | 71 | 0 |

*The buffer of pH 6.5 contains no detergent. Without detergent, its drops have the same volume as those of the buffer of pH 2.91.

After the total amount of the eluent has been pumped out of the mixer, after approximately 20 h, the mixer is switched back to the buffer reservoir (pH 2.91) with which the residue of the gradient is expelled from the column. The analysis is then completed. The total time of analysis is about 22 h. The column, after the completion of the analysis, is again equilibrated with the buffer of pH 2.91 and prepared for the application of the next sample. Hence it need not be regenerated.

*Technicon-type analyzer*

Technicon started the manufacture of the Type NC-1 amino acid analyzer, which comprised the chromatographic part and the basic type of AutoAnalyzer developed originally for clinical and similar routine analyses. The whole apparatus consisted of independent modules and therefore had great flexibility. It was equipped for a single-column system of analysis, but the elution conditions could be changed optionally. For the sake of comparison, the Piez and Morris method is described. The apparatus is shown in Fig. 32.6.

The eluent is pumped from the buffer reservoir or a Varigrad gradient mixer by a positive displacement pump into a jacketed column (140 × 0.9 cm), heated at 60°C and containing a 133 cm high column of packed Dowex 50W-X12, 20–30 μm. It is a column of common construction with semi-ball ground-glass joints, the lower end being provided with a sintered-glass disc and an exchangeable jacket. The effluent from the chromatographic column flows through a bubble flow meter into a stream splitter, from where it is pumped into the sample tube of the proportional pump pumping the ninhydrin reagent

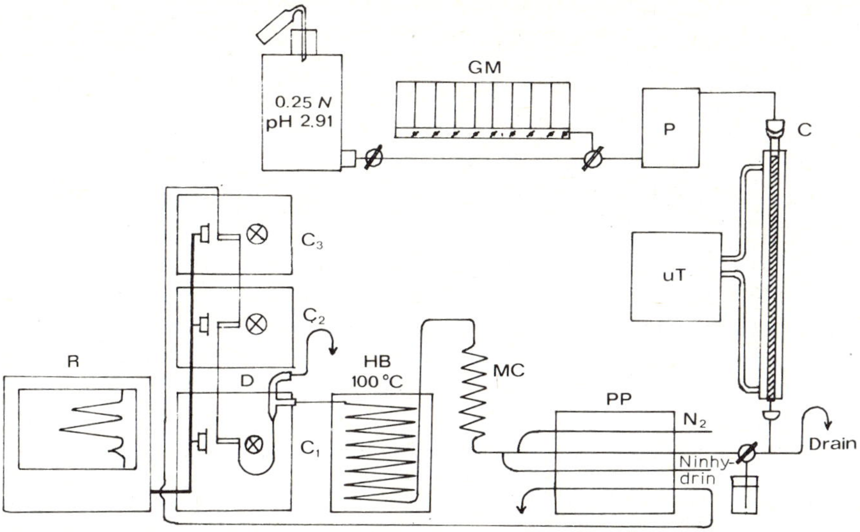


Fig. 32.6. Scheme of the Technicon amino acid analyzer with a single-column system. C = column; P = high-pressure pump; GM = Varigrad gradient mixer; uT = ultrathermostat; PP = proportioning peristaltic pump; MC = mixing coil; HB = reactor (10 m long glass capillary tubing, 1.6 mm I.D., in a 100°C oil-bath); $C_1$ = colorimeter (570 nm); $C_2$ = colorimeter (440 nm); $C_3$ = colorimeter (570 nm); D = debubbler; R = recorder (the signal from all colorimeters is printed as three dotted lines).

and nitrogen gas simultaneously. The flow-rate of the peristaltic pump should be lower than that of the effluent from the chromatographic column in order to prevent air from entering the pump. This, however, causes a partial loss of the sample and a decrease in the sensitivity of the analysis.

A mixture of the effluent with the ninhydrin reagent is separated into small volumes in the capillary tubing by nitrogen bubbles. The bubbles prevent the longitudinal spreading of the zones of single amino acids eluted from the chromatographic column during the passage of the solution through the tubing of the AutoAnalyzer.

The eluent and the reagent are pumped at the same rate and are combined in the proportions 1:1. Immediately after this mixing, the mixture is regularly separated by the nitrogen bubbles. It is further mixed in the mixing coil and is led into the thermostatting bath (100°C) through which it flows for 15 min and where the colour develops. Next it enters the debubbler, from which an aliquot of the coloured solution is pumped without bubbles through a capillary of 0.5 mm I.D. by means of another tube of the peristaltic pump into 15-mm cells of the photometer, in which the absorbance is measured at 570 nm. The photometers are connected in series and their cells are connected by capillaries of 0.5 mm I.D. One photometer measures the absorbance at 440 nm in a tubular cell while the second measures the absorbance at 570 nm in an 8-mm tubular cell. Finally, the measured solution is led via the peristaltic pump into the drain. The recorder registers the effluents from the colorimeters in the form of three dotted lines. During the debubbling, part of the sample is again lost in the form of waste of an aliquot of the reaction mixture and hence a further decrease in sensitivity takes place.

*References p. 710*

For the two-column system of analysis on a Technicon analyzer, the gradient mixer should be replaced with an automatic buffer change timer and temperature change timer and a switch-off timer.

*Procedure*

Before the sample is added to the chromatographic column, solutions are prepared in the Varigrad gradient mixer and the AutoAnalyzer is then set in operation. The sample tubing is switched over to pumping 0.25 *N* sodium citrate buffer solution of pH 2.91 and the baseline is allowed to stabilize. During this time, standard leucine solutions can be injected and the proper functioning of the apparatus checked.

The sample is introduced on to the chromatographic column in the conventional manner (see the section on the procedure with classical-type analyzers, p. 676). The eluent pump is connected to the reservoir of the pH 2.91 buffer and started, the start of the operation being indicated on the elution curve. When the pressure on the column reaches equilibrium, the flow-rate is measured and the sample tubing of the peristaltic pump is switched over to the effluent splitter. Simultaneously, the pumping of the elution buffer of pH 2.91 is switched over to the gradient mixer, which is already in operation with all chambers interconnected. After about 20 min, the values of the baseline are set on the recorder, as are the absorbance at 570 nm (to 0.020), reduced absorbance at 570 nm (to 0.000) and absorbance at 440 nm (to 0.050 unit of the extinction scale).

The consumption of all solutions in the gradient mixer is completed after about 20 h. The eluent pumping into the column is then switched over from the gradient mixer back to the buffer (pH 2.91) reservoir and pumping is continued for about a further 2 h in order to re-equilibrate the column. After elution of arginine, the analysis is terminated and the column is ready for the analysis of another sample. All tubing of the peristaltic pump is immersed in distilled water before it is stopped and the whole apparatus is rinsed.

The two-column system differs from the single-column system in its working programme, and is the same as in basic analyzers with a two-column system. The 15-cm column is operated during the day and the 150-cm column overnight. The change of buffer and the end of the analysis should be programmed using a timer switch. The 150-cm column should be regenerated after the analysis (see the section on the procedure with classical-type analyzers, p. 676). The procedure during the sample application and the termination of the analysis is identical with that given above for the single-column system of the Technicon analyzer.

*Commercial amino acid analyzers*

Beckman-Spinco (Palo Alto, Calif., U.S.A.) lists four amino acid analyzers. The Model 118 is the simplest and is capable of dealing with three hydrolyzates per day at a normal sensitivity of 4 nmole, using a 6-mm tubular cuvette or 2 nmole with a 12-mm tubular flow cell. The addition of a sampling device and a programmer converts this semi-automatic model into the fully automatic Model 119, which can analyze up to 30 samples in sequence at the rate of about six samples per day. Both models can be used in single- or double-column techniques and allow for automatic regeneration and selection of up to three buffers.

The Model 120 C, which is a direct descendent of the classical analyzer, is now a standard analytical instrument. It requires manual loading and some attention to buffer regeneration and equilibration, but is more versatile than the above models and has facilities for handling large-bore columns for preparative work. The addition of a manual sample injection kit permits an analyzer output of up to five accelerated hydrolyzates per day with an error of reproducibility of less than 2%.

The Beckman Model 121 is a fully automatic instrument with sample injection, which can analyze up to 72 samples in unattended operation over about 6 days. Sample volumes are pre-metered and the many functions of this instrument are adequately controlled by a 42-channel punched-tape programmer, each programme being written out on a separate continuous loop of paper tape. A digital integrator or an on-line data reduction computer is almost essential.

Beckman produces an extensive literature and bibliography of applications, and also has a large selection of optional accessories of high quality.

Technicon (Tarrytown, N.Y., U.S.A.) was the first company to introduce an automatic analyzer, which consisted of a series of independent modules in order to retain maximum flexibility. It was based on the single-column techniques previously mentioned and offered the small laboratory a high-resolution, extremely versatile instrument. The time of elution could be varied from 5.5 up to 21 h by appropriate choice of operating conditions. In order to accommodate further samples, variations were introduced, which incorporate three (NC-3) and five (NC-5) columns. For clinical application, the Model NC.S was devised for the rapid analysis of specific amino acids. All the above models had nitrogen segmentation, which gave enhanced resolution without any dead volume. Furthermore, it ensured that the ninhydrin reagent was not oxidized, nor did it need refrigeration.

The current NC-2 analyzer for amino acids retains the modular basis of the earlier models and single or multiple columns (75 × 0.5 cm) can be accommodated. A protein hydrolyzate can be analyzed in as little as 2.5 h, or a difficult biological sample may have its elution time extended to 36 h or more, depending on the volumes of buffers available. For the detection of the colour response, two colorimeters are used, one at 440 nm and the other at 570 nm; an alternative is to use one colorimeter set at a wavelength of 410 nm. The difference in sensitivity, it is claimed, can be more than offset by continuous range expansion of up to 10 times, which is a built-in feature of these colorimeters. Nitrogen segmentation, which is a special feature of Technicon analyzers, is still used in order to achieve maximum resolution, but the new formulation of the ninhydrin reagent using hydrazine sulphate does not require storage under nitrogen gas. The limit of detection is about 1 nmole.

The Technicon Sequential Multi-Sample Amino Acid Analyzer (TSM) is an automated instrument designed to separate, detect and determine amino acids present in plant or animal tissues and fluids. This instrument, which has been available since about 1968, can separate and analyze free acidic, neutral and basic amino acids from a protein hydrolyzate in 1¼ h, and from physiological fluids or tissue extracts in 5½ h. After the initial manual loading of up to 40 samples in eight cartridges, the analysis proceeds automatically. This model utilizes the two-column mode of operation and the sensitivity of detection is similar to that of the NC-2. A very wide range of columns, resins and other accessories is available and Technicon provides a good servicing scheme; they also hold

*References p. 710*

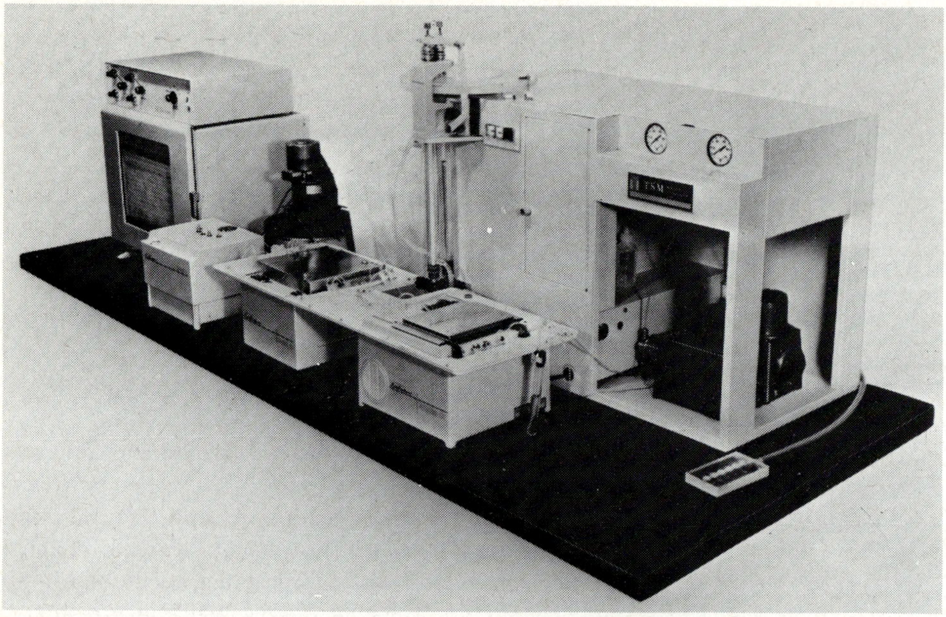

Fig. 32.7. Technicon TSM-1 automatic amino acid analyzer.

frequent colloquia for the benefit of their clients, at which new developments are discussed. The external characteristics of the TSM model are shown in Fig. 32.7.

LKB (Stockholm, Sweden) lists three automatic amino acid analyzers in the following order of increasing versatility; BC-100, BC-100L and BC-201. All three models are fitted with two long columns and one short column and can be operated in any of the conventional modes. An unusual and useful feature is the provision of a pre-wash column to guard against the contamination of buffers with ammonia. The simplest of these models (BC-100) is fitted with three flow cells linked to a conventional three-channel dotting recorder. The sensitivity range is claimed to be between 10 and 100 nmole. The next model (BC-100L) is fitted with a single linearized output high-sensitivity colorimeter, which monitors continuously at wavelengths of 440 and 570 nm using one cell of 15-mm path length and of small volume (14 $\mu$l). The sensitivity of this instrument, which also has a series of switched stable ranges built into the recorder, is such that 1 nmole of amino acid gives a peak height of 1 in.

The third model (BC-201) is also fitted with the single sensitive colorimeter but has an increased capability for controlling buffer and temperature changes. A unique feature of this model is the use of a "patchwork" programmer (see Fig. 32.8), which does not have the disadvantage of a paper loop which is easily damaged. With suitable accessories, up to 19 samples can be analyzed sequentially, each sample being stored in a PTFE loop until required.

Jeol (Tokyo, Japan) produces the Model JLC-6AH, which is a conventional automatic analyzer that can process up to 36 samples automatically by either the single- or double-column mode.

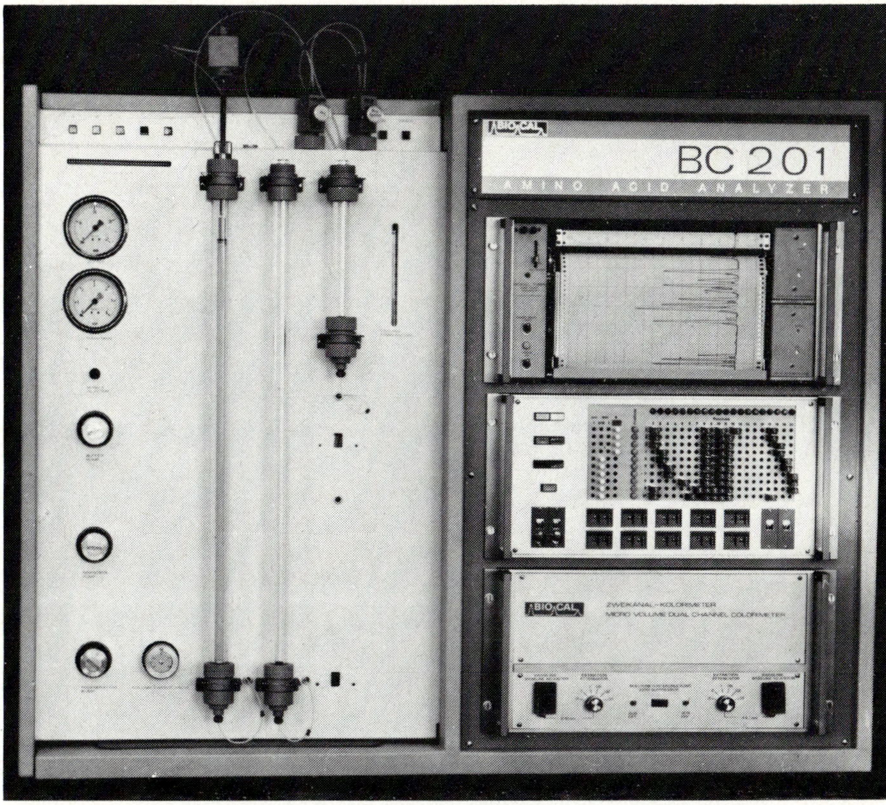


Fig. 32.8. LKB BC-201 automatic amino acid analyzer.

Virtis (Gardiner, N.Y., U.S.A.) are makers of the Phoenix amino acid analyzer, which can analyze up to eight samples sequentially by the single-column technique.

Perkin-Elmer (Beaconsfield, Great Britain) is the European agent for the Hitachi (Japanese) range of amino acid analyzers. Models KLA-3, KLA-5 and KLA-6 are all two-column instruments capable of automatic operation. The most recent model of this group (KLA-6) can deal with up to 12 samples in order.

A distinguishing feature of all these models is the existence of facilities for rapid conversion from ion-exchange chromatography to ligand-exchange chromatography, which, it is claimed, enhances resolution while simplifying the methodology. The detection limit is stated by the manufacturers to be $5 \times 10^{-10}$ mole.

The Mikrotechna (Prague, Czechoslovakia) Model AAA 881 is an automatic analyzer with two-column operation and manual sample injection (Fig. 32.9). It permits an output of five hydrolyzate analyses per day with an error of reproducibility of about 1%. The sensitivity range is 5–100 nmole. The addition of a sampling device makes it fully automatic.

There are many other manufacturers of amino acid analyzers producing mainly classical types of apparatus, for example Bender-Holbein (Munich, G.F.R.) and Carlo Erba (Milan, Italy).

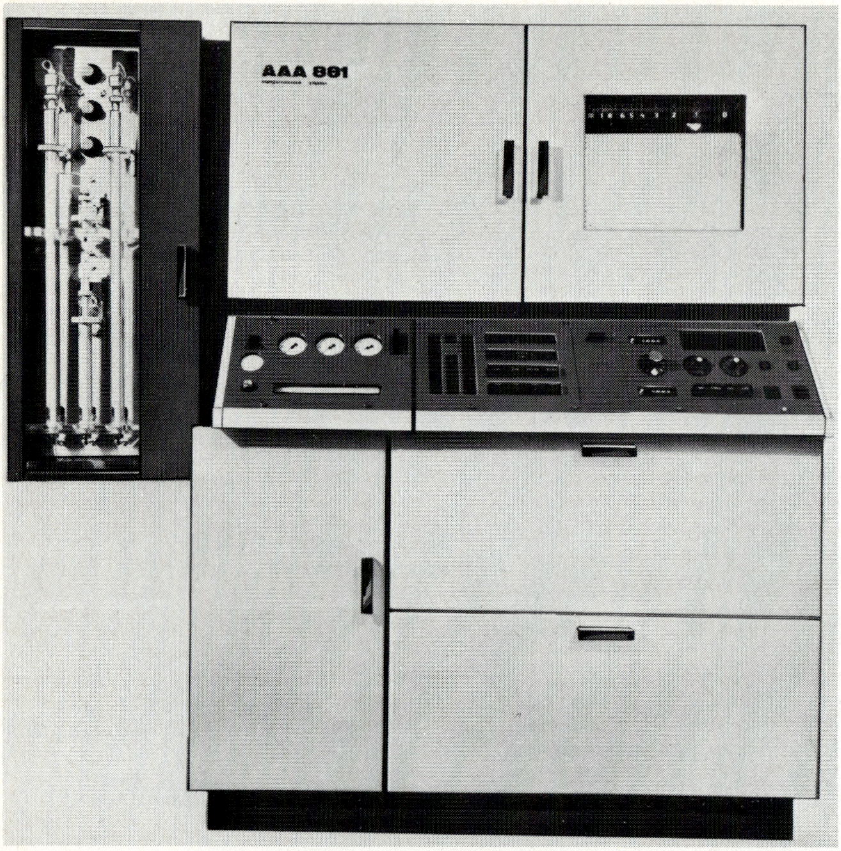

Fig. 32.9. Mikrotechna semi-automatic amino acid analyzer.

*Future trends in amino acid analysis*

Since the introduction of automated methods of amino acid analysis, the trend has been towards faster analyses on increasing numbers of samples. In this connection, clinical biochemistry has benefitted greatly from this rapid production of results, but the market has now become saturated with a variety of instruments that fulfil these needs. However, there remains a considerable demand for relatively simple, but versatile, analyzers to meet the occasional needs of the individual laboratory, rather than for a machine which is a factory for amino acid analysis.

*Ion-exchange chromatography*

A new type of automatic amino acid analyzer that combines the advantages of classical ion-exchange chromatography with the new engineering required for high-pressure liquid chromatography is the Durrum amino acid analyzer. This instrument (see Fig. 32.10) is produced by Durrum, Palo Alto, Calif., U.S.A. One of its main features is that the ion-

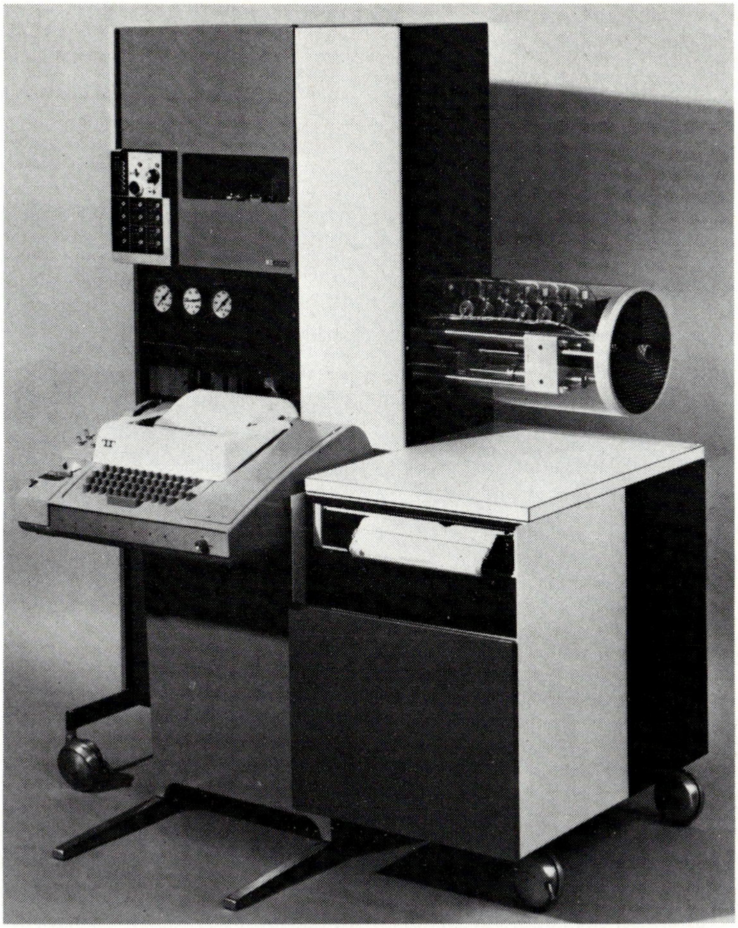

Fig. 32.10. Durrum fully automatic nanomole amino acid analyzer.

exchange resin used consists of very small spherical beads (8 μm in diameter) encased in a single stainless-steel column; the operating pressure for this column is 3000 p.s.i. A second important aspect is that a computer is an integral part of the whole instrument and not just an option, which in other instances is frequently produced by a different manufacturer. Another unusual feature is the photometer, which provides for rapid alternating wavelength scanning at two different wavelengths of 590 and 690 nm; the 690-nm scan is used as a standard that eliminates background interference. For the analysis of large numbers of proline samples, a filter of 440 nm can be used in place of that at 590 nm. The standard flow cell is 5 mm in path length but of extremely small volume (1.9 μl); other cells 1, 2, 10 and 20 mm in length can also be used. In each instance, the output from the photometer is fed directly to the computer and simultaneously to the visual chart. The computer prints out the concentration of each amino acid in nanomoles and 20 μl of serum is sufficient for each analysis; amounts of amino acid greater than

0.1 nmole are computed. One sample of protein hydrolyzate can be analyzed in 48 min and up to 80 samples can be analyzed without attention. The whole instrument costs £23,000 but clearly could be economic when very large numbers of samples need to be analyzed frequently.

In order to minimize capital outlay, there is a trend towards producing instruments that are capable of performing a number of different complicated analyses such as those required not only for amino acids, but also for peptides, nucleic acids, etc. Two such versatile instruments, which can be brought into operation by simple changes in the detector or the resin, are now commercially available; these are the Beckman Multichrom apparatus and the Hitachi 0.34 U. There is little doubt that such multi-purpose equipment has much to offer to those laboratories that require diverse analyses and are operating with a restricted capital expenditure.

In the field of ultramicroanalysis of amino acids, Udenfriend *et al.* and also Roth introduced a new technique in 1972, based on the interaction between an amino acid, ninhydrin and an aldehyde such as *o*-phthalaldehyde. Recently, a commercial phosphor, fluorescamine, has been used instead of ninhydrin for detecting amino acids at the picomole level. The American Instrument Co. (Silver Spring, Md., U.S.A.) markets a suitable adapter for converting the Technicon amino acid analyzer for quantitative fluorescence determinations.

## Packings for chromatographic columns

For the chromatography of amino acids, the most important carriers are those on which the substances are separated according to their charge, *i.e.*, according to their dissociation constants, because the sorption properties and solubilities of amino acids do not differ substantially. For this reason, attempts to separate amino acids by adsorption and partition chromatography were unsuccessful. However, the advantages of ion exchangers could be exploited only when synthetic ion exchangers were developed, whose properties could be suitably selected. The properties required were mechanical hardness, chemical stability and a perfectly reversible exchange and, as these conditions were not fulfilled by ion exchangers produced from cellulose or dextran gels, attention was therefore mainly devoted to polystyrene resins.

Owing to the low molecular weights of amino acids and the small differences between them, molecular sieves were also unsuitable means for efficient separations, even when small-pore types were developed. However, their adsorptive properties were made use of in the separation of amino acids by adsorption chromatography.

### *Sulphonated polystyrene cation exchangers*

The ion exchangers first produced were ground and graded on sieves in the dry state according to particle size. The limits of their particle size were roughly indicated by the range of sieves through which they would and would not pass, for example 200–400 mesh. The sieves were labelled by numbers indicating the number of meshes per square inch, and these values are recalculated as the diameter in millimetres according to the equation 16/ mesh = particle diameter (mm). Later, spherical ion exchangers were also classified in this manner.

For analytical purposes, the purity of the ion exchanger produced was of great importance. Ion exchangers contained the following two types of impurities. The admixture of ions was a consequence of their production in metallic apparatus, and this could be eliminated by washing. There were two sources of organic impurities: insufficiently pure starting monomers, especially divinylbenzene, and the products formed from impurities during sulphonation, especially the residues of wetting agents present in the polymer. The ion exchangers produced today are pure from this point of view. Almost all impurities could be also eliminated from ion exchangers by washing.

The moist ion exchanger is suspended in a 5-fold volume of 4 *N* hydrochloric acid (the dry ion exchanger is first allowed to swell in distilled water) and, if the supernatant after settling is coloured (yellow), decantation is repeated until the colour disappears. Then the ion exchanger sediment is stirred with a 10-fold amount of distilled water and again allowed to settle, and the same operation is repeated once more. The sediment is then suspended in a 5-fold volume of 2 *N* sodium hydroxide solution in an erlenmeyer flask, which is immersed in a boiling water-bath and heated to 100°C with occasional stirring for half an hour.

The settled ion exchanger is then decanted while still hot. If the supernatant is strongly coloured, the procedure with 2 *N* sodium hydroxide solution is repeated until the colour disappears. The supernatant may be weakly yellow. The sedimented ion exchanger is then decanted with 10-fold volumes of distilled water until neutral. This procedure eliminates from the ion exchanger impurities of heavy metal ions and organic substances from the production process. It was also found to be suitable for the elimination of denatured proteins adhering to the ion exchanger.

For the analysis of amino acids, the ground ion exchanger had to be freed from fines by decantation, which would otherwise cause a large increase in the backpressure of chromatographic columns. Later, a hydraulic method was developed (Hamilton, 1958) by which not only powder could be eliminated from an ion exchanger suspension, but particles of identical diameter were also gradually washed out, their diameters increasing with the increase in flow-rate. Chromatographic columns prepared from the fractions of ion exchangers obtained in this manner (*i.e.*, having a homogeneous particle size) had a substantially higher resolving power than the columns prepared from ungraded ion exchangers, and at the same time a lower back-pressure.

Experiments with ion exchangers of homogeneous particle size served as a basis for the deduction and checking of theoretical concepts in ion-exchange chromatography (Bogue; Hamilton, 1960) and also put new demands on the production of ion exchangers. The results of the solution of the linearized equation for the mass transfer during the transition of the concentration pulse at a given point of the chromatographic column showed that the efficiency of the column resolution expressed as the peak width of the chromatographed substance or by its fourth, $\sigma$, is directly proportional to the square of the particle diameter ($d_p$) of the ion exchanger used and indirectly proportional to the diffusion coefficient, $D_s$, of the chromatographed substance in the solid phase, *i.e.*, in the ion exchanger. The dependence is expressed by the approximate relationship $\sigma_{\text{exit}} >$

TABLE 32.2
ION EXCHANGERS FOR AMINO ACID ANALYSIS

| Ion exchanger | Cross-linkages (%) | Homogeneity | Purity | Particle form | Manufacturer |
|---|---|---|---|---|---|
| Dowex 50W | 8 | Mesh | Special | Beads | Dow Chem., |
| | 12 | Mesh | Special | Beads | Midland, Mich., U.S.A. |
| Amberlite IR-120/AS | | | | | |
| 2835 | 8 | μm | Laboratory | Powdered | Rohm & Haas, |
| 3545 | 8 | μm | Laboratory | Powdered | Philadelphia, Pa., |
| 4560 | 8 | μm | Laboratory | Powdered | U.S.A. |
| 80100 | 8 | μm | Laboratory | Powdered | |
| Zeo-Karb 225 | 8 | Mesh | Laboratory | Beads | Permutite, London, Great Britain |
| Ostion KS | | | | | |
| LG-0803 | 8 | μm | Special | Beads | United Chem. |
| LG-0802 | 8 | μm | Special | Beads | Metallurg. Works, Ústí, Czechoslovakia |
| Chromex UA-8 | 8 | μm | Laboratory | Beads | Reanal, Budapest, Hungary |
| Spherix | 8 | μm | Special | Beads | Phoenix, Gardiner, N.Y., U.S.A. |
| Custon Research Resin | | | | | |
| UR-30 | 8 | μm | Special | Beads | Beckman, Fullerton, |
| AA-15 | 8 | μm | Special | Beads | Calif., U.S.A. |
| AA-27 | 8 | μm | Special | Beads | |
| PA-28 | 8 | μm | Special | Beads | |
| PA-35 | 8 | μm | Special | Beads | |
| Resin H-70 | 8 | μm | Special | Beads | Hamilton, Reno, |
| HP-B80 | 7.75 | μm | Special | Beads | Nev., U.S.A. |
| HP-AN90 | 7 | μm | Special | Beads | |
| Aminex | | | | | |
| MS-C | 8 | μm | Laboratory | Powdered | Calbiochem, Los |
| MS-D | 8 | μm | Laboratory | Powdered | Angeles, Calif., |
| Blend Q-150S | 8 | μm | Laboratory | Powdered | U.S.A. |
| Q-50S | 8 | μm | Laboratory | Powdered | |
| Q-15S | 8 | μm | Laboratory | Powdered | |
| 50W | 12 | μm | Laboratory | Beads | |
| A-4 | 8 | μm | Special | Beads | |
| A-5 | 8 | μm | Special | Beads | |
| A-6 | 8 | μm | Special | Beads | |
| Chromobeads | | | | | |
| Type A | 8 | μm | Special | Beads | Technicon, Tarry- |
| Type B | 8 | μm | Special | Beads | town, N.Y., U.S.A. |
| Durrum | | | | | |
| DC-1A | 8 | μm | Special | Beads | Durrum, Palo Alto, |
| DC-6A | 8 | μm | Special | Beads | Calif., U.S.A. |

$\sim 0.1 \cdot d_p^2/D_s$. Hence, under constant conditions of diffusion, the separation efficiency depends on the particle size of the exchanger.

Attempts to increase the rate of analysis led, therefore, to the use of ion exchangers of ever smaller particle size, but this trend was limited by the increasing back-pressure of columns made of finer particles. Technically acceptable operating back-pressures (10–20 atm) were exerted by columns made of ion exchangers of 12–24 $\mu$m particle size, while finer granulated ion exchangers require large increases in pressure. For example, for an ion exchanger of 8 $\mu$m particle size, a pressure of about 200 atm is necessary. Also, the production of ion exchangers with such a small particle diameter and of such high homogeneity is very difficult and therefore these products are very expensive.

From the early days of ion exchanger production, it was known that the back-pressure in exchanger columns can be decreased to a certain extent by increasing the number of cross-linkages (divinylbenzene) in the polymer, *i.e.*, by making mechanically harder ion exchangers. For the separation of amino acids, ion exchangers with 8% of cross-linkages are the most suitable and this percentage cannot be increased without an adverse effect on the resolving power. Therefore, some workers have endeavoured to produce multicomponent copolymers by adding components that would produce mechanically hard polymers (Ertinghausen and Adler). The content of cross-linkages however, remains the same as in the original ion exchanger.

Sulphonated polystyrene cation exchangers of different quality for the chromatography of amino acids can be differentiated according to the above-mentioned stages of development. Some types can be identified according to their trade name, and in this respect, Dow Chem. (Midland, Mich., U.S.A.) chose a meaningful method of labelling its Dowex products. For example, Dowex 50 represents a strongly acidic sulphonated styrene–divinylbenzene cation exchanger, and the chemical purity is indicated by the letter W (white) and the percentage of the cross-linkages by X8, for example. The range of particle size for fractions graded by sieves is indicated by 200–400 mesh, for example, and fractions graded by the Hamilton hydraulic method are indicated by the particle diameter and the scattering, for example 40 ± 7 $\mu$m. The complete label is then: Dowex 50W-X8,200–400 mesh, or Dowex 50W-X8, 40 ± 7 $\mu$m, and it enables one to distinguish an ion exchanger from the point of view of its use in particular types of amino acid chromatography. Other producers of ion exchangers label ion exchangers with indications of the content of cross-linkages, and complete it with the range of particle sizes, as for example Amberlite IR-120, meaning that it is a sulphonated polystyrene–divinylbenzene with 8% of cross-linkages. In such instances the necessary information can be found in the producers' catalogues. Finally, special ion exchangers for certain new types of amino acid analyzers are indicated by trade names only. If the recommended ion exchangers are not used, these instruments do not achieve the declared quality of chromatographic separation.

A list of the main producers of ion exchangers for amino acid chromatography is given in Table 32.2.

### *Other ion exchangers*

As a result of the improvement and rapid development in the chromatography of amino acids on sulphonated polystyrene, the use of other ion exchangers has decreased.

The original development of chromatographic techniques meant that the weak cation exchanger Amberlite IRC-50 (Hannig; Ishii) was used, but only to a limited extent. Their disadvantage consisted in the difficult equilibration of the column and the necessity to adjust the pH of the sample accurately. Anion exchangers had a more important role. Strongly basic Dowex 2-X10 was used for the chromatography of strongly acidic amino acids and their derivatives (Schram *et al.*). Cysteic acid, phosphoserine and similar substances have appreciably larger elution volumes than other amino acids on an anion exchanger and therefore it is possible to separate them easily from other amino acids and among themselves. Elution is rapid because all basic and neutral amino acids can be eluted at once with the solvent front. The chromatographic separation ability of these ion exchangers is dependent on the same properties as in sulphonated polystyrenes, *i.e.*, on the particle diameter, its homogeneity, content of cross-linkages, etc.

As mentioned above, cellulose and dextran ion exchangers do not have suitable chromatographic and mechanical properties and therefore their use in the chromatography of amino acids is limited to special techniques. Carboxymethylcellulose was employed by Smith *et al.* for the separation of lysine from its oligomers and polymers. Munier and Drapier tried to modify the dextran anion exchangers DEAE- and QAE-Sephadex by cross-linking for the column chromatographic separation of cysteine and glutathione. The use of strongly basic Dowex 1-X8 and Sephadex G-10 represents a mechanical transition from ion-exchange chromatography to gel chromatography (Purdie and Hanafi).

*Molecular sieves*

The gel chromatography of amino acids was studied only after the development of the production of gels with small pores. However, the separation did not take place according to molecular weights, but according to differing adsorptivities. Conditions were found for adsorption chromatography on Sephadex G-10 (Eaker and Porath). The resolving ability for amino acids is not too high, with the exception of the separation of aromatic amino acids, which are bound firmly to the gel. This effect was put to further use by Ziska. The mechanical properties of swollen gels are not suitable for the rapid chromatography of amino acids, but molecular sieves of new types that are sufficiently hard have been produced.

## Preparation of eluents and reagents

*Buffers*

Analytical-grade chemicals are used for the preparation of all buffers necessary for amino acid analysis. Citrate buffers were originally prepared from citric acid and sodium hydroxide, but later it was found that it is more advantageous if sodium citrate is used because it does not absorb ammonia from the air and therefore there is no risk of an undefinable increase of the baseline. The rate of elution of substances from cation exchangers depends on the concentration of the cation in the eluent and therefore the elution buffers are prepared so that they have a defined normality of the cation. The compositions of the buffers for basic elution systems are given in Tables 32.3–32.6. Neutral detergent Brij-35 (polyethylene glycol lauryl ether) solution is added to all buffers so that they wash the surface of the ion exchanger well. They are prepared in large jars, 20–40 l, because it is important to adjust their pH value very accurately, which is time consuming. The adjustment and measurement of the pH is carried out on an accurate compensation pH meter. A definite setting is carried out after the first run of the standard mixture of amino acids. A buffer solution of pH 3.25 is set according to the position of

TABLE 32.3
COMPOSITION OF SODIUM CITRATE BUFFER SOLUTIONS FOR THE CHROMATOGRAPHY OF AMINO ACIDS BY THE TWO-COLUMN SYSTEM (MOORE *et al.*)

| Constituent | pH | | | | |
|---|---|---|---|---|---|
| | 2.2 ± 0.03 | 3.25 ± 0.01 | 4.25 ± 0.02 | 5.28 ± 0.02 | 4.26 ± 0.02 |
| | Sodium concentration (*N*) | | | | |
| | 0.2 | 0.2 | 0.2 | 0.35 | 0.38 |
| Sodium citrate dihydrate (g) | 9.8 | 784.3 | 784.3 | 1372.6 | 1489.2 |
| Conc. HCl (ml) | 15.5 | 498 | 335 | 260 | 614 |
| Thiodiglycol (ml) | 2.5 | 200 | 200 | – | – |
| Brij-35 solution, 50 g/100 ml (ml) | 1.0 | 80 | 80 | 80 | 80 |
| Caprylic acid (ml) | 0.05 | 4 | 4 | 4 | 4 |
| Final volume (l) | 0.5 | 40 | 40 | 40 | 40 |

TABLE 32.4
COMPOSITION OF LITHIUM CITRATE BUFFER SOLUTIONS FOR THE CHROMATOGRAPHY OF AMINO ACIDS BY THE TWO-COLUMN SYSTEM (BENSON *et al.*)

| Constituent | pH | | |
|---|---|---|---|
| | 2.20 ± 0.03 | 2.80 ± 0.01 (25°C) | 4.15 ± 0.01 |
| | Lithium concentration (*N*) | | |
| | 0.30 | 0.30 | 0.30 |
| | Citrate concentration (*M*) | | |
| | 0.10 | 0.153 | 0.10 |
| Lithium citrate tetrahydrate (g) | 14.1 | 150.5 | 282.0 |
| Lithium chloride (g) | – | 59.5 | – |
| Conc. HCl (ml) | 13.0 | 118.0 | 135.0 |
| Thiodiglycol (ml) | 1.25 | 25.0 | 25.0 |
| Caprylic acid (ml) | 0.05 | 1.0 | 1.0 |
| Brij-35 solution, 50 g/100 ml (ml) | – | 20.0 | 20.0 |
| Final volume (l) | 0.5 | 10 | 10 |

TABLE 32.5
COMPOSITION OF SODIUM CITRATE BUFFER SOLUTIONS FOR THE CHROMATOGRAPHY OF AMINO ACIDS BY THE SINGLE-COLUMN SYSTEM (PIEZ and MORRIS)

| Constituent | pH | |
|---|---|---|
| | 2.91 | 6.5 |
| | Sodium concentration (*N*) | |
| | 0.25 | 2.4 |
| | Citrate concentration (*M*) | |
| | 0.47 | 0.8 |
| Sodium citrate dihydrate (g) | 440.0 | 943.1 |
| Citric acid hydrate (g) | 1460.0 | – |
| Thiodiglycol | 90.0 | – |
| Brij-35 solution, 50 g/100 ml (ml) | 36.0 | – |
| Caprylic acid (ml) | 1.8 | – |
| Final volume (l) | 18 | 4 |

TABLE 32.6
COMPOSITION OF LITHIUM CITRATE BUFFER SOLUTIONS FOR THE CHROMATOGRAPHY OF AMINO ACIDS BY THE SINGLE COLUMN SYSTEM (PERRY *et al.*)

| Constituent | pH | | |
|---|---|---|---|
| | 2.80 ± 0.02 | 3.80 ± 0.02 | 6.10 ± 0.02 |
| | Lithium concentration (*N*) | | |
| | 0.2 | 0.2 | 1.2 |
| | Citrate concentration (*M*) | | |
| | 0.05 | 0.05 | 0.05 |
| Lithium citrate tetrahydrate (g) | 140.9 | 140.9 | 140.9 |
| 6 *N* HCl | Titrate to pH | Titrate to pH | Titrate to pH |
| Lithium hydroxide (g) | 12 | 12 | 12 |
| Lithium chloride (g) | – | – | 420.9 |
| Brij-35 solution, 50 g/100 ml (ml) | 20 | 20 | 20 |
| Thiodiglycol (ml) | 50 | 50 | – |
| Final volume (l) | 10 | 10 | 10 |

the cysteine peak, which should be mid-way between the peaks of alanine and valine. The buffer solution of pH 5.28 is adjusted according to the position of the peak of histidine, which should lie mid-way between the lysine and ammonia peaks. It is advisable to over-layer approximately 1 l of buffer solution of an accurately set pH with toluene and store it in a refrigerator as a standard for the measurement of the pH values of freshly prepared buffers.

In order to prevent the contamination of buffer solutions with microorganisms, a series of preservatives is employed. First phenol (0.1%) was used, but it turned red on ageing and then interfered in the detection of amino acids. Therefore, new preservatives, for example caprylic acid (0.01%), diethyl pyrocarbonate (0.01%) and pentachlorophenol ($10^{-5}$%) were introduced. Diethyl pyrocarbonate and pentachlorophenol are added in ethanolic solution. Some amino acids are oxidized during chromatography by the oxygen in the air and therefore an antioxidant is added to the buffers, *e.g.*, thiodiglycol, before the buffers are filled into the analyzed reservoir so that it is not oxidized during storage of the buffer. If some of the acidic buffer (pH < 5) gives an excessively high baseline on analysis, it can be purified on a Dowex 50-X8 (100–200 mesh) column or its equivalent. Approximately 2 l (2.5 kg) of ion exchanger is washed with 4 l of 2 *N* sodium hydroxide solution and 2 l of distilled water and is then eluted with the buffer until the pH of the effluent is identical with that of the original buffer. Further pure buffer is collected at a rate of several litres per hour. The capacity is approximately 18 l of buffer. Ready-prepared, specially pure buffers for amino acid analysis are now available from some producers of amino acid analyzers (Durrum, Technicon, Hamilton, etc.).

*Ninhydrin reagent*

The composition of the ninhydrin reagent has undergone many changes during the development of amino acid analysis. However, the basic principles applied in the original work of Moore and Stein (1948, 1954) have remained unchanged. The reagent contains a concentrated buffer, which adjusts the pH of the reaction mixture to about 5.0, which is optimum for the course of the reaction and the absorption spectrum of the product. A mixture of methyl Cellosolve with water ensures a good solubility of both organic compounds and inorganic salts. The reduced ninhydrin for the start of the colour reaction is obtained by addition of tin(II) chloride. Therefore, the preparation of the reagent must be carried out in the absence of oxygen from the air, *i.e.*, under nitrogen, and it is also stored under these conditions.

For the two-column system of amino acid analysis (Spackman *et al.*), the ninhydrin reagent is prepared as described below.

First, 4 *N* sodium acetate buffer solution of pH 5.5 is prepared as follows. Reagent-grade sodium acetate trihydrate (2720 g) is dissolved in 2 l of water on a steam-bath and, after cooling to room temperature, 500 ml of glacial acetic acid are added and the volume is made up to 5 l. The undiluted buffer should have a pH of 5.51 ± 0.03; if a final adjustment of the pH is necessary, 5 g of sodium hydroxide correspond to about 0.04 pH. The buffer can be stored at 4°C without a preservative and is filtered before use. The methyl Cellosolve peroxide content is checked before use by mixing 2 ml of methyl Cellosolve with 1 ml of 4% potassium iodide solution; the mixture should not be yellow.

The reagent solution is then prepared in a 4-l aspirator bottle, the bottom outlet of which is connected to pre-purified nitrogen supply. Three litres of methyl Cellosolve are stirred magnetically for 15 min while bubbling a slow stream of nitrogen through it. Reagent-grade ninhydrin (80 g) is then added and nitrogen is again bubbled through the mixture for 15 min. After the ninhydrin has dissolved, 1 l of the 4 *N* sodium acetate buffer solution of pH 5.5 is added and the solution is stirred for 30 min while bubbling

*References p. 710*

nitrogen through it. To the solution is then added 1.6 g of tin(II) chloride dihydrate (reagent-grade, not hydrolyzed) and, while it dissolves, the nitrogen line is closed and transferred to an inlet tubing of the bottle stopper. After dissolution is complete, the ninhydrin reagent is transferred by means of a Tygon tubing connector into the ninhydrin reservoir using the pressure of the nitrogen supply. Before the three-way stopcock connected to the ninhydrin reservoir is opened, all gas in the connecting line is allowed to flow out. The addition of ninhydrin solution should be complete in about 5 min.

For the single-column system of amino acid analysis (Piez and Morris), the reagent is prepared and filled into the reservoir in the same way as the above reagent, from which it differs in composition only. The constituents and the length of time nitrogen is bubbled through the solution after each addition are as follows:

| *Addition of* | *Time of $N_2$ bubbling (min)* |
|---|---|
| methyl Cellosolve (2500 ml) | 15 |
| ninhydrin (80 g) | 15 |
| 4 *N* sodium acetate buffer solution (1000 ml)<br>distilled water (500 ml) | 30 |
| tin(II) chloride dihydrate (1.333 g) | 15 |

Papers describing modifications of the ninhydrin reagent can be divided into two groups. In the 1950s (Troll and Cannan), modifications were studied that led to a change in solvents. Although such a modification increased the colour yield of the reaction, the method became impractical for the routine analysis of large number of samples. The second group of papers was devoted to the search for a more suitable reducing reagent than tin(II) chloride. A series of reducing substances was tested: ascorbic acid, potassium cyanide, sodium dithionite and titanium(III) chloride, and even electrolytic reduction. However, none of these methods has been used to any great extent.

### *Amino acid standards*

For the colorimetric determination of amino acids in fractions from chromatographic columns, chromatographically pure preparations of amino acids that gave reproducible colour yields closest to the theoretical values with ninhydrin reagent were used. These amino acids were leucine, isoleucine and alanine. The colour yields of all amino acids were referred to leucine as a unit, and these ratios were called leucine units and were used for the calculation of the true content of single amino acids from the values determined as leucine.

For amino acid analyzers, another method of standardization had to be chosen. The whole procedure was standardized by the determination of a mixture of all amino acid standards. The required accuracy in the preparation of the mixture was achieved by their large-scale preparation. Aliquots of the solution were then sealed under nitrogen in test-tubes containing smaller amounts used in current practice. Later, ready-for-use standard mixtures of amino acids became commercially available from Calbiochem (Los Angeles, Calif., U.S.A.), Serva (Heidelberg, G.F.R.), etc.

*Preparation of standard mixtures of amino acids*

All quantitatively recoverable amino acids either in protein hydrolyzates (hydrolyzate standard) or in physiological fluids and biological extracts (rare amino acids standard) are used to standardize the amino acid analyzer. The standards of amino acids must be of chromatographic purity, analyzed for total nitrogen, ash and dry weight.

In order to facilitate the weighings, the solution is prepared on a relatively large scale by transferring 1250 ± 2 μmole of each amino acid of a corresponding standard solution (half of this amount of cystine) into a 500-ml calibrated flask. The flask is half filled with water and 5 ml of concentrated hydrochloric acid are added in order to dissolve the less soluble amino acid. The final solution, containing 2.5 μmole of each amino acid per millilitre, was transferred in 5.2-ml amounts into 10-ml glass ampoules, which were sealed under nitrogen and stored in the cold.

A sample to be analyzed is prepared by diluting the standard stock solution with buffer solution of pH 2.2 containing Brij-35 and thiodiglycol.

## Chromatographic elution systems

*Two-column system* (*Spackman et al.*)

Amino acids can be separated from a protein hydrolyzate on a 150-cm column in two steps. With a 0.2 $N$ buffer solution of pH 3.25, acidic and some neutral amino acids are eluted. After the elution of cystine, no amino acid should be eluted with the same buffer for a long time. Therefore, after elution of glycine (250 ml; 8 h 20 min), the first buffer is switched over to a 0.2 $N$ buffer solution of pH 4.25 and other neutral amino acids are eluted with it. Finally, the aromatic amino acids tyrosine and phenylalanine are also eluted. According to their dissociation constants, aromatic amino acids should have substantially smaller elution volumes than they actually have. Their elution volumes are increased by the adsorption of the aromatic nucleus on the ion-exchange matrix. The remaining basic amino acids are eluted after a very long time and their elution can be enhanced by an appreciable increase in buffer concentration, but at the cost of higher baseline drift, increase in volume changes of the ion exchanger and a number of technical difficulties. Therefore, elution is stopped at this moment and the remaining amino acids are drained out of the column with alkali. The second aliquot of the same sample is then separated with a more concentrated and more basic 0.38 $N$ buffer solution of pH 5.28 on a 15-cm column. This elution is carried out so that acidic and neutral amino acids are eluted first and then basic amino acids are separated from tryptophan to arginine. After the elution of arginine, no ninhydrin-positive substance is present on the column. The elution on both columns lasts 21.5 h (5 and 16.5 h) at a flow-rate of eluent of 30 ml/h at 50°C. A chromatogram of the standard mixture of amino acids (hydrolyzate) is shown in Fig. 32.11.

Further development of the two-column system of elution of amino acids from hydrolyzates proceeded only by increasing the efficiency of ion exchangers and optimization of the dimensions and technical performance of chromatographic columns. In spite of its considerable acceleration, the elution system has not changed further (Benson and Patterson, Ertinghausen *et al.*).

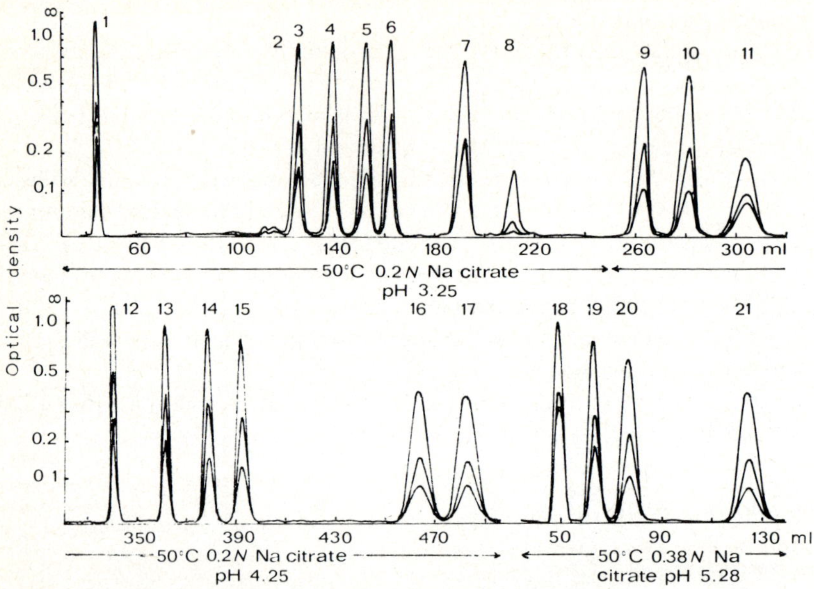


Fig. 32.11. Separation of amino acids from hydrolyzate on a two-column system. Standard mixture of amino acids. Amino acid analyzer: basic type. Columns: 150 × 0.9 cm and 15 × 0.9 cm. Ion exchanger: Amberlite IR-120, 40 ± 7 μm (150-cm column) or 25–30 μm (15-cm column). Operating conditions: 150-cm column operated at 50°C and 30 ml/h of 0.2 *N* sodium citrate buffer of pH 3.25 and 4.25; 15-cm column operated at 50°C and 30 ml/h of 0.38 *N* sodium citrate buffer of pH 5.28. Detection: ninhydrin. Peaks: 1 = cysteic acid; 2 = methionine sulphoxides; 3 = aspartic acid; 4 = methionine sulphone; 5 = threonine; 6 = serine; 7 = glutamic acid; 8 = proline; 9 = glycine; 10 = alanine; 11 = cystine; 12 = valine; 13 = methionine; 14 = isoleucine; 15 = leucine; 16 = tyrosine; 17 = phenylalanine; 18 = lysine; 19 = histidine; 20 = ammonia; 21 = arginine.

For the analysis of amino acids from physiological fluids (blood plasma, urine and mammalian tissue extracts), the elution conditions on a 150-cm column had to be adjusted. The separation efficiency is increased by elution with a buffer of pH 3.25 at 30°C and, of course, it lasts longer (330 ml; 11 h). In this manner, a separation of substances immediately behind the front is achieved as well as a better separation of the group between aspartic acid and glutamic acid. Elution is then continued at 50°C with a buffer solution of pH 4.25 (up to 670 ml; 22.5 h), until β-aminoisobutyric acid is eluted. The column is then regenerated (see the section on the procedure with amino acid analyzers, p. 676). For the separation of the basic components of the mixture, the separation capacity of the 15-cm column is insufficient and a 50-cm column is therefore used. Elution is carried out with the 0.38 *N* buffer solution of pH 4.26 at 30 and 50°C. The first three peaks belong to strongly acidic amino acids and acidic and neutral components. Then a combined peak of amino acids follows, moving on the 150-cm column at the end of the elution. Further basic components are separated up to carnosine. The very long time necessary for the elution of arginine, after carnosine, is shortened by increasing the temperature to 50°C after the elution of carnosine (400 ml; 13.3 h). The whole elution requires 21 h (630 ml). The chromatogram of the standard mixture of amino acids (physiological fluids) is shown in Figs. 32.12 and 32.13.

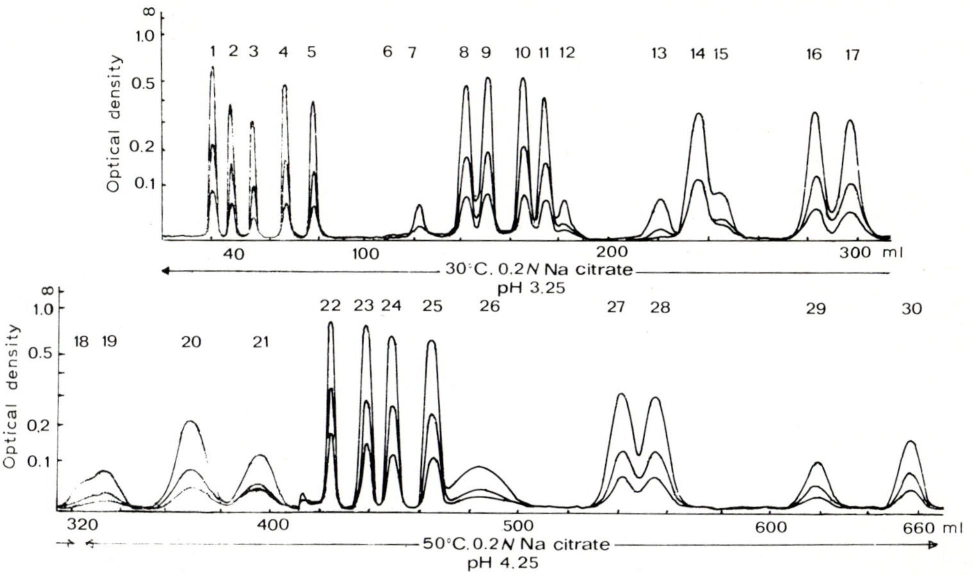


Fig. 32.12. Separation of acidic and neutral ninhydrin-positive compounds of blood plasma on a two-column system. Standard mixture. Amino acid analyzer: basic type. Column: 150 × 0.9 cm. Ion exchanger: Amberlite IR-120, 40 ± 7 μm, powdered. Operating conditions: operated at 30 and 50°C with a flow-rate of 30 ml/h of 0.2 *N* sodium citrate buffer solution of pH 3.25 and 4.25, respectively. Detection: ninhydrin. Peaks: 1 = phosphoserine; 2 = glycerophosphoethanolamine; 3 = phosphoethanolamine; 4 = taurine; 5 = urea; 6 = methionine sulphoxides; 7 = hydroxyproline; 8 = aspartic acid; 9 = threonine; 10 = serine; 11 = asparagine; 12 = sarcosine; 13 = proline; 14 = glutamic acid; 15 = citrulline; 16 = glycine; 17 = alanine; 18 = α-aminoadipic acid; 19 = α-amino-*n*-butyric acid; 20 = valine; 21 = cystine; 22 = cystathionine; 23 = methionine; 24 = isoleucine; 25 = leucine; 26 = glucosamine; 27 = tyrosine; 28 = phenylalanine; 29 = β-alanine; 30 = β-aminoisobutyric acid.

A disadvantage of this system of elution is that glutamine and asparagine cannot be separated on the 150-cm column and must be determined from the difference in the contents of aspartic and glutamic acids before and after hydrolysis and the total amount of amides. This disadvantage was eliminated by elution with lithium buffers (Benson *et al.*). In lithium buffers, the diffusion of the zones of amino acids is less and their separation is sharper, while the elution volumes of some amino acids also change and the total elution conditions must be modified. A 55 × 0.9 cm column is eluted with a 0.3 *N* buffer solution of pH 2.8 and a 0.3 *N* buffer solution of pH 4.15 at 37°C and a flow-rate of eluent of 70 ml/h. The change of buffer takes place after 150 min (175 ml) and the whole elution lasts 4 h 40 min. **α**-Aminoisobutyric acid is eluted last. The 50-cm column need not be eluted with a lithium buffer, but if such a buffer is used, better separations on the column can be achieved and there is no risk of changing the sodium or lithium buffers by mistake (Mondino *et al.*). The two-column system of elution of amino acids from biological materials has not been developed further and it has been changed to stepwise gradient elution from a single column.

*References p. 710*

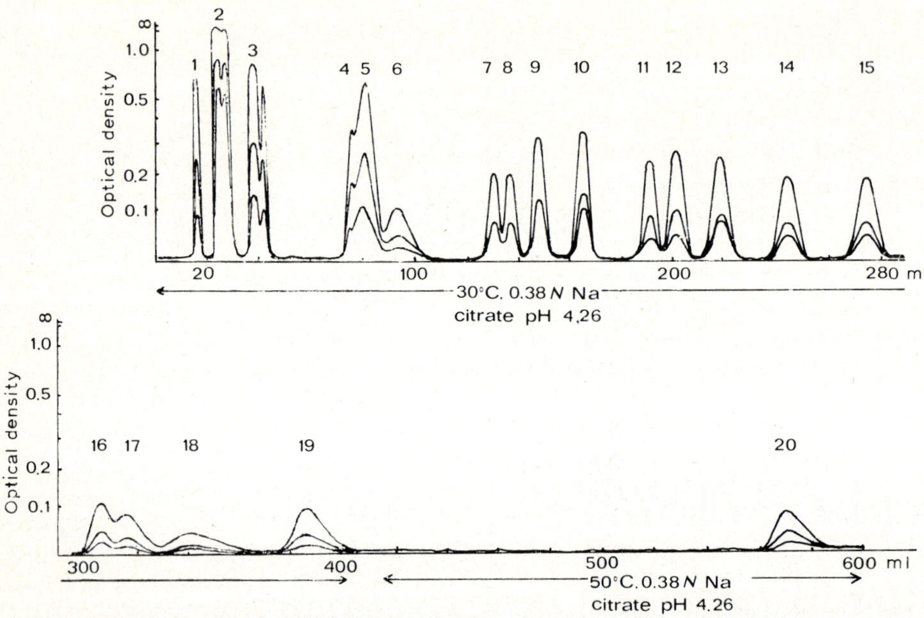


Fig. 32.13. Separation of basic ninhydrin-positive compounds of blood plasma on a single-column system. Standard mixture. Amino acid analyzer: basic type. Column: 50 × 0.9 cm. Ion exchanger: Amberlite IR-120, 25–30 μm. Operating conditions: operated at 30 and 50°C with a flow-rate of 30 ml/h of 0.38 *N* sodium citrate buffer solution of pH 4.26. Detection: ninhydrin. Peaks: 1 = phosphoserine; 2 and 3 = acidic and neutral components; 4 = β-alanine; 5 = tyrosine, phenylalanine, β-aminoisobutyric acid and glucosamine; 6 = galactosamine; 7 = hydroxylysine; 8 = *allo*-hydroxylysine; 9 = γ-amino-*n*-butyric acid; 10 = ornithine; 11 = ethanolamine; 12 = ammonia; 13 = lysine; 14 = 1-methylhistidine; 15 = histidine; 16 = 3-methylhistidine; 17 = anserine; 18 = tryptophan; 19 = creatinine and carnosine; 20 = arginine.

*Single-column system*

On a 133-cm column, amino acids from protein hydrolyzates and amino acids from physiological fluids and biological extracts are separated with the same elution system (Piez and Morris). For elution, a continuous gradient of the concentration of cations and of the pH is used, defined by starting buffers and the method of preparation (see Table 32.1) in a commercial nine-chamber Varigrad mixer (see Fig. 32.3); 0.25 *N* sodium citrate buffer solution of pH 2.91 is mixed with 2.4 *N* sodium citrate buffer solution of pH 6.5. The increase in salt concentration causes a gradual baseline drift (see Fig. 32.14), which is caused by increasing concentration of ammonia present in the chemicals from which buffers are prepared, but also by some cations (Brummel *et al.*) and other substances. For the analysis of amino acids in hydrolyzates, the separation efficiency of the column is sufficient (see Fig. 32.14), while for more complex mixtures of amino acids it is limited. The separation of the critical group of components of these mixtures in the region of the elution of amides and the pairs homocystine–β-alanine and γ-aminobutyric acid–ammonia can be improved by using lithium buffers (Meyer *et al.*, Perry *et al.*). The separa-

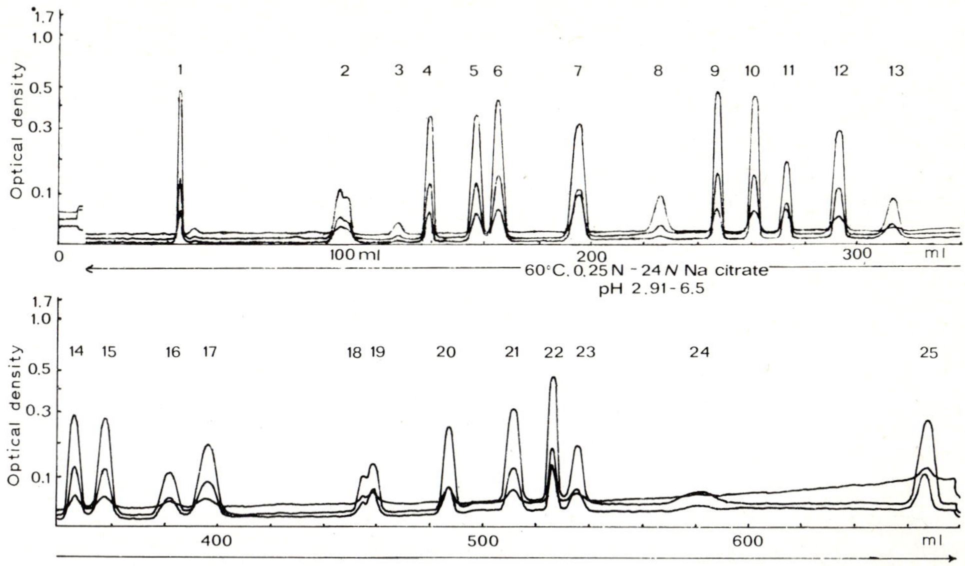


Fig. 32.14. Separation of amino acids from hydrolyzate on a single-column system. Standard mixture of amino acids. Amino acid analyzer: Technicon type. Column: 133 × 0.9 cm. Ion exchanger: Dowex 50W-X12, 20–30 μm, beads. Operating conditions: operated at 60°C with a flow-rate of 30 ml/h of a continuous gradient of concentration and pH (0.25–2.4 *N* sodium concentration and pH 2.91–6.5) formed in a Varigrad gradient mixer. Detection: ninhydrin. Peaks: 1 = cysteic acid; 2 = methionine sulphoxides; 3 = hydroxyproline; 4 = aspartic acid; 5 = threonine; 6 = serine; 7 = glutamic acid; 8 = proline; 9 = glycine; 10 = alanine; 11 = cystine; 12 = valine; 13 = methionine; 14 = isoleucine; 15 = leucine; 16 = tyrosine; 17 = phenylalanine; 18 = hydroxylysine; 19 = *allo*-hydroxylysine; 20 = ornithine; 21 = ammonia; 22 = lysine; 23 = histidine; 24 = tryptophan; 25 = arginine.

tion efficiency of the column can be increased further by decreasing the steepness of the gradient and by temperature changes. A mixture of up to 92 components was separated in this manner (Mardeus *et al.*).

Simultaneously with the single-column system of elution with a continuous gradient, a single-column system involving elution with a stepwise gradient was developed by Hamilton and Anderson. At first, the efficiency of this system was not comparable with that of the two-column system, especially in the analysis of hydrolyzates; the time of analysis was 38 h. The separation of amino acids was achieved by elution with buffers of pH 2.95, 4.15 and 5.0 at temperatures of 40 and 50°C on a 100 × 0.63 cm Amberlite IR-120 (20–40 μm) column. However, its advantage consisted in the satisfactory selection of the course of the gradient and other conditions of elution. As in all other instances, further development of this system was directed to shortening the elution time, *i.e.*, to increasing the efficiency, but also to increasing the separation effect in the chromatography of ninhydrin-positive substances from biological material. An increase in the elution rate was mainly important for the separation of hydrolyzates for which an elution time of 260 min was achieved with a standard analyzer (Power and Benet). The chromatography of a mixture of ninhydrin-positive substances from biological materials was aimed at the separation of all accessible substances on a single column, so that they might be

*References p. 710*

characterized by their elution volumes under identical elution conditions. Important work in this respect was described in a paper by Hamilton (1963), which contained the elution volumes of 180 substances. It can be used for the characterization of unidentified substances and for the choice of elution conditions for mixtures of ninhydrin-positive substances present in true natural mixtures. For improving the separation of amides and the critical pairs of amino acids, lithium buffers have been made use of even in this system of elution (Atkin and Ferdinand, Kedenburg).

*Special systems*

For some sources and groups of rare amino acids and their derivatives, special conditions of elution have been developed. Systems devised for the separation of mixtures of amino acids of animal origin had to be modified for use in the separation of amino acids extracted from plant materials, owing to their quantitative and qualitative differences in composition (Krishchenko and Gruzdev, Lorenz, Vega and Nunn). A map of elution volumes of single components from extracts from insects was also developed (Bonnot *et al.*). For control analyses in the foodstuffs industry, a procedure was developed for the analysis of amino acids in hydrolyzates of foods and meals (Jacobs, Roach) and of beer and malt (Mostek *et al.*). A method for the determination of the amino acid composition of grass and silage was worked out for the comparison of their fodder value (Macpherson and Wall). For industrial purposes, a procedure was developed for the separation of amino acids in wool hydrolyzates (Tasdhomme). Amino acid analyzers have often also been used for the separation and determination of non-amino acid substances, for example amines (Morris *et al.*, Rokushika *et al.*, Wall), uronic acids (Olson *et al.*), alcohols, aldehydes and aromatic acids (Lange and Hempel). Ion exchangers were utilized as ligand exchangers (Wagner and Shepherd) for the separation of amino sugars and cystine derivatives. Equipment for this type of chromatography was built into a Hitachi–Perkin-Elmer Type KLA-3B amino acid analyzer. The Jeol JLC-6AH analyzer can be adapted for the chromatography and analysis of sugars on anion exchanger LC-R-3 with borate buffers.

Hydrolyzates of collagen and elastin contain desmosine and isodesmosine and the elution conditions in a single-column system were modified for their separation (Miller and Piez, Starcher *et al.*). Later on conditions for elution in two-column systems were elaborated (Bandlow and Nordwig, Yu). Many papers have been devoted to the chromatography of sulphur-containing amino acids. Elution volumes of cysteine derivatives (Purdie *et al.*) and the products of cystine formed on substitution of proteins and their hydrolysis have been defined (Beale and Kent), and conditions have been established for the separation of the derivatives of lysine originating during protein substitution from native amino acids, and also conditions for accelerated elution (Ronca *et al.*, Seely *et al.*). Methylhistidine and some rare amino acids were separated on 15-cm columns by Kirkpatrick and Anderson. By decreasing the liquid flow-rate in the reactor by half, a 10–20-fold increase in sensitivity of detection of N-methylamino acids was achieved; the conditions for the elution were adjusted for this purpose (Coggins and Benoiton). Tryptophan and its derivatives were separated on Amberlite CG-50 (Contractor).

*Gel permeation chromatography*

The importance of the gel chromatography of amino acids increased after the introduction of the manufacture of gels of low porosity. However, with amino acids the separation is not based on the size of the molecules, but on adsorption. A Sephadex G-10 column is eluted with solvent systems consisting of acetic acid, salts and pyridine, citrate and Tris buffers, or even a weak solution of an alkali-metal hydroxide (Eaker and Porath). From the list of elution volumes of single amino acids in various solvent systems (Table 32.7), distinctly higher retention volumes of aromatic amino acids can be seen, caused by the adsorption of the aromatic ring on the gel. Sephadex G-10 can also be used for the separation of single amino acids from macromolecular substances, for example iodinated amino acids from serum (Mongey and Mason). A combination of molecular sieve and ion-exchange

TABLE 32.7

GEL FILTRATION OF AMINO ACIDS ON SEPHADEX G-10: DISTRIBUTION COEFFICIENTS IN TEN SOLVENTS

Packing: Sephadex G-10, water regain = 0.96, $<40\ \mu m$. Column: 142 × 1 cm, Perspex, void volume = 41 ml, internal volume of swollen gel = 43 ml. Operating conditions: 25°C, 10 ml/h. Solvents: $S_1$ = 0.2 *M* acetic acid, pH 2.7, freshly poured column; $S_2$ = 0.2 *M* acetic acid, pH 2.7, after exposure of gel to all other media; $S_3$ = 0.2 *M* acetic acid with 0.5 *M* sodium chloride; $S_4$ = 0.2 *M* acetic acid with 2.0 *M* sodium chloride; $S_5$ = 0.1 or 0.3 *M* pyridine–acetic acid (equimolar), pH 5.0; $S_6$ = 0.35 *M* sodium citrate, pH 5.28; $S_7$ = 1.0 *M* pyridine with 0.03 *M* acetic acid, pH 6.7; $S_8$ = 1.0 *M* pyridine with 0.5 *M* sodium chloride, pH 8.5; $S_9$ = 0.05 *M* Tris–sulphuric acid buffer, pH 8.07; $S_{10}$ = 0.01 *M* sodium hydroxide solution. Detection: ninhydrin colorimetry of the effluent. Sample: 0.5–0.8 μm of each amino acid in 1 ml.

| Substance | Solvent | | | | | | | | | |
|---|---|---|---|---|---|---|---|---|---|---|
| | $S_1$ | $S_2$ | $S_3$ | $S_4$ | $S_5$ | $S_6$ | $S_7$ | $S_8$ | $S_9$ | $S_{10}$ |
| Lysine | 0.20 | 0.06 | 0.29 | 0.35 | 0.19 | 0.17 | 0.16 | 0.19 | 0.14 | 0.24 |
| Histidine | 0.31 | 0.07 | 0.34 | 0.39 | – | – | 0.32 | – | – | – |
| Ornithine | – | 0.07 | – | 0.37 | – | – | – | – | – | – |
| Arginine | – | 0.08 | 0.48 | 0.60 | 0.27 | – | – | – | – | – |
| Threonine | – | 0.28 | – | 0.51 | – | – | – | – | – | – |
| Valine | 0.38 | 0.28 | 0.50 | 0.69 | – | – | – | – | – | – |
| Proline | – | 0.29 | 0.38 | 0.47 | – | – | – | – | – | – |
| Glycine | 0.44 | 0.29 | 0.48 | 0.54 | 0.34 | 0.36 | 0.35 | – | 0.35 | 0.32 |
| Serine | – | 0.30 | – | – | – | – | – | – | – | – |
| Isoleucine | – | 0.30 | 0.57 | 0.86 | – | – | – | – | – | – |
| Leucine | 0.40 | 0.32 | 0.58 | – | – | 0.39 | – | – | – | – |
| Norleucine | – | 0.32 | 0.59 | 0.90 | – | – | – | – | – | – |
| Methionine | 0.40 | – | – | 0.58 | – | – | – | – | – | – |
| Glutamic acid | 0.44 | 0.41 | 0.49 | 0.61 | – | – | 0.29 | 0.14 | – | 0.13 |
| Aspartic acid | – | 0.53 | – | 0.54 | – | – | – | – | – | 0.15 |
| Phenylalanine | 0.86 | 0.64 | 1.11 | 1.83 | 0.71 | 0.81 | 0.65 | 0.56 | 0.74 | 0.61 |
| Tyrosine | 1.28 | 0.93 | 1.73 | 2.85 | 0.96 | 1.16 | 0.88 | 0.61 | 1.11 | 0.38 |
| Tryptophan | – | 2.74 | – | – | – | – | – | – | – | 2.88 |
| Ammonia | 0.56 | 0.13 | 0.68 | 0.79 | 0.46 | 0.39 | 0.43 | 0.67 | 0.38 | 0.75 |
| Urea | 1.09 | 1.06 | 1.09 | 1.17 | 1.03 | 1.07 | 0.95 | 0.94 | 1.02 | 1.06 |
| Glucosamine | – | 0.07 | 0.49 | 0.51 | – | – | – | – | – | – |

*References p. 710*

effects was achieved during the separation of cysteine and glutathione derivatives on a mixed-bed column of Dowex 1 and Sephadex G-10 (Purdie and Hanafi). The same effect was utilized for the separation of amino acids and their dinitrophenyl derivatives on a cross-linked DEAE- and QAE-Sephadex (Munier and Drapier).

**Preparation of sample**

The hydrolysis of amino acids can be catalyzed by acids or hydroxides; acid hydrolysis is technically simpler and has been more thoroughly investigated than alkaline hydrolysis. During acid hydrolysis the destruction of amino acids takes place to an appreciably lower extent than during alkaline hydrolysis, and therefore mainly acid hydrolysis is used for the determination of the amino acid composition of proteins. This has been the subject of many studies (Tristram), but the basic procedure (Hirs *et al.*, 1954) has remained practically unchanged.

The hydrolysis of proteins consists of the following operations. Protein or an aliquot of a material containing a known amount of protein (4 mg; a dry sample or a solution that is evaporated directly in the hydrolysis tube) is introduced into a heavy-walled 180 × 20 mm test-tube, taking care that the walls are not wetted. Azeotropic hydrochloric acid distilled three times from glass (2 ml; 5.7 *N*) is then added and the test-tube is constricted with a small, sharp flame so that the final internal diameter is *ca.* 1 mm. After cooling, the contents are cooled with ice, the test-tube is evacuated with an oil pump (0.1 mmHg), and the tube is finally sealed with a sharp flame under evacuation.

The test-tube containing the mixture to be hydrolyzed is placed in an aluminium block heated at 110 ± 0.5°C for the required time. After the hydrolysis is completed, the test-tube is allowed to cool and is then opened with a small, sharp flame directed just below the sealed tip. The upper part of the test-tube is then cut off and the contents are transferred into a 100-ml flask. Both parts of the test-tube are rinsed quantitatively into the flask.

The hydrolyzate is mixed with distilled water and evaporated to dryness three times on a rotatory evaporator in order to eliminate hydrochloric acid. The residue is then dissolved in citrate buffer of pH 2.2 and made up to the required volume. The prepared hydrolyzate is allowed to stand for 4 h in air at room temperature before use in order to oxidize all cysteine to cystine.

If the protein does not contain any saccharides or heavy metal ions and the hydrolysis is carried out in the absence of air, the original content of the amino acids that are decomposed during hydrolysis can be calculated by using an equation for first-order reaction kinetics and the values of these amino acids determined after two suitably chosen periods of hydrolysis. The equation for the first-order kinetics is:

$$\log A_0 = \frac{t_2}{t_2 - t_1} \cdot \log A_1 - \frac{t_1}{t_2 - t_1} \cdot \log A_2$$

where $t_1$ and $t_2$ are the times of hydrolysis and $A_0, A_1$ and $A_2$ are the contents of a particular amino acid at the start and after $t_1$ and $t_2$, respectively. After 24 h of hydrolysis, labile amino acids are decomposed only to a small extent but strongly bound amino acids (such as isoleucine and valine) are not released completely. After 72 h of hydrolysis, the pattern is reversed.

During hydrolysis, hydroxyamino acids are partially decomposed, while tryptophan and amides are decomposed completely. Cysteine and cystine are oxidized to an undefined product. These amino acids must be determined by specific procedures (Schram *et al.*; Spies and Chambers; Zmrhal, 1968).

The content of labile amino acids cannot be calculated if the hydrolyzed protein was not pure. The crude protein extracts are therefore hydrolyzed for an average time (48 h) in order to keep the error of the determination of labile and firmly bound amino acids low. The hydrolysis of larger amounts of proteins is carried out by refluxing under nitrogen.

Free amino acids and ninhydrin-positive substances from physiological fluids (blood, plasma, urine and cerebrospinal fluid) are obtained by deproteinization either with picric or sulphosalicylic acid (Berridge *et al.*, Knipfel *et al.*, London). Solutions of various concentrations and in various ratios with respect to the analyzed fluid are used, depending on the type of fluid and its source. After removal of the precipitate, an aliquot of the supernatant is introduced on to the column. Physiological fluids can be also deproteinized with an ion exchanger in the $H^+$ form (Reid) or by centrifuging at high accelerations (Kedenburg). The extraction of amino acids from tissues is carried out by deproteinization of the centrifuged tissue homogenate with the same reagents, and also with trichloroacetic and perchloric acids (Saifer and Gerstenfeld). Reid's method can also be used successfully for most plant tissues (Lipke and Magnes).

The extraction of plant material is more difficult. With aqueous methanol, plant tissues can be extracted in a Soxhlet apparatus at 50°C for 10 h (Welte *et al.*). Repeated extraction with a mixture of phenol, acetic acid and water (1:1:1) and further working up of the extract results in 95% yield (Nguyen and Paquin). Borate buffer of pH 12.3 can also be used for extraction (Krishchenko and Gruzdev).

### Calculation of the elution curve

The result of column chromatography is an elution curve in which the contents of the separated components determined in the fractions are plotted against the effluent volume. Determined values in fractions containing the zone of some of the separated components form a characteristic symmetrical peak on the elution curve, while fractions that do not contain any separated component have the blank value, and form the so-called baseline. The content of a substance being determined in fractions containing some separated substance should have the baseline value subtracted from it. The average value of the blank is determined for all fractions of the peak by connecting the baseline before the peak with that after the peak by a straight line and by reading its value at the half-width of the peak.

The sum of the amounts of the substance determined in all fractions forming the peak, less the average blank value, multiplied by the number of added fractions gives the amount of the eluted substance in leucine equivalents. Applying the factor giving the ratio of the colour yield of the determined substance to the leucine yield for a given colorimetric method, leucine equivalents are converted into the true content of the substance in the

*References p. 710*

separated mixture. The factors are called leucine units and their values are given in Table 32.8.

TABLE 32.8
COLOUR YIELDS AND CONSTANTS OF AMINO ACIDS AND RELATED COMPOUNDS

Colour yields relative to leucine, called leucine units (Moore and Stein, 1954), were determined on 2-ml samples of 0.1 m*M* solutions in the buffers (pH 2.2–5.0), after heating for 15 min. The constants (see text) were determined on an amino acid analyzer (basic type) with a two-column system (Spackman *et al.*), with a heating time of 15 min, chart drive 3 in./h, 12 sec per dot on each curve, and flow-rate through the column 30 ml/h.

| Compound | Colour yield | Constant (50°C) |
|---|---|---|
| Cysteic acid | 0.99 | 25.5 |
| S-carboxymethyl-cysteine | – | 23.1 |
| Methionine sulphoxides | 0.98 | 25.0 |
| Aspartic acid | 0.94 | 25.2 |
| Methionine sulphone | 1.02 | 25.2 |
| Threonine | 0.94 | 25.4 |
| Serine | 0.95 | 26.1 |
| Glutamic acid | 0.99 | 26.3 |
| Proline (440 nm) | 0.225 | 6.37 |
| Glycine | 0.95 | 25.6 |
| Alanine | 0.97 | 26.0 |
| Half-cystine | 0.55 | 14.4 |
| Valine (pH 4.25) | 0.97 | 27.2 |
| Valine (pH 3.25) | 0.97 | 25.1 |
| Methionine | 1.02 | 25.6 |
| Isoleucine | 1.00 | 27.3 |
| Leucine | 1.00 | 27.6 |
| Tyrosine | 1.00 | 27.0 |
| Phenylalanine | 1.00 | 26.7 |
| Glucosamine | 1.03 | 26.6 |
| Galactosamine | – | *ca.* 23.2 |
| Tryptophan | 0.94 | 22.2 |
| Lysine | 1.10 | 30.7 |
| Histidine | 1.02 | 28.0 |
| Ammonia | 0.97 | 26.9 |
| Arginine | 1.01 | 27.5 |

| Compound | Colour yield | Constant (30–50°C) |
|---|---|---|
| Phosphoserine | – | 27.2 |
| Glycerophospho-ethanolamine | 0.50 | 20.6 |
| Phosphoethanol-amine | 0.43 | 15.7 |
| Taurine | 0.88 | 25.1 |
| Urea | 0.0314 | 0.97 |
| Hydroxyproline (440 nm) | 0.077 | 2.54 |
| Asparagine | 0.95 | 23.4 |
| Glutamine | 0.99 | *ca.* 21.8 |
| Sarcosine | 0.28 | 6.48 |
| Citrulline | 1.04 | 26.5 |
| α-Aminoadipic acid | 0.96 | 23.7 |
| α-Aminobutyric acid | 1.02 | 24.9 |
| Cystathionine (2 $NH_2$ groups) | – | 31.9 |
| Half-homocystine | – | 25.4 |
| β-Alanine | 0.50 | 11.4 |
| β-Aminoisobutyric acid | 0.44 | 12.5 |
| Hydroxylysine | 1.12 | 29.0 |
| γ-Amino-*n*-butyric acid | 1.01 | 27.3 |
| Ornithine | 1.12 | 29.9 |
| Ethanolamine | 0.91 | 20.5 |
| Ammonia | – | 24.7 |
| Lysine | 1.10 | 28.6 |
| 1-Methylhistidine | 0.88 | 22.5 |
| Histidine | 1.02 | 26.5 |
| 3-Methylhistidine | 0.86 | *ca.* 22.5 |
| Anserine | 0.78 | 18.2 |
| Tryptophan | 0.94 | 18.5 |
| Creatinine | 0.027 | 0.72 |
| Carnosine | 0.93 | 21.7 |
| Arginine | 1.01 | 25.4 |

In the same manner, the contents of the separated substances can be calculated by the integration method from the record of the elution curve from the amino acid analyzer. The peak area is divided by vertical lines into parts corresponding to the chart drive time necessary for the flow-through of an amount of the eluent equivalent to one fraction volume. For example, at an eluent flow-rate of 30 ml/h, chart drive 75 mm/h and fraction volume 1 ml, one division was 2.5 mm wide. In places where the vertical lines dissect the elution curve (peak), absorbances are read. However, they are not converted into leucine equivalents as in the fractional procedure because a standard leucine curve would be difficult to determine on the amino acid analyzer. The absorbance values are summed and the average blank value multiplied by the number of added values is subtracted from this sum; the result gives the value of the peak area. By analysis of a standard mixture of amino acids, unit peak areas of all amino acids, the so-called constants, $C_i^x$, are determined (see Table 32.8). The content of the analyzed substances ($\mu M$) is calculated by dividing the calculated areas of their peaks by these constants.

However, this method of calculation is tedious and lengthy and therefore a more simple calculation was developed. The shape of high, symmetrical peaks is very similar to a triangle. The peak is split by a horizontal line at the extinction half-height and this line crosses the sides of the peak at two points. If straight lines are drawn from the top of the peak through these cross-sections they will form a triangle with the baseline drawn under the peak. It was found by Spackman *et al.* that its area is 6% less than the peak area. This difference is constant and small and, under certain conditions (high, symmetrical peaks), it has no adverse effect on the accuracy of calculation. For the calculation of the area of the triangle, its height can be read from the graph (total height, in extinction units) and the average baseline value subtracted from it to give the "pure" height ($H$), and instead of the half-base of the triangle, the peak width at half of its "pure" height ($W$) can be measured. The peak width is measured by the number of time intervals between the points by which the elution curve is recorded. The points above the dividing line are added and 1 is subtracted to give the number of whole intervals. To this sum, part of the intervals from the limiting points to the halving line are added, measured with 10% accuracy. The magnitude of the peak area is calculated by multiplying the peak widths at half-height thus measured by its "pure" height. The peak area divided by the constant for a corresponding amino acid ($C_i$) obtained in the same manner with a known amount of standard indicates the content of the analyzed substance.

Because of its simplicity, this method of calculation is mainly used and the constants determined for it are also used for calculation by the integration method. The constants $C_i$ of the simplified calculation are converted into the constants $C_i^x$ of the calculation obtained by integration by multiplying by a factor:

$$C_i^x = FC_i \text{ and } F = f \cdot \frac{V_n}{V_f}$$

where $f$ (1.06) is the factor correcting the smaller area calculated by the simplified method, $V_f$ is the chart shift in the recorder (mm) per time of flow necessary for the fraction volume to pass through the column, *i.e.*, the width of the interval of the calculation by integration, and $V_n$ is the shift (mm) per time of the interval between the recording of two points in the recorder.

*References p. 710*

From the time when a measuring system with a linearized record was constructed (Orten *et al.*), the utilization of automatic integrators for the calculation of the contents of separated substances from elution curves was facilitated. Time losses of the technician caused by the necessary calculation of the results were thus eliminated. The same goal was aimed at by the manufacturers of analyzers when an automatic device recording the measured values on to computer tapes were connected to the analyzers. The programme for the calculation of the analysis of amino acids by computer was developed by a number of workers (Black *et al.*, Spitz *et al.*).

## PREPARATIVE CHROMATOGRAPHY

The aims of preparative amino acid chromatography are primarily the separation of the maximum amount of material, then the isolation of pure substances and finally the simplicity of the procedure. The losses of the separated substances are not the limiting factor, in contrast to analytical chromatography, provided that they are not too high.

Adsorption chromatography has the highest capacity and has been utilized in separations of amino acids on charcoal columns in all its forms (frontal and elution analysis and displacement chromatography). The development of these methods is due mainly to Tiselius (Williams *et al.*). Amino acids were separated into groups and less complex mixtures, and also into pure substances. The separation of aromatic amino acids from the mixture by this method was advantageous because they were strongly sorbed on charcoal (Schramm and Primosigh), but the properties of charcoal could not compete with those of ion exchangers. The use of charcoal may be of importance only in the isolation of aromatic amino acids from the complete mixture (Fromageot *et al.*).

A more effective technique for the preparative separation of amino acids is displacement chromatography on ion exchangers (Partridge and Brimley), which was developed on the basis of a series of studies with synthetic ion exchangers. It has the capacity of several hundred grams of hydrolyzate and phenylalanine and tyrosine are separated first on a charcoal column from hydrolyzates freed from humins and hydrochloric acid. Both amino acids are then gradually displaced from the column in a pure state with a mixture of phenol and acetic acid. The hydrolyzate is diluted to 40 l (pH 1.86) and allowed to run at the rate of 40 ml/min through a system of columns connected in series. The first is a 11 × 3.8 cm column of sulphonated polystyrene ion exchanger containing 4.5% of cross-links (nowadays Dowex 50-X4, 60–100 mesh). The subsequent three columns, 61 × 7.6 cm, 41 × 5.1 cm and 30 × 3.8 cm, contain Zeo-Karb 215, and this arrangement improves the sharpness of the separation. The effluent drains to waste. After the sample has been applied, the first column is disconnected and the remaining three columns are washed with 0.15 *N* ammonia solution (20 ml/min) and the effluent is collected into 250-ml fractions, which are tested for their amino acid contents by paper chromatography. The first column, containing bound basic amino acids, is then connected to two smaller columns, 20 × 2.5 cm and 13.6 × 1.7 cm, which also contain polystyrene, so that a sectional column is formed and the amino acids are displaced with 0.075 *N* sodium hydroxide solution (6 ml/min) into 250-ml fractions, which are tested for their amino acid contents by paper chromatography.

The scheme of the elution curve is constructed according to the chromatograms of the aliquots taken from the fraction from both columns. Bands are indicated on it in which single amino acids or mixtures of several of them are contained, the bands being chosen so that the mixtures formed are as simple as possible and contain the maximum amount of these amino acids. Following this resolution pattern, the fractions are combined and the simple mixtures obtained are re-chromatographed on Zeo-Karb 215 and a sulphonated polystyrene under different conditions or on Dowex 2 until pure amino acids are isolated. The transitional materials between the bands, containing small amounts of many amino acids, are discarded. The conditions for complementary chromatography should be chosen according to the quality of the primary separation of the mixtures. The yield of the isolated amino acids is about 50% of the initial weight of the protein. Amino acids can be isolated from protein hydrolyzates (egg albumin), or hydrolyzates of entire organisms (yeast) unless they contain large amounts of saccharides, and also from biological extracts (Westall).

From the preparative point of view, elution chromatography on ion exchangers is less effective as its capacity is about 100 times less than that of displacement chromatography. It has the advantage, however, of affording very pure substances; the degree of purity may be indicated as chromatographically pure. In 1950, a method for the separation of amino acids on Dowex 50-X8 was described that involved stepwise gradient elution with hydrochloric acid (Stein and Moore, 1950), which was ideally simple because the isolated amino acids could be obtained from the solution simply by evaporation of the acid. However, the separation was not complete in the regions of serine and threonine and of isoleucine and leucine. For the elution of basic amino acids, 4 *N* hydrochloric acid had to be used, but in spite of this it took a long time. Therefore, the authors developed a new system of elution with ammonium acetate and ammonium formate buffers with which the complete separation of all amino acids could be achieved. The procedure consisted of a primary elution with acetate buffers on a 15-cm Dowex 50-X8 column on which pure basic amino acids were isolated. The material from the first two peaks was re-chromatographed. The second peak, containing phenylalanine and tyrosine, was separated on a 60-cm Dowex 50-X8 column by elution with a buffer of a different composition. The first peak was re-chromatographed on a 15-cm Amberlite IR-4B column to afford glutamic and aspartic acids, each separately. All other amino acids present in the first peak of this column were further separated on a 120-cm Dowex 50 column with formate buffers, and on this column the remaining amino acids were separated, with the exception of serine and threonine, glycine and alanine, and isoleucine and methionine. The first two double peaks were combined and separated into single amino acids on a 60-cm Dowex 50 column, also using formate buffers. Then, using the same column, methionine and isoleucine were separated separately. Detection was carried out by ninhydrin colorimetry on aliquots of fractions. All columns were 7.5 cm I.D. The sample weight of the bovine serum albumin hydrolyzate was 2.5 g. The solutions of the isolated amino acids were evaporated and the ammonium salts sublimed off. The average yield was 66% of pure L-antipodes.

The elution systems utilizing volatile buffers were further developed for various purposes, for example for the detection of radioisotopes by liquid scintillation counting (Redford-Ellis and Kelson).

With the production of molecular sieves of low porosity, it became possible to carry

*References p. 710*

out preparative amino acid isolations on Sephadex G-10 (Eaker and Porath) in large segment columns of the KS-370 type (Pharmacia, Uppsala, Sweden). However, this method is still economically unsatisfactory.

## REFERENCES

Atkin, G. E. and Ferdinand, W., *Anal. Biochem.*, 38 (1970) 313.
Bandlow, W. and Nordwig, A., *J. Chromatogr.*, 39 (1969) 326.
Beale, D. and Kent, C. M., in D. I. Holy (Editor), *Sixth Colloquium on Amino Acid Analysis*, Technicon International Division, Geneva, 1968, p. 133.
Benson, Jr., J. V., Gordon, M. J. and Patterson, J. A., *Anal. Biochem.*, 18 (1967) 228.
Benson, Jr., J. V. and Patterson, J. A., *Anal. Chem.*, 37 (1965) 1108.
Berridge, Jr., B. J., Chao, W. R. and Peters, J. H., *Amer. J. Clin. Nutr.*, 24 (1971) 934.
Black, H. L., Buttery, P. J. and Gregson, K., *J. Chromatogr.*, 68 (1972) 103.
Bogue, D. C., *Anal. Chem.*, 32 (1960) 1777.
Bonnot, G., Gelobel, B. and Guilland, J., *Ann. Biol. Anim., Biochim. Biophys.*, 10 (1970) 357.
Brummel, M., Gerbeck, C. M. and Montgomery, R., *Anal. Biochem.*, 31 (1969) 331.
Coggins, J. R. and Benoiton, N. L., *J. Chromatogr.*, 52 (1970) 251.
Contractor, S. F., *J. Chromatogr.*, 11 (1963) 568.
Crestfield, A. M., *Anal. Chem.*, 35 (1963) 1762.
Eaker, D. and Porath, J., *Separ. Sci.*, 2 (1967) 518.
Ertinghausen, G. and Adler, H. J., *J. Chromatogr.*, 44 (1969) 620.
Ertinghausen, G., Adler, H. J. and Reichler, A. S., *J. Chromatogr.*, 42 (1969) 355.
Fromageot, C., Jutisz, M. and Lederer, E., *Biochim. Biophys. Acta*, 2 (1948) 487.
Giddings, J. C., *J. Chromatogr.*, 13 (1964) 301.
Hamilton, P. B., *Anal. Chem.*, 30 (1958) 914.
Hamilton, P. B., *Anal. Chem.*, 32 (1960) 1779.
Hamilton, P. B., *Anal. Chem.*, 35 (1963) 2055.
Hamilton, P. B. and Anderson, R. A., *Anal. Chem.*, 31 (1959) 1504.
Hamilton, P. B., Bogue, D. C. and Anderson, R. A., *Anal. Chem.*, 32 (1960) 1782.
Hannig, K., *Clin. Chim. Acta*, 4 (1959) 51.
Hirs, C. H. W., Stein, W. H. and Moore, S., *J. Biol. Chem.*, 211 (1954) 941.
Ishii, S. I., *J. Biochem. (Tokyo)*, 43 (1956) 531.
Jacobs, S., *Analyst (London)*, 95 (1970) 370.
Kedenburg, C. P., *Anal. Biochem.*, 40 (1971) 35.
Kirkpatrick, Ch. H. and Anderson, R. A, *J. Chromatogr.*, 14 (1964) 297.
Knipfel, J. E., Christensen, D. A. and Owen, B. D., *J. Ass. Offic. Anal. Chem.*, 52 (1969) 981.
Krishchenko, V. P. and Gruzdev, L. D., *Khim. Sel. Khoz.*, 9 (1971) 860.
Lange, H. W. and Hempel, K., *J. Chromatogr.*, 59 (1971) 53.
Lipke, W. G. and Magnes, L. J., *Advan. Autom. Anal., Technicon Int. Symp. 1969*, Mediad, New York, 1970, p. 339.
London, D. R., in D. I. Schmidt (Editor), *Technicon Amino Acid Analysis*, Technicon Instruments, Chertsey, 1966, p. 38.
Lorenz, H., *Phytochemistry*, 10 (1971) 63.
Macpherson, H. T. and Wall, R. A., *J. Sci. Food Agr.*, 21 (1970) 129.
Mardeus, Y., Van Sande, M. and Caers, J., *Anal. Lett.*, 4 (1971) 285.
Meyer, P. D., Stegink, L. D. and Shipton, H. W., *J. Chromatogr.*, 48 (1970) 538.
Miller, E. J. and Piez, K. A., *Anal. Biochem.*, 16 (1966) 320.
Mondino, A., Bongiovanni, G., Nol, V. and Raffaele, I. V., *J. Chromatogr.*, 63 (1971) 411.
Mongey, E. H. and Mason, J. W., *Anal. Biochem.*, 6 (1963) 223.
Moore, S., Spackman, D. H. and Stein, W. H., *Anal. Chem.*, 30 (1958) 1185.
Moore, S. and Stein, W. H., *J. Biol. Chem.*, 176 (1948) 367.
Moore, S. and Stein, W. H., *J. Biol. Chem.*, 192 (1951) 663.

Moore, S. and Stein, W. H., *J. Biol. Chem.*, 211 (1954) 893 and 907.
Morris, D. R., Koffron, K. L. and Okstein, C. J., *Anal. Biochem.*, 30 (1969) 449.
Mostek, J., Solinová, H. and Čepička, J., *Kvasný Prům.*, 17 (1971) 121.
Munier, R. L. and Drapier, A. M., *Chromatographia*, 10 (1969) 433.
Nguyen, S. T. and Paquin, R., *J. Chromatogr.*, 61 (1971) 349.
Northrop, J. H., Kunitz, M. and Herriott, R. M., *Crystalline Enzymes*, Columbia Univ. Press, New York, 2nd ed., 1948.
Olson, A. C., White, L. M. and Noma, A. T., *J. Chromatogr.*, 43 (1969) 399.
Orten, A. U., Doppke, H. J. and Spurrier, H. H., *Anal. Chem.*, 37 (1965) 623.
Partridge, S. M. and Brimley, R. C., *Biochem. J.*, 51 (1952) 628.
Perry, T. L., Stedman, D. and Hansen, S., *J. Chromatogr.*, 38 (1968) 460.
Peterson, E. A. and Sober, H. A., *Anal. Chem.*, 31 (1959) 857.
Piez, K. A. and Morris, L., *Anal. Biochem.*, 1 (1960) 187.
Power, T. F. and Benet, D. J., *Anal. Biochem.*, 36 (1970) 537.
Purdie, J. W., Gravelle, R. A. and Hanafi, D. E., *J. Chromatogr.*, 38 (1968) 346.
Purdie, J. W. and Hanafi, D. E., *J. Chromatogr.*, 59 (1971) 181.
Redford-Ellis, M. and Kelson, M. N., *J. Chromatogr.*, 59 (1971) 434.
Reid, R. H. P., in D. I. Schmidt (Editor), *Technicon Amino Acid Analysis*, Technicon Instruments, Chertsey, 1966, p. 43.
Roach, A. G., in D. I. Schmidt (Editor), *Technicon Amino Acid Analysis,* Technicon Instruments, Chertsey, 1966, p. 86.
Rokushika, S., Funakoshi, S., Murakami, F. and Hatano, H., *J. Chromatogr.*, 56 (1971) 137.
Ronca, G., Chiti, R. and Lucacchim, A., *J. Chromatogr.*, 47 (1970) 114.
Roth, M., *Anal. Chem.*, 43 (1972) 880.
Saifer, A. and Gerstenfeld, S., *Clin. Chem.*, 10/11 (1964) 970.
Schram, E., Moore, S. and Bigwood, E. J., *Biochem. J.*, 57 (1954) 33.
Schramm, G. and Primosigh, J., *Chem. Ber.*, 76 (1943) 373.
Seely, J. H., Edattel, S. R. and Benoiton, N. L., *J. Chromatogr.*, 44 (1969) 618.
Smith, M., Stahmann, M. A. and Gemenza, G., *J. Chromatogr.*, 18 (1965) 366.
Spackman, D. H., *Fed. Proc., Fed. Amer. Soc. Exp. Biol.*, 23 (1964) 371.
Spackman, D. H., Moore, S. and Stein, W. H., *Anal. Chem.*, 30 (1958) 1190.
Spies, J. R. and Chambers, D. C., *Anal. Chem.*, 20 (1948) 30.
Spitz, H. D., Henyon, G. and Silvertson, J. N., *J. Chromatogr.*, 68 (1972) 111.
Starcher, B. C., Weger, L. Y. and Johnson, L. D., *J. Chromatogr.*, 54 (1971) 425.
Stein, W. H. and Moore, S., *J. Biol. Chem.*, 176 (1948) 337.
Stein, W. H. and Moore, S., *Cold Spring Harbor Symp. Quant. Biol.*, 14 (1950) 179.
Synge, R. L. M., *Biochem. J.*, 38 (1944) 285.
Tasdhomme, M., *Bull. Inst. Text. Fr.*, 24 (1970) 237.
Thomson, A. R. and Eveleigh, J. W., *Anal. Chem.*, 41 (1969) 1073.
Tristram, G. R., in D. I. Schmidt (Editor), *Technicon Amino Acid Analysis*, Technicon Instruments, Chertsey, 1966, p. 61.
Troll, W. and Cannan, K. R., *J. Biol. Chem.*, 200 (1953) 803.
Udenfriend, S., Stein, S., Bohlen, P., Dairman, W., Leimgruber, W. and Weigle, M., *Science*, 178 (1972) 871.
Vega, A. and Nunn, P. B., *Anal. Biochem.*, 32 (1969) 446.
Wagner, S. L. and Shepherd, S. L., *Anal. Biochem.*, 41 (1971) 314.
Wall, R. A., *J. Chromatogr.*, 37 (1968) 549.
Welte, E., Przemek, E. and Nuh, M. C., *Z. Pflanzenernähr. Bodenk.*, 128 (1971) 243.
Westall, R. G., *J. Sci. Food Agric.*, 6 (1950) 191.
Williams, R. J. P., Hagdahl, L. and Tiselius, A., *Ark. Kemi*, 7 (1954) 1.
Yu, S. Y., *Anal. Biochem.*, 37 (1970) 212.
Ziska, P., *J. Chromatogr.*, 60 (1971) 139.
Zmrhal, Z., unpublished results, 1965.
Zmrhal, Z., *Rostl. Výroba*, 14 (1968) 517.

*Chapter 33*

# Amino acid derivatives

Z. DEYL and M. JUŘICOVÁ

## CONTENTS



## INTRODUCTION

The chromatography of amino acid derivatives is closely related to the elucidation of the primary structure of proteins. There is probably no other field in organic chemistry or biochemistry in which the advancement of understanding of the structure and function of a particular group of compounds has parallelled the development of a method so closely as has occurred in the concomitant development of protein structure and chromatographic techniques for proteins and amino acid derivatives. Much of the work carried out in this respect has been directed towards the identification of the N-terminal amino acid in proteins. No doubt flat-bed techniques play an important role in the separation of the corresponding derivatives. These techniques have been exhaustively surveyed in a recent two-part review by Rosmus and Deyl (1971, 1972). On the other hand, little remains in this field that can be studied by gas chromatography, but liquid column chromatography appears to be just beginning to be applied. There are several reasons that favour the advancement of liquid chromatographic techniques in the field of amino acid derivatives. Firstly, the number of modification reactions available is steadily increasing, which places further demands upon separation techniques. Secondly, rare proteins and protein fractions, which are frequently analyzed (and sequenced), require some kind of reliable quantitation, for which liquid column separations appear to be more applicable than diverse flat-bed techniques. Also, the isolation and preparation of unusual amino acids and attempts to avoid errors in protein sequencing may further increase the importance of column chromatographic techniques. The terms "modern" and "classical" separations are difficult to apply to amino acid derivatives because even very recent methods of separation still use the more or less standard arrangement of the chromatographic procedure: an automated version of the chromatographic separation of dinitrophenyl derivatives and of 5-dimethylaminonaphthalene-1-sulphonyl (Dns) derivatives may serve as a good example (Deyl and Rosmus, Kesner). Also, to our knowledge no attempt has been made to separate some types of amino acid derivatives by high-speed high-resolution liquid chromatography, although some derivatives, such as Dns-amino acids, seem to be very attractive from this viewpoint.

The use of different types of derivatives has shown very uneven development. It appears that dinitrophenylation has already passed its peak and is leaving the field free for different fluorescent derivatives, among which Dns derivatives are of prime importance. For double-checking the liberated N-terminal amino acid derivatives, the chromatography of hydantoins or modified hydantoins is useful. Of over 40 other types of derivatives, a few have been subjected to liquid chromatographic separations and the results are summarized in this chapter.

## 2,4-DINITROPHENYL (DNP) AMINO ACID DERIVATIVES

### Classical chromatographic procedures

The first attempts to separate DNP-amino acid derivatives were oriented towards different types of silica gel chromatography. In all of these procedures, the way in which the sorbent was prepared and hydrated was of decisive importance, which also holds true for more recent techniques such as the automated analysis of DNP derivatives on silica gel columns (see p. 720). A suitable method for standardization of the silica gel has been reported by Gordon *et al.* The overall scheme for the separation of DNP-amino acids in a silica gel column is rather complex. It has been described by Porter and Sanger and Fig. 33.1 makes the whole system, which is described below, easy to understand.

In the first step, a mixture of ether-soluble DNP-amino acids is subjected to elution with chloroform and an *n*-butanol–chloroform mixture. The original mixture is resolved into five bands, which, except for the third, are subjected to further fractionation. The third band is the DNP-derivative of hydroxyproline and can therefore be subjected to direct quantitation.

The first band from the chloroform and *n*-butanol–chloroform fractionation is eluted in three sub-fractions with an ethanol–ligroin mixture containing an increasing proportion of ligroin. By washing the column with chloroform–(ethanol–ligroin) (1:1), the mixture of DNP-leucine and DNP-isoleucine is eluted. A chloroform–(acetone–cyclohexane)–(ethanol–ligroin) (1:2:2) mixture elutes a complex band, which can be separated further with acetone–cyclohexane. If this complex band is treated with chloroform–(ethanol–ligroin)–(acetone–cyclohexane) (1:2:1), the DNP-valine band is eluted, while the same solvent system in a ratio of 1:2:2 elutes the DNP-derivative of phenylalanine.

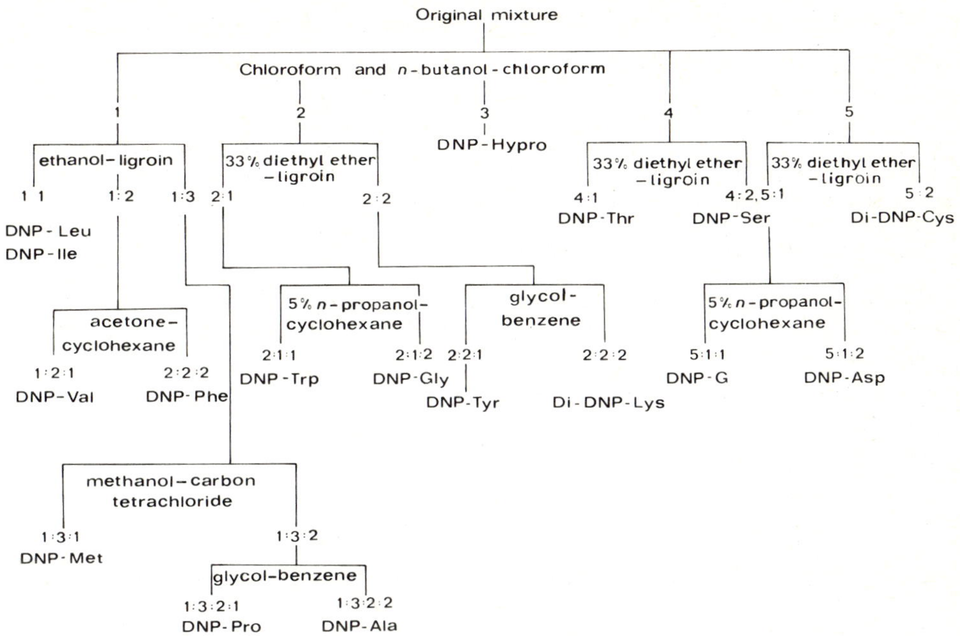


Fig. 33.1. Scheme for the fractionation of ether-soluble DNP-amino acids (Porter and Sanger).

The last fraction obtained from the chloroform–(ethanol–ligroin) fractionation is eluted with this mixture in a ratio of 1:3. The band eluted in this step is again complex, consisting of DNP-methionine, DNP-proline and DNP-alanine. If to the original chloroform–(ethanol–ligroin) (1:3) another portion of methanol–carbon tetrachloride solvent is added, making the overall proportions of the individual solvents 1:3:1, the band of DNP-methionine is eluted. If the proportion of methanol–carbon tetrachloride is further increased, making the overall proportion of chloroform–(ethanol–ligroin)–(methanol–carbon tetrachloride) 1:3:2, a slow-moving band consisting of DNP-proline and DNP-alanine is eluted. These last two DNP-amino acids can be separated by adding benzene to the eluent from the previous step. A ratio of chloroform–(ethanol–ligroin)–(methanol–carbon tetrachloride)–(ethylene glycol–benzene) of 1:3:2:1 elutes DNP-proline, while a further increase in the ratio of ethylene glycol to benzene elutes DNP-alanine. In this latter instance, the proportions of the components are chloroform–(ethanol–ligroin)–(methanol–carbon tetrachloride)–(ethylene glycol–benzene) 1:3:2:2.

The second band consists of four amino acid derivatives: DNP-tryptamine, DNP-glycine, DNP-tyrosine and bis-DNP-lysine. These derivatives are separated by eluting the column with an increasing proportion of 33% diethyl ether–ligroin solvent: if the ratio of chloroform–(33% diethyl ether–ligroin) is 2:1, the mixture of DNP-tryptamine and DNP-glycine is brought into solution. The further separation of these two amino acid derivatives is carried out by using a 5% solution of *n*-propanol in cyclohexane. A mixture of chloroform–(33% diethyl ether–ligroin)–(5% *n*-propanol–cyclohexane) (2:1:1) elutes DNP-tryptamine, whereas when the components are present in the ratio 2:1:2 DNP-glycine is eluted. The slower band of the 33% diethyl ether–ligroin fractionation eluted with chloroform–(33% diethyl ether–ligroin) (2:2) consisting of DNP-tyrosine and bis-DNP-lysine is separated by using an increasing proportion of ethylene glycol–benzene in the solute. Hence, chloroform–(33% diethyl ether–ligroin)–(ethylene glycol–benzene) (2:2:1) washes out the band of DNP-tyrosine, whereas when the components are present in the ratio 2:2:2 the band of di-DNP-lysine is washed out.

The third band as already mentioned, is pure DNP-hydroxyproline and does not require any further fractionation.

The fourth band is a mixture of DNP-threonine and DNP-serine, which is separated by washing first with chloroform–(33% diethyl ether–ligroin) (4:1) and then with the same mixture in the ratio 4:2. In the first instance, DNP-threonine is eluted, while in the second step DNP-serine is eluted.

The fifth band is washed with chloroform–(33% diethyl ether–ligroin) (5:1), which moves the complex band of DNP-aspartic acid and DNP-glutamic acid, while the same mixture in the ratio 5:2 elutes the last amino acid to be separated, *viz.*, di-DNP-cystine. The separation of the corresponding derivatives of aspartic and glutamic acids is achieved by using 5% *n*-propanol–cyclohexane as the solvent. The solvent system consisting of chloroform–(33% diethyl ether–ligroin)–(5% *n*-propanol–cyclohexane) elutes DNP-glutamic acid, while the same solvent system with a slightly different ratio of components (5:1:2) elutes DNP-aspartic acid.

In such a complex separation, one would not expect all of the separation steps to be carried out on a single column. Porter and Sanger suggested the use of four columns in parallel, thus permitting the separation of fractions that arise in individual steps of the separation in parallel. This operation prevents the whole procedure from being tedious and time consuming.

The following four columns should be used.

*(a) Ethanol–ligroin column.* One volume of water, 1 volume of ethanol and 10 volumes of ligroin (boiling range 80–100°C) are mixed, 1 ml of the lower aqueous phase is added for every 2 ml of silica gel and the upper ligroin phase is used as the mobile phase in the column.

*(b) Methanol–carbon tetrachloride column.* This column is prepared in a similar manner; 1 volume of water, 1 volume of methanol and 15 volumes of carbon tetrachloride are used.

*(c) Acetone–cyclohexane column.* This column is prepared by using 1 volume of water, 1 volume of acetone and 10 volumes of cyclohexane.

*(d) Ethylene glycol–benzene column.* Ethylene glycol and benzene are shaken together, 1 ml of the ethylene glycol layer is added for each 1 g of dry silica and the benzene layer is used as the mobile phase.

Blackburn was the first to introduce buffered silica gel columns into the chromatography of DNP-amino acid derivatives. He was also able to show that the difference observed by many workers between different batches of silica gel can be attributed to the variations in the pH of the aqueous phase that is in contact with the gel.

According to the method of Blackburn, silica gel is treated for 3 h with excess of hydrochloric acid after it has been precipitated from water-glass solution. The band rates (*R* values) of amino acid derivatives on buffered columns are summarized in Table 33.1. The buffer systems used were 0.2 *M* sodium dihydrogen phosphate, 0.5 *M* phosphate (pH 6.61) and 0.25 *M* phosphate (pH 5.95).

Acid-soluble DNP-amino acid derivatives may be subjected to fractionation in 0.2 *M* sodium dihydrogen phosphate using Sanger's original fractionation, described in the preceding section.

It appears that except for the elaborate automated analysis of DNP-amino acid derivatives on silica gel columns (see p. 720), all of the other procedures suffer from considerable disadvantages based mainly on the poor reproducibility of silica gel preparations. For this reason, Kieselguhr columns were introduced by Mills; acid-washed Kieselguhr is used in this separation.

Acid-soluble amino acid derivatives are first removed by extracting the sample into methyl ethyl ketone that has been acidified with 10% of 6 *N* hydrochloric acid. The residue after extraction is evaporated to dryness and dissolved in water-saturated methyl ethyl ketone–chloroform (3:1).

TABLE 33.1
BAND RATES (*R* VALUES) OF DNP-AMINO ACIDS IN CHLOROFORM AND *n*-BUTANOL–CHLOROFORM (BLACKBURN)

Buffer systems (saturated organic phase with the particular buffer): (A) 0.2 *M* sodium dihydrogen phosphate; (B) 0.5 *M* phosphate (pH 6.61); (C) 0.25 *M* phosphate (pH 5.95).

| DNP-amino acid | *R* value | | | |
|---|---|---|---|---|
| | $CHCl_3$, buffer A | $CHCl_3$, buffer B | 3% *n*-Butanol–$CHCl_3$, buffer C | 17% *n*-Butanol–$CHCl_3$, buffer B |
| DNP-leucine | Fast | 0.7 | – | – |
| DNP-valine | Fast | 0.5 | – | – |
| DNP-phenylalanine | Fast | 0.7 | – | – |
| DNP-methionine | Fast | 0.5 | – | – |
| DNP-proline | Fast | 0.3 | – | – |
| DNP-alanine | Fast | 0.2 | – | 0.5 |
| DNP-tryptophan | 0.6 | 0.4 | Fast | – |
| DNP-glycine | 0.9 | 0.03 | 0.5 | 0.2 |
| Di-DNP-lysine | 0.7 | 0.06 | Fast | Fast |
| Di-DNP-tyrosine | Fast | 0.07 | 0.9 | Fast |
| DNP-threonine | 0.3 | Slow | 0.18 | 0.15 |
| DNP-serine | 0.11 | Slow | 0.08 | 0.07 |
| DNP-glutamic acid | 0.17 | Slow | 0.20 | Slow |
| DNP-aspartic acid | 0.06 | Slow | Slow | Slow |

Silicic acid–Celite chromatography is one of the most rapid classical column procedures used for the separation of DNP-amino acid derivatives. Green and Kay reported the possibility of completing the analysis in less than 2 h. Most of the commercially available silicic acid preparations are suitable without any further pre-treatment. The only difficulty in the procedure is to ensure an adequate flow-rate, which is difficult to achieve with silicic acid alone. The sorbent is therefore mixed with half of its weight of Celite and the fraction that passes a 60-mesh filter is used for column preparation. The column size recommended by Green and Kay is 1–1.4 × 17 cm.

Schroeder reported that the intensity of sorption can be increased by pre-washing the column. It appears that the intensity of sorption increases in the following order: adsorbent pre-washed with alcohol < adsorbent not pre-washed < adsorbent pre-washed with diethyl ether < adsorbent pre-washed with acetone–diethyl ether. A similar procedure was described by Green and Kay.

## Recent chromatographic procedures

### *Separation on nylon columns*

Jellinek and Del Carmen Vara used non-stretched nylon 66 and Celite as sorbents for the separation of DNP-amino acid derivatives. The sorbent was prepared by mixing two volumes of an aqueous solution of non-stretched nylon 66 with one volume of an aqueous

*References p.738*

suspension of Celite 545. Acetic acid–ammonium acetate and ammonia–ammonium acetate buffer of ionic strength 0.2 (pH 4–10) were used as eluents, in some instances with addition of ethanol. The actual separation was carried out at 22°C with an elution rate of 1 ml per 8 min. The eluent was evaluated manually by measuring the optical density.

*Ion-exchange chromatography*

In column chromatographic techniques, ion exchange can also be used for the separation of water-soluble DNP-amino acids. A suitable procedure was introduced by Heinrich and Bugna, in which the cation-exchange resin Amberlite IRC-50 (Bio-Rex 70) ($Na^+$) (–400 mesh) was used with considerable success. The column used was maintained thermostatically at 50°C and was 15–20 cm in length. The flow-rate applied was *ca.* 0.5 ml/min. The eluting buffer was similar to the "pH 5" buffer specified for the Technicon amino acid analyzer.

The column effluent is recorded by using a suitable device in the recorded region (360 nm). The order of amino acid derivatives released from the column (0.5–1 h apart) is α-DNP-lysine, ε-DNP-lysine, DNP-arginine, bis-DNP-histidine, DNP-tryptamine and bis-DNP-lysine. The lower limits of this method are 0.01–0.05 μmole, provided that standard Technicon equipment is used.

Excellent results in the ion-exchange chromatography of water-soluble DNP-amino acids were achieved by Nishikawa *et al.*, using Dowex 50W-X2.

The chromatography of DNP-amino acids on Dowex 50W resins is difficult owing to strong non-electrostatic interactions between the DNP moiety of the amino acid derivatives and the aromatic constituents of the resin. These interactions result in highly diffused peaks, which are hard to integrate. A simple approach in order to avoid these interactions was to introduce aromatic compounds into the eluting buffers; *p*-hydroxybenzoic acid gave excellent results in this respect. A Dowex 50W column of dimensions 29 × 0.9 cm used in a Beckman-Spinco amino acid analyzer was applied. Resin filled to a height of 11 cm was used and elution of the column was carried out in stages. Two buffers containing *p*-hydroxybenzoic acid of pH 5.28 (buffer I) and 6.0 (buffer II) were used.

The column was operated at 54°C using a flow-rate of 60 ml/h while the ninhydrin flow-rate was 30 ml/h. The sample was loaded in the usual way and the column was eluted with buffer I for 60 min, after which it was replaced with buffer II. The yellow colour, together with the ninhydrin-positive reactions of amino acids, were recorded. The optimal resolution was obtained if the amount separated was 0.5 μmole per peak. The following order of maxima appeared on the chromatogram: the bulk of DNP-derivatives of neutral and acidic amino acids, Im-DNP-histidine, lysine, histidine, $NH_2$, ε-DNP-lysine, arginine and O-DNP-tyrosine.

*Liquid–liquid partition of DNP-amino acids (Hyflo Supercel chromatography)**

DNP-amino acid derivatives are partitioned between the organic and aqueous phase provided that they are undissociated; pH values affect the overall partition coefficient, as

*For different types of counter-current chromatography, see Chapter 8.

shown by Matheson (1963, 1965, 1966) by changing the equilibrium in the aqueous layer. However, according to Matheson and Sheltawy, in principle any of the ether-soluble DNP-amino acids can be partitioned between aqueous buffers and ethyl acetate in the ionic form. If ionized, the bands of DNP-amino acids are exceptionally narrow and the *R* values are only slightly pH-dependent.

For the liquid–liquid partition chromatography of DNP-amino acids, the following procedure was described in the series of papers by Matheson and his coworkers. The chromatographic column was 19 × 1 cm in size; ethyl acetate and the appropriate buffer were shaken together and the resulting phases were separated. Hyflo Supercel was slurried in the upper phase and 2.5 ml of the lower phase were added dropwise. The mixture was shaken vigorously until free from lumps. The column was eluted with some of the following buffers (flow-rate 1 ml/min): (a) 0.1 *M* Tris–maleic acid–salt buffer (pH 5.4); (b) 0.1 *M* sodium phosphate–salt buffer (pH 7.4); (c) 0.1 *M* Tris–hydrochloric acid–salt buffer (pH 7.4); and (d) 0.1 *M* sodium phosphate buffer (pH 12).

Optimum results were obtained in mixtures containing about 1.5 mg of each amino acid present in the form of the mono-DNP-derivative, while the optimum amount of bis-DNP derivatives was 2–3 mg. A special problem in this type of column chromatography is the separation of DNP-serine and DNP-lysine in complicated mixtures; for this purpose, the Tris–glycine–urea–salt buffer, 6 *N* in urea (the pH of 9.5, however, is not critical) was used by Matheson and Sheltawy.

For the column partition of DNP derivatives of methionine, phenylalanine, ornithine and lysine, buffers consisting of Tris–hydrochloric acid, 2-amino-2-methylpropane-1,3-diol hydrochloride and alternatively glycine–sodium hydroxide were used by Matheson and Sheltawy. For water-soluble amino acid derivatives, 0.2 *M* Tris–0.2 *M* glycine buffer of pH 9.1 gives satisfactory results; alternatively, methyl ethyl ketone–ethyl acetate–0.2 *M* Tris + 0.2 *M* glycine buffer (3:2:1 by volume) can be recommended for this purpose.

The sample can also be applied to a column equilibrated with a buffer of low pH, which makes it possible to resolve the slowly moving DNP-derivatives; the fast-moving derivatives are eluted and the eluate is applied to another column equilibrated to a higher pH. It is usually necessary to repeat the process several times until the complete separation of all components has been achieved.

The separation procedure reported by Portugal *et al.* for the separation of glutamic and aspartic acid derivatives is, in principle, the same as that reported above.

*Reversed-phase chromatography on chlorinated rubber*

The chlorinated rubber (150–200 mesh) was prepared (Partridge and Swain) by shaking it with a suspension of *n*-butanol (4 ml per 10 g of chlorinated rubber) in 0.2 *M* citrate–phosphate buffer previously saturated with *n*-butanol. The slurry obtained was used for packing the columns by filtration under slightly reduced pressure. The column was eluted with *n*-butanol at three different pH values (3, 4 and 5). The best results were obtained at pH 3, where DNP-lysine, DNP-aspartine, DNP-serine, DNP-aspartic acid, DNP-glycine, DNP-alanine, DNP-proline, DNP-valine and DNP-leucine can be separated (these amino acid derivatives are listed according to decreasing *R* values). With increasing pH, the fastest moving bands of DNP-amino acids, such as DNP-lysine, DNP-aspartine and DNP-serine, pass through the column virtually with the void volume.

*References p. 738*

## Automated chromatography of DNP-amino acids

### *Automated separation on silica gel columns*

The latest achievements in the column chromatography of DNP-amino acids are the automated procedures referred to in the papers by Kesner and co-workers (1963, 1964). Silica gel is used as the support and the procedure can easily be repeated in any laboratory that is equipped to carry out standard chromatographic procedures. An overall scheme of the apparatus used is presented in Fig. 33.2.

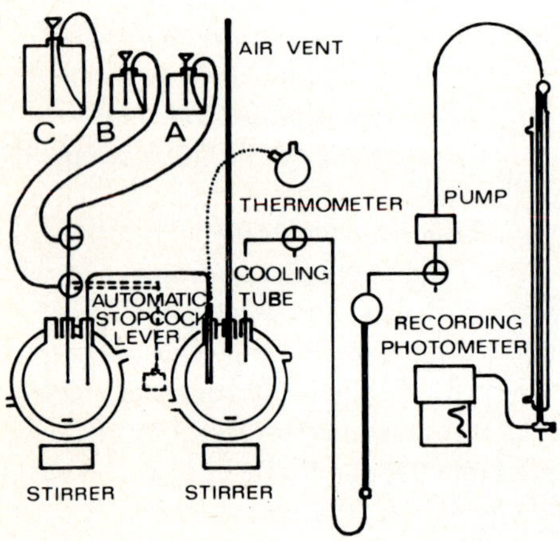


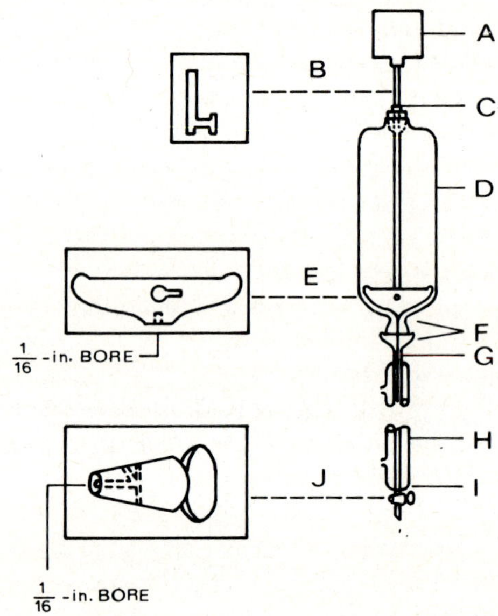


Fig. 33.2. Two-chamber gradient apparatus using solvent de-aeration, water saturation and continuous spectrophotometric recording. (Kesner *et al.*, 1963, 1964).

Fig. 33.3. Column pouring apparatus. A = stirrer with rheostat; B = button-tip glass stirring rod, 6 mm O.D.; C = stirring-rod guide; D = 700-ml infusion bottle, 12-mm outlet bore, attached to ball part; E = PTFE stirring blade (1/8 in. thick); F = ball-and-socket joint; G = PTFE tubing tail, 0.034 in. I.D., 0.058 in. O.D.; H = jacketed chromatographic column, 1 × 120 cm; I = sintered glass disk; J = PTFE stopcock, bored for connection to flow cell. (Kesner *et al.*, 1963, 1964).

As is usual with silica gel, the final separation is highly sensitive to the moisture content of the sorbent material. It is generally recommended that the sorbent is tested before the column is filled; on the other hand, by changing the activity of the sorbent it is possible to obtain resolutions of DNP derivatives that had moved together in a particular instance. In conclusion, it is stressed that constant conditions must be maintained during the automated analysis in order to obtain reproducible results.

A column of the specified dimensions (100 × 1 cm) is filled by using a simple device (see Fig. 33.3) which, in principle, consists of an infusion bottle clamped to the top of the column and provided with a stirrer blade which extends several inches below the ball-joint. The column stopcock is closed and the extra hole is plugged with a short piece of PTFE tubing into which a fine glass rod has been inserted. The whole system is filled with acid-washed *n*-heptane until about 50 ml are present in the infusion bottle. The silica gel sorbent is then transferred into the infusion bottle. Before this transfer is carried out, 0.3 and 0.5-g amounts of silica gel are retained in two paraffine foil sealed beakers. A further volume of acid-washed *n*-heptane is added to the infusion bottle until the mixture reaches the 600-ml mark on the bottle. Then a stopper bearing the stirrer guide is inserted, the stirrer speed is adjusted so as to maintain a clear *n*-heptane layer above the silica gel suspension, the lower stopcock of the column is opened and the silica gel suspension is allowed to sediment under gravity. The pouring of the column usually takes 30 min. After this period, the stopcock is closed, the PTFE tubing is attached to a suction flask and the excess of *n*-heptane is removed until its level falls to the ball-and-socket joint. The column is now ready to be clamped to the circulating water pump. If, during the pouring of the silica gel suspension, the silica gel particles stick to the column wall in the stirred area, it is recommended that these particles are removed with a narrow glass rod (by moving it back and forth several times) in order to prevent possible clogging of the column during packing under pressure.

The packing is carried out by passing *n*-heptane under pressure through the column. *n*-Heptane from a reservoir is passed through a stirred three-necked flask in which it is heated to 56°C, pumped through the acid-wash columns and delivered to the separation column. The socket area at the top of the separation column is freed from any particles of silica gel by filling it with hot acid-washed *n*-heptane and by allowing the particles to fall into the column. Excess of *n*-heptane is removed, the socket area is dried with a piece of filter-paper and the column is tightened to the inlet tubing. Care must be taken with the gaskets, which must be prevented from direct contact with *n*-heptane, which would cause them to swell excessively and cause leaking on the top of the column.

At this stage, the thermostatting bath is set to 35°C and both the column and the mixing chamber (three-necked flask) are heated to this temperature. The column outflow is connected to the photometer cell and the hot *n*-heptane is allowed to pass through the column at the rate of 180 ml/h. The passage of 45–60 ml of the solvent, which takes 15–20 min, should result in the column packing to a height of *ca.* 120–100 cm. Occasionally, the sorbent tends to become clogged during this operation, which can sometimes be prevented by applying a cushioned vibrator to the wall of the column.

The sample to be analyzed is mixed with the saved portion of the sorbent and transferred quantitatively in the dry form on to the top of the column. The DNP derivatives are protected from decomposition by light by wrapping the column with aluminium foil.

The gradient elution device consists of two heated mixing chambers, each of 500-ml volume. About 150 ml of de-aerated 0.1 *N* sulphuric acid are placed in the acid-washed column (60 × 1.2 cm) with a 150-ml bulb at the top. Then 450 ml of *n*-heptane are placed in the first mixing chamber and heated to 56°C, while the same amount of *n*-heptane placed in the other chamber is heated to 35°C. Reservoirs, connected to the latter mixing flask, contain:

(1) 3% *tert.*-amyl alcohol in *n*-heptane;

(2) 18% *tert.*-amyl alcohol in *n*-heptane;

(3) methyl ethyl ketone;

(4) *n*-heptane.

The separation of peptides is identical with that just described with the exception that the solutions in reservoirs are:

(1) 18% *tert*.-amyl alcohol in *n*-heptane;

(2) 50% methyl ethyl ketone in *n*-heptane;

(3) methyl ethyl ketone;

(4) *n*-heptane.

The apparatus usually has an additional *n*-heptane reservoir which is used solely for column packing.

The operation of the apparatus is carried out as follows. The stopcock, which allows the 3% *tert*.-amyl alcohol–*n*-heptane solution to pass down the second mixing chamber, is opened and the pump is set to deliver 180 ml/h of the solvent to the column. After 700 ml of the solvent have passed through the column, which takes almost 4 h (the elution of DNP-phenylalanine is completed by that time), the solvent is changed to 18% *tert*.-amyl alcohol–*n*-heptane. After a further 4 h (720 ml) the eluent is replaced with methyl ethyl ketone; about 2200 ml are allowed to pass through the column, by which time all amino acid derivatives are eluted.

Virtually the same procedure is applicable for DNP-peptides, and the solvents used in each of the three elution steps are those mentioned above.

The overall separation of DNP-amino acids can be seen from Fig. 33.4.

A further development in the technique described is based on the use of standard nine-chamber Technicon Varigrad equipment. A silicone rubber-insulated heating tape is wrapped round the bottom third of this chamber and is held in place by PTFE adhesive tape. Although the combinations can be adjusted according to the specific requirements

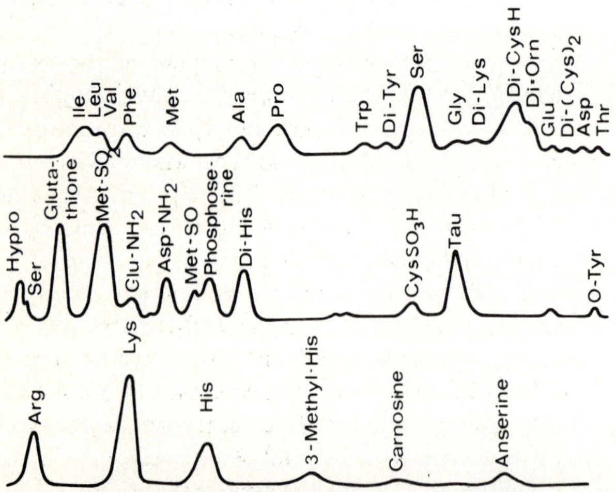


Fig. 33.4. Separation of a known mixture of DNP derivatives of amino acids and related substances. Peaks representing unreacted 2,4-dinitrofluorobenzene and dinitrophenol appear before the DNP-amino acids and are not shown. If dinitroaniline is not removed, its peak appears in the alanine–proline area. (Kesner *et al.*, 1963, 1964).

for a particular mixture to be separated, the following procedure appears to be generally applicable, according to Kesner and co-workers (1963, 1964). At a temperature of 45°C and a flow-rate of 100–150 ml/h, the following composition of solvents in individual chambers of the Varigrad system are used:

Chamber No. 1: 100 ml of 3% *tert.*-amyl alcohol in *n*-heptane;
No. 2: 100 ml of 3% *tert.*-amyl alcohol in *n*-heptane;
No. 3: 100 ml of 3% *tert.*-amyl alcohol in *n*-heptane;
No. 4: 100 ml of 10% *tert.*-amyl alcohol in *n*-heptane;
No. 5: 100 ml of 10% *tert.*-amyl alcohol in *n*-heptane;
No. 6: 100 ml of 3% *tert.*-amyl alcohol in *n*-heptane;
No. 7: 84 ml of methyl ethyl ketone;
No. 8: 84 ml of methyl ethyl ketone;
No. 9: 84 ml of methyl ethyl ketone.

The determination of DNP-amino acids on silica gel can also be carried out at a temperature of 25–28°C and a flow-rate of 250 ml/h. In this instance, chambers 1 to 6 contain 200 ml of 2, 2, 2, 7, 10 and 12% *tert.*-amyl alcohol in *n*-heptane, respectively, while the residual chambers contain 168 ml of methyl ethyl ketone, and sometimes it appears necessary to pass some additional methyl ethyl ketone in order to elute the slowly moving DNP-amino acid zone.

For mixtures that contain both DNP-amino acids and DNP-peptides, the gradient system should consist of two parts. In the first gradient, which consists of 2, 2, 2, 10, 10 and 15% *tert.*-amyl alcohol in *n*-heptane in chambers 1 to 6, all amino acids up to DNP-serine are eluted, the amount poured into each chamber is 150 ml and the column, which is operated at room temperature, is eluted at the rate of 150 ml/h. The second gradient elutes soluble DNP derivatives and DNP-peptides; it consists of 15% *tert.*-amyl alcohol in *n*-heptane (200 ml, chambers 1 and 2), 50% methyl ethyl ketone in *n*-heptane (183 ml, chambers 3 and 4) and pure methyl ethyl ketone (168 ml, chambers 5 to 9).

Kessner *et al.* (1964) successfully applied this procedure to the separation of DNP-amino acids and DNP-peptides obtained by the action of pronase upon native, oxidized, dinitrophenylated and dinitrophenylated + oxidized ribonuclease. They introduced slight modifications to the above procedure. In the original procedure, 2,4-dinitroaniline is eluted close to DNP-alanine; any confusion in this respect can be avoided by pre-extracting the mixture with diethyl ether at pH 8, which removes 2,4-dinitroaniline; alternatively, replacing 450 ml of *n*-heptane in the first mixing chamber with 450 ml of *n*-heptane–toluene makes the pre-extraction unnecessary. In the latter instance, the DNP-$NH_2$ peak is shifted towards the beginning of the chromatogram. In fact, any pre-extraction in this event is unnecessary as DNP-OH, DNP-$NH_2$ and unreacted 2,4-dinitrofluorobenzene are eluted prior to all amino acids. This buffer change is reported not to change the elution pattern of other DNP-amino acids.

Tris buffers, which are widely used in protein chemistry, are reported to cause a double peak of DNP-glutamine; however, this effect can be avoided by pre-extracting the sample to be analyzed with diethyl ether. It is also recommended that phosphate buffers are used instead of Tris, as they do not exhibit such effects.

The hydration of the silica gel sorbent appears to be critical for the separation of some DNP-derivatives, such as bis-DNP-cystine from DNP-glutamine and DNP-aspartic acid.

*References p. 738*

Failure to separate these DNP-amino acids is ascribed to the day-to-day-variations in atmospheric humidity that interfere with the oven-drying procedure. In general, in order to avoid such unwanted effects, it appears to be of great advantage to prepare large batches of partially hydrated silica gel. A suitable composition comprises 150 ml of deionized water mixed with 1000 g of dried silica gel and stored in air-tight bottles.

During drying, it is possible to use temperatures up to 250°C, as silica gel loses water progressively with increasing temperature. However, satisfactory results have been achieved by drying silica gel at 115°C for 1 week. This procedure results in a partially hydrated product, the water content of which is used as basis for calculating the amount of 0.75 *N* sulphuric acid to be added. However, Kesner and co-workers (1963, 1964) reported that even partially hydrated silica gel exhibits some variations with changes in atmospheric humidity. The optimum positioning of the di-DNP derivatives relative to mono-DNP derivatives is finally achieved by varying the final hydration of silica gel. Different degrees of hydration also change the order of elution of amino acids. Thus, for instance, at a high degree of hydration (5.7 ml of 0.75 *N* sulphuric acid per 8 g of silica gel), the order of eluted peaks runs as follows: tyrosine, glycine, lysine, glutamine, lysine, aspartic acid, threonine and serine, while at a low degree of hydration (4.7 ml of 0.75 *N* sulphuric acid per 8 g of silica gel) the order is glycine, tyrosine, lysine, glutamine, aspartic acid, threonine, lysine and serine.

Another system has been described by Tentori for the automated column chromatography of DNP derivatives in which a 120 × 0.9 cm column of silica gel is used. Elution is carried out with a complex gradient system composed of nine chambers. The gradient is formed by different proportions of *tert.*-amyl alcohol, *n*-heptane and methyl ethyl ether as follows:

| | *tert.-Amylalcohol* | *n-Heptane* | *Methyl ethyl ether* |
|---|---|---|---|
| Chamber No. 1 | 5 | 95 | – |
| 2 | 6 | 94 | – |
| 3 | 7 | 93 | – |
| 4 | 10 | 90 | – |
| 5 | 15 | 75 | 8.5 |
| 6 | 15 | 60 | 21 |
| 7–9 | – | – | 84 |

The column was maintained thermostatically at 47°C and the running time was about 5.5 h. Before being admitted to the column, the eluent was passed through a saturating chamber containing 0.1 *M* sulphuric acid.

In addition to the above gradient system, it is also possible to use the following system, with which the results are comparable as far as the quality of separation is concerned:

| | *tert.-Amyl alcohol* | *n-Heptane* | *Methyl ethyl ether* |
|---|---|---|---|
| Chamber No. 1 | 7.5 | 142.5 | – |
| 2 | 9 | 141 | – |
| 3 | 10.5 | 139.5 | – |
| 4 | 15 | 135 | – |
| 5 | 22.5 | 112.5 | 12.75 |
| 6 | 10 | 70 | 61 |
| 7–9 | – | – | 126 |

The column was maintained for the first 1.5 h at 47°C and then at −38°C. The separation time was 10 h.

The following method of filling the chambers permitted the separation of DNP derivatives soluble in water at the operating temperature of 28°C in 225 min:

| | *tert.-Amyl alcohol* | *n-Heptane* | *Methyl ethyl ether* |
|---|---|---|---|
| Chamber No. 1,2 | 15 | 85 | – |
| 3,4 | – | 46 | 46 |
| 5–9 | – | – | 84 |

*Automated separation on nylon columns*

Beyer and Schenk (1969a, b, 1971) applied column chromatography on nylon powder columns to DNP-amino acid determinations. The procedure is particularly suitable for water-soluble DNP derivatives and has been used successfully in the study of structural proteins such as keratin and collagen.

Columns of dimensions 50 × 1 cm appeared to be suitable for the separation. The nylon powder suspensions were boiled before use in order to prevent the formation of bubbles, and were then cooled to 30°C. At this temperature, the columns were poured with a 3 atm overpressure in order to ensure rapid and homogeneous setting. Finally, the columns were compressed with air and eluted twice with citrate buffer of pH 3.0. The buffer consisted of a mixture of 298.5 ml of 1 *N* hydrochloric acid, 42.33 g of citric acid and 16.12 g of sodium hydroxide, diluted to 5 l with distilled water.

The operation and sample application were carried out as follows. A 2-ml volume of hydrolysate containing about 200 mg of the sample was applied to the column with the usual precautions. The sample was washed into the column with two 0.5-ml portions of pH 3.0 buffer and the analysis was carried out using the same buffer at a flow-rate of 30 ml/h. When O-DNP-tyrosine had emerged from the column, the column was ready to be loaded with another sample. However, in practice, after two separations the column had to be regenerated by removing the nylon powder from the column and re-suspending it in 5% ammonia solution for 15 min. The alkaline suspension was then filtered and the residue washed until it was neutral. The cake was resuspended in 0.1 *N* hydrochloric acid,

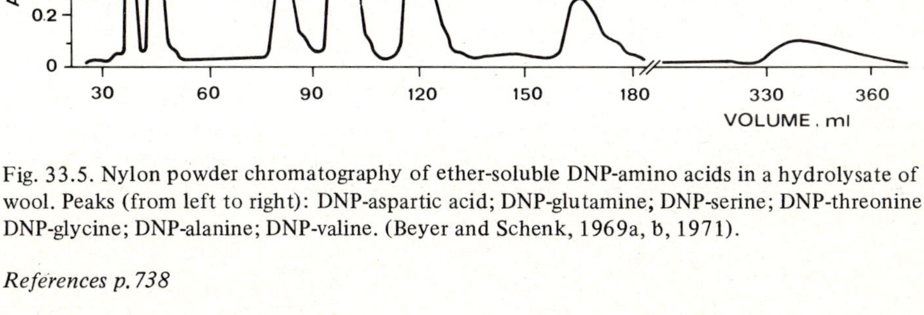


Fig. 33.5. Nylon powder chromatography of ether-soluble DNP-amino acids in a hydrolysate of DNP-wool. Peaks (from left to right): DNP-aspartic acid; DNP-glutamine; DNP-serine; DNP-threonine; DNP-glycine; DNP-alanine; DNP-valine. (Beyer and Schenk, 1969a, b, 1971).

stirred for 15 min, filtered, washed until neutral and preserved in 2.5 l of distilled water. After a further two runs it was discarded.

For the separation of ether-soluble DNP-amino acids, the chromatographic procedure is only slightly different. The preparation of the carrier material is identical with that described above. However, the columns are filled at 60°C and then eluted twice with phosphate buffer of pH 8.0. The buffer used for elution is prepared from 2.7 g of orthophosphoric acid and 55 g of disodium hydrogen orthophosphate dihydrate. This mixture is made up to 5 l and used for analysis.

Chromatograms obtained with ether-soluble and water-soluble amino acid derivatives on nylon powder columns are presented in Figs. 33.5 and 33.6, respectively.

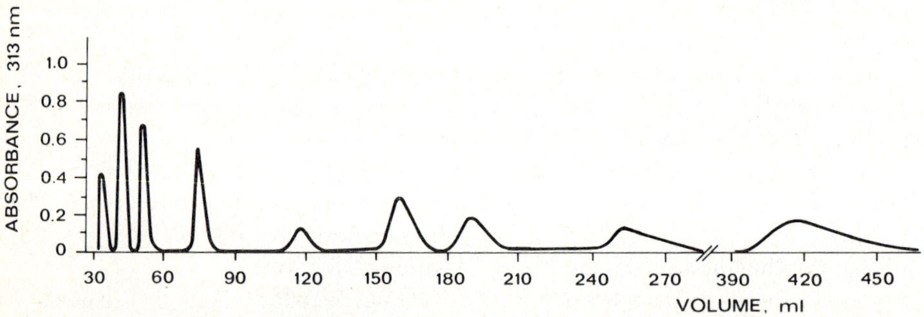


Fig. 33.6. Nylon powder column chromatography of water-soluble DNP-amino acids. Peaks (from left to right): DNP-histidine; DNP-serine; N-DNP-(2-amino-2-carboxyethyl)lysine; DNP-cystine; DNP-ornithine; DNP-lysine; DNP-tyrosine; DNP-arginine; DNP-cysteic acid. (Beyer and Schenk, 1969a, b, 1971).

## 5-DIMETHYLAMINONAPHTHALENE-1-SULPHONYL (Dns) AMINO ACIDS

### Separation in open capillary alkali-treated columns

Nota *et al.* applied open-glass alkali-treated capillary columns to the separation of Dns-amino acids. The success of these columns in gas chromatography made Nota *et al.* try these columns in submicro-scale separations of substances that are not amenable to gas chromatographic analysis. This method may have general applicability, but Dns-amino acids have been used because of the simplicity of detection; these compounds can be detected fluorimetrically, as they exhibit blue fluorescence (excitation maximum 340 nm and luminiscence maximum 500–550 nm, depending on the solvent used; for details see Rosmus and Deyl, 1971, 1972).

The capillary columns, according to the original description of Nota *et al.*, are prepared from Murano soft glass tubes, I.D. 2.0 mm, O.D. 6 mm, using a glass tube drawing apparatus. The columns are filled with 2.5 *N* sodium hydroxide solution and kept at 100°C for at least 2 h, washed with water until neutral, rinsed with acetone and dried in a stream of nitrogen. The injector is essentially the same as that used in gas chromatography.

For the separation of Dns-amino acids, the columns were equilibrated with benzene–pyridine–acetic acid (80:20:2) (Morse and Horecker). After equilibration, the needle valve was set to give a 1:30 ratio of the flow-rate in the column to the flow-rate in the valve. The flow-rate in the column was kept constant at 0.5 ml/h by applying a pressure of 20–100 torr.

Using a micro-syringe, 4 $\mu$l of Dns-amino acid mixture containing 6 $\mu$mole/ml of each derivative were then injected into the column. Each drop emerging from the column was examined by TLC on silica gel plates using the above-mentioned system, *i.e.*, benzene–pyridine–acetic acid (80:20:2) as the mobile phase. $R_F$ values were then plotted against the time after which the particular amino acid appeared in the effluent. In this way, the amino acids could be identified.

### Automated chromatography of Dns-amino acids

#### *Column preparation and operation*

The column used (Deyl and Rosmus) had dimensions of 100 × 1 cm and was adjusted for constant temperature operation (35°C); it was filled with Woelm polyamide (15 g). As the degree of separation obtained is considerably influenced by the method of column packing used, a special device developed by Kesner for uniform column filling (see p. 720) was used. The pouring device consisted of an infusion bottle fitted with a stirrer blade which extended below the ball-joint. The infusion bottle was filled with about 50 ml of distilled benzene and the portion of polyamide suspension was added. Adequate amounts of the sorbent were retained for sample application. The infusion bottle was then closed, the lower tap joining the column head with the infusion bottle was opened and polyamide particles were allowed to sediment under gravity. The stirrer speed was adjusted so as to maintain a clear benzene layer above the polyamide suspension. After the level of polyamide had reached the column head, the upper tap was closed, the filling device was removed and the excess of benzene was gently aspirated out. As the quality of the column packing influences the final separation to a great extent, the packing has to be carried out carefully so as to remove particles that might adhere to the column walls during the filling procedure. It is also advisable not to touch the sedimenting layer of polyamide in order to prevent uneven particle distribution and consequent deformation of bands during chromatography.

After the column had been filled, benzene was pumped through it for about 1 h in order to pack it. During the packing procedure, the flow-rate was kept at 2.5 ml/min and during operation of the column the flow-rate was decreased to 0.1 ml/min. During the packing procedure, the thermostat was set at 35°C and the column, the mixing chamber and the reservoirs were adjusted to this temperature. The outlet of the column was connected to an adapted Farrand spectrofluorimeter cell; as in most instances the fluorescence intensity was much too high for the recorder scale, a proportional pump was inserted. The excess of the outflow from the column was either discarded or retained in a fraction collector for further investigation by flat-bed techniques. The fluorescence wavelengths were set to 340 and 500 nm for excitation and luminescence, respectively.

The Farrand spectrofluorimeter was alternatively set for decreased sensitivity (1.0 position on the sensitivity scale), and the proportional pump was by-passed. The outlet flow was diluted with acetone or methyl Cellosolve from an additional reservoir. The individual parts used for the split-stream procedure and all the tubing used were parts of the Technicon amino acid analyzer. The overall assembly of the apparatus is shown schematically in Fig. 33.7.

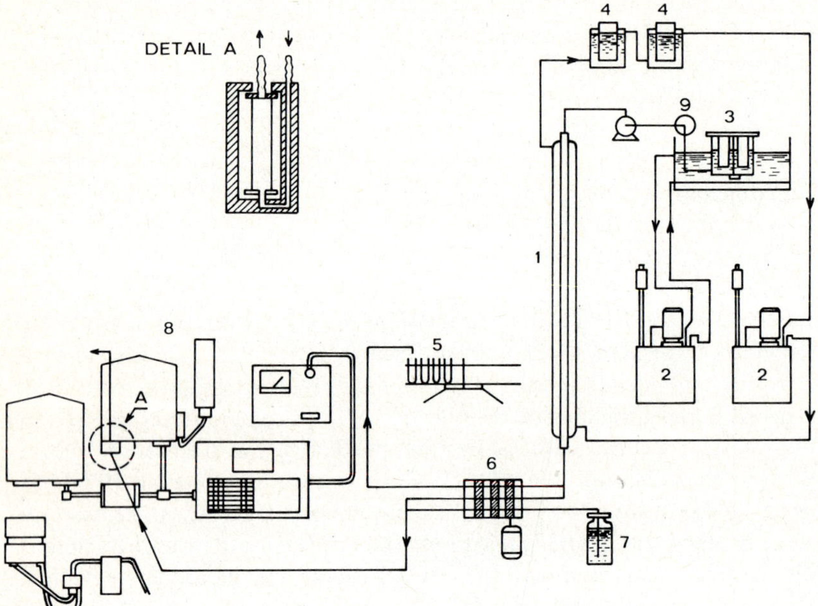


Fig. 33.7. The overall assembly of the chromatographic equipment. 1 = Separation column; 2 = thermostats; 3 = gradient device and reservoirs (benzene–acetic acid gradient); 4 = reservoirs for benzene–acetic acid (9:1 and 6:4 systems); 5 = fraction collector; 6 = proportional pump; 7 = acetone reservoir; 8 = Farrand spectrofluorimeter (detail A – flow-through cuvette); 9 = programmed three-way tap (Deyl and Rosmus).

The measuring cuvette was adapted from a 5-mm round-shaped quartz tube in the manner shown in Fig. 33.7. The spectrofluorimeter gear box was adapted to a lower speed (20 cm/h).

*Sample preparation*

The amount of sample analyzed varied from 50 to 500 $\mu$l. A peptide or standard amino acid mixture (0.5–5 mmole of each component to be detected in 0.1 *M* sodium hydrogen carbonate solution) was evaporated to dryness *in vacuo*. This step removes ammonia that might otherwise result in considerable problems when the peak of Dns-amide is being eluted during column regeneration. The dry residue was redissolved in 10–15 $\mu$l of ammonia-free water and an equal volume of the Dns reagent (a saturated solution of Dns chloride in acetone) was added. The mixture was incubated at 37°C for 1 h and the excess of the unreacted Dns chloride was extracted with 500 $\mu$l of toluene or ethyl acetate. The extraction step was repeated three times. The whole series of operations was carried out with protection against direct light. With proteins and peptides for which hydrolysis was necessary, this hydrolysis was carried out at 105°C for 18 h in a nitrogen atmosphere in a sealed tube. The sample was finally evaporated to dryness, redissolved in 1–2 ml of water and mixed with a portion of the retained polyamide to make an opaque suspension, and the mixture was evaporated to dryness at 40°C *in vacuo*. The dry residue was resuspended in benzene and loaded on to the separation column by several washings with 10-ml volumes of benzene.

*Elution*

As the solvent system used by Woods and Wang in thin-layer chromatography did not result in complete resolution of all the amino acid peaks, different proportions of benzene–acetic acid were examined. The most generally applicable mixture was benzene–acetic acid (90:5) in which the fast-moving peaks of leucine and isoleucine were not separated. In order to improve this situation, elution was started with a benzene–(benzene–acetic acid) (9:1) gradient, composed of two 200-ml reservoirs. After 300 min, the inlet was switched automatically to the benzene–acetic acid (9:1) mixture and elution was carried out for the next 800 min without a gradient. In the final stage, this eluent was suddenly switched to a benzene–acetic acid (6:4) mixture, which made it possible to elute asparagine, hydroxyproline, arginine, cysteine and cysteic acid. The bluish band of Dns-amide remained uneluted and was removed during the regeneration procedure.

*Column regeneration*

Before re-use, the column was washed with dried acetone (1.5 h was satisfactory). The flow-rate of the washing fluid was 1.5 ml/min. Acetone was then replaced with benzene, which was passed through the column for an additional 2 h. After this period, the column was ready for use in a new separation.

The gradient elution system exhibits several advantages compared with the widely used flat-bed techniques. Firstly, it minimizes the possibility of inducing errors, as the separation is very precise and can easily be completed with an additional flat-bed check-up by using the same material, which is therefore not lost and the demands on the amount to be analyzed are consequently reasonable. Another advantage is based on the fact that the column technique gives a good possibility of recovering unusual amino acids or hydrolysis-resistant peptides, which may be of considerable importance in special instances, such as in the analysis of complex peptide mixtures. As indicated in Fig. 33.8, this technique offers the possibility of separating almost all common amino acids in one run, and under standard conditions the technique can also be used for quantitative determinations. These advantages are, of course, obtained at the cost of using more complicated equipment and slightly larger samples for analysis (at least twice as much as in the flat-bed technique).

As in every separation of a complex mixture, there are pairs of Dns derivatives which are hard to separate, such as phenylalanine and a number of others with high chromato-

*References p. 738*

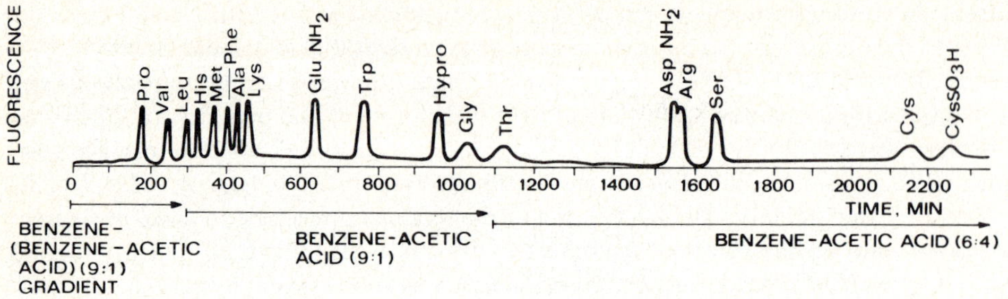


Fig. 33.8. Typical elution profile of Dns-amino acids on a poiyamide column (Deyl and Rosmus).

graphic mobilities. In order to achieve adequate separations, which may be subjected to quantitation by using the technique common in non-derivatized amino acid analysis, one has to work in the range of *ca.* 2000 theoretical plates. An improved separation has been obtained by introducing a gradient system at the beginning of the chromatographic run. The operating times and solvent systems used are as follows: 0–300 min – gradient of benzene–(benzene–acetic acid) (9:1), 200 ml of each solvent; 300–1100 min – benzene–acetic acid (9:1); and 1100–2500 min – benzene–acetic acid (6:4).

The chromatographic properties of the individual solvent systems used are summarized in Table 33.2.

TABLE 33.2
RETENTION VOLUMES AND RELATIVE RETENTION VOLUMES OF Dns-AMINO ACIDS

| Amino acid | Benzene–acetic acid (9:1) | | Benzene–acetic acid (6:4) | |
|---|---|---|---|---|
| | $V_e$* | $V_e/V_{pro}$** | $V_e$* | $V_e/V_{pro}$** |
| Leu | 138 | 1.53 | 80 | 1.02 |
| Val | 124 | 1.38 | 82.5 | 1.05 |
| His | 143 | 1.64 | – | – |
| Met | 152 | 1.69 | 81 | 1.04 |
| Ala | 163 | 1.81 | 92 | 1.08 |
| Lys | 196 | 2.18 | 108 | 1.38 |
| Gly | 765 | 8.39 | 117 | 1.50 |
| Trp | 500 | 5.55 | 134 | 1.71 |
| Thr | 865 | 9.60 | 142 | 1.82 |
| Ser | 2540 | 28.21 | 485 | 6.21 |
| Cys | – | – | 980 | 12.57 |
| Arg | – | – | 384 | 4.92 |
| Hyp | 700 | 7.76 | – | – |
| Pro | 90 | 1 | 78 | 1 |
| Phe | 155 | 1.72 | 85 | 1.09 |
| Gln | 370 | 4.12 | 126 | 1.61 |
| Asn | 1350 | 15.00 | 370 | 4.74 |

* $V_e$ = elution volume.
** $V_e/V_{pro}$ = elution volume relative to proline.

Although no precise rules for predicting chromatographic mobility can be formulated, there are some general features which, in the case of an unknown derivative, may serve as a guide. An increase in the number of carbon atoms in the amino acid side-chain decreases the retention times. Compared with a straight chain, the $-CH_2-$ difference in a branched side-chain has a much lower effect in decreasing the retention time. Hydroxylation, however, shifts retention times to much higher values and the differences in a homologous series are increased in hydroxylated amino acids. While the presence of a second amino group makes the amino acid move with a short retention time, guanidylation considerably retards the chromatographic mobility.

The column packing is able to withstand an almost unlimited number of separations provided that not too many impurities are loaded on to the column in each run and provided that they (mainly Dns-amides) are eluted adequately during the regeneration procedure. After several separations, a grey-brown ring appears at the top of the column packing; however, this does not disturb the separation.

## HYDANTOINS AND SUBSTITUTED HYDANTOINS

### Hydantoins

A very detailed method of chromatographic isolation and identification of hydantoins resulting from the following reaction is currently available:

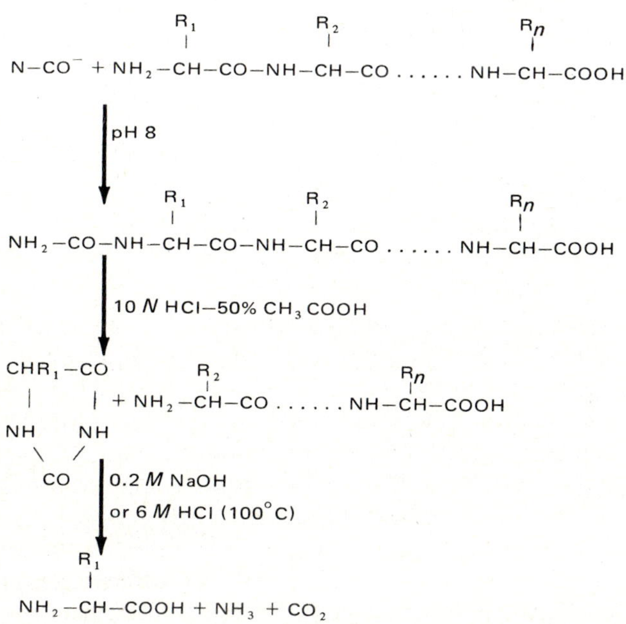


However, when only a single analysis is required and it is not worth building the automatic analyzer described below, a simplified procedure can be used with Dowex 50-X2 (200–400 mesh). In general, the separation follows the rules summarized in Fig. 33.9.

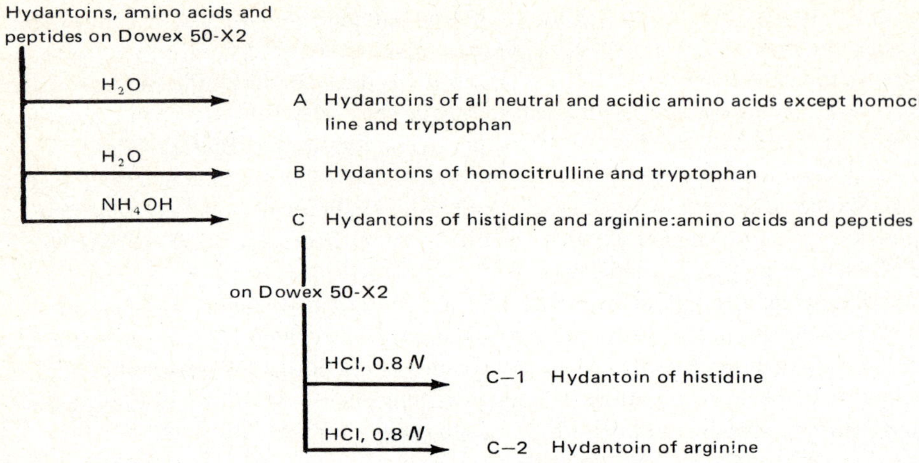


Fig. 33.9. Separation scheme for hydantoins, amino acids and peptides on Dowex 50-X2.

The solution is transferred quantitatively on to the Dowex 50-X2 column, the size 10 × 0.9 cm reportedly being adequate. Before use, the column is washed successively with 1 *M* sodium hydroxide solution, 6 *M* hydrochloric acid and water; the height of the column should be 10 cm after the last wash. The column is maintained thermostatically at 30–25°C and about 100 ml of water are allowed to pass through. The first 30 ml of this fraction are designated fraction A (see the scheme in Fig. 33.9), and the next 30–70 ml are designated fraction B. These solutions are evaporated to dryness and used for further assay. The column is eluted with 40 ml of ammonia solution, and fraction C is thereby obtained. Another column of Dowex 50-X2 of the same size is prepared by washing it with 1 *M* sodium hydroxide solution, 6 *M* hydrochloric acid and 0.8 *N* hydrochloric acid. The residue obtained by evaporating fraction C is dissolved in 1 ml of distilled water and placed quantitatively on this column, the temperature being maintained at 20–35°C. Elution is carried out with 0.8 *M* hydrochloric acid at a flow-rate of 60 ml/h. The first 35 ml are designated fraction C-1, and the next 75 ml fraction C-2.

The individual fractions are assayed for the presence of hydantoins by the usual amino acid analyzer. The practical procedures for fractions A to C differ slightly, as can be seen from the following descriptions.

*Fraction A.* After being taken to dryness, the residue is dissolved in 2 ml of freshly prepared sodium hydroxide solution (0.2 *M*). A 1.8-ml volume of this solution is transferred to a tube, which is sealed and then heated at 110°C for 24 h. After hydrolysis, the contents are neutralized with 1 ml of 1 *N* hydrochloric acid, the solution is evaporated to dryness, and 2 ml of citrate buffer (pH 2.2) are added. The solution is centrifuged so as to remove the precipitated silicates, and the supernatant solution is used for the determination of neutral and acidic amino acids. If glutamic acid is expected to be the N-terminal, it is recommended that the hydrolysis period should be extended to 96 h as excellent yields are thus obtained.

*Fraction B.* Fraction B is similarly processed. After evaporation, the residue is dissolved in 2 ml of 0.2 *N* sodium hydroxide solution. A 1.8-ml volume of this solution is transferred to a hydrolysis tube, which is sealed and heated as for fraction A. The hydrolyzate is analyzed only on the short column of a two-column amino acid analyzer as it has to be checked only for lysine.

*Fraction C.* Both fractions C-1 and C-2 are processed in a similar manner. The residues obtained after the respective fractions have been evaporated to dryness are dissolved in 2 ml of 6 *N* hydrochloric acid, and 1.8 ml of the solution are transferred into a hydrolysis tube and sealed. The hydrolytic cleavage is carried out at 110°C for 96 h. After the test-tube has been opened, excess of hydrochloric acid is removed by evaporation. The residue is dissolved in 11 ml of 0.2 *M* sodium hydroxide solution, and the samples are re-evaporated so as to remove excess of ammonia. An additional 1 ml of 1 *M* hydrochloric acid is added, the solution is evaporated to dryness and the dry residue is finally dissolved in 2 ml of buffer (pH 2.2). The sample is applied to the 15-cm column of the amino acid analyzer in order to evaluate the positions of histidine in fraction C-1 and arginine in fraction C-2. It should be noted that the first 10 ml of the effluent must be disconnected from the ninhydrin line as otherwise a large amount of neutral amino acids may precipitate and clog the apparatus.

Special attention to processing the samples must be paid if pyrrolidone carboxylic acid is expected to be present. In this event, fraction A from the Dowex column is evaporated to dryness and the residue is dissolved in 5 ml of 3 *M* hydrochloric acid and heated at 100°C for 30 min. The excess of acid is removed by evaporation, the residue is dissolved in 2 ml of distilled water, and the sample is quantitatively transferred to another Dowex column (5 × 0.4 cm) with an additional 1 ml of water. The column is eluted with 6 ml of water, and the effluent is evaporated to dryness and subjected to amino acid analysis as described above.

Recently, the chromatographic separation of hydantoin derivatives has been carried to a greatly advanced stage. Hagel and Gerding developed a method for column chromatographic separation using an automatic hydantoin analyzer with Bio-Rad AG 1-X8 ion-exchange resin. Sixteen hydantoins of the most common amino acids were separated. In the elution programme devised, a two-column system was eluted with three different ammonium chloride–ammonia buffers. The column lengths were 95 and 15 cm, respectively, and internal diameters were both 0.9 cm. The columns were maintained thermostatically at 20°C. The anion-exchange resin Bio-Rad AG 1-X8 ($Cl^-$,–400 mesh) (Bio-Rad Labs., Richmond, Calif., U.S.A.), was used for column packing. The resin to be used in the column was collected by fractional sedimentation in 2 *N* hydrochloric acid. It was washed with 1 *N* sodium hydroxide solution at 40°C so as to remove possible impurities and to achieve a low background and a high recovery of hydantoins, which is necessary if one takes into consideration the incomplete conversion of the parent amino acid into this derivative. The resin was washed with water so as to remove the excess of base. In the last stage, the chloride form was re-established washing the resin with 1 *N* hydrochloric acid. Two volumes of ammonium chloride–ammonia buffer were added per volume of settled resin.

The columns were filled in the usual manner for amino acid analyzers. In order to allow the resin sufficient time to settle and to yield reproducible results, buffers were pumped through the column over a period of several days. It has been reported that not

*References p. 738*

less than 1 month was required in order to obtain a column of perfect quality.

The following buffers were used for elution.

Buffer I: 0.3 *M* ammonium chloride + 5.5 ml of 25% ammonia solution per litre (pH 8.78);

Buffer II: 1.5 *M* ammonium chloride + 230 ml of 25% ammonia solution per litre (pH 9.88);

Buffer III: 2.5 *M* ammonium chloride + 260 ml of 25% ammonia solution per litre (pH 9.68).

The preparation of the buffers was given the appropriate care that is usual in automatic amino acid analysis. Further, a constant pressure of ammonia was maintained over the buffers in stock.

In the separation of a hydantoin mixture, the separating power of the column decreased sharply when the flow-rate was higher than 10 ml/h on a column with a surface area of 0.636 $cm^2$. Therefore, a very low back pressure was developed. The long column (95 cm) gave a back pressure of about 1 atm, while in the short column it was less than 0.1 atm. The long column was fed with buffer I for the first 13 h, and buffer II for 17 h. The small column was operated by using buffer III.

The order of hydantoins eluted from the long column was as follows: arginine, threonine, serine, glycine, alanine, aspartic acid, glutamine, histidine, lysine, valine, proline, isoleucine, methionine and leucine. On the short column, there appeared first two peaks of the amino acids resolved on the long column, and then the maxima of phenylalanine and tyrosine appear (in that order).

## Phenylthiohydantoins (PTH-amino acids)

The reaction of N-terminal amino acids with phenyl isothiocyanate occurs as follows:

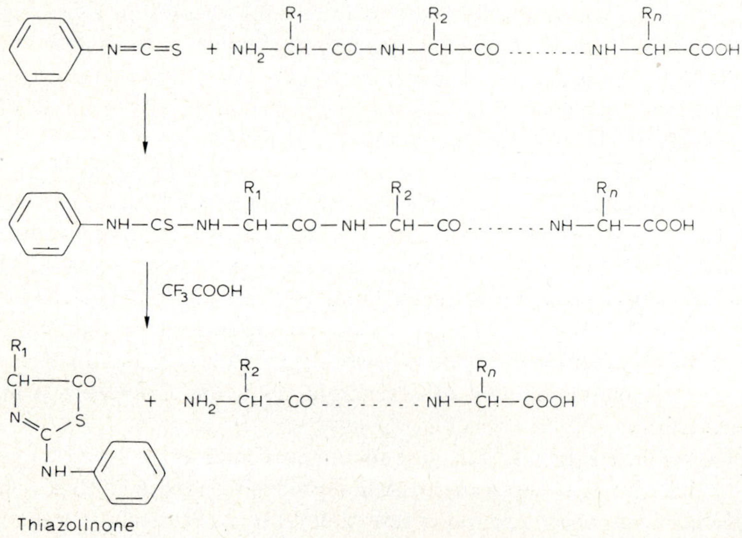

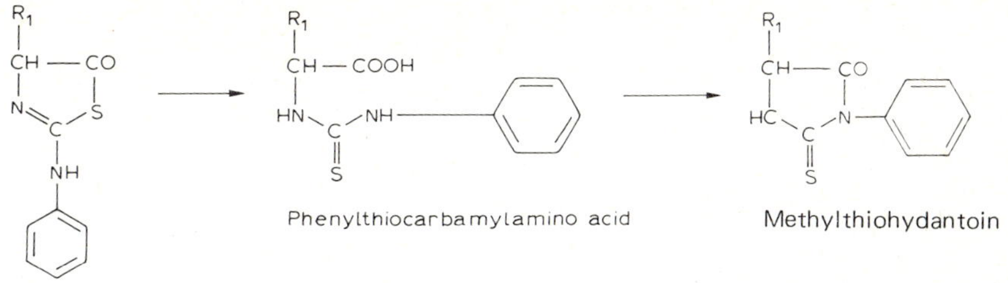


Up to the present time, only one column chromatographic separation of PTH-amino acids has been described, by Sjöquist (1955, 1957), who used Celite as the sorbent.

The experimental procedure is as follows. For the preparation of the column, 10 g of Celite is mixed with 6 ml of the stationary phase, the mixture is slurried with 100 ml of the mobile phase and the slurry is immediately added to the column (80 × 0.8 cm). The height of the filled space is *ca.* 50 cm.

Sample application is carried out by dissolving the sample (about 10 $\mu$g of each PTH-amino acid) in 0.1 ml of the stationary phase and transferring the solution with a capillary pipette to the top of the column. Two 0.1-ml portions of stationary phase and three 0.2-ml portions of mobile phase are used for washing the sample into the column. The space above the column is subsequently filled with the mobile phase.

Three different chromatographic operations for the separation of 20 PTH-amino acids are described below, which together form the analytical procedure. Two of the operations are carried out on columns, here called column I and column II, and these are supplemented by a paper chromatographic separation of PTH-arginine and PTH-histidine.

In column I, the preparation of which is described above, the following solvent systems are used.

(A) *n*-Heptane (250 ml) + propionic acid (100 ml) + ethylene chloride (5 ml) + water (30 ml); the lower layer (78 ml) is used as the stationary phase and the upper layer as the mobile phase.

(B) *n*-Heptane (85 ml) + propionic acid (5 ml) + ethylene chloride (10 ml); only one phase is obtained.

The flow-rate is adjusted to 15 ml/h. The elution is started with solvent A and solvent B is introduced when the peak of PTH-phenylalanine has left the column. The result of a typical separation is presented in Fig. 33.10.

Column II is operated in the same way as column I up to the stage where the PTH-phenylalanine has left the column. From that stage, a gradient elution is applied, consisting of 100 ml of solvent A to which is gradually added a mixture of *n*-heptane (58 ml) + propionic acid (17 ml) + ethylene chloride (25 ml).

The result of a typical separation is presented in Fig. 33.11. The only unresolved pair of PTH-amino acids is that of arginine and histidine phenylthiohydantoins. These PTH derivatives are separated by paper chromatography in *n*-heptane–*n*-butanol–75% formic acid (40:30:9). The $R_F$ value for PTH-histidine is 0.37 and for PTH-arginine 0.46. The spots are eluted with 70% ethanol and evaluated spectrophotometrically at 296 nm.

The recoveries in both column and paper separations are better than 95%, with the exception of PTH-serine and PTH-threonine, for which the recoveries are 40% and 70%, respectively.

*References p. 738*

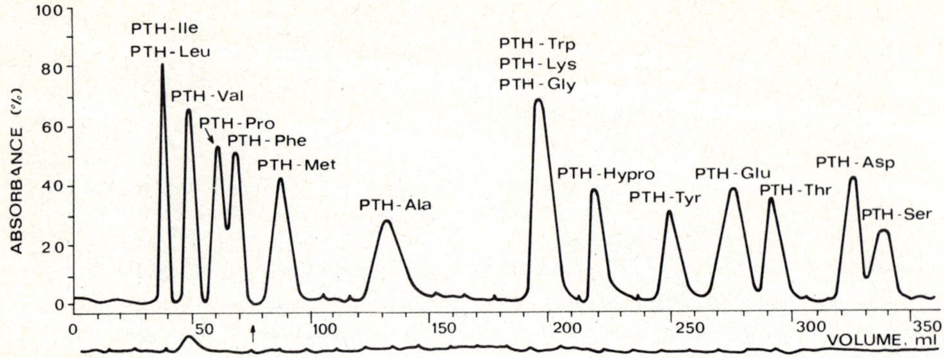


Fig. 33.10. Chromatographic separation of PTH-amino acids on a Celite column (column I). Optical density recorded at 269 nm. The arrow indicates the change of solvent. The lower curve is a trace of a blank without PTH derivatives. (Sjöquist, 1955, 1957).

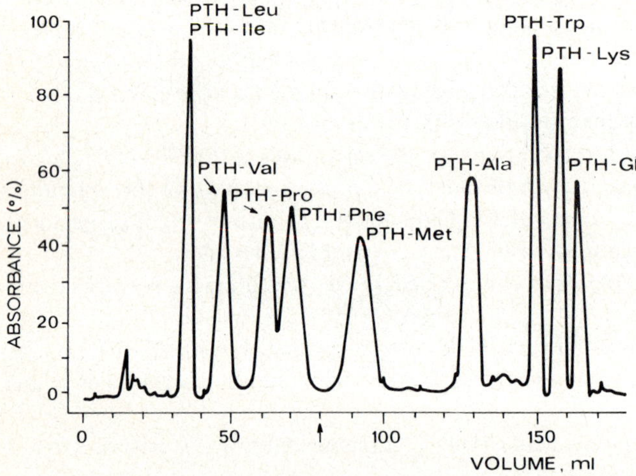


Fig. 33.11. Separation of PTH-amino acids on column II. The gradient began at the arrow. Optical density recorded at 269 nm (Sjöquist, 1955, 1957).

## MISCELLANEOUS DERIVATIVES

As already mentioned in the Introduction, there are over 40 different types of amino acid derivatives that can be used for the N-terminal amino acid analysis. However, only a few were chromatographed on columns, because flat-bed techniques appear to be far more simple and useful for qualitative analysis. Of the few which are suitable for liquid column chromatographic separations, those which are frequently used have been surveyed above. However, there are some others that are likely to be used in the future, and these are briefly outlined below.

The tendency of research workers to use fluorescent, coloured or radioactively labelled compounds, which removes the necessity for applying detection reactions, is reflected in the use of coloured hydantoins; 3,5-dinitrophenylthiohydantoins derived from 4-dimethylamino-3,5-dinitrophenyl isothiocyanate and 4-dimethylamino-3,5-dinitrophenylhydantoins derived from 4-dimethylamino-3,5-dinitrophenyl isocyanate are good examples (Evans and Reith, Nepluyev *et al.*, Reith and Waldron). Although the latter derivative was introduced over 20 years ago, its applications are still few (Chibnall and Spahr).

Levy (1950) and Levy (1957) have used isotopically labelled iodobenzene-*p*-sulphonyl chloride (pipsyl chloride) for quantitative amino acid analysis, N-terminal determinations and sequencing of protein structures. In spite of numerous studies, an extensive investigation of the properties of pipsyl amino acids, especially their chromatographic behaviour, has been investigated only recently by Fletcher. He suggested the use of column and/or paper chromatographic separation and identification of the resulting derivatives.

For the analysis of amino groups in proteins and peptides, the procedure of Velick and Udenfriend, applied to the N-terminal analysis of salmin, can be recommended. In this procedure Celite was used for column packing.

Alternatively, column chromatography, using a Dowex 1-X2 column in a $Cl^-$ cycle can be used for the same purpose. This technique was suggested by Levy (1957) and Levy and Carpenter, and compared with that mentioned above it is rather simple. The conditions are as follows. A total of 350 ml of an eluent comprising 10 ml of 95% ethanol, 10 ml of 1 *N* hydrochloric acid and water to 1 l is used, with a bed volume of 3 g of the ion exchanger, Dowex 1-X2, 200–400 mesh. The elution volumes of the most common amino acid derivatives are summarized in Table 33.3. Alternative possibilities for separations are presented in Table 33.4.

For detection, the above authors designed a special apparatus in which individual fractions were collected directly in trays, evaporated to dryness and the dry samples presented to the end-window of a Geiger-Müller counter. The ratio of count rate to mass of derivative was determined as follows. A portion of a standard solution of $[^{131}I]$pipsyl-

TABLE 33.3

RETARDATION OF PIPSYL COMPOUNDS ON A DOWEX 1-X2 ($Cl^-$) COLUMN (LEVY, 1957; LEVY AND CARPENTER)

Mobile phase: 350 ml of a solution comprising 10 ml of 95% ethanol, 10 ml of 1 *N* hydrochloric acid and water are mixed together to make the final volume 1 l, 3 g Dowex 1-X2 ($Cl^-$), 200–400 mesh.

| Pipsyl derivative | Volumes eluted at peak maximum (ml) |
|---|---|
| Ala | 66 |
| Asp | 155 |
| Glu | 107 |
| Gly | 93 |
| Pro-OH | 82 |
| Ile | 51 |
| Leu | 57 |
| Phe | 128 |
| Pro | 56 |
| Ser | 78 |
| N-Tyr | 210 |
| Gly–Ala–Gly–Ala | 53 |

*References p. 738*

TABLE 33.4
DETAILS OF CHROMATOGRAPHIC COLUMNS FOR THE SEPARATION OF PIPSYL AMINO ACIDS (LEVY, 1957; LEVY AND CARPENTER)

| Designation | Dimensions (cm) | pH of buffer | Eluting solvent | Fraction volume (ml) | Approx. flow-rate (ml/h) | Application |
|---|---|---|---|---|---|---|
| FR-7.2 | 20 × 0.9 | 7.2 | Methyl ethyl ketone–chloroform (1:1) followed by methyl ethyl ketone–chloroform (3:2) | 0.9 | 10–12 | Fast-running ether-soluble derivatives |
| G-5.2 | 10 × 0.45 | 5.2 | Chloroform–*n*-butanol gradient followed by *n*-butanol for pipsyl-cysteic acid | 0.5 | 8 | (a) Slow-running ether-soluble derivatives (b) Acid-soluble derivatives |
| G-7.2 | 10 × 0.45 | 7.2 | Chloroform–*n*-butanol gradient | 0.5 | 8 | Certain mixtures of ether-soluble derivatives |

glycine in 2% sodium hydrogen carbonate solution was acidified, and the derivative was extracted quantitatively into diethyl ether. The ethereal solution was evaporated, and the residue dissolved in a small, known volume of ethanol. Measured portions of this solution were evaporated on to filter disks. Three or four standards were run on the G-5.2 column, the first being put on at the start and the remainder in succession after the preceding peak had left the column. Standards were run on the FR-7.2 column, from which pipsylglycine is not eluted, by packing a column of the same dimensions with buffer of pH 5.2. In all standard runs, at least $10^4$ total counts were recorded. Of course, with unlabelled iodine, conventional procedures such as recording of optical density can be recommended, provided that enough peptidic material is available.

In order to make the enumeration of different types of N-terminal derivatives of amino acids used in liquid column chromatographic separations complete, it is necessary to mention methoxycarbonylamino acids, introduced by Chibnall and Spahr, and the hydrolysis of the N-terminal peptide bond with the β-hydroxyaquotriethylenetetramine-cobalt(III) ion, described by Buckingham *et al.*

## REFERENCES

Beyer, H. and Schenk, U., *J. Chromatogr.*, 39 (1969a) 428.
Beyer, H. and Schenk, U., *J. Chromatogr.*, 39 (1969b) 491.
Beyer, B. and Schenk, U., *J. Chromatogr.*, 61 (1971) 263.
Blackburn, S., *Biochem. J.*, 45 (1949) 579.
Buckingham, D. A., Collman, J. P., Happer, D. A. R. and Marzilli, L. G., *J. Amer. Chem. Soc.*, 89 (1967) 1082.
Chibnall, A. C. and Spahr, P. F., *Biochem. J.*, 68 (1958) 135.
Deyl, Z. and Rosmus, J., *J. Chromatogr.*, 69 (1972) 129.
Evans, G. G. and Reith, W. S., *Biochem. J.*, 56 (1954) 111.

Fletcher, J. C., *Biochem. J.*, 78 (1966) 34C.
Gordon, A. H., Martin, A. J. P. and Synge, R. L. M., *Biochem. J.*, 37 (1943) 79.
Green, F. G. and Kay, L. M., *Anal. Chem.*, 24 (1952) 726.
Hagel, P. and Gerding, J. J. T., *Anal. Biochem.*, 28 (1969) 463.
Heinrich, M. R. and Bugna, E., *Anal. Biochem.*, 28 (1969) 1.
Jellinek, M. R. E. and Del Carmen Vara, M., *Ann. Ass. Quím. Argent.*, 53 (1965) 203; *C.A.*, 66 (1967) 35291k.
Kesner, L., *Anal. Chem.*, 35 (1963) 83.
Kesner, L., Muntwyler, E. and Griffin, G. E., *Biochim. Biophys. Acta*, 85 (1964) 435.
Kesner, L., Muntwyler, E., Griffin, G. E. and Abratis, J., *Anal. Chem.*, 35 (1963) 83.
Levy, A. L., *J. Chem. Soc.*, (1950) 404.
Levy, D. and Carpenter, F. H., *Biochemistry*, 6 (1967) 3559.
Levy, M., *Methods Enzymol.*, 4 (1957) 238.
Matheson, N. A., *Biochem. J.*, 88 (1963) 146.
Matheson, N. A., *Biochem. J.*, 94 (1965) 513.
Matheson, N. A., *Biochem. J.*, 100 (1966) 389.
Matheson, N. A. and Sheltawy, *Biochem. J.*, 98 (1966) 297.
Mills, G. L., *Biochem. J.*, 50 (1952) 707.
Morse, D. and Horecker, B. L., *Anal. Biochem.*, 14 (1966) 429.
Nepluyev, V. M., Chernukhina, L. A. and Serebryanyi, S. B., *Biokhimiya*, 29 (1964) 51.
Nishikawa, A. H., Wu, L. H. L. and Becker, R. R., *Anal. Biochem.*, 18 (1967) 384.
Nota, G., Marino, G., Buonocore, V. and Ballio, A., *J. Chromatogr.*, 46 (1970) 103.
Partridge, S. M. and Swain, T., *Nature (London)*, 166 (1950) 272.
Porter, R. R. and Sanger, F., *Biochem. J.*, 42 (1948) 287.
Portugal, A. V., Green, R. and Sutherland, T., *J. Chromatogr.*, 12 (1963) 183.
Reith, W. S. and Waldron, N. M., *Biochem. J.*, 53 (1953) XXXV.
Rosmus, J. and Deyl, Z., *Chromatogr. Rev.*, 13 (1971) 163.
Rosmus, J. and Deyl, Z., *J. Chromatogr.*, 70 (1972) 221.
Schroeder, W. A., *Ann. N.Y. Acad. Sci.*, 49 (1948) 204.
Sjöquist, J., *Biochim. Biophys. Acta*, 16 (1955) 183.
Sjöquist, J., *Ark. Kemi*, 11 (1957) 129 and 151.
Tentori, L., *Ann. Ist. Sup. Sanità, 2*, Special No. 1 (1966) 187; *C.A.*, 66 (1967) 82195s.
Velick, S. F. and Udenfriend, S., *J. Biol. Chem.*, 191 (1951) 233.
Woods, K. R. and Wang, K. T., *Biochim. Biophys. Acta*, 133 (1967) 369.

*Chapter 34*

# Peptides

I. KLUH

## CONTENTS

## INTRODUCTION

Fundamental importance is attached to the chemical structures and the function of proteins and peptides in living organisms. The basis for further work on a protein is the isolation of that protein in a pure form and the exact determination of its chemical structure.

A major improvement in isolation techniques was the application of column chromatography to the separation of peptides from biological fluids and to the separation of peptides in the enzymatic and non-enzymatic hydrolyzates of proteins. The extent of recent progress with this technique can be judged not only by the large number of published papers, describing methods in which column chromatography was used, but predominantly by the decrease in the amount of starting material required for complete chemical analysis.

In spite of the high resolving power achieved in recent modifications of this technique, it is not likely that a single chromatographic step will separate a peptide mixture completely into its components. Some peptides can be obtained in a sufficiently pure state after a single chromatographic step, while other peptides are eluted in mixtures if they are similar in respect of size, net charge or binding strength to the matrix used. Re-chromatography of these mixtures under different conditions or on another ion exchanger or a different molecular sieve may effect a further separation. The selection of suitable conditions and media with a resolving power to give the performance required for the separation of known peptides is not difficult. However, initially there is usually an inadequate amount of information available about all of the constituents of the mixture and the choice of the separation procedure is a trial-and-error process. The availability of only a limited amount of material limits the number of trials that can be carried out as the recovery is frequently reduced owing to the irreversible binding of peptides to the matrix of the sorbent.

However, analysis of the amino acids in the mixture to be separated is usually possible and, by considering the type of hydrolysis or method of preparation of the mixture, some characteristics of the peptides can be predicted and many difficulties avoided.

This chapter cannot offer more than an account of some methods and a brief discussion of this topic. The choice of an actual method to be used depends solely on the experimentor, and it is hoped that this chapter will be of help to him in finding useful experimental data and references.

## METHODS FOR THE SEPARATION OF PEPTIDES

Methods for the separation of peptides can best be classified according to the nature of the properties responsible for the separation. These properties are of five types:

(1) Molecular or particle weight of peptides (gel permeation chromatography);

(2) Net electric charge (ion-exchange chromatography);

(3) Difference in solubility in two phases (partition chromatography);

(4) Adsorptive forces (adsorption chromatography);

(5) Highly selective affinity to different substances covalently bound to the solid matrix (affinity chromatography).

This list of properties is arranged in order of the number of papers in which the particular property is put to advantage in the separation of peptides published in the last 10 years.

### Gel permeation chromatography

The amount of work in which the molecular sieving effect of gel permeation is used has enormously increased from the time when several types of polymers (gels) with different degrees of cross-linking first became commercially available. This is understandable, as the molecular weights of peptides in the mixture are relatively easy to predict from the molecular weight of the starting protein, amino acid analyses and the type of hydrolysis used. The elution scheme often does not follow the order of molecular weights but rather the unpredictable particle weights of associated peptides. However, the ease of handling and the excellent (virtually 100%) recovery are good reasons why this type of separation is often used as the initial stage in the separation of peptides.

### Ion-exchange chromatography

Separations based on the net electric charge of the peptide depends on the formation of multiple electrostatic bonds between charged sites on the surface of the exchanger and the opposite charge sites of the peptide. The dipolar ionic character of the peptides allows the use of both cation and anion exchangers. The only restrictions are due to the limited solubility of peptides at different pH values. Sometimes, for a complex mixture of peptides, it can be very difficult to find conditions under which all of the peptides are soluble. The addition of a high concentration of urea or guanidine hydrochloride usually improves the solubility of peptides. The introduction of volatile buffers in ion-exchange chromatography avoids the need for subsequent desalting of the separated peptides. Nowadays, the ion-exchange chromatography of peptides is well developed, can be partially or fully automated and is so versatile that it can be used for the separation of virtually any peptide mixture.

### Partition chromatography

In partition chromatography, the aqueous phase can be anchored on a suitable chromatographic bed and packed in a chromatographic tube. The peptides to be separated are carried through the bed by the flow of liquid and peptides are retarded according to differences in their partition coefficients. Gels of different types and cellulose powder have recently been used as supports in the partition chromatography of peptides. The number of applications is unfortunately rather short.

*References p. 771*

### Adsorption and affinity chromatography

Adsorption chromatography using charcoal, starch or other classical materials is now hardly ever used for the separation of peptides. At the time of writing this chapter, a renaissance is occurring of adsorption chromatography in the more modern technique of affinity chromatography, which can be classified as an application of selective adsorbents, based on biological specificity.

Affinity chromatography, although not yet widely used for the isolation of peptides, offers considerable promise, especially for the study of highly physiologically active peptides.

## ANALYSIS OF THE EFFLUENT FROM THE CHROMATOGRAPHIC COLUMN

The location of the peptides in the fractions eluted from the chromatographic column can be estimated directly by colorimetry or spectrophotometry or, if the chromatography is carried out in volatile buffers, aliquots of the fractions can be evaporated to dryness and subjected to paper or thin-layer chromatography or electrophoresis. Colorimetric or spectrophotometric methods are rapid and are often used. Evaporation of aliquots and characterization by chromatography or electrophoresis is a time-consuming process but gives more information about the complexity of the sample and can serve as a guide for further purification. Further developments may lie in purely physical detector systems based on the differential refractive index measurement (see Chapter 8). Differential refractometers are in current use in the column chromatography of some compounds on a preparative scale. For the identification of peptides, it would be necessary to increase the sensitivity and decrease the noise of commercial refractometers.

### Spectrophotometric detection in ultraviolet light

A rapid estimation of the peptide concentration in the effluent of the column can be obtained by measuring the absorption of light in the ultraviolet region, particularly at 280 nm. Absorption at this wavelength occurs with the aromatic amino acids, tryptophan, tyrosine and phenylalanine. The advantage of this method is that it is rapid, simple to perform, non-destructive and suitable for continuous monitoring. In spite of its widespread use, however, the method has many disadvantages and while it can be recommended for use with proteins or large peptides, it cannot be recommended for the analysis of unknown mixtures of peptides. Only peptides that contain aromatic amino acids can be followed and many compounds interfere in the determination.

It is possible to determine the concentration of all peptides, regardless of their amino acid composition, if the absorbance is measured at 180–220 nm. Peptide bonds are responsible for a major part of the absorbtion in this region of the spectrum. The determination of the absorbance at 210 nm is 10–20 times more sensitive than that at 280 nm (Goldfarb *et al.*). This advantage is partly offset by the difficulty of making measurements in the far ultraviolet region, and a large number of different substances interfere in the measurement. Almost all commonly used buffer systems for the separation of peptides absorb too strongly in this region to permit their use, and they are restricted to non-volatile inorganic buffers.

Fluorescence measurement can also be used for the detection of peptides that contain aromatic amino acids. This method is more sensitive than the measurement of the absorbance at 280 nm but the limitation of the method to aromatic peptides is the same.

## Ninhydrin method

The most common analytical method used for the determination of peptide concentration in the effluent from the chromatographic column is the ninhydrin colorimetric method (Moore and Stein).

Ninhydrin colorimetry is one of the most sensitive colorimetric methods known and permits the use of very small samples. It is possible to use it in a manual or fully automated form for both peptides and amino acids, with the advantage that the ninhydrin reagent does not give rise to the corrosion problem and can be fed by means of a standard pump. The drawback of the ninhydrin method is that the free amino group of the peptide must be available to the reagent and that some peptides give low colour yields. Certain generalizations about the colour values of peptides can be made from the work of Callaham *et al.* on dipeptides. All dipeptides that contain arginine, threonine, serine, glutamic acid, glycine, phenylalanine, methionine, leucine and tyrosine as the N-terminal residue have colour values close to $1.6 \times 10^6$. For comparison, the colour value of leucine is $1.7 \times 10^6$. Dipeptides that contain N-terminal lysine and aspartic acid have average colour values which are 20% and 29% higher respectively, than $1.6 \times 10^6$, while N-terminal histidine and tryptophan dipeptides average 42% and 67% of that value, respectively. Dipeptides that contain N-terminal proline, valine and isoleucine have extremely low colour values, being 2.7%, 6.4% and 8.5% of $1.6 \times 10^6$, respectively.

These drawbacks to the ninhydrin reaction are overcome by following an alkaline hydrolysis introduced by Hirs *et al.* The procedure is based on the fact that in an alkaline medium virtually all peptides bonds are hydrolyzed and the subsequent ninhydrin reaction gives information about the concentration of free amino acids. As ammonia, which interferes in the ninhydrin reaction, is boiled off during the alkaline hydrolysis, the determination is not affected by the presence of ammonia in the sample. This hydrolysis ensures that peptides which give a low colour yield with ninhydrin will not be overlooked. From the increase in the ninhydrin colour of a peptide that has been subjected to alkaline hydrolysis in comparison with the value of an unhydrolyzed sample, it is possible to gain an idea of the approximate size of the peptide.

### *Ninhydrin colorimetry*

A 0.1–1.0-ml volume of the peptide solution is mixed in a tube with 1 ml of ninhydrin solution (see Chapter 32). The capped tube is briefly shaken and heated for 15 min in a covered boiling water-bath. After dilution with an appropriate volume of 50% ethanol, the tubes are thoroughly shaken, cooled to 30°C and the colour is read at 570 nm. The

*References p. 771*

buffer in the ninhydrin solution (see Chapter 32) is usually sufficiently concentrated so that adjustment of the pH of the sample solution to 5.5 is seldom necessary. The amount of diluent (50% ethanol) used depends on the concentration of the peptide and must be determined experimentally. Sometimes benzene is used as a denaturant in ethanol, in which event the concentration of ethanol used for dilution must be increased to 60% in order to prevent opalescence. Effluent concentration curves are constructed by plotting optical density at 570 nm against the elution volume, or are corrected to "leucine equivalents" (the number of micromoles of leucine which give, under the same conditions and dilution, the same absorbance at 570 nm).

*Ninhydrin reaction following alkaline hydrolysis*

Aliquots of 0.1–0.5 ml are transferred by pipette into Pyrex or polypropylene test-tubes, 1 ml of 2.5 *N* sodium hydroxide solution is added and the unstoppered tubes are placed in an open water-bath at 90°C for 2.5 h. To each cooled tube, 1 ml of 30% acetic acid is added and the tubes are shaken. The ninhydrin reaction is performed as described above.

Polypropylene test-tubes are preferred for alkaline hydrolysis as glass tubes gradually become etched by the alkali. In order to make the method more convenient, it is possible to deliver all reagents from a pipetting machine. If the water-bath is replaced with a steam-heated autoclave, the temperature of alkaline hydrolysis can be increased to 120°C and the time of hydrolysis decreased to 20 min (Hirs).

*Automated procedures*

In the automatic ninhydrin method, the effluent from the chromatographic column is split into three streams. Two of these streams are used for analysis, giving recorder traces corresponding to "direct" and "hydrolyzed" ninhydrin colours, while the third stream enters the fraction collector. For this purpose, any commercially available amino acid analyzer can be modified. The additional items required are mostly standard parts of analyzers. The original colorimetric unit can analyze one stream by the direct ninhydrin method. Alkaline hydrolysis, if needed, can be accomplished by means of commercially available modules of the Technicon AutoAnalyzers. A system using the Technicon AutoAnalyzer was described by Catravas. The separation of peptides on an amino acid analyzer with an automated detection system has a reproducibility which makes it very useful for the peptide mapping of enzymatic digests of proteins, which provides a means of distinguishing structural differences. Simultaneously, it permits the isolation of peptides required for amino acid analysis and sequence studies. Reference was made to a number of original papers in a review by Hill and Delanay.

Recently, the chromatographic behaviour of 140 different dipeptides in an amino acid analyzer was reported by Callaham *et al.* Column chromatography was performed on the Beckman-Spinco spherical resins AA-27 and AA-15. Standard citrate buffers and a 4-h procedure were used with the Beckman-Spinco Model 120 B amino acid analyzer in order to identify dipeptides resulting from the degrading action of dipeptidylaminopeptidase on proteins and peptides.

## Spectrophotometry of peptides with 2,4,6-trinitrobenzenesulphonic acid (TNBS)

In an alternative method, the ninhydrin reagent is replaced in both the manual or automated versions with TNBS, which was first used in the chemistry of peptides by Okuyma and Satake. The coloration of amino acids and peptides due to the formation of the Meisenheimer complex can be measured at 420 nm or the complex can be converted into TNP-peptide and measured at 340 nm (Satake *et al.*).

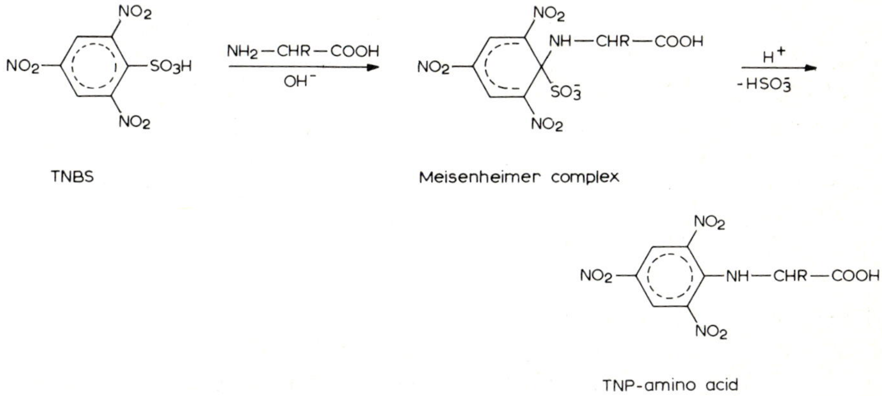


In comparison with ninhydrin colorimetry, the sensitivity is lower by 50–60%, but this method has the advantage that the peptide can be recovered after colorimetry. The TNP-group can be removed by the action of concentrated ammonia solution without appreciable splitting of the peptide bonds:

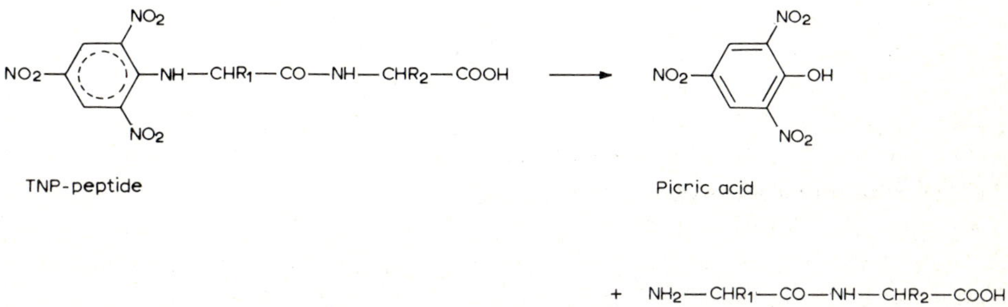


It is necessary to bear in mind that proline, in addition to peptides with proline at the N-terminal position, do not react with the reagent. The coloration due to ammonia is very small, but amines interfere in the reaction. The simplicity of the method and the similarity of the relative colour intensities of the various TNP-peptides make this spectrophotometric technique valuable in spite of its lower sensitivity. In the manual method, the coloration of TNP-peptides is preferred because of the lower blank value. The use of TNBS for the continuous detection of peptides or amino acids in the effluent from the column takes advantage of the orange colour of the Meisenheimer complex, with a higher colour yield (Machleidt *et al.*, Satake *et al.*).

*References p. 771*

The method was also used for the analysis of amino acids liberated from peptides by alkaline hydrolysis. The replacement of ninhydrin with TNBS in the automated procedure prevents the formation of methyl Cellosolve–ninhydrin breakdown products and clogging of the tubing (Delanay).

The manual procedure for the determination of the concentration of peptides is as follows (Satake *et al.*). Up to 1 ml of the sample solution to be analyzed (0.1–0.8 $\mu$mole of peptide), 1 ml of 4% sodium hydrogen carbonate solution and 1 ml of 0.1% TNBS solution are mixed and kept in the dark for 30 min at 40°C. The resulting solution is acidified with 1 ml of 1 *N* hydrochloric acid and the optical density is measured at 340 nm.

*Note:* It was reported by Satake *et al.* that some preparations of TNBS give a high blank value (0.4–0.5). TNBS can be purified as follows: the sodium salt of TNBS is dissolved in 1 *N* hydrochloric acid at a concentration of about 5% and the orange-coloured solution is passed through a column of Norite–Celite (1:1, w/w) in order to decolorize it. By concentrating the lemon-coloured filtrate *in vacuo*, TNBS is obtained as almost colourless crystals, which are stable for more than a year if kept in the dark.

### Fluorescent ninhydrin method

A very sensitive fluorimetric detection system for monitoring peptide separations involves the use of the ternary condensation product, containing one residue of peptide, phenylacetaldehyde and ninhydrin (Samejima *et al.*). The fluorescent ninhydrin procedure is 10–100 times more sensitive than the colorimetric ninhydrin procedure. The great advantage is that ammonia does not react and that at most pH values peptides yield more fluorescence than free amino acids. The future of chemical methods for the detection of peptides may lie in successful modifications and improved instrumentation of the fluorescent ninhydrin method.

### Folin–Lowry method

The concentration of peptides can also be determined by the Folin–Lowry method. The principle of the method is that a copper–tartrate complex is allowed to react with the peptide in an alkaline medium. The peptide–copper complex can reduce phosphotungstate to form a blue substance with a broad absorption peak at about 750 nm (see Chapter 35). The method was developed for and mostly used in the routine measurement of proteins in solution, and therefore it does not fall within the scope of this chapter.

### Specific detection of cystine-containing peptides

The principle of the method is the action of sulphite upon cystine, which is converted into one molecule of S-cysteinesulphonic acid and one molecule of cysteine. Cysteine is oxidized to cystine and the phospho-18-tungstic acid colour reagent acts as a hydrogen acceptor (Clarke; Kassel and Brand, 1938a, b; Lugg).

The procedure was modified and described by Spackman and Stein as follows. To 1 ml of the effluent from the column, the following reagents are added stepwise. (1) A 1-ml volume of buffer prepared by dissolving 105 g of citric acid and 52.5 g of sodium hydroxide in about 450 ml of water with cooling, then adding 110 g of solid sodium acetate ($C_2H_3O_2Na \cdot 3H_2O$) and 250 g of solid urea; separately prepared solutions of 13.6 g of zinc chloride in 20 ml of water and 26.8 g of ammonium chloride in 100 ml of water; the mixture is diluted to 1 l and preserved with toluene (it should be filtered before use). The pH of a mixture of 5 ml of buffer with 1 ml of colour reagent must be 5.7. (2) A 0.3-ml volume of water or an appropriate sodium hydroxide solution to keep the final reaction mixture at pH 5.6–5.8. As the pH and ionic strength, or both, of the column eluent is changing, it is necessary to determine the concentration of the sodium hydroxide solution experimentally. (3) A 0.2-ml volume of the phosphotungstic acid colour reagent (Folin). (4) A 0.5-ml volume of buffered sodium sulphite, made by dissolving 9.5 g of $Na_2S_2O_5$ in water, adding 15 ml of 4 *M* sodium acetate solution and diluting to 100 ml. This solution should be kept in a refrigerator and prepared freshly each week. The 3-ml fractions are allowed to stand for 15 mins, 5 ml of water is added to each tube and the tubes are read in a spectrophotometer at 700 nm.

*Note:* If pyridine-containing buffers are used for elution, it is necessary to evaporate aliquots from each fraction to dryness, which can be accomplished in a desiccator over sodium hydroxide and sulphuric acid, dissolve the residue in water and subject the solution to analysis. The procedure is standardized with a $1 \times 10^{-4} M$ solution of cystine hydrochloride.

## GEL PERMEATION CHROMATOGRAPHY

The use of granulated gels of different constitutions for the separation of peptides that differ in size is an important method. The chromatographic procedure in which the molecular sieve properties of gels are utilized has sometimes been called gel filtration. For the separation and desalting of peptides, dextran gels available from Pharmacia, Uppsala, Sweden, under the trade-name Sephadex and polyacrylamide gels from Bio-Rad Labs., Richmond, Calif., U.S.A., under the trade-name Bio-Gel P. Both gels are now produced as beads of different sizes and fractionation ranges (see Chapter 9).

The choice of the appropriate type of gel depends on the molecular weights of the peptides to be separated. Molecules with molecular weights above the upper limit of the range (the exclusion limit) are totally excluded from the gel and emerge from the column as a mixture in a void volume of the gel bed. Molecules with molecular weights below the fractionation range are eluted at an elution volume approximately equal to the total volume of the gel bed. Only peptides with molecular weights within the fractionation range are separated.

The ideal matrix for fractionation should act only by virtue of its molecular sieve properties. Both dextrans and polyacrylamide gels show non-ideal behaviour which must be taken into consideration. The first effect is the adsorption of peptides that contain aromatic amino acids (tryptophan, tyrosine and phenylalanine) on the matrix. Such peptides are usually retarded to varying extents compared with non-aromatic peptides of a similar size. The effect is particularly notable in tryptophan and tyrosine peptides, and less pronounced in phenylalanine peptides. The adsorption can be reduced if buffers that contain aromatic constituents are used or if the buffer contains urea or potassium thiocyanate, but the adsorption cannot be completely eliminated. On the other hand, it has been reported that the adsorption by the Sephadex matrix can be increased by increasing the ionic strength (Jansen). The interpretation that the increase in adsorption is due to the removal of water of hydration of the gel freeing more sites for interaction does not agree with the interpretation of the counter-effect of urea.

The second effect that must be considered is the presence of a few negatively charged ionized groups in the gel network. Positively charged peptides may be retarded or completely adsorbed, and negatively charged peptides may be excluded from the gel matrix, regardless of their molecular weights. The charge effect can be eliminated by the use of solvents with an ionic strength exceeding 0.02.

### Applications of gel permeation chromatography

The applications cited below and in other sections of this chapter are not intended to be comprehensive, but were chosen only in order to illustrate the method described.

#### *Desalting of peptides*

Gel permeation chromatography is the only universally applicable method for desalting all but the smallest or tryptophan-containing peptides. The method can be used not only for the removal of salts but also for the removal of other low-molecular-weight compounds such as sugars and urea.

The peptide to be desalted is dissolved in 0.1–0.2 *N* acetic or formic acid or in 0.2% ammonium carbonate or hydrogen carbonate solution and applied on a column of gel. The fractionation range of the gel must be chosen so that the peptidic material is eluted with or near the void volume. Peptides should be completely excluded from the gel matrix. Sephadex G-10, G-15 and G-25 and Bio-Gel P-2 and P-6 are well suited for the purpose. The gel bed volume of Sephadex G-25 should be chosen to be 3–4 times that of the sample to be applied and has to be equilibrated before the experiment with a volatile electrolyte that is used for elution. The choice of the electrolyte depends on the solubility of the peptides. Dilute organic acids or 0.1–0.2 *N* acetic or formic acid are usually used for peptides that are soluble in acidic media, and 0.2% ammonium carbonate or hydrogen carbonate for peptides that are soluble in alkaline or neutral media. If necessary, the desalting can be performed with distilled water as eluent, but tailing of the peptide zone is more pronounced so that the volume of the gel bed must be increased so as to avoid overlapping of peptides with salts. The flow-rate of the eluent is 7–20 ml/h $\cdot$ cm$^2$ and depends on the elution volume of the peptide, loading of the column and volume of the sample applied on the column. The volume of the gel bed must be increased if highly cross-linked gels are used because of the water regain of the gel used. The peptides can be detected in the eluate from the column by colorimetry or spectrophotometry and recovered by evaporation under reduced pressure or by freeze-drying.

*Note:* It has been reported that some batches of acetic acid are very low in but not entirely free from aldehydes. During the gel permeation chromatography and especially during evaporation of samples, N-terminal groups of peptides can be completely or partially blocked and it is therefore advisable to check each batch of acetic acid for the presence of reducing compounds: mix 5 ml of glacial acetic acid with 15 ml of water and 0.3 ml of 0.1% potassium permanganate solution; the purple colour of potassium permanganate must remain stable for 15 min, and if the colour fades the acetic acid must be refluxed with chromic acid, distilled twice on a Widmar column and the potassium permanganate test repeated.

### *Fractionation of peptides*

Gels are available in different particle size ranges. The super-fine grade is intended for column chromatography in which very high resolution is required. The flow-rate must be kept as low as possible, so that this technique can be used for analytical purposes only. For preparative chromatography, fine-grade gel beads (20–80 $\mu$ for Sephadex and 40–75 $\mu$ for Bio-Gel) are recommended.

In order to achieve an efficient fractionation, the gel must be chosen so that the sizes of the peptides fall within the fractionation range. The theoretical size of the peptides in the mixture is usually not difficult to predict, considering the molecular weight of the starting material and the type of hydrolysis with respect to amino acid analysis of the material.

In order to obtain a separation based on the size of the peptides, it is necessary to minimize the electrostatic peptide–gel interaction. This is the reason why the separation is never carried out in distilled water but always in a medium of high ionic strength at a suitable pH. In order to simplify the recovery of the material from the eluent, the same volatile electrolytes as used in the desalting procedure are recommended.

For the preparation of fragments obtained by maleylation of reduced and carboxymethylated human serum albumin, it was found suitable to use 0.1 *M* sodium hydrogen carbonate solution on a 250-cm column of Sephadex G-100. The peptides were separated strictly according to their size. As expected, the fastest peak contained the largest fragment with 162 amino acid residues, the second contained a peptide with 88 amino acid residues and the third contained a peptide with 36 amino acid residues (Fig. 34.1).

It is possible to find references in the literature to a number of successful separations of peptides according to their size. In spite of this, however, experience has shown that sometimes mixtures of certain peptides, especially the long-chain members, are separated only with great difficulty. The properties of such peptidic mixtures are to a large extent determined by cohesive forces, which remain intact after the disruption of primary linkages. Some components of the mixture produce more or less dissociable conformations. These interactions may either facilitate or, more frequently, greatly complicate the separation. Usually the process of aggregation is considered to be undesirable and to be avoided. Reagents known to disrupt hydrogen bonds, such as 8 *M* urea, 6 *M* guanidine hydrochloride, detergents or high concentrations of organic acids, are used as eluents to be on the safe side. This may be a short-sighted view, for aggregation can permit a

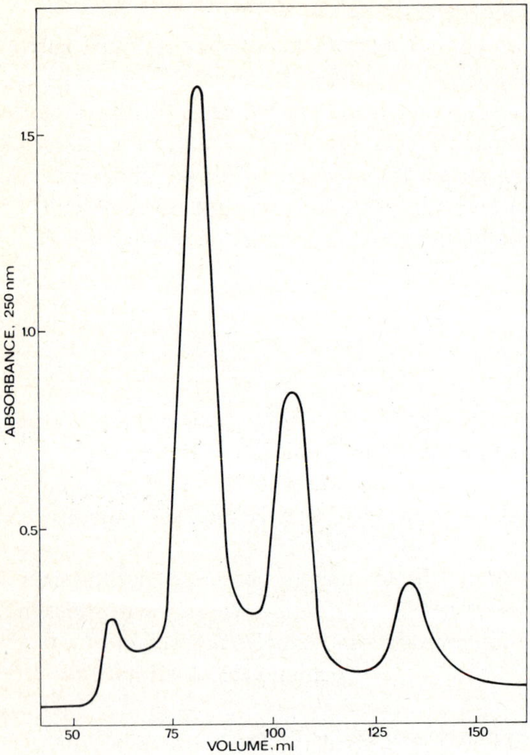


Fig. 34.1. Fractionation of a mixture of carboxymethylated and maleylated chains of Fragment N on Sephadex G-100 column in 0.1 *M* ammonium hydrogen carbonate solution (Kušnír and Meloun).

selective separation to be made. The C-terminal cyanogen bromide fragment of hog pepsin has been separated by taking advantage of selective aggregation (Kostka *et al.*) (Fig. 34.2).

The above example is also interesting from the point of view that a high concentration of urea completely failed to dissociate the tight conformation. As there are not many more powerful reagents, another approach to the problem of preventing aggregation of peptides must be considered. Reversible substitution of the basic groups in peptides by citraconylation, thioltrifluoroacetylation or maleylation of the $\epsilon$-amino groups of lysines seems to be a very promising approach to the problem.

## Side-effects of gel permeation chromatography

Advantage can be taken of the adsorption effect in addition to the charge effect of the gel matrix in order to achieve separations of peptides that cannot be distinguished by molecular sieving. An example of the utilization of the side-effect is the separation of the two hormones oxytocin and vasopressin on a column of Sephadex G-25 (Frankland *et al.*).

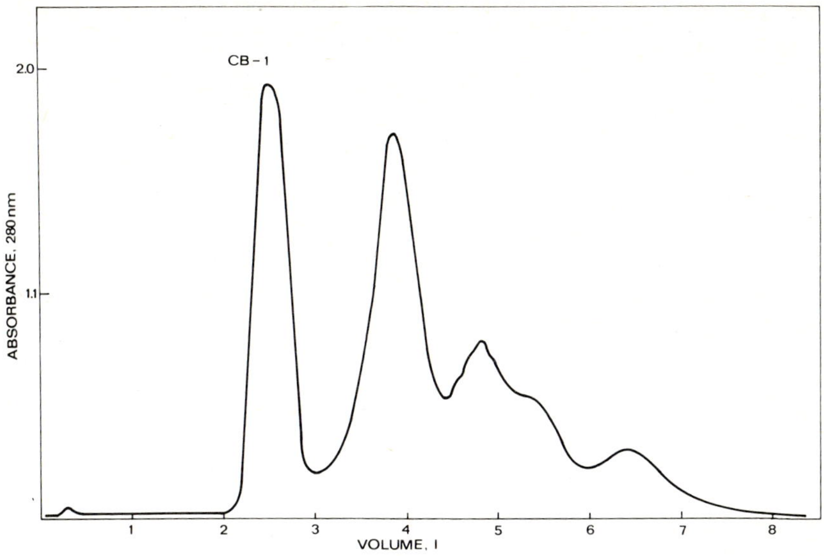


Fig. 34.2. Gel permeation chromatography of cyanogen bromide hydrolyzate of reduced and S-sulphonated hog pepsin (Kostka *et al.*). A 1-g sample of the hydrolyzate was dissolved in 200 ml of 0.3 *M* ammonium acetate solution, 8 *M* in urea at pH 6.0, and applied to a 70 × 11 cm column of Sephadex G-100 (40–120 μ) equilibrated with the same buffer at the rate of 190 ml/h, and 95-ml fractions were collected. Peptide CB-1, emerging with the void volume of the column, has 37 residues and is derived from the C-terminal part of the pepsin molecule.

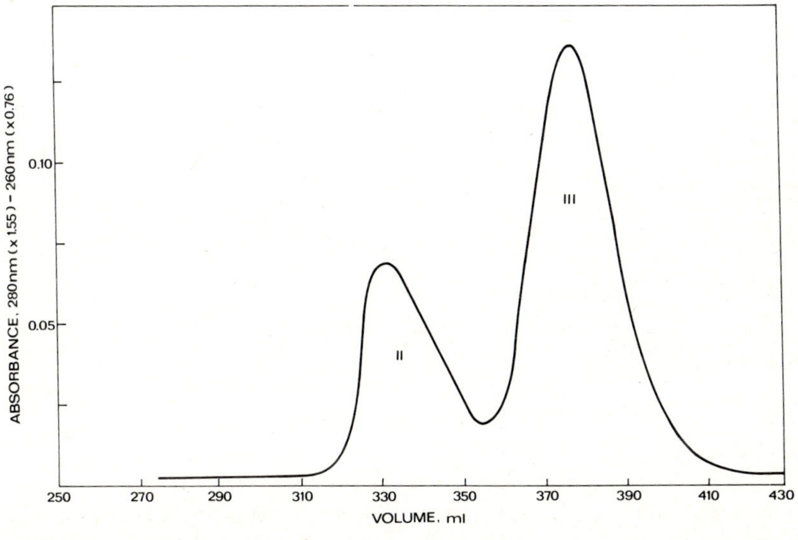


Fig. 34.3. Re-chromatography of the peptide components from the neurophysin–peptide complex on a 145 × 2 cm column of Sephadex G-25 (20–80 μ); the eluent was 0.1 *N* formic acid (Frankland *et al.*). Peak II, oxytocin; peak III, arginine–vasopressin.

*References p. 771*

The mixture of hormones is dissolved in 0.1 *N* acetic acid and applied on to a column of Sephadex G-25, equilibrated and eluted with 0.1 *N* acetic acid (Fig. 34.3). Both hormones were recovered by lyophilization and vasopressin was isolated in an adequate state of purity. The presence of serine in the oxytocin fraction has not been explained. From the study of the conformation of these hormones in aqueous solutions, the expected elution volume of hormones would be the reverse. The greater affinity of vasopressin for Sephadex G-25 may be due to the presence of an additional phenylalanine residue or to the electrostatic interaction between the guanidino group of vasopressin and the carboxyl groups of the gel matrix.

The use of a long Sephadex G-25 column equilibrated with 1% formic acid provided a good technique for the separation of more complex peptidic mixtures. The separation of peptic peptides of native lysozyme was based predominantly upon their molecular sizes. However, fragments containing aromatic amino acids were retarded by adsorption (Canfield and Liu) (Fig. 34.4).

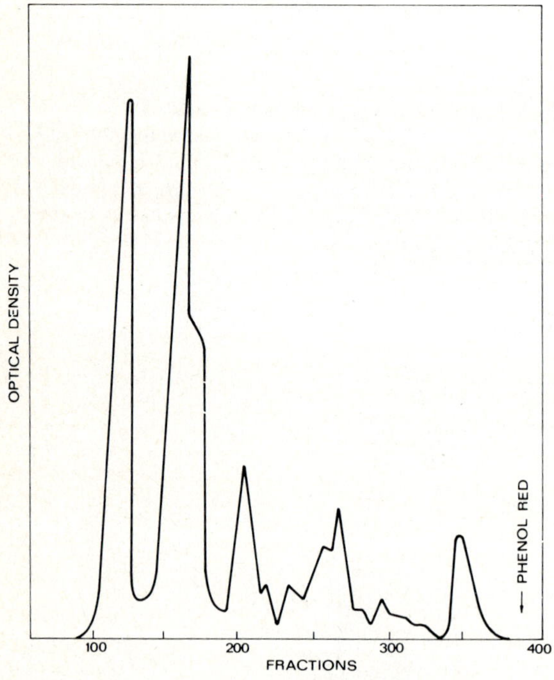


Fig. 34.4. Elution of peptides produced by peptic digestion of lysozyme from a 330 × 3.0 cm column of Sephadex G-25 (Canfield and Liu). The optical density was measured at 280 nm. The effluent was collected in 10-ml fractions. Phenol red was used as a reference marker.

The ion-exchange properties of dextran gel were utilized in a method for the reversible retention of toxins of scorpions (Miranda *et al.*) and adopted for the isolation of some peptides from the chymotryptic digest of reduced and S-methylated neurotoxin I from *Androctonus australis* Hector (Rochat *et al.*). A small proportion of carboxyl groups can exchange basic peptides if the gel bed is equilibrated and the sample applied in distilled water. Sephadex G-15 was selected for this purpose in preference to Sephadex G-25 because of the lower water regain. The mixture of chymotryptic peptides that emerges from Sephadex eluted with 30% acetic acid as one fraction was dissolved in water and applied on a 2 × 150 cm column of Sephadex G-15, equilibrated and eluted with water at the flow-rate of 13 ml/h. After 22 h, water was replaced with 1 *M* acetic acid (Fig. 34.5).

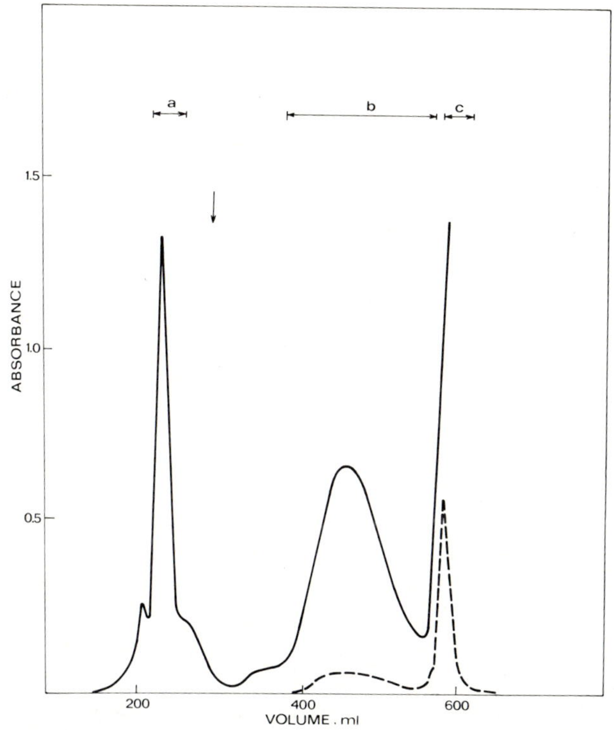


Fig. 34.5. Reversible adsorption on Sephadex (Rochat *et al.*). The fraction obtained by gel permeation chromatography of chymotryptic digest was dissolved in water and chromatographed through a 240 × 2 cm column of Sephadex G-15 in water. The vertical arrow indicates the change from water to 1.0 *M* acetic acid as eluent. The flow-rate was 13 ml/h. Full line, absorbance at 220 nm; broken line, absorbance at 280 nm. Ranges a, b and c indicated by double-headed arrows indicate pooling of fractions. The large increase in absorbance at 220 nm just before the elution of peptide c is due to the strong absorbance of acetic acid at this wavelength.

Three peptides were isolated in a pure state. Peptide a, containing serine, glycine, valine, and leucine, was eluted with the void volume of the column. Peptide b contained two asparagines, proline, two valines, isoleucine, S-methylcysteine and two tyrosines, and was retarded because of the content of the two tyrosine residues. Peptide c contained asparagine, glycine, tyrosine, lysine and arginine, and was strongly adsorbed to the gel. The application of 1 *N* acetic acid was necessary for elution.

*References p.771*

## ION-EXCHANGE CHROMATOGRAPHY

Ion-exchange chromatography has been widely used for the separation of peptides and the number of published papers involving the use of this technique is very large. The diversity of the material to be separated has led to an enormous number of modifications and nowadays it is difficult to rationalize all the information. This short survey could serve as a guide for making a choice of the most appropriate ion exchanger to be used for a particular material, but the extent of this section does not permit the inclusion of complete and detailed descriptions. Many techniques have been reviewed recently and the shortened descriptions here only provide brief summaries.

### Strong cation and anion exchangers with a polystyrene matrix

These types of exchangers have active groups attached to a cross-linked polystyrene matrix. Active groups can be strongly acidic ($SO_3H$, Dowex 50) or strongly basic ($N(CH_3)_3$, Dowex 1). The degree of cross-linking in exchangers commonly used for the separation of peptides is 2%, and less frequently 4% or 8%. These ion exchangers are the oldest types and have been used successfully many times for the fractionation of small peptides and amino acids. Non-specific adsorption, governed by the combination of Van der Waals forces and polar interactions, does not permit the elution of proteins or large peptidic fragments under conditions compatible with their stability. The main reason for the strong binding is the presence of many different charges that are capable of forming bonds with the fully ionized groups of the sorbent. The probability of finding conditions under which all bonds holding only one type of molecule would dissociate and the macromolecule then eluted is very low. Another reason why this type of ion exchanger was abandoned for use in the separation of high-molecular-weight solutes is their small capacity, because only exterior charges of the exchanger are accessible to macromolecules.

From the fact that even small peptides that contain aromatic amino acids, especially tryptophan, have not been isolated frequently from different hydrolyzates of proteins, it is possible to deduce that they are irreversibly adsorbed on the polystyrene matrix. The same is also likely to be true of cystine peptides, but the failure to isolate these peptides can be explained by the instability of the material and other reasons.

Ion exchangers of these types are supplied by different manufacturers and distributors with different designations relating to their degree of cross-linkage, commercial grade or particle size. In many instances ion exchangers of the same type can be used interchangeably, although they are made by different manufacturers and do not have completely identical physical properties.

For the separation of peptides, volatile buffers are usually used that can be evaporated or sublimed away, so avoiding the need for time-consuming desalting procedures. For analytical purposes when peptides are not isolated, non-volatile inorganic buffers are preferred. For the preparation of volatile buffers, the following bases and organic acids are usually used: 3-methylpiperidine, pyridine, α-picoline 2,4,6-lutidine, N-ethylmorpholine, acetic acid and formic acid. The purity of these substances varies considerably, depending on the source, and sometimes varies even between batches from the same supplier. The impurities can lead to the coloration of buffers and can interfere in the ninhydrin colorimetry.

### Purification of solvents

Pyridine (McDowall and Smith) is stirred vigorously under gentle reflux for 1 h with 2% (v/v) of concentrated sulphuric acid and then allowed to cool. The supernatant pyridine is decanted from the dark lower layer, which can be solidified, filtered through glass-wool so as to remove any suspended material, and then distilled through the fractionating column. The first 5% and the last 20% of the residue are rejected. The product should be colourless or faintly yellow and is then relatively free from ninhydrin-positive contaminants.

Pyridine, N-ethylmorpholine and other bases can be redistilled from a solution of 1 g of solid ninhydrin per litre before use so as to diminish the content of ninhydrin-positive contaminants. N-Ethylmorpholine and 3-methylpiperidine should be kept in the cold. For the purification of acetic acid, see the section on the gel permeation chromatography of peptides (p. 751). The purification of acetic acid is critical for fractionation on strong ion exchangers and for the desalting of large peptidic fragments by means of Amberlite IRC-50.

### Fractionation of peptides on strong ion exchangers

It is not possible to predict exactly the best sequence of fractionation procedures, even for a known peptidic mixture, without preliminary experiments. The resolution capacities of strong cation and anion exchangers for acidic and basic peptides under proper conditions can be compared. Neutral peptides are usually better resolved on the strong cation exchanger Dowex 50, because fractionation on the strong anion exchanger Dowex 1 depends more on the net charge of the peptide, whereas the elution from Dowex 50 is more complicated and advantage is taken of side-effects.

If it is intended to use both of these resins in combination, a Dowex 50 column is usually used for the initial fractionation rather than a Dowex 1 column, for purely practical reasons. It is not likely that a single chromatographic step will separate the peptide mixture into its components, so that the resulting simpler mixtures have to be re-chromatographed on a complementary ion exchanger. Dowex 50 columns have to be repacked after chromatography owing to shrinkage of the resin, whereas the Dowex 1 column does not need to be repacked but simply re-equilibrated, so that the simplest procedure is to prepare one column of Dowex 50 and one of Dowex 1, the former being used for the initial fractionation and the latter for repeated re-chromatographies. Dowex 1 has, of course, been used very successfully on many occasions for the initial fractionation of peptides.

*Fractionation of peptides on Dowex 50-X2*

The separation of peptides is accomplished by increasing together the pH and ionic strength of the elution buffer. A buffer system very frequently used consists of two pyridinium acetate buffers of different pH and concentration: (1) 0.2 *M* pyridinium

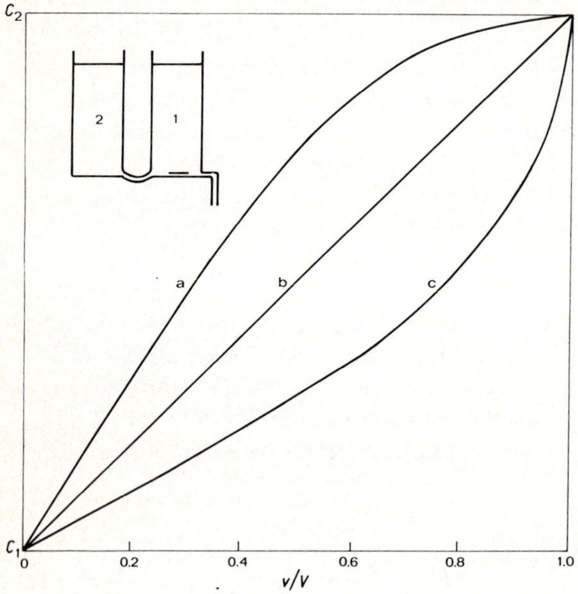


Fig. 34.6. Gradient-producing device consisting of two vertical-walled containers with cross-sectional areas $A_1$ and $A_2$ and solutions of initial concentration $C_1$ and $C_2$ (Bock and Nan-Sing Ling). Container 1 must be stirred continuously. This system delivers a solution of concentration $C = C_2 - (C_2 - C_1)$ $(1 - V) A_2/A_1$. Curves: a, $A_2 = 2A_1$; b, $A_2 = A_1$; c, $2A_2 = A_1$. $v$ is the fraction of the volume of the total system which has been delivered, $V$ is the volume of the mixing vessel (No. 1).

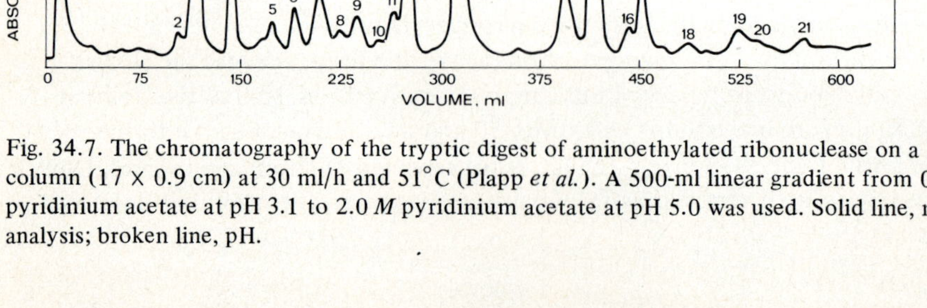


Fig. 34.7. The chromatography of the tryptic digest of aminoethylated ribonuclease on a Dowex 50 column (17 × 0.9 cm) at 30 ml/h and 51°C (Plapp *et al.*). A 500-ml linear gradient from 0.2 *M* pyridinium acetate at pH 3.1 to 2.0 *M* pyridinium acetate at pH 5.0 was used. Solid line, ninhydrin analysis; broken line, pH.

acetate, pH 3.1 (molarity in pyridine); and (2) 2.0 $M$ pyridinium acetate, pH 5.0. In order to produce the gradient, an assembly consisting of two vertical-walled containers with cross-sectional areas in the ratio 1:2 filled with solutions of initial concentrations 0.2 and 2.0 $M$ is commonly used. The container with the 0.2 $M$ buffer must be continuously stirred. This mixing system delivers a convex gradient. For the theory of this gradient, see Fig. 34.6 (Paar).

Another possibility for producing a gradient of the same type is to connect together three cylinders with the same diameter. The first is the mixer and contains the 0.2 $M$

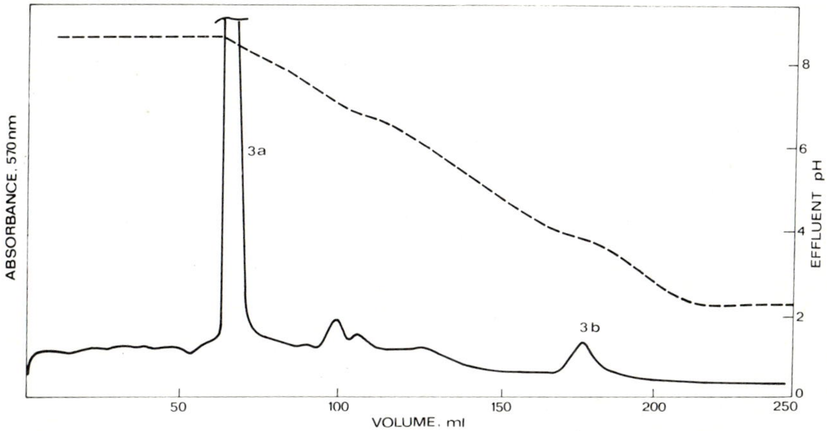


Fig. 34.8. Re-chromatography of fraction 3 (Fig. 34.7) on a Dowex 1 column (26 × 0.9 cm) at 40 ml/h and 35°C, with the use of a gradient from pH 9.4 to 2.4 (Plapp *et al.*).

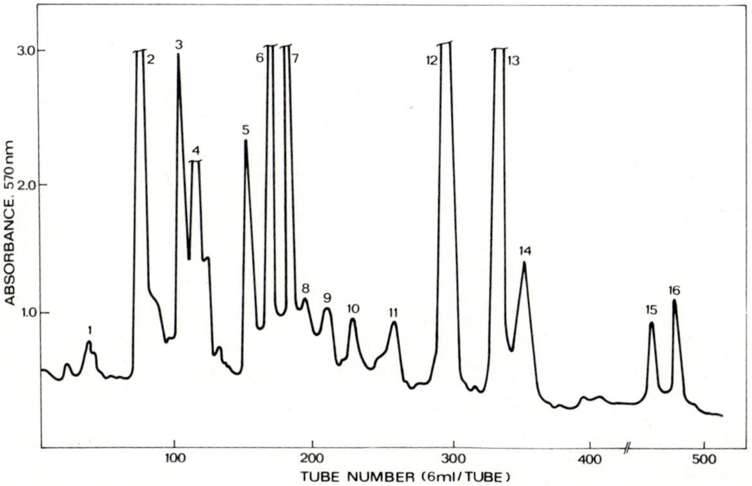


Fig. 34.9. Elution pattern of peptides from the tryptic digest of S-aminoethylated Bence-Jones protein Ag on a Dowex 1-X2 column (1.8 × 150 cm) at 35°C (Titani *et al.*, 1969a). The elution buffer was collected in 6-ml fractions at a flow-rate of 80 ml/h. The chromatogram was monitored by the ninhydrin reaction after alkaline hydrolysis.

buffer, while the other two contain 2.0 *M* buffer. The sample is applied on to the column as a solution in water or 0.1 *N* pyridinium acetate buffer, adjusted to pH 2.0 with hydrochloric acid. The volume of the sample is not critical but should not exceed about 2% of the volume of buffer needed for equilibration of the column, otherwise the performance is affected. An extensive review of the method, containing detailed procedures, was published by Schroeder (1972a).

An illustration of the above method is given by the fractionation of protein hydrolyzate on a Dowex 50-X2 column and the re-chromatography of one fraction on a column of Dowex 1 (Figs. 34.7 and 34.8). The opposite sequence of fractionation procedures is shown in Figs. 34.9 and 34.10.

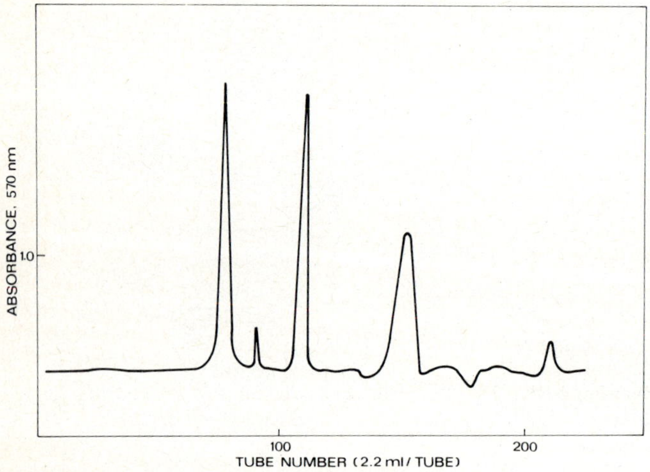


Fig. 34.10. Purification of fraction 3 obtained by Dowex 1 chromatography (Fig. 34.9) on a Dowex 50-X2 column (0.9 × 150 cm) (Titani *et al.*, 1969a). Peptides were eluted with 0.1 *M* pyridinium formate, pH 3.1 to 2.0 *M* pyridinium acetate, pH 5.0 in a linear gradient at 50°C. Fractions of 2.2 ml were collected at a flow-rate of 20 ml/h.

### *Fractionation of peptides on Dowex 1-X2*

The separation of peptides is accomplished by decreasing the pH and increasing the concentration of anions, usually acetic acid. Gradient elution is usually carried out by the use of a constant-volume mixing chamber and with a series of solvents introduced successively into the changing solution in the mixer. The buffer system recommended by Schroeder (1972b) consists of five solvents:

(1) pH 9.4: 60 ml of N-ethylmorpholine, 80 ml of α-picoline, 40 ml of pyridine and about 0.5 ml of acetic acid;

(2) pH 8.4: the same as in (1), but with about 3 ml of acetic acid;

(3) pH 6.5: the same as in (1), but with about 37 ml of acetic acid;

(4) 0.5 *N* acetic acid;

(5) 2.0 *N* acetic acid.

If one 100-ml column is to be used, the volume of the mixing chamber should be 135 ml and the volumes of the successively added buffers should be as follows: (1) 40 ml; (2) 120 ml; (3) 160 ml; (4) 240 ml; and (5) 500 ml.

The conditions of the gradient proposed above are suitable for the initial fractionation of the complete hydrolyzate of protein, containing basic, neutral and acidic peptides. Sometimes a Dowex 1-X2 column is used for re-chromatography of the fractions emerging from the Dowex 50 column. These fractions do not contain the whole pattern of peptides and the gradient can be restricted to the required area. A knowledge of the charge on the peptides to be separated, as determined by paper electrophoresis at pH 6.4, aids in choosing the appropriate conditions to be used for re-chromatography. Basic peptides leave the Dowex 1 column in the pH range 9.4–7.0, neutral peptides from pH 7.0 to 5.0 and acidic peptides below pH 5.0.

For neutral peptides, equilibration of the column with 1% pyridine can be recommended; the pH of our batches of pyridine is about 7.3. The first buffer in the mixing chamber is 1% pyridine, and the second solvent is 0.5 *N* acetic acid. For acidic peptides, the 0.5 *N* acetic acid is replaced with 2.0 *N* acetic acid.

For basic peptides, Nyman *et al.* used 0.1 *M* N-methylpiperidinium acetate of pH 11.2 for equilibration of the column and as the first buffer in the mixing chamber. The second buffer was 0.2 *M* methylpiperidinium acetate of pH 5.5. N-Methylpiperidine can be expected to be a useful constituent of buffers for extending the pH range in Dowex 1 columns when the resolution of basic peptides is desired. The p*K* value of N-methylpiperidine is 11, and it was used as a buffer constituent by Padieu and Maleknia for the separation of amino acids.

An excellent review of chromatography on Dowex 1 columns was given by Schroeder (1972b). There is only one recommendation to be added concerning the preparation of the exchanger before use. Dowex 1 is supplied in the chloride form, and it is usually not easy to remove the last traces of chlorides from the resin. The recommended procedure is to pour the exchanger into the wide chromatographic tube and to wash it with 0.5 *N* sodium hydroxide solution until the effluent is completely free from chlorides (checked with silver nitrate after acidification). If the chlorides are not completely removed before chromatography, they can cause the tailing of peptides, and the eluted peptides cannot be directly chromatographed on paper owing to the presence of pyridinium chloride.

### Weakly acidic cation exchanger Amberlite IRC-50

A weakly acidic, acrylic type of cation-exchange resin is Amberlite IRC-50. The sorption properties of the matrix, which is a polymer of methacrylic acid and divinylbenzene, are lower in comparison with the polystyrene-type matrix of strong cation and anion exchangers. Carboxyl groups of the resin are almost wholly undissociated below pH 5 and are almost fully dissociated at pH 7 and above. The extent of dissociation between pH 5 and 7 can be controlled by variation of the pH of the buffer.

Amberlite IRC-50 has been used successfully many times for the fractionation of basic peptides. The problem with other peptides is that in a buffer with a pH sufficiently low to bring about an appreciable degree of binding of non-basic peptides to the resin, the

*References p.771*

peptides are more or less irreversibly bound to the resin and cannot be eluted until the pH of the eluting buffer has been significantly increased. It is believed that the strong binding of peptides and especially of large fragments to the resin below pH 5 is due mainly to hydrogen bonding. Such bonds form multiple attachments to the resin which is very unfavourable for chromatographic separations as the peptides cannot be eluted from

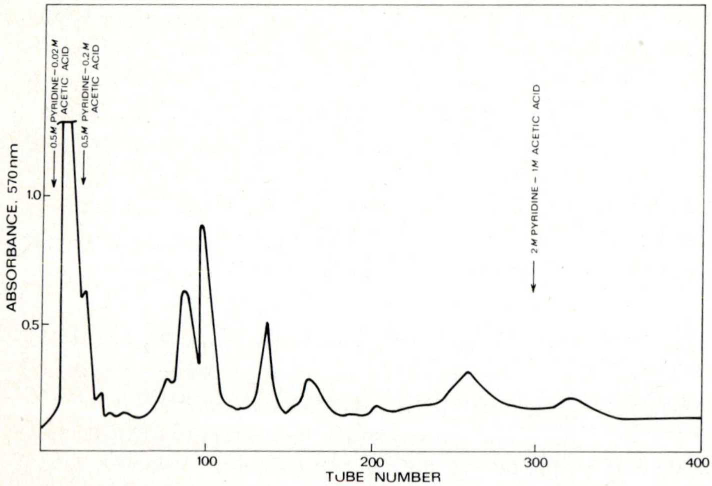


Fig. 34.11. Elution pattern of peptides from the tryptic digest of the κ-type Bence-Jones protein Ag on an Amberlite IRC-50 column with pyridine–acetate buffers (Titani *et al.*, 1969b). Fractions of 20 ml (tubes 1–35) and 10 ml (tubes 36 onwards) were collected at a flow-rate of 80 ml/h. The chromatogram was monitored by the ninhydrin reaction after alkaline hydrolysis.

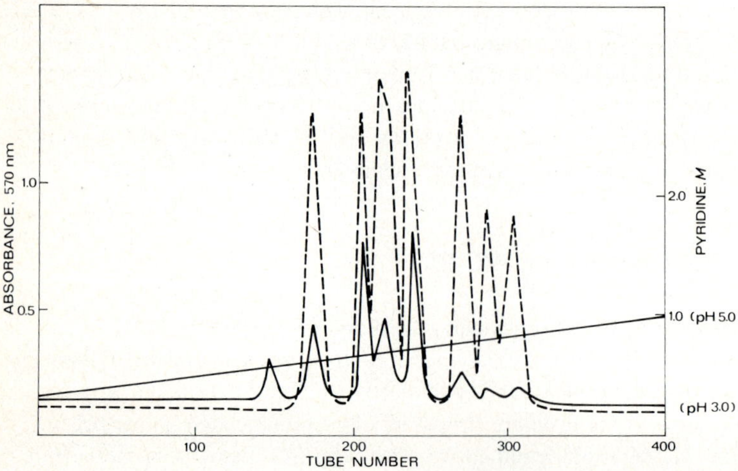


Fig. 34.12. Elution pattern of peptides from a tryptic digest of κ-type Bence-Jones protein Ag (first fraction eluted from the Amberlite IRC-50) on a Dowex 50-X2 column (150 × 1.8 cm) with pyridine–acetate buffers (see also Fig. 34.11) (Titani *et al.*, 1969b). The solid and broken lines represent the ninhydrin colour values before and after alkaline hydrolysis, respectively.

the column unless all of the bonds are broken. On the other hand, a buffer with a pH that is high enough to prevent hydrogen bonding renders most neutral and acidic peptides unsuitable for ion exchange on this type of resin.

For the fractionation of weakly basic peptides, it is helpful to use a buffer with a pH of about 6.5–7.0, when α-amino groups are partially charged. The high buffering capacity of the resin at this pH level and the long time required for equilibration make Amberlite IRC-50 more suitable for fractionation at a fixed pH by increasing the ionic strength. The pH level is well covered by phosphate buffers. The increase in the ionic strength of the buffer can be achieved by increasing the phosphate concentration or by adding sodium chloride to the buffer. The resolution probably depends on a combination of ion exchange and hydrogen bonding. This technique, when used with a Dowex 50-X2 column has been used for the complete fractionation of peptides from tryptic and chymotryptic digests of κ-type Bence-Jones protein by Titani *et al.* (1969b). Initial fractionation of peptides on an Amberlite CG-50 column permitted the isolation of basic peptides (Figs. 34.11 and 34.12). The first fraction eluted from the Amberlite column contained a mixture of acidic and neutral peptides, which was further fractionated on a column of Dowex 50-X2. The gradient system recommended for the Dowex 50 column was stopped at pH 4.5 (1 *M* pyridine–acetate). Peptides eluted in the pH range 4.5–5.0 from the Dowex column were retarded on the Amberlite IRC-50 column.

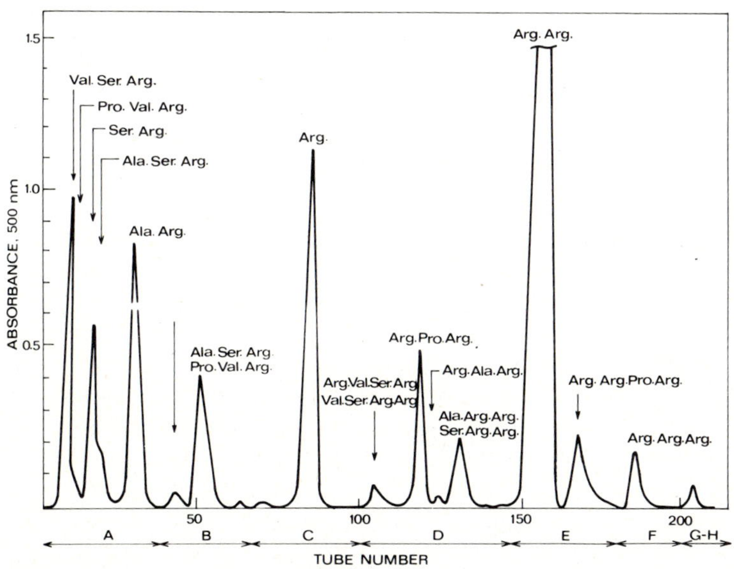


Fig. 34.13. Column separation of tryptic peptides of clupeine Z. The 20-h tryptic hydrolyzate was chromatographed on an Amberlite CG-50 column (30 × 1.0 cm) (Azegami *et al.*). Elution was performed with a sodium chloride concentration in 0.2 *M* sodium borate buffer increasing stepwise as indicated below at 30°C and at a flow-rate of 3 ml per tube per 45 min. Eluents: 0.2 *M* sodium borate containing:

A, 0.074 *N* NaCl (pH 8.10); B, 0.15 *N* NaCl (pH 8.10); C, 0.25 *N* NaCl (pH 8.10);
D, 0.6 *N* NaCl (pH 8.30); E, 1.0 *N* NaCl (pH 8.24); F, 1.5 *N* NaCl (pH 8.03);
G, 0.5 *N* acetic acid; H, 0.1 *N* HCl.

Strongly basic peptides can be fractionated above pH 7, which permits the exchanger to be fully ionized. This technique has often been used for the isolation of arginine-containing peptides. For an illustration of this method, see Fig. 34.13. The peptides were eluted in 16 chromatographic peaks. It has been found that the fractions are usually not homogeneous but contain groups of very similar peptides. Fractions 2–7 contained peptides with one arginine residue in the sequence. Fraction 8 was free arginine, fractions 9–12 contained two arginines in each peptide, fraction 13 was peptide arginyl–arginine, fraction 14 contained peptide with three arginines in the molecule and fraction 15 was arginyl–arginyl–arginine. The separation of peptides on Amberlite IRC-50 has been reviewed by Edmundson.

It has been mentioned above that many peptides, particularly large fragments, are strongly absorbed from aqueous solutions at low pH, owing to the hydrogen bonding by the IRC-50. This effect can be utilized for desalting large fragments; the elution of the peptidic material can be accomplished with 50% acetic acid.

**Ion exchangers with a cellulose, dextran or polyacrylamide matrix**

Ion exchangers of this type have found widespread application in the fractionation of proteins rather than in the fractionation of small peptides. They are also often employed in the fractionation of large peptidic fragments, often in buffer systems that contain 8 *M* urea or 6 *M* guanidine hydrochloride. The choice of exchanger and the conditions for the chromatography of these long-chain members follows the same rules as for the fractionation of proteins, and the reader should consult these general rules in Chapter 35.

It has been shown that some exchangers of this type are useful even for the separation of small peptides. The exchangers most frequently used in the chromatography of proteins are those that contain diethylaminoethyl or carboxymethyl groups, but these ion exchangers are not well suited for the fractionation of peptides. In general, all but the most acidic peptides are bound to the cation exchanger at a low ionic strength in the pH range 3.0–3.5 and can be eluted by increasing the concentration of the salts and the pH to 6–9. Exchangers that contain carboxylmethyl groups are only slightly ionized at pH 3, and thus have a reduced capacity. At the very low ionic strength needed for the sorption, the buffering capacity of the ion exchanger exceeds that of any practical volume of eluting buffer. In practice, by the abrupt change of the pH value of the eluting buffer, groups of peptides are eluted as a single peak. Ion exchangers with diethylaminoethyl groups have analogous disadvantages, especially when volatile buffers are used. Wide changes in ionic strength and pH during the chromatographic elution lead to a contraction of the column bed of this type of ion exchanger and to irregularities in the flow-rate and the elution profile. This effect is most pronounced in high-capacity and low-cross-linked modified gels, but modified celluloses also undergo considerable changes in volume. Useful exchangers of the type in question for the fractionation of peptides are cellulose phosphate and SE-, SP- and QAE-Sephadex.

*Cellulose phosphate*

Cellulose phosphate is a cation exchanger and a large proportion of phosphate groups attached to the cellulose gives a high ion-exchange capacity of 7 mequiv./g. The first p*K*

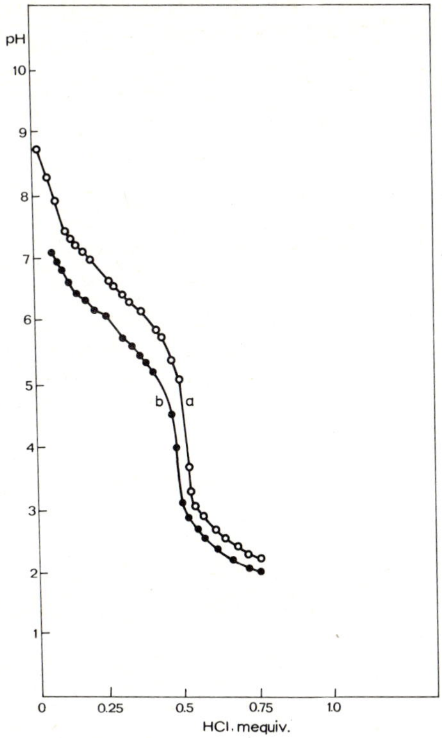


Fig. 34.14. Titration curves for cellulose phosphate (Whatman 325): (a) obtained in water alone; (b) obtained in 0.5 *M* sodium chloride solution (Peterson and Sober).

value of the phosphate group occurs at about pH 2.2 and the second at about pH 6.5 (Fig. 34.14). In the region of pH 2.5–5.5 the exchanger has little buffering capacity and does not alter the effluent pH appreciably even if the buffer is of low ionic strength. The use of buffers of low concentration is an advantage for further use with fractions from the column. Canfield and Anfinsen reported the successful separation of peptides from the peptic and chymotryptic hydrolyzates of egg-white lysozyme, using volatile buffers. The chromatography was accomplished by gradient elution with both the pH and the ionic strength varying.

A gradient of pyridinium acetate was used by Holmquist and Schroeder for the separation of tryptic peptides from $\alpha$- and $\beta$-chains of human haemoglobin. It was found that some peptides that were eluted together from a Dowex 50 column could be separated on cellulose phosphate, and *vice versa*. On the basis of the work of Holmquist and Schroeder, it is possible to correlate the structures of peptides with their chromatographic behaviour on Dowex 50 and cellulose phosphate. The chromatographic behaviour of peptides on a cellulose phosphate column is similar to but not identical with that on Dowex 50. With some exceptions, the longer, more negative, peptides emerge before the shorter, more positive, peptides. Peptides that contain more than six residues and are acidic with a charge no greater than 2 emerge before pH 4.7, while peptides that are less than seven amino acid residues in length and which bear a net charge positive emerge after pH 4.7. The retardation of histidine-containing peptides was reported by Canfield and Anfinsen and Holmquist and Schroeder.

Another advantage of the cellulose phosphate column is that the effluent is completely free from the dissolved resin. Fractions with a high concentration of pyridine from the Dowex 50 columns contain a certain amount of dissolved resin, which complicates the further handling of the fractions. These contaminants from the Dowex 50 column can easily be removed by chromatography on cellulose phosphate.

It is necessary to mention the general phenomenon that at the beginning of the ionic strength gradient the pH decreases. This initial pH drift is more pronounced in cellulose phosphate columns owing to the lower buffering capacity of the buffer in addition to that of the exchanger. The same effect can be found on Dowex 50 and other exchangers. The resolution of peptides is usually not badly affected. Typical separations and conditions are shown in Fig. 34.15 and Table 34.1.

Cellulose phosphate is routinely used for the separation of large fragments from cyanogen bromide digests of collagens. The procedure originally proposed by Bornstein *et al.* has been used many times without modifications. A 20 × 2.5 cm column of cellulose phosphate is equilibrated and eluted with 0.001 *M* sodium acetate solution of pH 3.8 on which is superimposed a linear gradient of sodium chloride from 0 to 0.3 *M* over a total volume of 820 ml. It is possible to follow the fractionation by continuous monitoring the absorbance at 230 nm.

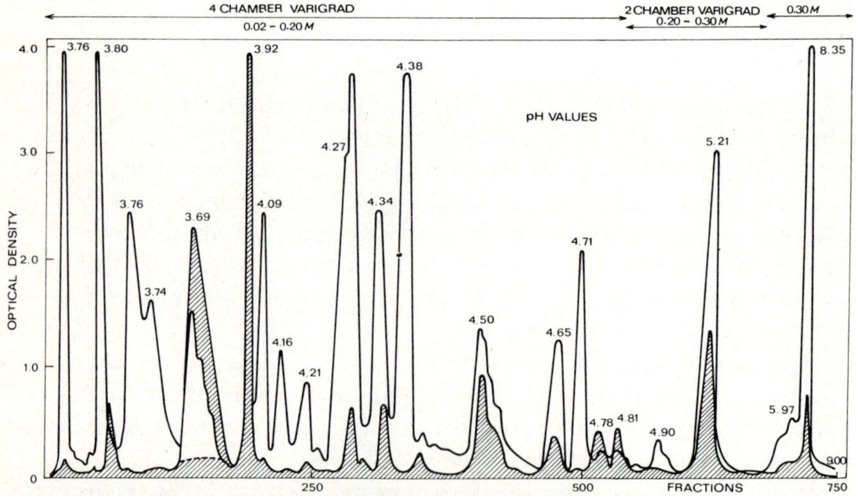


Fig. 34.15. The chromatographic separation of peptides produced by chymotrypsin digestion of carboxymethylated lysozyme (Canfield and Anfinsen). The solid line represents the optical density at 570 nm of aliquots subjected to ninhydrin analysis following alkaline hydrolysis. The shaded areas represent the optical density at 280 nm of the column effluent. The gradient used for the cellulose phosphate column is illustrated at the top of the figure.

TABLE 34.1
BUFFERS USED FOR ELUTION OF CELLULOSE PHOSPHATE COLUMNS BY CANFIELD AND ANFINSEN

A four-chamber Technicon Varigrad containing ammonium acetate buffer was used for fractionation on a 25 × 2.4 cm cellulose phosphate column. An eight-chamber Varigrad containing pyridinium acetate buffer was used with the same adsorbent and a column of the same dimensions.

| Buffer | Chamber | Acetate | | |
|---|---|---|---|---|
| | | Volume (ml) | Concentration (*M*) | pH |
| Ammonium acetate | 1 | 1100 | 0.02 | 3.95 |
| | 2 | 1100 | 0.07 | 4.03 |
| | 3 | 1100 | 0.13 | 4.30 |
| | 4 | 1100 | 0.20 | 5.06 |
| Pyridinium acetate | 1 | 500 | 0.05 | 3.91 |
| | 2 | 500 | 0.10 | 4.02 |
| | 3 | 500 | 0.15 | 4.21 |
| | 4 | 500 | 0.20 | 4.42 |
| | 5 | 500 | 0.25 | 4.63 |
| | 6 | 500 | 0.25 | 4.80 |
| | 7 | 500 | 0.30 | 5.01 |
| | 8 | 500 | 0.45 | 5.04 |

### *SE-Sephadex*

The strong cation exchanger SE-Sephadex is a sulphoethyl derivative of dextran gel, available in two porosity grades, designated C-25 and C-50. For the separation of peptides, the low porosity type C-25 is preferred, with a total capacity of 2.3 mequiv./g. This ion exchanger can serve advantageously in place of Dowex 50, owing to the similarity of the exchange groups and the lower non-specific adsorption. The swelling of the C-25 type of exchanger is essentially independent of pH over the pH range 2–12, thus allowing regeneration to be performed in the column.

Konigsberg and Hill recommended this exchanger particularly for the fractionation of basic peptides. In some instances Dowex 50-X2 is not adequate either alone or in combination with Dowex 1, as some basic peptides are eluted at the void volume of Dowex 1 and are completely retained on Dowex 50. Two peptides from the peptic digest of haemoglobin could not be eluted from Dowex 50 without using sodium hydroxide, but they were eluted without difficulty from SE-Sephadex C-25 with the use of the same gradient of pyridine–acetate buffers as for Dowex 50. The amino acid sequence of these peptides are Leu–Ala–His–Lys–Tyr–His and Tyr–Pro–Try–Thr–Glu($NH_2$)–Arg–Phe.

The fractionation of peptides obtained by tryptic digestion of Kazal-type trypsin inhibitor from porcine pancreas was reported by Tschesche and Wachter. For equilibration of the SE-Sephadex C-25 column, 0.05 *M* ammonium formate buffer of pH 4 was used, followed by stepwise or gradient elution with 0.05 *M* ammonium acetate solution of pH 7.0.

*References p. 771*

The chromatography of peptides on SE-Sephadex C-25 in 8 *M* urea-containing buffers was used for the fractionation of tryptic hydrolyzates of the chain of pig immunoglobulin by Franěk and Novotný. A buffer containing 0.005 *M* potassium formate, adjusted with formic acid to pH 3.0, was used for equilibration of the column. The column was developed with the same buffer, which was superimposed by a linear ionic strength gradient of potassium chloride solution. General conclusions about the chromatography, based on experimental work with the described type of gradient, were drawn by Novotný. The relationships derived were expressed as a simple rule permitting the determination of the column volume and the slope of the gradient most suitable for a particular mixture of peptides.

## AFFINITY CHROMATOGRAPHY*

There are many advantages in using affinity chromatography for the purification and isolation of peptides, which can be separated by highly specific and reversible bonding to the adsorbent. The column operation is very fast, peptides can be isolated from a large volume and the column can be used several times. The disadvantage that the investigator must prepare the adsorbent himself is compensated by the high specificity and ease of handling once the adsorbent has been prepared. The preparation of adsorbents has been discussed in Chapter 13. Applications are illustrated here with a few examples.

The isolation of nitrotyrosine-containing peptides from nitrated lysozyme by means of antibodies to nitrotyrosine attached to Sepharose was proposed by Helman and Givol (Fig. 34.16). The antibodies were prepared in goats or rabbits by injection of a nitrotyrosine–protein conjugate. Purification of antibodies from antisera was achieved by affinity chromatography on a column of nitro-γ-globulin–Sepharose conjugate by Wilchek *et al.* The adsorbed antibodies were eluted with 0.1 *M* acetic acid and used for coupling to Sepharose in order to prepare the antinitrotyrosine–Sepharose adsorbent. This adsorbent was used in the one-step column isolation of tryptic peptides containing nitrotyrosine residues from nitrated, reduced and carboxymethylated lysozyme.

The same useful technique was applied to the isolation of nitrotyrosyl-containing peptides from porcine carboxypeptidase B by Sokolovsky.

Affinity chromatography made possible the isolation of synthetic peptides on the basis of selective association with proteins attached to the solid matrix. This technique was used by Kato and Anfinsen for the isolation of RNase-S-peptide from the crude preparation resulting from the solid-phase synthesis.

Another use of affinity chromatography is for the one-step isolation of affinity-labelled peptides. Affinity-labelled peptides were isolated from the active sites of *Staphylococcus nuclease* and pancreatic ribonuclease by Wilchek.

The general method for the isolation of tryptophan-containing peptides is of special interest owing to the difficulties connected with the isolation of these peptides by ion-exchange chromatography. The method proposed by Wilchek and Miron was tested on the isolation of tryptophan-containing peptides from human serum albumin and horse cytochrome *c*. The tryptophan residues of these proteins were quantitatively modified with the highly specific reagent 2,4-dinitrophenylsulphenyl chloride, and the peptides

*For a more detailed description of this technique, see Chapters 7 and 14.

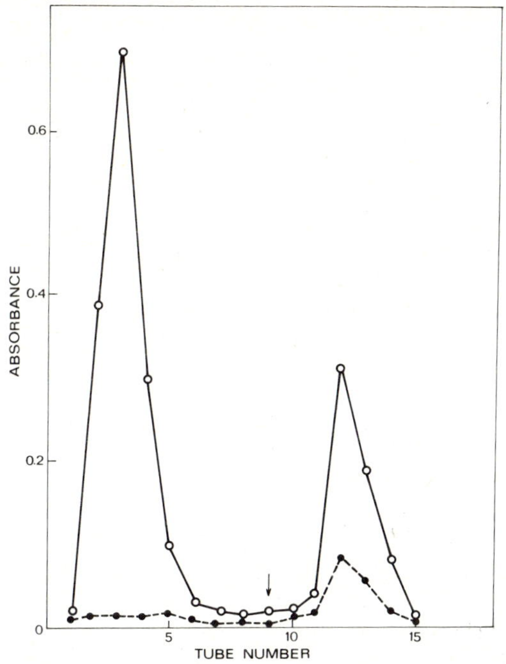


Fig. 34.16. Isolation of nitrotyrosyl peptides from nitrotyrosyl–lysozyme (Helman and Givol). A tryptic digest of 2 mg of reduced and alkylated nitrolysozyme was applied to a 6 × 1 cm column of antinitrotyrosyl antibody–Sepharose conjugate that contained 30 mg of antibodies. The column was washed with 0.1 *M* ammonium hydrogen carbonate solution and the yellow nitrotyrosyl peptides were eluted with 1 *M* ammonia solution (arrow). ○, $E_{280\ nm}$; ●, $E_{381\ nm}$.

containing the modified tryptophan residues were liberated by tryptic hydrolysis and selectively adsorbed and purified on an anti-DNP antibody column. The preparation of DNP–antibody adsorbent was described by Wilchek *et al.* The use of this method can be extended to the isolation of peptides containing methionine, cystine, cysteine, tyrosine or lysine, provided that the residues are specifically modified. Eluents for these strongly bound peptides must be carefully selected so as not to release the antibody from the solid matrix. It has been found that elution with 6 *M* guanidine hydrochloride released 2–3% of the antibody which was covalently bound to Sepharose.

### Isolation of cysteine-containing peptides

The isolation of cysteine-containing peptides by means of organomercurial adsorbents does not fall within the scope of the section on affinity chromatography as it does not exploit the unique property of macromolecules, namely their biological function. This method can be considered as an application of adsorption chromatography. The preparation of organomercurial derivatives of Sepharose was described by Cuatrecasas, Sluyterman and Wijdenes, of cellulose by Shainoff, of cross-linked dextran by Eldjarn and Jellum, and of maleic acid–ethylene copolymer by Liener.

*References p.771*

The last adsorbent was used by Liener and Chao to bind the SH-peptides of the peptic digest of insulin in which the disulphide bridges had been reduced with sodium borohydride. The release of peptides by elution with mercaptoethanol was followed by carboxymethylation of SH-peptides and their isolation by other methods.

## PARTITION CHROMATOGRAPHY

In spite of the fact that for a very long time paper chromatography was the method of choice for the fractionation of peptides, there were few attempts to transfer the experience with paper to chromatographic columns. The reason was probably the difficulty in finding a suitable material as the support for the anchored phase with good flow characteristics and low adsorption. Commercially available gels of different types and cellulose powder are nowadays most often used as the stationary phase support in the partition chromatography of peptides. Another advantage of these materials is that the suitability of solvent systems for the separation of the particular peptidic mixture can easily be tested by paper chromatography with the organic phase of the proposed solvent system being used as the developer. The most promising systems are those in which the $R_F$ values of peptides are different, and preferably around 0.5. Solvents systems with very low or very high $R_F$ values are not useful.

The complete cycle of column operation consists in:

(1) equilibration with the aqueous phase of the solvent system;
(2) equilibration with the organic phase of the solvent system;
(3) chromatography;
(4) discharge of the two-phase system and the material not eluted;
(5) washing of the column.

After the completion of the last stage, the column is ready to enter the first equilibration stage of the next cycle. A complete cycle of a column packed with Sephadex G-25 usually requires 5 or 6 days.

The procedure for the purification of oxytocin by column partition chromatography was developed by Yamashiro using a two-phase system similar to that used in the purification of oxytocin and vasopressin by counter-current distribution. The column was packed with Sephadex G-25 in 0.2 *N* acetic acid, and the volumes of solvents needed for equilibration were as shown in Table 34.2.

TABLE 34.2
SOLVENTS FOR EQUILIBRATION

| Stage | Influent solvent | Total volume | Flow-rate (ml/h·cm²) |
|---|---|---|---|
| 1 | Aqueous phase | 1.3 × bed volume | 10–15 |
| 2 | Organic phase | 0.3 × bed volume | 5–10 |
| 3 | Organic phase | 7–10 × bed volume | 5 |
| 4 | Pyridine–0.2 *N* acetic acid (*X*:*Y*) | 1.5 × bed volume | 5 |
| 5 | 0.2 *N* acetic acid | 1.3 × bed volume | 3–15 |

The void volume is measured during stage 2 after the emergence of the organic phase of the solvent system at the column exit. The washing solvent used in the fourth stage is expressed in volume terms. The ratio $X:Y$ was determined according to the solvent systems used in stages 1, 2 and 3. A washing system was found to be effective if not more than four volumes of it were required to obtain miscibility with one volume of the organic phase of the solvent system to be discharged.

Solvents for the partition chromatography of oxytocin on Sephadex G-25 column are listed in Table 34.3.

TABLE 34.3
SOLVENT SYSTEMS FOR THE PARTITION CHROMATOGRAPHY OF OXYTOCIN ON A SEPHADEX G-25 COLUMN (YAMASHIRO)

| Solvent system | pH | $R_F$ of oxytocin | $X:Y$ (stage 4) |
|---|---|---|---|
| (I) *n*-Butanol–*n*-propanol–0.2 *N* acetic acid (2:1:3) | 3.0 | 0.16–0.18 | 1:4 |
| (II) *n*-Butanol–benzene–pyridine–0.2 *N* acetic acid (6:1:1.5:9) | 5.5 | 0.24–0.27 | 3:5 |
| (III) *n*-Butanol–benzene–pyridine–0.1% acetic acid (6:2:1:9) | 6.2 | 0.20–0.25 | 3:5 |

The use of partition chromatography with Sephadex G-25 as the supporting phase was particularly useful in the isolation of S-DNP labelled peptides from ATP-creatine phosphotransferase (Mahovald). The technique used was the same as described above but with the following solvent systems:

(I) *n*-Butanol–*n*-propanol–3% pyridine in 3% aqueous acetic acid (2:1:3);

(II) *n*-Butanol–*n*-propanol–benzene–3% pyridine in 3% aqueous acetic acid (4:1:1:6);

(III) *n*-Butanol–*n*-propanol–benzene–3% pyridine in 3% aqueous acetic acid (8:1:3:12);

(IV) *n*-Butanol–benzene– 3% pyridine in 3% aqueous acetic acid (1:1:2).

The effluent from the column was followed by measuring the absorbance at 330 nm, fractions being evaporated on a rotary evaporator and subjected to sequence analysis.

## REFERENCES

Azegami, M., Ishi, S. and Ando, T., *J. Biochem.*, 67 (1970) 523.
Bock, M. R., Ling Nan-Sing, *Anal. Chem.*, 26 (1954) 1543.
Bornstein, P., Kang, A. H. and Piez, K., *Proc. Nat. Acad. Sci. U.S.*, 55 (1966) 417.
Callaham, P. X., McDonald, J. K. and Ellis, S., *Methods Enzymol.*, 25 (1972) 282.
Canfield, R. E. and Anfinsen, C. B., *J. Biol. Chem.*, 238 (1963) 2684.
Canfield, R. E. and Liu, A. K., *J. Biol. Chem.*, 240 (1965) 1997.
Catravas, G. N., *Anal. Chem.*, 36 (1964) 1146.
Clarke, H. T., *J. Biol. Chem.*, 97 (1932) 235.
Cuatrecasas, P., *J. Biol. Chem.*, 245 (1970) 3059.
Delanay, R., *Anal. Biochem.*, 46 (1972) 413.
Edmundson, A. B., *Methods Enzymol.*, 11 (1958) 369.
Eldjarn, L. and Jellum, E., *Acta Chem. Scand.*, 17 (1963) 2610.

Folin, O., *J. Biol. Chem.*, 106 (1934) 311.
Franěk, F. and Novotný, J., *Eur. J. Biochem.*, 11 (1968) 5591.
Frankland, B. T., Hollenberg, M. D., Hope, D. B. and Schachter, B. J., *Brit. J. Pharmacol.*, 26 (1966) 502.
Goldfarb, A. R., Saidel, L. J. and Mosovich, E., *J. Biol. Chem.*, 193 (1951) 397.
Helman, M. and Givol, D., *Biochem. J.*, 125 (1971) 971.
Hill, R. L. and Delanay, R., *Methods Enzymol.*, 11 (1958) 339.
Hirs, C. H. W., *Methods Enzymol.*, 11 (1958) 325.
Hirs, C. H. W., Moore, S. and Stein, W. H., *J. Biol. Chem.*, 211 (1954) 907.
Holmquist, W. R. and Schroeder, W. A., *J. Chromatogr.*, 26 (1967) 465.
Jansen, J. C., *J. Chromatogr.*, 28 (1967) 12.
Kassel, B. and Brand, E., *J. Biol. Chem.*, 125 (1938a) 115.
Kassel, B. and Brand, E., *J. Biol. Chem.*, 125 (1938b) 131.
Kato, I. and Anfinsen, C. B., *J. Biol. Chem.*, 244 (1969) 1004.
Konigsberg, W. and Hill, R. J., *J. Biol. Chem.*, 237 (1962) 2547.
Kostka, V., Morávek, L., Kluh, I. and Keil, B., *Biochem. Biophys. Acta*, 175 (1969) 459.
Kušnír, J. and Meloun, B., *Collect. Czech. Chem. Commun.*, 38 (1973) 143.
Liener, I. E., *Arch. Biochem. Biophys.*, 52 (1967) 67.
Liener, I. E., and Chao Li-Pen, *Anal. Biochem.*, 25 (1968) 317.
Lugg, J. W. H., *Biochem. J.*, 26 (1932) 2144.
McDowall, M. A. and Smith, E. L., *J. Biol. Chem.*, 240 (1965) 4635.
Machleidt, W., Kerner, W. and Otto, J., *Z. Anal. Chem.*, 252 (1970) 151.
Mahovald, T. A., *Biochemistry*, 4 (1965) 732.
Miranda, F., Rochat, H. and Lissitzky, S., *J. Chromatogr.*, 7 (1962) 142.
Moore, S. and Stein, W. H., *J. Biol. Chem.*, 211 (1954) 907.
Novotný, J., *FEBS Lett.*, 14 (1971) 7.
Nyman, P. O., Strid, L. and Westermark, G., *Eur. J. Biochem.*, 6 (1968) 172.
Okuyma, T. and Satake, K., *J. Biochem.*, 47 (1960) 454.
Paar, C. W., *Proc. Biochem. Soc., 324th Meeting,* XXVII.
Padieu, P. and Maleknia, N., *Bull. Soc. Chim. Biol.*, 47 (1965) 493.
Peterson, E. A. and Sober, H. A., *J. Amer. Chem. Soc.*, 78 (1956) 751.
Plapp, B. V., Raftery, M. A. and Cole, R. D., *J. Biol. Chem.*, 242 (1967) 265.
Rochat, H., Rochat, C., Lissitzky, S. and Edman, P., *Eur. J. Biochem.*, 17 (1970) 262.
Samejima, K., Dairman, W., Stone, J. and Udenfriend, S., *Anal. Biochem.*, 42 (1971) 237.
Satake, K., Take, T., Matsuo, A., Tazaki, K. and Hiraga, Y., *J. Biochem.*, 60 (1966) 12.
Schroeder, W. A., *Methods Enzymol.*, 25 (1972a) 203.
Schroeder, W. A., *Methods Enzymol.*, 25 (1972b) 214.
Shainoff, J. R., *J. Immunol.*, 100 (1968) 187.
Sluyterman, L. A. and Wijdenes, J., *Biochim. Biophys. Acta*, 200 (1970) 593.
Sokolovsky, M., *Eur. J. Biochem.*, 25 (1972) 267.
Spackman, D. H. and Stein, W. H., *J. Biol. Chem.*, 235 (1960) 648.
Titani, K., Shinoda, T. and Putnam, F. W., *J. Biol. Chem.*, 244 (1969a) 3550.
Titani, K., Whitley, E. J. and Putnam, F., *J. Biol. Chem.*, 244 (1969b) 3521.
Tschesche, H. and Wachter, E., *Hoppe-Seyler's Z. Physiol. Chem.*, 351 (1970) 1449.
Wilchek, M., *FEBS Lett.*, 7 (1970) 161.
Wilchek, M. and Bocchini, V., Becker, M. and Givol, D., *Biochemistry*, 10 (1971) 2828.
Wilchek, M. and Miron, T., *Biochim. Biophys. Acta*, 278 (1972) 1.
Yamashiro, D., *Nature (London)*, 201 (1964) 76.

*Chapter 35*

# Proteins

Z. PRUSÍK

## CONTENTS



## INTRODUCTION

Proteins are essential components of living matter and the range of their concentrations and molecular weights is wide. In contrast to other components of macromolecular character, which are present in every organism, proteins are extremely variable in their electrochemical properties. The many possible combinations of occurrence, and the sequence of amino acids in the polypeptide chain and of polypeptide chains in the protein also cause considerable variations in their solubilities in water and other strongly polar solvents. The many structural variations in proteins also mean that proteins may occur together in mixtures that have very different structures but are still similar with respect to their physico-chemical properties. The amphoteric character of proteins and the large variations in their molecular weights are the main factors that permit an effective separation by means of column processes, mainly in aqueous media. The specific binding properties of proteins became a factor that extended the availability of highly effective and specific separation processes by the method of affinity chromatography. On the other hand, the protease and peptidase activity of proteins that are regularly present in complex mixtures of starting material impose certain limitations on separation processes from the point of view of temperature and/or acidity of the eluent.

An important factor in the separation is also the tendency of proteins to become aggregated or denatured. In the study of living matter, poorly soluble proteins that form the structure of cell membranes also cannot be ignored. Recently, requirements for efficient separations of lipoproteins were laid down, and attention was also devoted to the proteins of cell nucleus, where the main stress was laid not only on the resolving power of column methods but also on rapid procedures accompanied by small losses during the separation process.

The aim of this chapter is to give a review of separation and detection methods, and to emphasize, as far as possible, some general principles that enable an experimenter provided with basic information on the starting material to suggest a rational separation procedure, although the trial-and-error method cannot be excluded completely during the search for the most suitable procedure for the isolation of proteins.

## GENERAL RULES FOR THE SEPARATION OF PROTEINS

### Selection of the separation procedure

The first requirement for the use of column separation methods is perfect solubilization of the proteins, which can be achieved during the extraction of biological preparations by a suitable choice of salt concentration in the solution, by detergents and by pH adjustment. After the elimination of mechanical impurities, the required protein is concentrated, either by precipitation procedures or by sorption on ion exchangers and subsequent desorption at a suitable ionic strength and pH of the solution. The elimination of salts and the transfer of the proteins into a buffer of the required composition can then be carried out by GPC; the same method can also be used for the basic separation according to the size and shape of the protein molecules. As the next step, the sorption method is usually applied, during which use is made of the charge or specific binding properties of proteins; see also the chapters on Enzymes (Chapter 36) and Affinity chromatography (Chapter 7). In the final part of the separation, GPC is again often applied in view of the fact that most proteins are usually transformed into a dry state by freeze-drying. In favourable instances, for example if the proteins possess extreme properties of solubility, charge or binding strength, some parts of the procedure can be omitted.

Although this chapter deals with column operations, it should be mentioned that operations on columns are often controlled or even completed by electrophoretic procedures, for example electrophoresis in a polyacrylamide gel.

### Choice of temperature in column separations of proteins

With increasing temperature, the viscosity of the mobile phase decreases and the hydrodynamic resistance of the column and the time necessary for equilibration in the separation process are correspondingly shortened. Theoretically, the upper limit of the temperatures that can be used in column separations of proteins is that above which thermal denaturation of proteins occurs. This temperature depends on the individual properties of the proteins, whether they are more resistent to the formation of polymeric forms at higher or at lower temperatures. In practice, the temperature limit is often shifted downwards, especially if hydrolases of the protease type or other destructively acting enzymes are present in the separated mixture, which often occurs in work with tissue extracts. In such instances, the work must be carried out at temperatures close to or even below 0°C, with the addition of several per cent of isopropanol. The necessity for this procedure can be avoided only if a medium is chosen the pH of which is far from the optimum pH of the interfering enzymatic activities, or a medium that contains, for example, inhibitors of proteolytic activities; if the activity is eliminated specifically, for example by affinity chromatography (Carey and Wells); or if a medium is selected that blocks the enzyme activity by its random-coil forming effect, *e.g.*, highly concentrated solutions of guanidine hydrochloride or urea.

### Separation according to molecular size

#### *Gel permeation chromatography*

The principle of the GPC of proteins is the differing retardation of the molecules during the flow of their solution through a column consisting of gel particles, smaller molecules that are capable of penetrating easily into the gel pores being retarded more than larger molecules. The molecules that are unable to penetrate into the gel are eluted in the retained volume, $V_0$, of a given column. In addition to this prevailing process, ion exchange also may be involved to a small extent, as may the sorption properties of the gel matrix and also the partition effect of the mobile phase (eluent) provided that it is immiscible with the phase fixed on the gel. The principles of the GPC method are explained in Chapter 5. As proteins especially have a vast range of molecular weights, ($10^3$–$10^6$ daltons and above), the GPC method is used very extensively for the characterization and separation of protein mixtures. A detailed treatment of GPC techniques with a review of gels that are suitable for protein separations was given by Reiland.

#### *Characterization of proteins by distribution constants*

On the supposition that no interactions take place between the gel and the substances passing the gel, column GPC can be considered to be an example of liquid–liquid chromatography. During the passage through the column of a total volume $V_t$, the substances in solution are distributed between the liquid volume, $V_0$, around the gel particles and the volume, $V_i$, which is the volume of the liquid inside the gel particles. After passing through the column, the substances are washed out in a volume $V_e$. The properties of proteins can be characterized either by the distribution constant, $K_d$, according to the relationship

$$K_d = \frac{V_e - V_0}{V_i}$$

or by means of the distribution constant, $K_{av}$, which is more useful in practice, defined as

$$K_{av} = \frac{V_e - V_0}{V_t - V_0}$$

Both $K_d$ and $K_{av}$ are independent of the column geometry and can be used in the characterization of the molecular weights of proteins. As $V_i$ is determined less easily than $V_t$, the characterization of proteins by $K_{av}$ is more advantageous. Both $K_d$ and $K_{av}$ have values from 0 to 1 unless an interaction of the gel matrix with the protein passing through takes place.

## Group separations – desalting

GPC is used for group separations when a fine separation is not imperative; however, the separation usually requires a large amount of preparation. For example, a common

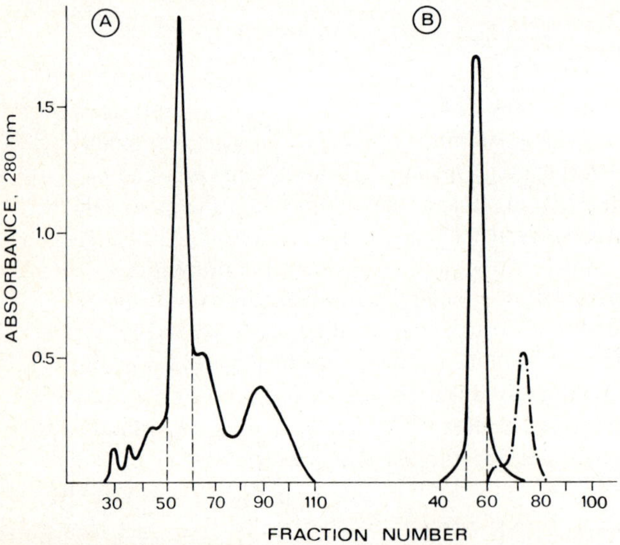


Fig. 35.1. Last step in the purification of luteinizing hormone by gel permeation chromatography on Sephadex G-75 (Hennen *et al.*). Column: 96 × 2.5 cm. Sorbent: Sephadex G-75. Buffer: 0.05 *M* ammonium hydrogen carbonate solution. Operating conditions: flow-rate, 14 ml/h; fraction volume, 3.4 ml; sample A, crude porcine luteinizing hormone, 160 mg in 3 ml of buffer; sample B, material from A after selection, 100 mg in 3 ml of buffer. The GPC pattern of the α-sub-unit of luteinizing hormone is indicated by the broken line.

problem consists in the separation of protein components from the components of low molecular weight in the solution, *i.e.*, desalting. For such separations gels with a low exclusion limit are suitable, for example Sephadex G-25 and Bio-Gel P-6. The proteins are excluded ($K_{av} = 0$), while the substances of low molecular weight are strongly retarded and their $K_{av}$ values are close to unity. The sample volume, $V_s$, should not exceed one quarter of the column volume, $V_t$ ($V_t \geqslant 0.25\ V_s$).

### Fine separations of protein mixtures

GPC is used for finer separations of proteins, where the distribution constant $K_{av}$ is usually between 0.1 and 0.3. Therefore, a suitable type of gel should be chosen for separation which would not cause either protein exclusion or an unnecessarily weak differentiation of the components of a mixture if a gel was chosen with an excessive exclusion limit. Agarose gels, polydextran and polyacrylamide gels are now available for a large range of molecular weights of proteins that occur in nature. During the separation of protein mixtures, attention must be paid to the small difference in viscosity between the sample solution and the eluent; the quality of the resolution is influenced by the sample volume, $V_s$, which should be at least 25–100 times less than the column volume, $V_t$ ($V_t \geqslant 25{-}100\ V_s$). An example is shown in Fig. 35.1.

### Determination of molecular weights

The shape and the size of the protein molecule are correlated with the molecular weight, so that basic information on the molecular weight or shape can be obtained by GPC. The method described by Andrews (1964, 1965, 1970) permits the molecular weight or the Stokes' radius of the molecule of the investigated protein to be determined by graphical interpolation if the elution volumes, $V_e$, of several proteins of known molecular weight are plotted against the molecular weight. In contrast to some other methods of molecular weight determination, the investigated protein need not be free from all contaminating components. When the protein possesses a specific property, such as activity or characteristic absorbance, its molecular weight can be determined even in mixtures in which it is a minor component.

The correct determination of molecular weight requires a suitable choice of protein standards for the construction of a calibration graph of elution volume, $V_e$, *versus* log (mol. wt.) or of the distribution coefficient, $K_{av}$, *versus* log (mol. wt.) (see Table 35.1). This means that for globular-type proteins, standards should be chosen among globular proteins, thus avoiding the errors that follow from the different behaviour of linear and globular proteins during GPC. The determination of molecular weight within a broad range of molecular weights on a single agarose gel column can be carried out with advantage in a dissociating medium (Bryce and Crichton). For the determination of the molecular weight and the separation of proteins with molecular weights ranging from $10^4$ to $10^5$ daltons, polydextran gels with a lower matrix density are suitable, for example Sephadex G-75, G-100, G-150 or G-200, or synthetic gels of corresponding properties, for example the polyacrylamide gels Bio-Gel P-30 to P-300. For higher molecular weights, ranging from $10^5$ to $10^7$ daltons, agarose gels of the type Sepharose 2B to 6B or Bio-Gel A-0.5 to A-0.150 can be used, for example. Detailed characteristics of suitable gels in relation to different ranges of molecular weight are given in the chapter on GPC (Chapter 5).

### Choice of eluent

All types of aqueous buffers can be used as eluents for the GPC of proteins unless their properties impair the stability of the gel matrix and do not interact with either the gel matrix or the proteins. Hence in some instances borate buffers are not desirable, nor are aqueous eluents with a pH less than 2–2.5, because they disintegrate the structure of the cross-linked polydextran gels in a similar manner to solutions with oxidative properties. Water also is not a good eluent, because polydextran gels at extremely low ionic strengths ($< 0.02$) behave as weak cation exchangers with extremely low capacities. For the desalting of proteins and for their separations, 0.05 to 0.10 $M$ ammonium hydrogen carbonate solution is very suitable, because it can be eliminated by lyophilisation. Because they are weakly alkaline, solutions of ammonium hydrogen carbonate solubilize proteins well, and another advantage of its use lies in its resistance to microorganisms, which is especially important during prolonged work with cross-linked polydextran gels and agarose gels.

## GEL PERMEATION CHROMATOGRAPHY

### Gel permeation chromatography with recycling

During the separation of a mixture of proteins by GPC, the components of the investigated mixture may have closely similar distribution constants, $K_{av}$, even on a suitable type of gel. If such protein components must be separated, the length of the column can be increased, of course, but such an increase has practical limits. An increase in the column length, *i.e.*, in the bed height, leads to an increase in the hydrodynamic resistance of the column and thus a decrease in the elution rate. The pressure gradient is unfortunately limited, especially in more deformable gels with larger pores, which are suitable for the separations of proteins.

Porath and Bennich therefore proposed the use of recycling chromatography for such materials, and this method can generally be used to give an increase in the effective bed height in all columnar processes in which the separation is independent of the concentration of the substances in solution. The method was found to be especially advantageous in the GPC isolation of proteins and protein fragments with small differences in $K_{av}$.

The principle of the recycling method consists in repetition of the GPC process on the same gel column. The effluent from the column passes through the UV absorptiometer and is then re-injected into the column from below by means of a peristaltic pump, and by the repetition of the process the column length is effectively increased. The maximum number of cycles is limited by the requirement that the proteins with the lowest elution volume, $V_e$, in the separated mixture do not penetrate into the zone of stronger retarded proteins with the highest $V_e$ during a repeating cycle. The more similar are the elution volumes of the proteins, the greater is the number of times the cycle can be repeated without interfering with the process. In order to prevent the mixing of the least retarded proteins with the most retarded proteins, the required fractions can be separated from the recycling system by bleeding the effluent, and recycling can be continued with the other components of the mixture. Opening and closing the recycling circuit at suitable stages permits the elution of the interfering component. The amount of effluent necessary for the collection of fractions is replaced with an equivalent amount of eluent through the recycling valve.

If unknown protein mixtures are being studied, it is desirable that the separation should first be carried out with an open system, and the programme for the recycling process is then deduced from the elution data.

Porath and Bennich demonstrated the efficiency of recycling in the separation of 5.2 ml of a 3% solution of minor components of human ceruloplasmin on a 93 × 3.2 cm column of polydextran gel (Sephadex G-100), equilibrated in 0.1 *M* Tris + 0.5 *M* sodium chloride buffer adjusted to pH 8.0 with hydrochloric acid, at a flow-rate of 20 ml/h at 10°C. The cycle lasted 10 h and all material passed through the column in 14 h, so that three cycles could be allowed to take place without remixing. During the third cycle, the tailing material with the largest elution volume was separated. After this cycle, the dominant component B, with a higher $V_e$, could be eliminated completely by bleeding. The effective bed height corresponded to 370 cm after the fourth cycle. After the sixth and the seventh cycles, the residue of component D was removed completely. Component A, which already had a symmetrical peak, was isolated after the eighth cycle, at an effective bed height of 7.4 m. This separation is shown in Fig. 35.2.

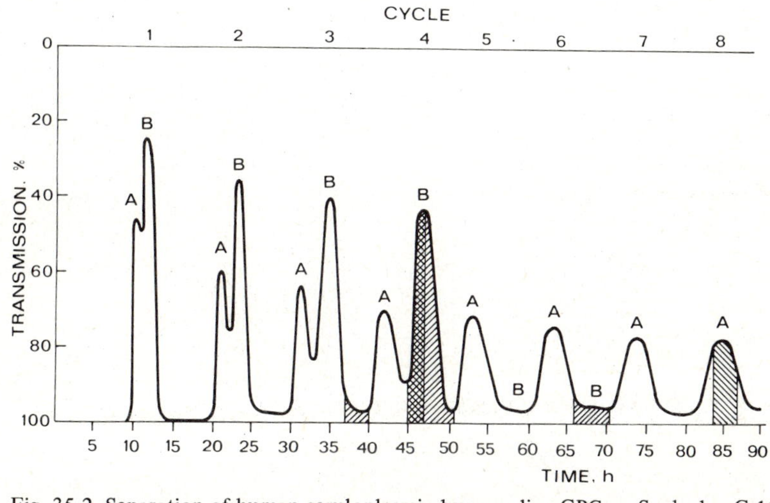


Fig. 35.2. Separation of human ceruloplasmin by recycling GPC on Sephadex G-100 (Porath and Bennich). Column: 93 × 3.2 cm. Buffer: 0.1 *M* Tris containing 0.5 *M* sodium chloride and adjusted to pH 8.0 with hydrochloric acid. Operating conditions: flow-rate, 20 ml/h; temperature, 10°C; sample, 5.2 ml of a 3% solution of ceruloplasmin. Detection: transmittance at 354 nm. Bleeding of the column is indicated by hatched areas.

*References p.805*

When a more detailed analysis of single components in the effluent is necessary before recycling, the fractions can be stored frozen and submitted to analysis gradually. In the next phase, the so-called discontinuous recycling according to Morávek, the fractions are again injected into the column in order of their elution volumes. By this method, a higher degree of resolution of single components than in simple re-chromatography can be achieved, even when applied to preparative fractionation that requires a longer time.

### Gel permeation chromatography of proteins in dissociating media and the solubilization of proteins

Anomalies in the distribution constants, $K_{av}$, or elution volumes, $V_e$, in relation to the molecular weights of proteins occur in GPC in aqueous solutions for molecules that differ in shape and density. These anomalies disappear to an appreciable extent or completely if an eluent is used for GPC that contains highly concentrated urea or guanidine hydrochloride solutions. Under the effect of urea and especially guanidine hydrochloride, the polypeptide chains of proteins are transformed into a random coil configuration, which causes the disappearance of the variation in $V_e$ that resulted from the different shapes of the protein molecules in their native form. While solutions of urea of 8–9 *M* concentration often do not give a sufficient dissociating effect, guanidine hydrochloride causes strong dissociation and the formation of random coils at a concentration of 4–6 *M*. On a Sepharose 6B column equilibrated with 6 *M* guanidine hydrochloride, Bryce and Crichton separated proteins with molecular weights ranging from $80 \cdot 10^3$ to $1.4 \cdot 10^3$ daltons (see Table 35.1) and determined their molecular weights. The resolution of proteins was very

TABLE 35.1
GPC OF PROTEINS AND PEPTIDES IN 6 *M* GUANIDINE HYDROCHLORIDE ON SEPHAROSE 6B (BRYCE AND CRICHTON)

$K_{av}$ values of several proteins and peptides compared with their molecular weights. $K_{av} = (V_e - V_0)/(V_t - V_0)$, where $V_0$ = exclusion volume estimated with Dextran Blue 2000, $V_t$ = total volume obtained by co-chromatography of sample with tryptophan and $V_e$ = elution volume of the sample.

| Protein | Molecular weight | $K_{av}$ |
|---|---|---|
| Transferrin (horse) | 76,600 | 0.0758 |
| Serum albumin (bovine) | 68,000 | 0.1024 |
| Immunoglobulin H-chain (porcine) | 55,000 | 0.1343 |
| Ovalbumin | 43,000 | 0.1769 |
| Chymotrypsinogen A | 25,700 | 0.2380 |
| Immunoglobulin L-chain (porcine) | 23,500 | 0.2673 |
| Myoglobin (horse heart) | 17,200 | 0.3218 |
| Lima bean trypsin inhibitor | 8,400 | 0.5141 |
| Glucagon | 3,480 | 0.7021 |
| Insulin (B chain) | 3,400 | 0.7127 |
| Insulin (A chain) | 2.340 | 0.7739 |
| Bacitracin | 1,411 | 0.8563 |

good even at a molecular weight of $10^3$ daltons. From the empirical equation

$$\log M = A - B \cdot K_{av}$$

Bryce and Crichton found that on GPC on Sepharose 6B in 6 *M* guanidine hydrochloride, two zones of linearity appeared, the first for the molecular weight range $11 \cdot 10^3$–$80 \cdot 10^3$ daltons, where $A = 5.091$ and $B = 2.652$, and the second for the molecular weight range $1.4 \cdot 10^3$–$8 \cdot 10^3$ daltons, where $A = 5.176$ and $B = 2.357$.

While guanidine hydrochloride has the highest dissociating effect and is also suitable for histones with a strong tendency to aggregate, urea is still used for preparative separations by GPC because it is cheaper. Ammonium isocyanate, which is present in urea, represents a risk, however, owing to its ability to substitute the amino groups of the prepared proteins, especially if the pH is higher than 5.

For the separation of strongly aggregated protein oligomers, complexes of proteins and peptides, lipoprotein complexes and other poorly solubilizable aggregates the protein component of which is the object of the separation, protein solubilization based on the use of detergents in an aqueous medium is also an effective method. Detergents of the ionic type, such as sodium dodecyl sulphate (SDS) and sodium deoxycholate (DOC), or of the non-ionic type, such as Triton X-100, represent a group of dissociating reagents that can be used for GPC on all types of gels. The application of detergents is especially valuable for the solubilization and the separation of lipoprotein complexes and membrane proteins, which can otherwise be extracted only with difficulty. For extraction and GPC, 0.1–0.5% SDS is most often used. Triton X-100 and DOC are also used at lower concentrations, mostly from 0.05 to 0.5%.

Detergents were also used at concentrations of 3–4% (SDS and taurodeoxycholate). However, some proteins, for example cytochrome *c*, show a greater tendency to dimerize when the concentration of the detergent is increasing. Bovine serum albumin and ovalbumin remained unaffected by 10 m*M* taurodeoxycholate, as had already been found by Morgan *et al.* Although the solubilization of membrane proteins by detergents is not complete, nevertheless most apolipoproteins can be kept in solution during the separation by means of detergents and thus a fair separation of apolipoproteins, phospholipids and neutral lipids can be achieved.

Variations in the solubilities of proteins produced by varying detergent concentrations are made use of in gradient zonal precipitation according to Swanljung. After the elimination of the detergent from the protein preparation, an appreciable decrease in solubility often takes place. In spite of this disadvantage, the damage that occurs to the preparation is often less than that during delipidation of the starting material by organic solvents (Zahler and Wallach).

## CHROMATOGRAPHY ON GLASS WITH CONTROLLED PORE SIZE

Chromatography on glass of controlled pore size (CPG), also called steric chromatography, differs from GPC mainly as regards the matrix, which is finely granulated glass with a large number of pores the size of which can be regulated within a broad range during production. In addition, the distribution of the pore diameters is much smaller

*References p.805*

than in materials used for GPC and does not exceed ± 20% of the mean pore diameter. The use of this method for the separation of proteins is not very widespread, and the available data show that it is most useful when speed of separation is essential. Mechanical properties of CPG permit the use of extremely high flow-rates, larger by an order of magnitude than in the conventional GPC. Even at high flow-rates and corresponding high-pressure gradients, a change in column volume does not take place. The excellent chemical stability and thermal resistance of CPG permits the regeneration of CPG columns even with powerful reagents without risking a change in the column properties, so that thermal sterilization of the CPG column is also possible.

This method was used by Haller *et al.* for the separation of the immunoglobulin fraction of serum, and especially macroglobulins; they also desalted tobacco mosaic virus (TMV) on the same CPG column. On a 100 × 1.2 cm CPG column with particles of average pore size 17.5 nm chromatography of 3 ml of serum was completed within 40 min at a flow-rate of 120 ml/h. The authors considered the method to be capable of industrial application, especially in the production of macroglobulin serum fractions.

The use of the method in protein separations is limited only by the risk of pressure aggregation of proteins, which was observed at extreme pressures between 2800 and 3500 kp/cm$^2$ used for the separation of serum albumin according to Bidlingmeyer and Rogers.

Today, the range of porosities of commercially available CPG has increased in the direction of smaller pore diameters so that the application of CPG chromatography to proteins with lower molecular weights may be expected in the near future.

## ION-EXCHANGE CHROMATOGRAPHY

The binding of proteins to ion exchangers is based on the formation of multiple ionic bonds between the charged groups of the ion exchanger and inversely charged groups of the protein. The affinity of an ion exchanger towards the protein, and also its capacity, is determined by the salt concentration and pH. The separation of bond-forming proteins is based on changes in the protein charge caused by variations in pH, or by affecting the bonds by means of substances that compete with the protein for the charges on the ion exchanger. In order to define the separation process well, the pH and the salt concentration, and especially the concentration of ions with a charge opposite to that of the ion exchanger, should be specified.

### Sorption and the choice of ion exchanger according to the protein charge

Proteins are bound by electrostatic bonds if their charge is opposite to that of the dissociated ion exchanger groups. Hence, a protein is bound to a cation exchanger at a pH lower than that which corresponds to its isoelectric point; binding to an anion exchanger does not take place under the same conditions. On the contrary, if the pH is above the isoelectric point of the protein, a bond is formed between the protein and the anion exchanger. A suitable choice of the type of ion exchanger depends on the charge of the protein and the pH range over which it is stable. If this pH range for stability of the protein is less than the value of the isoelectric point, ion-exchange chromatography should be carried out on a cation exchanger, while a protein that is stable above the isoelectric point should be chromatographed on an anion exchanger. If the proteins are sufficiently stable at least in the range of ± 1 pH unit from the isoelectric point, either an anion or a cation exchanger can be used.

Although the method of isoelectric fractionation of proteins may permit the determination of the isoelectric point of the investigated protein with sufficient accuracy (in a density gradient), or at least approximately, the isoelectric point of the protein need not necessarily be known. The suitability of an ion exchanger can be determined by a simple batch operation. When testing the binding to an anion exchanger, the pH is chosen so as to be in the upper region of the stability interval of the protein, while for binding to a cation exchanger, the lower region of the stability interval should be chosen. The protein, freed from salts, is dissolved in a suitable buffer and after gentle agitation with an ion exchanger stabilized with the corresponding buffer (no foam should be formed), the concentration of the protein in the supernatant is determined.

The properties of the matrix and the functional groups of ion exchangers are discussed below.

## Range of pH used in the ion-exchange chromatography of proteins

For the separation of proteins, ion exchangers based on cellulose, cross-linked polydextran gels, agarose gels on polyacrylates and, less often, polyacrylamide gels, are commonly used. Hydroxyapatite of sufficiently fine grain and to a lesser extent also magnesium silicate (Florisil) are also used because of their advantageous properties in particular instances.

Proteins are frequently stable in their anionic form, and if their isoelectric points are not extremely high they may be chromatographed on an anion exchanger at about pH 7. For groups of weakly dissociated ion exchangers, the degree of dissociation of which changes in the middle of the pH interval, the operation is carried out in the pH region where the groups are completely ionized. This principle, intended to ensure full utilization of the ion-exchange capacity, has often been neglected when working with CM-cellulose and CM-Sephadex, although a finer separation can be achieved by neutralizing proteins than by neutralizing ion exchangers.

The anion exchangers most frequently used for the separation of proteins are those of medium strength containing diethylaminoethyl groups. If bound to a cellulose or polydextran matrix, their working range at lower pH values is limited by the lower stability of the matrix towards the medium at a pH lower than 2.7–3.0, below which level many proteins are no longer sufficiently stable. At higher pH values, the application of anion exchangers is limited by the point at which the dissociation of the anionic group of the exchanger is suppressed.

The stability of ion exchangers with a polyacrylamide matrix is overlapped by the stability interval of proteins, although slight splitting of the amide groups (with a simultaneous increase in the number of carboxyl groups) in polyacrylamide gels in alkaline solutions has been observed. The limitation of the working range of ion exchangers in alkaline conditions concerns anion exchangers with aminoethyl and diethylaminoethyl groups and other weak anion exchangers. The working range of strong anion exchangers, for example those which contain triethylaminoethyl and diethyl-(2-hydroxypropyl)aminoethyl groups, is limited only by the range of protein stability. At pH values higher than 10, irreversible changes are to be expected for most proteins, especially due to changes in the tyrosine present. While for weak cation exchangers with carboxymethyl groups the limitation given by the degree of dissociation of the carboxyl groups applies, medium-strong cation exchangers such as cellulose phosphate have their working range limited only by the stability of the matrix, the same as for strong cation exchangers such as matrices with sulphoethyl or sulphopropyl groups.

Examples of the working pH ranges for polydextran-type ion exchangers, given by the producer (Pharmacia, Uppsala, Sweden), are given in Table 35.2. Similar values can be assumed to be approximately valid for cellulose ion exchangers also, although for the isolation of glycoproteins on DEAE-cellulose for example, even pH 9.9 was used. With cellulose-type ion exchangers, the separation properties improve with decreasing particle

TABLE 35.2
RECOMMENDED pH RANGES FOR ION-EXCHANGE CHROMATOGRAPHY OF PROTEINS ON EXCHANGERS OF THE SEPHADEX TYPE (PHARMACIA, 1969)

| Type of ion exchanger | Recommended pH range |
|---|---|
| QAE-Sephadex | 2–3 to 10 |
| DEAE-Sephadex | 2–3 to 9.5 |
| CM-Sephadex | 6 to 10 |
| SE-Sephadex | 2–3 to 10 |
| SP-Sephadex | 2–3 to 10 |

TABLE 35.3
RECOMMENDED TYPES OF CROSS-LINKING OF POLYDEXTRAN ION EXCHANGERS IN RELATION TO THE MOLECULAR WEIGHT OF THE SEPARATED PROTEINS (PHARMACIA, 1969)

| Molecular weight of proteins | Sephadex ion exchanger | |
|---|---|---|
| | Cross-linking type* | Ionogenic group |
| $\leqslant 3 \cdot 10^4$ | A-25<br>C-25 | DEAE, QAE<br>CM, SE, SP |
| $3 \cdot 10^4 - 2 \cdot 10^5$ | A-50<br>C-50 | DEAE, QAE<br>CM, SE, SP |
| $> 2 \cdot 10^5$ or mixture | A-25<br>C-25 | DEAE, QAE<br>CM, SE, SP |

*A = Anion exchanger, C = cation exchanger

size, and microgranular celluloses give better resolutions than fibrous types, especially if they are kept humid. The matrix is formed by particles that are impenetrable to proteins. For steric reasons, the capacity of cellulose ion exchangers is several times less for macromolecular proteins in comparison with the titratable capacity of ion exchangers for small ions. In a similar manner, the capacity is also changed in ion exchangers with a gel matrix, but the dependence of the capacity on the dimensions of the macromolecules is much more distinct because larger protein molecules can be bound only to those ion exchanger groups which are on the surface of the gel particles. Examples of recommended types of more or less cross-linked polydextran gel ion exchangers in relation to the molecular weights of the separated proteins are given in Table 35.3.

From Table 35.3, it follows that polydextran ion exchangers with a denser matrix have a larger field of applications. The advantage of the higher capacity of a looser matrix of a polydextran ion exchanger for proteins with medium molecular weights is most striking if the capacities of the dry weight of ion exchangers are compared. As the initial ionic strength of the elution buffers is low, the same weight of ion exchangers with a looser matrix (Sephadex A-50 and C-50) of polydextran gels assumes a substantially larger volume. In fact, this decreases to a certain extent the differences in capacity between ion exchangers with a matrix of the Sephadex G-25 and G-50 type. The ratios of the capacities for haemoglobin and ion exchangers derived from the polydextran gels Sephadex G-25 and Sephadex G-50 are given in Tables 35.4 and 35.5.

TABLE 35.4
BINDING CAPACITY OF ION EXCHANGERS OF THE SEPHADEX TYPE FOR PROTEINS IN BUFFERS OF IONIC STRENGTH 0.01 (PHARMACIA, 1970)

| Type of exchanger | Capacity of haemoglobin (w/w in grams) | pH of buffer |
|---|---|---|
| QAE-Sephadex A-25 | 0.3 | 8.0 |
| QAE-Sephadex A-50 | 6 | |
| DEAE-Sephadex A-25 | 0.5 | 8.0 |
| DEAE-Sephadex A-50 | 5 | |
| CM-Sephadex C-25 | 0.4 | 5.0 |
| CM-Sephadex C-50 | 9 | |
| SP-Sephadex C-25 | 0.2 | 5.0 |
| SP-Sephadex C-50 | 7 | |

The capacity of an ion exchanger cannot be utilized to its full extent in separations on columns; in the sorption of proteins, 5–10% of the ion exchanger capacity at most is utilized. The sample should be introduced on to the ion-exchange column in a small volume only when the stabilizing buffer is simultaneously the solvent for the sample and the eluent, without subsequent elution with a gradient of ionic strength and/or pH. As the samples are usually complex mixtures, which are most suitably separated by gradient elution, dilute protein solutions may be applied in a large volume. If the flow-rate is not

*References p.805*

TABLE 35.5
CAPACITY OF DEAE- AND CM-CELLULOSE (WHATMAN) FOR INSULIN AND LYSOZYME (REEVE ANGEL AND CO.)

| Type of ion exchanger | Designation | Capacity (w/w) | | | Remarks |
|---|---|---|---|---|---|
| | | Protein | Amount (g) | pH | |
| DEAE-cellulose | DE-22 | Insulin | 0.75 | 8.5 | free base |
| | DE-23 | | 0.75 | | |
| | DE-32 | | 0.85 | | |
| | DE-52 | | 0.85 | | |
| CM-cellulose | CM-22 | Lysozyme | 0.60 | 5.0 | $Na^+$ |
| | CM-23 | | 0.60 | | |
| | CM-32 | | 1.26 | | |
| | CM-52 | | 1.26 | | |

excessive, the sample remains sorbed on the upper part of the column, independent of the extent of dilution of the sample.

## Elution

Buffers with a sufficient buffering capacity are used for elution so that proteins will not affect the pH of the eluent by their own buffering capacity. The buffering ion in the eluent should, if possible, have the same charge sign as the functional group of the exchanger. Elution of proteins is carried out either by increasing the ionic strength of the buffer stepwise, or by a stepwise change in the pH of the buffer in the direction of the isoelectric point of the protein; continuous gradient elution can also be used with advantage, because the risk of the formation of artificial peaks is less than with a stepwise change in the eluent properties.

Elution without a gradient, where the eluent is identical with the buffer that stabilizes the exchanger and also serves for the dissolution of the sample, can be used for very fine separations of closely related proteins. The separation of various types of haemoglobins, described by Dozy and Huisman, is an example. The establishment of suitable conditions for elution without a gradient is a lengthy task and the requirement of the proximity of the isoelectric point and the pH of the eluent can serve as the only guideline. Optimum ionic strength and pH values for particular separations must be sought empirically. A further disadvantage is that the sample volume decides the peak widths obtained after elution.

For gradient elution, the proteins are either bound strongly to the ion exchanger or they do not form any bonds. In gradient elution, only the values 0 or 1 for the $R_F$ value can be taken into consideration (Porath and Fryklund). Lampson and Tytell observed that a series of proteins is desorbed at a pH that differs from the isoelectric point by 0.4–0.6 pH units at an ionic strength of approximately 0.1 on an exchanger containing cationic carboxymethyl groups, and this relationship may help in the selection of optimum conditions for desorption.

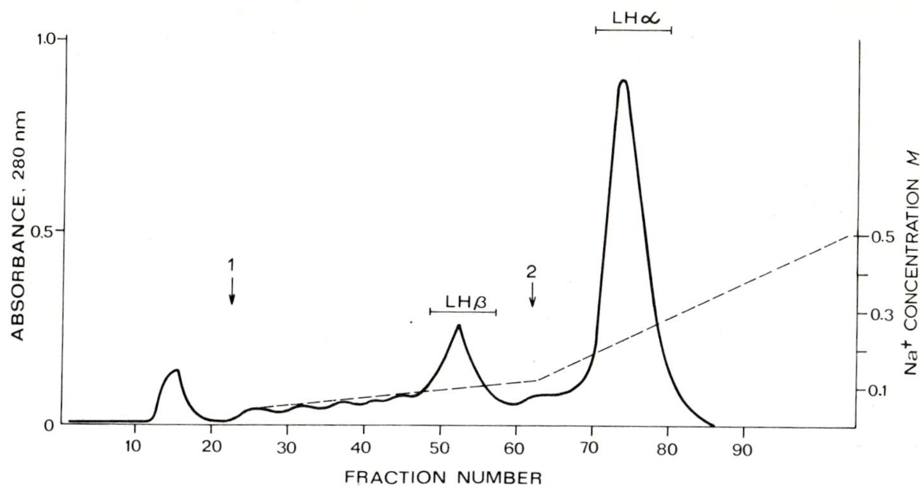


Fig. 35.3. Separation of sub-units of porcine luteinizing hormone by ion-exchange chromatography on SE-Sephadex C-25 in 8 *M* urea (Hennen *et al.*). Column: 10 × 0.9 cm. Buffer: equilibrating buffer– 0.0025 *M* sodium acetate buffer, pH 4.9, in 8 *M* urea; two linear gradients of 0.025–0.10 *M* (150 ml; start: arrow 1) and 0.10–0.50 *M* (100 ml; start: arrow 2) of sodium acetate buffer, both in 8 *M* urea. Operating conditions: flow-rate, 11 ml/h; fraction volume, 3.7 ml; sample, 80 mg of porcine luteinizing hormone. Detection: absorbance at 280 nm (solid line). The gradient was applied after collection of the non-adsorbed fraction, the slope of the $Na^+$ concentration being indicated by the broken line.

An ionic strength gradient or the concentration of ions capable of binding with the functional groups of the exchanger with an opposite charge should always be increasing (Fig. 35.3). For an anion exchanger, the pH gradient decreases, but for a cation exchanger it increases. Most often ionic strength gradients are used. Within broad concentration limits, sodium chloride or potassium chloride gradients are used in the presence of buffer.

The optimum resolution of proteins and peptides can be achieved by decreasing the steepness of the gradient only if the concentration of the eluted protein or peptide is equal to, or greater than, approximately $10^{-3}$ *M*. However, in practice, the concentration of the eluted protein is more commonly $10^{-4}$ *M* or less. In such instances, the quality of the separation apparently improves when a steeper gradient is applied; this was shown by Novotný for the separation of light immunoglobulin chains on SE-Sephadex and QAE-Sephadex.

*Selective elution from ion exchangers*

Sometimes, if the proteins are distinguished in mixtures by their extreme charge, the ionic strength and the pH of the eluting buffer may be set so that the required substance is eluted individually, without being sorbed on the exchanger, while all others are retained by the exchanger. An example of such a procedure, which is also applicable to larger scale work, is the purification of human immunoglobulin, IgG, on a QAE-Sephadex A-50 column, described by Joustra and Lundgren and outlined below. The method involves the use of buffers with a constant ionic strength during the elution and the regeneration. Regeneration is achieved only by a pH change, which, in the case of a strong anion exchanger and under the conditions mentioned, does not cause a change in gel volume. This enables the separation and regeneration process to be repeated many times in the same column.

The serum is freed from β-lipoproteins by the addition of Aerosil (Degussa, Frankfurt am Main, G.F.R.). Two grams of Aerosil 380 are added to 100 ml of serum and the mixture is stirred at room temperature for 4 h. After centrifugation at 12,000 *g* for 30 min, the supernatant is equilibrated by GPC on a Sephadex G-25 column with the elution buffer A (specified below). The serum is then diluted in a 1:2 ratio with elution buffer A and applied on to an ion-exchange column. If the β-lipoproteins are eliminated, a volume can be applied which may be up to the bed volume ($V_t$) of the column. If the sample volume is further increased, transferring appears in the eluate. After the elution of IgG with buffer A, the ion-exchange column is regenerated with buffer B. The eluate of IgG is concentrated 10-fold by ultrafiltration and immediately lyophilized. The yield is approximately 70% of IgG, depending on the type of serum.

The two buffers have the following compositions:

*Buffer A:* ethylenediamine–acetic acid, pH 7.0, ionic strength $I = 0.1$.

Ethylenediamine (2.88 g; distilled under reduced pressure) is dissolved in 73 ml of 1 *M* acetic acid and the volume is made up to 1 litre with distilled water.

*Buffer B:* sodium acetate–acetic acid, pH 4.0, ionic strength $I = 0.1$.

A 435-ml volume of 0.6 *M* acetic acid plus 130 ml of 0.6 *M* sodium acetate are diluted to 1 litre with distilled water.

An example of the separation is as follows. On a 11 × 1.5 cm column of QAE-Sephadex A-50 equilibrated with buffer A, a 10-ml sample of human serum was applied, equilibrated and diluted with buffer A. At a flow-rate of 8 ml/h · cm$^2$, elution of IgG with 65 ml of buffer A was carried out and the column was regenerated with buffer B. The regeneration and the elution of IgG were controlled by UV absorption of the eluate at 254 nm. The first peak contained approximately 0.3% of IgG.

## CHROMATOGRAPHY ON HYDROXYAPATITE AND ON CALCIUM PHOSPHATE

The principle of the separation of proteins on hydroxyapatite (HA) (introduced by Swingle and Tiselius in 1951 and developed by Tiselius *et al.*) and calcium phosphate gel (according to Price and Greenfield) consists in the interaction of negatively and positively charged groups of protein molecules with phosphate and calcium ions in HA, and it is therefore a special case of ion-exchange chromatography of proteins. Either micro-particular crystalline HA of composition $Ca_{10}(PO_4)_6(OH)_2$ or calcium phosphate gel anchored on an inert carrier, for example cellulose, serves as the support.

HA is amphoteric and the isoelectric point of the carrier is strongly affected by the method of preparation; its value changes in the range from 6.5 to 10.2. The binding of positively charged groups on HA is strongly influenced by salts, for example sodium and potassium chlorides, while with negatively charged groups in proteins the salt concentration does not affect the bond strength significantly. In addition, the effect of salt concentration on the sorption of proteins is also a function of their molecular weight. The mechanism of binding, the application of HA chromatography to the separation of proteins and the operating technique for use with HA columns were described in a review by Bernardi. The use of HA chromatography seems to be particularly suitable for basic proteins and polypeptides, among which histones can be easily released by higher salt concentrations.

The separation of proteins, especially enzymes, including a newer method for coating cellulose with calcium phosphate gel, was described by Koike and Hamada. It seems that the separation properties of HA and calcium phosphate are not yet sufficiently appreciated, owing to the polyfunctional character of the bond, originating from the amphoteric properties of the carrier, and from the few results available for the complete characterization of the effect of the molecular weight of the separated proteins on the course of the separation process.

A survey of selected protein separations is presented in Table 35.6

## SOLUBILITY CHROMATOGRAPHY

The principle of solubility chromatography consists in the gradual precipitation of proteins and their elution from a column usually formed by a suitable inert carrier. Single proteins differ substantially in their solubilities in relation to various salt concentrations or ionic strengths, concentrations of organic solvents and detergents. The wide range of protein solubilities permits their separation on the basis of gradual solubilization of the protein precipitate by changes in salt concentration, organic solvents and detergents in suitable solutions; this method is analogous to classical fractional precipitation procedures used for the isolation of proteins. In contrast to batch fractional procedures, the use of columns permits a better separation of fractions because the application of a continuous gradient of the eluent permits the attainment of the optimum eluent concentration, which just suffices for elution but does not cause occlusion of the proteins from the solution. Keil *et al.* proposed a method which they called column gradient extraction of proteins, in which the different solubilities of proteins in the presence of ammonium sulphate are made use of. First, proteins are precipitated from a solution in concentric layers, on to a Kieselguhr carrier (Hyflo Supercel) using an increasing concentration of ammonium sulphate. When the proteins have been precipitated, the carrier with the layer of protein is introduced into a tube and in the second phase of the operation a solution of ammonium sulphate of decreasing concentration is pumped into the column. Elution of proteins takes place gradually down to zero concentration of the salt. Keil *et al.* demonstrated the efficiency of the method with several examples of the isolation of proteins. Fig. 35.4 illustrates the separation of the proteins of equine serum.

Saffran *et al.* extracted hypophysis tissue on a column with a concentration gradient of solvents of increasing polarity, in the system ethanol–water–acetic acid, and achieved an appreciable enrichment of the fraction with corticotropic activity (ACTH). Diatomaceous earth was used as the carrier because it prevents an excessive increase in the hydrodynamic resistance of the column during gradual swelling of the tissue. The method of Saffran *et al.* is particularly suitable when a minor component of tissue preparations, such as acetone-dried residues, has to be isolated. Its advantage is that the eluent can be eliminated completely by evaporation.

*References p.805*

TABLE 35.6
SELECTED PROTEIN SEPARATIONS ON ION EXCHANGERS

| Protein | Source | Ion exchanger | Equilibration buffer | Sorption–desorption region | Note | Reference |
|---|---|---|---|---|---|---|
| *Hormones* | | | | | | |
| γ-Lipotropic hormone | Bovine pituitary gland | CM-cellulose | 0.01 *M* Ammonium acetate, pH 4.6 | (a) Equilibration buffer<br>(b) 0.2 *M* Ammonium acetate, pH 6.7 | Gradient<br>Re-chromatography | Gráf *et al.* |
| | | SE-Sephadex C-25 | 0.01 *M* Ammonium formate, pH 3.5 | (a) 0.01 *M* Ammonium formate pH 3.5<br>(b) 0.05 *M* Ammonium acetate, pH 5.5 | Gradient | |
| α-MSH (melanocyte stimulating hormone)<br>β-MSH<br>ACTH (adrenocorticotropic hormone) | Porcine and bovine pituitary gland | CM-cellulose | 0.01 *M* Ammonium acetate, pH 4.6 | (a) Equilibration buffer<br>(b) 0.2 *M* Ammonium acetate, pH 6.7<br>Elution order α-MSH, β-MSH, ACTH | Continuous gradient | Hussa *et al.* |
| TCT (thyrocalcitonin) | Porcine thyroid gland | CM-cellulose | 0.2 *M* Pyridine acetate, pH 4.0 | (a) Equilibration buffer<br>(b) 1.1 *M* Pyridine acetate, pH 4.0 | Linear gradient | Hawker *et al.* |
| TCT | Porcine | CM-cellulose | 0.062 *M* Pyridine acetate, pH 4.4 | (a) Equilibration buffer<br>(b) 1.86 *M* Pyridine acetate, pH 5.7 | Linear gradient | Bell *et al.* |
| FSH (follicle stimulating hormone) | Human pituitary gland | DEAE-Sephadex A-50 | 0.1 *M* phosphate, pH 6.8 | Equilibration buffer | Without gradient | Peckham and Parlow |

| | | | | | | |
|---|---|---|---|---|---|---|
| FSH<br>LH + TSH (luteinizing and thyroid stimulating hormone) | Human pituitary gland | CM-Sephadex C-50 | 0.004 *M* Ammonium acetate, pH 5.5 | (a) Equilibration buffer<br>(b) 0.1 *M* Ammonium acetate, pH 6.7 (elution of FSH)<br>(c) 0.1 *M* Ammonium acetate, pH 9.5 | | Rathnam and Saxena |
| TSH | Bovine pituitary gland | DEAE-cellulose | 0.005 *M* Sodium glycinate, pH 9.5 | (a) Equilibration buffer<br>(b) 0.005 *M* Sodium glycinate, pH 9.5, containing 0.02 *M* NaCl<br>(c) As (b) containing 0.5 *M* NaCl | Linear gradient | Lindsay *et al.* |
| β-Lipotropic hormone | Bovine | CM-cellulose | 0.01 *M* Ammonium acetate, pH 4.6 | (a) Equilibration buffer<br>(b) 0.1 *M* Ammonium acetate, pH 6.8<br>(c) 0.2 *M* Ammonium acetate, pH 6.8 | Continuous gradient | Lohmar and Li |
| Light-adapting hormone | Crustacean eyestalks, *Pandalus borealis* | CM-Sephadex C-25 | 0.075 *M* Ammonium acetate, pH 4.92, containing $2 \cdot 10^{-4}$ volumes of thiodiglycol | (a) Equilibration buffer<br>(b) 0.1 *M* Ammonium acetate, pH 9.51 | | Fernlund |
| | | | 0.02 *M* $NH_4HCO_3$ | Equilibration buffer | Re-chromatography without gradient | |
| Red pigment concentrating hormone | *Pandalus borealis* | Dowex 50W-X2 | 0.1 *M* Triethylamine formate equilibrated to pH 3.7 | (a) Equilibration buffer<br>(b) 0.1 *M* Triethylamine acetate, pH 5.5<br>(c) 0.1 *M* Triethylamine, pH 11.4 | Stepwise | Josefsson |

(Continued on p. 792)

TABLE 35.6 *(continued)*

| Protein | Source | Ion exchanger | Equilibration buffer | Sorption–desorption region | Note | Reference |
|---|---|---|---|---|---|---|
| Prolactin | Human pituitary gland | DEAE-cellulose | 0.012 *M* Tris–HCl, pH 7.6 | (a) Equilibration buffer<br>(b) As (a), in 0.05 *M* NaCl<br>(c) As (a), in 0.5 *M* NaCl | Stepwise | Hwang *et al.* |
| | | CM-cellulose | 0.01 *M* Ammonium acetate, pH 5.6 | (a) Equilibration buffer<br>(b) As (a), in 0.2 *M* NaCl<br>(c) As (a), in 0.5 *M* NaCl | Prolactin eluted at 0.2 *M* NaCl | |
| *Blood proteins* | | | | | | |
| Fibrinogen | Human blood platelets | DEAE-cellulose | 0.05 *M* Tris–HCl, pH 7.2 | (a) Equilibration buffer<br>(b) Equilibration buffer containing 0.5 *M* NaCl | Linear gradient | Ganguly |
| Fibrinogen | Human | DEAE-cellulose | 0.005 *M* Tris–phosphate, pH 8.6 | (a) Equilibration buffer<br>(b) 0.02 *M* Tris–phosphate, pH 4.3<br>(c)–(h) Increasing molarity to 0.5 *M* Tris–phosphate, decreasing pH to 4.1 | Seven-chamber mixer, Tris–phosphate and pH gradient | Mosesson *et al.* |
| Prothrombin | Bovine | DEAE-Sephadex A-50 | 0.1 *M* sodium phosphate, pH 6.0 | (a) Equilibration buffer<br>(b) Equilibration buffer in 1.0 *M* NaCl | Linear gradient | Ingwall and Scheraga |
| Plasminogen | Human | DEAE-Sephadex A-50 | 0.02 *M* Sodium phosphate, pH 8.0, containing 0.04 *M* NaCl | (a) Equilibration buffer<br>(b) Equilibration buffer in 0.05 *M* $\epsilon$-aminocaproic acid | | Wallen and Wiman |
| Plasminogen | Bovine | CM-Sephadex C-50 | 0.1 *M* Sodium acetate, pH 5.0, containing 0.3 *M* NaCl | (a) Equilibration buffer<br>(b) Equilibration buffer in 0.6 *M* NaCl | Linear gradient of NaCl | Nagasawa and Suzuki |

| | | | | | | |
|---|---|---|---|---|---|---|
| *Metalloproteins* | | | | | | |
| Haemocyanin | Decapod crustaceans: *Callinectes sapidus*, *Libinia emarginata*, *Gecarcinus lateralis* | DEAE-cellulose | 0.05 *M* Ammonium hydrogen carbonate, pH 7.8 | (a) Equilibration buffer<br>(b) Equilibration buffer containing 0.5 *M* NaCl, 0.002 *M* $MgCl_2$ | Concave gradient | Kerr |
| Stellacyanin | *Rhus vernicifera* | Amberlite CG-50 | 0.05 *M* potassium phosphate, pH 5.5 | (a) Equilibration buffer<br>(b) 0.2 *M* $KH_2PO_4$ | Stepwise | Peisaca *et al.* |
| Umecyanin | Horse radish | CM-cellulose | 0.004 *M* Sodium acetate, pH 4.7 | (a) Equilibration buffer<br>(b) 0.03 *M* Sodium acetate, pH 5.7 | Re-chromatographed stepwise | Paul and Stigbrand |
| | | DEAE-cellulose | 0.05 *M* Sodium phosphate, pH 5.5 | | | |
| Myoglobin | Mollusc, *Busycon caricum* | DEAE-Sephadex A-50 | 0.04 *M* $NH_4HCO_3$ | Equilibration buffer | | Read |
| | | SE-Sephadex C-50 | Tris-cacodylate, pH 6.5, ionic strength $I = 0.1$ | Equilibration buffer | Re-chromatographed | |
| Cytochrome $c_{555}$ | *Crithidia fasciculata* | Amberlite CG-50 | 0.01 *M* Sodium phosphate, pH 8.0 | (a) 0.01 *M* Sodium phosphate, pH 8.0<br>(b) As (a), containing 0.4 *M* NaCl | | Kusel *et al.* |
| | | | 0.01 *M* Sodium phosphate, pH 8.0, containing 0.01 *M* $K_3Fe(CN)_6$ | (a) 0.01 *M* Sodium phosphate, pH 8.0 | Re-chromatographed Linear gradient $Na^+$ | |
| | | | | (b) As (a), containing 0.4 *M* NaCl | Cytochrome eluted at 0.19 *M* $Na^+$ | |

*(Continued on p. 794)*

TABLE 35.6 *(continued)*

| Protein | Source | Ion exchanger | Equilibration buffer | Sorption–desorption region | Note | Reference |
|---|---|---|---|---|---|---|
| Cytochrome *c* | *Spirillum itersonii* | DEAE-cellulose | 0.003 *M* Tris–HCl, pH 8.7 | Equilibration buffer | Cytochrome not adsorbed | Clark-Walker and Lascelles |
| | | CM-cellulose | 0.002 *M* Potassium phosphate, pH 6.2 | (a) Equilibration buffer<br>(b) 0.05 *M* Potassium phosphate, pH 6.2 | Re-chromatographed stepwise | |
| Cytochrome *c* | Horse heart | CM-cellulose | 0.05 *M* $Na_2HPO_4$ containing 0.05 *M* $NaH_2PO_4$, pH 6.75 | Equilibration buffer | Without gradient | Dixon and Thompson |
| Cytochrome $b_5$ | Rabbit liver | DEAE-Sephadex A-50 | 0.02 *M* Tris–HCl, containing 0.5% Triton X-100, 0.002 *M* EDTA, 0.1 *M* KCl | (a) Equilibration buffer<br>(b) Equilibration buffer in 0.3 *M* KCl | Linear gradient cytochrome $b_5$ eluted at 0.25 *M* KCl | Ito and Sato |
| Transferrin | Human | DEAE-Sephadex A-50 | 0.05 *M* Tris–HCl, pH 8.0 | (a) Equilibration buffer<br>(b) 0.07 *M* Tris–HCl, pH 8.0<br>(c) 0.15 *M* Tris–HCl, pH 8.0 | | Aisen *et al.* |
| Haemoglobin | Human | DEAE-cellulose or DEAE-Sephadex | 0.05 *M* Tris–HCl, pH 8.4–7.2 for different Hb types | 0.05 *M* Tris–HCl, pH 8.3–7.1 | Stepwise pH gradients | Horton and Chernoff |
| Haemoglobin | *Ascaris lumbricoides* body walls | DEAE-cellulose | 0.001 *M* Potassium phosphate, pH 7.0 | (a) Equilibration buffer<br>(b) 0.03 *M* Potassium phosphate, pH 7.0<br>(c) 0.3 *M* Potassium phosphate, pH 7.0 | Two linear gradients | Okazaki *et al.* |

| | | | | | | |
|---|---|---|---|---|---|---|
| *Histones* | | | | | | |
| Histones | Calf thymus | CM-cellulose | 0.04 *M* Sodium acetate, pH 4.6, in 3.7 *M* urea | (a) Equilibration buffer<br>(b)–(i) Increasing molarity of sodium acetate, 0.1–0.6 *M* and increasing pH, 5.7–8.6 | Stepwise | Yang |
| | | | 0.005 *M* Potassium phosphate, pH 6.8, containing 3.7 *M* urea | (a) Equilibration buffer<br>(b)–(h) Equilibration buffer containing NaCl of increasing molarity, 0.15–0.60 *M*<br>(i) 0.05 *N* HCl<br>(j) 0.10 *N* HCl | | |
| Histone | HeLa cells | Amberlite CG-50 | 0.1 *M* Sodium phosphate, pH 6.8, in 8% guanidine hydrochloride | (a) Equilibration buffer<br>(b) 0.1 *M* Sodium phosphate, pH 6.8, in 8.5% guanidine hydrochloride<br>(c) 0.1 *M* Sodium phosphate, pH 6.8, in 13% guanidine hydrochloride (elution of lysine-rich histones) | | Sadgopal and Bonner |
| | | | | (d) 0.1 *M* Sodium phosphate, pH 6.8, in 40% guanidine hydrochloride (elution of arginine-rich histones) | Non-linear gradient | |
| Histone | Calf thymocytes | Amberlite CG-50 | 0.1 *M* Sodium phosphate, pH 6.8, in 8% guanidine hydrochloride | (a) Equilibration buffer<br>(b) 0.1 *M* Sodium phosphate, pH 6.8, containing 8% guanidine hydrochloride<br>(c) 0.1 *M* Sodium phosphate, pH 6.8, containing 13% guanidine hydrochloride<br>(d) 0.1 *M* Sodium phosphate, pH 6.8, containing 40% guanidine hydrochloride | Linear gradient, 8–13% guanidine hydrochloride<br>Linear gradient, 13–40% guanidine hydrochloride | Pallotta and Berlowitz |

*(Continued on p. 796)*

TABLE 35.6 *(continued)*

| Protein | Source | Ion exchanger | Equilibration buffer | Sorption–desorption region | Note | Reference |
|---|---|---|---|---|---|---|
| *Inhibitors* | | | | | | |
| Trypsin inhibitor | Potato | DEAE-cellulose | 0.02 *M* Sodium borate, pH 8.3 | (a) Equilibration buffer<br>(b) Equilibration buffer in 0.2 *M* NaCl | Linear gradient | Hochstrasser *et al.* |
| | | Cellulose phosphate | 0.02 *M* Sodium borate, pH 7.0 or 6.7 | Equilibration buffer | Re-chromatographed | |
| Trypsin inhibitor | Soya bean | DEAE-cellulose | 0.05 *M* Ammonium acetate, pH 5.0 | (a) Equilibration buffer<br>(b) 0.5 *M* Ammonium acetate, pH 6.5 | Gradient produced by rectangular Varigrad | Frattali and Steiner |
| Trypsin inhibitor B | Cow colostrum | DEAE-cellulose | 0.02 *M* Sodium phosphate, pH 7.2 | (a) Equilibration buffer<br>(b) Equilibration buffer in 0.01 *M* NaCl<br>(c) Equilibration buffer in 0.02 *M* NaCl | Two linear gradients | Čechová *et al.* |
| Antiplasmin | Human plasma $\alpha_2$-macroglobulin | DEAE-Sephadex A-50 | 0.02 *M* Tris–HCl in 0.1 *M* NaCl, pH 7.7 | (a) Equilibration buffer<br>(b) 0.02 *M* Tris–HCl, pH 7.7, in NaCl of increasing molarity | Linear gradient of NaCl | Iwamoto and Abiko |
| *Toxins* | | | | | | |
| Viscotoxins | Mistletoe, *Viscum album* L. | SE-Sephadex C-25 | 0.067 *M* Phosphate, pH 5.0 | (a) 0.033 *M* $KH_2PO_4$ (sample)<br>(b) 0.067 *M* Phosphate, pH 5.0<br>(c) 0.067 *M* Phosphate, pH 8.0, NaCl to 0.125 *M* ($Na^+$) | Linear gradient | Samuelsson and Pettersson |

| | | | | | | |
|---|---|---|---|---|---|---|
| Neurotoxin | Scorpion venoms | | | | Without gradient | Miranda *et al.* |
| | *Andronoctus australis* | DEAE-Sephadex A-25 | 0.1 *M* Ammonium acetate, pH 8.5 | Equilibration buffer | | |
| | *Buthus occitanus* | Amberlite CG-50 | 0.2 *M* Ammonium acetate, pH 6.7, 6.3 or 6.15 | 0.2 *M* Ammonium acetate, pH 6.7, 6.3 or 6.15 | | |
| | *Leiurus quinques-triatus* | CM-Sephadex C-50 | 0.2 *M* Ammonium acetate, pH 6.7 | 0.02 *M* Ammonium acetate, pH 6.7 | | |
| Anaphylatoxin | Hog serum | CM-Sephadex C-50 | 0.02 *M* Sodium acetate, pH 5.6 | (a) 0.02 *M* Ammonium formate, pH 5.5<br>(b) 0.5 *M* Ammonium formate, pH 6.5 | Linear gradient of ammonium formate | Vogt |
| A-toxin | *Clostridium botulinum* | DEAE-Sephadex A-50 | 0.15 *M* Tris–HCl, pH 8.0 | (a) Equilibration buffer<br>(b) 0.15 *M* Tris–HCl, pH 8.0, in 0.3 *M* NaCl | Linear gradient of $Cl^-$ concentration | Dasgupta *et al.* |
| B-toxin | *Clostridium botulinum* | DEAE-cellulose | 0.07 *M* Tris–HCl, pH 8.0 | (a) Equilibration buffer<br>(b) 0.07 *M* Tris–HCl, pH 8.0, in 0.5 *M* NaCl | Linear gradient, toxin eluted at 0.14 *M* $Cl^-$ | Beers and Reich |
| E-toxin | *Clostridium botulinum* | CM-Sephadex C-50 | 0.02 *M* Sodium acetate, pH 6.0 | (a) Equilibration buffer<br>(b) 0.02 *M* Sodium acetate, pH 6.0, in 0.5 *M* NaCl | Linear gradient | Kitamura *et al.* |

*References p.805*

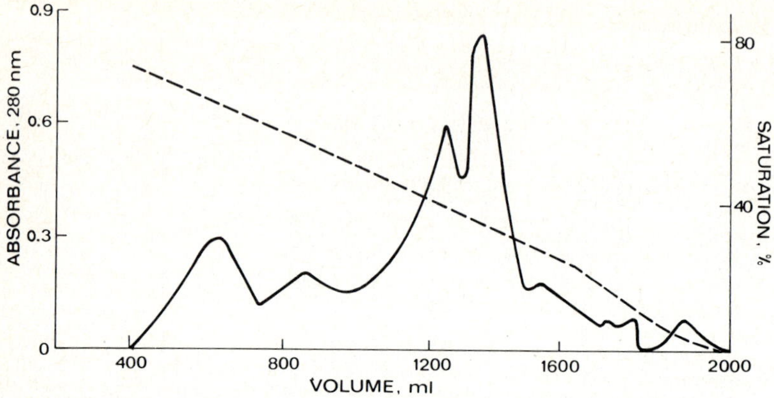


Fig. 35.4. Separation of horse serum by solubility chromatography (Keil *et al.*). Column: 22 × 2.7 cm. Sorbent: Kieselguhr (Hyflo Supercel). Eluent: gradient of ammonium sulphate in water. Operating conditions: flow-rate, 44 ml/h; fraction volume, 22 ml; sample, 50 ml of horse serum; mixer diameter, 10 cm; height of solution, 12.5 cm, containing ammonium sulphate of 85% saturation; reservoir diameter, 10 cm; height, 15.5 cm, containing water. Solid line, absorbance at 280 nm; broken line, percentage of saturation of ammonium sulphate gradient.

## TECHNIQUE OF GEL PERMEATION CHROMATOGRAPHY IN A DETERGENT GRADIENT

If a constant concentration of detergent in the eluent is used for the GPC of proteins solubilized by the detergent, then the ratio of detergent to protein is usually higher than necessary. The effect of the detergent manifests itself in reversible and often also irreversible changes. The ratio of detergent to protein is critical from the point of view of the preservation of the original protein properties and therefore it is advisable to keep the lowest practical detergent concentration during the separation. The condition of minimum detergent concentration is fulfilled if the proteins are separated on a column in a concentration gradient of the detergent. The procedure according to Swanljung was used for the purification of the crude ATPase concentrate. The separation has four steps: in the first, a Sepharose 6B column is stabilised with a detergent-free buffer, in the second the column is washed with a buffer in which the detergent concentration is gradually increased, in the third the sample is applied on to the column dissolved in the buffer the concentration of which corresponds to the highest detergent concentration in the column, and in the fourth the column is washed with a buffer of constant (highest) detergent concentration. If water-insoluble proteins attain the lowest critical value of detergent concentration in the column, *i.e.*, if the volume of the buffer with the detergent gradient is not much less than that of the column, then precipitation occurs and the proteins move further with the lowest detergent concentration at which they enter the solution. In a similar manner to ATPase, a series of flavoproteins and the $b$–$c_1$ cytochrome complex were also separated. This method of separation is an inverted variant of zone precipitation gel chromatography described by Porath. In this case also a concentration gradient is formed, which moves slowly down the column. Faster moving proteins begin to precipitate at the critical ammonium sulphate concentration and then move with the same speed as the gradient.

## AFFINITY CHROMATOGRAPHY

The principle of affinity chromatography was defined by Cuatrecasas (1972) as a method by which a substance or a group of substances is separated specifically from a mixture on the basis of the affinity given by the biological bonding function of macromolecules, and the technique was rapidly accepted for use in protein separations. This promising selective separation method is described in general terms in the chapter on affinity chromatography (Chapter 7). In this chapter, the possibilities for the separation of proteins with a non-enzymatic character will only be briefly mentioned as the isolation of enzymes is described in Chapter 36.

In addition to the purification of a series of enzymes and their inhibitors or substrates, which may be alternatively bound on the matrix, the purification of antibodies and antigens, hormone-transport and vitamin-binding proteins, and group purifications of proteins that contain a glycidic component, also belong to the group of separations by affinity chromatography. On the border of the definition of affinity chromatography is the method of isolation of SH-proteins on agarose–organomercury derivatives. The fixation of antibodies offers wide possibilities, and of the carriers available agarose was found to be the most suitable. In contrast to the commonly used method of the reaction of activated agarose with proteins, it is advantageous if the pH value is shifted to lower values during the preparation of the carrier with bound antibodies; the optimum pH is 6–7. Cuatrecasas (1970) found that at this pH the capacity of the adsorbent is one order of magnitude higher than when the reaction takes place in a very alkaline medium, although the amount of the bound antibodies remains unchanged. Agarose with bonded antibodies against human somato-mammotropin (HCS) was used by Weintraub for the preparation of HCS, labelled and non-labelled, for radioimmunoassays. The same sorbent was used by Guyda and Friesen for the elimination of 99% of growth hormone activity from the homogenate of simian hypophyses.

An example of group isolation is the isolation of glycoproteins according to Aspberg and Porath. Glycoproteins undergo binding with concanavalin A and agarose is used as the support, and today, commercial preparations of agarose-bound concanavalin A are available for affinity chromatography. The binding site of concanavalin A is specific for α-D-glucosyl, α-D-mannosyl and other sterically similar residues that are usually present in glycoproteins. Allan *et al.* isolated, by affinity chromatography on concanavalin A–sorbent, glycoprotein receptors of concanavalin from the membranes of porcine lymphocytes of plasma in a medium containing sodium deoxycholate. Instead of the preparation of adsorbent according to Allan *et al.*, it is also possible to use commercially available preparations, for example Con A–Sepharose, (Pharmacia, Uppsala, Sweden) and Glycosylex A (Miles-Seravac, Lausanne, Switzerland).

The separation of glycoproteinic receptors of concanavalin A according to Allan *et al.* is carried out as follows. Porcine lymphocyte plasma membranes (20 mg of protein) are extracted with 5 ml of 1% sodium deoxycholate and the soluble fraction, forming approximately 85% of the membrane protein, is washed through the adsorbent column and eluted with 1% deoxycholate until the extinction at 280 nm reassumes the value of the blank. Approximately 75% of the added proteins are eluted. The receptor glycoproteins (5% of the content of membrane proteins) are eluted with a solution of methyl-α-D-gluco-

*References p.805*

pyranoside (20 mg/ml in 1% deoxycholate). The eluted material is precipitated by addition of one tenth of the volume of 2% acetic acid and the precipitate is extracted three times with 10-ml portions of 95% ethanol in order to eliminate the sugar and deoxycholic acid. The preparation, dried in a current of air, contains five components of glycoproteins that can be resolved by polyacrylamide gel electrophoresis in 0.1% sodium dodecylsulphate.

## DETECTION OF PROTEINS IN THE EFFLUENT

### Colorimetric detection

The modification by Lowry *et al.* of the colorimetric determination of proteins with the phenolic reagent described by Folin and Ciocalteau and the biuret reaction is a sensitive and universal colorimetric detection method that is very suitable for following the protein concentration in the effluent; it permits the determination of down to 5–100 $\mu$g of protein in 1 ml. The blue-violet coloration formed is sufficiently stable and the reaction can be carried out not only on single fractions but also in a continuous flow-through system with adjusted reagent concentrations. The detection method of Lowry *et al.* is about 100 times more sensitive than the biuret reaction, and the ninhydrin colorimetric method also does not attain the sensitivity of the method of Lowry *et al.* without previous alkaline hydrolysis of the sample. Thus, in a 0.05-ml micro-cell an amount of less than 0.02 $\mu$g can be determined. The colour intensity depends on the tyrosine content, as in the original Folin reaction. The method is empirical and for quantitative determinations calibration with a protein, the concentration of which is determined by some other method, is necessary. The calibration is carried out in the same system as that used for the actual determination.

The colorimetry of soluble proteins according to Lowry *et al.* is carried out as follows. A 1-ml volume of protein solution and 5 ml of solution C are mixed and allowed to stand at room temperature for at least 10 min. Solution D (0.5 ml) is added rapidly and the mixture stirred immediately (within 1 to 2 sec). After standing for 30 min, colorimetry is carried out either at a wavelength of 500 nm if the protein concentration is high, or at 750 nm if the concentration is low. Bovine or human serum albumin at concentrations from 0.02 to 0.5 mg/ml can be used as a standard. Solution A is 2% of sodium carbonate in 0.1 *N* sodium hydroxide; B is 0.5% of copper(II) sulphate ($CuSO_4 \cdot 5\,H_2O$) in 1% sodium tartrate (or 1% sodium potassium tartrate); C is 50 ml of A plus 1 ml of B; and D is 1 ml of concentrated Folin–Ciocalteau reagent (diluted to a concentration corresponding to 1 *N* acid).

The phenol reagent is prepared according to Folin and Ciocalteau as follows. Ammonium molybdate ($(NH_4)_2MoO_4 \cdot 2H_2O$; 25 g) and 100 g of sodium tungstate ($Na_2WO_4 \cdot 2H_2O$) are mixed with 700 ml of water in a 1500-ml flask, 50 ml of 85% phosphoric acid and 100 ml of concentrated hydrochloric acid are added and the mixture is refluxed for 10 h. Then 150 g of lithium sulphate are added to the boiling solution, followed by a few drops of bromine solution and 50 ml of water. The mixture is boiled for a further 15 min without a reflux condenser. When all of the excess of bromine has been boiled off, the reaction mixture is cooled, made up to 1 l and filtered. The filtrate should not be greenish, as interference in the colorimetry will result.

In the biuret reaction, on reaction of $Cu^{2+}$ with the –NH–CO– group, which is characteristic of proteins and peptides, a complex is formed the absorption maximum of which is about 555 nm. The colour intensity is not affected by ammonium salts and for all proteins it is dependent only on the number of peptide groups. The biuret method is suitable for the determination of higher concentrations of proteins (0.25–25 mg/ml). The colour yield is standardized with pure protein. The biuret reaction can also be carried out in a continuous flow-through system. The procedure is as follows: 0.1–4 ml protein solution containing 1–5 mg of protein is made up to 5 ml with solution A, then 5 ml of solution B are added and the mixture is heated on a water-bath at 32°C for 30 min, the resulting blue-violet coloration being measured at 555 nm. Solution A is 0.85% sodium chloride solution. Solution B comprises 45 g of sodium potassium tartrate, 15 g of copper(II) sulphate ($CuSO_4 \cdot 5H_2O$), 5 g of potassium iodide, and 0.2 *M* sodium hydroxide solution (without carbonate), prepared as follows. After the dissolution of the sodium potassium tartrate, copper(II) sulphate is added while stirring until it is dissolved. Finally, potassium iodide is added and the mixture is made up to 1 l with 0.2 *M* sodium hydroxide solution. Sodium hydroxide solution without carbonate can be prepared by heating 50% sodium hydroxide solution at 90°C for 24 h and, after sedimention of sodium carbonate, diluting the clear solution with boiled water.

## Spectrophotometric detection

The principle of the spectrophotometric detection of protein and peptide solutions consists in the measurement of the absorbance of the solutions in the region of the absorption maximum. Most proteins contain tyrosine, phenylalanine and, at lower concentrations, also tryptophan, in the form of amino acid residues. The absorption maxima of these residues appear in the short-wave region of the UV spectrum, and absorptions characteristic of the peptide bond, the helical structure of the peptide chain, disulphide bonds of cystine residues, etc., also appear in the same part of the spectrum. For practical determinations of the concentration of proteins, the absorption maxima of tyrosine and tryptophan are the most important. The relatively broad maximum enables the determination of the absorbance at a wavelength of 275–280 nm. Above 280 nm, substantial changes in absorbance may take place with proteins as a result of the effect of pH, because the ionization of the tyrosine hydroxyl group in strongly alkaline conditions causes both a change in the extinction coefficient and a shift of the absorption maximum to higher wavelengths. The changes in the extinction coefficient of tyrosine as a function of pH are evident from Table 35.7.

The shift in the absorption maximum of tryptophan in the same pH interval is substantially less and so is the change of the absolute molar extinction coefficient $\epsilon$. For 0.1 *N* hydrochloric acid, the $\lambda_{max.}$ of tryptophan is 278 nm ($\epsilon$ = 5450), while for 0.1 *N* sodium hydroxide solution $\lambda_{max.}$ is 280.5 nm ($\epsilon$ = 5250). In the direction of shorter wavelengths, a minimum appears in the spectrum of proteins, followed by a steep increase in absorbance. If an eluent with a low absorbance is used, the sensitivity of detection may be increased several times when measuring below 235 nm in comparison with measurement at 280 nm. Measurements at these wavelengths can be made on spectrophotometers of average quality. An example is shown in Fig. 35.5.

*References p.805*

TABLE 35.7
VARIATION OF THE MOLAR EXTINCTION COEFFICIENT (ϵ) OF TYROSINE WITH pH

| ϵ | pH of medium | $\lambda_{max.}$ (nm) |
|---|---|---|
| 1500 | 1.09* | 277.7 |
| 1300 | 8.0 | 275 |
| 2600 | 12 | 293 |

*0.1 *N* HCl

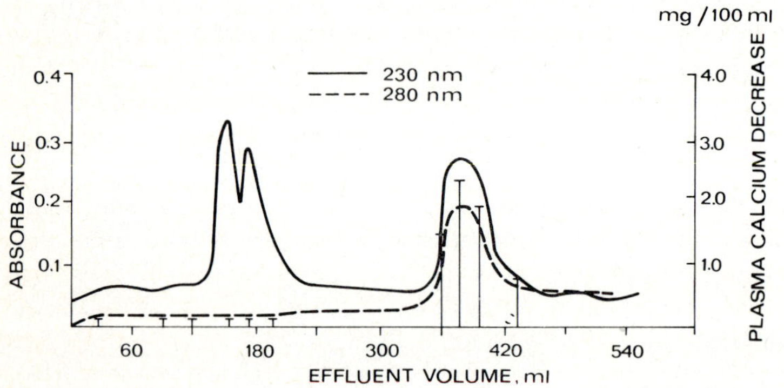


Fig. 35.5. Separation of thyrocalcitonin concentrate by partition chromatography on Sephadex G-25 (Hawker *et al.*). Example of differences in absorbance at 280 and 230 nm. Vertical lines are the bioassay responses. Eluent: *n*-butanol–acetic acid–water (7:1:9).

At 210–220 nm, the detection of proteins is very specific and sensitive. However, buffers in which the components contain carbonyl groups are unsuitable for use in the 210–220 nm region and the choice of components for the buffered eluents is limited. For the 192–194 nm region, in which the presence of substances that contain peptide bonds can be determined specifically, only aqueous solutions of alkali metal fluorides can be used. The importance of the spectrophotometric determination of proteins at wavelengths below 210 nm is limited to special cases of GPC. The sensitivity of the spectrophotometric determination at 210 nm is comparable with the sensitivity of the colorimetric detection of Lowry *et al.* Tombs *et al.* mentioned a sensitivity of 2 μg/ml for serum proteins. The absorbance is due mainly to the peptide bond and therefore proteins have a similar extinction coefficient (see Table 35.8).

In the presence of UV-absorbing impurities which interfere at 260–280 nm, *i.e.*, substances with a typical spectrum of nucleic acids, the absorbance at 220 nm is almost identical with that in the colorimetric determination of proteins according to Lowry *et al.* Wrigley and Webster have shown that at 220 nm, succinate, phthalate and barbiturate buffers cannot be used, while sodium hydroxide solution, acetate, glycine and Tris buffers can be used up to a concentration of 0.01 *M*. Sodium chloride solution, cacodylate, borate, phosphate and ammonium sulphate can be used even at concentrations above 0.1 *M*.

For proteins that contain, in addition to aromatic amino acids, further groups with characteristic absorption maximum bands in the long-wave UV region or bands in the

TABLE 35.8
COMPARISON OF EXTINCTION COEFFICIENTS ($E^{1\%}_{1\,cm}$) OF PROTEINS AT WAVELENGTHS OF 280 AND 210 nm (TOMBS *et al.*)

| Protein | $E^{1\%}_{1\,cm}$ | |
|---|---|---|
| | 280 nm | 210 nm |
| Human serum albumin | 6 | 203 |
| Human immunoglobulin | 15 | 213 |
| Bovine serum albumin | 6.8 | 204 |
| Human siderophilin | 14 | 200 |

visible region are utilized for detection. The requirement is that the groups with distinct light absorption should remain firmly bound to apoprotein in the course of the separation process. The measurement of absorbance in the long-wave UV region is used for the sensitive detection of some enzymes with firmly bound prosthetic groups.

In the visible region, some types of metalloproteins may be detected by absorption spectrophotometry, especially those which contain complexed iron, copper, chromium, vanadium, etc. Characteristic absorption maxima of some metalloproteins are listed in Table 35.9.

TABLE 35.9
CHARACTERISTIC ABSORPTION MAXIMA OF SOME METALLOPROTEINS IN THE VISIBLE REGION OF THE SPECTRUM

| Metalloproteins | $\lambda_{max.}$ (nm) |
|---|---|
| Haemoproteins | 412 |
| Haemoglobin (human) | 415 |
| Haemovanadin | 425 |
| Cytochromes | 550–560 |
| Haemocyanins | 563–580 |
| Caeruloplasmin | 605–610 |
| Plastocyanin | 597 |
| Erythrocuprein | 655 |

Extinction coefficients and wavelengths of maximum absorption are dependent on the oxidation state of the metal. The ratio of absorbance in the visible and the UV regions, *i.e.*, $A_{visible}/A_{280}$, is used as a criterion for the purity of metalloproteins. If the solutions are turbid, the true value of the optical density at 280 nm may be determined by means of optical density values in the visible region of the spectrum, by extrapolating the plot of log O.D. *versus* log λ. The contribution of light scattering may represent 50–60% of the effective optical density at 280 nm.

## Detection by ultraviolet fluorescence

The principle of this method consists in the measurement of the ultraviolet radiation emitted at 340–350 nm due to the excitation of tryptophan or tyrosine in the protein

*References p.805*

molecule by radiation of a shorter wavelength (about 280–290 nm). The greatest part of the emission is due to the tryptophan residue, in relation to the molecular structure and the ionization state of the protein. The fluorescence of the phenylalanine residue is less important, especially in the presence of tyrosine and tryptophan in the protein molecule. The sensitivity of the detection by fluorescence is approximately 1000 times higher than that based on absorption at 280 nm; however, the method is much more demanding with respect to the purity and properties of the eluent. A complication in the application of the method arises as a result of individual changes in the dependence of the fluorescence intensity on pH in various types of proteins.

Another method of making use of the fluorescence measurements and thus increasing the sensitivity of detection independently of the presence of tryptophan or tyrosine and phenylalanine in the protein consists in labelling the proteins with fluorescent reagents, such as fluorescein isothiocyanate or 5-dimethylaminonaphthalene-1-sulphonyl chloride (Dns chloride). This method was used for labelling antibodies, for example, by Goldstein *et al.* and Rinderknecht. Antibodies labelled with a fluorescent group can be further utilized for the detection of antigens.

In all methods in which UV light or UV fluorescence is used for measurements, it should be borne in mind that short-wave UV radiation, if sufficiently intense, may damage the protein molecule, which may also undergo photo-oxidation. The energy of the absorbed light should therefore be as low as possible and the time of exposure as short as possible.

**Automation of spectrophotometric detection**

Automation of the detection process can be based either on the less often used fractional principle as proposed, for example, by Groulade *et al.*, or on the flow-through principle. The most commonly used apparatus for continuous control of the process of protein separation is the absorptiometer for the UV region of the spectrum. Its function is based mainly on the presence of tyrosine, tryptophan and phenylalanine in proteins. Currently used apparatus is of simpler single-beam construction and transmittance recording, serving as semi-quantitative indicators of the presence of proteins, while less commonly used are the more efficient double-beam absorptiometers, which permit higher amplification with lower noise at the zero line.

Current types of the UV absorptiometers are provided with low-pressure mercury lamps emitting at 253.7 nm. The dose of the UV radiation energy is about $10^{-11}$ Einstein/min, *i.e.*, approximately 0.09 $\mu$W. For specific protein detection, the radiation of 280 nm obtained by means of a fluorescent transformer (lead-glass or crystal) is most convenient. If the sensitivity requirements, and especially the requirements placed on the specificity of detection, are lower, absorptiometers for 253.7 nm can also be used.

If the recording is carried out simultaneously at 253.7 and 280 nm, it is possible to estimate, according to Thacker *et al.*, the purity of the protein eluates on the basis of the ratio of absorbances, $A_{280}/A_{253.7}$, from the point of view of contamination with the nucleic acid components that often accompany complex mixtures of proteins.

A simple apparatus for the flow-through detection of proteins at 280 nm, fitted with a magnesium spectral lamp, was proposed by Bennett *et al.* More efficient types of UV absorptiometers, provided with a deuterium lamp that affords a continuous spectrum in the UV region, or with a monochromator and a logarithmic amplifier of the transmittance signal (Hoffmann), are almost as useful as UV spectrophotometers, because they give a direct record of the optical density.

Although the capacity of flow-through cells represents a volume of only $10^{-1}$–$10^{-2}$ ml, the volume of the quartz flow-through cell and the necessary hydraulic system always causes broadening (tailing) of the protein peak. In the region of the maximum protein peak, especially at higher protein concentrations, the density of the eluent changes abruptly. Hence, flow-through detection on the effluent is suitable in instances when a direct fractional determination of the protein concentration is not possible for reasons of volume or time. Flow-through detection is indispensable in the recycling process of GPC.

## REFERENCES

Aisen, P., Leibman, A. and Reich, H. A., *J. Biol. Chem.*, 241 (1966) 1666.
Allan, D., Auger, J. and Crumpton, M. J., *Nature (London)*, 236 (1972) 23.
Andrews, P., *Biochem. J.*, 91 (1964) 222.
Andrews, P., *Biochem. J.*, 96 (1965) 595.
Andrews, P., *Methods Biochem. Anal.*, 18 (1970) 1.
Aspberg, K. and Porath, J., *Acta Chem. Scand.*, 24 (1970) 1839.
Beers, W. H. and Reich, E., *J. Biol. Chem.*, 244 (1969) 4473.
Bell, P. H., Colucci, D. F., Dziobkowski, C., Snedeker, E. H., Barg, Jr., W. F. and Paul, R., *Biochemistry*, 9 (1970) 1665.
Bennett, P. A., Sullivan, J. V. and Walsh, A., *Anal. Biochem.*, 36 (1970) 123.
Bernardi, G., *Methods Enzymol.*, 22 (1971) 325.
Bidlingmeyer, B. A. and Rogers, L. B., *Anal. Chem.*, 43 (1971) 1882.
Bryce, C. F. A. and Crichton, R. R., *J. Chromatogr.*, 63 (1971) 267.
Çarey, W. F. and Wells, J. R., *Biochem. Biophys. Res. Commun.*, 41 (1970) 574.
Čechová, D., Jonáková-Švestková, V. and Šorm, F., *Collect. Czech. Chem. Commun.*, 35 (1970) 3085.
Clark-Walker, G. D. and Lascelles, J., *Arch. Biochem. Biophys.*, 136 (1970) 153.
Cuatrecasas, P., *J. Biol. Chem.*, 245 (1970) 3059.
Cuatrecasas, P., *Advan. Enzymol.*, 36 (1972) 30.
Dasgupta, B. R., Berry, L. J. and Borroff, D. A., *Biochim. Biophys. Acta*, 214 (1970) 343.
Dixon, H. B. F. and Thompson, C. M., *Biochem. J.*, 107 (1968) 427.
Dozy, A. M. and Huisman, H. J., *J. Chromatogr.*, 40 (1969) 62.
Fernlund, P., *Biochim. Biophys. Acta*, 237 (1971) 519.
Folin, O. and Ciocalteau, V., *J. Biol. Chem.*, 73 (1927) 627.
Frattali, V. and Steiner, R. F., *Biochemistry*, 7 (1968) 521.
Ganguly, P., *J. Biol. Chem.*, 247 (1972) 1809.
Goldstein, G., Slizyo, I. S. and Chase, M. W., *J. Exp. Med.*, 114 (1961) 89.
Gráf, L., Cseh, G. and Medzihradszky-Schweiger, H., *Biochim. Biophys. Acta*, 175 (1969) 444.
Groulade, J., Chicault, M. and Waltzinger, W., *Bull. Soc. Chim. Biol.*, 49 (1967) 1609.
Guyda, H. and Friesen, H., *Biochem. Biophys. Res. Commun.*, 41 (1971) 1068.
Haller, W., Tympner, K. D. and Hannig, K., *Anal. Biochem.*, 35 (1970) 23.
Hawker, C. D., Rasmussen, H. and Glass, J. D., *Proc. Nat. Acad. Sci. U.S.*, 58 (1967) 1535.
Hennen, G., Prusík, Z. and Maghuin-Rogister, G., *Eur. J. Biochem.*, 18 (1971) 376.
Hochstrasser, K., Werle, E., Siegelmann, R. and Schwarz, S., *Hoppe-Seyler's Z. Physiol. Chem.*, 350 (1969) 897.
Hoffmann, L. G., *J. Chromatogr.*, 40 (1969) 39.
Horton, B. F. and Chernoff, A. I., *J. Chromatogr.*, 47 (1970) 493.
Hussa, R. O., Landon, J. and Winnick, T., *Biochem. J.*, 114 (1969) 519.
Hwang, P., Guyda, H. and Friesen, H., *J. Biol. Chem.*, 247 (1972) 1955.
Ingwall, J. S. and Scheraga, H. A., *Biochemistry*, 8 (1969) 1860.
Ito, A. and Sato, R., *J. Biol. Chem.*, 243 (1968) 4922.

Iwamoto, M. and Abiko, Y., *Biochim. Biophys. Acta*, 214 (1970) 402.
Josefsson, L., *Biochim. Biophys. Acta*, 148 (1967) 300.
Joustra, M. and Lundgren, H., *Annual Colloquium, Protides of Biological Fluids, Brugge*, 17 (1969) 511.
Keil, B., Keilová, H. and Bartošek, I., *Collect. Czech. Chem. Commun.*, 27 (1962) 2940.
Kerr, M. S., *Comp. Biochem. Physiol.*, 34 (1970) 301.
Kitamura, M., Sakaguchi, S. and Sakaguchi, G., *Biochim. Biophys. Acta*, 168 (1968) 207.
Koike, M. and Hamada, M., *Methods Enzymol.* 22 (1971) 339.
Kusel, J. P., Suriano, J. R. and Weber, M. M., *Arch. Biochem. Biophys.*, 133 (1969) 293.
Lampson, G. P. and Tytell, A. A., *Anal. Biochem.*, 11 (1965) 374.
Lindsay, R. H., Starnes, W. R., Hershman, J. M. and Pittmann, J. A., *Proc. Soc. Exp. Biol. Med.*, 131 (1969) 875.
Lohmar, P. and Chom Hao Li, *Biochim. Biophys. Acta*, 147 (1967) 381.
Lowry, H. O., Rosebrough, N. J., Farr, A. L. and Randall, R. J., *J. Biol. Chem.*, 193 (1951) 265.
Miranda, F., Kupeyan, Ch., Rochat, H., Rochat, C. and Lissitzky, S., *Eur. J. Biochem.*, 16 (1970) 514.
Morávek, L., *J. Chromatogr.*, 59 (1971) 343.
Morgan, R. G. H., Barrowman, J. and Borgström, B., *Biochim. Biophys. Acta*, 175 (1969) 65.
Mosesson, M. W., Alkjaersig, N., Sweet, B. and Sherry, S., *Biochemistry*, 6 (1967) 3279.
Nagasawa, S. and Suzuki, T., *J. Biochem. (Tokyo)*, 66 (1969) 273.
Novotný, J., *FEBS Lett.*, 14 (1971) 7.
Okazaki, T., Wittenberg, B. A., Briehl, R. W. and Wittenberg, J. B., *Biochim. Biophys. Acta*, 140 (1967) 258.
Pallotta, D. and Berlowitz, L., *Biochim. Biophys. Acta*, 200 (1970) 538.
Paul, K. G. and Stigbrand, T., *Biochim. Biophys. Acta*, 221 (1970) 255.
Peckham, W. D. and Parlow, A. F., *Endocrinology*, 84 (1969) 953.
Peisaca, J., Levine, W. G. and Blumberg, W. E., *J. Biol. Chem.*, 242 (1967) 2847.
Pharmacia, *Supplement to Sephadex Ion Exchanger Booklet*, Pharmacia, Uppsala, Sweden, 1969, pp. 2–3.
Pharmacia, *Sephadex-Ionenaustauscher (Leitfaden zur Ionenaustausch-Chromatographie)*, Pharmacia, Uppsala, Sweden, 1970, p. 12.
Porath, J., *Nature (London)*, 196 (1962) 47.
Porath, J. and Bennich, H., *Arch. Biochem. Biophys., Suppl.*, 1 (1962) 152.
Porath, J. and Fryklund, L., *Nature (London)*, 226 (1970) 1169.
Price, V. E. and Greenfield, R. E., *J. Biol. Chem.*, 209 (1954) 365.
Rathnam, P. and Saxena, B. B., *J. Biol. Chem.*, 245 (1970) 3725.
Read, K. R. H., *Comp. Biochem. Physiol.*, 22 (1967) 1.
Reeve Angel & Co., *Technical Bulletin 1032, 2000 and 15 M*, H. Reeve Angel & Co., Clifton, N.J., U.S.A.
Reiland, J., *Methods Enzymol.* 22 (1971) 287.
Rinderknecht, H., *Nature (London)*, 193 (1962) 167.
Sadgopal, A. and Bonner, J., *Biochim. Biophys. Acta*, 207 (1970) 206.
Saffran, M., Muhlstock, B. and Caplan, B., *Can. J. Biochem. Physiol.*, 39 (1961) 653.
Samuelsson, G. and Pettersson, B., *Acta Chim. Scand.*, 24 (1970) 2751.
Shipman, W. H. and Cole, L. J., *Anal. Biochem.*, 29 (1969) 490.
Swanljung, P., *Anal. Biochem.*, 43 (1971) 382.
Swingle, F. M. and Tiselius, A., *Biochem. J.*, 48 (1951) 171.
Thacker, L. H., Scott, C. D. and Pitt, W. W., *J. Chromatogr.*, 51 (1970) 175.
Tiselius, A., Hjertén, S. and Levin, O., *Arch. Biochem. Biophys.*, 65 (1956) 132.
Tombs, M. P., Souter, F. and Maclagan, N. F., *Biochem. J.*, 73 (1959) 167.
Vogt, W., *Biochem. Pharmacol.*, 17 (1968) 727.
Wallen, P. and Wiman, B., *Biochim. Biophys. Acta*, 221 (1970) 20.
Weintraub, B., *Biochem. Biophys. Res. Commun.*, 39 (1970) 83.
Wrigley, C. V. and Webster, H. L., *J. Chromatogr.*, 33 (1968) 534.
Yang, H., *J. Biol. Chem.*, 245 (1970) 6404.
Zahler, P. H. and Wallach, D. F. H., *Biochim. Biophys. Acta*, 135 (1967) 371.

*Chapter 36*

# Enzymes

O. MIKEŠ

## CONTENTS



## SPECIAL REQUIREMENTS FOR THE CHROMATOGRAPHY OF ENZYMES

Enzymes differ from many other natural substances in their fragile macromolecular polyvalent amphoteric nature and from other proteins in their specific topography, reflecting their particular activities . The special, very susceptible and complicated structure of enzymes must be kept in mind when methods are selected for their isolation and fractionation. Some types of enzymes very easily change their tertiary structure, which is accompanied by a loss in activity, often irreversible. The ever-present risk of the denaturation of enzymes is generally known. The surface denaturation of enzymes in contact with unsuitable chromatographic sorbents that allow strong hydrophobic contacts has often been observed; adsorbents of a hydrophilic nature are preferably used.

The researcher will certainly not expose enzyme preparations to extreme conditions of heat, pH, oxidizing or reducing agents, detergents or unsuitable organic solvents. However, what may be neglected are slight changes in conformation of enzymes which are not connected with a loss of activity, but which modify the enzymes so that they then differ from the products of proteosynthesis. For example, Sipos and Merkel found that exposure of protease from marine bacteria to an increase in temperature in the presence of $Ca^{2+}$ changes the optimum temperature of this enzyme, and the same was later found to be true of bovine trypsin and chymotrypsin. A similar influence of some ions on amylase activity is also known. Mikeš *et al.* found changes in the optimum pH of digestion of haemoglobin with alkaline protease from *Aspergillus flavus* exposed for a certain time to an elevated temperature in the presence of various additives. These transformations, which lead to polymorphism of certain enzymes, also take place to a certain extent at ambient temperature but not at 4°C. In most instances, it is important when treating enzyme preparations to work in a laboratory at a temperature of just above 0°C, as the risks of

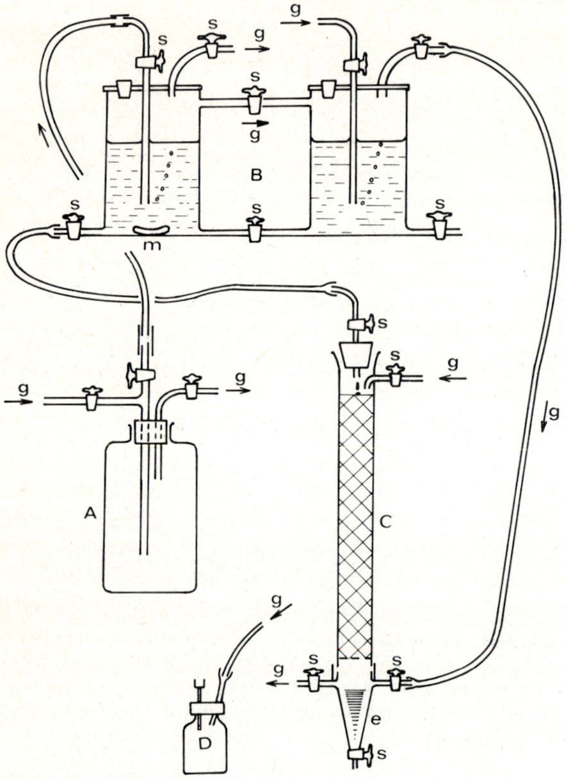


Fig. 36.1. Principle of the anaerobic column system (Repaske). A, plastic bottle as a reservoir of gas; B, buffer reservoir for stepwise elution or a gradient device; C, chromatographic column; D, serum bottle with hypodermic needles for collecting fractions; e, calibrated end unit; g, stream of oxygen-free gas; m, magnetic stirrer; s, stopcocks.

enzyme autolysis, of proteolytic attack on raw enzymic preparations or of microbial contamination are minimized at this temperature. Another general rule is to work as rapidly as possible.

Other risks arise from the sensitivity of many enzymes to the ions of heavy metals. Therefore deionized water or water distilled from glass should be used for all extractions and in the preparation of all buffers. Special ion-exchange treatment is necessary in some instances. Ions extracted from the walls of the vessels used sometimes influence the results and bottles coated with silicone films should be used in such instances.

Some enzymes are sensitive to the method of concentration and cannot even be lyophilized. The stability of the enzyme studied should certainly be examined at the beginning of any work.

Specific operating procedures must be followed in special instances, for example, contact with oxygen in the air must be avoided with some enzymic systems (hydrogenases, enzymes of methane bacteria, for the study of enzymic reduction and oxidation of various substrates, enzymic systems of anaerobes and of anaerobic mutants, experiments involving the reduction of disulphide bridges in enzymes, etc.). All of these procedures require special conditions, *e.g.*, anaerobic glove-boxes.

Column chromatography of oxygen-sensitive enzymes requires a strict anaerobic technique. Repaske described equipment for the chromatography of hydrogenases with a 90% recovery (Fig. 36.1). All spaces and lines are carefully washed with a suitable oxygen-free gas, introduced into the system through bottle A, the same gas is bubbled through the buffers used before application to the column and the slurry of the sorbent is carefully introduced into the column using an anaerobic atmosphere. Also, the collection and preservation of fractions is performed under oxygen-free conditions. This or similar equipment cannot be replaced by the usual chromatographic apparatus in which reducing agents in elution buffers are used, as very low recoveries result.

For longer systematic experiments with oxygen-sensitive enzymes, an anaerobic laboratory has many advantages, allowing greater versatility and the easy utilization of a long sequence of different methods without contact with the outer atmosphere. The construction of such an anaerobic chamber with all of the necessary equipment was described by Poston *et al.*, who discussed the anaerobic facilities at the National Institute of Health, Bethesda, Md., U.S.A. The laboratory is systematically washed with nitrogen and the last traces of oxygen are removed by reaction with hydrogen on a catalyst bed. Persons who work in the laboratory, of course, must wear masks supplying them with air and are connected to a vacuum line for removing exhaled air. Normal laboratory equipment and methods can be used in such a chamber.

## TECHNIQUES AND AUTOMATED ANALYSES

Many effective methods have been developed for the separation of enzymes from various components of living matter and for their fractionation and concentration. These methods include the following: crystallization (Davies and Segal, Jakoby, Zeppezauer); dialysis, ultrafiltration and hollow-fibre techniques (Blatt, McPhie, Rony); electrophoretic methods (Prusík, Shuster, Vesterberg, and Chapter 35); extraction (Penefsky and Tzagoloff, Tzagoloff and Penefsky); lyophilization (Everse and Stolzenbach); solvent precipitation and methods involving the use of soluble non-ionic polymers (Fried and Chun, Kaufman); and zonal centrifugation (Cline and Ryel).

The most important and most often used fractionation method for the preparation and analysis of enzymes is chromatography. A bibliography on column chromatography by Deyl *et al.* covers the period from 1967 to 1970 and contains 1687 references to papers dealing with enzymes; this was extended to 1972 by 131 references to papers using affinity chromatography for the isolation of enzymes by Turková in Chapters 7 and 14, and elsewhere. In papers cited by both authors, 3288 different column chromatographic experiments were described. It may be interesting for the reader to compare the survey of the different methods in Table 36.1, but it should be noted that in fact the proportion of individual methods changes with time. In recent years, great progress has been made in the use of affinity chromatography. Recent references to particular methods are as follows: affinity chromatography (Cuatrecasas, Cuatrecasas and Anfinsen (1971a, b), Feinstein, Porath, Turková, and Chapters 7 and 14); chromatography on hydroxyapatite

TABLE 36.1
SURVEY OF USE OF DIFFERENT CHROMATOGRAPHIC METHODS FOR THE FRACTIONATION OF ENZYMES

Calculated from the bibliographic data collected by Deyl *et al.*, Turková, and those given in Chapters 7 and 14.

| Method | Percentage |
|---|---|
| Liquid–solid chromatography<br>(aluminium oxide, brushite, calcium phosphate modifications, hydroxyapatite) | 7.2 |
| Liquid–liquid chromatography<br>(Celite, cellulose, silica gel, starch) | 0.4 |
| Gel permeation chromatography<br>(agarose, polyacrylamide gel, polydextran) | 41.5 |
| Ion-exchanging derivatives of cellulose | 30.5 |
| Ion-exchanging derivatives of polydextran | 11.8 |
| Ion-exchange resins | 4.3 |
| Affinity chromatography | 4.0 |
| Other methods | 0.3 |

(Bernardi, and Chapter 35); chromatography on calcium phosphate gel (Koike and Hamada, and Chapter 35); detergent gradient gel chromatography – technique for the purification of membrane-bound enzymes (Swanljung, Swanljung and Frigeri); gel permeation chromatography (Reiland, and Chapters 5 and 12); electrochromatographic methods (Prusík); ion-exchange chromatography (Himmelhoch, Mikeš, and Chapters 6 and 13); and substrate-specific elution (Pogel and Sarngadharan).

The last method, specific elution with substrate, is a combination of classic chromatographic sorption with specific desorption, which has the benefit of specific enzyme–substrate, enzyme–inhibitor or enzyme–effector interactions. The enzyme is bound, for example, on an ion exchanger, in the usual way. The desorption is effected by a gradient of the substrate concentration in addition to the usual ionic strength gradient. Only the particular enzyme that is capable of specific interaction with the substrate changes its tertiary structure a little, while other proteins and enzymes are not influenced. The specific interaction modifies topographic parameters (*e.g.*, the number of charged groups on the surface of the enzyme and the number of hydrophobic contacts), which is reflected in changes in binding forces to the ion exchanger or other sorbent. The final effect is specific desorption.

Examples of this modern method for the specific separation of enzymes are the purification of chicken pancreatic ribonuclease by elution from cellulose phosphate with ribonucleic acid (Eley) and the purification of rabbit liver fructose-1,6-diphosphatase on CM-cellulose (Sarngadharan *et al.*). The efficiency of the method can be seen from the next example: in the purification of glucose-6-phosphate dehydrogenase by elution of the enzyme from CM-Sephadex with 2 m*M* glucose-6-phosphate (Rattazzi), a 51-fold increase in specific activity was achieved with 91% recovery of the enzyme activity.

The need to repeat individual enzyme experiments using standard procedures led to the development of the automation of enzyme analysis, which has great importance primarily for clinical biochemists. However, research biochemists often must repeat large

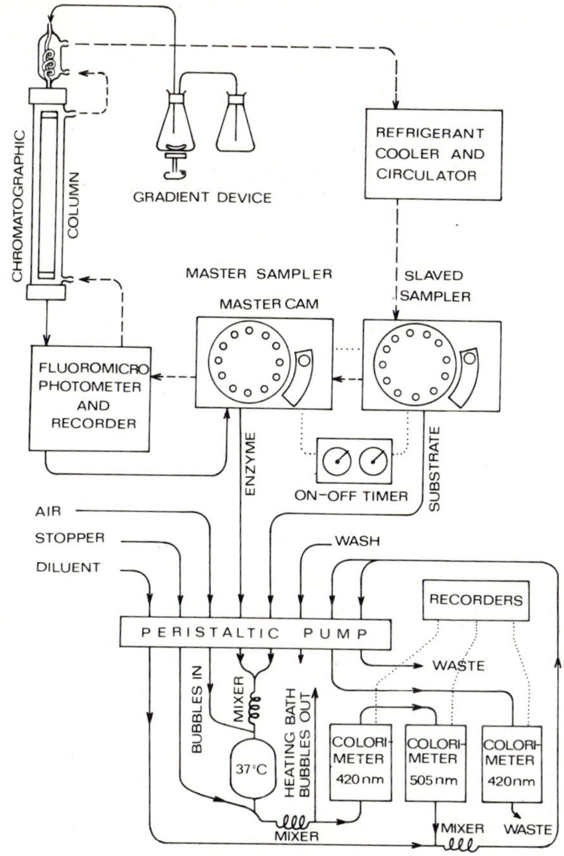


Fig. 36.2. Schematic diagram of single flow path autoanalytical system for enzymes based on Technicon AutoAnalyzer components and used for the monitoring of column chromatograph effluents (example compiled according to Beck and Tappel, Eveleigh *et al.* and Tappel). Solid lines, liquid flow; broken lines, refrigerant flow; dotted lines, electrical control. The slaved sampler can be replaced with a substrate refrigerator, programmed multichannel valve and distributor, which allow increased substrate capacity.

numbers of experiments, which can hardly be realized without automation. The first approaches to this problem were made several years ago (*cf.*, Moss, Schwartz *et al.*, Večerek *et al.*, and others). The present apparatus (still being developed) allows the complete automation of processes, including calculation and presentation of results and feedback control of instruments by computer. A recent survey of automated instruments, allowing continuous flow analysis, discrete sample analysis and continuous kinetic experiments, was given by Schwartz (1971). A survey of instruments allowing multiple enzyme analyses was presented by Tappel. Instruments available for automated analysis of discrete samples have been reviewed by Alpert. In most automated enzyme experiments that have been described, the Technicon AutoAnalyzer (Schwartz, 1968) was used, which can be adapted for the analysis of all types of enzymes (*cf.*, Wacker and Coombs) and, by using a sample splitter, also for the simultaneous analysis of several enzymes in the same sample (Smythe *et al.*).

The systems for automated multiple enzyme analysis are applicable to continuous monitoring of column chromatograph effluents (*cf.*, Beck and Tappel) in addition to other functions. The principle of such apparatus (*cf.*, Bradley and Tappel, Eveleigh *et al.*) is illustrated by Fig. 36.2. The master sampler acts as a column fraction collector. The on–off timer distributes the appropriate amounts of enzymes and substrates to the mixer. The air bubbles split the stream of solution into well mixed segments, which are then heated in the heating bath. After mixing with a suitable reagent, which terminates the enzyme reaction, and developing the chromatotropic product, and after deaeration, the streams flow through colorimeters. In order to measure peak absorbances accurately, part of the concentrated stream effluent is then diluted with water (or with a suitable diluent) in a fixed proportion and measured again. This system was used for the simultaneous determination of six enzyme activities (acid phosphatase, α-D-galactosidase, β-D-galactosidase, β-D-glucuronidase, aryl sulphatase and N-acetyl-β-D-glucosaminidase) in the soluble enzyme fraction of rat liver lysozymes chromatographed on a CM-cellulose column. Only about one quarter of each 7.7-ml fraction was consumed in this determination (Beck and Tappel). The method was improved (Tappel) in order to allow further simultaneous determinations (additional activities: α-L-fucosidase, 5′-phosphodiesterase I, 5′-phosphodiesterase IV, α-D-glucosidase, β-D-glucosidase and α-D-mannosidase) directly in the soluble fraction of rat liver lysozomes, omitting chromatography. Fig. 36.3 illustrates the results obtained.

Examples of purification or fractionation of particular enzymes using the various methods mentioned above are described below in greater detail. As an introduction

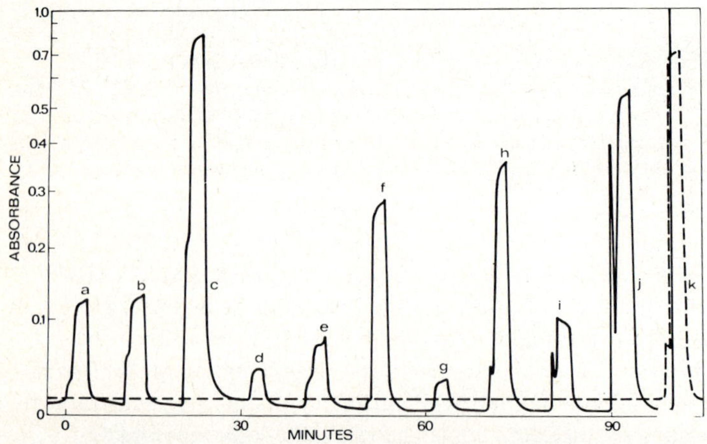


Fig. 36.3. Automatic determination of the activities of 11 hydrolytic enzymes from rat liver lysozymes (Tappel). Solid line, absorbance at 420 nm; broken line, absorbance at 505 nm. a, α-L-fucosidase; b, 5′-phosphodiesterase I; c, 5′-phosphodiesterase IV; d, α-D-glucosidase; e, α-D-galactosidase; f, β-D-galactosidase; g, β-D-glucosidase; h, α-D-mannosidase; i, β-D-glucuronidase; j, N-acetyl-β-D-glucosaminidase; k, arylsulphatase.

TABLE 36.2
SURVEY OF DIFFERENT TYPES OF ENZYMES FRACTIONATED BY COLUMN CHROMATOGRAPHIC METHODS (DEYL *et al.*)

| Enzyme type | Proportion of published papers (%) |
|---|---|
| E.C.1: oxidoreductases | 19.0 |
| E.C.2: transferases | 22.4 |
| E.C.3: hydrolases | 36.8 |
| E.C.4: lyases | 7.3 |
| E.C.5: isomerases | 1.4 |
| E.C.6: ligases | 5.6 |
| Other enzymic preparations | 7.5 |

Table 36.2 surveys the proportions of published papers that describe column chromatographic methods for the treatment of the main classes of enzymes.

## OXIDOREDUCTASES

The preparation of 3-hydroxy-3-methylglutaryl coenzyme A reductase from *Pseudomonas* using a column of hydroxyapatite gel and stepwise elution was described by Bensch and Rodwell. Their procedure is an example of a successful fractionation using reducing additives in the preparation of a reductase. The cells of *Pseudomonas* were grown aerobically at 30°C and pH 7 in a medium containing 12 m*M* ammonium DL-mevalonate and collected from the stationary phase. All other manipulations were carried out at 0–5°C. The washed, centrifuged and frozen cells retained full activity for several months. The turbid supernatant of sonically disrupted cells (50 m*M* Tris–hydrochloric acid, pH 7.1; 1.2 m*M* EDTA) was centrifuged at 60,000 *g* and fractionated using ammonium sulphate (25–35% of ammonium sulphate precipitate contains the enzyme). This preparation can be stored frozen for several weeks. After dissolving the preparation in 10 m*M* potassium phosphate solution (pH 6.8), 10 μ*M* in dithiothreitol, the solution was dialyzed overnight against this buffer (the tubing was previously boiled in 1 m*M* EDTA) and subjected to chromatography.

A 6.0 × 2.5 cm column of Bio-Gel HT hydroxyapatite gel was equilibrated with 10 m*M* potassium phosphate solution (pH 6.8), 10 μ*M* in dithiothreitol. After the application of the sample, the column was washed with 30 ml of the above buffer (one column volume) and eluted successively with 60-ml portions of potassium phosphate solution (pH 6.8), 10 μ*M* in dithiothreitol and 20, 28 and 40 m*M* in total phosphate. Fractions of 3 ml were collected and combined according to the scheme illustrated in Table 36.3. The combined fractions 42–48 were made 2.5 m*M* in EDTA and brought to 40% saturation in ammonium sulphate. The precipitate was dissolved in 100 m*M* Tris–hydrochloric acid (pH 7.1), 2.5 m*M* in EDTA and 10μ*M* in dithiothreitol and dialyzed against the same buffer. This purified oxidoreductase loses activity rapidly in the absence of EDTA or dithiothreitol.

Fractions 49–70 can be dialyzed and re-chromatographed on hydroxyapatite. Gradient elution gave an improved yield but only half of the final specific activity.

*References p.829*

TABLE 36.3
SURVEY OF PURIFICATION OF *Pseudomonas* 3-HYDROXY-3-METHYLGLUTARYL COENZYME A REDUCTASE ON A HYDROXYAPATITE COLUMN (BENSCH AND RODWELL)

Equivalent of 760 mg dry weight cells = 100%. The activity was tested using mevalonate oxidation.

| Fraction | Volume (ml) | Total activity (I.U.) | Total protein (mg) | Specific activity (I.U./mg) | Enrichment | Yield (%) |
|---|---|---|---|---|---|---|
| Crude extract after sonic treatment | 12.8 | 1070 | 575 | 1.86 | 1.0 | 100 |
| Fraction after ammonium sulphate precipitation and dialysis | 2.0 | 570 | 34.8 | 16.4 | 8.8 | 53 |
| Chromatographic fractions 49–70 | 61.1 | 436 | 23.0 | 18.9 | 10.2 | 40 |
| Chromatographic fractions 42–48 | 32.9 | 77 | 1.98 | 38.9 | 21.0 | 7.2 |

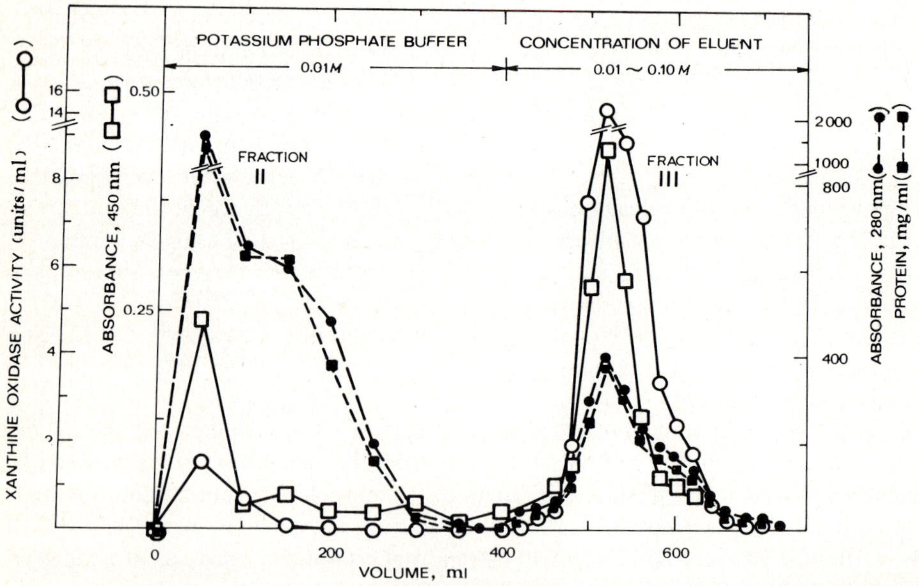


Fig. 36.4. Chromatography of milk xanthine oxidase preparation on DEAE-Sephadex using gradient elution (Roussos and Morrow). Column dimensions: 1.6 cm × 4.9 cm². Load: 201 mg of protein. The ion exchanger was equilibrated with 10 m*M* potassium phosphate buffer (pH 7.4), 1 m*M* in EDTA (starting buffer). After washing off all enzyme activities and proteins, the linear gradient elution (400 ml of 100 m*M* potassium phosphate buffer (pH 7.4), 10 μ*M* in EDTA, to 400 ml of 10 m*M* potassium phosphate) was started. Protein was determined by the method of Lowry *et al.* Enzymic activity was tested at 25°C and the unit was defined according to Fridovich (*cf.* also Avis *et al.*).

The purification of milk xanthine oxidase by a combination of chromatography on anion-exchange polydextran and chelating resin was described by Roussos and Morrow. Trace amounts of some heavy metals often play an important role when present in the purified enzyme preparation because they can be considered to be an integral part of the enzyme. Bovine milk xanthine oxidase serves as an example. Therefore Roussos and Morrow made an effort to prepare enzyme preparations free from molybdenum and very low in iron, and favoured the conclusion that molybdenum and probably also iron are not essential in the xanthine–oxygen oxidoreductase activity.

All procedures were carried out at 0–4°C. Commercial xanthine oxidase (Worthington Biochemicals Corp., Freehold, N.J., U.S.A.) was dialyzed twice against 10 m*M* potassium phosphate solution (pH 7.4), 1 m*M* in EDTA. The supernatant was applied to a pre-equilibrated DEAE-Sephadex column (Fig. 36.4). The eluates with the highest specific activity (fraction III) were pooled and applied on to a chelating resin column.

The Chelex 100 column (2 cm × 4.2 cm$^2$) was pre-equilibrated with 20 m*M* phosphate buffer (pH 7.4), 1 m*M* in EDTA. After the application of the sample, the column was washed with the same buffer (5 ml) and the combined effluent was dialyzed against 100 $\mu M$ phosphate buffer (pH 7.4) 10 $\mu M$ in EDTA. Electrophoretic examination indicated the presence of homogeneous protein. The enzyme prepared in this manner possessed a specific activity that was significantly higher than those hitherto reported (about 3-fold),

TABLE 36.4
RESULTS OF THE PURIFICATION OF TRIPHOSPHOPYRIDINE NUCLEOTIDE ISOCITRATE DEHYDROGENASE FROM *Bacillus stearothermophilus* USING A COMBINATION OF METHODS (HOWARD AND BECKER)

| Fraction | | Total activity (I.U.) | Total $A_{280}$ | $A_{280}/A_{260}$ | Specific activity |
|---|---|---|---|---|---|
| (1) | Combined supernatant fractions after sonic disruption of cell suspension | 1488 | 40,000 | 0.55 | 0.037 |
| (2) | Supernatant of the mixture of the preceding fraction and sodium DL-isocitrate after the pH was adjusted to 4.9–5.0 | 1300 | 11,400 | 0.56 | 0.11 |
| (3) | Solution of the pellet after ammonium sulphate precipitation at pH 7.5 | 1270 | 3280 | 0.745 | 0.39 |
| (4) | Desalted protein (Sephadex G-25) treated with ammonium sulphate and chromatographed on an agarose (Bio-Gel A-0.5 m) column in the presence of a small amount of ammonium sulphate | 968 | 341 | 1.6 | 2.84 |
| (5) | Solution of combined and reprecipitated fractions chromatographed on hydroxyapatite (Bio-Gel HTP) | 645 | 52.4 | 1.61 | 12.3 |
| (6) | Active pooled fractions after chromatography on ECTEOLA-cellulose (Cellex E) | 248 | 8.16 | 1.66 | 30.4 |
| (7) | Combined homogeneous fractions after chromatography of concentrated solution (ultrafiltration; Amicon UM-2 membrane) on a Sephadex G-75 column | 128 | 3.73 | 1.70 | 34.2 |

*References p.829*

it was active in the absence of molybdenum and the content of iron was 3–4-fold lower than that in previous preparations. The minimum molecular weight calculated on the basis of the FAD content was 354,000 ± 94,000, *i.e.*, 2.3-fold higher than the value published previously for the crystalline preparation.

In many instances, the application of one or two chromatographic or other purification steps is not sufficient to obtain a homogeneous enzyme from a natural source and a combination of several methods must be used. The isolation of isocitrate dehydrogenase using successive chromatographic and other methods (Howard and Becker) is an example. The efficiency of the procedure is illustrated by the survey in Table 36.4 (in which the details are omitted). All solutions used in the preparation of isocitrate dehydrogenase contained 1 m*M* EDTA and no antioxidants were used. A 1000-fold purification (based on the absorbance at 280 nm) was achieved. The yield of total activity was 8.6%.

On the other hand, the method of affinity chromatography does not require as many operations. Newbold and Harding described a single-step procedure that gave a 4000-fold purification of dihydrofolate reductase from mammalian skin. This comparison illustrates the efficiency of the latter method.

## TRANSFERASES

The purification of ornithine carbamoyl-transferase from *Halobacterium salinarium* using gel filtration, sucrose gradient centrifugation and chromatography on calcium phosphate gels was described by Dundas. This is an example of the processing of a typical extremely halophilic enzyme, which shows a high activity in 4 *M* sodium chloride solution and is rapidly and irreversibly inactivated in a salt-free environment. Therefore, all purification procedures were carried out on 4.3 *M* sodium chloride solutions. A 0.1 *M* solution of L-ornithine stabilizes the enzyme and was included throughout the purification. Owing to the necessarily high ionic strength, the use of ion-exchange purification techniques is impossible in this instance.

The broken cells of *Halobacterium salinarium* were resuspended in 4 *M* sodium chloride–0.1 *M* ornithine solution and the supernatant was fractionally precipitated with acetone at 0°C. The dissolved precipitate was dialyzed against 4.3 *M* sodium chloride–0.1 *M* ornithine and stored at –30°C. The specific activity increased 5–15-fold.

Gel filtration was carried out on a Sephadex G-200 column previously equilibrated with 4.3 *M* sodium chloride–0.1 *M* ornithine solution. The same solution was used for elution. This procedure resulted in an approximately 2-fold increase in specific activity.

Volumes of 1 ml of acetone-purified fractions were centrifuged using 30-ml gradients made with 25% (w/v) sodium chloride and 0.1 *M* ornithine and a decreasing sucrose concentration from 30% (w/w) to 10% (w/w). After centrifugation (Beckman 60 Ti angle rotor, 5°C, 60,000 rpm for 6–12 h), the tubes were collected, giving 0.7-ml fractions. The pooled runs gave a 2-fold increase in specific activity.

The best result (a 6-fold increase in activity) was obtained by calcium phosphate gel chromatography (Fig. 36.5) or by its combination with prior gel filtration. The sorbent was prepared according to Tiselius *et al.* as described by Levin. The combined active fractions were dialyzed in the presence of urea and mercaptoethanol and the electrophoretic and sedimentation pattern obtained showed the homogeneity of the preparation.

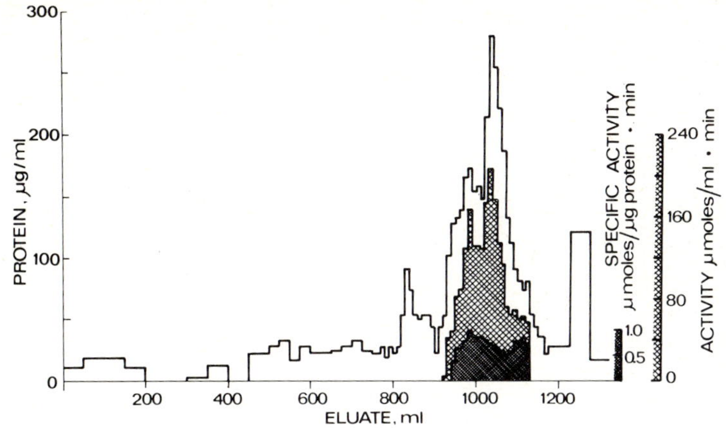


Fig. 36.5. Calcium phosphate gel chromatography of ornithine carbamoyl transferase (Dundas). Column dimensions: 20 × 2.5 cm. Washing: 500 ml of 5 m*M* phosphate buffer (pH 6.8), 500 ml of 4.5 *M* sodium chloride, and 100 ml of 4.3 *M* sodium chloride, 0.1 *M* in ornithine. Then the sample (60 mg of protein) was applied. Eluent: 4.3 *M* sodium chloride, 0.1 *M* in ornithine solution, with stepwise increasing phosphate concentration, 5–300 m*M* (input phosphate concentration to 50, 100, 150, 200, 250 and 300 m*M* after 300, 550, 750, 850, 950 and 1150 ml, respectively). Fractions: 10 ml (or 25 ml).

The isolation of three multiple forms of aminoacyl transferase I of rat liver using hydroxyapatite and polydextran chromatography was described by Schneir and Moldave. This method permitted the preparation of three forms of aminoacyl transferase I (differing in molecular weight) from rat liver mitochondria and proved the transformation of high-molecular-weight species to lower-molecular-weight enzyme. This fractionation is briefly described below.

Homogenized excised livers were centrifuged at 10,000 *g* and the supernatant, after precipitation by adjustment to pH 5.0, was freed from low-molecular-weight components by passing it through Sephadex G-25 (0.05 *M* Tris–hydrochloric acid, pH 8). This and all other treatments were performed at +4°C. The effluent (500 ml), containing transferases I and II, was made 1 m*M* in dithiothreitol (all other solutions used for washing also contained this sulphydryl-activating agent) and mixed with 166 ml of well suspended hydroxyapatite (Clarkson Chemical Co., Inc.) at 4°C for 1 h. The supernatant after centrifugation (100,000 *g*, 10 min, 4°C) with washings of the sediment was discarded. The residue was eluted successively with three 300-ml volumes of potassium phosphate solutions at pH 6.8 to obtain fractions extracted with 0.125, 0.15, 0.175, 0.25 and 0.50 *M* potassium phosphate solution. The active fraction (0.25 *M*) was precipitated by ammonium sulphate to a final concentration of 70% (pH 7) and concentrated by dialysis under vacuum against Tris–hydrochloric acid buffer.

Gel filtration of this sample is illustrated by Fig. 36.6a. The combined fractions 31–37, 42–48 and 56–72 were precipitated with ammonium sulphate and, after vacuum dialysis, applied individually to the same column. The results are illustrated in Fig. 36.6b. Form A of transferase I can be converted into form B when exposed briefly to the influence of 1 *M* ammonium chloride solution; prolonged incubation leads to loss of activity. Form B seems

*References p.829*

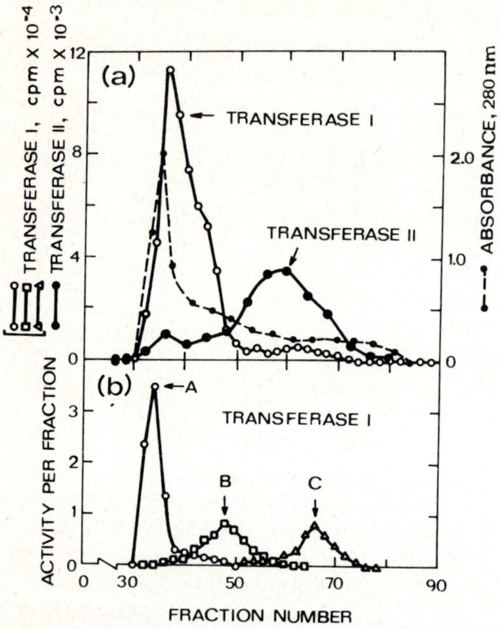


Fig. 36.6. Chromatography of transferase I on Sephadex G-200 (Schneir and Moldave). Column dimensions: 90 × 1.5 cm. Eluent: 0.05 *M* Tris–hydrochloric acid pH 8, 1 m*M* in dithiothreitol. Flow-rate: 8 ml/h. Fractions: 1.1 ml. Temperature: 4°C. Activity tests are described in the original paper. (a) Load: 1 ml of solution after hydroxyapatite treatment (see text) containing 25 mg of protein. (b) Survey of re-chromatography of pooled and dialyzed fractions 31–37, 42–48 and 56–72 from the preceding chromatography. These three samples were applied individually. (b) represents a composite of three columns. A, B and C: three forms of transferase I.

to be a natural enzyme, while form C could represent active sub-units, and the existence of form A can be explained by the binding of B to large molecular weight materials that have no enzymic activity.

The purification of tyrosine aminotransferase by affinity chromatography was described by Miller *et al.* They purified L-tyrosine–2-oxoglutarate aminotransferase using affinity adsorbents containing pyridoxamine phosphate covalently bound to agarose (Sepharose 4B). The sorbed enzyme can be released by changing the buffer, with 125-fold purification. Further purification can be achieved by gel permeation chromatography on Sephadex G-200 columns. A mixture of different isozymes is obtained by this method, because affinity chromatography binds and releases all molecules with the same specific activity.

## HYDROLASES

During the isolation of enzymes, sometimes multiple forms are prepared with the same type of activity but differing in their chromatographic or electrophoretic properties. There may be several causes of this phenomenon, and some examples are given below.

TABLE 36.5
PURIFICATION OF ARYLSULPHATASE FROM *P. aeruginosa* (DELISLE AND MILAZZO)

Enzyme assays after Dodgson and Spencer.

| Fraction | Total protein (mg) | Specific activity (u/mg) | | Recovery (%) | |
|---|---|---|---|---|---|
| | | *p*-Nitrophenol sulphate | Nitrocatechol sulphate | *p*-Nitrophenol sulphate | Nitrocatechol sulphate |
| Acetone powder extract | 1539 | 0.0224 | 0.0451 | 100 | 100 |
| DEAE-cellulose | 432.8 | 0.0941 | 0.1249 | 118* | 78 |
| DEAE-Sephadex | 199.2 | 0.137 | 0.239 | 79 | 68 |
| $(NH_4)_2SO_4$ precipitate | 112.0 | 0.1862 | 0.3686 | 60 | 59 |
| DEAE-Sephadex (pH 8.5) | 26.5 | 0.7592 | 1.444 | 58 | 55 |
| Acrylamide gel electrophoresis | 6.6 | 2.49 | 4.464 | 48 | 42 |

*Endogenous inhibitor was present in the initial extract.

The isolation of arylsulphatase isoenzymes from *Pseudomonas aeruginosa* using cellulose or polydextran ion-exchange chromatography was described by Delisle and Milazzo. This is an example of a simple method leading to the isolation of an identical pair (charge-isomers) of microbial arylsulphate sulphohydrolases, formed by the micro-organism in duplicate.

The enzyme was extracted from an acetone powder of the micro-organism in 0.05 *M* Tris–hydrochloric acid buffer (pH 7.5). Sorption on DEAE-cellulose or DEAE-Sephadex and elution using 0.01 *M* Tris-hydrochloric acid buffer (pH 7.5) and linear sodium chloride gradients 0–1 *M* and 0.1–0.6 *M*, respectively, led to the first purification shown in Table 36.5. Further purification was achieved by precipitation with ammonium sulphate. The enzyme was present in the fraction salted out between 35–75% of saturation. The fraction after preparative disc electrophoresis had the highest enrichment (about 100-fold) with an approximately 45% yield. The enzyme, when examined by gel electrophoresis, indicated a high purity; only two zones were present, both hydrolyzing both substrates. The presence of two isoenzymes in the microorganism was proved by repeated cross-experiments. The method of Hedrick and Smith was used to establish whether these two enzymes were size- or charge-isomers or perhaps both. The results are shown in Fig. 36.7. Parallel lines demonstrate that the two enzymes are similar in size but that they differ in charge; they are therefore charge-isomers.

The isolation of two active forms of lipase from *Rhizopus arrhizus* using a cation exchanger was described by Sémériva *et al.* This is an example of the isolation of two forms of an enzyme, the first being a direct product of the microorganism and the second being formed from the first by a slow conversion of an unknown type.

The lyophilized powder (5 g; activity 40 units/mg = specific activity 400), prepared according to Laboureur and Labrousse, was dissolved in 800 ml of water at 2°C and the solution was pumped (300 ml/h) into a 27 × 1.7 cm column of Amberlite IRC-50 (Type

*References p.829*

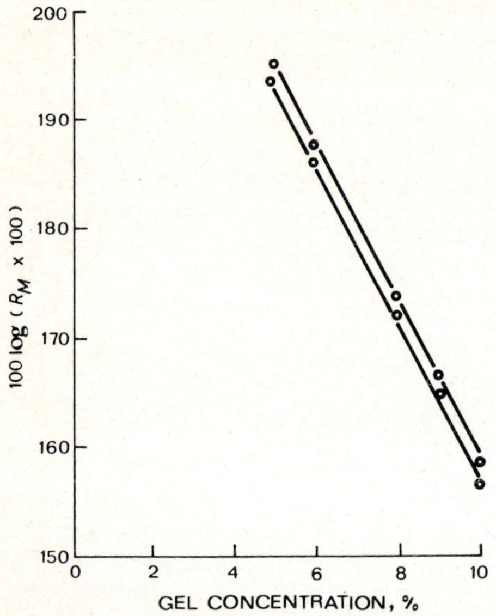


Fig. 36.7. Application of Hedrick and Smith's method for distinguishing types of arylsulphatase enzyme isomers (Delisle and Milazzo). The effect of different polyacrylamide gel concentrations on electrophoretic mobility of the sample was examined. Points represent the average of triplicate determinations on each gel.

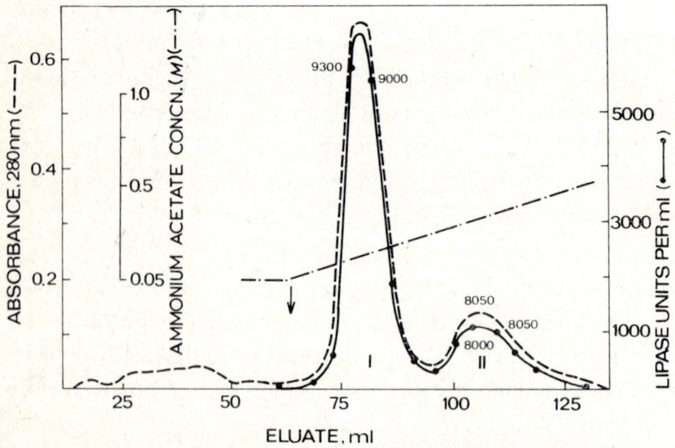


Fig. 36.8. Chromatography of a concentrate of a lipase from *Rhizopus arrhizus* on a weakly acidic cation-exchange resin, Amberlite IRC-50 (Sémériva *et al.*). Column dimensions: 20 × 0.9 cm. Buffer: the column was equilibrated with buffer B (see text). Sample: solution after Sephadex filtration ($10^5$ units). Washing: 30 ml of buffer B (10 ml/h). Desorption: a linear gradient of 50 ml of buffer B to 50 ml of buffer A. Solid line, lipase activity; broken line, total proteins; dotted line, buffer concentration in the eluate. I and II: different lipase forms. For the measurement of the lipase activity and definition of the unit, see original paper.

IRF-97, Rohm and Haas, Philadelphia, Pa., U.S.A.) equilibrated with 20 m*M* calcium acetate solution at pH 4.7. The column was washed (300 ml/h) with 400 ml of 5 m*M* calcium acetate solution and eluted (20 ml/h) with 1 *M* ammonium acetate buffer (pH 5.7) 5 m*M* in $Ca^{2+}$ (buffer A). The yield was 95% and the specific activity about 5000. The pooled active fractions were concentrated to 6.5 ml by vacuum dialysis against 50 m*M* ammonium acetate buffer (pH 5.7), 5 m*M* in calcium acetate (buffer B). The solution was freed from non-active contaminants on a 32 × 3.2 cm Sephadex G-75 column equilibrated and eluted with buffer B. Lipase emerged as a symmetrical peak (in 30 ml; 1.6 $V_0$; average specific activity 8000).

This fraction was re-chromatographed on an Amberlite IRC-50 column (Fig. 36.8). Two active peaks (I and II) were obtained with a specific activity of 9200 and 8000, respectively. The yield was 95%. The re-chromatography and disc electrophoresis of peak I immediately and after storage for 16 or 92 days in the cold indicated the conversion of form I into II, the latter being stable.

The chromatographic resolution of two forms of alkaline proteinase from *Aspergillus flavus* arising by conversion of the native enzyme when exposed to conditions near the transition state was described by Mikeš *et al.* The polymorphism of alkaline aspergillopeptidase, caused by transformations of the native enzyme when exposed either to high temperatures (near the transition temperature) or to treatment with 8 *M* urea at room temperature, was described. Various substances ($Ca^{2+}$, EDTA or ϵ-aminocaproic acid) added at a concentration of 0.02 *M* specifically influenced the forms that arise. The new forms of enzyme differed in the optimum pH of cleavage of haemoglobin, and in the specificity of cleavage of the B-chain of oxidized insulin.

In the example given in Fig. 36.9, an aqueous solution of the native enzyme was heated for 20 min at 45°C without any additives. The figure illustrates the influence of pH on the separation of the forms that arise. The best results were obtained only at pH 5.9, because at lower and higher pH peaks I and II coalesced.

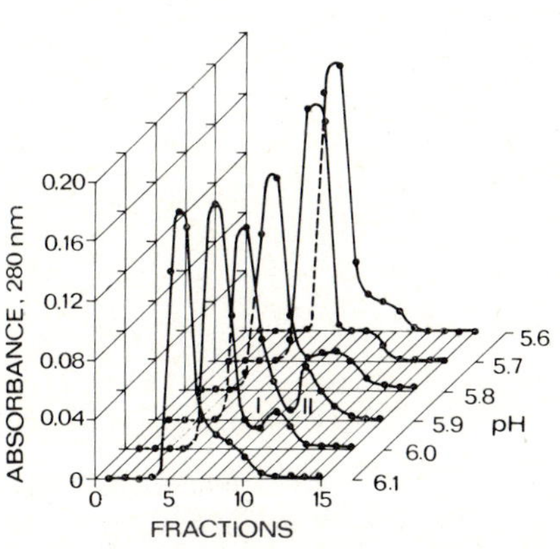


Fig. 36.9. Establishment of the optimum pH for the chromatographic separation of two forms of enzyme formed by thermal treatment of alkaline proteinase from *Aspergillus flavus* (Mikeš *et al.*). Sorbent: DEAE-Sephadex. Column dimensions: 20 × 1 cm. Buffers for equilibration and elution: the same concentration of phosphate was used (0.01 *M*), but the buffers differed in the pH indicated for particular runs. Simple elution was used. Fractions: 1.5 ml/h. Temperature: 4°C.

*References p.829*

The purification of acetylcholinesterase using affinity chromatography was described by Kalderon *et al.* Affinity chromatography of enzymes can utilize their specific interaction with fixed antibodies, substrates, effectors or inhibitors. The last case is illustrated by the example given below. The phenyltrimethylammonium ion is a good competitive inhibitor of acetylcholinesterase (Wilson and Alexander) and its affinity for the enzyme decreases with increase in ionic strength (Changeux). Hence there are good conditions for the possibility of selective desorption of the enzyme by changing the salt concentration in the eluent.

Kalderon *et al.* prepared this inhibitor in the form of [N-(ε-aminocaproyl)-*p*-aminophenyl]-trimethylammonium bromide hydrobromide, $[H_3\overset{+}{N}(CH_2)_5CO{\cdot}NH{\cdot}C_6H_4{\cdot}\overset{+}{N}(CH_3)_3]\ {\cdot}2\ Br$ (abbreviation: ε-aminocaproyl-PTA), the ε-amino group of which served to link the inhibitor to the BrCN-activated Sepharose 2B (Axén *et al.*). Also, ε-aminocaproyl-PTA was shown to be a good inhibitor of the enzyme (the inhibition constant $K_i$ is 6 $\mu M$).

A crude preparation of the enzyme was prepared from toluene-treated tissue from the electrical organ of the electric eel by extraction and fractionation with ammonium sulphate according to Leuzinger and Baker. The fraction between 15 and 35% ammonium sulphate was dissolved in 0.1 *M* sodium chloride–0.01 *M* phosphate (pH 7) and dialyzed against the same buffer for 12 h at 4°C. The enzyme prepared as described had a specific activity of 240 units per milligram of protein (1 unit being the amount of enzyme that hydrolyzes 1 μmole of acetylcholine per minute) and was applied on the column (*cf.*, Fig. 36.10). The first peak represents the unsorbed enzyme accompanied by other proteins in the washings. The second peak was found to have a specific activity of 4100 units per milligram of protein, *i.e.*, 17 times greater than the specific activity of the enzyme applied to the column. Higher concentrations of the inhibitor bound to Sepharose sorbed a larger

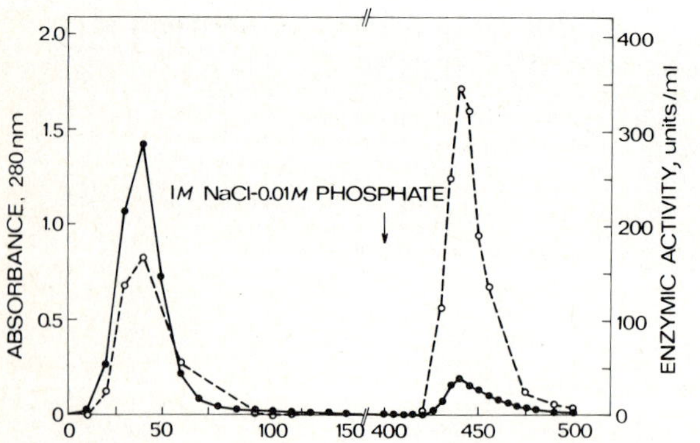


Fig. 36.10. Elution pattern of acetylcholinesterase from an affinity chromatography column (Kalderon *et al.*). Adsorbents: ε-aminocaproyl-PTA–Sepharose (0.16 μmole/ml of inhibitor). Column dimensions: 40 × 1.1 cm. Load applied on the column: 10 ml of a solution of crude enzyme preparation containing 15,000 units of esterase activity in 0.1 *M* sodium chloride–0.01 *M* phosphate (pH 7). Time of equilibration: 1 h. Washing: 0.1 *M* sodium chloride–0.01 *M* phosphate (pH 7). Elution: 1 *M* sodium chloride–0.01 *M* phosphate (pH 7). Fractions: 5 ml. Evaluation of fractions: Absorbance at 280 nm and determination of activity according to Kremzner and Wilson (pH-stat method at pH 7 and 25°C). Full line, protein ($A_{280}$); dashed line, enzymic activity.

amount of enzyme but the desorbate had a lower specific activity (*e.g.*, about 2000 units per milligram of protein with Sepharose containing 1.4 μmole/ml of inhibitor). The probable explanation is the non-specific sorption of proteins on ion-exchange quaternary ammonium groups which cannot be suppressed by the effect of higher ionic strength during the sorption.

The isolated enzyme displayed only one band on disc electrophoresis but contained aggregates owing to the known lability of acetylcholineesterase to lower ionic strength. They decreased the specific activity to 40% in comparison with the purest enzyme.

## LYASES

A simple method for the purification of L-glutamate 1-decarboxylase from *Escherichia coli* using a DEAE-Sephadex column and crystallization of the enzyme was described by Strausbauch *et al.* The starting material was a by-product after purification of pyruvate oxidase from *E. coli* described by Williams and Hager. The procedure of the latter authors was interrupted before the last protamine sulphate fractionation step, which was replaced with DEAE-Sephadex column chromatography, as illustrated in Fig. 36.11. The second peak was found to contain glutamate decarboxylase (60% purity). The active fractions were collected, concentrated by precipitation in ammonium sulphate and frozen.

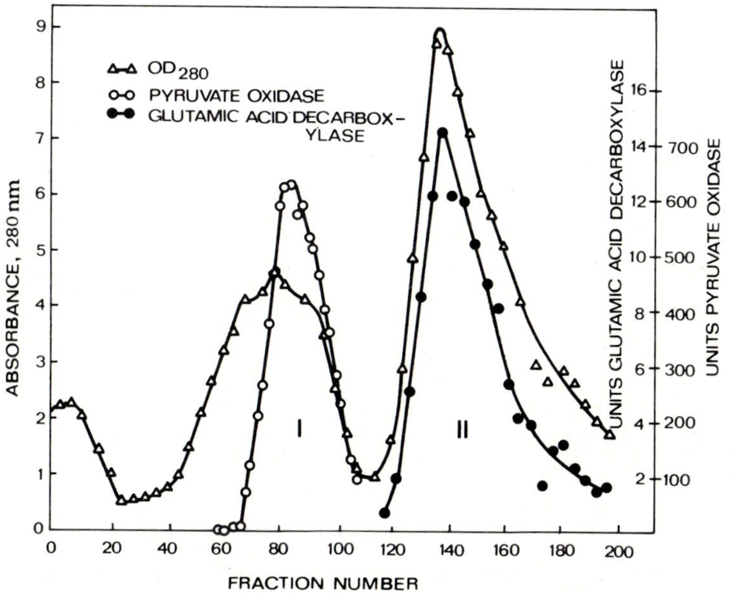


Fig. 36.11. Chromatographic purification of glutamate decarboxylase (*Escherichia coli*) on a DEAE-Sephadex A-50 column (Strausbauch *et al.*). Elution: linear gradient 0.02 *M* potassium phosphate buffer (pH 5.7)–0.3 *M* phosphate (pH 5.3). The pooled fractions of the second peak were further processed by crystallization. For the determinations of the enzyme activities and the definition of the units, see the original paper.

*References p.829*

For crystallization, the thawed fraction (40 mg/ml of protein) was adjusted to pH 6.5 with sodium phosphate buffer (0.05 *M* final concentration) and solid ammonium sulphate was added in small portions in the cold over a period of 5 days. Thin needles or flat plates appeared at a 15% (w/v) concentration of ammonium sulphate, which was then added up to 20% (w/v). Crystallization was repeated twice after dissolving the crystals in a minimum amount of 0.05 *M* phosphate buffer of pH 6.5. When examined by various methods, the crystals represent a homogeneous enzyme.

The isolation of chicken breast muscle aldolase by a combination of ion-exchange chromatography on cellulose and ammonium sulphate precipitation was described by Marquardt. Pure homogeneous aldolase (fructose 1,6-diphosphate D-glyceraldehyde-3-phosphate lyase) was prepared in six steps. All operations were carried out at 0–5°C. (I) *Extraction.* Breast muscles of freshly killed mature female chickens were chilled on ice and frozen in 200-g amounts. The thawed preparations were homogenized with 600 ml of 1 m*M* EDTA–5 m*M* 2-mercaptoethanol (pH 7.6) and the supernatant and the filtrate were used for further processing. (II) *First precipitation.* Solid ammonium sulphate was added to 50% saturation over 1 h and the supernatant after centrifugation (25,000 *g*, 30 min) was saturated to 63% over 1 h. After 3 h, the suspension was centrifuged again and the precipitate dissolved in the minimum volume of 0.1 *M* Tris–50 m*M* EDTA–10 m*M* 2-mercaptoethanol, pH 7.5. (III) *TEAE-cellulose chromatography.* The enzyme was

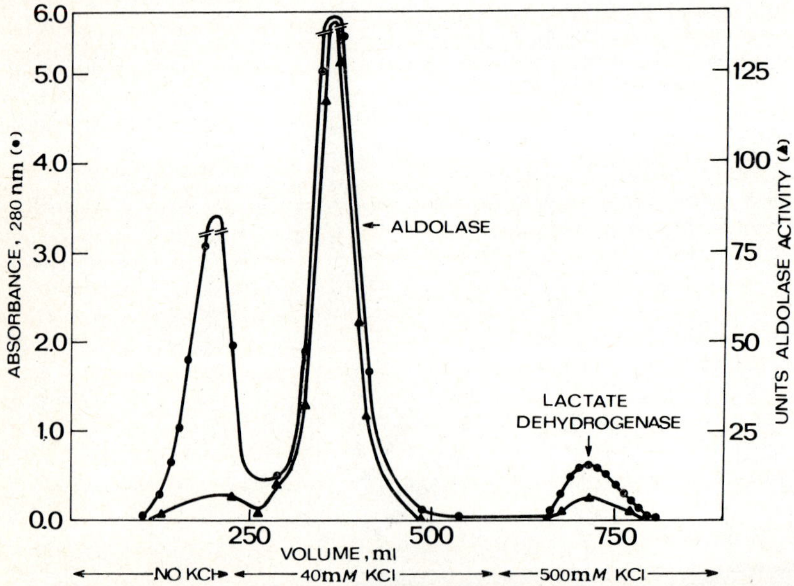


Fig. 36.12. Chromatography of chicken muscle aldolase on a CM-cellulose column (Marquardt). Ion exchanger: Cellex CM. Column dimensions: 22 × 4 cm. The column was equilibrated with pH 6.5 buffer (see text) and the unsorbed protein of the sample was flushed with 150 ml of the pH 6.5 buffer. Elution: stepwise, 40 m*M* potassium chloride in a pH 6.5 buffer and 500 m*M* potassium chloride in the same buffer. Flow-rate: 2.6 ml/min. Second peak: aldolase. Third peak: lactate dehydrogenase. Enzyme tests for these two enzymes were those of Rajmakur *et al.* and Kornberg, respectively.

dialyzed for 16 h against 12 l of 50 m$M$ Tris–10 m$M$ sodium phosphate–10 m$M$ EDTA–5 m$M$ 2-mercaptoethanol, pH 9.2. After adjustment to pH 9.3, the centrifuged solution (50,000 $g$, 20 min) was chromatographed on a 33 × 4.5 cm TEAE-cellulose (Cellex T, Bio-Rad Labs, Richmond, Calif., U.S.A.) column equilibrated with pH 9.2 dialyzing buffer. The column was washed with the same buffer (5 ml/min) and the void volume (identified by the coloured solution) emerged slightly before aldolase (approximately 30 ml). (IV) *Second precipitation.* The active pooled fractions were saturated with ammonium sulphate to 70% over 1 h and centrifuged (25,000 $g$, 30 min). The sediment was dissolved in 75 ml of 10 m$M$ sodium phosphate–5 m$M$ EDTA–5 m$M$ 2-mercapto-ethanol (pH 6.5) and dialyzed against 12 l of the pH 6.5 buffer. (V) *Chromatography on CM-cellulose.* The centrifuged supernatant (50,000 $g$, 30 min) of dialyzed enzyme was sorbed on to a CM-cellulose column (Cellex CM, Bio-Rad Labs.) and chromatographed (*cf.*, Fig. 36.12). All fractions with a specific activity of 20 and better were combined. (VI) *Crystallization.* The combined fractions were saturated with solid ammonium sulphate to 50% and centrifuged at 50,000 $g$ for 30 min. Solid ammonium sulphate was added until the first turbidity appeared. After 12 days at 4°C, the enzyme began to crystallize and the process was completed by adding 4% of ammonium sulphate. The crystals were collected after 1 day by centrifugation. Recrystallization was carried out as described above except that the buffer used was 0.1 $M$ sodium phosphate–1 m$M$ EDTA, pH 7.6. The purification was 4.4-fold and the yield 50%. Various methods proved the homogeneity of the preparation.

## ISOMERASES

Only 24 papers describing the use of column chromatographic methods for the fractionation of isomerases were found in the bibliography by Deyl *et al.* Therefore, only one example of this class of enzymes will be given here as an illustration of an *in vitro* hybridization of mouse phosphoglucose isomerase variants (Carter and Yoshida).

D-Glucose-6-phosphate ketol-isomerase has been found to exist in three genetically determined electrophoretic phenotypes in various mouse tissues:

Phenotype I: associated with F type (fast cathode-migrating enzyme);

Phenotype II: associated with S type (slow migrating enzyme);

Phenotype III: possesses three enzyme components, F, FS and S (FS being an enzyme of intermediate mobility).

When F and S types of mice were cross-bred, the offspring showed three enzyme bands; two of them were identical with the F and S bands of the parents, but the third seemed to be a hybrid of F and S. To verify this possibility, Carter and Yoshida tried to prove that the recombination occurred also *in vitro*. Hybridization with the crude extract was not possible and therefore the purification of enzymes was necessary.

The muscle tissue from three phenotypes of laboratory mouse was selected as a source of enzymes and was pooled. After homogenization in 0.01 $M$ potassium chloride solution and fractionation with 0.03 $M$ zinc acetate and ammonium sulphate (Noltmann), the enzyme was dialyzed against 0.005 $M$ sodium phosphate buffer (pH 6.8) and purified on a 35 × 1.5 cm column of calcium phosphate gel equilibrated with the same buffer. The enzyme was eluted with a linear gradient comprising 300 ml of 0.005 $M$ sodium phosphate

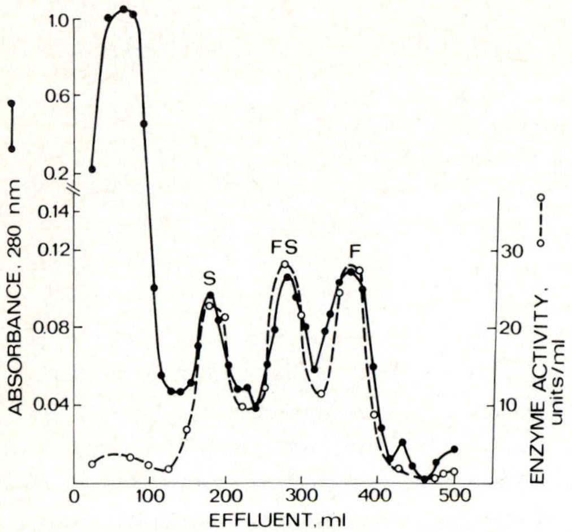


Fig. 36.13. Chromatography of mouse phosphoglucose isomerase on a CM-Sephadex column (Carter and Yoshida). Column dimensions: 35 × 1.5 cm. Equilibration: 0.005 *M* sodium phosphate buffer, pH 6.8. Elution: linear gradient, 300 ml of the same buffer–300 ml of 0.025 *M* sodium phosphate buffer, pH 6.8. For enzyme assay and definition of the unit, see the original paper. S, FS and F are enzymes of the three phenotypes (see text), distinct in starch gel electrophoresis.

to 300 ml of 0.025 *M* sodium phosphate (pH 6.8). The active fractions were pooled and concentrated by vacuum dialysis and the enzyme preparation was then dialyzed against the more diluted buffer of pH 6.8 and chromatographed on CM-Sephadex (Fig. 36.13). The material of the active enzyme peak was used for further experiments.

For hybridization, equal volumes (0.1 ml) of solution of peaks F and S of the same activity were mixed, and a similar aliquot of the FS peak was prepared. These two solutions were made 2 *M* in guanidine hydrochloride, 50 m*M* in 2-mercaptoethanol, 35 m*M* in EDTA and 25 m*M* in sodium phosphate buffer, pH 6.8. The solutions were kept on ice for 3 min and then dialyzed against 2-1 volumes of 0.005 *M* sodium phosphate, pH 6.8. Then the solution was concentrated to the initial volume by vacuum dialysis. A control sample without the guanidine hydrochloride treatment was also prepared. The samples were examined electrophoretically. The results can be summarized as follows:

F + S + guanidine hydrochloride →
FS + guanidine hydrochloride → F + FS + S
Offspring from the mating of F and S phenotypes →

Control samples were without effect on the hybridization.

The three enzymes in the three phenotypes are composed of peptide chains (sub-units) f and s, controlled by two different alleles and associated in the enzymes in question, ff, fs and ss.

## LIGASES

The isolation of two methionyl-tRNA synthetases from *Escherichia coli* and the evidence for their specific interaction with particular tRNA using chromatography on methylated albumin was described by Cerhová and Rychlík. A problem in proteosynthesis has been to decide whether the amino acid is attached to the respective tRNAs by only one enzyme or whether there are as many activating enzymes as there are tRNAs for the same amino acid. Cerhová and Rychlík isolated two methionyl tRNAs and studied their specific interaction.

The crude extract, after sonic disintegration of *E. coli* cells (100 g in 500 ml of 0.025 *M* Tris–hydrochloric acid buffer, pH 8.0), was incubated at 37°C for 2 h. The autolyzate was precipitated with ammonium sulphate, the enzyme being separated between 35 and 55% saturation. The sediment, after centrifugation at 30,000 *g* for 10 min, was dissolved in 0.02 *M* potassium phosphate, pH 7.2 (100 mg/ml of protein) and chromatographed on a Sephadex G-75 column (50 × 4 cm), equilibrated by and eluted with the same buffer. The active 5-ml fractions were pooled and stirred for 10 min with alumina $C_\gamma$ gel (12 mg/ml) at pH 6.5. The centrifugation gel was washed with 200 ml of water. The enzyme was eluted with 100 ml of 0.1 *M* potassium phosphate solution of pH 7.0 and dialyzed against 0.02 *M* phosphate buffer of pH 7.5.

Chromatography on a 20 × 2.5 cm column of DEAE-cellulose (gradient: 0.07 to 0.2 *M* phosphate buffer; 5-ml fractions resolved the preparation into two peaks, I and II, both charging tRNA with methionine. On re-chromatography, the positions of the peaks remained unchanged. Methionyl-tRNA synthetase I differs from II: the double-labelling experiments showed that each of them acylates with methionine a different species of methonine-tRNA.

tRNA was charged with [U-$^{14}$C] methionine (4.7 μCi/μmole) by methionyl-tRNA synthetase I or with [U-$^{3}$H] methionine (153 μCi/μmole) by methionyl-tRNA synthetase II. Both samples were mixed and chromatographed on a methylated albumin–Kieselguhr column (Fig. 36.14). Methionyl-tRNA of *E. coli* was resolved into two components (*cf.* also Černá *et al.*); $^{3}$H was found to be attached to the first and $^{14}$C to the second component. Methionyl-tRNA synthetase I selectively acylated the second component of methionine-tRNA, and methionyl-tRNA synthetase II the first component. Hence the two different methionyl-tRNA synthetases correspond to two different species of methionyl-tRNA.

The mechanism of pyruvate carboxylase formation from apoenzyme and biotin in a thermophilic bacillus was described by Cazzulo *et al.* (1969, 1970). This is another example of the use of chromatographic methods for the solution of biochemical problems. Pyruvate carboxylase is formed by the attachment of biotin to its inactive protein precursor, apopyruvate carboxylase. This reaction is catalyzed by holoenzyme synthetase. The reconstruction of the pyruvate carboxylase activity in cell-free extract requires acetyl-CoA, $Mg^{2+}$ and ATP in addition to biotin. An experiment was designed to demonstrate the effect of acetyl-CoA on the incorporation of labelled biotin into the apoenzyme.

First the apoenzyme was purified from a culture of *Bacillus coagulans* (a variant of *B. stearothermophilus*). The washed cells were digested with lysozyme and the extract was fractionated with ammonium sulphate. Apoenzyme and holoenzyme synthetase can be resolved from this preparation by chromatography on Sephadex G-200 (*cf.*, Fig. 36.15a), the apoenzyme being eluted well before holoenzyme synthetase. The following incubation mixture was prepared: 2.2 mg of apoprotein and 3.1 mg of holoenzyme synthetase obtained after chromatography on Sephadex G-200, plus 12.5 μmoles of

*References p.829*

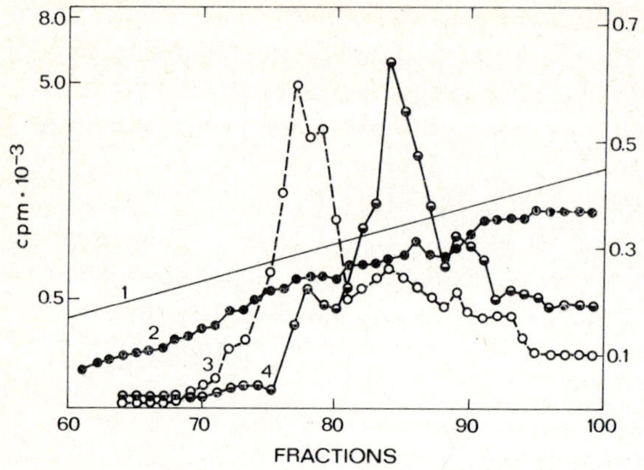


Fig. 36.14. Chromatography of methionyl-tRNA on methylated albumin–Kieselguhr using specific double labelling (Cerhová and Rychík). The 20 × 2.5 cm column of methylated albumin–Kieselguhr was prepared according to Yamane and Sueoka. Load: 10 mg of [U-$^{14}$C] methionyl-tRNA (specific activity 1.4 $\mu$Ci/$\mu$mole) and 3 mg of [U-$^{3}$H] methionyl-tRNA (specific activity 4.9 $\mu$Ci/$\mu$mole). 1, gradient 0.25 $M$–0.35 $M$ sodium chloride solution; 2, $A_{280}$ values; 3, number of impulses from methionyl-tRNA linked through methionyl-tRNA synthetase II; 4, number of impulses from methionyl-tRNA linked through methionyl-tRNA synthetase I.

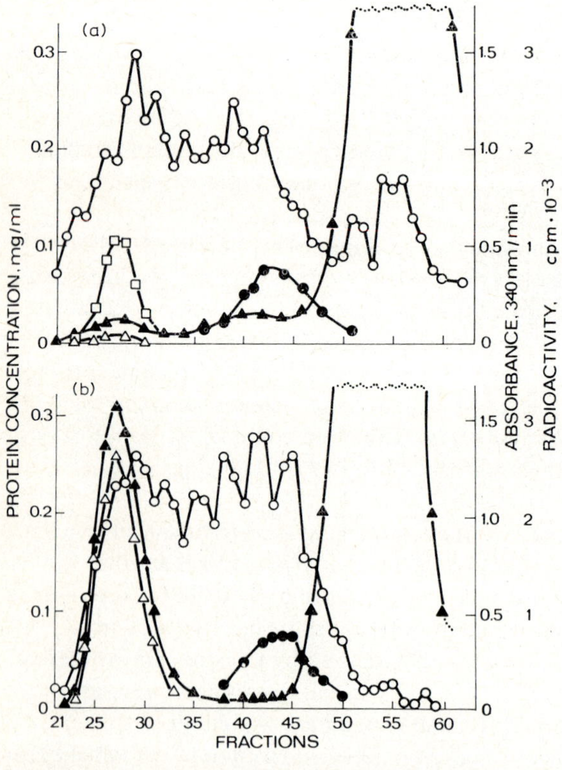


Fig. 36.15. Chromatographic evidence of the role of acetyl-CoA on the synthesis of pyruvate carboxylase using a Sephadex G-200 column (Cazzulo *et al.*, 1970). Column dimensions: 30 × 1.7 cm. Equilibration: 50 m$M$ Tris–hydrochloric acid (pH 7.6), 1 m$M$ in EDTA and 0.4 $M$ in ammonium sulphate. The solution of precipitated incubation mixture (see text) was applied to the column. The proteins were eluted with the same buffer as above. Fractions: 1 ml. Assays: ○, protein spectrophotometrically; ▲, radioactivity; △, holopyruvate carboxylase spectrophotometrically (Cazzulo *et al.*, 1969). Apopyruvate carboxylase (□) or holoenzyme synthetase (●) were tested after incubation with other additive substances (details of the procedures are given in the original paper). a, Experiment without acetyl-CoA; b, experiment with acetyl-CoA.

magnesium chloride, 3.1 $\mu$moles of ATP, 0.17 $\mu$moles of (+)-[$^{14}$C] biotin (10 $\mu$Ci) and 37.5 $\mu$moles of Tris–hydrochloric acid (pH 7.6), in a final volume of 7 ml. After incubation for 30 min at 45°C and cooling to 0°C, 20 $\mu$moles of EDTA were added and the proteins precipitated with 20 ml of saturated ammonium sulphate solution. The precipitate was dissolved in 0.5 ml of 50 m$M$ Tris–hydrochloric acid (pH 7.6, 1 m$M$ in EDTA and 0.4 $M$ in ammonium sulphate, and the solution was applied to the column (Fig. 36.15b).

When acetyl-CoA was not included in the incubation mixture (Fig. 36.15a), there was little incorporation of (+)-[$^{14}$C] biotin into apoprotein and no formation of pyruvate carboxylase activity was observed. When 2.5 $\mu$moles of acetyl-CoA were included in the incubation mixture in a duplicate experiment (Fig. 36.15b), a large peak of radioactivity was associated with a protein that possessed pyruvate carboxylase activity. This peak was eluted at the same place as that occupied in the above experiment by the apoprotein.

On the basis of these and other experiments, Cazzulo *et al.* (1970) considered the acetyl-CoA to be an allosteric effector of the reconstitution process.

## REFERENCES


Alpert, N. L., *Clin. Chem.*, 15 (1969) 1198.
Avis, P. G., Bergel, F. and Bray, R. C., *J. Chem. Soc., London*, (1955) 1100.
Axén, R., Porath, J. and Ernback, S., *Nature (London)*, 214 (1967) 1302.
Beck, C. and Tappel, A. L., *Anal. Biochem.*, 21 (1967) 208.
Bensch, W. R. and Rodwell, V. W., *J. Biol. Chem.*, 245 (1970) 3755.
Bernardi, G., *Methods Enzymol.*, 22 (1971) 325.
Blatt, W. F., *Methods Enzymol.*, 22 (1971) 39.
Bradley, D. W. and Tappel, A. L., *Anal. Biochem.*, 33 (1970) 400.
Carter, N. D. and Yoshida, A., *Biochim. Biophys. Acta*, 181 (1969) 468.
Cazzulo, J. J., Sundaram, T. K. and Kornberg, H. L., *Nature (London)*, 223 (1969) 1137.
Cazzulo, J. J., Sundaram, T. K. and Kornberg, H. L., *Nature (London)*, 227 (1970) 1103.
Čerhová, M. and Rychlík, I., *Collect. Czech. Chem. Commun.*, 32 (1967) 3808.
Černá, J., Rychlík, I. and Šorm, F., *Collect. Czech. Chem. Commun.*, 31 (1966) 336.
Changeux, J. P., *Mol. Pharmacol.*, 2 (1966) 369.
Cline, G. B. and Ryel, R. B., *Methods Enzymol.*, 22 (1971) 39.
Cuatrecasas, P., *Advan. Enzymol.*, 36 (1972) 29.
Cuatrecasas, P. and Anfinsen, C. B., *Annu. Rev. Biochem.*, 40 (1971a) 259.
Cuatrecasas, P. and Anfinsen, C. B., *Methods Enzymol.*, 22 (1971b) 345.
Davies, D. R. and Segal, D. M., *Methods Enzymol.*, 22 (1971) 266.
Delisle, G. and Milazzo, F. H., *Biochim. Biophys. Acta*, 212 (1970) 505.
Deyl, Z., Rosmus, J., Juřicová, M. and Kopecký, J., *Bibliography of Column Chromatography 1967–70*, Elsevier, Amsterdam, London, New York, 1973.
Dodgson, K. S. and Spencer, B., *Methods Biochem. Anal.*, 4 (1957) 246.
Dundas, I. E. D., *Eur. J. Biochem.*, 16 (1970) 393.
Eley, J., *Biochemistry*, 8 (1969) 1502.
Eveleigh, J. W., Adler, H. J. and Reichler, A. S., *Automat. Anal. Chem., Technicon Symp., 1967*, Vol. 1, Mediad Inc., White Plains, N.Y., 1968, p. 311.
Everse, J. and Stolzenbach, F. E., *Methods Enzymol.*, 22 (1971) 33.
Feinstein, G., *Naturwissenschaften*, 58 (1971) 389.
Fridovich, I., *J. Biol. Chem.*, 237 (1962) 584.
Fried, M. and Chun, P. W., *Methods Enzymol.*, 22 (1971) 238.
Hedrick, J. L. and Smith, A. J., *Arch. Biochem. Biophys.*, 126 (1968) 155.
Himmelhoch, S. R., *Methods Enzymol.*, 22 (1971) 273.
Howard, R. L. and Becker, R. R., *J. Biol. Chem.*, 245 (1970) 3186.
Jakoby, W. B., *Methods Enzymol.*, 22 (1971) 248.
Kalderon, N., Silman, I., Blumberg, S. and Dudai, Y., *Biochim. Biophys. Acta*, 207 (1970) 560.
Kaufman, S., *Methods Enzymol.*, 22 (1971) 233.
Koike, M. and Hamada, M., *Methods Enzymol.*, 22 (1971) 339.
Kornberg, A., *Methods Enzymol.*, 1 (1955) 491.
Kremzner, L. T. and Wilson, I. B., *J. Biol. Chem.*, 238 (1963) 1714.

Laboureur, P. and Labrousse, M., *C.R. Acad. Sci., Paris,* 259 (1964) 4394; *Bull. Soc. Chim. Biol.*, 48 (1966) 747.

Leuzinger, W. and Baker, A. L., *Proc. Nat. Acad. Sci. U.S.*, 57 (1967) 446.

Levin, O., *Methods Enzymol.*, 5 (1962) 27.

Lowry, O. H., Rosebrough, N. J., Farr, A. L. and Randall, R. J., *J. Biol. Chem.*, 193 (1951) 265.

McPhie, P., *Methods Enzymol.*, 22 (1971) 23.

Marquardt, R. R., *Can. J. Biochem.*, 47 (1969) 517.

Mikeš, O. (Editor), *Laboratory Chromatographic and Other Separation Methods*, Ellis Horwood Publ. Co., Chichester, and SNTL, Prague (Czech version), 1975, in press.

Mikeš, O., Worowski, K. and Turková, J., *Collect. Czech. Chem. Commun.*, 38 (1973) 3339.

Miller, J., Cuatrecasas, P. and Thompson, E. B., *Biochim. Biophys. Acta*, 276 (1972) 407.

Moss, D. W., *Med. Electron. Biol. Eng.*, 3 (1965) 327.

Newbold, P. C. H. and Harding, N. G. L., *Biochem. J.*, 124 (1971) 1.

Noltmann, E. A., *J. Biol. Chem.*, 239 (1964) 1545.

Penefsky, H. and Tzagoloff, A., *Methods Enzymol.*, 22 (1971) 204.

Pogel, B. M. and Sarngadharan, M. G., *Methods Enzymol.*, 22 (1971) 379.

Porath, J., *Biotechnol. Bioeng. Symp.*, No. 3, 1972, p. 145.

Poston, J. M., Stadtman, T. C. and Stadtman, E. R., *Methods Enzymol.*, 22 (1971) 49.

Prusík, Z., in Mikeš, O. (Editor), *Laboratory Chromatographic and Other Separation Methods*, Ellis Horwood Publ. Co., Chichester, and SNTL, Prague (Czech version), 1975, in press.

Rajmakur, T. V., Woodfin, B. M. and Rutter, W. J., *Methods Enzymol.*, 9 (1966) 491.

Rattazzi, M. C., *Biochim. Biophys. Acta*, 181 (1969) 1.

Reiland, J., *Methods Enzymol.*, 22 (1971) 287.

Repaske, R., *Methods Enzymol.*, 22 (1971) 322.

Rony, P. P., *Biotechnol. Bioeng.*, 13 (1971) 431.

Roussos, G. G. and Morrow, B. H., *Biochem. Biophys. Res. Commun.*, 29 (1967) 388.

Sarngadharan, M. G., Watanabe, A. and Pogell, B. M., *J. Biol. Chem.*, 245 (1970) 1926.

Schneir, M. and Moldave, K., *Biochim. Biophys. Acta*, 166 (1968) 58.

Schwartz, M. K., *Automat. Anal. Chem., Technicon Symp., 1967*, Vol. 1, Mediad Inc., White Plains, N.Y., 1968, p. 587.

Schwartz, M. K., *Methods Enzymol.*, 22 (1971) 5.

Schwartz, M. K., Kessler, G. and Bodansky, O., *Ann. N.Y. Acad. Sci.*, 87 (1960) 616.

Sémériva, M., Benzonana, G. and Desnuelle, P., *Biochim. Biophys. Acta*, 144 (1967) 703.

Shuster, L., *Methods Enzymol.*, 22 (1971) 412 and 434.

Sipos, T. and Merkel, J., *Biochem. Biophys. Res. Commun.*, 31 (1968) 522; *Biochemistry*, 9 (1970) 2766.

Smythe, W. J., Shamos, M. M., Morgesern, S. and Skeggs, L. T., in *Automat. Anal. Chem., Technicon Symp., 1967,* Vol. 1, Mediad Inc., White Plains, N.Y., 1968, p. 105.

Strausbauch, P. H., Fischer, E. H., Cunningham, C. and Hager, L. P., *Biochem. Biophys. Res. Commun.*, 28 (1967) 525.

Swanljung, P., *Anal. Biochem.*, 43 (1971) 382.

Swanljung, P. and Frigeri, L., *Biochim. Biophys. Acta*, 283 (1972) 391.

Tappel, A. L., *Methods Enzymol.*, 22 (1971) 219.

Tiselius, A., Hjertén, S. and Levin, Ö., *Arch. Biochem. Biophys.*, 65 (1956) 132.

Turková, J., in Mikeš, O., (Editor), *Laboratory Chromatographic and Other Separation Methods,* Ellis Horwood Publ. Co., Chichester, and SNTL, Prague (Czech version), 1975, in press.

Tzagoloff, A. and Penefsky, H. S., *Methods Enzymol.*, 22 (1971) 219.

Večerek, B., Kácl, K., Kolařík, L., Chundela, B. and Večerková, J., *Chem. Listy*, 53 (1959) 279.

Vesterberg, O., *Methods Enzymol.*, 22 (1971) 389.

Wacker, W. E. C. and Coombs, T. C., *Annu. Rev. Biochem.*, 38 (1969) 539.

Williams, F. R. and Hager, L. P., *Biochim. Biophys. Acta*, 116 (1966) 168.

Wilson, I. B. and Alexander, J., *J. Biol. Chem.*, 237 (1962) 1323.

Yamane, T. and Sueoka, N., *Proc. Nat. Acad. Sci. U.S.*, 50 (1963) 1093.

Zeppezauer, M., *Methods Enzymol.*, 22 (1971) 253.

*Chapter 37*

# Low-molecular-weight constituents of nucleic acids

# Nucleosides, nucleotides and their analogues

S. ZADRAŽIL

## CONTENTS



## INTRODUCTION

Nucleic acids are composed of phosphoric acid, a sugar component (deoxyribose or ribose) and purine and pyrimidine bases (adenine, guanine and cytosine, thymine or uracil). These basic components can be isolated from total hydrolyzates of polymers, while partial hydrolysis of the polynucleotide chain leads to fragments in the form of nucleosides, nucleotides and oligonucleotides (Table 37.1). These components can also be isolated from the cell pool, the composition of which is the result of an equilibrium between biosynthetic and catabolic cell processes (Table 37.2). In addition to the above substances, cellular pools also contain more highly phosphorylated nucleoside derivatives (diphosphates and triphosphates) and some nucleotide-type coenzymes (DPN, TPN, NAD, FAD, UDPG, coenzyme A, etc.) (Hutchinson).

In addition to the four fundamental nucleosides for each type of nucleic acid, six so-called minor components were found in DNA (mostly from bacteriophages) and about 35 in RNA (mainly tRNA). A survey of these components, which can be isolated only as cleavage products of natural polymers because they are formed by modification of the fundamental components on the macromolecular level, is shown in Table 37.3. Obviously, when studying the structure and function of nucleic acids, many synthetic intermediates are encountered, *e.g.*, nucleosides with protective groups in organic synthesis, and analogues of bases and nucleosides, *e.g.*, azapyrimidines (Škoda), the separation of which must therefore also be taken into account in experimental work (Zadražil, 1972).

TABLE 37.1
HYDROLYTIC PROCEDURES FOR THE ISOLATION OF NUCLEIC ACID COMPONENTS

| Conditions | Substrate(s) | Final products | Reference |
|---|---|---|---|
| 72% $HClO_4$ (12 *N*), 100°C, 1–2 h | DNA, RNA, oligonucleotides | Purine and pyrimidine bases | Littlefield and Dunn, Marshak and Vogel, Sluyser and Bosch |
| 98–100% HCOOH, 175°C, 1 h | DNA and oligonucleotides | Purine and pyrimidine bases | Wyat, Wyat and Cohen |
| 0.3 *M* KOH, 37°C, 16 h | RNA and oligonucleotides | 3′(2′)-Mono-nucleotides (96%) | Bock, Singh and Lane |
| *Crotalus adamanteus* venom (1:10 enzyme: substrate) and bacterial alkaline phosphatase (1:30), 37°C, 20–24 h | tRNA and RNA in buffer of pH 8.6 with 5 m*M* $MgCl_2$ | Nucleosides | Hall (1965) |
| DNase I (1:100), 37°C, 2 h, pH 6–7, and *Naja naja* venom (1:50), 37°C, 5 h, pH 8–9 | Denatured DNA in a buffer with 30 m*M* $MgSO_4$ | Nucleosides | Piťhová *et al.* |
| Specific nucleases (various conditions of complete and partial hydrolyses) | DNA, RNA and oligonucleotides | Mono- and oligonucleotides | Zadražil (1973) |

TABLE 37.2
EXTRACTION OF CELL NUCLEOTIDE POOL (HUTCHISON AND MUNRO)

| Extraction agent | Conditions | Removal of extraction agent |
|---|---|---|
| Trichloroacetic acid | 5–10% at 4°C, repeated four times | Extraction with diethyl ether (several times) |
| Perchloric acid | 1.2–6% at 4°C, repeated 3–4 times | As $KClO_4$ by centrifugation for 10 min at 2000 *g* below 4°C |

## GENERAL TECHNIQUES IN THE SEPARATION OF LOW-MOLECULAR-WEIGHT COMPONENTS OF NUCLEIC ACIDS

In order to separate purine and pyrimidine bases, nucleosides and nucleotides, all types of chromatographic techniques are employed in practical laboratory work, including paper chromatography (paper electrophoresis and fingerprinting techniques), thin-layer chromatography (Randerath) and all types of column chromatography (adsorption, partition, ion exchange and gel permeation). It is usually stated, from the quantitative point of view, that the optimum amounts of nucleotides for separation by means of the above techniques are 0.2–30 μg for TLC, 10–200 μg for paper chromatography and 100–500 μg for paper electrophoresis, while for columns the optimum amount is between 50 μg and several hundred milligrams. Evidently, these ranges may be considerably

TABLE 37.3
LIST OF NUCLEOSIDES OCCURRING IN NATURAL NUCLEIC ACIDS (HALL, 1971)

| Compound | Compound |
|---|---|
| Adenosine | Uridine |
| 1-Methyl- | 3-Methyl- |
| 2-Methyl- | 5-Methyl- |
| $N^6$-Methyl- | 2′-O-Methyl- |
| $N^6$, $N^6$-Dimethyl- | 2-Thio-5-carboxymethyl- |
| 2′-O-Methyl- | (methyl ester) |
| $N^6$-($\Delta^2$-Isopentenyl)- | 5-Hydroxy- |
| $N^6$-(*cis*-4-Hydroxy-3-methylbut-2-enyl)- | 5-Carboxymethyl- |
| $N^6$-($\Delta^2$-Isopentenyl)-2-methylthio- | 5,6-Dihydro- |
| 2′ (3′)-O-Ribosyl- | 4-Thio- |
| N-[9-(β-D-Ribofuranosyl-9H-purin-6-yl) | 2-Thio-5-(N-methylaminomethyl)- |
| carbamoyl]-L-threonine-[N-(nebularin- | Pseudo- |
| 6-ylcarbamoyl)]-L-threonine | [5-(β-D-Ribofuranosyl)uracil] |
| | 2′-O-Methylpseudo- |
| Inosine | [5-(2′-O-Methylribosyl)uracil] |
| 1-Methyl- | |
| | Deoxyadenosine |
| Guanosine | $N^6$-Methyl- |
| 1-Methyl- | |
| 7-Methyl- | Deoxyguanosine |
| $N^2$-Methyl- | |
| 2′-O-Methyl- | Deoxycytidine |
| $N^2$, $N^2$-Dimethyl- | 5-Methyl- |
| | 5-Hydroxymethyl- |
| Cytidine | |
| 3-Methyl- | Deoxyuridine |
| 5-Methyl- | 5-Methyl- |
| 2′-O-Methyl- | (thymidine) |
| $N^4$, $O^2$-Dimethyl- | 5-Hydroxymethyl- |
| $N^4$-Acetyl- | 5-(4′, 5′-dihydroxypentyl)- |
| 2-Thio- | |

exceeded in special cases of analytical or preparative separations.

In order to prepare samples for chromatography and further treatment, it is usually necessary to remove excessive concentrations of salts and low-molecular-weight substances (*e.g.*, urea), which would have an unfavourable influence on later separation and analysis. As mixtures of low-molecular-weight substances are involved, the use of dialysis and normal exclusion in gel filtration (the main methods of desalting nucleic acids) is greatly limited.

In general, chemical and adsorption methods can be recommended, *e.g.*, extraction with an acetone–alcohol mixture (Blumson and Baddiley, Christianson *et al.*) and adsorption on activated carbon (Rudner *et al.*; Zadražil, 1972), which can also be used in a column arrangement (adsorption at pH 4–5 and elution with ammonia-containing alcohol). With column techniques, anion-exchange carriers on a cellulose base (DEAE-

*References p.855*

cellulose with elution with carbonate and hydrogen carbonate buffers; Cohn and Bollum; Rushizky and Sober, 1962) are mainly used for nucleotides and their polyphosphates. Most salts used in chromatographic gradients can be separated from nucleic acid components, particularly nucleosides and bases, by means of gel filtration on Bio-Gel P-2 (Table 37.4; Uziel; Uziel and Cohn, 1965a, b) and Sephadex columns (Flodin). Hence the two main types of gel carriers employed can also, in general, be used in order to desalt bases, particularly due to adsorption of the bases on the column material. Most salts (chlorides, iodides, acetates, formates, phosphates, etc.) are eluted earlier in the form of sharp, distinct peaks (Khym and Uziel, 1970; Simkin).

TABLE 37.4
DESALTING OF NUCLEIC ACID COMPONENTS ON A COLUMN OF BIO-GEL P-2 (UZIEL; UZIEL AND COHN, 1965b)

| Compound | $K_d$ (at room temperature) |
|---|---|
| Oligonucleotides (RNA included) at pH 8 | 0 |
| Adenosine-5′-monophosphate | 0.24 |
| Cytidine- and uridine-5′-monophosphates | 0.3 |
| Guanosine-5′-monophosphate | 0.7 |
| Acetate and Tris | 0.8 |
| $HPO_4^{2-}$ | 0.82 |
| Formate | 0.84 |
| $H_2PO_4^-$ and ammonium hydrogen carbonate | 0.85 |
| Tris $H^+$ (as $Cl^-$) and $Cl^-$ | 1.0 |
| Formic and acetic acids | 1.1 |
| Urea and $Br^-$ | 1.4 |
| Cytidine, uridine and thiouridine | 1.6 |
| Cytosine, thymine and uracil | 1.8 |
| Adenosine and guanosine | 2.1 |
| Adenine and guanine | 3.1 |
| Oligonucleotides (RNA included) at pH 3 | 8 |

Insoluble poly-N-vinylpyrrolidone is a recent material for desalting all nucleic acid components; when used in a column it separates nucleotides, nucleosides, pyrimidines and purines in that order (Lerner *et al.*) and can be eluted with water (Dougherty and Schepartz, 1969a). In this instance also, as with Bio-Gels, salts pass through the column unimpeded while nucleoside and bases are delayed (interaction by hydrogen bonds; Dougherty and Schepartz, 1969b). Lithium chloride and sodium chloride can easily be separated from all components, while ammonium sulphate is eluted together with the nucleotide fraction. For nucleotides, it is therefore advantageous to use the above procedures with substituted cellulose or gels for desalting.

With respect to the wide variations in the chemical compositions of the substances involved, the differences in the dissociation of the various substituents (for p$K$ values, see Table 37.5), distribution coefficients, electron density, size and shape of molecules, etc., may all be utilized in separation processes. The sorbents most widely used in separating bases, nucleosides and nucleotides are synthetic ion exchangers, which were introduced by Cohn (1949a, b), and which together with gradient elution (Hurlbert *et al.*, Schmitz *et al.*) are the most generally applicable separation techniques. For oligonucleotides, classical ion-exchange resins were replaced mostly with substituted celluloses and dextran gels (Rushizky and Sober, 1968; Staehelin), which, used with urea-containing gradient solutions (Tomlinson and Tener), are the main tool for use in column separations of enzyme hydrolyzates of nucleic acids in sequence analysis.

TABLE 37.5
$pK'_a$ VALUES OF PURINES, PYRIMIDINES, NUCLEOSIDES AND NUCLEOTIDES* (SMITH)

| Compound | Primary phosphate | Amino (basic) | Secondary phosphate | Enol (acidic) |
|---|---|---|---|---|
| Adenine | – | 4.22 | – | 9.8 |
| 6-Methylaminopurine | – | 4.18 | – | 9.99 |
| 6-Dimethylaminopurine | – | 3.87 | – | 10.5 |
| Adenosine | – | 3.45 | – | – |
| 2′-phosphate | 0.89 | 3.80 | 6.15 | – |
| 3′-phosphate | 0.89 | 3.65 | 5.88 | – |
| 5′-phosphate | | 3.74 | 6.05 | – |
| 5′-pyrophosphate | | 3.95 | 6.26 | – |
| 5′-triphosphate | | 4.0 | 6.48 | – |
| Guanine | – | 3.3 | – | 9.2, 12.3 |
| Guanosine | – | 1.6 | – | 9.16 |
| 2′-phosphate | 0.7 | 2.3 | 5.9 | 9.7 |
| 3′-phosphate | 0.7 | 2.3 | 5.9 | 9.7 |
| 5′-phosphate | | 2.4 | 6.1 | 9.4 |
| 5′-pyrophosphate | | 2.9 | 6.3 | |
| 5′-triphosphate | | 3.3 | 6.5 | |
| Hypoxanthine | – | 1.98 | – | 8.9, 12.1 |
| Inosine | – | – | – | 8.75 |
| 5′-phosphate | 1.54 | – | 6.0 | 8.9 |
| Xanthine | – | – | – | 7.4, 11.1 |
| Xanthosine | – | – | – | 5.75 |
| Cytosine | – | 4.45 | – | 12.2 |
| 5-Methylcytosine | – | 4.6 | – | 12.4 |
| 5-Methylcytidine | – | 4.28 | – | – |
| Cytidine | – | 4.22 | – | – |
| 2′-phosphate | 0.8 | 4.36 | 6.17 | |
| 3′-phosphate | 0.8 | 4.28 | 6.0 | |
| 5′-phosphate | | 4.5 | 6.3 | |
| 5′-pyrophosphate | | 4.6 | 6.4 | |
| 5′-triphosphate | | 4.8 | 6.6 | |
| Uracil | – | – | – | 9.5 |
| Uridine | – | – | – | 9.17 |
| 2′-phosphate | 1.0 | – | 5.9 | 9.4 |
| 3′-phosphate | 1.0 | – | 5.9 | 9.4 |
| 5′-phosphate | | – | 6.4 | 9.5 |
| 5′-pyrophosphate | | – | 6.5 | 9.4 |
| 5′-triphosphate | | – | 6.6 | 9.5 |
| Thymine | – | – | – | 9.8 |
| Thymidine-5′-phosphate | 1.6 | – | 6.5 | 10.0 |

*The deoxyribonucleotides have p*K* values similar to those of the ribonucleotides.

*References p.855*

Partition and adsorption column chromatography are mainly carried out with Kieselguhr (Celite; Hall, 1962, 1964), cellulose (Randerath and Struck), silica gel (Žemlička *et al.*), hydrophilic gels (Kull and Soodak) and aluminium oxide (Prystaš and Šorm, 1964a, b), usually of commercial purity grade. Before column preparation, Kieselguhr is generally purified by washing with acid (2–3 *M*) and water to remove UV-absorbing impurities, as UV absorbance measurements at different wavelengths is the only practical detection method available. Eluents applied in this field are organic solvents (their purity with respect to polar admixtures is very important) and their mixtures, the elution capacity of which increases with the polarity of the mixture and also depends to some extent on the carrier employed (Trappe). Binary and ternary elution mixtures, similar to those used in paper chromatography, are used in partition chromatography, particularly for anomalous and minor components of nucleic acids and for synthetic intermediates.

Sorbents available for the gel permeation chromatography of low-molecular-weight substances include cross-linked dextran gels (Sephadex, Pharmacia, Uppsala, Sweden) and polyacrylamide gels (Bio-Gel, Bio-Rad Labs. Richmond, Calif., U.S.A.). The gels mostly work on the molecular-sieve principle (group-wise separation of nucleotides, nucleosides and bases), but they also have some adsorption and ion-exchange properties (greater affinity to aromatic and heterocyclic compounds). For this reason, dilute neutral buffer solutions are used for elution instead of water (De Bersaques, Zadražil *et al.*). pH changes of the elution solution can obviously be utilized to separate those substances which are more strongly bound to the column bed (Khym and Uziel, 1970).

## AUTOMATED PROCEDURES FOR THE ANALYSIS OF NUCLEIC ACID COMPONENTS

Most column-type fractionation techniques are nowadays carried out with the use of automatic fraction collectors and instruments recording the substance analyzed. In an attempt to shorten as much as possible the time needed for separation, to enhance the sensitivity and resolution of the column (nanomoles of substances separated), etc., special chromatographic instruments and systems were developed even for the separation of nucleic acid components (in analogy with amino acid analyzers).

One of the systems for the analysis of nucleoprotein components (ninhydrin-positive and UV-absorbing substances) is based on the principle of the amino acid analyzer with a column of Amberlite IR-120, amplified with an LKB Uvicord Model II instrument with recording facilities for absorbance measurements at 254 nm (Ženíšek *et al.*). The entire process is shown diagrammatically in Fig. 37.1. High-pressure cation-exchange chromatography was employed in a Varian Aerograph Model 4100 liquid chromatograph system with a dynamically filled column of the cation exchanger VC-10 (10X; giving a high degree of homogeneity of the column formed; Scott and Lee) for automatic analysis of the nucleotide composition of RNA and DNA (Burtis, 1970a). This fully automated system with recording facilities separates on the column hydrolyzate corresponding to 0.25 $\mu$g of RNA within 2–4 min (Fig. 37.2). Uziel *et al.* recommend a similar microanalytical method for separating UV-absorbing substances on a Dowex 50 column and related cation-exchange resins, applying it to the separation of a mixture of bases obtained by the gradual degradation of the polynucleotide chain by periodate oxidation (Khym

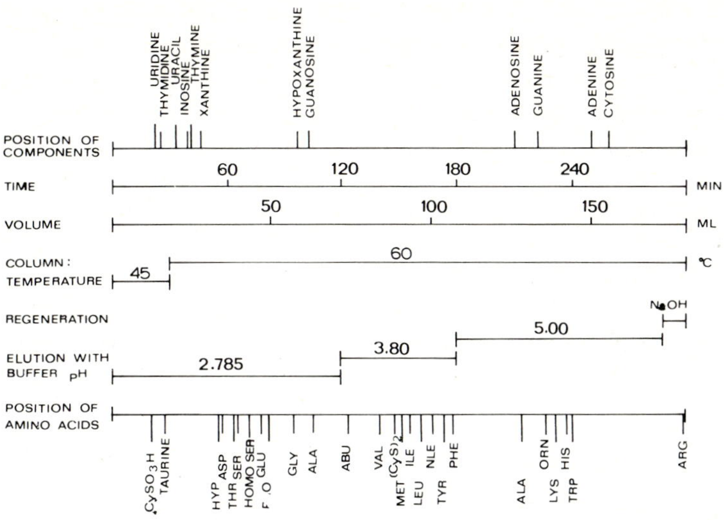


Fig. 37.1. Diagrammatic representation of a separation of ninhydrin-positive and UV-absorbing substances in the amino acid analyzer with a column of Amberlite IR-120 (Ženíšek *et al.*).

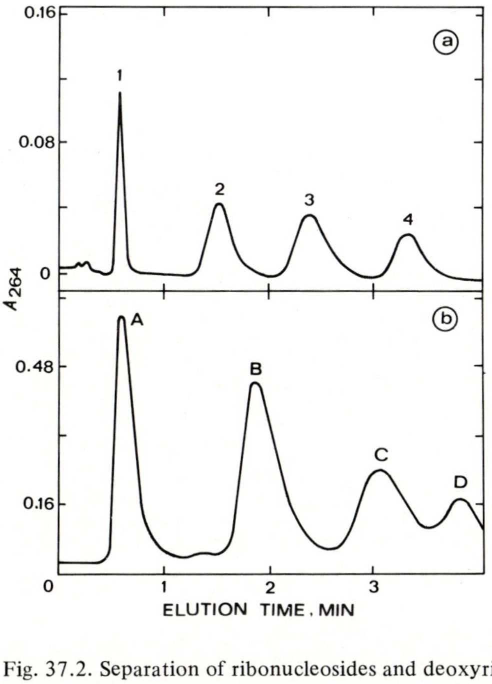


Fig. 37.2. Separation of ribonucleosides and deoxyribonucleosides by high-pressure cation-exchange chromatography on a Varian Aerograph Model 4100 liquid chromatograph with a column of VC-10 cation exchanger (Burtis, 1970a). Column: 25 × 0.24 cm. Elution: 0.4 *M* ammonium formate solution, pH 4.0. Flow-rate: 50 ml/h. Pressure: 3000 p.s.i. Temperature: 75° C. Sample: (a), mixture containing uridine (1), guanosine (2), adenosine (3), and cytidine (4), 0.1 μg of each; and (b), mixture containing thymidine (A), deoxyguanosine (B), deoxyadenosine(C), and deoxycytidine (D), 0.4 μg of each.

and Uziel, 1968). An ion-exchange technique of high efficiency (a column of the cation-exchange resin Aminex A-4 eluted with a citrate complex gradient made in a Technicon Autograd instrument) was used in the analysis of a mixture of nucleotides, nucleosides and bases (Murakami *et al.*) and polyphosphates, nicotinamide and flavine nucleotides (Drobishev *et al.*) in an automatic system with a Hitachi Model 034 liquid chromatograph (with detection at 260, 270 and 280 nm). Thirteen components (0.1–0.8 $\mu$moles) of the mixture were separated with good resolution in quantitative (± 3%) and qualitative analyses (Murakami *et al.*). A similar separation of deoxyribonucleotides on a Zipax SAX column was used for determinations of the oligomers nucleotide composition (Gabriel and Michalewsky).

A fully automated system with a Dowex 2 column and two continuous-flow measuring

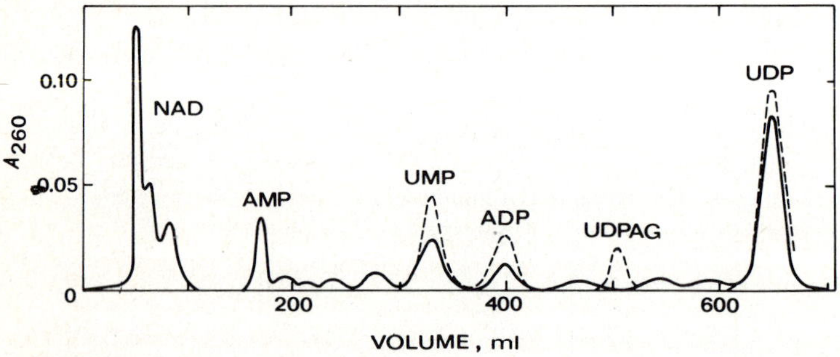


Fig. 37.3. Separation of a *Mycoplasma* acidic extract and 100 nmole of UMP, UDP, UDPAG and ADP (broken lines) added to, by an automated anion-exchange column chromatography (Virkola). Column: Dowex 2-X8 (200–400 mesh), 12 × 0.8 cm. Elution: a gradient prepared in a Varigrad mixer with nine compartments containing 200 ml each of the following solutions: compartment 1, water; 2, 0.5 *M* formic acid; 3 and 4, 4 *M* formic acid; 5–7, 4 *M* formic acid with 1 *M* ammonium formate. Flow-rate: 29.2 ml/h. Temperature: 22°C.

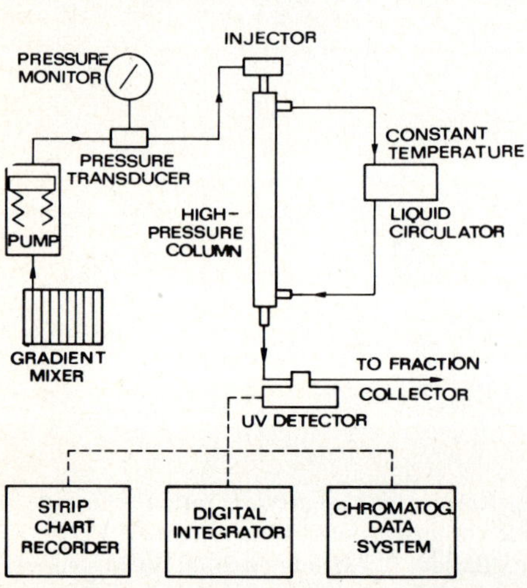


Fig. 37.4. Schematic representation of an automated liquid chromatograph based on the Varian Aerograph model (Burtis, 1970a).

cells (ensuring a stable background line) was employed by Virkola to separate an extract of acid-soluble substances in *Mycoplasma laidlawii* A. The applicability of this analytical system is shown in Fig. 37.3. Burtis (1970b) employed the same principle in the analysis of urine, using a Varian Aerograph Model LCS-4010 urine analyzer. Dinucleotides (Kennedy and Lee) and adenosine polyphosphates (Schmukler) can also be analyzed with a commercial nucleic acid analyzer (*e.g.*, the Picker-Nuclear LCS-1000 analyzer) with a column of anion-exchange resin.

In all of these instances, the analytical application of ion-exchange columns combined with control and detection systems is involved, and precise preparation of homogeneous column fillings, buffer solutions and elution gradients is also required, together with, in many instances, special instrumentation for corresponding results to be achieved (Anderson, Thacker *et al.*, Uziel *et al.*). A general diagram of a liquid chromatograph based on the Varian Aerograph system is shown in Fig. 37.4.

## INDIVIDUAL TYPES OF NUCLEIC ACID CONSTITUENTS

### Purine and pyrimidine bases and their analogues

In laboratory practice, these substances are encountered as components of whole nucleic acid hydrolyzates or as synthetic substances. The main separation technique is paper chromatography, the main advantage of which is simplicity. Most of the bases involved contain at least one substituent that is capable of ionization, which thus gives the molecule a positive or negative charge (Table 37.5). These properties permit the use of both ion-exchange resins and hydrophilic gels containing a certain proportion of ionizable groups (dextran). On the other hand, even slight differences in the structure of the substituents have a substantial influence on the behaviour of bases in the course of elution in partition column chromatography (analogous to paper chromatography).

A good example of the most common ion-exchange separation process is the use of a Dowex 50 ($H^+$) column, which, when eluted with 2 $N$ hydrochloric acid, gives distinct peaks for all fundamental bases of RNA (Cohn, 1949b) in the order uracil, cytosine, guanine and adenine. The same column eluted with a linear gradient of 1–4 $M$ hydrochloric acid has been used for the isolation of methylated bases, mainly guanine derivatives, from the perchloric acid hydrolyzate of the total RNA (Craddock *et al.*). Similar ion-exchange resins can be used to analyze bases split off in the gradual periodate oxidation of polyribonucleotides (Khym and Uziel, 1968). An anion-exchange column of Dowex 1, which also serves as an example of the first application of ion-exchange columns to nucleic acid components, combines the advantages of elution at constant solution concentration with good resolution of the sorbent (Fig. 37.5). Synthetic anion exchangers can be replaced with DEAE-cellulose (Weith and Gilham), which, however, is mainly employed to separate higher components (mono- and oligonucleotides) and rather serves in the instances mentioned to separate groups of bases for preliminary purification.

The method of partition or adsorption chromatography with Kieselguhr as sorbent in the field of nucleic acid components has achieved the widest application in separating nucleosides (see p. 843), for which it has been worked out in great detail (Hall, 1971). This "nucleoside" procedure was also used by Hall (1967) to separate the main bases of

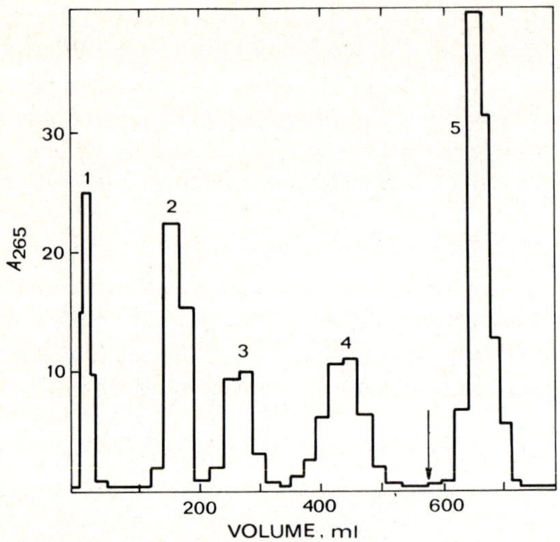


Fig. 37.5. Separation of purines and pyrimidines on a Dowex 1 column (Cohn, 1949b). Column: 8.5 × 0.49 cm. Elution: 0.2 *M* ammonia solution with 0.025 *M* ammonium chloride (pH 10.6) changed, as indicated by the arrow, for 0.1 *M* ammonium chloride (pH 10.0). Flow-rate: 60 ml/h. Sample: cytosine (1), uracil (2), thymine (3), guanine (4), and adenine (5), 1–2 mg of each base.

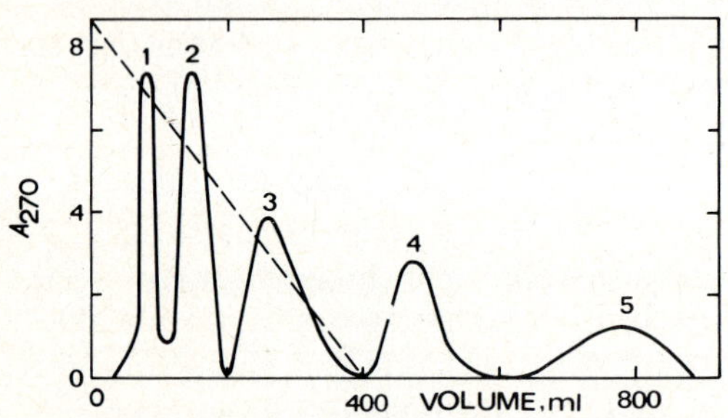


Fig. 37.6. Separation of purine and pyrimidine bases by partition chromatography on Kieselguhr (Hall, 1967). Column: 42 × 1.9 cm column of 50 g of Celite 545–Microcel E (9:1) mixture in a lower phase of the solvent system ethyl acetate–2-ethoxyethanol–10% formic acid (4:1:2). Elution: 400-ml linear gradient of upper phase with a concentration of formic acid decreasing to zero, followed by an upper phase of ethyl acetate–1-butanol–water (1:1:1). Flow-rate: 60 ml/h. Sample: 2 ml of lower phase of the starting solvent system containing 5 mg of each of the bases thymine (1), uracil (2), adenine (3), guanine (4), and cytosine (5), applied in the form of a suspension with 4 g of the carrier mixture.

nucleic acids, as shown in Fig. 37.6. The separation of bases on poly-N-vinylpyrrolidone, based on hydrogen bonding of the bases with the sorbent, likewise belongs to this group of methods (Dougherty and Schepartz, 1969b).

Adsorption of purine bases on the sorbent material during gel permeation chromatography has already been mentioned in the section on desalting. This interaction of bases, which is actually an ion-exchange process, allows the mutual separation of substances within this group to take place together with separation of the group of bases from higher components (Table 37.6). Sweetman and Nyhan (1968, 1971) have shown that the

TABLE 37.6
DISTRIBUTION CONSTANTS ($K_d$ VALUES) OF SOME PURINES, PYRIMIDINES, NUCLEOSIDES AND NUCLEOTIDES ON VARIOUS SEPHADEX COLUMNS

Procedures: A, column 25 × 1.5 cm eluted with 130 m*M* ammonium formate of pH 6 (De Bersaques); B, column 100 × 1 cm eluted with 50 m*M* sodium dihydrogen phosphate of pH 7 (Sweetman and Nyhan, 1968); C, column 150 × 1.3 cm eluted with 10 m*M* ammonium carbonate of pH 9 (Gorbach and Henke); D, column 35 × 3.5 cm eluted with water (Gelotte); E, column 90 × 2.2 cm eluted with 0.005% aqueous ammonium carbonate of pH 7.6 (Zadražil *et al.*); F, column 180 × 3.5 cm eluted with 10 m*M* ammonium carbonate of pH 9 (Hohn and Pollman).

| Compound | $K_d$ value on Sephadex column | | | | | | |
|---|---|---|---|---|---|---|---|
| | G-10 | | G-15 | | | G-25 | |
| | A | B | A | C | D | E | F |
| Adenine | 6.00 | 7.66 | 4.26 | 4.62 | 2.2 | 3.38 | 3.62 |
| Adenosine | 3.36 | 4.23 | 2.92 | 2.79 | 1.7 | 2.59 | 2.50 |
| 2′-Deoxy- | 3.23 | 4.23 | 2.81 | – | – | – | – |
| 5′-phosphate | 0.91 | 0.93 | 1.37 | 0.58 | 0.1 | 0.85 | 1.27 (3′) |
| 5′-pyrophosphate | 0.41 | – | 0.74 | – | – | – | – |
| 5′-triphosphate | 0.29 | – | 0.50 | – | – | – | – |
| Guanine | 3.35 | 5.84 | 2.80 | 3.22 | – | 3.40 | 3.23 |
| Guanosine | 2.43 | 3.14 | 2.29 | 2.09 | 1.6 | 2.60 | 2.30 |
| 2′-Deoxy- | 2.66 | 3.28 | 2.40 | – | – | – | – |
| 5′-phosphate | 0.8 | 0.82 | 1.07 | 0.43 | 0.4 | 0.91 | 1.04 |
| Hypoxanthine | 2.17 | 2.83 | 1.89 | 2.20 | 1.6 | – | – |
| Inosine | 1.26 | 1.56 | 1.31 | 1.50 | 1.2 | – | – |
| 2′-Deoxy- | 1.26 | 1.60 | 1.30 | – | – | – | – |
| Xanthine | 3.15 | 4.34 | 2.89 | 2.20 | 1.8 | – | – |
| Xanthosine | 1.83 | 2.51 | 1.95 | 1.50 | – | – | – |
| Cytosine | 1.43 | 1.84 | 1.26 | 1.50 | 1.6 | 1.80 | 1.69 |
| Cytidine | 0.99 | 1.32 | 1.06 | 1.18 | 1.2 | 1.60 | 1.42 |
| 2′-Deoxy- | 1.09 | – | 1.16 | – | – | – | – |
| 5′-phosphate | 0.4 | 0.40 | 0.54 | 0.37 | 0.1 | 0.66 | 0.69 |
| Uracil | 1.56 | 1.91 | 1.50 | 1.18 | 1.1 | 1.82 | 1.54 |
| Uridine | 1.07 | 1.28 | 1.21 | 0.85 | 1.0 | 1.66 | 1.27 |
| 2′-Deoxy- | 1.09 | – | 1.15 | – | – | – | – |
| 5′-phosphate | 0.45 | 0.44 | 0.74 | 0.10 | 0.1 | 0.66 | 0.69 |
| Thymine | 1.86 | 2.23 | 1.63 | 1.45 | – | – | 1.54 |
| 1-β-D-Ribofuranosyl- | 1.00 | – | 1.15 | – | – | – | – |
| Thymidine | 1.24 | 2.23 | 1.29 | 1.02 | – | – | 1.23 |
| 5′-phosphate | 0.52 | – | 0.73 | – | – | – | 0.89 |
| 5′-pyrophosphate | 0.28 | – | 0.44 | – | – | – | – |
| 5′-triphosphate | 0.19 | – | 0.32 | – | – | – | – |

adsorption of purines, the mechanism of which they investigated, may serve not only for separation but also to predict the elution volume of a molecule of a given structure on Sephadex G-10, and *vice versa*. As the elution volume of purines are very high when elution is carried out with water, dilute phosphate or volatile buffers (ammonium carbonate and hydrogen carbonate) are used when separating bases, for practical reasons. Khym

*References p.855*

and Uziel (1970) used columns of Sephadex G-10, equilibrated with 0.01 *M* hydrochloric acid or ammonia solution and ammonium chloride buffer of pH 9.7 to separate purine and pyrimidine bases in the order uracil, cytosine, adenine, guanine and uracil, cytosine, guanine, adenine, respectively. The variation of the pH between 9 and 10 considerably altered the elution profile of the mixture (a dependence on the ionization of the individual bases).

**Nucleosides**

As the sugar component of a nucleoside does not contribute greatly to changes in the formation of the charge of the molecule (with the exception of the use of borate buffers with ribose derivatives; Khym and Cohn, Khym and Zill), the same applies to the separation of nucleosides as to that of bases. The main separation method in this field is adsorption and partition chromatography, although ion-exchange and gel permeation chromatography are also utilized.

Natural nucleosides of both types can be separated by a similar ion-exchange method to that described for bases [Dowex 1 in the formate cycle with elution by ammonium formate at pH 10.2 (Andersen *et al.*; Cohn, 1950) or Dowex 50 used in the "nucleoside

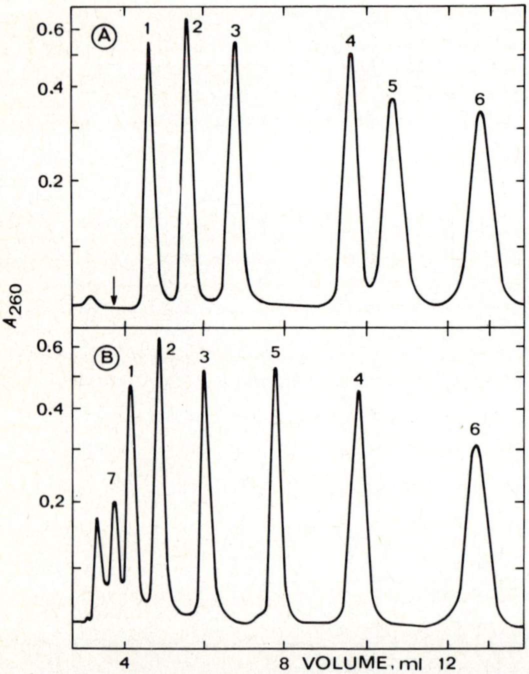


Fig. 37.7. Separation of ribonucleosides on a sulphonated polystyrene cation-exchange column (Singhal and Cohn). Column: Bio-Rad Aminex A-6, 50 × 0.5 cm with void and inner volumes of 3 and 5 ml, respectively. Elution: 0.02 *M* ammonium carbonate with ammonia solution of (A) pH 9.3 or (B) 9.85. Flow-rate: 0.98 ml/cm$^2$ and 0.194 ml/min. Temperature: 50°C. Sample: 10 $\mu$l of about 100 nmole mixture of pseudouridine (1), uridine (2), thymidine (ribo-, 3), cytidine (4), guanosine (5), adenosine (6) and 4-thiouridine (7).

analyzer" according to Uziel *et al.*]. Using the ion-exclusion or ion-repulsion techniques on a column of the cation exchanger Bio-Rad Aminex A-6, Singhal and Cohn achieved one of the best nucleoside separations known (Fig. 37.7). When sodium or potassium borate ($10^{-2}$ to $10^{-5}$ *M*) in an elution solution of pH 8–10 is used, substances containing the *cis*-diol group obtain another negative charge (borate diols are formed), which facilitates ion-exchange separation. Jaenicke and Von Dahl used the borate form of a strong anion-exchange resin to separate ribonucleosides (in the order cytidine, adenosine, uridine and guanosine), eluting with increasing concentrations of borate buffer of pH 9.2 and 0.01–0.1 *M* sodium chloride solution.

Ion-exchange chromatography, however, is more important for the separation and purification of nucleoside products from reaction mixtures of organic synthesis. Ziff and Fresco isolated uridine and cytidine derivatives from the mixture obtained by oxidation of 4-thiouridine (Dowex 1 ($Cl^-$ or $CH_3COO^-$) and Dowex 50 ($H^+$), both eluted by salt gradients); Bobek *et al.* (1969a, b) isolated 6-azauridine and 6-azapseudouridine derivatives and Bobek and Šorm isolated homouridine and homocytidine on the same types of ion exchangers. The separation of different nucleoside mixtures, differing also in their sugar components, was effected on a Dowex 1 column by Dekker and by Gin and Dekker. Metabolic conversions of isopentenyladenosine were studied using a DEAE-cellulose column (Hall *et al.*).

Partition chromatography on a Celite 545 column, when combined with paper chromatography (Fig. 37.8), allows nearly all of the nucleoside components known so far to occur in tRNA enzyme hydrolyzates to be separated (Table 37.3). Because of the general importance of this technique, a detailed description of the column preparation and the sample application is given below (Hall, 1962, 1967, 1971).

A suspension of Celite 545 in 3 *N* hydrochloric acid is washed on a suction funnel with the same acid until a clear filtrate is obtained. After washing with distilled water until the filtrate is neutral, the Celite is washed with ethanol and dried in a thin layer for 16 h at 100°C. Microcel E is washed in the same manner. The column is best prepared by means of the "dry pack" technique. To 160 g of a dry mixture of Celite 545–Microcel E (9:1), roughly two parts of the aqueous (lower) phase of the solvent system employed (see p. 845) are added so as to obtain a free-floating powder (the bonding capacity of the sorbent for the liquid is just saturated); in this form, the powder is suitable for filling a thick-walled glass tube (height 80 cm, I.D. 2.54 cm, fitted with a ground-glass joint) by means of a rod with a flat end cut at a right-angle, moving closely in the column (in a similar manner to the piston of a syringe). Small portions of the above moist Celite are compressed into compact layers by means of the rod, the height of these layers being equal to the column diameter. In this way, a very compact carrier column is obtained with no chanelling effect on elution.

A column of this type is suitable for separating 1–1.5 g of RNA hydrolyzed to nucleosides. The best method of sample application is to dissolve the lyophilized enzyme RNA hydrolyzate in 9 ml of the aqueous phase of the solvent system and, after centrifuging for 15 min at 15,000 *g*, and mixing with 18 g of dry Celite 545–Microcel E (9:1), to fill the moist suspension obtained on to the top of the column prepared as above. After application of the sample, the column can be eluted with the appropriate solvent system (see Fig. 37.8).

*References p.855*

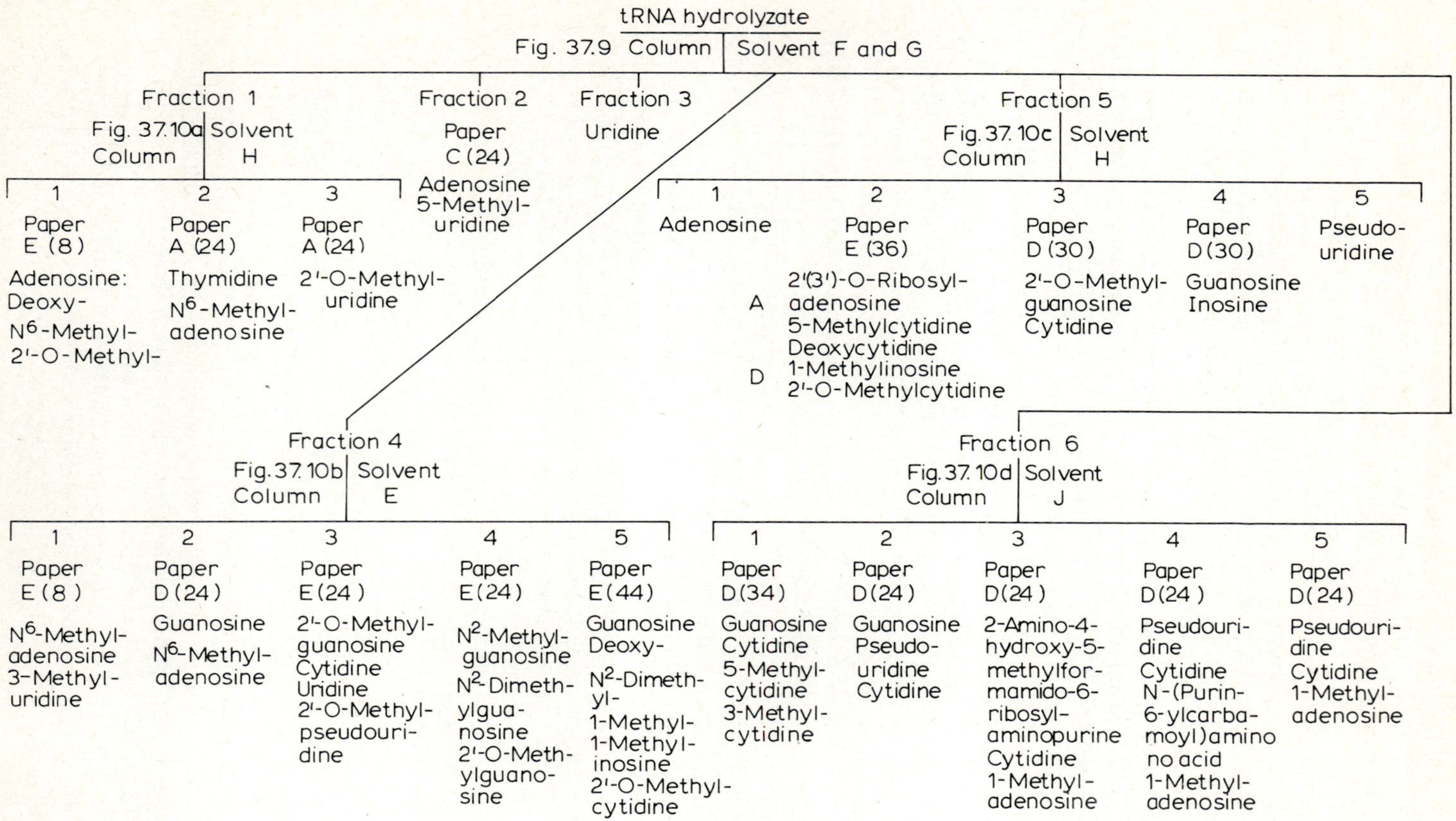


Fig. 37.8. General procedure for the complete separation of nucleosides from a tRNA enzymic hydrolyzate (Hall, 1967). A–J: solvent systems used (see p. 845). Numbers in parentheses represent the developing time of paper chromatography in the given solvent system.

Figs. 37.9 and 37.10 show the distribution of the total hydrolyzate into individual fractions and their re-chromatography on smaller columns of the same sorbent. The peaks obtained after re-chromatography were divided into individual components by chromatography on Whatman No. 3MM paper (Fig. 37.8). The solvent mixtures used for chromatography were: (A) 1-butanol–water–concentrated ammonia solution (86:14:5); (B) ethyl acetate–2-ethoxyethanol–2% aqueous formic acid (4:1:2); (C) 2-propanol–concentrated hydrochloric acid–water (680:176:144); (D) 2-propanol–water–concentrated ammonia solution (7:2:1); (E) ethyl acetate–1-propanol–water (4:1:2); (F) ethyl acetate–2-ethoxyethanol–water (4:1:2); (G) ethyl acetate–1-butanol–ligroin (b.p. 66–75°C)–water (1:2:1:1); (H) 1-butanol–water–concentrated ammonia solution (3:1:0.05); and (J) ethyl acetate–1-butanol–water (1:1:1). Hall (1962) separated deoxyribonucleosides on a Celite 545 column in solvent system B in the order thymidine, deoxyadenosine, deoxyguanosine and deoxycytidine.

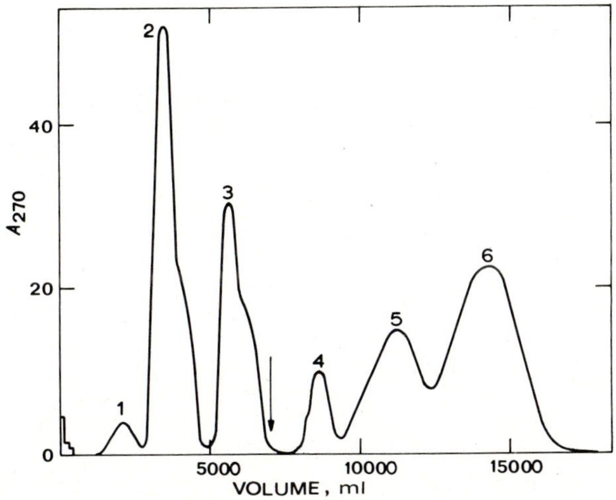


Fig. 37.9. Fractionation of an enzymic hydrolyzate of tRNA by partition chromatography on a column of Celite 545–Microcel E (Hall, 1965, 1967). Column: 80 × 5.08 cm column made of 690 g of carrier mixture (9:1) in the aqueous phase of solvent system F. Elution: upper phase of solvent system F changed for an upper phase of solvent system G at the point marked by the arrow. Flow-rate: 600 ml/h. Sample: 35 ml of aqueous phase of F containing nucleosides of the enzymic hydrolyzate of tRNA (5.4 g), applied as a suspension with 80 g of carrier mixture; fractions 1–6, containing $N^6$-methyladenosine (1), adenosine (2), uridine (3), methylated guanosines (4), guanosine (5), and cytidine (6) as their main components, were used for further separations (see Figs. 37.8 and 37.10).

Neutral aluminium oxide can be used with advantage as sorbent in the adsorption separation of protected nucleosides, *e.g.*, anomeric ribofuranosyl derivatives of uracil (elution with a benzene-ethyl acetate mixture in various proportions; Prystaš and Šorm, 1964a, b, 1965, 1966).

The $K_d$ values in Table 37.6 characterize the properties of the main nucleosides on Sephadex columns and the possibility of their mutual separation by gel permeation chromatography with the use of eluents of different concentrations and pH values (De

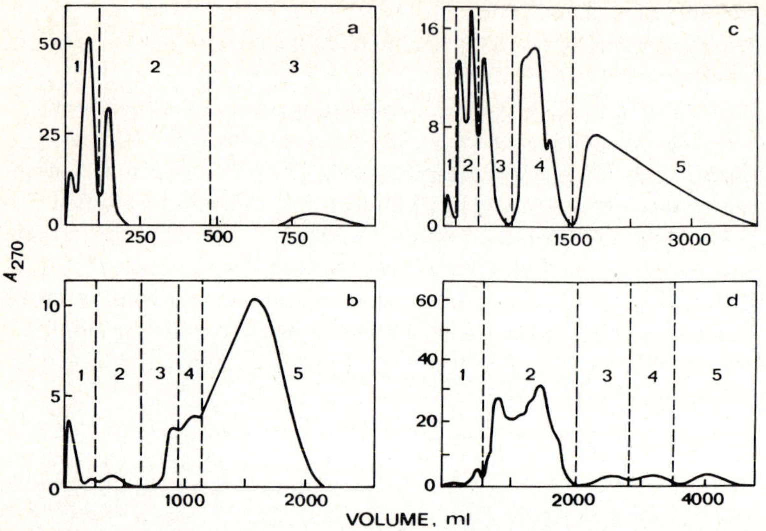


Fig. 37.10. Re-chromatography of fractions from Fig. 37.9 on a column of Celite 545–Microcel E (Hall, 1965, 1967). Column: 80 × 2.54 cm column made of 150 g of carrier mixture. Flow-rate: 150 ml/h. (a) fraction 1 of Fig. 37.9 eluted with solvent system H; (b) fraction 4 of Fig. 37.9 eluted with solvent system E; (c) fraction 5 of Fig. 37.9 eluted with solvent system H; (d) fraction 6 of Fig. 37.9 eluted with solvent system J. Numbers in each figure indicate regions the fractions of which are used for further separation by paper chromatography (see Fig. 37.8).

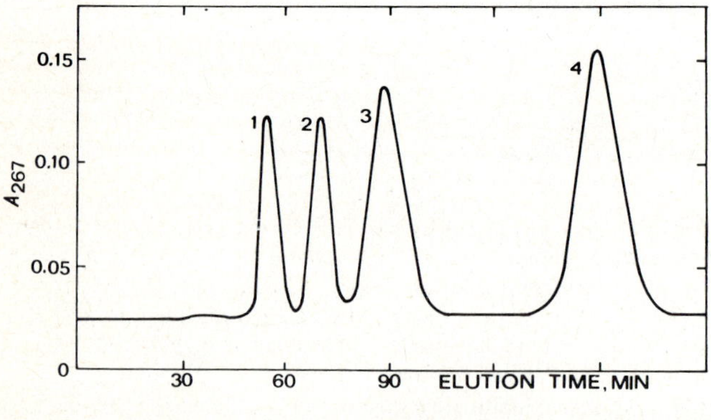


Fig. 37.11. Separation of deoxyribonucleosides on a column of fractionated Sephadex G-10 (Ehrlich *et al.*, 1971a). Column: 25 × 0.5 cm. Elution: 0.025 *M* ammonium carbonate of pH 10.4. Flow-rate: 3 ml/h. Sample: 50 μl containing 0.25 $A_{260}$ unit of each deoxyribonucleoside: thymidine (1), deoxycytidine (2), deoxyguanosine (3) and deoxyadenosine (4).

Bersaques; Gelotte; Gorbach and Henke; Hohn and Pollman; Sweetman and Nyhan, 1968; Zadražil *et al.*). The most successful separation of nucleosides was carried out by Bernardi and co-workers on commercial and fractionated gels (to obtain a more homogeneous column filling), *viz.* Bio-Gel P-2 (Carrara and Bernardi, Piperno and Bernardi) and Sephadex G-10 (Ehrlich *et al.*, 1971a). It was found that both types of gel can be used for precise determinations of the base composition of nucleic acids even with a different elution order of the components to be separated (Fig. 37.11). Bio-Gel P-2 can also be used to separate nucleosides in the presence of a larger amount of nucleotides, which are eluted considerably earlier than on Sephadex G-10. As a microanalytical method, this procedure has been applied in many instances to determine the nucleoside composition of enzyme hydrolyzates of DNAs (Bernardi *et al.*, 1968, 1970; Corneo *et al.*) and of terminal nucleotides (Ehrlich *et al.*, 1971b). Both ribonucleosides and deoxyribonucleosides can be separated on the same column of fractionated Bio-Gel P-2 in a single operation in the presence of sodium tetraborate in the elution solution at pH 10.1 (Fig. 37.12; Piperno and Bernardi). A Sephadex G-10 column eluted with a citric acid–phosphate buffer of pH 3.5 can be recommended for the separation of thymidine from its analogues 5-bromo- and 5-iododeoxyuridine (Braun, Visser).

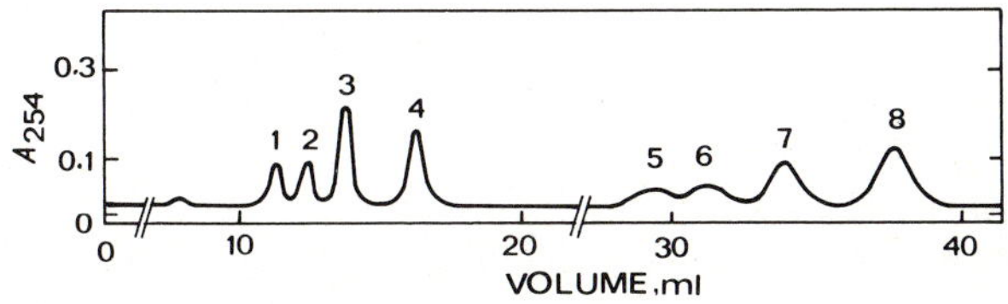


Fig. 37.12. Separation of ribonucleosides and deoxyribonucleosides on a column of fractionated Bio-Gel P-2 (Piperno and Bernardi). Column: 60 × 0.8 cm. Elution: 0.1 m*M* borate in 2 m*M* ammonium carbonate of pH 10.2. Flow-rate: 2.8 ml/h. Sample: 50 µl containing 1 $A_{260}$ unit of each ribunocleoside: (1) uridine, (2) guanosine, (5) cytidine and (6) adenosine; and 2 $A_{260}$ units of each deoxyribonucleoside: (3) deoxyguanosine, (4) thymidine, (7) deoxycytidine and (8) deoxyadenosine.

## Nucleotides and oligonucleotides

In contrast to bases and nucleosides, nucleotides bear a strongly acidic phosphate group, so that they are mainly ionized as anions, although their behaviour to some extent also depends on the substituents of the base present (Table 37.4). They cannot be separated by means of paper chromatography, but with respect to their charge, paper electrophoresis is frequently employed (Markham and Smith, Sanger and Brownlee, Smith). For the same reasons, the main column separation technique is ion-exchange chromatography on polystyrene resins (Dowex), as well as on substituted celluloses and dextrans. In order to separate oligonucleotide mixtures on the basis of the molecular sizes of the components, DEAE-substituted carriers eluted with urea-containing gradients (Tener) can be recommended in addition to gel permeation chromatography (Stanley).

A column of anion-exchange resin in the chloride cycle (Dowex 1), successively eluted with water and diluted hydrochloric acid, was first employed by Cohn (1950) and was found to be an ideal tool for separating RNA alkaline hydrolyzates (elution order: bases and nucleosides, CMP, AMP, UMP and GMP). For a routine method of investigating the base composition of RNA Katz and Comb recommend a Dowex 50 ($H^+$) column for separating UMP (0.05 *M* hydrochloric acid), GMP and mixtures of AMP and CMP, which can be further separated, if necessary (*e.g.*, in order to measure their specific radioactivity), on a Dowex 1 column. Dowex 1 with stepwise concentration elution with acetate at pH

*References p.855*

4.7 also served to separate the nucleotides in a DNA enzyme hydrolyzate (Sinsheimer and Koerner) and, with ammonium acetate of pH 4.3, to resolve 5-hydroxymethylcytidylic acid from its mono- and diglycosylated derivatives (DNA from T-even phages; Lehman and Pratt). Similarly, 5,6-dihydrouridylic acid was isolated from the total RNA hydrolyzate after treatment with RNase T1 on a Dowex 1 column with a complex ammonium formate gradient (pH 3.5), and re-chromatography of the minor peak between cytidylic and pseudouridylic acid was carried out on a DEAE-Sephadex column (linear gradient of 0.1–0.7 *M* ammonium carbonate; Madison, Madison and Holley).

Anion-exchange resins of the Dowex 1 type were also used in the separation of oligonucleotides, serving as sorbents in the first attempts to characterize the primary structure of nucleic acids (Volkin and Cohn, RNA enzyme hydrolyzate after treatment with pancreatic RNase; Fig. 37.13). Using gradient elution with formic acid and formate, oligonucleotides from a similar hydrolyzate were isolated on Dowex 1-X2 columns (Zadražil and Šormová). Oligonucleotides from a DNA enzyme hydrolyzate were separated in similar manner by Sinsheimer. The separation of isomeric 3′,5′- and 2′,5′-dinucleoside monophosphates was carried out by Taylor and Hall.

A widely used anion-exchange sorbent is DEAE-cellulose or its Sephadex analogue, which, in addition to being used for the fractionation of reaction mixtures in a synthetic laboratory (Holý and Šorm, separation of isomeric monophosphites on a column in borate buffer; Holý and Žemlička, fractionation of a mixture of nucleoside and mono- and

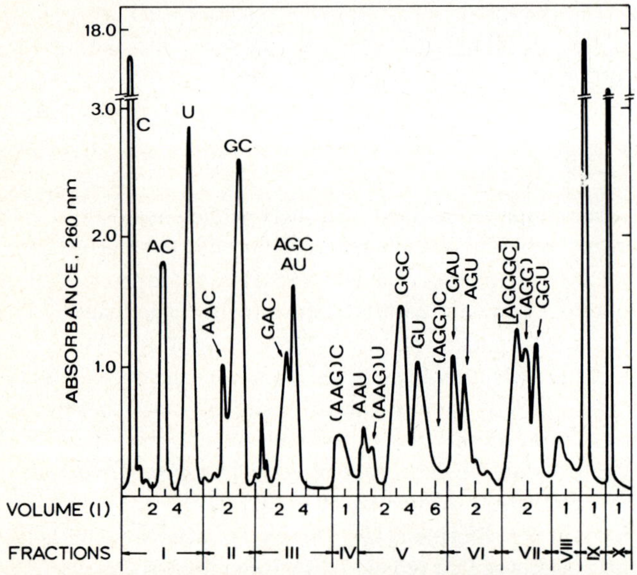


Fig. 37.13. Preparative fractionation of a pancreatic RNase digest of calf liver RNA on a Dowex 1 column (Volkin and Cohn). Column: 15 × 1.09 cm, Dowex 1-X2 (400 mesh). Elution: for fractions I to X, the following solutions were used in a stepwise manner: 0.005 *N* hydrochloric acid; 0.01 *N* hydrochloric acid; then 0.0125 *M*; 0.025 *M*; 0.05 *M*; 0.1 *M*; 0.2 *M*; 0.3 *M*; 1.0 *M*; and 2.0 *M* sodium chloride added to 0.01 *N* hydrochloric acid in each instance. Sample: 700 mg of RNA hydrolyzed with 10 mg of RNase. Parentheses indicate an unknown sequence or a mixture of sequences. Brackets indicate empirical composition.

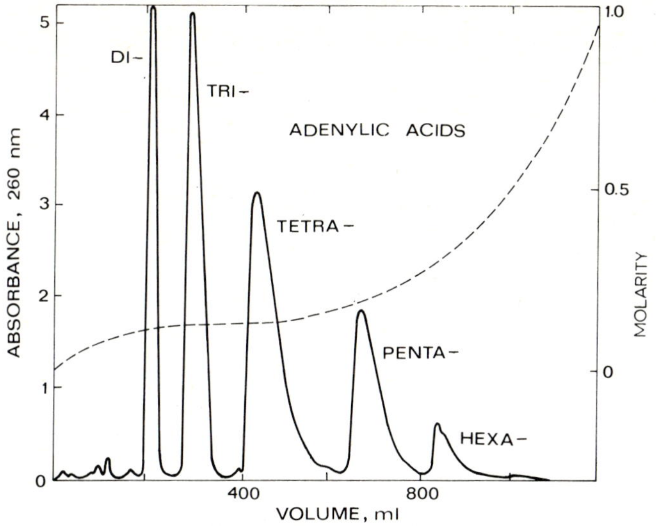


Fig. 37.14. Separation of oligoadenylic acids from an *Azatobacter* nuclease digest of poly-A on a column of DEAE-cellulose (Staehelin *et al.*, 1959). Column: 30 × 0.9 cm. Elution: a gradient prepared in a Varigrad mixer with six compartments containing ammonium hydrogen carbonate–carbonate buffer of pH 8.6 with the following concentrations: 1, 0.01 *M*; 2, 0.24 *M*; 3, 0.01 *M*; 4 and 5, 0.20 *M*; 6, 1 *M*.

dinucleotide by a linear triethylammonium hydrogen carbonate gradient of pH 7.5; Tener *et al.*, fractionation of oligomers in synthetic polymerization of thymidylate) also find a particularly important field of application in separating oligonucleotides in enzyme hydrolyzates of nucleic acids and polynucleotides (Staehelin *et al.*, 1959, separation of homologous oligoadenylic acids from poly-A enzyme hydrolyzate, Fig. 37.14; and a similar fractionation of RNA hydrolyzate obtained with pancreatic RNase, Fig. 37.15).

It can be seen from the latter two examples, that the form of the gradients to which this column fractionation is always bound is decisive for the degree of separation. In general, for DEAE-substituted cellulose and dextran, the homologous oligonucleotides are separated in accordance with their chain lengths (Fig. 37.14), while the prevailing base of the oligonucleotide increases the elution volume in the series C, U, A and G (Fig. 37.15). The introduction of buffers containing 7 *M* urea (Tener, Tomlinson and Tener) simplifies the situation in such a way that at a neutral pH (5.4–8.0) the oligonucleotide mixture is separated in accordance with the magnitude of the charge only, *i.e.*, according to the number of phosphate groups or degree of polymerization (Fig. 37.16), even when a simple linear gradient is applied. Subfractionation according to the composition of the bases in fragments bearing the same charge can be carried out on the same column simply by omitting the urea (the above-mentioned order of influence of bases) or by elution at lower pH (2.7–4.0) in the presence of urea (protonization of adenine and cytosine), and also by elution at higher pH (8.5–10.0 with carbonate or hydrogen carbonate buffers) with no urea (ionization of uracil and guanine). Urea-containing buffers have been found to be very useful, particularly for separating purine and pyrimidine "isopliths" after specific chemical degradation of nucleic acids, apyrimidinic and apurinic acids being

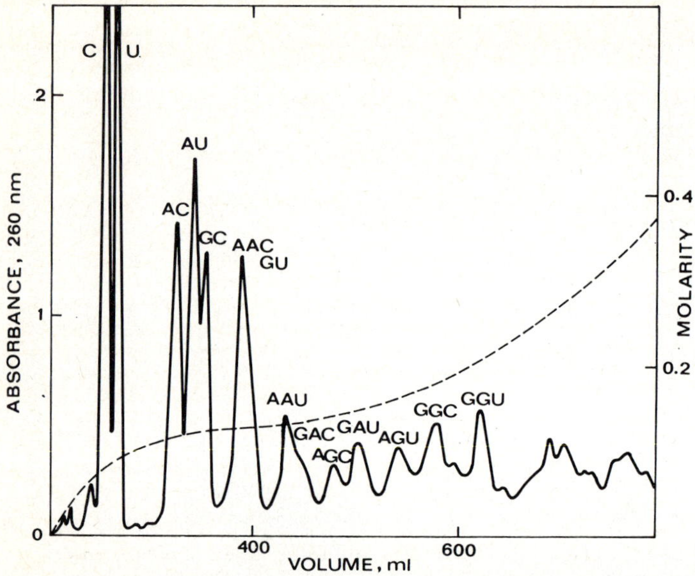


Fig. 37.15. Separation of a pancreatic RNase digest of yeast RNA on a column of DEAE-cellulose (Staehelin *et al.*, 1959). Conditions as in Fig. 37.14, except for the last Varigrad compartment, which contained 0.4 *M* buffer.

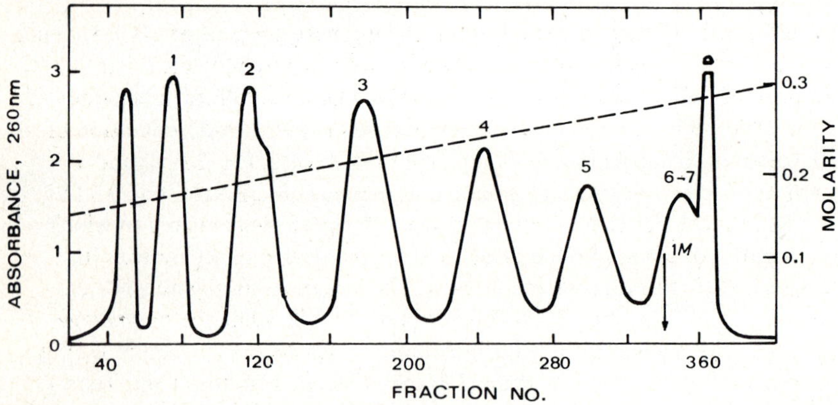


Fig. 37.16. Fractionation of an RNase T1 digest of yeast RNA on a column of DEAE-Sephadex in the presence of 7 *M* urea (Tener). Column: 50 × 4 cm. Elution: a linear gradient of sodium chloride in 0.02 *M* Tris–hydrochloric acid buffer of pH 7.6 with 7 *M* urea. Flow-rate: 100 ml/h. Fractions: 20 ml. The peaks are numbered according to the chain-length of the compounds present.

intermediates in the process (Habermann; Petersen and Reeves; Vanyushin and Bur'yanov, 1969a, b).

Mononucleotides are not usually separated in the course of fractionation on molecular sieves and, therefore, Sephadex G-10 and Bio-Gel P-2 are used particularly to separate adenosine and thymidine polyphosphates, which fractionate according to their molecular weights when eluted with formate at pH 6 (De Bersaques), and to study, for example, nucleotide–metal interactions (Colman). The separation of oligothymidylic acids and a study of the influence of their terminal phosphate group on the separation was likewise successful (Haynes *et al.*; Hohn and Pollman).

Stanley verified the possibility of using columns of Sephadex G-25, G-50, G-75 and G-100 to separate substances by their degree of polymerization, finding that 0.5 *M* ammonium hydrogen carbonate of pH 8.6 with 8 *M* urea is needed for an elution process that is independent of the purine bases content, and that the elution volume of oligonucleotides may also be influenced by their secondary structure. Combinations of different sorbent columns, such as Sephadex, Bio-Gel, DEAE-cellulose and hydroxyapatite, were used during a study of oligodeoxyribonucleotide biosynthesis in pulse-labelled mammalian cells (Schandl and Taylor).

### Complex mixtures

In order to separate complex mixtures containing components of different types (bases, nucleosides, nucleotides and their polyphosphates, or oligonucleotides together with their sequence isomers, etc.), encountered mainly when isolating the acid-soluble pool of cells and tissues (Table 37.2) or fractionating enzyme hydrolyzates of nucleic

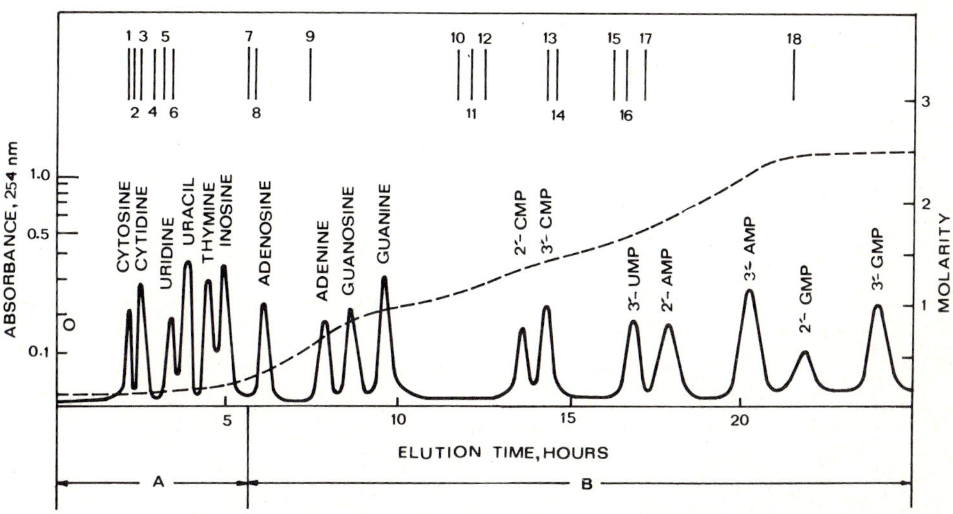


Fig. 37.17. Separation of RNA components and the position of some bases, nucleosides and nucleotides on a column of Dowex 1 (Hori). Column: Dowex 1-X8 (200–400 mesh), 150 × 0.9 cm. Elution: a gradient prepared in a Varigrad mixer with nine compartments containing 135 ml of acetate buffer of pH 4.4 of the following concentrations: 1, 0.1 *M*; 2, water; 3, 1 *M*; 4, 1.2 *M*; 5, 2 *M*; 6, 0.4 *M*; 7, 3 *M*; 8 and 9, 2.5 *M*. Flow-rates: 0.445 ml/min (A) and 0.89 ml/min (B). Temperature: 35°C (A) and 45°C (B). Components: 1 = 5-methylcytosine; 2 = pyrimidine; 3 = pseudouridine; 4 = 2-aminopyrimidine; 5 = thymidine; 6 = 5-hydroxymethyluracil; 7 = purine; 8 = hypoxanthine; 9 = xanthine; 10 = cytidine-5'-phosphate; 11 = xanthosine; 12 = pseudouridine-monophosphate; 13 = thymidylic acid; 14 = uridine-5'-phosphate; 15 = uridine-2'-phosphate; 16 = inosine-5'-phosphate; 17 = adenosine-5'-phosphate; 18 = guanosine-5'-phosphate.

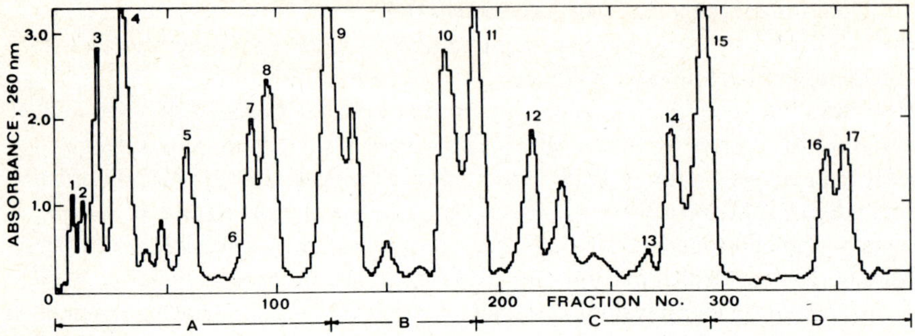


Fig. 37.18. Separation of a yeast nucleotide extract on a Dowex 1 column (Schmitz). Column: 19 × 0.8 cm, Dowex 1-X8. Elution: linear gradients of 0–4 *M* formic acid (A) and 0–0.2 *N* (B), 0.2–0.4 *N* (C) and 0.4–0.8 *N* (D) ammonium formate in 4 *N* formic acid. Flow-rate: 60 ml/h. Fractions: 4 ml. Sample: a neutralized perchloric acid extract of 40 g of yeast. 1 = Adenosine; 2 = cytidine-5′-phosphate; 3 = diphosphopyridinenucleotide; 4 = adenosine-5′-phosphate; 5 = guanosine-5′-phosphate and triphosphopyridinenucleotide; 6 = cytidine-5′-pyrophosphate; 7 = inosine-5′-phosphate; 8 = uridine-5′-phosphate; 9 = adenosine-5′-pyrophosphate; 10 = uridinediphosphoaminosugar peptide; 11 = uridinediphosphoglucose or -galactose; 12 = guanosine-5′-pyrophosphate; 13 = cytidine-triphosphate; 14 = uridine-5′-pyrophosphate and uridinediphosphoglucuronic acid; 15 = adenosine-5′-triphosphate; 16 = guanosine-5′-triphosphate; 17 = uridine-5′-triphosphate.

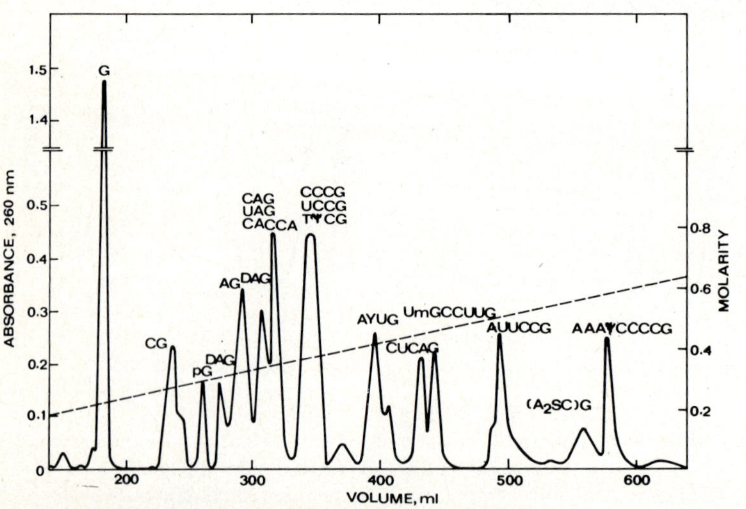


Fig. 37.19. Separation of oligonucleotides from an RNase T1 digest of $tRNA^{Phe}_{E. coli}$ on a column of DEAE-cellulose (Uziel and Gassen). Column: Whatman DE-32, 120 × 0.5 cm. Elution: a linear gradient of 0.05–0.75 *M* ammonium acetate of pH 8.3 (total volume 760 ml). Flow-rate: 12 ml/h. Sample: RNase T1 hydrolyzate of 47 $A_{260}$ units in 0.2 ml of 0.2 *M* Tris (pH 7.3). Letters A, G, C and U correspond to basic nucleotides; Ψ = pseudouridylic acid; T = ribothymidylic acid; S = 4-thiouridylic acid; D = 5,6-dihydrouridylic acid (all with 3′-phosphate end in oligonucleotide).

acids, ion-exchange chromatography is used exclusively, and, if necessary, subfractionation is carried out on several columns under different conditions, as stated in the preceding section on nucleotide fractionation. Thus the same rules apply for the separation of these mixtures and, therefore, the discussion here is limited to a few examples.

The possibilities of separating acid extracts containing UV-absorbing substances of different origins have already been mentioned in conjunction with the description of automatic "nucleoside and nucleotide analyzers", in which use is made of Dowex-type anion and cation exchangers (see p. 836). Fig. 37.17 shows one such example of the separation of a synthetic mixture of low-molecular-weight RNA components on a Dowex 1-X8 column (Hori) with an acetate buffer gradient, prepared in a Varigrad instrument (Peterson and Sober). Lesser demands on the separation of yeast nucleotide extract, with good resolution, are illustrated in Fig. 37.18 (Schmitz), where the separation of polyphosphates and nucleotide coenzymes is also involved.

The successful separation of complicated mixtures of polynucleotides and oligonucleotides is an essential condition for elucidating the primary structure of polynucleotides (Brownlee). Ion-exchange chromatography, particularly on DEAE-cellulose and DEAE-Sephadex, using gradient elution in the presence of 7 *M* urea (Rushizky *et al.*, Tomlinson and Tener) as well as at elevated temperatures (Penswick and Holley), which decrease the secondary interaction with the carrier, completely satisfies the demands of successive separation (see p. 849) of the components of total and partial hydrolyzates of nucleic acids.

Uziel and Gassen, who carried out sequence studies of $tRNA^{Phe}_{E.coli}$, mentioned several illustrative examples (Figs. 37.19–37.21). Table 37.7 shows several completed tRNA structures, in the investigation of which various column techniques were employed. For

TABLE 37.7
SOME PRIMARY STRUCTURES OF tRNAs ELUCIDATED USING COLUMN FRACTIONATION TECHNIQUES

| tRNA | Source | Number of bases | Reference |
|---|---|---|---|
| Alanine I | Yeast | 77 | Holley *et al.*, Merrill |
| Serine I and II | Yeast | 85 | Zachau *et al.* (1966a, b) |
| Serine | Rat liver | 85 | Delihas and Staehelin, Staehelin *et al.* (1968) |
| Tyrosine | Yeast | 78 | Madison *et al.* |
| Tyrosine | *Torulopsis utilis* | 78 | Hashimoto *et al.* |
| Phenylalanine | Yeast | 76 | Raj-Bhandary *et al.* |
| Phenylalanine | *Escherichia coli* | 76 | Uziel and Gassen |
| Phenylalanine | Wheat germ | 76 | Dudock *et al.* |
| Valine | Yeast | 77 | Bayev *et al.* |
| Valine I | *Torulopsis utilis* | 75 | Mizutani *et al.* |
| Isoleucine | *Torulopsis utilis* | 77 | Takemura *et al.* |
| Aspartic acid | Yeast | 75 | Gangloff *et al.* |
| Tryptophane | Yeast | 75 | Keith *et al.* |
| Glutamic acid II | *Escherichia coli* | 76 | Ohashi *et al.* |
| Arginine III | Yeast | 75 | Kuntzel *et al.* |

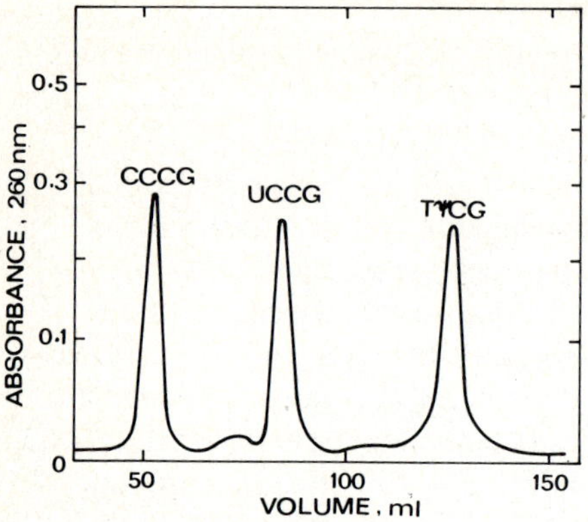


Fig. 37.20. Re-chromatography of tetranucleotide mixture from Fig. 37.19 on a column of DEAE-cellulose at pH 3.7 (Uziel and Gassen). Column: Whatman DE-32, 140 × 0.6 cm. Elution: a linear gradient of 0.1–0.75 *M* ammonium formate of pH 3.7. Flow-rate: 39.6 ml/h. Sample: mixture of $C_3G$, $UC_2G$ and TΨCG from Fig. 37.19.

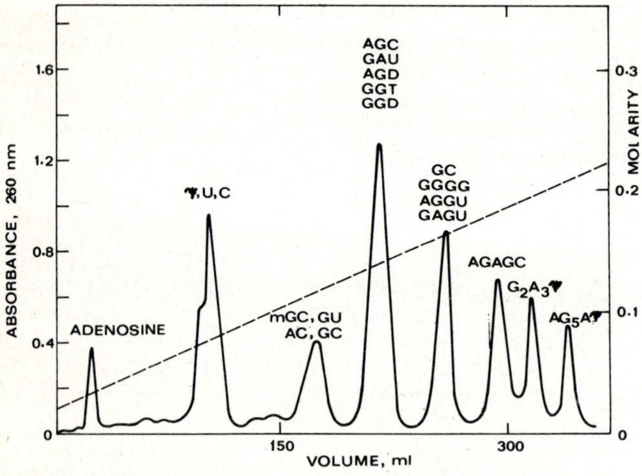


Fig. 37.21. Separation of oligonucleotides from an RNase A digest of $tRNA^{Phe}_{E.\ coli}$ on a column of DEAE-cellulose (Uziel and Gassen). Column: Whatman DE-32, 100 × 0.5 cm. Elution: a linear gradient of 0.02–0.3 *M* sodium chloride in 0.02 *M* Tris–hydrochloric acid (pH 7.8) with 7 *M* urea (total volume 500 ml). Flow-rate: 12 ml/h. Sample: RNase A hydrolyzate of 75 $A_{260}$ units of tRNA. Symbols as in Figs. 37.19 and 37.20.

similar examples of oligonucleotide separations, the comprehensive work of Professor Khorana's laboratory, published in well known biochemical journals under the over-all title of "Studies on polynucleotides" (about 100 papers have been published so far) is to be recommended; these papers contain a large amount of information of a similar type in the fields of both natural and synthetic polymers.

## REFERENCES

Andersen, W., Dekker, C. A. and Todd, A. R., *J. Chem. Soc.*, (1952) 2721.
Anderson, N. G., *Anal. Biochem.*, 4 (1962) 269.
Bayev, A. A., Venkstern, T. V., Mirzabekov, A. D., Krutilina, A. I., Li, L. and Axelrod, V. D., *Mol. Biol. (USSR)*, 1 (1967) 754.
Bernardi, G., Carnevali, F., Nicolaieff, A., Piperno, G. and Tecce, G., *J. Mol. Biol.*, 37 (1968) 493.
Bernardi, G., Faures, M., Piperno, G. and Slonimski, P. P., *J. Mol. Biol.*, 48 (1970) 23.
Blumson, M. L. and Baddiley, J., *Biochem. J.*, 81 (1961) 114.
Bobek, M., Farkaš, J. and Šorm, F., *Collect. Czech. Chem. Commun.*, 34 (1969a) 1673.
Bobek, M., Farkaš, J. and Šorm, F., *Collect. Czech. Chem. Commun.*, 34 (1969b) 1690.
Bobek, M. and Šorm, F., *Collect. Czech. Chem. Commun.*, 34 (1969) 1684.
Bock, R. M., *Methods Enzymol.*, 12A (1967) 224.
Braun, R., *Biochim. Biophys. Acta*, 149 (1967) 601.
Brownlee, G. G., in T. S. Work and E. Work (Editors), *Laboratory Techniques in Biochemistry and Molecular Biology*, Vol. III, North-Holland Co., Amsterdam, 1972, p. 1.
Burtis, C. A., *J. Chromatogr.*, 51 (1970a) 183.
Burtis, C. A., *J. Chromatogr.*, 52 (1970b) 97.
Carrara, M. and Bernardi, G., *Biochim. Biophys. Acta*, 155 (1968) 1.
Christianson, D. D., Paulis, J. W. and Wall, J. S., *Anal. Biochem.*, 22 (1968) 35.
Cohn, W. E., *J. Amer. Chem. Soc.*, 71 (1949a) 2275.
Cohn, W. E., *Science*, 109 (1949b) 377.
Cohn, W. E., *J. Amer. Chem. Soc.*, 72 (1950) 1471.
Cohn, W. E. and Bollum, F. J., *Biochim. Biophys. Acta*, 48 (1961) 588.
Colman, R. F., *Anal. Biochem.*, 46 (1972) 358.
Corneo, G., Ginelli, E., Soave, C. and Bernardi, G., *Biochemistry*, 7 (1968) 4373.
Craddock, V. M., Villa-Trevino, S. and Magee, P. N., *Biochem. J.*, 107 (1968) 179.
De Bersaques, J., *J. Chromatogr.*, 31 (1967) 222.
Dekker, C. A., *J. Amer. Chem. Soc.*, 87 (1965) 4027.
Delihas, N. and Staehelin, M., *Biochim. Biophys. Acta*, 119 (1966) 385.
Dougherty, T. M. and Schepartz, A. I., *J. Chromatogr.*, 40 (1969a) 299.
Dougherty, T. M. and Schepartz, A. I., *J. Chromatogr.*, 43 (1969b) 397.
Drobishev, V. I., Mansurova, S. E. and Kulaev, I. S., *J. Chromatogr.*, 69 (1972) 317.
Dudock, B. S., Katz, G., Taylor, E. K. and Holley, R. W. *Proc. Nat. Acad. Sci. U.S.*, 62 (1969) 941.
Ehrlich, S. D., Thiery, J. P. and Bernardi, G., *Biochim. Biophys. Acta.* 246 (1971a) 161.
Ehrlich, S. D., Torti, G. and Bernardi, G., *Biochemistry*, 10 (1971b) 2000.
Flodin, P., *J. Chromatogr.*, 5 (1961) 103.
Gabriel, T. F. and Michalewsky, J., *J. Chromatogr.*, 67 (1972) 309.
Gangloff, J., Keith, G., Ebel, J. P. and Dirheimer, G., *Biochim. Biophys. Acta*, 259 (1972) 198 and 210.
Gelotte, B., *J. Chromatogr.*, 3 (1960) 330.
Gin, J. B. and Dekker, C. A., in W. W. Zorbach and R. S. Tipson (Editors), *Synthetic Procedures in Nucleic Acid Chemistry*, Vol. 1, Wiley, New York, 1968, p. 208.
Gorbach, G. and Henke, J., *J. Chromatogr.*, 37 (1968) 225.
Habermann, V., *Collect. Czech. Chem. Commun.*, 28 (1963) 510.
Hall, R. H., *J. Biol. Chem.*, 237 (1962) 2283.
Hall, R. H., *Biochemistry*, 3 (1964) 876.
Hall, R. H., *Biochemistry*, 4 (1965) 661.
Hall, R. H., *Methods Enzymol.*, 12A (1967) 305.
Hall, R. H., *The Modified Nucleosides in Nucleic Acids*, Columbia Univ. Press, New York, 1971.
Hall, R. H., Alam, S. M., McLennan, B. D., Terrine, C. and Guern, J., *Can. J. Biochem.*, 49 (1971) 623.
Hashimoto, S., Miyazaki, M. and Takemura, S., *J. Biochem (Tokyo)*, 65 (1969) 659.
Haynes, F. N., Hansbury, E. and Mitchell, W. E., *J. Chromatogr.*, 16 (1964) 410.
Hohn, T. and Pollman, W., *Z. Naturforsch.*, 18B (1963) 919.

Holley, R. W., Apgar, J., Everett, G. A., Madison, J. T., Marquisee, M., Merril, S. H., Penswick, J. R. and Zamir, H., *Science*, 147 (1965) 1462.
Holý, A. and Šorm, F., *Collect. Czech. Chem. Commun.*, 34 (1969) 1929.
Holý, A. and Žemlička, J., *Collect. Czech. Chem. Commun.*, 34 (1969) 3921.
Hori, M., *Methods Enzymol.*, 12A (1967) 381.
Hurlbert, R. B., Schmitz, A., Brumm, A. F. and Potter, V. R., *J. Biol. Chem.*, 209 (1954) 23.
Hutchinson, D. W., *Nucleotides and Coenzymes*, Methuen, London, 1964.
Hutchison, W. C. and Munro, H. N., *Analyst (London)*, 86 (1961) 768.
Jaenicke, L. and Von Dahl, K., *Naturwissenschaften*, 39 (1952) 87.
Katz, S. and Comb, D. G., *J. Mol. Biol.*, 238 (1963) 3065.
Keith, G., Roy, A., Ebel, J. P. and Dirheimer, G., *FEBS Lett.*, 17 (1971) 306.
Kennedy, W. P. and Lee, J. C., *J. Chromatogr.*, 151 (1970) 203.
Khym, J. X. and Cohn, W. E., *J. Amer. Chem. Soc.*, 75 (1953) 1153.
Khym, J. X. and Uziel, M., *Biochemistry*, 7 (1968) 422.
Khym, J. X. and Uziel, M., *J. Chromatogr.*, 47 (1970) 9.
Khym, J. X. and Zill, L. P., *J. Amer. Chem. Soc.*, 74 (1952) 2090.
Kull, F. J. and Soodak, M., *Anal. Biochem.*, 32 (1969) 10.
Kuntzel, B., Weissenbach, J. and Dirheimer, G., *FEBS Lett.*, 25 (1972) 189.
Lehman, I. R. and Pratt, E. A., *J. Biol. Chem.*, 235 (1960) 3254.
Lerner, J., Dougherty, T. M. and Schepartz, A. I., *J. Chromatogr.*, 37 (1968) 453.
Littlefield, J. W. and Dunn, D. B., *Biochem. J.*, 70 (1959) 642.
Madison, J. T., *Methods Enzymol.*, 12A (1967) 137.
Madison, J. T., Everett, G. A. and Kung, H., *Science*, 153 (1966) 531.
Madison, J. T. and Holley, R. W., *Biochem. Biophys. Res. Commun.*, 18 (1965) 153.
Markham, R. and Smith, J. D., *Biochem. J.*, 52 (1952) 552.
Marshak, A. and Vogel, H. J., *Fed. Proc., Fed. Amer. Soc. Exp. Biol.*, 9 (1950) 85.
Merrill, C. R., *Biopolymers*, 6 (1968) 1727.
Mizutani, T., Miyazaki, M. and Takemura, S., *J. Biochem. (Tokyo)*, 64 (1968) 839.
Murakami, F., Rokushika, S. and Hatano, H., *J. Chromatogr.*, 53 (1970) 584.
Ohashi, Z., Harada, F. and Nishimira, S., *FEBS Lett.*, 20 (1972) 239.
Penswick, J. R. and Holley, R. W., *Proc. Nat. Acad. Sci. U.S.*, 53 (1965) 543.
Petersen, G. B. and Reeves, J. M., *Biochim. Biophys. Acta*, 129 (1966) 438.
Peterson, E. A. and Sober, H. A., *Anal. Chem.*, 31 (1959) 857.
Piperno, G. and Bernardi, G., *Biochim. Biophys. Acta*, 238 (1971) 388.
Pithová, P., Fučík, V., Zadražil, S., Šormová, Z. and Šorm, F., *Collect. Czech. Chem. Commun.*, 30 (1965) 2879.
Prystaš, M. and Šorm, F., *Collect. Czech. Chem. Commun.*, 29 (1964a) 121.
Prystaš, M. and Šorm, F., *Collect. Czech. Chem. Commun.*, 29 (1964b) 131.
Prystaš, M. and Šorm, F., *Collect. Czech. Chem. Commun.*, 30 (1965) 2960.
Prystaš, M. and Šorm, F., *Collect. Czech. Chem. Commun.*, 31 (1966) 1053.
Raj-Bhandary, U. L., Chang, S. H., Stuart, A., Faulkner, R. D., Hoskinson, R. M. and Khorana, H. G., *Proc. Nat. Acad. Sci. U. S.*, 57 (1967) 751.
Randerath, K., *Thin Layer Chromatography*, Verlag Chemie, Weinheim, and Academic Press, New York, 1969.
Randerath, K. and Struck, H., *J. Chromatogr.*, 6 (1961) 365.
Rudner, R., Shapiro, H. S. and Chargaff, E., *Biochim. Biophys. Acta*, 129 (1966) 85.
Rushizky, G. W., Mozejko, J. H., Woodford, P. C. and Sober, H. A., *Anal. Biochem.*, 46 (1972) 443.
Rushizky, G. W. and Sober, H. A., *Biochim. Biophys. Acta*, 55 (1962) 217.
Rushizky, G. W. and Sober, H. A., *Progr. Nucl. Acid Res. Mol. Biol.*, 8 (1968) 171.
Sanger, F. and Brownlee, G. G., *Methods Enzymol.*, 12A (1967) 361.
Schandl, E. K. and Taylor, J. H., *Biochim. Biophys. Acta*, 228 (1971) 595.
Schmitz, H., *Biochem. Z.*, 325 (1954) 554.
Schmitz, H., Hurlbert, R. B. and Potter, V. R., *J. Biol. Chem.*, 209 (1954) 41.

Schmukler, H. W., *J. Chromatogr. Sci.*, 8 (1970) 653.
Scott, C. D. and Lee, N. E., *J. Chromatogr.*, 42 (1969) 263.
Simkin, P. A., *J. Chromatogr.*, 47 (1970) 103.
Singh, H. and Lane, B. G., *Can. J. Biochem.*, 42 (1964) 1011.
Singhal, R. P. and Cohn, W. E., *Biochim. Biophys. Acta*, 262 (1972) 565.
Sinsheimer, R. L., *J. Biol. Chem.*, 208 (1954) 445.
Sinsheimer, R. L. and Koerner, J. F., *Science*, 114 (1951) 42.
Škoda, J., *Progr. Nucl. Acid Res.*, 2 (1963) 197.
Sluyser, M. and Bosch, L., *Biochim. Biophys. Acta*, 22 (1962) 479.
Smith, J. D., *Methods Enzymol.*, 12A (1967) 350.
Staehelin, M., *Progr. Nucl. Acid Res.*, 2 (1963) 169.
Staehelin, M., Peterson, E. A. and Sober, H. A., *Arch. Biochem. Biophys.*, 85 (1959) 289.
Staehelin, M., Rogg, H., Baguley, B. C., Ginsberg, T. and Wehrli, J., *Nature (London)*, 219 (1968) 1363.
Stanley, W. M., *Methods Enzymol.*, 12A (1967) 404.
Sweetman, L. and Nyhan, W. L., *J. Chromatogr.*, 32 (1968) 662.
Sweetman, L. and Nyhan, W. L., *J. Chromatogr.*, 59 (1971) 349.
Takemura, S., Murakami, M. and Miyazaki, M., *J. Biochem. (Tokyo)*, 65 (1969) 489.
Taylor, P. R. and Hall, R. H., *J. Org. Chem.*, 29 (1965) 1078.
Tener, G. M., *Methods Enzymol.*, 12A (1967) 398.
Tener, G. M., Khorana, H. G., Markham, R. and Pol, E. M., *J. Amer. Chem. Soc.*, 80 (1958) 6223.
Thacker, L. H., Scott, C. D. and Pitt, W. W., Jr., *J. Chromatogr.*, 51 (1970) 175.
Tomlinson, R. V. and Tener, G. M., *Biochemistry*, 2 (1963) 697 and 703.
Trappe, W., *Biochem. Z.*, 305 (1940) 150.
Uziel, M., *Methods Enzymol.*, 12A (1967) 407.
Uziel, M. and Cohn, W. E., *Fed. Proc., Fed. Amer. Soc. Exp. Biol.*, 24 (1965a) 668.
Uziel, M. and Cohn, W. E., *Biochim. Biophys. Acta*, 103 (1965b) 539.
Uziel, M., and Gassen, H. G., *Biochemistry*, 8 (1969) 1643.
Uziel, M., Koh, C. K. and Cohn, W. E., *Anal. Biochem.*, 25 (1968) 77.
Vanyushin, B. F. and Bur'yanov, Y. I., *Biokhimiya*, 34 (1969a) 546.
Vanyushin, B. F. and Bur'yanov, Y. I., *Biokhimiya*, 34 (1969b) 718.
Virkola, P., *J. Chromatogr.*, 51 (1970) 195.
Visser, D. W., in W. W. Zorbach and R. S. Tipson (Editors), *Synthetic Procedures in Nucleic Acid Chemistry*, Vol. 1, Wiley, New York, 1968, p. 408.
Volkin, E. and Cohn, W. E., *J. Biol. Chem.*, 205 (1953) 767.
Weith, H. L. and Gilham, P. T., *Science*, 166 (1969) 1004.
Wyat, G. R., *Biochem. J.*, 48 (1951) 581.
Wyat, G. R. and Cohen, S. S., *Biochem. J.*, 55 (1953) 774.
Zachau, H. L., Düttig, D. and Feldmann, H., *Hoppe-Seyler's Z. Physiol. Chem.*, 347 (1966a) 212.
Zachau, H. L., Düttig, D. and Feldmann, H., *Angew. Chem.*, 78 (1966b) 392.
Zadražil, S., in W. W. Zorbach and R. S. Tipson (Editors), *Synthetic Procedures in Nucleic Acid Chemistry*, Vol. 2, Wiley, New York, 1973, p. 533.
Zadražil, S., *An Introduction to the Structural Studies of Nucleic Acids (Sequence Determination)*, Koch-Light Labs., Colnbrook, 1974.
Zadražil, S. and Šormová, Z., *Coll. Czech. Chem. Commun.*, 26 (1961) 2651.
Zadražil, S., Šormová, Z. and Šorm, F., *Collect. Czech. Chem. Commun.*, 26 (1961) 2643.
Žemlička, J., Chládek, S., Haladová, Z. and Rychlík, I., *Collect. Czech. Chem. Commun.*, 34 (1969) 3755.
Ženíšek, Z., Laštovková, J. and Varhaník, J., *Sci. Tools*, 16 (1969) 39.
Ziff, E. B. and Fresco, J. R., *J. Amer. Chem. Soc.*, 90 (1968) 7338.

*Chapter 38*

# Nucleic acids

S. ZADRAŽIL

## CONTENTS



## INTRODUCTION AND GENERAL TECHNIQUES IN NUCLEIC ACID SEPARATIONS

Nucleic acids are the fundamental genetic material of all living organisms. The study of their structure and of the molecular mechanism of their functions necessitates a wide range of methodical approaches, by means of which nucleic acids are isolated (liberated from cells, subcellular particles and bonds to other cell components), purified (to remove low- and high-molecular-weight admixtures, both natural and those added in the isolation process) and fractionated (to decrease the heterogeneity of the final preparation with respect to chemical composition, molecular weight and higher structures). In these processes, the various techniques of modern column chromatography play a substantial role. With regard to the complex chemical composition of the macromolecule (nitrogenous bases and the sugar–phosphate chain, giving the polymer a negative charge at neutral pH) and the widely varied arrangements of individual types of nucleic acids (native and single-stranded DNA, circular and supercoiled DNA, viral and phage RNAs, rRNA, mRNA, tRNA, DNA–RNA hybrids, natural and synthetic mono- and polythematic polymers, etc.), ion-exchange, adsorption and partition chromatographic techniques and gel permeation chromatography can be used to separate nucleic acids.

The procedures used for isolating nucleic acids are generally complex processes that differ according to the source of the material to be isolated. Some general steps are illustrated in Table 38.1. Reviews by Cantoni and Davies and Kirby (1964) serve as introduction to the detailed study of isolation methods adapted to individual nucleic acid sources. The technique of gel filtration on a column of Sephadex G-150 can, however, also be used for the direct isolation of high-molecular-weight DNA obtained from nuclei lysate in guanidine hydrochloride (Pivec and Štokrová). Similarly, Sepharose 4B can be used with animal DNA (Loeb and Chauveau) and for transforming DNA from bacterial lysate (Šatava *et al.*), elution of the column in both instances being carried out with citrate containing 2 *M* sodium chloride. An example of such an elution is shown in Fig. 38.1.

TABLE 38.1
BASIC STEPS OF ISOLATION PROCEDURES FOR NUCLEIC ACIDS

| Isolation step | Methods used | References |
|---|---|---|
| Tissue and cell disruption | Mechanical (alumina, pressure, glass beads)<br>Chemical – detergents<br>Enzymic – lysozyme, pronase | Fraenkel-Conrat and Singer (1954a, b), Kay *et al.*, Nomoto *et al.*, Marmur, Berns and Thomas |
| Isolation of subcellular particles | Differential centrifugation, density gradients, zonal centrifugation | Brakke, Reid |
| Nuclease inhibitors | EDTA, citrate, polyvinyl sulphate, bentonite, sodium dodecylsulphate, phenol | Brownhill *et al.*, Singer and Fraenkel-Conrat, Littauer and Sela, Keller *et al.* |
| Deproteinization | Phenol, organic solvents (chloroform), sodium dodecyl-sulphate | Kay *et al.*, Marmur, Kirby (1964) |
| Enzymic degradation | Specific nucleases (DNase, RNase) | Marmur, Fleissner and Borek |
| Precipitation | Ethanol, methoxy- and ethoxyethanol, isopropanol | Kirby (1956), Kay *et al.*, Marmur |
| Further separation and fractionation | Density gradients (sucrose, CsCl) | Vinograd and Hearst, Zadražil and Mach, Schildkraut *et al.*, Williamson |
| | Counter-current distribution | Kirby *et al.*, Kidson and Kirby (1963, 1964), Doctor *et al.*, Zachau *et al.* (1961) |
| | Aqueous polymer two-phase system | Albertsson |
| | Chemical methods | Stephenson and Zamecnik, Katchalsky *et al.* |
| | Chromatography | |

Most of the known types of chromatographic sorbents can be used for the separation of nucleic acids (mixtures of polyanions). A brief review of these sorbents, together with their general separation mechanisms, is given in Table 38.2. However, all sorbents are not equally effective with different types of nucleic acids.

The method most generally used to detect the separated macromolecules is UV absorbance measurement at 260 nm. Differences in types of higher structures are most easily studied by observing the hyperchromic effect after thermal or chemical denaturation, or chemical or enzymic hydrolysis. The spectrophotometric method may be accompanied by the more sensitive measurement of radioactivity (incorporation of $[^{32}P]$ phosphate or $^{3}H$- and $^{14}C$-specific precursors, *e.g.*, thymidine or uridine). Qualitative and quantitative analysis of the fractions obtained can be carried out in most instances by means of colour reactions with the sugar component, such as the orcinol reaction for RNA (Ceriotti) and the diphenylamine reaction for DNA (Burton), which are used most

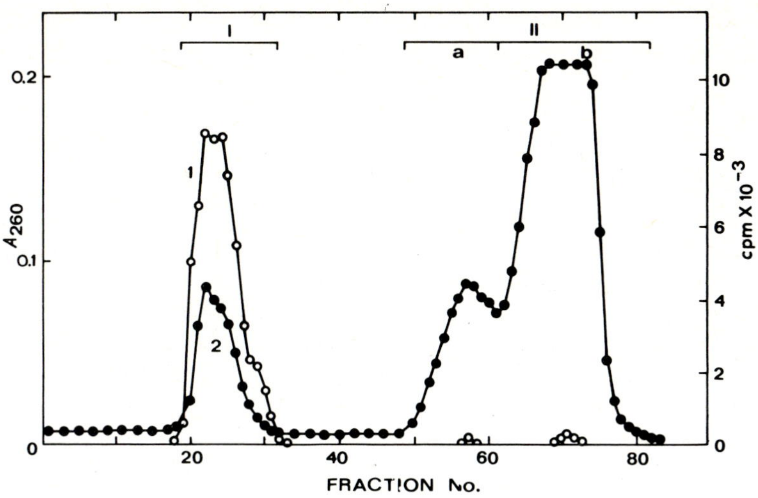


Fig. 38.1. Isolation of a *Bacillus subtilis* DNA on a Sepharose 4B column (Šatava *et al.*). Column: 80 × 3 cm. Elution: 2 *M* sodium chloride in 0.01 *M* sodium citrate solution. Flow-rate: 24 ml/h. Fractions: 10 ml. Sample: 15 ml of a lysozyme lysate of 2 g of wet bacterial cells labelled with [$^3$H] thymidine. I, DNA; IIa, cellular proteins; IIb, mainly lysozyme. 1, Radioactivity; 2, absorbance at 260 nm.

TABLE 38.2
SOME CHROMATOGRAPHIC MATERIALS USED IN NUCLEIC ACID FRACTIONATION

| Type of chromatography | Separation forces involved | Materials |
|---|---|---|
| Ion exchange | Ion exchange and weak secondary interactions (aromatic residues, etc.) | ECTEOLA-, DEAE-, BD-, BND-celluloses; DEAE- and QAE-Sephadexes |
| Adsorption | Electrostatic and complex interactions, charge density in molecules (nucleic acid–protein interactions) | MAK, PLK and histone–Kieselguhr (Celite); hydroxyapatite; nitrocellulose; Amberlite (with $Mg^{2+}$ or $Al^{3+}$) |
| Partition | Hydrophilic and hydrophobic affinity (differences in distribution coefficients) | Sephadex G-25; Sephadex LH-20; diatomaceous earth (Chromosorb W); Kieselguhr (Celite); aluminium oxide |
| Gel permeation | Sieving effects (differences in molecule size and conformation) | Dextran gels (Sephadexes); polyacrylamide gels (Bio-Gel P); agarose gels (Sepharoses; Bio-Gel A); agar; porous glass |

frequently to determine nucleic acids in tissue extracts and in isolated preparations (Hutchison and Munro). Specific nucleases (DNase, RNase, *Neurospora* nuclease, Lehman enzyme, RNase H, etc.) can be used in a similar manner. Important information may also be obtained from other analytical methods, *e.g.*, velocity and equilibrium sedimentation in an ultracentrifuge (molecular-weight heterogeneity, chemical composition and molecular shapes; Schildkraut *et al.*, Smith and Levine), hybridization (sequential relationships; Kennell, Walker), electron microscopy (molecular length and shape; Lang *et al.*, Lang and Mitani), etc. Of biological and biochemical methods, bacterial transformation (Spizizen) and determination of the acceptor activity for amino acids in tRNA (Cherayil and Bock, Von Ehrenstein) deserve mention.

## DEOXYRIBONUCLEIC ACIDS

The primary structure of DNA consists of a long, non-branching deoxyribosophosphate chain with four fundamental nitrogenous bases (adenine, guanine, cytosine and thymine). The elementary higher structure of DNA is the Watson–Crick right-handed double helix, consisting of two polynucleotide chains winding round the same axis and held together by their base interaction A–T and G–C (Watson and Crick). The molecular weight of different samples varies within a wide range, from $10^6$ to $3 \cdot 10^9$ daltons, being generally the result of degradation of the molecule during isolation, because extremely large molecules (*e.g.*, circular DNA of *Escherichia coli*, $2.8 \cdot 10^9$ daltons; Bleeken *et al.*, Cairns), of the order of $10^6$ nucleotide pairs, are very sensitive to mechanical degradation. Therefore, most DNA samples are mixtures of fragments of different sizes, the mean molecular weight of which is about $10^7$ daltons. In addition to the linear polymer, the structure of a covalent closed circle was also found in many instances, *e.g.*, single-stranded circular DNA of the $\varphi$X-174 phage (Kleinschmidt *et al.*) and its double-stranded replicative form (Rush and Warner), which was also found in the $\lambda$ phage (Caro, MacHattie and Thomas, Radloff *et al.*, Weissbach *et al.*) and mitochondrial DNA. All these properties of DNA molecules may affect their behaviour during column fractionation and can be utilized for their separation.

Adsorption chromatography on hydroxyapatite, methylated albumin–Kieselguhr and polylysine–Kieselguhr columns or nitrocellulose is the best method for separating DNA from RNA and higher oligonucleotides, and for removing proteins. These sorbents are also used with advantage for separating native (double-stranded) and denatured (single-stranded) DNA.

Of the many methods available for preparing hydroxyapatite, the method according to Tiselius *et al.* has found the widest use, being preferred even over commercial preparations:

Hydroxyapatite is prepared as a precipitate by gradual mixing of equal volumes (2 l) of 0.5 *M* calcium chloride solution and 0.5 *M* disodium hydrogen orthophosphate solution dropwise (120 drops/min) from separating funnels in a beaker. The precipitate obtained is decanted four times by washing with distilled water (4 l each time). Finally, distilled water is added so as to give a final volume of 4 l, and 100 ml 40% sodium hydroxide solution are added. The mixture is boiled for 1 h with agitation. The precipitate is again decanted four times by washing with water, then 0.01 *M* phosphate buffer of pH 6.8 is added, the suspension is heated to boiling and immediately cooled, then boiled for a further 5 min after decanting and addition of the same buffer. This operation is repeated, with boiling for 15 min, and then twice more with 0.001 *M* phosphate buffer, boiling for 15 min each time. The hydroxyapatite thus obtained can be stored in 0.001 *M* phosphate buffer of pH 6.8. In order to prepare the column, the adsorbent is introduced into the column, the lower end of which is stopped with glass-wool, at atmospheric pressure. When the column has been filled, the adsorbent is washed from the walls and the uppermost adsorbent layer is stirred in order to create an even surface for sample application. The water content of the column is about 70%.

Methylated albumin-coated Kieselguhr (MAK) is usually prepared in a similar manner in the laboratory (Mandell and Hershey). As it can be used universally to fractionate nucleic acids, a detailed description of the preparation of this column is given below.

A 4.2-ml volume of 12 *N* hydrochloric acid is added in small portions to 5 g of bovine serum albumin (fraction V) in 500 ml of absolute methanol, by which process the protein is dissolved and re-precipitated. The suspension is allowed to stand at room temperature in the dark for 3 days with occasional stirring and, after centrifugation at 4000 *g* for 10 min, the sediment is washed at least three times with absolute methanol and twice with diethyl ether. It is then dried *in vacuo* over potassium hydroxide, and likewise stored in the form of an easily water-soluble powder.

Hyflo Supercel (Kieselguhr) is washed before preparation of the MAK column with 200 ml of 1 *N* sodium hydroxide solution per 100 g of carrier, filtered and washed on the filter with another 300 ml of 1 *N* sodium hydroxide solution. Further washing is carried out with 500 ml of 1 *N* hydrochloric acid and then a large volume of distilled water so as to remove the acidic reaction of the filtrate. The washed Kieselguhr is then dried in an oven.

A 20-g amount of washed Kieselguhr is suspended in 100 ml of 0.1 *M* sodium chloride solution in 0.05 *M* phosphate buffer of pH 6.7 and boiled briefly in order to remove air. After cooling, 5 ml of 1% aqueous methylated albumin solution is added with stirring, the mixture is stirred for a further 15 min, then 20 ml of the same buffer with 0.1 *M* sodium chloride solution are added and stirring is continued for 5 min. The suspension is centrifuged for 10 min at 2000 *g* and washed with 0.4 *M* sodium chloride solution in the same buffer. The final sediment is suspended in 125 ml of buffered 0.4 *M* sodium chloride solution, and can be stored in a cool place for several weeks.

In order to prepare a 400 × 20 mm column with three layers, 8 g of Kieselguhr and 2 ml of 1% methylated albumin solution are taken for the first layer, 6 g of Kieselguhr in 40 ml of buffered 0.4 *M* sodium chloride solution and 10 ml of MAK suspension from the first layer (lower protein content) for the second, and 1 g of Kieselguhr in 10 ml of buffered 0.4 *M* sodium chloride solution (protein-free – covering layer) for the third layer. The three suspensions described are introduced into the column in the above order on to a 2-cm layer of cotton-wool or cellulose, and the column is washed with 150 ml of buffered 0.1 *M* sodium chloride solution. Amounts of 1.2–7 mg of nucleic acids in 10–50 ml of buffered 0.1 *M* sodium chloride solution can be applied to such a column. The initial buffer concentration depends on the composition of the nucleic acid mixture (lower concentration in the presence of tRNA, which is gradually increased for native DNA, rRNA and denatured DNA). A typical example of the results obtained with this column is shown later in Fig. 38.5.

Poly-L-lysine-coated Kieselguhr (PLK), prepared in a similar manner to MAK, with a ratio of 1 mg of polylysine to 1 g of Kieselguhr can also be used in a similar manner (Ayad and Blamire, 1968, 1969). Separation on a nitrocellulose column (prepared from a washed carrier, which, being gradually compacted in a chromatographic tube with a glass rod, makes a homogeneous column; Armstrong and Boezi), analogous to the use of nitrocellulose membrane filters in hybridization (Gillespie and Spiegelman, Nygaard and Hall), is a very rapid method for separating nucleic acids.

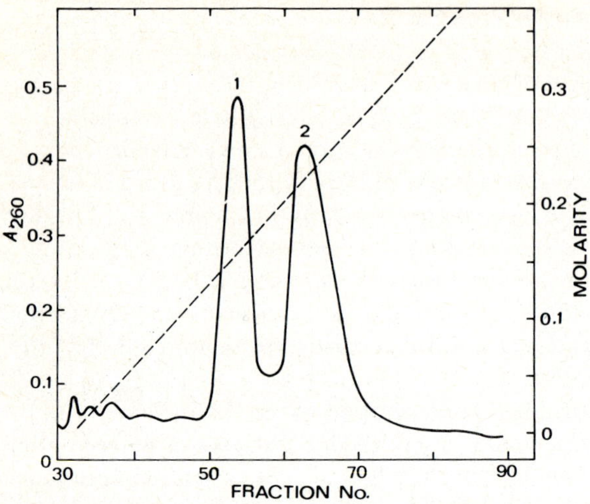


Fig. 38.2. Separation of a mixture of native and heat-denatured transforming DNA from *H. influenzae* on a hydroxyapatite column (Chevalier and Bernardi). Column: 15 × 2 cm. Elution: a linear gradient of 0.001–0.5 *M* potassium phosphate, pH 6.8 (100 ml of each). Fractions: 2.7 ml. Sample: 36 ml of a mixture of 775 μg of heat-denatured DNA and 375 μg of native DNA. 1, Denatured DNA; 2, native DNA.

TABLE 38.3
PROPERTIES AND CHROMATOGRAPHIC BEHAVIOUR OF NUCLEIC ACIDS ON A HYDROXYAPATITE COLUMN

| Nucleic acid | Structural characteristics | Buffer molarity at elution peak | Reference |
|---|---|---|---|
| *DNA* | | | |
| Calf thymus | Native | 0.20–0.22 | Bernardi (1961, |
| | | 0.20, 0.25* | 1969b) |
| | Denatured | 0.10, 0.15* | |
| *Saccharomyces cerevisiae* | | | |
| Mitochondria | Circular | 0.27 | Bernardi *et al.* |
| Nuclei | Linear | 0.25 | |
| Polyoma virus | Linear and circular | 0.29 | Bourgaux-Ramoisy |
| | Superhelix | 0.26 | *et al.*, Bourgaux and Bourgaux-Ramoisy |
| Phage T2 | Unglucosylated | 0.22 | Bernardi (1969a, b) |
| | Glucosylated | 0.25–0.27 | |
| Phage φX-174 | Single-stranded circle | 0.10, 0.15* | Bernardi (1969b, c) |
| *RNA* | | | |
| *E. coli* tRNA | Clover-leaf structure | 0.13 | Bernardi (1969c) |
| Plant rRNA | | 0.15 | Pinck *et al.* |
| Ehrlich ascites Tumour | High-molecular | 0.15, 0.20* | Bernardi and Timasheff |
| Plant virus | Replicative form | 0.20 | Pinck *et al.* |
| *Polynucleotides* | Random-coiled | 0.10–0.15 | Bernardi (1969c) |
| | Double-stranded | 0.20–0.22 | |
| | Triple-stranded | 0.45 | |

*Stepwise elution.

The principle of separation and fractionation on hydroxyapatite consists in electrostatic interaction between the negative phosphate groups of the polymer and the positive charges of calcium ions in the sorbent crystals (Bernardi, 1965). Thus, an elution can be carried out with a concentration gradient of phosphate buffer solution or by increasing the column temperature with a constant concentration and pH of the eluent (Miyazawa and Thomas). This mechanism is clearly demonstrated by the different behaviour of native and denatured DNA (Fig. 38.2) and RNA (Table 38.3, Bernadi, 1965, 1969a, b, c; Chevalier and Bernardi). No changes in the DNA molecule were observed during the chromatographic process and fractionation with respect to base composition or molecular weight did not occur. The different behaviour of some nucleic acids, evidently caused by conformational differences in the molecules from the form of native double-stranded DNA (dependence on molecule volumes: Bernardi, 1969a) is summarized in Table 38.3.

The hydroxyapatite column was also used to obtain evidence of the fact that the residual transforming activity of heat-denatured DNA is related to the native-like (eluted in the same position, Fig. 38.3), probably cross-linked, sample fraction (Chevalier and Bernardi). The determination of the number of sex factors on *E. coli* chromosomes has been carried out by the same method (Frame and Bishop). When stepwise elution is used, however, artefact peaks may be formed (Fig. 38.4); this formation evidently depends on unsuitable DNA:carrier ratios or on the length of the column (Bernardi, 1961).

Elution at a constant but elevated temperature (60–70°C) is used to isolate rapidly denaturing DNA fractions (satellite DNA, Votavová *et al.*; renaturation studies, McGallum and Walker). When variable temperatures are used, with a phosphate buffer of constant concentration, DNA can be fractionated by the base composition, denatured fractions being obtained (Miyazawa and Thomas). The method can be modified for use with large amounts of material and larger DNA sample series by using a heated centrifuge, which

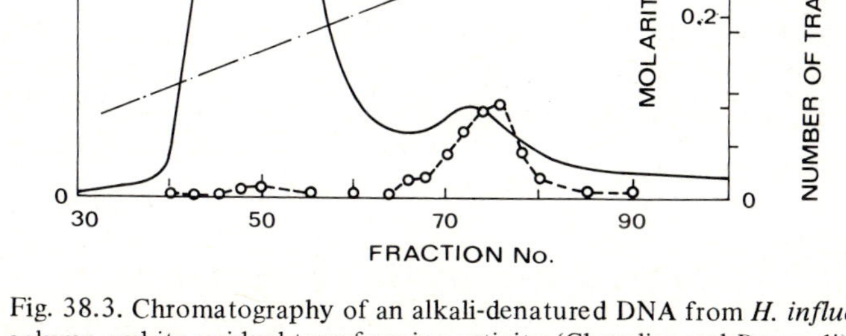


Fig. 38.3. Chromatography of an alkali-denatured DNA from *H. influenzae* on a hydroxyapatite column and its residual transforming activity (Chevalier and Bernardi). Column: 20 × 1.2 cm. Elution: a linear gradient of 0.001–0.5 *M* potassium phosphate, pH 6.8 (150 ml of each). Fractions: 2.4 ml. Sample: 1.8 mg of alkali-denatured transforming DNA. Circles indicate the number of transformed cells (cathomycin marker). The specific biological activity of fraction 76 was 43% of untreated DNA.

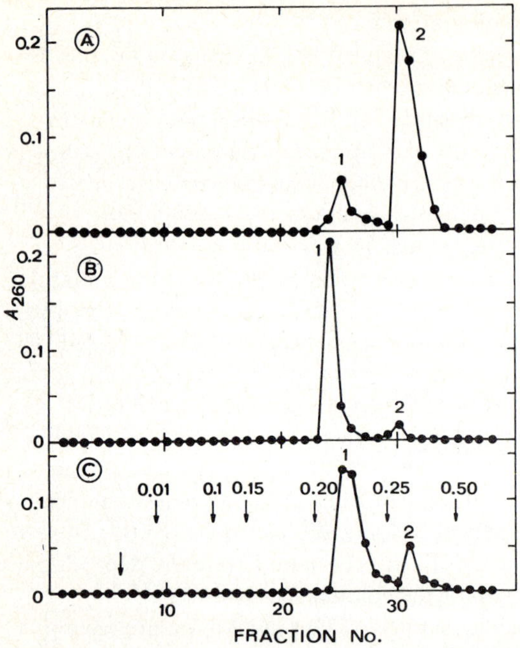


Fig. 38.4. Chromatography of native calf thymus DNA on a column of hydroxyapatite with stepwise elution (Bernardi, 1961). Column: A, 5 × 1.3 cm; B and C, 3 × 1.3 cm. Elution: stepwise with phosphate buffer of indicated concentrations. Fractions: 3 ml. Sample: 1.28 mg of DNA. The first arrow indicates the point of DNA application. A, separation of the original DNA sample; B and C, re-chromatography of the material from peaks 1 and 2, respectively, of A. The whole procedure shows the artifact origin of the second peak.

speeds up the process considerably (Brenner *et al.*, Flamm *et al.*). In most instances the yield obtained by chromatography on a hydroxyapatite column is 90–100%.

Native DNA can be separated by stepwise elution from denatured DNA and DNA–RNA complex on nitrocellulose (Armstrong and Boezi, Klamerth), being eluted in the opposite order to that on hydroxyapatite. Riggsby, who modified the nitrocellulose column technique used by Bautz and Reilly, achieved approximately 100-fold purification of genetically labelled DNA.

Differing from the preceding sorbents, the MAK column, which operates mainly on the basis of electrostatic interaction between DNA and the basic protein, is able not only to separate DNA from RNA and the native form from the denatured form of DNA, but also to fractionate double-stranded DNA by its composition and molecular weight. However, with insufficiently deproteinated samples, the fractionation may be influenced to a large extent by the presence of residual proteins.

A three-layer column (Mandell and Hershey), the preparation of which is described above, can be used to separate at room temperature the entire extract of nucleic acids from *E. coli* (Fig. 38.5) into four fractions, which are successively eluted at a sodium chloride concentration of about 0.4 *M* for tRNA, 0.6 *M* for DNA, 0.75 *M* for 16S rRNA and 0.85 *M* for 23S rRNA, in 0.05 *M* phosphate buffer of pH 6.7. It is generally advantageous to separate about 1 mg of DNA on 10 ml of MAK suspension, the column height being irrelevant. Transformation DNA from *B. subtilis* (Pivec *et al.*, Zadražil *et al.*) was repeatedly fractionated on a similar column (Šponar *et al.*), the fractions obtained differing in terms of mean base composition and molecular weight, and also in biological activity for various markers (Fig. 38.6). For elution from the MAK column, native DNA with a mean molecular weight of $10^7$ to $2 \cdot 10^7$ is eluted in the sodium chloride concentration range 0.60–0.75 *M* and denatured DNA at 0.80–0.95 *M*. A decrease in the molecular weight leads to a decrease in the concentration of the eluting salt, while the G–C content is indirectly proportional to the concentration of the eluting solution (Cheng and Sueoka, Mandell and Hershey). The recovery, particularly of denatured DNA, is highest with an elution temperature of less than 10°C, decreasing with increase in temperature and becoming practically zero at about 40°C (Roger). Stepwise elution gives a better recovery of the sample applied, while gradient elution is better for analytical purposes as it is more sensitive to differences in the desorption of individual molecules.

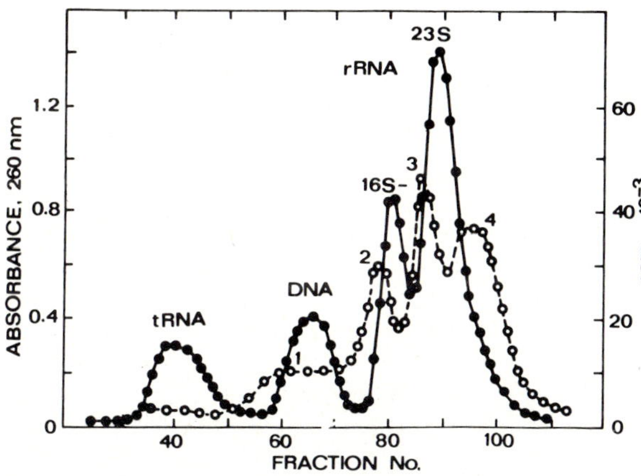


Fig. 38.5. Chromatography of $^{32}P$-labelled nucleic acids extracted from *E. coli* on a methylated serum albumin column (Takai *et al.*). Column: according to Mandell and Hershey (see text). Elution: a linear gradient of 0.2–1.0 *M* sodium chloride in phosphate buffer. Flow-rate: 24 ml/h. Fractions: 4 ml. Sample: a phenol-extracted and alcohol-precipitated nucleic acid mixture from 200 ml of a bacterial culture labelled with $^{32}P$. Closed circles, absorbance at 260 nm; open circles, four fractions of labelled mRNA (1–4).

The different cytosine contents of individual DNA strands from bacteria have enabled the MAK column with an intermittent gradient (Rudner *et al.*, 1968a, b, 1969) to be used to separate the complementary strands of thermally or alkaline-denatured DNA (Fig. 38.7).

Poly-L-lysine is a further protein sorbent used with Kieselguhr (Ayad and Blamire, 1968). An extract of whole nucleic acid from *B. subtilis* (Fig. 38.8) and a DNA preparation from the same source (Fig. 38.9) was reproducibly fractionated, particularly by the base composition (Ayad and Blamire, 1968, 1969), as sonication of the DNA sample or thermal denaturation did not markedly influence the elution profile (Table 38.4). In spite of some advantages (reproducibility, affinity for the base composition used to separate plasmide

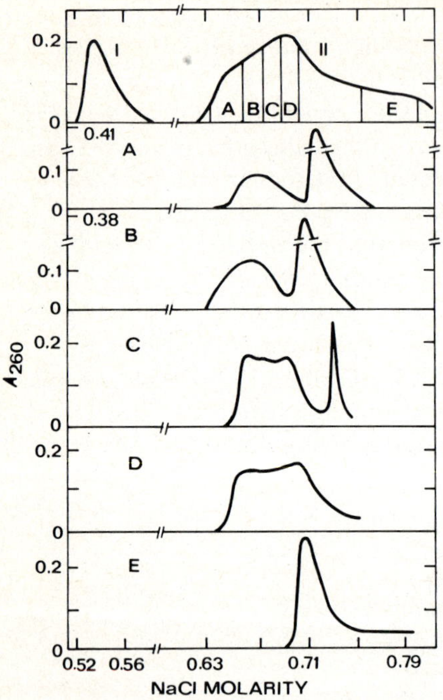


Fig. 38.6. Repeated fractionation of *B. subtilis* DNA on an MAK column (Pivec *et al.*). Top curve – Column: 25 × 3.5 cm prepared according to Mandell and Hershey (see text). Elution: a linear gradient of 0.5–1 *M* sodium chloride in 0.013 *M* phosphate buffer of pH 7. Flow-rate: 40 ml/h. Fractions: 7 ml. Sample: 20 mg of transforming DNA. I, oligonucleotides and residual protein; II, DNA distribution peak. Fractions A to E were used for re-chromatography. Curves A to E – Column: 15 × 2 cm, prepared and eluted as mentioned above. Flow-rate: 20 ml/h. Fractions: 5 ml. Samples: material of fractions A to E, each about 2.5 mg.

from chromosomal DNA (Cannon and Dunican), known fractionation mechanism, etc.), PLK chromatography is not yet being widely used. This applies to a far greater extent to other carriers of a similar nature, such as histone (Ayad and Wilkinson, Brown and Martin, Brown and Watson, Tichonenko), protamine (Légault-Démare *et al.*) etc., and sometimes even chemically bound to cellulose instead of Kieselguhr.

In order to fractionate bacterial and animal nucleic acids, a method of interaction between the mixture to be separated and the hexamine cobalt(II) salt of synthetic or natural polynucleotides bound to Kieselguhr has been developed (Lin). The sorbents used are denatured DNA, poly-A, poly-I, poly-C and poly-U, and the mutual bond depends on the presence of dioxane, so that there is no need for the polymers involved to be sequentially complementary. Elution is carried out with a linear dioxane gradient mixed with a buffer. On the column with bound poly-I, complementary strands of denatured DNA from *E. coli* were separated, while partial fractionation of tRNA and rRNA was achieved on bound animal DNA (Lin).

Experiments with DNA fractionation on Amberlite IRC-50 in the presence of $Mg^{2+}$ (Frankel and Crampton) and on IR-120 with $Al^{3+}$ (Kothari, 1970a, b, 1972) appear to lie

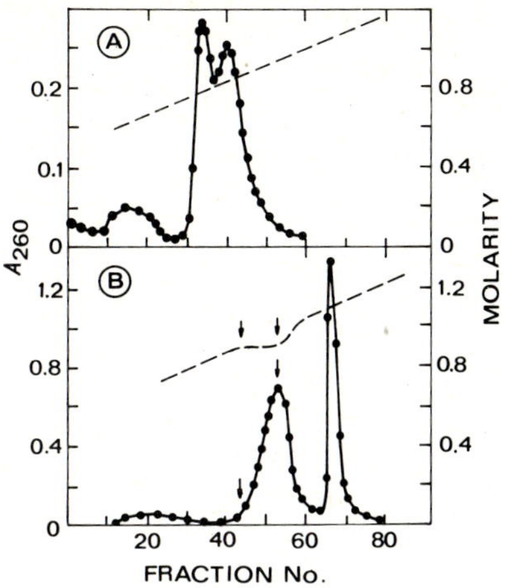


Fig. 38.7. Denatured *B. subtilis* DNA eluted from an MAK column with the use of linear gradient (A) and by the intermittent gradient technique (B) for the separation of complementary strands (Rudner *et al.*, 1968a). Column: 15.5 × 1.9 cm. Elution: a linear sodium chloride gradient in 0.05 *M* sodium phosphate of pH 6.7 (A) or the same used with the intermittent technique (B). Fractions: 5 ml. Sample: 2 mg of denatured DNA. A, heat-denatured DNA with a total gradient volume of 400 ml (0.6–1.2 *M* sodium chloride); recovery 64%. B, alkali-denatured DNA with a total gradient volume of 450 ml (0.7–1.4 *M* sodium chloride); recovery 80%. As indicated by arrows, the gradient was cut at tube 44 and reconnected at tube 53.

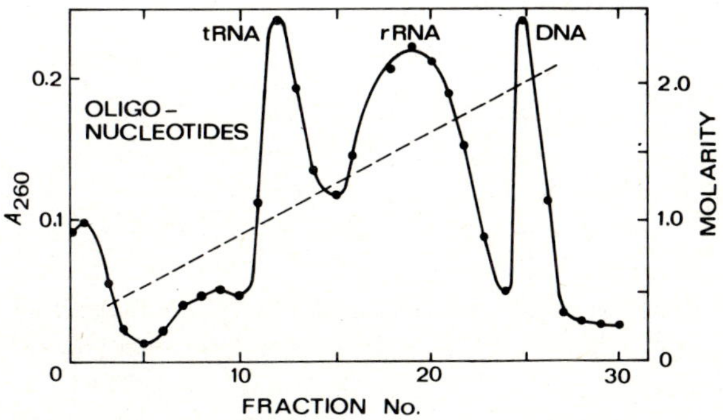


Fig. 38.8. Separation of a nucleic acid mixture extracted from *B. subtilis* on a PLK column (Ayad and Blamire, 1969). Column: 5 g of PLK material. Elution: a linear gradient of 0.4–4 *M* sodium chloride in 0.02 *M* phosphate buffer of pH 6.7 (150 ml). Flow-rate: 20 ml/h. Fractions: 4 ml.

on the borderline with ion-exchange chromatography. As weak cation exchangers are involved, DNA fractionation is a matter of adsorption (formation of chelate complexes between the carboxyl groups of the resin, metal ions and the phosphate groups of DNA). This means that the sorbent used fractionates by base composition, A + T-rich molecules being more firmly bound to the carrier (Mindich and Hotchkiss, 1964b; Pullman and Pullman). Transforming DNA from *H. influenzae* was fractionated in this way in order to

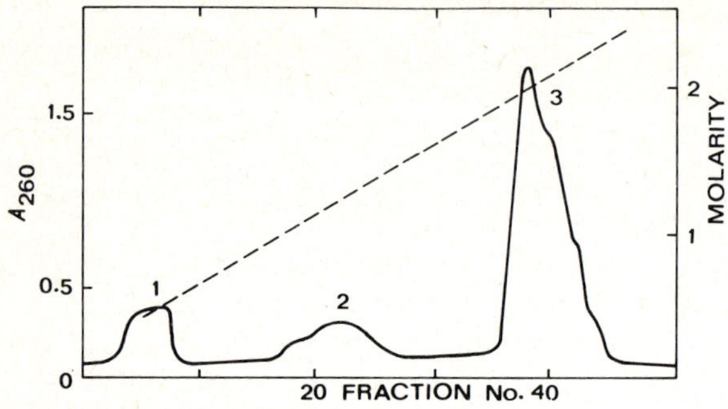


Fig. 38.9. Fractionation of *B. subtilis* native DNA on a PLK column (Ayad and Blamire, 1968). Column: 10 g of PLK material. Elution: a linear gradient of 0.4–4 *M* sodium chloride in 0.02 *M* potassium phosphate buffer of pH 6.7 (150 ml). Flow-rate: 20 ml/h. Fractions: 4 ml. Sample: 1.5 mg of DNA in 15 ml of 0.4 *M* buffered sodium chloride. 1, Low-molecular-weight DNA admixtures; 2, heterogeneous fraction of DNA of no definite conformation; 3, high-molecular-weight native DNA (each fraction of different G + C content).

TABLE 38.4
BEHAVIOUR OF *Bacillus subtilis* DNA ON A PLK COLUMN (AYAD AND BLAMIRE, 1968)

| Structural characteristics | Molarity of elution buffer | | |
|---|---|---|---|
| | Peak I | Peak II | Peak III |
| Native | 0.5 | 1.20 | 2.0 |
| Sonicated | 0.6 | 1.25 | 1.84 |
| Heat-denatured | 0.6 | 1.35 | 2.07 |

cumulate biological activity to certain markers (Mindich and Hotchkiss, 1964a).

Initially, there was an intensive study of the ion-exchange fractionation of DNA, particularly on substituted cellulose types (Bendich *et al.*, 1955, 1958; Davila *et al.*, 1965a, b; Kit, 1960a, b; Otaka *et al.*; Rosenkranz and Bendich), which are more suitable than synthetic ion exchangers because of their hydrophilic nature, better permeability and enormous surface area and, therefore, greater capacity for macromolecules. It was found, however, that this field is more suitable for fractionating complex mixtures of oligonucleotides and tRNA than for high-molecular-weight DNA. Recently, benzoylated and naphthoylated DEAE-cellulose, functioning on the basis of differentiation of the secondary structure of molecules, was used to separate an extract of nucleic acids from *E. coli* infected with MS2 phage. The separation, however, was only partial, *viz.*, of rRNA from the mixture of DNA and tRNA (Sedat *et al.*). This type of carrier was also used in order to fractionate replicating DNA of λ phage (Kiger and Sinsheimer). The physical basis of separation has not yet been fully clarified, but these columns have been used successfully for the separation of the native and partially denatured DNAs (Iyer and Rupp, Pyeritz *et al.*) and, being treated with deoxycholate, also for the purification of the bacterial genes for rRNAs (Udvardy and Venetianer).

DNA itself can be covalently bound to a cellulose and serves as an adsorbent for isolating complementary DNA strands and RNA (according to the sequence) from a mixture of different nucleic acid molecules (Bautz and Hall). The ideal conditions for adsorption are comparable with those which favour renaturation of nucleic acid strands. By decreasing the ionic strength and increasing the column temperature the adsorbed polynucleotide can be eluted.

As with ion-exchange resins, the preparation of a gel column for nucleic acid fractionation is the same as for the chromatography of low-molecular-weight substances (the manufacturers supply detailed instructions for column preparation with all gel types). Originally, gel columns (sorbents with higher degrees of cross-linking, see Fig. 38.10) were used in order to replace and speed up dialysis (desalting of high-molecular-weight samples and removal of phenol after isolation, Shepferd and Petersen), which means purification of the macromolecule in the course of its isolation (Bauer and Johanson) or removal of superfluous components in studies of interactions with DNA (Attardi *et al.*, Hanson, Sekine *et al.*).

The actual separation of macromolecules was achieved by separating DNA and RNA on more porous gels of the dextran type, such as Sephadex G-200 (Fig. 38.11, Bartoli

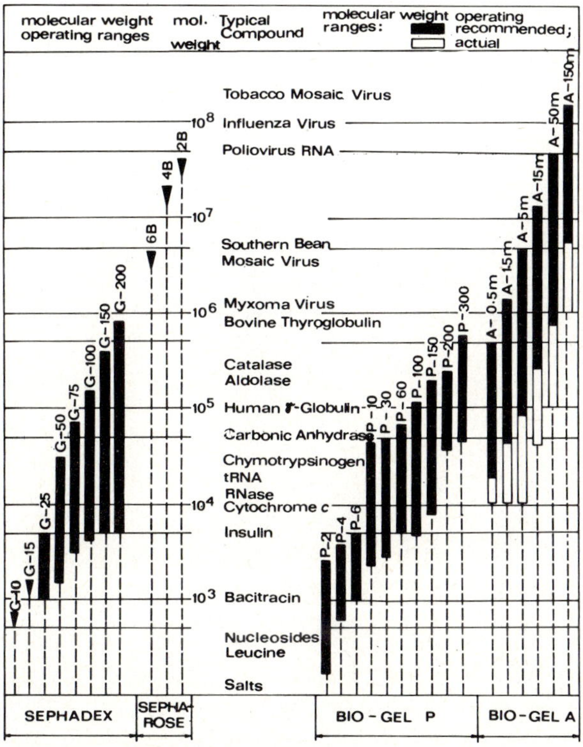


Fig. 38.10. Relationship between operating range and molecular weights for gel filtration materials (Pharmacia and Calbiochem leaflets).

*References p.883*

and Rossi), or of the agarose type, such as Sepharose (Fig. 38.12, Öberg and Philipson), the particles of which are capable of absorbing and delaying the flow of macromolecules. The above examples show that separations on gels of both types are based on differences in molecular weight and shape. As mentioned already (p. 861), Sepharose is also used in order to separate DNA from admixtures that are difficult to remove (Cozzarelli *et al.*, Young and Jackson) and this is therefore one of the mildest separation methods (Loeb and Chauveau). The relatively recent introduction of agarose carriers into laboratory use and the possibility of simple elution with gradient-free solutions indicate that these gels should find wider application in the future. The resolution of the column depends, as with low-molecular-weight substances, on the ratio between the volume of the sample and that of the column filling.

Gels are also used as stationary phases, serving to anchor DNA as the fractionation

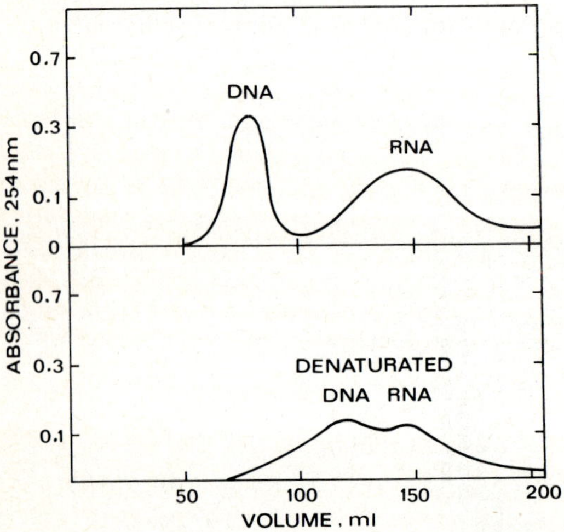


Fig. 38.11. Separation of liver DNA and RNA by gel filtration on a Sephadex G-200 column (Bartoli and Rossi). Column: 25 × 2.5 cm. Elution: 1 *M* sodium chloride in 0.1 *M* Tris–hydrochloric acid buffer of pH 7.2. Flow-rate: 30 ml/h.

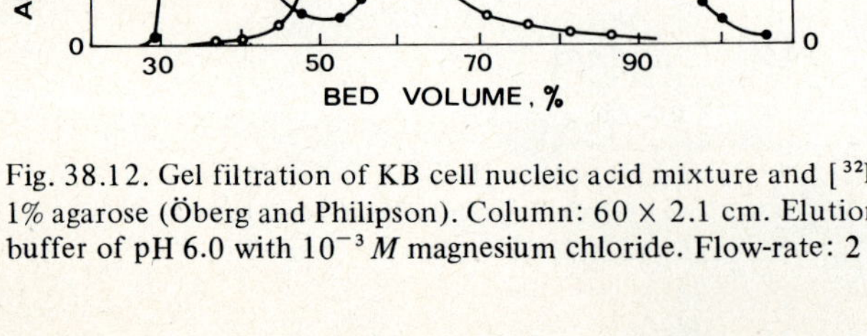


Fig. 38.12. Gel filtration of KB cell nucleic acid mixture and [$^{32}$P] RNA of poliovirus on a column of 1% agarose (Öberg and Philipson). Column: 60 × 2.1 cm. Elution: $2 \cdot 10^{-3}$ *M* sodium phosphate buffer of pH 6.0 with $10^{-3}$ *M* magnesium chloride. Flow-rate: 2 ml/h·cm$^2$. Sample volume: 1.5 ml.

agent (hybridization on an agar column, Bendich and Bolton) where it served, for example, to isolate the DNA anticodon strand (Doskočil and Hochmannová). When bound to various gel carriers, DNA can also be used for the purification of DNases (Naber *et al.*, Poonian *et al.*, Schabort) or of other enzymes that take part in DNA metabolism and biosynthesis (polymerases, ligase, etc.).

The use of partition chromatography (a continuation of the now little used technique of counter-current distribution, Kidson and Kirby, 1963, 1964) is limited in the case of DNA to a modified gel carrier, such as hydrophobic Sephadex LH-20 (Kidson, 1969). A column prepared in the organic phase of a multicomponent solvent system (amyl alcohol, 2-methoxyethanol, 2-butoxyethanol, tripentylamine, acetic acid and trilithium citrate) could distinguish the native and denatured DNAs by elution with a linear gradient of citrate in the aqueous phase (used as the mobile phase). This method, the separation mechanism of which is due to a stronger interaction of the denatured DNA phosphate groups with the amine, with exchange of $Li^+$, was used successfully to study the structure of *E. coli* DNA in the vicinity of the replicating point (Kidson, 1968).

## RIBONUCLEIC ACIDS

In comparison with DNA, RNAs are a far more heterogeneous group with macromolecules that differ distinctly in structure and function. RNAs are found in all living cells: rRNA with a molecular weight between $0.5 \cdot 10^6$ and $2 \cdot 10^6$ daltons (80% of total RNA), tRNA ($2.5 \cdot 10^4$ daltons; 15% of total DNA) and mRNA (very heterogeneous and labile, mean molecular weight $10^5$ daltons; 5% of total DNA). In the primary structure, a large number of minor modified components are added to the four fundamental bases, particularly in the case of tRNA (Hall). The secondary structure is likewise very varied (Cox), from totally double-stranded viral RNA (Gomatos and Tamm), through molecules with extensive hydrogen bonding (rRNA and tRNA, Cox, Levitt) up to linear single-stranded structures of mRNA (in a translation process, Matthaei and Nirenberg).

As most of the techniques that are capable of separating RNA from DNA have already been mentioned in the preceding section on DNA, we shall now limit the discussion to fractionations that differentiate between individual types of RNA molecules. Owing to the great variations in the structures of the molecules all column chromatographic techniques are used.

From the field of the adsorption chromatography, the MAK column separates all RNA types (Fig. 38.5) including mRNA, the heterogeneity of which can be studied in this way (Monier *et al.*, Oravec). It can also be used in order to fractionate infectious RNA from poliomyelitis virus (Cocito *et al.*). Most attention, however, has been devoted to the most complex group of tRNA. Using $^{14}C$-labelled amino acids, Sueoka and Yamane found that sub-fractionation takes place in the tRNA peak (Fig. 38.13), even allowing a search for modified tRNA after infection of bacteria by a phage (Kano-Sueoka and Sueoka), after transformation of animal cells by oncornaviruses (Trávníček and Říman) or after incorporation of fluorouracil (Lowrie and Bergquist). When Kieselguhr is replaced with silicic acid, the capacity of the methylated albumin column for fractionation of tRNA from *E. coli* increases up to 100-fold (Stern and Littauer, Ziv *et al.*). A situation similar to that

*References p.883*

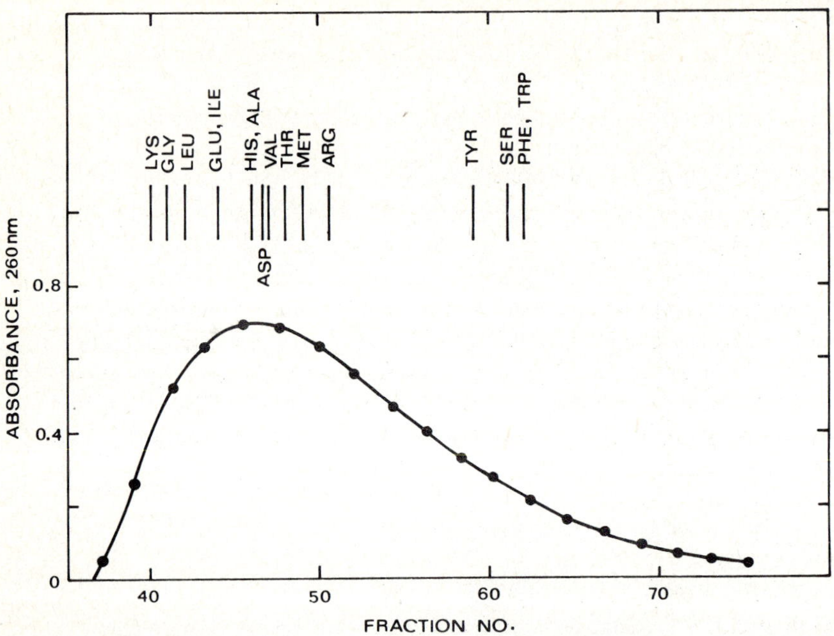


Fig. 38.13. Chromatography of a mixture of 16 aminoacyl tRNAs from *E. coli* on an MAK column (Sueoka and Yamane). Column: 8 × 1.8 cm. Elution: a linear gradient of 0.2–1.1 *M* sodium chloride in 0.05 *M* phosphate buffer of pH 6.7 (110 ml of each solution). Flow-rate: 60 ml/h. Fractions: 2 ml. Sample: 2.5 mg of tRNA incubated with a mixture of one radioactively labelled amino acid and the 19 remaining non-radioactive amino acids. The vertical lines indicate the positions of the main radioactive peaks of the individual aminoacyl tRNAs (multiform not being marked).

for MAK also applies to the use of PLK (Fig. 38.8). Gradient elution successively liberates fractions according to their increasing molecular weight, but for double-stranded RNA and 16S and 23S rRNA the secondary structure and conformation may also be important (Ayad and Blamire, 1970).

Differing from its widespread application with DNA, the hydroxyapatite column cannot be used with high-molecular-weight rRNA because it is degraded during chromatography (Bernardi and Timasheff). Owing to their structural analogy with DNA, however, RNA replicative forms and intermediates can be fractionated (Pinck *et al.*). In spite of a detailed study of the fractionation of different tRNAs, viral RNA and polynucleotides (Bernardi, 1969c), chromatography of RNA on hydroxyapatite has not been widely used.

Of greater importance for tRNA are ion-exchange procedures on substituted celluloses and Sephadex, which permit the isolation of pure specific tRNAs that can be used for sequential analysis and for protein biosynthesis. The existence of weak bonding forces (secondary interactions dependent on RNA base composition) is of decisive importance in fractionation owing to the similarity of the separated molecules in terms of size and charge. Control of these forces, together with changes in pH and the concentration gradient, may be decisive for satisfactory separations (urea, temperature, BD and BND substitution, etc.). Detailed studies of these problems were made by Bock and Cherayil and Cherayil and Bock, who used DEAE-cellulose and DEAE-Sephadex under different conditions so as to

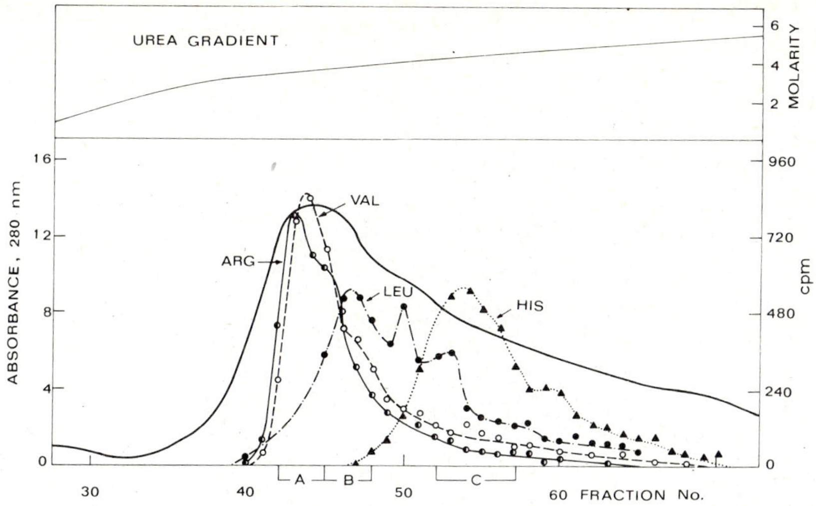


Fig. 38.14. Fractionation of tRNA on a DEAE-cellulose column with a urea gradient (Cherayil and Bock). Column: 100 × 1.5 cm. Elution: an exponential gradient of 0–7 *M* urea in 0.02 *M* Tris–hydrochloric acid buffer of pH 7.5 with 0.34 *M* sodium chloride prepared in a constant-volume mixer containing 300 ml of a starting solution. Flow-rate: 15 ml/h. Fractions: 7.5 ml. Sample: 200 mg of yeast tRNA. Arg, Val, Leu and His represent the acceptor activity of tRNA fractions for the given amino acids. Material of regions A, B and C was used for further re-chromatography (Fig. 38.15). Solid line, absorbance at 280 nm.

permit the isolation or enrichment of most specific tRNAs (Figs. 38.14 and 38.15). General conclusions concerning this fractionation process can be drawn: (a) secondary interaction of the polynucleotide with cellulose is considerably greater than with dextran (mainly purine bases) and can be eliminated with the aid of urea; (b) separation by length of the polyanion chain can be achieved on both cellulose and dextran ion exchangers in the presence of urea; (c) re-chromatography of the peaks at lower pH values facilitates the further separation of tRNA by the base composition (partial protonation of adenine and cytosine residues, Fig. 38.15) and (d) the use of a urea gradient considerably increases the resolution of the column for tRNA compared with a sodium chloride gradient. Similarly, a change of column temperature in the 20–65°C range (Baguley *et al.*) causes, on elution, the tRNA peak to widen and to shift towards a higher concentration of the eluting gradient (a change in molecular conformation takes place).

Secondary interactions, particularly of the lipophilic part of the chain, may be enhanced by the introduction of benzoyl (BD), and benzoyl and naphthoyl (BND) groups into DEAE-cellulose (Gillam *et al.*, 1967). This leads to stronger binding of the polynucleotide, in spite of the simultaneous decrease in capacity of the ion exchanger. tRNAs specific for the aromatic amino acids phenylalanine, tyrosine and tryptophan can therefore be separated more easily from other tRNAs in the form of aminoacyl tRNA (Gillam *et al.*, 1968; Fink *et al.*; Maxwell *et al.*), the affinity of which to the column is greater (they are eluted with 1 *M* sodium chloride solution only after the addition of ethanol). In this instance, however, the decisive factor for successful separation is the purity of the amino

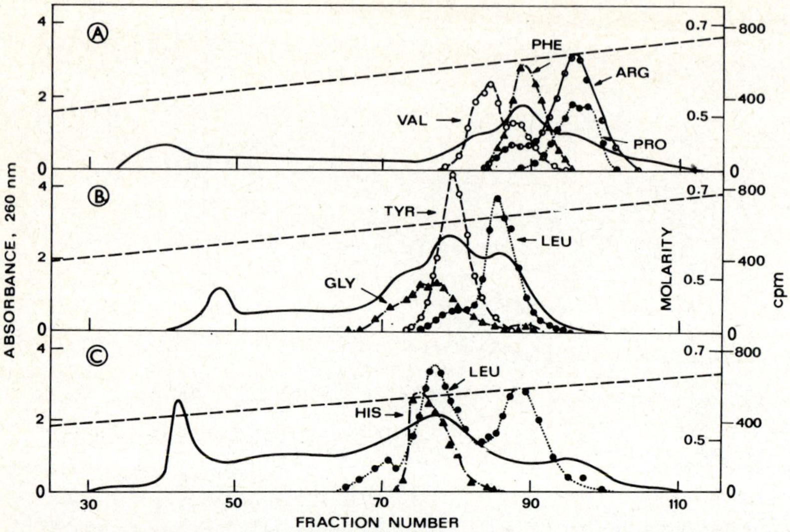


Fig. 38.15. Re-chromatography of fractions A, B and C of tRNA from Fig. 38.14 on a DEAE-Sephadex column at pH 4.5 (Cherayil and Bock). Column: 100 × 1.2 cm. Elution: a linear gradient of 0.52–0.7 *M* sodium chloride (the total volume 500 ml) in the presence of 0.03 *M* acetate buffer of pH 4.5 and 7 *M* urea. Flow-rate: 8 ml/h. Fractions: 4 ml. Sample: fractions A, B and C from the previous column (Fig. 38.14) adjusted to 7 *M* urea and pH 4.5. Val, Phe, Arg, Pro, Gly, Tyr, Leu and His represent the acceptor activity of tRNA fractions for a given labelled amino acid. Solid line, absorbance at 260 nm.

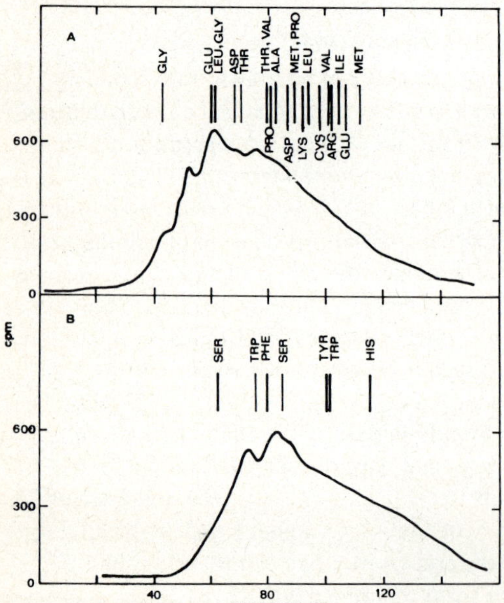


Fig. 38.16. Chromatography of aminoacyl tRNA on a BD-cellulose column (according to Dejesus and Gray). Column: 90 × 1.5 cm. Elution: a linear gradient of 0.55–0.9 *M* sodium chloride (A) or 0.75–1.2 *M* sodium chloride with 10% ethanol (B), both in 0.01 *M* sodium acetate of pH 5 with 0.002 *M* magnesium chloride. Fractions: 10 ml. Sample: 100 μg in 10 ml of aminoacyl tRNA; each radioactively labelled aminoacyl tRNA was prepared separately.

acid and of the synthetase preparation used for aminoacylation. An example of an aminoacyl tRNA separation and a BD-cellulose column (Dejesus and Gray) is given in Fig. 39.16. The separation of both free and aminoacylated $tRNA_f^{Met}$ and $tRNA_m^{Met}$ can be achieved and the effect of formylation can be studied by this technique (Samuel and Rabinowitz, Stanley, White and Bayley).

Partition chromatography is used almost exclusively to fractionate tRNA and is a continuation of experience gained with counter-current distribution. The best known variant is reversed-phase chromatography (Kelmers *et al.*), in which the carrier of the stationary phase (4% dimethyllaurylammonium chloride in isoamyl acetate) was Chromosorb W (diatomaceous earth) and elution was effected with a salt gradient in the presence of 0.01 *M* magnesium chloride. The column served to differentiate 16 tRNAs, multiforms being observed with $tRNA^{Ser}$ and $tRNA^{Arg}$. Further modifications have been made by the same group (Pearson *et al.*, Weiss and Kelmers) and widely used by others (Gallagher *et al.*, Müller *et al.*). Muench and Berg achieved a 24-fold enrichment of the specific activity of tRNA in single-stage partition chromatography on Sephadex G-25 in which a column with an aqueous phase of a solvent system containing potassium phosphate buffer of pH 6.88, ethoxyethanol, butoxyethanol, mercaptoethanol and triethylamine was eluted with a linear gradient of triethylamine in the organic phase.

Differing from DNA, RNAs of lower molecular weight can also be separated on gels

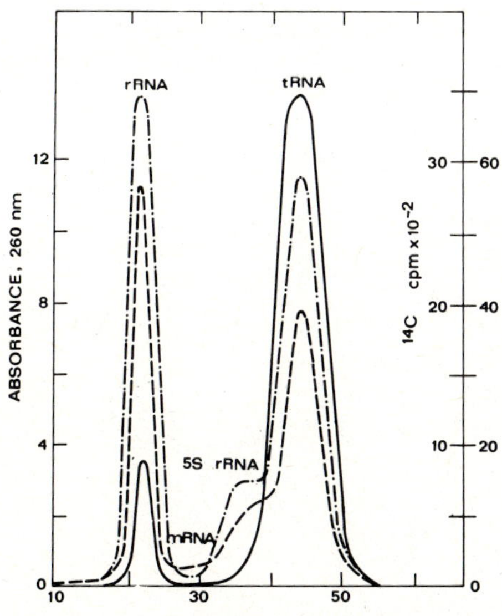


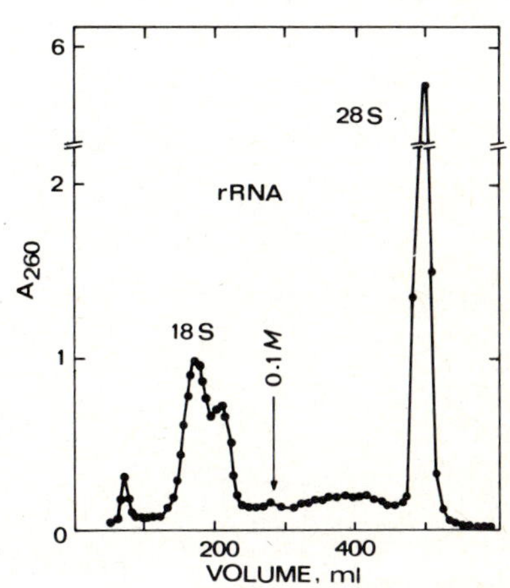


Fig. 38.17. Gel permeation chromatography of [$^{14}$C]methyl and [$^{3}$H]uracil-labelled tRNA of *E. coli* on a Sephadex G-100 column (Schleich and Goldstein). Column: 150 × 2 cm. Elution: 1 *M* sodium chloride. Sample: tRNA isolated on a DEAE-cellulose column in a volume of 0.5 ml. – – – – –, Absorbance at 260 nm; –·–·–, $^{3}$H; ———, $^{14}$C. The tRNA peak represents 75% of the total material.

Fig. 38.18. Fractionation of rat liver ribosomal RNA by Sepharose 2B gel filtration (Petrović *et al.*). Column: 200-ml bed. Elution: gradually with 0.5 and 0.1 *M* sodium chloride (arrow) in 0.02 *M* Tris–hydrochloric acid buffer of pH 7.5 containing 0.1% of sodium dodecylsulphate and 0.0025 *M* EDTA. Flow-rate: 5 ml/h. Temperature: 21°C. Sample: 6.3 mg of rRNA.

*References p.883*

with a higher degree of cross-linking by gel permeation chromatography. A preparation of whole tRNA from *E. coli*, isolated on DEAE-cellulose, was fractionated on a Sephadex G-100 column (Fig. 38.17), on which residues of rRNA, mRNA and even of 5S rRNA were found to be well separated (Schleich and Goldstein). Agarose gels can also be used in order to separate and purify high-molecular-weight rRNAs (Fig. 38.18, Novaković and Petrović, Petrović *et al.*), and viral RNA (*e.g.*, poliomyelitis virus, Öberg *et al.*). In the latter instance, with 2% agarose, the elution volume of RNA with a molecular weight of $2 \cdot 10^6$ depends greatly on conformation changes caused by the presence of $Mg^{2+}$: $10^{-3}$ *M* lithium sulphate solution elutes RNA at the void volume, while $10^{-2}$ *M* lithium sulphate solution with $10^{-3}$ *M* magnesium sulphate elutes RNA at 60% of the bed volume.

A $tRNA^{Ile}$ covalently bound to Sephadex may be mentioned as an example of affinity chromatography using RNA for the isolation and purification of specific tRNA synthetases (Bartkowiak and Pawelkiewicz).

## POLYNUCLEOTIDES AND LARGE OLIGONUCLEOTIDES

Chemically, polynucleotides do not differ from natural nucleic acids and they can therefore be fractionated by a similar or even identical procedure. Homologous oligo- or polynucleotides of synthetic origin, derived from both the ribo- and deoxyribo- types, can be fractionated relatively easily by size on columns of DEAE-cellulose and DEAE-

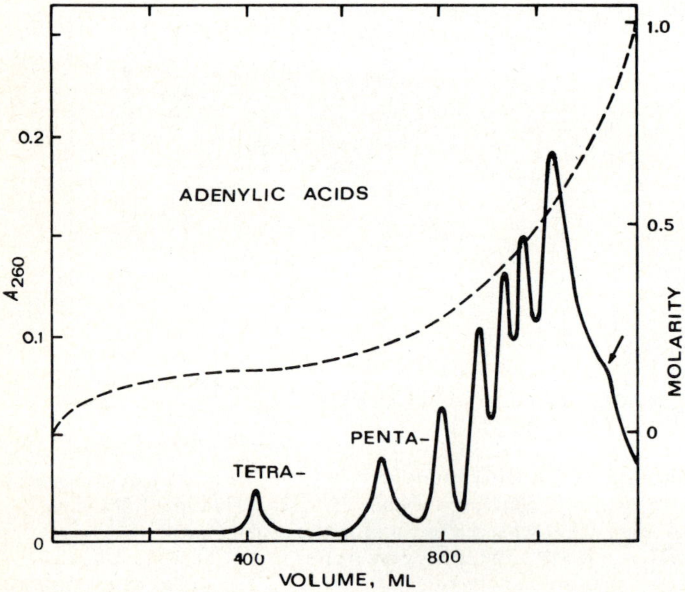


Fig. 38.19. Separation of oligonucleotides from an *Azotobacter* nuclease digest of poly-A on a DEAE-cellulose column (Stevens and Hilmoe). Column and elution gradient as in Fig. 37.14. Sample: 17.5 μmole of poly-A hydrolyzed for 30 min at pH 7.7.

Sephadex (Hall and Sinsheimer, Narang *et al.*, Tener *et al.*). Fig. 38.19 shows the elution profile of the partial enzyme hydrolyzate poly-A obtained from a DEAE-cellulose column by using a complex gradient of volatile buffers (Stevens and Hilmoe). In order to resolve the oligonucleotide mixture in the enzyme hydrolyzate of natural polynucleotides, however, buffers containing 7 *M* urea must be used (see the separation of nucleic acid components, p. 831), which eliminates secondary interactions with the carrier.

Gel permeation chromatography with dextran and agarose has been widely applied in this field, having now become the main fractionation technique. In the synthetic preparation of polythymidylic acid containing up to 24 monomer units, gel filtration on Sephadex G-15, G-25 and G-75 was used to isolate the intermediates and final product; this technique resulted in complete separation on elution with 0.1 *M* triethylammonium hydrogen carbonate of pH 7.5. The products can be desalted by direct lyophilization (Narang *et al.*). The separation of oligonucleotides on these gels can also be used in the study of the influence of terminal groups of oligonucleotides on elution (Haynes *et al.*, Hohn and Pollman, Hohn and Schaller).

A combination of chromatography on an MAK column with thermal chromatography on a hydroxyapatite column (Fig. 38.20) was used in order to isolate the natural polymers poly d(A–T) in an extract of nucleic acids from the testes of *Cancer borealis* (Brzezinski *et al.*). This isolation is an example of the possibility of using the resolution of the MAK column according to base composition and polymer size in order to fractionate polynucleotides, combined with the different re-naturation rates of the fractions on a hydroxyapatite column at elevated temperature.

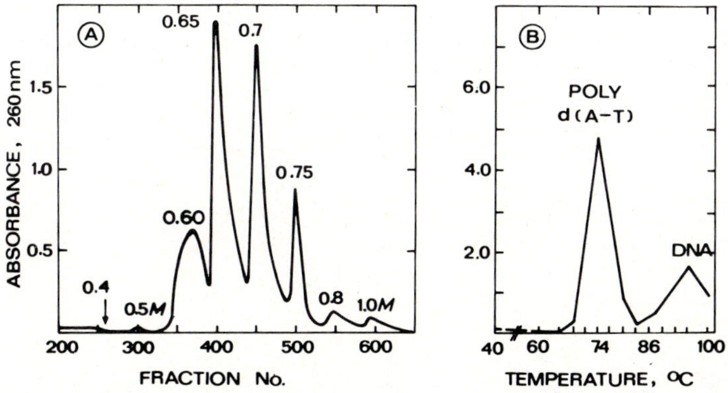


Fig. 38.20. Isolation of poly d(A–T) from purified crab DNA on an MAK column and by thermal chromatography on hydroxyapatite (Brzezinski *et al.*). (A) Column: MAK, 9.6 × 2.5 cm. Elution: stepwise with the indicated concentrations of sodium chloride in 0.05 *M* phosphate buffer of pH 6.7. Flow-rate: 6.5 ml/min. Fractions: 19.6 ml. Sample: 2800 $A_{260}$ units of crab DNA in 0.4 *M* buffered sodium chloride. Each peak contains both poly d(A–T) and the major DNA component in different proportions. (B) Column: hydroxyapatite, 42 × 2.5 cm. Elution: 0.14 *M* phosphate buffer of pH 6.7 with varying temperature (40°C; 60°C; 65°C; from 68°C to 95°C in 3°C intervals; 100°C) and 5-min equilibrating time. Flow-rate: 3 ml/min. Fractions: 15 ml. Sample: peak material from (A) eluted with 0.7 *M* sodium chloride, 81% of which is poly d(A–T). Only the fractions collected at 71°C, 74°C and 77°C were pooled for the isolation.

*References p.883*

## AUTOMATED PROCEDURES AND POLYNUCLEOTIDE SEQUENCE ANALYSIS

In modern sequential analysis, the separation of complex mixtures of oligonucleotides is achieved mainly by fingerprint techniques (Murray, Sanger *et al.*, Southern). In spite of this the first complete analysis of sequences in $tRNA^{Ala}_{yeast}$ and $tRNA^{Ser}_{yeast}$ (Table 37.7; p. 853) was carried out by classical methods of column separation on ion exchangers (Holley *et al.*; Zachau *et al.*, 1966a, b), and suitable combinations of DEAE-cellulose and DEAE-Sephadex columns were mostly involved (reviews by Rushizky and Sober, Staehelin).

The major difficulties are associated with the separation of large fragments after partial

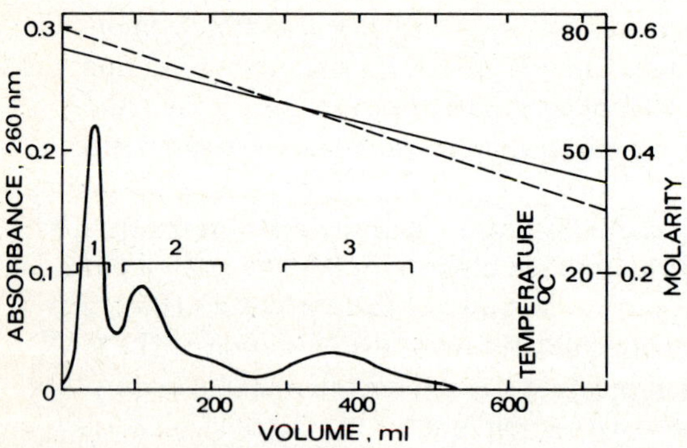


Fig. 38.21. Separation of two halves of $tRNA^{Val}_{yeast}$ after its partial digestion with RNase T1 (Mirzabekov). Column: DEAE-cellulose, 18 × 1.1 cm. Elution: a linear gradient with decreasing concentration of sodium chloride from 0.56 to 0.38 *M* in 0.02 *M* Tris buffer of pH 8, simultaneously under conditions of decreasing temperature from 80 to 36°C (at 2°C/h); total gradient volume, 750 ml. Flow-rate: 36 ml/h. Sample: 160 $A_{260}$ units of the tRNA digest. 1 = 5′-half of the molecule; 2 = 3′-half of the molecule; 3 = whole tRNA molecule (see Fig. 38.23).

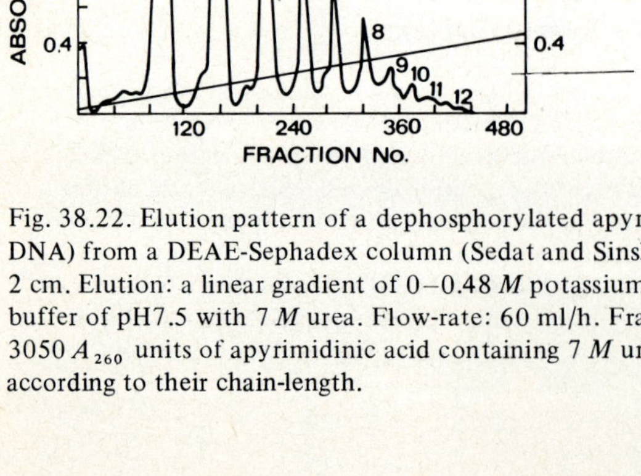


Fig. 38.22. Elution pattern of a dephosphorylated apyrimidinic acid (purine tracts from calf thymus DNA) from a DEAE-Sephadex column (Sedat and Sinsheimer). Column: DEAE-Sephadex A-25, 30 × 2 cm. Elution: a linear gradient of 0–0.48 *M* potassium chloride in 0.005 *M* Tris–hydrochloric acid buffer of pH7.5 with 7 *M* urea. Flow-rate: 60 ml/h. Fractions: 10 ml. Temperature: 55°C. Sample: 3050 $A_{260}$ units of apyrimidinic acid containing 7 *M* urea in the starting buffer. Fractions are numbered according to their chain-length.

enzymic hydrolysis, the structure of which is important for the final reconstruction of the entire molecule from small fragments after complete hydrolysis by specific nucleases. Penswick and Holley used a column of DEAE-cellulose, on which they isolated the two large fragments covering the whole structure of $tRNA^{Ala}_{yeast}$, after mild hydrolysis with RNase T1. Mirzabekov also isolated two halves of $tRNA^{Val}_{yeast}$ using the same column eluted with decreasing gradients of both temperature and salt concentration, as shown in Fig. 38.21.

With DNA of calf thymus, an entire spectrum of purine oligonucleotides (up to a size of 12 monomer units) was separated, after isolation of apyrimidinic acid, on DEAE-Sephadex in 7 *M* urea at 55°C (Fig. 38.22, Sedat and Sinsheimer).

It is often the aim to achieve gradual degradation of the chain starting at one end (analogous to sequential analysis of proteins), but the required standard of results for proteins could not be achieved because of a lack of suitable specific procedures. The main sequential method for nucleic acids, *viz.*, block cleavage of polynucleotides by specific enzymes, an example of which is represented in Fig. 38.23 (reviewed by Brownlee and by Zadražil) and fractionation of the oligonucleotide mixture obtained (reviewed by Staehelin and by Rushizky and Sober), may therefore be amplified only by the method of gradual degradation of oligoribonucleotides by periodate oxidation and β-elimination (Neu and Heppel; Khym and Uziel; Weith and Gilham, 1967) with the possibility of implementing a maximum of 26 steps (in the case of $tRNA^{Phe}_{E.coli}$, Uziel and Khym, 1969). In the study cited, the liberated bases and sugar fragments were separated by re-precipitation of the reaction mixture with cetyltrimethylammonium bromide (the respective tRNA salt is soluble in 1 *M* sodium chloride solution, but it is insoluble at salt concentrations lower than 0.3 *M*), permitting repeated dissolution and precipitation of the tRNA studied in the process, with no influence on the phosphomonoesterase activity removing the phosphate group formed after β-elimination. The supernatant containing the liberated bases after centrifugation of the precipitate at 4000 *g* for 10 min can be separated, directly or after partial desalting and concentration, on a Dowex 50 column with automatic measurement by means of a UV analyzer (Khym and Uziel, Uziel *et al.*). Repetition of the 10 steps of this degradation process was achieved in a study of a terminal sequence of phage Qβ RNA, in which the polynucleotide was separated from the liberated bases on a DEAE-cellulose column by elution with water and a triethylammonium hydrogen carbonate gradient (Weith and Gilham, 1969). It seems that the method may be further improved by binding the polyribonucleotide to cellulose (through a 5′-phosphate group to Cellex N-1, Calbiochem, Los Angeles, Calif., U.S.A.), enabling reactions to be carried out quantitatively, and stabilizing the polynucleotide molecule against non-specific cleavage on oxidation at higher temperatures (noise bases, Wagner *et al.*).

Also, with DNA, attempts were made to gradually degrade the model trinucleotide after oxidation of the 3′-hydroxyl group and separation of the products obtained on DEAE-cellulose with a triethylammonium hydrogen carbonate gradient (Gabriel *et al.*). At present, the method can be used only for low-molecular-weight model oligodeoxyribonucleotides.

*References p.883*

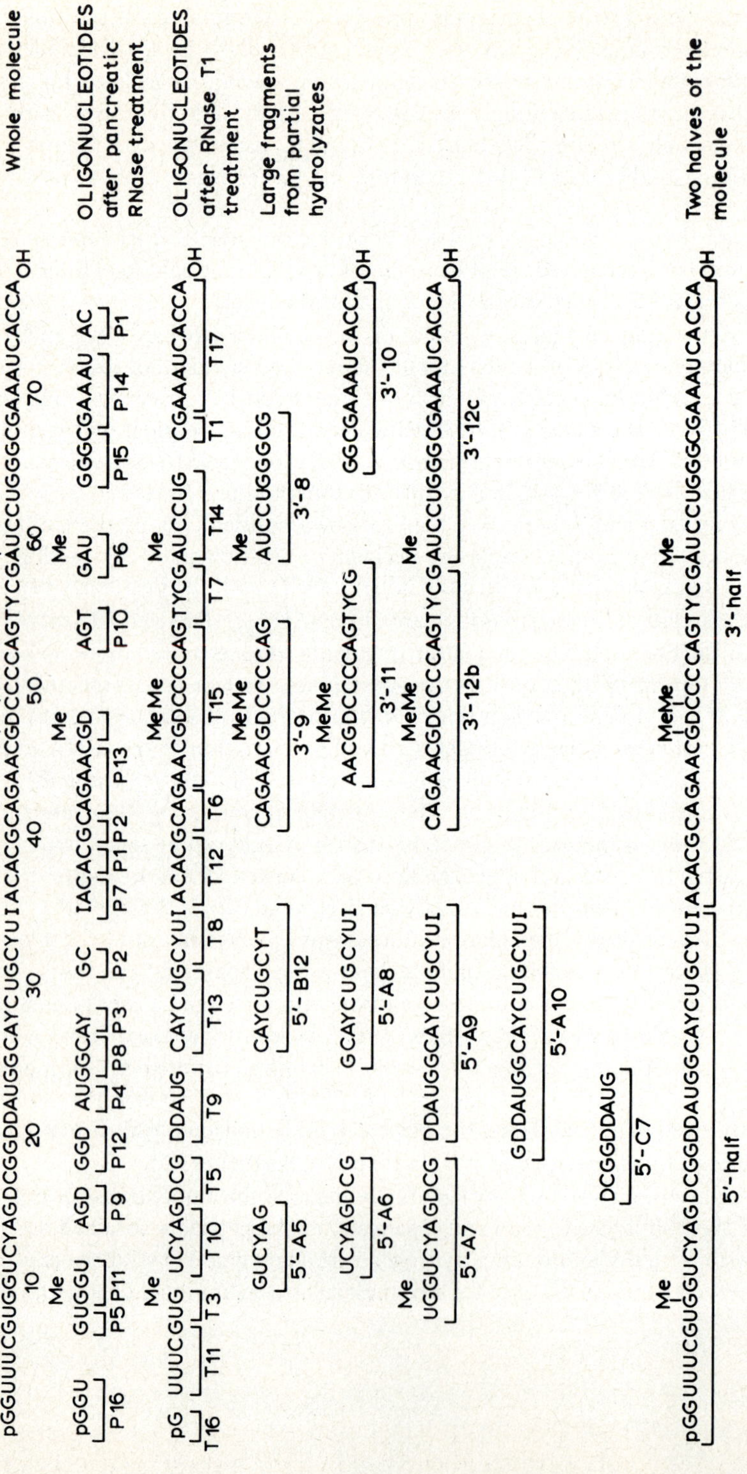


Fig. 38.23. Reconstruction of a primary structure of $tRNA^{Val}_{yeast}$ from fragments obtained by the block-cleavage sequencing technique with specific enzymes (Mirzabekov).

It is hardly possible to speak of true automatic sequential analysis yet, as the available methods and instrumentation relate only to control and detection in separation processes (see the separation of low-molecular-weight components of nucleic acids, p. 831).

## REFERENCES


Albertsson, P. A., *Partition of Cell Particles and Macromolecules*, Wiley, New York, 1960.
Armstrong, R. L. and Boezi, J. A., *Biochim. Biophys. Acta*, 103 (1965) 60.
Attardi, G., Huang, P. C. and Kabat, S., *Proc. Nat. Acad. Sci. U.S.*, 53 (1965) 1490.
Ayad, S. R. and Blamire, J., *Biochem. Biophys. Res. Commun.*, 30 (1968) 207.
Ayad, S. R. and Blamire, J., *J. Chromatogr.*, 42 (1969) 248.
Ayad, S. R. and Blamire, J., *J. Chromatogr.*, 48 (1970) 456.
Ayad, S. R. and Wilkinson, A. E., *J. Chromatogr.*, 66 (1972) 277.
Baguley, B. C., Bergquist, P. R. and Ralph, R. K., *Biochim. Biophys. Acta*, 95 (1965) 510.
Bartkowiak, S. and Pawelkiewicz, J., *Biochim. Biophys. Acta*, 272 (1972) 137.
Bartoli, F. and Rossi, C., *J. Chromatogr.*, 28 (1967) 30.
Bauer, R. D. and Johanson, R., *Biochim. Biophys. Acta*, 119 (1966) 418.
Bautz, E. K. F. and Hall, B. D., *Proc. Nat. Acad. Sci. U.S.*, 48 (1962) 400.
Bautz, E. K. F. and Reilly, E., *Science*, 151 (1966) 328.
Bendich, A. J. and Bolton, E. T., *Methods Enzymol.*, 12B (1968) 635.
Bendich, A., Fresco, J. R., Rosenkranz, H. S. and Beiser, S. M., *J. Amer. Chem. Soc.*, 77 (1955) 3671.
Bendich, A., Pahl, H. B., Korngold, G. C., Rosenkranz, H. S. and Fresco, J. R., *J. Amer. Chem. Soc.*, 80 (1958) 3949.
Bernardi, G., *Biochem. Biophys. Res. Commun.*, 6 (1961) 54.
Bernardi, G., *Nature (London)*, 206 (1965) 779.
Bernardi, G., *Biochim. Biophys. Acta*, 174 (1969a) 423.
Bernardi, G., *Biochim. Biophys. Acta*, 174 (1969b) 435.
Bernardi, G., *Biochim. Biophys. Acta*, 174 (1969c) 449.
Bernardi, G., Carnevali, F., Nicolaieff, A., Piperno, G. and Tecce, G., *J. Mol. Biol.*, 37 (1968) 493.
Bernardi, G. and Timasheff, S. N., *Biochem. Biophys. Res. Commun.*, 6 (1961) 58.
Berns, K. I. and Thomas, C. A., *J. Mol. Biol.*, 11 (1965) 476.
Bleeken, S., Strohbach, G. and Sarfert, E., *Alg. Mikrobiol.*, 6 (1966) 121.
Bock, H. M. and Cherayil, J. D., *Methods Enzymol.*, 12A (1967) 638.
Bourgaux, P. and Bourgaux-Ramoisy, D., *J. Gen. Virol.*, 1 (1967) 323.
Bourgaux-Ramoisy, D., Van Tieghem, M. and Bourgaux, P., *J. Gen. Virol.*, 1 (1967) 589.
Brakke, M. K., *J. Amer. Chem. Soc.*, 73 (1951) 1847.
Brenner, D. J., Fanning, G. R., Rake, A. V. and Johnson, K. E., *Anal. Biochem.*, 28 (1969) 447.
Brown, G. L. and Martin, A. V., *Nature (London)*, 176 (1955) 971.
Brown, G. L. and Watson, M., *Nature (London)*, 172 (1953) 339.
Brownhill, T. J., Jones, A. S. and Stacey, M., *Biochem. J.*, 73 (1959) 434.
Brownlee, G. G., in T. S. Work and E. Work (Editors), *Laboratory Techniques in Biochemistry and Molecular Biology*, Vol. 3, North-Holland, Amsterdam, 1972, p. 1.
Brzezinski, A., Szafranski, P., Johnson, P. H. and Laskowski, M., Sr., *Biochemistry*, 8 (1969) 1228.
Burton, K., *Biochem. J.*, 62 (1956) 315.
Cairns, J., *Cold Spring Harbor Symp. Quant. Biol.*, 28 (1963) 43.
Cannon, M. C. and Dunican, L. K., *Biochem. Biophys. Res. Commun.*, 39 (1970) 423.
Cantoni, G. L. and Davies, D. R. (Editors), *Procedures Nucleic Acid Res. 1966*, Harper & Row, New York, London, 1966.
Caro, L., *Virology*, 25 (1965) 226.
Ceriotti, G., *J. Biol. Chem.*, 214 (1955) 59.
Cheng, T. Y. and Sueoka, N., *Science*, 141 (1963) 1194.
Cherayil, J. D. and Bock, H. M., *Biochemistry*, 4 (1965) 1174.

Chevalier, M. R. and Bernardi, G., *J. Mol. Biol.*, 32 (1968) 437.
Cocito, C., Prinze, A. and De Somer, P., *Nature (London)*, 191 (1961) 573.
Cox, R. A., *Quart. Rev., Chem. Soc.*, 22 (1968) 499.
Cozzarelli, N. R., Kelly, R. B. and Kornberg, A., *Proc. Nat. Acad. Sci. U.S.*, 60 (1968) 992.
Davila, C., Charles, P. and Ledoux, L., *J. Chromatogr.*, 19 (1965a) 382.
Davila, C., Charles, P. and Ledoux, L., *J. Chromatogr.*, 19 (1965b) 396.
Dejesus, T. G. S. and Gray, E. D., *Biochim. Biophys. Acta*, 254 (1971) 419.
Doctor, B. P., Apgar, J. and Holley, R. W., *J. Biol. Chem.*, 236 (1961) 1117.
Doskočil, J. and Hochmannová, J., *Biochim. Biophys. Acta*, 108 (1965) 504.
Fink, L. M., Lanks, K. W., Goto, T. and Weinstein, I. B., *Biochemistry*, 10 (1971) 1873.
Flamm, W. G., Walker, P. M. B. and McGallum, M., *J. Mol. Biol.*, 40 (1969) 423.
Fleissner, E. and Borek, E., in G. L. Cantoni and D. R. Davies (Editors), *Procedures Nucleic Acid Res. 1966*, Harper & Row, New York, London, 1966, p. 461.
Fraenkel-Conrat, H. and Singer, B., *J. Amer. Chem. Soc.*, 76 (1954a) 180.
Fraenkel-Conrat, H. and Singer, B., *Biochim. Biophys. Acta*, 24 (1954b) 540.
Frame, R. and Bishop, J. O., *Biochem. J.*, 121 (1970) 93.
Frankel, F. L. and Crampton, C. F., *J. Biol. Chem.*, 237 (1962) 3200.
Gabriel, T., Chen, W. Y. and Nussbaum, A. L., *J. Amer. Chem. Soc.*, 90 (1968) 6833.
Gallagher, R. E., Ting, R. C. and Gallo, R. C., *Biochim. Biophys. Acta*, 272 (1972) 568.
Gillam, I., Blew, E., Warrington, R. C., Von Tigerstrom, M. and Tener, G. M., *Biochemistry*, 7 (1968) 3459.
Gillam, I., Milward, S., Blew, D., Von Tigerstrom, M., Wimmer, E. and Tener, G. M., *Biochemistry*, 6 (1967) 3043.
Gillespie, D. and Spiegelman, S., *J. Mol. Biol.*, 12 (1965) 829.
Gomatos, P. J. and Tamm, I., *Proc. Nat. Acad. Sci. U.S.*, 50 (1963) 878.
Hall, J. B. and Sinsheimer, R. L., *J. Mol. Biol.*, 6 (1963) 115.
Hall, R. N., *The Modified Nucleosides in Nucleic Acids*, Columbia Univ. Press, New York, 1971.
Hanson, C. V., *Anal. Biochem.*, 32 (1969) 303.
Haynes, F. N., Hansbury, E. and Mitchell, W. E., *J. Chromatogr.*, 16 (1964) 410.
Hohn, T. and Pollman, W., *Z. Naturforsch.*, 18B (1963) 919.
Hohn, T. and Schaller, H., *Biochim. Biophys. Acta*, 138 (1967) 466.
Holley, R. W., Apgar, J., Everett, G. A., Madison, J. T., Marquisee, M., Merrill, S. H., Penswick, J. R. and Zamir, H., *Science*, 147 (1965) 1462.
Hutchison, W. C. and Munro, H. N., *Analyst (London)*, 86 (1961) 768.
Iyer, Y. N. and Rupp, W. D., *Biochim. Biophys. Acta*, 228 (1971) 117.
Kano-Sueoka, T. and Sueoka, N., *J. Mol. Biol.*, 20 (1966) 183.
Katchalsky, E., Yankofsky, S., Novogradsky, A., Galanter, Y. and Littauer, U. Z., *Biochim. Biophys. Acta*, 123 (1966) 641.
Kay, E. P. M., Simmons, M. S. and Dounce, A. L., *J. Amer. Chem. Soc.*, 74 (1952) 1724.
Keller, P. J., Cohen, E. and Wade, R. D., *J. Biol. Chem.*, 239 (1964) 3292.
Kelmers, A. D., Novelli, G. D. and Stulberg, M. P., *J. Biol. Chem.*, 240 (1965) 3979.
Kennell, D. E., *Progr. Nucl. Acid Res. Mol. Biol.*, 11 (1971) 259.
Khym, J. X. and Uziel, M., *Biochemistry*, 7 (1968) 422.
Kidson, C., *Cold Spring Harbor Symp. Quant. Biol.*, 33 (1968) 179.
Kidson, C., *Biochemistry*, 8 (1969) 4376.
Kidson, C. and Kirby, K. S., *Biochim. Biophys. Acta*, 76 (1963) 624.
Kidson, C. and Kirby, K. S., *Biochim. Biophys. Acta*, 91 (1964) 627.
Kiger, J. A. and Sinsheimer, R. L., *J. Mol. Biol.*, 40 (1969) 467.
Kirby, K. S., *Biochem. J.*, 64 (1956) 405.
Kirby, K. S., *Progr. Nucl. Acid Res.*, 3 (1964) 1.
Kirby, K. S., Hastings, J. R. B. and O'Sullivan, M. A., *Biochim. Biophys. Acta*, 61 (1962) 978.
Kit, S., *Arch. Biochem. Biophys.*, 87 (1960a) 318.
Kit, S., *Arch. Biochem. Biophys.*, 87 (1960b) 330.
Klamerth, O., *Nature (London)*, 208 (1965) 1318.
Kleinschmidt, A., Burton, A. K. and Sinsheimer, R. L., *Science*, 142 (1963) 961.
Kothari, R. M., *J. Chromatogr.*, 52 (1970a) 119.

Kothari, R. M., *J. Chromatogr.*, 53 (1970b) 580.
Kothari, R. M., *J. Chromatogr.*, 64 (1972) 85.
Lang, D., Bujard, H., Wolff, B. and Russell, D., *J. Mol. Biol.*, 23 (1967) 163.
Lang, D. and Mitani, M., *Biopolymers*, 9 (1970) 373.
Légault-Démare, J., Rebeyrotte, R., Leprieur, A. and Rossaux, J., *Biochim. Biophys. Acta*, 87 (1964) 165.
Levitt, M., *Nature (London)*, 224 (1969) 759.
Lin, H. J., *Biochim. Biophys. Acta*, 217 (1970) 232.
Littauer, U. Z. and Sela, M., *Biochim. Biophys. Acta*, 61 (1962) 609.
Loeb, J. E. and Chauveau, J., *Biochim. Biophys. Acta*, 182 (1969) 225.
Lowrie, R. J. and Bergquist, P. L., *Biochemistry*, 7 (1968) 1761.
McGallum, M. and Walker, P. M. B., *Biochem. J.*, 105 (1967) 163.
MacHattie, L. A. and Thomas, C. A., Jr., *Science*, 144 (1964) 1124.
Mandell, J. D. and Hershey, A. D., *Anal. Biochem.*, 1 (1960) 66.
Marmur, J., *J. Mol. Biol.*, 3 (1961) 208.
Matthaei, J. H. and Nirenberg, M. W., *Proc. Nat. Acad. Sci. U.S.*, 47 (1961) 1580 and 1588.
Maxwell, I. H., Wimmer, E. and Tener, G. M., *Biochemistry*, 7 (1968) 2629.
Mindich, L. and Hotchkiss, R. D., *Biochim. Biophys. Acta*, 80 (1964a) 73.
Mindich, L. and Hotchkiss, R. D., *Biochim. Biophys. Acta*, 80 (1964b) 93.
Mirzabekov, A. D., *Valine tRNA I – Structural Basis of Recognition, Dissertation Thesis,* U.S.S.R. Academy of Sciences, Moscow, 1972.
Miyazawa, Y. and Thomas, C. A., *J. Mol. Biol.*, 11 (1965) 223.
Monier, R., Naono, S., Hayes, D., Hayes, F. and Gros, F., *J. Mol. Biol.*, 5 (1962) 311.
Müller, P., Wehrli, W. and Staehelin, M., *Biochemistry*, 10 (1971) 1885.
Muench, K. H. and Berg, P., *Biochemistry*, 5 (1966) 1970.
Murray, K., *Biochem. J.*, 118 (1970) 831.
Naber, J. E., Schepman, A. M. J. and Rörsch, A., *Biochim. Biophys. Acta*, 14 (1966) 326.
Narang, S. A., Michniewicz, J. J. and Dheer, S. K., *J. Amer. Chem. Soc.*, 91 (1969) 936.
Neu, H. C. and Heppel, L. A., *J. Biol. Chem.*, 239 (1964) 2927.
Nomoto, M., Narahashi, Y. and Murakami, M., *J. Biochem. (Tokyo)*, 48 (1968) 453.
Novaković, M. B. and Petrović, S. L., *Anal. Biochem.*, 49 (1972) 367.
Nygaard, A. P. and Hall, B. D., *Biochem. Biophys. Res. Commun.*, 12 (1963) 68.
Öberg, B., Bengtsson, S. and Philipson, L., *Biochem. Biophys. Res. Commun.*, 20 (1965) 36.
Öberg, B. and Philipson, L., *Arch. Biochem. Biophys.*, 119 (1967) 504.
Oravec, M., *Anal. Biochem.*, 49 (1972) 58.
Otaka, E., Osawa, S., Oota, Y., Ishihama, A. and Mitsui, H., *Biochim. Biophys. Acta*, 55 (1962) 310.
Pearson, R. L., Weiss, J. F. and Kelmers, A. D., *Biochim. Biophys. Acta*, 228 (1971) 770.
Penswick, J. R. and Holley, R. W., *Proc. Nat. Acad. Sci. U.S.*, 53 (1965) 543.
Petrović, S., Novaković, M. and Petrović, J., *Biochim. Biophys. Acta*, 254 (1971) 493.
Pinck, L., Hirth, L. and Bernardi, G., *Biochem. Biophys. Res. Commun.*, 31 (1968) 481.
Pivec, L., Šrogl, M., Zadražil, S., Fučík, V. and Šormová, Z., in M. Kohoutová and J. Hubáček (Editors), *The Physiology of Gene and Mutation Expression*, Academia, Prague, 1965, p. 205.
Pivec, L. and Štokrová, J., *FEBS Lett.*, 14 (1971) 157.
Poonian, M. S., Schlabach, A. J. and Weissbach, A., *Biochemistry*, 10 (1971) 424.
Pullman, B. and Pullman, A., *Biochim. Biophys. Acta*, 36 (1959) 343.
Pyeritz, R. E., Schlegel, R. A. and Thomas, C. A., Jr., *Biochim. Biophys. Acta*, 272 (1972) 504.
Radloff, R., Bauer, W. and Vinograd, J., *Proc. Nat. Acad. Sci. U.S.*, 57 (1967) 1514.
Reid, E. (Editor), *Separations with Zonal Rotors*, Wolfson Bioanalytical Centre, University of Surrey, Guildford, 1971.
Riggsby, W. S., *Fed. Proc., Fed. Amer. Soc. Exp. Biol.*, 27 (1968) 804.
Roger, M., *Proc. Nat. Acad. Sci. U.S.*, 59 (1968) 200.
Rosenkranz, H. S. and Bendich, A., *J. Amer. Chem. Soc.*, 81 (1959) 902.
Rudner, R., Karkas, J. D. and Chargaff, E., *Proc. Nat. Acad. Sci. U.S.*, 60 (1968a) 630.
Rudner, R., Karkas, J. D. and Chargaff, E., *Proc. Nat. Acad. Sci. U.S.*, 60 (1968b) 921.
Rudner, R., Karkas, J. D. and Chargaff, E., *Proc. Nat. Acad. Sci. U.S.*, 63 (1969) 152.
Rush, M. G. and Warner, R. C., *Cold Spring Harbor Symp. Quant. Biol.*, 33 (1968) 459.

Rushizky, G. W. and Sober, H. A., *Progr. Nucl. Acid Res. Mol. Biol.*, 8 (1968) 171.
Samuel, C. E. and Rabinowitz, J. C., *Anal. Biochem.*, 47 (1972) 244.
Sanger, F., Brownlee, G. G. and Barrell, B. G., *J. Mol. Biol.*, 13 (1965) 373.
Šatava, J., Zadražil, S. and Šormová, Z., *Collect. Czech. Chem. Commun.*, 38 (1973) 2167.
Schabort, J. C., *J. Chromatogr.*, 73 (1972) 253.
Schildkraut, C. L., Marmur, J. and Doty, P., *J. Mol. Biol.*, 4 (1962) 430.
Schleich, T. and Goldstein, J., *J. Mol. Biol.*, 15 (1966) 136.
Sedat, J. and Sinsheimer, R. L., *J. Mol. Biol.*, 9 (1964) 489.
Sedat, J. W., Kelly, R. B. and Sinsheimer, R. L., *J. Mol. Biol.*, 26 (1967) 537.
Sekine, H., Nakano, E. and Sakaguchi, K., *Biochim. Biophys. Acta*, 174 (1969) 202.
Shepferd, G. R. and Petersen, D. F., *J. Chromatogr.*, 9 (1962) 445.
Singer, B. and Fraenkel-Conrat, H., *Virology*, 14 (1961) 59.
Smith, H. O. and Levine, M., *Methods Enzymol.*, 12A (1967) 557.
Southern, E. M., *Nature (London)*, 227 (1970) 794.
Spizizen, J., *Proc. Nat. Acad. Sci. U.S.*, 44 (1958) 1302.
Šponar, J., Pivec, L. and Šormová, Z., *Collect. Czech. Chem. Commun.*, 29 (1964) 2077.
Staehelin, M., *Progr. Nucl. Acid Res.*, 2 (1963) 169.
Stanley, W. M., Jr., *Anal. Biochem.*, 48 (1972) 202.
Stephenson, M. L. and Zamecnik, P. C., *Proc. Nat. Acad. Sci. U.S.*, 7 (1962) 91.
Stern, R. and Littauer, U. Z., *Biochemistry*, 7 (1968) 3469.
Stevens, A. and Hilmoe, R. J., *J. Biol. Chem.*, 235 (1960) 3016.
Sueoka, N. and Yamane, T., *Proc. Nat. Acad. Sci. U.S.*, 48 (1962) 1254.
Takai, M., Kondo, N. and Osawa, S., *Biochim. Biophys. Acta*, 55 (1962) 418.
Tener, G. M., Khorana, H. G., Markham, R. and Pol, E. H., *J. Amer. Chem. Soc.*, 80 (1958) 6223.
Tichonenko, T. I., *Biochimiya*, 27 (1962) 131.
Tiselius, A., Hjertén, S. and Levine, O., *Arch. Biochem. Biophys.*, 65 (1956) 132.
Trávníček, M. and Říman, J., *Biochim. Biophys. Acta*, 199 (1970) 283.
Udvardy, A. and Venetianer, P., *Eur. J. Biochem.*, 20 (1971) 513.
Uziel, M. and Khym, J. X., *Biochemistry*, 8 (1969) 3254.
Uziel, M., Koh, C. K. and Cohn, W. E., *Anal. Biochem.*, 25 (1968) 77.
Vinograd, J. and Hearst, J. E., *Progr. Chem. Org. Nat. Prod.*, 20 (1962) 372.
Von Ehrenstein, G., *Methods Enzymol.*, 12A (1967) 592.
Votavová, H., Šponar, J. and Šormová, Z., *Eur. J. Biochem.*, 12 (1970) 208.
Walker, P. B. M., *Progr. Nucl. Acid Res. Mol. Biol.*, 9 (1969) 301.
Wagner, T. E., Chai, H. G. and Warfield, A. S., *J. Amer. Chem. Soc.*, 91 (1969) 2388.
Watson, J. D. and Crick, F. H. C., *Nature (London)*, 171 (1953) 737 and 964.
Weiss, J. F. and Kelmers, A. D., *Biochemistry*, 6 (1967) 2507.
Weissbach, A., Bartl, P. and Salzman, L. A., *Cold Spring Harbor Symp. Quant. Biol.*, 33 (1968) 525.
Weith, H. L. and Gilham, P. T., *J. Amer. Chem. Soc.*, 89 (1967) 5473.
Weith, H. L. and Gilham, P. T., *Science*, 166 (1969) 1004.
White, B. N. and Bayley, S. T., *Biochim. Biophys. Acta*, 272 (1972) 583.
Williamson, R., in E. Reid (Editor), *Separations with Zonal Rotors*, Wolfson Bioanalytical Centre, University of Surrey, Guildford, 1971, p. 83.
Young, F. E. and Jackson, A. P., *Biochem. Biophys. Res. Commun.*, 23 (1966) 490.
Zachau, H. G., Düttig, D. and Feldmann, H., *Hoppe-Seyler's Z. Physiol. Chem.*, 347 (1966a) 212.
Zachau, H. G., Düttig, D. and Feldmann, H., *Angew. Chem.*, 78 (1966b) 392.
Zachau, H. G., Tada, M., Lawson, T. B. and Schwieger, M., *Biochim. Biophys. Acta*, 53 (1961) 221.
Zadražil, S., *An Introduction to Structural Studies of Nucleic Acids (Sequence Determination)*, Technical Booklet, Koch-Light, Colnbrook, 1974.
Zadražil, S., Fučík, V. and Šormová, Z., *Biochem. Genet.*, 7 (1972) 57.
Zadražil, S. and Mach. O., in E. Reid (Editor), *Separations with Zonal Rotors*, Wolfson Bioanalytical Centre, University of Surrey, Guildford, 1971, p. 91.
Ziv, E., De Groot, N. and Lapidot, Y., *Biochim. Biophys. Acta*, 228 (1971) 135.

*Chapter 39*

# Alkaloids

K. MACEK

## CONTENTS



## INTRODUCTION

Alkaloids are a heterogeneous group of organic compounds containing a tertiary or quaternary amino group in the molecule; only a few alkaloids contain a secondary amino group. In the past, alkaloids were isolated almost exclusively from vegetable material, while at present attempts are being made to prepare many of them synthetically or by partial syntheses, or to prepare derivatives that are unknown in natural material. Owing to their physiological effects, alkaloids are important therapeutics, although some are abused as narcotics and hallucinogenic drugs and are also used for doping and for suicidal purposes. The detection of these substances therefore has an important position in toxicological analysis.

Chromatographic procedures have been used in alkaloid chemistry for a long time; if some investigations in which ion exchange was used for the purification of alkaloids are disregarded, chromatography was used for the first time in 1937 by Merz and Franck for the separation of tinctures of belladonna, quinine, ipecacuanha and strychnine with alumina as the sorbent. Chromatographic procedures were first used for the purification of single alkaloids or whole groups, for separations from ballasts or other drugs, followed by isolation or determination by classical methods of analysis. The introduction of flat-bed procedures (paper and thin-layer chromatography) brought about a revolution in alkaloid analysis, especially in the identification and determination of structurally related alkaloids (ergot, opium and rauwolfia alkaloids, etc.). Of column techniques, only the application of gas chromatography in the 1960s achieved similar success. In view of the chemical nature of alkaloids, it is now more probable that in the near future high-resolution column chromatography will take over a series of tasks, which today are solved by flat-bed techniques.

Extensive literature exists on classical column chromatography (*cf.*, Graf, Higuchi and Bodin, Janot and Le Hir, Lederer and Lederer). Literature references to more recent work can be found in the bibliography by Deyl *et al.* and in reviews published in the journal *Analytical Chemistry*.

## PREPARATION OF SAMPLES

The preparation of samples for chromatography is governed by the material that is to be chromatographed and also by the sorbent on which the separation should be carried out. For adsorption or partition chromatography, alkaloids are applied on to the column mostly as bases dissolved in common organic solvents of polar or less polar character (acetone, ethyl acetate, diethyl ether, chloroform). For ion-exchange chromatography, alkaloids can be applied in the form of salts either in aqueous solution or in mixtures of lower aliphatic alcohols (methanol, ethanol).

Except when they occur as solutions of the pure substances, alkaloids submitted to chromatography are mostly present in vegetable material, various medicinal preparations, reaction mixtures after alkaloid synthesis, or biological material. In almost all instances attempts are made to carry out a pre-separation of the basic substances from the ballast by a suitable extraction procedure. Inorganic salts and substances of lipophilic character are among the most commonly used ballasts. When vegetable material is analyzed, the plant is dried, then finely ground and extracted first with light petroleum in order to remove lipids. The material is then made alkaline and the alkaloids are extracted with diethyl ether, chloroform or some other solvent; alkaloid bases, but not inorganic salts or acidic substances, pass into the extract. When solutions are analyzed (for example injection solutions, reaction mixtures in syntheses, or biological fluids), it usually suffices to make the solution alkaline and to extract the alkaloids with a solvent (chloroform, diethyl ether, etc.). In some instances, the pre-purification can be carried out directly by ion exchange. Aqueous solutions of mineral acids are used for extraction. Extracts that contain alkaloid salts in addition to ballasts are poured through a suitable cation exchanger; the alkaloids are retained and when the acids have been eluted from the column, the alkaloids are eluted with mixtures of lower aliphatic alcohols with ammonia.

## TECHNIQUES

### Classical column chromatography

*Sorbents and mobile phases*

For column chromatography, virtually all available sorbents have been used for separations based on adsorption, partition and ion-exchange mechanisms. Adsorption is advantageous mainly for purification and crude fractionation, partition chromatography for finer separations of alkaloids, and ion exchange both for crude fractionation and for the determination of alkaloids in the form of salts as they occur in various pharmaceutical preparations.

One of the oldest sorbents, which is still much used today, is alumina. Alumina is also one of the most complex sorbents, with which it is not a simple task to achieve reproducibility in separations. In addition to the generally known effect of particle size and activity (see Chapter 9), the pH of alumina also has a very important effect. The common basic alumina is most often used, but for a number of separations neutral or acidic alumina may be more suitable. Depending on the nature of the mobile phase, alumina may function either as an adsorbent or as an ion exchanger. As an adsorbent, alumina is used mainly in non-aqueous non-polar or weakly polar solvents, for example, light petroleum and benzene. Under these conditions, alkaloids are relatively strongly adsorbed, while some interfering substances (for example, plant pigments) pass into the eluate. For the desorption of alkaloids from the column, more polar solvents (lower alcohols, acetone) or their mixtures with water have to be used. Ion exchange takes place mainly in aqueous alcoholic solutions. On basic alumina, the following reaction takes place:

$$\text{Al–ONa} + (\text{BH})^+\,\text{Cl}^- \rightarrow \text{Al–OH} + \text{B} + \text{Na}^+ + \text{Cl}^-$$

and alumina functions as a cation exchanger, while on acidic alumina the reaction

$$\text{Al–Cl} + (\text{BH})^+\,\text{OH}^- \rightarrow \text{Al–OH} + (\text{BH})^+\,\text{Cl}^-$$

occurs and alumina functions as an anion exchanger. With neutral alumina, both types of reactions may take place, depending on the conditions used. Hence, basic alumina may be used for the liberation of alkaloid bases from their salts, while plant pigments are simultaneously retained if the column is eluted with aqueous alcohols.

Other adsorbents used for chromatography include various earths, such as franconite, florisin, kaolin, talc, Kieselguhr and Celite. Today, silica gel is used predominantly as a carrier for partition chromatography. Charcoal has found only limited use.

Sorbents for partition chromatography serve mainly as carriers of the liquid stationary phase, mostly water and aqueous buffer solutions of pH 3–7 (depending on the nature of the alkaloids). Polar organic solvents, such as formamide, are less often anchored on the carrier. In classical column chromatography, cellulose is most often used as the sorbent, followed by diatomaceous earth (under various commercial names, such as Hyflo Supercel and Celite 545). Silica gel is less distributed, owing to its weak cation-exchange properties.

As mobile phases for sorbents with fixed buffer solutions, diethyl ether or chloroform are most commonly used, saturated with the corresponding buffer. For chromatography in systems with a stationary polar organic solvent (formamide, ethylene glycol, etc.) chloroform (for more hydrophilic alkaloids), benzene and cyclohexane (for more hydrophobic alkaloids), or their mixtures serve as mobile phases.

The third large group of sorbents consists of ion exchangers. For the separation of alkaloids, cation exchangers can be used that retain bases from the solutions of alkaloid salts, or ion exchangers that, on the contrary, retain anions, and free bases flow through the column. In the first instance, aqueous solutions must be used in order to achieve the retention of alkaloids, and elution is carried out with alcoholic solutions suitably alkaline (in order to liberate alkaloid bases, but to prevent simultaneous precipitation), while in the second instance alkaloids are introduced on to the column directly in alcoholic

*References p.894*

solutions. Most of the synthetic resins may be used as anion exchangers, the choice being governed by the purpose of the analysis and the type of alkaloids to be separated. An example of the effect of the structure of the ion exchanger and effects of the ionic forms and of the solvent medium in the *Cinchona* alkaloids group can be found in the papers by Bhat *et al.* and Kanhere *et al.*

Interesting separations of opium alkaloids were achieved on SE-Sephadex C-25 (a coarse cation exchanger) (Broich *et al.*; Gladyshev *et al.*, 1969, 1971). Ion exchangers based on cellulose (CM-cellulose) have been used only exceptionally for separations of alkaloids (McMartin *et al.*).

*Detection and determination*

Recently, the detection of substances directly on the column has been used in exceptional cases: in long-wave ultraviolet light, some alkaloids fluoresce and in short-wave light they quench fluorescence. The sorbent can also be impregnated with a suitable indicator (methyl red, Tunmann and Hudmann; dimethylaminobenzene, Trautner and Roberts; etc.) and the amount of the alkaloid can be determined from the width of the zones formed.

In classical column chromatography for the determination of alkaloids, fractions are usually collected which are then analyzed by spectrophotometry, colorimetry, acid–base or potentiometric titration, weighing of the dry residue, etc. In more recent work, the eluate was conducted into a flow-through cell of the photometer and the concentration registered continuously with a recorder. For automatic analysis, for example in toxicology,

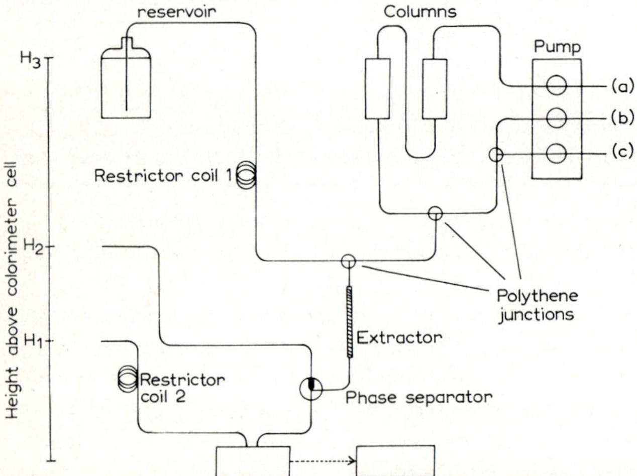


Fig. 39.1. Apparatus used for CM-cellulose chromatography combined with a dye-extraction assay procedure (McMartin *et al.*). Reagents and their flow-rates were as follows: (a) buffer used to elute columns at 0.33 ml/min; (b) dye solution (1% bromocresol green in 0.032 *M* sodium hydroxide solution) at 0.07 ml/min; (c) 1 *M* citrate buffer, pH 3.7, to control the pH in the extractor, at 0.07 ml/min. Resistance coils 1 and 2 were 232-cm and 198-cm length of P.P. 25 polyethylene cannula (from Portex Plastics). The connections to and from the pump to the columns, from the columns to the extractor and from the separator to the colorimeter were P.P. 25 polyethylene cannula. All other connections were P.P. 100 polyethylene cannula.

an interesting arrangement was developed by McMartin *et al.*, who combined column chromatography with a dye-extraction assay procedure. The principle of this method can be seen in Fig. 39.1.

If column chromatography is used as a preparative method, the presence of alkaloids can be checked in single fractions, for example, by thin-layer chromatography.

### High-resolution column chromatography

It is surprising that so few high-resolution column chromatographic procedures have been described so far. Up to now, the greatest attention has been paid to natural and semi-synthetic opium alkaloids, especially in combined preparations, and also to purine alkaloids.

#### *Sorbents and mobile phases*

For column chromatography, all of the main types of sorbents have been used. For the liquid–solid chromatography of some opium alkaloids, basic alumina can be used, or Corasil, SIL-X, and other sorbents. Due to the high adsorptivity of many alkaloids it is sometimes inevitable to deactivate the adsorbent with a small amount of water (0.5–2.0%) (Heacock *et al.*). For the separation and the determination of caffeine, codeine and some other analgesics, Henry and Schmit used ion-exchange chromatography on Zipax coated with a strong anion-exchange resin (SAX) with an aqueous buffer of pH 9.2 as the mobile phase. Other ion exchangers can also be used, such as WBAX (Waters Ass., Framingham, Mass., U.S.A.), which is a weakly basic ion exchanger on a silica gel support. Citric acid (0.1 *M*, pH 3.0) can be used as the mobile phase.

For purine alkaloids and for the separations of brucine and strychnine, Wu and Siggia made use of liquid–liquid chromatography. Corasil II was found to be a suitable support, while Corasil I gave more elongated peaks under the same conditions, evidently because an adsorption mechanism interfered. Poly-G 300 was used as the stationary phase at a concentration of 1.1%; for coating, the Poly-G 300 should be dissolved in dichloromethane. A heptane–ethanol mixture (10:1) was used as the mobile phase (Fig. 39.2).

#### *Detection and determination*

For the detection of alkaloids, common detectors that are supplied with commercial chromatographs can be used, for example, a differential refractometer or an ultraviolet spectrophotometer. The ultraviolet spectrophotometer is especially suitable for alkaloids with maximum absorption in the short-wave region (for example, for purine alkaloids at 270 nm). In this instance, the detector sensitivity is in the range of tens of nanograms.

For fluorescing alkaloids it is of great advantage to use a fluorescence detector. For ergot alkaloids the recommended excitation wavelength, 350 nm, and fluorescence wavelength, 390 to 410 nm, were used by Heacock *et al.* The excitation and emission

*References p.894*

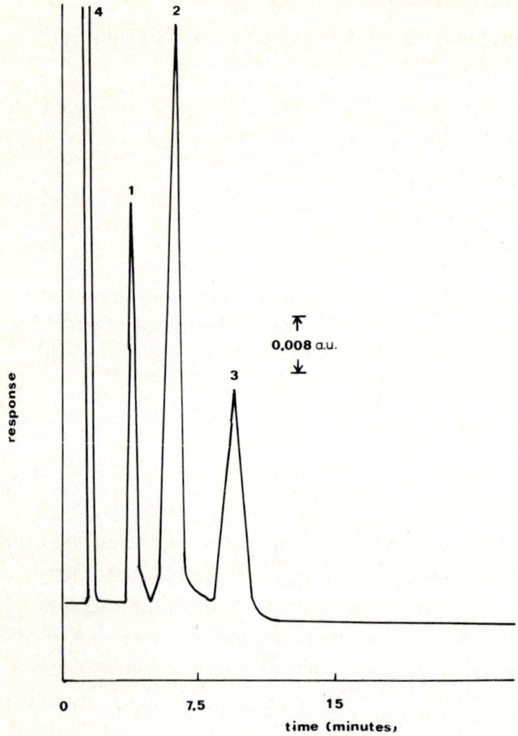


Fig. 39.2. Separation of purine alkaloids (Wu and Siggia). Column: Corasil II coated with 1.1% Poly-G 300; 1 m × 1 mm. Mobile phase: *n*-heptane–ethanol (10:1). Flow-rate: 0.27 ml/min. Column input pressure: 300 p.s.i. Detection: UV spectrophotometer, 270 nm. Alkaloids: 2 μl of solution containing 160 ng of caffeine (1), 250 ng of theophylline (2) and 83 ng of theobromine (3); peak 4, solvent.

TABLE 39.1
EXAMPLES OF SEPARATIONS OF ALKALOIDS

| Alkaloids | Stationary phase | Mobile phase | Remarks | Reference |
|---|---|---|---|---|
| Caffeine and some other analgesics | Zipax coated with anion-exchange resin | Buffer of pH 9.2 (ionic strength increased with ammonium nitrate) | High-resolution CC | Henry and Schmit |
| Caffeine and some other analgesics | WBAX anion exchanger | 0.1 *M* citric acid, pH 3.0 | High-resolution CC | Anonymous |
| Curare | Cellulose | Ethyl acetate–pyridine–water (10:6:3) | | Wieland and Merz |
| Ephedrine and some amphetamines | Copper–Bio-Rex 70 | 0.1 *M* ammonia in 33% ethanol | High-resolution CC | De Hernandez and Walton |
| Ergot | Cellulose with citrate–phosphate buffer, pH 3 | Diethyl ether | | Carless |

TABLE 39.1 *(continued)*

| Alkaloids | Stationary phase | Mobile phase | Remarks | Reference |
|---|---|---|---|---|
| Ergot | Corasil or Zipax | Chloroform–methanol–ethyl acetate–acetic acid (60:20:50:3) or various chloroform–methanol mixtures | High-resolution CC | Heacock *et al.* |
| Ergot | SIL-X | Acetonitrile–isopropyl ether (40:60) | High-resolution CC | Wittwer and Kluckhohn |
| | Corasil II | Acetonitrile–isopropyl ether (25:75) | High-resolution CC | |
| Morphine and some other basic drugs | CM-cellulose | 0.014 *M* borate, pH 8.5 | Automated assay for detection and determination | McMartin *et al.* |
| Opium | Two columns: IMAC C-22 ($H^+$) and Amberlite IRA-400 ($OH^-$) | 1 *M* ammonia; alcoholic 1 *M* ammonia; on Amberlite, acetate buffers | Separation of 6 principal opium alkaloids | Kamp |
| Opium | SE-Sephadex C-25 | Phosphate buffer, pH 4.6 | Opium alkaloids together with strychnine and quinine | Broich *et al.*; Gladyshev *et al.* (1969, 1971) |
| Oxindole | Corasil $C_{18}$ | Methanol–water (4:1) at 60°C | High-resolution CC | Jolliffe and Shellard |
| Papaver | Alumina | Heptane–diethyl ether | Stepwise gradient | Pfeifer and Döhnert |
| Papaver | Silica gel | Benzene–acetone–methanol (7:2:1) | | Pfeifer and Kühn |
| Piperine isomers | Silica gel H | Ethyl acetate | High-resolution CC | De Cleyn and Verzele |
| Purine | Corasil + Poly-G 300 | Heptane–ethanol (10:1) | High-resolution CC | Wu and Siggia |
| Rauwolfia | Alumina | Chloroform–ethanol | Stepwise gradient from 0.5 to 5% | Hofmann |
| Strychnos | Corasil + Poly-G 300 | Heptane–ethanol (10:1) | High-resolution CC | Wu and Siggia |
| Tropine | Kieselguhr + 0.5 *M* phosphate buffer, pH 7.3 | Chloroform | | Evans and Partridge |
| Tropine | SIL-X | Ammonia–tetrahydrofuran (1:100) | High-resolution CC | Stutz and Sass |
| Veratrum | Celite + buffer (pH 3.0)–ethylene glycol (2:1) | Chloroform, ethylene chloride | Stepwise gradient | Levine and Fishbach |
| Yohimbine isomers | Alumina | Benzene, diethyl ether, diethyl ether + 2% methanol | Stepwise elution | Le Hir *et al.* |

*References p.894*

wavelength were selected by using appropriate filters. However, it is necessary to keep in mind that some common solvents used as components of the mobile phase like chloroform may have quenching effects.

## APPLICATIONS

Most published papers have been devoted to purification procedures and to the isolation of one or a group of alkaloids (unresolved). These procedures do not represent a chromatographic separation in the usual sense, and therefore they will not be discussed here. Examples of some more typical separations are given in Table 39.1. For the separation of further alkaloids, see Deyl *et al.*

## REFERENCES


Anonymous, *Chromatogr. Notes*, 1, No. 6 (1971).

Bhat, C. V., Kamath, B. R., Shah, R. S., Kanhere, S. S. and Bafna, S. L., *J. Pharm. Sci.*, 57 (1968) 1195.

Broich, J. R., De Mayo, M. M. and Dal Cortivo, L. A., *J. Chromatogr.*, 33 (1968) 526.

Carless, J. E., *J. Pharm. Pharmacol.*, 5 (1953) 883.

De Cleyn, R. and Verzele, M., *Chromatographia*, 5 (1972) 346.

De Hernandez, C. M. and Walton, M. F., *Anal. Chem.*, 44 (1972) 890.

Deyl, Z., Rosmus, J., Juřicová, M. and Kopecký, J., *Bibliography of Column Chromatography 1967–1970*, Elsevier, Amsterdam, London, New York, 1973.

Evans, W. C. and Partridge, M. W., *J. Pharm. Pharmacol.*, 1 (1949) 593.

Gladyshev, P. P., Goryaev, M. I. and Baigalieva, A. N., *Izv. Akad. Nauk Kaz. SSR, Ser. Khim.*, 19 (1969) 57; *C.A.*, 71 (1969) 64112m.

Gladyshev, P. P., Goryaev, M. I. and Baigalieva, A. N., *Izv. Akad. Nauk Kaz. SSR, Ser. Khim.*, 21, No. 4 (1971) 43; *C.A.*, 75 (1971) 122380t.

Graf, E., *Arzneim.-Forsch.*, 1 (1951) 257.

Heacock, R. A., Langille, K. R., MacNeil, J. D. and Frei, R. W., *J. Chromatogr.*, 77 (1973) 425.

Henry, R. A. and Schmit, J. A., *Chromatographia*, 3 (1970) 116.

Higuchi, T. and Bodin, J. I., in T. Higuchi and E. Brochmann-Hanssen (Editors), *Pharmaceutical Analysis*, Interscience, New York, 1961, pp. 468–543.

Hofmann, A., *Helv. Chim. Acta*, 37 (1954) 314.

Janot, M. M. and Le Hir, A., in E. Lederer (Editor), *Chromatographie*, Vol. I, Masson, Paris, 1959, pp. 505–530.

Jolliffe, G. H. and Shellard, E. J., *J. Chromatogr.*, 81 (1973) 150.

Kamp, W., *Pharm. Weekbl.*, 92 (1957) 1.

Kanhere, S. S., Shah, R. S. and Bafna, S. L., *J. Pharm. Sci.*, 57 (1968) 342.

Lederer, E. and Lederer, M., *Chromatography*, Elsevier, Amsterdam, 2nd ed., 1959.

Le Hir, A., Goutarel, R. and Janot, M. M., *Ann. Pharm. Fr.*, 11 (1953) 546.

Levine, J. and Fishbach, H., *J. Amer. Pharm. Ass., Sci. Ed.*, 44 (1955) 713; 46 (1957) 191.

McMartin, C., Simpson, P. and Thorpe, N., *J. Chromatogr.*, 43 (1969) 72.

Merz, K. W. and Franck, R., *Arch. Pharm. (Weinheim)*, 275 (1937) 345.

Pfeifer, S. and Döhnert, H., *Pharmazie*, 23 (1968) 585.

Pfeifer, S. and Kühn, L., *Pharmazie*, 23 (1968) 267.

Stutz, M. H. and Sass, S., *Anal. Chem.*, 45 (1973) 2134.

Trautner, E. M. and Roberts, M., *Analyst (London)*, 73 (1948) 140.

Tunmann, P. and Hudmann, W., *Arch. Pharm. (Weinheim)*, 287 (1954) 281.

Wieland, T. and Merz, H., *Chem. Ber.*, 85 (1952) 731.

Wittwer, J. D., Jr. and Kluckhohn, J. H., *J. Chromatogr. Sci.*, 11 (1973) 1.

Wu, Ch. Y. and Siggia, S., *Anal. Chem.*, 44 (1972) 1499.

*Chapter 40*

# Other heterocyclic compounds

J. DAVÍDEK, M. JANDA and I. STIBOR

## CONTENTS



## INTRODUCTION

At present, there is a lack of published information giving a general insight into the dependence of the chromatographic behaviour of heterocyclic compounds on their structure. Demetriou *et al.* investigated the gel chromatography on Sephadex of some indoles in comparison with the analogous aromatic (phenyl) derivatives, showing that the nature of the ring structure and the nuclear substitution have only a minimal effect on the chromatographic behaviour of monocarboxylic acids. The effect of the chemical nature of the aliphatic portion of a heterocyclic or aromatic compound has been found to be stronger in the adsorption of carboxy derivatives in comparison with amine analogues. With regard to substitution of the nucleus, it has been shown that planarity of the molecule may result in either ion exclusion or an increase in adsorption. These effects are attributed to the electron-donating properties of the substituent group. Woof and Pierce showed that the presence of carboxy and nitro groups decreases the adsorption of aromatic compounds, whereas the effects of methoxy and hydroxy groups in phenolic and indolic compounds were noted also in the studies of Demetriou *et al.*

The properties and identification of simple derivatives of γ-pyrone are described in textbooks on organic qualitative analysis. The distribution, taxonomic significance, genetics, metabolism, biosynthesis, enzymology, function, isolation and identification of the derivatives of γ-pyrone, mostly flavonoid compounds, have been excellently reviewed in the last few years (Mabry *et al.*; Robinson; Seshadri; Weinges *et al.*, 1968, 1969). The methods for the quantitative analysis of these compounds are mostly based on spectrophotometric measurements.

## DERIVATIVES OF γ-PYRONE

### Isolation procedures

Many compounds in this group are water soluble, especially in the glycoside form, and are present in aqueous plant extracts. Even those which are only slightly soluble in water are sufficiently polar to be extracted well with methanol, ethanol or acetone, which are the solvents most frequently used for the extraction of flavonoid compounds. In general, however, alcohols are the solvents of choice. Methanol, because of its lower boiling point, is often preferred to ethanol, propanols and higher alcohols being used only rarely. For fresh plant material, absolute solvents may be used, while for dried material the use of aqueous alcohol is often an advantage. When only one solvent is used, its choice depends not only on the compounds that are being examined, but also on the plant tissue in which they occur and the nature of any other substances which may be present. It is often useful to remove lipids from certain tissues (*e.g.*, seeds) with light petroleum before proceeding to extract flavonoids with alcohol. Re-extraction of an aqueous solution with polar organic solvents is frequently of value in separating this group from more polar compounds, such as carbohydrates. Ethyl acetate is a suitable solvent for the cleaning of extracts of catechins and leucoanthocyanidins. Benzene can be used for benzophenones and stilbenes. Amyl alcohol has been used extensively for anthocyanins. *sec.*-Butanol is the most polar alcohol to be incompletely miscible with water, and if the aqueous extract is saturated with sodium chloride or magnesium sulphate, it is very successful for removing compounds of this group. Whichever solvent is used for the extraction, the next step before separation is concentration of the extract using the lowest possible temperature, preferably under nitrogen or in a vacuum. Polyphenolic substances are also sensitive to air oxidation in neutral and basic solutions so that it is good practice to prepare extracts in the presence of a dilute acid (*e.g.*, hydrochloric acid). On the other hand, the use of hot acid or standing for a long period with cold acid may cause hydrolysis of glycosides.

Classically, various precipitating agents have been used for these compounds, and neutral and basic lead acetate have been particularly recommended. Picric acid, barium hydroxide, pyridine, etc., have also been used.

Charcoal is useful for the preliminary purification of mixtures of flavonoids, particularly flavonoid glycosides, which are usually present in crude material. The charcoal procedure separates flavonoids from non-aromatic constituents such as the common carbohydrates.

**General chromatographic techniques**

Column chromatography does not, in most instances, separate complex mixtures of flavonoids that may be present in crude plant extracts, but when larger amounts of the flavonoids are required, column chromatography may be the method of choice. The general use of very polar adsorbents, such as alumina, precluded their successful application to compounds such as the flavonoids. The introduction of partition chromatography on silica gel led to its application in the separation of flavonoid compounds being first used for tea catechins (Bradfield *et al.*). Gage *et al.* used a polyacrylic-exchange resin and Ice and Wender the mild inorganic adsorbent magnesol (hydrated magnesium silicate) for the isolation and separation of flavonol glycosides. Magnesol columns are developed and eluted with wet ethyl acetate, followed by aqueous ethanol. Glycosides are more strongly adsorbed than aglycones (the reverse is true for polyamide). Forsyth applied cellulose, while Neu used CM-cellulose. For powdered cellulose columns, all of the chromatographic systems employed in paper chromatography can be used. More recently, powdered polyamide has come into increasing use because of its high loading capacity (Carelli *et al.*; Davídek; Hörhammer *et al.*, 1961). Polyamide and CM-cellulose columns are developed with water followed by aqueous alcohol. Flavanones, which are difficult to separate from chalcone impurities by means of recrystallization, are readily purified using polyamide. Polyamide column separation often yields pure flavonoids or simple mixtures, which can be further separated by additional column or paper chromatography. Johnston *et al.* (1968) described a procedure for the separation of flavonoids (aglycones and glycosides) on Sephadex LH-20 columns using methanol as eluent. The degree of adsorption of flavonoid aglycones on to Sephadex generally depends on the number of hydroxyl groups but not on their acidity, while with flavonoid glycosides of higher molecular weights both gel sieving and adsorption are important. Sephadex appears to be an efficient, high-capacity medium for both analytical and preparative work.

**Chromones**

Chromones are structurally related to the coumarins, but occur much less frequently. Eugenin, a typical chromone, was isolated from *Eugenia aromatica.*

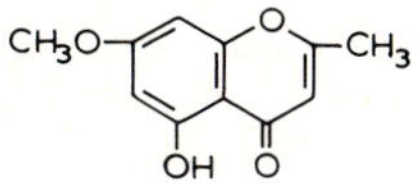


Eugenin

Relatively simple mixtures of some chromone derivatives resulting during the synthesis of these compounds can be separated by using different types of adsorbents according to the nature of the compound to be separated.

Silicic acid columns were used for the partial purification of 5,7-dihydroxy-2,2-dimethylchromanone resulting from Friedel-Crafts acylation using light petroleum–diethyl ether (6:4) as eluent. 5-Hydroxy-2,2,8,8-tetramethylbenzo[1,2-6:3,4-*b′*] dipyran-

4(3H),10(9H)-dione was eluted with light petroleum–diethyl ether (45:50). 3,4,9,10-Tetrahydro-2,2,8,8-tetramethylbenzo[1,2-*b*:3,4-*b*′]dipyran-5-ol resulting from sodium borohydride reduction of the former compound can be separated using a silicic acid column and elution with a mixture of light petroleum–diethyl ether (35:65). Crude 3,4,9,10-tetrahydro-6-isobutyl-2,2,8,8-tetramethylbenzo[1,2-*b*:3,4-*b*′]dipyran-5-ol was obtained from the same column using light petroleum–diethyl ether (85:15) (Byrne and Shannon). The mixture of 3-methylisothiochroman-4-one and 3,3-dimethylisothiochroman-4-one can be separated on a silicic acid column using diethyl ether–*n*-hexane (10:90 and 20:80) (Berchtold and Lumma).

Alumina was used for the isolation of products resulting from the isomerization of 7-methoxychromano-(3,4-*d*)-isooxazole (Kasturi *et al.*). Blickenstaff and Tao used alumina for the separation of products resulting from the reduction of 7-methoxychroman, which can be isolated as a product of 7-methoxy-4-chromanone by reduction with Raney nickel using light petroleum (b.p. 30–60°C) as the eluent. The dihydro derivative of 7-methoxychroman could be isolated on Florisil by elution with benzene–light petroleum (1:1) and by benzene. Bhalerao and Thyagarajan used silica gel for the separation of the Schmidt rearrangement products in the conversion of chromanones into 1,4- and 1,5-benzoxazepines. The chromans formed by cyclization of the isoprenylphenol can be isolated on silica gel. Dihydro*allo*evodionol and tetrahydrofranklinone can be separated by elution with *n*-hexane–ethyl acetate (99:1). *allo*Evodionol resulting from the reduction of dihydro-(4′-methylpent-3′-enyl)-5-hydroxy-6-carbethoxy-7-pentylchromene and D,L-cannabichromene, a constituent of hashish, can be separated using the above procedure. Nuclear prenylation of 2-methyl-5,7-dihydroxychromone and column chromatography of the products employing silica gel and elution with benzene–light petroleum (40:60 and 50:50) and benzene gives the following three fractions: 2-methyl-6,8-di-C-prenyl-5,7-dihydroxychromone, 2-methyl-7-prenyloxy-5-hydroxychromone and peucenin. 2,2-Dimethyl-5-hydroxy-6-acetylchromene can be prepared from a reaction mixture by elution with benzene–light petroleum (1:3), and isobavachin by elution with benzene–ethyl acetate (95:5) (Cardillo *et al.*). The isolation of siccanochromenes can be carried out using silver nitrate–silica gel column chromatography of their acetates (Nozoe *et al.*).

Kieselgel was employed for the isolation of metabolites from *Panus rudis*. By elution with chloroform–methanol (98:2) and by further chromatography, *cis*-3,4-dihydroxy-6-methoxy-2,2-dimethylchroman and *trans*-3,4-dihydro-6-methoxy-2,2-dimethylchroman were obtained. From *Panus conchatus*, 6-hydroxy-2,2-dimethylchroman-4-one together with small amounts of panepoxydione can be isolated (Kis *et al.*).

## Flavonoids

Flavonoids belong to the large group of polyphenolic compounds. The occurrence of flavonoids (2-phenylchromone derivatives) and related substances is widespread in plants. Examples are shown below.

For the separation of flavonoids from plant materials, different chromatographic procedures have been used. The variety of flavonoid compounds and the presence of interfering substances do not allow the use of one simple procedure, and therefore the most suitable methods for different types of plant materials are listed in Table 40.1 and those for synthetic mixtures in Table 40.2.

*Separation of flavonoids on Amberlite XAD-2*

A 16 × 0.9 cm column of Amberlite XAD-2 (200–400 mesh) maintained at 45°C was used for the separation of many flavonoids (Fig. 40.1).

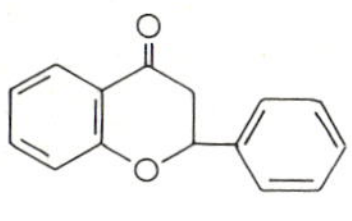

Flavanone

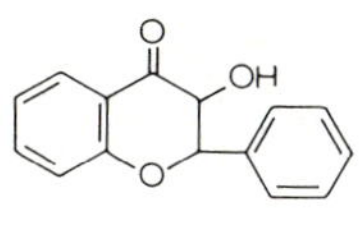

Dihydroflavonol

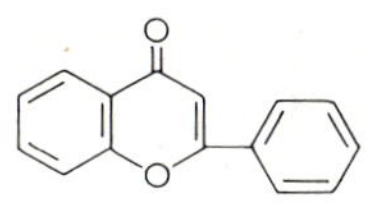

Flavone

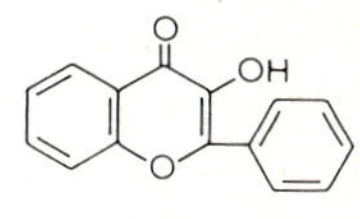

Flavonol

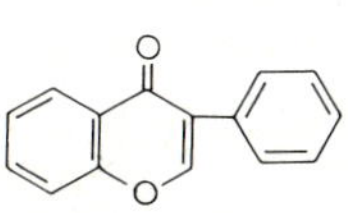

Isoflavone

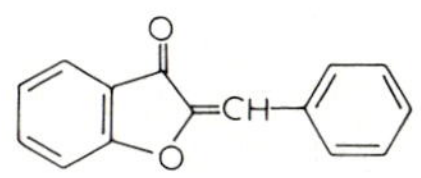

Aurone

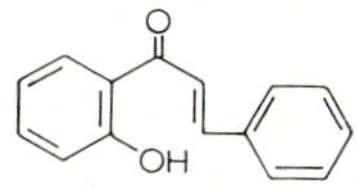

Chalcone

The column is first equilibrated with 20% ethanol at a flow-rate of 60 ml/h and a solution or suspension of flavonoids in 20% ethanol (each containing less than 500 μg/ml) is placed on the top of the column. A 100-ml volume of 20% ethanol is run first, followed by linear gradient elution with a total volume of 1 l, the ethanol concentration increasing from 20 to 90%. The flavonoids are eluted from the column in the order sugar esters, glycosides and aglycones. Of the aglycones, the compounds that have more hydroxyl groups are eluted faster than those with fewer hydroxyl groups. The elution time for the glycosides varies depending on the position at which the sugar moiety is attached, even if the sugar and aglycone are the same. This procedure was also applied to the determination of flavonoids in crude methanolic extracts from plants (Hori).

Johnston *et al.* (1969) described the separation of polyphenolic glycosides extracted from plant material on dextran gel (Sephadex G-10 and G-15). The adsorption properties of the Sephadex may be suitably modified using aqueous methanol eluents of varying methanol content. This procedure is suitable for the separation of aesculin, chlorogenic acid, arbutin, quercitrin-4′-glucoside, phloridzin, rhoifolin, apiin, rutin, apigenin-7-glucoside and quercitrin.

Nordström separated 20 flavones, 7 flavanones and *d*-catechin on Sephadex LH-20. The degree of adsorption of the aglycones generally depends on the number of hydroxyl groups in the molecule, especially in positions 3 and 5. The flavones are eluted faster than their flavonol analogues.

*References p.924*

TABLE 40.1
SEPARATION OF FLAVONOIDS FROM PLANT MATERIALS

| Plant material | Isolation procedure | Adsorbent | Solvent | Flavonoids | References |
|---|---|---|---|---|---|
| *Hymenoxys scaposa* flower heads | Extd. with light petroleum, methylene chloride; methanol, evapd. | | | | |
| | Evapd., aq. extract chromatographed on | Polyamide | Water, water–methanol (4:1) Water–methanol (1:1) | Patulitrin, quercetagitrin | Thomas and Mabry (1968a) |
| | Residue from methanol extract chromatographed on | Polyamide | Water–methanol (1:1) Water–methanol (4:1) | Patuletin-3-rutinoside Patuletin-3-O-glucoside | |
| *Hymenoxys scaposa* leaves | Extd. with cold light petroleum, methylene chloride | Silica gel | Chloroform–methanol (99.5:0.5) | Kymenoxin, scaposin (5,5′, 7-trihydroxy-3′,4′,6,8-tetramethoxyflavone) | Thomas and Mabry (1968b) |
| *Polypodium vulgare* | Extd. with ethanol, evapd., water added; extd. with light petroleum (b.p. 80–100°C) | Polyamide | Water, water–methanol (3:1), Benzene–ethyl acetate–acetic acid (3:4:3) | | Weinges and Wild |
| | Eluate re-chromatographed on | Sephadex LH-20 | Ethanol | (+)-Catechin-7-L-arabinoside | |
| *Madotheca platyphylla* | *n*-Pentane, ethanol, concd., extd. with water at 40°C, evapd., extd. with chloroform | Polyamide | Water Water–methanol (1:1) Water–methanol (2:3) | Saponarin (C-β-D-glucopyranosyl-(6)-O-mono-β-D-glucoside-(7)-5,7,4′-trihydroxyflavone) | Tjukavkina *et al.* |
| | Eluate re-chromatographed on | Polyamide | | Saponaretin (6-C-β-D-glucopyranosyl-5,7,4′-trihydroxyflavone) | |
| *Trifolium pratense* | Methanol, evapd., extd. with *n*-hexane, aq. phase with ethyl acetate at pH 3.0, organic phase with aq. $NaHCO_3$, acidified with HCl to pH 3, extd. with ethyl acetate | Silicic acid | Ethyl acetate–*n*-butanol | | |
| | Eluate re-chromatographed on | Celite | Chloroform–60% ethanol–ethyl acetate (2:2:1) | 5-Malonyl-7-D-β-glucosyl-5,7-dihydroxy-4′-methoxyisoflavone | Tamura *et al.* |

| | | | | | |
|---|---|---|---|---|---|
| | Organic phase (neutral fraction) after stirring with aq. $NaHCO_3$ | Charcoal | Acetone (aq.)<br>Acetone, acetone–aq. ammonia | 7-D-β-Glucosyl-5,7-dihydroxy-4′-methoxyisoflavone, trifolirhizin, ononin, formononetin, daidzein, daidzein glucoside | |
| *Salvia triloba* | Extd. with 70% ethanol, evapd. | Silicic acid–diatomaceous earth (3:1) | Light petroleum, light petroleum–benzene (9:1, 1:1), benzene, benzene–chloroform (9:1, 1:1), chloroform–diethyl ether (9:1, 1:1), diethyl ether, ethanol | Salvigenin (5-hydroxy-6,7,4′-trimethoxyflavone) | Ulubelen *et al.* |
| *Prunus spinosa* leaves | Extd. with 70% ethanol, evapd. aq. residue extracted with chloroform | Polyamide | 25–35% ethanol | Ternozide (3-α-L-arabofuranosido-7-α-L-rhamnofuranoside) | Makarov *et al.* |
| *Artemisia porrecta* var. coerula stems | Extd. with hot water, chloroform | Acidic alumina | Light petroleum–diethyl ether (8:2) | Gerniarin | Maksyutina |
| *Teclea sudanica* leaves | Extd. with light petroleum, chloroform, ethyl acetate (methanol) | Polyamide | Gradient elution, 0–50% ethanol | Vitexin, homovitexin, saponaretin, homoorientin, orientin | Paris and Etchepare |
| *Calophyllum blancoi* bark | Extd. with *n*-pentane, 5% $Na_2CO_3$, aq. layer acidified with HCl, extd. with diethyl ether, evapd. | Silica gel | *n*-Hexane–ethyl acetate (20:1) | Blancoic acid | Stout *et al.* (1968) |
| *Calophyllum papuanum* bark | Extd. with *n*-pentane, 5% $Na_2CO_3$, aq. layer acidified with HCl, extd. with diethyl ether, evapd. | Silica gel | *n*-Hexane–ethyl acetate (7:1) | Papuanic acid, isopapuanic acid | Stout *et al.* (1968) |

*(Continued on p. 902)*

TABLE 40.1 *(continued)*

| Plant material | Isolation procedure | Adsorbent | Solvent | Flavonoids | References |
|---|---|---|---|---|---|
| *Citrus aurantium* fruits | Extd. with acetone, benzene | Neutral alumina | Chloroform | | |
| | Eluate chromatographed on | Neutral alumina | Benzene–ethyl acetate (95:5) | 3,3′,4′,5,6,7,8-Heptamethoxyflavone, 3′,4′,5,6,7,8-hexamethoxyflavone, tangeretin (4′,5,6,7,8-pentamethoxyflavone), sinensetin (3′,4′,5,6,7-pentamethoxyflavone) isosinensetin (3′,4′,5,7,8-pentamethoxyflavone), isoscutallarein (4′,5,7,8-tetramethoxyflavone) | Schneider *et al.* |
| *Citrus reticulata* var. Cleopatra, var. Avana fruits | Extd. with diethyl ether, chloroform, methanol | Acidic alumina | Benzene, benzene–diethyl ether (4:6) | Tangeritin, nobiletin | Pinkas *et al.* |
| | Extd. with chloroform | Silica gel | Benzene–ethyl acetate (increasing concentrttion) | | |
| | Eluate re-chromatographed on | Silica gel H | Benzene–ethyl acetate (4:6) | 5-O-Desmethylnobiletin, 4,5-dihydroxy-6,7,8-trimethoxyflavone, 4,5-dihydroxy-3′,6,7,8-tetramethoxyflavone, tangeritin, nobiletin | |
| *Prunus persica* bark | Extd. with light petroleum (b.p. 40–60°C), diethyl ether, ethyl acetate | | | | |
| | Light petroleum extract chromatographed on | Neutral alumina | Various | | Rahman and Bhatmagar |
| | Flavanone-containing fraction from light petroleum and ether extract chromatographed on | Magnesium silicate | | Persicogenin (3′,5-dihydroxy-4′,7-dimethoxyflavone) naringenin, aromadendrin | |
| *Viscum album* var. coloratum leaves | Extd. with boiling ethanol, chloroform, aq. ethyl acetate | Cellulose | Ethyl acetate satd. with water, ethanol | Flavoyladorinin-β-(7,3′di-O-methylluteolin-4′-O-mono-D-glucoside) homoflavoyladorinin (7,3′-di-O-D-glucosapioside) | Ohta and Yagishita |

| | | | | | |
|---|---|---|---|---|---|
| *Sparattosperma vernicosum* fruits | Extd. with ethanol, evapd. refluxed with Norite in methanol, filtered, evapd. | Silica gel | Gradient elution with chloroform–methanol (5–30%) | Pinocembrin-7-β-neohesperidoside | Kutney *et al.* |
| *Baptisia lecontei* | Extd. with 25% methanol | Polyclar AT (polyvinylpyrrolidone) | Water, increasing concns. of methanol up to 100%, 0.3 *N* HCl in methanol, 4.5 *N* HCl in methaonl | Purified extract for further identification | Markham and Mabry |
| *Cupressus funebris* leaves | Extd. with hot ethanol, evapd, extd. with light petroleum | Silica gel | Toluene–pyridine–acetic acid (10:1:1, 20:1:1) | Isocryptomerin, hinokiflavone, enpressuflavone, amentoflavone | Natarajan *et al.* |
| *Citrus paradisi* leaves | | Polyclar AT (polyvinylpyrrolidone) | Water | Apigenin-7-β-rutinoside, naringenin-7-β-rutinoside, naringin | Nordby *et al.* |
| *Brassica napus* leaves | Extd. with methanol, pH 6.0 (with $NaHCO_3$), extd. with chloroform, diethyl ether, ethyl acetate | Perlon | 90% methanol | | Hörhammer *et al.* (1967) |
| | Extract re-chromatographed on | Kieselgel | Ethyl acetate–methanol–water (100:16.5:13.5) | Brassicosid | |
| *Sinapsis arvensis* leaves | Extracted with methanol | Perlon | 90% methanol | | Hörhammer *et al.* (1967) |
| | Extract re-chromatographed on | Kieselgel | Ethyl acetate–methanol–water (100:16.5:13.5) | Brassidin (isorhamnetin-3-mono-β-D-glucoside) | |
| Green tea | Water | Cellulose | 2% acetic acid | 6,8-Di-D-C-glucopyranosyl apigenins | Sakamoto |
| *Iris Nertshinskia Loddiges f. alubiflora* Honda | Ethanol, benzene, aq. soln. satd. with ethyl acetate | Nylon | Methanol | Swertisin, swertijaponin, genkwanin-6-C-β-D-glucopyranosyl-X″-O-mono-D-glucoside | Kawase |

(Continued on p. 904)

References p. 924

TABLE 40.1 *(continued)*

| Plant material | Isolation procedure | Adsorbent | Solvent | Flavonoids | References |
|---|---|---|---|---|---|
| *Metasequoia glyptostroboides* | Extd. with light petroleum, diethyl ether, methyl ethyl ketone–water (azeotropic mixture), concd., dissolved in acetone | Polyamide | 10–90% acetone | Myricetin-3-rhamnoside with quercetin-3-rhamnoside, kämpferol-3-rhamnoside, myricetin-3-rhamnoside, quercetin-3-rhamnoside, apigenin-7-glucoside, luteolin-7-glucoside, tricetin-7-glucoside | Clark-Lewis and Dainis |
| | Further sepn. on | Cellulose | Acetone–water–acetic acid (20:80:5) | | |
| | Further sepn. on | Sephadex LH-20 | Acetone–water–methanol (2:1:1) | | |
| *Mallotus phillipinensis* | Kamala powder (fruits from *M. phillipinensis*) extd. with light petroleum (b.p. 40–60°C) carbon tetrachloride, dichloromethane, acetone | Silica gel | Benzene–chloroform | C-Methylated cinnamoyl chromene, rearranged flavonone chromone | Crombie *et al.* |
| *Veronicastrum sibiricum* leaves and stems | Extd. with *n*-butanol, aq. phase chromatographed on | Charcoal–Celite (1:1, w/w) | Water | | Inouye *et al.* |
| | Re-chromatographed on | Charcoal | Methanol–water | | |
| | Organic phase evapd., dissolved in water, extd. with diethyl ether, aq. phase chromatographed on | Perlon | | Luteolin-7-β-D-glucopyranoside, luteolin-7-β-neohesperidoside | |
| *Petroselimum sativum* stems | Extd. with water, light petroleum (b.p. 60–80°C), methanol | Sephadex LH-20 | 90% methanol | Derivatives of apigenin | Tomás *et al.* |
| *Trifolium pratense*, protein concentrate from leaves | Extd. with boiling ethanol, evapd., dissolved in 70% ethanol, extd. with light petroleum (b.p. 60–80°C), isoflavones extd. with diethyl ether | Sephadex G-25 | 0.1 *M* ammonia | Biochanin A, formononetin, daidzein, genistein | Glencross *et al.* |

TABLE 40.2
SEPARATION OF FLAVONOIDS IN SYNTHETIC MIXTURES

| Flavonoid | Adsorbent | Solvent | References |
|---|---|---|---|
| Flavan-(4α-ylthio) acetic acid, 7-(methoxyflavan-4α-ylthio) acetic acid, methyl-(3-hydroxyflavan-4-ylthio) acetate, methyl-(3-O-benzoyloxyphenyl-1-*p*-methoxyphenylpropylthio) acetate | Silica gel | Light petroleum (b.p. 40–60°C)–diethyl ether (4:1) | Brown *et al.* |
| 4′-γ,γ-Dimethylallylnaringenin | Silica gel | Chloroform | Chari *et al.* |
| Isoelliptol isoflavone, hydrogenated 8,9-dimethoxy-furano[3′,2′: 2,3]pterocarpan | Silica gel | Chloroform–*n*-hexane (1:1) | Fukui *et al.* |
| 3,5,3′-Trihydroxy-7,4′-dibenzyloxy-flavon-3-[β-D-galactopyranoside tetraacetate], 2,3,4,6-tetra-O-acetyl-α-D-galactopyranosyl bromide | Silica gel | Toluene–ethyl acetate (5:4) or ethyl acetate–methanol–water (100:20:14) | Hörhammer *et al.* (1969) |
| Poncirin, naringin | Silica gel | Ethyl acetate–methanol–water (80:14:10) | Wagner *et al.* (1969b) |
| 5,7,3′,4′-Tetrametyl-(±)-[2-$^{14}$C]epicatechin | Silica gel | Benzene–acetone (9:1) | Weinges *et al.* (1969) |
| Cumarone | Silica gel | Benzene–acetone (95:5) | Weinges *et al.* (1969) |
| 7,4′-Dibenzylquercetin-3-glucoside | Silica gel | Ethyl acetate–methanol–water (100:20:15) | Hörhammer *et al.* (1968a) |
| 2-methyl-2′,4,4′,6′-tetrabenzoyl-3-methoxychalcone | Polyamide | Cyclohexane–dioxane–acetic acid (50:3:1) | Hänsel *et al.* |
| *trans*-3-Methyl-3′-methoxy-4′,5,7-trihydroxyflavanone | Polyamide | Ethanol–water (6:4) | Hänsel *et al.* |
| Naringenin-7-β-[cellobioside heptaacetate] | Silica gel | Diethyl ether–benzene (3:8) | Wagner *et al.* (1969a) |
| Naringenin-7-β-[rutinoside hexaacetate] | Polyamide | 10% methanol | Wagner *et al.* (1969b) |
| 7,4′-Dibenzylkämpferol-3-β-sophoroside | Silica gel | Toluene–ethyl acetate (5:4) | Wagner *et al.* (1968) |

(Continued on p. 906)

References p.924

TABLE 40.2 *(continued)*

| Flavonoid | Adsorbent | Solvent | References |
|---|---|---|---|
| Kämpferol-3-β-sophoroside | Cellulose | 2% acetic acid | Wagner *et al.* (1968) |
| Degradation product of (+)-catechin | Silica gel, Silicic acid–Celite (5:1) | Benzene–acetone–methanol (90:7:3) | Weinges *et al.* (1969) |
| Hydrogenated products of 3-veratrylidene-7-methoxyflavone | Alumina | Benzene–light petroleum (30:70) | Morsingh |
| 3-Methoxyrisnagin | Neutral alumina | Ethyl acetate–benzene | Mukorjee *et al.* |
| Oxidation products of apigenin | Sephadex LH-20 | Methanol | Molyneux *et al.* |
| Enzymatic degradation products of (+)-catechin | Sephadex LH-20 | Ethanol | Weinges and Huthwelker |
| Methylated talboflavone | Neutral alumina | Benzene, chloroform | Joshi *et al.* |
| 3,5,3′-Trihydroxy-7,4′-dimethoxyflavon-3-β-[6-O-α-L-rhamnopyranosyl-D-glucopyranoside] ombuoside | Silica gel | Ethyl acetate–methanol–water (100:20:15) | Hörhammer *et al.* (1968b) |
| Luteolin-7-β-D-glucopyranoside, luteolin-7-β-neohesperidoside | Silica gel | Chloroform–methanol (9:1) | Inouye *et al.* |

TABLE 40.3
CHROMATOGRAPHIC SYSTEMS FOR THE SEPARATION OF XANTHONES

| Compound | Adsorbent | Solvent | References |
|---|---|---|---|
| 2,4-Di-C-prenyl-1,3-dihydroxy-7-methoxyxanthone, 1-hydroxy-3-prenyloxy-7-methoxyxanthone, 2-C-prenyl-1,3-dihydroxy-7-methoxyxanthone | Silica gel | Benzene–light petroleum (30:70, 50:50, 60:40) | Jain *et al.* (1969a) |

| | | | |
|---|---|---|---|
| 2-C-Allyl-3-allyloxy-1-hydroxyxanthone, 4-C-allyl-3-allyloxy-1-hydroxyxanthone, 3-allyloxy-1-hydroxyxanthone | Silica gel | Light petroleum (b.p. 60–80°C)–benzene (9:1, 9:3) | Jain *et al.* (1969b) |
| 1,3-Dihydroxyxanthone, 2,4-di-C-allyl-1,3-dihydroxyxanthone, 2-C-allyl-1,3-dihydroxyxanthone, 4-C-allyl-1,3-dihydroxyxanthone | Silica gel | Light petroleum (b.p. 60–80°C)–benzene (1:1) | Jain *et al.* (1969b) |
| 3-Hydroxy-1,2-dimethoxyxanthone, 3-hydroxy-1,5,6-trimethoxyxanthone, 4-hydroxy-2,3-dimethoxyxanthone, 1,5-dihydroxy-3,3-methoxyxanthone, 5-hydroxy-1,3-dimethoxyxanthone (from *Kielmeyera rupestris*) ($Na_2CO_3$-soluble chloroform extract) | Silica gel | Chloroform, chloroform–methanol (99:1, 98:2) | De Barros Correa *et al.* |
| 2-Hydroxy-1-methoxyxanthone, 3-methoxy-1,5,6-trimethoxyxanthone, 4-methoxy-2,3′-methylenedioxyxanthone, 3-hydroxy-1,2-dimethoxyxanthone, 5-hydroxy-1,3-dimethoxyxanthone (from *Kielmeyera rupestris*) ($Na_2CO_3$-insoluble portion) | Silica gel | Light petroleum (b.p. 60–70°C)–benzene (1:1)<br>Benzene<br>Benzene–chloroform (1:1), chloroform | De Barros Correa *et al.* |
| 1-Hydroxy-3,7-dimethoxyxanthone, 8-hydroxy-1,2,6-trimethoxyxanthone (from *Macrocarpaea glabra*) (*n*-pentane-soluble fraction) | Silica gel | Light petroleum–ethyl acetate in different proportions | Stout *et al.* (1969) |
| Crude osajaxanthone (from *Kielmeyera corymbosa*) | Silica gel | Chloroform–methanol (98:2) | Gottlieb *et al.* |
| 6-(3,3-Dimethylallyl)-1,5-dihydroxyxanthone (from *Colophyllum inophyllum*) | Silica gel | *n*-Hexane<br>*n*-Hexane–benzene (1:1)<br>Benzene–ethyl acetate (20:1) | Govindachari *et al.* |

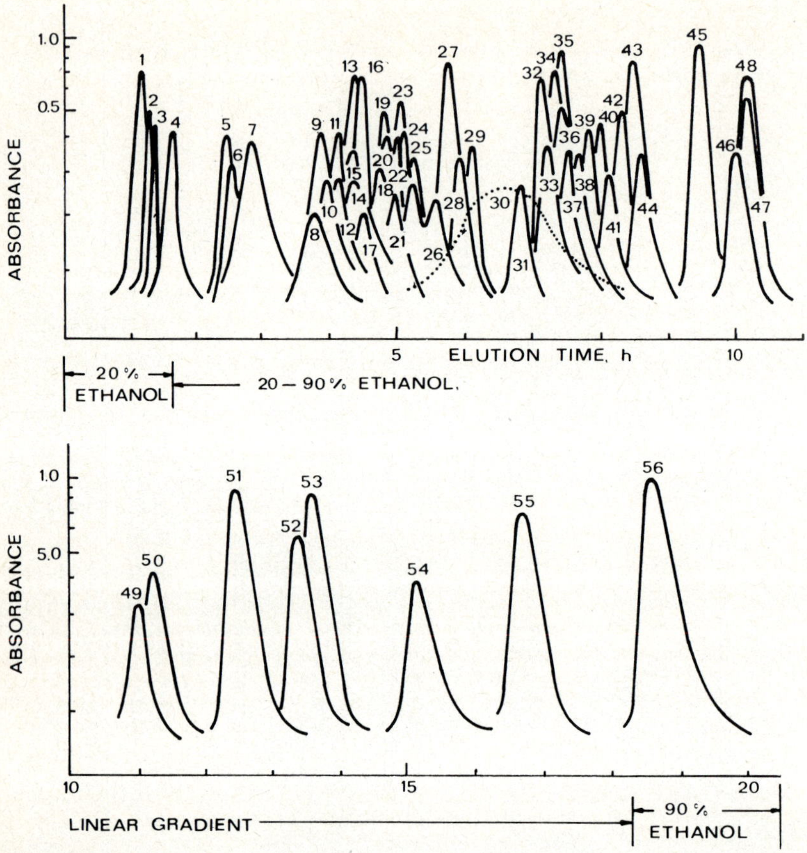


Fig. 40.1. Separation of flavonoids (Hori). Column: Amberlite XAD-2, 200–400 mesh, 16 × 0.9 cm, 45°C. Solvent: aqueous ethanol. Flow-rate: 60 ml/h. 1 = Apigenine-7-glucuronide; 2 = scutellarin; 3 = nelumboside; 4 = baicalin; 5 = homoorientin; 6 = saponarin; 7 = robinin; 8 = rutin; 9 = scoparin; 10 = lonicerin; 11 = rhoifolin; 12 = hyperin; 13 = liquiritin; 14 = tagetin; 15 = pedaliin; 16 = plantaginin; 17 = quercitrin; 18 = luteolin-7-glucoside; 19 = trifolin; 20 = narcissin; 21 = avicularin; 22 = kaempheritrin; 23 = bakkoside; 24 = reynoutrin; 25 = hesperidin; 26 = apiin; 27 = cosmosiin; 28 = juglanin; 29 = phellamurin; 30 = morin; 31 = aromadendrin; 32 = myricetin; 33 = linarin; 34 = pectoliuarin; 35 = cirsimarin; 36 = pedalitin; 37 = chrysosplenin; 38 = fisetin; 39 = fukugetin; 40 = chrysosplenetin; 41 = quercetin; 42 = amurensin; 43 = luteolin; 44 = neolinarin; 45 = geinstein; 46 = naringrnin; 47 = kaempherol; 48 = apigenin; 49 = hesperetin; 50 = baicalein; 51 = cirsimaritin; 52 = 2-methylthiochromone; 53 = wogonin; 54 = swertianol; 55 = flavone; 56 = euparin.

## Xanthones

These yellow compounds (illustrated below) of which about 20 are known, have been isolated from the flowering plants, fungi and lichens.

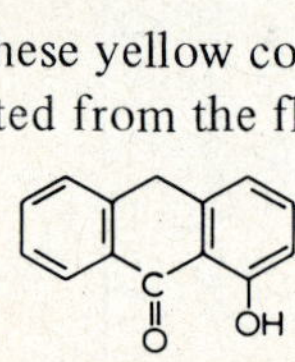

*Isolation procedures*

Xanthones from plants, fungi and lichens can be isolated by extraction with non-polar solvents. De Barros Correa *et al.* used extraction with hot benzene and with hot ethanol, while Stout *et al.* (1969) used *n*-hexane, *n*-pentane and methylene chloride. The wood of *Kielmeyera corymbosa* is extracted with ethanol, benzene and chloroform (Gottlieb *et al.*). Acetone is used for the extraction of xanthones of *Frasera caroliniensis* and the concentrated acetone solution is then extracted with *n*-pentane–methylene chloride (4:1) (Stout and Balkenhol). The roots of *Frasera albicaulis* are extracted with methanol, and the concentrated extract is re-extracted with 20% methylene chloride in *n*-pentane (Stout *et al.*, 1969). Xanthones of *Callophyllum fragrans* are extracted with boiling chloroform and then with boiling acetone (Locksley and Murray). Cold *n*-hexane is used for the extraction of xanthones of *Callophyllum inophyllum* combined with re-extraction with acetone–*n*-hexane (1:4) (Govindachari *et al.*).

Xanthones are best purified on silica gel. The chromatographic systems most often used are listed in Table 40.3.

## ANTHOCYANS

About 20 different classes of anthocyanidin glycosides are known and the pattern that emerges is very similar to that of flavonols.

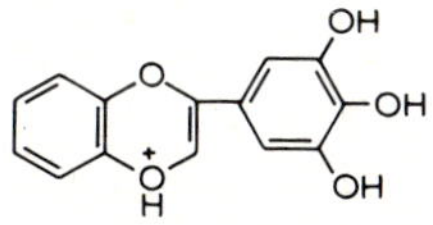

Delphinidin cation

### Isolation procedures

In general, hydroxylic solvents such as water or alcohols are used for the extraction of these substances, as with other classes of flavonoid compounds. However, just sufficient hydrochloric acid or another volatile acid to prevent their conversion into the pseudo-base form should be present. The classical method of isolating anthocyanin pigments is to precipitate them from alcoholic extracts by adding an organic solvent, such as diethyl ether, or to precipitate them as lead salts or by adsorption on a cation-exchange resin with subsequent elution from the washed resin with acidified methanol. Both of these methods have some disadvantages. The use of insoluble polyvinylpyrrolidone, which forms unusually strong hydrogen bonds with the proton of phenolic hydroxyl groups, in order to isolate anthocyanin pigments, was developed by Wrolstad and Putnam. The following techniques can be used for the isolation and preliminary purification of anthocyanin pigments. Care should be taken during the concentration of the extract because of the loss of labile components. The procedure is best carried out at low temperature and under nitrogen or *in vacuo*.

*References p.924*

The isolation of anthocyanin pigments from blackcurrant was described by Chandler and Harper. The fruit (500 g) is macerated in a blender with methanol (1 l) containing 1% of 10 N hydrochloric acid and the slurry allowed to stand for 30 min at room temperature before filtering it. The pulp is extracted in this way several times and the combined extracts are evaporated to 1 l at 40°C. Excess of a saturated solution of lead acetate is then added, and the lead–anthocyanin complex is removed by centrifuging and washed with 500 ml of methanol. The precipitate is again dissolved in methanol and the process repeated until a fine powder is obtained.

The isolation of anthocyanins from red cherries (*Prunus cerasus* L. var. Montmorency) was described by Dekazos. The anthocyanins are extracted from 50 g of de-stoned cherries by macerating them for 90 sec at 0°C in a blender with 1% of cold hydrochloric acid in methanol (100 ml). The macerate is filtered through filter-paper and the residue is re-extracted with the same solvent several times. The combined extracts are centrifuged at 10,000 *g* for 5 min and filtered through Celite. The pigment solution is mixed with Bio-Rad AG 50W-X4 ($H^+$) cation-exchange resin, 100–200 mesh, and after 2 h at 0°C the solution is filtered off and the resin washed several times with water. This procedure separates the pigments from the free sugars. Then the resin is transferred to a 50 × 4 cm column and washed with pure methanol. The pigments are eluted by successive extractions with 0.1, 0.5 and 1% of hydrochloric acid in methanol, as described by Smith and Luh. The eluates are combined, filtered and concentrated at 40°C.

The isolation of strawberry anthocyanin pigments was described by Wrolstad and Putnam. The strawberry homogenate (10 g) is extracted three times with 50-ml portions of water at 100°C. The extracts are filtered, the filtrate is shaken with 10 g of insoluble polyvinylpyrrolidone for 15 min, the mixture is centrifuged at 600 *g* for 4 min and the supernatant discarded. The pellet is washed with water and with the same amount of methanol. The pigments are recovered from the polyvinylpyrrolidone–pigment adsorbate by shaking the pellet with 0.01% of hydrochloric acid in methanol (100 ml) four or five times for 30 sec. The extracts are combined and concentrated to a small volume at 40°C.

### General chromatographic techniques

The polar adsorbent alumina was used by Karrer and Strong for the final purification of crude peonin chloride, but the method was not successful with total plant extracts. The successful separation of small amounts of a mixture of the synthetic anthocyanidins malvidin, petunidin and delphinidin on columns of silicic acid with 10% of phosphoric acid as the stationary phase and a mixture of phenol and toluene as the non-aqueous phase was used by Spaeth and Rosenblatt. Silicic acid and Celite columns were used for separations of various proanthocyanidins (Weinges *et al.*, 1968, 1970a,b). Polyamide was used in the column chromatography of blackcurrant fruit extracts by Chandler and Harper. Anthocyanin pigments purified on the column were identified as cyanidin and delphinidin and their 3-glucosides and 3-rutinosides. Chromatography on a 100 × 5.5 cm Perlon column and elution with a mixture of ethanol–dimethylformamide (8:2) enables a mixture of dimeric proanthocyanidins from plant extracts to be obtained (Weinges *et al.*, 1968). Column chromatography on cellulose powder has been used to separate anthocyanin mixtures (Bendz *et al.*). The cellulose column retards anthocyanins that are stronger than sugars and acids but less strong than other flavonoids. Garber *et al.* used a cellulose column (65 × 3 cm) and elution with the upper phase of the solvent system *n*-butanol–acetic acid–water (5:4:1) for the separation of a methanol–hydrochloric acid extract from flowers of *Collinsia heterophylla* and obtained five well defined bands of anthocyanins.

Leucoanthocyanidins of flowers of *Camellia japonica* can be separated from anthocyanins on ion-exchange columns (12 × 1.3 cm) of Amberlite IRC-120 ($H^+$). The anthocyanin is quantitatively adsorbed on the resin while leucoanthocyanidin can be obtained from the neutralized filtrate (0.1 *N* sodium hydroxide solution) by extraction with ethyl acetate (Endo).

Gel filtration on Sephadex G-25 with aqueous alcoholic hydrochloric acid (Somers, 1966) and aqueous acetone (Somers, 1968) was used for the separation of condensed wine pigments from the monomers. The condensed pigment fraction is sharply separated; however, the anthocyanins are only partially resolved, apparently because of adsorptive partition. Good resolution of anthocyanins by using partition chromatography on Sephadex columns with the organic phase of *n*-butanol–acetic acid–water (4:1:5) as the eluent was reported (Gombkötö). This method has the same disadvantage as that with cellulose powder if macro-scale work has to be carried out. Sephadex LH-20 columns (30 × 1.5 cm) and 1% of methanolic hydrochloric acid at a flow-rate of 1 ml/min were used for the purification of the anthocyanin obtained by dialysis of the pigment of the Prof. Blaauw iris (Asen *et al.*).

Chromatography on polyvinylpyrrolidone shows promise for the separation of anthocyanins in larger amounts, and is convenient because it permits the direct use of plant material extracts without the need for preliminary and time-consuming purification procedures.

### *Application to separations of natural mixtures from plants*

Using column chromatography on the insoluble polyvinylpyrrolidone Polyclar AT, Hrazdina successfully separated the major pigments present in grapes: delphinidin-, petunidin-, malvidin- and peonidin-3,5-diglucosides.

A 60 × 2.5 cm column packed with Polyclar AT was washed with water (1 l) and grape juice, prepared by pressing and filtering, was percolated through the column, followed by washing with water until the effluent was tasteless (800 ml). The pigment material was then eluted from the column with 30% of aqueous ethanol containing 1 ml/l of 1 *N* hydrochloric acid and the elution was followed spectrophotometrically at 254 nm. Fractions containing the same pigment were pooled, concentrated to a small volume (30 ml) at 40°C under nitrogen, except with delphinidin-3,5-diglucoside, and re-chromatographed on a Polyclar AT column (50 × 2.5 cm) using the same solvent for elution. The main pigment fractions after re-chromatography showed no contamination on TLC and were concentrated to 20 ml under the same conditions and after addition of 1 ml of concentrated hydrochloric acid were placed in a refrigerator, where the pigment crystallized. The condensed pigment material was irreversibly adsorbed on the top of the column and was not eluted.

*References p.924*

Wrolstad and Struthers reported the conditions used in the Polyclar AT column separation of strawberry, rhubarb and raspberry anthocyanins that resulted in improved separation, eliminating the use of re-chromatography as in the experiments of Hrazdina. Very good resolution was obtained with strawberry pigments (diethyl ether precipitate) on a 40 × 5 cm column prepared by Hrazdina using 0.1% hydrochloric acid in methanol, and with rhubarb pigments on 35 × 1.5 and 70 × 1.5 cm columns and elution with 0.1% of hydrochloric acid in 30% ethanol and 0.1% of hydrochloric acid in methanol, respectively. Good resolution of raspberry pigments was achieved on a 30 × 1.5 cm column using 0.1% of hydrochloric acid in 70% methanol for elution.

## AFLATOXINS AND MYCOTOXINS

Aflatoxins are toxic materials produced by certain strains of *Aspergillus flavus* and *A. parasiticus.* The presence of aflatoxins has been reported in many agricultural commodities, groundnuts, cottonseed, soyabeans, maize, rice, wheat, millet, sorghum, sesame, barley, peas, etc. The best known are aflatoxins $B_1$, $B_2$, $G_1$ and $G_2$:

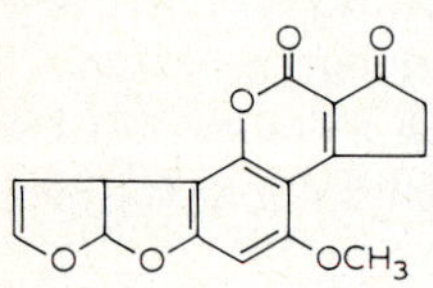

$B_1$

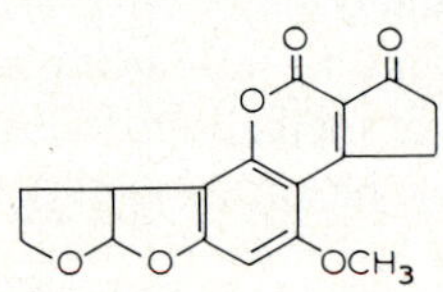

$B_2$

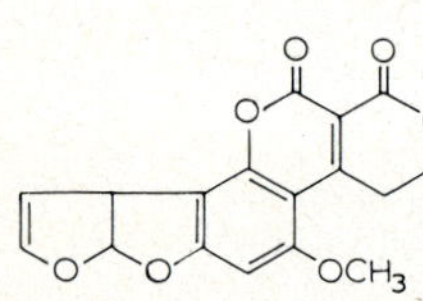

$G_1$

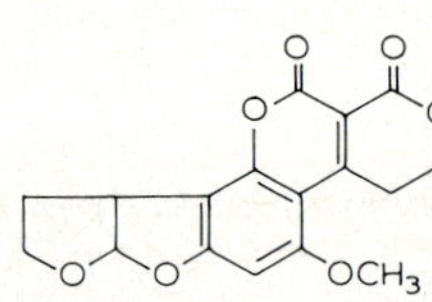

$G_2$

### Isolation procedures

In general, non-polar organic solvents are used for the extraction of these substances. Chloroform (Stubblefield *et al.*, 1970; Wiley and Waiss) is used for the extraction of toxins from *Aspergillus flavus*, and ethyl acetate for the isolation of toxins from *Fusarium tricinctum* and *Trichoderma liguorum* cultures (Bamburg and Strong). Chloroform–methanol (97:3) is used for the preparation of an extract of groundnuts (Holaday), while chloroform–methanol–*n*-hexane (8:2:1) extracts toxins of maize, groundnuts and sorghum that have been inoculated with mould spores of *Aspergillus flavus, A. ochraceus* and *A. nidulans* (Vorster). Cottonseed products are extracted with acetone–water–glacial acetic acid (850:150:8) (Pons *et al.*).

A significant proportion of the fat content of some samples is extracted with the solvent used for mycotoxin extraction. Extraction with *n*-hexane, Skellysolve B or a similar solvent is then recommended. Vorster extracted the fat-containing mixture with 0.1 *M* sodium hydrogen carbonate solution. Interferences from pigments and fluorescent substances, produced either by fungi or naturally present in the substrate, have been reported in the literature, together with the procedures for their removal. Wiseman *et al.* used solid copper(II) carbonate to remove quinonoid pigments from chloroform extracts of aflatoxin cultures, while lead acetate was used with cottonseed products (Pons *et al.*).

A crude aflatoxin can be prepared by precipitation of the defatted chloroform extract with Skellysolve B (Wiley and Waiss) and then used for further purification and fractionation.

### General chromatographic techniques

For further purification, fractionation and concentration of the toxin extracts prior TLC or PC analysis, column chromatographic methods are often used. Pons *et al.* use a 40 × 2 cm column packed with 15 g of silica gel, which is washed with a mixture of diethyl ether–*n*-hexane (3:1) in order to remove impurities, and aflatoxins of cottonseed products are then eluted with chloroform–acetone–2-propanol (34:5:1). The extract of maize, groundnuts and sorghum can be cleaned up on a 30 × 2.2 cm column packed with 10 g of silica gel (0.05–0.2 mm). Elution with a mixture of light petroleum–diethyl ether (3:1) at a flow-rate of 15 ml/min followed by chloroform–methanol (97:3) and finally benzene–acetic acid (9:1) gives three fractions which contain sterigmatocystin, aflatoxins and ochratoxins, respectively (Vorster). A method for detecting and quantifying aflatoxins, based on column chromatography, has been tested by Holaday. Columns of dimensions 4.5 × 0.4 cm are used with the chloroform–methanol (97:3) extract of a groundnut sample. If aflatoxins are present, a blue fluorescent band is observed when the column is exposed to UV light. For the isolation of toxins from *Fusarium tricinctum* and *Trychoderma liguorum* cultures, 70 × 2 cm and 70 × 1.5 cm columns of silica gel (0.05–0.2 mm) and elution with the solvent system toluene–ethyl acetate (1:3) is used. Also, 30 × 1.5 cm columns packed with neutral alumina, eluted with ethanol–ethyl acetate–acetone (1:4:4) and Skellysolve B–benzene–acetone (2:1:1) can be employed (Bamburg and Strong).

The removal of interfering substances from crude extracts of *A. flavus* has been accomplished by rapid chromatography through an acidic alumina column with benzene–chloroform (5:1) and benzene–chloroform (1:1) as the eluting solvents. Silica gel columns can be used for further fractionation of the toxins (Rodricks). Steyn has developed a chromatographic method for the preparation of pure aflatoxins from crude aflatoxin-containing extracts. A 60 × 3–5 cm column containing 50–70 g of the mixed adsorbent (basic alumina containing 6.25% of oxalic acid dihydrate) is eluted with each of the solvents benzene–chloroform (1:1), chloroform, chloroform–methanol (98:2), chloroform–methanol (95:5) and chloroform–methanol (90:10) at a flow-rate of 2–8 ml/min, or a single-phase elution with benzene or chloroform can be used. Aflatoxins $B_1$, $B_2$, $G_1$ and $G_2$ have been separated on a series of columns packed with silicic acid (100 mesh),

*References p.924*

washed with chloroform–ethanol (99:1) (separation of relatively pure $B_1$), followed by fractionation on a 30 × 2 cm silica gel (10–40 μm) column with chloroform–acetone–ethanol (97.3:2:0.75) (Stubblefield *et al.*, 1968).

Parasiticol has been separated together with aflatoxins and purified by chromatography on columns of silicic acid (elution with chloroform–ethanol, 99:1 and 85:5), silica gel (elution with chloroform–acetone–ethanol, 97.3:2:0.75) and alumina (30 × 0.9 cm column, elution with chloroform containing 0.75% of ethanol) (Stubblefield *et al.*, 1970). In a simplified procedure for the detection of aflatoxin $B_1$ in cottonseed meals, a single chromatographic column of Celite (45 × 3.5 cm) was used and elution was carried out with chloroform (Velasco). Aflatoxins have been purified on a 30 × 2.2 cm column containing 10 g of washed and deactivated Florisil, the column being washed with tetrahydrofuran and aflatoxins eluted with acetone containing 1% of methanol (Levi). Wiley and Waiss used a Sephadex LH-20 column prepared and eluted with chloroform for the further purification of aflatoxin $M_1$ from a chloroform extract of *A. flavus* cultures on rice. 6-Hydroxyramulosin, a metabolite of *Pestalotia ramulosa*, was extracted from the cultures with chloroform, crystallized from *n*-hexane and purified on a column of Florisil by elution with chloroform (Tannenbaum *et al.*).

*Application to separations of natural mixtures*

The extraction and chromatography of aflatoxins of *A. flavus* cultures on rice has been described by Wiley and Waiss.

A 4.5-kg amount of the cultures was defatted by extraction for 8 h with 4.5 l of Skellysolve B. The cultures were then extracted three times for 8 h each with 4.5 l of chloroform and the combined extracts were evaporated, the residue dissolved in *ca.* 100 ml of chloroform and the aflatoxins precipitated by the addition of *ca.* 1 l of Skellysolve B. The crude aflatoxin (7.55 g) was first chromatographed on 100 × 5 cm columns of silica gel (Silic AR, CC-7, 100–200 mesh) using ethyl acetate as the eluent. On this column there was no separation of aflatoxins from each other, but most of the brown, oily material was removed. Next, the crude aflatoxin was chromatographed on the same column prepared with silica gel, with a loading of 1 g per kilogram of silica gel, eluted with chloroform until $B_1$ started to appear, and then the solvent was changed to 5% methanol–chloroform with gradient elution until the fractions containing $M_1$ had been eluted (the fractions were monitored by TLC). This procedure thus allowed large amounts of $B_1$ to be separated from the remainder of the metabolites. From this column, pure $B_1$ (3.41 g) after crystallization from chloroform–methanol and about 2 g of mixed B and G were obtained. Also eluted was 18 mg of pure $G_2$, but, for the most part, $B_2$, $G_1$ and $G_2$ were not separated. The $M_1$ fractions, containing a yellow oil, were chromatographed on columns of Sephadex LH-20 prepared and eluted with chloroform. Then 38 mg of $M_1$ was crystallized from chloroform. The complete separation of aflatoxins $B_1$, $B_2$, $G_1$ and $G_2$ could be achieved by TLC, but the method was too tedious for a reasonable quantity of material to be obtained. Excellent resolution was obtained on 100 × 2.5 cm columns of silica gel H for TLC (E. Merck, Darmstadt, G.F.R.) prepared with chloroform and eluted as before with chloroform and chloroform–methanol. On these columns, virtually complete resolution of the four aflatoxins B and G was obtained.

## OTHER COMPOUNDS CONTAINING HETEROCYCLIC OXYGEN

Most of the furans discussed are associated with naturally occurring compounds that have already been reviewed by Dean. Column chromatography is in most instances the most suitable method by which the furan consituents of natural material could be isolated. The separation of a number of compounds containing the furan ring from Indian Yam Beans (*Pachyrrhizus erosus*) described by Krishnamurti *et al.* serves as an example of these applications.

The ground seeds (2.5 kg) were extracted repeatedly with light petroleum (b.p. 60–80°C) (70 h) to yield an oil. A subsequent extraction (70 h) with diethyl ether yielded fraction A (43.5 g) and a further continuous extraction (50 h) with diethyl ether yielded fraction B (6.2 g). Fraction A was refluxed for 45 min with light petroleum (b.p. 60–80°C) (2 X

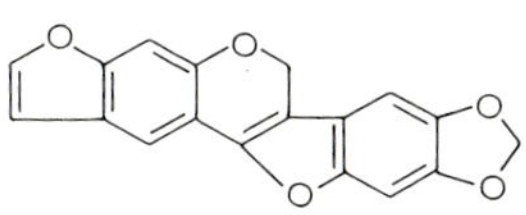

Erosnin

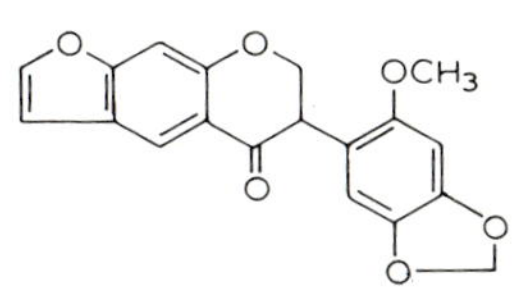


Neotenone

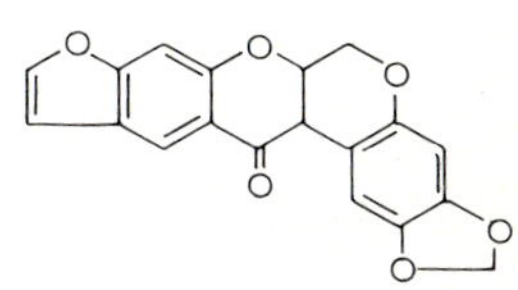

Dolineone

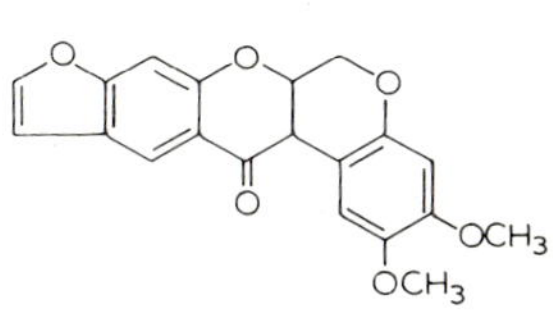


Erosone

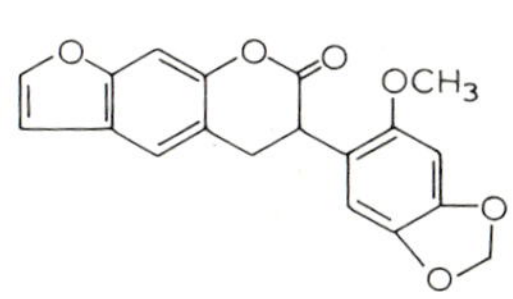


Pachyrrhizin

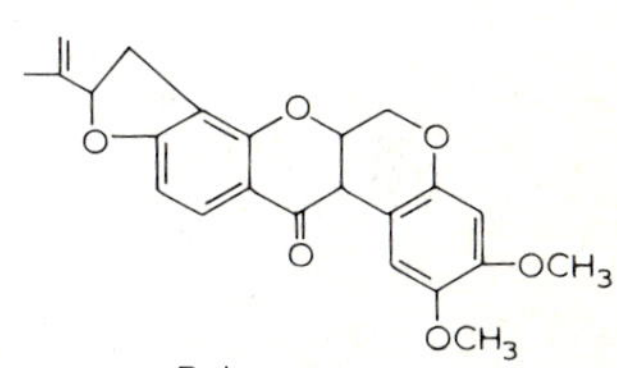


Rotenone

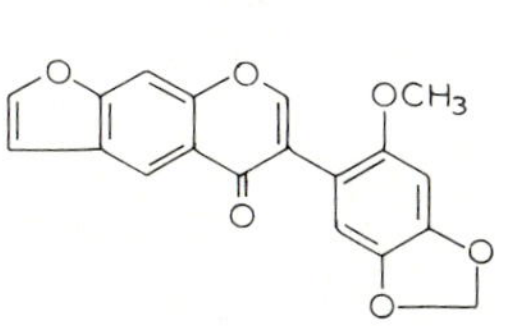


Dehydroneotenone

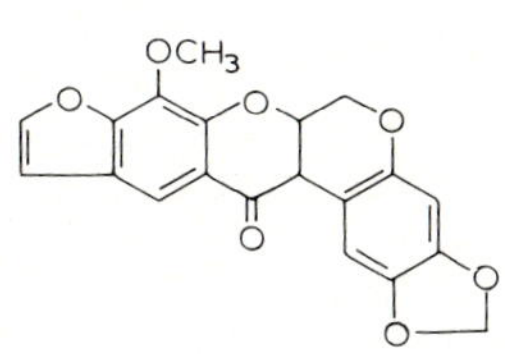


Pachyrrhizone

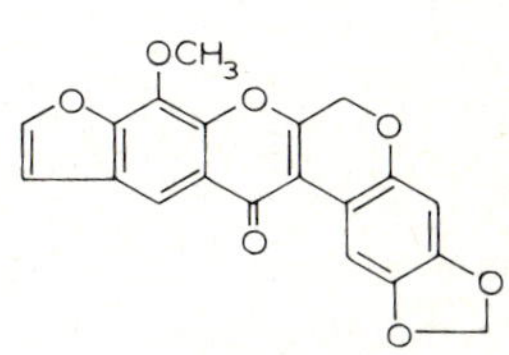


Dehydropachyrrhizone

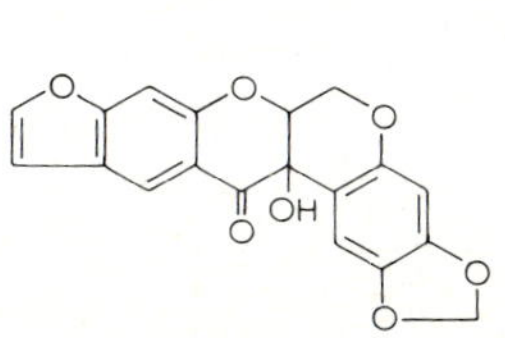


12a - Hydroxydolineone

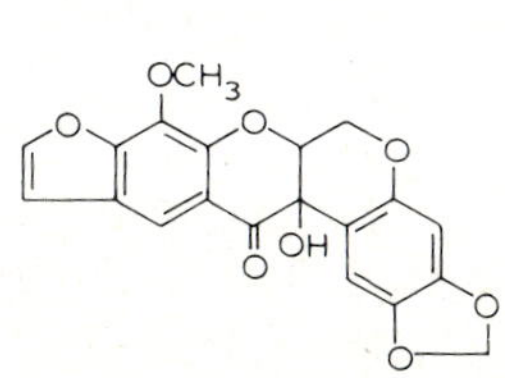


12a - Hydroxypachyrrhizone

*References p.924*

200 ml) in order to remove fatty material. The remaining solid (33.2 g) was dissolved in 250 ml of dry benzene and chromatographed over alumina (neutral, 1 kg) using successively benzene, benzene plus 10, 25 and 50% of chloroform and chloroform plus 10, 20 and 40% of acetone as eluting mixtures and collecting 400-ml fractions. These fractions were studied by thin-layer chromatography, and similar fractions were combined and the mixture separated by re-chromatography and fractional crystallization. Elution with benzene gave the following fractions: 1–4, erosnin; 5–21, neotenone, dolineone and erosone; 22–40, dolineone and pachyrrhizin; 41–50, pachyrrhizin and rotenone. Benzene–chloroform (9:1) yielded pachyrrhizin and rotenone, while pachyrrhizin, rotenone and dehydroneotenone were isolated from benzene–chloroform (3:1). The residue obtained from benzene–chloroform (1:1) was combined with fraction B and this mixture was separated into dehydroneotenone and pachyrrhizone. The chloroform and chloroform–acetone fractions yielded a brownish residue (2.15 g). This residue was dissolved in benzene (10 ml) and chromatographed on silica gel (75 g) using light petroleum (b.p. 60–80°C) with increasing amounts of benzene as eluent. Three compounds were isolated: dehydropachyrrhizone, 12*a*-hydroxydolineone and 12*a*-hydroxypachyrrhizone.

Wojtowicz and Diliberto have described a very useful application of column chromatography to furan-based drugs. They developed the method for the separation of furazolidone and nifuroxime in suppositories on magnesium silicate. The accuracy of the proposed method was based on the results of 10 replicate analyses of synthetic mixtures; more than 97% of each ingredient was recovered.

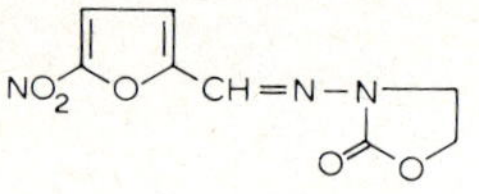


Furazolidone

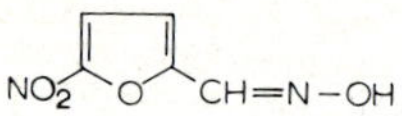


Nifuroxime

The separation was carried out in a 250 × 10 mm column, packed with Florisil (60–100 mesh), attached to a UV–visible range spectrophotometer. Standard solutions were prepared from reference standards: furazolidone (5 μg/ml) was dissolved in acetic acid–dimethylformamide solution (10 ml of glacial acetic acid dissolved in 200 ml of formamide) and nifuroxime (7.5 μg/ml) was dissolved in methanol–chloroform (5:1). Diluted samples of standards were transferred on to the column (5 ml of each compound) by pipette. The top of the column was rinsed with chloroform and elution was carried out with chloroform at a rate of 1–2 drops/sec. This eluent was followed by acetic acid–dimethylformamide and 5 ml of eluate were collected. Then elution was continued with the same mobile phase until 48 ml were collected; the exact volume of eluting solvents used was not crucial. The elution patterns indicated that 100% of the nifuroxime present was recovered from the column in the first 10 ml of chloroform and 100% of furazolidone in the first 25 ml of acetic acid–dimethylformamide solution. For quantification, measurement of the optical density at 338 nm (for nifuroxime with 1:9 methanol–chloroform as reference standard) or at 370 nm (for furazolidone with acetic acid–dimethylformamide as reference standard) appears to be the most advantageous procedure.

## PORPHYRINS AND RELATED COMPOUNDS

The chromatography of porphyrins has recently been reviewed by Falk. The nomenclature is preferably based on the following numbering of the porphyrin nucleus, so that each compound is defined only by the substituents on each position:

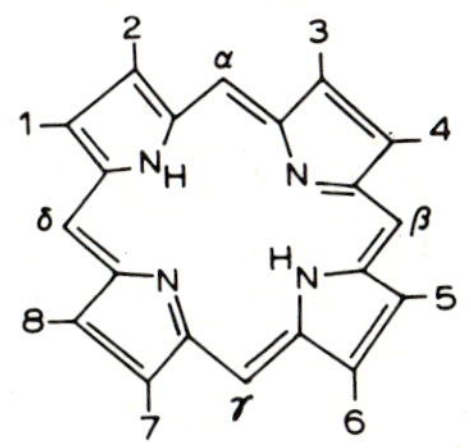

Brockmann *et al.* separated seven substituted porphyrins (A–G), using a 500-fold excess of stationary phase. The method of separation and the conditions and results are summarized in Fig. 40.2 and Table 40.4.

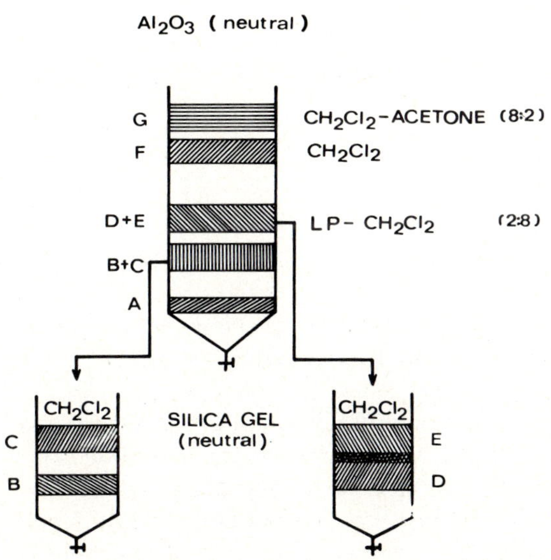


Fig. 40.2. Separation of a complex mixture of porphyrins (A–G) (Brockmann *et al.*). LP = light petroleum (b.p. 40–60°C).

Baum and Ellsworth separated a mixture containing protoporphyrin dimethyl ester and its magnesium and zinc haems, using a 22 × 2 cm column of sucrose (Domini 10X containing 3% (w/w) of corn starch) and elution with light petroleum (b.p. 35–60°C)–benzene (3:1). Wolf *et al.* described the separation of compounds with different configurations in the side-chain, using silica gel and the solvent system indicated overleaf.

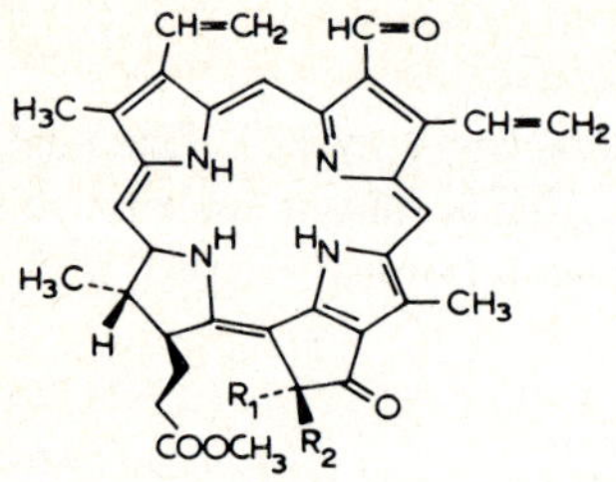


| $R_1$ | $R_2$ | Eluent |
|---|---|---|
| $COOCH_3$<br>$OCH_3$ | $OCH_3$<br>$COOCH_3$ | $CCl_4$ |
| $COOCH_3$<br>$OCOCH_3$ | $OCOCH_3$<br>$COOCH_3$ | $CCl_4$–acetone (96:4) |

TABLE 40.4
IDENTIFICATION OF SUBSTITUTED PORPHYRINS (FOR SEPARATION SEE FIG. 40.2) (BROCKMANN *et al.*)

P = $CH_2CH_2COOCH_3$

| Position | Porphyrin | | | | | | |
|---|---|---|---|---|---|---|---|
| | A | B | C | D | E | F | G |
| 1 | $CH_3$ | $CH_3$ | $CH_3$ | $CH_3$ | $CH_3$ | $CH_3$ | $CH_3$ |
| 2 | H | H | H | H | CHO | CHO | CHO |
| 3 | $CH_3$ | $CH_3$ | $CH_3$ | $CH_3$ | $CH_3$ | $CH_3$ | $CH_3$ |
| 4 | H | H | H | CHO | H | H | CHO |
| 5 | $CH_3$ | $CH_3$ | $CH_3$ | $CH_3$ | $CH_3$ | $CH_3$ | $CH_3$ |
| 6 | P | P | P | P | P | P | P |
| 7 | P | P | P | P | P | P | P |
| 8 | $CH_3$ | $CH_3$<br>α-CHO | $CH_3$<br>β-CHO | $CH_3$ | $CH_3$ | $CH_3$ | $CH_3$<br>β-CHO |

Ion-exchange chromatography (mainly on Sephadex dextran gels) is a very convenient method for the chromatographic separation of porphyrins and haems, and some examples of these applications are given below.

Porphyrins can be separated on columns of Sephadex G-25 gel using sodium borate buffer of pH 8.6. Alteration of the molarity of the buffer influences the separation in a predictable manner; for most applications 0.01 *M* is optimal, but for special separations other appropriate concentrations can be used. The recovery is quantitative and the technique is suitable for preparative purposes. These methods have already been applied successfully to the separation of complex mixtures of porphyrins derived from natural sources. Separation techniques employing Sephadex gels and an automatic fraction collecting device have the advantages of requiring minimum attention, having high flow-rates and being easily reproducible. The amount of material handled is limited virtually only by the size of the column (Rimington and Belcher).

Porphyrins can also be separated by extraction chromatography on a tri-*n*-butyl phosphate column by elution with a pH gradient. The compounds are eluted at discrete pH values according to the number of carboxyl groups in the side-chain. A very complex mixture isolated from urine could be resolved by a combination of the above method with thin-layer chromatography and absorption spectrophotometry (Mundschenk, 1968a, b, 1969). Uroporphyrins I and III were separated by this technique (Mundschenk, 1968a, b, 1969). Uroporphyrin I: (1 = 3 = 5 = $CH_2$ COOH, 2 = 4 = 6 = 7 = 8 = $CH_2$ $CH_2$ COOH); uroporphyrin III: (1 = 3 = 5 = 8 = $CH_2$ COOH, 2 = 4 = 6 = 7 = $CH_2$ $CH_2$ COOH) (The numbers refer to the formula on p. 917).

## INDOLES

For the preliminary fractionation of naturally occurring compounds, separation into acidic, basic, neutral and amphoteric groups is one of the most useful methods. Among heterocyclic compounds, ion-exchange chromatography has been developed for the separation of indoles (Raj and Hutzinger). A mixture of indoles is passed through a Dowex 50 ($(C_2H_5)_3N^+$) resin column, which retains the basic and amphoteric indoles, and the effluent is passed through a Sephadex A-25 ($CH_3COO^-$) column, where the acidic indoles are retained and the neutral fraction is obtained in the effluent. Basic and amphoteric indoles could be eluted with triethylamine solution and further separated on a Dowex 1 ($HCOO^-$) column. Acidic indoles were eluted from the Sephadex A-25 column with acetic acid or ammonium acetate.

The application of gradient elution chromatography to the separation of acidic indoles (converted into esters using diazomethane) was evaluated by Grunwald *et al.* Generally, this type of chromatography, as thin-layer chromatography, (*n*-hexane–*n*-butanol, 3:1) is a useful pre-purification method that gives highly purified samples which could be examined directly by gas–liquid chromatography. On the other hand, Änggård *et al.* described an excellent separation of acidic and neutral compounds derived from indoleamine on Sephadex LH-20 in 1,2-dichloroethane–methanol (various ratios). The following method was found to be useful in *in vitro* and *in vivo* studies on the metabolism, and also for the isolation of metabolites.

The corresponding acids or acidic extracts are esterified by the use of freshly prepared diazomethane. The sample is dissolved in a small amount of ethyl acetate and sufficient methanol added, and, after the addition of excess of diazomethane in cold diethyl ether (–20°C), the reagent is immediately removed with a stream of nitrogen so as to prevent the methylation of carboxy groups. In the preparation of the column, Sephadex is allowed to swell in methanol and is then refluxed three times for several hours with a large excess of methanol. After drying, a suitable amount is equilibrated with the solvents (1,2-dichloroethane–methanol, 9:1, 8:2 or 7:3) for a minimum of 2 h in a filtration flask. The gel is de-aerated by the brief application of a 15 mm Hg vacuum and then immediately used for packing the column. A glass container serving as a reservoir for the solvent is fitted to the top of the column (750 × 10 mm), while the bottom of the column has a narrow stopcock. The gel is poured into the column on a pad of a tightly packed glass-wool and allowed to settle under gravity. A small circular plug of porous PTFE is placed on the top of the gel surface so as to prevent disturbances to the fragile gel surface. The columns are conditioned by running the solvent mixture through the column for 6–12 h. The samples are applied in the smallest possible volume of the solvent mixture and allowed to sink into the gel, and the remaining sample solution is rinsed from the PTFE disc with several 0.5-ml volumes of solvent mixture. The column and top reservoir are filled with about 300 ml of solvent mixture. Under these conditions, the bed volumes of the columns are about 45 ml, and the flow-rate is about 0.15 ml/min · cm$^2$; fractions of about 0.8 ml are collected. In Fig. 40.3, the chromatography of 5-hydroxytryptophol and the methyl esters of 5-hydroxy-3-indoleacetic and indoleacetic acids is demonstrated.

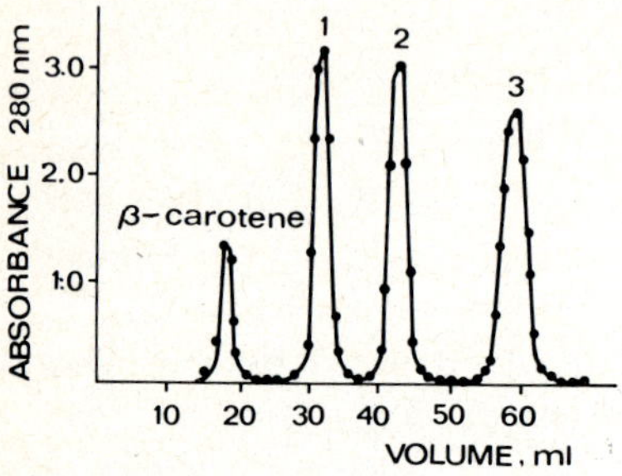


Fig. 40.3. Chromatography of 5-hydroxytryptophol (3) and the methyl esters of 5-hydroxy-3-indoleacetic acid (2) and indoleacetic acid (1) on Sephadex LH-20 (Änggård *et al.*). Column: 750 × 10 mm. Flow-rate: 0.15 ml/min · cm$^2$. Solvent system: 1,2-dichloroethane–methanol (7:3). β-Carotene was used as an internal standard.

## PYRIDINE AND RELATED COMPOUNDS

Brook and Robertson achieved an excellent separation of a mixture of pyridine derivatives by ion-exchange chromatography. This method has a high efficiency for pyridines and related heterocyclic systems. A Sephadex G-10 column was used, with distilled water and 0.1 *N* sodium chloride solution as eluents. About 1 mg of the synthetic

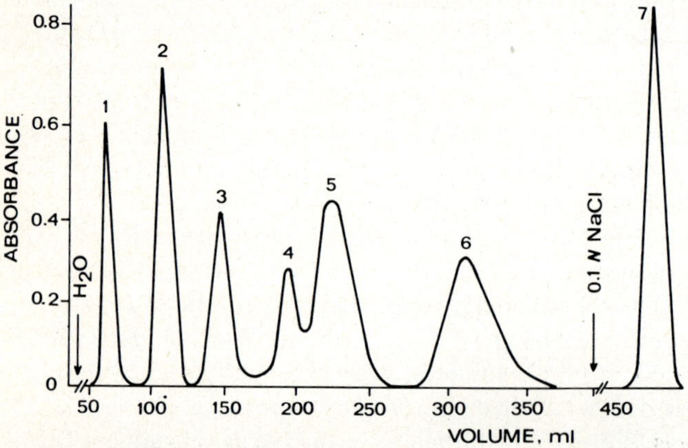


Fig. 40.4. Separation of a synthetic mixture of substituted pyridine compounds (Brook and Robertson). Column: 87 × 1.5 cm, Sephadex G-10. Peaks: 1 = pyridine-2-carboxylic acid; 2 = N-methylpyridone; 3 = C-(2-pyridyl)-N-methylaldonitrone; 4 = pyridine-2-carboxamide; 5 = 2-(methoxyiminomethyl) pyridine; 6 = pyridine-2-aldoxime; 7 = 2-hydroxyiminomethyl-N-methylpyridinium methanesulphonate.

mixture was taken for analysis, and successive 3-ml volumes of column effluent were scanned on an Optica CF4R spectrophotometer, the absorption maxima being plotted against the volume of eluent. The components were eluted in the order indicated in Fig. 40.4.

## POLYNUCLEAR AZA-HETEROCYCLICS AND COMPLEX MIXTURES OF HETEROCYCLIC COMPOUNDS

The application of high-speed liquid chromatography to heterocyclic compounds has begun only recently. In an important paper, Ray and Frei reported the separation of a number of polynuclear aza-heterocyclic compounds using a column packing that was prepared by the reaction of *p*-nitrophenyl isocyanate with the surface silanol groups of Corasil I. The method, owing to its sensitivity, has a practical application in air pollution problems.

The material used for the column packing has been called chromatographic brush, and can be characterized by the following method of preparation. Corasil I (Waters Ass., Framingham, Mass., U.S.A.) was treated with a 1:1 mixture of concentrated nitric and sulphuric acids, washed free from acid with distilled water and finally dried overnight at 110°C. To 18 g of Corasil I in 50 ml of benzene, 7.0 g of *p*-nitrophenyl isocyanate were added and the mixture was refluxed for 24 h in an atmosphere of nitrogen. Then the residue was decanted several times with benzene and extracted successively in a Soxhlet apparatus with benzene, *n*-hexane and methylene chloride. The extraction was finished when the ultraviolet absorbance of the extractant was zero. At this stage, the brush material was dried at 60°C and used for column preparation (1000 × 2.2 mm), using 1% of acetonitrile in *n*-hexane as the mobile phase at 1000 p.s.i. A typical example of a separation is presented in Fig. 40.5.

From the results obtained by Ray and Frei, a number of conclusions about factors that affect the separation under the specified conditions can be drawn. The elution order of compounds can be explained on the basis of the formation of a donor–acceptor complex and therefore also on the basis of electronic and steric affects of the molecules. The brush material formed by the reaction of *p*-nitrophenyl isocyanate with the silanol groups of Corasil I obviously behaves as a good charge acceptor, while the aza-heterocyclic compounds should serve as good donors owing to the presence of non-bonded electrons that are available for coordination. However, according to Ray and Frei, this expected behaviour is affected by the lone pairs of electrons on the nitrogen atoms, which are inaccessible owing to the stereochemical arrangement of the molecule.

During separation, the largest molecule is eluted first as the nitrogen atoms are the most sterically hindered, thus preventing effective complex formation. Identical steric hindrance is apparent in the benzo[*h*]quinoline molecule, which is eluted immediately after dibenzo-[*a,c*]phenazine. In the absence of steric hindrance, relatively strong donor–acceptor complexes are formed and the retention time is high. Low retentions, such as that of phenazine, can be attributed to a lower basicity (compared with acridine).

The separation of complex mixtures of nitrogen-containing heterocyclic compounds has been reported by Popl *et al.*, who measured the adsorption energies of three acidic

*References p.924*

TABLE 40.5
EQUIVALENT RETENTION VOLUMES OF NITROGEN-CONTAINING HETEROCYCLIC COMPOUNDS ON ALUMINA

| Compound | Equivalent retention volume | | | | | |
|---|---|---|---|---|---|---|
| | Acidic alumina | | Neutral alumina | | Basic alumina | |
| | *n*-Pentane + 10% $CH_2Cl_2$ | *n*-Pentane + 10% $(CH_3CH_2)_2O$ | *n*-Pentane + 10% $CH_2Cl_2$ | *n*-Pentane + 10% $(CH_3CH_2)_2O$ | *n*-Pentane + 10% $CH_2Cl_2$ | *n*-Pentane + 10% $(CH_3CH_2)_2O$ |
| Pyrrole | 2.70 | 1.32 | 3.36 | 1.78 | 8.57 | 3.38 |
| Indole | 4.78 | 2.59 | 6.65 | 3.55 | 14.43 | 6.81 |
| Carbazole | 6.86 | 2.82 | 8.18 | 3.68 | 17.40 | 9.40 |
| Pyridine | 23.2 | 4.60 | 12.36 | 3.14 | 14.5 | 5.15 |
| Quinoline | 21.40 | 3.42 | 11.68 | 2.58 | 17.44 | 4.12 |
| Acridine | 12.1 | 3.61 | 8.0 | 2.86 | 10.46 | 5.00 |
| Isoquinoline | 39.0 | 7.80 | 22.64 | 4.85 | 28.93 | 8.15 |
| 7,8-Benzoquinoline | 4.62 | 0.88 | 3.37 | 0.77 | 4.76 | 1.60 |

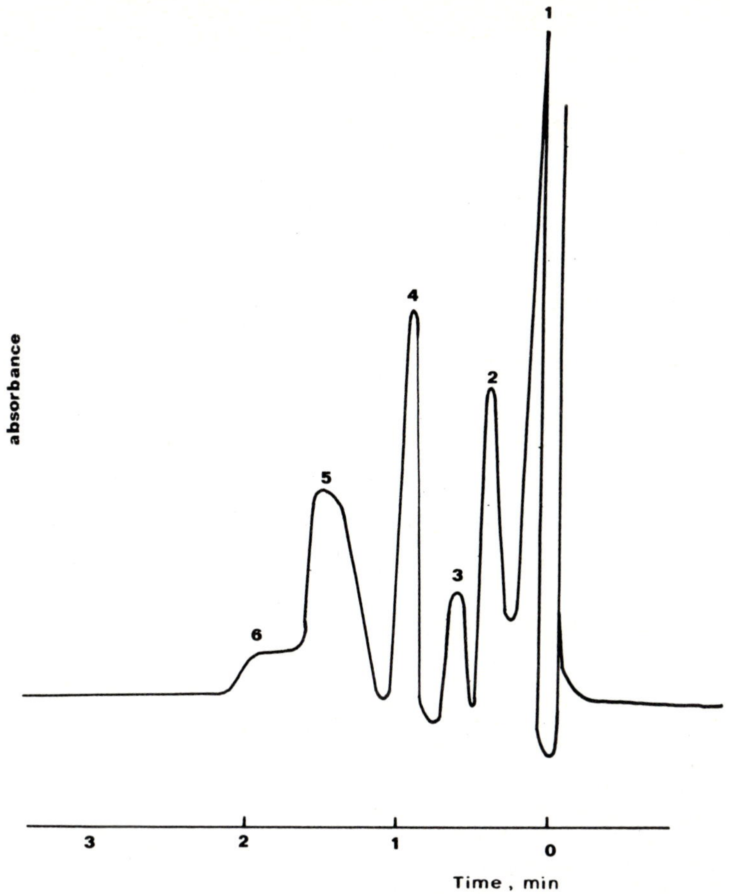


Fig. 40.5. Separation of polynuclear aza-aromatics on a chemically bonded stationary phase column (Ray and Frei). Column: 1 m × 22 mm. Support: *p*-nitrophenyl isocyanate bonded to Corasil I. Mobile phase: 1% of acetonitrile in *n*-hexane. Pressure: 1000 p.s.i. Peaks in order of elution: (1) dibenzo[*a,c*]phenazine; (2) benzo[*h*]quinoline; (3) phenazine; (4) acridine; (5) benzo[*c*]quinoline; (6) benzo[*f*]quinoline.

and five basic compounds of this type on alumina with surface pH values of 4.0, 7.5 and 10.0. Mixtures of 10% diethyl ether and 10% methylene chloride with *n*-pentane were used for elution. The calculated values of the adsorption energies of the basic nitrogen compounds depend on the adsorbent surface pH and show the advantage of using acidic alumina for the separation of nitrogen bases from aromatic hydrocarbons (see also Chapter 17). It was observed that the use of diethyl ether as eluent considerably decreased the effect of the acidic surface of the adsorbent on basic substances. Equivalent retention volumes of nitrogen-containing heterocyclic compounds on alumina are summarized in Table 40.5.

## REFERENCES

Änggård, E., Sjöquist, B. and Sjöström, R., *J. Chromatogr.*, 50 (1970) 251.

Asen, S., Stewart, R. N., Norris, K. H. and Masie, D. R., *Phytochemistry*, 9 (1970) 619.

Bamburg, J. R. and Strong, F. M., *Phytochemistry*, 8 (1969) 2405.

Baum, S. J. and Ellsworth, R. K., *J. Chromatogr.*, 47 (1970) 503.

Bendz, G., Martensson, O. and Terenius, L., *Acta Chem. Scand.*, 16 (1962) 1183

Berchtold, G. A. and Lumma, W. C., *J. Org. Chem.*, 34 (1969) 1566.

Bhalerao, U. T. and Thyagarajan, G., *Can. J. Chem.*, 46 (1968) 3367.

Blickenstaff, R. T. and Tao, I. Y. C., *Tetrahedron*, 24 (1968) 2495.

Bradfield, A. E., Penney, M. and Wright, W. B., *J. Chem. Soc.*, (1947) 32.

Brockmann, H., Jr., Bliesender, K.-M. and Inhoffen, H. H., *Justus Liebigs Ann. Chem.*, 718 (1968) 148.

Brook, A. J. W. and Robertson, R. K., *Chem. Ind. (London)*, (1967) 2110.

Brown, B. R., Betts, M. J. and Shaw, M. R., *J. Chem. Soc., C*, (1969) 1178.

Byrne, E. and Shannon, P. V. R., *J. Chem. Soc., C*, (1969) 1540.

Cardillo, G., Cricchio, R. and Merlini, L., *Tetrahedron*, 24 (1968) 4825.

Carelli, V., Liquori, A. M. and Mele, A., *Nature (London)*, 176 (1955) 70.

Chandler, B. V. and Harper, K. A., *Aust. J. Chem.*, 15 (1962) 114.

Chari, V. M., Aurnhammer, G. and Wagner, H., *Tetrahedron Lett.*, (1970) 3079.

Clark-Lewis, J. W. and Dainis, I., *Aust. J. Chem.*, 21 (1968) 425.

Crombie, L., Green, C. L., Tuck, B. and Whiting, D. A., *J. Chem. Soc., C*, (1968) 2625.

Davídek, J., *Nature (London)*, 189 (1961) 487.

Dean, F. M., *Naturally Occurring Oxygen Ring Compounds*, Butterworths, London, 1963.

De Barros Correa, D., Fonseca E. Silva, L. G., Gottlieb, O. R. and Goncalves, S. J., *Phytochemistry*, 9 (1970) 447.

Dekazos, E. D., *J. Food Sci.*, 35 (1970) 242.

Demetriou, J. A., Macias, F. M., McArthur, M. J. and Battie, J. M., *J. Chromatogr.*, 34 (1968) 342.

Endo, T., *Nature (London)*, 182 (1952) 801.

Falk, J. E., *J. Chromatogr.*, 5 (1961) 277.

Forsyth, W. G. C., *Biochem. J.*, 51 (1952) 511.

Fukui, K., Nakayama, M. and Harano, T., *Bull. Chem. Soc. Jap.*, 42 (1969) 1693.

Gage, T. B., Morris, Q. L., Detty, W. E. and Wender, S. H., *Science*, 113 (1951) 522.

Garber, E. D., Redding, W. F. and Chorney, W., *Nature (London)*, 193 (1962) 801.

Glencross, R. G., Festenstein, G. N. and King, H. G. C., *J. Sci. Food Agr.*, 23 (1972) 371.

Gombkötö, G., *Kertesz. Szolesz. Foiskola Kozlem.*, 31 (1967) 121.

Gottlieb, O. R., Taviera Magalhaes, M., Ottoni Da Silva Pereira, M., Lins Mesquita, A. A., De Barros Correa, D. and De Oliveira, G. G., *Tetrahedron*, 24 (1968) 1601.

Govindachari, T. R., Pai, B. R., Muthukumaraswamy, N., Rao, U. R. and Rao, N. N., *Indian J. Chem.*, 6 (1968) 57.

Grunwald, C., Vendrell, M. and Stowe, B. B., *Anal. Biochem.*, 20 (1967) 484.

Hänsel, R., Barthel, L. and Rimpler, H., *Arch. Pharm. (Weinheim)*, 303 (1970) 860.

Holaday, C. E., *J. Amer. Oil Chem. Soc.*, 45 (1968) 680.

Hörhammer, L., Stich, L. and Wagner, H., *Arch. Pharm. (Weinheim)*, 294 (1961) 685.

Hörhammer, L., Wagner, H., Arndt, H.-G., Dirscherl, R. and Farkas, L., *Chem. Ber.*, 101 (1968a) 450.

Hörhammer, L., Wagner, H., Arndt, H.-G., Hitzler, G. and Farkas, L., *Chem. Ber.*, 101 (1968b) 1183.

Hörhammer, L., Wagner, H., Hitzler, G., Farkas, L., Wolfner, A. and Nógrádi, M., *Chem. Ber.*, 102 (1969) 792.

Hörhammer, L., Wagner, H., Krämer, H. and Farkas, L., *Chem. Ber.*, 100 (1967) 2301.

Hori, M., *Bull. Chem. Soc. Jap.*, 42 (1969) 2333.

Hrazdina, G., *J. Agr. Food Chem.*, 18 (1970) 243.

Ice, O. H. and Wender, S. H., *Anal. Chem*, 24 (1952) 1616.

Inouye, H., Aoki, Y., Wagner, H., Hörhammer, L., Aurnhammer, G. and Budweg, W., *Chem. Ber.*, 102 (1969) 5009.

Jain, A. C., Khanna, V. K. and Seshadri, T. R., *Tetrahedron*, 25 (1969a) 2787.
Jain, A. C., Khanna, V. K. and Seshadri, T. R., *Indian J. Chem.*, 7 (1969b) 1182.
Johnston, K. M., Stern, D. J. and Waiss, A. C., Jr., *J. Chromatogr.*, 33 (1968) 539.
Joshi, B. S., Kamat, V. N. and Viswanathan, N., *Phytochemistry*, 9 (1970) 881.
Karrer, P. and Strong, F. M., *Helv. Chim. Acta*, 19 (1936) 25.
Kasturi, T. R., Damedaran, K. M., Subrahmanyam, G., Brown, P. and Pettit, G. R., *Chem. Commun.*, (1968) 794.
Kawase, A., *Agr. Biol. Chem.*, 32 (1968) 1028.
Kis, Z., Closse, A., Sigg, H. P., Hraban, L. and Snatzke, G., *Helv. Chim. Acta*, 53 (1970) 1577.
Krishnamurti, M., Sambhy, Y. R. and Seshadri, T. R., *Tetrahedron*, 26 (1970) 3023.
Kutney, J. P., Warnock, W. D. C. and Gilvert, B., *Phytochemistry*, 9 (1970) 1877.
Levi, C. P., *J. Ass. Offic. Anal. Chem.*, 52 (1969) 1300.
Locksley, H. D. and Murray, I. G., *J. Chem. Soc., C*, (1969) 1567.
Mabry, T. J., Markham, K. R. and Thomas, M. B., *The Systematic Identification of Flavonoids,* Springer, Berlin, Heidelberg, New York, 1970.
Makarov, V. A., Shinkarenko, A. L., Litvinenko, V. I. and Kovalev, I. P., *Khim. Prir. Soedin.*, (1969) 345.
Maksyutina, N. P., *Khim. Prir. Soedin.*, (1970) 201.
Markham, K. R. and Mabry, T. J., *Phytochemistry*, 7 (1968) 791.
Molyneux, R. J., Waiss, A. C. and Haddon, W. F., *Tetrahedron*, 26 (1970) 1409.
Morsingh, F., *Tetrahedron*, 25 (1969) 355.
Mukerjee, S. K., Rao, M. M. and Seshadri, T. R., *Indian J. Chem.*, 6 (1968) 1.
Mundschenk, H., *J. Chromatogr.*, 38 (1968a) 106.
Mundschenk, H., *J. Chromatogr.*, 37 (1968b) 431.
Mundschenk, H., *J. Chromatogr.*, 40 (1969) 393.
Natarajan, S., Murti, V. V. S. and Seshadri, T. R., *Phytochemistry*, 9 (1970) 575.
Neu, R., *J. Chromatogr.*, 4 (1960) 489.
Nordby, H. E., Fisher, J. F. and Kew, T. J., *Phytochemistry*, 7 (1968) 1653.
Nordström, G. Q., *Acta Chem. Scand.*, 21 (1967) 2885.
Nozoe, S., Suzuki, K. T. and Okuda, S., *Tetrahedron Lett.*, (1968) 3643.
Ohta, N. and Yagishita, K., *Agr. Biol. Chem.*, 34 (1970) 900.
Paris, R. and Etchepare, S., *Ann. Pharm. Fr.*, 26 (1968) 51.
Pinkas, J., Lavie, D. and Chorin, M., *Phytochemistry*, 7 (1968) 169.
Pons, W. A., Jr., Cucullu, A. F., Franz, A. O., Jr. and Goldblatt, L. A., *J. Amer. Oil Chem. Soc.*, 45 (1968) 694.
Popl, M., Dolanský, V. and Mostecký, J., *J. Chromatogr.*, 74 (1972) 51.
Rahman, W. and Bhatmagar, S. P., *Aust. J. Chem.*, 21 (1968) 539.
Raj, R. K. and Hutzinger, O., *Anal. Biochem.*, 33 (1970) 43.
Ray, S. and Frei, R. W., *J. Chromatogr.*, 71 (1972) 451.
Rimington, C. and Belcher, R. V., *J. Chromatogr.*, 28 (1967) 112.
Robinson, T., *The Organic Constituents of Higher Plants*, Burgess Publishing Co., Minneapolis, 1967.
Rodricks, J. V., *J. Amer. Oil Chem. Soc.*, 46 (1969) 149.
Sakamoto, Y., *Agr. Biol. Chem.*, 34 (1970) 919.
Schneider, G., Unkrich, G. and Pfaender, P., *Arch. Pharm. (Weinheim)*, 301 (1968) 785.
Seshadri, T. R., in T. A. Geissman (Editor), *The Chemistry of Flavonoid Compounds*, Pergamon Press, New York, 1962.
Smith, R. M. and Luh, B. S., *J. Food Sci.*, 30 (1965) 995.
Somers, T. C., *Nature (London)*, 209 (1966) 368.
Somers, T. C., *Vitis*, 7 (1968) 303.
Spaeth, E. C. and Rosenblatt, D. H., *Anal. Chem.*, 22 (1950) 1321.
Steyn, M., *J. Ass. Offic. Anal. Chem.*, 53 (1970) 619.
Stout, G. H. and Balkenhol, W. J., *Tetrahedron*, 25 (1969) 1947.
Stout, G. H., Hickernell, G. K. and Sears, K. D., *J. Org. Chem.*, 33 (1968) 4191.
Stout, G. H., Reid, B. J. and Breck, G. D., *Phytochemistry*, 8 (1969) 2417.

Stubblefield, R. D., Shotwell, O. L. and Shannon, G. M., *J. Amer. Oil Chem. Soc.*, 45 (1968) 686.
Stubblefield, R. D., Shotwell, O. L., Shannon, G. M., Weisleder, D. and Rhowedded, W. K., *J. Agr. Food Chem.*, 18 (1970) 391.
Tamura, S., Chang, C. -F., Suzuki, A. and Kumai, S., *Agr. Biol. Chem.*, 33 (1969) 291.
Tannenbaum, S. W., Agarwal, S. C., Williams, T. and Pitcher, R. G., *Tetrahedron Lett.*, (1970) 2377.
Thomas, M. B. and Mabry, T. J., *Phytochemistry*, 7 (1968a) 787.
Thomas, M. B. and Mabry, T. J., *Tetrahedron*, 24 (1968b) 3675.
Tjukavkina, N. A., Benešová, V. and Herout, V., *Collect. Czech. Chem. Commun.*, 35 (1970) 1306.
Tomás, F., Mataix, J. J. and Carpena, O., *Rev. Agroquim. Tecnol. Aliment.*, 12 (1972) 263.
Ulubelen, A., Öztür, S. and Isildatici, S., *J. Pharm. Sci.*, 57 (1968) 1037.
Velasco, J., *J. Amer. Oil Chem. Soc.*, 46 (1969) 105.
Vorster, L. J., *Analyst (London)*, 94 (1969) 136.
Wagner, H., Aurnhammer, G., Hörhammer, L. and Farkas, L., *Chem. Ber.*, 102 (1969a) 2089.
Wagner, H., Aurnhammer, G., Hörhammer, L., Farkas, L. and Nógrádi, M., *Chem. Ber.*, 102 (1969b) 785.
Wagner, H., Hörhammer, L., Dirscherl, R., Farkas, L. and Nógrádi, K., *Chem. Ber.*, 101 (1968) 1186.
Weinges, K., Ebert, W., Huthwelker, D., Mattauch, H. and Perner, J., *Justus Liebigs Ann. Chem.*, 726 (1969) 114.
Weinges, K. and Huthwelker, D., *Justus Liebigs Ann. Chem.*, 731 (1970) 161.
Weinges, K., Kaltenhaeuser, W., Mar, H. D., Nader, E., Nader, F., Perner, J. and Seiler, D., *Justus Liebigs Ann. Chem.*, 711 (1968) 184.
Weinges, K., Marx, H. D. and Goritz, K., *Chem. Ber.*, 103 (1970a) 2336.
Weinges, K., Perner, J. and Marx, H. D., *Chem. Ber.*, 103 (1970b) 2344.
Weinges, K. and Wild, R., *Justus Liebigs Ann. Chem.*, 734 (1970) 46.
Wiley, M. and Waiss, A. C., Jr., *J. Amer. Oil Chem. Soc.*, 45 (1968) 870.
Wiseman, H. G., Jacobson, W. C. and Harmeyer, W. C., *J. Ass. Offic. Anal. Chem.*, 50 (1967) 982.
Wojtowicz, P. F. and Diliberto, J. J., *J. Pharm. Sci.*, 60 (1971) 1387.
Wolf, H., Richter, I. and Inhoffen, H. H., *Justus Liebigs Ann. Chem.*, 725 (1969) 177.
Woof, J. B. and Pierce, J. S., *J. Chromatogr.*, 28 (1967) 94.
Wrolstad, R. E. and Putnam, T. B., *J. Food Sci.*, 34 (1969) 154.
Wrolstad, R. E. and Struthers, B. J., *J. Chromatogr.*, 55 (1971) 405.

*Chapter 41*

# Organic sulphur compounds

J. CHURÁČEK

## CONTENTS



## INTRODUCTION

This chapter includes sulphur compounds of the sulphide, sulphoxide, sulphone and sulphonic acid types and their derivatives. All of these compounds generally contain a strongly polar functional group containing sulphur in various oxidation states, which enables their different properties to be used for selective sorption on various materials. Among adsorbents, silica gel is more convenient to use because on alumina such a strong adsorption of substances may take place that some of them, such as disulphonic acids, can be desorbed only with great difficulty. However, ion exchangers are most often used in this field and are suitable even for the separation of some neutral sulphur compounds. Their importance is greatest in the analysis of sulphonic acids and their salts.

The column chromatography of sulphur compounds, particularly sulphonic acids and sulphonamides, is less common than paper and thin-layer chromatography. At present, attempts are being made to automate some analytical control procedures in the production of dyestuff intermediates, *i.e.*, primarily aromatic sulphonic acids.

High-speed liquid chromatography has great potential in the separation of sulphonic acids. Already the use of highly efficient solid phases with controlled porosity of the Zipax type in combination with a suitable ion exchanger permits very rapid and usually highly efficient separations. For other sulphur compounds, column chromatography is employed predominantly for purification and preparative purposes; the good separations achieved with PC and TLC guarantee these methods wide application in the future.

## SULPHONIC ACIDS

Sulphonic acids are sorbed on ion-exchange resins both by ionic forces and by molecular sorption, which is sometimes very strong. In a study of the sorption of aromatic sulphonic acids (*p*-toluenesulphonic, $\beta$-naphthalenesulphonic, anthraquinonesulphonic and other acids) on weakly basic anion-exchange resins (Lewatit M-II; Lewatit MIH;

Amberlite IR-45), it was found that Van der Waals forces between the aromatic nuclei and the resinous structure of the exchanger are so strong that they even sometimes prevail over the coulombic attraction forces between oppositely charged ions (Narebska and Kostecka).

Molecular sorption of sulphonic acids on ion exchangers of the styrene type was utilized by Scoggins and Miller for the rapid analytical separation of alkyl- and arylmonosulphonic acids from disulphonic acids by column chromatography on Amberlite XAD-2 on a 25 × 1.25 cm column. Mineral acids and the more acidic disulphonic acids were eluted first with water or sodium chloride solution, and monosulphonic acids were eluted with methanol. In this manner the following mixtures were separated: tetradecanedisulphonic and tetradecanesulphonic acids; sulphuric acid, dodecanedisulphonic and dodecanesulphonic acids; 2-methylnaphthalenedisulphonic and 2-methylnaphthalenesulphonic acids; 1,5-naphthalenedisulphonic and 2-naphthalenesulphonic acids; 4,4′-biphenyldicarboxylic and 4-biphenylcarboxylic acids; 4,4′-biphenylsulphonic and 4-biphenylsulphonic acids. The separation of hexadecanedisulphonic and hexadecanesulphonic acids was unsuccessful.

Funasaka *et al.* (1961, 1962, 1963, 1968) studied the salting-out chromatography of aromatic sulphonic acids, which in acidified sodium chloride solution, are easily sorbed on the weakly acidic cation exchanger Amberlite CG-50. The distribution coefficients increase with increasing concentration of sodium chloride and with decreasing pH in the range 1.0–5.0. On a 225 × 11 mm column of this cation exchanger (200–400 mesh), a mixture of $\alpha$- and $\beta$-anthraquinonesulphonic acids was separated successfully using 0.1 *N* sodium chloride solution of pH 2.7 for elution. On a 270 × 17 mm column, a mixture of the following acids was separated into individual components with 4 *N* sodium chloride solution of pH 1.3 as eluent at 40°C: $\alpha$-anthraquinonesulphonic acid and 1,5-, 1,6-, 1,7- and 1,8-anthraquinonedisulphonic acids. The individual acids were determined in the eluate by UV spectrophotometry.

Mixtures of sulphates and alkane mono-, di- and polysulphonates were separated on columns of Bio-Rad AG 50W-X8 ($H^+$) cation exchanger and the dextran anion exchanger DEAE-Sephadex A-25 ($OH^-$) (Mutter). For the elution of the substances from these columns, aqueous ethanol solutions of various concentrations were used, and also ammonium hydrogen carbonate solutions in water and aqueous ethanol and propanol. The eluate was evaporated, the residue extracted with alcohol, and the acids were titrated with potassium hydroxide solution with Bromocresol red as indicator.

Fudano and Konishi described the separation and determination of linear alkylbenzene sulphonates (LAS) and alkylsulphates (AS) by salting-out chromatography on Amberlite CG-50 (100–200 mesh). The resin was washed with methanol in a Soxhlet extractor and then treated with 3 *N* sodium hydroxide solution and 3 *N* hydrochloric acid. After treatment with 30 successive volumes of 3 *N* sodium chloride solution followed by washing with five volumes of water, the resin was dried at 105°C for 5 h. The sodium chloride–methanol system was selected for elution on the basis of preliminary experiments. The effects of methanol and sodium chloride concentrations on the distribution coefficients of LAS and AS, using the $H^+$ form of the resin, are shown in Table 41.1. In the case of the resin in the purely $Na^+$ form, which was conditioned with 3 *N* sodium hydroxide solution instead of sodium chloride solution, the distribution

TABLE 41.1
EFFECT OF CONCENTRATION OF METHANOL AND SODIUM CHLORIDE ON DISTRIBUTION COEFFICIENTS OF LINEAR ALKYLBENZENE SULPHONATES AND ALKYLSULPHATES AT 40°C (FUDANO AND KONISHI)

| Concentration of methanol | Concentration of NaCl (*M*) | | | | | | | |
|---|---|---|---|---|---|---|---|---|
| | 0.5 | | 1.0 | | 1.5 | | 2.0 | |
| (%) | LAS | AS | LAS | AS | LAS | AS | LAS | AS |
| 30 | 98.7 | 34.8 | 187.7 | 65.7 | 257.9 | 91.3 | 269.6 | 119.9 |
| 35 | 65.5 | 24.7 | 78.0 | 39.1 | 159.5 | 52.0 | 131.4 | 83.2 |
| 40 | 34.2 | 14.6 | 70.6 | 22.8 | 121.4 | 25.7 | 185.2 | 44.8 |
| 45 | 25.2 | 7.8 | 29.5 | 11.4 | 35.3 | 14.4 | 53.0 | 24.6 |

coefficients of LAS and AS were much lower than the values in Table 41.1, probably because in the purely $Na^+$ form of the resin the electrical repulsion forces act between the solute and the carboxyl groups of the exchanger, thereby reducing the distribution coefficients of LAS and AS. On the other hand, as the dissociation of resin in the $H^+$ form is suppressed, the Van der Waals interaction between the solute and the resin is strong and the distribution coefficients are correspondingly large.

On the basis of the results in Table 41.1, 40% methanol–0.5 *M* sodium chloride solution was selected as the most suitable eluent. A 100 × 25 mm column and this eluent were used at 40°C. Although the AS were eluted completely, the LAS were mainly adsorbed on the resin and the elution was difficult; this effect is probably caused by Van der Waals interactions between the LAS molecules and the skeleton of the resin, which were stronger in this instance owing to the presence of an aromatic ring in the molecule of LAS whereas the AS contain only alkyl groups.

When the concentration of methanol in the eluent was increased, the elution volume of LAS approached that of AS and the separation of both was incomplete. By increasing the column temperature the same result was obtained. Therefore, after the AS were eluted, 40% aqueous methanol without sodium chloride was used for the elution of LAS.

The separation and determination of aromatic sulphonates by ion-exchange column chromatography was described in detail by Stehl. The use of a cross-linked polyalkylene-amine anion-exchange resin minimizes the usual absorption of these compounds. Using a mixed aqueous organic solvent system (water–acetonitrile–methanol) and lithium chloride gradients of 0–0.5 *M*, the elution of mono- and disulphonates and their resolution were satisfactory. By utilizing relatively small columns (500 × 2 mm) and detection by monochromated UV absorption, the sensitivity and the specificity for these materials can be high. Under these conditions, mono- and disulphonates are baseline resolved within 30–45 min. The separation of isomeric or homologous sulphonates (mono- or di-) was reported for benzenesulphonates, phenolsulphonates, biphenylsulphonates and diphenyl oxide sulphonates.

The results of the separation of some benzenesulphonic and naphthalenesulphonic acids are represented in Figs. 41.1–41.5.

The separation of the two aminobenzenesulphonates shown in Fig. 41.4 demonstrates the ability of this technique to resolve closely related homologues. The equivalent separation of the same two compounds at a constant eluent strength is shown in Fig. 41.3.

An indication of the resolving ability of this resin for isomers of aminobenzenesulphonates is shown in Fig. 41.5. Although the separation is not complete ($R$ = 0.30), this

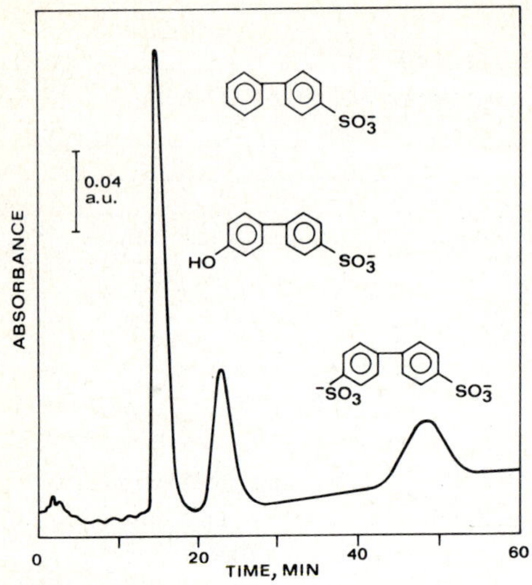


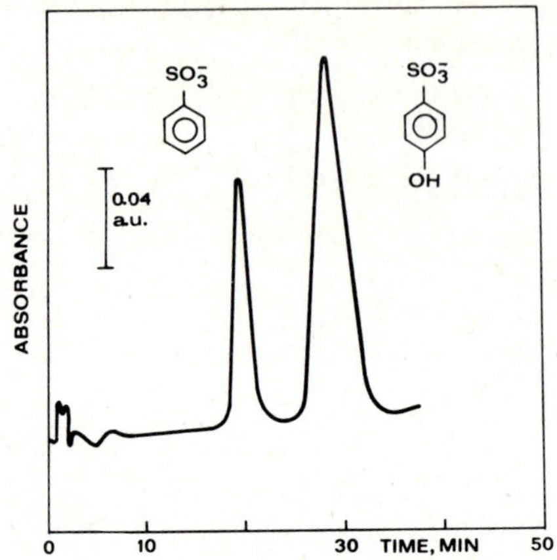


Fig. 41.1. Separation of biphenylsulphonates (Stehl). Column 500 × 2 mm. Ion exchanger: Bio-Rex 5 ($Cl^-$). Mobile phase: water–acetonitrile–methanol (1:1:1); gradient of 1 *M* lithium chloride into 40 ml of solvent. Flow-rate: 0.5 ml/min. Detection: spectrophotometric, λ = 250 nm. Sample: 10 μg of each component.

Fig. 41.2. Separation of sulphonic acids (Stehl). Column: 500 × 2 mm. Sorbent: Bio-Rex 5 ($Cl^-$). Mobile phase: water–acetonitrile–methanol (1:1:1); gradient of 1 *M* lithium chloride into 50 ml of solvent. Flow-rate: 1 ml/min. Detection: spectrophotometric, λ = 250 nm. Sample: 10 μg of each component.

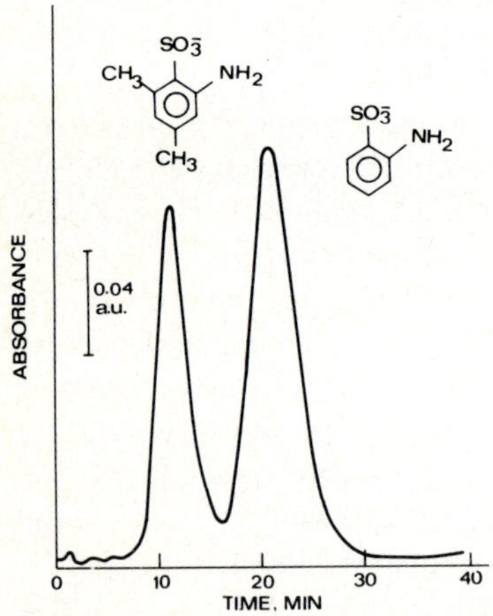


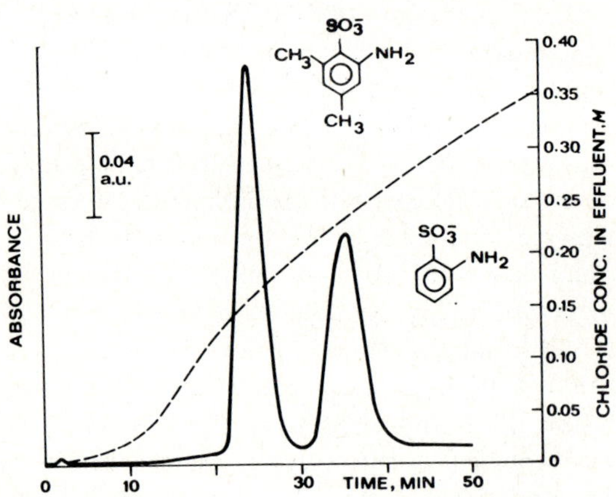


Fig. 41.3. Separation of aminobenzenesulphonates (Stehl). Column: 200 × 2 mm. Ion exchanger: Bio-Rex 5 ($Cl^-$). Mobile phase: 0.1 *M* lithium chloride in water–acetonitrile–methanol (1:1:1) (no gradient). Flow-rate: 1 ml/min. Detection: spectrophotometric, λ = 250 nm. Sample: 10 μg of each component.

Fig. 41.4. Separation of aminobenzenesulphonates (Stehl). Column: 500 × 2 mm. Ion exchanger: Bio-Rex 5 ($Cl^-$). Mobile phase: gradient of 0.66 *M* lithium chloride into 50 ml of water–acetonitrile–methanol (1:1:1). Flow-rate: 1 ml/min. Detection: spectrophotometric, λ = 250 nm. Sample: 10 μg of orthanilic acid + 20 μg of dimethylorthanilic acid.

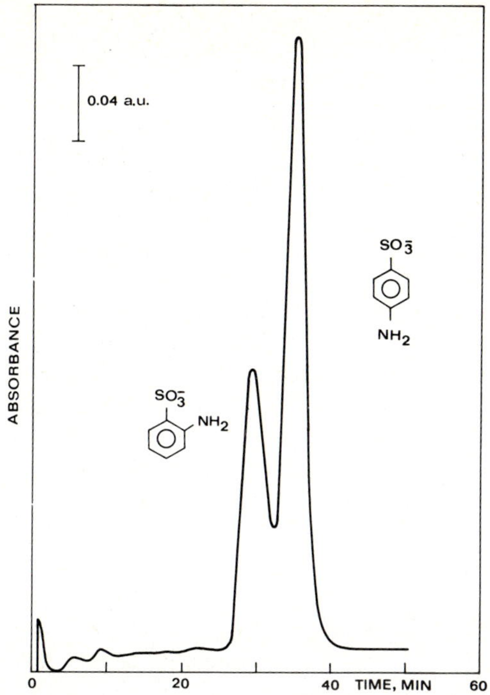


Fig. 41.5. Separation of aminobenzenesulphonate isomers (Stehl). Column: 500 × 2 mm. Ion exchanger: Bio-Rex 5 ($Cl^-$). Mobile phase: gradient of 1 *M* lithium chloride into 75 ml of water–acetonitrile–methanol (1:1:1). Flow-rate: 1 ml/min. Detection: spectrophotometric, λ = 250 nm. Sample: 20 μg.

again demonstrates the potential of the method for the separation of aromatic sulphonates.

A comparison of this method using the cross-linked polyalkyleneamine resin and the elution of the same materials from the conventional styrenedivinylbenzene matrix resin is shown in Table 41.2, where the elution of a mono- and disulphonated, chlorinated, alkylated diphenyl oxide species (Dowex 6A1) in terms of the volume and concentration of eluent is given.

For the separation of orthanilic and sulphanilic acids from aqueous solutions, Spillane and Scott used an Amberlite IRA-400 ($Cl^-$) column eluted with hydrochloric acid of various concentrations (0.01–0.1 *N*).

The possibilities of using liquid ion exchangers for the separation of aromatic sulphonic acids have also been investigated (Oi *et al.*). On columns packed with a support

TABLE 41.2
COMPARISON OF BIO-REX 5 ($Cl^-$) AND DOWEX 1-X2 ($Cl^-$) (STEHL)

| Component | Retention volume (ml) | | LiCl molarity | |
|---|---|---|---|---|
| | Bio-Rex 5 | Dowex 1-X2 | Bio-Rex 5 | Dowex 1-X2 |
| Monosulphonate | 5 | 25 | 0.03 | 0.28 |
| Disulphonate | 15 | 70 | 0.16 | 0.58 |

(Chromosorb or PTFE) coated with a liquid anion exchanger (Alamine 336, a high-molecular-weight amine), some aromatic sulphonic acids were separated quantitatively by gradual elution beginning with 0.5 *M* hydrochloric acid and 1.0 *M* perchloric acid and continuing with 1.0 *M* hydrochloric acid. This method was found useful for the separation of aminosulphonic acids from other sulphonic acids. The following mixtures were separated successfully; (a) sulphanilic and 2-naphthol-3,6-disulphonic acids; (b) sulphanilic and 4,5-dihydroxy-*m*-benzenedisulphonic acids; (c) *o*-aminobenzenesulphonic and 6,7-dihydroxy-naphthalene sulphonic acids; and (d) *o*-aminobenzenesulphonic and 6,7-dihydroxynaphthalene-2-sulphonic acids, on a column with Chromosorb as support.

For the separation of binary mixtures of 2-naphthol-6-sulphonic and 2-naphthol-8-sulphonic acids, 2-naphthol-3,6-disulphonic and 2-naphthol-8-sulphonic acids, and 2-naphthol-3,6-disulphonic and 2-naphthol-8-sulphonic acids, 2 *M* hydrochloric acid was used for the elution of the first component, while the second component was eluted with 0.1 *M* perchloric acid (Fritz and Gillette).

It is probable that the industrial importance of these substances, especially sulphonates, will lead to the development of other types of specially treated sorbents, either silica gels or special types of ion-exchange resins that will have the specific properties necessary for complete separations of even isomeric compounds. Although most types of ion exchangers can be used for the separation of sulphonic acids, experiments were made to separate these substances on specially treated silica gel (Wirzing). The results obtained in the separation of 1- and 2-naphthalenesulphonic acids showed that a silica gel can be prepared that will permit the separation of these isomeric compounds. When silica gel is prepared, it is important to precipitate it from the solution correctly. It was found that naphthalenesulphonic acid added during the precipitation has a decisive effect on the properties of silica gel. The shapes of the elution curves indicate, however, that the desorption of the sulphonic acids is impaired, which is reflected in a strong tendency for tailing to occur.

## OTHER SULPHUR COMPOUNDS

Aliphatic sulphoxides are bound selectively to strongly acid polystyrene cation exchangers in the $H^+$ form, evidently in consequence of hydrogen bond formation between the sulphonic acid groups of the exchanger and the sulphonyl groups of the sulphoxides. Aromatic sulphoxides are not sorbed and can therefore be separated from aliphatic sulphoxides. This fact was utilized in the isolation and purification of crude aliphatic sulphoxides on Dowex 50. The sulphoxides were sorbed from benzene solution, and ethanol was found to be the most suitable solvent for desorption (Horák and Pecka). On the same cation exchanger, dimethyl sulphoxide was separated from a mixture of amino acids, peptides and similar compounds, and it was determined quantitatively (Krull and Friedman). Sulphoxides formed in petroleum fractions by air oxidation could be retained together with other stronger bases on Duolite C-10 cation exchanger. Weak bases were then eluted with *n*-pentane, benzene and methanol, while for the elution of stronger bases 10% isopropylamine in methanol had to be used. The methanolic fraction containing the major portion of the sulphoxides was analyzed by gas chromatography (Okuno *et al.*). Ishibashi *et al*. have shown that ligand exchange can be used for the chromatographic separation of mercaptans and sulphides. Mercaptans and dialkyl sulphides are strongly sorbed from toluene, methanol or *n*-hexane solutions on to the macroporous cation exchanger Amberlyst 15 ($Ag^+$ or $Cu^{2+}$), while on the macroporous anion exchanger Amberlyst A-27 ($OH^-$) they are bound only weakly. Mercaptans are not sorbed on the cation exchanger in the $Ag^+$ form, but diethyl sulphide is bound more strongly by the $Ag^+$ form than by the $Cu^{2+}$ form of the cation exchanger.

For the separation of diaminodiphenyl sulphones, microbore column chromatography was used, silica gel being used as the sorbent (Gordon and Peters). It was found that the separation achieved by thin-layer chromatography can evidently be adapted for the conditions of microbore column chromatography (Fig. 41.6).

A mixture of sulphamide and its condensation products can be separated on the strongly basic anion exchanger Dowex 1-X8 ($Cl^-$) by gradient elution with water and sodum chloride solutions (Masuda and Ito).

Some neutral or weakly basic sulphur compounds are sorbed on ion exchangers of the polystyrene type. In addition to molecular sorption, specific interactions may also

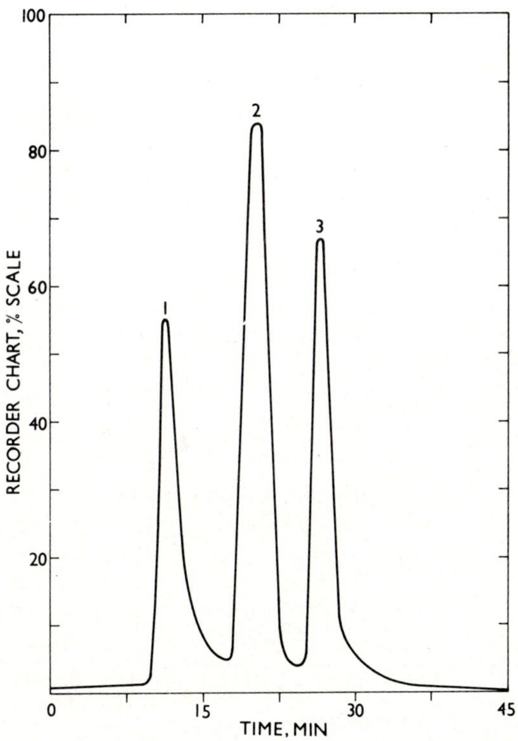


Fig. 41.6. Separation of sulphones (Gordon and Peters). Column: 300 × 2.8 mm. Sorbent: dry silica gel (silica TLC-7, Mallinckrodt, St. Louis, Mo., U.S.A.). Mobile phase: anhydrous ethyl acetate. Flow-rate: 0.25 ml/min, eluent pumped through at 100 lb./in.$^2$. Detection: spectrophotometric, $\lambda$ = 280 nm. 1 = 15 $\mu$g of 4,4'-diaminodiphenyl sulphone; 2 = 10 $\mu$g of 4-amino-4'-acetamidodiphenyl sulphone; 3 = 15 $\mu$g of 4,4'-diacetamidodiphenyl sulphone.

play a role. For example, hydrogen sulphide is strongly sorbed on anion exchangers in the $OH^-$ form, with which it evidently forms addition compounds of the xanthate type, on reaction with the amino groups of the resin and the $OH^-$ (Caplino and Davenkov, Dolby and Samuelson).

Some papers on applications in which liquid chromatography was used mainly for the purification and preparation of products are reviewed in Table 41.3.

TABLE 41.3
SURVEY OF DIFFERENT PROCEDURES APPLICABLE TO THE SEPARATION OF SULPHUR COMPOUNDS

| Compounds chromatographed | Sorbent | Mobile phase | References |
|---|---|---|---|
| Preparation of sulphides and their derivatives | Silica gel | Light petroleum, ethyl acetate and others | Russell and Ochromowycz |
| Benzyl chloromethyl sulphide – preparation | Silica gel | Benzene–chloroform (4:1) | Paquette *et al.* |
| Disulphides – preparation | Florisil | Cyclohexane–benzene (1:1), followed by benzene and finally chloroform | Stoffey |
| Methioxy ketoxime | Acidic alumina, activity I | Diethyl ether–methanol (150:1) Benzene–diethyl ether (25:1) | Autrey and Scullard |
| Purification and preparation of sulphenyl derivatives | Alumina with 1% $AgNO_3$ | Tetrachloromethane | Schmid and Heimola |
| Preparation of benzyl *p*-tolylsulphoxide and *p*-toluenesulphenate | Florisil | Chloroform, ethyl acetate–benzene (1:9) | Miller *et al.* |
| 2,4-Dinitrosulphenyl derivatives | Alumina | Benzene Benzene–light petroleum (1:1) | Nakano *et al.* |
| Methyl *p*-tolylsulphoxide | Silica gel | Dichloromethane–diethyl ether (9:1) or (1:1) | Wudl *et al.* |
| Diazosulphonate purification and preparation | Alumina | Chloroform | Hodson *et al.* |

## HIGH-SPEED LIQUID CHROMATOGRAPHY

Fig. 41.7 shows the separation of a synthetic mixture of aromatic sulphonic acids on a Zipax SAX column operated with perchloric acid of concentration 0.0025 *M*, later changed to 0.005 *M*. This change was made after 18 min in order to elute the last component more rapidly (Kirkland). A good separation of sulphonamide drugs was achieved on Corasil $C_{18}$ (reversed phase) (Waters Ass., Framingham, Mass., U.S.A.) in a 2 ft. × 2.3 mm column, with acetonitrile–water (5:95) as the mobile phase (Fig. 41.8).

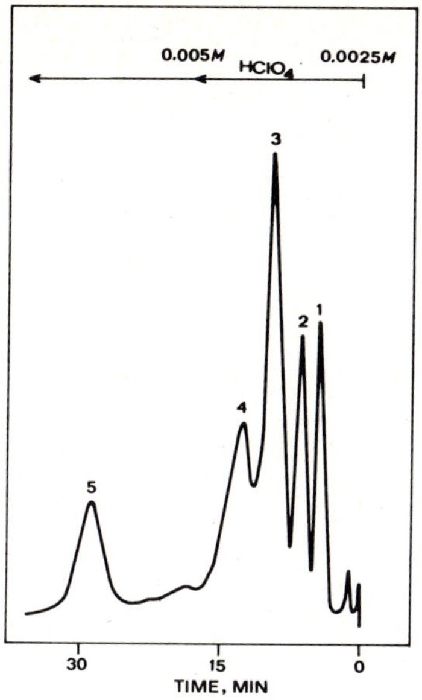


Fig. 41.7. Separation of aromatic sulphonic acids (Kirkland). Column: 1 m × 2.1 mm. Sorbent: Zipax SAX. Mobile phase: initial eluent 0.0025 *M* perchloric acid, followed by 0.005 *M* perchloric acid after 18 min. Flow-rate: 0.54 ml/min. Column pressure: 370 p.s.i.g. Temperature: 66°C. Detection: Spectrophotometric, 0.1 absorbance full-scale. 1 = Sodium benzenesulphonate; 2 = sodium *p*-toluenesulphonate; 3 = sodium 2,5-dimethylbenzenesulphonate; 4 = sodium *p*-chlorobenzenesulphonate; 5 = sodium naphthalene-β-sulphonate.

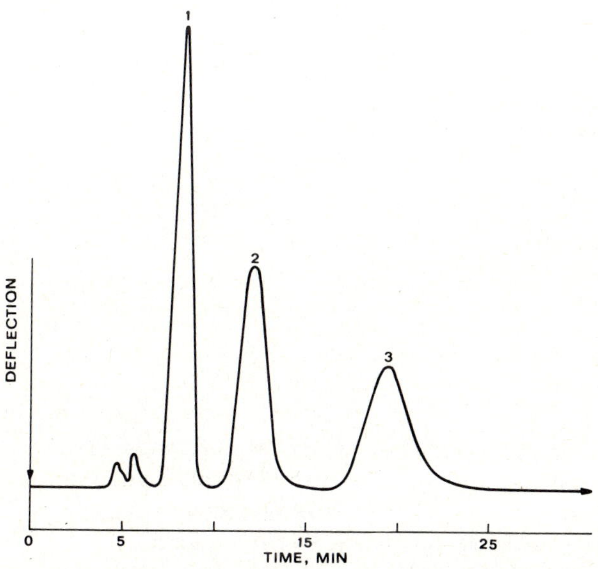


Fig. 41.8. Separation of sulphonamides (Waters Ass.). Column: 2 ft. × 2.3 mm. Sorbent: Corasil $C_{18}$ (reversed phase). Mobile phase: acetonitrile–water (5:95). Flow-rate: 0.25 ml/min. Detection: spectrophotometric. Sample: 10 μl containing 0.33 μg of each sulphonamide. 1 = Sulphadiazine; 2 = sulphamerazine; 3 = sulphamethazine.

*References p.937*

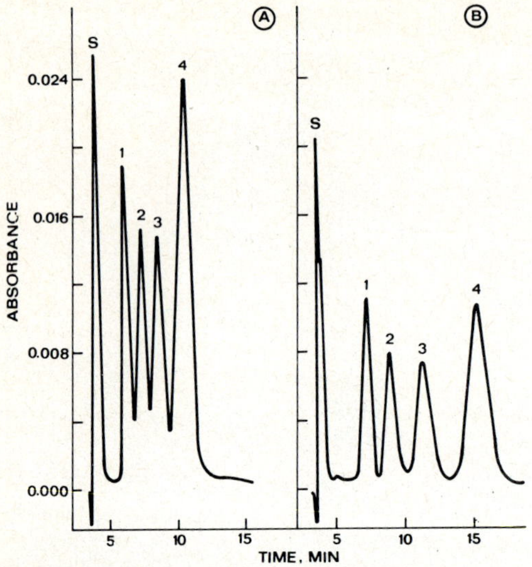


Fig. 41.9. Separation of sulphonylureas (Beyer). Column 1000 × 2.1 mm. Sorbent: 1% ethylene-propylene copolymer on Zipax. Mobile phase: (A) 0.01 *M* monobasic sodium citrate containing 15% of methanol; (B) 0.01 *M* monobasic sodium citrate containing 10% of methanol. Flow-rate: 0.36 ml/min. Detection: Spectrophotometric, λ = 254 nm. S = solvent; 1 = chlorpropamide, 1.50 μg; 2 = tolazamide, 2.50 μg; 3 = tolbutamide, 2.50 μg; 4 = acetohexamide, 0.25 μg. Structural formulae of separated substances:

Chlorpropamide

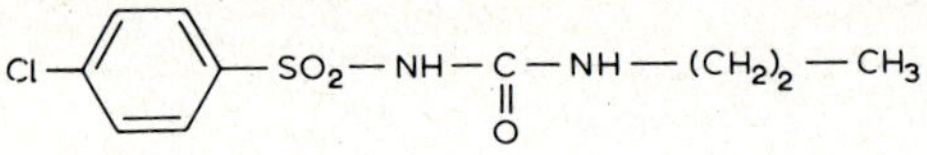

Tolazamide

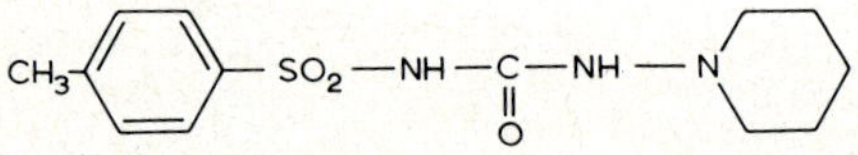

Tolbutamide

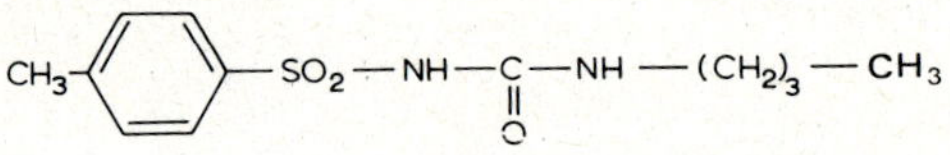

Acetohexamide

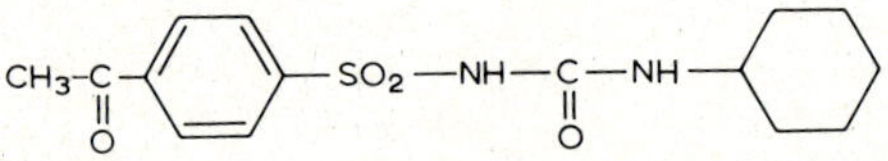

Beyer described a quantitative liquid chromatographic separation of sulphonylureas in pharmaceutical products. A 1000 × 2.1 mm column filled with Zipax and 1% ethylene–propylene copolymer was used. Two different mobile phases were used: 0.01 *M* sodium borate + 27.5% (v/v) methanol and citrate buffer consisting of 0.01 *M* monobasic sodium citrate + 10 or 15% (v/v) methanol. A typical separation is shown in Fig. 41.9.

It is to be expected that the use of these sorbents and their future development will permit the rapid and effective separation of the broad spectrum of intermediates containing sulpho groups, which are important in the dyestuffs industry.

## REFERENCES


Autrey, R. L. and Scullard, P. W., *J. Amer. Chem. Soc.*, 90 (1968) 4924.
Beyer, W. F., *Anal. Chem.*, 44 (1972) 1312.
Caplino, L. A. and Davenkov, A. B., *Zh. Prikl. Khim.*, 39 (1966) 608.
Dolby, L. and Samuelson O., *Acta Chem. Scand.*, 20 (1966) 892.
Fritz, J. S. and Gillette, R. K., *Anal. Chem.*, 40 (1968) 1777.
Fudano, S. and Konishi, K., *J. Chromatogr.*, 66 (1972) 153.
Funasaka, W., Kojima, T. and Fujimura, K., *Bunseki Kagaku (Jap. Anal.)*, 10 (1961) 374.
Funasaka, W., Kojima, T. and Fujimura, K., *Bunseki Kagaku (Jap. Anal.)*, 11 (1962) 936.
Funasaka, W., Kojima, T. and Fujimura, K., *Bunseki Kagaku (Jap. Anal.)*, 17 (1968) 48.
Funasaka, W., Kojima, T., Fujimura, K. and Kustrida, S., *Bunseki Kagaku (Jap. Anal.)*, 12 (1963) 1170.
Gordon, G. R. and Peters, J. H., *J. Chromatogr.*, 47 (1970) 269.
Hodson, D., Holt, G. and Wall, D. K., *J. Chem. Soc., C.*, (1968) 2201.
Horák, V. and Pecka, J., *J. Chromatogr.*, 14 (1964) 97.
Ishibashi, N., Kamata, S. and Matsuura, M., *Kogyo Kagaku Zasshi*, 70 (1967) 1036; *C.A.* 68 (1968) 16437n.
Kirkland, J. J., *Anal. Chem.*, 43 (1971) 37A.
Krull, L. H. and Friedman, M., *J. Chromatogr.*, 26 (1967) 336.
Masuda, E. and Ito, Y., *J. Chromatogr.*, 31 (1967) 650.
Miller, E. G., Rayner, D. R., Thomas, H. T. and Mislow, K., *J. Amer. Chem. Soc.*, 90 (1968) 4861.
Mutter, M., *Tenside*, 5 (1968) 138.
Nakano, T., Barton, D. H. R. and Sammes, P. G., *J. Chem. Soc., C.*, (1968) 322.
Narebska, A. and Kostecka, I., *Rocz. Chem.*, 39 (1965) 1305.
Oi, N., Miyazaki, K. and Shizaga, N., *Bunseki Kagaku (Jap. Anal.)*, 16 (1967) 607.
Okuno, I., Latham, D. R. and Haines, W. E., *Anal. Chem.*, 39 (1967) 1830.
Paquette, L. A., Wittenbrook, L. S. and Schreiber, K., *J. Org. Chem.*, 33 (1968) 1080.
Russell, G. A. and Ochromowycz, L. A., *J. Org. Chem.*, 35 (1970) 764.
Schmid, G. H. and Heimola, M., *J. Amer. Chem. Soc.*, 90 (1968) 3466.
Scoggins, M. W. and Miller, J. W., *Anal. Chem.*, 40 (1968) 1155.
Spillane, W. J. and Scott, F. L., *Lab. Pract.*, 17 (1968) 352; *Anal. Abstr.*, 17 (1969) 287.
Stehl, R. H., *Anal. Chem.*, 42 (1970) 1802.
Stoffey, D. G., *J. Org. Chem.*, 33 (1968) 1651.
Waters Ass., *Application High-lights*, No. 10, Waters Ass., Framingham, Mass.
Wirzing, G., *Chromatographia*, 3 (1970) 19.
Wudl, F., Lightner, D. A. and Cram, D. J., *J. Amer. Chem. Soc.*, 89 (1967) 4099.

*Chapter 42*

# Organic phosphorus compounds

J. ZABRANSKÝ

## CONTENTS



This chapter deals with the application of column chromatography to the separation of various types of organic compounds that contain a phosphorus atom and which were not discussed in the chapters on sugars, lipids, nucleotides and pesticides.

## APPLICATION OF COLUMN CHROMATOGRAPHY

Column chromatography is a useful method for separating organic phosphorus compounds from crude reaction mixtures when investigating their synthesis or reactions with other compounds. The choice of the technique depends on the nature of the compounds and the material to be extracted. In organic chemistry, liquid–solid chromatography is the most widely applied technique, while ion-exchange chromatography is preferably used in biological chemistry. No special operations are needed in order to separate organic phosphorus compounds by liquid–solid or ion-exchange chromatography, but these compounds have widely differing natures so that it is necessary to find the optimal conditions for the separation of each of the compounds being investigated. Some of the applications of column chromatography in the separation of various types of organic phosphorus compounds are reported in Table 42.1. The distribution coefficients of various inositol polyphosphates on different Sephadex gels are listed in Table 42.2 (Steward and Tate).

The analysis of the eluate is usually carried out by gas–liquid, thin-layer or paper chromatography and the structure of the separated compounds is established on the basis of elemental analysis and UV, IR, NMR or mass spectra.

TABLE 42.1
APPLICATIONS OF COLUMN CHROMATOGRAPHY IN THE SEPARATION OF ORGANIC PHOSPHORUS COMPOUNDS

| Technique | Compounds separated | Sorbent or ion exchanger | Eluent | Notes | References |
|---|---|---|---|---|---|
| LSC | S,S-Di-(N-alkylcarbamoylmethyl) tri- and tetrathiophosphates | Silica gel (KSK), 75–100 mesh, 120 g | Light petroleum–acetone mixtures | Separation from reaction mixture; column 3 cm O.D. | Mandelbaum *et al.* |
| LSC | Dialkylamino-O-alkyl-S-(N-alkylcarbamoylmethyl) dithiophosphates | Silica gel (KSK), 75–100 mesh, 120 g | Light petroleum–acetone mixtures | Separation from reaction mixture; column 3 cm O.D.; flow-rate 1.6 ml/min | Soifer *et al.* |
| LSC | Bis-*p*-nitrobenzyl-(2-palmitoyloxy-octadecyl) phosphate | Silicic acid | *n*-Hexane–diethyl ether (1:3) | Separation from reaction mixture | Parfenov *et al.* |
| LSC | Products of irradiation of dialkyl α-ketophosphonates | Silica gel, 100 mesh | 10% acetone in benzene with increasing amounts of acetone | Separation from reaction mixture after treatment with diazomethane | Ogata and Tomioka |
| LSC | Diphenyl ethylphosphonate | Silica gel, 100 mesh, 40 g | Benzene | Separation from reaction mixture; column 50 × 2.5 cm | Koketsu and Ishii |
| LSC | Dialkyl α-hydroxy-β-ethoxyethyl-phosphonates; diethyl β-hydroxy-γ-ethoxypropylphosphonate | Silicic acid, 50 g | Benzene (250 ml) Chloroform (400–500 ml) | Separation from reaction mixture | Griffin and Kundu |
| LSC | Menthyl phosphinates<br>Phospine oxides | Silica gel<br>Silica gel | Benzene<br>Benzene–chloroform (1:1) | Separation from reaction mixture | Korpiun *et al.* |
| LSC | Phosphatidic acid | Silicic acid, activated at 100°C, 5 g | Diethyl ether–benzene mixtures; diethyl ether; methanol–chloroform mixtures | Separation from a solubilized fraction of rat brain | Martensson and Kanfer |
| LSC | Substituted phosphinimines | Alumina | Chloroform | Separation from reaction mixture | Wiegraebe and Bock |

| | | | | | |
|---|---|---|---|---|---|
| LSC | Triphenylphosphines | Alumina | Benzene | Separation from reaction mixture | Hands and Mercer |
| | Triphenylphosphine oxides | Alumina | Diethyl ether | | |
| LSC | Hexaphenylcyclotriphosphazatriene | Alumina | Benzene | Separation from reaction mixture | Biddlestone and Shaw |
| IEC | Alkyl dihydrogen phosphates | Amberlite IR-120 ($H^+$) | 50% ethanol | Separation from reaction mixture; column 20 × 1.2 cm | Hata *et al.* |
| IEC | 3-Hydroxy-2-pyridylmethyl and 6-methyl-2-pyridylmethyl phosphates | Dowex 50W-X8 ($H^+$) | Mineral-free water | Separation from reaction mixture; column 34 × 3 cm; flow-rate 300 ml/h | Murakami *et al.* |
| IEC | Dialkyl esters of alkylphosphonic acids | AV-16 ($OH^-$) | Water (50 ml) | Separation from reaction mixture; flow-rate 0.5–0.7 ml/min | Beresnev and Vlasova |
| | Alkyl esters of alkylphosphonic acids | | 0.025 *N* KOH (100 ml) | | |
| | Alkylphosphonic acids | | 1 *N* KOH (50 ml) | | |
| IEC | Trimetaphosphinic acid | KV-2 ($H^+$), 0.4–1.0 mm | Water | Column 52 × 1 cm; flow-rate 2.5–3.0 ml/min | Nikolaev *et al.* |
| IEC | Phosphoryl–Tris | Dowex 1-X4 ($HCOO^-$) | 0–0.5 *N* formic acid gradient | Column 33 × 3.8 cm | Henderson *et al.* |
| | | Dowex 50W-X4 ($H^+$) | 0–0.03 *N* HCl gradient | Column 40 × 3.8 cm; biological application | |

TABLE 42.2
GEL CHROMATOGRAPHY OF INOSITOL POLYPHOSPHATES AND RELATED COMPOUNDS (STEWARD AND TATE)

| Packing: | | | | | | |
|---|---|---|---|---|---|---|
| Sephadex: grade | | G-15 | G-25 | G-50 | G-100 | G-200 |
| size ($\mu$m) | | 40–120 | 100–300 | 100–300 | 40–120 | 140–400 |
| Column parameters: | | | | | | |
| length (cm) | | 46 | 52 | 58 | 78 | 36 |
| volume of the gel column (ml) | | 300 | 500 | 710 | 770 | 1020 |
| void volume (ml) | | 120 | 210 | 270 | 260 | 290 |
| volume of the stationary phase (ml) | | 130 | 250 | 410 | 470 | 700 |
| Compound | Eluent (LiCl) concentration (*M*) | Distribution coefficient ($K_d \pm 0.05$) | | | | |
| Phosphitin | 0.01 | 0.00 | 0.00 | 0.00 | 0.03 | 0.21 |
| | 0.10 | | | 0.00 | | |
| | 0.20 | 0.00 | | | | |
| | 2.00 | | 0.00 | | | |
| Myoinositol hexaphosphate | 0.01 | 0.00 | 0.06 | 0.45 | 0.79 | 0.90 |
| | 0.10 | | | 0.61 | | |
| | 0.20 | 0.06 | 0.19 | | | |
| | 2.00 | | 0.30 | | | |
| Pyrophosphate | 0.01 | 0.04 | 0.31 | 0.81 | 0.97 | |
| | 0.10 | | | 0.83 | | |
| | 0.20 | 0.25 | | | | |
| | 2.00 | | 0.67 | | | |
| Adenosine triphosphate | 0.01 | 0.16 | 0.60 | 0.91 | | |
| | 0.10 | | | 0.93 | | |
| | 0.20 | 0.35 | | | | |
| | 2.00 | | 0.93 | | | |
| Orthophosphate | 0.01 | 0.26 | 0.56 | 0.91 | 0.97 | 0.98 |
| | 0.10 | | | 0.89 | | |
| | 0.20 | 0.40 | | | | |
| | 2.00 | | 0.77 | | | |

| | | | | | | |
|---|---|---|---|---|---|---|
| Fructose | 0.01 | 0.63 | 0.75 | 0.89 | 0.97 | 0.98 |
| | 0.10 | | | 0.93 | | |
| | 0.20 | 0.65 | | | | |
| | 2.00 | | 0.81 | | | |
| Adenosine monophosphate | 0.01 | 0.74 | 0.91 | 1.14 | 1.07 | |
| | 0.10 | | | 1.14 | | |
| | 0.20 | 1.00 | | | | |
| | 2.00 | | 1.30 | | | |
| Myoinositol tripyrophosphate | 0.01 | | 0.06 | 0.48 | | |
| | 0.10 | | | 0.61 | | |
| | 0.20 | | 0.19 | | | |
| Alkaline hydrolysis myoinositol pentaphosphate | 0.01 | | 0.06 | 0.56 | | |
| | 0.10 | | | 0.62 | | |
| | 0.20 | | 0.19 | | | |
| Chicken blood myoinositol pentaphosphate | 0.01 | | 0.06 | 0.54 | | |
| | 0.10 | | | 0.64 | | |
| | 0.20 | | 0.19 | | | |
| Myoinositol tetraphosphate | 0.01 | | 0.12 | 0.57 | | |
| | 0.10 | | | 0.68 | | |
| | 0.20 | | 0.25 | | | |
| Myoinositol triphosphate | 0.01 | | 0.19 | 0.64 | | |
| | 0.10 | | | 0.75 | | |
| | 0.20 | | 0.33 | | | |
| Myoinositol diphosphate | 0.01 | | 0.31 | 0.70 | | |
| | 0.10 | | | 0.79 | | |
| | 0.20 | | 0.44 | | | |
| Myoinositol monophosphate | 0.01 | | 0.44 | 0.86 | | |
| | 0.10 | | | 0.86 | | |
| | 0.20 | | 0.59 | | | |
| Glycerol phosphoryl myoinositol | 0.01 | | 0.53 | 0.97 | | |
| | 0.10 | | | 0.97 | | |
| | 0.20 | | 0.56 | | | |
| Myoinositol | 0.01 | | 0.75 | 0.83 | | |
| | 0.10 | | | 0.86 | | |
| | 0.20 | | 0.75 | | | |

*References p.944*

## REFERENCES

Beresnev, A. N. and Vlasova, T. E., *Zh. Prikl. Khim.*, 42 (1969) 410.
Biddlestone, M. and Shaw, R. A., *J. Chem. Soc., A*, (1969) 178.
Griffin, C. E. and Kundu, S. K., *J. Org. Chem.*, 34 (1969) 1532.
Hands, A. R. and Mercer, J. H., *J. Chem. Soc., C*, (1968) 1331.
Hata, T., Mushika, Y. and Mukaiyama, T., *J. Amer. Chem. Soc.*, 91 (1969) 4532.
Henderson, R. J., Jr., Hill, F. L. and Mills, G. C., *Arch. Biochem. Biophys.*, 139 (1970) 311.
Koketsu, J. and Ishii, Y., *Bull. Chem. Soc. Jap.*, 43 (1970) 2527.
Korpiun, O., Lewis, R. A., Chickos, J. and Mislow, K., *J. Amer. Chem. Soc.*, 90 (1968) 4842.
Mandelbaum, Y. A., Soifer, R. S., Melnikov, N. N. and Belova, L. A., *Zh. Obshch. Khim.*, 37 (1967) 2287.
Martensson, E. and Kanfer, J., *J. Biol. Chem.*, 243 (1968) 497.
Murakami, Y., Sunamoto, J., Sadamori, H., Kondo, H. and Takagi, M., *Bull. Chem. Soc. Jap.*, 43 (1970) 2518.
Nikolaev, A. F., Dreiman, N. A. and Zyryanova, T. A., *Zh. Obshch. Khim.*, 40 (1970) 937.
Ogata, Y. and Tomioka, H., *J. Org. Chem.*, 35 (1970) 596.
Parfenov, E. A., Serebrennikova, G. A. and Preobrazhenskii, N. A., *Zh. Obshch. Khim.*, 37 (1967) 2363.
Soifer, R. S., Mandelbaum, Y. A., Melnikov, N. N. and Belova, L. A., *Zh. Obshch. Khim.*, 37 (1967) 2291.
Steward, J. H. and Tate, M. E., *J. Chromatogr.*, 45 (1969) 400.
Wiegraebe, W. and Bock, H., *Chem. Ber.*, 101 (1968) 1414.

*Chapter 43*

# Boron compounds

S. HEŘMÁNEK

## CONTENTS



## GENERAL TECHNIQUES

The separation of rigid boron compounds is based on differences in the sizes and shapes of the molecules, in dipole moments, Brønsted or Lewis acidities and in the characters of the substituents. In some instances, the use of liquid–solid chromatography is limited by the reactivity of the compounds to be separated towards water, air or some solvents. However, a careful choice of chromatographic conditions can decrease this sensitivity to an acceptable level.

Up to now, classical column chromatography on silica gel or aluminium oxide has been the most widely used technique, but recently dry column chromatography has been introduced into borane chemistry. The main advantages of the latter technique are good separations, rapidity, negligible destruction of the solid-phase column by hydrogen (due to undesirable hydrolysis) and, in particular, the easy selection of conditions by means of thin-layer chromatography. Very good results were obtained not only in separations of coloured compounds (sandwich-type metallocarboranes), but also in separations of UV-detectable compounds (see below), chromatographed in a quartz column or in a column with polyethylene or polypropylene walls. Attempts to use gel permeation chromatography (Čoupek and Heřmánek) and especially high-efficiency liquid chromatography (Krejčí and Heřmánek) are promising, but more detailed studies are necessary. It would appear that other kinds of column chromatography have not been used.

**Stationary phase**

Water- and air-stable borane compounds (most ligand derivatives, *closo*-carboranes and metallocarboranes) are chromatographed on silica gel or aluminium oxide. With hydride compounds, which are hydrolyzed by water, it is recommended that an adsorbent of activity I should be used so as to prevent evolution of hydrogen and destruction of the adsorbent column.

Acidic *nido*-boranes and *nido*-carboranes can be separated on silica gel activated *in vacuo* (100–150°C) and deactivated after cooling by the injection of an eluent containing 3–7% of acetic or trifluoroacetic acid into the evacuated flask (Stuchlík *et al.*).

Some of the air-sensitive boron compounds can be separated in an inert atmosphere on an activated and gas-free stationary phase. An adsorbent is activated *in vacuo*, the flask filled with nitrogen, the eluent is added, a small part of it is evaporated *in vacuo* at ambient temperature (so as to remove the last traces of oxygen from the active surface) and the flask is filled with nitrogen. The suspension is then transferred through a connecting tube to the chromatographic column, which has previously been flushed with nitrogen. All of the succeeding operations, *i.e.*, the addition of the mixture to be chromatographed, the stepwise addition of the gas-free eluent and the collection of fractions, are performed in an inert atmosphere in a simple apparatus (Fig. 43.1).

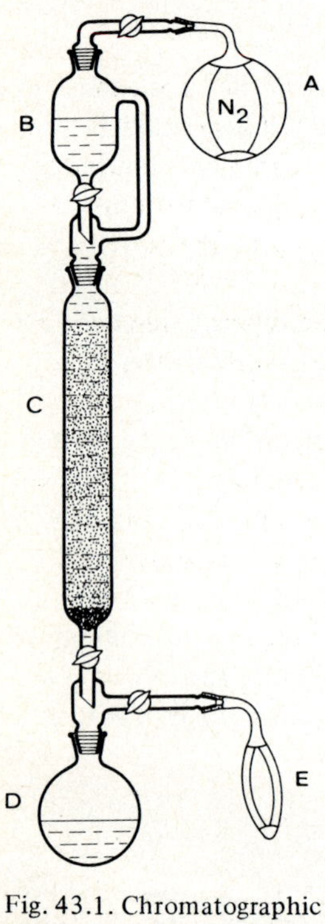


Fig. 43.1. Chromatographic apparatus for the separation of air-sensitive compounds. A, E = football bladders containing nitrogen; B = dropping funnel containing eluent; C = column containing stationary phase; D = collecting flask.

### Mobile phase

The best information on eluents is obtained by thin-layer chromatography on appropriate stationary phases. In classical elution chromatography, solvents in which a distinct separation proceeds at an $R_F$ value of 0.1–0.2 are to be preferred. In dry column chromatography, solvents are used that give the best separations within the full length of the plate. *Caution*: tetrachloromethane is not recommended for the separation of larger amounts of those boron hydride compounds which reduce silver nitrate, because of the danger of an explosion.

### Detection

Some groups of borane compounds are deeply coloured (metallocarboranes), while others are visible in ultraviolet light (higher boranes, *nido*-carboranes, ligand derivatives). Most *nido*-compounds absorb in the region of 200–350 nm and reduce aqueous silver nitrate solution, but the detection of *closo*-carboranes is difficult owing to the transmittance of ultraviolet light and low reactivity towards chemical reagents. According to Plešek *et al.* (1972), fractions that contain *closo*-carboranes in amounts above 1 μg can be monitored by thin-layer chromatography on starch-bound silica gel, followed by detection with iodine vapour. In high-efficiency liquid chromatography, virtually all types of boron compounds were found to be detectable by refractometric monitoring.

## BORANES AND SUBSTITUTED BORANES

Higher boranes and their substituted derivatives are mildly sensitive to air and show different sensitivities towards water and heat. They have a relatively high dipole moment (*ca.* 3 Debye) and various Brønsted acidities ($B_{10}H_{14}$, p$K_a$ *ca.* 5; *n*-$B_{18}H_{22}$, p$K_a$ <1). Some of them ($B_{10}H_{14}$, $XB_{10}H_{13}$) react with water, forming the strongly acidic $H_2O^-$ borane adducts (p$K_a$ 1.8). The separation of these compounds was carried out on silica

TABLE 43.1
ELUTION CHROMATOGRAPHY OF BORANES AND THEIR DERIVATIVES

| Compounds separated | Sorbent* | Eluent | Reference |
|---|---|---|---|
| Decaborane < alkyldecaboranes | Silica gel | *n*-Hexane | Siegel and Mack |
| Halodecaboranes:<br>1-Br > 2-Br; 6-Br > 5-Br;<br>1-Cl > 2-Cl | Silica gel<br>(i) | *n*-Hexane, benzene | Stuchlík *et al.* |
| *n*-$B_{18}H_{22}$ > iso-$B_{18}H_{22}$ | Silica gel<br>(activity II) | *n*-Hexane | Heřmánek and Plešek |
| *n*-$B_{18}H_{22}$ < 6-$C_6H_5CH_2$–$B_{18}H_{21}$ | Silica gel<br>(i) | *n*-Hexane | Heřmánek *et al.*, unpublished |

*(i) = Impregnated with $CH_3COOH$ or $CF_3COOH$.

*References p. 951*

TABLE 43.2
CHROMATOGRAPHY OF CARBORANES AND THEIR DERIVATIVES

| Compounds separated | Sorbent | Eluent | Reference |
|---|---|---|---|
| 1,2- < 1,6- < 1,10-$C_2B_8H_{10}$ | Silica gel | *n*-Hexane | Štíbr *et al.* |
| *o*- < *m*- < *p*-carborane | $Al_2O_3$ (basic) | *n*-Pentane | Sieckhaus *et al.* |
| *o*- < *m*- < *p*-carborane | | Tetrahydrofuran | Čoupek *et al.* |
| *o*-Carborane > 4-bromo-*o*-carborane | Silica gel | *n*-Hexane | Plešek *et al.* (1970) |
| Mono-, di- and trihalogeno-*o*-carboranes | $Al_2O_3$ | *n*-Hexane | Stanko and Goltjapin; Stanko and Anorova; Zakharkin and Kalinin; Zakharkin *et al.* (1966) |
| Halogeno-*m*- and -*p*-carboranes | $Al_2O_3$ | *n*-Hexane | Stanko and Goltjapin |
| *o*-Carborane > 1-ethoxy-*o*-carborane | Silica gel | *n*-Hexane | Gregor *et al.* |
| 3-Hydroxy-, 3-acetoxy-*o*-carborane | $Al_2O_3$ | *n*-Hexane–chloroform (3:2) | Zakharkin *et al.* (1970a) |
| 1-Acyl-*o*-carborane, reaction products | $Al_2O_3$ (acid-washed) | *n*-Hexane | Stanko *et al.* |
| 1-Phenyl-*o*-, 1-phenyl-*m*-carborane | Silica gel (activity I) | *n*-Pentane | Hawthorne *et al.* (1968b) |
| 3-Phenyl-*o*-carborane, reaction products | Silica gel | *n*-Hexane–benzene (1:1) | Hawthorne and Wegner |
| 1- and 3-fluorphenyl-*o*-carborane; purification | $Al_2O_3$ | *n*-Hexane | Zakharkin *et al.* (1969) |
| 1-Fluorphenyl-*o*- and -*m*-carboranes, reaction products | Silica gel (activity II) | *n*-Hexane, cyclohexane | Hawthorne *et al.* (1965) |
| 1-Ferrocenyl-*o*-carborane, reaction products | $Al_2O_3$ | *n*-Hexane | Zakharkin *et al.* (1970b) |
| *o*-, *m*- and *p*-$CPB_{10}H_{12}$ | $Al_2O_3$ (activity II) | *n*-Hexane | Zakharkin and Kysin |

TABLE 43.3
CHROMATOGRAPHY OF LIGAND DERIVATIVES OF BORANES AND CARBORANES

| Compounds separated | Sorbent | Eluent | Reference |
|---|---|---|---|
| $B_{10}H_{14} > B_9H_{13} \cdot THF$ | Silica gel | Benzene | Plešek *et al.* (1967) |
| $B_9H_{13}L > B_{10}H_{12}L_2$ ($L = p\text{-}X\text{–}C_6H_4NH_2$; $C_5H_5N$) | Silica gel (activity II) | Dichloromethane, diethyl ether | Heřmánek *et al.* (1968) |
| $B_9H_{13}L$, $B_9H_{13}L'$ ($L = (CH_3CH_2)_2S > X\text{–}C_5H_4N$) | $Al_2O_3$ (neutral) | Dichloromethane | Graybill *et al.* |
| $((CH_3)_3N)_2B_{10}H_8$ (2,7- > 2,3- + 2,4-) | $Al_2O_3$ | Dichloroethane | Hertler and Raasch |
| $((CH_3)_2S)_2B_{10}H_8$ | $Al_2O_3$ (acidic) | Benzene–dichloroethane<br>Acetonitrile–dichloroethane | Knoth *et al.* |
| 7,8-$C_2B_9H_{11}L$ (*sym.* > *asym.*) | Silica gel | Benzene ($L = C_4H_8O$)<br>Acetonitrile–ethyl acetate ($L = CH_3CN$)<br>Benzene–*n*-pentane ($L = (CH_3CH_2)_2S$) | Young *et al* |
| 3-($BrC_5H_4N$)-7,8-$CPB_9H_{10}$ | Silica gel | Benzene | Beer and Todd |

TABLE 43.4
CHROMATOGRAPHY OF METALLOCARBORANES

| Compounds separated | Sorbent | Eluent | Reference |
|---|---|---|---|
| $X\text{–}C_2B_7H_8\text{–}Co\text{–}C_5H_5$ ($X = H$, $\text{–}COCH_3$) | Silica gel | *n*-Hexane–dichloromethane | Graybill and Hawthorne |
| 1,10-$[(C_5H_5)Fe(CO)_2]_2\text{–}C_2B_8H_8$ purification | Silica gel | *n*-Hexane–benzene (1:1) | Smart *et al.* |
| $(R\text{–}C_2B_9H_{10})_2Ni$, $[(R\text{–}C_2B_9H_{10})_2Ni]^-$ | Silica gel | *n*-Hexane–benzene (1:1) | Hawthorne *et al.* (1968a); Warren and Hawthorne |
| Isomeric $(R_2C_2B_9H_9)_2Ni$, isomeric $[(R_2C_2B_9H_9)_2Ni]^-$ | Silica gel | *n*-Hexane–dichloromethane (2:1) | Paxson *et al.* |
| $(C_2B_9H_{11})Co\text{–}C_5H_5$, reaction products | Silica gel | *n*-Hexane–benzene (1:1) | Hawthorne *et al.* (1968a) |
| $(R_2C_2B_9H_9)\text{–}Pd(C_4(C_6H_5)_4)$, reaction products | Silica gel | *n*-Hexane–benzene (4:6) | Hawthorne *et al.* (1968a) |
| $[(C_2B_9H_{11})_2Fe]^-$, $(C_2B_9H_{10}S(CH_2CH_3)_2)_2Fe$ and $(C_2B_9H_{11})Fe(C_2B_9H_{10}S(CH_2CH_3)_2)$ | Silica gel | *n*-Hexane–benzene (7:3) | Hawthorne *et al.* (1971) |
| 1-$[C_5H_5)Fe(CO)_2]$-2-$CH_3\text{–}C_2B_{10}H_{11}$, reaction products | Silica gel | *n*-Hexane–benzene (5:1) | Smart *et al.* |
| $(B_{10}H_{10}S)Co\text{–}C_5H_5$, reaction products | $Al_2O_3$ | Benzene–dichloroethane (5:1) | Hertler *et al.* |

gel (Heřmánek and Plešek, Siegel and Mack) or silica gel deactivated with acetic acid (Stuchlík *et al.*) or with trifluoroacetic acid (Heřmánek *et al.*, unpublished). The separation of boranes and their substituted derivatives is summarized in Table 43.1.

In accordance with our experience, we propose that the separation of these sensitive compounds should be carried out by elution chromatography on silanized and air-free silica gel in an inert atmosphere. Our latest results also show the advantage of the use of high-efficiency liquid chromatography in the identification and separation of borane compounds (Krejčí and Heřmánek).

## CARBORANES

The chromatographic behaviour of various carboranes is based on their dipole moments (dependent on the character of the skeleton, the location of hetero-atoms or the different locations of substituents) and on the character and number of substituents. The most stable class is the *closo*-carboranes, $C_2B_nH_{n+2}$, which are generally stable enough towards heat, oxidation and acid hydrolysis and are separated satisfactorily on silica gel or aluminium oxide. It is of interest that a good separation of *o*-, *m*- and *p*-carborane, *i.e.*, compounds that differ significantly in their dipole moments but not in the sizes and shapes of the molecules, was achieved by gel permeation chromatography (Čoupek and Heřmánek), which is believed not to be influenced by the dipole moment. Some examples of the column chromatography of *closo*-carboranes and their derivatives are reported in Table 43.2.

A theoretically possible but rarely used technique is the column chromatography of more sensitive intermediate and higher *nido*-carboranes (Plešek *et al.*, unpublished). The optimum separation conditions for these compounds are similar to those for boranes.

## LIGAND DERIVATIVES OF BORANES AND CARBORANES

The chromatographic character of these relatively air- or hydrolysis-stable compounds is mainly influenced by their high dipole moments, owing to the dative bond between the ligand and the borane framework. Generally, the mobility decreases in following order: mono-ligand $\gg$ di-ligand derivatives and $(CH_3CH_2CH_2CH_2)_2S > (CH_3CH_2)_2S > (CH_3)_2S > (C_6H_5)_3P > CH_3CN$ = pyridine $\gg (C_6H_5)NH_2$ (approximately).

Examples of chromatographed mixtures are listed in Table 43.3.

## METALLOCARBORANES

Column chromatography on silica gel is the most useful method for the purification of many types of σ-bonded or sandwich-type metallocarboranes, most of which are relatively stable and of low volatility. The chromatographic behaviour of these compounds is determined predominantly by their ionic charge or dipole moment.

Good results were achieved in separations of neutral metallocenes from ionic metallocenes (*i.e.*, compounds with differently charged metal atoms), substituted from unsubstituted compounds, metallocenes with different ligands, or in separations of isomers, with different positions of the carbon atoms. Mixtures of these compounds are often intensively coloured and therefore well separated by the dry column technique. Typical examples of chromatographed mixtures and experimental conditions are collected in Table 43.4.

## REFERENCES


Beer, D. C. and Todd, L. J., *J. Organometal. Chem.*, 36 (1972) 77.
Čoupek, J. and Heřmánek, S., unpublished results.
Čoupek, J., Heřmánek, S., Plešek, J. and Pokorný, S., unpublished results.
Graybill, B. M. and Hawthorne, M. F., *Inorg. Chem.*, 8 (1969) 1799.
Graybill, B. M., Pitochelli, A. R. and Hawthorne, M. F., *Inorg. Chem.*, 1 (1962) 626.
Gregor, V., Heřmánek, S. and Plešek, J., *Collect. Czech. Chem. Commun.*, 33 (1968) 980.
Hawthorne, M. F., Berry, T. E. and Wegner, P. A., *J. Amer. Chem. Soc.*, 87 (1965) 4746.
Hawthorne, M. F., Warren, L. F., Callahan, K. P. and Travers, N. F., *J. Amer. Chem. Soc.*, 93 (1971) 2407.
Hawthorne, M. F. and Wegner, P. A., *J. Amer. Chem. Soc.*, 90 (1968) 896.
Hawthorne, M. F., Young, D. C., Andrews, T. D., Howe, D. V., Pilling, R. L., Pitts, A. D., Reintjes, M., Warren, L. F., Jr. and Wegner, P. A., *J. Amer. Chem. Soc.*, 90 (1968a) 879.
Hawthorne, M. F., Young, D. C., Garret, P. M., Owen, D. A., Schwerin, S. G., Tebbe, F. N. and Wegner, P. A., *J. Amer. Chem. Soc.*, 90 (1968b) 862.
Heřmánek, S. and Plešek, J., *Collect. Czech. Chem. Commun.*, 35 (1970) 2488.
Heřmánek, S., Plešek, J. and Fetter, K., unpublished results.
Heřmánek, S., Plešek, J., Štíbr, B. and Hanousek, F., *Collect. Czech. Chem. Commun.*, 33 (1968) 2177.
Hertler, W. R., Klanberg, F. and Muetterties, E. L., *Inorg. Chem.*, 6 (1967) 1696.
Hertler, W. R. and Raasch, M. S., *J. Amer. Chem. Soc.*, 86 (1964) 3661.
Knoth, W. H., Hertler, W. R. and Muetterties, E. L., *Inorg. Chem.*, 4 (1965) 280.
Krejčí, M. and Heřmánek, S., unpublished results.
Paxson, T. E., Kaloustian, M. K., Tom, G. M., Wiersema, R. J. and Hawthorne, M. F., *J. Amer. Chem. Soc.*, 94 (1972) 4882.
Plešek, J., Gregor, V. and Heřmánek, S., *Collect. Czech. Chem. Commun.*, 35 (1970) 346.
Plešek, J., Gregor, V. and Heřmánek, S., *J. Chromatogr.*, 74 (1972) 149.
Plešek, J., Heřmánek, S. and Štíbr, B., *Collect. Czech. Chem. Commun.*, 33 (1967) 691.
Plešek, J., Štíbr, B. and Heřmánek, S., unpublished results.
Sieckhaus, J. F., Semenuk, N. S., Knowles, T. A. and Schroeder, H., *Inorg. Chem.*, 8 (1969) 2452; *C.A.*, 69 (1968) 87047.
Siegel, B. and Mack, J. L., *Phys. Chem.*, 63 (1959) 1212.
Smart, J. C., Garrett, P. M. and Hawthorne, M. F., *J. Amer. Chem. Soc.*, 91 (1969) 1031.
Stanko, V. I. and Anorova, G. A., *Zh. Obshch. Khim.*, 41 (1971) 1521.
Stanko, V. I. and Goltjapin, Ju. V., *Zh. Obshch. Khim.*, 40 (1970) 127.
Stanko, V. I., Klimova, A. I., Tschapovskij, Ju. A. and Klimova, T. P., *Zh. Obshch. Khim.*, 36 (1966) 1779.
Štíbr, B., Plešek, J. and Heřmánek, S., unpublished results.
Stuchlík, J., Heřmánek, S., Plešek, J. and Štíbr, B., *Collect. Czech. Chem. Commun.*, 35 (1970) 339.
Warren, L. F., Jr. and Hawthorne, M. F., *J. Amer. Chem. Soc.*, 92 (1970) 1157.
Young, D. C., Howe, D. V. and Hawthorne, M. F., *J. Amer. Chem. Soc.*, 91 (1969) 859.
Zakharkin, L. I. and Kalinin, V. N., *Zh. Obshch. Khim.*, 37 (1967) 939.
Zakharkin, L. I., Kalinin, V. N. and Gedymin, V. V., *Zh. Obshch. Khim.*, 40 (1970a) 2653.
Zakharkin, L. I., Kalinin, V. N. and Snyakin, A. P., *Zh. Obshch. Khim.*, 40 (1970b) 2246.
Zakharkin, L. I., Kalinin, V. N., Snyakin, A. P. and Kvasov, B. A., *J. Organometal. Chem.*, 18 (1969) 19.
Zakharkin, L. I. and Kysin, V. I., *Zh. Obshch. Khim.*, 40 (1970) 2234.
Zakharkin, L. I., Stanko, V. I. and Klimova, A. I., *Izv. Akad. Nauk SSSR, Ser. Khim.*, (1966) 1946.

*Chapter 44*

# Vitamins

J. DAVÍDEK

## CONTENTS

## INTRODUCTION

The application of physicochemical methods for the determination of vitamins in mixtures requires the isolation of the vitamins from substances that might interfere in the analysis. The samples are purified by chromatography; for routine analyses, column chromatography is the most suitable.

TABLE 44.1
SEPARATION OF FAT-SOLUBLE VITAMINS USING HIGH-SPEED LIQUID CHROMATOGRAPHY (DUPONT)

| Vitamin | Column | Mobile Phase | Temperature (°C) |
|---|---|---|---|
| K | Permaphase ODS | Methanol–water (1:4) | 50 |
| $K_1$ | Permaphase ODS | Methanol–water (85:15) | 50 |
| A acetate, $D_2$, $D_3$ | Permaphase ODS | Methanol–water (4:1) | 50 |
| E and E acetate | Permaphase ODS | Methanol–water (85:15) | 50 |
| A | Permaphase ODS | Methanol–water (95:5) | 50 |
| A palmitate | Permaphase ODS | Isopropanol–water (85:15) | 50 |
| E succinate | Zipax HCP | Methanol–water (78:22) plus 0.1% $H_3PO_4$ | Ambient |

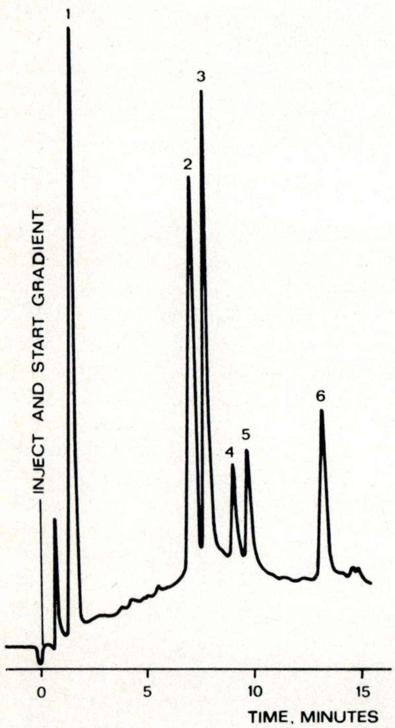


Fig. 44.1. Separation of fat-soluble vitamins (DuPont). Instrument: DuPont 820 liquid chromatograph. Mobile phase: gradient from water to methanol at 5%/min. Column temperature: 70°C. Column pressure: 1200 p.s.i. Flow-rate: 2 ml/min. Detection: UV spectrophotometrically. Peaks: 1 = vitamin K; 2 = vitamin A acetate; 3 = vitamin $D_2$; 4 = vitamin E; 5 = vitamin E acetate; 6 = vitamin A.

Most vitamins are sensitive to light and some also to atmospheric oxidation, and therefore all analytical procedures should be carried out with care (constant temperature; subdued light; pure or redistilled solvents). Owing to the solubility of vitamins (water-soluble and fat-soluble), generally two different approaches have to be applied, and even under these conditions it is not possible to separate all water- or fat-soluble vitamins in a single system. The behaviour and separation of individual vitamins are therefore discussed in this chapter.

## FAT-SOLUBLE VITAMINS

High-speed liquid chromatography has been used for the separation of fat soluble vitamins, and the conditions required for individual vitamins are summarized in Table 44.1 (DuPont). Mixtures of fat-soluble vitamins can be separated using Permaphase ODS and a gradient from water to methanol at 5%/min (Fig. 44.1).

### Vitamin A group

The vitamin A group consists of vitamins $A_1$ and $A_2$ in the alcohol, aldehyde, acid and ester forms and of a large group of carotenoid pigments. Substances of this type are generally not very stable, and exposure to light, oxygen and acids should therefore be avoided. Most methods utilize the different adsorption affinities to alumina, silicic acid or cellulose for the separation of substances of this group. Molecular sieves and ion exchangers have also been used for some special purposes.

The procedures for the separation of vitamins $A_1$ and $A_2$ in natural material include extraction, separation on a column of a suitable adsorbent and the determination of the separated substances.

Protein denaturation, saponification and digestion may be carried out in some instances prior to extraction. Denaturation of proteins is achieved by adding ethanol to an analyzed tissue with thorough mixing, while saponification is usually performed using alcoholic potassium hydroxide solution under controlled conditions. The loss of vitamins by oxidation during the heating process must be minimized by adding a small amount of an antioxidant. Nevertheless, it is possible to hydrolyze any esters of the vitamin A and some isomerization is to be expected. Digestion is usually performed with an alkali solution more concentrated than that used for saponification, with an increased risk of loss or isomerization of vitamins. Most hydrocarbons are suitable for the extraction of vitamin A from saponificated samples; light petroleum and *n*-hexane are almost universally used for this purpose.

The most commonly used adsorbent for the separation of these substances is alumina (Bertram and Krisch, Eremina, McLaren *et al.*) and silicic acid (Bertram and Krisch). For the elution of the separated substances from alumina, a mixture of light petroleum and diethyl ether (Bertram and Krisch) is used, while for the separation and elution from a sodium hydroxide-treated silicic acid column, Crain *et al.* used a gradient of the mixture

*References p.976*

diethyl ether–*n*-hexane–methanol. Zile and De Luca applied a Skelly-Solve B gradient (150 ml in the mixing chamber) to a 15% (v/v) mixture of diethyl ether with Skelly-Solve B (250 ml in the holding chamber) and others.

Shichi *et al.* used cellulose in studies on rhodopsin from cattle retina. Roberts and De Luca studied the oxidative decarboxylation of retinoic acid in microsomes of rat liver and kidney using DEAE-cellulose and silicic acid.

A study of the protein complex of vitamin A and its characterization was performed using molecular sieves of the Sephadex type (G-50, G-100 and G-200) and molecular sieves with ion-exchange character, such as DEAE-Sephadex A-50 and SE-Sephadex C-50 (Peterson, 1971a,b; Peterson and Berggård).

For rendering the vitamin A group visible and for their determination, absorption in the UV region can be used (Bertram and Krisch, McLaren *et al.*), and also colour reactions with antimony trichloride or pentachloride or with trifluoroacetic acid or its anhydride at 620 nm (McLaren *et al.*) and their fluorescence after activation at 325 and 470 nm (Bertram and Krisch).

*Determination of vitamin A in serum on alumina*

Into a 50 × 6 mm test-tube are placed 100 μl of serum, 100 μl of absolute ethanol and 150 μl of light petroleum (b.p. 40–60°C). After thoroughly mixing, the light petroleum layer is separated by centrifuging (2000–3000 rpm, 10 min) and transferred into another test-tube, where the light petroleum is removed by means of a stream of nitrogen. The residue is dissolved in 5 μl of light petroleum and applied to the micro-column (capillary glass column, 150 × 1.5 mm I.D., alumina of greater than 200 mesh containing 5% of water), the tube is washed with 5 μl of light petroleum and the liquid similarly applied to the column. The column is developed with cyclohexane–acetone (95:5) in the dark. A similar column to which has been applied a known mixture of retinol and retinyl palmitate in light petroleum is developed under the same conditions. After development, the compounds must be rendered visible in UV light. The upper zone contains retinol and the lower retinyl palmitate; the two zones are cut out and eluted with 150 μl of chloroform. After centrifugation (2000–3000 rpm, 10 min), 100 μl of the supernatant are transferred into a small tube and 100 μl of trifluoroacetic acid (diluted 1:1 with chloroform) are added. The absorbance is read at 620 nm 30 sec after the addition of the trifluoroacetic acid reagent (McLaren *et al.*).

*Chromatography of vitamin A compounds on a silicic acid column*

All column chromatography is carried out in a 60-cm column consisting of four 15-cm sections, adjacent sections decreasing in diameter by a factor of approximately 1.4 (the diameter of the uppermost section is 13 mm and that of the lowest is 4.5 mm). The column is packed with 15 g of silicic acid (activated for 12 h at 120°C), slurried with Skelly-Solve B and developed using convex gradient elution. The first gradient consists of Skelly-Solve B to 15% (v/v) diethyl ether in Skelly-Solve B (150 ml in the mixing chamber and 250 ml in the holding chamber). The second gradient is developed subsequent to the completion of the first gradient by placing 250–300 ml of 35% (v/v) diethyl ether in

Skelly-Solve B in the holding chamber. Chromatography is carried out in a cold room under pressure with a flow-rate of about 1 ml/min. The synthetic mixtures (all-*trans*-retinoic acid, 9,13-di-*cis*-retinoic acid, all-*trans*-retinol, 13-*cis*-retinol, 9,13-di-*cis*-retinol, 9-*cis*-retinol, 11,13-di-*cis*-retinol, all-*trans*-retinene, 13-*cis*-retinene, all-*trans*-methyl retinoate, 13-*cis*-methyl retinoate, 9,13-di-*cis*-methyl retinoate, all-*trans*-retinyl acetate and all-*trans*-retinyl palmitate) are applied to the column in a small volume of Skelly-Solve B under a slight pressure of nitrogen. Usually an excess (1–10 mg) of *dl*-*a*-tocopherol is added before separation in order to minimize oxidation. After elution from the column, the fractions are evaporated under a stream of nitrogen and redissolved in ethanol or benzene. The identification of compounds is based on their UV absorption characteristics (Fig. 44.2) (Zile and De Luca).

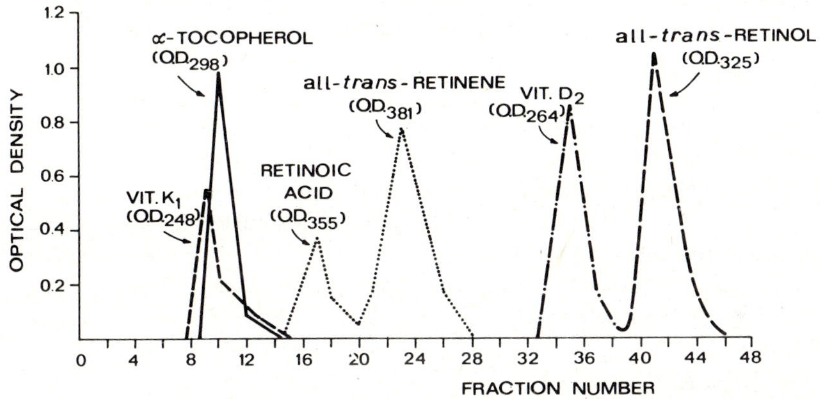


Fig. 44.2. Separation of a synthetic mixture of various fat-soluble vitamins (Zile and De Luca). Column: 60-cm long consisting of four 15-cm sections, adjacent sections decreasing in diameter by a factor of approximately 1.4; the uppermost section had an I.D. of 13 mm, the lowest 4.5 mm. Sorbent: silicic acid. Eluent: gradient of a mixture of diethyl ether in Skelly-Solve B (0–*ca.* 30%). Detection: UV spectrophotometrically according to the spectral characteristics of the individual components. Fractions of 10.6 ml collected.

## Calciferols

Calciferols are substances with antirachitic efficiency, in terms of chemical structure they are the steroids and from the analytical point of view one must note their instability towards light and in the presence of oxygen.

For the separation of calciferols, their different adsorption activities towards various adsorbents are utilized. Silicic acid is the most commonly used polar adsorbent (Bell and Kodicek, De Luca, Haussler *et al.*, Myrtle *et al.*, Suda *et al.*), often in combination with subsequent re-chromatography on Celite with methanol–water as stationary phase (De Luca, Haussler *et al.*, Myrtle *et al.*, Suda *et al.*). Partially deactivated alumina (Mariani and Mariani-Vicari, Takahashi and Yamamoto) and silica gel (Lawson *et al.*) have also been used for the separation of the vitamin D group. Dollwet and Norman used Factise (a polymerized and vulcanized vegetable oil) as the stationary phase.

*References p.976*

For the elution of substances separated on the polar adsorbants silical gel, silicic acid and alumina, gradient elution with the mixtures light petroleum (b.p. 40–60°C or 60–80°C) –diethyl ether–methanol (Bell and Kodicek, Haussler *et al.*, Lawson *et al.*, Myrtle *et al.*), diethyl ether–1,2-dichloroethane–methanol (Haussler *et al.*, Myrtle *et al.*), diethyl ether –Skelly-Solve B (Suda *et al.*) and a non-gradient system using 16.5% of diethyl ether in light petroleum (b.p. 40–60°C or 60–80°C) (Takahashi and Yamamoto) or 20% of 1,2-dichloroethane in diethyl ether (Haussler *et al.*) have been used. From the stationary phase system of methanol–water on Celite, calciferols were eluted with a gradient of the mixture 1,2-dichloroethane in light petroleum–1,2-dichloroethane (Haussler *et al.*) or Skelly-Solve B (De Luca, Suda *et al.*). With the reversed phase of Factise, a mixture of water and acetone was used for the elution.

Calciferols can be detected in eluents by utilizing their absorption maxima in UV light (300–220 nm) (Dollwet and Norman), spectrophotometrically after reaction with antimony trichloride (Mariani and Mariani-Vicari, Takahashi and Yamamoto) or radiometrically (Dollwet and Norman).

### *Chromatographic separation of vitamins $D_2$ and $D_3$ and related compounds using Factise*

For the preparation of the reversed phase (Dollwet and Norman), Factise is first triturated in acetone, and then forced with the bottom of an erlenmeyer flask through a 60-mesh sieve and collected below on a 100-mesh sieve. The Factise is kept moist with a small amount of acetone during this screening procedure, and the screened Factise is then stored in acetone. Later the Factise polymer is washed several times with glacial acetic

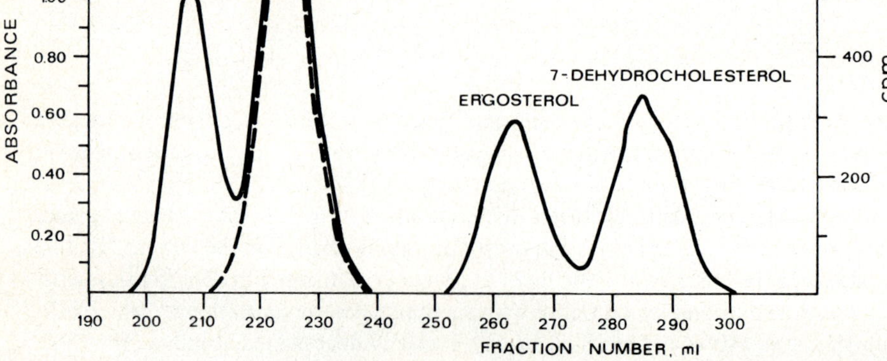


Fig. 44.3. Separation of [$^3$H]vitamin $D_3$, vitamin $D_3$, vitamin $D_2$, ergosterol and 7-dehydrocholesterol (Dollwet and Norman). Column: 214 × 0.8 cm. Stationary phase: Factise. Eluent: 5% of water in acetone. Detection: solid line, absorbance at 264 and 282 nm; broken line, radioactivity.

acid and acetone (Hirsch). Before filling the columns, the Factise is slurried several times in the solvent system to be used, 50 ml of the slurry are then forced into the columns, the Factise is allowed to settle and the eluent is then pumped through the column for 30 min. After completing the preparation of the column, the eluent is pumped out overnight at a flow-rate of *ca.* 10 ml/h so as to compact the column. The sample in 1.5–2 ml of acetone is placed on the top of the column, then rinsed into the column with at least four or five 0.5-ml portions of the eluent (5% of water in acetone) before the eluent is pumped through the column. Fractions of 1 ml are collected at a flow-rate of 10 ml/h and evaporated under vacuum. Then 5 ml of absolute ethanol are added, the contents stirred and the UV spectrum is scanned between 300 and 220 nm. This procedure was used for the separation of (a) vitamin $D_2$, $D_3$ and ergosterol; (b) cholesterol and 7-dehydrocholesterol; (c) dihydrotachysterol-2; (d) tritiated vitamin $D_3$ and $D_2$; (e) [4-$^{14}$C]vitamin $D_3$; and (f) [1,2-$^{3}$H] cholesterol (Fig. 44.3) (Dollwet and Norman).

*Isolation and identification of 25-hydroxyergocalciferol in blood*

Blood is collected and mixed immediately with one tenth of its volume of 0.1 *M* sodium oxalate solution so as to prevent clotting. Plasma is separated from the cells by means of a De Laval blood separator and is made up to 70% saturation with ammonium sulphate and allowed to stand at 4°C for 7 days. The precipitate of the protein is collected by centrifugation (25,000 rpm for 25 min; Sharples AS-16-P centrifuge) and extracted with methanol–chloroform (2:1). After further addition of chloroform, the denatured proteins are removed by filtration through glass-wool and re-extracted with a further portion of methanol–chloroform (2:1). The phases are allowed to separate and the aqueous phase is drawn off and re-extracted with chloroform. The combined chloroform layers are washed with tap water and allowed to stand for 24 h. The chloroform phase is concentrated to 50 ml, washed with saturated sodium chloride solution and dried over anhydrous magnesium sulphate. The solvent is then evaporated to dryness and the residue dissolved in 100 ml of Skelly-Solve B (a petroleum fraction, b.p. 65–67°C) (Suda *et al.*).

The first chromatography is carried out on a column of silicic acid (60 × 1 cm). The column is eluted with a diethyl ether–Skelly-Solve B gradient, obtained by running 400 ml of 85% diethyl ether in Skelly-Solve B from a holding chamber into a 250-ml mixing chamber initially containing 250 ml of Skelly-Solve B. Then diethyl ether is placed in the holding chamber and finally methanol is applied directly on to the column. From the fractions obtained, the fraction with the greatest biological activity is collected, evaporated to dryness and re-chromatographed on a multibore silicic acid column. The mixing chamber contains 250 ml of Skelly-Solve B and the holding chamber 400 ml of 85% of diethyl ether in Skelly-Solve B. As soon as the holding chamber has been emptied, it is filled with 300 ml of diethyl ether. The eluted fractions of the main peak are collected, combined and chromatographed on a Celite partition column. For this purpose, 200 ml of Skelly-Solve B is equilibrated at 10°C with an equal volume of methanol–water (8:2), 15 ml of the methanolic phase is mixed with 20 g of Celite and dry packed into a 60 × 1 cm column in 2-cm portions. The upper phase is used as the mobile phase. The residue obtained from the silicic acid column chromatography (after evaporation) is applied in a small volume (1–3 ml) of mobile phase to the column and is developed with mobile phase,

*References p.976*

5-ml fractions being collected. Tubes 11–17 are combined, evaporated to dryness as before and re-chromatographed on another partition column. The separated substances are determined radiometrically or spectrophotometrically (UV region) or by GLC (Suda *et al.*).

## Tocopherols

The preparation of a sample of natural material includes extraction and saponification. Owing to the high sensitivity of tocopherols to oxidation, antioxidants (ascorbic acid, pyrogallol, etc.) are added before saponification. Otherwise, the saponification is carried out in the manner described, for example, for vitamin A.

For the separation of tocopherols by column chromatography, different adsorbents are used. The chromatographic procedure utilizes the different adsorption affinities of the tocopherols and interfering substances to the adsorbent used. Silica gel (Cassagne and Baraud, Skinner *et al.*), silica gel with Celite (Williams), silicic acid (Dicks-Bushnell), silicic acid with diatomaceous earth (Cinquina), Florisil hydrated to different degrees (Dicks-Bushnell, Skinner *et al.*), magnesium hydrogen orthophosphate (Dicks-Bushnell) and a mixture of alumina, zinc carbonate and Celite (Millar and Caravaggi) were found to be suitable for this purpose.

The elution from the column of silica gel was carried out with a gradient of *n*-hexane–benzene–diethyl ether–methanol (Cassagne and Baraud) or by stepwise elution with 0.5% 3%, 10% and 20% of diethyl ether in *n*-hexane. Stepwise elution using a mixture of isooctane, 22% of chloroform in isooctane and 50% of chloroform in isooctane from the column of silicic acid (Cinquina) or a mixture of Skelly-Solve F and diethyl ether (98:2) (Dicks-Bushnell) has been used. Stepwise elution was applied to a Florisil column using Skelly-Solve F–diethyl ether (199:1, 99:1 and 39:1) (Dicks-Bushnell). With an alumina–zinc carbonate–Celite column, a mixture of benzene and cyclohexane (1:4) was used.

The detection and determination of the separated substances is mostly carried out by UV spectroscopy (246 and 292 nm) (Cinquina) before and after reduction with potassium borohydride (Williams) or using the iron(III) chloride–dipyridyl reaction at 520 nm (Dicks-Bushnell, Williams).

### *Separation and determination of quinones and α-tocopherol*

Approximately 30 g of leaves of the *Vicia faba* (broad bean) plant are frozen in liquid air and ground with 40–50 g of anhydrous disodium hydrogen orthophosphate. The finely ground leaves are lyophilized overnight at 4°C in the dark and the lyophilized leaves are re-ground, extracted with acetone and filtered. The residue is washed with acetone until colourless. The liquid extract is dried and the green residue is dissolved in chloroform, again evaporated, the residue re-dissolved in *n*-hexane (b.p. 67–70°C) and a known volume of the solution applied to the top of a Kieselgel G–Celite (1:1, w/w) column. The mixture of Celite and Kieselgel G is thoroughly washed with diethyl ether, which removes a yellow contaminant, and dried at 100°C for 1 h. The column material is prepared as a slurry with *n*-hexane, packed into the column (6 × 1.2 cm) and washed with a small volume (10 ml) of $n$-hexane. The sample of quinones is eluted with 60 ml of 0.5%, 60 ml of 3.0%, 60 ml of 10% and finally with 20% of diethyl ether in $n$-hexane. Fractions of approximately 3.5 ml are collected and all of the fractions are evaporated to dryness and the residues re-dissolved in 3 ml of 95% ethanol and identified (Williams).

*Determination of α-tocopherol in serum*

To 4 ml of serum are added 2 ml of water and 6 ml of ethanol containing 1% of pyrogallol. The mixture, containing 0.3 ml of 11 $N$ potassium hydroxide, is refluxed for 15 min at 90°C and, after cooling, is extracted with three 20-ml volumes of diethyl ether. The combined ethereal fractions are dried with 5 g of anhydrous sodium sulphate and evaporated to dryness at 40°C under a stream of nitrogen. The residue is dissolved in 0.1–0.5 ml of $n$-hexane and the solution is applied to a column of 0.6 g of alumina, 0.6 g of basic zinc carbonate and 0.3 g of Celite 545. The column is developed with benzene–cyclohexane (1:4). The first 3 ml of eluate contains β-carotene, the next 10 ml α-tocopherol, and ubichromenol remains on the column (Millar and Caravaggi).

**Vitamin K group**

There are many homologues of vitamin K, such as $K_1$, $K_2$ and $K_3$. Vitamins $K_1$ and $K_2$ also have homologues with side-chains of various lengths at the 3-position. The K vitamins are extremely sensitive to light and special care must be taken to avoid their photolysis during analytical operations.

Column chromatography has been used for the preliminary purification of analyzed material and for the isolation of the pure substances on a preparative scale. The most common adsorbents used were alumina, Decalso, Florisil, Permutit, zinc carbonate, silicic acid and occasionally magnesium oxide. Synthetic K vitamins were separated on alumina deactivated with dilute acetic acid. For the separation of homologues of vitamin K, a methylated Sephadex (Nyström and Sjövall) and solvent mixture comprising chloroform–methanol–$n$-heptane (1:1:2) have been used.

Column chromatography has in most instances been used especially for the preliminary purification of test material from interfering substances.

*Separation of vitamin $K_2$ on capillary columns of methylated Sephadex*

This method is suitable for the separation of the isoprenologue homologues of vitamin $K_2$ ($K_{2(10)}$–$K_{2(40)}$).

Columns with a diameter of 2 cm are prepared with about 25 g of methylated Sephadex G-25 superfine. The samples (0.2–1 mg) are applied to the columns in 0.5–1 ml of solvent.

Capillary columns are prepared in the following way. A small piece of glass-wool and a 2-cm length of stainless-steel capillary tube (O.D. 1/16 in., I.D. 0.25 mm, cut to a tip at the distal end) are inserted into the distal end of a PTFE tube about 2 m in length and this tube is filled with the solvent to be used for chromatography. The proximal end is connected with a stainless-steel tube (O.D. 1/16 in., I.D. 0.6 mm, length 5 cm) silver-soldered to a stainless-steel cylindrical reservoir (O.D. 30 mm, length 100 mm) that contains a slurry of methylated Sephadex G-25 superfine in the same solvent. The upper end of the cylindrical reservoir is connected to a nitrogen tank and a pressure of about 1–2 kp/cm$^2$ is applied. The slurry passes slowly through the capillary into the PTFE tube. Clogging is prevented by vibrating the reservoir. When the PTFE tube is completely filled with the gel, the pressure is released and the tubing is disconnected from the reservoir. An injection port is attached to the proximal end of the PTFE tubing and connected to another cylindrical reservoir (300 X 38 mm) that contains the solvent to be used (chloroform–methanol–*n*-heptane, 1:1:2). A pressure of 1–3 kp/cm$^2$ is applied. A 5-$\mu$l volume of a solution of the sample in the solvent is injected into the column. The elution of vitamin $K_2$ derivatives with chloroform–methanol–*n*-heptane (1:1:2) is followed by measurement of the absorption at 270 nm or by using a platinum chain-flame ionization detector. The compounds are eluted in order of decreasing molecular weight and in order of increasing polarity (Nyström and Sjövall).

## WATER-SOLUBLE VITAMINS

It is not possible to separate this large group of vitamins, which consist of different chemical constituents, in one universal procedure and the behaviour and separation of individual water-soluble vitamins will therefore be described.

Of new techniques, high-speed liquid chromatography has been used for the separation of water-soluble vitamins. The chromatographic conditions required for single component samples are summarized in Table 44.2 (DuPont).

TABLE 44.2
SEPARATION OF WATER-SOLUBLE VITAMINS USING HIGH-SPEED LIQUID CHROMATOGRAPHY (DUPONT)

| Vitamin | Column | Mobile phase |
|---|---|---|
| Niacinamide | Zipax SCX | pH 3, 0.07 *M* $NaClO_4$ |
| Riboflavin | Zipax SCX | pH 4, 0.14 *M* $NaClO_4$ |
| Pyridoxine | Zipax SCX | pH 4, 0.14 *M* $NaClO_4$ |
| Corrinoide | Zipax SCX | pH 9, 0.2 *M* $NaClO_4$ |
| Thiamine mononitrate | Zipax SCX | pH 9, 0.6 *M* $NaClO_4$ |
| Ascorbic acid | Zipax SAX | pH 7, no modifier |
| Niacin | Zipax SAX | pH 7, 0.002 *M* $NaNO_3$ |
| Folic acid | Zipax SAX | pH 7, 0.02 *M* $NaNO_3$ |

### Thiamine

During all analytical operations with thiamine and its derivatives, their instability, especially in neutral and alkali media, must be kept in mind.

For the separation of thiamine and its derivatives and metabolites, zeolite ion exchangers have most often been used. If the concentration of thiamine in a sample is low, the best results can be achieved using Permutit T (Decalso). For higher concentrations of thiamine, some synthetic cation exchangers such as Amberlite GC-50 and Dowex have most often been applied (Amos and Neal; Diorio and Lewin; Matsuo and Suzuoki; Neal, 1968, 1969; Suzuoki *et al.*). Dowex 1-X4 was found to be advisable for the separation of thiamine, monophosphothiamine, diphosphothiamine and triphosphothiamine (Koike *et al.*). Diorio and Lewin used Dowex AG 50W-X8 for the separation of pyrimidine precursors of thiamine.

Thiamine can be eluted from a Decalso column using an acidic solution of potassium chloride. Koike *et al.* eluted thiamine and its derivatives from Dowex 1-X4 with water followed by an acetate buffer. Neal (1968, 1969) eluted thiamine from Amberlite GC-50 with water followed by pyridine–acetic acid–water (7.5:1.5:91). Amos and Neal used a linear gradient of distilled water to 0.35 *M* pyridinium acetate for Amberlite GC-50 and 0.01 *N* hydrochloric acid for Sephadex G-10. Diorio and Lewin used 2 *M* ammonium hydroxide solution for Dowex AG 50W-X8.

*Separation of thiamine from interfering substances on Decalso*

The acidic extract, or the enzymatic hydrolyzate in the case of thiamine phosphate, is transferred by pipette into the column of Decalso, and thiamine in the acidic extract is bound on the ion exchanger. A 1-g amount of normally grained Decalso binds quantitatively 40 $\mu$g of thiamine. For isolation from natural materials, there is a maximum amount of about 10 $\mu$g, owing to the decreasing activity of Decalso in the presence of other compounds with the same binding activity. After the application of a sample, the column is washed with hot distilled water and thiamine is eluted with an acidic solution of potassium chloride.

*Separation of thiamine, monophosphothiamine (MPT), diphosphothiamine (DPT) and triphosphothiamine (TPT)*

The sample (5 ml, pH adjusted to 4.5 with 0.1 *N* hydrochloric acid) is passed through a 190 × 6 mm column of Dowex 1-X4 ($CH_3COO^-$), 200–400 mesh. Thiamine and MPT are eluted with 14 ml of water, DPT with 24 ml of 0.1 *M* sodium acetate buffer of pH 4.5 and TPT with 24 ml of 1 *M* sodium acetate buffer of pH 4.5. The separation of thiamine plus MPT, DPT and TPT is sharp (Fig. 44.4), but thiamine and MPT must be distinguished by the thiochrome method, before and after acid phosphatase digestion (Koike *et al.*).

*Separation of thiamine from its metabolites in urine and cells*

Metabolites of thiamine from urine and cells are chromatographed on a column of Amberlite GC-50 ($H^+$), eluted with water followed by pyridine–acetic acid–water (7.5:1.5:91). Thiamineacetic acid (4-methylthiazole-5-acetic acid) is the main metabolite, followed by the fraction corresponding to thiamine (Neal, 1968, 1969). Amos and Neal in addition separated 2-methyl-4-amino-5-formylaminomethylpyrimidine on the same column.

*References p.976*

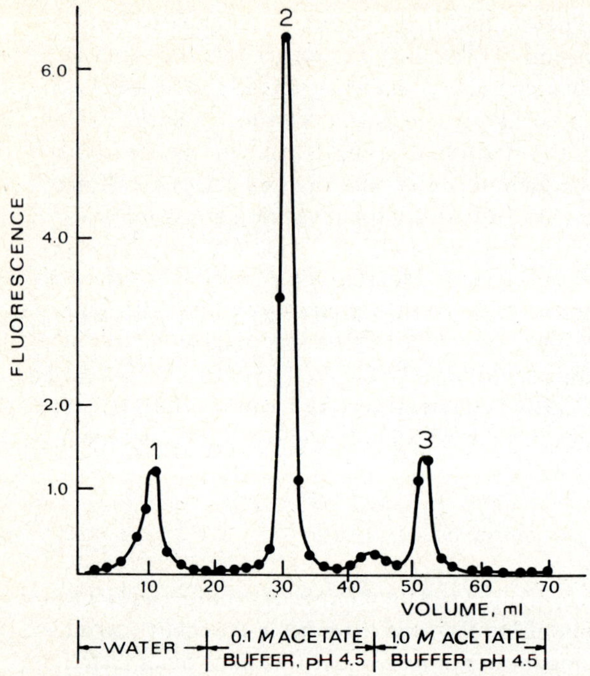


Fig. 44.4. Chromatography of thiamine and its phosphoric acid esters from biological materials (Koike *et al.*). Column: Dowex 1-X4 ($CH_3COO^-$). Eluent: water followed by acetate buffer of pH 4.5. Peaks: 1 = thiamine + MPT; 2 = DPT; 3 = TPT.

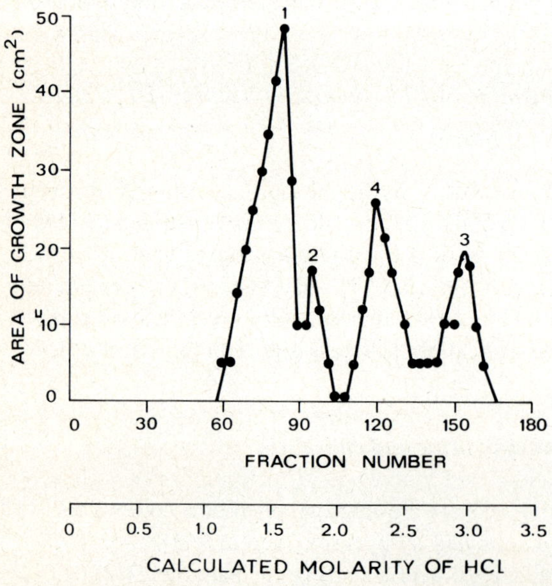


Fig. 44.5. Chromatography of pyridine precursors on a Dowex AG 50W-X8 ($H^+$) column (Diorio and Lewin). Eluent: linear gradient of hydrochloric acid (0.4 *N* in 2 l). Peaks: 1 = 2-methyl-4-amino-5-hydroxymethylpyridine; 2 = 2-methyl-4-amino-5-formylpyrimidine; 3 = 2-methyl-4-aminomethylpyrimidine; 4 = 2-methyl-4-amino-5-methoxymethylpyrimidine.

*Separation of pyrimidine precursors of thiamine on Dowex AG 50W-X8*

This procedure has been used for the separation of precursors from *Neurospora crassa* (Diorio and Lewin).

Dowex AG 50W-X8 ($H^+$) is mixed with broth on which *Neurospora crassa* has grown. The mixture is stirred for 1 h before filtering and discarding the filtrate. The resin is then washed with distilled water and eluted with 2 *M* ammonium hydroxide solution. The eluate containing the pyrimidine precursors is applied to a column of washed Dowex 50W-X8 ($H^+$) (70 × 1.5 cm) and the column is then washed with distilled water until the effluent is at pH 5–6. Elution of pyrimidine precursor compounds is accomplished with a linear gradient of hydrochloric acid (0–4 *N* in 2 l). With this procedure, four compounds can be separated in the above material (Fig. 44.5). It is possible to recover 90–95% of biologically active precursors in the analyzed medium.

## Riboflavin and other flavins

Substances of this type are generally not very stable when exposed to light, so that all analytical procedures, including the chromatographic separation, must be carried out in darkness or subdued red light. The determination of riboflavin and flavins is generally most often based on their fluorescense. In natural materials, however, many interfering fluorescent compounds occur, which must be removed prior to its determination. For this purpose, chromatographic procedures have most often been used.

The separation of riboflavin from an interfering fluorescent compound was first made on several activated clays. Klatzkin *et al.* used Florisil, while Strohecker and Henning used Permutit. Nowadays ion exchangers are most often used for these purposes (Koziołowa and Kozioł). Good results were obtained with Zeo-Karb 215, Wofatit P, F and KS, Lewatit PN and KSN, Staionit FN and F extra, MSF resin and phenolcarboxylic resins such as Lewatit CNS and Zeo-Karb 216. All types of phenolic resins gave completely quantitative sorption of riboflavin, lumiflavin and lumichrom from aqueous solutions, dilute acids (up to 0.5 *N*) and salt solutions (up to 2 *N*) over a pH range from 1 to 8. The capacity of the resins was found to be virtually unaffected by the form of the resin ($H^+$, $Li^+$, $Na^+$, $K^+$ or $NH_4^+$) and closely related to the number of free phenolic OH groups. For the separation of riboflavin, Lammi and Lerner used poly-N-vinylpyrrolidone (Polyclar AT). Recently, Lerner substituted Polyclar AT for Bioclar G. Carletti *et al.* achieved the satisfactory separation of all pure flavins on cellulose. Gupta *et al.* (1967b) used DEAE-cellulose for the same purpose.

Riboflavin can be eluted from Permutit T (Decalso) with water and from phenolic resins with alkaline solutions such as sodium, potassium and ammonium hydroxide, sodium carbonate and sodium tetraborate. Riboflavine 5′-phosphate (FMN) and flavine–adenine dinucleotide (FAD) in aqueous solutions are only partly bound by resorcinol-type resins and are not sorbed on sulphonic- and carboxylic-type resins. They can be eluted from resorcinol resins using a 10% aqueous solution of acetone, and riboflavin with a 1:1 acetone–water mixture. For Polyclar AT, elution of riboflavin with water or salt solution is most suitable (Lammi and Lerner). *tert.*-Butanol–0.01 *N* hydrochloric acid–

*References p.976*

water (50:2.5:40) followed by water is a suitable elution procedure for cellulose (Carletti *et al.*) and 0.01 *M* sodium carbonate solution for DEAE-cellulose (Gupta *et al.*, 1967a, b).

### *Separation of riboflavin from thiamine and vitamin $B_6$ on Permutit*

A neutral or weakly acidic sample solution (pH 4–6) containing 5–50 μg of riboflavin is applied on a column of Permutit T (Decalso). Riboflavin is not bound on Permutit and can be eluted easily with water. In this way, riboflavin can be separated from thiamine and vitamin $B_6$, which remain bound on the column of Permutit (Strohecker and Henning).

### *Separation of riboflavin from vitamin $B_{12}$ and cytochrome c on Bioclar G*

Bioclar G is slurried in a minimal volume of deionized water and the slurry poured into a 40 × 0.8 cm column. The sample containing vitamin $B_{12}$ (red), riboflavin (yellow) and horse cytochrome *c* (orange) is applied on the column, which is eluted with water. Vitamin $B_{12}$ begins to appear in the effluent at 5 min and is completely removed after 7 min, riboflavin follows immediately after this and its elution is completed in 14 min. At this point, water remaining above the bed is removed and replaced with 1% sodium sulphate solution. After 3 min, cytochrome *c* begins to emerge from the column and is completely recovered in 8 min (Lerner).

### *Separation of flavins and lumazines on cellulose*

This method is suitable for the determination of flavins and lumazines in flavinogenic microbes. A good separation of FMN, FAD, 6,7-dimethyl-8-ribityllumazine and 6-methyl-7-hydroxy-8-ribityllumazine is achieved on a 20 × 0.75 cm column packed with cellulose powder. The eluent is *tert.*-butanol–0.01 *N* hydrochloric acid–water (50:2.5:40) followed, after elution of FMN, by water at a flow-rate of 15 ml/h (Carletti *et al.*).

### *Separation of riboflavin from flavin nucleotides*

The separation of riboflavin from other flavin nucleotides is successful with all types of phenolic resins (Zeo-Karb 215, Wofatit F, P and KS, Lewatit PN) using a two-step elution. FMN and FAD can be eluted with water and pure riboflavin in the next step with acetone–water (1:1). With a resorcinol-type resin, the nucleotides can be eluted with a 5–10% aqueous solution of acetone and riboflavin with a 50% acetone–water mixture. For the determination of total flavins in biological materials, the crude extract obtained from samples by hydrolysis with 0.1 *N* sulphuric acid is neutralized with ammonia solution and then passed through the column of resorcinol resin. The column is washed with 0.1 *N* ammonium sulphate in order to remove the non-flavin compounds.

**Nicotinic acid and its derivatives**

No specific procedure is available for the determination of nicotinic acid and its amide and therefore their separation prior to determination is necessary. Nicotinic acid and its derivatives are stable under normal conditions and therefore none of the chromatographic procedures need special conditions.

The methods originally used for the isolation of nicotinic acid from biological materials were based on its selective adsorption on aluminium silicate (Perlzweig *et al.*). Finholt and Higuchi separated nicotinic acid from nicotinamide on the ion exchanger Amberlite IRA-400. McDonald and Stewart separated nicotinamide from nicotinamide mononucleotide using Dowex 2 ($HCOO^-$) (anionic constituents were separated by this step) and nicotinamide from nicotin nucleotide using Amberlite IRC-50. Kahn and Blum used Dowex 1 ($HCOO^-$) and Lee *et al.* Dowex 50 ($H^+$) in combination with Dowex 1. Nicotinic acid formed by the degradation of several nicotinic acid esters could be satisfactorily separated from other degraded components using the anion-exchange resin Amberlite CG-4B (Suzuki and Tanimura). The separation of small amounts of thionicotinamide and selenonicotinamide from the corresponding NADP analogues was achieved by chromatography on Sephadex G-10 and G-25 (Christ *et al.*).

The separation of nicotinic acid on Amberlite IRA-400 is based on its binding on the resin from pH 4.5 to 5.0; nicotinamide can be eluted with water at this pH (Finholt and Higuchi). Kahn and Blum eluted nicotinic acid and its derivatives from Dowex 1 using water followed by a concave gradient of formic acid. Lee *et al.* separated metabolites of nicotinic acid on Dowex 50 with water as eluent followed by their separation on Dowex 1 again with water as eluent.

*Separation of nicotinic acid from its degradation products*

Amberlite CG-4B (50 g) is purified by soaking it in distilled water, washing it with 50 ml of 0.5 *N* hydrochloric acid and subsequently washing it with 500 ml of 0.5 *N* sodium hydroxide solution in a column after removal of the acidic solution with distilled water. The resin is converted from its hydroxide form into its chloride form again with 0.5 *N* hydrochloric acid. This product is washed with distilled water and poured into a chromatographic tube, making a column of dimensions 4 × 0.9 cm. A 5-ml volume of 1 *M* acetate buffer (pH 4.9) is passed through the resin before addition of the sample solution. A 5-ml volume of a sample solution of degraded nicotinic acid esters of polyhydric alcohols is adjusted to pH 2.0–2.3 by addition of 5 ml of sodium hydroxide solution at an appropriate concentration and then the mixture is shaken with an equal volume of chloroform in a glass-stoppered tube for 10 min so as to remove slightly soluble compounds, and then centrifuged for 5 min. Then 5 ml of the aqueous layer and 1 ml of 1 *M* acetate buffer (pH 4.9) are transferred by pipette into a reservoir connected to the column and passed slowly through the column at a rate of 0.7 ml/min. The partly solvolyzed nicotinic acid esters of polyhydric alcohols are eluted with distilled water into a 100-ml calibrated flask at a rate of 1.4 ml/min. The nicotinic acid absorbed to the resin is eluted with 0.3 *N* hydrochloric acid and exactly 50 ml of effluent are collected in a 50-ml calibrated flask at a rate of 1.4 ml/min (Suzuki and Tanimura)

*References p. 976*

*Separation of nicotinic acid from its derivatives on Dowex 1*

This method was used by Kahn and Blum for the separation of nicotinic acid and its derivatives from *Astasia longa* cultivated in a synthetic medium containing [7-$^{14}$C] nicotinic acid. A trichloroacetic acid extract is applied to a Dowex 1 ($HCOO^-$) column (50 × 0.9 cm), which is washed with water. After about 100 fractions of 10 ml each have been collected, a concave gradient of formic acid is applied to the column and fractions of 5 ml are collected.

## Pyridoxine group

This group consists of pyridoxol, pyridoxal and pyridoxamine. Before their determination, it is necessary to extract these components using acidic or enzymatic hydrolysis and separate them using column, thin-layer or paper chromatography.

The original simple method for the separation of substances that interfere in the determination of pyridoxol was adsorption on activated clay. At present, the use of ion-exchange resins based on weakly acidic cation exchangers has been developed. Strohecker and Henning recommend the ion exchanger IV, while Platzer and Roberts used Dowex AG W-X8. Takanashi *et al.* carried out the separation on Dowex followed by Amberlite, while Johansson and Lindstedt used Dowex AG 50W-X8. Contractor and Shane followed the metabolism of pyridoxol on rats using cellulose phosphate, charcoal and DEAE-cellulose.

From ion exchanger IV (Merck), pyridoxol can be eluted with 1 *N* hydrochloric acid, from Dowex AG W-X8 with a hot mixture of 0.6 *M* potassium chloride and 0.1 *M* potassium dihydrogen orthophosphate (Platzer and Roberts). Takanashi *et al.* eluted pyridoxol from a Dowex column with acetate buffer of pH 4 and from an Amberlite CG-120 column with 0.4 *M* phosphate buffer of pH 7.5. Johansson and Lindstedt used Dowex 50W-X8 with 0.05 *M* ammonium formate solution of pH 4.25 followed by gradient elution with 100 ml of 0.05 *M* ammonium formate of pH 4.25 in a closed chamber to which 0.5 *M* ammonium formate of pH 7.5 was added.

*Separation of vitamin $B_6$ from interfering materials on a Dowex column*

A 10-ml volume of wet settled Dowex AG W-X8 ($K^+$, 100–200 mesh), is placed in a 25.5 × 1.0 cm glass column and adjusted to pH 4.5 by washing with 50 ml of 0.01 *M* potassium acetate solution of pH 4.5. An acid extract of 500 mg of tissue or vitamin $B_6$ standards adjusted to pH 4.5 is placed on the column, which is then washed with 50 ml of 0.02 *M* potassium acetate solution of pH 5.45. The vitamin $B_6$ components are eluted in one step with 30 ml of a boiling mixture of 0.6 *M* potassium chloride and 0.1 *M* potassium dihydrogen orthophosphate of pH 8.0. Then 4.5 ml of 1 *M* calcium chloride solution are added to the eluate and the volume is adjusted to 35 ml with water. The mixture is centrifuged at 700 *g* for 15 min and the supernatant fluid is assayed for total vitamin $B_6$. A convex gradient of increasing pH and molarity of potassium acetate, pH

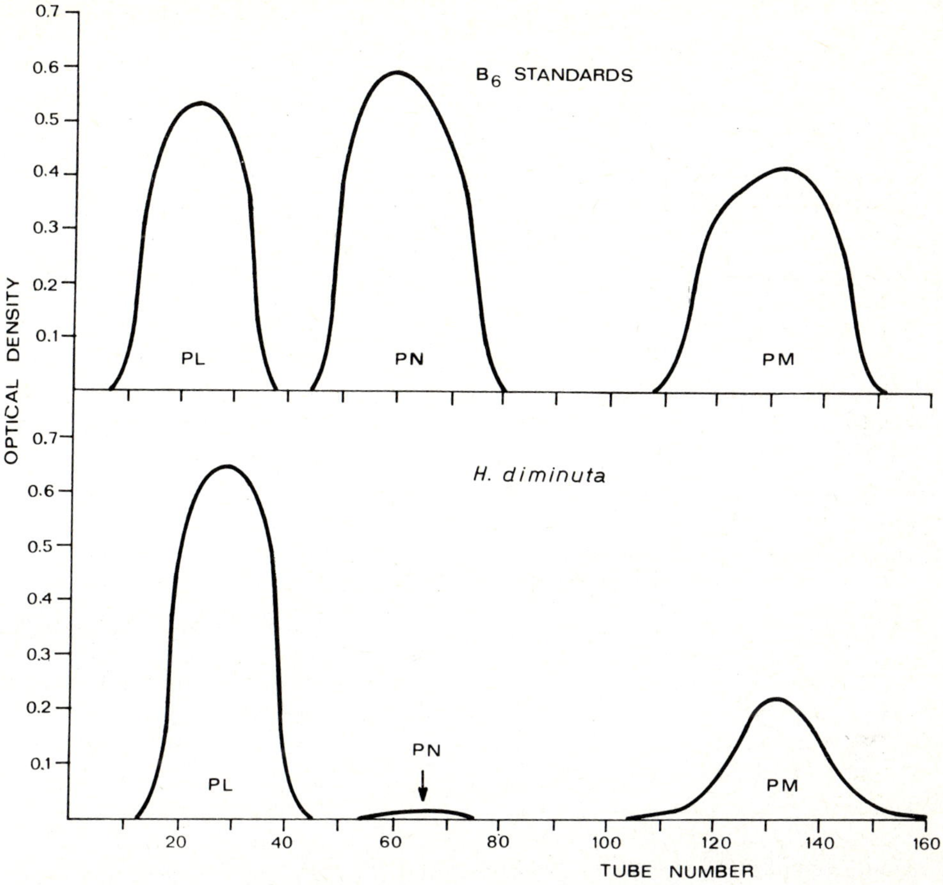


Fig. 44.6. Chromatography of vitamins $B_6$ (Platzer and Roberts). Column: Dowex AG W-X8 ($K^+$). Eluent: boiling 0.6 *M* potassium chloride and 0.1 *M* potassium dihydrogen orthophosphate (pH 8.0) and convex gradient of increasing pH and molarity of potassium acetate, pH 5.45 and 0.02 *M* to pH 7.0 and 0.1 M. Peaks: PL = pyridoxal; PN = pyridoxine; PM = pyridoxamine.

5.45 and 0.02 *M* to pH 7.0 and 0.1 *M*, is used for the separation of the vitamin $B_6$ group (Fig. 44.6) (Platzer and Roberts).

### *Separation of vitamin $B_6$ compounds on Dowex 50W-X8*

A perchloric acid extract of muscle tissue is prepared. A suitable portion is dissolved in 1–2 ml of 0.05 *M* ammonium formate buffer of pH 4.25 and placed on a 40 × 0.9 cm column of Dowex 50W-X8. The column is eluted first with 100 ml of 0.05 *M* ammonium formate of pH 4.25, then a gradient is started with 100 ml of 0.05 *M* ammonium formate of pH 4.25 in a closed chamber to which 0.5 *M* ammonium formate of pH 7.5 is added (Fig. 44.7) (Johansson and Lindstedt).

*References p.976*

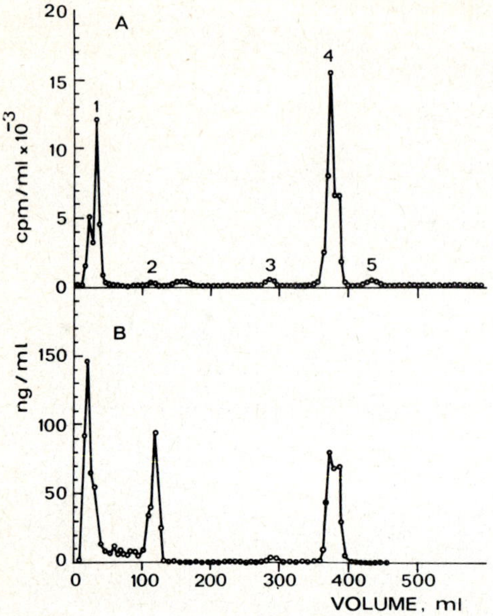


Fig. 44.7. Chromatography of vitamin $B_6$ from mouse liver (Johansson and Lindstedt). Column: Dowex 50W-X8. Eluent: ammonium formate, pH 4.25, and gradient of ammonium formate. A, unhydrolyzed extract; B, same extract after hydrolysis. Peaks: 1 = pyridoxine-5′-phosphate and pyridoxal-5′-phosphate; 2 = pyridoxalamine-5′-phosphate; 3 = pyridoxal; 4 = pyridoxine; 5 = pyridoxamine.

## Biotin

Chemical and physico-chemical methods for the determination of biotin are used only rarely; the biological activity of biotin is most often tested microbiologically. In special instances, column chromatography can also be used.

Column chromatography has been applied to the study of metabolites during the cultivation of microorganisms on culture media containing biotin or during the biosynthesis of biotin or its vitamers. For the binding of biotin or biotin-containing peptides, Sepharose–avidin columns are efficient (Bodanszky and Bodanszky). Avidin can be coupled with Sepharose 4B activated with cyanogen bromide. Enzymatically synthesized [$^{14}$C] desthiobiotin was purified by anion-exchange chromatography on Bio-Rad AG 1-X8; the desthiobiotin–avidin complex was isolated on Sephadex G-25F (Eisenberg and Krell).

### *Separation of biotin from interfering compounds on Dowex*

Iwahara *et al.* studied the products isolated from culture filtrates of *Pseudomonades* grown on a biotin–salts medium to which [$^{14}$C] carbonyl-labelled biotin was added.

The bacterial culture is centrifuged for 20 min at 16,000 *g* in order to remove the cells. Dowex 50W-X8 (100–200 mesh) is stirred into a total of 9 l of supernatant solution which has been filtered through paper. The resin is washed with 6 l of 70% ethanol, combined with the first filtrate and concentrated to 600 ml. After removal of the final precipitate, the solution (pH 7.0) is divided into three equal portions and each portion is poured over a 70 × 1.5 cm column of Dowex 1-X2 ($HCOO^-$), 100–200 mesh. The radioactive materials are eluted from the column by successive treatments with water and 0.01 *M* and 0.1 *M* formic acid.

Yang *et al.*, in studies on the changes to biotin due to *Rhodotorula, Penicillium* and *Endomycopsis* also investigated its separation on Dowex 1-X2. Eisenberg and Maseda followed the biotin derivatives from *Penicillium chrysogenum* on Dowex 50W-X8.

## Pantothenic acid and coenzyme A

In addition to the separation of pantothenic acid and its derivatives from natural materials, column chromatography has also been applied to prepare some analogues of pantothenic acid and, especially, to prepare coenzyme A.

Saccharides and riboflavin, during alkaline hydrolysis of analyzed materials, form a brown colour that interferes in the spectrophotometric determination. The colour can be removed on ion exchanger V (Merck) (Strohecker and Henning). Nagase *et al.*, in the preparation of some pantothenic acid analogues, used different types of Amberlite. Yoshioka *et al.* separated D-pantothenic acid 4-phosphate from the reaction mixture on DEAE-Sephadex. Cha *et al.* and Zahler and Cleland used DEAE-cellulose for the separation of coenzyme A. DEAE-cellulose was also used for the separation of some analogues of coenzyme A (Shimizu *et al.*, 1968) and this procedure was improved using DEAE-Sephadex A-25 (Shimizu *et al.*, 1970a) and Dowex 50 (Shimizu *et al.*, 1970b). Stearylcoenzyme A synthesized by Haeffner was isolated using Sephadex G-15.

Pantothenol can be separated from pantothenic acid on an anion exchanger, the elution being carried out with water. Pantothenic acid is eluted with 0.1 *N* hydrochloric acid (Knobloch). Yoshioka *et al.* eluted pantothenic acid 4-phosphate from DEAE-Sephadex with an increasing gradient of ammonium carbonate (0–1.0 *M*). In order to separate coenzyme A, Cha *et al.* used DEAE-cellulose and gradient elution with triethylamine hydrogen carbonate (pH about 7.5) from 0 to 0.8 *M* in a total volume of 1 l. For a similar purpose, DEAE-cellulose and a linear gradient of lithium chloride from 0.05 to 0.5 *M* was used by Zahler and Cleland and by Shimizu *et al.* (1968). Stearylcoenzyme A can be eluted from Sephadex G-15 using 0.5% mercaptoethanol.

### *Separation of coenzyme A analogues*

Some coenzyme A analogues, such as α-methyl-, β-methyl- and α-carboxycoenzyme A, can be separated on a DEAE-cellulose column (Shimizu *et al.*, 1968). The reaction mixture is passed through a 30 × 2 cm column of DEAE-cellulose ($Cl^-$) and elution is carried out using a linear salt gradient. The reservoir contains 0.3 *M* lithium chloride in 0.003 *N* hydrochloric acid (1 l) and the mixing vessel contains 0.003 *N* hydrochloric acid (1 l). The same workers (Shimizu *et al.*, 1970a) improved this method by using DEAE-Sephadex A-25 ($Cl^-$) with the same eluent.

*References p.976*

**Folic acid and other pteridine derivatives**

During the preparation of samples, their instability in light and in the presence of air must be considered. Hence all preparative procedures, including the chromatographic separation, must be carried out in the dark (or subdued red light) and under an atmosphere of nitrogen.

For the separation of a mixture of pteridine derivatives, columns of cellulosic ion exchangers are mostly used. Columns of cellulose phosphate are utilised mainly for the preparation of synthetic mixtures of pteridines (Chippel and Scrimgeour, Guroff and Rhoads, Jones and Brown, Mitsuda and Suzuki); cellulose phosphate was also used for the separation of substances of this type from natural material (Fukushima, Rembold *et al.*). For the isolation of pteridines from natural material, DEAE-cellulose (Chippel and Scrimgeour; Gupta and Huennekens; Gupta *et al.*, 1967b; Ho and Jones; Zakrzewski *et al.*; Zakrzewski and Sansone) or Dowex 1-X2 (Kaufman) and Dowex 1-X8 (Rembold *et al.*, Sugiura and Goto) can also be used. For the differentiation of isomers of methyl tetrahydropteridines, Whiteley *et al.* applied the cation exchanger Dowex 50, while Guroff and Rhoads used CM-Sephadex. The modified celluloses used for this purpose are represented by ECTEOLA-cellulose (Fukushima, Rembold *et al.*).

For the separation of pteridines from natural material, it is possible to use a molecular sieve of the Sephadex type. Sephadex G-10 and G-25 (Dewey and Kidder, Fukushima, Sugiura and Goto), suitable for the separation of low-molecular-weight substances, were applied for this purpose.

In some special instances, Florisil (Guroff and Rhoads), alumina (Hla-Pe and Aung-Than-Batu) and cellulose (Sugiura and Goto) were used for the separation of pteridine derivatives from other interfering ballast components.

Combinations of the above methods have been often used for the separation of natural mixtures In a study of the biosynthesis of pteridines in the tadpole of the bullfrog, Fukushima used separations on ECTEOLA-cellulose, cellulose phosphate and Sephadex G-25. Guroff and Rhoads applied Florisil and Sephadex in studies on the hydroxylation of phenylalanine by *Pseudomonas* species. To study the properties and metabolisms of pteridine derivatives in the rat liver, Rembold *et al.* applied Dowex 1-X8, ECTEOLA-cellulose and cellulose phosphate.

When using these ion exchangers, various eluents have been used according to the procedure and the properties of the separated components and the ion exchanger used. The elution of pteridines from a column of CM-Sephadex was carried out with a gradient of a mixture of sodium acetate in Cleland's reagent (a solution of dithiothreitol) (Guroff and Rhoads). For the separation of synthetic standards of pteridines on cellulose phosphate, elution in two steps was used, acidic and neutral substances being eluted with water and basic substances with 5% formic acid and buffer solution (Rembold *et al.*). Using Dowex 1-X2 ($CH_3COO^-$), the elution of reduced biopterine derivatives was carried out with 0.028 *M* 2-mercaptoethanol (Kaufman). For the elution of pteridines from a column of DEAE-cellulose, a gradient of ammonium acetate (Chippel and Scrimgeour, Gupta and Huennekens, Ho and Jones) in admixture with mercaptomethanol, or a phosphate gradient in admixture with potassium ascorbate (Rohringer *et al.*), was used.

*Separation of pteridines in the tadpole of the bullfrog*

The skin (10–15 g) is boiled in water (15 ml) for 5 min and homogenized in a Waring blender for 1 min. The homogenate is mixed with 80 ml of ethanol and centrifuged. This procedure is repeated and the combined supernatant fluid is concentrated by evaporation. A concentrated extract is applied to a pH 7 cellulose column (15 × 3 cm). Yellow fluorescent substances are eluted with water first, followed by blue fluorescent substances. As isoxanthopterin and pterin-6-carboxylic acid are not eluted readily with water, these compounds are eluted with 0.1 *N* hydrochloric acid. The yellow fluorescent fraction is concentrated and applied to a 20 × 1.8 cm Sephadex G-25F column and the yellow fluorescent fraction from this column is collected, concentrated and applied to a 20 × 1.8 cm cellulose phosphate column. The fraction is resolved into three zones; the first eluted material consists mainly of riboflavin and the second of sepiapterin. The concentrated blue fluorescent fraction from the first column is applied to a pH 7 cellulose column (35 × 1.8 cm), and biopterin and 2-amino-4-hydroxypteridine are separated from each other in this step. The biopterin fraction is purified using a 20 × 1.8 cm cellulose phosphate column. Chromatographic data for the normal pteridines on the columns used are presented in Table 44.3.

TABLE 44.3
COLUMN CHROMATOGRAPHY OF PTERIDINES (FUKUSHIMA)

| Compound | Relative elution volume | | $K_d$ |
|---|---|---|---|
| | pH 7 ECTEOLA-cellulose column | Cellulose phosphate column | Sephadex G-25F column |
| Biopterin | 100 | 100 | 1.6 |
| Neopterin | 103 | 97 | 1.6 |
| Dihydrobiopterin | 60 | –* | 2.0 |
| Sepiapterin | 67 | 26 | 2.5 |
| 2-Amino-4-hydroxypteridine | 126 | 112 | 1.8 |
| 6-Hydroxymethylpterin | 126 | 121 | 1.8 |
| Isoxanthopterin | 350** | 15 | 2.2 |
| Pterin-6-carboxylic acid | –*** | 10 | 1.0 |
| Lumazine | 107 | 13 | 1.7 |

*The compound is not stable on this column.
**Rough estimation.
***Not eluted with water.

## Corrinoids

Corrinoids are substances derived from simple corrine with centrally bound cobalt within complexes of these compounds with proteins.

For the separation of compounds of this type, ion exchangers and molecular sieves have been used, particularly Amberlite XAD-1 and XAD-2 (Kamikubo and Narahara), DEAE-cellulose (Gräsbeck *et al.*, Wolff *et al.*) and DEAE-Sephadex A-50 (Ellenbogen and Highley, Gräsbeck *et al.*). Cation exchangers used for the separation of corrinoids include

CM-cellulose (Finkler *et al.*; Tortolani *et al.*, 1970a) Dowex 50W-X2 (Tortolani *et al.*, 1970a), SP-Sephadex C-25 (Tortolani *et al.*, 1970b) and CM-Sephadex C-50 in combination with other molecular sieves (Gräsbeck *et al.*, Simons and Weber).

Molecular sieves of the Sephadex type play an important role in studies on the qualitative and quantitative relationships between cobalamins and their protein carriers. Using Sephadex G-50 (Gullberg, 1970a), G-25 (Hom), G-100 (Finkler *et al.*, Garrido-Pinson *et al.*, Simons and Weber), G-150 (Gullberg, 1970b) and G-200 (Gräsbeck *et al.*, Hom, Olesen *et al.*, Simons and Weber), the protein carriers of cobalamins were isolated from different sources (human plasma, human gastric juice, human leukocytes, etc.). Yurkevich *et al.* used Sephadex G-15 for the separation of cobalamins in the form of triphenylphosphine complexes.

Simons and Weber and Wolff *et al.* eluted corrinoids from DEAE-cellulose using a gradient of pH and molarity of a phosphate buffer while Simons and Weber used a gradient of sodium chloride and phosphate buffer for elution from DEAE-Sephadex A-50. With Amberlite, Kamikubo and Narahara used ethanol and acetone for the elution. For molecular sieves, Tris–sodium chloride buffer (Gulberg, 1970a,b) and phosphate buffer (Garrido-Pinson *et al.*) were mostly used.

*Separation and determination of corrinoids on an SP-Sephadex column*

SP-Sephadex is dispersed in a beaker with a 0.05 *M* sodium acetate buffer (pH 5.0) and poured into the chromatographic column (20 × 0.9 cm). The column is washed repeatedly with distilled water in order to remove excess ions and the eluate is checked for spectrophotometric purity in the wavelength range 260–400 nm. Then 2 ml of a solution containing 50 μg/ml of each corrinoid are applied on the column. The first elution is carried out with 40 ml of water, followed by 70 ml of 0.05 *M* acetate buffer (pH 5.0),

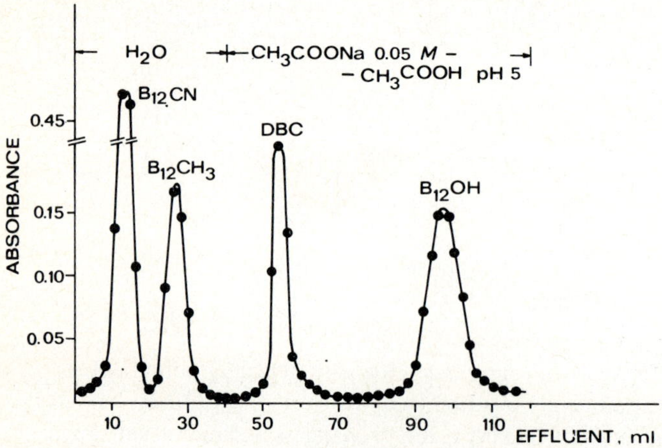


Fig. 44.8. Separation of cyanocobalamin ($B_{12}CN$), methylcobalamin ($B_{12}CH_3$), hydroxycobalamin ($B_{12}OH$) and cobamamide (DBC) (Tortolani *et al.*, 1970b). Column: SP-Sephadex C-25 ($Na^+$) Eluent: distilled water followed by 0.05 *M* sodium acetate buffer of pH 5.0.

and 2-ml fractions are collected. Cyanocobalamin and methylcobalamin are eluted separately with distilled water, while cobalamide and hydroxycobalamin remain fixed at the top of the column (Fig. 44.8); they can be separated by increasing the ionic strength of the eluent. Each 2-ml fraction is read at the corresponding wavelength. All operations must be conducted in subdued red light so as to prevent photolytic degradation of the cobalamins (Tortolani *et al.*, 1970b).

### L-Ascorbic and L-dehydroascorbic acids

Column chromatography is not often used in analyses of vitamin C, paper chromatography more often being employed. Ascorbic and dehydroascorbic acids behave in a similar manner to sugars, and in many instances the chromatographic procedures for sugars can be applied to these substances.

Friberg *et al.* separated osazones of dehydroascorbic acid using chromatography on charcoal, Sephadex G-50 and DEAE-Sephadex A-25 while Fiddick and Heath separated bound ascorbic acid from rat adrenals by gel filtration using Sephadex G-50 with 0.05 $N$ sodium chloride solution as eluent (Fig. 44.9).

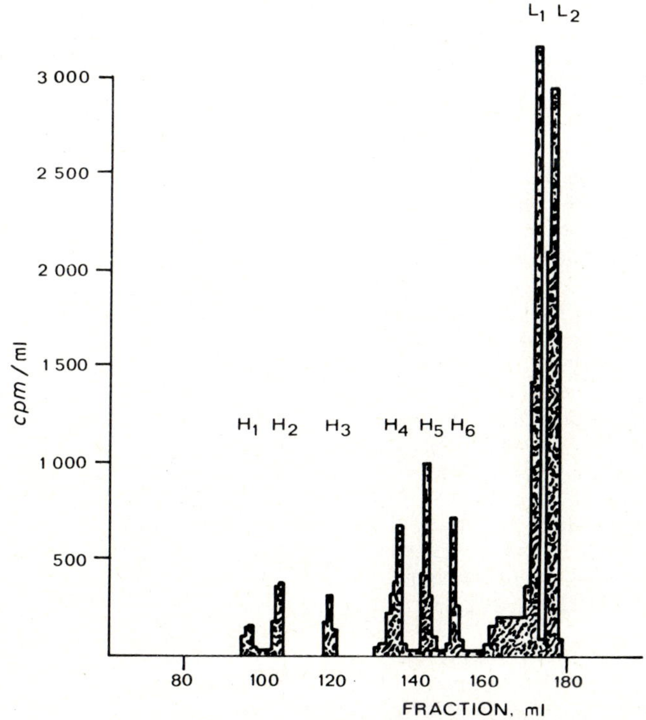


Fig. 44.9. Separation of ascorbic and dehydroascorbic acids from rat adrenals (Fiddick and Heath). Column: Sephadex G-50. Eluent: 0.1 N sodium chloride solution. Peaks: $H_1$ –$H_6$ represent probably complexes of amino acids with polypeptides; $L_1$ and $L_2$ represent ascorbic acid and dehydroascorbic acid, respectively.

*References p.976*

In some instances, chromatographic techniques can be applied to separate ascorbic and dehydroascorbic acid from compounds that might interfere in their determination. Davídek *et al.* and Grundová *et al.* separated anthocyanine pigments, which interfere in the polarographic determination of dehydroascorbic acid, on polyamide or Dowex 50W. Anthocyanine pigments are bound on the column and ascorbic and dehydroascorbic acids together with non-interfering substances can be eluted with water or 2% oxalic acid.

*Separation of L-ascorbic and L-dehydroascorbic acids on polyamide powder*

Polyamide powder (54 mesh) is mixed with water and allowed to stand for about 2 h and then a 20 × 1 cm column is prepared. A 20-ml volume of analyzed extract (the extract of ascorbic and dehydroascorbic acids is prepared with 2% of oxalic acid) is added to the top of the column. Ascorbic and dehydroascorbic acids are eluted into a 50-ml calibrated flask with 2% oxalic acid and the eluate is used for analysis. Dowex 50W-X4 ($Na^+$) can be used instead of polyamide (Davídek *et al.*).

## REFERENCES

Amos, W. H. and Neal, R. A., *J. Biol. Chem.*, 245 (1970) 5643.
Bell, P. A. and Kodicek, E., *Biochem. J.*, 115 (1969) 663.
Bertram, S. and Krisch, K., *Eur. J. Biochem.*, 11 (1969) 122.
Bodanszky, A. and Bodanszky, M., *Experientia*, 26 (1970) 327.
Carletti, P., Strom, R., Giovenco, S., Barra, D. and Giovenco, M. A., *J. Chromatogr.*, 29 (1967) 182.
Cassagne, C. and Baraud, J., *Bull. Soc. Chim. Fr.*, (1968) 1470.
Cha, S., Cha, M., Chung-Ja. and Parks, Jr., R., *J. Biol. Chem.*, 242 (1967) 2582.
Chippel, D. and Scrimgeour, K. G., *Can. J. Biochem.*, 48 (1970) 999.
Christ, W., Schmidt, D. and Coper, H., *J. Chromatogr.*, 51 (1970) 537.
Cinquina, C. L., *J. Bacteriol.*, 95 (1968) 2436.
Contractor, S. F. and Shane, B., *Biochem. Biophys. Res. Commun.*, 39 (1970) 1175.
Crain, F. D., Lotspeich, F. J. and Krause, R. F., *J. Lipid Res.*, 8 (1967) 249.
Davídek, J., Grundová, K., Velíšek, J. and Janíček, G., *Lebensm. -Wiss. Technol.*, 5 (1972) 213.
De Luca, H. F., *J. Agr. Food Chem.*, 17 (1969) 778.
Dewey, V. C. and Kidder, G. W., *J. Chromatogr.*, 31 (1967) 326.
Dicks-Bushnell, M. W., *J. Chromatogr.*, 27 (1967) 96.
Diorio, A. F. and Lewin, L. M., *J. Biol. Chem.*, 243 (1968) 4006.
Dollwet, H. H. A. and Norman, A. W., *Anal. Biochem.*, 25 (1968) 297.
DuPont, *Liquid Chromatography Methods Bulletin*, DuPont, Wilmington, Del., March, 1972.
Eisenberg, M. A. and Krell, K., *J. Biol. Chem.*, 244 (1969) 5503.
Eisenberg, M. A. and Maseda, R., *Biochemistry*, 9 (1970) 108.
Ellenbogen, L. and Highley, D. R., *J. Biol. Chem.*, 242 (1967) 1004.
Eremina, G. V., *Lab. Delo*, (1968) 81.
Fiddick, R. and Heath, H., *Biochim. Biophys. Acta*, 136 (1967) 206.
Finholt, P. and Higuchi, T., *J. Pharm. Sci.*, 51 (1962) 655.
Finkler, A. E., Green, P. D. and Hall, Ch. A., *Biochim. Biophys. Acta*, 200 (1970) 151.
Friberg, V., Lohmander, S. and Carlsson, G., *Acta Chem. Scand.*, 221 (1968) 3037.
Fukushima, T., *Arch. Biochem. Biophys.*, 139 (1970) 361.
Garrido-Pinson, G. C., Turner, M. D., Miller, L. L. and Segal, H. L., *Biochim. Biophys. Acta*, 127 (1966) 478.
Gräsbeck, R., Simons, K. and Sinkkonen, I., *Biochim. Biophys. Acta*, 127 (1966) 47.

Grundová, K., Davídek, J., Velíšek, J. and Janíček, G., *Lebensm. Wiss. Technol.*, 6 (1973) 11.
Gullberg, R., *Clin. Chim. Acta*, 27 (1970a) 251.
Gullberg, R., *Clin. Chim. Acta*, 29 (1970b) 97.
Gupta, V. S. and Huennekens, F. M., *Biochemistry*, 6 (1967) 2168.
Gupta, V. S., Kraft, S. C. and Samuelson, J. S., *J. Chromatogr.*, 26 (1967a) 158.
Gupta, V. S., Ozols, J. G. and Huennekens, F. M., *Biochemistry*, 6 (1967b) 2159.
Guroff, G. and Rhoads, C. A., *J. Biol. Chem.*, 244 (1969) 142.
Haeffner, E. W., *J. Chromatogr.*, 50 (1970) 140.
Haussler, M. R., Myrtle, J. F. and Norman, A. W., *J. Biol. Chem.*, 243 (1968) 4055.
Hirsch, J., *J. Lipid Res.*, 4 (1963) 1.
Hla-Pe, U. and Aung-Than-Batu, *Clin. Chim. Acta*, 24 (1969) 381.
Ho, P. P. K. and Jones, L., *Biochim. Biophys. Acta*, 148 (1967) 662.
Hom, B. L., *Biochim. Biophys. Acta*, 175 (1969) 20.
Iwahara, S., McCormick, D. B., Wright, L. D., and Lih, Ch., *J. Biol. Chem.*, 244 (1969) 1393.
Johansson, S. and Lindstedt, S., *Biochemistry*, 7 (1968) 2327.
Jones, T. H. D. and Brown, G. M., *J. Biol. Chem.*, 242 (1967) 3989.
Kahn, V. and Blum, J. J., *J. Biol. Chem.*, 243 (1968) 1441.
Kamikubo, T. and Narahara, H., *Vitamins*, 37 (1968) 225; *C.A.*, 68 (1968) 93131h.
Kaufman, S., *J. Biol. Chem.*, 242 (1967) 3989.
Klatzkin, Ch., Norris, F. W. and Wokes, F., *J. Pharm. Pharmacol.*, 1 (1949) 915.
Knobloch, E., *Fyzikálně Chemické Metody Stanovení Vitaminů*, Czechoslovak Academy of Sciences, Prague, 1956.
Koike, H., Wada, T. and Minakami, H., *J. Biochem.*, 62 (1967) 492.
Koziołowa, A. and Kozioł, J., *J. Chromatogr.*, 34 (1968) 216.
Lammi, C. J. and Lerner, J., *J. Chromatogr.*, 43 (1969) 395.
Lawson, D. E. M., Wilson, P. W. and Kodicek, E., *Biochem. J.*, 115 (1969) 269.
Lee, Y. C., Gholson, R. K. and Raica, N., *J. Biol. Chem.*, 244 (1969) 3277.
Lerner, J., *J. Chem. Educ.*, 47 (1970) 32.
McDonald, J. W. D. and Stewart, H. B., *Can. J. Biochem.*, 45 (1967) 363.
McLaren, D. S., Read, W. W. C., Awdeh, Z. L. and Tchalian, M., *Methods Biochem. Anal.*, 15 (1967) 1.
Mariani, A. and Mariani-Vicari, C., *Ann. Ist. Super. Sanita*, 4 (1968) 90.
Matsuo, T. and Suzuoki, Z., *J. Biochem.*, 65 (1969) 953.
Millar, K. R. and Caravaggi, C., *N. Z. J. Sci.*, 13 (1970) 329.
Mitsuda, H. and Suzuki, Y., *Biochem. Biophys. Res. Commun.*, 36 (1969) 1.
Myrtle, J. F., Haussler, M. R. and Norman, A. W., *J. Biol. Chem.*, 245 (1970) 1190.
Nagase, O., Tagawa, H. and Shimizu, M., *Chem. Pharm. Bull.*, 16 (1968) 977.
Neal, R. A., *J. Biol. Chem.*, 243 (1968) 4634.
Neal, R. A., *J. Biol. Chem.*, 244 (1969) 5201.
Nyström, E. and Sjövall, J., *J. Chromatogr.*, 24 (1966) 212.
Olesen, H., Rehfeld, J., Hom, B. L. and Hippe, E., *Biochim. Biophys. Acta*, 194 (1969) 67.
Perlzweig, W. A., Levy, E. D. and Sarett, H. P., *J. Biol. Chem.*, 136 (1940) 729.
Peterson, P. A., *J. Biol. Chem.*, 246 (1971a) 34.
Peterson, P. A., *J. Biol. Chem.*, 246 (1971b) 44.
Peterson, P. A. and Berggård, I., *J. Biol. Chem.*, 246 (1971) 25.
Platzer, E. G. and Roberts, L. S., *Comp. Biochem. Physiol.*, 35 (1970) 535.
Rembold, H., Metzger, H., Sudershan, P. and Gutensohn, W., *Biochim. Biophys. Acta*, 184 (1969) 386.
Roberts, A. B. and De Luca, H. F., *J. Lipid Res.*, 9 (1968) 501.
Rohringer, R., Kim, W. K. and Samborski, D. J., *Can. J. Biochem.*, 47 (1969) 1161.
Shichi, H., Lewis, M. S., Irreverre, F. and Stone, A. L., *J. Biol. Chem.*, 244 (1969) 529.
Shimizu, M., Nagase, O., Hosokawa, Y. and Tagawa, Y., *Tetrahedron*, 24 (1968) 5241.
Shimizu, M., Nagase, O., Hosokawa, Y., Tagawa, H. and Yotsui, Y., *Chem. Pharm. Bull.*, 18 (1970a) 838.
Shimizu, M., Nagase, O., Okada, S. and Hosokawa, Y., *Chem. Pharm. Bull.*, 18 (1970b) 313.
Simons, K. and Weber, T., *Biochim. Biophys. Acta*, 117 (1966) 201.

Skinner, W. A., Parkhurst, R. M., Scholler, J. and Schwarz, K., *J. Med. Chem.*, 12 (1969) 64.
Strohecker, R. and Henning, H. M., *Vitamin-Bestimmungen*, Verlag Chemie, Weinheim, 1963.
Suda, T., De Luca, H. T., Schnoes, H. K. and Blunt, J. W., *Biochemistry*, 8 (1969) 3515.
Sugiura, K. and Goto, M., *J. Biochem.*, 64 (1968) 657.
Suzuki, T. and Tanimura, Y., *Chem. Pharm. Bull.*, 17 (1969) 1422.
Suzuoki, N. Z., Matsuo, T. and Tominaga, F., *J. Biochem.*, 63 (1968) 792.
Takahashi, T. and Yamamoto, R., *Yakugaku Zasshi,* 89 (1969) 993; *C.A.*, 71 (1969) 84510t.
Takanashi, S., Matsunaga, J. and Tamura, Z., *J. Vitamin.*, 16 (1970) 132.
Tortolani, G., Bianchini, P. and Mantovani, V., *Farmaco, Ed. Prat.*, 25 (1970a) 772.
Tortolani, G., Bianchini, P. and Mantovani, V., *J. Chromatogr.*, 53 (1970b) 577.
Whiteley, J. M., Drais, J. H. and Huennekens, F. M., *Arch. Biochem. Biophys.*, 133 (1969) 436.
Williams, J. P., *J. Chromatogr.*, 36 (1968) 504.
Wolff, R., Linden, G. and Nicolas, J. P., *Bull. Soc. Chim. Biol.*, 51 (1969) 191.
Yang, H. C., Kusumoto, M., Iwahara, S., Tochikura, T. and Ogata, K., *Agr. Biol. Chem.*, 33 (1969) 1730.
Yoshioka, M., Samejima, K. and Tamura, Z., *Chem. Pharm. Bull.*, 17 (1969) 1265.
Yurkevich, A. M., Rudakova, I. P. and Pospelova, T. A., *Zh. Obshch. Khim.*, 39 (1969) 425.
Zahler, W. L. and Cleland, W. W., *J. Biol. Chem.*, 243 (1968) 716.
Zakrzewski, S. F., Evans, E. A. and Phillips, R. F., *Anal. Biochem.*, 36 (1970) 197.
Zakrzewski, S. F. and Sansone, A., *J. Biol. Chem.*, 242 (1971) 5661.
Zile, M. and De Luca, H. F., *Anal. Biochem.*, 25 (1968) 307.

*Chapter 45*

# Antibiotics

V. BETINA

## CONTENTS



## INTRODUCTION

Antibiotics are microbial metabolites possessing antimicrobial, cytotoxic, antitumour, antinematodal or insecticidal properties. Within the three decades from 1937 to 1966, discoveries of 1630 antibiotics were reported but 427 of them were rediscoveries (Neelameghan *et al.*). Some tens of new antibiotics have been discovered since 1966. Antibiotics belong to very different groups of organic compounds but until now a classification from a chemical point of view has been virtually non-existent. The classification of antibiotics used in the *"Bibliography of Column Chromatography"* (Deyl *et al.*) was accepted in writing the present chapter.

Classical and modern techniques of column chromatography are indispensable tools in the study and preparation of antibiotics. They are used mainly for the following purposes:

(i) laboratory and commercial isolation and purification of antibiotics; (ii) separation of mixtures of antibiotics; (iii) structural studies; (iv) studies of the biogenesis and preparation of labelled antibiotics; (v) studies of chemical modifications and biotransformations of antibiotics; and (vi) purity control.

Samples of various qualities can be applied on to chromatographic columns. They include filtrates and extracts of fermentation broths, crude concentrates of antibiotics, partially purified substances and pharmaceutical preparations.

Liquid–solid chromatography is widely used in separations of antibiotics belonging to different chemical groups. Applications of gel chromatography in the field of penicillins, tetracyclines and peptidic antibiotics are included in this chapter. Ion-exchange chromatography is of a great value for the isolation and separation of carbohydrate and peptidic antibiotics. It is also helpful in the separation of amino acids as building units of the latter group of compounds, which is important in structural studies.

In addition to physical and chemical methods used in qualitative and quantitative determinations of compounds separated by means of CC, assays of antibiotics are also based on their biological activity. Aliquots of fractions can be tested for inhibitory effects on sensitive microorganisms, and the zones of inhibition and their diameters then indicate the presence and amounts of active components in eluates.

As paper chromatography and thin-layer chromatography are usually combined with CC in order to check the purity of eluates, bioautography must be mentioned as a general method for the detection of antibiotics. The bioautography of antibiotics in PC and TLC is based on their inhibitory effects. Using sensitive microorganisms, the positions of antibiotics on paper or thin-layer chromatograms are rendered visible by the zones of inhibition. Technical details and references concerning the bioautography of antibiotics can be found elsewhere (Betina, 1972, 1973).

This chapter deals mainly with antibiotics which are or could be used pharmaceutically, in agriculture or in basic research. In some instances, chemically related antibiotics, semisynthetic and synthetic derivatives and analogues are also included.

## PENICILLINS AND CEPHALOSPORINS

### Penicillins and 6-APA

Penicillins are N-acyl derivatives of 6-aminopenicillanic acid (6-APA). Cephalosporin N is also a derivative of 6-APA and is now named penicillin N. Its side-chain is identical with that of cephalosporin C, which is an N-acyl derivative of 7-aminocephalosporanic acid (7-ACA). Cephalosporin $P_1$ is a steroid compound. Natural, biosynthetic, semisynthetic and synthetic derivatives of 6-APA and 7-ACA are often named as $\beta$-lactam antibiotics.

Isolation of 6-APA at the end of the fifties made possible the preparation of large numbers of N-substituted derivatives, the "semisynthetic penicillins". Since that time some 1800 different penicillins have been prepared and studied by a pharmaceutical company (Nayler) and the number examined throughout the world must amount to several thousands.

In recent years, CC was used in three fields of studies of penicillins: a) in chemical transformations of penicillins and 6-APA; b) in studies of penicillin allergy in humans; c) in isolation of polymers formed during storage of aqueous solutions of penicillins and 6-APA.

Attempts to convert penicillins into other β-lactam systems have been fruitful. Morin *et al.* (1969a) described transformations of penicillin sulphoxide into cephalosporin compounds. A $\Delta^3$-cephem derivative was prepared from phenoxymethylpenicillin sulphoxide methyl ester and was purified by CC. The conversion was (schematically) as follows:

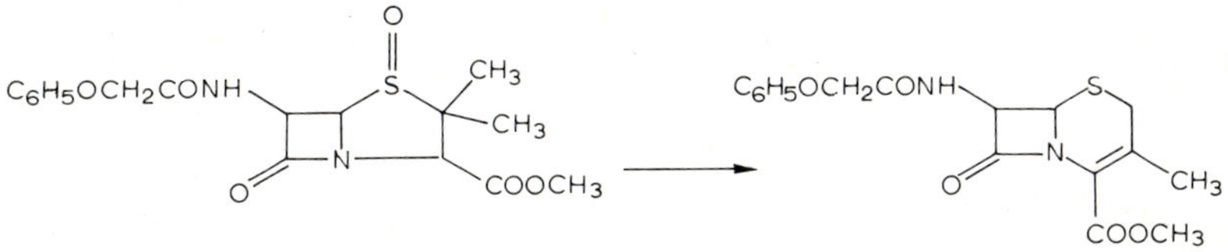


Phenoxymethylpenicillin sulphoxide methyl ester

Methyl 3-methyl-7-(2-phenoxyacetamido)-3-cephem-4-carboxylate

Chromatography on columns of silica gel was used extensively in transformations of penicillins into new β-lactam systems by Heusler and by Fechtig *et al.* A new intermediate, 7-aminocephalocillanic acid, was prepared by Scartazzini *et al.* and contains structural characteristics of both 6-APA and 7-ACA. Its N-acyl derivatives were named "cephalocillins", and two of the cephalocillins prepared were purified on columns of silica gel using toluene–acetic acid ester (4:1) as the eluent.

High-speed LLC was used successfully for the analysis of the methyl benzyl esters of penicillins. A Zipax column was developed with an ethanol–5% *n*-hexane solution which served as mobile phase. The elution time of the methyl benzyl esters is governed by the percentage of *n*-hexane in the mobile phase. The separation of the methyl benzyl ester of penicillin G from its impurities was achieved within 4 min. As the penicillins vary widely in solubility and in the substitutions on the molecule, it is doubtful that any single chromatographic system will be applicable to all of the penicillin derivatives (Schmit).

6-APA and benzylpenicillin were found to contain small but significant amounts of a high-molecular-weight protein that stimulated the formation of antibodies of penicilloyl specificity (Weston). Protein impurities were isolated by fractionation of the sodium salt of 6-APA on columns of Sephadex G-25 (Fig. 45.1).

β-Lactam antibiotics form polymers during storage as aqueous solutions. 6-APA also forms a dimer, presumably through nucleophilic attack of the amino group of one molecule to the β-lactam of another to form a penicillin. Dennen separated penicillins, 6-APA and a degradation mixture of 6-APA by gel chromatography (Fig. 45.2).

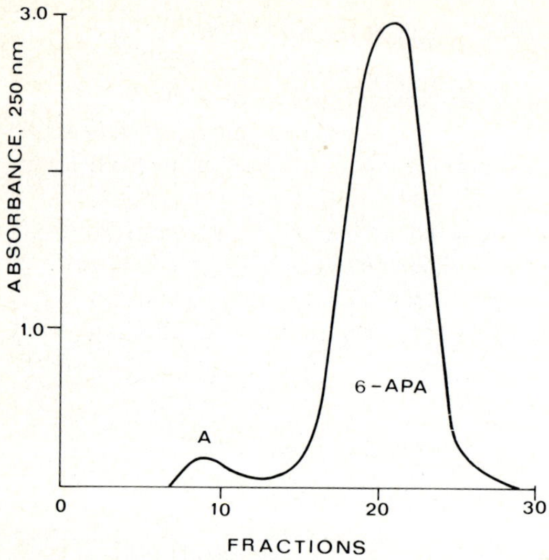


Fig. 45.1. Separation of protein impurities from the sodium salt of 6-aminopenicillanic acid (Weston). Sorbent: Sephadex G-25. Columns equilibrated and eluted with 0.5% sodium chloride solution. Detection: eluates were scanned automatically at 254 nm with an LKB Uvicord or manually at 280 nm with a Uvispek 700 spectrophotometer. A = impurities; 6-APA = 6-aminopenicillanic acid.

TABLE 45.1
ION-EXCHANGE CHROMATOGRAPHY OF D-PENICILLAMINE AND RELATED COMPOUNDS (PURDIE *et al.*)

| Compound | Time (min) |
|---|---|
| L-Cysteine | 249 |
| Glutathione (ox.) | 275 |
| DL- + *meso*-Lanthionine | 296, 319* |
| D-Penicillamine | 331 |
| L-Cysteine-glutathione disulphide | 394 |
| L-Cystine | 473 |
| L-Cysteine-D-penicillamine disulphide | 504 |
| D-Penicillamine disulphide | 545 |
| DL- + *allo*-Cystathionine | 548, 554* |
| L-Isoleucine | 566 |
| L-Cysteine trisulphide | 595 |
| L-Cysteine-D-penicillamine trisulphide | 616 |
| L-Cysteine-L-homocysteine disulphide | 618 |
| L-Homocysteine-D-penicillamine disulphide | 640 |
| D-Penicillamine trisulphide | 655 |
| L-Cysteine-L-homocysteine trisulphide | 675 |
| L-Homocysteine | 723 |
| D-Penicillamine tetrasulphide | 735 |
| Cysteamine-glutathione disulphide | 737 |
| L-Homocysteine trisulphide | 796 |
| L-Cysteine-cysteamine disulphide | 976 |

*Twin peaks.

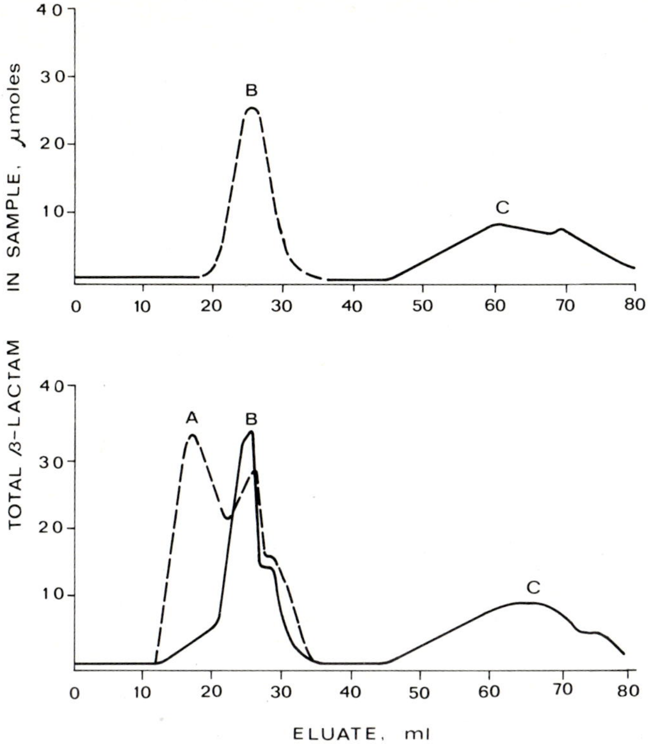


Fig. 45.2. Separation of penicillins, 6-APA and a degradation mixture of 6-APA by gel chromatography (Dennen). Column: 39 × 1.0 cm. Sorbent: Sephadex G-10. Detection: measurement of β-lactam content was performed by reacting intact β-lactam with neutral hydroxylamine to give the hydroxamate, followed by colour development with iron(III) chloride. The absorbance at 304 nm represented polymeric material. Above: 14.9 μmoles of penicillin and 11.50 μmoles of 6-APA applied to column; broken line = phenoxymethylpenicillin or benzylpenicillin; solid line = 6-APA; recovery of total β-lactam = 88%. Below: the starting material was 6-APA, which was degraded and then applied on to the column; solid line = intact β-lactam of 6-APA and polymeric material; broken line = 304-nm absorbing material. Peaks A, B and C represent polymeric material, penicillin or 6-APA dimer, and 6-APA, respectively.

Smith and Marshall separated high-molecular-weight materials, formed in aqueous solutions of 6-APA, benzylpenicillin, ampicillin and hetacillin, by repeated chromatography through Sephadex G-25 fine with water and 0.5% sodium chloride solution as eluents.

Of the degradation products of penicillin, D-penicillamine is of practical importance. D-penicillamine, its derivatives and sulphur-containing amino acids were separated, using a Technicon Model NC-1 amino acid analyzer with a column of Chromobeads A cation-exchange resin (150 × 0.6 cm), by Purdie *et al.* The column was maintained at 60°C throughout the runs as this temperature was satisfactory for the compounds being studied. An Autograd gradient elution device was used with 75 ml of buffer solution in each chamber as follows: 1–4, pH 2.875; 5, pH 3.80; 6, pH 3.80 + pH 5.0 (1:2); 7–9, pH 5.0. A buffer flow-rate of 35 ml/h was used. The results obtained are presented in Table 45.1.

*References p.1005*

## Cephalosporins and 7-ACA

Cephalosporin C has been used directly in medicine and also for the preparation of 7-ACA, the intermediate of "semi-synthetic" cephalosporins. Woodward *et al.* achieved the total synthesis of cephalosporin C, using CC on silica gel for the separation of the last of 18 intermediates in the series of reactions leading to cephalosporin C.

Nagarajan *et al.* reported the isolation of penicillin N and of three new β-lactam antibiotics of the cephalosporin C type from two species of *Streptomyces*. The antibiotics present in the broth filtrate were concentrated by carbon and anion-exchange CC. The final purification was achieved by CC on cellulose and silica gel to yield the purified antibiotics. The structures of the three new cephalosporins and that of cephalosporin C are as follows:

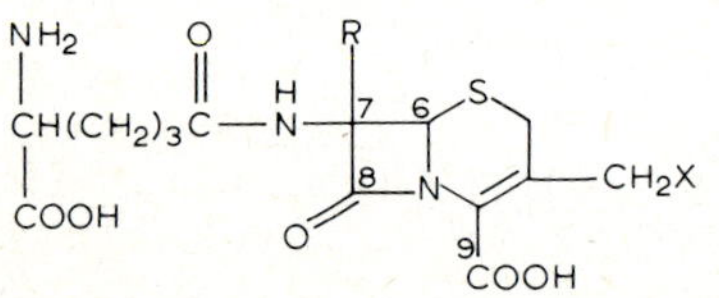


| | | |
|---|---|---|
| Cephalosporin C | R = H | X = $OC(=O)CH_3$ |
| A16886A | R = H | X = $OC(=O)NH_2$ |
| A16886B | R = $OCH_3$ | X = $OC(=O)NH_2$ |
| A16884A | R = $OCH_3$ | X = $OC(=O)CH_3$ |

Upon acid degradation, antibiotic A16886B and A16884A yielded, in addition to α-aminoadipic acid, approximately three times more glycine than did the antibiotic A16886A or cephalosporin C. Brannon *et al.* determined the origin of glycine obtained upon acid hydrolysis of A16886B-$^{14}$C-6 and A16886B-$^{14}$C-8. The results of these chromatographic analyses are presented in Fig. 45.3. Thus the higher yield of radioactive glycine obtained from the 6-$^{14}$C labelling compared with that from the 8-$^{14}$C labelling indicates that only the presence of the methoxy group at $C_7$ must be responsible for the significant increase of glycine and, therefore, for a mechanism of hydrolysis different from that with cephalosporin C.

7-ACA is prepared commercially by reaction of cephalosporin C with nitrosyl chloride in formic acid as solvent. Morin *et al.* (1969a) studied the reaction conditions in order to obtain higher yields of 7-ACA. Using CC, they found that the reaction mixture also contained other biologically active substances, which were purified and characterized.

Desacetylcephaloglycine is known to be a metabolic product present in the blood and urine after treatment with cephaloglycin. In an acidic medium, cephaloglycin (I) is also converted into desacetylcephaloglycin (II, having antibiotic activity) and desacetylcephalo-

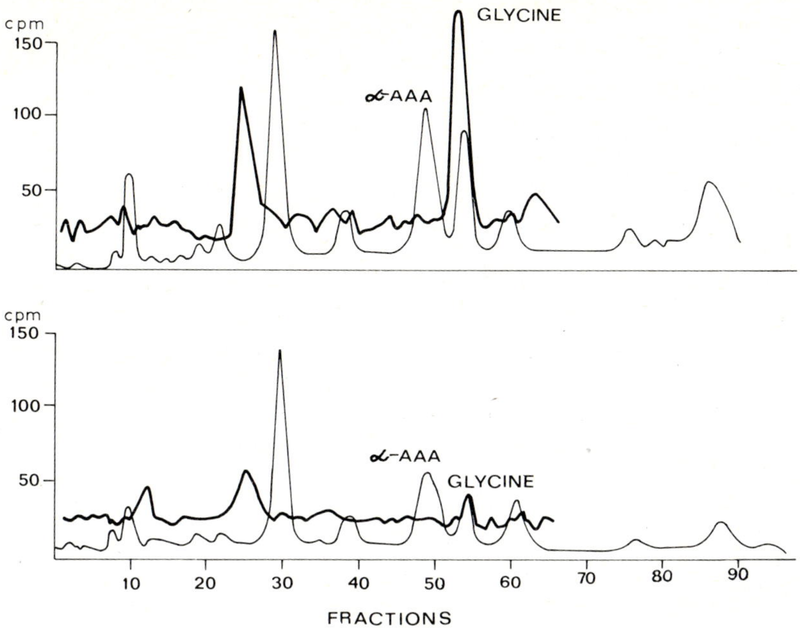


Fig. 45.3. Ninhydrin response and radioactivity profiles of column elution fractions of the acid hydrolyzates of β-lactam antibiotics [6-$^{14}$C] A16886B and [8-$^{14}$C] A16886B (Brannon *et al.*). Column: 52 × 0.9 cm. Ion exchanger: Durrum DC-1A resin. Buffers: (1) 2% thiodiglycol, 0.2 *N* $Na^+$, pH 2.2 for application; (2) 0.5% thiodiglycol, 0.2 *N* $Na^+$, pH 3.23. Operating conditions: Beckman-Spinco Model 120 C amino acid analyzer equipped with a BioCal Model BC501 automatic sample injector was used; flow-rate, 70 ml/h; temperature, 56.5°C. Detection: A ninhydrin flow of 35 ml/h was used when a ninhydrin response curve was desired; for radioactive scintillation counting, 1-min fractions were collected directly from the column. α-AAA = α-aminoadipic acid; thick line = radioactivity; thin line = ninhydrin response curve.

sporin lactone (III, considerably less active). Kukolja separated the reaction mixture using a cellulose column, collecting the fractions with an automatic fraction collector and following the progress of the separation by TLC (detection with UV light and a ninhydrin spray). Fractions containing I and trace amounts of III were followed by fractions containing the required II.

## CARBOHYDRATE ANTIBIOTICS

This section includes antibiotics known as aminoglycosidic or aminocyclitol antibiotics.

### Streptomycins

Pharmaceutical preparations of streptomycin may contain salts of alkali metals as impurities. The behaviour of mixtures consisting of streptomycin sulphate and alkali metal salts in gel filtration on Sephadex G-10 was investigated by Störl. Elution with deionized water resulted in a virtually salt-free product and the yield was about 95%. Streptomycin is eluted in two fractions with deionized water. The normal molecular sieve effect was obtained using buffer solutions (0.2 *M* ammonium acetate solution, pH 6.5) for elution.

Streptomycin is inactivated (by adenylation or phosphorylation) by resistant bacteria, and experiments of Umezawa *et al.* (1968c) illustrated the use of IEC in similar studies. A sample of streptomycin sulphate was inactivated by a cell-free system obtained from *Escherichia coli* carrying R factor. The reaction mixture was then passed through a column of Amberlite IRC-50 ($Na^+$) and the inactivated streptomycin was eluted with 0.5 *N* hydrochloric acid. The fraction that gave a positive Sakaguchi reaction and showed no antibacterial activity was collected and passed through a column of active carbon and the inactivated streptomycin was eluted with 0.2 *N* hydrochloric acid–methanol (1:1). The fraction that gave a positive Sakaguchi reaction and negative biological activity was neutralized with Dowex 44 ($OH^-$), and, after concentration under vacuum, the inactivated streptomycin was precipitated by the addition of 14 volumes of acetone and dried *in vacuo*. Part of the inactivated streptomycin in the reaction mixture passed through the column of IRC-50 resin, but it was adsorbed by another column of the same resin and purified by the same procedure as above. In total, 250 mg of inactivated streptomycin were obtained from the original sample (500 mg). It was characterized by PC and other analyses as adenylylstreptomycin.

De Fabrizio developed a single and sensitive method for the determination of dihydrostreptomycin in complex pharmaceutical preparations (Sulpec and sulphaguanidine cum dihydrostreptomycin). The method entails the recovery of dihydrostreptomycin by cation-exchange CC with subsequent fluorimetric determination.

**Neomycins**

The term neomycin usually refers to the amino sugar antibiotic complex, composed of two stereoisomers, neomycin B and neomycin C, and their degradation product, neamine (neomycin A). Commercial preparations contain neomycin B as the major constituent. "Extra" neomycins appear in varying amounts (always less than 1%) as constituents of many commercial neomycin samples. Three "extra" neomycins, neomycins D, E and F, were identified by Hessler *et al.* as paromamine, paromomycin I and paromomycin II, respectively. A commercial sample of neomycin sulphate was neutralized and subjected to IEC on Amberlite CG-50 ($NH_4^+$) resin. Gradient elution with ammonia solution yielded two fractions, the first of which contained neamine, neomycin D and neomycin F, and the second neamine and neomycins D, F, B and E, in order of elution. The two fractions were then further separated by ion-exclusion chromatography on Dowex 1-X2 ($OH^-$). Elution with water gave the individual components, which were identified as mentioned above.

Majumdar and Majumdar (1970) isolated and characterized three phosphoamidoneomycins, which are converted into neomycin by *Streptomyces fradiae* in the later stages of the fermentation. The compounds were isolated and separated using Amberlite IRC-50 ($NH_4^+$) and were characterized as neomycin B pyrophosphate, neomycin C

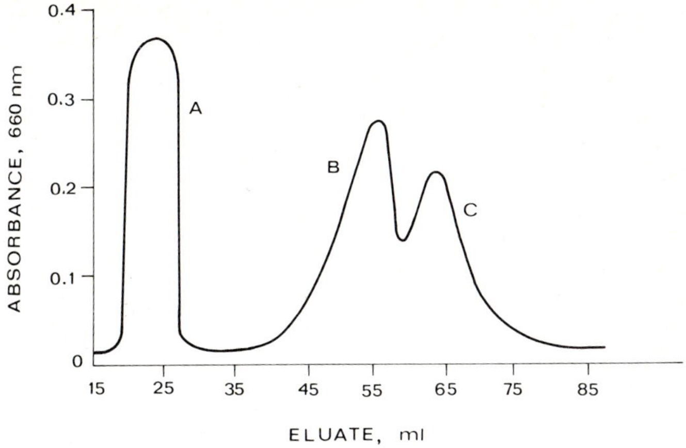


Fig. 45.4. Separation of phosphorylated neomycin intermediates (Majumdar and Majumdar, 1970). Column: 65 × 1.4 cm. Ion exchanger: Amberlite IRC-50 ($NH_4^+$), 200–400 mesh, equilibrated with 1500 ml of 18 m*M* ammonia solution. Buffer: 90 m*M* ammonia solution. Operating conditions: elution volumes are apparent from the figure; 1.6-ml fractions were collected; flow-rate, 0.15 ml/min. Detection: the ammonium molybdate reagent of Hanes and Isherwood was used and absorbances at 660 nm were measured. A = Neomycin C dipyrophosphate complex; B = neomycin C pyrophosphate; C = neomycin B pyrophosphate.

pyrophosphate and neomycin C dipyrophosphate complex. Fig. 45.4 shows a typical pattern of their separation. A method for the quantitative analysis of the neomycin components in fermentation broths was devised by Majumdar and Majumdar (1967), and involves a combination of IEC with PC as described below.

Reagents R-1 to R-5 were as follows:

*R-1:* commercial carbon tetrachloride is washed with concentrated sodium hydroxide solution, layer-separated, treated with fused calcium chloride and distilled. Methanol–acetic anhydride–carbon tetrachloride reagent (3:2:95) is prepared just before use.

*R-2: chlorine–carbon tetrachloride reagent.* Chlorine is prepared from 100 ml of concentrated hydrochloric acid and 50 g of potassium permanganate, then passed through water and concentrated sulphuric acid and absorbed in 1 l of carbon tetrachloride. About 5 g of barium carbonate and 5 g of fused calcium chloride are added and stored in a glass-stoppered amber-coloured bottle.

*R-3: starch–iodine–pyridine reagent.* Starch (1 g) and 0.25 g of potassium iodide are dissolved by heating in 3.5 ml of water and then 0.5 ml of this solution is added quickly to 50 ml of pyridine. Freshly prepared reagent is used.

*R-4: starch–iodine–barium carbonate reagent.* Starch (1 g) and 0.25 g of potassium iodide are dissolved by heating in 100 ml of water, then 2 g of barium carbonate are added to the cold solution and thoroughly shaken. After the solid particles have settled, the upper turbid portion is used for spraying the chromatograms. This reagent can be used for 5 to 6 days.

*R-5: starch–iodine–hydrochloric acid reagent.* This reagent is prepared freshly by boiling 1 g of starch, 0.25 g of potassium iodide and 1 ml of 5 *N* hydrochloric acid in 100 ml of water.

The resin is prepared as follows. Amberlite IRC-50, 200–400 mesh, is converted into the $NH_4^+$ form by shaking with 1 *N* ammonia solution. The resin is suspended in water and allowed to settle and the upper turbid portion is decanted; this process is repeated until all finer particles are removed. The resin is then suspended in an equal volume of water, and 2 ml of the suspension are transferred to a funnel with its stem (I.D. 8 mm; column volume 1 ml) plugged with glass-wool and then washed with water until the washings are neutral. Such a column can be used several times. After each use, it is washed with 1 *N* hydrochloric acid and then converted into the ammonium form.

N-Acetylneomycins are prepared in the following manner. The fermentation broth or the solution of the neomycin sample (containing about 4 mg of neomycins) is passed through an Amberlite IRC-50 ($NH_4^+$) column, which is washed with 5 ml of water and eluted with 5 ml of 1 *N* ammonia solution. This eluate is spotted at points 20 mm apart along the baseline (7.5 cm from the end) of Whatman No. 4 paper (43 × 16 cm) in a volume sufficient to contain 4–12 μg of neomycin. Standard solutions of neomycin B and C and neamine-free bases are applied in the same concentration range. For N-acetylation, the papers are rolled and soaked in R-1 for at least 12 h, then air-dried for 1 h in order to remove acetic anhydride.

The papers are developed for 24–36 h at 28°C with *n*-butanol–water–piperidine (84:16:2) by the descending technique. The chromatograms are air-dried and placed for 1–2 h in a chamber with a water-saturated atmosphere at 37°C for humification and complete removal of piperidine from the papers by the same subsequent step. The papers are next rolled inside the chamber and quickly transferred to a cylinder and the filtered R-2 is poured in. The apparatus is kept in the dark for 20 min at room temperature, and the chromatograms are air-dried for 0.5 h at 4–6°C in the dark in order to remove excess of chlorine. The papers are divided into a number of parallel strips each 20 mm wide. One of a pair of strips is sprayed with R-3 and bluish pink spots are obtained by transferring it to a water-saturated atmosphere. For rapid work, detection with R-4 and marking of the wet spots with a ball-point pen is preferred.

Rectangular areas corresponding to N-acetylneomycin B and C and N-acetylneamine are removed from the untreated paper strip with the help of the guide chromatogram and then cut in small pieces. Rectangles are cut from spot-free areas as a blank. The spots are then extracted for 30 min with 4 ml of water and 1 ml of R-5 in clean test-tubes and the absorbances of the coloured solutions are measured at 570 nm. The amount of neomycins is determined by reference to standard curves; all three neomycins can be determined by reference to a single standard curve. Theoretically, 2.56 μg of neomycin B or C are equivalent to 2.01 μg of neamine. The positions of the neomycins on the chromatograms relative to N-acetylneamine are about 0.69–0.70 (N-acetylneomycin B), 0.35 (N-acetylneomycin C) and 1.0 (N-acetylneamine). As little as 1 μg of neomycin base can be determined with an accuracy of ± 0.7% and a 60-μg sample can be analyzed if the amount of the minor component present is very low.

Pénasse *et al.* prepared several mono-N-alcoyl derivatives of neomycin B and paromomycin. Sulphates of the derivatives were successfully purified by IEC on Amberlite IRC-50 ($H^+$) with 0.066–1.0 *N* ammonia solution as the buffer.

## Kanamycins

Kanamycins A, B and C contain 3-amino-3-deoxy-D-glucose (3AG) in their molecules. 3AG exhibits antibacterial activity and was also isolated from *Bacillus aminoglucosidicus* by Umezawa *et al.* (1967a, b). The isolation procedure included IEC.

Total syntheses of kanamycins were reported by Umezawa *et al.* (1968b, d, 1969a, b, c). Synthetic hepta-O-acetyltetra-N-(2,4-dinitrophenyl) derivatives of the three kanamycins were purified on column of silica gel, deacetylated and the dinitrophenyl groups were removed by IEC using Dowex 1-X2 ($OH^-$).

## Gentamicins

Wagman *et al.* described the preparative separation of gentamicins $C_1$, $C_{1a}$ and $C_2$ by use of the following cellulose and Chromosorb W chromatographic column procedures.

In the cellulose column chromatography, Whatman No. 1 cellulose powder is packed in small segments in a column (I.D. 2.4 cm) to a height of 30 cm. The upper phase of the solvent system, consisting of chloroform–methanol–17% ammonia (2:1:1), is run through the column until a yellow band of impurities emerges, and the column is allowed to drain. A 200-mg amount of gentamicin sulphate is mixed with 2 g of cellulose powder, packed on top of the cellulose in the column and wetted with a small amount of the upper phase. The lower phase is allowed to run through the column at the rate of 2 ml/min; 16-ml fractions are collected every 8 min. Aliquots of each fraction are spotted on filter-paper and tested with ninhydrin reagent in order to determine the presence or absence of antibiotic. Like fractions, converted into their free bases on the column, are pooled and evaporated to dryness.

In the Chromosorb W column chromatography, 60–100 mesh Chromosorb W (Johns-Manville, Denver, Colo., U.S.A.) is slurried with the upper phase of chloroform–methanol–17% ammonia (2:1:1) and filtered using suction on a büchner funnel until excess of solvent is removed. The Chromosorb is packed into a column (I.D. 3 cm) to a height of 50 cm in 5-cm segments (*ca.* 150 g as dry Chromosorb). An alternative method is to pour the slurry into the column and to remove excess of solvent by suction from the bottom of the column. One litre of lower phase is run through the column in order to wash the Chromosorb. A 3-g portion of gentamicin base is dissolved in 10 ml of methanol, adsorbed on to the smallest possible amount of Chromosorb and dried under vacuum using a film evaporator. This mixture is packed on top of the column and wetted with a small amount of the lower phase. The column is eluted with the lower phase at the rate of 1 ml/min, collecting five fractions per hour. The fractions are tested as described for cellulose column chromatography. The component peaks are located, pooled, decolourized on IRA-401S ($OH^-$) resin and evaporated to dryness.

The results of the separation using a cellulose column are presented in Table 45.2. PC of the bases and bioautography against *Staphylococcus aureus* showed that the $C_1$ and $C_2$ components were free from impurities and that $C_{1a}$ contained approximately 5% of $C_2$. Re-chromatography of the $C_{1a}$ fraction using the described column resulted in the isolation of this component free of $C_2$. With Chromosorb W, the separation was as shown

TABLE 45.2
CHROMATOGRAPHIC SEPARATION OF THE GENTAMICIN COMPLEX USING A CELLULOSE COLUMN (WAGMAN *et al.*)

Starting material: 200 mg of the gentamicin complex sulphate (129 mg base equivalent). Total yield = 121.5 mg (94%).

| Component | Fraction Nos. | Weight (mg) |
|---|---|---|
| $C_1$ | 12–19 | 58.2 |
| $C_2$ | 23–33 | 50.8 |
| $C_{1a}$ | 38–49 | 12.5 |

TABLE 45.3
CHROMATOGRAPHIC SEPARATION OF THE GENTAMICIN COMPLEX USING A CHROMOSORB W COLUMN (WAGMAN *et al.*)

Starting material: 3 g of gentamicin complex base.

| Component | Fraction Nos. | Weight of purified fractions (g) | Antibiotic activity* (μg/mg) |
|---|---|---|---|
| $C_1$ | 35–60 | 1.01 | 833 |
| $C_2$ | 95–140 | 0.74 | 1050 |
| $C_{1a}$ | 165–231 | 0.31 | 1050 |

*Average of four assays against complex standard.

in Table 45.3. PC of the bases and bioautography indicated that all components were free from microbiologically active impurities.

IEC was used for the preparative separation of the entire gentamicin complex by Maehr and Schaffner. The strongly basic resin Dowex 1-X2, 100–200 mesh, was used without further purification or sizing. The resin was converted into the hydroxide form with 10 resin-bed volumes of 8% carbonate-free sodium hydroxide solution. All washings and elutions were made with distilled, carbon dioxide-free water. Typically, 35 g of gentamicin complex were dissolved in 30 ml of water and washed into the resin bed, the column flow-rate not exceeding 0.7 cm/min; manual fraction collection, according to the continuously recorded conductivity of the effluent (using an Industrial Instruments Model RC-16B conductivity bridge, operated at a frequency of 1000 Hz, and a Leeds & Northrup Sidomax H instrument recording the 0–10 mV range).

Of the 10 fractions collected, the first appeared as a sharp peak in the conductivity diagram, exhibiting a high ash content with only traces of biological activity. All of the following fractions displayed high activity against *Staphylococcus aureus*. Fraction aliquots were analyzed by TLC on non-activated layers of Brinkmann (Westbury, N.Y., U.S.A.) silica gel G, 0.75 mm thickness, developed with the solvent system chloroform–methanol–28% ammonia (2:1:1). In parallel experiments, the spots were located by bioautography and with ninhydrin, the correlation affording qualitative agreement between ninhydrin

TABLE 45.4
SUMMARY OF CC AND TLC CHARACTERISTICS OF GENTAMICIN ANTIBIOTICS (ADAPTED FROM MAEHR AND SCHAFFNER)

Figures are TLC $R_F \times 100$ values.

| Column effluent volume (ml) | | | | | | | | |
|---|---|---|---|---|---|---|---|---|
| A* | | | | | | | B** | |
| 1950–2670 | 2670–3155 | 3155–3485 | 3485–4175 | 4175–4770 | 4770–5110 | 5110–5550 | 5550–9340 | 9340–10120 |
| 42<br>54<br>58 | 18<br>24 | 9<br>17<br>26<br>64 | 13<br>39 | 22 | – | 26<br>45 | 18 | 59 |

*A = Mixture of the three major antibiotics, gentamicins $C_1$, $C_2$ and D.
**B contained gentamicin A as the major constituent.

colour and zones of inhibition. Sixteen different active components of the entire gentamicin complex were found (Table 45.4).

## Lincomycin group

IEC was used by Thomas *et al.* for the purification of tritiated lincomycin, and it seems that this procedure is applicable to the purification of all lincomycins.

Clindamycin is 7-chloro-7-deoxylincomycin prepared by chemical modification of the sugar moiety of lincomycin. Argoudelis *et al.* (1969) described the microbial transformation of clindamycin into N-demethylclindamycin. A mixture of the two substances was separated by CC on silica gel (Merck, Catalogue No. 7734). A column was prepared from 600 g of the sorbent in chloroform–methanol (6:1), and 6 g of the crude mixture were chromatographed using the same eluent. The fractions obtained were analyzed by TLC; clindamycin was eluted first, followed by N-demethylclindamycin.

During studies on the possibility of *Streptomyces lincolnensis* methylating N-demethylclindamycin to give clindamycin, it was found that N-demethylclindamycin had been transformed into a new bioactive material, compound A. When CM-Sephadex was used, N-demethylclindamycin and lincomycin (the latter being produced by *S. lincolnensis*) were adsorbed, while compound A passed through the column. It was found to be N-demethyl-N-hydroxymethylclindamycin (Argoudelis *et al.*, 1972). Recently, urinary excretion products of clindamycin in rats and dogs were isolated and purified by means of LCC, TLC and counter-current distribution (Sun).

Celesticetins are antibiotics produced by *S. caelestis*. Four members of this group, desalicetin and celesticetins B, C and D, were separated by counter-current distribution and/or silica gel CC (Argoudelis and Brodasky).

*References p.1005*

**Other aminoglycosides**

Aminoglycosidic antibiotics were separated by high-speed LLC on a Carbowax 750 column with *n*-hexane–isopropanol as the mobile phase. The hydroxy groups of the antibiotics interact with the Carbowax to retain the sample. The antibiotics are sufficiently soluble in isopropanol that a small amount in the mobile phase will elute the compounds satisfactorily. Retention times are governed by the concentration of isopropanol in the mobile phase (Schmit).

Fukagawa *et al.* (1968a, b) devised procedures, including IEC, for the isolation and purification of kasugamycin and [$^{14}$C] kasugamycin from fermentation broths. Various Dowex ion exchangers were used for the purification of semi-synthetic derivatives of kasugamycin by Cron *et al.*

On the basis of PC and bioautography, nebramycin was differentiated from the related antibiotics neomycin, paromomycin, kanamycin and gentamicin.

Thompson and Presti separated the nebramycin complex as follows. A preparation of nebramycin complex was dissolved in deionized water and, after adjusting the pH of the solution to 4.5 with sulphuric acid, partial decolourization was achieved by treatment for 1 h with Darco G-60 carbon. The carbon was removed by filtration and the filtrate was charged on 7 l of Amberlite IRC-50 ($NH_4^+$) resin in a 105 × 9.2 cm column. After the column had been thoroughly washed with deionized water, the resin was eluted with 0.1 *N* ammonia solution at a flow-rate of 20 ml/min. The eluate was collected in portions of 900 ml. Following the elution of components 1–5, the concentration of the eluent was increased to 0.3 *N* in order to elute component 6 more efficiently. The eluates were pooled on the basis of bioautographic purity and potency and the pooled eluates were concentrated and dried *in vacuo* to yield the amorphous free bases of the respective nebramycin components. Four components of the complex nebramycins 2, 4, 5 and 6 were characterized satisfactorily. In studies of the structure of nebramycin factor 6 (now called tobramycin), Koch and Rhoades isolated the amino sugar nebramine by CC on Bio-Rad 1-X4 ($OH^-$) with water as the eluent.

Lividomycins have been investigated by Mori *et al.* Two of them, containing 2-amino-2,3-dideoxy-D-glucose, were named lividomycins A and B. One further compound, a new member of the paromomycin group, was provisionally designated as antibiotic No. 2230-C, while the fourth compound, No. 2230-D, was identified as paromomycin I. Two methods were used for their separation.

(*a*) Crude lividomycins, *ca.* 25 mg, are dissolved in 5 ml of water and the solution is adsorbed on a 40 × 1 cm CM-Sephadex C-25 ($NH_4^+$) column. After the column has been thoroughly washed with deionized water, active portions are obtained by gradient elution between 200 ml of 0.12 *N* and 25 ml of 0.35 *N* ammonia solution at a flow-rate of 25 ml/h at 27°C. All fractions, each containing 3 ml, are assayed by the paper disc method against *Bacillus subtilis* ATCC 6633. The active fractions are lyophilized and weighed. Four peaks of the active fractions are designated as No. 2230-C, lividomycin A, No. 2230-D and lividomycin B, respectively.

(*b*) A solution of about 6 g of crude powder in 600 ml of deionized water is adsorbed on an Amberlite CG-50 Type I ($NH_4^+$) column (30 × 3 cm). After the column has been thoroughly washed with deionized water and 0.08 *N* ammonia solution, the compounds

No. 2230-C, lividomycin A, No. 2230-D and lividomycin B are eluted stepwise from the column with 0.1 *N*, 0.12 *N*, 0.15 *N* and 0.17 *N* ammonia solution, respectively. The eluates are pooled on the basis of biological activity as in (*a*) and concentrated under reduced pressure. Further purification of the antibiotics is carried out by CC on a column of Dowex 1-X2 ($OH^-$), 200–400 mesh, using deionized water for development. After detection, each active fraction is collected, concentrated at below 40°C and finally lyophilized. The free bases are obtained as amorphous powders.

A new antibiotic complex, butirosin, was isolated from a strain of *Bacillus circulans* by Dion *et al.* The complex was separated into butirosins A and B by means of Dowex 1-X1 or 1-X2 ($OH^-$) resin. The 3-l amount of resin ($Cl^-$, 50–100 mesh) in a 50 × 2 cm column was converted into the hydroxide form with 16 l of 2 *N* sodium hydroxide solution, washed with *ca.* 9.5 l of water, treated with 17.5 l of 5% boric acid, and washed with several hold-ups of water. Butirosin (6.01 g) in 15 ml of water was added to the prepared resin column. After percolation of the sample, the column was washed with 6.42 l of water at a flow-rate of *ca.* 420 ml/h. The column was then developed with 4 l of 1% boric acid, 4 l of 2% boric acid, and finally with several hold-ups of 5% boric acid. Portions of the eluate fractions were lyophilized, and the residues were analyzed as the N-acetyl derivatives for butirosin A and B content by PC. Most of the butirosin A (*ca.* 4.5 g), free from B, was found in fractions in the initial effluent volume of 1.5–4.95 l. Most of the butirosin B (0.6 g) was found in the effluent volume of 2.64–4 l after application of 5% boric acid to the column. The butirosin A and B, present in the effluent volumes from the Dowex 1 columns, were purified by means of Amberlite IRC-50 ($NH_4^+$), XE-243 (free base form) and sometimes Dowex 1-X16 resin columns.

TABLE 45.5
APPLICATIONS OF LIQUID COLUMN CHROMATOGRAPHY IN SEPARATIONS OF AMINO-GLYCOSIDIC ANTIBIOTICS

| Compounds | Sorbent or ion exchanger | Eluent | Reference |
|---|---|---|---|
| *Liquid-solid chromatography* | | | |
| α-Lipomycin | Oxalic acid–Silica gel | Benzene–diethyl ether–acetone (1:1:1) | Kunze *et al.* |
| Moneomycin D | Silica gel | Isopropanol–2 *N* ammonia (5:1) | Schacht and Huber |
| Ribostamycin | Silica gel | Not given | Ito *et al.* |
| Yazumycin | Cellulose | *n*-Propanol–pyridine–acetic acid–water (10:15:3:10) | Akasaki and Abe |
| *Ion-exchange chromatography* | | | |
| Hygromycin B | Bio-Rad AG 1-X2 | Water | Neuss *et al.* |
| Moneomycins | Dowex 1 ($Cl^-$) | 0.6% KCl in methanol–water (4:1) | Schacht and Huber |
| Nojirimycin | Dowex 1-X2 ($OH^-$) | Water | Inouye *et al.* |
| SF-701 | Amberlite IRC-50 ($Na^+$) | 0.5 *N* HCl | Tsuruoka *et al.* |
| Yazumycin | Amberlite IRC-50 ($H^+$) | 0.3 *N* HCl | Akasaki and Abe |

Khokhlov and Reshetov developed an effective general procedure for the separation of streptothricin mixtures based on IEC on CM- cellulose with a sodium chloride gradient. By means of this method, they demonstrated that all of the available preparations of antibiotics of this type were mixtures of six streptothricins differing in the number of L-β-lysine residues. Taniyama *et al.* (1971a, b) modified the method of IEC on CM-cellulose by replacing sodium chloride with volatile pyridine–acetic acid buffer, and also developed gel chromatographic techniques on Sephadex LH-20. Using these procedures they separated individual racemomycins A, C, B and D, and demonstrated their identity as streptothricins F, E, D and C, respectively. They also showed by the same methods the identity of the antibiotics racemomycins A and C as yazumycins A and C (Taniyama *et al.*, 1971c). The above techniques, in combination with TLC on cellulose, were used by Khokhlov and Shutova in order to confirm the chemical structure of streptothricins.

Applications of CC in the isolation, purification and separation of some other carbohydrate antibiotics are given in Table 45.5.

### Biotransformations of aminoglycosidic antibiotics

Strains of *Escherichia coli* that are resistant to various aminoglycosidic antibiotics are known to carry R factors with genetic information to produce enzymes that transform these antibiotics. Three kinds of transformations are known: adenylation, acetylation and phosphorylation. In studies on transformation by *E. coli* carrying R factors, IEC is widely used in order to isolate and to purify the transformed compounds from reaction mixtures (Benveniste and Davies; Umezawa *et al.*, 1968a, c). In addition to IEC, cellulose phosphate paper binding assays were developed and used to monitor the phosphorylation (Ozanne *et al.*), adenylation (Benveniste *et al.*) and acetylation (Benveniste and Davies). These assays can be combined with IEC analysis.

## MACROCYCLIC ANTIBIOTICS

### Macrolides

Banaszek *et al.* described the separation of erythromycins on a preparative scale as follows. Silica gel, $<0.08$ mm (Merck), 1 part by weight, is treated with 0.5 part of formamide in 1.5 parts of acetone. Acetone is removed in a rotatory evaporator, and the gel is suspended in benzene and introduced into a column. The substance to be separated, preferably in benzene solution, is placed on the column in the proportion of 1:100 with reference to the gel. The eluents are (I) *n*-hexane or benzene–chloroform or methylene chloride–ethanol (30–40:50–60:5); and (II) *n*-hexane–chloroform or methylene chloride–ethyl acetate–ethanol (35:30:30:5). The chloroform used must be free from phosgene and hydrogen chloride. For detection, *Bacillus subtilis* ATCC 6633 can be used for testing the biological activity using a paper disc soaked with aliquots of the fractions. When the column is packed with similarly impregnated aluminium oxide, the eluent used is *n*-hexane–chloroform or methylene chloride–ethanol (70:20:5). The small amounts of

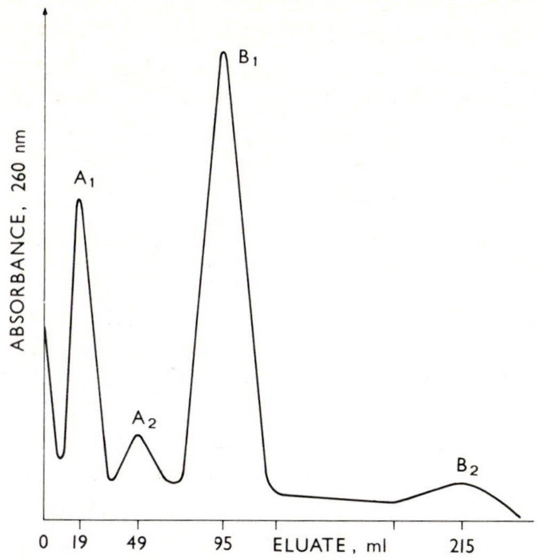


Fig. 45.5. Separation of pristinamycins $I_A$, $I_B$, $II_A$, $II_B$ (Preud'homme *et al.*). Column: 15 × 1.0 cm. Adsorbent: Celite 545 (Johns-Manville) mixed with the bottom layer of the eluent. Eluent: Cyclohexane–dioxane–distilled water (3:4:3), upper layer. Operating conditions: elution volumes are apparent from the figure; the flow-rate of 1 ml/min was controlled by nitrogen pressure. Detection: Spectrophotometric at 260 nm. $A_1$ = Pristinamycin $I_A$; $A_2$ = pristinamycin $I_B$; $B_1$ = pristinamycin $II_A$; $B_2$ = pristinamycin $II_B$. To separate a 5-mg sample, 5 g of Celite were wetted with 2.3 ml of the stationary phase. Elution was carried out with a total volume of 250 ml of eluent.

formamide in the fractions from the column are removed by washing with water.

Pristinamycin was fractionated into five components belonging to two different chemical groups. The three components of group I are amphoteric cyclopeptides and the two components of group II are neutral macrolides (Preud'homme *et al.*). Except for pristinamycin $I_C$, the other pristinamycins were separated by means of CC on Celite (Fig. 45.5).

Megalomicin complex was partially purified on a column of Sephadex LH-20 with aqueous ethanol as the eluent. TLC on silica gel G with the solvent system chloroform–methanol (3:2) indicated that the complex is composed of four major components, A, B, $C_1$ and $C_2$. Separation of $C_1$ and $C_2$ from the complex was accomplished by CC on silica gel developed with the above solvent system (TLC). Components A and B, eluted together, were resolved by CC on Florisil. Elution was carried out with ethyl acetate followed by increasing amounts of acetone in ethyl acetate. Megalomicin was eluted first, followed by component A (Weinstein *et al.*).

Sixteen-membered macrolides, 18-dihydroleucomycin $A_3$, 9-dehydro-18-dihydroleucomycin $A_3$ and tetrahydrotylosin were separated on columns of silica gel using benzene–acetone (5:1; 6–3:1 for tetrahydrotylosin) as the eluents (Ōmura *et al.*, 1972). The same adsorbent and eluent (10:1) were used for acetyl-leucomycins (Ōmura *et al.*, 1968).

TLC and CC showed that the antifungal antibiotic filipin contains a minimum of eight filipin-like pentaenes. Three of the components, which seemed to constitute 96% of the complex, were crystallized after partition CC on siliceous earth and adsorption CC on silica gel (Bergy and Eble).

*References p.1005*

[$^{14}$C] Brefeldin A, which is identical with cyanein (Betina *et al.*), was purified on a column of silica gel with chloroform–acetic acid esters (1:1) as the eluents (Handschin *et al.*). The fungal macrolides phomin and 5-dehydrophomin were separated on columns of alumina (activity II) with methylene chloride–methanol (998:2) as the eluent, or on a silica gel column with the above solvents (98:2 and 98:1) as the eluents (Rothweiler and Tamm).

### Other macrocyclic antibiotics

Rifamycin SV was isolated from a mutant of *Streptomyces mediterranei* strain and was purified by CC on silica gel by Lancini and Hengeller.

Rifampicin undergoes deacetylation on alkaline treatment, affording the corresponding deacetyl derivative without substantial loss of antibacterial activity. Under milder alkaline conditions, rifampicin yields two additional products in which the acetyl group migrates to two different positions in the molecule. The reaction products, after extracting the rifamycins from the reaction mixture with chloroform, and the remaining rifampicin were recovered by CC on a pH 6.0 (McIlvaine) buffered silica gel by stepwise elution with chloroform containing 1–4% of methanol (Maggi *et al.*).

LLC was used for the separation of some impurities in a 3-formylrifampin sample. The optimum partition ratio range was obtained conveniently by adjusting the carrier polarity. It was found that the optimum carrier for this separation with a Zipax polyamide column was *n*-hexane–ethanol (3:1). This combination permits the low-level detection of the desired impurities (a quinone, rifampin and X32) and at the same time provides a good measurement of the very polar, relatively high-molecular-weight major constituent, 3-formylrifampin (Kirkland).

Kinoshita and Umezawa reported the total synthesis of dehexyldeisovaleryloxyantimycin $A_1$, which established a possible general synthetic pathway to the members of antimycin A. The final product was purified by CC on silica gel with *n*-hexane–ethyl acetate (4:3) as the eluent.

## TETRACYCLINES AND RELATED ANTIBIOTICS

### Tetracyclines

Griffiths studied the adsorption characteristics of tetracycline (TC), 4-*epi*-anhydrotetracycline (EATC) and anhydrotetracycline (ATC) on Sephadex gel under various conditions of pH and salt concentration. The adsorption of the compounds was strong in acidic solvents but diminished as the pH was increased. In the alkaline region of pH 8.5–9.5, the ATC and EATC epimer were eluted at different rates from the column. An increase in the salt concentration resulted in stronger adsorption of the compounds to the gel, but did not influence their separation efficiency.

In Fig. 45.6 are shown the elution diagrams of mixtures of TC, ATC and EATC at various pH values. At pH 2.5, ATC and EATC displayed equal and maximal retardation on

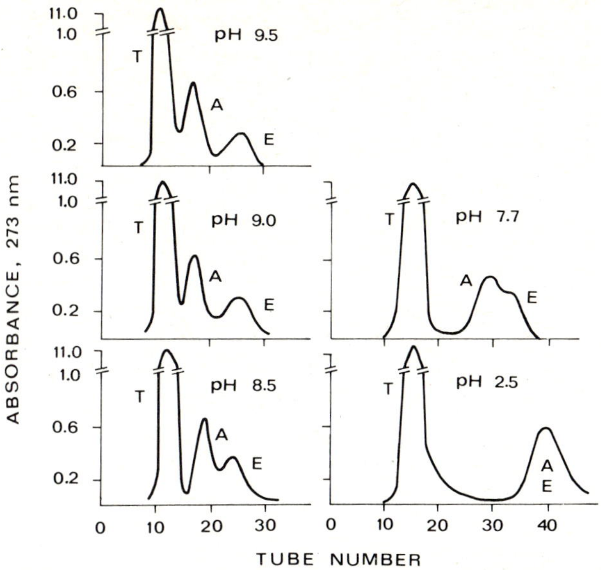


Fig. 45.6. Gel chromatography of mixtures of tetracycline, anhydrotetracycline and 4-*epi*-anhydrotetracycline (Griffiths). Column: 25 × 1.8 cm. Sorbent: Sephadex G-25 F (particle size 20–80 μm). Buffers: (1) dilute hydrochloric acid, pH 2.5; (2) 0.04 *M* phosphate buffer, pH 7.7; (3) 0.05 Tris buffer, pH 8.5, 9.0 and 9.5. Collection of samples was carried out with an LKB Radi Rac fraction collector and 4.4-ml aliquots were collected. Detection: the absorbance at 273 nm was measured. The columns were equilibrated with the above buffers, and the column with the Tris buffer was re-equilibrated with several solvents containing Tris buffer at various pH values. T = Tetracycline; A = anhydrotetracycline; E = 4-*epi*-anhydrotetracycline.

the column but did not show evidence of separation from each other. At pH 9.0 and 9.5, TC and the two derivatives produced constant elution patterns and optimal separation between ATC and EATC.

Fig. 45.7 depicts the elution diagram for the TC and the derivatives eluted separately on Sephadex columns equilibrated with phosphate buffer of pH 7.7 and Tris buffer of pH 9.0. It can be seen that the elution positions of the separate compounds agree closely with the admixture elution in Fig. 45.6. The sensitivity of the separation of the components is evident, since trace amounts of ATC impurities could be detected in the TC and EATC products.

With a low salt molarity and a constant pH of 9.0, ATC and EATC showed weak adsorption to the column. As the salt concentration was increased, the ATC and EATC exhibited stronger adsorption, as evidenced by their delayed elution. Evidence of broadening of the TC peak in Tris–sodium chloride solvent was observed (Fig. 45.8c).

The findings on the separation of the ATC and EATC epimer on Sephadex are of both theoretical and practical importance because they reveal a refined chromatographic mechanism operating under optimal solvent conditions. The results are of immediate practical interest because the toxic nature of EATC requires its analytical determination in pharmaceutical preparations for human use. Recently, an automated LCC method for

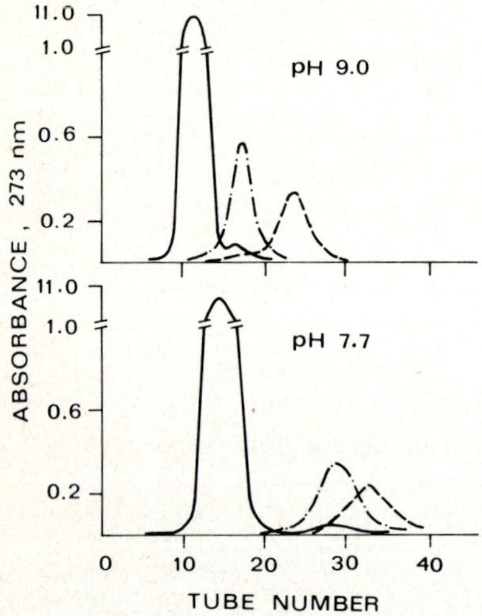


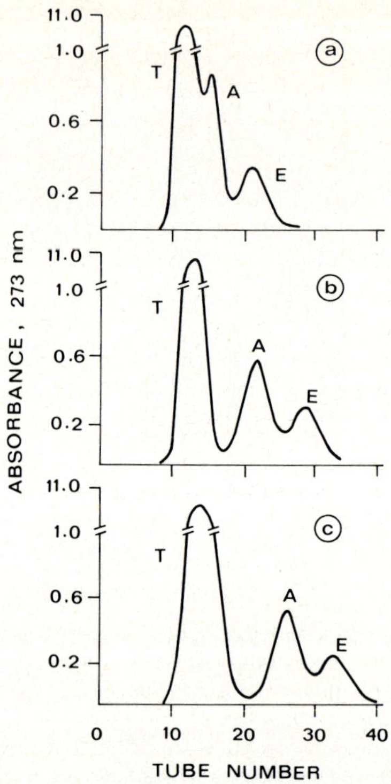


Fig. 45.7. Gel chromatography of tetracycline, anhydrotetracycline and 4-*epi*-anhydrotetracycline eluted separately on Sephadex at pH 7.7 and 9.0 (Griffiths). Column, sorbent, buffers, operating conditions and detection as in Fig. 45.6. ———, Tetracycline; —·—·—·, anhydrotetracycline; – – – – –, 4-*epi*-anhydrotetracycline.

Fig. 45.8. Gel chromatography of tetracycline; anhydrotetracycline and 4-*epi*-anhydrotetracycline in solvents of different salt concentrations (Griffiths). Column, sorbent, operating conditions and detection as in Fig. 45.6. T = Tetracycline; A = anhydrotetracycline; E = 4-*epi*-anhydrotetracycline. a, 0.025 *M* Tris buffer, pH 9.0; b, 0.050 *M* Tris buffer in 0.025 *M* sodium chloride solution, pH 9.0; c, 0.050 *M* Tris buffer in 0.2 *M* sodium chloride solution, pH 9.0.

determining the tetracycline antibiotics was reported by Ascione *et al.* The method was applicable to crystalline tetracyclines and their various pharmaceutical dosage forms.

## Anthracyclines

Several *Streptomyces* strains produce yellowish red pigments of the anthracyclinone group and their glycosides, anthracycline antibiotics. Silica gel (Merck), modified with acids or alkali salts, is mostly used for the CC of these antibiotics and their aglycones. The modifications of the sorbent according to Brockmann *et al.* (1965) are as follows: oxalic acid–silica gel, 1 kg of silica gel ($<$ 0.08 mm) is mixed with 2 l of 0.5 *N* oxalic acid solution and dried at 110°C; sodium hydrogen carbonate–silica gel, 1 kg of silica gel

(< 0.08 mm) is mixed with 2 l of 0.5 *N* sodium hydrogen carbonate solution and dried at 110°C.

The separation of β-rhodomycinone and β-isorhodomycinone after hydrolysis of their glycosides was achieved by the same workers. Similar techniques were used in the separation of aglycones of anthracyclines from *S. purpurascens* (Brockmann and Niemeyer, 1968a) and of galirubins A and B (Eckardt), in studies of the degradation products of aklavinone and 7-deoxyaklavinone (Brockmann and Niemeyer, 1968b) and in configuration and conformation studies on rhodomycinones (Brockmann and Niemeyer, 1967).

CC was also used in structural studies on cinerubin A (Keller-Schierlein and Richle), cinerubin B (Richle *et al.*) and beromycins B and C (Kudinova *et al.*).

## NUCLEOSIDE ANTIBIOTICS INCLUDING POLYOXINS

In studies of puromycin biosynthesis, enzymatic methylation of O-demethylpuromycin was investigated by Rao *et al.* The latter compound was synthesized and purified by CC on silica gel. Puromycin obtained from its demethyl precursor was isolated by TLC on cellulose and characterized by TLC and paper electrophoresis. Shelton and Clark described gel filtration on Bio-Gel P-2 (polyacrylamide gel) with 0.01 *M* ammonium carbonate–0.05% sodium dodecyl sulphate for the purification of puromycin.

Sangivamycin was recovered from a fermentation broth by adsorbtion on Darco G-60 (1.5%) followed by several elutions with 0.05 *N* methanolic hydrochloric acid (Rao).

The anion-exchange CC of Dekker was used by Uematsu and Suhadolnik to convert toyocamycin quantitatively into sangivamycin, as follows. An aqueous concentrate (30 ml) containing toyocamycin was added to a column of Bio-Rad AG 1-X8 ($OH^-$), (200–400 mesh, 6 g) in water–methanol (85:15). The column was washed with 200 ml of water–methanol (70:30), and sangivamycin was eluted with 500 ml of water–methanol (30:70) and evaporated to 1 ml. The antibiotic crystallized overnight at 0°C.

Mammalian cells convert formycin A into formycin B and the latter into oxoformycin B, which can be isolated from urine by the IEC method of Ishizuka *et al.* A strain of *Nocardia interforma* having low productivity of formycin A was found to be able to convert radioactive formycin B into formycin A and oxoformycin B. Sawa *et al.* separated the three compounds as follows. The concentrated mixture (2 ml) was charged on to a column of Dowex 50-X20 ($H^+$) (4 ml wet volume) and eluted thoroughly with water. [$^3$H] Oxoformycin B appeared at the front followed by [$^3$H] formycin B, with satisfactory separation; [$^3$H] formycin A was eluted with 0.5 *N* ammonia solution.

Polyoxins A, B, D, E, F, G and H, after previous purification on Dowex 50W-X8 and Sephadex, were fractionated on Amberlite IR-4B ($Cl^-$) followed by CC on cellulose (Suzuki *et al.*, 1970a, b).

## PEPTIDES AND RELATED ANTIBIOTICS

### Analogues of amino acids

Stapley *et al.* isolated cycloserine from broth samples by adsorption on Dowex 50-X2 ($Na^+$) resin at pH 3 and elution with 1% aqueous pyridine. IEC was used in the isolation and purification of three new antimetabolites: L-2-amino-4-methoxy-*trans*-3-butenoic acid (Scannell *et al.*, 1972b), an iminomethyl derivative of ornithine (Scannell *et al.*, 1972a) and 2-amino-4-methyl-5-hexenoic acid (Kelly *et al.*).

### Actinomycins

Brockmann and Lackner (1968a, b) synthesized actinomycins via actinomycin acids. Procedures used in the separation of actinomycin acids and of actinomycins, obtained by cyclization of the former, include CC on cellulose, silica gel and alumina.

In their investigation of actinomycin biosynthesis by *Streptomyces antibioticus* and *S. chrysomalus*, Yajima *et al.* analyzed amino acids in hydrolyzates of actinomycins. Typical results are presented in Fig. 45.9.

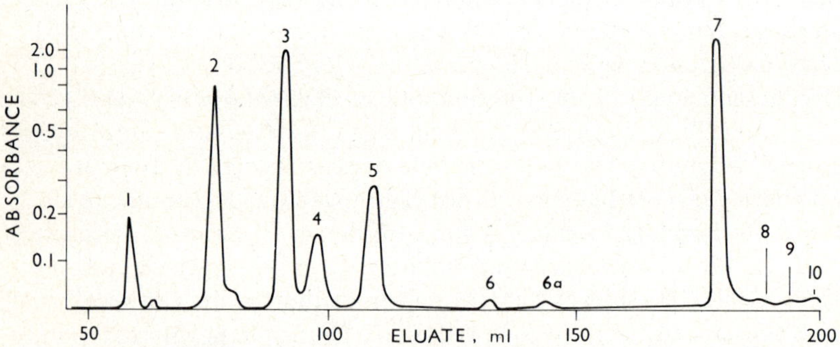


Fig. 45.9. Separation of the amino acid hydrolyzate obtained from an actinomycin mixture synthesized by *S. antibioticus* (Yajima *et al.*). Column: 69 × 0.9 cm. Ion exchanger: Beckman-Spinco PA-28 resin. Buffers: (1) 0.2 *M* sodium citrate solution, pH 3.05; (2) 0.2 *M* sodium citrate solution, pH 4.25. The separation was carried out with a Beckman-Spinco Model 120 C amino acid analyzer. Eluted volumes are apparent from the figure. Flow-rate of the buffer and ninhydrin: 34 ml/h. Temperature: 56°C. Retention times of the amino acids employed were as follows: 1, hydroxy-L-proline, 110; 2, L-threonine, 137; 3, sarcosine, 163; 4, N-methyl-D,L-valine, 175; 5, L-proline, 193; 6, N-methyl-D,L-*allo*isoleucine, 235; 6a, N-methyl-D,L-isoleucine, 279; 7, D-valine, 320; 8, D-*allo*isoleucine, 337; 9, D-isoleucine, 344 min; and 10, leucine, retention time not given. The buffer was changed from pH 3.05 to pH 4.25 at 230 min. Detection: the absorbance was measured at 570 nm, except for the imino acids, which were measured at 440 nm.

## Other peptides

Sulphur-containing thiopeptins were separated by CC into a major component, thiopeptin B, and four minor components, thiopeptins $A_1$–$A_4$ (Miyairi *et al.*). The separation procedures used were as follows. A 10-g amount of crude crystalline thiopeptin was dissolved in 20 ml of chloroform and subjected to chromatography by use of a 40 × 5 cm column packed with silica gel (Merck). After washing the column thoroughly with chloroform, elution was carried out with chloroform–methanol (19:1). The fraction of thiopeptin $A_1$ was eluted first, and a mixture of thiopeptins $A_1$, $A_2$, $A_3$ and $A_4$ followed. After these processes, thiopeptin B was eluted by using chloroform–methanol (9:1).

The 1-g mixture of thiopeptins $A_1$–$A_4$ was dissolved in 5 ml of chloroform and chromatographed on a 100 × 3 cm column containing silica gel (Merck) using chloroform–methanol (50:1). By this process, $A_1$, the $A_1$–$A_2$–$A_3$ mixture, $A_3$, the $A_3$–$A_4$ mixture, and $A_4$ could be separated. The $A_1$–$A_2$–$A_3$ mixture was further separated by the same process on an 80 × 2 cm silica gel (Merck) column.

The remaining mixture of $A_1$, $A_2$ and $A_3$ (30 mg) was dissolved in 1 ml of chloroform and the solution was spotted on five thin-layer plates (20 cm × 20 cm × 0.5 mm) of silica gel $GF_{254}$ (Merck) and was developed with chloroform–methanol (20:1). The spot of the $A_2$ fraction ($R_F$ 0.42) was scraped off and collected under UV light in a dark-room. The collected silica gel powder was extracted with chloroform–methanol (4:1), after which the extract was concentrated *in vacuo* to yield $A_2$ crystals. Each component of thiopeptin thus prepared was dissolved in warm acetone and crystallized to obtain the pure crystalline form.

Phleomycin, an antibacterial and antitumour antibiotic, was separated into ten components (Fig. 45.10) by CC on CM-Sephadex C-25 (Ikekawa *et al.*). When bleomycin complex, designated as bleomycin Cu–Bt, was separated by gradient CC on the same

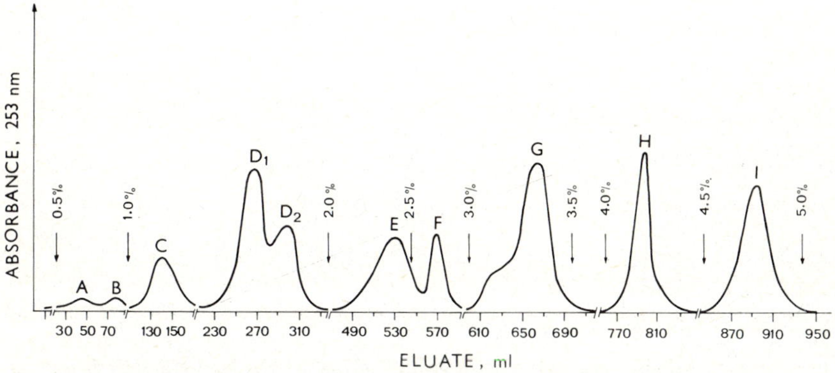


Fig. 45.10. Separation of phleomycins (Ikekawa *et al.*). Column: 20 × 2.5 cm. Sorbent: CM-Sephadex C-25. Eluents: ammonium formate, 0.5–5.0%. Eluted volumes are apparent from the figure. Flow-rate: 15 g per 10 min per tube. Reading the optical density curve, the concentration of ammonium formate was increased stepwise as indicated in the figure. Detection: the absorbance of the eluate at 253.7 nm was continuously recorded with an LKB Uvicord instrument. The individual phleomycins are designated in alphabetical order.

sorbent (Fig. 45.11), its components Cu–Bt2 and Cu–Bt5 were not differentiated from phleomycin C and F, respectively (Umezawa *et al.*, 1966).

Tuberactinomycins were separated by IEC (Fig. 45.12) and tuberactinomycin B was found to be identical with viomycin (Izumi *et al.*).

In Table 45.6, further applications of CC in studies of peptidic antibiotics are given.

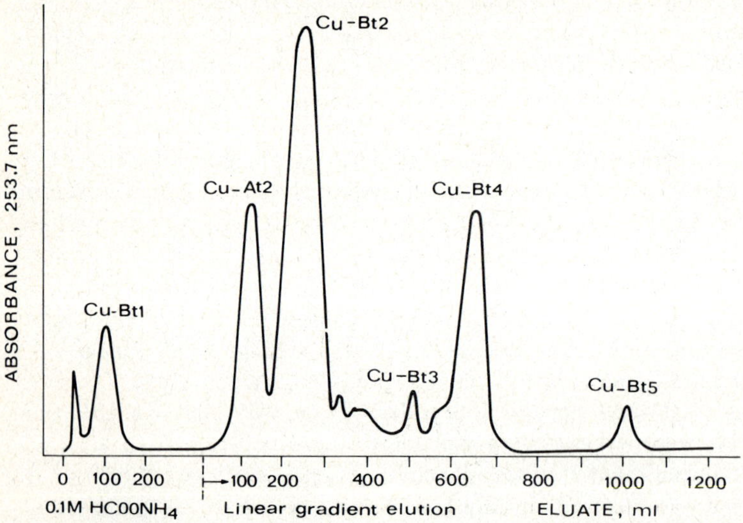


Fig. 45.11. Separation of bleomycin Cu–Bt complex (Umezawa *et al.*, 1966). Column: 21.5 × 1.5 cm. Sorbent: CM-Sephadex C-25. Eluents: (1) 0.1 *M* ammonium formate; (2) linear gradient of 0.1–1.0 *M* ammonium formate. Eluted volumes are apparent from the figure. Volume of mixing chamber: 1000 ml. Flow-rate: 0.5 ml/min. Temperature: 15–20°C. Detection: the absorbance of the eluate at 253.7 nm was continuously recorded with an LKB Uvicord instrument. The individual bleomycins were as designated in the figure.

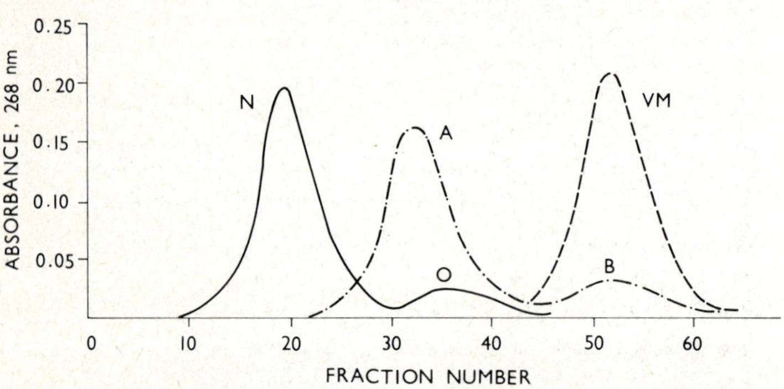


Fig. 45.12. Elution of tuberactinomycins and viomycin on an Amberlite CG-50 column (Izumi *et al.*). Column: 50 × 1.0 cm. Ion exchanger: Amberlite CG-50 ($NH_4^+$), 100–200 mesh. Buffer: 0.4 *M* ammonium acetate solution, pH 9.0. The eluate was fractionated into 20-ml portions. Flow-rate: 40 ml/h. Detection: all fractions were assayed for biological activity against *Bacillus subtilis* and also by measuring the UV absorption at 268 nm using a Hitachi Model 101 spectrophotometer. A = Tuberactinomycin A; B = tuberactinomycin B (identical with viomycin); VM = viomycin; N = tuberactinomycin N; O = tuberactinomycin O.

TABLE 45.6
APPLICATIONS OF LIQUID COLUMN CHROMATOGRAPHY IN SEPARATIONS OF PEPTIDIC AND RELATED ANTIBIOTICS

| Compound | Sorbent or ion exchanger | Eluent | Reference |
|---|---|---|---|
| *Liquid–solid chromatography* | | | |
| Enniatin B | Alumina (activity I) | Cyclohexane–methylene chloride–isopropanol* | Losse and Raue |
| Laspartomycin | Alumina Woelm (neutral) | 0.1 *N* Ammonia–methanol (1:4) | Naganawa *et al.* |
| Lindebein | Alumina Woelm | Water | Molloy *et al.* |
| Melinacidins | Silica gel | Toluene–ethyl acetate (3:1) | Argoudelis |
| Sulphonomycins | Kieselgel G | Chloroform–methanol (100:3) | Egawa *et al.* |
| Tsushimycin | Silica gel | Ethanol–1.4% ammonia (4:1) | Shoji *et al.* |
| *Gel chromatography* | | | |
| Neotelomycins | Sephadex G-25 | 0.1 *N* Acetic acid | Silaev *et al.* |
| Perhydroviomycin | Sephadex G-15 | 0.01 *N* Formic acid | Dyer *et al.* |
| *Ion-exchange chromatography* | | | |
| Amphomycin** | Dowex 1-X8 and 50W-X12 | 2 *N* Acetic acid | Bodanszky *et al.* |
| | | Water followed by 1% ammonia | Bodanszky *et al.* |
| Bacilysin | Zeo-Karb 225, SRC-5 | 1 *M* aq. Trimethylamine | Walker and Abraham |
| Bleomycin $A_2$ | Dowex 50W-X4 | Not given | Takita *et al.* |
| Enduracidin** | Amberlite IR-120 | 1 *N* Ammonia | Asai *et al.* |
| Macromomycin | Dowex 1-X2 | Not given | Chimura *et al.* |
| Polymyxin P | Amberlite IRC-50 | 0.05 *N* Hydrochloric acid | Kimura *et al.* |

*Ratio not specified
**Amino acids in hydrolyzates of the antibiotic.

## MISCELLANEOUS ANTIBIOTICS

Two autoantibiotics, inhibiting their own producer, *Candida albicans*, were isolated by Lingappa *et al.* A crude extract was separated into components by CC on silica gel. Compound I (phenethyl alcohol) was obtained by elution with light petroleum–diethyl ether (4:1) as the major component. Compound II (3-β-hydroxyethylindole) was obtained by eluting the column with a 3:2 mixture of the same solvents.

Vancomycin is not a pure, homogeneous substance and two components were demonstrated by PC (Betina, 1964). Best *et al.* separated vancomycin into three biologically active fractions by gradient elution from CM-Sephadex G-50 with 0.0–0.13 *M* ammonium hydrogen carbonate solution (pH 7.8). The three fractions, in the order of their elution, were designated CM-1, CM-2 and CM-3. CM-3 was 14 times more active than CM-2 (Fig. 45.13).

*References p.1005*

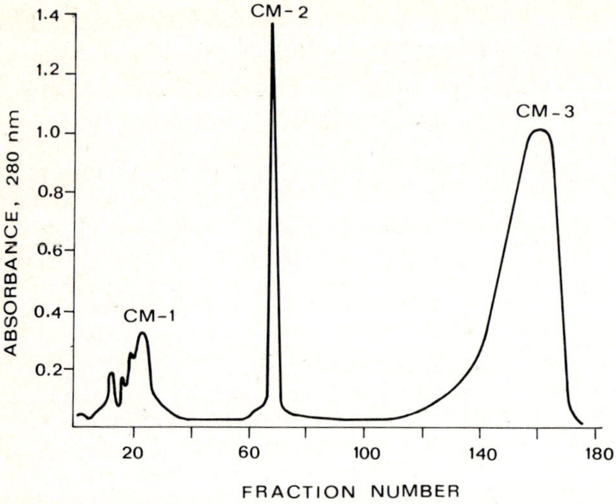


Fig. 45.13. Separation of vancomycin by gel chromatography (Best *et al.*). Column: 55 × 3 cm. Sorbent: CM-Sephadex C-50. Buffer: 0.0–0.13 *M* ammonium hydrogen carbonate solution, pH 7.8. Elution was accomplished with a linear gradient of the buffer with the use of an initial mixing chamber containing 500 ml of distilled water. Fractions (3.8 ml) were collected at 3-min intervals (initial rate). Detection: the absorbance of each fraction at 280 nm was measured. Fractions CM-1, CM-2 and CM-3 comprised about 10, 15 and 75% of the complex, respectively.

TABLE 45.7

APPLICATIONS OF LIQUID COLUMN CHROMATOGRAPHY IN SEPARATIONS OF MISCELLANEOUS ANTIBIOTICS

| Compound | Sorbent or ion exchanger | Eluent | Reference |
|---|---|---|---|
| *Liquid–solid chromatography* | | | |
| Aquamycin | Silica gel | Butyl acetate satd. with water | Sezaki *et al.* |
| Ascochlorin | Silica gel | Benzene–methanol (97:3) | Tamura *et al.* |
| Atrovenetin | Silica gel | Benzene–chloroform (1:1) | Narasimachari and Vining |
| Dehydrogriseofulvin | Florisil | Chloroform–methanol (1–3%) | Segal and Taylor |
| Kalafungin | Silica gel | Ethyl acetate–cyclohexane (1:9, 1:3) | Bergy |
| Paecilomycerol | Silica gel | Benzene–methanol (9:1) | Kato *et al.* |
| Resistomycin | Cellulose | Xylene–acetic acid–water (10:8:2), upper phase | Brockmann *et al.* (1969) |
| *Gel chromatography* | | | |
| Desacetylantimycin A | Sephadex LH-20 | Methanol | Singh *et al.* |
| *Ion-exchange chromatography* | | | |
| Ablastmycin | Dowex 50W-X8 | 0.2 *M* Pyridinium acetate buffer, pH 5.25 | Hashimoto *et al.* |
| Aristeromycin | Amberlite IRC-50 | 1 *N* Ammonia | Kusaka *et al.* |
| Pseudomonic acid | Amberlite XAD-2 | Methanolic and aq. ammonia; linear gradient | Fuller *et al.* |

Most recently, antibiotics from *Pseudomonas reptilivora* were fractionated by gel filtration with the use of Sephadex G-25 Superfine. One of these antibiotics seems to be a chelate compound of copper (Del Rio *et al.*).

Some other applications of CC in studies of miscellaneous antibiotics are given in Table 45.7.

## REFERENCES


Akasaki, K. and Abe, H., *J. Antibiot.*, 21 (1968) 98.

Argoudelis, A. D., *J. Antibiot.*, 25 (1972) 171.

Argoudelis, A. D. and Brodasky, T. F., *J. Antibiot.*, 25 (1972) 194.

Argoudelis, A. D., Coats, J. H. and Magerlein, B. J., *J. Antibiot.*, 25 (1972) 191.

Argoudelis, A. D., Coats, J. H., Mason, D. J. and Sebek, O. K., *J. Antibiot.*, 22 (1969) 309.

Asai, M., Muroi, M., Sugita, N., Kawashima, H., Mizuno, K. and Miyake, A., *J. Antibiot.*, 21 (1968) 138.

Ascione, P. P., Zagar, J. B. and Chrekian, G. P., *J. Chromatogr.*, 65 (1972) 377.

Banaszek, A., Krowicki, K. and Zamojski, A., *J. Chromatogr.*, 32 (1968) 581.

Benveniste, R. and Davies, J., *Biochemistry*, 10 (1971) 1787.

Benveniste, R., Yamada, T. and Davies, J., *Infection Immunity*, 1 (1970) 109.

Bergy, M. E., *J. Antibiot.*, 21 (1968) 454.

Bergy, M. E. and Eble, T. E., *Biochemistry*, 7 (1968) 653.

Best, G. K., Best, N. H. and Durham, N. N., *Antimicrob. Ag. Chemother. 1968*, (1969) 115.

Betina, V., *J. Chromatogr.*, 15 (1964) 379.

Betina, V., in K. Macek (Editor), *Pharmaceutical Applications of Thin-Layer and Paper Chromatography*, Elsevier, Amsterdam, London, New York, 1972, p. 503.

Betina, V., *J. Chromatogr.*, 78 (1973) 41.

Betina, V., Nemec, P., Kováč, Š., Kjaer, A. and Shapiro, R. H., *Acta Chem. Scand.*, 19 (1965) 519.

Bodanszky, M., Bodanszky, A. A., Ralofsky, C. A., Strong, R. C. and Foltz, R. L., *J. Antibiot.*, 24 (1971) 294.

Brannon, D. R., Mabe, J. A., Ellis, R., Whitney, J. G. and Nagarajan, R., *Antimicrob. Ag. Chemother.*, 1 (1972) 242.

Brockmann, H. and Lackner, H., *Chem. Ber.*, 101 (1968a) 1312.

Brockmann, H. and Lackner, H., *Chem. Ber.*, 101 (1968b) 2231.

Brockmann, H., Meyer, E., Schrempp, K., Reiners, F. and Reschke, T., *Chem. Ber.*, 102 (1969) 1224.

Brockmann, H. and Niemeyer, J., *Chem. Ber.*, 100 (1967) 3578.

Brockmann, H. and Niemeyer, J., *Chem. Ber.*, 101 (1968a) 1341.

Brockmann, H. and Niemeyer, J., *Chem. Ber.*, 101 (1968b) 2409.

Brockmann, H., Niemeyer, J. and Rode, W., *Chem. Ber.*, 98 (1965) 3145.

Chimura, H., Ishizuka, M., Hamada, M., Hori, S., Kimura, K., Iwanaga, J., Takeuchi, T. and Umezawa, H., *J. Antibiot.*, 21 (1968) 44.

Cron, M. J., Smith, R. E., Hooper, I. R., Keil, J. G., Ragan, E. A., Schreiber, R. H., Schwab, G. and Godfrey, J. C., *Antimicrob. Ag. Chemother. 1969*, (1970) 219.

De Fabrizio, F., *J. Pharm. Sci.*, 58 (1969) 136.

Dekker, C. A., *J. Amer. Chem. Soc.*, 87 (1965) 4027.

Del Rio, L. A., Gorgé, J. L., Olivares, J. and Mayor, F., *Antimicrob. Ag. Chemother.*, 2 (1972) 189.

Dennen, D., *J. Pharm. Sci.*, 56 (1967) 1273.

Deyl, Z., Rosmus, J., Juřicová, M. and Kopecký, J. (Editors), *Bibliography of Column Chromatography 1967–1970*, Elsevier, Amsterdam, London, New York, 1973.

Dion, H. W., Woo, P. W. K., Willmer, N. E., Kern, D. L., Onaga, J. and Fusari, S. A., *Antimicrob. Ag. Chemother.*, 2 (1972) 84.

Dyer, J. R., Carter, J. H. and Van Wyk, P. J., *J. Med. Chem.*, 14 (1971) 1120.

Eckardt, K., *Chem. Ber.*, 100 (1967) 2561.

Egawa, Y., Umino, K., Tamura, Y., Shimizu, M., Kaneko, K., Sakurazawa, M., Awataguchi, S. and Okuda, T., *J. Antibiot.*, 22 (1969) 12.
Fechtig, B., Bickel, H. and Heusler, K., *Helv. Chim. Acta*, 55 (1972) 417.
Fukagawa, Y., Sawa, T., Homma, I., Takeuchi, T. and Umezawa, H., *J. Antibiot.*, 21 (1968a) 410.
Fukagawa, Y., Sawa, T., Takeuchi, T. and Umezawa, H., *J. Antibiot.*, 21 (1968b) 450.
Fuller, A. T., Mellows, G., Woolford, M., Banks, G. T., Barrow, K. D. and Chain, E. B., *Nature (London)*, 234 (1971) 416.
Griffiths, B. W., *J. Chromatogr.*, 38 (1968) 41.
Handschin, U., Sigg, H. P. and Tamm, Ch., *Helv. Chim. Acta*, 51 (1968) 1943.
Hanes, C. S. and Isherwood, F. A., *Nature, (London)*, 164 (1949) 1107.
Hashimoto, T., Kito, M., Takeuchi, T., Hamada, M., Maeda, K., Okami, Y. and Umezawa, H., *J. Antibiot.*, 21 (1968) 37.
Hessler, E. J., Jahnke, H. K., Robertson, J. H., Tsuji, K., Rinehart, Jr., K. L., and Shier, T., *J. Antibiot.*, 23 (1970) 464.
Heusler, K., *Helv. Chim. Acta*, 55 (1972) 388.
Ikekawa, T., Iwami, F., Hiranaka, H. and Umezawa, H., *J. Antibiot., Ser. A*, 17 (1964) 194.
Inouye, S., Tsuruoka, T., Ito, T. and Niida, T., *Tetrahedron*, 23 (1968) 2125.
Ishizuka, M., Sawa, T., Koyama, G., Takeuchi, T. and Umezawa, H., *J. Antibiot.*, 21 (1968) 1.
Ito, T., Akita, E., Tsuruoka, T. and Niida, T., *Antimicrob. Ag. Chemother. 1970*, (1971) 33.
Izumi, R., Noda, T., Ando, T., Take, T. and Nagata, A., *J. Antibiot.*, 25 (1972) 201.
Kato, A., Ando, K., Kimura, T., Tamura, G. and Arima, K., *J. Antibiot.*, 22 (1969) 419.
Keller-Schierlein, W. and Richle, W., *Antimicrob. Ag. Chemother. 1970*, (1971) 68.
Kelly, R. B., Martin, D. G. and Hanka, L. J., *Can. J. Chem.*, 47 (1969) 2504.
Khokhlov, A. S. and Reshetov, P. D., *J. Chromatogr.*, 14 (1964) 495.
Khokhlov, A. S. and Shutova, K. I., *J. Antibiot.*, 25 (1972) 501.
Kimura, Y., Murai, E., Fujisawa, M., Tatsuki, T. and Nobue, F., *J. Antibiot.*, 22 (1969) 449.
Kinoshita, M. and Umezawa, S., *Bull. Chem. Soc. Jap.*, 43 (1970) 897.
Kirkland, J. J., in Kirkland, J. J. (Editor), *Modern Practice of Liquid Chromatography*, Wiley-Interscience, New York, 1971, p. 161.
Koch, K. F. and Rhoades, J. A., *Antimicrob. Ag. Chemother. 1970*, (1971) 309.
Kudinova, M. K., Borisova, V. N., Petukhova, N. M. and Brazhnikova, M. G., *Antibiotiki*, 18 (1972) 689.
Kukolja, S., *J. Med. Chem.*, 11 (1968) 1067.
Kunze, B., Schabacher, K., Zähner, H. and Zeeck, A., *Arch. Mikrobiol.*, 86 (1972) 147.
Kusaka, T., Yamamoto, H., Shibata, M., Muroi, M., Kishi, T. and Mizuno, K., *J. Antibiot.*, 21 (1968) 255.
Lancini, G. and Hengeller, C., *J. Antibiot.*, 22 (1969) 637.
Lingappa, B. T., Prasad, M. and Lingappa, Y., *Science*, 163 (1969) 192.
Losse, G. and Raue, H., *Chem. Ber.*, 101 (1968) 1532.
Maehr, H. and Schaffner, C. P., *J. Chromatogr.*, 30 (1967) 572.
Maggi, N., Vigevani, A., Gallo, G. G. and Pasqualucci, C. R., *J. Med. Chem.*, 11 (1968) 936.
Majumdar, M. K. and Majumdar, S. K., *Anal. Chem.*, 39 (1967) 215.
Majumdar, M. K. and Majumdar, S. K., *Biochem. J.*, 120 (1970) 271.
Miyairi, M., Miyoshi, T., Aoki, H., Kohsaka, M., Ikushima, H., Kunugita, K., Sakai, H. and Imanaka, H., *Antimicrob. Ag. Chemother.*, 1 (1972) 192.
Molloy, B. B., Lively, D. H., Gale, R. M., Gorman, M. and Boeck, L. D. , *J. Antibiot.*, 25 (1972) 137.
Mori, T., Ichiyanagi, T., Kondō, H., Tokunaga, K., Oda, T. and Munakata, K., *J. Antibiot.*, 24 (1971) 339.
Morin, R. B., Jackson, B. G., Flynn, E. H., Roeske, R. W. and Andrews, S. L., *J. Amer. Chem. Soc.*, 91 (1969a) 1396.
Morin, R. B., Jackson, B. G., Mueller, R. A., Lavagnino, E. R., Scanlon, W. B. and Andrews, S. L., *J. Amer. Chem. Soc.*, 91 (1969b) 1401.
Naganawa, H., Hamada, M., Maeda, K., Okami, Y., Takeuchi, T. and Umezawa, H., *J. Antibiot.*, 21 (1968) 55.
Nagarajan, R., Boeck, L. D., Gorman, M., Hamill, R. L., Higgens, C. E., Hoehn, M. M., Stark, W. M. and Whitney, J. G., *J. Amer. Chem. Soc.*, 93 (1971) 2308.
Narasimachari, N. and Vining, L. C., *J. Antibiot.*, 25 (1972) 155.

Nayler, J. H. C., *Proc. Roy. Soc., Ser. B*, 179 (1971) 357.
Neelameghan, A., Ravichandra Rao, I. K. and Gupta, B. S. S., *Life Sci.*, 9 (1972) 139.
Neuss, N., Koch, K. F., Molloy, B. B., Day, W., Huckstep, L. L., Dorman, D. E. and Roberts, J. D., *Helv. Chim. Acta*, 53 (1970) 2314.
Ōmura, S., Katagiri, M. and Hata, T., *J. Antibiot.*, 21 (1968) 272.
Ōmura, S., Tischler, M., Nakagawa, A., Hironaka, Y. and Hata, T., *J. Med. Chem.*, 15 (1972) 1011.
Ozanne, B., Benveniste, R., Tipper, D. and Davies, J., *J. Bacteriol.*, 100 (1969) 1144.
Pénasse, L., Barthélémy, P. and Nominé, G., *Bull. Soc. Chim. Fr.*, (1969) 2391.
Preud'homme, J., Tarridec, P. and Belloc, A., *Bull. Soc. Chim. Fr.*, (1968) 585.
Purdie, J. W., Gravelle, R. A. and Hanafi, D. E., *J. Chromatogr.*, 38 (1968) 346.
Rao, K., *J. Med. Chem.*, 11 (1968) 939.
Rao, M. M., Rebello, P. F. and Pogell, B. M., *J. Biol. Chem.*, 244 (1969) 112.
Richle, W., Winkler, E. K., Hawley, D. M., Dobler, M. and Keller-Schierlein, W., *Helv. Chim. Acta*, 54 (1972) 467.
Rothweiler, W. and Tamm, Ch., *Helv. Chim. Acta*, 53 (1970) 696.
Sawa, T., Fukagawa, Y., Homma, I., Wakashiro, T., Takeuchi, T. and Hori, M., *J. Antibiot.*, 21 (1968) 334.
Scannell, J. P., Ax, H. A., Pruess, D. L., Williams, T., Demny, T. C. and Stempel, A., *J. Antibiot.*, 25 (1972a) 179.
Scannell, J. P., Pruess, D. L., Demny, T. C., Sello, L. H., Williams, T. and Stempel, A., *J. Antibiot.*, 25 (1972b) 122.
Scartazzini, R., Peter, H., Bickel, H., Heusler, K. and Woodward, R. B., *Helv. Chim. Acta*, 55 (1972) 408.
Schacht, U. and Huber, G., *J. Antibiot.*, 22 (1969) 597.
Schmit, J. A., in J. J. Kirkland (Editor), *Modern Practice of Liquid Chromatography,* Wiley-Interscience, New York, 1971, p. 375.
Segal, A. and Taylor, E. H., *J. Pharm. Sci.*, 57 (1968) 874.
Sezaki, M., Hara, T., Ayukawa, S., Takeuchi, T., Okami, Y., Hamada, M., Nagatsu, T. and Umezawa, H., *J. Antibiot.*, 21 (1968) 91.
Shelton, K. R. and Clark, J. M., Jr., *Biochemistry*, 6 (1967) 2735.
Shoji, J. -I., Kozuki, S., Okamoto, S., Sakazaki, R. and Otsuka, H., *J. Antibiot.*, 21 (1968) 439.
Silaev, A. B., Katrukha, G. S., Trifonova, Zh. P. and Sinyavskaya, I. G., *Antibiotiki*, 18 (1968) 13.
Singh, K., Schillings, G., Rakhit, S. and Vézina, C., *J. Antibiot.*, 25 (1972) 141.
Smith, H. and Marshall, A. C., *Nature (London)*, 232 (1971) 45.
Stapley, E. O., Miller, T. W. and Jackson, M., *Antimicrob. Ag. Chemother. 1968*, (1969) 268.
Störl, H. J., *J. Chromatogr.*, 40 (1969) 71.
Sun, F. F., *J. Pharm. Sci.*, 62 (1973) 1657.
Suzuki, S., Isono, K. and Nagatsu, J., *Jap. Pat.*, 70 00,957; *C.A.*, 72 (1970a) 120075m.
Suzuki, S., Isono, K. and Nagatsu, J., *Jap. Pat.*, 70 00,954; *C.A.*, 72 (1970b) 120076n.
Takita, T., Muraoka, Y., Maeda, K. and Umezawa, H., *J. Antibiot.*, 21 (1968) 79.
Tamura, G., Suzuki, S., Takatsuki, A., Ando, K. and Arima, K., *J. Antibiot.*, 21 (1968) 539.
Taniyama, H., Sawada, J. and Kitagawa, T., *Chem. Pharm. Bull.*, 19 (1971a) 1627.
Taniyama, H., Sawada, J. and Kitagawa, T., *J. Chromatogr.*, 56 (1971b) 360.
Taniyama, H., Sawada, J. and Kitagawa, T., *J. Antibiot.*, 24 (1971c) 390.
Thomas, R. C., Ikeda, G. J. and Harpootlian, H., *J. Pharm. Sci.*, 56 (1967) 862.
Thompson, R. Q. and Presti, E. A., *Antimicrob. Ag. Chemother. 1967*, (1968) 332.
Tsuruoka, T., Shoumura, T., Azaki, N., Niwa, T. and Niida, T., *J. Antibiot.*, 21 (1968) 237.
Uematsu, T. and Suhadolnik, R. J., *Biochemistry*, 9 (1970) 1260.
Umezawa, H., Doi, O., Ogura, M., Kondo, S. and Tanaka, N., *J. Antibiot.*, 21 (1968a) 154.
Umezawa, S., Koto, S., Tatsuta, K., Hineno, H., Nishimura, Y. and Tsumura, T., *J. Antibiot.*, 21 (1968b) 424.
Umezawa, S., Koto, S., Tatsuta, K., Hineno, H., Nishimura, Y. and Tsumura, T., *Bull. Chem. Soc. Jap.*, 42 (1969a) 537.
Umezawa, S., Koto, S., Tatsuta, K. and Tsumura, T., *Bull. Chem. Soc. Jap.*, 42 (1969b) 529.

Umezawa, H., Okanishi, M., Kondo, S., Hamana, K., Utahara, R., Maeda, K. and Mitsuhashi, S., *Science*, 157 (1967a) 1559.
Umezawa, H., Suhara, Y., Takita, T. and Maeda, K., *J. Antibiot.*, 19 (1966) 210.
Umezawa, H., Takasawa, S., Okanishi, M. and Utahara, R., *J. Antibiot.*, 21 (1968c) 81.
Umezawa, S., Tatsuta, K. and Koto, S., *J. Antibiot.*, 21 (1968d) 367.
Umezawa, S., Tatsuta, K. and Koto, S., *Bull. Chem. Soc. Jap.*, 42 (1969c) 533.
Umezawa, S., Umino, K., Shibahara, S. and Omoto, S., *Bull. Chem. Soc. Jap.*, 40 (1967b) 2419.
Wagman, G. H., Marquez, J. A. and Weinstein, M. J., *J. Chromatogr.*, 34 (1968) 210.
Walker, J. E. and Abraham, E. P., *Biochem. J.*, 118 (1970) 557.
Weinstein, M. J., Wagman, G. H., Marquez, J., Oden, E., Testa, R. T. and Waitz, J. A., *Antimicrob. Ag. Chemother. 1968* (1969) 260.
Weston, R. D., *Antimicrob. Ag. Chemother. 1967* (1968) 553.
Woodward, R. B., Heusler, K., Gosteli, J., Naegeli, P., Oppolzer, W., Ramage, R., Ranganathan, S. and Vorbrüggen, H., *J. Amer. Chem. Soc.*, 88 (1966) 852.
Yajima, T., Grigg, M. A. and Katz, E., *Arch. Biochem. Biophys.*, 151 (1972) 565.

*Chapter 46*

# Pesticides

J. ZABRANSKÝ and J. CHURÁČEK

## CONTENTS



## INTRODUCTION AND GENERAL TECHNIQUES

Pesticides are substances with various basic chemical structures, used in agriculture for the protection of plants against pests, diseases and weeds and for the protection of animals against ectoparasites. Recently, great attention has been paid to the analysis of pesticides, because in many instances they are highly toxic substances and their residues can pass into food and menace human health.

Liquid column chromatography was found to be suitable as a clean-up procedure during the analysis of pesticide residues and also as a separation method in the analysis of pesticide preparations and in the study of their metabolism.

The chromatographic separation of pesticides is usually preceded by extraction with a suitable solvent. Particularly in the application of liquid column chromatography to the analysis of pesticide residues in biological material, the sensitivity and the accuracy of determination are dependent to a considerable degree on the extraction procedure employed. The choice of a suitable solvent is dependent primarily on the character of the pesticide and of the material being extracted and therefore a general extraction procedure, valid for all types of pesticides, cannot be proposed.

The most widespread use of liquid column chromatography is the application of liquid–solid chromatography in the determination of pesticides in biological material for the elimination of fats, pigments and other interfering substances. For this purpose, magnesium silicate (Florisil) is a suitable sorbent. However, it was observed that, for example, dieldrin and endrin are sorbed on Florisil so strongly that even a strongly polar solvent cannot elute them without some losses. It is possible, however, to apply other sorbents, such as silica gel, Kieselguhr and Celite. The use of alumina is not recommended because with polar substances desorption does not take place quantitatively.

Recent developments in the field of instrumentation and column technology enables liquid–liquid chromatography to be explored as a valuable adjunct to gas chromatography

*References p.1030*

TABLE 46.1
PERCENTAGE RECOVERIES OF CHLORINATED PESTICIDES FROM FLORISIL COLUMNS USING THE REYNOLDS PROCEDURE (BEVENUE AND OGATA)

| Adsorbent | Eluting agent | Lindane | Heptachlor | Aldrin | Heptachlor epoxide | DDE | DDD | *p,p'*-DDT | Dieldrin |
|---|---|---|---|---|---|---|---|---|---|
| Florisil, PR grade* | 200 ml of | 89 | 91 | 91 | 0 | 87 | 83 | 88 | 0 |
| Florisil, regular grade | *n*-hexane | 7 | 94 | 95 | 0 | 93 | 0 | 100 | 0 |
| Florisil, PR grade | 250 ml of 20% | 6 | 0 | 0 | 96 | 0 | 0 | 0 | 97 |
| Florisil, regular grade | diethyl ether in *n*-hexane | 92 | 0 | 0 | 67 | 0 | 94 | 0 | 94 |

*The columns were packed with PR and regular grades of Florisil which had been desiccated for 24 h after treatment for 16 h at 130°C. All of the other columns were packed with hot Florisil immediately after its removal from a 130°C oven.

for the rapid analysis of pesticides, particularly those which require the synthesis of derivatives for analysis or which cannot be handled by gas chromatography.

New porous support materials such as Zipax (DuPont, Wilmington, Del., U.S.A.) and Corasil (Waters Ass., Framington, Mass., U.S.A.) can be used in highly efficient columns which give high-speed separations. Pesticides can be chromatographed using a liquid–liquid partition system consisting of a $\beta,\beta'$-oxydipropionitrile or trimethylene glycol stationary phase and *n*-heptane or *n*-heptane with 5–10% of a polar modifier such as tetrahydrofuran, dioxane or chloroform as the mobile phase.

Gel chromatography is a useful addition to the other chromatographic techniques when used to separate pesticides into groups on the basis of their elution sequence. This technique is of little use as a clean-up procedure because much of the pigment material is eluted in the range over which pesticides begin to emerge. It is an advantage that the column can be used repeatedly over long periods of time without any detectable changes in the elution volumes of the pesticides.

Ion-exchange chromatography was found to be a suitable method for the separation of metabolites of pesticides.

The identification and determination of pesticides is very often carried out by some other chromatographic method, for example GLC, using selective, highly sensitive detectors (Thornburg and Beckman), or by spectrometric methods.

## Liquid–solid chromatography as a clean-up technique

When extracting insecticides from biological material, a considerable amount of pigments, waxes and other materials are extracted simultaneously, and these materials may interfere in the subsequent separation and identification of the insecticides and their metabolites. The nature of these co-extractives differs according to the material being extracted and no comprehensive clean-up procedure can be recommended. In general, the interfering compounds can be removed by partition between two solvents or by column chromatography.

One of the most widely used methods for the clean-up of fatty or oily samples is to extract a solution of the sample in a hexane or a light petroleum with an immiscible solvent that selectively dissolves the pesticides. The solvents most often used for this purpose are acetonitrile (Jones and Riddick), dimethylformamide (Burchfield and Storrs) and dimethyl sulphoxide (Eidelman, 1962, 1963, 1967; Haenni *et al.*). Of these, dimethylformamide is the best solvent for chlorinated pesticide residues; dimethyl sulphoxide is almost as good as dimethylformamide, but acetonitrile is inferior to the other two (Thornburg). However, dimethyl sulphoxide dissolves much less oil or fat than either dimethylformamide or acetonitrile. Experiments on the partitioning of butter fat between *n*-hexane and various solvents have indicated that the solubility of the fat in dimethyl sulphoxide is only about 5% of that in acetonitrile and 3% of that in dimethylformamide (Thornburg). These facts indicate that the best clean-up should be achieved with dimethyl sulphoxide although, in general, analysts have preferred to use acetonitrile or dimethylformamide.

Florisil is the sorbent most often used for the separation of chlorinated pesticides (Bevenue and Ogata, Burke and Malone, Reynolds). The influence of the quality of Florisil on the recovery of chlorinated pesticides is shown in Table 46.1.

*References p.1030*

Moubry *et al.* described a method in which minimum amounts of adsorbent and eluent are required in the analysis. Florisil is activated at 650°C and a critical amount of water is added in the range 0.5–2.0%; the exact amount of water required to obtain the most favourable results from each batch of adsorbent is determined by a series of trial experiments. Moubry *et al.* claim that if the correct amount of water is added to the Florisil, not more than 10% of the sample is eluted from the adsorbent.

Florisil is of little use for organophosphorus pesticides. Beckman and Garber investigated the recovery of 65 phosphorus pesticides from Florisil with a solvent–elution system of benzene and diethyl ether–benzene (1:2), acetone and methanol as eluents. Most of the compounds are recovered in the eluting mixture or acetone. Of all the investigated pesticides, only three were eluted with benzene.

The clean-up of aqueous and *n*-hexane extracts from vegetables can be performed on a column of alumina of activity V. *n*-Hexane-soluble insecticides such as phorate, bromophos, diazinon, fenitrothion, malathion and parathion are eluted with *n*-hexane or a 2% solution of acetone in *n*-hexane (Sissons and Telling, 1970a), whereas the complete elution of all of the water-soluble insecticides is obtained in 30 ml of chloroform (Sissons and Telling, 1970b). Elution of *n*-hexane-soluble insecticides from a column containing 8 g of alumina of activity V using various eluting agents is shown in Table 46.2.

TABLE 46.2
ELUTION OF *n*-HEXANE-SOLUBLE INSECTICIDES FROM ALUMINA COLUMN (SISSONS AND TELLING, 1970a)

| Eluting agent | | |
|---|---|---|
| *n*-Hexane | 2% Acetone in *n*-hexane | |
| 30-ml fraction | 30-ml fraction | 50-ml fraction |
| Aldrin | Chlorfenson | Azinphos-methyl |
| Chlordane | Endosulfan B | Chlorfenvinphos |
| Chlorbenside | Methoxychlor | |
| *p,p*′-DDT | Tetradifon | |
| *o,p*′-DDT | Diazinon | |
| DDE | Dioxathion | |
| TDE | Fenitrothion | |
| Endosulfan A | Malathion | |
| Dieldrin | Parathion | |
| Endrin | Thionazin | |
| Heptachlor | | |
| Heptachlor epoxide | | |
| Isodrin | | |
| Lindane | | |
| Telodrin | | |
| Toxaphene | | |
| Phorate | | |
| Bromophos | | |
| Disulfoton | | |

In investigations on environmental polution, Leoni recommends the separation of complex mixtures of pesticides into groups on a silica gel micro-column before GLC analysis. The micro-column is made of Pyrex glass and consists of two sections. The lower section, 30 × 0.42 cm I.D., is filled with 1 g of silica gel, grade 950 (60–200 mesh) (Grace Davison, Baltimore, Md., U.S.A.) to a height of 10 cm (the height is important for the reproducibility of the separation). The upper section serves as a reservoir for the eluents. The four eluents are 20 ml of *n*-hexane, 8 ml of 60% benzene in *n*-hexane, 8 ml of benzene and 14 ml of 50% ethyl acetate in benzene. The flow-rate must be about 1 ml/min and can be obtained by exerting a slight pressure upon the column. The silica gel is activated for 2 h before use with pre-warmed air at 130°C in an oven. After cooling, the product is weighed and partially deactivated by adding distilled water (5%, w/w).

The results of the separation of pesticides and polychlorobiphenyls on the silica gel micro-column are shown in Table 46.3. Some pesticides may be present in two successive elution groups, such as α- and γ-chlordane or ethion, methoxychlor and others.

This technique was used for the investigation of pesticides in samples of surface and drinking waters, but it can also be usefully applied to the analysis of pesticides in samples other than water.

TABLE 46.3

SEPARATION OF PESTICIDES AND POLYCHLORINATED BIPHENYLS IN GROUPS BY SILICA GEL MICRO-COLUMN CHROMATOGRAPHY (LEONI)

| Solvent system | | | |
|---|---|---|---|
| *n*-Hexane (20 ml) | 60% Benzene in *n*-hexane (8 ml) | Benzene (8 ml) | 50% Ethyl acetate in benzene (14 ml) |
| Hexachlorobenzene | *o,p'*-TDE | Parathion | Captan |
| Aldrin | *p,p'*-TDE | Parathion-methyl | Malathion |
| *p,p'*-DDE | α-BHC | Photodieldrin | Diazinon |
| Heptachlor | β-BHC | 2,4-D (methyl ester) | Paraoxon |
| *o,p'*-DDE | γ-BHC | | Paraoxon-methyl |
| Polychlorinated biphenyls | δ-BHC | | Guthion-methyl |
| *o,p'*-DDT | Perthane | | Dimethoate |
| *p,p'*-DDT | Trithion | | |
| | Bromophos | | |
| | Dursban | | |
| | Kelthane | | |
| | Heptachlor epoxide | | |
| | Endosulfan A | | |
| | Ronnel | | |
| | Trithion-methyl | | |
| | Endrin | | |
| | Vegadex | | |
| | Dieldrin | | |

*References p.1030*

## CHLORINATED PESTICIDES AND THEIR METABOLITES

Chlorinated pesticides are a large group of substances with widely varying characters, and therefore a single, universal method for their separation cannot be devised. A substantial role is played not only by the type of the sorbent used but also by its purity.

Great attention has been devoted to the metabolism of the most frequently used substances, for example dieldrin. The isolation of metabolites and the following of the metabolic changes of dieldrin in living organisms is not a simple task. Hedde *et al.* described a method for the analysis of the distribution and isolation of urinary metabolites of dieldrin in sheep. In order to be able to follow dieldrin and its metabolites more easily both in the body of sheep and during the isolation and the separation of metabolites, a preparation labelled with $^{14}C$ was used.

The metabolites of [$^{14}C$]dieldrin extracted with *n*-hexane were separated on a Sephadex G-10 gel column (Feil *et al.*). The urine extract was reduced to dryness by flash evaporation and the residue was dissolved in a minimum amount of acetone and applied on the Sephadex G-10 column. The column was eluted with about 20 column volumes of water followed by methanol at the rate of 0.1 ml/min. The metabolites remained on the column during elution with water, while many urinary pigments and salts were removed. The metabolites were removed in a broad band by elution with methanol.

After the methanol had been evaporated, the metabolites were dissolved in acetone ($<$ 0.5 ml) and applied on to a 100 × 2 cm Sephadex LH-20 column which had been washed with acetone. This column, when eluted with acetone at a rate of 0.1–0.2 ml/min, separated the *n*-hexane-soluble metabolites into four radioactive fractions (Fig. 46.1).

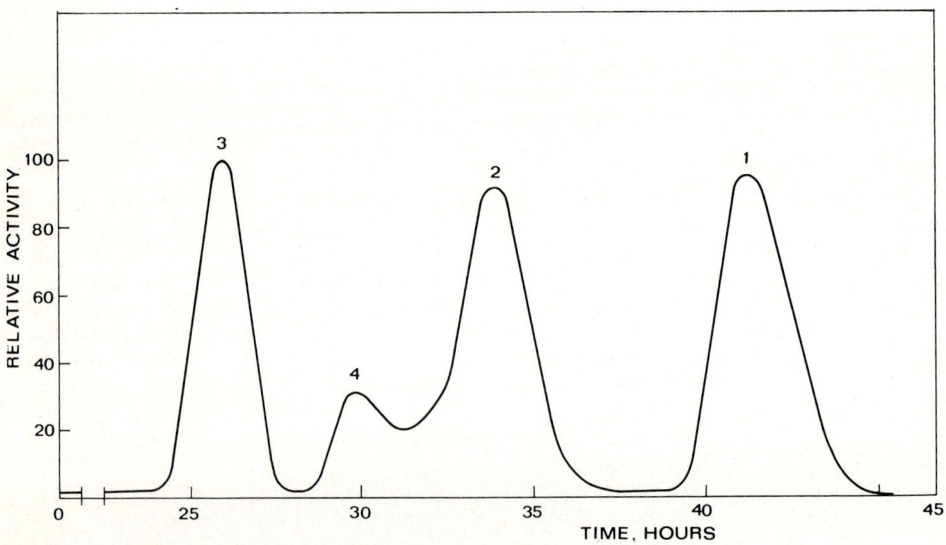


Fig. 46.1. Separation of [$^{14}C$] dieldrin metabolites (Feil *et al.*). Column: 100 × 2 cm. Sorbent: Sephadex LH-20. Mobile phase: acetone. Flow-rate: 0.22 ml/min. Detection: radioactivity measurement. 1, *trans*-6,7-Dihydroaldrin; 2, not identified; 3, 5-hydroxy-1,2,3,4,10,10-hexachloro-6,7-epoxy-1,4,4α, 5,6,7,8,8α-octahydro-1,4-endo-5,8-exodimethanonaphthalene; 4, not identified.

An interesting procedure for the separation and determination of dieldrin in a mixture with pentachlorophenol was described by Wilson *et al.* Removal of the pentachlorophenol by extraction with aqueous alkali is difficult and leads to inaccurate results because of entrainment and sometimes the formation of emulsions. These difficulties were overcome by the use, under non-aqueous conditions, of a macroreticular strong anion-exchange resin, such as De Acidite K or Amberlyst 29. A column of the resin is first generated in the hydroxide form by passing four bed volumes of 2 *N* sodium hydroxide solution through the column. After copious washing with deionized water in order to remove excess of alkali and all traces of chloride (the resins are supplied in the stable chloride form), water is removed by passing dry methanol through the column. The sample is diluted with dry methanol and percolated through the column. The pentachlorophenol is retained on the resin, while the dieldrin is found in the effluent. The column is washed with four bed volumes of methanol, the washings being added to the effluent. After removal of methanol on a water-bath, the residue is assayed for chlorine by the oxygen-flask technique.

Liquid column chromatography is very often used for the purity control of technical preparations. For example, Ďulák *et al.* chromatographed the herbicide pyrazone on an alumina column and so determined the impurities present (Fig. 46.2).

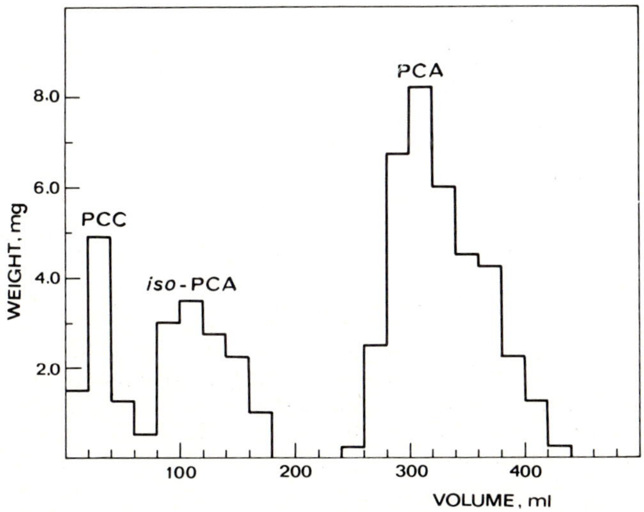


Fig. 46.2. Separation of pyrazone (Ďulák *et al.*). Column: 80 × 0.9 cm. Sorbent: alumina (6 g). Mobile phase: benzene, 5% chloroform in benzene, 10% chloroform in benzene. Flow-rate; 1.3–1.6 ml/min. Single components appear in the order PCC, iso-PCA and PCA. The amount of the components (in mg) corresponds to 20-ml fractions. PCA = 5-Amino-4-chloro-2-phenyl-3(2*H*)-pyridazinone (pyrazone); iso-PCA = 4-amino-5-chloro-2-phenyl-3(2*H*)-pyridazinone; PCC = 4,5-dichloro-2-phenyl-3(2*H*)-pyridazinone.

High-speed liquid chromatography has also been used successfully for the analysis of chlorinated pesticides. For example, using classical liquid–liquid partition chromatography, on a porous support, 10 pesticides belonging to different groups could be separated successfully (Fig. 46.3; Waters Ass., a). Under the conditions of liquid–solid chromato-

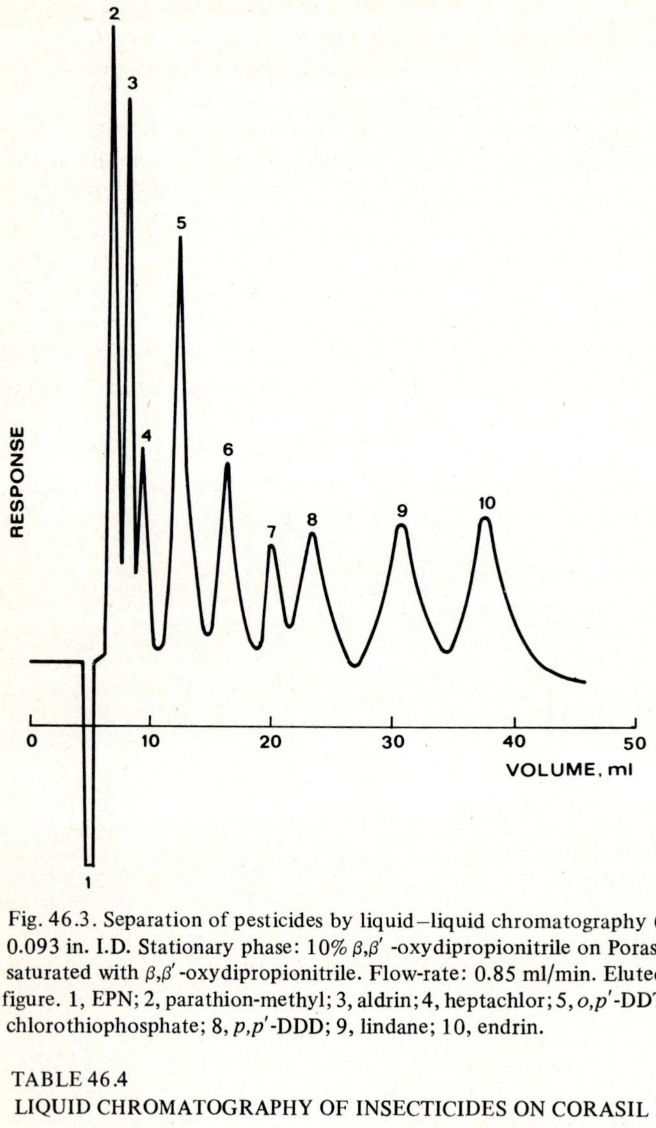


Fig. 46.3. Separation of pesticides by liquid–liquid chromatography (Waters Ass., a). Column: 4 ft × 0.093 in. I.D. Stationary phase: 10% β,β′-oxydipropionitrile on Porasil 60. Mobile phase: isooctane saturated with β,β′-oxydipropionitrile. Flow-rate: 0.85 ml/min. Eluted volumes are apparent from the figure. 1, EPN; 2, parathion-methyl; 3, aldrin; 4, heptachlor; 5, *o,p′*-DDT; 6, *p,p′*-DDT; 7, O,O′-dimethyl chlorothiophosphate; 8, *p,p′*-DDD; 9, lindane; 10, endrin.

TABLE 46.4
LIQUID CHROMATOGRAPHY OF INSECTICIDES ON CORASIL II (BOMBAUGH *et al.*)

Column: 50 cm × 2.3 mm. Sorbent: Corasil II. Mobile phase: *n*-hexane. Flow-rate: 1.5 ml/min. Pressure: 280–320 p.s.i. Temperature: ambient. Detection: refractive index. $k'$ = Phase ratio of the solute.

| Compound | Elution volume (ml) | $k'$ | HETP | Plates |
|---|---|---|---|---|
| Aldrin | 1.17 | 0.17 | 1.0 | 495 |
| *p,p′*-DDT | 1.64 | 0.64 | 1.0 | 512 |
| DDD | 2.32 | 1.32 | 1.2 | 406 |
| Lindane | 3.72 | 2.72 | 1.3 | 385 |
| Endrin | 10.20 | 9.20 | 1.2 | 400 |

graphy, Bombaugh *et al.* separated five pesticides on a Corasil II column in 8 min. Elution was carried out with *n*-hexane at *ca.* 300 p.s.i. The separating power of Corasil II is evident from Table 46.4. Still more effective and rapid is the separation on Corasil I (Waters Ass., 1970). Using *n*-hexane and a pressure of 600 p.s.i., a mixture of five pesticides can be separated within 45 sec (Fig. 46.4).

A very good separation was also achieved on a column packed with SIL-X adsorbent (Fig. 46.5; Nester-Faust Corp.).

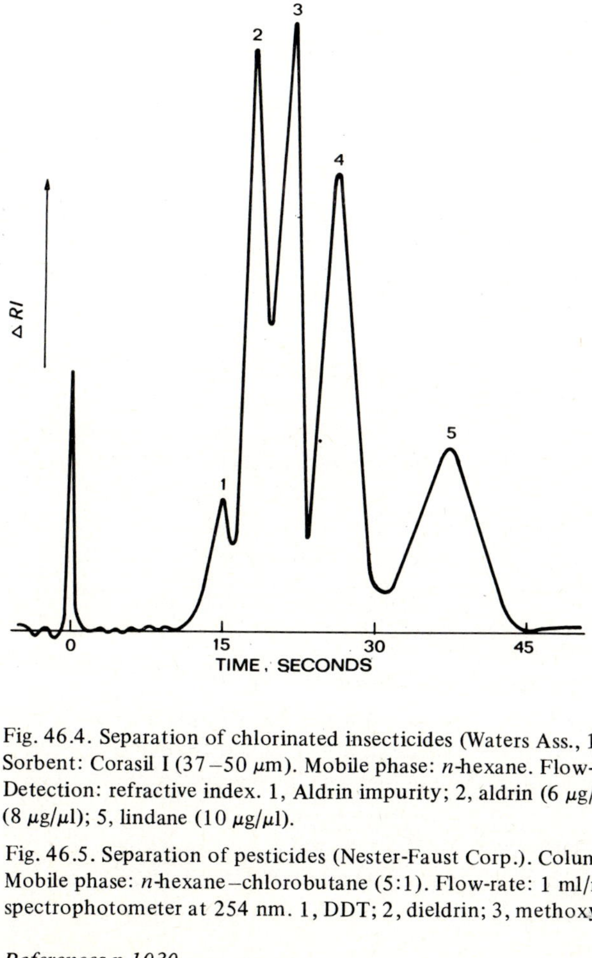


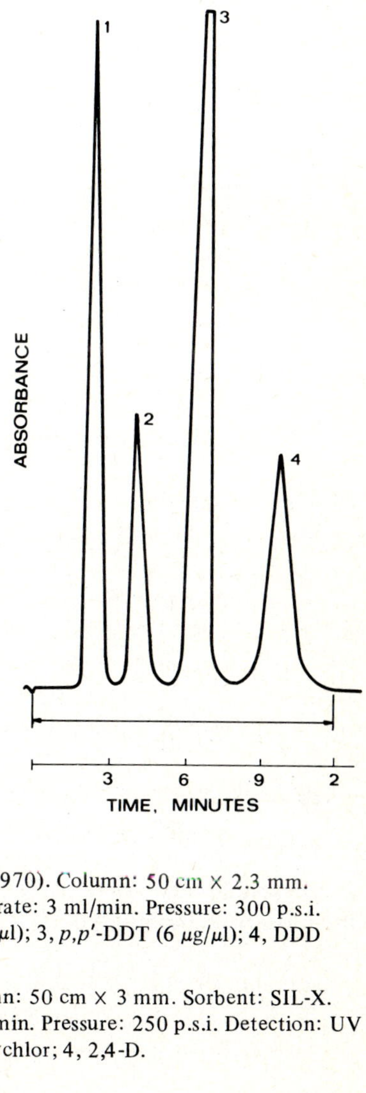


Fig. 46.4. Separation of chlorinated insecticides (Waters Ass., 1970). Column: 50 cm × 2.3 mm. Sorbent: Corasil I (37–50 μm). Mobile phase: *n*-hexane. Flow-rate: 3 ml/min. Pressure: 300 p.s.i. Detection: refractive index. 1, Aldrin impurity; 2, aldrin (6 μg/μl); 3, *p,p'*-DDT (6 μg/μl); 4, DDD (8 μg/μl); 5, lindane (10 μg/μl).

Fig. 46.5. Separation of pesticides (Nester-Faust Corp.). Column: 50 cm × 3 mm. Sorbent: SIL-X. Mobile phase: *n*-hexane–chlorobutane (5:1). Flow-rate: 1 ml/min. Pressure: 250 p.s.i. Detection: UV spectrophotometer at 254 nm. 1, DDT; 2, dieldrin; 3, methoxychlor; 4, 2,4-D.

*References p.1030*

## PHOSPHORUS PESTICIDES

Most organophosphorus pesticides are esters of phosphonic, phosphorothionic, phosphorothiolic or phosphorothiolothionic acids of the general structure

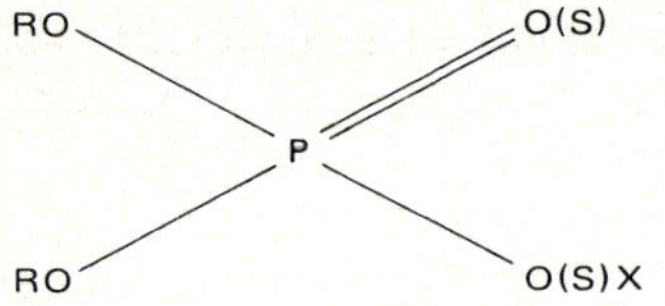


where R is alkyl and X is an organic radical. On the basis of this structure, a large number of compounds with different chemical properties and biological effects can be derived.

In liquid column chromatography, good results may be achieved by liquid–solid chromatography, which is the most widely used column chromatographic technique in the analysis of organophosphorus insecticides. Silica gel is preferably used for the purpose of separation while charcoal is used for the clean-up of organophosphorus insecticide residues.

Organophosphorus pesticides and their metabolites vary greatly in their polarities and the extent of their extraction is markedly dependent upon the solubility of the insecticides in the solvent used and the nature of the material to be extracted. For instance, chloroform is the most widely applicable polar solvent, applied both to the non-polar parent organophosphorus insecticides and the more polar metabolites. When the insecticides and the metabolites studied cover a wide range of polarities, it is recommended that they should be separated into water-soluble and non-polar solvent-soluble groups either by using two different solvents for extraction (Sissons and Telling, 1970a,b) or by partition of the extracted compounds into water and light petroleum (b.p. 40–60°C) (Laws and Webley).

The most frequently used procedure involves the use of a column containing charcoal or a mixture of charcoal and other sorbents. For the separation of water-soluble and oxidized organophosphorus insecticides from tissue components, it is recommended that a charcoal column and elution with chloroform should be used (Askew *et al.*, Laws and Webley).

Active carbon can be obtained by heating 14–22 mesh carbon at 600°C in closed crucibles for 2 h and by boiling it twice with concentrated hydrochloric acid for 30 min each time. Carbon is washed with water in order to remove acid and dried in an oven at 100–110°C.

A general method of cleaning up the organophosphorus insecticides from non-fatty food consists in applying a mixture of acid-treated charcoal, hydrated magnesium oxide (Sea Sorb 43, Fisher Scientific, Pittsburgh, Pa., U.S.A.) and Celite 545 and an eluting solution composed of acetonitrile and benzene (Storherr *et al.*), which results in less plant extractives than a solution of 25% of ethyl acetate in benzene on the same column (Watts *et al.*).

For the purification of non-polar phosphates, such as demeton, diazinon, malathion and parathion, it is recommended that a chromatographic column packed with layers of acid-washed alumina, activated charcoal mixed with Celite 545 and acid-washed alumina and developed with a 25% solution of benzene in ethyl acetate should be used (Getz, 1962a, b). Some other systems involving elution through a multiple sorbent column containing carbon can be used as a clean-up procedure for phorate (Archer *et al.*, Manuel) parathion and parathion-methyl (Camoni *et al.*) and commonly for non-polar organophosphorus compounds (Bates).

For example, the clean-up of non-polar organophosphorus pesticides can be achieved on a 1.5-cm O.D. column prepared from a slurry of 0.5 g of Nuchar carbon (Eastman-Kodak, Rochester, N.Y., U.S.A.), 2.0 g of magnesium oxide and 1.5 g of powdered cellulose in benzene–chloroform (1:1). The column is washed with 50 ml of the same solvent and the washings are discarded. The chloroform extract from vegetables is added to the top of the column and the column is washed with the same solvent until 150 ml of eluate have been collected. The eluate is evaporated nearly to dryness on a water-bath and then the volume is made up to 10 ml with chloroform in a stoppered measuring cylinder. Pesticides are determined by two-dimensional paper chromatography.

Liquid–solid chromatography is a suitable technique for the investigation of the metabolism of organophosphorus pesticides in plants and animal tissues. Organophosphorus insecticides are metabolized readily into compounds that are more polar and toxic than

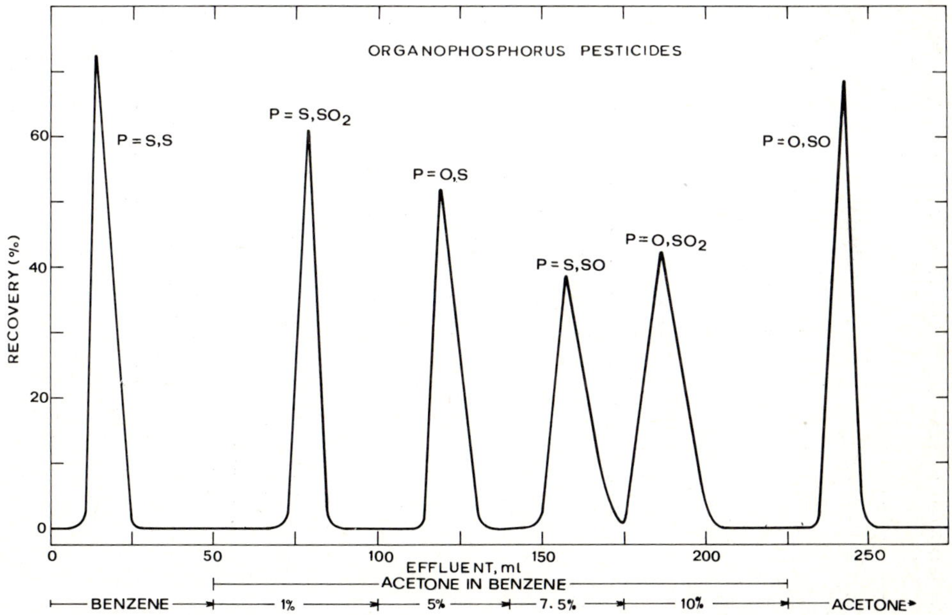


Fig. 46.6. Separation of fenthion and five of its metabolites (Bowman and Beroza). Column: 1.2 cm I.D. Sorbent: silica gel (4 g). Eluents: (1) benzene; (2) 1% of acetone in benzene; (3) 5% of acetone in benzene; (4) 7.5% of acetone in benzene; (5) 10% of acetone in benzene; (6) acetone. Eluted volumes are apparent from the figure. Temperature: ambient. Detection: GLC using the following conditions: column, glass, 90 × 0.4 cm I.D.; packing, DC-200, 10% (w/w) on 80–100 mesh Gas-Chrom Q; carrier gas, nitrogen, 160 ml/min; other gases, oxygen, 40 ml/min, and hydrogen, 200 ml/min; temperatures, column 210°C, injection port 225°C, detector 200°C. P=S,S = Fenthion; P=S,SO = fenthion sulphoxide; P=S,$SO_2$ = fenthion sulphone; P=O,S = fenthion O-analogue; P=O,SO = sulphoxide of fenthion O-analogue; P=O,$SO_2$ = sulphone of fenthion O-analogue.

*References p.1030*

TABLE 46.5
APPLICATIONS OF LIQUID–SOLID CHROMATOGRAPHY TO THE SEPARATION OF ORGANOPHOSPHORUS PESTICIDES AND THEIR METABOLITES

| Compounds separated | Sorbent | Eluent | Notes | Reference |
|---|---|---|---|---|
| Disulfoton and its metabolites | Silica gel | 1% Acetone in benzene (50 ml) | Disulfoton and its sulphone | Bowman *et al.* |
| | | 20% Acetone in benzene (50 ml) | Suphoxide of disulfoton, sulphide and sulphone of O-analogue | |
| | | Acetone (50 ml) | Sulphoxide of O-analogue | |
| Dimethoate and its metabolites | Silica gel | *n*-Hexane–chloroform mixtures with increasing polarity | Water-soluble metabolites separated by IEC on Dowex 1-X8 | Zayed *et al.* |
| Dimethoate and its metabolites | Celite | *n*-Hexane, *n*-hexane–chloroform mixtures, chloroform | Investigation of metabolisms in rats and plants | Lucier and Menzer |
| Phosphamidon and its metabolites | Celite | *n*-Hexane, *n*-hexane–chloroform mixtures, chloroform | Investigation of metabolism in rats and goats. Column 50 × 1.9 cm | Clemons and Menzer |
| Parathion, parathion-methyl, fenitrothion and their metabolites | Florisil deactivated with 5–8% of water | Benzene (180 ml) | Thionates | Moellhoff |
| | | Acetone (150 ml) | Oxidized metabolites and S-isomers. Column 30 × 1.8 cm, 8 g of sorbent. Flow-rate 1.5 ml/min. | |

the parent insecticides. Conversion of P=S to P=O and oxidation of the sulphur atom in the side-chain into sulphoxide and sulphone constitutes a major metabolic pathway for the insecticides. These compounds can be separated on the basis of their different polarities by using column chromatography. In most of the procedures, silica gel is used as the sorbent, together with an eluting solvent with increasing polarity. For example, in the metabolic investigation of fenthion in grass, maize and milk, the complete separation of fenthion and five of its metabolites was achieved by using a silica gel column and elution with benzene followed by varying ratios of benzene and acetone in order to give an increase in the polarity of the solvent (Fig. 46.6) (Bowman and Beroza).

Some other applications of liquid–solid chromatography in the separation of organophosphorus pesticides and their metabolites are given in Table 46.5.

Practical applications of high-speed liquid chromatography in organophosphorus pesticide analysis are still relatively few. An example of the analysis of the technical-grade larvicide Abate was described by Schmit. The separation is performed on a 1 m $\times$ 2.1 mm column using 1% $\beta,\beta'$-oxydipropionitrile on Zipax as the stationary phase and $n$-hexane as the mobile phase. The column pressure is 600 p.s.i. and the flow-rate 1 ml/min. Other related organophosphorus compounds can also be chromatographed by employing similar systems.

The application of gel chromatography to the separation of organophosphorus insecticides on Sephadex LH-20 swollen in acetone, tetrahydrofuran and ethanol was investigated by Ruzicka *et al.* (1968).

Sephadex LH-20 was allowed to swell for approximately 24 h in the solvents acetone, tetrahydrofuran and ethanol. The swollen gel slurries were poured into glass chromatographic columns of 2 cm diameter to give a packed bed volume of approximately 75 ml. One millilitre of a mixture of pesticides containing 20 $\mu$g of each compound was placed on the top of each column in the corresponding solvent, and eluted at the rate of 1 ml/min. Each mixture contained up to five pesticides and always included parathion as an internal standard. Fractions of the eluted solutions were collected and examined for the presence of the component pesticides using gas chromatography (Ruzicka *et al.*, 1967). Certain oxidation products that could not have been detected on the GC columns used were examined by carrying out a colorimetric phosphorus determination on each fraction. The molecular weights and elution volumes of pesticides relative to parathion established on the gel swollen in ethanol and acetone are shown in Table 46.6. The elution volumes of parathion eluted on this column with ethanol and acetone were found to be 75.5 and 43.5 ml, respectively. The elution sequences in acetone and tetrahydrofuran were identical and the data for the latter solvent have been omitted. Some differences in separation on gels swollen in ethanol and acetone were found in relation to the variations in the pesticide structure.

Askew *et al.* separated organophosphorus pesticides from river water extracts, prior to gas or thin-layer chromatography, on a Sephadex LH-20 column of 75-ml bed volume, into three groups, as shown in Table 46.7. The gel was swollen and eluted with ethanol and the fractions eluting in the ranges 40–60, 61–70 and 71–110 ml were collected.

Another technique used to establish the analysis of organophosphorus pesticides is ion-exchange chromatography. An initial study of the separation of mono- and diesters of phosphoric and phosphorothioic acids on the anion-exchange resin

TABLE 46.6
ELUTION VOLUME OF ORGANOPHOSPHORUS PESTICIDES ON SEPHADEX LH-20 RELATIVE TO PARATHION (RUZICKA *et al.*, 1968)

| Pesticide | Molecular weight | Gel swollen and eluted with | |
|---|---|---|---|
| | | Acetone | Ethanol |
| Aspon | 378 | 87 | 66 |
| Ethion | 384 | 92 | 82 |
| Sulfotep | 322 | 93 | 75 |
| Malathion | 330 | 95 | 86 |
| Mecarbam | 329 | 96 | 85 |
| Diazinon | 304 | 98 | 77 |
| Demeton-S | 258 | 99 | 79 |
| Pyrimithate | 333 | 99 | 81 |
| Disulfoton | 274 | 100 | 87 |
| Parathion | 291 | 100 | 100 |
| Fenitrothion | 277 | 101 | 114 |
| Phorate | 260 | 103 | 89 |
| Demeton-O-methyl | 230 | 103 | 93 |
| Phorate sulphone | 309 | 104 | 98 |
| Malathion O-analogue | 314 | 104 | 76 |
| Chlorthion | 283 | 107 | 118 |
| Fenthion | 278 | 108 | 112 |
| Thionazin | 248 | 108 | 89 |
| Thiometon | 248 | 108 | 100 |
| Parathion-methyl | 249 | 109 | 114 |
| Fenchlorphos | 321 | 109 | 101 |
| Parathion O-analogue | 275 | 112 | 86 |
| Chlorfenvinphos | 359 | 114 | 77 |
| Bromophos | 366 | 115 | 100 |
| Mevinphos | 224 | 116 | 79 |
| Phosphamidon | 264 | 117 | 72 |
| Phorate O-analogue | 244 | 118 | 74 |
| Trichlorfon | 257 | 121 | 80 |
| Dichlorvos | 221 | 122 | 83 |
| Morphothion | 285 | 124 | 106 |
| Demeton-S-methyl sulphone | 262 | 124 | 99 |
| Demeton-S-methyl | 230 | 130 | 86 |
| Thionazin O-analogue | 232 | 133 | 79 |
| TEPP | 290 | 134 | 73 |
| Bidrin | 237 | 141 | 77 |
| Phorate sulphoxide | 293 | 147 | 82 |
| Dimethoate | 229 | 148 | 95 |
| Dimefox | 154 | 150 | 71 |
| Demeton-S-methyl sulphoxide | 246 | 200 | 79 |

Dowex 1-X8 was carried out by Plapp and Casida. In these investigations phosphoric, phosphorothioic and mono- and dialkylphosphoric acids were eluted with hydrochloric acid gradients. During elution of dialkylphosphorothioic, dialkylphosphorodithioic, monoalkylphenylphosphoric and monoalkylphenylphosphorothioic acid, it is necessary to use methanol and acetone as co-solvents with acid gradients (Fig. 46.7).

TABLE 46.7
SEPARATION OF ORGANOPHOSPHORUS PESTICIDES ON SEPHADEX LH-20 (ASKEW *et al.*)

| Elution volume (ml) | | |
|---|---|---|
| 40–60 | 61–70 | 71–110 |
| Chlorfenvinphos | Demeton-O-methyl | Azinphos-ethyl |
| Crufomate | Demeton-S-methyl | Azinphos-methyl |
| Demeton-S | Dibrom | Bromophos |
| Diazinon | Dichlorofenthion | Carbophenothion |
| Dimefox | Dichlorvos | Coumaphos |
| Mevinphos | Disulfoton | Dimethoate |
| Oxydemeton-methyl | Ethion | Fenchlorphos |
| Phosphamidon | Ethoate-methyl | Fenitrothion |
| Schradan | Malathion | Haloxon |
| Sulfotep | Mecarbam | Morphothion |
| TEPP | Phorate | Parathion |
| Vamidothion | Pyrimithate | Phenkapton |
| Trimethyl phosphate | Thionazin | Phosalone |
| Triethyl phosphate | Trichlorphon | |
| Tributyl phosphate | | |

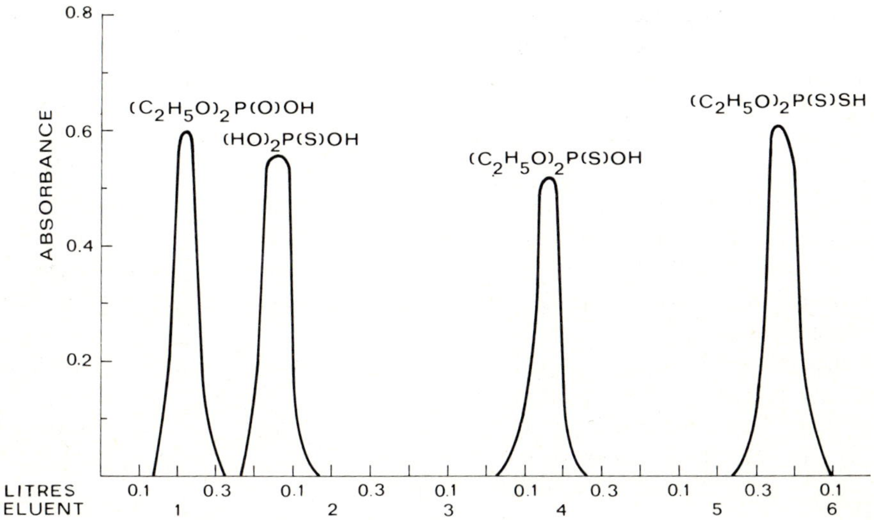


Fig. 46.7. Ion-exchange separation of metabolites of an O,O-dialkyl phosphorodithioate insecticide (Plapp and Casida). Column: 29 × 2.5 cm. Ion exchanger: Dowex 1-X8 ($Cl^-$), 100–200 mesh (50 g). Eluents: 1, elution gradient, pH 2 to pH 1 HCl; 2, elution gradient, pH 1 to 1 *N* HCl; 3, water to bring eluate above pH 1; 4, elution gradient, pH 1 HCl + methanol (1:3) to 1 *N* HCl + methanol (1:3); 5, elution gradient, 1 *N* HCl + acetone (1:3) to concentrated HCl + water + acetone (1:1:6); 6, concentrated HCl + water + acetone (1:1:6). Eluted volumes are apparent from the figure; 20-ml fractions were collected at the rate of about 3.5 min per fraction. Temperature: ambient. Detection: each fraction was analyzed for total phosphorus by the Allen method.

*References p.1030*

The gradient elution was performed by placing the weaker acid solution in the separating funnel directly above the column. The stronger acid solution was located in another funnel of identical dimensions, positioned at the same level, and the two funnels were connected by a siphon. As the weaker acid entered the column, the stronger acid was siphoned into the first funnel, producing a continuous increase in the eluent acid concentration. A stream of air was passed over the siphon inlet so as to ensure mixing of the two solutions.

This chromatographic technique is applicable to investigations of the *in vivo* and *in vitro* degradation of organophosphorus insecticides and related compounds. It was modified for the separation of anionic metabolites of $^{32}P$- and $^{35}S$-labelled parathion in urine samples from rats (Nakatsugawa *et al.*), of water-soluble metabolites of $^{32}P$-labelled trichlorfon (Bull and Ridgway) and of water-soluble metabolites of dimethoate in plants (Dauterman *et al.*).

The compounds eluted can be tested in the column fractions by thin-layer chromatography using some of the chromogenic reagents discussed by Watts or by paper chromatography using, for example, the procedure developed by Bates.

Askew *et al.* recommend the separation and identification of organophosphorus pesticides on 250-$\mu$m thick layers of silica gel G (E. Merck, Darmstadt, G.F.R.) on 20 × 20 cm glass carrier plates activated by heating at 120°C for at least 2 h. Generally, the plate is developed in the solvent system *n*-hexane–acetone (5:1) by the ascending technique.

Two-dimensional paper chromatography was successfully applied by Bates to the analysis of residues of organophosphorus pesticides in foodstuffs. For non-polar pesticides, it is recommended that 30% dimethylformamide in acetone as the stationary phase and *n*-hexane as the mobile phase should be used for development in the first direction, while in the second direction, the chromatogram is developed using reversed-phase systems of 5% of liquid paraffin (B.P. grade) in diethyl ether and a mixture of dimethylformamide and water (1:1). The stationary phase for developing polar compounds consists of 20% of formamide in acetone. The paper is developed twice in the first direction with *n*-hexane and then in the second direction with benzene–chloroform (9:1) as the mobile solvents.

The detection of organophosphorus pesticides on chromatograms can be carried out in UV light or using spray solutions of 2,6-dibromo-N-chloro-*p*-quinonimine, blue tetrazolium, 4-methylumbelliferone or silver nitrate–bromophenol blue.

Quantitative procedures involve the colorimetric determination of pesticidal phosphorus and, more recently, gas–liquid chromatography using a variety of detectors and columns. The structures of the separated compounds can be determined on the basis of elemental analysis and UV, IR, NMR and mass spectra.

## CARBAMATE PESTICIDES AND THEIR METABOLITES

The metabolism of carbamate pesticides was investigated using the milk and urine of lactating cows and goats (Robbins *et al.*, 1970) and the urine of rats (Robbins *et al.*, 1969). A $^{14}C$-labelled preparation of benzo-[*b*]-thien-4-yl methylcarbamate (mobam) was employed. Fig. 46.8 presents the general scheme for the isolation of mobam. A liquid anion-exchange column for extracting the metabolites from urine was prepared by coating

50 g of Porapak Q with 10 ml of trioctylamine. The fractions were further purified on cellulose ion-exchange columns. A cation-exchange cellulose column (40 × 2.2 cm), packed with 25 g of Cellex P, was converted into the $H^+$ form with hydrochloric acid. An anion-exchange cellulose column (48 × 2.3 cm) containing 25 g of Cellex AE was converted into the $OH^-$ form with 1 *N* ammonia solution. Each fraction from the trioctylamine–Porapak Q (ammonia solution and methanol eluates) was placed on the Cellex P column and eluted with water in order to remove cations and amphoteric compounds, then each fraction from the Cellex P column, without reduction in volume, was chromatographed on the Cellex AE column. Fig. 46.9 shows the elution sequence from the Cellex AE column eluted with water, 0.5 *M* formic acid, 0.1 *M* ammonium formate (pH 4), and 1 *N* ammonia solution. Single fractions were chromatographed on paper.

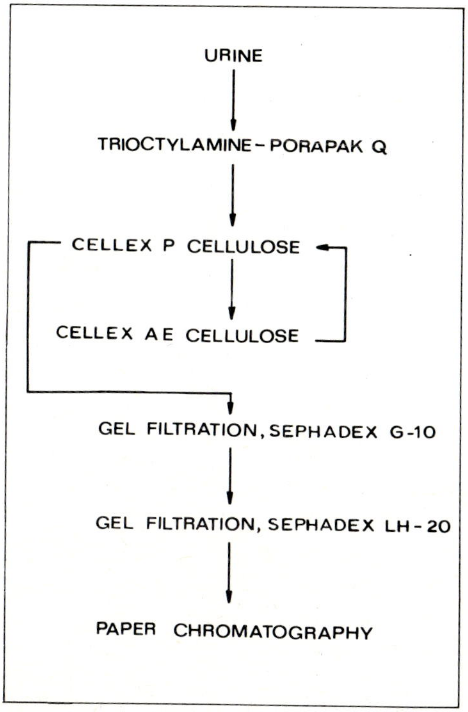


Fig. 46.8. Scheme for the isolation of water-soluble metabolites of [$^{14}$C]mobam from urine (Robbins *et al.*, 1969).

The metabolism of N-[$^{14}$C]methyl- and [$^{14}$C]benzyl-labelled 3,5-dichlorobenzyl N-methylcarbamate was investigated in the rat by Knaak and Sullivan. Fig. 46.10 shows the chromatographic results for the metabolites on an analytical column (24 × 1.5 cm) of DEAE-cellulose.

Except for the neutral compounds (A), the N-[$^{14}$C]methyl-labelled metabolites of the carbamate did not chromatograph identically with the [$^{14}$C]benzyl-labelled metabolites. The N-[$^{14}$C]methyl-labelled neutral compounds (A) amounted to 56% of the radioactivity

*References p.1030*

TABLE 46.8
CHROMATOGRAPHIC TECHNIQUES USED IN PESTICIDE ANALYSIS

| Technique | Compounds separated | Sorbent | Mobile phase | Reference |
|---|---|---|---|---|
| LSC | Polychlorinated biphenyl, DDT and its analogue | Silicic acid, Celite | Light petroleum, acetonitrile, *n*-hexane | Armour and Burke |
| LSC | Effects of pesticide-grade *n*-hexane on silicic acid chromatography of organochlorine pesticides and polychlorinated biphenyls | Silicic acid: Silicar ® (100–200 mesh) activated at 130°C | *n*-Hexane | Zitko |
| LSC | Pesticides in plant extracts | Acidic alumina, activity I and II | Light petroleum (b.p. 30–40°C)–diethyl ether (7:3) | Diemair *et al.* |
| LSC | Chlorinated pesticides, residues in foodstuffs: pre-separation and concentration | Silica gel G – $Na_2SO_4$ (9:1) | *n*-Hexane–benzene (3:2) | Krasnodebski and Kubacki |
| LSC | Separation of DDT and DDE in vegetable material, purification | Florisil | 20% of dichloromethane in light petroleum, 15% of diethyl ether in *n*-pentane | Ware and Dee |
| LSC | Determination of prophan and chloroprophan in potatoes: pre-separation on a column | Carbon–Florisil–Celite (1:2:2) | Methylene chloride | Reinhard |
| LSC | Determination of 2,4,5-trichlorophenoxyacetic acid in animal tissue, blood and urine | Florisil washed with 2.5% $H_3PO_4$ | 20% diethyl ether in light petroleum, 0.5% $H_3PO_4$ in diethyl ether | Clark |

| | | | | |
|---|---|---|---|---|
| LSC | Determination of toxaphene residues in clover | Florisil | Light petroleum (b.p. 40–60°C) | Adamovic and Hus |
| LSC | Determination of carbamate pesticides in grass and fruits | Florisil | Water-saturated dichloromethane | Argauer |
| LSC | Metabolites of furadane (carbamate pesticide) in plants | Silicic acid, Florisil | Diethyl ether–*n*-hexane, chloroform, ethyl acetate, methanol, stepwise elution | Metcalf *et al.* |
| IEC | Determination of sodium 2,4-dichlorophenoxyacetate in soil | Anion exchanger AB-17 ($Cl^-$) | 4 *N* HCl | Tsitovich and Kuzmenko |
| IEC | Metabolites of atrazine (2-chloro-4-ethylamino-*s*-triazine) | Dowex 50W-X8, DEAE-cellulose | 30% methanol, water, 2 *N* ammonia solution (4:1:3) acetic acid gradient | Lamoureux *et al.* |
| GPC | Metabolites of carbamate insecticides in chicken urine | For pre-separation Sephadex G-10 and LH-20 | Water<br>Methanol | Paulson *et al.* |
| LSC | Natural insecticides, pyrethrum analysis | Alumina<br>Silicic acid–Celite (4:1) | Diethyl ether–light petroleum (1:3)<br>Light petroleum (b.p. 60–68°C) | Wachs and Hanley<br>Doskotch and El-Feraly |

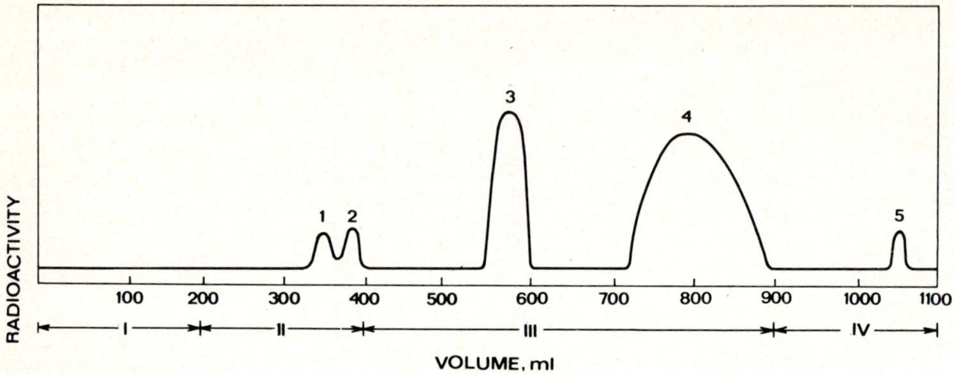


Fig. 46.9. Separation of pesticides (Robbins *et al.*, 1969). Column: 48 × 2.2 cm. Sorbent: Cellex AE (Bio-Rad Labs., Richmond, Calif., U.S.A.), 25 g. Mobile phase: I, $H_2O$; II, 0.5 *M* formic acid; III, 0.1 *M* ammonium formate (pH 4); IV, 1 *N* ammonia solution. Detection: radioactivity measurement. Peaks 1–5, metabolites (not identified).

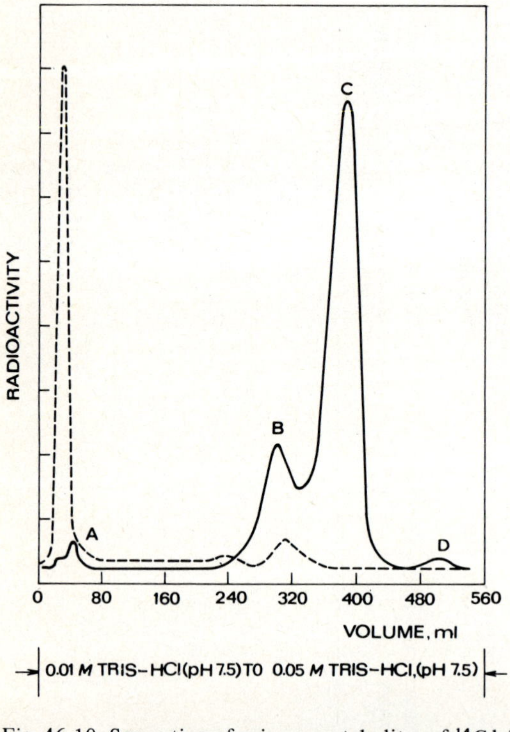


Fig. 46.10. Separation of urinary metabolites of $^{14}$C-labelled 3,4-dichlorobenzyl N-methylcarbamate (Knaak and Sullivan). Column: 24 × 1.5 cm. Sorbent: DEAE-cellulose. Mobile phase: 0.01 *M* Tris–HCl (pH 7.5) to 0.05 *M* Tris–HCl (pH 7.5). Detection: radioactivity measurement. A, neutral compounds; B, 3,4-dichlorobenzylglucuronide and 3,4-dichlorobenzoylglucuronide; C, 3,4-dichlorohippuric acid; D, 3,4-dichlorobenzoic acid. ——, [$^{14}$C] Benzyl-labelled metabolites; ----, N-[$^{14}$C] methyl-labelled metabolites.

recovered from the column. The two small peaks chromatographing in the region of B represent an additional 20% of the recovered radioactivity. Gas chromatography was used in combination with anion-exchange and silica gel chromatography to identify and quantitate the metabolic products.

High-speed LC has been applied to some thermally labile pesticides of the carbamate series using a column packed with Zipax impregnated with β,β′-oxydipropionitrile. Elution was carried out with di-*n*-butyl ether at 300 p.s.i. pressure (DuPont). Fig. 46.11 shows the separation of some substituted urea herbicides. Other papers mainly of the application type from the field of residue analysis are listed in Table 46.8.

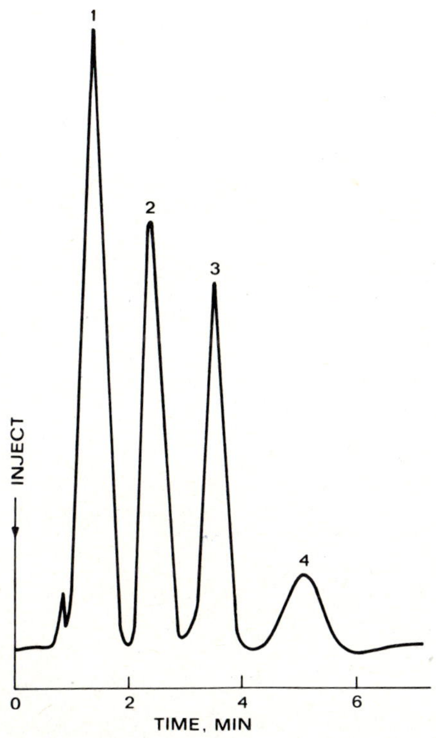


Fig. 46.11. Separation of substituted urea herbicides (DuPont). Column: 1 m × 2.1 mm. Sorbent: β,β′-oxydipropionitrile on Zipax. Mobile phase: di-*n*-butyl ether. Pressure: 300 p.s.i. Detection: UV spectrophotometer at 254 nm. 1, Linuron; 2, diuron; 3, monuron; 4, fenuron.

## PYRETHRINS

Of natural insecticides, pyrethrins in particular have been analyzed by column chromatography. A concentrated solution of pyrethrins was obtained by the clean-up of pyrethrum oleoresin on a column of silica gel, the pyrethroid fraction being eluted with

*References p.1030*

diethyl ether–*n*-hexane (1:3). A 0.5-ml volume of this solution, containing *ca.* 130 mg of pyrethrins, was injected on to the column (450 × 25 mm). The chromatography was carried out on the column using nitromethane, supported on Celite, as the stationary phase, and *n*-hexane saturated with nitromethane and, subsequently, carbon tetrachloride–*n*-hexane (1:3) as the mobile phase (Rickett).

## REFERENCES


Adamovic, V. M. and Hus, M., *Michochim. Acta*, (1969) 12.
Allen, R. J. L., *Biochem. J.*, 34 (1940) 858.
Archer, T. E., Zweig, G., Winterlin, W. L. and Francis, E. K., *J. Agr. Food Chem.*, 11 (1963) 58.
Argauer, R. J., *J. Agr. Food Chem.*, 17 (1969) 888.
Armour, J. A. and Burke, J. A., *J. Ass. Offic. Anal. Chem.*, 53 (1970) 761.
Askew, J., Ruzicka, J. H. and Wheals, B. B., *Analyst (London)*, 94 (1969) 275.
Bates, J. A. R., *Analyst (London)*, 90 (1965) 453.
Beckman, H. and Garber, D., *J. Ass. Offic. Anal. Chem.*, 52 (1969) 286.
Bevenue, A. and Ogata, J. N., *J. Chromatogr.*, 50 (1970) 142.
Bombaugh, K. J., Levangie, R. F., King, R. N. and Abrahams, L., *J. Chromatogr. Sci.*, 8 (1970) 660.
Bowman, M. C. and Beroza, M., *J. Agr. Food Chem.*, 16 (1968) 399.
Bowman, M. C., Beroza, M. and Gentry, C. R., *J. Ass. Offic. Anal. Chem.*, 52 (1969) 157.
Bull, D. L. and Ridgway, R. L., *J. Agr. Food Chem.*, 17 (1969) 837.
Burchfield, H. P. and Storrs, E. E., *Contrib. Boyce Thompson Inst.*, 17 (1953) 333.
Burke, J. A. and Malone, B., *J. Ass. Offic. Anal. Chem.*, 49 (1966) 1003.
Camoni, I., Gandolfo, N., Ramell, G. C., Sampaolo, A. and Binetti, L., *Boll. Lab. Chim. Prov.*, 19 (1968) 615.
Clark, D. E., *J. Agr. Food Chem.*, 17 (1969) 1168.
Clemons, G. P. and Menzer, R. E., *J. Agr. Food Chem.*, 16 (1968) 312.
Dauterman, W. C., Viado, G. B., Casida, J. E. and O'Brien, R. D., *J. Agr. Food Chem.*, 8 (1960) 115.
Diemair, W., Maier, G., Pfeilsticker, K. and Schlögel, K., *Z. Lebensm.-Unters.-Forsch.*, 139 (1969) 67.
Doskotch, R. W. and El-Feraly, F. S., *Can. J. Chem.*, 47 (1969) 1139.
Ďulák, K., Kováč, J. and Rapoš, P., *J. Chromatogr.*, 31 (1967) 354.
DuPont, *Liquid Chromatography, Laboratory Report,* 820 M6, DuPont, Wilmington, Del., March 30, 1970.
Eidelman, M., *J. Ass. Offic. Agr. Chem.*, 45 (1962) 672.
Eidelman, M., *J. Ass. Offic. Agr. Chem.*, 46 (1963) 182.
Eidelman, M., *J. Ass. Offic. Anal. Chem.*, 50 (1967) 591.
Feil, V. J., Hedde, R. D., Zaylskie, R. G. and Zachrison, C. H., *J. Agr. Food Chem.*, 18 (1970) 120.
Getz, M., *J. Ass. Offic. Agr. Chem.*, 45 (1962a) 393.
Getz, M., *J. Ass. Offic. Agr. Chem.*, 45 (1962b) 397.
Haenni, E. O., Howard, J. W. and Joe, F. L., *J. Ass. Offic. Agr. Chem.*, 45 (1962) 67.
Hedde, R. D., Davison, K. L. and Robbins, J. D., *J. Agr. Food Chem.*, 18 (1970) 116.
Jones, L. R. and Riddick, J. A., *Anal. Chem.*, 24 (1952) 569.
Knaak, J. B. and Sullivan, L. J., *J. Agr. Food Chem.*, 16 (1968) 454.
Krasnodebski, P. and Kubacki, S. J., *Tluszcze Jadalne*, 12 (1968) 107.
Lamoureux, G. L., Shimabukuro, R. H., Swanson, H. R. and Frear, D. S., *J. Agr. Food Chem.*, 18 (1970) 81.
Laws, E. Q. and Webley, D. J., *Analyst (London)*, 86 (1961) 249.
Leoni, V., *J. Chromatogr.*, 62 (1971) 63.
Lucier, G. W. and Menzer, R. E., *J. Agr. Food Chem.*, 18 (1970) 698.
Manuel, A. J., *J. Agr. Food Chem.*, 16 (1968) 57.

Metcalf, R. L., Fukuto, T. R., Collins, C., Borok, K., Abdel-Aziz, S., Munoz, R. and Cassil, C. C., *J. Agr. Food Chem.*, 16 (1968) 300.
Moellhoff, E., *Pflanzenschutz-Nachr. Bayer*, 21 (1968) 331.
Moubry, R. P., Myrdal, G. R. and Jensen, H. P., *J. Ass. Offic. Anal. Chem.*, 50 (1967) 885.
Nakatsugawa, T., Tolman, N. M. and Dahm, P. A., *Biochem. Pharmacol.*, 18 (1969) 1103.
Nester-Faust Corp., *Applications Bulletin*, Nester-Faust Corp., Newark, Del., 1971.
Paulson, G. D., Zaylskie, R. G., Zehr, M. V., Portnoy, C. E. and Feil, V. J., *J. Agr. Food Chem.*, 18 (1970) 110.
Plapp, F. W. and Casida, J. E., *Anal. Chem.*, 30 (1958) 1622.
Reinhard, C., *Deut. Lebensm.-Rundsch.*, 63 (1967) 340.
Reynolds, L. M., *Bull. Environ. Contam. Toxicol.*, 4 (1969) 128.
Rickett, F. E., *J. Chromatogr.*, 66 (1972) 356.
Robbins, J. D., Bakke, J. E. and Feil, V. J., *J. Agr. Food Chem.*, 17 (1969) 236.
Robbins, J. D., Bakke, J. E. and Feil, V. J., *J. Agr. Food Chem.*, 18 (1970) 130.
Ruzicka, J. H., Thomson, J. and Wheals, B. B., *J. Chromatogr.*, 30 (1967) 92.
Ruzicka, J. H., Thomson, J., Wheals, B. B. and Wood, N. F., *J. Chromatogr.*, 34 (1968) 14.
Schmit, J. A., in J. J. Kirkland (Editor), *Modern Practice of Liquid Chromatography*, Wiley-Interscience, New York, 1971, p. 403.
Sissons, D. J. and Telling, G. M., *J. Chromatogr.*, 47 (1970a) 328.
Sissons, D. J. and Telling, G. M., *J. Chromatogr.*, 48 (1970b) 468.
Storherr, R. W., Ott, P. and Watts, R. R., *J. Ass. Offic. Anal. Chem.*, 54 (1971) 513.
Thornburg, W., *J. Ass. Offic. Agr. Chem.*, 48 (1965) 1023.
Thornburg, W. and Beckman, H., *Anal. Chem.*, 41 (1969) 140R.
Tsitovich, I. K. and Kuzmenko, E. A., *Khim. Sel. Khoz.*, 6 (1968) 468.
Wachs, H. and Hanley, A. V., *Soap Chem. Spec.*, 43 (1967) 119.
Ware, G. W. and Dee, M. K., *Bull. Environ. Contam. Toxicol.*, 3 (1968) 375.
Waters Ass., *Product Bulletin*, ALC-1001, Waters Ass., Framingham, Mass., (a).
Waters Ass., *Technical Information Bulletin*, No. 19919, Waters Ass., Framingham, Mass., 1970.
Watts, R. R., *Residue Rev.*, 18 (1967) 105.
Watts, R. R., Storherr, R. W., Pardue, J. R. and Osgood, T., *J. Ass. Offic. Anal. Chem.*, 52 (1969) 522.
Wilson, J. N., Franks, M. C. and Sherlock, D. R., *Analyst (London)*, 92 (1967) 782.
Zayed, S. M. A. D., Hassan, A. and Fakhr, I. M. I., *Biochem. Pharmacol.*, 17 (1968) 1339.
Zitko, V., *J. Chromatogr.*, 59 (1971) 444.

*Chapter 47*

# Synthetic dyes

J. CHURÁČEK and J. GASPARIČ

## CONTENTS



## INTRODUCTION

In the field of synthetic dyes, liquid chromatography is used mainly for isolation and purification, usually preceded by extraction of the dyes from the original material (foods, cosmetic products, etc.) or crystallization (in the case of commercial dyes). They are then concentrated on a column and separated from ballasts, which makes them suitable for further analytical treatment, for example by paper or thin-layer chromatography or spectrophotometry. A common alternative consists in the determination of impurities (additives, salts) in commercial types of dyes, which must be separated on a column. In other instances, single components of a mixture of dyes have to be separated. It must be remembered that today synthetic dyes comprise a number of vast groups of substances with very different chemical and physical properties, and the differences in the tasks and the dyes require a careful choice of the chromatographic method and the use of virtually all of the commonly used inorganic adsorbents, ion exchangers, powdered cellulose, polyamides and dextran gels.

In this field, high-resolution chromatography also has a promising future. The possibility of the application of liquid–liquid chromatography to the determination of impurities in anthraquinone dyes was demonstrated by Schmit. The dyes were chromatographed in a reversed-phase system with Permaphase ODS as the stationary and methanol–water (15:85) as the mobile phase.

## GENERAL TECHNIQUES

Synthetic dyes can be divided into several groups, depending on their solubility – water soluble, soluble in organic solvents, poorly soluble. Representatives of the first group are acidic and direct dyes; the acidic dyes can be chromatographed by virtually any method, while with the direct dyes their affinity for cellulose (substantivity), on to which they are strongly adsorbed, should be taken into account. For fat-soluble dyes (disperse dyes), adsorption systems (alumina, silica gel) should be considered. The poorly soluble dyes include some pigments and mainly vat dyes; these dyes require the use of special conditions.

## CHROMATOGRAPHY ON ADSORBENTS

The choice of the adsorbent–mobile phase system depends on the character of the dye and the strength of its interaction with the adsorbent, *i.e.*, on the presence and the number of polar groups in the molecule, as well as on its solubility. Dyes of low or medium polarity can be chromatographed on alumina columns; for weakly polar types (Sudan I type), an adsorbent of activity I to II should be used with *n*-hexane or its mixtures with benzene, while for dyes of medium polarity, such as disperse azo dyes and anthraquinone derivatives containing hydroxy and amino groups, adsorbents of activity II to III and a mixture of benzene with acetone are more suitable (Bridgeman and Peters, Egerton *et al.*). If the dyes are adsorbed too strongly on to alumina, a weaker adsorbent should be used, such as silica gel, magnesium carbonate (Egerton *et al.*) or a polyamide. On a cellulose column, these dyes are virtually not adsorbed and therefore they can be easily eluted with benzene or acetone, while salts, dispersing agents and other auxiliary additives are retained on the column and can be eluted later with *n*-propanol–ammonia (2:1) (Gasparič and Kadlecová).

For the separation of foodstuff dyes, Davídek and Davídková used polyamide powder with water and aqueous alcohol as eluents. For dyes that contain several hydroxy groups in the molecule (especially those of the resorcinol type, which are more strongly bound to polyamide), a mixture of methanol with dilute ammonia must be employed (Davídek). Some dyes used as foodstuff colouring (Yellow 5 type) have been chromatographed by high-speed chromatography on SIL-X adsorbent and a methanol–tetrahydrofuran gradient 1:6 to 1:3 (Nester-Faust Corp.).

For dyes that are soluble only at elevated temperatures in higher boiling solvents (pigments and vat dyes), alumina can be used in combination with higher boiling solvents at 100–200°C in a jacketted column with thermostatted circulating paraffin. *o*-Dichlorobenzene or 1,2,4-trichlorobenzene (sometimes with the addition of 1–3% of phenol or *m*-cresol) was employed as the mobile phase (Unni and Venkataraman).

Water-soluble dyes containing sulphonic and carboxylic groups, *i.e.*, direct and acidic dyes, can be chromatographed on deactivated alumina, as shown in 1935 by Ruggli and Jensen and more recently by Raban and Gregora and by Schweppe (1963), or on silica gel or polyamide (Schweppe, 1970). Elution was carried out with water, aqueous ammonia or pyridine, ammoniacal ethanol, a mixture of *n*-propanol and ammonia (2:1), etc. The choice of the eluent is determined by the properties of the dye and also by the batch of adsorbent. If the dye in its free sulphonic acid form is introduced on to the column of alumina, the appearance of an ammonium or sodium salt in the eluate must be considered (originating from the alumina).

Metal complex dyes have also been separated on alumina (Schetty). The mechanism of the adsorption of water-soluble dyes on alumina was investigated by Giles *et al.*

Cellulose columns can be used for separating acidic dyes; direct dyes are adsorbed so strongly that they cannot be eluted. For example, dyes present in Sulphonazo III (Slovák *et al.*) and in xylenol orange (Murakami *et al.*) have been successfully purified on a cellulose column.

The chromatography of vat dyes, after their transformation to a vat, on a wick, *i.e.*, on a column composed of a bunch of cotton fibres introduced into a glass tube, is a special case of chromatography on cellulose material (Janicka and Kacprzak).

## CHROMATOGRAPHY ON HYDROPHILIC GELS

For the separation of azo and azomethine dyes, Reeves *et al.* described a method based on sorption on Sephadex G-25 or G-50. Azo dyes are eluted with 0.01 *N* potassium hydroxide solution and azomethine dyes with 0.01 *M* potassium chloride solution.

Sephadex G-10 was found to be suitable for the separation of complex mixtures of azo dyes used as fluorimetric chelating agents. The compounds involved were derived from *o,o'*-dihydroxyazobenzene by the introduction of one and of two methyleneiminodiacetic acid groups (Baetz and Diehl). Sephadex G-25 was used for the chromatography of some food dyes (Parrish), and both column and thin-layer techniques (the latter on microscope slides) were applied. The results are compared in Table 47.1. The relative $R_F$ values of the dyes tartrazine, indigo carmine and Orange G in 0.1% sodium sulphate solution corresponded to the relative distance travelled by these dyes on a column of Sephadex, and a mixture of them was completely separated on a 6-cm column when eluted with 0.1% sodium sulphate solution. The recovery of a pure dye from such a column was better than 98%.

TABLE 47.1
$R_F$ VALUES OF DYES ON SEPHADEX G-25 (PARRISH)

Eluents: I = water; II = 0.1% sodium sulphate solution; III = 4% sodium sulphate solution.

| Dye (Colour Index No.) | Eluent | | |
|---|---|---|---|
| | I | II | III |
| Blue VRS (42045) | 0.48 | 0.41 | 0.31 |
| Ponceau SX (14700) | 0.47 | 0.27 | 0.20 |
| Ponceau 4R (16255) | 0.46 | 0.27 | 0.21 |
| Tartrazine (19140) | 0.42 | 0.27 | 0.13 |
| Ponceau 3R (16155) | 0.40 | 0.12 | 0.06 |
| Indigo carmine (73015) | 0.36 | 0.13 | 0.06 |
| Amaranth (16185) | 0.34 | 0.15 | 0.06 |
| Naphthol yellow S (10316) | 0.33 | 0.29 | 0.16 |
| Carmoisine (14720) | 0.27 | 0.08 | 0.03 |
| Orange G (16230) | 0.25 | 0.07 | 0.04 |

The effect of the molecular weight of dyes on their liberation from Sephadex with aqueous acetone was investigated by King and Pruden. The correlation between the swelling characteristics and elution parameters of several food dyes during gel filtration were examined by Matsumoto and co-workers (1970a,b,c).

Polydextran gels can be further utilized for the separation and determination of

TABLE 47.2
APPLICATION-TYPE PAPERS ON THE CHROMATOGRAPHY OF DYES ISOLATED FROM VARIOUS NATURAL AND SYNTHETIC MATERIALS

| Type of dye | Adsorbent | Mobile phase | Reference |
|---|---|---|---|
| Water-soluble food dyes | Polyamide; CM-cellulose; Bentonite; Fuller's earth; Celite 545 | Methanol–ammonia (95:5); 0.1% NaOH in 70% methanol; pyridine–acetic acid–water (20:2:80) | Lehmann *et al.* (1970c) |
| Fat-soluble food dyes in coating materials and lipsticks | Polyamide | 0.1% NaOH in 70% methanol; methanol–ammonia (95:5) | Lehmann *et al.* (1970b) |
| Water-soluble acidic and basic food dyes in wines and fruit-juices | Polyamide | Formic acid–methanol (4:6); methanol–ammonia (95:5); 2.5% aq. sodium citrate–ammonia–methanol (80:20:12) | Lehmann *et al.* (1970a) |
| Food dyes in egg liqueur | Polyamide | Methanol–ammonia (95:5) | Lehmann and Collet (1970a) |
| Water-soluble and insoluble food dyes | Alumina at different pH; silica gel | Water; methanol–ammonia (99:1); water–ammonia (99:1); water-saturated *n*-butanol | Yanuka *et al.* |
| Water-soluble artificial food dyes | Alumina | Water–ammonia (9:1) | Stanley and Kirk |
| Water-soluble food dyes | Alumina | Water; ammonia | Sadini |
| Synthetic dyes in drugs | Polyamide | | Lehmann and Collet (1970b) |
| Synthetic food dyes | Alumina | | Mathew *et al.* |
| Water-soluble food colours | Polyamide | Methanol–ammonia–water (90:5:5); acetone–ammonia–water (40:9:1) | Gilhooley *et al.* |
| Water-soluble food dyes | Polyamide | 0.05% NaOH in 70% methanol | Lehmann and Hahn |

electrolytes (salts) in crude or commercial dyes. For example, Sephadex was employed for the desalting of sulphophthalein dyes (Jirsa and Hykeš). Further papers on the application of these methods are listed in Table 47.2.

## REFERENCES


Baetz, A. L. and Diehl, H., *J. Chromatogr.*, 34 (1968) 534.
Bridgeman, I. and Peters, A. T., *J. Chromatogr.*, 51 (1970) 534.
Davídek, J., *Z. Lebensm.-Unters.-Forsch.*, 132 (1967) 285.
Davídek, J. and Davídková, E., *Z. Lebensm.-Unters.-Forsch.*, 131 (1966) 99.
Egerton, G. S., Gleadle, J. M. and Uffindell, N. D., *J. Chromatogr.*, 26 (1967) 62.
Gasparič, J. and Kadlecová, J., *Chem. Listy*, 66 (1972) 1090.
Giles, C. H., Easton, I. A. and McKay, R. B., *J. Chem. Soc.*, (1964) 4495.
Gilhooley, R. A., Hoodless, R. A., Pitman, K. G. and Thomson, J., *J. Chromatogr.*, 72 (1972) 325.
Janicka, K. and Kacprzak, F., *Chem. Anal. (Warsaw)*, 4 (1959) 915.

Jirsa, M. and Hykeš, P., *J. Chromatogr.*, 21 (1966) 122.
King, H. G. C. and Pruden, G., *J. Chromatogr.*, 52 (1970) 285.
Lehmann, G. and Collet, P., *Z. Lebensm-Unters.-Forsch.*, 143 (1970a) 348.
Lehmann, G. and Collet, P., *Arch. Pharm. (Weinheim)*, 303 (1970b) 855.
Lehmann, G., Collet, P. and Morán, M., *Z. Lebensm.-Unters.-Forsch.*, 143 (1970a) 191.
Lehmann, G., Einschütz, H. and Collet, P., *Z. Lebensm.-Unters.-Forsch.*, 143 (1970b) 187.
Lehmann, G. and Hahn, H.-G., *Z. Anal. Chem.*, 238 (1968) 445.
Lehmann, G., Hahn, H.-G., Collet, P., Seiffert-Eistert, B. and Morán, M., *Z. Lebensm.-Unters.-Forsch.*, 143 (1970c) 256.
Mathew, T. V., Banerjee, S. K., Mukherjee, A. K. and Mitra, S. N., *Res. Ind.*, 14 (1969) 140; *C.A.*, 72 (1970) 109878e.
Matsumoto, U., Nagase, Y. and Tanada, S., *Yakugaku Zasshi,* 90 (1970a) 230; *C.A.*, 72 (1970) 104218e.
Matsumoto, U., Nagase, Y. and Tanada, S., *Yakugaku Zasshi*, 90 (1970b) 236; *C.A.*, 72 (1970) 104317d.
Matsumoto, U., Tanada, S. and Nagase, Y., *Yakugaku Zasshi*, 90 (1970c) 1316; *C.A.*, 74 (1971) 6419t.
Murakami, M., Yoshino, T. and Harasawa, S., *Talanta*, 14 (1967) 1293.
Nester-Faust Corp., Newark, Del., booklet.
Parrish, J. R., *J. Chromatogr.*, 33 (1968) 542.
Raban, P. and Gregora, V., *Chem. Prům.*, 14 (1964) 359.
Reeves, R. L., Kaiser, R. S. and Finley, K. T., *J. Chromatogr.*, 47 (1970) 217.
Ruggli, P. and Jensen, P., *Helv. Chim. Acta*, 18 (1935) 624; 19 (1936) 64.
Sadini, V., *Proc. 16th Int. Dairy Congr., Copenhagen, 1962,* Vol. 3. Section C, p. 474; *C.A.*, 59 (1963) 15850 h.
Schetty, G., *Helv. Chim. Acta*, 52 (1969) 1016.
Schmit, J. A., in J. J. Kirkland (Editor), *Modern Practice of Liquid Chromatography*, Wiley-Interscience, New York, 1971, p. 409.
Schweppe, H., *Paint Technol.*, 27, No. 8 (1963) 12; *Anal. Abstr.*, 12 (1965) 792.
Schweppe, H., in E. S. Kováts (Editor), *5th Int. Symp. Separ. Methods, 1969, Section Column Chromatography,* Sauerländer AG, Aarau, 1970, p. 56; *C.A.*, 76 (1972) 73699p.
Slovák, Z., Borák, J. and Fischer, J., *Chem. Prům.*, 18 (1968) 142.
Stanley, R. L. and Kirk. P. L., *Agr. Food Chem.*, 11 (1963) 492.
Unni, M. K. and Venkataraman, K., *J. Sci. Ind. Res.*, 19B (1960) 355; *C.A.*, 55 (1961) 8864i.
Yanuka, Y., Shalon, Y., Weissenberg, E. and Nir-Grosfeld, I., *Analyst (London)*, 88 (1963) 872.

*Chapter 48*

# Pigments of plastids and photosynthetic chromatophores

Z. ŠESTÁK

## CONTENTS



## INTRODUCTION

Special, photosynthetically active pigments occur in unique structural units of autotrophic plants and are present in the coloured plastids (the chloroplasts) of higher plants and various algae. They are found in the thylakoid structures of the chromatoplasm of blue-green algae and photosynthetic bacteria and also occur in non-photosynthetic structures as the chromoplasts and their precursors, usually called etioplasts.

These special chloroplast pigments are divided into three principal groups: the green fat-soluble chlorophylls; the yellow fat-soluble carotenoids; and the red and blue water-soluble phycobilins. The carotenoids are usually subdivided into two groups: the polyene hydrocarbons or carotenes, and their oxy-derivatives, called xanthophylls. One to several chlorophylls, one to three carotenes, and several xanthophylls occur in the chloroplasts of autotrophic plants belonging to various taxonomic groups. The pigments phycobilins occur in the form of protein complexes called biliproteins in the red and blue-green algae, together with the chlorophylls and carotenoids.

The separations of natural mixtures of the pigments, mainly for their subsequent spectroscopic determination, usually depend upon differential solubility or differential sorption, properties that are related to the polarity of the pigments themselves. As a rule, phycobilins are the most polar and carotenes are the least polar, and some chlorophylls and certain xanthophylls exhibit similar polarities. All the pigments are very sensitive to heat, light, oxygen and corrosive reagents, and consequently their extraction and separation should be carried out under mild conditions, such as at reduced temperature (about 0°C), in a nitrogen atmosphere, in dim light and with neutral solvents.

The separation of the chlorophylls and carotenoids was the object of the classical chromatographic studies of Tswett, who examined the suitability of many adsorbents for this purpose. His method has been improved many times, and the successful modifications are used routinely in many laboratories (see, for example, reviews by Holden; Strain; Strain and Svec, 1969), which explains why only a few new procedures for liquid column chromatography have been published recently.

## SAMPLE PREPARATION

In principle, only fresh, turgid plant material should be analyzed, as storage may result in altered pigments. If the material cannot be analyzed immediately after sampling, storage at temperatures below 0°C for 1 or 2 days in polyethylene bags or rapid freezing in a mixture of acetone and dry ice or in liquid air or liquid nitrogen can be used.

Leaves or other plant materials are either ground by hand in a mortar with an abrasive (quartz sand or carborundum powder; one third of the volume of the plant material) and a cold extraction solvent (in an amount that just saturates the slurry), or are blended. Blending in a chilled Waring blender or a similar device, in models with an overhead motor, or in macerators of double cylinder-knife systems, requires a larger amount of plant material and of solvent (about 20 times the weight of the tissue). Centrifuged uni- or oligocellular algae and photosynthetic bacteria can be treated in a piston homogenizer or in vibration homogenizers with small glass beads.

A mixture of acetone and water (85:15) is most frequently used as the extraction solvent. Pure methanol or methanol–light petroleum (b.p. 40–60°C or 20–40°C) (2:1) (good for carotenoids) is preferable for some plants and algae, but unfortunately there is the possibility of chlorophyll allomerization. Tissue acids that may alter the chlorophylls during extraction are neutralized with a small amount of magnesium carbonate or a few drops of fresh dimethylaniline (these procedures are not satisfactory with strong acids). Chlorophyllase activity and chlorophyll oxidation can be arrested by preliminary immersion for 30 sec–1 min in neutral boiling water, followed immediately by cooling with running water or ice.

After completion of the extraction of pigments with additional portions of the cold solvent and filtration through a sintered-glass filter or centrifugation of the turbid extracts, the chlorophylls and carotenoids (together with other extracted lipoidal substances) are transferred to a non-polar solvent. This solvent, usually light petroleum (b.p. 20–40°C or higher), is added to the extract in a separating funnel, together with an excess of sodium chloride solution. The light petroleum layer is washed carefully about five times with distilled water (without shaking). In order to avoid the occasional formation of heavy emulsions in light petroleum, the pigments are often transferred to diethyl ether, which is then evaporated at reduced pressure, and the pigments are dissolved immediately in a small volume of light petroleum or another solvent suitable for loading on the column. (For details of the extraction procedure, see Davies, 1965, or Šesták, 1971.)

The detailed analysis of carotenoids is often facilitated by a preliminary saponification of unwanted lipids and chlorophylls in the acetone or methanol extracts by addition of an excess of a methanolic solution of potassium hydroxide. The mixture is left for 20 min–12 h in the dark under nitrogen, and the carotenoids are then transferred to light petroleum (b.p. 20–40°C or higher) by partition as described above; saponified chlorophylls remain in the aqueous phase. Extracts from brown algae and diatoms, whose specific xanthophylls are altered by alkalis, should not be saponified.

Phycobilin–protein complexes (biliproteins) are leached from cells of red and blue-green algae disrupted by grinding with water or dilute salt solutions to which coarse powdered material is eventually added, by blending with solid carbon dioxide, repeated freezing and thawing, or by treating with methanol or concentrated hydrochloric acid. The biliproteins in the aqueous extracts are usually precipitated with 40–75% of ammonium sulphate, centrifuged and dialyzed prior to chromatography.

## CHROMATOGRAPHIC PROCEDURES

### Chlorophylls and their mixtures with carotenoids

The chemical nature of the chlorophylls determines the liquid–solid systems that can be used for their chromatographic separation. All known chromatographic sorbents and a wide variety of solvents and their mixtures have already been tested by various workers, but no single solvent–sorbent combination can serve for the separation of all of the chlorophylls and carotenoids from a single plant species. The lability of chlorophylls necessitates the use of the mildest sorbents, such as powdered sugar, cellulose powder or starch.

Commercial powdered sugar varies a great deal in its composition. Some products (confectioner's sugar) contain 3% of maize starch, while others (icing sugar) contain 1.5% of calcium phosphate in order to reduce caking. An increase in the percentage of starch reduces the caking effect, but sometimes it lowers the quality of the separations. There are significant differences in the suitability of different brands of powdered sugar, involving different origins (from sugar cane or sugar beet), size of grains, content of cations (calcium), the presence of additives to increase the icing quality, etc. A practice in large laboratories is to test several brands of sugar (and solvents) and then to buy a supply of the most appropriate brand(s) for a few years. Unfortunately, the sugar cakes on standing and must be reground and sieved before use.

Resolutions on maize and potato starch are similar to those on sucrose. As the preparation of uniform starch columns is difficult and the prices of sugar and starch are similar, sugar is usually preferred to starch. Columns of cellulose powder give separations similar to those on paper or on cellulose thin layers (for reviews, see Šesták 1958, 1965, 1967). Some preparations of cellulose powder (Whatman CF-11, W. & R. Balston, Maidstone, Great Britain) may be acidic and require preliminary washing with neutral distilled water and air drying.

Columns of length 100–600 mm and diameter 7–100 mm are uniformly packed by compressing small portions of dry adsorbent with a plunger (such as an inverted stopper or a plastic disc) on a bar. Metallic plungers or packers operated by compressed air are also used (Strain and Svec, 1969).

Uniform loading of the column with an appropriate amount of pigments is necessary for the effective separation of pigment zones (the higher the loading, the poorer is the resolution). For example, loading of columns 10 mm in diameter with *ca.* 200 $\mu$l of

*References p.1048*

TABLE 48.1

CHROMATOGRAPHIC SEQUENCE (FROM TOP TO BOTTOM OF THE COLUMN) OF CHLOROPHYLLS AND CAROTENOIDS AND SOME OF THEIR PRECURSORS AND DEGRADATION PRODUCTS FROM VARIOUS PLANT MATERIALS IN COLUMNS OF POWDERED SUGAR (CONTAINING 3% OF STARCH) WITH LIGHT PETROLEUM (b.p. 20–40°C) + 0.5–2.0% OF *n*-PROPANOL AS ELUENT (AFTER STRAIN AND SVEC, 1969)

Water-soluble pigments (*e.g.*, chlorophyllides) are adsorbed at the top of the column. Pheophytin *a* is adsorbed between chlorophyll *a* and the carotenes.

| Chlorophylls and their derivatives* | Leaves of higher plants, green algae** | Green sulphur bacterium | Blue-green algae | Brown algae§ | Yellow-green algae (*Heterokontae*)§§ | Red algae |
|---|---|---|---|---|---|---|
| Concentration of *n*-propanol in light petroleum (%) | | | | | | |
| 0.5–2.0 | 0.5 | 1–2.5 | 0.5 | 0.5–2.0 | 0.5 | 0.5 |
| Chlorophyll *c* | Neoxanthin | Two yellow zones | Myxoxanthophyll | Chlorophyll *c* | Xanthophyll 600 | Chlorophyll *d* |
| Pyrochlorophyll *b* | Violaxanthin | *Chlorobium* chlorophyll | Zeaxanthin ± lutein | Neofucoxanthin *a* | Vaucheriaxanthin ester 766 | Lutein ± zeaxanthin |
| Chlorophyll *b* | Chlorophyll *b* | Bacteriochlorophyll | Chlorophyll *a* | Neofucoxanthin *b* | Xanthophyll 566 | Chlorophyll *a* |
| Chlorophyll *d* | Zeaxanthin + lutein | Green zone | Myxoxanthin | Fucoxanthin | Xanthophyll 582 | Carotenes ($\beta \pm \alpha$)*** |
| Chlorophyll *b*′ | Chlorophyll *a* | Yellow zone | β-Carotene | Violaxanthin | Chlorophyll *a* | |
| Pyrochlorophyll *a* | Cryptoxanthin | Chlorobactene | | Chlorophyll *a* | β-Carotene | |
| Chlorophyll *a* | Carotenes ($\beta + \alpha$)*** | | | Carotenes ($\beta + \alpha$)*** | | |
| Chlorophyll *a*′ | | | | | | |
| Protochlorophyll(ide) | | | | | | |
| Pheophytin *a* | | | | | | |
| Pyropheophytin *a* | | | | | | |

*Chlorophylls *a*′ and *b*′ are isomers of chlorophylls *a* and *b*. Pyrochlorophylls and pyropheophytins originate by heating of pyridine solutions of chlorophylls and their magnesium-free derivatives pheophytins (produced from chlorophylls by the action of acids). Deuterated chlorophylls are not separated from the ordinary chlorophylls. For the sequence of other derivatives, see Strain and Svec (1966).

**In the alga *Euglena*, violaxanthin and α-carotene are missing and the zone of zeaxanthin + lutein is replaced by diadinoxanthin.

***Carotenes do not separate by this procedure; they move fastest and can wash along with the solvent front.

§In diatoms, the zone of violaxanthin is replaced by two zones of diadinoxanthin and diatoxanthin.

§§The xanthophylls of this group are characterized by their molecular weights (*e.g.*, xanthophyll 600).

extract from less than 1 g of plant material, of 30–50 mm columns with 2–5 ml of extract from less than 10 g of plant material, and of 80–100 mm columns with 10–20 ml of extract from less than 100 g of plant material is recommended (Strain and Svec, 1969).

Solvent systems consist mostly of a non-polar liquid (*e.g.*, light petroleum) with a small portion of a polar liquid (often an alcohol, *e.g.*, *n*-propanol). One must use solvents that permit selective sorption of the pigments and that do not cause decomposition of the pigments.

The separation system recently used with extracts of many plants consists of powdered sugar as sorbent and light petroleum (b.p. 20–40°C) plus 0.5–2.0% of *n*-propanol as eluent (Strain and Svec, 1969). The sequences of pigments separated on the column are shown in Table 48.1. An alternative eluent is acetone (gradient from 5–25%) in light petroleum (Sweeney and Martin).

Starch and cellulose powder sorbents yield a sequence of pigments similar to that given by powdered sugar, but these sorbents require a higher proportion of the polar component in the eluent (*e.g.*, 1% of *n*-propanol instead of 0.5%). With cellulose powder (Whatman CC-41) columns, toluene also seems to be an appropriate eluent. After the elution of all the carotenoids (apart from neoxanthin), the chlorophyll zones are separated further using 2% of *n*-propanol or isopropanol in toluene. Cellulose powder yields more diffuse pigment zones than powdered sugar, but the results on parallel columns are usually more comparable.

For the complete separation of all chlorophylls and carotenoids present in an unsaponified plant extract, subsequent separations on columns of different sorbents with various elution solvents are essential: Deroche, for example, used a preliminary elution of $\beta$-carotene from the cellulose (Whatman CF-11) column (17 × 100 mm) with light petroleum. The acetone eluate of the chlorophylls plus xanthophylls was then separated into two fractions on a polyethylene (Hoechst-Peralta, grains 0.1–0.5mm) column (140 × 23 mm) by elution with acetone containing 30% and 15% of water. The xanthophyll and chlorophyll fractions were then divided into individual components on two cellulose columns (100 × 13 or 17 mm) by gradient elution with 3–10% of acetone in light petroleum.

For the purification of individual chlorophylls, several systems involving the use of sugar, cellulose and other sorbents (powdered polyamide and polyethylene, polymethacrylate and Sephadex gels, etc.) have been proposed. A comparison of these methods is presented in Table 48.2.

The separation of chlorophylls from other pigments on powdered polyethylene (a suitable brand has to be chosen) does not require the transfer of pigments to a non-polar solvent (Anderson and Calvin). An 80% aqueous acetone extract is applied to a sorbent washed with 70% acetone and the column (I.D. 5 cm) is developed with the same solvent. The sequence of pigments from the top of the column is carotenes, pheophytins, chlorophyll *a*, chlorophyll *b* and xanthophylls. The purification of chlorophylls *a* and *b* requires further separation on a powdered cellulose column, which reverses the order of adsorption of the chlorophylls.

This method was modified by Houssier and Sauer for the purification of the immediate chlorophyll precursor protochlorophyll(ide) *a* and some allied pigments (4-vinylprotochlorophyll *a*; protopheophytin *a*). A 400 × 100 mm column of polyethylene powder is eluted with acetone–water (7:10 to 9:1). In the subsequent chromatography on a

*References p.1048*

TABLE 48.2
PROCEDURES FOR THE SEPARATION OF VARIOUS CHLOROPHYLLS AND THEIR DERIVATIVES AND PRECURSORS

| Compound separated | Sorbent and dimensions of the column | Eluent | Notes | Reference |
|---|---|---|---|---|
| Chlorophyll *a* (pure) | Polyamide powder, 100 × 6 mm | Benzene–chloroform (1:1) after preliminary washing of carotenes with *n*-hexane and pheophytin + xanthophylls with benzene | From plant leaves; flow-rate 1 ml/min | Frič and Haspel-Horvatovič |
| Chlorophyll *b* | Polymethacrylate gel, 600 × 7 mm | Benzene | Last fractions contain α-carotene | Wieland *et al.* |
| Chlorophyll *c** (pure) | (a) Cellulose powder (Whatman CF-11 or Schleicher and Schüll 123, Dassel, G.F.R.), 350–800 × 35 mm | 0.5% *n*-propanol in light petroleum (b.p. 60–80° C), after elution with 5% methanol in diethyl ether or light petroleum re-chromatographed with the later solvent system | | Jeffrey**, Sagromsky and Rieth |
| | (b) Sephadex LH-20 in methanol | Methanol–chloroform (1:1) | This final purification step may be replaced by separation between methanol and ligroine (Sagromsky and Rieth) | Croft and Howden |
| Chlorophylls $c_1$ and $c_2$ | Polyethylene powder, 350 × 50 mm | Acetone–tetrahydrofuran (30:1, warm) and subsequent acetone washing for $c_1$ and impurities; $c_2$ washed with tetrahydrofuran | From brown algae | Strain *et al.* (1971) |
| Chlorophyll *d* | Powdered sugar | Light petroleum (b.p. 40–70°C)–benzene (1:0, 4:1, 3:2, 2:3, 1:4, 0:1 in sequence) | From red algae after preliminary elimination of most of the chlorophyll *a*; repeat 5–6 times for complete purification | Evstigneev and Cherkashina |
| Bacteriochlorophyll | Kieselguhr G*** (Brinkmann, Westbury, N.Y., U.S.A.) impregnated with triolein, 200 × 25 mm | Methanol–acetone–water (6:2:1) | Bacteriochlorophyll eluted first | Kim |

TABLE 48.2 *(continued)*

| Compound separated | Sorbent and dimensions of the column | Eluent | Notes | Reference |
|---|---|---|---|---|
| *Chlorobium* chlorophylls 650 and 660 | Powdered sugar | 40 and 50% diethyl ether in light petroleum | From green sulphur bacteria | Stanier and Smith |
| *Chlorobium* chlorophyll precursors | Polyethylene powder, 300–400 × 35–75 mm | Phosphate buffer with 2,6-lutidine (0–10%) | | Richards and Rapoport |
| Protoporphyrin IX dimethyl esters (chlorophyll precursors) | Powdered sugar, 250 × 20 mm | Benzene–light petroleum (b.p. 35–60°C) (1:3), then 0.5% pyridine in light petroleum (b.p. 35–60°C) | | Baum and Ellsworth |

*An alternative procedure (Seely *et al.*) involves a dimethyl sulphoxide extract from brown algae. The pigments are dissolved in *n*-hexane, and chlorophyll *a* and xanthophylls are eluted from a sucrose column with 0–20% acetone in *n*-hexane. Chlorophyll *c* is then resolved into 2–3 bands and eluted with *n*-hexane–ethyl acetate–acetone (6:4:1).

**In the original procedure of Jeffrey, the separation step (a) was followed by removal of lipids on a silicic acid–Hyflo Supercel (5:2) column with 5% methanol in chloroform and final chromatography on neutral aluminium oxide [successive eluents: methanol–chloroform (9:1), ethanol–chloroform–water (5:2:1 and 5:2:2.5)].

***Kieselguhr and some other diatomaceous earths as well as silicic acid sorbents may induce isomerization of some xanthophylls (Strain *et al.*, 1967).

400 × 100 mm column of powdered sugar, the concentration of *n*-propanol in isooctane is increased from 0 to 0.5%. Another procedure (Ellsworth) for the precursors combines separation on a 100–150 × 25 mm column of silica gel G (E. Merck, Darmstadt, G.F.R.) with diethyl ether in light petroleum (b.p. 35–60°C) (successively 0 and 30%) and on a sucrose column.

Chlorophylls (*a* + *b*) can be separated from methyl chlorophyllides (*a* + *b*), or their magnesium-free derivatives pheophytins (*a* + *b*) from pheophorbides (*a* + *b*), by elution (flow-rate 32 ml/h) with methanol–chloroform (1:1) from a column of Sephadex LH-20 swollen in the same solvent system (Schenk and Dässler).

The separation of pheophytins can also be accomplished on the ion-exchange resin Dowex 50W-X4 ($H^+$), which converts chlorophylls into pheophytins (Wilson and Nutting). The resin in the column (200–240 × 10 mm, with water-jacket) is dehydrated with acetone, and the pigment solution in acetone is added immediately. The acetone elutes pheophytin *b* and the contaminating carotenoids (flow-rate 2–3 ml/min). Pheophytin *a* is then eluted with nearly boiling 85% acetone.

## Chlorophyll–protein complexes

Column chromatography can also be used for separating the natural pigment–protein complexes of chlorophylls or protochlorophyllides extracted from plants. For example,

*References p.1048*

Argyroudi-Akoyunoglou *et al.* separated two chlorophyll–protein complexes in a 55 × 10 mm hydroxyapatite column (pre-washed with 5 ml of 1% sodium dodecylsulphate in 0.1 *M* phosphate buffer, pH 7.0) by stepwise elution with 0.1 *M* phosphate buffer (pH 7.0) followed by the 0.35 *M* phosphate buffer (pH 7.0). Schopfer and Siegelman used a 150 × 75 mm hydroxyapatite column for the separation, with 0.2 *M* potassium chloride in tricine buffer followed by 0.25 *M* potassium phosphate in tricine buffer (pH 8.0) at a flow-rate of 20 ml/min, as a step in the purification of the protochlorophyllide–protein complex (protochlorophyllide holochrome).

## Carotenoids

As carotenoids are more resistant than chlorophylls to inorganic sorbents, there is a wider choice of procedures for the extensive resolution of these pigments (for reviews, see Davies, 1965; Strain and Svec, 1969). Nevertheless, some sorbents (*e.g.*, silicic acid and some diatomaceous earths) induce changes in carotenoids, namely isomerization of neoxanthin and violaxanthin (Strain *et al.*, 1967), and should not be used. In general, there are five types of chromatographic procedures for carotenoids: (1) separation of unsaponified plant extracts either (A) directly on a single column or (B) after elimination of unwanted pigments by (a) a combination of several columns, (b) their preliminary extraction (*e.g.*, analysis of hypophasic carotenoids) or (c) their degradation directly on the sorbent; and (2) separation of saponified extracts.

Hypophasic carotenoids can be separated on a 250 × 25 mm calcium carbonate–Hyflo Supercel (1:1) column by stepwise elution with 5, 7.5, 10 and 15% acetone in light petroleum (b.p. 60–80°C) (Jensen). Columns of this and other inorganic sorbents can be packed with the dry sorbent or with a slurry in the first eluent.

For separations of types 1Ac and 2 above, Strain *et al.* (1968) and Strain and Svec (1969) recommended the use of aluminium oxide as sorbent and benzene + 36% acetone as eluent, or a mixture of special activated magnesium oxide (Sea Sorb 43, Fisher Scientific, Pittsburgh, Pa., U.S.A.) with Celite 545 (1:1–1:2, slurried, dried at 20°C in air for 16 h and reactivated at 110°C for 0.5 h) and development with 1,2-dichloroethane or with light petroleum (b.p. 65–110°C) plus 40% of acetone or 5–10% of *n*-propanol. The magnesium oxide–Celite sorbent also serves for the separation of $\alpha$- and $\beta$-carotenes, when washed with 1% acetone in light petroleum (b.p. 65–110°C).

For carotenoids in saponified extracts, columns of sugar, starch or cellulose powder can also be used. They are eluted with light petroleum + 0.5–1% of *n*-propanol (Strain and Svec, 1969) or stepwise with light petroleum (b.p. 40–60°C) containing 0–10% of acetone (Costes). The eluents that contain acetone are more appropriate for use with cellulose sorbents. The separation of the main carotenoids is complete, but their zones are usually not as sharp as on the above-mentioned sorbents. The adsorption capacity and selectivity of powdered cellulose for the separation of carotenoids can be increased by grinding with polar solvents, *e.g.*, acetone (Miller *et al.*). Automated devices for carotenoid analysis with photometric measurement of the eluates were described by Monéger.

The relative sorbability of carotenoids varies not only with the sorbent and eluent, but also with the structure of the carotenoid molecules, *i.e.*, with the number of double bonds and their conjugation, the number of hydroxyl and epoxy groups, etc. Hence the usual sequence of the main chloroplast carotenoids from the top of the column is: neoxanthin, violaxanthin, zeaxanthin, lutein, cryptoxanthin and carotenes. The relative positions of the xanthophylls often vary owing to the composition of the eluent, the type of plant material, etc. For comparisons of the sequences of elution of carotenoids see, for example, Table 3 in Subbarayan *et al.*, or Table V in Davies (1965). Methods for the isolation of individual carotenoids are summarized in Table 14 in Goodwin and in comprehensive tables in Foppen.

A continuous flow separation in a 280 × 4 mm glass column or 100–200 × 6 mm stainless-steel tubing with standard Swagelok fittings, filled with magnesium oxide (Sea Sorb 43, Fisher Scientific, for carotenes) or precipitated zinc carbonate (Fisher Scientific, for xanthophylls) was recommended by Stewart and Wheaton. α- and β-carotenes were separated with *n*-hexane containing 5% *tert.*-pentanol (flow-rate 1 ml/min), and xanthophylls with a gradient of *tert.*-pentanol in *n*-hexane (flow-rate 0.5 ml/min). The pumps used were operated at pressures of 800–2000 p.s.i. ($5.5-13.8 \cdot 10^6$ N/m$^2$). Isomerization of carotenoids was restricted by incorporating 1% of butylated hydroxytoluene (Eastman-Kodak, Rochester, N.Y., U.S.A.) in the solvent. The elution sequence of pigments in a tangerine extract was (from the top of the column): *cis*-violaxanthin, *cis*-antheraxanthin, violaxanthin, antheraxanthin, lutein, zeaxanthin, β-citraurin, cryptoxanthin and carotenes (β-, α-). In another automated continuous procedure (Davies, 1967), a column of silica gel and Celite (1:1) was used, with a continuously increasing concentration of ethyl acetate–methanol (5:1) in benzene as the eluent.

## Biliproteins

Classical methods of biliprotein chromatography involve separations on columns of tricalcium phosphate gel or calcium phosphate gel mixed with Hyflo Supercel or Celite filter aid, of brushite ($CaHPO_4 \cdot 2\,H_2O$) or hydroxyapatite [$Ca_5(PO_4)_3OH$]. The eluents consist of increasing concentrations of acetate buffers or phosphate buffers, 2.5–100 m*M* (pH 6.0–6.5). Phycoerythrins usually precede phycocyanins. These methods exist in many modifications (see, for example, Haxo *et al.*, Ó Carra, Swingle and Tiselius, Tiselius *et al.*, Troxler and Lester).

Biliprotein separations are also carried out with solutions of ammonium sulphate of various concentration. Precipitated biliproteins are added to coarse diatomaceous earth (Celite 545), the slurry is poured into a 1000 × 150 mm column and eluted with a linear gradient of 75 to 5% ammonium sulphate solution (flow-rate 20 ml/min). The order of eluted chromoproteins is ferredoxin, cytochromes, *allo*phycocyanin and phycocyanin (Siegelman *et al.*).

Gel filtration on a Sephadex G-150 column (Pharmacia, Uppsala, Sweden) was used by Erokhina and Krasnovskiĭ for the isolation of phycocyanin. Water-cooled columns of Sephadex G-100 and G-200 can also be used for the determination of the molecular weights of phycoerythrins and phycocyanins from various algae (Nolan and Ó hEocha); the eluent is 0.05 *M* Tris–hydrochloric acid buffer (pH 7.5) containing 0.5% of sodium azide.

*References p.1048*

Ericksson and Halldal separated red algae biliproteins by a combination of gel filtration on a column of Sephadex G-25 (370 × 62 mm) and ion-exchange chromatography on a 280 × 20 mm column of DEAE-cellulose (pre-washed with 1% sodium chloride solution and the starting eluent). The eluent for the gel column and the starting eluent for the DEAE-cellulose column was 0.01 *M* Tris–hydrochloric acid (pH 7.2), and its ionic strength was then gradually increased to 0.5 *M* sodium chloride. Complete separation of phycoerythrin and phycocyanin was achieved on a second DEAE-cellulose column with an eluent 0.42 *M* sodium chloride in the above buffer. Neufeld and Riggs purified dialyzed phycocyanin on a column of DEAE-cellulose (460 × 19 mm) equilibrated with 0.1 *M* phosphate buffer (pH 7.0) and eluted with 0.08 *M* phosphate buffer (pH 7.0).

## DETECTION

As chlorophylls, carotenoids and phycobilins are intensely coloured substances, their zones can be detected visually. Small concentrations of chlorophylls and their derivatives are detected by their red fluorescence under UV radiation. 5,6-Monoepoxycarotenoids give a greenish blue colour reaction with trace amounts of hydrochloric acid. The eluted pigments are transferred into pure solvents and usually identified spectrophotometrically; for the spectral characteristics of chlorophylls see, for example, Šesták (1971), of carotenoids, Foppen or Davies (1965), and of phycobilins, Ó hEocha.

## REFERENCES


Anderson, A. F. H. and Calvin, M., *Nature (London)*, 194 (1962) 285.

Argyroudi-Akoyunoglou, J. H., Feleki, Z. and Akoyunoglou, G., *Biochem. Biophys. Res. Commun.*, 45 (1971) 606.

Baum, S. J. and Ellsworth, R. K., *J. Chromatogr.*, 47 (1970) 503.

Costes, C., *Bull. Soc. Fr. Physiol. Vég.*, 15 (1969) 55.

Croft, J. A. and Howden, M. E. H., *Phytochemistry*, 9 (1970) 901.

Davies, B. H., in T. W. Goodwin (Editor), *Chemistry and Biochemistry of Plant Pigments*, Academic Press, New York, London, 1965, p. 489.

Davies, B. H., *Biochem. J.*, 103 (1967) 51P.

Deroche, M.-E., *Chim. Anal. (Paris)*, 53 (1971) 704.

Ellsworth, R. K., *Anal. Biochem.*, 39 (1971) 540.

Eriksson, C. E. A. and Halldal, P., *Physiol. Plant.*, 18 (1965) 146.

Erokhina, L. G. and Krasnovskiĭ, A. A., *Mol. Biol.*, 5 (1971) 399.

Evstigneev, V. B. and Cherkashina, N. A., *Biokhimiya*, 35 (1970) 48.

Foppen, F. H., *Chromatogr. Rev.*, 14 (1971) 133.

Frič, F. and Haspel-Horvatovič, E., *J. Chromatogr.*, 68 (1972) 264.

Goodwin, T. W., in K. Paech and M. V. Tracey (Editors), *Modern Methods of Plant Analysis*, Vol. III, Springer, Berlin, Göttingen, Heidelberg, 1955, p. 272.

Haxo, F., Ó hEocha, C. and Norris, P., *Arch. Biochem. Biophys.*, 54 (1955) 162.

Holden, M., in T. W. Goodwin (Editor), *Chemistry and Biochemistry of Plant Pigments*, Academic Press, New York, London, 1965, p. 461.

Houssier, C. and Sauer, K., *Biochim. Biophys. Acta*, 172 (1969) 476.

Jeffrey, S. W., *Biochem. J.*, 86 (1963) 313.

Jensen, A., *Norsk Inst. Tang-Tareforskning Report*, No. 31, 1966, p. 1.

Kim, W. S., *Z. Naturforsch.*, 22b (1967) 1054.
Miller, C. K., Steffenson, D., Frame, H. D., Jr., and Strain, H. H., *Anal. Chem.*, 35 (1963) 93.
Monéger, R., *Rev. Gen. Froid*, 1968 (1968) 425.
Neufeld, G. J. and Riggs, A. F., *Biochim. Biophys. Acta*, 181 (1969) 234.
Nolan, D. N. and Ó hEocha, C., *Biochem. J.*, 103 (1967) 39P.
Ó Carra, P., *Biochem. J.*, 94 (1965) 171.
Ó hEocha, C., in T. W. Goodwin (Editor), *Chemistry and Biochemistry of Plant Pigments*, Academic Press, New York, London, 1965, p. 175.
Richards, W. R. and Rapoport, H., *Biochemistry*, 5 (1966) 1079.
Sagromsky, H. and Rieth, A., *Arch. Protistenk.*, 114 (1972) 46.
Schenk, J. and Dässler, H. -G., *Pharmazie*, 24 (1969) 419.
Schopfer, P. and Siegelman, H. W., *Plant Physiol.*, 43 (1968) 990.
Seely, G. R., Duncan, M. J. and Vidaver, W. E., *Mar. Biol.*, 12 (1972) 184.
Šesták, Z., *J. Chromatogr.*, 1 (1958) 293; *Chromatogr. Rev.*, 1 (1959) 193.
Šesták, Z., *Chromatogr. Rev.*, 7 (1965) 65.
Šesták, Z., *Photosynthetica*, 1 (1967) 269.
Šesták, Z., in Z. Šesták, J. Čatský and P. G. Jarvis (Editors), *Plant Photosynthetic Production. Manual of Methods*, Dr. W. Junk Publishers, The Hague, 1971, p. 672.
Siegelman, H. W., Chapman, D. J. and Cole, W. J., in T. W. Goodwin (Editor), *Porphyrins and Related Compounds*, Academic Press, New York, London, 1968, p. 107.
Stanier, R. Y. and Smith, J. H. C., *Biochim. Biophys. Acta*, 41 (1960) 478.
Stewart, I. and Wheaton, T. A., *J. Chromatogr.*, 55 (1971) 325.
Strain, H. H., *Chloroplast Pigments and Chromatographic Analysis*, Pennsylvania State University, University Park, Pa., 1958.
Strain, H. H., Cope, B. T., McDonald, G. N., Svec, W. A. and Katz, J. J., *Phytochemistry*, 10 (1971) 1109.
Strain, H. H., Sherma, J. and Grandolfo, M., *Anal. Chem.*, 39 (1967) 926.
Strain, H. H., Sherma, J. and Grandolfo, M., *Anal. Biochem.*, 24 (1968) 54.
Strain, H. H. and Svec, W. A., in L. P. Vernon and G. R. Seely (Editors), *The Chlorophylls*, Academic Press, New York, London, 1966, p. 21.
Strain, H. H. and Svec, W. A., *Advan. Chromatogr.*, 8 (1969) 119.
Subbarayan, C., Jungalwala, F. B. and Cama, H. R., *Anal. Biochem.*, 12 (1965) 275.
Sweeney, J. P. and Martin, M. E., *Food Technol.*, 15 (1961) 263.
Swingle, S. M. and Tiselius, A., *Biochem. J.*, 48 (1951) 171.
Tiselius, A., Hjertén, S. and Levin, Ö., *Arch. Biochem. Biophys.*, 65 (1956) 132.
Troxler, R. F. and Lester, R., *Biochemistry*, 6 (1967) 3840.
Tswett, M. S., *Ber. Deut. Bot. Ges.*, 24 (1906) 384.
Wieland, T., Lüben, G. and Determann, H., *Naturwissenschaften*, 51 (1964) 138.
Wilson, J. R. and Nutting, M. -D., *Anal. Chem.*, 35 (1963) 144.

*Chapter 49*

# Macromolecular substances and plastics

M. KUBÍN and J. ČOUPEK

## CONTENTS



## INTRODUCTION

Since 1964, gel permeation chromatography has become widely used in polymer science and technology as a rapid and reliable method for determining molecular weights and molecular-weight distributions of plastics, resins, rubbers, etc., and superseded almost completely the older, time-consuming methods of polymer fractionation. In industry, GPC serves as a method for characterizing and analyzing new polymeric products and for controlling the quality of standard products (Harmon). Not only can the method rapidly indicate out-of-specification batches, but sometimes it can indicate directly the source of a particular problem, as the molecular-weight distribution directly reflects the conditions of preparation, either in polymerization or polycondensation reactions or in blending of polymer compositions from individual, pre-synthesized products (see, for example, Little *et al.*). In these instances, it is often unnecessary to convert the chromatogram into the true distribution curve, as the direct comparison of GPC records obtained under standard conditions yields enough information about the product quality and specification. Reliable chromatograms can be obtained within 3–4 h and consecutive samples can be injected into the column at intervals much shorter than the elution time of a single sample. As a method for separating solutes according to their size, GPC is valuable even for studying the concentration and type of low-molecular-weight additives present in polymers, such as solvents, plasticizers and antioxidants. Packings that are capable of effecting separations only in the low-molecular-weight region are commercially available and GPC has been successfully applied to lubricants, polyglycols, tars, asphaltenes and a series of other oligomeric substances.

GPC can also be used in fundamental studies of the mechanism and kinetics of polymerization and polycondensation processes, in elucidating the course of degradation and depolymerization reactions and in investigations concerned with the relationship between the structure and properties of complex macromolecular substances and systems.

By increasing the inner diameter of GPC columns, it is possible to fractionate much larger samples with excellent efficiency; thus, by repeatedly collecting the corresponding cuts, virtually monodisperse fractions can be obtained on a preparative scale, not only for calibration purposes but also for basic studies of the effect of molecular weight and molecular-weight distribution on the physical properties of polymers. The polymer chemist now has at his disposal the means of preparing, by blending suitable monodisperse fractions, polymer samples with any desired molecular-weight distribution and can reliably correlate their properties with the shape and position of the corresponding distribution curve.

Except probably in the analysis of oligomeric substances, the concept of baseline separation is of relatively minor importance in the gel permeation chromatography of plastics and synthetic high-molecular-weight substances in general. This is because virtually all synthetic polymers are mixtures of individual species (mers) covering a wide range of molecular weights and it is unreasonable to expect that any separation method could resolve two successive mers with a total chain-length corresponding to hundreds or even thousands of monomeric units. The chromatogram of a polymer in most instances is therefore a single broad zone, representing the sum of excessively overlapping, unresolved peaks. Fig. 49.1 shows a typical chromatogram of a polystyrene sample with broad molecular-weight distribution.

The problem is to derive from the GPC record the continuous molecular-weight distribution curve and/or to calculate the individual molecular weight averages (the number average, $\bar{M}_n$, the weight average, $\bar{M}_w$, the *z*-average, $\bar{M}_z$, etc.). Methods for solving this problem are discussed in Chapter 5 and the prerequisite calibration of the sets of GPC columns is treated in Chapter 12. Most of the work on the gel permeation chromatography of synthetic high polymers has been carried out on a commercial instrument

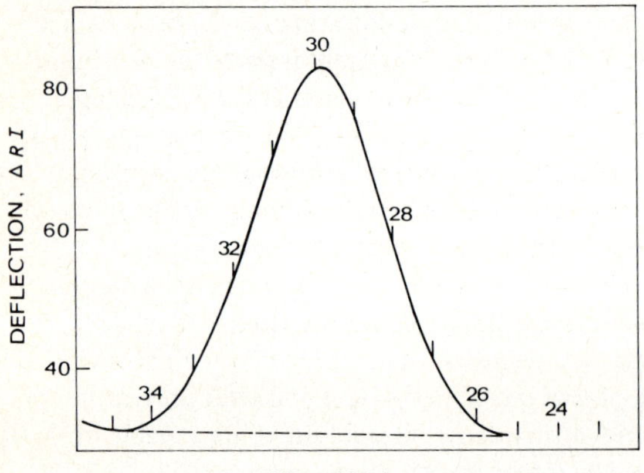


Fig. 49.1. Typical gel permeation chromatogram of a polystyrene sample (Huang *et al.*). Styragel packing (four columns designated $10^6$, $10^5$, $10^4$, and $10^3$ Å) with tetrahydrofuran as solvent, 25°C, 1 ml/min; 2 ml of a 0.25% solution injected.

(Model 200 gel permeation chromatograph, Waters Ass., Framingham, Mass., U.S.A.).

For these reasons, this chapter differs from the other chapters in this part of the book and represents a survey, hopefully reasonably complete, of the numerous applications of GPC in the field of polymer science and technology.

## VINYL POLYMERS

Polystyrene is by far the most extensively studied substance. This is only partly due to its unquestionable importance as a polymer; its ready solubility in most GPC solvents, its relatively simple and well known solution behaviour and, last but not least, the commercial availability of narrow polystyrene fractions for calibration purposes, are the reasons why nearly all of the fundamental studies concerned with the mechanism of GPC separations and the influence of operational variables on the separation efficiency have been performed with polystyrene. These aspects have been covered in the first part of this book and will not be reviewed here.

Several studies have been described that deal with the kinetics and mechanism of styrene polymerization determined on the basis of distribution curves from GPC data. Thus, Huang and his collaborators (Huang and Chandramouli; Huang and Westlake, 1969, 1970; Huang *et al.*) examined the polymerization of styrene induced by $\gamma$-irradiation and the influence of temperature and the presence of water and methanol in the reaction mixture on the resulting distribution. Fig. 49.2 shows schematically the changes in the experimental chromatograms of radiation-polymerized styrene, brought about by changes in the polymerization conditions (decreasing the temperature, Fig. 49.2A; increasing the conversion, Fig. 49.2B).

May and Smith, and also Hsieh and McKinney, used GPC on Styragel packings in studies of the kinetics and mechanism of styrene polymerization.

Polystyrene was used in several studies in which the GPC results were compared with those obtained by other methods (Alliet; Meyerhoff, 1967, 1968; Screaton and Seemann; Smith *et al.*; Yamada *et al.*, 1966d). Peaker and Patel obtained a high efficiency in GPC by running a mixture of a high-molecular-weight, inactive polystyrene fraction with a lower fraction labelled with $^{14}C$ and measuring the degree of contamination of the former.

Changes in the molecular weight and distribution curves of polystyrenes, brought about by degradation, were followed by Hendrickson and also by Smith and Temple. Preparative fractionation of polystyrene by GPC has also been described (Hattori and Hamashima).

The above studies were performed on Styragel packing (cross-linked, macroporous polystyrene–divinylbenzene beads). Other packings have been tested for the fractionation of polystyrene: silica gel (MacCallum, Unger *et al.*), porous glass (Cantow and Johnson; Cooper and Johnson, 1969, 1971); specially prepared beads based on methacrylates (Determann *et al.*, Heufer and Braun).

The operating conditions employed in the analysis of polystyrene by GPC are summarized in Table 49.1.

Bagby *et al.* examined the thermal degradation of poly(methyl methacrylate) (PMMA)

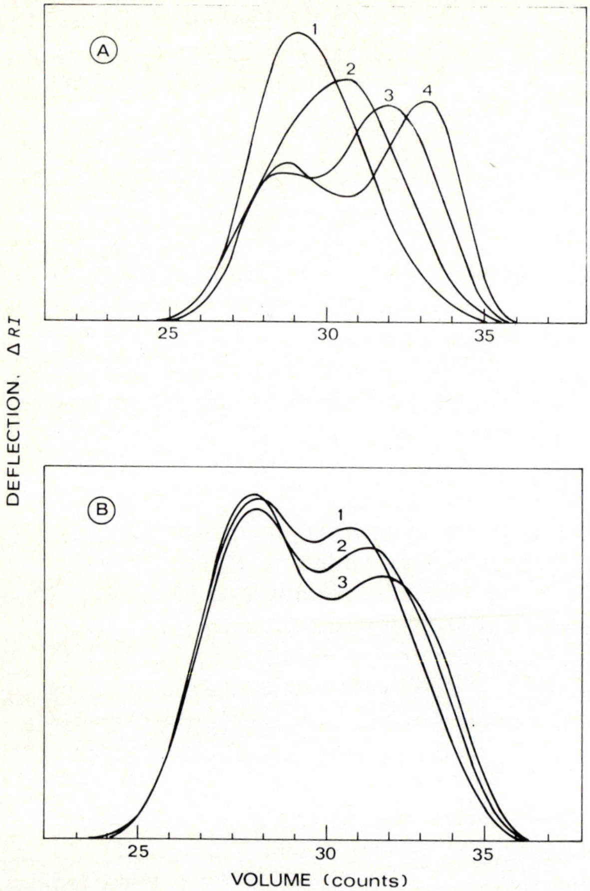


Fig. 49.2. Changes in the molecular-weight distribution of polystyrene, as reflected in the shape of the GPC chromatograms, brought about by (A) changes in temperature (1=30°C; 2=15°C; 3=0°C; 4=−10°C) and (B) changes in conversion (1=1.36%; 2=2.52%; 3=5.29%) in the radiation-induced polymerization of styrene (Huang and Westlake, 1970). Styragel packing (four columns designated $10^7$, $10^5$, $10^4$ and $10^3$ Å) with tetrahydrofuran as solvent, 1 ml/min, 25°C.

and fractions prepared therefrom, analyzing the initial samples and the residue after degradation by GPC on Styragel with THF as a solvent at ambient temperature. PMMA was also used by several workers in tests of the reliability of GPC by comparing the results with those obtained by other methods. Berger and Schulz (1970, 1971) employed PMMA fractions characterized by osmometry, light scattering and viscometry for correcting molecular weights and polydispersity as determined by GPC. Smith *et al.* compared the molecular-weight distributions of non-fractionated PMMA samples, determined by GPC on Styragel with THF at 50°C with curves calculated from known polymerization kinetics and mechanisms, and found good agreement. Yamada *et al.* (1967), using similar experimental conditions (Styragel, THF at 30°C), reported that the GPC results were in accordance with the fractionation of the same PMMA sample by elution chromatography.

TABLE 49.1
OPERATING CONDITIONS FOR THE GEL PERMEATION CHROMATOGRAPHY OF POLYSTYRENE

| Solvent* | Packing | Temperature (°C) | References |
|---|---|---|---|
| THF | Styragel | 50 | May and Smith |
| THF | Styragel | 25 | Huang and Westlake (1970) |
| THF | Styragel | Ambient | Alliet, Barson and Robb, Goedhart and Opschoor, Goetze *et al.* |
| TCB | Styragel | 130 | Hsieh and McKinney |
| THF | Styragel | 35 | Hattori and Hamashima |
| Toluene | Styragel | 20 | Pannell |
| Benzene | Silica gel | Ambient | MacCallum |
| $CCl_4$ | Porous glass | 25 | Cooper and Johnson (1969, 1971) |
| Toluene | Porous glass | 23 | Cantow and Johnson |

*THF = tetrahydrofuran; TCB = 1,2,4-trichlorobenzene.

TABLE 49.2
OPERATING CONDITIONS FOR THE GEL PERMEATION CHROMATOGRAPHY OF POLY(METHYL METHACRYLATE)

| Solvent | Packing | Temperature (°C) | References |
|---|---|---|---|
| THF | Styragel | 50 | Smith *et al.* |
| THF | Styragel | Ambient | Berger and Schulz (1970, 1971), Bagby *et al.* |
| THF | Styragel | 30 | Yamada *et al.* (1967) |

Table 49.2 summarizes the operating conditions employed in the GPC analysis of PMMA.

Lyngaa-Jorgensen studied poly(vinyl chloride) (PVC) by GPC on different combinations of Styragel columns and was able to confirm experimentally the hypothesis that PVC prepared by heterogeneous polymerization has a distribution curve close to the "most probable" curve. Pedersen and Lyngaa-Jorgensen examined the molecular-weight distribution of PVC polymerized in suspension. Rudin and Benshop-Hendrychova verified the applicability of GPC to the measurement of molecular-weight averages of PVC by comparing the results with those obtained by osmometry and light scattering. Chan proposed an empirical procedure for the interpretation of GPC data on PVC with regard to chromatogram skewness due to the solute concentration effect and found that the number- and weight-average molecular weights, as determined by GPC, agreed with the corresponding results obtained by conventional methods. Rohn found a good correlation between the first moment of the elution curve of non-fractionated PVC and its intrinsic viscosity. Abdel-Alim and Hamielec developed a novel technique, based on GPC measurements, for the determination of molecular aggregation in solutions of PVC and its dependence on the tacticity of the polymer.

Table 49.3 summarizes the operating conditions employed by these investigators in the GPC analysis of PVC.

Fritsche used Sephadex G-75 and G-200 in an attempt to fractionate polyacrylonitrile by GPC.

TABLE 49.3
OPERATING CONDITIONS FOR THE GEL PERMEATION CHROMATOGRAPHY OF POLY(VINYL CHLORIDE)

| Solvent | Packing | Temperature (°C) | References |
|---|---|---|---|
| THF | Styragel | 25 | Rohn |
| THF | Styragel | Ambient | Goedhart and Opschoor, Lyngaa-Jorgensen, Rudin and Benshop-Hendrychova |
| Benzene–THF (3:2) | Silica gel | Ambient | MacCallum |
| THF | Styragel | 30 | Chan |
| THF | Porous glass and Styragel | 25 | Abdel-Alim and Hamielec |

## RUBBERS

Several workers tested the applicability of GPC to the determination of molecular weights and molecular-weight distributions of elastomers. Adams *et al.* compared the results of GPC on polybutadiene with conventional fractional precipitation and also with osmometry and light scattering; they found that GPC indicates systematically higher polydispersity. Gamble *et al.* performed a similar analysis and obtained comparable results for butyl rubber. Cantow *et al.* (1967b) found good agreement between GPC and gradient elution in a study of broad- and narrow-distribution polyisobutenes in the molecular-weight range $2 \cdot 10^3 - 1 \cdot 10^6$. MacCallum found comparable efficiency between GPC and fractional precipitation in the fractionation of polychloroprene. Cantow *et al.* (1967a) studied the effect of temperature on the GPC separation of fractions of polyisobutene. Funt and Hornof discussed the criteria for establishing a simplified universal GPC calibration for poly-(1,2-butadiene) on the basis of polystyrene fractions.

Dawkins *et al.* used polyisoprene as a model substance to test the validity of a universal GPC calibration based on unperturbed dimensions of the polymer coil in solution. Runyon proposed a direct calibration procedure, employing dual detectors (UV and differential refractometer), especially suited for low-molecular-weight polybutadienes prepared by initiation with the α-methylstyrene tetramer–disodium compound. Cooper and Johnson (1971) employed polyisobutene fractions in studying the effect of pore-size distribution on the separation efficiency of porous glass beads. Boháčková *et al.* analyzed fractions of cyclic polyisoprene, obtained by a modified column precipitation technique, by means of GPC in order to test the efficiency of the precipitation method.

Hsieh and McKinney, using GPC on Styragel with 1,2,4-trichlorobenzene as solvent at 130°C, found that a single master curve describes the correlation of polymer polydispersity (expressed as the ratio of weight- and number-average molecular weights determined by GPC) with the kinetic ratio $k_i/k_p$ (where $k_i$ and $k_p$ are the rate constants of initiation and propagation, respectively) for anionically polymerized polystyrene, polybutadiene and polyisoprene. Nishida *et al.* used GPC on Styragel in THF at 25°C to measure the molecular-weight distribution in samples of polyisobutene.

Kraus and Stacy proposed a method for determining long-chain branching in polybutadiene, based on preparative fractionation and subsequent analysis of the fractions by GPC, using a Styragel packing and THF as a solvent at ambient temperature, and by viscometry. Menin and Roux analyzed mixtures of high-molecular-weight polybutadiene ($\bar{M}_w$ = 4.5 · $10^5$) and low-molecular-weight polyisobutene ($\bar{M}_n = 10^3$) and found that GPC can be used for the quantitative determination of polymer blend compositions. Mate and Lundstrom determined the amount of oil extenders in stereospecific polybutadienes by GPC; with proper selection of column combinations, the oil and polymer can be separated completely. Thus, the molecular-weight distribution of elastomers and the oil content could be measured simultaneously.

Table 49.4 gives the operating conditions of GPC employed by various investigators for the analysis of elastomers.

Silicon-containing elastomers are discussed separately in the section Miscellaneous polymers (p. 1064).

TABLE 49.4
OPERATING CONDITIONS FOR THE GEL PERMEATION CHROMATOGRAPHY OF ELASTOMERS

| Elastomer | Solvent* | Packing | Temperature (°C) | References |
|---|---|---|---|---|
| Polybutadiene | THF | Styragel | 30 | Menin and Roux |
| | THF | Styragel | Ambient | Kraus and Stacy |
| | THF | Styragel | 20 | Funt and Hornof, Yamada *et al.* (1966a, b, c) |
| | TCB | Styragel | 130 | Hsieh and McKinney |
| | Toluene | Styragel | 70 | Mate and Lundstrom |
| Polyisobutylene | TCB | Styragel | 150 | Cantow *et al.* (1967c) |
| | TCB | Styragel | 35–150 | Cantow *et al.* (1967a) |
| | $CCl_4$ | Porous glass | 25 | Cooper and Johnson (1971) |
| Polyisoprene | TCB | Styragel | 130 | Hsieh and McKinney |
| | THF | Styragel | 42 | Crammond *et al.* |
| Natural rubber | TCB | Styragel | 130 | Harmon and Jacobs |

*THF = tetrahydrofuran; TCB = 1,2,4-trichlorobenzene.

## POLYOLEFINS

The fractionation and distribution analysis of polyolefin homopolymers and copolymers are among the most difficult problems in polymer chemistry. Polyolefins can be dissolved only at relatively high temperatures (above 100°C) and the interpretation of the results is usually complicated by peculiarities of their structure, such as long- and short-chain branching, heterogeneity and crystallinity. It is therefore not surprising that since the advent of gel permeation chromatography, much work has been devoted to the application of this promising method in this difficult field. It is now definitely established that Styragel packings can be used at temperatures well above 100°C, provided that the solvent is continuously purged with an inert gas and enough antioxidant is added to prevent degradation of the sample and column packing. Nevertheless, some loss of column efficiency is always observed and it is recommended that the columns should be recalibrated from time to time. Some workers therefore prefer porous glass or silica gel as a packing for GPC analyses at elevated temperatures, but most of the work reported so far has been carried out on styrene–divinylbenzene beads.

Nakajima (1966) compared GPC as a method for determining the molecular-weight distribution of polyethylene with the conventional fractional elution technique and found good agreement between the results of the two methods. He used 1,2,4-trichlorobenzene as solvent at 135°C on commercial Styragel packings. Columns with increased resolution for molecular weights above $10^6$ are necessary for the analysis of polyethylene, as its distribution curve is often very asymmetric and extends to the region of extremely high molecular weights (Nakajima, 1971a), well beyond the highest calibration standards available. A typical chromatogram of a polyethylene sample with a very broad molecular-weight distribution is shown in Fig. 49.3. In order to obtain reliable information about the molecular-weight distribution of linear polyethylene, Nakajima (1971a) recommended the construction and comparison of cumulative distribution curves, which are free of errors (arising from the uncertain calibration and resolution of GPC in the high-molecular-weight region) up to at least 95% cumulative weight. Nakajima (1971b) also reported on the reproducibility of GPC data on linear polyethylene with the same set of Styragel columns over a period of 2.5 years. In order to account for the change in column properties, a procedure was suggested in which a polyethylene sample with a broad distribution

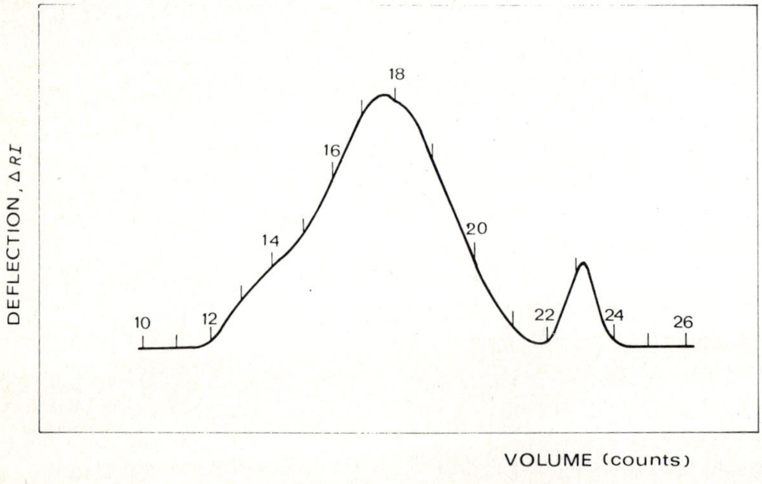


Fig. 49.3. Typical chromatogram of a polyethylene sample with a very broad molecular-weight distribution (Nakajima, 1966). Styragel packing with 1,2,4-trichlorobenzene as solvent, 1 ml/min, 135°C.

is used as a standard, which is run whenever necessary and the GPC set is recalibrated by plotting the cumulative fraction of the standard sample against the retention volume.

Crouzet *et al.* used silica beads (Spherosil) and 1,2,4-trichlorobenzene as solvent at 135°C in their study of the applicability of GPC to the distribution analysis of polypropylene. Good separations in the molecular-weight range $5 \cdot 10^3 - 1.5 \cdot 10^6$ were reported. The number- and weight-average molecular weights calculated from the GPC results were in reasonable agreement with the corresponding values calculated on the basis of gradient elution fractionation, although somewhat higher values of the ratio $\bar{M}_w/\bar{M}_n$ were found by the former method. A similar study on polypropylene with *o*-dichlorobenzene at 135°C and a Styragel packing has been described by Ogawa *et al.* By comparing the molecular-weight distributions obtained by GPC and gradient elution by means of calculated $\bar{M}_n$, $\bar{M}_w$, standard deviation, skewness and kurtosis, and also from the experimentally determined values of the number- and weight-averages, it was concluded that GPC is the more reliable of the two methods.

Much attention has been paid to the applicability of "universal calibration" concepts for polyolefins; the sharp fractions of polyethylene and polypropylene required for calibration are difficult to prepare, and therefore the possibility of calibrating GPC columns for the analysis of polyolefins by means of well characterized, commercially available polystyrene fractions is very attractive. Coll and Gilding discussed the theoretical basis of universal calibration based on the hydrodynamic volume theory; they were able to transform the calibration curve based on polystyrene fractions into curves for polyethylene and polypropylene; the agreement lay within the limits of experimental error. Williams and Ward also confirmed the possibility of using polystyrene calibration for polyethylene. The validity of universal calibration based on hydrodynamic volume was experimentally confirmed for polyethylene by many other investigators (Boni *et al.*, Holmström and Sörvig, Troth, Wild and Guliana, Williamson and Červenka). Dawkins and Maddock transformed the calibration curve determined by means of polystyrene fractions into a calibration curve for linear polyethylene by using (a) the hydrodynamic volume concept and (b) the unperturbed mean-square end-to-end distance as universal parameters. Both procedures predict correct calibration for polyethylene within the limits of experimental error for this case (*o*-dichlorobenzene at 138°C) where the two systems have similar polymer–solvent interactions.

Ross and Frolen examined carefully a standard polyethylene sample with a broad molecular-weight distribution by GPC at 135°C with 1,2,4-trichlorobenzene as solvent and Styragel columns, and discussed in detail the calibration procedure used. The reliability of the distribution curve was tested in the following way. The original sample was fractionated by gradient elution into 16 fractions, which were then analyzed by GPC and the distribution curves thus obtained were summed (after proper normalization with respect to the weights of the individual fractions). From the summed molecular-weight distribution, the number- and weight-average molecular weights were calculated as $\bar{M}_n$ = 17,400 and $\bar{M}_w$ = 53,300, in excellent agreement with the values of 18,300 and 53,100 obtained from the true experimental chromatogram of the parent sample.

Salovey and Hellman examined long-chain branching as a possible source of error in determining the molecular-weight distribution of commercial polyethylene samples. Ram and Miltz proposed a method for analyzing the polydispersity of branched polyethylenes by GPC with 1,2,4-trichlorobenzene as solvent at 130°C on Styragel packings. Another method for determining the molecular-weight distribution of branched polyethylene is due to Wild *et al.* (1971a, 1971b), who suggested a procedure in which fractions obtained by gradient elution (or some other preparative fractionation method) are analyzed by GPC.

The problem of evaluating long-chain branching in polydisperse polymers, especially polyolefins, has attracted much attention. In addition to the contributions discussed in Chapter 12 several papers have dealt with this question with special reference to low-density polyethylene (Cote and Shida, Miltz and Ram, Otocka *et al.*, Prechner *et al.*). Drott and Mendelson used their previously described method and presented evidence (based on combined GPC and viscosity measurements) about the presence of long-chain branches even in the so-called "linear" high-density polyethylene.

Holmström and Sörvig used GPC on Styragel with 1,2,4-trichlorobenzene as solvent at 135°C for determining the changes in molecular-weight distribution, brought about by thermal degradation of low-density polyethylene in a controlled atmosphere. Ross and Casto (1967, 1968) recognized some limitations connected with the use of Styragel packings and a differential refractometer as the detector at the high temperatures needed in

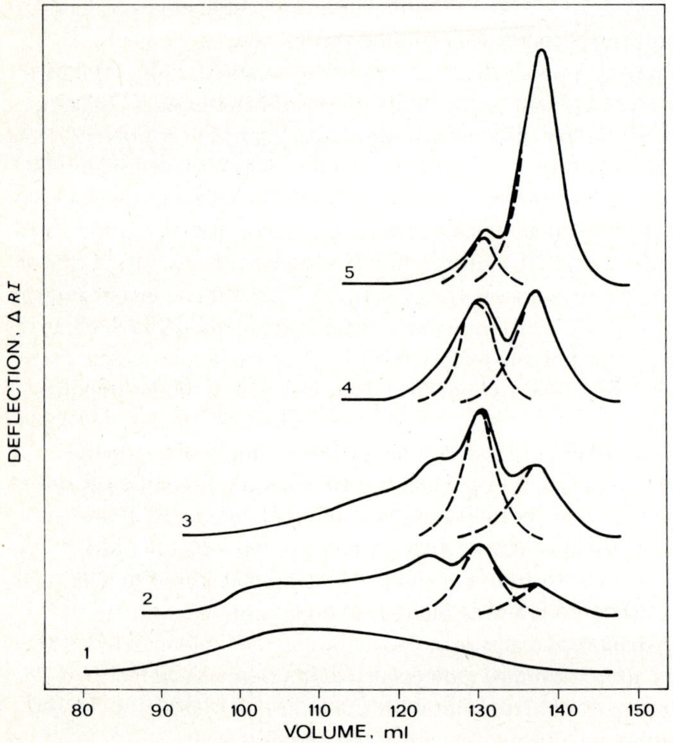


Fig. 49.4. Changes in the distribution of polyethylene single crystals, brought about by degradation in fuming nitric acid at 60°C, as reflected in the GPC chromatograms (Williams *et al.*, 1968a). Time of degradation (hours): (1) 0 (parent sample); (2) 10; (3) 24; (4) 50; (5) 126.

the analysis of polyolefins and recommended the use of porous glass beads as the column packing and an infrared detector for monitoring the eluent; they used perchloroethylene as solvent at 110°C on Bio-Rad porous glass packing.

An interesting application of GPC was found in the investigation of the morphology of polyethylene crystals. Blundell *et al.* degraded monolayer single crystals of polyethylene with fuming nitric acid, which caused chain scission at the basal surfaces of the crystal lamellae, and examined the degradation products by GPC with 1,2,4-trichlorobenzene as solvent at 135°C on a Styragel packing. A regular pattern was observed in the chromatograms, which showed several very sharp peaks; it was suggested that these correspond to single, double and multiple traverses of the polymer chain through the lamella. In a later paper, Williams *et al.* (1968a) investigated a series of polyethylene crystals of various types by this technique. In all instances, chromatograms were observed that corresponded to single and double traverses of the folding chains. Typical results are shown in Fig. 49.4, which shows the changes in the chromatograms with increasing time of exposure to fuming nitric acid at 60°C. The first two peaks were found to correspond to molecular weights of 1260 ± 90 and 2530 ± 150, respectively, and the third peak, which could be observed only after short treatment times, corresponds to a molecular weight of about 4500. The same group of investigators (Williams *et al.*, 1968b) then examined bulk polyethylene (see also Ward and Williams) and polyethylene fibrous crystals (Willmouth *et al.*) by this method. Williams *et al.* (1970) reported on the increased width of peaks in gel permeation chromatography of nitric acid-degraded polyethylene as compared with pure paraffins; this spreading was attributed to the polar groups attached to the chain ends.

The operating conditions employed by the various investigators in analyzing polyolefins by GPC are summarized in Table 49.5.

TABLE 49.5
OPERATING CONDITIONS FOR THE GEL PERMEATION CHROMATOGRAPHY OF POLYOLEFINS

| Polyolefin | Solvent* | Packing | Temperature (°C) | References |
|---|---|---|---|---|
| Polyethylene | TCB | Styragel | 130 | Miltz and Ram |
| | TCB | Styragel | 135 | Cote and Shida, Prechner *et al.*, Ross and Frolen, Salovey and Hellman |
| | TCB | Styragel | 140 | Wild *et al.* (1971b) |
| | DCB | Styragel | 130 | Williams *et al.* (1970) |
| | DCB | Styragel | 138 | Dawkins and Maddock |
| | DCB | Styragel | 140 | Troth |
| | PCE | Porous glass | 110 | Ross and Casto (1968) |
| | $C_{10}H_{12}$ | Silica gel | 125 | MacCallum |
| Polypropylene | TCB | Styragel | 135 | Coll and Gilding |
| | TCB | Silica gel | 135 | Crouzet *et al.* |
| | DCB | Styragel | 135 | Ogawa *et al.* |
| Poly(l-butene) | TCB | Styragel | 135 | Ring and Holtrup |

*TCB = 1,2,4-trichlorobenzene; DCB = *o*-dichlorobenzene; PCE = perchloroethylene; $C_{10}H_{12}$ = tetralin.

## POLYCONDENSATES

Hoare and Hillman, using methylene chloride as the solvent at room temperature, analyzed polycarbonates based on bisphenol A by GPC on Styragel; they calibrated the columns with prepared fractions of the same polymer and ascertained that GPC is a suitable method for the characterization of polycarbonates. Edwards published the results obtained in a study of the molecular-weight distribution of three epichlorohydrin–bisphenol A polymers. The values of $\bar{M}_n$ and $\bar{M}_w$ from GPC gave a good straight-line correlation with the results obtained by ebulliometry and light scattering. Biesenberger *et al.* (1971a), using a previously described method of recycling gel permeation chromatography (Biesenberger *et al.*, 1971b, Duvdevani *et al.*), studied the composition of epoxy resins of different origin.

Wagner and Greff analyzed resol type resins and examined the effect of catalyst type, reaction temperature and time on the total composition of the product by GPC on Styragel with THF as solvent at 25°C. Other studies on phenol–formaldehyde condensation products have been reported (Montague *et al.*, Quinn *et al.*, Yoshikawa *et al.*); tetrahydrofuran was the solvent of choice in these investigations.

Scholtan and Kranz compared the elution pattern of low-molecular-weight poly(propylene glycol)s in two solvents (THF and toluene) on a Styragel packing. Feist *et al.* separated the lower homologues of poly(ethylene glycol)s by gel permeation chromatography on Sephadex G-50 and G-100. Blanchard and Baijal measured the polydispersity of poly(propylene glycol)s and poly(ethylene glycol)s by determining the ratio $\bar{M}_w/\bar{M}_n$ by GPC and also by light scattering and osmometry, and found good agreement between the results.

Goebel and also Prince and Stapelfeldt discussed the problems associated with the use of *m*-cresol as eluent in the gel permeation chromatography of polyamides. Dark *et al.*

TABLE 49.6
OPERATING CONDITIONS FOR THE GEL PERMEATION CHROMATOGRAPHY OF POLYCONDENSATES

| Polycondensate | Solvent* | Packing | Temperature (°C) | References |
|---|---|---|---|---|
| Phenol–formaldehyde resins | THF | Styragel | 25 | Wagner and Greff |
| | THF | Styragel | Ambient | Montague *et al.*, Quinn *et al.* |
| | THF | Styragel | 50 | Yoshikawa *et al.* |
| Polyamides | *m*-Cresol | Styragel | 135 | Goebel |
| | TFE | Styragel | Ambient | Dark *et al.* |
| | HMPT | Styragel | 85 | Panaris and Pallas |
| Polyesters | *m*-Cresol | Styragel | 120 | Moore *et al.*, Shaw |
| | THF | Styragel | 37 | Billmeyer and Katz |
| Polyglycols | THF; toluene | Styragel | Ambient | Scholtan and Kranz |
| | $H_2O$ | Sephadex | Ambient | Feist *et al.* |

*THF = tetrahydrofuran; HMPT = hexamethylphosphorotriamide; TFE = trifluoroethanol.

recommended trifluoroethanol as a solvent for use at room temperature with nylon 66. Ede developed a method for determining the molecular-weight distribution of nylon 6 by GPC on Styragel, using chlorobenzene–*m*-cresol (1:1) as eluent. Dudley, working with *m*-cresol at 130°C, characterized several samples of nylon 66 having different polydispersities and showed that the hydrodynamic volume is a universal calibration parameter for this system, so that columns can be calibrated for nylon 66 by means of polystyrene fractions. As shown by Panaris and Pallas, polyamides 11 and 12, which are not soluble in trifluoroethanol and show a very small refractive index difference in *m*-cresol, can be studied by GPC in hexamethylphosphorotriamide.

Table 49.6 summarizes the experimental conditions employed in the GPC analysis of various polycondensates.

## COPOLYMERS

In comparison with the numerous applications of GPC in the characterization of homopolymers, relatively few attempts have been made to employ this method for determining the molecular weights and their distribution in copolymers of various types.

In 1966, Benoit *et al.* and Grubisic *et al.* proved that the hydrodynamic volume concept of universal calibration is also valid for some copolymers. Crammond *et al.* analyzed three isoprene–styrene block copolymers by GPC on a Styragel packing in THF as solvent at 42°C and were able to prove that no homopolymer is formed under the copolymerization conditions employed (low-temperature initiation by butyl-lithium) and that the copolymers prepared have a very narrow molecular-weight distribution. They also attempted to determine the heterogeneity in chemical composition by collecting samples of the effluent and analyzing them by UV spectrometry for the content of styrene units; insignificant differences in chemical composition were found between the front and back halves of the elution curve. Good agreement between the $\bar{M}_n$ values calculated from the distribution curve determined by GPC and measured directly by means of high-speed osmometry indicated the reliability of the GPC data.

Blanchard and Baijal, in an attempt to explain the discrepancy observed in number-average molecular weights measured by two independent methods (vapour pressure osmometry and end-group analysis of terminal hydroxyl groups) in tetrahydrofuran–propylene oxide copolymers, determined their molecular-weight distribution by GPC on Styragel in THF as solvent at 25°C and found that under the polymerization conditions employed, a low-molecular-weight species was formed that did not contain terminal hydroxyl groups.

Heller *et al.* studied 4-vinylbiphenyl–isoprene copolymers of the type ABA (prepared by coupling the AB-type living anions with phosgene) by GPC in toluene at 70°C on a Styragel packing. Fig. 49.5 shows the differences in the GPC chromatograms of (a) a single segment A (poly-4-vinylbiphenyl), (b) a segment of the type AB; and (c) a coupled ABBA product.

Williamson and Červenka included fractions of an ethylene–1-butene copolymer in their study of the validity of universal calibration of the Benoit *et al.* type for polyolefins. Kraus and Stacy applied a method of determining long-chain branching, based on a preparative-scale fractionation and subsequent analysis of the fractions by GPC, to poly(styrene–*co*–butadiene) rubbers.

*References p.1071*

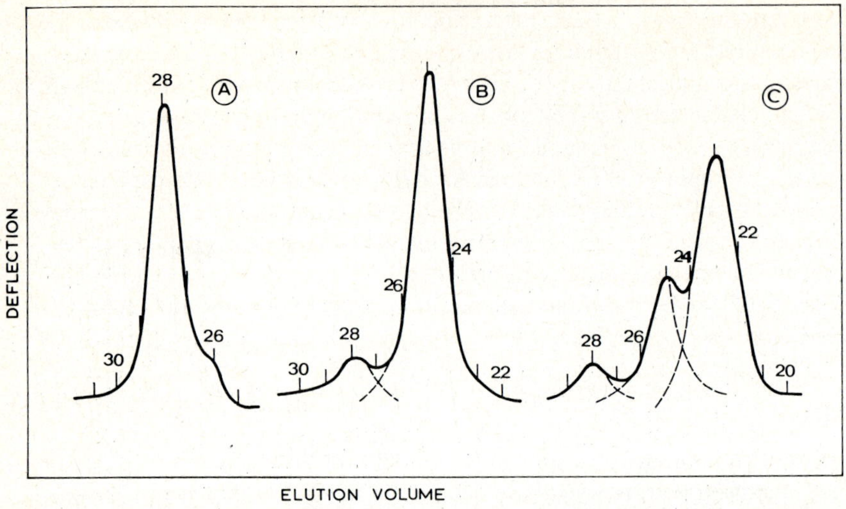


Fig. 49.5. GPC used in the control of the extent of coupling between two polymeric segments: (A) the starting segment A (poly-4-vinylbiphenyl); (B) the AB segment (poly-4-vinylbiphenyl–polyisoprene); (C) the final product, coupled ABBA block copolymer. Composition of the final product: homopolymer A, 4.7%; AB block, 26%; ABBA block, 69% (Heller *et al.*). Styragel packing ($1 \cdot 10^6$, $1.5 \cdot 10^5$, $3 \cdot 10^4$ and $8 \cdot 10^3$ Å), with toluene as solvent, 1 ml/min, 70°C.

Applications of gel permeation chromatography to the simultaneous determination of molecular-weight distributions and chemical heterogeneity in copolymers are briefly mentioned in Chapter 5.

## MISCELLANEOUS POLYMERS

### Silicones

Rodriguez *et al.*, using trichloroethylene as a solvent at room temperature, Styragel-packed columns and an infrared flow cell, studied dimethylsilicones and an equilibrium polymer prepared by potassium hydroxide-initiated polymerization of octamethyltetrasiloxane. Levin and Carmichael examined the distribution curves of polydimethylsilmethylenes prepared under different polymerization conditions by GPC on Styragel with toluene as eluent at ambient temperature. Kendrick compared the distributions of three polydimethylsiloxane samples, determined by GPC, with the results of fractional precipitation and gas–liquid chromatography. He employed Styragel columns at 20°C and toluene as the eluent and found very good agreement between the results of the different methods, provided that a correction for zone broadening was applied to the gel permeation chromatograms of narrow-distribution polymers. Andrianov *et al.* recommended GPC as a suitable method for controlling the composition of the reaction mixture during the polymerization reaction leading to high-boiling organosiloxanes.

**Cellulose derivatives**

Meyerhoff (1965) included cellulose nitrate, among other polymers, in his search for a universal calibration parameter for GPC. In a later paper (Meyerhoff, 1970), he found significant differences between cellulose nitrate and some vinyl polymers when the hydrodynamic volume parameter $[\eta] \cdot M$ was plotted against retention volume, measured on styrene–divinylbenzene beads in acetone. Muller and Alexander tested GPC on Styragel with THF as solvent at 25°C as a method for characterizing the chain-length distribution of cellulose nitrate. Brewer *et al.* and also Brewer and Tanghe found that fractional precipitation and GPC on Styragel with THF as solvent at ambient temperature yield comparable information on the molecular-weight distributions of various cellulose esters (including acetates, tricarbanilates and tripropionates). The two methods can be combined, GPC being used as an analytical method for the characterization of samples obtained by fractional precipitation; this combination is claimed to give considerably more information than either of the two methods used separately.

Segal *et al.* recognized the inadequacy of calculating molecular weights of cellulose derivatives on the basis of polystyrene calibration and extended chain-length, and used the universal calibration of the Benoit *et al.* type. Okunev *et al.* determined the molecular-weight distribution of secondary cellulose acetates by GPC on cross-linked polystyrene gels in acetone and tetrahydrofuran. Rinaudo and Merle examined by GPC on a Styragel packing with THF as solvent at ambient temperature the degradation of cellulose by the nitration solution and also the degradation of the cellulose backbone by the enzymatic attack of a cell-free cellulase. Ouano *et al.* compared the molecular weights of cellulose nitrates, measured by viscometry, with values calculated from gel permeation chromatograms on the basis of polystyrene calibration. The assumption that the extended chain-length is the proper link between the behaviour of the two polymers led to very poor agreement, whereas with the universal calibration based on the hydrodynamic volume, the calculated and experimental values agreed within 10% except for the highest molecular weights; these were determined, however, by extrapolating the linear calibration far beyond the value corresponding to the highest polystyrene standard available.

Altgelt (1965a) fractionated asphaltenes obtained from three asphalts of different origin and properties by GPC on columns packed with cross-linked polystyrene–divinylbenzene beads, using a benzene–methanol mixture as eluent at room temperature. A wide molecular-weight distribution was found in all of the samples studied, the molecular weight ranging from about 6000 to the upper limit of 40,000. Altgelt (1965b) also studied a series of asphaltenes and maltenes together with the parent asphalts by GPC under the same operating conditions.

Screaton and Seemann analyzed a polysulphone sample in four solvents to show the inadequacy of the extended chain-length as a parameter for converting GPC data into molecular-weight distributions. Ho-Duc *et al.* found that poly-2-vinylpyridine is irreversibly adsorbed on both Styragel and Porasil (porous silica gel) packings if THF is used as the eluent. They were able to obtain reasonable chromatograms when THF was replaced with dimethylformamide at 50°C on Porasil packings, but a specific interaction of poly-2-vinylpyridine with the packing still interfered in the separation even in this highly polar solvent.

*References p.1071*

## OLIGOMERS

The gel chromatography of oligomeric compounds makes possible an elegant separation of mixtures that in most instances are insufficiently volatile to be analyzed by gas chromatography. The individual members of a homologous series of oligomers often differ only very slightly in their chemical properties, thus making separation by liquid–liquid chromatography very difficult. So far, a comparatively small number of papers have been published in this field, although the interest in the gel chromatography of oligomers has continued to increase during the last 3 years.

In contrast to the gel chromatography of high-molecular-weight compounds, a high separation efficiency and resolution of the system are considered to be important in the separation of oligomers, with the aim of attaining separations to the baseline, thus permitting the isolation and identification of the individual oligomers by independent methods. It has been demonstrated experimentally that the effects of the end- and side-groups are more pronounced during the elution of oligomeric compounds compared with polymers, which provides evidence for specific interactions of the compounds being separated with the gel. The differences in the form and size of the structural units of oligomers also play a considerably more important role in the formation of an average molecule conformation than is the case with polymer molecules. All these facts make the construction of a universal calibration curve very difficult, if not impossible. If, however, model compounds are used for calibration (for instance, isolated members of a homologous series of oligomers), gel chromatography allows a rapid and comparatively very precise analysis of even very complicated mixtures.

The gels used for the separation of oligomeric compounds differ considerably in their internal structures from those used in the gel chromatography of high polymers. They are always homogeneous and relatively densely cross-linked, their molecular-weight exclusion limit being roughly one third higher than the molecular weights of the highest separated members of the homologous series of oligomers. The use of homogeneous gels is limited, however, by their mechanical stability, because with decreasing density of the gel network the equilibrium degree of swelling increases in proportion, and the column easily becomes plugged during high-speed gel chromatography. For the gel chromatography of oligomeric compounds in an arrangement usual in the analyses of polymers, homogeneous gels with an exclusion limit up to 5000–6000 molecular-weight units are used with advantage. If oligomeric systems with higher molecular weights are to be separated, a semi-heterogeneous or macroporous gel must be applied.

The gel chromatography of oligomers has been systematically studied by Heitz and Kern, who used a series of oligophenylenes for the calibration of the packings. By means of these substances, they investigated the chromatographic properties of homogeneous gel structures. Heitz and Čoupek made use of the fact that, owing to the well defined rigid backbone of the individual homologues, it is possible to study in more detail the behaviour of the system as a function of the experimental conditions. Using experiments involving model systems, Čoupek and Heitz stressed the importance of the choice of a gel with an appropriate pore distribution for the optimum separation of oligomeric mixtures. The efficiency of separation of homogeneous gels is a function not only of the network density, but to a certain extent also of the ratio of the pore volume to the size of molecules of the compounds to be separated. If the effective volume of solutes approaches the volume of the accessible pores, the separation efficiency falls drastically. Recyclization with continuous concentration of the eluate was used by Heitz and Ullner and this method separated oligomeric mixtures in gram amounts successfully. As for the efficiency and degree of resolution, the latter procedure can be compared with the best analytical techniques. The optimum resolution is achieved after triple recyclization of the eluate. The use of findings obtained in the optimization of conditions for preparative gel chromatographic separations enabled Heitz *et al.*, on using styrene–divinylbenzene (2%) copolymer, to achieve a complete separation up to $P_n = 15$ ($P_n$ is the number-average degree of polymerization). The model mixtures used were oligomeric polystyrenes (Fig. 49.6), poly(methyl methacrylate)s (Fig.49.7), polyglycols and several types of surfactants.

Bömer *et al.* used the perfect separation technique with discontinuous homologous series of oligomers of poly(ethylene glycol)s obtained by condensation of triethylene glycol and nonaethylene glycol for the isolation and characterization of the homologues up to $P_n = 45$. A polystyrene gel cross-linked with 2% divinylbenzene was used as packing. Up to 5 g of the reaction mixture were separated on a 200 X 5 cm column during a single elution with tetrahydrofuran. As higher oligo(ethylene glycol)s are soluble in tetrahydrofuran only to a limited extent, their separation was carried out on Merckogel OR-20,000 in methanol. The latter arrangement enabled oligo(ethylene glycol)s to be separated up to $P_n = 72$.

Polystyrene gels cross-linked with divinylbenzene were used by Barson and Robb in an investigation of the mechanism of photoinitiated reactions of bromotrichloromethane and carbon tetrachloride with styrene and of bromotrichloromethane with methyl methacrylate. The telomeric compounds arising during the reaction were followed quantitatively by gel chromatography in tetrahydrofuran, owing to the good resolution of the gel. The authors determined the transfer constants for the first few polymerization steps and suggested the possibility of the gel chromatographic separation of telomers on a preparative scale.

Hammond *et al.* used a standard set of columns of the Waters Ass. Model 200 gel chromatograph for the separation of cyclic oligomers arising as side-products during the polymerization of propylene oxide, 1,2-butene oxide and *n*-propyl glycidyl ether and during the copolymerization of these monomers with tetrahydrofuran. A bimodal distribution formed by cyclic and linear oligomers could be observed. The separation efficiency of the chosen system of columns was not too high for work in the oligomeric region, so that the results could be given only a more or less qualitative evaluation.

A mixture of oligostyrenes prepared by cationic polymerization was separated by Braun and Meier into individual oligomers by gel chromatography on a gel obtained by copolymerization of methyl methacrylate with ethylene dimethacrylate. The separations were followed quantitatively by gas chromatography of the eluate. The structure of the separated oligomers was determined by means of other analytical methods.

Modified dextran and acrylamide gels proved to be excellent materials for the separation of oligomeric mixtures in organic solvents. Mulder and Buytenhuys developed a technique for the separation of hydrocarbons and triglycerides on Bio-Beads SX-1 and SX-2 gels in benzene and achieved an outstanding separation of further low-molecular-

*References p.1071*

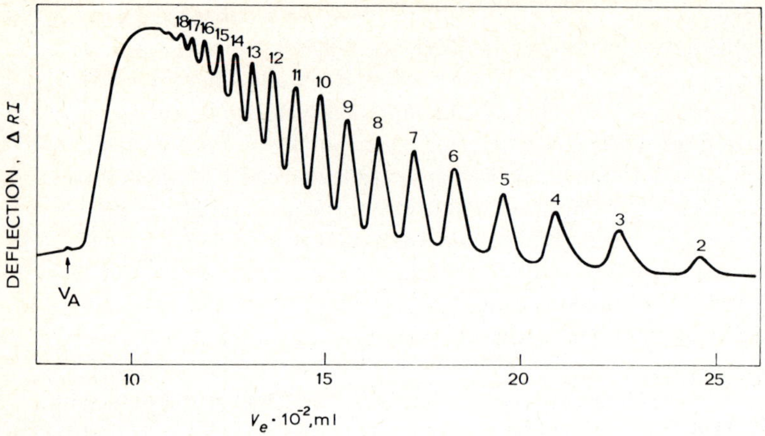


Fig. 49.6. Gel chromatographic separation of oligostyrenes (Heitz *et al.*). The numbers over the peaks correspond to the degree of polymerization. Column, 200 × 5 cm; flow-rate, 200 ml/h; solvent, tetrahydrofuran; $V_e$ = elution volume.

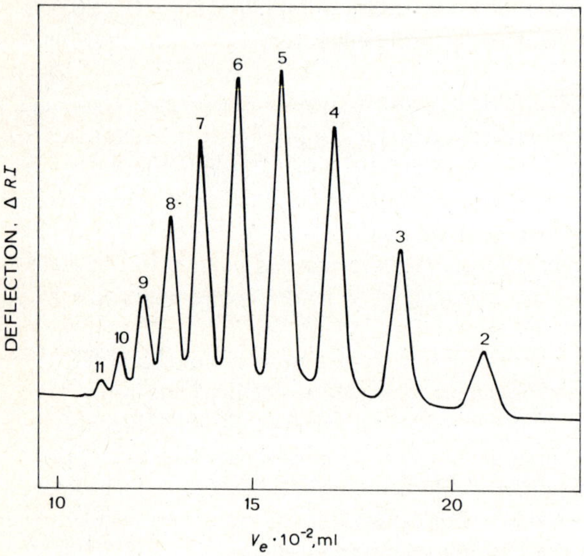


Fig. 49.7. Gel chromatographic separation of oligomeric methyl methacrylates (Heitz *et al.*). Column, 200 × 5 cm; flow-rate, 200 ml/h; solvent, tetrahydrofuran; $V_e$ = elution volume.

weight compounds on Bio-Beads SX-8 gel in the same solvent. By using Sephadex LH-20, a very good separation was attained in methanol for cyclic oligomers extracted from nylon 6 (Fig. 49.8). A series of three columns was used for quantitative analysis and the complete separation of all oligomers present in the extract was obtained. Similarly, excellent resolutions of the individual compounds of the mixture were obtained in the analyses of oligomeric esters.

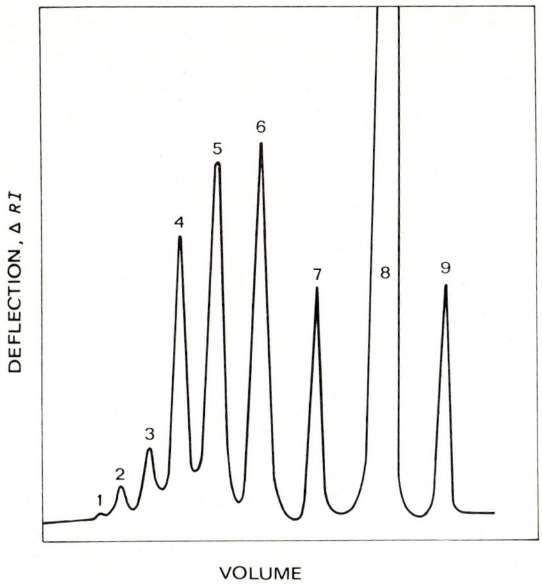


Fig. 49.8. Separation of cyclic oligomers of caprolactam (Mulder and Buytenhuys). Sephadex LH-20 in methanol; two columns. Sample application, 30 mg; flow-rate, 30 ml/h. 1–7, octamer up to dimer; 8, caprolactam; 9, water.

Mori and Takeuchi (1970a) determined monomers and oligomers extracted from nylon 6 and nylon 66 with ethanol. The materials used in the separation were Sephadex G-15 and G-25 and Bio-Gel P-4 with 0.1 *N* hydrochloric acid as the eluting agent, with spectrophotometric detection at 210 nm. The linear dependence of the elution volumes of the individual components on the logarithm of their molecular weights was confirmed for cyclic oligomers of nylon 6. The deviation from linearity observed for the cyclic dimer of nylon 66 was explained by its interaction with the solvent. The linear oligomers were separated by Mori and Takeuchi (1970b) in the form of their 2,4-dinitrophenyl derivatives by gel chromatography on Sephadex LH-20 in 0.05 *N* hydrochloric acid–methanol solution. If the contribution of the dinitrophenyl group is subtracted when calculating the molecular weight of the individual modified oligomers, the linear dependence of the elution volume is obtained once again. The cyclic monomers and oligomers do not interfere in the separation and determination of linear monomers and oligomers during detection in the range 370–450 nm.

Gel chromatography of oligomers is widely used in the analysis of complex systems that arise during the preparation of various resins by polycondensation or polyaddition reactions. Considerable attention has been devoted to the study of the products of the reaction of phenol with formaldehyde. Resols formed by alkali-catalyzed condensation were analyzed by Wagner and Greff by gel chromatography. The low-molecular-weight fractions consisting of mono- and binuclear methylolphenols were separated into individual compounds and identified by means of model compounds. As a differential refractometer was used for detection, quantitative analysis was rendered difficult because of the different refractive index values of the different compounds present in the resol. Gel chromatography was used together with NMR spectrometry in order to demonstrate the effect of various catalysts, temperature and reaction time on the final composition of the resol. Aldersley and Hope reported that during the catalytic methylolation of phenol in dilute aqueous solutions, the initial reaction rate is directly proportional to the concentrations of sodium hydroxide and formaldehyde. By integrating the rate equations for the methylolation of phenol, 2-methylphenol and 4-methylphenol obtained by gel chromatography, it was possible to simulate the experimental rate curves and to calculate the optimum values of the rate constants for all stages of the methylolation reaction. In this way, the unequivocal superiority of gel chromatography over the other methods so far used for studying the reactions of phenol with formaldehyde could be demonstrated.

The products of an acid-catalyzed reaction of formaldehyde with phenol (Novolacs) were analyzed by Montague *et al.* The first step of condensation proceeds by addition of formaldehyde to the 2- or 4-position of the phenol molecule, so that three different dimers can be formed that differ in their molecular dimensions. Various types of catalysts can, to a certain extent, affect the character of the dimer formed by condensation and which determines the final properties of the resin. The gel chromatograms of four resins obtained by means of different catalysts confirmed this finding. The weight- and number-average molecular weights of the Novolacs were also calculated from the chromatographic data.

The use of gel chromatography for the quantitative analysis of polyurethane oligomers was described by Kornbau and Ziegler. The detection was carried out refractometrically, and the individual fractions were identified spectrometrically. The polymerization of furfuryl alcohol followed by gel chromatography was described by Wewerka (1968a, b). A resin obtained by the condensation of furfuryl alcohol and containing a high portion of low-molecular-weight oligomers was divided into 12 compounds by Wallon *et al.* The separated compounds were characterized and used for the calibration proper that is required for the quantitative and qualitative description of the products of the condensation of furfuryl alcohol.

Biesenberger *et al.* (1971a) used recyclization gel chromatography to analyze epoxide resins. By using the effective carbon number defined by Hendrickson and Moore, they calibrated a gel chromatographic system consisting of a standard set of columns and compared the products of several important producers of epoxide resins.

Čoupek and Bouchal chose high-resolution gel chromatography for the analysis of the products of the Diels-Alder reaction of chloroprene. The separation of three isomeric chloroprene dimers in addition to the polymer was achieved on a styrene–divinylbenzene copolymer with a low value of the molecular-weight exclusion limit. The kinetics of the thermal transformations of chloroprene in the presence of free radical inhibitors gave some more data that enabled the reaction mechanism to be suggested (Bouchal *et al.*). In this way, gel chromatography enabled several elementary reactions to be measured simultaneously by a single analysis. The same gel packing was used by Hrabák and Čoupek for the analysis of a series of homologous oligomeric compounds arising by the reaction of diacyl peroxides with tertiary aromatic amines and for the determination of the mechanism of their formation.

Gel chromatography is important in the analysis of hydrocarbons and other components present in the products of the petrochemicals industry, as these compounds are similar in character, to a certain extent, to homologous series of oligomers. Oelert described the chromatographic behaviour of 40 hydrocarbons during their elution with cyclohexane, methylene chloride and isopropanol while using a vinyl acetate gel (Merckogel OR-500). He obtained a linear relationship between the logarithm of the molecular volume and the elution volume, but the curve was strongly influenced by the type of hydrocarbon and by the character of the mobile phase. Successful separations of complicated industrial mixtures have been reported by Cogswell *et al.*, Haley, Hillman, Hsieh *et al.* and other workers.

Evreinov *et al.* used ASK silica gel with methyl ethyl ketone as solvent for the separation of polyethylene glycol adipates with molecular weights of 370, 740, 980 and 2240, while alumina was applied successfully by Pyl and Wuntke to the separation of oligomeric polyethylene glycol terephthalates by means of the gradient technique. Schoellner and Hellwig separated polyesters modified with oil by using column and thin-layer chromatography. Sephadex LH-20 was used in the gel chromatographic separation of oligomeric model compounds from diphenyl carbonate and bisphenol A by Sotobayashi *et al.*

By using column chromatography of phenylpropionitrile oligomers, Janič *et al.* achieved not only a separation according to the molecular weight, but also the separation of *cis–trans* isomers. Calcium hydroxide and magnesium oxide were used as sorbents. In order to improve the flow-rate through the column, Kieselguhr was added to the calcium hydroxide (1:1, w/w). The elution was performed by using the gradient technique with benzene–ethanol.

## REFERENCES


Abdel-Alim, A. H. and Hamielec, A. E., *J. Appl. Polym. Sci.*, 16 (1972) 1093.

Adams, H. E., Farhat, K. and Johnson, B. L., *Ind. Eng. Chem., Prod. Res. Develop.*, 5 (1966) 126.

Aldersley, J. W. and Hope, P., *Angew. Makromol. Chem.*, 24 (1972) 137.

Alliet, F. D., *Appl. Polym. Symp.*, 8 (1969) 39.

Altgelt, K. H., *J. Appl. Polym. Sci.*, 9 (1965a) 3389.

Altgelt, K. H., *Makromol. Chem.*, 88 (1965b) 75.

Andrianov, K. A., Zhdanov, A. A., Zavin, B. G. and Sunekantz, T. I., *Vysokomol. Soedin., Ser. A*, 12 (1970) 20.

Bagby, G., Lehrle, R. S. and Robb, J. C., *Makromol. Chem.*, 119 (1968) 122.

Barson, C. A. and Robb, J. C., *Brit. Polym. J.*, 3 (1971) 53.

Benoit, H., Grubisic, Z., Rempp, P., Decker, D. and Zilliox, J., *J. Chim. Phys. Physicochim. Biol.*, 63 (1966) 1507.

Berger, K. C. and Schultz, G. V., *Makromol. Chem.*, 136 (1970) 221.

Berger, K. C. and Schultz, G. V., *Separ. Sci.*, 6 (1971) 297.

Biesenberger, J. A., Tan, M. and Duvdevani, I., *J. Appl. Polym. Sci.*, 15 (1971a) 1549.

Biesenberger, J. A., Tan, M., Duvdevani, I. and Maurer, T., *J. Polym. Sci., Part B*, 9 (1971b) 353.

Billmeyer, Jr., F. W. and Katz, I., *Macromolecules*, 2 (1969) 105.

Blanchard, L. P. and Baijal, M. D., *J. Polym. Sci., Part A-1*, 5 (1967) 2045.

Blundell, D. J., Keller, A., Ward, I. M. and Grant, I. J., *J. Polym. Sci., Part B*, 4 (1966) 781.

Bömer, B., Heitz, W. and Kern, W., *J. Chromatogr.*, 53 (1970) 51.

Boháčková, V., Poláček, J., Grubisic, Z. and Benoit, H., *J. Chim. Phys. Physicochim. Biol.*, 66 (1969) 207.

Boni, K. A., Sliemers, F. A. and Stickney, P. B., *J. Polym. Sci., Part A-2*, 6 (1968) 1579.
Bouchal, K., Čoupek, J., Pokorný, S. and Hrabák, F., *Makromol. Chem.*, 137 (1970) 95.
Braun, D. and Meier, W., *Kolloid-Z. Z. Polym.*, 224 (1968) 11.
Brewer, R. J. and Tanghe, L. J., *J. Polym. Sci., Part A-1*, 7 (1969) 1635.
Brewer, R. J., Tanghe, L. J., Bailey, S. and Burr, J. T., *J. Polym. Sci., Part A-1*, 6 (1968) 1697.
Cantow, M. J. R. and Johnson, J. F., *J. Appl. Polym. Sci.*, 11 (1967) 1851.
Cantow, M. J. R., Porter, R. S. and Johnson, J. F., *J. Polym. Sci., Part A-1*, 5 (1967a) 987.
Cantow, M. J. R., Porter, R. S. and Johnson, J. F., *J. Polym. Sci., Part C*, 16 (1967b) 13.
Cantow, M. J. R., Porter, R. S. and Johnson, J. F., *J. Appl. Polym. Sci.*, 11 (1967c) 335.
Chan, R. K. S., *Polym. Eng. Sci.*, 11 (1971) 152.
Cogswell, T. E., McKay, F. F. and Latham, D. R., *Anal. Chem.*, 43 (1971) 645.
Coll, H. and Gilding, D. K., *J. Polym. Sci., Part A-2*, 8 (1970) 89.
Cooper, A. R. and Johnson, J. F., *J. Appl. Polym. Sci.*, 13 (1969) 1487.
Cooper, A. R. and Johnson, J. F., *J. Appl. Polym. Sci.*, 15 (1971) 2293.
Cote, J. A. and Shida, M., *J. Polym. Sci., Part A-2*, 9 (1971) 421.
Čoupek, J. and Bouchal, K., *Makromol. Chem.*, 135 (1970) 69.
Čoupek, J. and Heitz, W., *Makromol. Chem.*, 112 (1968) 286.
Crammond, D. N., Hammond, J. M. and Urwin, J. R., *Eur. Polym. J.*, 4 (1968) 451.
Crouzet, P., Fine, F. and Mangin, P., *J. Appl. Polym. Sci.*, 13 (1969) 205.
Dark, W. A., Levangie, R. F. and Bombaugh, K. J., *6th International Gel Permeation Chromatography Seminar, Miami Beach, 1968*, Waters Ass., Framingham, Mass., 1968.
Dawkins, J. V., Denyer, R. and Maddock, J. W., *Polymer*, 10 (1969) 154.
Dawkins, J. V. and Maddock, J. W., *Eur. Polym. J.*, 7 (1971) 1537.
Determann, H., Lüben, G. and Wieland, T., *Makromol. Chem.*, 73 (1964) 168.
Drott, E. E. and Mendelson, R. A., *Polym. Prepr., Amer. Chem. Soc., Div. Polym. Chem.*, 12 (1971) 277.
Dudley, M. A., *J. Appl. Polym. Sci.*, 16 (1972) 493.
Duvdevani, I., Biesenberger, J. A. and Tan, M., *J. Polym. Sci., Part B*, 9 (1971) 129.
Ede, P. S., *J. Chromatogr. Sci.*, 9 (1971) 275.
Edwards, G. D., *J. Appl. Polym. Sci.*, 9 (1965) 3845.
Evreinov, V. V., Gerbin, V. I., Surynina, L. I. and Entelis, S. G., *Vysokomol. Soedin., Ser. A*, 12 (1970) 829.
Feist, W. C., Southerland, C. F. and Tarkow, H., *J. Appl. Polym. Sci.*, 11 (1967) 149.
Fritsche, P., *Faserforsch. Textiltech.*, 18 (1967) 21.
Funt, B. L. and Hornof, V., *J. Appl. Polym. Sci.*, 15 (1971) 2439.
Gamble, L. W., Westerman, L. and Knipp, E. A., *Rubber Chem. Technol.*, 38 (1965) 823.
Goebel, C. V., *4th International Gel Permeation Chromatography Seminar, 1967*, Waters Ass., Framingham, Mass., 1967.
Goedhart, D. and Opschoor, A., *J. Polym. Sci., Part A-2*, 8 (1970) 1227.
Goetze, K. P., Porter, R. S. and Johnson, J. F., *J. Polym. Sci., Part A-2*, 9 (1971) 2255.
Grubisic, Z., Rempp, P. and Benoit, H., *J. Polym. Sci., Part B*, 5 (1967) 753.
Haley, G. A., *Anal. Chem.*, 43 (1971) 371.
Hammond, J. M., Hooper, J. F. and Robertson, W. G. P., *J. Polym. Sci., Part A-1*, 9 (1971) 281.
Harmon, D. J., *J. Appl. Polym. Sci.*, 11 (1967) 1333.
Harmon, D. J. and Jacobs, H. L., *J. Appl. Polym. Sci.*, 10 (1966) 253.
Hattori, S. and Hamashima, M., *Kobunshi Kagaku (Chem. High Polym.)*, 27 (1970) 775.
Heitz, W., Bömer, B. and Ullner, H., *Makromol. Chem.*, 121 (1969) 102.
Heitz, W. and Čoupek, J., *J. Chromatogr.*, 36 (1968) 290.
Heitz, W. and Kern, W., *Angew. Makromol. Chem.*, 1 (1967) 150.
Heitz, W. and Ullner, H., *Makromol. Chem.*, 120 (1968) 58.
Heller, J., Schimscheimer, J. F., Pasternak, R. A. and Kingsley, C. B., *J. Polym. Sci., Part A-1*, 7 (1969) 73.
Hendrickson, J. G., *J. Appl. Polym. Sci.*, 11 (1967) 1419.
Hendrickson, J. G. and Moore, J. C., *J. Polym. Sci., Part A*, 4 (1966) 167.

Heufer, G. and Braun, D., *J. Polym. Sci., Part B*, 3 (1965) 495.
Hillman, D. E., *Anal. Chem.*, 43 (1971) 1007.
Hoare, H. C. and Hillman, D. E., *Brit. Polym. J.*, 3 (1971) 259.
Ho-Duc, N., Daoust, H. and Gourdenne, A., *Polym. Prepr., Amer. Chem. Soc., Div. Polym. Chem.*, 12 (1971) 639.
Holmström, A. and Sörvig, E., *J. Chromatogr.*, 53 (1970) 95.
Hrabák, F. and Čoupek, J., *Makromol. Chem.*, 145 (1971) 289.
Hsieh, B. C. B., Wood, R. E., Anderson, L. L. and Hill, G. R., *Anal. Chem.*, 41 (1969) 1066.
Hsieh, H. L. and McKinney, O. F., *J. Polym. Sci., Part B*, 4 (1966) 843.
Huang, R. Y. M. and Chandramouli, P., *J. Polym. Sci., Part B*, 7 (1969) 245.
Huang, R. Y. M. and Westlake, J. F., *J. Polym. Sci., Part B*, 7 (1969) 713.
Huang, R. Y. M. and Westlake, J. F., *J. Polym. Sci., Part A-1*, 8 (1970) 49.
Huang, R. Y. M., Westlake, J. F. and Sharma, S. C., *J. Polym. Sci., Part A-1*, 7 (1969) 1729.
Janič, N., Beneš, M. J. and Peška, J., *Makromol. Chem.*, 138 (1970) 99.
Kendrick, T. C., *J. Polym. Sci., Part A-2*, 7 (1969) 297.
Kornbau, N. D. and Ziegler, D. C., *Anal. Chem.*, 42 (1970) 1290.
Kraus, G. and Stacy, C. J., *J. Polym. Sci., Part A-2*, 10 (1972) 657.
Levin, G. and Carmichael, J. B., *J. Polym. Sci., Part A-1*, 6 (1968) 1.
Little, J. N., Waters, J. L. and Dark, W. A., *Polym. Prepr., Amer. Chem. Soc., Div. Polym. Chem.*, 12 (1971) 840.
Lyngaa-Jorgensen, J., *J. Polym. Sci., Part C*, 33 (1971) 39.
MacCallum, D., *Makromol. Chem.*, 100 (1967) 117.
Mate, R. D. and Lundstrom, H. S., *J. Polym. Sci., Part C*, 21 (1968) 317.
May, Jr., J. A. and Smith, W. B., *J. Phys. Chem.*, 72 (1968) 216 and 2993.
Menin, J. P. and Roux, R., *J. Polym. Sci., Part A-1*, 10 (1972) 855.
Meyerhoff, G., *Makromol. Chem.*, 89 (1965) 282.
Meyerhoff, G., *Polym. Prepr., Amer. Chem. Soc., Div. Polym. Chem.*, 8 (1967) 1295.
Meyerhoff, G., *J. Polym. Sci., Part C*, 21 (1968) 31.
Meyerhoff, G., *Makromol. Chem.*, 134 (1970) 129.
Miltz, J. and Ram, A., *Polymer*, 12 (1971) 685.
Montague, P. G., Peaker, F. W., Bosworth, P. and Lemon, P., *Brit. Polym. J.*, 3 (1971) 93.
Moore, L. D., Overton, J. R. and Rash, J., *6th International Gel Permeation Chromatography Seminar, Miami Beach, 1968*, Waters Ass., Framingham, Mass., 1968.
Mori, S. and Takeuchi, T., *J. Chromatogr.*, 49 (1970a) 230.
Mori, S. and Takeuchi, T., *J. Chromatogr.*, 50 (1970b) 419.
Mulder, J. L. and Buytenhuys, F. A., *J. Chromatogr.*, 51 (1970) 459.
Muller, T. E. and Alexander, W. J., *J. Polym. Sci., Part C*, 21 (1968) 283.
Nakajima, N., *J. Polym. Sci., Part A-2*, 4 (1966) 101.
Nakajima, N., *Separ. Sci.*, 6 (1971a) 275.
Nakajima, N., *J. Appl. Polym. Sci.*, 15 (1971b) 3089.
Nishida, N., Salladay, D. G. and White, J. L., *Polym. Prepr., Amer. Chem. Soc., Div. Polym. Chem.*, 12 (1971) 522.
Oelert, H. H., *J. Chromatogr.*, 53 (1970) 241.
Ogawa, T., Suzuki, Y. and Inaba, T., *J. Polym. Sci., Part A-1*, 10 (1972) 737.
Okunev, P. A., Dorofeev, C. P., Kundryavtzeva, Z. A. and Tarakanov, O. G., *Vysokomol. Soedin., Ser. A*, 13 (1971) 1680.
Otocka, E. P., Roe, R. J., Hellman, M. Y. and Muglia, P. M., *Polym. Prepr., Amer. Chem. Soc., Div. Polym. Chem.*, 12 (1971) 274.
Ouano, A. C., Broido, A., Barrall, E. M. and Javier-Son, A. C., *Polym. Prepr., Amer. Chem. Soc., Div. Polym. Chem.*, 12 (1971) 859.
Panaris, R. and Pallas, G., *J. Polym. Sci., Part B*, 8 (1970) 441.
Pannell, J., *Polymer*, 13 (1972) 277.
Peaker, F. W. and Patel, J. M., *Appl. Polym. Symp.*, 8 (1969) 125.
Pedersen, H. L. and Lyngaa-Jorgensen, J., *Brit. Polym. J.*, 1 (1969) 138.

Prechner, R., Panaris, R. and Benoit, H., *Makromol. Chem.*, 156 (1972) 39.
Prince, L. A. and Stapelfeldt, H. E., *6th International Gel Permeation Chromatography Seminar, Miami Beach, 1968*, Waters Ass., Framingham, Mass., 1968.
Pyl, T. and Wuntke, K., *Plaste Kaut.*, 15 (1968) 274.
Quinn, E. J., Osterhoudt, H. W., Heckles, J. S. and Ziegler, D. C., *Anal. Chem.*, 40 (1968) 547.
Ram, A. and Miltz, J., *J. Appl. Polym. Sci.*, 15 (1971) 2639.
Rinaudo, M. and Merle, J. P., *Eur. Polym. J.*, 6 (1970) 41.
Ring, W. and Holtrup, W., *Makromol. Chem.*, 103 (1967) 83.
Rodriguez, F., Kulakowski, R. A. and Clark, O. K., *Ind. Eng. Chem., Prod. Res. Develop.*, 5 (1966) 121.
Rohn, C. L., *J. Polym. Sci., Part A-2*, 5 (1967) 547.
Ross, G. and Frolen, L., *J. Res. Nat. Bur. Stand. Sect. A*, 76 (1972) 163.
Ross, J. H. and Casto, M. E., *Polym. Prepr., Amer. Chem. Soc., Div. Polym. Chem.*, 8 (1967) 1278.
Ross, J. H. and Casto, M. E., *J. Polym. Sci., Part C*, 21 (1968) 143.
Rudin, A. and Benshop-Hendrychova, I., *J. Appl. Polym. Sci.*, 15 (1971) 2881.
Runyon, J. R., *Separ. Sci.*, 6 (1971) 249.
Salovey, R. and Hellman, M. Y., *J. Polym. Sci., Part A-2*, 5 (1967) 333.
Schoellner, R. and Hellwig, J., *Fette, Seifen, Anstrichm.*, 70 (1968) 770.
Scholtan, W. and Kranz, D., *Makromol. Chem.*, 110 (1967) 150.
Screaton, R. M. and Seemann, R. W., *Appl. Polym. Symp.*, 8 (1969) 81.
Segal, L., Timpa, J. D. and Wadsworth, J. I., *J. Polym. Sci., Part A-1*, 8 (1970) 3577.
Shaw, G., *7th International Gel Permeation Chromatography Seminar, 1969*, Waters Ass., Framingham, Mass., 1969.
Smith, W. B., May, J. A. and Kim, Ch. W., *J. Polym. Sci., Part A-2*, 4 (1966) 365.
Smith, W. B. and Temple, H. W., *J. Phys. Chem.*, 72 (1968) 4613.
Sotobayashi, H., Lie, S. L., Springer, J. and Ueberreiter, K., *Makromol. Chem.*, 111 (1968) 172.
Troth, H. G., *5th International Gel Permeation Chromatography Seminar, London, 1968*, Waters Ass., Framingham, Mass., 1968.
Unger, K., Vogel, K. and Kohlschütter, H. W., *Z. Naturforsch.*, B22 (1967) 8.
Wagner, E. R. and Greff, R. J., *J. Polym. Sci., Part A-1*, 9 (1971) 2193.
Wallon, S. B., Barrand, J. B. and Petro, B. A., *J. Chromatogr.*, 54 (1971) 33.
Ward, I. M. and Williams, T., *J. Macromol. Sci.*, B5 (1971) 693.
Wewerka, E. M., *Carbon (Oxford)*, 6 (1968a) 93.
Wewerka, E. M., *J. Appl. Polym. Sci.*, 12 (1968b) 1671.
Wild, L. and Guliana, R., *J. Polym. Sci., Part A-2*, 5 (1967) 1087.
Wild, L., Ranganath, R. and Ryle, T., *Polym. Prepr., Amer. Chem. Soc., Div. Polym. Chem.*, 12 (1971a) 266.
Wild, L., Ranganath, R. and Ryle, T., *J. Polym. Sci., Part A-2*, 9 (1971b) 2137.
Williams, T., Blundell, D. J., Keller, A. and Ward, I. M., *J. Polym. Sci., Part A-2*, 6 (1968a) 1613.
Williams, T., Keller, A. and Ward, I. M., *J. Polym. Sci., Part A-2*, 6 (1968b) 1621.
Williams, T., Udagawa, Y., Keller, A. and Ward, I. M., *J. Polym. Sci., Part A-2*, 8 (1970) 35.
Williams, T. and Ward, I. M., *J. Polym. Sci., Part B*, 6 (1968) 621.
Williamson, G. R. and Červenka, A., *Eur. Polym. J.*, 8 (1972) 1009.
Willmouth, F. M., Keller, A., Ward, I. M. and Williams, T., *J. Polym. Sci., Part A-2*, 6 (1968) 1627.
Yamada, S., Imai, S. and Kitahara, S., *Kobunshi Kagaku (Chem. High Polym.)*, 23 (1966a) 400.
Yamada, S., Imai, S. and Kitahara, S., *Kobunshi Kagaku (Chem. High Polym.)*, 23 (1966b) 577.
Yamada, S., Imai, S. and Kitahara, S., *Kobunshi Kagaku (Chem. High Polym.)*, 23 (1966c) 505.
Yamada, S., Kitahara, S. and Hattori, Y., *Kobunshi Kagaku (Chem. High Polym.)*, 23 (1966d) 683.
Yamada, S., Kitahara, S., Hattori, Y. and Konahara, Y., *Kobunshi Kagaku (Chem. High Polym.)*, 24 (1967) 97.
Yoshikawa, T., Kimura, K. and Fujimura, S., *J. Appl. Polym. Sci.*, 15 (1971) 2513.

*Chapter 50*

# Cells and subcellular particles

M. JUŘICOVÁ and Z. DEYL

## CONTENTS



## INTRODUCTION

In the previous chapters in this book, a number of specific aspects of liquid column chromatography has been outlined. However, at least one has not yet been discussed, namely the capability of this technique to separate intact cells and cell particles, a task that cannot be undertaken by any of the comparable chromatographic techniques. For subcellular particles and viruses, glass beads, cellulose, different types of gels and ion exchangers have been successfully applied by a number of investigators, while glass beads have been the most widely used in the chromatography of cells. The future in this respect presumably lies in affinity chromatography, a highly selective technique that has been discussed in more detail elsewhere (see the respective chapter). Most of the chromatographic separations of cells can be considered as adherence chromatography, *i.e.*, a specified type of cell is more or less selectively adsorbed on the bed used and eluted after removal of contaminating substances. However, the second step is frequently omitted and replaced with a batch procedure, during which the column bed is destroyed and adhering cells are extracted from the material and removed from the column. It is with this aspect that it is extremely hard to draw a border-line between strictly chromatographic procedures and batch techniques. Some of the batch operations may be easily adapted into a chromatographic procedure and for this reason they are mentioned in the following survey.

In general, the chromatographic procedures used for cells and subcellular particles are used for strictly preparative purposes. Some of these techniques resemble frontal analysis and can be seen to be related to chromatography, while others will perhaps seem unusual to a chromatographer.

These procedures are helpful in the preparation of particular types of cells and subcellular structures, and their present status simply reflects the diverse possibilities that are available to any rapidly developing branch of chromatography in its early days.

## RIBOSOMES

In general, the chromatography of ribosomes resembles to a certain extent the chromatography of nucleic acids. However, the size of these structures introduces some specific features into the chromatographic separations; ion-exchange chromatography on modified celluloses offers the best results, such as DEAE-cellulose and ECTHAM-cellulose (Peterson and Kuff, Stanley and Wahba), but purification on unmodified cellulose or chromatography on Bio-Gel can also be applied (Kliffen, Stenesh and Yang).

The purification of ribosomes on a DEAE-cellulose bed has been carried out for the preparation of ribosomes free from ribonuclease. In principle, this method is based on the fact that most monovalent cations at moderate concentrations (0.1–1.0 *M*) are capable of activating ribonuclease I, which, in its inactive form, is bound to ribosomes, with concomitant depolymerization of ribosomal RNA. In the presence of low concentrations of magnesium acetate, however, ribosomes are stable over a wide concentration range of ammonium chloride, which allows chromatographic separations on DEAE-cellulose without loss of ribosomal integrity. In 0.5 *M* ammonium chloride solution, ribonuclease I is dissociated from the other type of ribonuclease present in ribosomes, ribonuclease II, which is separated during DEAE-cellulose chromatography as it is eluted at lower concentrations of ammonium chloride than the ribosomes themselves. The first ribonuclease, ribonculease I, is removed by differential centrifugation in a preceding step.

Unmodified cellulose in combination with sand can be recommended for the isolation and purification of ribosomes from plant material, as a good separation from plant pigments is achieved (Kliffen).

### Chromatography on ECTHAM-cellulose (epichlorohydrin- and Tris-treated cellulose)

Peterson and Kuff introduced a special type of cellulose, ECTHAM-cellulose, into the chromatography of ribosomes. Into a chilled solution of 20 g of Tris and 60 g of sodium hydroxide dissolved in 175 ml of distilled water, 60 g of cellulose (100–230 mesh) were mixed. The dry, crumby mixture was immersed in an ice-bath for 30 min with occasional stirring, then 30 ml of epichlorohydrin were added in small portions and the reaction mixture was allowed to stand overnight in a fume-hood. Peterson and Kuff reported that the temperature of the mixture increased steadily and reached a maximum 2–3 h after the reaction has begun. A 500-ml volume of 2 *M* sodium chloride solution was added to the product, the excess moisture was removed by gentle filtration on a coarse sintered-glass filter and the filter cake was washed with 1 *N* sodium hydroxide solution, 1 *N* hydrochloric acid and 1 *N* sodium hydroxide solution, with additional water rinses between each wash. The modified cellulose was then decanted in 10 l of distilled water and the fine material was removed. The residue was filtered, washed with ethanol and filtered to dryness. Excess of ethanol was evaporated under vacuum in a rotating evaporator fitted with a coarse sintered-glass filter and heated to 40°C. Then 10 g of this product were immersed in 150 ml of 0.5 *N* sodium hydroxide solution, and after a few minutes, the mixture was diluted several times with water and allowed to settle. The sediment was filtered again on a coarse sintered-glass filter and all fines were discarded. The residue on the filter was washed free from alkali and then conditioned by washing it with two 100-ml portions and one 500-ml portion of the starting buffer. This suspension was poured into a 1.8 cm I.D. column. Pressure was applied immediately and gradually increased from 5 to 15 p.s.i. If a high pressure is applied from the beginning of the sedimentation, the column becomes clogged and must be repacked. The final column height was 24 cm. Up to this stage, this column could be operated at room temperature, but henceforth it had to be transferred to a cold-room and maintained below 5°C. Here it was first washed with an additional portion of chilled starting buffer, consisting of 0.625 *M* Tris, 0.005 *M* in magnesium chloride with the pH varying from 7.4 to 7.5. Elution was carried out with a salt gradient starting with zero concentration in the starting buffer and ending with 2 *M* saline in the limiting buffer. Each separation was terminated with a wash with 0.1 *M* trisodium orthophosphate solution. Regeneration was carried out with water. When the trisodium orthophosphate had been removed, starting buffer was pumped on to the column again until it was equilibrated. The progress of washing and equilibration was followed by conductivity and pH measurements. The columns were operated under 15 p.s.i. over-pressure. Samples were applied as suspensions in the starting buffer and the effluent was monitored at 260, 280 and 415 nm.

### Chromatography on Bio-Gel A-15

A procedure for the chromatographic separation of ribosomes by gel filtration was worked out by Stenesh and Yang. Ribosomes, obtained in this particular instance from *Bacillus licheniformis*, were suspended in 0.01 *M* Tris buffer of pH 7.0 that was 0.01 *M* in magnesium acetate and 0.06 *M* in ammonium chloride. Before use, this buffer was made 0.006 *M* in spermidine and 0.006 *M* in 2-mercaptoethanol. This buffer was also used for elution (buffer I). Alternatively, another buffer consisting of 0.005 *M* Tris of pH 7.4 that was 0.002 *M* in magnesium acetate was used for elution (buffer II). Chromatography was carried out with a 100 × 2.5 cm column. The column was equipped with a flow adaptor and was packed with Bio-Gel A-15 (100–200 mesh) (Bio-Rad Labs., Richmond, Calif., U.S.A.) and *ca.* 0.4–0.8 ml of ribosome solution (with an absorbance of about 300 at 260 nm) was applied to the top of the column. The column was maintained thermostattically at 4°C, and measurement of the optical density at 260 nm was the most suitable detection method. The flow-rate was 30 ml/h. A typical example of this separation is presented in Fig. 50.1.

## VIRUSES

For the preparative chromatography of viruses, cellulose, hydroxyapatite and gel filtration with Sephadex G-25 can be used. As viruses cover a whole category of particles,

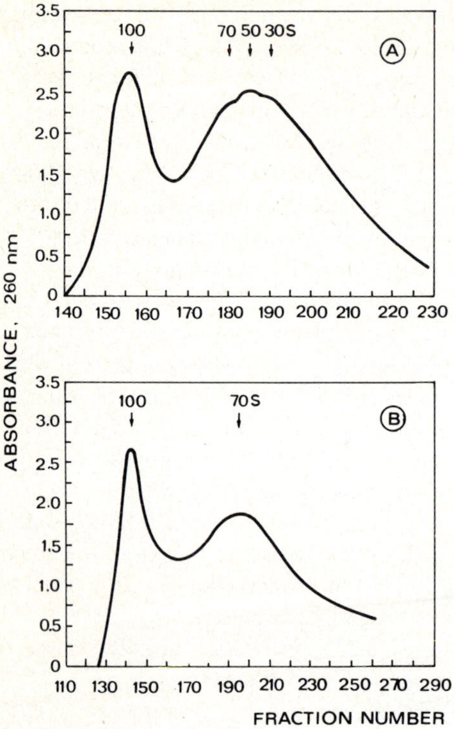


Fig. 50.1. Gel filtration fractionation of ribosomes (Stenesh and Yang). (A) buffer I, flow-rate 39 ml/h; (B) buffer II, flow-rate 30 ml/h.

it is difficult to outline general rules for their chromatography; detailed separation procedures depend on the nature of the particular virus to be separated. Therefore, different chromatographic procedures are mentioned without making a special attempt to generalize.

Chromatography on cellulose for preparative purposes was introduced by Venekamp and Mosch and used for the purification of potato virus X, potato virus Y, tobacco mosaic virus and potato stem mottle virus. For virus extraction, a two-step procedure was applied. Cellulose (10 g) was mixed with 5% polyethylene glycol that contained 0.05% Dextran 500 (Pharmacia, Uppsala, Sweden; viscosity number *dl/g* 0.48), 4.5% glucose, 2% sodium chloride, 0.004 *M* in magnesium chloride and 0.01 *M* in phosphate buffer of pH 7.0, to form a slurry. This slurry was poured into a chromatographic column 3 cm wide to make a bed 5 cm high located between sand layers 2 cm high. Alternatively, it was also possible to use buffers of the same composition in which the concentration of sodium chloride was decreased to 0.1%. The columns were jacketted and the separation was carried out at 2°C.

Homogenate prepared from fresh potato leaves (25 g in 70 ml of 0.5 *M* sodium acetate solution that was 0.004 *M* in magnesium chloride) was applied to the column and incubated for 30 min. During this period chloroplasts were precipitated. Clean sand was mixed with the chloroplasts until a flow-rate of 2 ml/min was achieved. The initial brown effluent was monitored colorimetrically. Elution was carried out with the buffer system used for preparing the column until a volume of 250 ml had been used, and then the buffer was suddenly replaced with another buffer that contained no polyethylene glycol or sodium chloride and 150 ml of this modified mobile phase was passed through the column. Then, polyethylene glycol and salt were added to the effluent and the whole volume was passed through a second chromatographic column similar to the first. The viruses were again retained on the cellulose. The column was then eluted with a series of buffers in which the concentrations of polyethylene glycol, dextran, glucose, magnesium chloride and phosphate buffer of pH 7.0 were kept constant while the sodium chloride concentration was reduced step-wise from 2.0 to 1.5, 1.0, 0.5, 0.1 and 0.0 *M*; 100-ml portions of each of these solvents were used. These percolations were continued by passage of 100-ml amounts of solvent mixtures in which the dextran, glucose, magnesium chloride and pH 7 phosphate buffer were held at a constant level while the polyethylene glycol concentration was reduced step-wise to 3 and 0%.

In a simplified version of this procedure, the homogenate was added on to the top of the column and incubated for 30 min before it was drained into the column. The brownish coloured effluent was removed by passing 250 ml of the initial buffer. The percolation

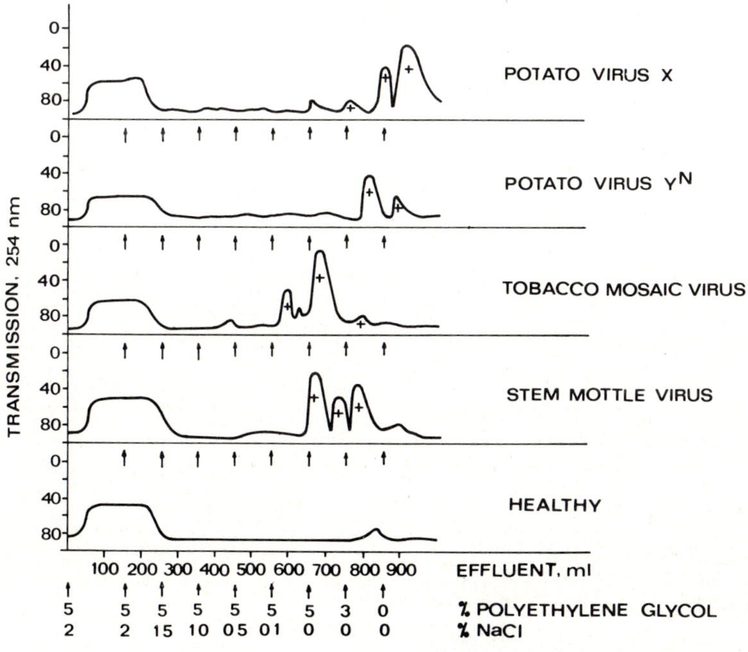


Fig. 50.2. Fractionation of viral nucleoproteins and other UV-absorbing substances from tomato plant homogenates on cellulose columns (Venekamp and Mosch). Nucleoproteins from homogenates were obtained by previous elution through cellulose with a solution containing 0.05% dextran, 4.5% glucose, 0.004 *M* in magnesium chloride and 0.01 *M* in phosphate buffer of pH 7. Cellulose column: 10 g dry weight, 5 × 3 cm. The percentages of polyethylene glycol and sodium chloride in the eluting solvent are indicated by arrows. The absorption of the effluents was recorded on an LKB Uvicord absorptiometer at 254 nm. The presence of an infectious virus is indicated by a plus sign. Corresponding fractions from healthy tomato are recorded in the bottom graph.

was continued by passage of 200-ml portions of solvent mixtures in which the dextran, glucose, magnesium chloride and pH 7 phosphate buffer concentrations were kept constant while the polyethylene glycol concentration was reduced stepwise to 3.2, 1 and 0%. The presence of viruses in the effluent was monitored by the UV absorbance at 254 nm and confirmed by a biological assay. A typical example of chromatographic profiles is presented in Fig. 50.2.

A chromatographic procedure suitable for the purification and isolation of S, M and L variants of *Mengo* virus was described by Scraba *et al.*, in which hydroxyapatite was used as the column packing. Methanol-precipitated virus-rich material was subjected to sonication, cleaved with α-chymotrypsin (enzyme concentration 0.8 mg/ml) and incubated for 45 min at 37°C. An equal volume of 0.2 *M* sodium pyrophosphate solution of pH 8.0 was added together with ribonuclease (0.08 mg/ml) and incubation was continued for an additional 30 min. The mixture was then chilled and spun at 7000 *g* for 30 min. The virus was sedimented at 100,000 *g* for 60 min and then re-suspended in a few millilitres of 0.02 *M* potassium phosphate buffer of pH 7.1 (for the S and M variants) or pH 6.0 (for the L variant). The mixture was then added to a 22 × 1.6 column of hydroxyapatite and eluted with a concave gradient of potassium phosphate solution of pH 7.1. Fractions of 5 ml

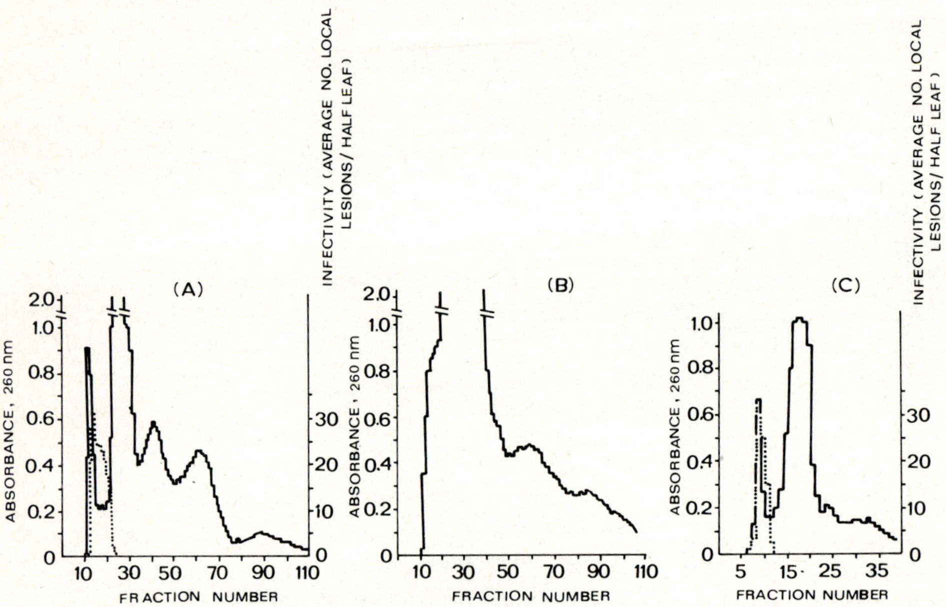


Fig. 50.3. (A) Elution diagram of the crude preparation of turnip mosaic virus chromatographed on a 40 × 2.5 cm column of Sephadex G-25 (medium mesh). Mobile phase, water; flow-rate, 100 ml/h; 4-ml fractions collected. Solid line, absorbance at 260 nm; broken line, infectivity. (B) Gel filtration of the crude preparation of turnip mosaic virus on Sephadex G-25 equilibrated with 0.1 *M* sodium chloride solution. Eluent, 0.1 *M* sodium chloride solution; flow-rate, 100 ml/h; 4-ml fractions collected. Absorbance at 260 nm plotted against volume of eluate (fraction number). (C) Chromatography of pooled fractions 12–22 from (A) on a 40 × 2.5 cm column of 4% agarose gel. Mobile phase, 0.05 *M* phosphate buffer (pH 7.5); flow-rate, 25 ml/h; 4-ml fractions collected. Solid line, absorbance at 260 nm; broken line, infectivity (Jacoli).

were collected and assayed for their optical density at 260 and 280 nm and infectivity by a special test.

For the purification of turnip mosaic virus, chromatography on Sephadex G-25 was used by Jacoli. Sephadex G-25 (medium mesh) was filled into a 40 × 2.5 cm column to give a total bed volume of 200 ml. The column was equipped with a flow adaptor and a cooling jacket. After the gel had been equilibrated with water, the sample of turnip mosaic virus was applied in small amounts using a peristaltic pump. The sample was eluted with distilled water at a flow-rate of 100 ml/h and the eluate was monitored at 254 nm. The location of the virus along the elution pattern was established in the following ways: (1) by measuring the double refraction flow in low and intermediate velocity gradients using a flow birefringence apparatus, and (2) serologically by a tube precipitin test. An example of this separation according to Jacoli is presented in Fig. 50.3.

## BACTERIOPHAGES

Several bacteriophages, namely $\varphi X_{174}$, M 13, M 12, Q$\beta$ and T 4, of *Enterobacteria* were effectively purified from crude extracts by chromatography on glass of controlled pore size. The advantage of this method is based on the fact that during the separation procedure no decrease in the toxicity and invasivity of the bacteriophage was observed. The method was developed by Gschwender *et al.*

Columns of dimensions 100 × 1 cm were filled with glass grains with a narrow pore-size distribution (average pore size 280 Å). The pore volume of the glass used averaged 0.7 ml/g. The column packing was poured in as an aqueous slurry and the column was made to settle by gentle vibration. The slurry was prepared by adding boiling water to dry glass of controlled pore size. With most samples of this glass, no evacuation appeared to be necessary, except when the starting material was not clean for some reason. The slurry was allowed to settle, decanted and re-suspended in deaerated water. After two to three decantations, the slurry was poured into the column and, as the glass grains settled down rapidly, excess of water was aspirated off from above. The bed was packed by vibration and the column was then ready for operation.

It was recommended that the column should be equilibrated by pumping several column volumes of water through the bed. If column back-pressure increases, which happens frequently shortly after the column has been set in operation, it is advised that the direction of flow of the water through the column should be reversed. Alternatively, it is also possible to fill the column with dry porous glass and then to pump water through the bed. It is believed that the former method of filling the column gives better results. The void volume of the column coincides with the exclusion volume of the tobacco mosaic virus suspension.

The column was used over a long period of time without the need for regeneration. If some contamination occurred, the column was rinsed by injecting 5–10 ml of dilute ammonia solution (1 part of concentrated ammonia solution plus 5 parts of water), which was flushed into the column with water (the addition of buffer at this stage should be avoided) until the effluent was free from ammonia. Columns of this type can be sterilized in an autoclave. Nitric acid can be used instead of ammonia for washing the column.

Another advantage of this procedure is that any buffer of pH 7.0 and above can be used as the mobile phase. The optimal flow-rate is about 2 ml/min and a peristaltic pump allowing a variation of flow-rate between 0.5 and 10.0 ml/min is usually used. The column effluent is monitored at 254 nm and special additional tests such as infectivity, expressed as plaque-forming units, were used for detection. Owing to the high flow-rates, no refrigeration of the column was necessary. The chromatographic system may be tested before operation by injecting 0.5 ml of 0.1% tobacco mosaic virus (void volume) and 1% benzyl alcohol (salt volume). A typical example of this separation is presented in Fig. 50.4.

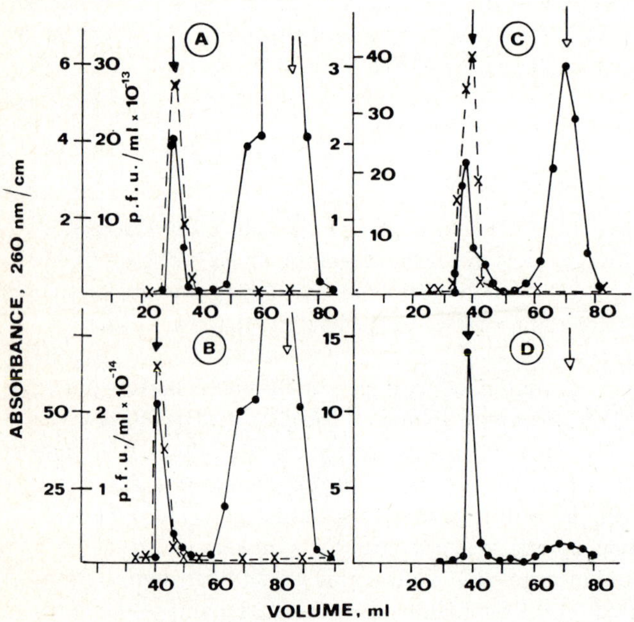


Fig. 50.4. Elution profiles of phage concentrates purified on columns containing glass of controlled pore size (Gschwender *et al.*). Flow-rate of eluent, 2 ml/min; fractions were collected every 1 or 2 min. Arrows mark exclusion (closed arrows) and salt (open arrows) volumes of the columns as determined by calibration with tobacco mosaic virus and benzyl alcohol. •——•, absorbance at 260 nm; X-----X, infectivity (p.f.u. = plaque-forming units). (A) 3 ml of $\varphi X_{174}$ phage concentrate, containing $3.5 \cdot 10^3$ infective particles per millilitre were purified on a controlled pore size (280 Å) glass column. (B) 2.5 ml of M 12 concentrate containing $2.5 \cdot 10^{14}$ infective particles per millilitre were chromatographed on a controlled pore size (240 Å) glass column. Q $\beta$ phage concentrates gave the same elution pattern on this column. (C) 4 ml of M 13 concentrate containing $6 \cdot 10^{13}$ infective particles per millilitre were purified on a controlled pore size (250 Å) glass column. (D) T 4 concentrate was chromatographed on a controlled pore size (250 Å) glass column.

## BLOOD CELLS

The separation of lymphocytes and polymorphonuclear neutrophils was carried out by Garvin using 16 × 1 cm columns filled with siliconized Superbrite glass beads No. 100

(3M Co., St. Paul, Minn., U.S.A.). The glass beads were siliconized using Dow Corning No. 1208 vapour phase siliconizing mixture. The vapour of mixed chlorosilanes was obtained by passing air through the above reagent. From the reaction flask, it was led through vigorously agitated glass beads for exactly 1 min, and flushed with air for a further 5 min. Finally, the bead was heated to 100°C for 30–60 min. The flow-rate was maintained by a motor-driven syringe and kept at 0.1 ml/min, but it can be increased up to 1 ml/min in specific experiments. For some experiments, the bed height was decreased to 4 cm. Recoveries of polymorphonuclear neutrophils and lymphocytes in successive millilitres of effluent blood are given in Table 50.1. The column in this instance was operated at 24–27°C.

TABLE 50.1
RECOVERIES OF POLYMORPHONUCLEAR NEUTROPHILS AND LYMPHOCYTES IN SUCCESSIVE MILLILITRES OF EFFLUENT BLOOD (GARVIN)

| Millilitre analyzed | Recovery in effluent blood (%) | |
|---|---|---|
| | Polymorphonuclear neutrophils | Lymphocytes |
| 1st | 15 ± 10 | 86 ± 15 |
| 2nd | 5 ± 4 | 79 ± 14 |
| 3rd | 5 ± 5 | 69 ± 16 |
| 4th | 4 ± 6 | 73 ± 38 |

Blood for the separation experiment was prepared as follows. The blood was drawn with sterile precautions and was gently squirted into a bottle containing 0.1 ml of heparin for each 10 ml of blood. It was mixed by swirling for 1 min and transferred into test-tubes (5 ml each portion). Suitable additions were made, the blood was incubated for 15 min at 37°C in corked tubes, mixed by inversion for 3 min, and 5-ml portions were applied to the column.

Lymphocytes can be also separated on antigen-coated plastic beads (Wigzeil and Mäkelä). This procedure, however, is a combination of column and batch separation and is mentioned here only for the sake of completeness. Columns of dimensions 80–150 X 1.5 cm were used at a flow-rate of 1–2 ml/min. Degalan V-26 particles (Deguss Wolfgang, Hanau am Main, G.F.R.) were used for antigen binding. The procedure was described in great detail by Wigzell and Andersson. Separation was carried out at 4°C and Eagle's minimal essential medium was used as the mobile phase. During the separation, about 50% of the cells applied to the column were retained for unspecified reasons. The retained cells were eluted by vigorous mechanical shaking of the extruded beads in a glass vessel. Attempts at specific elution with free hapten or complete immunogen were not always successful.

## CELLS FROM THE SPLEEN

Plotz and Talal modified the original procedure of Garvin and of Rabinowitz for the separation of splenic antibody-forming cells. Columns of dimensions 15 X 0.75 in. or 8.1 X 9/16 in. were used, filled to three quarters of their height with glass beads (Superbrite

*References p.1085*

glass beads, 3M Co., Type 100–5005). The column was filled with a glass bead slurry in fuming nitric acid and after 10 min which were allowed in order to clean and condition the glass filling of the column, the bed was washed until free from acid, for which 20–30 bed volumes of distilled water were needed. Air was aspirated through the column overnight and next morning the column was gently shaken so as to loosen the glass particles. The columns were operated at 37°C. Elution was carried out with Eagle's minimal essential medium with divalent cations (0.0018 *M* $Ca^{2+}$ and 0.001 *M* $Mg^{2+}$) and 10% foetal bovine serum albumin (first 90 ml), followed by 90 ml of the same medium without divalent cations (0.01 *M* EDTA and 10% foetal bovine serum albumin). A typical example of this separation is presented in Fig. 50.5.

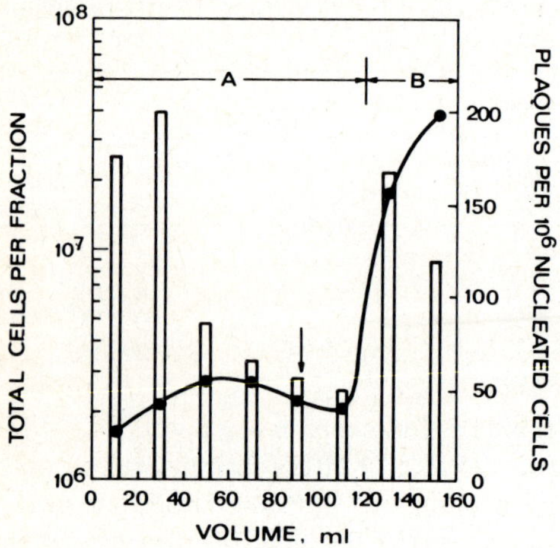


Fig. 50.5. Elution of a suspension of spleen cells from fine Fischer rats injected 4 days previously with sheep red cells with 90 ml of 10% foetal bovine serum albumin followed by 90 ml of the same buffer without divalent cations and 0.01 *M* EDTA, as indicated by the arrow (Plotz and Talal). The arrows at the top of the graph indicate the fractions pooled as A and B. The bars represent total cells per fraction, and the circles represent plagues per $10^6$ nucleated cells.

## BONE MARROW CELLS

The separation of different types of bone marrow cells was developed by Přistoupil *et al.* Cells were prepared in the basal Eagle's medium (BEM) and chromatographed on glass beads. Chromatographic columns were 30 × 1.3 cm in size and were maintained thermostattically at 37°C. Before use, glass beads, 0.2 mm in size, were washed either with boiling 65% nitric acid or with a mixture of potassium bichromate and sulphuric acid at 95°C for 20 min, and then rinsed until the pH of the effluent decreased to 6.0. The slurry was poured into a chromatographic column up to a height of 10 cm. Before use, the columns were washed with distilled water or, preferably, with 0.9% saline (200 ml). Then 30–50 ml of BEM were introduced. All mobile phases were adjusted to pH 7.1–7.2.

Then 1 ml of cell suspension (containing about $7.5 \cdot 10^6$ cells) was drained into the column at a rate about 1 ml/min · $cm^2$, BEM containing 10% of inactivated serum was

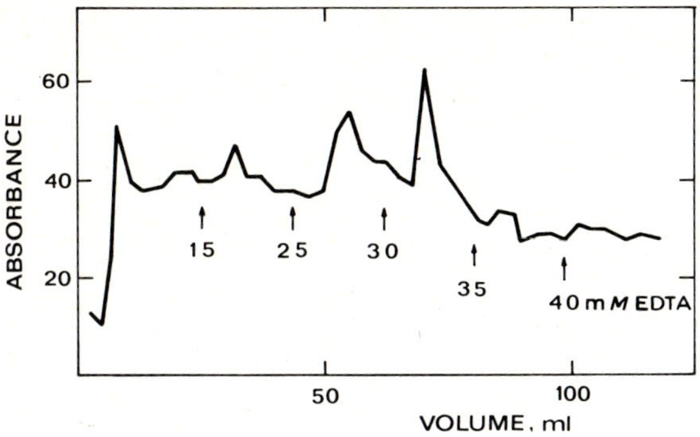


Fig. 50.6. Adherence chromatography of rat bone marrow cells (Přistoupil *et al.*) The 10 × 1.3 cm column of glass beads, 0.2 mm diameter, were treated with nitric acid, then washed with water followed by BEM; 75 · $10^6$ cells suspended in 1 ml of BEM containing 20% of calf serum were applied, pH 7.1–7.2, 37°C. Elution was carried out with BEM + 20% of calf serum containing an increasing concentration of EDTA.

applied in two 1-ml portions, the flow-rate was adjusted to 0.4–0.5 ml/min · $cm^2$ and fractions of 2–3 ml were collected. Then 17 ml of the same medium were added and passed through the column at a lower flow-rate until 10 ml of eluate had been collected. The flow-rate was then increased to the original level and was maintained at this level until the end of the experiment. Successive portions of BEM containing 10 ml of calf serum and 1.5–10 mmoles of EDTA were then added step-wise just at the moment when the previous portion had drained into the bed. All mobile phases were adjusted to pH 7.1–7.2 with 7.5% sodium hydrogen carbonate solution. The tip of the applicator (a 20-ml pipette) was bent upwards in order to avoid turbulence in the upper layer of the glass beads. The effluent was evaluated turbidimetrically. A typical example of this separation is presented in Fig. 50.6.

## REFERENCES

Garvin, J. E., *J. Exp. Med.*, 114 (1961) 51.
Gschwender, H. H., Haller, W. and Hofschneider, P. H., *Biochim. Biophys. Acta*, 190 (1969) 460.
Jacoli, G. G., *Biochim. Biophys. Acta*, 165 (1968) 299.
Kliffen, C., *J. Chromatogr.*, 53 (1970) 531.
Peterson, E. A. and Kuff, E. L., *Biochemistry*, 8 (1969) 2916.
Plotz, P. H. and Talal, N., *J. Immunol.*, 99 (1967) 1236.
Přistoupil, T. I., Fričová, V., Hrubá, A. and Kramlová, M., *J. Chromatogr.*, 67 (1972) 63.
Rabinowitz, Y., *Blood*, 23 (1964) 811.
Scraba, D. G., Hostvedt, P. and Colter, J. S., *Can. J. Biochem.*, 47 (1969) 165.
Stanley, Jr., W. M. and Wahba, A. J., *Methods Enzymol.*, 12 (1967) 524.
Stenesh, J. and Yang, K., *J. Chromatogr.*, 47 (1970) 108.
Venekamp, J. H. and Mosch, W. H. M., *Virology*, 23 (1964) 394.
Wigzell, H. L. R. and Andersson, B., *J. Exp. Med.*, 129 (1969) 23.
Wigzell, H. and Mäkelä, O., *J. Exp. Med.*, 132 (1970) 110.

*Chapter 51*

# Inorganic, coordination and organometallic compounds

F. JURSÍK

## CONTENTS



## INTRODUCTION

Chromatography was first applied in inorganic chemistry in 1937 by Schwab and Jockers, who separated cations on an alumina column. However, this method soon found a wider application after the discovery of partition chromatography by Martin and Synge. At present, there are several methods for the separation of cations and anions (for reviews, see Marcus and Kertes; Muzarelli; Nickless, 1967, 1968), the development of which is connected with the pioneer work of Lederer, Pollard, Burstall, Linstead, Wells and Lacourt (for reviews, see Nickless, 1967, 1968).

Chromatography has been used not only for analytical purposes, but also for the investigation of reaction mechanisms, the preparation of pure inorganic compounds, the study of the behaviour of inorganic species in aqueous solution and for solving stereochemical problems. The last aspect especially stimulated the application of chromatography in coordination chemistry, because since Linhard *et al.* separated geometrical isomers of $[Co(NH_3)_4(N_3)_2]^+$ on alumina, chromatography has proved to be a useful means for the study of problems in coordination chemistry (for reviews, see Druding and Kauffman, Garunchio and Strazza). Also, Lederer and colleagues showed the possibility of using chromatography for the study of metal complexes (see, for example, Kertes and Lederer).

This chapter includes the separation of cations and anions, but of the great number of applications that have been described only a few examples will be given here. These examples represent groups of cations the separation of which is somewhat difficult and which also serves as an example of a particular chromatographic technique. For these reasons, a general scheme for the separation of a multicomponent mixture of cations is also included (see p. 1094).

Coordination chemistry is represented here by stereochemical applications. The utilization of chromatography in synthetic coordination chemistry is described in the mentioned reviews.

## SIMPLE INORGANIC COMPOUNDS

For the separation of simple inorganic ions, several methods are applied, as described below. Their choice is influenced by the possible relationships between the solute, sorbent (ion exchanger) and eluent. If, for example, there are considerable size differences between the ions to be separated, gel chromatography or separation on inorganic ion exchangers can be used with advantage, because the sieving effect can be utilized. On the other hand, different charge densities of the ions will make ion-exchange or adsorption chromatography more suitable. However, in instances when these principal differences are negligible, methods in which several types of interactions (formation of chelates, extraction, etc.) occur, are preferable. In addition, one must take into account that during chromatography different interactions take place to different extents. For example, when the eluent contains an organic solvent, the retention of some cations on an ion exchanger can be altered owing to the extraction effect (Korkisch; Korkisch and Klakl, 1969). Further, typical sorbents such as alumina and silica gel behave simultaneously as weak ion exchangers with the possibility of forming hydrogen bonds, etc.

### Cations

*Gel permeation chromatography*

According to Saunders, cations are only weakly adsorbed on polyacrylamide gel, but interactions between the anions and amide protons of the gel occur. This effect was demonstrated by Egan, who observed that different elution volumes are necessary for the elution of sodium chloride and nitrate. The order of cations eluted on Bio-Gel P-2 (or P-100) is $K^+ > Na^+ > Li^+ > Mg^{2+} > Ca^{2+}$ (water as eluent), while from Sephadex G-10, $Mg^{2+}$ is eluted first followed by a mixture of $Li^+$, $Na^+$ and $Ca^{2+}$. For this reason, in addition to size differences among the ions, which appear to be the main determining factor, side-effects such as adsorption (Egan, Lindqvist), ion exclusion (Neddermeyer and Rogers, Pecsok and Saunders), restricted diffusion (Ackers) and ion exchange (Ortner and Pacher) must be also taken into account. Another phenomenon is the dependence of the elution volumes on the nature of the background electrolytes employed (Ueno *et al.*, 1970a).

For the elution of both anions and cations, water or aqueous solutions of different salts are used. Yoza and Ohashi found that elution with salt solutions does not lead to complete separation owing to the formation of tails. These tails can be depressed to a minimum by using an acidic eluent, which can change the order of elution of the ions. The eluent also affects the symmetry of the elution curves (Yoza and Ohashi), which is ascribed both to the multicomponent character of metal ions in solution and to the polyfunctional character of the gel.

As a typical example of application of gel chromatography, the following method for the separation of alkaline earths on Sephadex G-15 (Ueno *et al.*, 1970a) is described. For preparation of the column, Sephadex G-15 (particle size 40–120 $\mu$m) suspended in the eluent is allowed to swell for 2 days and undesirable fine particles are removed by decantation. A deaerated (under reduced pressure) suspension of the swollen gel is then poured into a column (60 × 1.5 or 90 × 1.5 cm) partially filled with the eluent and, after the gel bed has reached a height of about 5 cm, the outlet at the bottom is opened so as to allow the eluent to flow at a rate of approximately 30 ml/h. The addition of the gel is continued until the gel bed reaches the desired height. The bed volume is adjusted to 100 or 150 ml. In order to protect the surface of the gel from disturbance, a disc of filter-paper is placed on the top of the bed. After packing, 500 ml of the eluent are passed through the column in order to settle the gel bed.

The separation procedure is as follows. A 1-ml volume of a 0.01 *M* sample solution (chlorides) is placed on the column bed just as the layer of the eluent (0.1 *M* potassium chloride–0.01 *M* hydrochloric acid) is soaking into the bed. As soon as the solution has entered the bed, about 4 ml of the eluent are added and elution is started at a constant flow-rate of 25–35 ml/h. The effluent is collected in fractions of 1 ml with a drop count fraction collector (see Fig. 51.1). The amounts of the samples in the effluents are determined by EDTA titration.

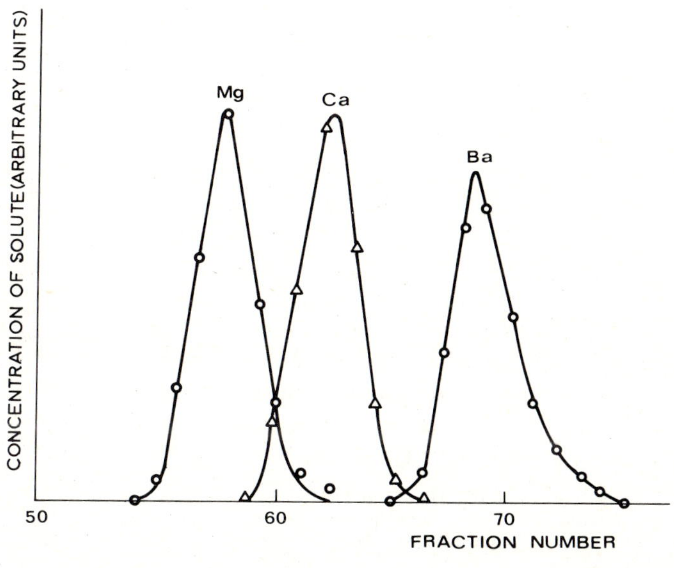


Fig. 51.1. Gel permeation chromatographic separation of magnesium, calcium and barium (Ueno *et al.*, 1970a).

*Ion-exchange chromatography*

In order to achieve efficient separations, the following general requirements must be met:

(1) the concentration of the sample applied on the column (this amount depends on the column capacity) should be low ($<0.2\,M$);

(2) the rate of elution (which depends on the particle size) to establish equilibrium should be low (0.2–2.2 ml/min · $cm^2$);

(3) the longer the column, the better is the quality of separation.

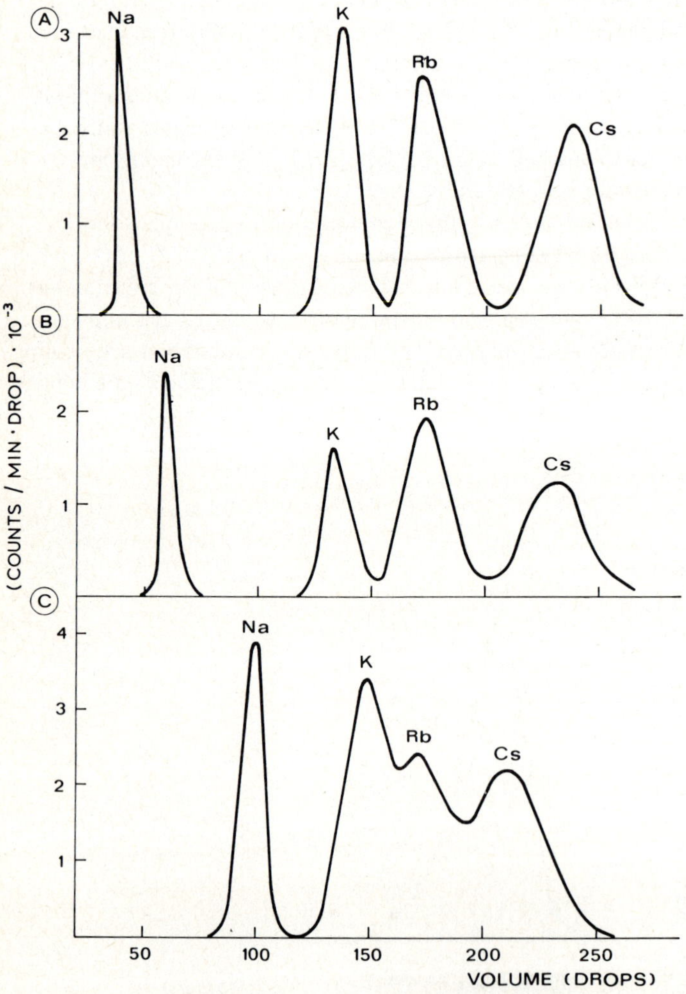


Fig. 51.2. Effect of degree of resin cross-linking on the separation of alkali metals (Dybczyński). Resin: A, Dowex 50W-X16; B, Dowex 50W-X8; C, Dowex 50W-X2.

Considering the above conditions, sufficient information should also be available in order to be able to predict the behaviour of the elements. These conditions can be predicted from the exchange constants or by using distribution coefficients (Tsitovich). Similarly, adsorption maxima in gradient elution can also be predicted (Massart and Bossaert). The possibility of predicting the resin cross-linking effect on the quality of ion-exchange separations was shown by Dybczyňski, who found that the best separation of alkali metals is achieved at the highest degree of cross-linking (see Fig. 51.2). To some extent, results obtained from paper ion-exchange chromatography can also be used for the determination of optimal conditions, but these results are applicable only if it is assumed that the same solutes or resins will be used (Lederer and Ossicini). The differences observed are caused by the different techniques used (Ossicini and Lederer) or by the presence of some binder in the finished ion-exchange paper (Sherma).

Cations are often separated as metal complexes of either the cationic or anionic type with different charges. This technique considerably increases the separability of individual cations and also allows the cations originally present in the solution on anion exchangers to be separated (see, for example, Coleman *et al.*, 1969; Faris; Ishida *et al.*; Korkisch and Klakl, 1968; Marcus and Eyal; Molnár *et al.*; Neumann *et al.*). This separation can be achieved by using one of the following methods:

(1) by chromatography of the prepared metal complexes (Vanderdeelen);

(2) by elution with a complex-forming agent (De Bruyne; Faris; Orlandini; Shibata *et al.*, 1968; Winget and Lindstrom);

(3) by the use of specific ion exchangers containing complexing agents as ionogenic groups (Birney *et al.*, Kühn and Hoyer, Wódkiewicz and Dybczyňski). These groups can be retained on cellulose or polystyrene supports. According to Schmitt and Fritz, cellulose-based ion exchangers are better than those based on polystyrene, because the former exhibit a higher rate of ion exchange.

In several instances, especially for the separation of alkali metals, the use of phosphomolybdate (Coetzee and Rohwer), heteropolyacids (Cziboly *et al.*), zirconium phosphate (Araki *et al.*) or hydrated antimonic acids (Abe, Abe and Ito) as ion exchangers has been reported. The results of Abe indicate that the separability (alkali metals) on antimonic acid depends both on the lattice geometry and on the lattice constants.

As an example of the application of specific ion exchangers, the separation of the rare earth elements on Dowex 1-X4 (Wódkiewicz and Dybczyňski) is described. For preparation of the column, Dowex 1-X4 ($Cl^-$) resin is placed in a column and conditioned by passing through 1 *M* potassium hydroxide solution, water and 1 *M* hydrochloric acid in large excess. Finally, the resin is converted into the 1,2-diaminocyclohexane-N,N′-tetraacetate form by passing a 0.1 *M* solution of the disodium salt of the acid until no chloride can be detected in the effluent, then the ion exchanger is washed with deionized water and air-dried. The column (I.D. 2 mm) is filled with a slurry of the converted Dowex 1-X4 resin and, after the exchanger has settled in the column, it is rinsed with a few millilitres of eluent.

The procedure is as follows. The load (a solution of radioactively labelled rare earths in hydrochloric acid together with a sufficient amount of complexing agent) is forced through the column by applying a slight over-pressure. Then a burette containing the eluent (a 0.0011 *M* solution of disodium 1,2-diaminocyclohexane-N,N′-tetraacetate) is connected to the column and a flow-rate of 0.5–1 ml/min is maintained constant under a mercury column pressure. The effluent is collected in drops on a moving paper band and the spots are dried, cut from the paper and their radioactive count rates are measured. Using this procedure, thulium, dysprosium, lanthanum and praseodymium are separated (see Fig. 51.3).

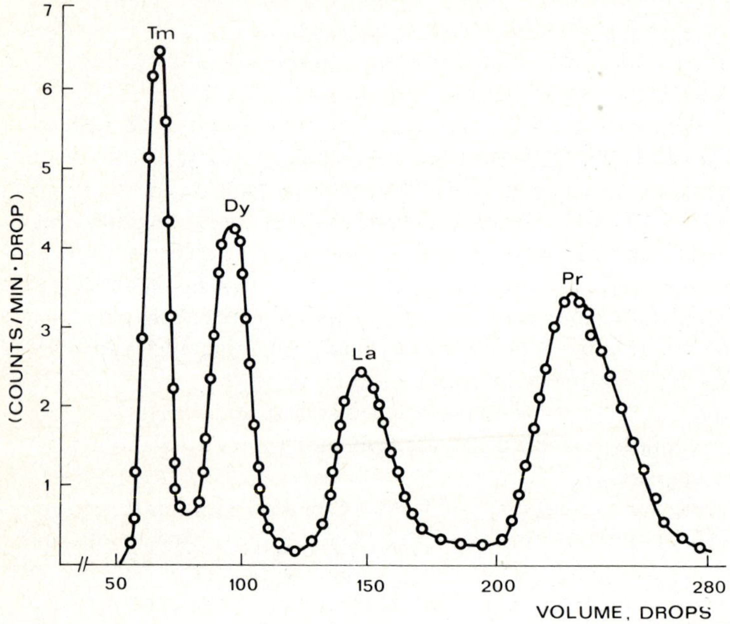


Fig. 51.3. Ion-exchange chromatographic separation of thulium, dysprosium, lanthanum and praseodymium (Wódkiewicz and Dybczyński).

*Reversed-phase chromatography*

This method is, in principle, an improved extraction, in which the stationary phase is formed by various organic compounds. Such a stationary phase can have exchange properties, with cations forming ionic association complexes or metal chelates. For effective column separations, two factors are important:

(1) the ratio of the partition coefficients (separation factor);

(2) the height equivalent to a theoretical plate (HETP).

These factors may be affected by the flow-rate, temperature, particle size, etc. According to O'Laughlin and Jensen, both of these factors are relatively insensitive to slight changes in flow-rate and temperature. Further, assuming the particle size of the supporting material to be constant, the particular support used as extractant has little effect on the HETP. As was found by Fidelis and Siekierski, tri-*n*-butyl phosphate used as stationary phase shows less sensitivity to the amount of carrier (Kieselguhr). An acidic stationary phase (*e.g.*, di-(2-ethylhexyl) orthophosphoric acid, 2-ethylhexylphenylphosphonic acid) is to be preferred to a neutral stationary phase (*e.g.*, tri-*n*-octylphosphine oxide) because it has considerably greater separation factors and requires a lower acid concentration for the mobile phase (Fidelis and Siekierski).

As several factors (*e.g.*, extraction, adsorption, formation of metal chelates) contribute to the overall separation effect, reversed-phase chromatography can be used with advantage for the separation of elements that have similar chemical properties and for which only a significant difference in extractability is needed. Examples are the separation of rare earth elements (O'Laughlin and Jensen), alkaline earth metals (Akaza, 1966a) and alkali metals (Akaza, 1966b), and, as a practical example of this method, the separation of Ga(III), In(III) and Tl(III) by reversed-phase chromatography (Frizt *et al.*) is described. For preparation of the column, Amberlyst XAD-2 (80–100 mesh, washed with 6 *M* hydrochloric acid and methanol and air-dried) is slurried, equilibrated with diisopropyl ether and allowed to remain in contact with this solvent for 1 h. The slurry is then transferred to the column (120 × 10 mm) and a small plug of glass-wool is used to prevent the top of the column from being disturbed by the addition of the sample or eluent. Approximately three column volumes of equilibrated acid are passed through the column so as to displace the interstitial organic solvent from the column. The organic phase is run down to the level of the column before the equilibrated acid is added. The sample (containing 300 μmole of elements dissolved in 5 ml of the appropriate eluent) is then transferred into the column and eluted. For details, see Fig. 51.4.

In conclusion, many separations of cations have been reported. However, these include systems that usually can be used only for the separation of certain cations (see Table 51.1). For this reason, Fritz and Latwesen combined two methods that are

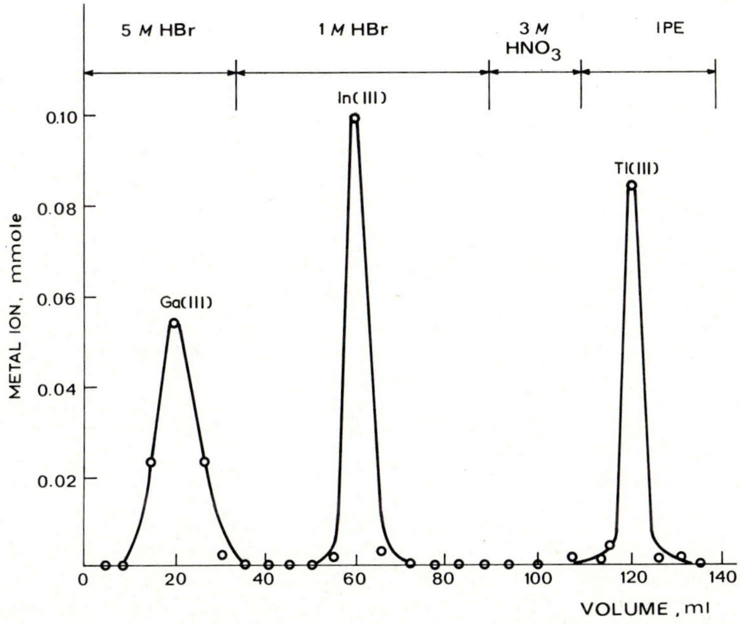


Fig. 51.4. Reversed-phase chromatographic separation of gallium, indium and thallium (Fritz *et al.*). Packing: Amberlyst XAD-2 impregnated with diisopropyl ether (IPE). Flow-rate: 1.0–1.5 ml/min.

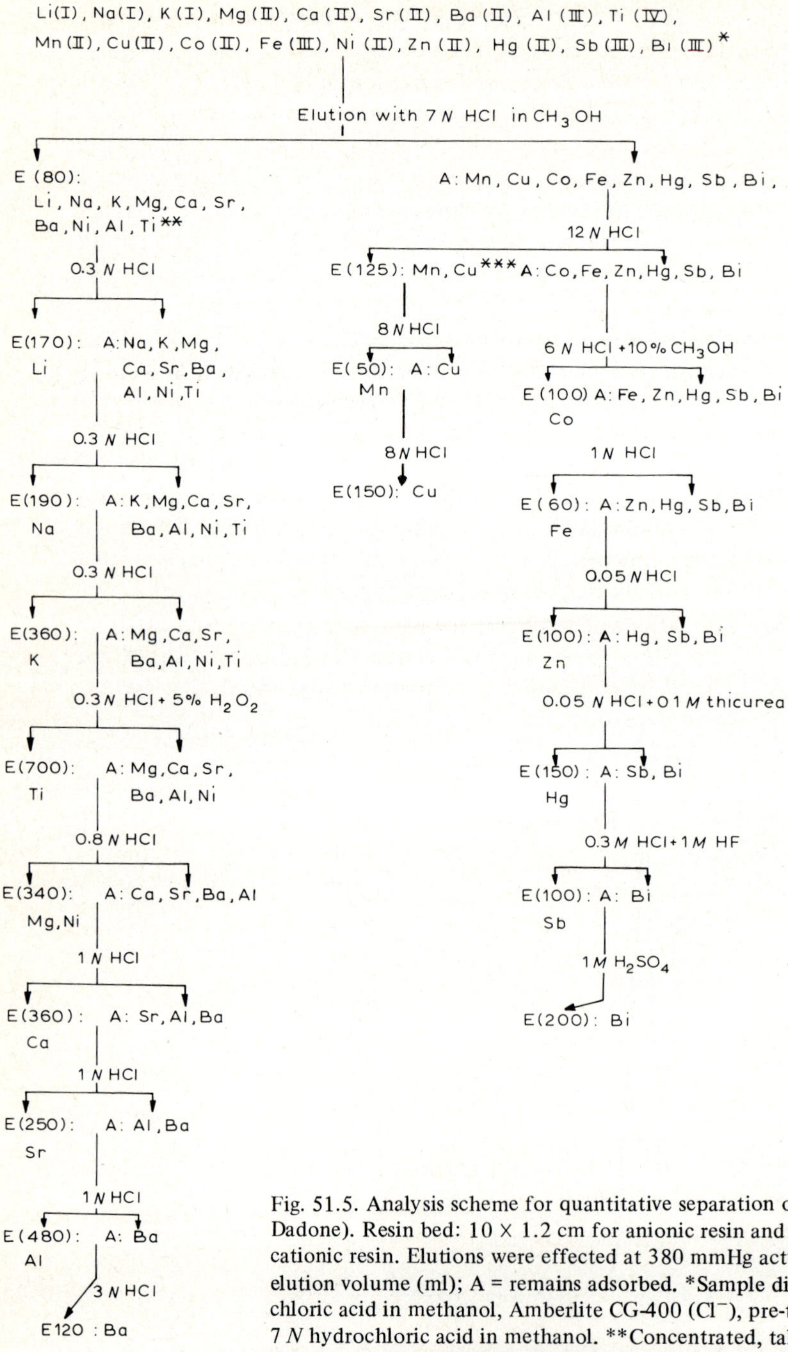


Fig. 51.5. Analysis scheme for quantitative separation of cations (Frache and Dadone). Resin bed: 10 × 1.2 cm for anionic resin and 18 × 1.2 cm for cationic resin. Elutions were effected at 380 mmHg actual pressure. E = elution volume (ml); A = remains adsorbed. *Sample dissolved in 7 *N* hydrochloric acid in methanol, Amberlite CG-400 ($Cl^-$), pre-treated with 20 ml of 7 *N* hydrochloric acid in methanol. **Concentrated, taken up in 0.3 *N* hydrochloric acid, Amberlite CG-120 ($H^+$). ***Concentrated, taken up in 7 *N* hydrochloric acid in methanol, Amberlite CG-400 ($Cl^-$), pre-treated with 20 ml of 7 *N* hydrochloric acid in methanol.

especially suitable for this purpose (ion-exchange and reversed-phase chromatography) and attempted to find a general scheme for the analysis of multi-component samples. More recently, Frache and Dadone described a quantitative separation scheme for common cations. This separation (Fig. 51.5) is based in principle on the ability of different cations to form anionic chloride complexes with different degrees of adsorption on the anion exchanger and on the differences in adsorptivity of the remaining cations on the cation exchanger.

TABLE 51.1
SEPARATION OF DIFFERENT GROUPS OF CATIONS

Abbreviations used: EtOH = ethanol; IPE = diisopropyl ether; MeOH = methanol; MHDPO = methylenebis(di-*n*-hexylphosphine oxide); TBP = tri-*n*-butyl phosphate; TMP = 2,2,4-trimethylpentane; TOPO = tri-*n*-octylphosphine oxide.

| Cations* | Ion exchanger (sorbent) | Eluent | References |
|---|---|---|---|
| Na(I) > K(I) > Rb(I) > Cs(I) > Be(II) > Mg(II) | Bio-Rad AG 50W-X8, column 75 ml | 0.6 *M* $HNO_3$ flow-rate 2 ml/min | Strelow *et al.* (1968) |
| Fe(III) > U(VI) > Ca(II) > La(III)<br>In(III) > Ga(III) > Be(II) > Al(III) > Y(III)<br>Cd(II) > Zn(II) > Fe(III) > Ca(II) > Ba(II) | Bio-Rad AG 50W-X8, 10 × 1.9 cm | §<br><br>3 *M* $HNO_3$ | Strelow *et al.* (1969) |
| La(III) > Pr(III) > Nd(III) > Sm(III) > Eu(III) > Gd(III) > Tb(III) > Dy(III) > Ho(III) > Er(III) > Lu(III) > Tm(III) | Hyflo Supercel + 10% MHDPO, 20.8 × 0.9 cm | 16 *M* $HNO_3$ flow-rate 0.2 ml/min · cm² | O'Laughlin and Jensen |
| Sm(III) > Th(IV) > U(VI)<br>As(III) > Sn(IV) > Sb(III)**<br>As(III) > Te(IV) > Cd(II) > Bi(III)***<br>Al(III) > Fe(III) > As(III) > Mo(VI)<br>Cr(III) > Fe(III)<br>Fe(III) > Re(VII)**<br>Ge(IV) > Sn(IV)** | Amberlite CG-4B, 14 × 1 cm | 0.1 – 11.4 *M* HCl | Kuroda *et al.* |
| Ga(III) | Amberlyst XAD-2 + IPE, 6 × 1.2 cm | § | Fritz and Latwesen |
| Pb(II) > Zn(II) | Amberlyst A-26, 6 × 1 cm | | |
| Ni(II) > Co(II) > Cu(II) > Zn(II) > Cd(II) | Amberlyst A-26, 14 × 1 cm | | |
| Ti(IV) | Amberlyst XAD-2 + TOPO, 11 × 1.4 cm | § | |
| V(IV) > Al(III) | Dowex 50W-X8, 6 × 1.2 cm | | |
| Ni(II) > Ti(IV) > Th(IV) | Amberlyst XAD-2 + TOPO, 10 × 1.3 cm | | |

(*Continued on p. 1096*)

TABLE 51.1 (*continued*)

| Cations* | Ion exchanger (sorbent) | Eluent | References |
|---|---|---|---|
| La(III) > Ti(IV) > Th(IV) > Zr(IV) | Amberlyst XAD-2 + TOPO, 13 × 1.2 cm | | |
| Ni(II) > Mg(II) > Ca(II) | Dowex 50W-X8, 7 × 1.2 cm | | |
| Pd(II) > Pt(IV) > Au | Daiflon M-300, + TBP, 30 × 0.92 cm | HCl, $HNO_3$ | Akaza *et al.* (1970) |
| Cd(II) > Zn(II) > Fe(III) > Cu(II) > Co(II) > Mn(II) > Tl(III) > In(III) > Ga(III) > Al(III) > Yb(III)<br>V(V) > Fe(III) > U(VI) > Ti(IV) > Ca(II) > Ba(II) | Bio-Rad AG 50W-X8, 17 × 2.1 cm | § | Strelow *et al.* (1971) |
| Be(II) > Fe(III) > Cr(III) > Co(III) §§, §§§<br>Be(II) > Cu(II) > Ru(III) > Co(III) §§,†<br>Be(II) > Cu(II) > Al(III) > Cr(III) > Ru(III) > Co(III) §§,†† | Kieselguhr, 50 × 0.27 cm; stationary phase, $H_2O$–EtOH–TMP (0.343: 0.641:0.016, w/w) | $H_2O$–EtOH–TMP (0.0007:0.022: 0.977, w/w) | Huber *et al.* |

*Cations listed in order of elution.
**Cations eluted with 1 *M* NaOH.
***Cations eluted with 3 *M* $H_2SO_4$.
§For details, see original paper.
§§High-speed chromatography (cations separated as acetylacetonates), ligand present in mobile phase.
§§§Flow-rate, 1.7 ml/min.
†Flow-rate, 1.9 ml/min.
††Flow-rate, 1.1 ml/min.

## Anions

### *Inorganic phosphorus compounds*

The application of chromatography to the inorganic chemistry of phosphorus includes in principle either the separation of compounds that have phosphorus atoms in different oxidation states or the separation of phosphates that contain a different number of phosphorus atoms in a molecule.

Lower oxo anions of phosphorus are advantageously separated on Dowex 1-X8 (Benz and Paiaxo; Pollard *et al.*, 1962, 1963). For the complete separation, gradient elution is necessary (Koguchi *et al.*; Pollard *et al.*, 1962, 1963). Due to the nature of the compounds to be separated, changes in pH alter the order of elution. For example, at pH 6 the order of elution was found to be $HPO_4^{2-} > HPO_3^{2-}$, whereas at pH 11 the $HPO_3^{2-}$ anion was eluted before $HPO_4^{2-}$ (Pollard *et al.*, 1962). The adsorbability of phosphorus oxo anions also depends on the concentration of the eluting agent used. This dependence (logarithm of concentration), which has a linear character, allows the calculation of the position of the elution peak in gradient elution (Koguchi *et al.*).

For the separation of lower oxo anions of phosphorus, Pollard *et al.* (1967) used gradient elution chromatography on cellulose. The results obtained by TLC can be used in column chromatography, but the sample must be applied as a solid (adsorbed on cellulose) and, in addition, the water content in the solvent must be lower than that used in TLC. In order to obtain a narrow band, the minimum volume of sample must be applied.

Condensed phosphates that differ in molecular weight can also be separated using Dowex 1-X4 (Ohashi *et al.*), but gel chromatography, in which the molecular-sieve effect appears to be the main factor, is more convenient. Ueno *et al.* (1970b) studied the separation of linear polyphosphates in detail and found:

(1) elution with water results in low distribution coefficients with ortho, di- and triphosphates;

(2) electrostatic repulsions between oxo anions and the gel matrix can be reduced by elution with potassium bromide solution;

(3) the distribution coefficients of oxo anions increase with increasing concentration

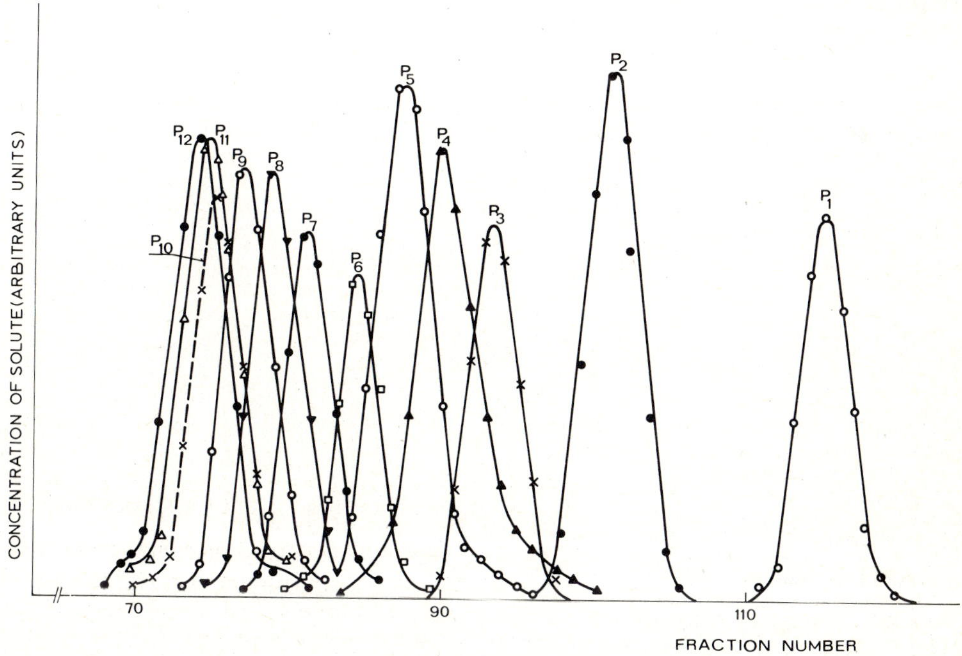


Fig. 51.6. Elution curves for linear phosphates ($P_1 - P_{12}$ = number of phosphorus atoms) (Ueno *et al.*, 1970b). Column: 150 ml (volume of bed). Sorbent: Sephadex G-25. Eluent: 0.1 *M* KCl, pH 7.0. Phosphorus concentration in sample solution = $3 \cdot 10^{-3} - 5 \cdot 10^{-3} M$ for individual phosphates and $0.01 - 0.02\,M$ for the polyphosphate fractions. One fraction = 1.08 ml.

*References p.1111*

TABLE 51.2
SEPARATION OF LOWER PHOSPHORUS OXO ANIONS (POLLARD *et al.*, 1963)

Column: 50 × 1.5 cm. Ion-exchanger: Dowex 1-X8 ($Cl^-$), 100–200 mesh. Elution: potassium chloride buffer solution, pH 6.8. Potassium chloride concentration: (a) in mixing vessel, 0.05 *M* (750 ml); (b) in reservoir, 0.2 *M*. Flow-rate: 60 ml/h. Temperature: 18°C.

| Compounds separated* | Retention volume (ml) |
|---|---|
| $H_2PO_2^-$ | 150 |
| $HPO_4^{2-}$ | 260 |
| $HPO_3^{2-}$ | 330 |
| $H_2P_2O_6^{2-}$ | 590 |
| $HP_2O_5^{3-}$ | 680 |
| $H_2P_2O_5^{2-}$ | 720 |
| $HP_2O_6^{3-}$ | 890 |
| $P_2O_7^{4-}$ | 990 |
| $P^4-P^3-P^4$ ** | 1180 |

* Listed in order of elution.
** $Na_5P_3O_8$.

TABLE 51.3
SEPARATION OF DIFFERENT ANIONS

| Anion* | Sorbent (ion exchanger) | Eluent | References |
|---|---|---|---|
| $I^- > Br^- > Cl^-$ | Hydrous zirconium oxide 12 cm × 0.075 cm² | 0.15–1 *M* $KNO_3$, flow-rate 110 cm/h | Tustanowski (1967a) |
| $I^- > Br^- > Cl^-$ | Alumina 29.5 cm × 0.075 cm² | 0.2 *M* $KNO_3$, flow-rate 20–70 ml/h | Tustanowski (1967b) |
| $F^- > Cl^- > Br^- > I^- > SO_4^{2-}$ | Cellulose-20** | Water | Brown and Chitumbo |
| $[Fe(CN)_6]^{3-} > [Fe(CN)_6]^{4-}$ | Kel-F + tri-*n*-butyl-phosphate | $HCl-H_2O$ | Akaza *et al.* (1969) |
| $(Mo_xO_y)^{n-}$*** | Sephadex G-10 and LH-20, Bio-Gel P-2 | $H_2SO_4$ | Streuli and Rogers |
| $ReO_4^- > MoO_4^{2-} > WO_4^{2-}$ § | DEAE-cellulose | 0.02–0.1 *M* $NH_4CNS$ §§ | Ishida and Kuroda |

*Anions listed in order of elution.
**For preparation, see original paper.
***Different types of molybdates, the composition of which depends on pH.
§Eluted with 0.1 *M* NaOH–0.1 *M* NaCl.
§§pH 3 for $ReO_4^-$, pH 5 for $MoO_4^{2-}$.

of the eluting agent;

(4) the higher distribution coefficients of both $K^+$ and $Cl^-$ compared with those of oxo anions cause the phosphates to be accompanied by background electrolyte;

(5) only at phosphorus concentrations lower than 0.01 *M* are the elution curves symmetrical (see Fig. 51.6);

(6) the pH does not affect the distribution coefficients;

(7) the elution volume increases with increasing gel porosity.

As expected, polyphosphates can be also separated using a Bio-Gel P-2 column, from which they are eluted in order of decreasing molecular weight (Neddermeyer).

As an example of this type of separation, the separation of oxo anions on Dowex 1-X8 (Pollard *et al.*, 1963) is described. A 1-ml volume of a mixture of anions containing 2 mg of phosphorus per anion per millilitre is applied on the column (50 × 1.5 cm). As soon as no liquid remains above the bed of Dowex 1-X8 ($Cl^-$) resin (100–200 mesh), 2 ml of a 0.2 *M* buffered solution of potassium chloride (the pH of which is adjusted to 6.8 by the addition of 25 ml of a 2 *M* solution of ammonium acetate to 1 litre of potassium chloride solution) are added. Anions are further eluted by gradient elution (see Table 51.2) according to Grande and Beukenkamp. Fractions of 10 ml are collected and their phosphorus contents determined.

Examples of the use of chromatography for the separation of some other anions are shown in Table 51.3.

## COORDINATION AND ORGANOMETALLIC COMPOUNDS

### General survey

The choice of the best chromatographic method is affected by the character of the compounds to be separated. While the use of ion-exchangers is very effective for charged complexes, the separation of non-ionic metal complexes is difficult and different sorbents are employed. However, some of these materials behave simultaneously as weak ion exchangers. A typical example is silica gel, the silanol sites (≡ Si–OH) of which act as a strong hydrogen-bond donor (Basila). For this reason, this material will strongly adsorb species that are hydrogen-bond acceptors and therefore, in such instances, the use of solvents capable of forming strong hydrogen bonds is necessary (Jursík).

According to Burwell *et al.*, competition for the coordination sphere between the silanol or siloxane sites and ligands occurs during the chromatography of metal complexes. From this it follows that any effect that increases the association between the complex and the ≡ Si–OH centre will increase the tendency of the silanol (siloxane) group to penetrate into the coordination sphere of the central atom. This means on the one hand that solvents with low dielectric constants will increase the adsorption of solutes and on the other hand such solvents (or those with a low flow-rate) will make the adsorption irreversible (Bradley and Pantony).

The adsorption of metal complexes on silica gel is also affected by its activity. In general, greater mobility is observed on hydrated than on dehydrated silica gel (Hathaway and Lewis, Jursík).

*References p.1111*

Apart from silica gel, alumina is most frequently used as a sorbent. As far as its adsorption mechanism is concerned, it is similar to that of silica gel, because the surface of alumina also contains hydroxyl groups. However, stronger adsorption of metal complexes can be expected on alumina owing to the different coordination numbers of aluminium and silicon.

**Geometrical isomers**

Geometrical isomerism of coordination compounds is the consequence of the co-ordination of different kinds of donor atoms, which leads to a different orientation of molecules in an electric or magnetic field. Such metal complexes are characterized by a permanent dipole moment, the value of which indicates the polarity of the metal complex. In addition, the different orientation of alkyl groups of a symmetrical bidentate ligand coordinated to a metal ion also gives rise to geometrical isomerism.

The results show that it is only the polarity of the metal complex or the character of the ligands that influences the chromatographic behaviour of geometrical isomers, and that the different dipole moments of geometrical isomers result in *trans*-isomers having a higher mobility than *cis*-isomers (King and Walters). The observed lower mobility of *cis*-isomers is due to their ability to bind two sorbent sites (*e.g.*, silanol centres of silica gel or functional groups of an ion exchanger), while *trans*-isomers bind only one sorbent site (Druding and Hagel).

The different chromatographic behaviour of geometrical isomers is not only related to their different dipole moments. Legg, for example, described the separation of $[Co(EDDA)(pn)]^+$ isomers, (EDDA = ethylenediamine-N,N′-diacetate; pn = propylenediamine) for which the geometrical isomerism arises only as a result of the different orientation of the methyl groups in propylenediamine.

The chromatography of isomers is accompanied by a series of negative effects, such as substitution of labile bonded ligands for the hydroxyl groups of silica gel. Isomerization promoted by the sorbent must also be taken into account (Kauffman *et al.*).

For the efficient separation of isomeric pairs, which depends on the interaction of solvent, sorbent and solute, the following ideal conditions can be derived:

(1) isomers must have maximally different dipole moments;

(2) isomers must be soluble in both polar and non-polar solvents;

(3) an ideal solvent is one which does not promote isomerization; for example, the following equilibrium:

$$cis\text{-}[Co(en)_2Cl_2]^+ \rightleftharpoons trans\text{-}[Co(en)_2Cl_2]^+$$

(en = ethylenediamine) is strongly influenced by solvents (Tobe and Watts, 1962, 1964), and this observed influence of polarity has been attributed to the dipolar character of the *cis*-isomer (Fitzgerald and Watts; Pearson *et al.*; Tobe and Watts, 1962, 1964);

(4) an ideal sorbent binds the *cis*-isomer sufficiently firmly, but there must still be the possibility of eluting it with a polar solvent;

(5) the sorbent must be isomerically inert; in this sense, it must be noted that silica gel is a sorbent that sometimes catalyzes isomerization (Kauffman *et al.*);

(6) faster flow-rates (>2 ml/min) give diffuse bands (Girgis and Fay).

As far as the practical use of column chromatography for the separation of geometrical isomers is concerned, two examples are given below.

(1) Separation of isomers of $[Co(EDDA)(en)]^+$ (Legg and Cooke, 1965). To a column containing 90 ml of Dowex 50W-X8 ($Na^+$), 50–100 mesh, 3 mmoles of this complex dissolved in 100 ml of water are added. The column is then allowed to swell with water and the adsorbed complex is eluted with 0.5 *M* sodium perchlorate solution at the rate of 0.1–0.5 ml/min. The *trans*-isomer is eluted first.

(2) Separation of isomers of tris(benzoylacetonato)chromium(III) (Girgis and Fay). A column (84 × 2.8 cm) equipped with an electric vibrator is packed with a slurry of Florisil (180 g) in *n*-hexane and allowed to drain under continuous vibration. Then 0.5 g of the complex dissolved in the minimum volume of benzene–*n*-hexane (1:1) is added and elution is carried out with benzene–diethyl ether (19:1) at a flow-rate of 2 ml/min. After eluting the first band from the column (*trans*-isomer), the flow-rate is increased to 8 ml/min and the *cis*-isomer is obtained.

For other examples of the separation of geometrical isomers, see Table 51.4.

## Optical isomers and diastereoisomers

Neglecting the origin of the metal complex chirality, these compounds may contain one or more centres of chirality, which affects the choice of both solvent and sorbent.

Considering the separation of optical antipodes, according to the classical method (fractional crystallization of diastereoisomers), another centre of chirality is introduced into the racemate molecule so that enantiomers are transformed into diastereoisomers that differ from each other in their physico-chemical properties. In chromatography, this can be achieved either by using a chiral sorbent or a chiral solvent. The first instance, which is the more frequently used, is based on the fact that both enantiomers, *e.g.* (+)-M and (–)-M (M = metal atom), are bonded on the chiral sorbent, which can be expressed by (+)-Ads, with the formation of both (+)-Ads–(–)-M and (+)-Ads–(+)-M. However, diastereoisomers formed in such a manner are not equally stable, so that one enantiomer may be adsorbed more strongly than the other.

As typical examples of the separation of enantiomeric metal complexes on chiral sorbents (cellulose, starch, etc.), four methods of resolution are described below.

(1) Resolution of $[Co(EDDA)(B)]^+$ (B = N-methyl, N-ethyl, N,N′-dimethyl or N,N′-diethylenediamine) (Legg and Douglas). For the preparation of the column, 20 g of Cellex CM-cellulose (Bio-Rad Labs., Richmond, Calif., U.S.A.) are stirred with 400 ml of 0.01 *M* sodium perchlorate solution, allowed to settle for 30 min and then the process is repeated. A further 100 ml of 0.1 *M* sodium perchlorate solution are added and this suspension is quickly stirred and immediately and continuously poured into the column (50–70 × 2.5 cm). The cellulose is allowed to settle for about 30 min and a plug of glass-wool is inserted in order to protect the cellulose surface. The column is washed with 100 ml of 0.01 *M* sodium perchlorate solution and then with 100 ml of water at a flow-rate of 2–4 ml/min. Then 0.2–0.3 g of the complex dissolved in 100 ml of water are loaded on to the column at a rate of 1–2 ml/min. The 3–4-cm layer of adsorbed

*References p.1111*

complex is eluted with 0.01 *M* sodium perchlorate solution and the eluate is collected in 15-ml fractions.

(2) Resolution of triaquotribenzo(*b,f,j*)(1,5,9)triazacycloduodecinenickel(II) nitrate on cellulose (Taylor and Busch). A suspension of microcrystalline cellulose (Avicel, technical grade) in water is poured into a column (60 × 2.2, 65 × 2.2 or 100 × 2.0 cm) and water is allowed to filter through the bed at a rate of approximately 0.2 ml/min. The column is washed with water for 24–26 h at a flow-rate of 10 ml/min. A saturated aqueous solution of the complex containing a few drops of 0.01 *M* nitric acid is then loaded on to the column. The adsorbed complex is eluted with water as one broad band at a rate of 0.5 ml/min and collected into several fractions. The best separation was achieved with a column size of 65 × 2.2 cm.

(3) Resolution of the same complex as in (2) on starch (Taylor and Busch). Before use, the starch is stirred with 0.01 *M* nitric acid and decanted or filtered. A starch suspension in 0.01 *M* nitric acid is added to the column and washed with 0.01 *M* nitric acid for 48 h. In order to achieve a flow-rate of 0.2 ml/min (suitable flow), air pressure is applied. A 10-ml volume of a saturated 0.01 *M* nitric acid solution of the complex is then placed on the column and eluted with 0.01 *M* nitric acid as one broad band, which is separated into fractions.

(4) Resolution of cobalt(III) or chromium(III) tris-($\beta$-diketonates) on D-(+)-mannitol or D-(+)-sorbitol (Marković and Schweitzer). A thin slurry of D-(+)-mannitol (or D-(+)-sorbitol) (sieved to 80 mesh) in a benzene–ligroine (1:6) mixture is poured into a column so as to form a 124 × 5 cm bed. After removing the excess of solvent by flow through the column, 0.102–0.120 g of sample dissolved in 15 ml of benzene is applied to the column. The adsorbed complex is then eluted with benzene–ligroine (1:6) solvent and the eluate is separated into several fractions.

Chromatography on the chiral sorbents usually does not lead to the total resolution of enantiomers (Girgis and Fay; Jursík *et al.*, 1973; Marković and Schweitzer), with the exception of the cobalt(III) trinuclear complex (Brubaker *et al.*). Also, comparison of the sorbents shows that the separation on cellulose is not as efficient as that on starch (Taylor and Busch). The most effective resolution on starch, which also depends on the amount of sorbent used, can be attained with a water-soluble racemate (Krebs and Diewald). Furthermore, the presence of a larger number of polar groups in the metal complex is necessary in order to achieve a multi-site adsorption (Krebs and Rasche).

Resolution of enantiomers, as mentioned above, can also be achieved in principle by using a chiral solvent, in which racemic metal complexes anti-racemize (Bosnich). As the mechanism of this anti-racemization includes a stereospecific substitution reaction (Smith and Haines) with the solvent being coordinated as a bidentate ligand, it is possible to assume preferential adsorption of one of the enantiomers on the achiral sorbent. However, the application of this method, which seems to be promising, has not yet been reported.

In addition to the described chiral sorbents, alumina or silica gel can also be employed, but before using it, it must be covered with a layer of optically active material, *e.g.* (+)-tartaric acid (Piper). In other instances, Amberlite IRA-400 with chiral ionogenic groups was used (Gillard and Mitchel). By modifying the latter method, the racemate is transformed into diastereoisomers, which are then separated chromatographically (Dhar *et al.*, 1958, 1959).

As can be seen from the two examples given below, for the resolution of diastereoisomeric metal complexes, where differences in non-bonding interactions in individual stereoisomers are utilized, alumina, silica gel and ion exchangers are most frequently used.

(1) Resolution of tris-[(+)-3-acetylcamphorato]cobalt(III) on alumina (Springer *et al.*). Approximately 500 g of acid-washed alumina suspended in benzene are poured into a column (bed size 40 X 4 cm) and, as soon as the benzene has soaked into the alumina, 7 g of the complex dissolved in 25 ml of benzene are added to the column. When all of the complex has been adsorbed, the column is filled with benzene so as to form a column of 115 cm above the bottom of the alumina (this benzene layer maintains a flow-rate of approximately 10 ml/min). Elution with benzene separates the complex into two bands (geometrical isomers). Each band is collected in two fractions, each of which contains two isomers and, by repeated chromatography under the same conditions (on a smaller scale), pure optical isomers are obtained.

(2) Resolution of $[Co(gly)_2(pn)]^+$ and $[Co(gly)(pn)_2]^{2+}$ (gly = glycine; pn = propylenediamine) on Dowex 50W-X8 (Kojima and Shibata, 1970). The reaction mixture, containing both complexes dissolved in water, is added to the column of Dowex 50W-X8 ($Na^+$), 14.5 X 8 cm, and the adsorbed complexes form a compact band at the top of the column. The column is allowed to swell with water. The adsorbed bands are eluted with a 0.5 *M* solution of sodium chloride at a rate of about 0.6 ml/min and, after elution for about 24 days, five bands develop, four of which are collected separately. The band remaining at the top of the column is then eluted with a 2 *M* solution of sodium chloride for about 36 days, so that the fifth band is separated into two more bands, which are eluted separately.

The resolution of both enantiomers and diastereoisomers naturally depends on their minimum rate of isomerization and racemization.

**Relationship between chromatographic behaviour and configuration of optical isomers**

The resolution of enantiomeric metal complexes on chiral sorbents resembles the method of the "less soluble diastereoisomers". It seems that here also there exists a relationship between the adsorptivity of individual enantiomers (diastereoisomers) and their configurations, so that, for example, the less adsorbed enantiomers (diastereoisomers) will have the same configuration. However, this would be valid only for sorbents or metal complexes with the same adsorption sites. In addition, the dependence of the order of elution of isomers on the kind of eluent (Krebs and Diewald) and kind of sorbent (Marković and Schweitzer) must be also taken into account. The order of elution of isomers is also affected by the ligand configuration, as was demonstrated by Schoenberg *et al.*

As an example, Piper observed that on alumina modified by (+)-tartaric acid, (+)-isomers of Co(III) and Cr(III) acetylacetonates are less adsorbed. Similar results were obtained when these complexes were resolved on a D-lactose column (Fay *et al.*). Several racemic complexes, *e.g.* $[Co(en)_3]^{3+}$, $[Co(pn)_3]^{3+}$, $[Cr(en)_3]^{3+}$ and $[Rh(en)_3]^{3+}$, were separated into enantiomers on Amberlite IRA-400 in the (+)-tartarate form. The first isomers eluted were: (–)-$[Co(en)_3]^{3+}$, (–)-$[Co(pn)_3]^{3+}$, (–)-$[Cr(en)_3]^{3+}$ and (+)-$[Rh(en)_3]^{3+}$. Because of the similarity of the adsorption sites in these complexes, it can

TABLE 51.4
SEPARATION OF GEOMETRICAL ISOMERS

| Type of compound | Compounds separated* | Sorbent (ion-exchanger) | Eluent | Isomer eluted first | References |
|---|---|---|---|---|---|
| $MA_6$ | $[Co(bzacac)_3]$ | Florisil | Benzene–diethyl ether (19:1) | *trans* | Girgis and Fay |
| | $[Cr(bzacac)_3]$ | Florisil | Benzene–diethyl ether (19:1) | *trans* | Girgis and Fay |
| | $[Co(tfacac)_3]$ | Florisil | Benzene–*n*-hexane (1:1) | *trans* | Girgis and Fay |
| | $[Cr(tfacac)_3]$ | Florisil | Benzene–*n*-hexane (1:1) | *trans* | Girgis and Fay |
| | $[Co(tmb)_3]$ | Florisil | Benzene–*n*-hexane (3:1) | *trans* | Girgis and Fay |
| | $[Cr(tmb)_3]$ | Florisil | Benzene–*n*-hexane (3:1) | *trans* | Girgis and Fay |
| | $[Co(MHDTH)_3]$ | Alumina | Benzene–*n*-hexane (2:3) | *trans* | Gordon and Holm |
| | $[Co(dien)_2]^{3+}$ | Cellulose phosphate | *n*-Butanol–HCl–$H_2O$ (200:15:15) | *trans* | Keene *et al.* |
| | $[Co(dien)_2]^{3+}$ | SE-Sephadex C-25 | 0.3 *M* sodium tartrate | s-*cis*** | Keene and Searle |
| $MA_4B_2$ | $[Co(ox)(gly)_2]^-$ | Dowex 1-X10 | 0.5 *M* $NaClO_4$ | *trans* | Matsuoka *et al.* (1967) |
| | $[Co(en)(gly)_2]^+$ | Dowex 50W-X8 | 0.5 *M* KBr | *trans* | Matsuoka *et al.* (1967) |
| | $[Co(NH_3)_2(EDDA)]^+$ | Dowex 50W-X8 | 0.5 *M* $NaClO_4$ | *trans* | Legg and Cooke (1965) |
| | $[Co(en)(EDDA)]^+$ | Dowex 50W-X8 | 0.5 *M* $NaClO_4$ | *trans* | Legg and Cooke (1965) |
| | $[Co(dien)IDA)]^+$ | Dowex 50W-X8 | 0.5 *M* $NaClO_4$ | *trans* | Legg and Cooke (1966) |
| | $[Co(dien)(MIDA)]^+$ | Dowex 50W-X8 | 0.5 *M* $NaClO_4$ | *trans* | Legg and Cooke (1966) |
| | $[Co(dap)_2]^+$ | Dowex 50W-X8 | 0.25 *M* $NH_4Br$ | *trans* | Freeman and Liu |
| | $[Co(gly)_2(pn)]^+$ | Dowex 50W-X8 | 0.5 *M* NaCl | *trans* | Kojima and Shibata (1970) |
| | $[Co(L\text{-}asp)_2]^-$ | Dowex 1-X2 | 0.1 *M* NaCl | *trans* | Yamada *et al.* |
| | $[Co(en)(N,N'\text{-}DMEEN)(NO_2)_2]^+$ | Dowex 50W-X8 | 0.5–0.075 *M* $NaClO_4$ | *trans* | Tiethof and Cooke |
| | $[Ru(NH_3)_4(py)_2]^{3+}$ | Bio-Rad AG 50W-X2 | 2.2 and 2.5 *M* NaCl | *trans* | Ford and Sutton |
| | $[Rh(H_2O)_2Cl_4]^-$ | Dowex 1-X8 | 0.3 *M* $HNO_3$ | *trans* | Bridges and Chang |
| $MA_3B_3$ | $[Co(L\text{-}val)_3]$ | Alumina | Ethanol–water (85:15) | *trans* | Gillard *et al.* |
| | $[Co(\beta\text{-}ala)_3]$ | Alumina | Water | *trans* | Ćelap *et al.* |

| | | | | | |
|---|---|---|---|---|---|
| $MA_2B_2$ | $[Pt\{(n\text{-}Bu)_3P\}_2Cl_2]$ | Alumina | Benzene | *trans* | Kauffman *et al.* |
| | | | Acetone | *cis* | |
| | | Silica gel | Benzene | *trans* | |
| | | | Acetone | *cis* | |
| | $[Pt\{(Et)_2S\}_2Cl_2]$ | Alumina | Benzene | *trans* | |
| | | | Acetone | *cis* | |
| | | Silica gel | Acetone–benzene (0.5:100) | *trans* | |
| | | | Acetone | *cis* | |
| | $[Pt(py)_2Cl_2]$ | Alumina | Acetone–benzene (7:100) | *trans* | |
| | | | Acetone | *cis* | |
| | | Silica gel | Acetone–benzene (17:100) | *trans* | |
| | | | Acetone | *cis* | |

* bzacac = benzoylacetonate; tfacac = trifluoroacetylacetonate; tmb = 1,1,1-trifluoro-4-*p*-methoxyphenyl-2,4-butanedionate; MHDTH = 5-methylhexane-2,4-dithionate; dien = diethylenetriamine; ox = oxalate; gly = glycinate; IDA = iminodiacetate; MIDA = methyliminodiacetate; dap = L- or DL-$\alpha,\beta$-diaminopropionate; asp = aspartate; N,N′-DMEEN = N,N′-dimethylethylenediamine; py = pyridine; val = valinate; ala = alaninate; *n*-Bu = *n*-butyl; Et = ethyl.

** Symmetrical-*cis*.

TABLE 51.5

RELATIONSHIP BETWEEN THE STRUCTURE OF COBALT COMPLEXES AND THE ORDER OF ELUTION OF ISOMERS USING SORBENTS

| Compounds separated* | Sorbent | Eluent | Sign of rotation of less adsorbed isomer | References |
|---|---|---|---|---|
| *cis*-$[Co(gly)_3]$ | Starch | 10% KCl | + | Douglas and Yamada |
| *trans*-$[Co(gly)_3]$ | | Water | + | Douglas and Yamada |
| *trans*-$[Co(sarc)_3]$ | | Water | – | Jursík *et al.* (1972) |
| *trans*-$[Co(\alpha\text{-isobut})_3]$ | | Water | – | Jursík *et al.* (1973) |

(*Continued on p. 1106*)

*References p.1111*

TABLE 51.5 (*continued*)

| Compounds separated* | Sorbent | Eluent | Sign of rotation of less adsorbed isomer | References |
|---|---|---|---|---|
| *trans*-[Co(L-val)$_3$] | Alumina | Ethanol–acetone (1:2) | – | Shibata *et al.* (1966) |
| *trans*-[Co(L-leu)$_3$] | | Ethanol–water (95:5) | – | Gillard and Payne |
| *trans*-[Co(L-norleu)$_3$] | | Isopropanol–water (4:1) | – | Jursík and Hájek |
| *trans*-[Co {(2L,3L)-ileu}$_3$] | | Isopropanol–water (4:1) | – | Jursík and Hájek |
| *trans*-[Co(L-leu)$_2$(gly)] | | Isopropanol–water (3:2) | – | Jursík *et al.* (1970) |
| *trans*-[Co(L-leu)(gly)$_2$] | | Isopropanol–water (3:2) | – | Jursík *et al.* (1970) |
| *trans*-[Co(L-γ-GluOMe)$_3$] | | Methanol | + | Gillard and Payne |
| [Co(pn)$_3$]$^{3+}$ | Cellulose | *n*-Butanol**–10 *M* HCl (97:2) | – | Dwyer *et al.* |
| [Co(pn)$_2$(en)]$^{3+}$ | | *n*-Butanol**–10 *M* HCl (97:2) | – | Dwyer *et al.* |
| [Co(pn)(en)$_2$]$^{3+}$ | | *n*-Butanol**–10 *M* HCl (97:2) | – | Dwyer *et al.* |

* Sarc = sarcosinate; α-isobut = α-aminoisobutyrate; leu = leucinate; norleu = norleucinate; ileu = isoleucinate.
** Saturated with water.

TABLE 51.6
RELATIONSHIP BETWEEN THE STRUCTURE OF COBALT COMPLEXES AND THE ORDER OF ELUTION OF ISOMERS ON ION EXCHANGERS

| Compounds separated* | Ion exchanger | Elution | Sign of rotation of less adsorbed isomer | References |
|---|---|---|---|---|
| [$Co_3N_6S_6$] | Cellex CM | 0.1 *M* NaCl | – | Brubaker *et al.* |
| [Co(EDDP)(en)]$^+$ | Dowex 50W-X8 | 0.35 *M* $NaClO_4$ | + | Schoenberg *et al.* |
| *cis*-[Co(IDA)$_2$]$^-$ | Cellex AE | 0.01 *M* NaCl | – | Legg and Douglas |
| [Co(EDTA)]$^-$ | Cellex AE | 0.01 *M* NaCl | – | Legg and Douglas |

| | | | | |
|---|---|---|---|---|
| $[Co(TRDTA)]^-$ | Cellex AE | 0.01 *M* NaCl | – | Legg and Douglas |
| $[Co(ox)(L\text{-}ser)_2]^-$ | Dowex 1-X10 | 0.07 *M* KCl | + | Matsuoka *et al.* (1970) |
| $[Co(gly)_2(\text{-}pn)]^+$ | Dowex 50W-X8 | 0.5 *M* NaCl | + | Kojima and Shibata (1970) |
| $[Co(gly)(\text{-}pn)_2]^+$ | Dowex 50W-X8 | 2 *M* NaCl | + | Kojima and Shibata (1970) |
| $[Co(L\text{-}aspH)_2(l\text{-}pn)]^+$ | Dowex 50W-X8 | 0.1–0.4 *M* $NaClO_4$ | + | Kojima and Shibata (1971) |
| $[Co(L\text{-}asp)(l\text{-}pn)_2]^+$ | Dowex 50W-X8 | 0.1–0.4 *M* $NaClO_4$ | + | Kojima and Shibata (1971) |
| $[Co(L\text{-}asp)_2]^-$ | Dowex 1-X2 | 0.1 *M* NaCl | + | Yamada *et al.* |
| $[M(phen)_2X_2]^+$** | CM-cellulose | 0.4 *M* $CH_3COOH$ | + | Broomhead and Grumley |

* $[Co_3N_6S_6]$ = hexakis(2-aminoethanethiolo)tricobalt(III) bromide; EDDP = ethylenediaminediaminopropionate; IDA = iminodiacetate; TRDTA = trimethylenediaminetetraacetate; ox = oxalate; ser = serinate; aspH = monoanion of aspartic acid; asp = aspartate; phen = 1,10-phenanthroline.
** M = Ir(III), Rh(III), Cr(III), Co(III); X = Cl or Br.

TABLE 51.7
RELATION BETWEEN THE STRUCTURE OF FERROCENES AND THE ORDER OF ELUTION OF ISOMERS

| Compounds separated* | Sorbent | Eluent | References |
|---|---|---|---|
| 2-Acetyl-1,1′-dimethylferrocene<br>3-Acetyl-1,1′-dimethylferrocene | Alumina | Benzene | Reinehart |
| 1′-Acetyl-2-ethylferrocene<br>1-Acetyl-1′-ethylferrocene<br>3-Acetyl-3-ethylferrocene | Alumina | Benzene | Rosenblum and Woodward |
| 1,1′-Dimethylferrocene-2-carboxylic acid** | Alumina | Benzene | Westman and Reinehart |
| (RR)-α-[2-Trimethylsilylferrocenyl]ethyldimethylarsine<br>(RS)-α-[2-Trimethylsilylferrocenyl]ethyldimethylarsine | Alumina*** | Ligroine–isopropanol (20:1) | Marquarding *et al.* |

* Listed in order of elution.
** Separated as *p*-brombenzyl esters.
*** Modified with a 2% aqueous solution of ammonia.

be concluded that these isomers will have the same configuration (Gillard and Mitchel).

With the exception of tris-($\beta$-diketonato)metal complexes, adsorptivity can be correlated with the different alkyl group chelate ring conformation. This is apparent for Co(III) complexes of C-substituted diamines and amino acids. In all instances (except for both $Co(gly)_3$ and $Co(\gamma\text{-gluOMe})_3$ (where gluOMe is the methyl ester of glutamic acid) the (−)-isomer is less strongly adsorbed (see Table 51.5). The sign of the Cotton effect indicates isomers with axial (pseudo-axial) arrangements of alkyl groups. The exception of $Co(gly)_3$ is due to the absence of alkyl groups. This fact, however, does not reflect the influence of the chiral sorbent because the (−)-isomer of $Co(\alpha\text{-isobut})_3$ (isobut = amino isobutyric acid) is the first to be eluted from a starch column (Jursík *et al.*, 1973). On the other hand, the unusual order of elution of $Co(\gamma\text{-gluOMe})_3$ (see Table 51.5) may be due to the presence of a polar side-chain in which, for example, hydrogen bonding may occur. The effect of the latter on the order of elution of isomers, the importance of which was observed by Matsuoka *et al.* (1970), must not be overlooked.

The above examples indicate that an equatorial disposition of alkyl groups ((+)-isomers) facilitates the adsorption of the corresponding isomer on both chiral and achiral sorbents (Jursík and Hájek). However, the resolution of optical isomers using ion exchangers shows the opposite behaviour (see Table 51.6). Schoenberg *et al.* and Kojima and Shibata (1970) ascribed this effect to the possible steric interactions of ligand alkyl groups with the resin.

## Ferrocenes

Most of the chromatographic applications concern the separation of reaction mixtures (see, for example, Dormond and Decombe, Hughes *et al.*, Nesmeyanov *et al.*, Shiga *et al.*). In several investigations, chromatography was used for the purification of the reaction products (Combs *et al.*, King *et al.*, Reich-Rohrwig and Schlögl).

Another very useful application includes the stereochemistry of ferrocenes. Falk *et al.* (1969), for example, separated isomeric ferrocenophan carboxylic acid diphenylamides on a silica gel column (30 × 4 cm). Elution with benzene gave the $\alpha$-isomer, while the $\beta$-isomer was eluted with benzene–ethanol (10:1). Other examples include the separation of *cis*- and *trans*-isomers of bis($\alpha$-ketotetramethylene)ferrocene on alumina (the *cis*-isomer is eluted first) (Falk and Schlögl, 1965), isopropylferrocene carboxylic acid amides (Schlögl and Fried) and isomers of acetylmethylferrocene (Benkeser *et al.*, Hill and Richards, Nagai *et al.*). On the other hand, attempts to separate a mixture of 1-ferrocenyl-2-propyl acetate and 2-ferrocenyl-1-propyl acetate using an alumina column proved unsuccessful (Nugent *et al.*).

Table 51.7 lists a few applications showing the relationship between the structure and chromatographic behaviour of ferrocene derivatives. From Table 51.7, it follows that $\alpha$-isomers are less strongly adsorbed than $\beta$-isomers. This fact can be utilized for the study of the structure of ferrocene derivatives.

Both chiral ferrocenes and ferrocenes with the chirality centre outside the ferrocene part can be partially resolved, as shown in the resolution of $\alpha$-acetylmethylferrocene on acetylcellulose by Falk and Schlögl (1966). A 300-mg amount of racemic complex dissolved in benzene is added to the column (400 × 3 cm) containing 700 g of partially acetylated cellulose. The adsorbed complex is then eluted with benzene and fractions of effluent are collected (the (+)-isomer is eluted first).

Similarly, N,N′-diferrocenylcarbodiimide possessing axial chirality can be resolved (Schlögl and Mechtler).

**Metallocenes**

The application of chromatography to other metallocenes is similar to its application to ferrocenes. An example is the separation of the α- and β-isomers of methylcymanthreonylpropionic acid on alumina (activity II–III). When elution is carried out with a benzene–light petroleum mixture, the α-isomer is eluted first (Gowal and Schlögl, 1968a). On alumina, the α- and β-isomers of aminoacetylcymanthrene were also separated (Egger and Nikiforov).

As far as resolution of racemic complexes of "metallocene asymmetry" is concerned, the following procedure is a typical example, involving the resolution of (1-tetralone)-chromium tricarbonyl on acetylcellulose (Falk *et al.*, 1966). To the glass column (400 × 3 cm), 700 g of partially acetylated cellulose suspended in benzene are added. The cellulose bed thus formed is then washed with benzene and 250 mg of racemate is applied.

TABLE 51.8
APPLICATIONS OF COLUMN CHROMATOGRAPHY IN METALLOCENE CHEMISTRY

| Compounds separated* | Sorbent | Elution | References |
|---|---|---|---|
| $C_5H_5V(CO)_4$<br>$(C_5H_5)_2V_2(CO)_5$<br>$(C_5H_5)V(CO)_3PH_3$<br>$(C_5H_5)V(CO)_2(PH_3)_2$ | Alumina (5% $H_2O$) | *n*-Hexane | Fischer *et al.* (1969) |
| $C_5H_5ReC_6H_6$<br>$CH_3COC_6H_5ReC_5H_5$<br>$CH_3COC_5H_4ReC_6H_6$ | Alumina (4% $H_2O$) 1500 × 2 cm | Benzene<br>Benzene–diethyl ether (1:1) | Fischer and Wehner |
| $C_5H_5Mo(0)C_7H_7$<br>$(C_5H_5)_2Mo(IV)H_2$ | Alumina (4% $H_2O$) 1500 × 2 cm | Benzene | Wehner *et al.* |
| $C_5H_5W(0)C_7H_7$<br>$(C_5H_5)_2W(IV)H_2$ | Alumina (6% $H_2O$) 1500 × 2 cm | Benzene | Wehner *et al.* |
| $C_5H_5V(CO)_4$<br>$C_5H_5V(I)(CO)_2[(n\text{-Bu})_3P]_2$<br>$C_5H_5V(I)(CO)_3[(n\text{-Bu})_3P]$ | Alumina (activity II) | *n*-Hexane | Fischer and Schneider |
| $Na_2[C_5H_5V(CO)_2(CN)_2]$<br>$Na[C_5H_5V(CO)_3(CN)]$ | Sephadex G-10 40 × 2 cm | Water | Fischer and Schneider |

* Listed in order of elution.

TABLE 51.9
APPLICATIONS OF COLUMN CHROMATOGRAPHY IN THE CHEMISTRY OF METAL CARBONYLS AND ORGANOMETALLIC COMPOUNDS

| Compounds separated* | Sorbent | Eluent | References |
|---|---|---|---|
| *trans*-[$(CO)_4ML_2$] **<br>*cis*-[$(CO)_4ML_2$] ** | Alumina<br>70 × 3 cm | Light petroleum<br>(b.p. 30–60°C) | Grim and Wheatland |
| *trans*-[$(CO)_3$-Cr-(1-acetoxytetralin)] | Alumina<br>(deactivated) | Diethyl ether–light petroleum<br>(b.p. 40–60°C)<br>(1:12) | Jackson and Mitchell |
| *cis*-[$(CO)_3$-Cr-(1-acetoxytelin)] | | (1:1) | |
| $C_5HF_6(2H)Mn(CO)_5$<br>$C_5HF_6(4H)Mn(CO)_5$<br>$C_5HF_6(5H)Mn(CO)_5$<br>$C_5HF_4O(4H)Mn(CO)_5$ | Alumina<br>(deactivated)<br>80 × 2.5 cm | *n*-Pentane | Haszeldine *et al.* |
| *anti*-[$CH_3S$-Fe-$(CO)_3$]$_2$<br>*syn*-[$CH_3S$-Fe-$(CO)_3$]$_2$<br>[$CH_3S$-$Fe_2$-$(CO)_6$]$_2$S | Acid alumina<br>(activity III)<br>60 × 2.5 cm | *n*-Hexane | Coleman *et al.* (1967) |
| 2,7-DOX-Fe-$(CO)_3$<br>2,7-DOX-[Fe-$(CO)_3$]$_2$ | Alumina<br>(activity III) | Light petroleum | Fischer *et al.* (1967) |
| $(CO)_6$-$Os_2$-$I_2$<br>$(CO)_8$-$Os_2$-$I_2$ | Florisil | *n*-Hexane<br>Dichloromethane–*n*-hexane (1:1) | Green *et al.* |
| MT-Cr-$(CO)_3$<br>MXL-Cr-$(CO)_3$<br>TO-Cr-$(CO)_3$ | Carbowax 400 on Porasil C | 2,2,4-Trimethylpentane | Veening *et al.* |
| *trans*-[$PtCl_2(Et_3P)_2$]<br>*trans*-[Pt(*n*-$C_3H_7$)Cl$(Et_3P)_2$]<br>*trans*-[PtH$(Et_3P)_2$Cl]<br>*cis*-[Pt(n-$C_3H_7)_2(Et_3P)_2$] | Alumina<br>35 × 2 cm | Diethyl ether–light petroleum<br>(b.p. 40–60°C)<br>(20:80) | Chatt *et al.* |

* Listed in order of elution. DOX = dimethyloxepin; MT = mesitylene; MXL = *m*-xylene; TO = toluene.
** M = W or Mo; L = alkyl, arylalkyl or diarylphosphine.

The adsorbed complex eluted with benzene forms a band 1 m in length. The effluent is collected in fractions and these are further purified by TLC.

Other applications, with the exception of purification processes (see Fischer *et al.*, 1969; Gowal and Schlögl, 1968b; Herberich and Michelbring), concerning the separation of the reaction mixtures are listed in Table 51.8.

Applications of column chromatography in the chemistry of organometallic compounds and metal carbonyls are summarized in Table 51.9.

## REFERENCES

Abe, M., *Bull. Chem. Soc. Jap.*, 42 (1969) 3701.
Abe, M. and Ito, T., *Bull. Chem. Soc. Jap.*, 40 (1967) 1013.
Ackers, G. K., *Biochemistry*, 3 (1964) 723.
Akaza, I., *Bull. Chem. Soc. Jap.*, 39 (1966a) 980.
Akaza, I., *Bull. Chem. Soc. Jap.*, 39 (1966b) 585.
Akaza, I., Kiba, T. and Kiba, T., *Bull. Chem. Soc. Jap.*, 43 (1970) 2063.
Akaza, I., Kiba, T. and Taba, M., *Bull. Chem. Soc. Jap.*, 42 (1969) 1291.
Araki, S., Suzuki, S., Hobo, T., Yoshida, T., Yoshizaki, K. and Yamada, M., *Bunseki Kagaku (Jap. Anal.)*, 17 (1968) 847.
Basila, M. R., *J. Chem. Phys.*, 35 (1961) 1151.
Benkeser, R. A., Nagai, Y. and Hooz, J., *Bull. Chem. Soc. Jap.*, 36 (1963) 482.
Benz, C. and Paiaxo, L. M., *Chim. Anal. (Paris)*, 50 (1968) 247.
Birney, D. G., Blake, W. E. and Meldrum, P. R., *Talanta*, 15 (1968) 557.
Bosnich, B., *J. Amer. Chem. Soc.*, 89 (1967) 6143.
Bradley, M. P. T. and Pantony, D. A., *Talanta,* 16 (1969) 473.
Bridges, K. L. and Chang, J. C., *Inorg. Chem.*, 6 (1967) 619.
Broomhead, J. A. and Grumley, W., *Inorg. Chem.*, 10 (1971) 2002.
Brown, W. and Chitumbo, K., *J. Chromatogr.*, 63 (1971) 478.
Brubaker, Jr., G. R., Legg, J. I. and Douglas, B. E., *J. Amer. Chem. Soc.*, 88 (1966) 3446.
Burwell, Jr., R. L., Pearson, R. G., Haller, G. L., Tjok, P. B. and Chock, S. P., *Inorg. Chem.*, 4 (1965) 1123.
Ćelap, M. B., Niketić, S. R., Janjić, T. J. and Nikolić, V. N., *Inorg. Chem.*, 6 (1967) 2063.
Chatt, J., Coffey, R. S., Gough, A. and Thompson, D. T., *J. Chem. Soc., A*, (1968) 190.
Coetzee, C. J. and Rohwer, E. F. C. H., *Anal. Chim. Acta*, 44 (1969) 293.
Coleman, J. M., Wojcicki, A., Pollick, P. J. and Dahl, L. F., *Inorg. Chem.*, 6 (1967) 1236.
Coleman, J. S., Asprey, L. B. and Chisholm, R. C., *J. Inorg. Nucl. Chem.*, 31 (1969) 1167.
Combs, C. S., Ashmore, C. I., Bridges, A. F., Swanson, C. R. and Stephens, W. O., *J. Org. Chem.*, 34 (1969) 1511.
Cziboly, C., Zsinka, L. and Szirtes, L., *Magy. Kem. Lapja,* 24 (1969) 470; *Anal. Abstr.*, 19 (1970) 3700.
De Bruyne, R. C., *J. Inorg. Nucl. Chem.*, 32 (1970) 348.
Dhar, S. K., Doron, V. and Kirschner, S., *J. Amer. Chem. Soc.*, 80 (1958) 753.
Dhar, S. K., Doron, V. and Kirschner, S., *J. Amer. Chem. Soc.*, 81 (1959) 6372.
Dormon, A. and Decombe, J., *Bull. Soc. Chim. Fr.*, (1968) 3673.
Douglas, B. E. and Yamada, S., *Inorg. Chem.*, 4 (1965) 1561.
Druding, L. F. and Hagel, R. B., *Anal. Chem.*, 38 (1966) 478.
Druding, L. F. and Kauffman, G. B., *Coord. Chem. Rev.*, 3 (1968) 408.
Dwyer, F. P., MacDermott, T. E. and Sargeson, A. M., *J. Amer. Chem. Soc.*, 85 (1963) 2913.
Dybczyński, R., *J. Chromatogr.*, 71 (1972) 507.
Egan, B. Z., *J. Chromatogr.*, 34 (1968) 382.
Egger, H. and Nikiforov, A., *Monatsh. Chem.*, 99 (1968) 2296.
Falk, H., Hofr, O. and Schlögl, K., *Monatsh. Chem.*, 100 (1969) 624.
Falk, H. and Schlögl, K., *Monatsh. Chem.*, 95 (1965) 266.
Falk, H. and Schlögl, K., *Tetrahedron,* 22 (1966) 3047.
Falk, H., Schlögl, K. and Steyrer, W., *Monatsh. Chem.*, 97 (1966) 1029.
Faris, J. P., *J. Chromatogr.*, 26 (1967) 232.
Fay, R. C., Girgis, A. Y. and Klabunde, U., *J. Amer. Chem. Soc.*, 92 (1970) 7056.
Fidelis, I. and Siekierski, S., *J. Chromatogr.*, 17 (1965) 542.
Fischer, E. O., Kreiter, C. G., Rühle, H. and Schwarzhaus, K. E., *Chem. Ber.*, 100 (1967) 1905.
Fischer, E. O., Louis, E., Bathelt, W. and Müller, J., *Chem. Ber.*, 102 (1969) 2547.
Fischer, E. O. and Schneider, R. J. J., *Chem. Ber.*, 103 (1970) 3684.

Fischer, E. O. and Wehner, H. W., *Chem. Ber.*, 101 (1968) 454.
Fitzgerald, W. R. and Watts, D. W., *J. Amer. Chem. Soc.*, 89 (1967) 821.
Ford, D. C. and Sutton, C., *Inorg. Chem.*, 8 (1969) 1544.
Frache, R. and Dadone, A., *Chromatographia*, 5 (1972) 581.
Freeman, W. A. and Liu, C. F., *Inorg. Chem.*, 7 (1968) 764.
Fritz, J. S., Frazee, R. T. and Latwesen, G., *Talanta*, 17 (1970) 857.
Fritz, J. S. and Latwesen, G., *Talanta*, 17 (1970) 81.
Garunchio, V. and Strazza, G. G., *Chromatogr. Rev.*, 8 (1966) 260.
Gillard, R. D. and Mitchel, P. R., *Struct. Bonding (Berlin)*, 7 (1970) 49.
Gillard, R. D. and Payne, N. C., *J. Chem. Soc., A*, (1967) 1197.
Gillard, R. D., Payne, N. C. and Phillips, D. C., *J. Chem. Soc., A*, (1968) 973.
Girgis, A. Y. and Fay, R. C., *J. Amer. Chem. Soc.*, 92 (1970) 7061.
Gordon, II, J. G. and Holm, R. H., *J. Amer. Chem. Soc.*, 92 (1970) 5219.
Gowal, H. and Schlögl, K., *Monatsh. Chem.*, 99 (1968a) 267.
Gowal, H. and Schlögl, K., *Monatsh. Chem.*, 99 (1968b) 972.
Grande, J. E. and Beukenkamp, J., *Anal. Chem.*, 28 (1956) 1497.
Green, M., Bruce, M. I., Cooke, M. and Westlake, D. J., *J. Chem. Soc., A*, (1969) 987.
Grim, S. O. and Wheatland, D. A., *Inorg. Chem.*, 8 (1969) 1716.
Haszeldine, R. N., Fields, R., Green, M., Harrison, T., Jones, A. and Lever, A. B. P., *J. Chem. Soc., A*, (1970) 49.
Hathaway, B. J. and Lewis, C. E., *J. Chem. Soc., A*, (1969) 1176.
Herberich, G. E. and Michelbring, R., *Chem. Ber.*, 103 (1970) 3615.
Hill, E. A. and Richards, J. H., *J. Amer. Chem. Soc.*, 83 (1961) 4216.
Huber, J. F. K., Kraak, J. C. and Veening, H., *Anal. Chem.*, 44 (1972) 1554.
Hughes, R. E., Vuuren, P. J. V., Fletterick, R. J. and Heinvald, J., *Chem. Commun.*, (1970) 883.
Ishida, K., Kiriyama, T. and Kuroda, R., *Anal. Chim. Acta*, 41 (1968) 537.
Ishida, K. and Kuroda, R., *Anal. Chem.*, 39 (1967) 212.
Jackson, W. R. and Jennings, W. B., *J. Chem. Soc., B*, (1969) 1221.
Jursík, F., *Separ. Sci.*, 3 (1968) 235.
Jursík, F. and Hájek, B., *Collect. Czech. Chem. Commun.*, 38 (1973) 907.
Jursík, F., Petrů, F. and Hájek, B., *Proc. 3rd Conf. on Coordination Chemistry, Debrecen, 1970*, Vol. 2, Akadémiai Kiadó, Budapest, p. 251.
Jursík, F., Sýkorová, D. and Hájek, B., unpublished results, 1972.
Jursík, F., Wollmanová, D. and Hájek, B., *Collect. Czech. Chem. Commun.*, 38 (1973) 3627.
Kauffman, G. B., Pinnell, R. P. and Takahashi, L. F., *Inorg. Chem.*, 1 (1962) 544.
Keene, F. R. and Searle, G. H., *Inorg. Chem.*, 11 (1972) 148.
Keene, F. R., Searle, G. H., Yoshikawa, Y., Imai, A. and Yamasaki, K., *Chem. Commun.*, (1970) 784.
Kertes, S. and Lederer, M., *Anal. Chim. Acta*, 15 (1956) 543.
King, E. L. and Walters, R. R., *J. Amer. Chem. Soc.*, 74 (1952) 4471.
King, R. B., Houk, L. W. and Pannell, K. H., *Inorg. Chem.*, 8 (1969) 1042.
Koguchi, K., Waki, H. and Ohashi, S., *J. Chromatogr.*, 25 (1966) 398.
Kojima, Y. and Shibata, M., *Inorg. Chem.*, 9 (1970) 238.
Kojima, Y. and Shibata, M., *Inorg. Chem.*, 10 (1971) 2382.
Korkisch, J., *Separ. Sci.*, 1 (1966) 159.
Korkisch, J. and Klakl, H., *Talanta*, 15 (1968) 339.
Korkisch, J. and Klakl, H., *Talanta*, 16 (1969) 377.
Krebs, H. and Diewald, J., *Z. Anorg. Allg. Chem.*, 287 (1956) 98.
Krebs, H. and Rasche, R., *Naturwissenschaften*, 41 (1954) 63.
Kühn, G. and Hoyer, E., *Z. Anal. Chem.*, 228 (1967) 166.
Kuroda, R., Ishida, K. and Kiriyama, T., *Anal. Chem.*, 40 (1968) 1502.
Lederer, M. and Ossicini, L., *J. Chromatogr.*, 13 (1964) 188.
Legg, J. I., *Chem. Commun.*, (1967) 675.
Legg, J. I. and Cooke, D. W., *Inorg. Chem.*, 4 (1965) 1576.
Legg, J. I. and Cooke, D. W., *Inorg. Chem.*, 5 (1966) 594

Legg, J. I. and Douglas, B. E., *Inorg. Chem.*, 7 (1968) 1452.
Lindqvist, B., *Acta Chem. Scand.*, 16 (1962) 1794.
Linhard, M., Weigel, M. and Flygare, H., *Z. Anorg. Allg. Chem.*, 263 (1950) 233.
Marcus, Y. and Eyal, E., *J. Inorg. Nucl. Chem.*, 32 (1970) 2040.
Marcus, Y. and Kertes, A. S., *Ion Exchange and Solvent Extraction of Metal Complexes*, Wiley, New York, 1969.
Marković, V. and Schweitzer, G. K., *J. Inorg. Nucl. Chem.*, 33 (1971) 3197.
Marquarding, D., Klusacek, H., Gokel, G., Hoffmann, P. and Ugi, I., *J. Amer. Chem. Soc.*, 92 (1970) 5389.
Martin, A. J. P. and Synge, R. L. M., *Biochem. J.*, 35 (1941) 1358.
Massart, D. L. and Bossaert, W., *J. Chromatogr.*, 32 (1968) 195.
Matsuoka, N., Hidaka, J. and Shimura, Y., *Bull. Chem. Soc. Jap.*, 40 (1967) 1868.
Matsuoka, N., Hidaka, J. and Shimura, Y., *Inorg. Chem.*, 9 (1970) 719.
Molnár, F., Horwáth, A. and Khalkin, V. A., *J. Chromatogr.*, 26 (1967) 225.
Muzarelli, R. A. A., *Advan. Chromatogr.*, 5 (1968) 127.
Nagai, Y., Hooz, J. and Benkeser, R. A., *Bull. Chem. Soc. Jap*, 37 (1964) 53.
Neddermeyer, P. A., *Diss. Abstr., B*, 29 (1968) 2782.
Neddermeyer, P. A. and Rogers, L. B., *Anal. Chem.*, 41 (1969) 94.
Nesmeyanov, A. N., Polovyanyuk, I. V., Lokshin, B. V., Makarova, L. G. and Chapovskij, V. A., *Zh. Obshch. Khim.*, 37 (1967) 2015.
Neumann, J. F., Paxson, J. R. and Cummishey, C. J., *J. Inorg. Nucl. Chem.*, 30 (1968) 2243.
Nickless, G., *Lab. Pract.*, 16 (1967) 1238.
Nickless, G., *Advan. Chromatogr.*, 7 (1968) 121.
Nugent, M. J., Kummer, R. and Richards, J. H., *J. Amer. Chem. Soc.*, 91 (1969) 6141.
Ohashi, S., Tsuji, N., Ueno, Y., Takeshita, B. and Muto, M., *J. Chromatogr.*, 50 (1970) 349.
O'Laughlin, J. W. and Jensen, D. F., *J. Chromatogr.*, 32 (1968) 567.
Orlandini, K. A., *Inorg. Nucl. Chem. Lett.*, 5 (1969) 325.
Ortner, H. M. and Pacher, O., *J. Chromatogr.*, 71 (1972) 55.
Ossicini, L. and Lederer, M., *J. Chromatogr.*, 17 (1965) 387.
Pearson, R. G., Henry, P. M. and Basolo, F., *J. Amer. Chem. Soc.*, 79 (1957) 5379, 5382.
Pecsok, R. L. and Saunders, D., *Separ. Sci.*, 3 (1968) 325.
Piper, T. S., *J. Amer. Chem. Soc.*, 83 (1961) 3908.
Pollard, F. H., Nickless, G. and Murray, J. D., *J. Chromatogr.*, 27 (1967) 271.
Pollard, F. H., Nickless, G. and Rothwell, M. T., *J. Chromatogr.*, 10 (1963) 212.
Pollard, F. H., Rogers, D. E., Rothwell, M. T. and Nickless, G., *J. Chromatogr.*, 9 (1962) 227.
Reich-Rohrwig, P. and Schlögl, K., *Monatsh. Chem.*, 99 (1968) 1752.
Reinehart, A. B., *J. Amer. Chem. Soc.*, 79 (1957) 5749.
Rosenblum, M. and Woodward, R. B., *J. Amer. Chem. Soc.*, 80 (1958) 5443.
Saunders, D. L., *Diss. Abstr., B*, 29 (1969) 2783.
Schlögl, K. and Fried, M., *Monatsh. Chem.*, 95 (1964) 558.
Schlögl, K. and Mechtler, H., *Angew. Chem.*, 78 (1966) 606.
Schmitt, D. H. and Fritz, J. S., *Talanta,* 15 (1968) 515.
Schoenberg, L. N., Cooke, D. W. and Liu, C. F., *Inorg. Chem.,* 7 (1968) 2386.
Schwab, G. M. and Jockers, K., *Angew. Chem.*, 50 (1937) 546.
Sherma, J., *J. Chromatogr.*, 26 (1967) 327.
Shibata, M., Nishikawa, J. and Nishida, Y., *Bull. Chem. Soc. Jap.*, 39 (1966) 2310.
Shibata, M., Tanabe, I., Okuda, K. and Kadota, K., *Bull. Chem. Soc. Jap.*, 41 (1968) 2627.
Shiga, M., Tsunashima, M., Kono, H., Motoyama, I. and Hata, K., *Bull. Chem. Soc. Jap.*, 43 (1970) 841.
Smith, A. A. and Haines, R. A., *J. Amer. Chem. Soc.*, 91 (1969) 6280.
Springer, Jr., C. S., Feibush, B. and Sievers, R. E., *Inorg. Chem.,* 10 (1971) 1242.
Strelow, F. W. E., Coetzee, J. H. J. and Van Zyl, C. R., *Anal. Chem.*, 40 (1968) 197.
Strelow, F. W. E., Van Zyl, C. R. and Bothma, C. J. C., *Anal. Chim. Acta*, 45 (1969) 81.
Strelow, F. W. E., Victor, A. H., Van Zyl, C. R. and Eloff, C., *Anal. Chem.*, 43 (1971) 870.

Streuli, C. A. and Rogers, L. B., *Anal. Chem.*, 40 (1968) 653.
Taylor, L. F. and Busch, D. H., *J. Amer. Chem. Soc.*, 89 (1967) 5372.
Tiethof, J. A. and Cooke, D. W., *Inorg. Chem.*, 11 (1972) 315.
Tobe, M. L. and Watts, D. W., *J. Chem. Soc., London*, (1962) 4614.
Tobe, M. L. and Watts, D. W., *J. Chem. Soc., London*, (1964) 2891.
Tsitovich, J. K., *Izv. Vyssh. Ucheb. Zaved., Khim. Khim. Tekhnol.*, 11 (1968) 758; C.A., 69 (1968) 99730x.
Tustanowski, S., *J. Chromatogr.*, 31 (1967a) 268.
Tustanowski, S., *J. Chromatogr.*, 31 (1967b) 270.
Ueno, Y., Yoza, N. and Ohashi, S., *J. Chromatogr.*, 52 (1970a) 321.
Ueno, Y., Yoza, N. and Ohashi, S., *J. Chromatogr.*, 52 (1970b) 469.
Vanderdeelen, J., *Anal. Chim. Acta*, 49 (1970) 360.
Veening, H., Greenwood, J. M., Shanks, W. H. and Willeford, B. R., *Chem. Commun.*, (1969) 1305.
Wehner, H. W., Fischer, E. O. and Müller, J., *Chem. Ber.*, 103 (1970) 2258.
Westman, B. C. and Reinehart, A. B., *Acta Chem. Scand.*, 16 (1961) 1199.
Winget, J. O. and Lindstrom, R. S., *Separ. Sci.*, 4 (1969) 209.
Wódkiewicz, L. and Dybczyński, R., *J. Chromatogr.*, 32 (1968) 394.
Yamada, S., Hidaka, J. and Douglas, B. E., *Inorg. Chem.*, 10 (1971) 2187.
Yoza, N. and Ohashi, S., *J. Chromatogr.*, 41 (1969) 429.

Chapter 52

# Isotopes and radioactive compounds

J. DRŠATA and I. M. HAIS

## CONTENTS



## ISOTOPIC EFFECTS IN LIQUID COLUMN CHROMATOGRAPHY

There have been relatively few reports in recent years describing even the partial separation of chemically identical substances that differ only in atomic or molecular weight. Thus Knyazev *et al.* applied the "stationary zone" method of Spedding *et al.* to the separation of the isotopes of iron-54 to iron-58 on a column of the cation-exchange resin Dowex 50-X4 ($H^+$). When 15% citric acid buffered with ammonia to pH 3.5 was used, the head fraction was enriched with iron-54. With 0.75 *N* hydrofluoric acid and 0.75 *N* ammonium fluoride solution, the head fraction was enriched with iron-58. The isotopes in question have been differentiated by mass spectroscopy.

The effect of the isotopic composition of the solutes on their ion-exchange behaviour has been noted for $^3$H-labelled compounds (Gottschling and Freese) and $^{14}$C-labelled compounds (Piez and Eagle 1955, 1956). This observation was confirmed by Gaitonde and Nixey, who reported the enrichment of $^{14}$C-labelled amino acids in the tail fraction on

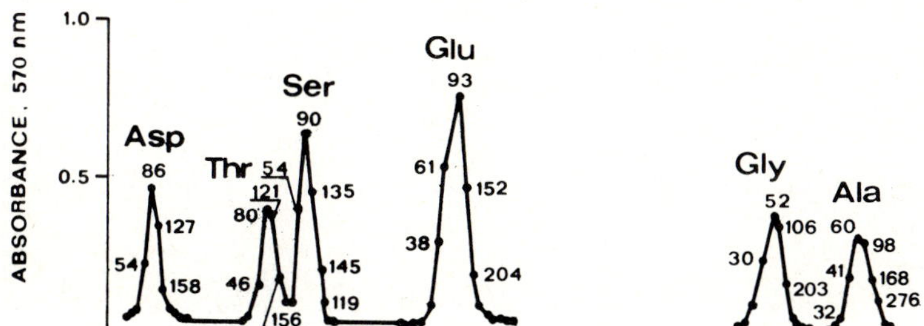


Fig. 52.1. Resolution of $^{14}$C- and $^{12}$C-labelled amino acids on Chromobeads. A cation-exchange resin by elution with lithium citrate buffer, pH 3.0 (Gaitonde and Nixey). The numbers of the elution profile indicate the specific radioactivity of the amino acid in the fraction as a percentage of the mean specific radioactivity of that amino acid. The mean specific radioactivities (dpm/$\mu$mole) of each amino acid determined by pooling all fractions representing that amino acid were as follows: aspartate, 24,790; threonine, 13,820; serine, 7830; glutamate, 14,820; glycine, 13,260; and alanine, 26,210.

TABLE 52.1
SEPARATION OF RADIOACTIVE SUBSTANCES

NAA = neutron activation analysis.

| Separated substances or nuclides | Purpose | Column | Mobile phase | Notes on detection etc. | References |
|---|---|---|---|---|---|
| $^{64}$Cu, $^{66}$Cu, $^{65}$Zn, $^{68}$Zn, $^{69m}$Zn | Cu and Zn in plants by NAA | Dowex 1-X10 ($Cl^-$) | 3 *N* HCl (Cu); 0.005 *N* HCl (Zn) | NaI(Tl) crystal, multi-channel analyzer | Souliotis |
| $^{110m}$Ag, $^{60}$Co, $^{24}$Na, $^{51}$Cr, $^{59}$Fe, $^{64}$Cu, $^{67}$Cu | Trace impurities in Zn by NAA | Dowex 1-X8 ($Cl^-$) | 12 *N* HCl (Cr, Na); 10 *N* HCl (Ag); 4 *N* HCl (Co); 2.5 *N* HCl (Cu); 0.5 *N* HCl (Fe) | | Machiroux and Mousty |
| $^{51}$Cr, $^{64}$Cu, $^{58}$Co, $^{60}$Co, $^{59}$Fe, $^{187}$W, $^{99}$Mo, $^{99}$Tc, $^{65}$Zn, $^{92}$Ga | Cr by NAA | Dowex 1-X4 | HF; HCl–HF; $NH_4Cl$–$NH_4F$ | γ-spectra | Dugain *et al.* |
| $^{56}$Mn | $^{56}$Mn from neutron-irradiated $KMnO_4$ | A block of porous PTFE supporting bis(di-2-ethylhexyl)phosphoric acid | 4% solution of neutron-irradiated $KMnO_4$ | β–γ coincidence | Zhukovskii *et al.* |
| $^{99m}$Tc, $^{186}$Re | Mo and Re in $ZnSO_4$ by NAA | Dowex 1-X8 ($NO_3^-$) | 0.5 *N* $HNO_3$ ($^{239}$Nb, $^{187}$W, $^{203}$Hg, $^{22}$Na); 0.2 *N* $HClO_4$ (Re); 2 *N* $HClO_4$ (Tc) | γ-spectra | Dams and Hoste |
| $^{181}$Hf, $^{182}$Ta, $^{233}$Pa, $^{95}$Zr | Ta and Hf in silicates by NAA | Dowex 1-X8 | 1 *M* HF–0.25 *M* HCl; 5 *M* HF–2.5 *M* HCl (Hf); 2.6 *M* $NH_4Cl$–1 *M* HF, $NH_3$ to pH 5.5 (Ta) | $^{182}$Ta: γ, single-channel analyzer; $^{181}$Hf: multi-channel analyzer; $^{233}$Pa, $^{95}$Zr: γ-spectrometry | Greenland |

| | | | | | |
|---|---|---|---|---|---|
| $^{179m}Hf$, $^{46m}Sc$ | Hf in Sc by NAA | Dowex 1-X8 | 0.4 *M* $H_2SO_4$ (Sc); 2 *M* $H_2SO_4$ (Hf) | Separation and determination takes 1 h | Balsenc *et al.* |
| U, rare earths | KU-2 separation of rare earths from U, their neutron activation, coprecipitation with $Fe(OH)_3$, HCl solution on Dowex 1, individual rare earths on KU-2 | KU-2 (cation-exchange resin) | Ascorbic acid (U); 4–5 *N* HCl (rare earths) | Drops on paper strip, dried (IR), each drop counted separately | Alimarin *et al.* |
| Fe, rare earths | | Dowex 1-X8 ($Cl^-$) | | | |
| $^{175}Yb$, $^{166}Ho$, $^{165}Dy$, Gd, Eu, Sm, $^{140}La$ | | KU-2 ($NH_4^+$) | Ammonium α-hydroxyisobutyrate | | |
| $^{160}Tb$, $^{169}Yb$, $^{175}Yb$, $^{177}Lu$ | Lu, Yb and Tb in rocks by NAA | Dowex 1-X8 ($NO_3^-$) | 7 *M* $HNO_3$–methanol (1:9) (Yb and Lu); 7 *M* $HNO_3$–methanol (2:8) (Tb) | 400-channel pulse height analyzer with a well-type NaI(Tl) crystal | Brunfelt and Steinnes |
| $^{169}Yb$, $^{175}Yb$, $^{177}Lu$ | Lu in gadolinite by NAA | Dowex 50W-X8 | 0.5 *M* α-hydroxyisobutyric acid | γ-spectroscopy | Massart and Hoste |
| $^{153}Gd$, $^{159}Gd$, $^{161}Tb$, $^{152m}Eu$, $^{153}Sm$, $^{60}Co$ | Rare earth impurities in $Gd_2O_3$ by NAA | Bio-Rad AG 50W-X12 (cation-exchange resin) | 0.4 *M* α-hydroxyisobutyric acid with $NH_3$ to pH 4 (first chromatography to remove Eu, second to purify Pm) | High-resolution γ-spectrometry with Ge(Li) crystal | Pinajian and Raman |
| $^{51}Ti$, $^{46}Sc$, $^{24}Na$, $^{51}Cr$, $^{76}As$, $^{122}Sb$, $^{124}Sb$, $^{187}W$, $^{64}Cu$, $^{182}Ta$, $^{65}Zn$, $^{60}Co$, $^{59}Fe$, $^{198}Au$, $^{117m}Sn$, $^{99}Mo$, $^{110m}Ag$ | 16 trace elements in Ti by NAA | Dowex 1-X8 | 1 *M* HF, 9 *M* HCl | High-resolution Ge(Li) γ-spectrometer | Neirinckx *et al.* (1968) |

(Continued on p. 1118)

TABLE 52.1 *(continued)*

| Separated substances or nuclides | Purpose | Column | Mobile phase | Notes on detection etc. | References |
|---|---|---|---|---|---|
| 27 elements | Trace constituents in $TiO_2$ by NAA | Dowex 1-X8 | 1 *M* HF; 22 *M* HF; 6 *N* HCl | γ-spectroscopy with a high-resolution Ge(Li) detector | Neirinckx *et al.* (1969a) |
| 21 metals | Trace impurities in $TiO_2$ single crystals by NAA | Dowex 1-X8 | 12 *M* HCl; 9 *M* HCl; HF | γ-spectrometry with Ge(Li) crystal | Neirinckx *et al.* (1969b) |
| $^{69m}Zn$, $^{65}Zn$, $^{24}Na$, $^{56}Mn$, $^{64}Cu$, $^{67}Cu$, $^{60}Co$, $^{115m}Cd$, $^{115}Cd$, $^{114m}In$, $^{115m}In$, $^{116m}In$, $^{55}Fe$, $^{59}Fe$ | Co, Cd, Fe, and In in electrolytic $ZnSO_4$ by NAA | Dowex 1-X8 | 2 *N* HCl (Co, Na, Cu, Mn); 0.3 *N* HCl (In, Fe); 0.1 *N* HCl (Zn); $H_2O$ (Cd) | | Liessens *et al.* |
| $^{233}Th$ | Th in geological materials by NAA | Dowex 1-X8 ($NO_3^-$) | 8 *N* $HNO_3$; 6 *N* $HNO_3$ (rare earths); 0.1 *N* $HNO_3$ (Th) | β-counting of air-dried samples | Aruscavage and Millard |
| $^{233}Th$ | Th in rocks and minerals by NAA | Dowex 1-X8 ($NO_3^-$) | 8 *M* $HNO_3$; 2 *M* HCl (Th) | β-activity counted in gas-flow proportional counters | Norton and Stoenner |
| | | Dowex 1-X8 ($Cl^-$) | Conc. HCl (Th) | | |
| $^{109}Cd$, $^{65}Zn$ | Cyclotron-produced $^{109}Cd$ from the Ag matrix | Dowex 1-X8 ($Cl^-$) | 0.02 *M* HCl (Zn, Cu); 3 *M* $HNO_3$ (Cd) | Chromatography following precipitation of Ag | Neirinckx |

| | | | | | |
|---|---|---|---|---|---|
| $^{51}Cr$ | Cr(VI), Cr(III)-monomer, -dimer and -polymer species by NAA | Bio-Rad AG 50W-X8 or Dowex 50W-X8 ($H^+$) | 0.05 *M* $HClO_4$ [Cr(VI)]; 4 *M* $HClO_4$ [Cr(III)-monomer]; 5 *M* $HClO_4$ [Cr(III)-dimer]; then batch treatment | Well-type NaI(Tl) scintillation counting system | Collins *et al.* |
| $^{149}Pm$, $^{147}Pm$ | Pm from neutron-irradiated $Nd_2O_3$ (enriched in $^{148}Nd$) | Hydrophobised Kieselguhr supporting di-2-ethylhexyl-phosphoric acid | 3 *N* $HNO_3$ (Nd); 4 *N* $HNO_3$ (Pm) | γ-spectra; liquid scintillation counting of $^{147}Pm$ | Joon *et al.* |
| $^{59}Co$, $^{60}Co$, $^{51}Cr$, $^{46}Sc$, $^{59}Fe$ | Impurities in single crystals of MgO by NAA | Dowex 1 ($Cl^-$) | 12 *N* HCl (Cr, Sc); 90% aqueous acetone (Fe); 1 *N* HCl (Co) | γ-spectrometry | Lee |
| $^{128}I$ | Routine production of $^{128}I$ by neutron irradiation | Dowex 50-X8 ($H^+$ and $Na^+$ in series) | Solution of irradiated KI | | Domondon |
| $Cl^-$, $Br^-$, $I^-$ ($^{131}I$, $^{82}Br$ in elaboration of the method) | Separation of $I^-$ (from urine), which is then subjected to NAA | Amberlite IRA-400 ($Cl^-$), 14–52 mesh | 0.2 *N* $NH_4NO_3$ ($Br^-$, $Cl^-$); $H_2O$ ($I^-$) | γ-activities of air-dried samples | Morgan *et al.* |
| $^{99}Mo$ | Separation of Mo from U fission products | $Al_2O_3$ | 3 *M* $HNO_3$; $H_2O$; 0.01 *M* ammonia solution; 15 *M* ammonia solution (Mo) | | Kulesza |
| $^{99m}Tc$ | $^{99m}Tc$ (after equilibrium) from $^{99}Mo$ | | 0.1 *M* $HNO_3$ | | |

(Continued on p. 1120)

TABLE 52.1 *(continued)*

| Separated substances or nuclides | Purpose | Column | Mobile phase | Notes on detection etc. | References |
|---|---|---|---|---|---|
| $^{152}Pm$, $^{152}Nd$, rare earths | Rapid isolation of individual rare earths from $^{235}U$ fission, identification of $^{152}Pm$ produced from $^{152}Nd$ | Dowex 50W-X12 | 0.23 *M* α-hydroxyisobutyric acid with ammonia solution to pH 2.5 | | Wakat and Griffin |
| Ce | Radio-cerium in fission products | Siliconized Kieselguhr supporting bis(2-ethylhexyl) hydrogen phosphate | 5 *M* $HNO_3$, 0.5 *M* $NaBrO_3$; 5 *M* $HNO_3$; 5 *M* $HNO_3$, 0.1 *M* ascorbic acid (Ce) | β-activities (GM tube), γ-spectra (200-channel analyzer) | Krtil *et al.* |
| $^{147}Pm$ | $^{147}Pm$ from other fission product lanthanides | Exchange resin | | Pressure to increase flow-rate | Lowe |
| $^{95}Zr$, $^{95}Nb$; U fission products | $^{95}Zr$ and $^{95}Nb$ | Silica gel (A.S.T.M.) | 6 *N* HCl + 0.005 *N* $HNO_3$; 0.5% oxalic acid (Zr, Nd) | γ-spectra [NaI(TI) detector, 400-channel analyzer] | El-Garhy *et al.* |
| U fission products | Sequential preparation of fission products | Diaion SK-1 ($NH_4^+$, $H^+$), Diaion SA-100 ($Cl^-$) (similar to Dowex 50W-X8 and Dowex 1-X8, respectively) | HCl; oxalic acid; $NH_4Cl$; ammonium citrate in ammonia solution; ammonium citrate–citric acid; ammonium α-hydroxyisobutyrate–α-hydroxyisobutyric acid | β–γ coincidence | Natsume *et al.* |

| | | | | | |
|---|---|---|---|---|---|
| $^{90}Sr$, $^{89}Sr$ | Radio-Sr after chromatographic removal of anions and EDTA separation | Cation-exchange resin ($NH_4^+$) | EDTA, pH 6 (Sr) | $^{90}Sr$ calculated from $^{90}Y$ activity | Bouquiaux and Gillard-Baruh |
| $^{90}Sr$ | $^{90}Sr$ in milk and bones | Amberlite IR-120 | 0.05 *M* 1,2-diaminocyclohexane-N,N,N′, N′-tetraacetic acid | | Zima and Giacintov |
| $^{140}Ba$, $^{140}La$, $^{228}Ac$ | Separation of La and Ac | Ftoroplast-4 supporting tri-*n*-butyl phosphate | 10 *M* $NH_4NO_3$ + 0.1 *M* $HNO_3$ (Ba, then Ac, then La) | | Ziv *et al.* |
| $^{137}Cs$, $^{86}Rb$ | Extraction chromatography of alkali metals | Teflex (perfluorochloropolyethylene) or Ftoroplast (perfluoropolyethylene) supporting dipicrylamine in nitrobenzene | 0.15 *M* $(NH_4)_2CO_3$ (Rb); 0.5 *M* $(NH_4)_2CO_3$ (Cs) | Dried drops on Al planchettes GM counted | Kyrš and Kadlecová |
| $^{24}Na$, $^{42}K$, $^{86}Rb$, $^{134}Cs$ | Rapid separation of alkali metals | Aminex Q-150 S ($H^+$) | 1.61 *M* HCl | Pressure to increase flow-rate | Huber and Van Urk-Schoen |
| $^{7}Be$, $^{10}Be$ | Be $(OH)_2$ from rainwater and soil | Sodium diallyl phosphate complexing resin | 0.5 *M* $NH_4F$ | γ-spectrum ($^{7}Be$). The β-emitter $^{10}Be$ does not interfere | Möller |
| $^{60}Co$, $^{59}Fe$ | Co from Fe isotopes by extraction chromatography | Styrene–divinylbenzene copolymer supporting 0.018% dithizone | 0.2 *N* sodium acetate–0.2 *N* acetic acid buffer, pH 5.6 ($Fe^{2+}$) | γ-spectroscopy ($Fe^{3+}$ reduced with ascorbic acid before chromatography) | Spěváčková and Křivánek |

*(Continued on p. 1122)*

TABLE 52.1 *(continued)*

| Separated substances or nuclides | Purpose | Column | Mobile phase | Notes on detection etc. | References |
|---|---|---|---|---|---|
| $^{65}$Zn, $^{115m}$Cd, $^{110m}$Ag, $^{203}$Hg, $^{210}$Pb (RaD), $^{210}$Bi (RaE), $^{210}$Po (RaF) | Zn from Cd (applied in $HClO_4$–$NaClO_4$, pH 4) | Hydrophobized Celite 545 supporting $5 \times 10^{-4}$ *M* dithizone in $CCl_4$ | 0.1 *M* oxalic acid–sodium oxalate, pH 4.1 (Zn); oxalic acid (Cd) | Scintillation with well-type NaI(Tl) crystal; end-window GM | Šebesta |
| | Ag from Hg | | 2% KCNS in 0.1 *N* $H_2SO_4$ (Ag); 6% KI in acetate, pH 4 (Hg) | | |
| | Cd from Ag | | 0.1 *N* $HNO_3$ (Cd); 6% KI in acetate, pH 4 (Ag) | | |
| | Pb, Bi and Po | | 0.1 *N* HCl (Pb); 0.3 *N* HCl (Bi); 1.5 *N* HCl (Po) | | |
| Natural Th; enriched U; $^{131}$I | Th, U and Y from urine, I from foodstuffs | Microthene-710 supporting tri-*n*-octylphosphine oxide in cyclohexane | 4 *M* $HNO_3$; 3 *M* $H_2SO_4$ (Th); 1 *M* HF (U) | NaI(Tl) crystal with 400-channel analyzer; low-background $\beta$-detector | Testa |
| | | Microthene supporting benzene | 1 *M* $H_2SO_4$; $H_2O$; 0.1 *M* ascorbic acid | | |
| Pu | Pu in sea water | Bio-Rad AG 1-X8 ($Cl^-$) | 12 *M* HCl; 12 *M* HCl–1 *M* $NH_4I$ (20:1) (Pu) | $\alpha$-spectrometry (5.76 MeV, $^{236}$Pu) | Wong |
| Pu; rare earths ($^{152-154}$Eu and $^{177}$Lu as tracers) | Rare earths in Pu | Dowex 1-X4 ($NO_3^-$) | 7.2 *M* $HNO_3$ | $\alpha$ activity (Pu), NaI detector for $\gamma$-activity | Joshi *et al.* |

| | | | | | |
|---|---|---|---|---|---|
| Pu | Pu in environmental and biological samples | Bio-Rad AG 1-X2 ($CL^-$) or Dowex 1-X2 ($Cl^-$) | 9 *M* HCl; 7.2 *M* $HNO_3$; 1.2 *M* HCl–30% $H_2O_2$ (50:1) (Pu) | α-spectrometry for $^{236}Pu$ | Talvitie |
| $^{237,238,239}Np$, $^{233,235,238}U$, $^{232,234}Th$ | Separation of Th, U, Np | Ammonium 12-molybdophosphate + asbestos | 0.025 *M* $HNO_3$ [Np(V)]; 0.2 *M* HF (U); 12 *M* HCl [Th, Np(IV)] | α- and γ-spectra | Ganzerli-Valentini *et al.* |
| [1, 2-$^{14}C$]phosphorylethanolamine | Separation and determination of synthesized phosphorylethanolamine | Zeo-Karb 225 ($Na^+$) | 0.2 *M* sodium citrate, pH 3.1 | Automatic scintillation counting of the effluent passing through anthracene crystals | Porcellati and Di Jeso |
| $^{32}P$-labelled nucleoside triphosphates | Preparation of $^{32}P$-labelled nucleoside triphosphates | Dowex 1 ($HCOO^-$) | 4 *N* formic acid–0.1 *M* ammonium formate (diphosphates); 4 *N* formic acid–2 *M* ammonium formate (triphosphates) | | Keenan *et al.* |
| [$^{14}C$]ethidium bromide | Control of synthesized [$^{14}C$]ethidium bromide | $Al_2O_3$ | Acetone–methanol (92:8); acetone–methanol (5:1) (ethidium bromide) | Liquid scintillation counting in toluene-based scintillator | Loomeijer and Kroon |
| $^{125}ICl$, [$^{125}I$]immunoglobulin | Radioiodination of IgG; excess of reagent (ICl) is removed on Sephadex | Sephadex G-25 | Borate, pH 8 | γ-counting | Zappacosta and Rossi |

(Continued on p. 1124)

References p.1125

TABLE 52.1 *(continued)*

| Separated substances or nuclides | Purpose | Column | Mobile phase | Notes on detection etc. | References |
|---|---|---|---|---|---|
| $^{32}P$-labelled RNA | Labelling and separation of $^{32}P$-labelled RNA from Ehrlich ascites cells | Methylated albumin–Kieselguhr | Linear NaCl gradient from 0.4 to 1.0 *M* in 0.05 *M* sodium phosphate, pH 6.7 | Liquid scintillation counting ("NCS" in toluene–dioxane-based scintillator) | Hellung-Larsen and Frederiksen |
| $^{14}C$-labelled glutamic acid and glutamine | L-Glutamic acid for glutamate decarboxylase assay | Dowex 1-X2 ($Cl^-$) | $H_2O$ (glutamine); $H_2O$; 0.01 *N* HCl (glutamic acid) | Liquid scintillation counting | Wood *et al.* |
| [$^{14}C$]citrullin | Assay of carbamyl phosphate synthetase in presence of $H^{14}CO_3^-$, glutamine, ornithine and ornithine transcarbamylase | Dowex 50 ($H^+$) | $H_2O$; 0.5 *M* ammonium acetate | Liquid scintillation counting of fractions | Hager and Jones |
| $^{14}C$- and $^3H$-labelled δ-aminolaevulinic acid, $^{14}C$- and $^3H$-labelled succinic acid | Radiochemical assay of aminolaevulinate synthetase | Dowex 1 ($CH_3COO^-$) placed in tandem on Dowex 50 ($H^+$) | $H_2O$; then the acetate column was removed; 0.1 *N* HCl; 1 *M* sodium acetate | Liquid scintillation counting | Strand *et al.* |

cation-exchange resins (Fig. 52.1) and in the head fractions on anion-exchange resins. If this phenomenon is not borne in mind and specific radioactivities are estimated from individual fractions, gross errors may ensue. It is therefore necessary to pool all of the fractions under the peak for the estimation of specific radioactivity.

It is possible that the gain in resolving power (increase in the number of theoretical plates per column) that has been achieved in liquid column chromatography recently and is still increasing, will, within a short time, be reflected by more success in separations based on minute differences in distribution coefficients between substances that differ in isotope composition.

## SEPARATION OF RADIOACTIVE SUBSTANCES

In this book, most of the separations of radioactively labelled compounds have been covered in the various chapters in the special part, according to the chemical nature of the compounds (mostly organic). Applications to organic compounds and biochemicals are illustrated by several examples at the end of Table 52.1. Most of the other examples included in Table 52.1 belong to the realm of radiochemistry.

The products of nuclear fission have been separated by means of LCC. Neutron activation analysis (NAA) has often been supplemented by chromatographic separation, either before activation (Morgan *et al.*) in order to remove components (often those which occur in bulk) that might interfere in the interpretation of the NAA results or, more frequently, after activation. In some instances, chromatography has been used to separate unstable products that have then been subjected to further chromatographic analysis. NAA is involved in a substantial part of Table 52.1. In experiments with $^{3}H$- or $^{14}C$-labelled substances, the position of $^{3}H_2O$ or $^{14}CO_2$ ($HCO_3^-$), respectively, on the elution curves should be established in order to avoid misinterpretation of the chromatograms.

## REFERENCES

Alimarin, I. P., Miklishanskii, A. Z. and Yakovlev, Yu., *J. Radioanal. Chem.*, 4 (1970) 45.
Aruscavage, P. J. and Millard, Jr., H. T., *J. Radioanal. Chem.*, 11 (1972) 67.
Balsenc, L., Haerdi, W. and Monnier, D., *Anal. Chim. Acta*, 48 (1969) 213.
Bouquiaux, J. and Gillard-Baruh, J. H. C., *Radiochim. Acta*, 9 (1968) 153.
Brunfelt, A. O. and Steinnes, E., *Analyst (London)*, 94 (1969) 979.
Collins, C. H., Collins, K. E. and Ackerhalt, R. E., *J. Radioanal. Chem.*, 8 (1971) 263.
Dams, R. and Hoste, J., *Anal. Chim. Acta*, 41 (1968) 197.
Domondon, D. B., *Philipp. Nucl. J.*, 11 (1969) 129; *C.A.*, 72 (1970) 117060x.
Dugain, F., Cestre, C. and Beyssier, B., *Anal. Chim. Acta*, 42 (1968) 39.
El-Garhy, M., Shehata, M. K. K. and El-Bayoumy, S., *J. Radioanal. Chem.*, 11 (1972) 175.
Gaitonde, M. K. and Nixey, R. W. K., *Anal. Biochem.*, 50 (1972) 416.
Ganzerli-Valentini, M. T., Maxia, V., Meloni, S., Martinelli, A. and Rollier, M. A., *J. Radioanal. Chem.*, 11 (1972) 179.
Gottschling, H. and Freese, E., *Nature (London)*, 196 (1962) 829.
Greenland, L. P., *Anal. Chim. Acta*, 42 (1968) 365.
Hager, S. E. and Jones, M. E., *J. Biol. Chem.*, 242 (1967) 5667.

Hellung-Larsen, P. and Frederiksen, S., *Anal. Biochem.*, 40 (1971) 227.
Huber, J. F. K. and Van Urk-Schoen, A. M., *Anal. Chim. Acta*, 58 (1972) 395.
Joon, K., den Boef, R. and de Wit, R., *J. Radioanal. Chem.*, 8 (1971) 101.
Joshi, B. D., Patel, B. M. and Page, A. G., *Anal. Chim. Acta*, 57 (1971) 379.
Keenan, R. W., Zishka, M. K. and Nishimura, J. S., *Anal. Biochem.*, 47 (1972) 601.
Knyazev, D. A., Dobizha, E. V. and Klinskii, G. D., *Isotopenpraxis*, 6 (1970) 130.
Krtil, J., Bezděk, M. and Mencl, J., *Radioanal. Chem.*, 1 (1968) 369.
Kulesza, A., *Nukleonika*, 14 (1969) 216.
Kyrš, M. and Kadlecová, D., *J. Radioanal. Chem.*, 1 (1968) 103; *C.A.*, 68 (1968) 111002a.
Lee, H. M., *Anal. Chim. Acta*, 41 (1968) 431.
Liessens, J., Dams, R. and Hoste, J.. *Anal. Chim. Acta*, 45 (1969) 213.
Loomeijer, F. J. and Kroon, A. M., *Anal. Biochem.*, 49 (1972) 455.
Lowe, J. T., *U.S. At. Energy Comm.*, 1969, Rep. No. DP-1194, 29 pp.; *C.A.*, 72 (1970) 27327k.
Machiroux, R. and Mousty, F., *Anal. Chim. Acta*, 42 (1968) 371.
Massart, D. L. and Hoste, J., *Anal. Chim. Acta*, 42 (1968) 15.
Möller, P., *J. Inorg. Nucl. Chem.*, 32 (1970) 2473.
Morgan, D. J., Black, A. and Mitchell, G. R., *Analyst (London)*, 94 (1969) 740.
Natsume, H., Umezawa, H., Suzuki, T., Ichikawa, F., Sato, T., Baba, S. and Amano, H., *J. Radioanal. Chem.*, 7 (1971) 189.
Neirinckx, R., Adams, F. and Hoste, J., *Anal. Chim. Acta*, 43 (1968) 369.
Neirinckx, R., Adams, F. and Hoste, J., *Anal. Chim. Acta*, 46 (1969a) 165.
Neirinckx, R., Adams, F. and Hoste, J., *Anal. Chim. Acta*, 48 (1969b) 1.
Neirinckx, R. D., *Anal. Chim. Acta*, 58 (1972) 237.
Norton, E. F. and Stoenner, R. W., *Anal. Chim. Acta*, 55 (1971) 1.
Piez, K. A. and Eagle, H., *Science*, 122 (1955) 968.
Piez, K. A. and Eagle, H., *J. Amer. Chem. Soc.*, 78 (1956) 5284.
Pinajian, J. J. and Raman, S., *J. Inorg. Nucl. Chem.*, 30 (1968) 3151.
Porcellati, G. and Di Jeso, F., *J. Label. Compounds*, 3 (1967) 206.
Šebesta, F., *J. Radioanal. Chem.*, 7 (1971) 41.
Souliotis, A. G., *Analyst (London)*, 94 (1969) 359.
Spedding, F. H., Powell, J. E. and Svec, H. J., *J. Amer. Chem. Soc.*, 77 (1955) 6125.
Spěváčková, V. and Křivánek, M., *Radiochem. Radioanal. Lett.*, 3 (1970) 63.
Strand, L. J., Swanson, A. L., Manning, J., Branch, S. and Marver, H. S., *Anal. Biochem.*, 47 (1972) 457.
Talvitie, N. A., *Anal. Chem.*, 43 (1971) 1827.
Testa, C., *Anal. Chim. Acta*, 50 (1970) 447.
Wakat, A. and Griffin, C., *Radiochem. Radioanal. Lett.*, 2 (1969) 351.
Wong, K. M., *Anal. Chim. Acta*, 56 (1971) 355.
Wood, A. W., McCrea, M. E. and Seegmiller, J. E., *Anal. Biochem.*, 48 (1972) 581.
Zappacosta, S. and Rossi, G., *Immunochemistry*, 4 (1967) 122.
Zhukovskii, Yu. G., Katykhin, G. S., Martynov, A. L. and Nikitin, M. K., *Radiokhimiya*, 10 (1968) 252.
Zima, S. and Giacintov, P., *J. Radioanal. Chem.*, 7 (1971) 19.
Ziv, D. M., Shestakov, B. I. and Shestakova, I. A., *Radiokhimiya*, 10 (1968) 738.

# Subject index*,**

## A



*Compiled by H. Bečvářová.

**For some general terms which occur throughout the book, reference is given only to the page where the particular term is explained or to the chapter or paragraph dealing with the term in question. Sorbents are referred to without further specification (particle size, degree of cross-linking, etc.). All specifications available to us are summarized in tables in the part entitled Practice of liquid chromatography.

**B**

## C

## D

## E

## H



## I

## K

## L



## M

## N



## O



## P

## Q



## R

## S

## T

## List of compounds chromatographed*



*Compiled by H. Bečvářová.

**B**

## C

## D

## E

## F

## G

## H

## I

## J

## K



## L

## M

## N

## O



## P

## Q



## R

## S

## T

## U



## V

## W



## X

## Y



## Z